Contemporary Logic and Computing

edited by

Adrian Rezuş

ISBN 978-1-84890-340-1

College Publications
Scientific Director: Dov Gabbay
Managing Director: Jane Spurr

http://www.collegepublications.co.uk

Cover design by Laraine Welch

Contents

Foreword

Adrian Rezuş
Nijmegen
equivalences.org@gmail.com

The present volume stems from a book-proposal made about two years ago to College Publications, London. The main idea was that of *illustrating the interplay between the contemporary work in logic and the mainstream mathematics.*

The outcome is somewhat variegated and, as expected, it lacks conceptual unity. This made the classification of the subjects rather difficult.

The division of the volume in two sections – topics in 'logic' vs topics in 'computing' – is more or less conventional.

Some contributions are focussed on historical and technical details meant to put in perspective the impact of the work of some outstanding mathematicians and philosophers on the contemporary research in logic and computing science. Cases in point are Jan Łukasiewicz (1878–1956), Luitzen Egbertus Jan Brouwer (1881–1966), Wilhelm Ackermann (1896–1962), Frank Plumpton Ramsay (1903–1930), Bolesław Sobociński (1906–1980), Alan Mathison Turing (1912–1954), Johan J. de Iongh (1915–1999), Corrado Böhm (1923–2017), Julius Richard Büchi (1924–1984), and Errett Albert Bishop (1928–1983). Some other papers, also with a historical flavour, were supposed to evidentiate punctual methods of research and specific concepts or topics, as, e.g., decidability, computability, randomness, and computational or descriptive complexity.

In general, the papers were intended as *specific surveys of results*, so I have deliberately left out contributions with a generalistic orientation.

I am, first of all, indebted to the authors – about three dozen – who managed to find time in order to step back, reflect on, and sort out their – and their colleagues – achievements during the last decades, for a larger audience.

On the practical side, I am grateful to Dov Gabbay, the scientific advisor of College Publications, who accepted the somewhat adventurous idea of hosting such a vast project, as well as to those many friends and colleagues – including Mark van Atten, Cristian Calude, Hristo Ganchev, William Gasarch, Christine Gaßner, Ulrich Kohlenbach, Edwin Mares, Klaus Meer, Anil Nerode, Eike Neumann, Dimiter Skordev, Theodor A. Slaman, and Wim Veldman – who have gently agreed to act as referees for the volume, oft under a short notice.

Other volumes – to be issued subsequently in the same series – will hopefully delineate aspects of *the contemporary logic landscape* that have not been illustrated here.

The intended audience of the book includes graduate students in mathematical logic, foundations of mathematics, and computing science, as well as philosophers, mathematicians, and, possibly, other scientists interested in the recent research on logic and computing.

The volume has been generated with software provided by the publisher. I am particularly indebted to Jane Spurr, the managing director of College Publications, for technical assistance, advice, and suggestions.

Adrian Rezuş

Nijmegen, 23 June 2020

I Selected Topics in Contemporary Logic

THE DEVELOPMENT OF DECIDABILITY PROOFS BASED ON SEQUENT CALCULI

KATALIN BIMBÓ
Department of Philosophy, University of Alberta, Canada
bimbo@ualberta.ca

ABSTRACT Some logics are decidable, that is, there is an algorithm to determine whether a formula is provable in a calculus formalizing the logic. This paper aims to depict the evolution of decidability proofs based on sequent calculi from the first such proof for intuitionist logic to some of the latest results that use—in an essential way—the *Curry–Kripke technique*.

KEYWORDS Curry's lemma, decidability, intuitionist logic, Kripke's lemma, modal logic, relevance logic, sequent calculus

Introduction

A desire to know if a sentence is true or not may be the original motivation for human inquiry. Famously, Hilbert, in his Paris lecture, in 1900, claimed that there are no unknowable mathematical theorems. Of course, the tools to gain knowledge, in general, or to prove mathematical claims, in particular, has to be narrowed down to make sense of claims concerning the possibility or impossibility of attaining knowledge.

Decidability is a much more specific problem than the informal idea of somehow establishing or refuting mathematical claims. First of all, a fixed language and a formal theory is assumed. Furthermore, the question is simplified to whether a given formula is provable from the formal theory. Lastly, a solution cannot use arbitrary means; it should involve an effective procedure. Notably, the first formal notions of computability were not introduced until the 1920s.

Our focus in this paper is on formal theories that are propositional logics, some of which are sufficiently complex to be undecidable (e.g., the major relevance logics, T (ticket entailment), E (entailment) and R (relevant implication)). Quantification, as a rule, leads to undecidability; indeed, it is well known that even 2-valued predicate logic with predicates of arbitrary finite arity is undecidable. This is so despite the fact that 2-valued propositional logic is decidable, and 2-valued monadic predicate logic is also decidable. For 2-valued (or, in general, a finitely valued) logic, a quick decidability argument can appeal to truth tables, which can be effectively constructed for any n ($n \in \mathbb{N}$). Then it is trivial to scan the final column for all T's (or distinguished values). Of course, this argument relies on the adequacy of the truth assignment interpretation of 2-valued logic or on the claim that a logic is finitely valued. Sometimes, assuming a suitable semantics for a logic, the decidability problem is explicitly phrased as the question if a formula is valid. This may be labeled as the *semantic decidability problem.* There are techniques beyond truth tables (e.g., algebraic methods and filtration) that can be used to answer the semantic version of the question (e.g., for logics that have the finite model property). However, we limit our considerations here to the *syntactic* question. More specifically, we look at decision procedures that use a *sequent calculus.*

The next section presents an outline of Gentzen's original proof of the decidability of propositional intuitionist logic. Section 2 is an application of a similar method by Lambek to his two calculi. Section 3 describes "Curry's turn," which literally, changes the direction of the proof search (in relation to a sequent calculus proof). In Section 4, we briefly recall Kleene's $G3$ approach together with his influential concept of cognate sequents. In Section 5, we outline how Kripke solved, in three short summer months, the decidability problem for $E_{\rightarrow}$, implicational entailment. Kripke introduced new ideas into Curry's framework, which led to a new group of logics shown to be decidable. Then, we devote Section 6 to ways in which the Curry–Kripke method has been used to obtain further decidability results. Finally, we draw some conclusions in the last section.

1 The decidability of propositional intuitionist logic

Intuitionist logic was formulated by Heyting [32], in 1930. Gödel [31] showed that intuitionist logic is not finitely valued. In other words, there is no hope to obtain a truth-table like semantics for it, and then to check validity through a semantical interpretation.

Gentzen [27, 28] provided new formalizations for intuitionist logic in the form of NJ (a natural deduction calculus) and LJ (a logistic calculus).[1] We are interested in the *propositional* part of LJ here, hence, we will use the label to refer to that logic without the quantifiers; furthermore, we will not consider NJ here at all.

[1] [29, 30] are translations of [27, 28], and they may also be found in [43].

Gentzen's LJ (and LK, his logistic calculus for 2-valued logic) has served as the original blueprint for many later sequent calculi, which motivates us to give a full definition of (propositional) LJ.[2]

Definition 1.1. The language of LJ contains *four connectives*, namely, $\neg$ (negation), $\wedge$ (conjunction), $\vee$ (disjunction) and $\supset$ (intuitionistic implication), as well as a denumerable set of *propositional variables*, that we denote by $p_0, p_1, \ldots, p_n, \ldots$.

The set of formulas is given as the set of strings that can be generated by the following context-free grammar, where $\mathbf{P}$ rewrites to a propositional variable.

$$A := \mathbf{P} \mid \neg A \mid (A \wedge A) \mid (A \vee A) \mid (A \supset A)$$

Remark 1.2. There are other connectives that we could introduce, for instance, $\supset\subset$ (co-implication) or $\mathbf{F}$ (falsity constant), which are, indeed, mentioned in [29, p. 289]. However, they are definable using the connectives already introduced and they are not used in LJ.

Definition 1.3. A *sequent* is a pair of finite sequences of formulas $\langle \Gamma, \Theta \rangle$, where Θ contains at most one element. We use the symbol $\Vdash$ to separate the two sequences writing $\Gamma \Vdash \Theta$.

All the sequents in a proof in LJ have to satisfy the size restriction on Θ, hence, we will not keep repeating it. As a notational convention, we use A, Γ to indicate the concatenation of the formula A and the sequence Γ (in that order). Similarly, Γ, A and Γ, Θ indicate concatenations too.

The intended informal interpretation of a provable sequent $\Gamma \Vdash \Theta$ is that the conjunction of the elements of Γ implies θ, where θ is either the only element in Θ or it is $\mathbf{F}$, if Θ is empty.

Definition 1.4. The sequent calculus LJ consists of an *axiom* and *rules*, which are divided into three groups: connective rules, structural rules and the cut rule.

$$A \Vdash A \ \text{(Id)} \qquad \frac{\Gamma \Vdash A}{\neg A, \Gamma \Vdash} \ (\neg\Vdash) \qquad \frac{A, \Gamma \Vdash}{\Gamma \Vdash \neg A} \ (\Vdash\neg)$$

$$\frac{A, \Gamma \Vdash \Theta}{A \wedge B, \Gamma \Vdash \Theta} \ (\wedge_1\Vdash) \qquad \frac{B, \Gamma \Vdash \Theta}{A \wedge B, \Gamma \Vdash \Theta} \ (\wedge_2\Vdash) \qquad \frac{\Gamma \Vdash A \quad \Gamma \Vdash B}{\Gamma \Vdash A \wedge B} \ (\Vdash\wedge)$$

$$\frac{A, \Gamma \Vdash \Theta \quad B, \Gamma \Vdash \Theta}{A \vee B, \Gamma \Vdash \Theta} \ (\vee\Vdash) \qquad \frac{\Gamma \Vdash A}{\Gamma \Vdash A \vee B} \ (\Vdash\vee_1) \qquad \frac{\Gamma \Vdash B}{\Gamma \Vdash A \vee B} \ (\Vdash\vee_2)$$

[2] We will not follow Gentzen's notation as to the use of particular symbols such as &.

$$\frac{\Gamma \Vdash A \quad B, \Lambda \Vdash \Theta}{A \supset B, \Gamma, \Lambda \Vdash \Theta} (\supset \Vdash) \qquad \frac{A, \Gamma \Vdash B}{\Gamma \Vdash A \supset B} (\Vdash \supset)$$

$$\frac{A, A, \Gamma \Vdash \Theta}{A, \Gamma \Vdash \Theta} (W \Vdash) \qquad \frac{\Gamma, A, B, \Lambda \Vdash \Theta}{\Gamma, B, A, \Lambda \Vdash \Theta} (C \Vdash)$$

$$\frac{\Gamma \Vdash \Theta}{A, \Gamma \Vdash \Theta} (K \Vdash) \qquad \frac{\Gamma \Vdash}{\Gamma \Vdash A} (\Vdash K) \qquad \frac{\Gamma \Vdash A \quad A, \Lambda \Vdash \Theta}{\Gamma, \Lambda \Vdash \Theta} (\text{cut})$$

A *proof* in LJ is a tree, in which the nodes are occurrences of sequents, specifically, the leaves are occurrences of instances of the axiom, and all other nodes result by applications of rules. The root of the tree is the sequent that is *proved.* (Every sequent in a proof tree is the root of a subtree in the proof tree, and every sequent in a proof tree is a provable sequent.)

Remark 1.5. There are four two-premise rules in LJ, and the $(\Vdash \wedge)$ and $(\vee \Vdash)$ rules differ from the $(\supset \Vdash)$ and the (cut) rules in that the former two require the two premises to be the same save the formulas A and B. The following two rules would do just as well.

$$\frac{\Gamma \Vdash A \quad B, \Gamma \Vdash \Theta}{A \supset B, \Gamma \Vdash \Theta} (\supset \Vdash') \qquad \frac{\Gamma \Vdash A \quad A, \Gamma \Vdash \Theta}{\Gamma \Vdash \Theta} (\text{cut}')$$

Here is a quick argument for the equivalence of the versions of the rules. If Γ happens to be Λ in the $(\supset \Vdash)$ or the (cut) rule, then the lower sequent will contain Γ, Γ, which may be reduced to Γ by finitely many applications of the $(C \Vdash)$ and $(W \Vdash)$ rules. Conversely, Γ and Λ have a (possibly, empty) common part, thus, Γ and Λ may be depicted—with appeal to $(C \Vdash)$—as Γ', Ξ and Ξ, Λ'. Then both can be beefed up (using $(K \Vdash)$ and $(C \Vdash)$) to Γ', Ξ, Λ'. Then, the primed rules yield lower sequents that are missing a Ξ, which may be inserted by finitely many applications of $(K \Vdash)$ and $(C \Vdash)$.

Remark 1.6. The cut rule is an *admissible rule*, hence, LJ may be formulated without it. In [29], an inductive proof was used to show that the *mix* rule is admissible, which is in turn equivalent to the cut rule. The mix rule has the following form:

$$\frac{\Gamma \Vdash A \quad A, \Lambda \Vdash \Theta}{\Gamma, \Lambda^{-A} \Vdash \Theta} (\text{mix}),$$

where Λ^{-A} is the sequence of formulas Λ with *all* the occurrences of A deleted. The proof uses an induction on the number of applications of the mix rule in a proof, as well as a double induction on two parameters that characterize the size of the mix formula and the place of the mix within a proof tree. The lower sequent of the cut $(\Gamma, \Lambda \Vdash \Theta)$ can be easily restored from $\Gamma, \Lambda^{-A} \Vdash \Theta$ by applications of $(K\Vdash)$ and $(C\Vdash)$. On the other hand, all the occurrences of A is Λ can be reduced to a single occurrence by (finitely many) uses of $(C\Vdash)$ and $(W\Vdash)$, and then one cut yields Λ^{-A}.

We are interested in the decidability proof for LJ, hence, we will not dwell on the proof of the admissibility of the cut rule. The decidability of LJ was an open problem in 1935, moreover, as we already mentioned, it was known from Gödel's [31] that there was no hope to find "intuitionistic truth tables." We turn to presenting the decidability proof for LJ along the lines of [30]. The flexibility of LJ illustrated by the use of the structural rules in Remarks 1.5 and 1.6 plays a crucial role in the proof.

Definition 1.7. A *reduced* sequent is a sequent in which no formula has more than *three* occurrences in the antecedent.

Obviously, for any sequent with a non-empty antecedent there are at least two reduced sequents, which do not contain occurrences of a formula with no occurrences in the starting sequent, that are equivalent to it. (Equivalence means provability from one another.) We call these sequents *reduced versions.*

Lemma 1.8. *If* $\Gamma \Vdash \Theta$ *is provable in* LJ *and* $\Gamma^\rho \Vdash \Theta$ *is a reduced version of* $\Gamma \Vdash \Theta$, *then there is a proof of* $\Gamma^\rho \Vdash \Theta$, *in which* all *sequents are* reduced.

Proof. (Sketch) The proof utilizes the ideas mentioned in Remarks 1.5 and 1.6. If $\Gamma \Vdash \Theta$ is provable, then it is provable without applications of the cut rule. Hence, continuing any such proof, if necessary, with applications of $(C\Vdash)$ and $(W\Vdash)$, a cut-free proof of $\Gamma^\rho \Vdash \Theta$ can be obtained.

Then it remains to show that if there is a sequent in the latter proof that is not reduced, then the proof can be transformed into one that has only reduced sequents. Instances of the axiom are reduced sequents, hence, any sequent that does not meet the condition in Definition 1.7, must have resulted by an application of a rule. As a critical case, we consider the $(\supset\Vdash)$ rule, which is the most capable rule to produce multiple occurrences of a formula in the lower sequent. Here is a chunk of a proof that we will consider.

$$\dfrac{\begin{matrix}\vdots \\ \Gamma^{-(A\supset B)}, (A\supset B)^m \Vdash A\end{matrix} \qquad \begin{matrix}\vdots \\ B, (A\supset B)^n, \Lambda^{-(A\supset B)} \Vdash \Theta\end{matrix}}{\Gamma^{-(A\supset B)}, (A\supset B)^{m+n+1}, \Lambda^{-(A\supset B)} \Vdash \Theta}\ (\supset\Vdash)$$

With the superscripts, we made explicit the number of occurrences of $A\supset B$. (Since $(C\Vdash)$ is a rule, we may assume that all the $A\supset B$'s have been shepherded together.)

The lowest value for $m + n + 1$ is 1, when $m = n = 0$. If $m \geq 1$ or $n \geq 1$, then applications of $(W \Vdash)$ can reduce those numbers to 1, and of course, $1 + 1 + 1 = 3$, the number in Definition 1.7. It is easy to see that the insertion of sufficiently many $(W \Vdash)$ and $(C \Vdash)$ steps will create a proof from a proof. Applications of other rules may be dealt with similarly.

The number of occurrences of a formula can be always increased by applications of the rule $(K \Vdash)$. However, we should note that no connective rule requires more that one occurrence of any formula. It is not so for the structural rules. But if the occurrences of A have already been reduced to 1, then a subsequent application of the $(W \Vdash)$ rule may be simply omitted. Similarly, if $(C \Vdash)$ was applied to $\Gamma, A, A, \Lambda \Vdash \Theta$, then that application could have been omitted in the first place, and if there is only one A left, then $(C \Vdash)$ definitely should be omitted. In other words, applications of rules remain applications of the same rule when the upper sequents are replaced by their reduced versions, or if the application of a rule became superfluous, then it is omitted altogether. □

Remark 1.9. A way to think about proofs that comprise reduced sequents throughout is that there is no need to make detours in a proof, not only via applications of the cut rule, but via accumulating many copies of one and the same formula either. Of course, it is not true that all provable sequents have cut-free proofs comprising reduced sequents only—simply, because non-reduced sequents are provable too. However, it if trivial to prove $\Gamma \Vdash \Theta$ from $\Gamma^{\rho} \Vdash \Theta$.

A glance at the rules of LJ helps to establish that a proof without cut has the subformula property.

Definition 1.10. A proof has the *subformula property*, when every formula in any sequent is either a formula or a subformula of a formula in the sequent proved.

Formulas can be viewed as *types*, so to speak, rather that *tokens* or occurrences. For example, an application of $(W \Vdash)$ discards a token (an occurrence), but not a type (a formula).

Definition 1.11. Let $\Gamma \Vdash \Theta$ be a given sequent, and let $\Gamma^{\rho} \Vdash \Theta$ be a reduced version of it. There are finitely many subformulas in $\Gamma^{\rho} \Vdash \Theta$, the set of which we denote by Σ. There are finitely many instances of (Id) that can be constructed from the elements of Σ, we denote the set of these sequents by Π. The decision procedure is given as the following three steps.

1. The elements of Π can be used as upper sequents in rules. We generate all the possible lower sequents that satisfy the conditions (1) all the formulas in the sequents are elements of Σ, and (2) the sequents are reduced.

2. We update Π with the newly obtained sequents, and return to step 1. If no new sequents were generated, then we proceed to step 3.

3. We check whether $\Gamma^\rho \Vdash \Theta$ is or is not an element of Π, accordingly, $\Gamma \Vdash \Theta$ is or is not provable.

Remark 1.12. We stress that this decision procedure is essentially Gentzen's. It is a *top-down* procedure in the sense that it starts with instances of (Id). An inessential difference from the original in [30] is that we limited the generation of the lower sequents by (1) and (2) instead of outright generating all the reduced sequents from Σ, and then moving sequents into Π. Perhaps, in the spirit of intuitionist logic, it is "more constructive" to generate a sequent when it is (known to be) provable.

The decidability proof for LJ could be adapted to obtain a decidability proof for the propositional part of LK. A key modification would be to define reduced sequents to limit the number of occurrences of a formula in the succedent to *two.* Indeed, Gentzen defined reduced sequents, but he permitted three occurrences for any formula in the succedent. If we glance at the rules for LK, then we see that no rule behaves with respect to the succedent as $(\supset \Vdash)$ does with respect to the antecedent. Namely, $(\supset \Vdash)$ combines Γ, Λ and $A \supset B$. We may conclude that Gentzen's three is not optimal, and at the same time we may note that any natural number larger than three in place of three would do just as well. Although the decidability of propositional LK is not a new result, it hints at other potential applications.

2 The decidability of Lambek's calculi

Joachim Lambek published two influential papers [37] and [38] in which he introduced what afterward became to be known as the *associative Lambek calculus* and the *non-associative Lambek calculus.* He can be seen to continue a long tradition, which goes back (at least) to Frege, where sentences are decomposed into a function and its arguments. Semantic and syntactic types had been investigated by Ajdukiewicz [2], Church [19] and Curry [21] before Lambek's work, however, it seems that previously nobody considered using a sequent calculus to specify derivations of compound types.

We will denote the non-associative Lambek calculus by LN and the associative one by LG. Our definitions will closely resemble Lambek's original definitions.[3]

[3]In [14], two calculi, which are labeled LQ and LA, are essentially Lambek's calculi (LN and LG here)—except that the left-hand side of the turnstile was not permitted to be empty, which harmonized well with some sequent calculi there. Lambek's own label for one of his sequent calculi

Definition 2.1. The language of LN and LG comprises *three binary connectives* $\cdot$, / and \ together with denumerably many *propositional variables.*

The set of formulas is given as the set of strings generated by the following CFG, where $\mathbf{P}$ may be rewritten as any of the propositional variables.

$$A := \mathbf{P} \mid (A \cdot A) \mid (A \,/\, A) \mid (A \setminus A)$$

Definition 2.2. The set of *ropes* is inductively defined by (1)–(3).

(1) The empty rope is a rope;

(2) if A is a formula, then A is a rope;

(3) if Γ and Θ are ropes, then (Γ, Θ) is a rope.

Remark 2.3. We will assume that the empty rope, which may be denoted by a space, behaves as follows: $(\Gamma, \ ) = (\ , \Gamma)$ and $(\ , \Gamma) = \Gamma$. Also, if Θ is a rope that occurs in Γ, then by $\Gamma[\Theta]$ we denote the rope Γ, in which one particular occurrence of Θ has been chosen. If Θ is replaced by another rope, let us say, Λ, then $\Gamma[\Lambda]$ will be the shorthand for the usual $\Gamma[\Theta/\Lambda]$ within the context of a rule.

The way we think of ropes is that they are like *strings* (or finite sequences) but bulkier, due to the presence of parentheses. Thus, if we omit all the parentheses that indicate groupings in a rope we get a string.

Definition 2.4. A *sequent* is a pair $\langle \Gamma, A \rangle$ where Γ is a rope and A is a formula. As before, we use the notation $\Gamma \Vdash A$ for this pair.

Definition 2.5. The sequent calculus LN consists of an *axiom* and seven *rules.*

$$A \Vdash A \quad \text{(Id)}$$

$$\frac{\Gamma \Vdash A \quad \Theta[B] \Vdash C}{\Theta[B \,/\, A, \Gamma] \Vdash C} \ (/ \Vdash) \qquad \frac{\Gamma, A \Vdash B}{\Gamma \Vdash B \,/\, A} \ (\Vdash /)$$

$$\frac{\Gamma \Vdash A \quad \Theta[B] \Vdash C}{\Theta[\Gamma, A \setminus B] \Vdash C} \ (\setminus \Vdash) \qquad \frac{A, \Gamma \Vdash B}{\Gamma \Vdash A \setminus B} \ (\Vdash \setminus)$$

$$\frac{\Gamma[A, B] \Vdash C}{\Gamma[A \cdot B] \Vdash C} \ (\cdot \Vdash) \qquad \frac{\Gamma \Vdash A \quad \Theta \Vdash B}{\Gamma, \Theta \Vdash A \cdot B} \ (\Vdash \cdot)$$

$$\frac{\Gamma \Vdash A \quad \Theta[A] \Vdash B}{\Theta[\Gamma] \Vdash B} \ \text{(cut)}$$

is Σ_G, while the other has no label. We use here the labels LN and LG in order to retain some "L-labels" introduced by others, e.g., Curry's LA.

Definition 2.6. The sequent calculus LG is the result of replacing ropes with *sequences* (or strings) of formulas in the antecedent of a sequent.

Remark 2.7. It seems that the introduction of grouping into the antecedent of a sequent is original with Lambek. This idea proved extremely useful later on. Lambek's motivations seem to have come from algebra (where non-associative binary operations are a common place), and from linguistics (where certain groupings of words are preferred over others in the grammatical analysis of sentences).

We illustrate the latter by an example from [38, p. 158]. The English sentence "John likes fresh milk" in phrase structure grammar is viewed as "John (likes (fresh milk))" rather than the four other possible ways to parse the sentence (without changing the order of the words). (E.g., "(John (likes fresh)) milk" is pretty weird.)

The main influential works dealing with sequent calculi in the 1950s, especially, [22] and [33], focused on 2-valued and intuitionist logics. In that context, it is not necessary, indeed, it would be a nuisance to distinguish sequences of formulas with different groupings. Even the order and the (positive) number of occurrences is more than what is needed in a sequent calculus capturing those logics. Lambek took the bold steps of omitting all the structural rules, of using Ketonen's rule for $(\cdot \Vdash)$ and of discarding the assumption that the antecedent of a sequent is a string of formulas. As a result of these changes, there are two residuals to $\cdot$, which is a well-known fact from algebra. Accordingly, Lambek introduced two versions of Gentzen's $(\supset \Vdash)$ rule—with situating the formulas (potentially) affected inside of a sequent. We also have to point out that the $(\Vdash \cdot)$ rule is the appropriate generalization of the $(\Vdash \wedge)$ rule when there is no $(K\Vdash)$ rule in the calculus.

Theorem 2.8. *The cut rule is* admissible *in LN and in LG without the cut rule.*

Proof. (Idea.) Lambek stated this theorem and outlined the proof in his [37] and [38]. His proof is *not* a direct adaptation of Gentzen's proof; mix could not be shown to be equivalent to the cut rule. Nor his proof goes along the lines of Curry's in [22]. The total absence of structural rules implies that every application of every rule (save cut) introduces a new occurrence of a connective into the sequent. Then, the whole (cut-free) proof above a cut can be characterized in terms of the number of occurrences of connectives in a sequent. Lambek calls the sum of the number of occurrences of connectives in the premises of the cut rule the *degree of the cut.*[4]

The induction is on the degree of the cut, which is reduced through modifications of the proof. The latter are straightforward, because there are no structural rules.

[4]This usage differs from usage by others in the literature, where the degree of the cut (or more precisely, of the cut formula) is a number that characterizes the *complexity of the cut formula* per se in terms of its logical components.

Then another induction on the number of cuts in a proof completes the demonstration of the cut theorem. □

Theorem 2.9. *LG and LN are* decidable.

Proof. (Idea.) Lambek sketched the proof for his calculi, and in both cases he outlined a proof search starting from the bottom. By the cut theorem, it is sufficient to look for cut-free proofs. Then, each step in a proof is an application of a connective rule, hence, the search for a proof can be seen as a decomposition of the given sequent in every possible way. There are finitely many ways to decompose a sequent. Thus, either a proof is found or there is no proof.[5] □

Remark 2.10. To complete our quick overview of Lambek's LG and LN, we wish to emphasize that the careful selection of the rules is the key to the cut theorem being provable. In the case of these calculi, it is true (what is often a misunderstanding for other sequent calculi) that the admissibility of the cut rule delivers decidability.

Gentzen's LK and LJ, on one hand, and Lambek's LG and LN, on the other, represent two extremes: *all* structural rules and *no* structural rules are available. However, there are interesting logics that are situated in the middle, so to speak. And to prove their decidability—if they are decidable—requires further new ideas. Next we look at how Curry turned proof search around.

3 Curry's lemma

The concept of a formal system was of primary interest in the 1940s. Curry gave a series of lectures in 1949, which became the basis for his [22]. The systems that are considered by him range from sequent calculi for the positive fragments of 2-valued logic and of intuitionist logic to modal systems through type-assignment systems.

Curry introduced two sequent calculi LA and LC, which formalized the negation-free fragments of intuitionist and 2-valued logic, respectively.[6] We focus on these calculi and their variants. The language and the formulas are as in Definition 1.1 except that $\neg$ is omitted.

[5]Lambek [38, p. 155, p. 165] suggests that his proof "follows" Gentzen's, which is, of course, true in the sense that finding a cut-free proof in a sequent calculus is the target. However, the direction of the proof search is similar to Curry's (whose [22] Lambek referenced). Describing a top-down proof search, which would more closely resemble Gentzen's procedure in its direction would be unproblematic.

[6]We present LA and LC as sequent calculi without stressing their epitheoretic part, and we do not follow Curry's terminology everywhere.

Definition 3.1. A *sequent* in LA is $\langle \Gamma, A \rangle$, where Γ is a (finite) sequence of formulas, and A is a formula. A *sequent* in LC is $\langle \Gamma, \Theta \rangle$, where Γ and Θ are (finite) sequences of formulas.

The difference in the definition of a sequent for LJ and LA is explained by the lack of a rule (namely, of the $(\neg \Vdash)$ rule) that could empty the succedent of a sequent. The single succedent restriction on sequents applies to all sequents in LA—even if we do not repeat it every time.

Definition 3.2. The *axioms* and *rules* for LC (for LA) are the following. (We adopt Lambek's notation to indicate an occurrence of a formula. Θ' is *contained* in Θ (in rule $(\supset \Vdash)$) in the sense that every occurrence of every formula in Θ' is in Θ with no requirement that their order be preserved.)

$$\Gamma[A] \Vdash \Theta[A] \quad (\mathrm{Id})$$

$$\frac{\Gamma, A \Vdash \Theta}{\Gamma, A \wedge B \Vdash \Theta}\ (\wedge_1 \Vdash) \qquad \frac{\Gamma, B \Vdash \Theta}{\Gamma, A \wedge B \Vdash \Theta}\ (\wedge_2 \Vdash) \qquad \frac{\Gamma \Vdash A, \Theta \quad \Gamma \Vdash B, \Theta}{\Gamma \Vdash A \wedge B, \Theta}\ (\Vdash \wedge)$$

$$\frac{\Gamma, A \Vdash \Theta \quad \Gamma, B \Vdash \Theta}{\Gamma, A \vee B \Vdash \Theta}\ (\vee \Vdash) \qquad \frac{\Gamma \Vdash A, \Theta}{\Gamma \Vdash A \vee B, \Theta}\ (\Vdash \vee_1) \qquad \frac{\Gamma \Vdash B, \Theta}{\Gamma \Vdash A \vee B, \Theta}\ (\Vdash \vee_2)$$

$$\frac{\Gamma \Vdash A, \Theta' \quad \Gamma, B \Vdash \Theta}{\Gamma, A \supset B \Vdash \Theta}\ (\supset \Vdash) \qquad \frac{\Gamma, A \Vdash B, \Theta}{\Gamma \Vdash A \supset B, \Theta}\ (\Vdash \supset)$$

$$\frac{\Gamma \Vdash \Theta}{\Xi \Vdash \Theta}\ (C \Vdash) \qquad \frac{\Theta \Vdash \Gamma}{\Theta \Vdash \Xi}\ (\Vdash C)^7$$

$$\frac{\Gamma[A], A \Vdash \Theta}{\Gamma[A] \Vdash \Theta}\ (W \Vdash) \qquad \frac{\Gamma \Vdash A, \Theta[A]}{\Gamma \Vdash \Theta[A]}\ (\Vdash W)$$

Remark 3.3. Curry considered axiomatic extensions too; we omitted an axiom and a rule that pertain to those components. Looking at the systems above, we should note several interesting details.

First, there is no thinning rule, either on the left- or on the right-hand side of the $\Vdash$ (i.e., $(K \Vdash)$ and $(\Vdash K)$ are dropped). Instead, a more general version of (Id) is postulated, which allows for the inclusion of formulas on either side in LC (on the left in LA), in addition to A appearing on both sides. This is a step toward *dispensing* with the structural rules, which goes against Gentzen's original aim to separate the operational rules from the structural ones. However, Curry is clearly

[7] Ξ is any permutation of Γ in these two rules.

intrigued by possibilities for changing the structural rules. And some variations on the structural rules—introduced by Curry and by others—proved exceptionally fruitful. (Lambek's calculi from the previous section are just the first examples.) The permutation rules ($C \Vdash$) and ($\Vdash C$) are more general than the similar rules in LK in the sense that several LK (LJ) steps can be combined into one step in LC (LA). Indeed, Curry quickly introduced the convention [22, p. 33] that permutation steps are left tacit, which we could view as him using multisets rather than sequences of formulas. The contraction rules here generalize Gentzen's contraction rules in a different way. Several permutation steps together with one contraction step in a proof in LK (LJ) can be performed as one step in LC (LA).

Second, Curry did not introduce any notation for an occurrence of a formula in a sequence of formulas, though this would be useful for the formulation of (Id) and the contraction rules as shown above. Lastly, we may point out that the ($\supset \Vdash$) rule is a blend of Gentzen's rule and Ketonen's rule. Like Gentzen's rule, Curry's rule permits the formulation of a 2-valued and an intuitionistic calculus in one fell swoop (unlike Ketonen's rule does), but Curry's rule is more restrictive than Gentzen's by requiring Γ in both premises.

Curry proved that the *thinning* rules are *admissible* [22, Theorem 2, p. 35] in the following more general forms. (The primed sequences of formulas are contained in their unprimed companions.)

$$\frac{\Gamma' \Vdash \Theta}{\Gamma \Vdash \Theta}\ (K\Vdash) \qquad\qquad \frac{\Gamma \Vdash \Theta'}{\Gamma \Vdash \Theta}\ (\Vdash K)$$

For us, the most interesting development in [22] is Theorem 3, which is the origin of what is called "Curry's lemma" in some of the literature, and what is called "height-preserving admissibility of contraction" in some other publications. To stress the importance of this theorem, we quote it.

> Theorem 3. If the rules Ol and, in LC, Or and Er, are so modified as to require the principal constituents to appear in all the premises on the same side as in the conclusion, then the rules W are redundant. — [22, p. 36]

The modifications affect eight rules in LC (four of which are impossible in LA, just as ($\Vdash W$) is impossible); we restate them all.

$$\frac{\Gamma[A \wedge B], B \Vdash \Theta}{\Gamma[A \wedge B] \Vdash \Theta}\ (\wedge_2\Vdash)^+ \qquad \frac{\Gamma \Vdash A, \Theta[A \wedge B] \quad \Gamma \Vdash B, \Theta[A \wedge B]}{\Gamma \Vdash \Theta[A \wedge B]}\ (\Vdash \wedge)^+$$

$$\frac{\Gamma[A \wedge B], A \Vdash \Theta}{\Gamma[A \wedge B] \Vdash \Theta}\ (\wedge_1\Vdash)^+ \qquad \frac{\Gamma \Vdash B, \Theta[A \vee B]}{\Gamma \Vdash \Theta[A \vee B]}\ (\Vdash \vee_2)^+$$

$$\frac{\Gamma[A \vee B], A \Vdash \Theta \quad \Gamma[A \vee B], B \Vdash \Theta}{\Gamma[A \vee B] \Vdash \Theta} (\vee \Vdash)^+ \qquad \frac{\Gamma \Vdash A, \Theta[A \vee B]}{\Gamma \Vdash \Theta[A \vee B]} (\Vdash \vee_1)^+$$

$$\frac{\Gamma[A \supset B] \Vdash A, \Theta' \quad \Gamma[A \supset B], B \Vdash \Theta}{\Gamma[A \supset B] \Vdash \Theta} (\supset \Vdash)^+ \qquad \frac{\Gamma, A \Vdash B, \Theta[A \supset B]}{\Gamma \Vdash \Theta[A \supset B]} (\Vdash \supset)^+$$

An initial reaction to these rules could be a surprise along the lines that if we already have the formula that is being introduced by the rule (i.e., the principal formula), then why to apply the rule at all? Of course, this riposte overlooks that some formulas *disappear* due to the application of the rule. Indeed, Curry notes that now "the essential function of the rules is ... to eliminate components." [22, p. 37]

Lemma 3.4. *If* ($\natural$) *is a connective rule of* LC*, then it is* derivable *from* $(\natural)^+$ *and thinning.*

Proof. Assuming the premise(s) of ($\natural$), one application of the appropriate thinning rule, $(K\Vdash)$ or $(\Vdash K)$, to each premise yields the premise(s) of $(\natural)^+$. An application of $(\natural)^+$ gives the same lower sequent that ($\natural$) has. □

Remark 3.5. If we denote by LC^+ the calculus obtained by replacing the ($\natural$) rules with the $(\natural)^+$ ones, then Lemma 3.4 shows that everything provable in LC^+ is provable in LC. Unquestionably, this is the "easy direction" in showing the equivalence of the two calculi. (By omission, we have also shown half of the equivalence between LA and LA^+, where the latter is LA with $(\wedge_1\Vdash)$, $(\wedge_2\Vdash)$, $(\vee\Vdash)$ and $(\supset\Vdash)$ replaced by their plussed versions.)

Lemma 3.6 (Admissibility of contraction). *If* $\Gamma[A], A \Vdash \Theta$ *(or* $\Gamma \Vdash A, \Theta[A]$*) is provable in* LC*, then* $\Gamma[A] \Vdash \Theta$ *(or* $\Gamma \Vdash \Theta[A]$*) is provable in* LC^+*.*

Proof. The proof is by induction on the structure of a proof (or equivalently, on the height of the proof tree).
1. If $\Gamma[A], A \Vdash \Theta$ or $\Gamma \Vdash A, \Theta[A]$ is an instance of (Id), then so is $\Gamma[A] \Vdash \Theta$ and $\Gamma \Vdash \Theta[A]$, because A must occur in Θ or Γ, respectively.
2. If the claim holds for the premise of $(C\Vdash)$ or $(\Vdash C)$, then it holds for the lower sequent. (In some cases, the application of the permutation rule may be omitted altogether.)

As Curry noted earlier, the permutation steps may be left tacit everywhere. Indeed, counting them is cumbersome. From now one, we use a generalized version of the contraction rules, which are similar to Curry's generalized thinning and generalized permutation rules. These rules are:

$$\frac{\Gamma[A][A] \Vdash \Theta}{\Gamma[A] \Vdash \Theta} (W\Vdash) \qquad \frac{\Gamma \Vdash \Theta[A][A]}{\Gamma \Vdash \Theta[A]} (\Vdash W)$$

These rules better capture what we want to show about LC^+ (than the earlier rules do). (**1.** and **2.** above remain true.)

3. Let us assume that the $(\wedge\Vdash)^+$ rule has been applied in a proof. We consider whether $\Gamma[A]\,[A]\,[B\wedge C]\Vdash\Theta$ or $\Gamma[B\wedge C]\,[B\wedge C]\Vdash\Theta$ is the provable sequent of LC. In the first case, by the hypothesis of the induction, we have that $\Gamma[A]\,[B\wedge C], B\Vdash\Theta$ or $\Gamma[A]\,[B\wedge C], C\Vdash\Theta$ is provable, whichever was the premise. An application of $(\wedge\Vdash)^+$ gives $\Gamma[A]\,[B\wedge C]\Vdash\Theta$. In the second case, the upper sequent must have had *at least two* occurrences of $B\wedge C$ (e.g., as in $\Gamma[B\wedge C]\,[B\wedge C], B\Vdash\Theta$), and that number can be reduced by one. That is, the inductive hypothesis supplies the provability of $\Gamma[B\wedge C], B\Vdash\Theta$ or $\Gamma[B\wedge C], C\Vdash\Theta$. Hence, by an application of the rule $(\wedge\Vdash)^+$, we get $\Gamma[B\wedge C]\Vdash\Theta$.

If A occurs on the right-hand side of $\Vdash$, then the lower sequent after the application of the rule $(\wedge\Vdash)$ is $\Gamma[B\wedge C]\Vdash\Theta[A]\,[A]$, but the upper sequents must have been $\Gamma[B\wedge C], B\Vdash\Theta[A]\,[A]$ or $\Gamma[B\wedge C], C\Vdash\Theta[A]\,[A]$. By the hypothesis of the induction, $\Gamma[B\wedge C], B\Vdash\Theta[A]$ and $\Gamma[B\wedge C], C\Vdash\Theta[A]$ are provable. The application of the $(\wedge\Vdash)^+$ rule does not depend on the shape of the succedent, nor does the rule affect any change in it.

The other conjunction rules, all the disjunction rules, and $(\Vdash\supset)^+$ follow the same pattern of argument.

4. We consider the $(\supset\Vdash)^+$ rule. First, if A occurs in the antecedent and it is distinct from $B\supset C$ and C, then we only need to appeal to the inductive hypothesis. Second, if A is $B\supset C$, then we may appeal to the inductive hypothesis and an application of the $(\supset\Vdash)^+$ rule. Third, we scrutinize what happens when A occurs in the succedent, since Θ' and Θ may not coincide. If $\Gamma[B\supset C]\Vdash\Theta[A]\,[A]$ is provable in LC, then one of the premises of the rule is $B, \Gamma[B\supset C]\Vdash\Theta[A]\,[A]$, and we may appeal to the hypothesis of induction to have $\Theta[A]$ as the succedent of the sequent. We have to consider the other premise of $(\supset\Vdash)^+$. The largest Θ' is Θ, hence, we consider this case. (If Θ' is lacking some occurrences of formulas, then that surely does not create a possibility for an application of $(\Vdash W)$; at most some steps become superfluous.) If A is B, then by hypothesis, we have $\Gamma[A\supset B]\Vdash B, \Theta[\ \ ]$ (where $[\ \ ]$ indicates the place where one of the occurrences of B was). If A is not B, then similarly, we have $\Gamma[A\supset B]\Vdash B, \Theta[A]$. Either way, an application of $(\supset\Vdash)^+$ yields the sequent $\Gamma[B\supset C]\Vdash\Theta[A]$ as we needed. $\square$

Remark 3.7. Curry did not list all the rules in their new form (that we indicated by a $^+$). However, presenting further modifications, he listed the principal formulas in upper sequents next to the immediate subformulas. An advantage of using Lambek's notation and generalized versions of the contraction rules is that we can avoid the insertion of a permutation step in the modified proofs, which occasionally would be necessary, and we may obtain the following corollary.

Corollary 3.8 (No height increase). *If the height of the proof of* $\Gamma \Vdash \Theta$ *in* LC *was* n*, then there is a proof of the same sequent in* LC^+ *with height* m *st* $m \leq n$.

Proof. (Idea) Notice that due to the context sharing in the two-premise rules in LC, contractions that are not on formulas introduced by connective rules can be eliminated (by adjusting the numbers of formulas in (Id)). If an application of contraction is on formulas that are introduced by rules, then the matching rules with $^+$'s eliminate the contraction without an extra step in the proof. □

As an illustration of the above sketch, we give an example, namely, the proof of $(A \supset B) \wedge A \Vdash B$ in LC and LC^+. (φ stands in for the formula on which it is a subscript—to shorten some sequents.)

$$
(W\Vdash)\,\dfrac{(\wedge_2\Vdash)\,\dfrac{(\wedge_1\Vdash)\,\dfrac{(\supset\Vdash)\,\dfrac{A \Vdash A, B \qquad A, B \Vdash B}{A \supset B, A \Vdash B}}{A \supset B, (A \supset B) \wedge A \Vdash B}}{(A \supset B) \wedge A, (A \supset B) \wedge A \Vdash B}}{(A \supset B) \wedge A \Vdash B}
\qquad
\dfrac{\dfrac{\dfrac{\varphi, A \supset B, A \Vdash A \qquad \varphi, A \supset B, A, B \Vdash B}{\varphi, A \supset B, A \Vdash B}\,(\supset\Vdash)^+}{((A \supset B) \wedge A)_{\varphi}, A \Vdash B}\,(\wedge_2\Vdash)^+}{(A \supset B) \wedge A \Vdash B}\,(\wedge_1\Vdash)^+
$$

Corollary 3.9. *If* $\Gamma[A]\,[A] \Vdash \Theta$ $(\Gamma \Vdash \Theta[A]\,[A])$ *is provable in* LC^+ *with the height of the proof tree* n*, then* $\Gamma[A] \Vdash \Theta$ $(\Gamma \Vdash \Theta[A])$ *is provable in* LC^+ *with the height of the proof tree* m*, where* $m \leq n$.

Proof. It is easy to check that the steps in the inductive proof of Lemma 3.6 do not increase the height of the proof, because we avoided permutations. Thus, a sequent that results by contraction from a provable formula does not require a longer proof. □

Remark 3.10. *Curry's Lemma*, in the contemporary literature, often refers to Lemma 3.6 and Corollary 3.9 together. A, perhaps, useful way to think about Curry's lemma is that if a formula could be contracted in a proof at a step, then it could have been contracted when it was introduced. Hence, there is no need to accumulate multiple occurrences of a formula for no reason, and then, in later steps contract them (which is, in general, not possible).

Lemma 3.6, Corollary 3.8 and Corollary 3.9 have their analogs for LA and LA^+. (We will not state them separately.)

Curry went on to consider further modifications of the connective rules. In particular, he considered rules where not only the principal formula occurs in the premises, but the immediate subformulas may occur too. He gives as an example the four versions of the ($\Vdash\wedge$) rule. The conclusion that he draws from the latter modifications [22, Rem. 6, p. 38] is that the premises may be considered to be *pairs of sets of formulas* rather than pairs of sequences of formulas. We will denote these systems by LA^{++} and LC^{++}.

Remark 3.11. We note that LC^{++} is just one short step from *analytic tableaux*, especially, from a version where each node carries the set of all formulas from the previous step. Signs prefixed to formulas—such as $\mathfrak{t}$ and $\mathfrak{f}$—could indicate whether the formula occurs on the left of on the right from $\Vdash$.

In [22, §6], Curry states the decidability of LA and LC. Of course, these are not new results, but the *proof method* Curry uses to establish them is *new*. We quote his reasoning first.

> ... certain decidability properties should be expected ... each of the rules O derives a more complex statement from simpler ones, and the complexity once introduced can never be got rid of at a later stage. ... It ought to be possible by examining an elementary statement to determine what rules it could be a consequence of and from what premises; ... — [22, p. 39]

The sequent calculi LA and LC, as well as their versions are formulated without a cut rule (so far). Thus, once the rules are inspected, it is *immediate* that each calculus has the subformula property.

Lemma 3.12 (Subformula property). *In a proof of $\Gamma \Vdash \Theta$ in LA (LC, LA^+, LC^+, LA^{++} and LC^{++}), if A occurs in a sequent in the proof, then A occurs as a formula or as a subformula of a formula in $\Gamma \Vdash \Theta$.*

Theorem 3.13 (Decidability). *LA and LC are* decidable.

Proof. This is, essentially, Curry's Theorem 7 in [22, §6]. Given the equivalent formulations of the systems, the subformula property and the finiteness of the set of subformulas of all formulas in any sequent, we can start to search for a proof in a *bottom-up* fashion. Curry gives some examples where he relies on the modified versions of the systems, and he also suggests amending the proof search to shorten it. Using the $^{++}$ formulations, it is obvious that after finitely many steps either all the premises are instances of (Id), or no new sequent can be obtained, hence, the search has failed. □

We give a detailed example with a simple formula. (This is not Curry's example.)

Example 3.14. $\Vdash ((A \lor B) \land (C \lor B)) \supset ((A \land C) \lor B)$ is a theorem of LC; it expresses the distributivity of $\lor$ over $\land$ (in lattices). Here is a proof in LC.

$$\dfrac{\dfrac{\dfrac{\dfrac{\dfrac{\dfrac{\dfrac{\dfrac{A \Vdash A, B \quad B \Vdash A, B}{A \lor B \Vdash A, B}\,(\lor \Vdash)}{(A \lor B) \land (C \lor B) \Vdash A, B}\,(\land_1 \Vdash) \qquad \dfrac{\dfrac{C \Vdash C, B \quad B \Vdash C, B}{C \lor B \Vdash C, B}\,(\lor \Vdash)}{(A \lor B) \land (C \lor B) \Vdash C, B}\,(\land_2 \Vdash)}{(A \lor B) \land (C \lor B) \Vdash A \land C, B}\,(\Vdash \land)}{(A \lor B) \land (C \lor B) \Vdash (A \land C) \lor B, B}\,(\Vdash \lor_1)}{(A \lor B) \land (C \lor B) \Vdash B, (A \land C) \lor B}\,(\Vdash C)}{(A \lor B) \land (C \lor B) \Vdash (A \land C) \lor B, (A \land C) \lor B}\,(\Vdash \lor_2)}{(A \lor B) \land (C \lor B) \Vdash (A \land C) \lor B}\,(\Vdash W)}{\Vdash ((A \lor B) \land (C \lor B)) \supset ((A \land C) \lor B)}\,(\Vdash \supset)$$

For the purposes of the proof search, we may use those rules from LC^{++} that permit the largest amount of contraction. Since we "use" the rules in the reverse direction, this means that we will retain as many formulas as possible. This is unproblematic, because (Id) may contain spurious formulas.

For instance, we want to use the left-hand side rule, rather than the right-hand side rule.

$$\dfrac{\Gamma, A, A \land B, B \Vdash \Theta}{\Gamma, A \land B \Vdash \Theta} \qquad\qquad \dfrac{\Gamma, A \land B, B \Vdash \Theta}{\Gamma, A \land B \Vdash \Theta}$$

Now, we give some of the branches of the proof-search tree, which is constructed from its *bottom* toward the top. In order to keep the sequences on the page, we will subscript formulas with φ_n, and repeat only these φ's in lieu of the whole formula in the sequents above. (Π is the proof of $(A \lor B) \land (C \lor B) \Vdash C, \varphi_2, B, \varphi_1, \varphi_0$ which is similar to the left chunk above the line.)

$$\dfrac{\dfrac{\dfrac{\dfrac{\dfrac{\varphi_3, \varphi_4, \varphi_5, A \Vdash A, \varphi_2, B, \varphi_1, \varphi_0 \quad \varphi_3, \varphi_4, \varphi_5, B \Vdash A, \varphi_2, B, \varphi_1, \varphi_0}{\varphi_3, (A \lor B)_{\varphi_4}, (C \lor B)_{\varphi_5} \Vdash A, \varphi_2, B, \varphi_1, \varphi_0}\,(\text{Id})}{((A \lor B) \land (C \lor B))_{\varphi_3} \Vdash A, \varphi_2, B, \varphi_1, \varphi_0}\,(\lor \Vdash) \qquad \Pi}{(A \lor B) \land (C \lor B) \Vdash (A \land C)_{\varphi_2}, B, \varphi_1, \varphi_0}\,(\Vdash \land)}{(A \lor B) \land (C \lor B) \Vdash ((A \land C) \lor B)_{\varphi_1}, \varphi_0}\,(\Vdash \lor)}{\Vdash (((A \lor B) \land (C \lor B)) \supset ((A \land C) \lor B))_{\varphi_0}}\,(\Vdash \supset)$$

The example illustrates "safe choices" for 2-valued logic, because retaining the φ formulas does not preclude getting to an axiom (if that is possible at all). With some squinting, one might find the above proof—again, starting at the root and moving upward—a slightly uneconomical version of an analytic tableaux.

Remark 3.15. LA and LC were formulated without a cut rule. However, the cut rule is important for any reasonable sequent calculus. Arguably, Curry aimed at providing a general theory of formal systems, and as part of it, he invented a *new method* to prove the admissibility of the cut rule in [22, §7]. The elimination of the mix for LJ and LK typically uses very localized modifications on a given proof such as swapping the application of a pair of rules, or completely eliminating an application of the cut rule, when one of its premises is an axiom. Curry's idea is to formally characterize categories of formulas with respect to their role in a proof, as well as certain kinds of rules, and then, perform modifications in sequents throughout the subproof. The proof still proceeds by induction, but it is, perhaps, fair to say that Curry's proof is more abstract than Gentzen's. We will not go into the details of Curry's proof of ET (the Elimination Theorem [22, p. 45]), because it is somewhat tangential to our goals in this paper (as the details of previous cut theorems were).

4 Kleene's G calculi

Kleene wrote a textbook [33], which appeared in 1952 and turned out to be a very influential introductory text for 2-valued and intuitionist logics. Kleene did not concentrate on sequent calculi, yet in Chapter XV he introduced sequent calculi for 2-valued and intuitionist logics. As it was fashionable—following Gentzen—he defined sequent calculi for the two logics so that intuitionist logic turn out to be the logic obtained by some restriction on 2-valued logic.[8]

Curry already "officially" generalized the exchange rule into a permutation rule that can permute several formulas at once. The idea of permutations is quite old; in group theory it has been used in the 19th century. However, the notion of a *multiset*, which can be thought of as an equivalence class of permutations, seem not to have been considered as a legitimate data type until the 1970s or so.[9] Kleene almost (but not quite) invented the notion of a multiset, and Curry de facto used it when he suppressed permutation steps.

Definition 4.1. Let us assume the structural rules of LK, in particular, the rules $(W \Vdash)$, $(\Vdash W)$, $(C \Vdash)$ and $(\Vdash C)$.[10] The sequents $\Gamma_1 \Vdash \Theta_1$ and $\Gamma_2 \Vdash \Theta_2$ are *cognate sequents* if there is a sequent $\Phi \Vdash \Psi$ $\widehat{\text{st}}$ it is derivable from both $\Gamma_1 \Vdash \Theta_1$ and $\Gamma_2 \Vdash \Theta_2$ by applications of the contraction and exchange rules.

[8] An unfortunate side effect of this strategy is that Kleene's labels (e.g., $G3$) refer ambiguously to a particular version of either the 2-valued or intuitionist logic; cf. [33, p. 480]. We add subscripts to $G3$ to distinguish these two sequent calculi "in $G3$-style," which are related but formalize different logics.

[9] See Blizard's [18] for historical remarks.

[10] We did not list $(\Vdash W)$ and $(\Vdash C)$ in LJ, but these rules in LK are just like the $(W \Vdash)$ and $(C \Vdash)$ rules in LJ but operating on the right-hand side of the $\Vdash$, which may contain any finite number of formulas in LK.

Remark 4.2. A way to look at cognate sequents is that they are strings or multisets that setify into the same set. It is easy to see that given $\Gamma \Vdash \Theta$, which is a pair of finite sequences or multisets of formulas, there are infinitely many $\Gamma' \Vdash \Theta'$ that are cognate with $\Gamma \Vdash \Theta$. Also, the shortest sequents that are cognate with $\Gamma \Vdash \Theta$ contain exactly one occurrence of every formula in Γ (on the left-hand side of the $\Vdash$) and exactly one occurrence of every formula in Θ (on the right-hand side of the $\Vdash$).

Although Curry has already noted [22, Remark 6, p. 38] that sets of formulas can be used instead of sequences of formulas in a sequent, Kleene did not change sequences to sets, rather he allowed a certain ambiguity using the concept of cognate sequents.

Definition 4.3. The sequent calculi $G3_k$ and $G3_j$ are defined by the following *axiom* and *rules*, with certain provisos for $G3_j$. First, there is a general restriction that there is at most one formula on the right-hand side of $\Vdash$. Second, the principal formulas of the rules are not required to occur in the premise(s) of the right-introduction rules. Specifically, in $(\Vdash\neg)$, the premise is $A, \Gamma \Vdash \ $. Lastly, in the $(\neg\Vdash)$ rule, the lower sequent is $\neg A, \Gamma \Vdash C$ (instead of $\neg A, \Gamma \Vdash \Theta$), where C is an arbitrary formula or the empty multiset.

$$A, \Gamma \Vdash \Theta, A \quad (\mathrm{Id})$$

$$\frac{\neg A, \Gamma \Vdash \Theta, A}{\neg A, \Gamma \Vdash \Theta}\ (\neg\Vdash) \qquad \frac{A, \Gamma \Vdash \Theta, \neg A}{\Gamma \Vdash \Theta, \neg A}\ (\Vdash\neg)$$

$$\frac{A, A \wedge B, \Gamma \Vdash \Theta}{A \wedge B, \Gamma \Vdash \Theta} \quad \frac{B, A \wedge B, \Gamma \Vdash \Theta}{A \wedge B, \Vdash \Theta}\ (\wedge\Vdash) \qquad \frac{\Gamma \Vdash \Theta, A \wedge B, A \quad \Gamma \Vdash \Theta, A \wedge B, B}{\Gamma \Vdash \Theta, A \wedge B}\ (\Vdash\wedge)$$

$$\frac{A, A \vee B, \Gamma \Vdash \Theta \quad B, A \vee B, \Gamma \Vdash \Theta}{A \vee B, \Gamma \Vdash \Theta}\ (\vee\Vdash) \qquad \frac{\Gamma \Vdash \Theta, A \vee B, A}{\Gamma \Vdash \Theta, A \vee B} \quad \frac{\Gamma \Vdash \Theta, A \vee B, B}{\Gamma \Vdash \Theta, A \vee B}\ (\Vdash\vee)$$

$$\frac{A \supset B, \Gamma \Vdash \Theta, A \quad B, A \supset B, \Gamma \Vdash \Theta}{A \supset B, \Gamma \Vdash \Theta}\ (\supset\Vdash) \qquad \frac{A, \Gamma \Vdash \Theta, A \supset B, B}{\Gamma \Vdash \Theta, A \supset B}\ (\Vdash\supset)$$

Remark 4.4. The list above, however, is not the full specification of the $G3$ calculi, because Kleene permits the *replacement* of any sequent with another sequent that is *cognate* with the former. This move means that the $G3$ calculi contain *generalized thinning* as part of the axiom, and generalized versions of three structural rules, namely, of *permutation*, *contraction* and *expansion*. (Expansion is a special version of thinning, where new occurrences of old formulas may be introduced.)[11]

[11] Kleene mentions a further variation of the $G3$ calculi; arbitrary formulas may be omitted from

Definition 4.5. A proof is *irredundant* if and only if there are no cognate sequents on any branch.

The notion of an irredundant proof is intended to counterbalance the extremely lenient way of keeping track of the number of copies of formulas. This notion of irredundancy is compatible with logics that contain both thinning and contraction rules.

Theorem 4.6. *The logic $G3_j$ is* decidable.

Proof. Kleene's proof combines several steps. First, another sequent calculus ($G2_j$) already has been shown to correspond to intuitionist logic in the sense that every intuitionistic theorem has a cut-free proof in it. Second, there is a match between the rules in the two calculi, since the explicit structural rules of $G2_j$ are built into the connective rules in $G3_j$ (and do not need to be paired). Third, every proof in $G3_j$ that is not irredundant simplifies to one that is, hence, every theorem has an irredundant proof in $G3_j$. The argument for the elimination of redundancies is based on the observation that if $\Gamma \Vdash \Theta$ occurs above $\Gamma' \Vdash \Theta'$, where the two sequents are cognate, then any sequents between these two sequents as well as $\Gamma' \Vdash \Theta'$ itself may be omitted together with any branches rooted in the omitted sequents. The proof remains a proof, because $\Gamma' \Vdash \Theta'$ can be replaced by $\Gamma \Vdash \Theta$ in an application of a rule. Lastly, a decision procedure results from a bottom-up proof search in $G3_j$, when the search is limited to irredundant proofs. It is sufficient to note that there are finitely many formulas in a sequent of the form $\Vdash A$, where A is the purported theorem. Then, there are finitely many sequents that can be obtained from the set of subformulas of A s.t. the sequents are pairwise incognate, or in other words, they belong to different cognation classes.[12] Then the proof-search tree is, obviously, finite. □

Remark 4.7. Kleene's decision procedure does not use Curry's lemma, which focuses on the admissibility of the contraction rules. Rather, Kleene allows using any cognate sequents in place of each other in applications of rules, hence, he simply excludes proofs in which cognate sequents are above each other.

A similar claim and a similar decision procedure can be fabricated for $G3_k$ too. However, the notion of cognate sequences relies on both contraction and thinning (in addition to permutations). It is, perhaps, fair to say that cognate sequents of formulas already have a name—they are called pairs of finite sets of formulas. It is

the premise(s) in application of the rules. The omissions, presumably, should exclude the subalterns, which license an application of the rule itself, otherwise, soundness is lost.

[12]"Incognate" and "cognation class" are Kleene's terms, possibly, his neologisms.

possible that using sets in the formulation of the rules would appear less plausible informally. E.g., from the premise $\{A, A \wedge B\} \cup \Gamma \Vdash \Theta$ and application of $(\wedge_1 \Vdash)$ could yield $\{A, A \wedge B\} \cup \Gamma \Vdash \Theta$ and $\{A \wedge B\} \cup \Gamma \Vdash \Theta$, which would need some motivation. (Of course, the free use of cognate sequents produces the same effect, but perhaps in a less obvious manner.) The thinning rules are, typically, not part of *relevance logics*. Hence, new ideas were required to prove the decidability of some relevance logics, which were the *new logics* introduced in the 1950s.

5 Kripke's lemma

The implicational fragment of the logic of relevant implication, $R_{\to}$ was introduced by Church [20], in 1951. However, Anderson and Belnap focused on the logic of entailment, E in the later part of the decade. The implicational fragment of E (i.e., $E_{\to}$) was denoted at the time by I, and its decision problem was of great interest for several reasons.

Remark 5.1. We use $\to$ for the binary connective that is *entailment* in E and *relevant implication* in R. The $\to$ connective has some similarities to $\supset$ (material implication or intuitionistic implication), but also to Lambek's $/$.

There was no finite characteristic matrix for $E_{\to}$ at hand to bypass the problem via truth tables; indeed, $E_{\to}$ is not finitely valued. A difference between $R_{\to}$ and $E_{\to}$ is that $(A \to (B \to C)) \to (B \to (A \to C))$ is a theorem of $R_{\to}$, but not of $E_{\to}$. The latter has only a restricted version of this formula as its theorem, namely, $(A \to ((B \to D) \to C)) \to ((B \to D) \to (A \to C))$. Although both logics have $(A \to B) \to ((B \to C) \to (A \to C))$ as their theorem, $(A \circ (B \circ C)) \to ((A \circ B) \circ C)$ is not a theorem of $E_{\to}$. The connective $\circ$ is to $\to$ as $\cdot$ is to $/$ in Lambek's calculi, and it is known that $\circ$ may be added to $E_{\to}$ conservatively. So far all calculi—save LN—contained a sequence of formulas (or something even less structured) in the antecedent of a sequent. Given that $E_{\to}$—unlike LN—has $(A \to (A \to B)) \to (A \to B)$ as its theorem, it is not immediately clear how to define a sequent calculus for $E_{\to}$. (See [10] and [11, Ch. 5] for sequent calculi for fragments of E.) Assuming that the problems stemming from the lack of associativity for $\circ$ can be overcome in a sequent calculus formulation, another problem emerges, namely, $A \to (B \to A)$ and $((A \to B) \to A) \to A$ are not theorems, that is, the thinning rules cannot be postulated. Hence, the more relaxed form of the axiom, that is, $A, \Gamma \Vdash \Theta, A$ cannot be included into the system.

Kripke defined sequent calculi for $E_{\to}$ and $S4_{\to}$ too. Moreover, he designed a decision procedure not only for $S4_{\to}$, but also for $E_{\to}$ and $R_{\to}$, as he reported in [35].

5.1 Kripke's correspondence around 1959

According to mail preserved in the Kripke Archive at CUNY, Kripke was in contact (through mail) with several logicians around this time. In particular, Alan R. Anderson, Nuel D. Belnap and Timothy J. Smiley each had written to Kripke (responding to communication from him). In a letter, dated January 8, 1958, Anderson discusses his ideas about adding deontic modalities to some of Lewis's modal logics ($S3$, $S4$ and $S5$). However, Anderson and Belnap's interests seem to had been focused, at the time, on their logic of entailment E (and Ackermann's Π', which is one rule away)—as [3] and [5] show. In Π' and in E, necessity can be defined as $\Box A := (A \to A) \to A$, and this gives an $S4$-like modality. The idea that entailment—as distinguished from a mere conditional—has a modal component goes back to at least Lewis and his work on strict implication.

The axiomatic system Π' was defined by Wilhelm Ackermann in [1], and it has four rules. The implicational axioms with detachment give implicational ticket entailment, $T_{\to}$, and this logic T (including other usual connectives) was defined by Anderson in [4]. Ackermann's third rule (γ) is omitted both from T and E, but the latter retains the fourth rule (δ). Anderson and Belnap re-axiomatized E to replace (δ) with an axiom, which strongly suggests that the permutation of the antecedents of an implication (i.e., $(A \to (B \to C)) \to (B \to (A \to C)))$ is not a theorem of E, and—indeed—it is not. In a natural deduction system, dependencies between assumptions and further formulas can be expressed using some sort of indexing on the formulas (see e.g., [3]). Let us reiterate that in the absence of permutation, indeed, even of the associativity for fusion (which was not a connective in Π'), a sequent calculus formulation for E or $E_{\to}$ does not seem to be forthcoming.

In a letter to Kripke, written on May 31, 1959, Belnap describes LI, which is the merge formulation of the implicational fragment of E.[13] It seems that Kripke must have been experimenting with combining the idea of subscripting formulas (as is done in some natural deduction systems) with a sequent calculus. Belnap writes in his letter:

> I find your use of subscripts very interesting, and I think fruitful. It is quite different from anything I have tried, inasmuch as it never occurred to me to assign sets of subscripts to the consequent. My original essays along the lines of subscripting were in the nature of permutation-controls in the antecedent (under a nesting-of-entailments interpretation). I subsequently abandoned these in favor of the notion of the merge, chiefly with an eye on simplicity of proof

[13] A closely related calculus $L_{\mu}E_{\to}$ is in [6, §7.3], together with merge calculi for two other relevance logics ($T_{\to}$ and $R_{\to}$, as well as for $S4_{\to}$ and $J_{\to}$).

of ET.

In the same letter, Belnap called the decidability problem for LI (that is, $E_{\rightarrow}$) to Kripke's attention. We quote him again:

> I have been able to find no decision procedure for LI. Gentzen's techniques seem to require that we have A→B→.(A→.A→B), a theorem which fails for I and LI: contraction is the stumbling block. ... Of peripheral interest is the fact that one can get the L-formulation of the pure implicational fragment of S4 by permitting primes of the form α, A ⊩ A. For this system we would have a decision procedure, as also for the system like LI but without W.

By mid June, 1959, Kripke had a sequent calculus (with no subscripts) for the pure strict implication part of $S4$. In a letter to Belnap (June 13th, 1959), Kripke specifies the rule Pr with a side condition that X comprises implicational formulas (both for $S4$ and I). That is, the right introduction rule for $\rightarrow$ is

$$\frac{X, A \Vdash B}{X \Vdash A \rightarrow B}\ \text{Pr}, \text{ where all elements of } X \text{ have the form } C \rightarrow D.$$

Imposing side conditions on a connective rule in a sequent calculus is not unheard of (though it is not very common). For example, Curry's formulation of the absolute calculus (i.e., J_+) with multiple right-hand sides (LA_m in [23, 5C4]) has one side condition, and it pertains to Pr.

Kripke's letter also shows that he was thinking about a sequent calculus for E_+. His attention was on the problem of the distributivity of $\wedge$ and $\vee$, and how different variants of the $\wedge$ and $\vee$ rules interact.

Some three months later, Kripke had the *decidability proof for I*. In a letter to Belnap, dated September 21, 1959, he outlined the proof. The letter, which is merely 2 pages long, contains the crucial component in the proof that has become known as "Kripke's lemma." Before we leave this brief historical excursion into the contents of the Kripke Archive, we should mention that Kripke was aware that the lemma was a more general statement (not specific to sequent calculi) and could be stated in terms of positive integers. He also mentioned that the lemma does not permit (in general) the calculation of an upper bound on the length of irredundant sequences of sequents, though it *guarantees finiteness.* As a result, Kripke conjectured, that the whole decision procedure is general recursive, rather than primitive recursive. The lemma is stated and its proof is presented in [6, pp. 138–139], where it is attributed to Kripke with a mention of a letter, which must be the letter we just mentioned.[14]

[14] As far as I know, Kripke did not publish anything about his decidability results after the

5.2 Church's weak implication is decidable

Church who preferred his λI-calculus over the λK-calculus, introduced, what he called "weak implication" by replacing $A \to (B \to A)$ by $A \to A$ in an axiomatization of $J_\to$ in [20]. Nowadays, this logic is better known as the *implicational fragment of the logic of relevant implication* or $R_\to$. We will illustrate the use of Kripke's lemma (together with the new notion of irredundancy) on this logic. First, $R_\to$ extends $E_\to$ with the permutation of the antecedents of an implication (which does not spoil the connection in information between premises and conclusions). Second, $R_\to$ can be formulated as a sequent calculus without side conditions on $(\Vdash\to)$, which allows us to focus on the essence in Kripke's decidability proofs. Third, the decidability of $R_\to$ led to further decidability results, some of which we will mention in Section 6. Fourth, although Belnap pointed Kripke's attention to $E_\to$, which was his and Anderson's favorite logic, Kripke's decidability proof for $E_\to$ is easily seen to simplify to a decidability proof for $R_\to$—cf. [35].

Definition 5.2. The logic $R_\to$ is defined by (A1)–(A4) and the rule of detachment (from A and $A \to B$, infer B). The axioms are (A1) $A \to A$, (A2) $(A \to B) \to ((C \to A) \to (C \to B))$, (A3) $(A \to (B \to C)) \to (B \to (A \to C))$ and (A4) $(A \to (A \to B)) \to (A \to B)$.

Remark 5.3. The notion of a proof is the usual one for axiomatic calculi, that is, there is no requirement that all the formulas in a proof—save the last formula—be used in the proof. However, the axioms do not include the principal type schema of the combinator K. The above axioms can be conceptualized by saying that $R_\to$ is implicational IBCW.[15]

Definition 5.4. The sequent calculus $LR_\to$ is defined by the following axiom and four rules. (A sequent is a pair of a multiset and a formula, that is, it is of the form $\Gamma \Vdash A$.)

$$A \Vdash A \ \text{(Id)} \qquad \frac{\Gamma \Vdash A \quad B, \Theta \Vdash C}{A \to B, \Gamma, \Theta \Vdash C} \ (\to\Vdash) \qquad \frac{A, \Gamma \Vdash B}{\Gamma \Vdash A \to B} \ (\Vdash\to)$$

$$\frac{A, A, \Gamma \Vdash B}{A, \Gamma \Vdash B} \ (W\Vdash) \qquad \frac{\Gamma \Vdash C \quad C, \Theta \Vdash B}{\Gamma, \Theta \Vdash B} \ \text{(cut)}$$

abstract [35], which appeared in the *Journal of Symbolic Logic* in 1959. Belnap and Wallace in [7] (published as [8]) applied Kripke's method to prove the decidability of the negation–implication fragment of E.

[15] The axioms for these combinators are $\mathsf{K}xy \rhd x$, $\mathsf{I}x \rhd x$, $\mathsf{B}xyz \rhd x(yz)$, $\mathsf{C}xyz \rhd xzy$ and $\mathsf{W}xy \rhd xyy$.

Remark 5.5. The notion of a proof in this sequent calculus is usual. The axioms of $R_{\to}$ are provable in the sense that if A is an axiom, then $\Vdash A$ has a proof.

The cut rule helps to emulate applications of the detachment rule in a proof in the axiomatic system. To prove that if $LR_{\to}$ proves a formula, then so does $R_{\to}$, it is useful to start with noting that the Elimination Theorem (ET) holds. Then it remains to show that the moves sanctioned by the three remaining rules can be mimicked in $R_{\to}$. A sequent $\Gamma \Vdash A$ can be taken to be $\overrightarrow{\Gamma} \to A$, where the elements of Γ are the antecedents of the implication in an arbitrary order of the formulas.

For the proof of ET, it is convenient (though not necessary) to use a version of the cut that is intermediate between the single cut rule and the mix rule; let us call this multi-cut. Namely, in the lower sequent of the cut rule, an arbitrary number of occurrences of C (the cut formula) may be omitted from Θ. The equivalence of the single cut rule and of the multi-cut rule is immediate.

As Belnap stressed in his letter (quoted above), the issue for a decision procedure is the rule $(W \Vdash)$. However, the only rule that allows an increase in the number of formulas in the antecedent (once cut is counted out) is $(\to \Vdash)$. Thus, if contraction can be built into that rule cautiously, then one could hope to be able to control the size of a bottom-up proof-search tree.

Definition 5.6. The sequent calculus $[LR_{\to}]$ is defined with the *axiom* (Id), and the *rule* $(\Vdash \to)$ as in 5.4, with the following rules added.

$$\frac{\Gamma \Vdash A \quad B, \Theta \Vdash C}{[A \to B, \Gamma, \Theta] \Vdash C}\ ([\to \Vdash]) \qquad\qquad \frac{\Gamma \Vdash C \quad C, \Theta \Vdash B}{[\Gamma, \Theta^-] \Vdash B}\ ([\text{cut}]),$$

where Θ^- is Θ from which an arbitrary number of occurrences of C have been omitted.

The bracketing in the rules indicates that some contractions are *permitted* (but not prescribed). For $([\to \Vdash])$ we have the following: if D is not $A \to B$, and D occurs both in Γ and Θ, then $[A \to B, \Gamma, \Theta]$ may omit an occurrence of D. If $A \to B$ occurs in Γ or Δ, then $[A \to B, \Gamma, \Theta]$ may omit an occurrence of $A \to B$. Lastly, if $A \to B$ occurs both in Γ and Θ, then two occurrences of $A \to B$ may be omitted from $[A \to B, \Gamma, \Theta]$. For ([cut]), if D occurs in Γ and Θ, then the number of occurrences of D in $[\Gamma, \Theta]$ may be the sup of those in Γ and Θ separately.

Remark 5.7. The amount of permitted contractions in $([\to \Vdash])$ may be specified slightly differently than above, where we followed Dunn [25, §3.6]. Belnap preferred a more relaxed version of the rules, which he suggested to Kripke in his letter of September 28, 1959 (Kripke Archive, CUNY). Let D occur m times in Γ and n times in Θ. Then the number of occurrences of D in $[\Gamma, \Theta]$ is p, where $\max(m, n) \leq p \leq m + n$. (See also [8, p. 279].)

The latter form of the side condition (sometimes) allows one to "postpone" contractions. On the other hand, the former condition rhymes with the slogan that formulas may be contracted if they could not have been contracted before.

The calculus $[LR_{\to}]$ is intended as a formalization of $R_{\to}$. The cut theorem holds, and so one could consider establishing the equivalence of the logics directly. However, it is perhaps easier to show the equivalence of the sequent calculi. It is obvious that if $[LR_{\to}]$ proves A, then so does $LR_{\to}$ (without cut). To prove the other direction, we can use Curry's lemma.

Lemma 5.8 (Curry's lemma for $[LR_{\to}]$). *If $\Gamma \Vdash A$ is provable in $[LR_{\to}]$ with the height of the proof being n, and $\Gamma' \Vdash A$ could be obtained from $\Gamma \Vdash A$ by zero or more applications of the rule $(W \Vdash)$, then $\Gamma' \Vdash A$ is provable in $[LR_{\to}]$ and it has a proof with height less than or equal to n.*

Proof. We proceed by induction on the height of the proof, considering at each step how the sequent has been obtained.

The case of the axiom is vacuous, and the case of $(\Vdash \to)$ is by an application of the hypothesis of the induction. Thus, we turn to the next case, which is the crucial one.

3. If the last rule applied in a proof is $([\to \Vdash])$, then if $\Gamma' \Vdash A$ is by zero contraction, then the claim obviously holds. There are two possibilities that we consider, namely, a contraction is on the principal formula of the rule or on another formula. (The combination of these can be handled by the combination of the two subcases.)

Let the principal formula be $B \to C$. If it has more than one occurrence in the antecedent of $[B \to C, \Gamma, \Theta] \Vdash A$, we may consider the origin of the non-principal occurrences. The inductive hypothesis ensures that multiple occurrences of $B \to C$ in Γ as well as in Θ can be reduced to a single occurrence. Then by an application of the rule $([\to \Vdash])$, the number of occurrences of $B \to C$ may be reduced to *one*.

If let us say D is not $B \to C$, and D has multiple occurrences in $[B \to C, \Gamma, \Theta]$ in the sequent, then by inductive hypothesis, we can have the number of occurrences of D be reduced to (at most) one in Γ and in Θ. If D did not occur in Γ or Θ, then this suffices for the truth of the claim. If D occurred in the antecedent of both premises, then an application of $([\to \Vdash])$ can further reduce the number from two to *one*. □

Remark 5.9. The proof is very similar to the proof of Lemma 3.6. However, a difference is that the principal formula is *not required* to occur in the premises and the contractions are *not mandatory.* That is, if we have five $B \to C$'s in Γ, and four in Θ, then we still can prove—using the rule ($[\to \Vdash]$)—the sequent $(B \to C)^n, \Gamma^{-(B\to C)}, \Theta^{-(B\to C)} \Vdash A$, where $n = 10$ (or any other n st $1 \leq n < 10$).

For the decision procedure we need Kripke's notion of irredundancy, which is different than the earlier notion.

Definition 5.10. A sequence of cognate sequents is *irredundant* when earlier elements of the sequence are neither identical to later ones nor can be obtained by contractions from later elements in the sequence.

Remark 5.11. It may be useful to consider what irredundancy would mean in a *proof tree.* Let us assume that the root is $\Gamma \Vdash A$, and we have a branch that comprises sequents some of which may be cognate. If we select a particular cognation class of sequents, then the subsequence in the branch will be *redundant* if there is a pair of sequents $\Theta_1 \Vdash B$ and $\Theta_2 \Vdash B$ st $\Theta_1 \Vdash B$ is *above* $\Theta_2 \Vdash B$ in the proof tree and $\Theta_2 \Vdash B$ is the result of contractions on Θ_1 (or Θ_1 and Θ_2 are identical). In other words, contractions may not be postponed in a proof if the branches are to be irredundant with respect to every cognation class exemplified in the proof.

The *proof-search tree,* however, is built from the root sequent $\Gamma \Vdash A$ upward. And the order of sequents in a sequence (in Definition 5.10) should be the order in the proof-search tree, that is, $\Theta_2 \Vdash B$ precedes $\Theta_1 \Vdash B$. So if the sequence is irredundant, then $\Theta_2 \Vdash B$ cannot be a contracted version of $\Theta_1 \Vdash B$.

Lemma 5.12 (Kripke's lemma). *An irredundant sequence of cognate sequents is* finite.

We do not repeat the proof of this lemma; a readily available detailed proof by induction is in [6, p. 139] (which is credited to Kripke).

Theorem 5.13. $R_\to$ *is* decidable.

Proof. (Sketch) We outline the components of the proof, so that it can be generalized to other logics.

1. $R_\to$ has a sequent calculus formulation, $LR_\to$ in which the cut rule is admissible.
2. $LR_\to$ has a modified version $[LR_\to]$, in which the cut rule is admissible in $[LR_\to]$.
3. Curry's lemma holds for $[LR_\to]$, that is, contraction is admissible with no increase in the height of the proof tree.
4. Given a sequent $\Gamma \Vdash A$, there is a finite proof-search tree, which either contains a proof of the sequent (if the sequent is provable), or it does not (if it is not provable).

The proof-search tree is built from the given sequent upward by iteratively adding sequents that could be premises of rules that could result in the sequent. Branches are discontinued if they result in an instance of the axiom, if they would become irredundant or if there is no way to obtain the top sequent from some premises by the rules.

(1) The cut theorem guarantees that it is sufficient to search for cut-free proofs. (2) Sequents are finite, hence, contain finitely many formulas. (3) Finitely many formulas have finitely many subformulas (because there are no quantifiers). (4) Each rule has finitely many premises (because there are no rules similar to an ω-rule). (5) There are finitely many cognation classes that appear in the proof-search tree.

Given (1)–(5), the only way the proof-search tree could be infinite—according to Kőnig's lemma about trees—is by having an infinite branch. This possibility is what is excluded by Kripke's lemma. □

Remark 5.14. Some components of the proof are easily transferable to other logics, including the use of Kripke's lemma. To put it succinctly, given a sequent calculus formalization of a (propositional) logic, one needs to define an equivalent sequent calculus in which contractions that were part of the first calculus are admissible. The lemmas that have to be proved for each such calculus is the *admissibility of cut* and *Curry's lemma*.

The fact that Kripke's lemma is equivalent to some lemmas, and in this sense, it is not specific to each sequent calculus in itself, does not diminish the ground breaking role it played in the expansion of the range of logics that could be proved decidable using sequent calculi.

6 Extension of the Curry–Kripke method

The Curry–Kripke method has been extended into several directions. We briefly indicate three of them.

Additional connectives. We mentioned above that Belnap and Wallace proved the implication–negation fragment of E decidable in the early 1960s. The addition of $\wedge$ and $\vee$ to $LR_{\rightarrow}$ is not difficult, however, the usual rules without thinning do not permit the proof of the R theorem $(A \wedge (B \vee C)) \rightarrow ((A \wedge B) \vee (A \wedge C))$. The latter problem was known to Belnap and to Kripke. It was solved by Dunn in [24], where two structural connectives were used together with $\boldsymbol{t}$ (intensional truth). A careful use of $\boldsymbol{t}$ prevents proofs of sequents that express irrelevancies to be constructed by a detour through cut. It was a hope of some relevance logicians in the mid 1960s that the solution to the decidability problem of R (and perhaps, E too) was to be found through the decidability of their non-distributive versions. Robert K. Meyer added

$\wedge$, $\vee$, $\sim$ (negation) and N (necessity) to R and defined a sequent calculus LR^N in his dissertation [39]. Meyer proved his logic, what he called lattice-R, decidable.

The addition of $\boldsymbol{t}$ to $LR_{\rightarrow}$ is relatively easy, and it was carried out in [15], where $LR^{\boldsymbol{t}}_{\rightarrow}$ was proved decidable (as an auxiliary step toward a proof of the decidability of $T_{\rightarrow}$). The addition of $\boldsymbol{t}$ as well as of $\boldsymbol{f}$ to LR was carried out in [17], together with proofs of decidability $LR^{\boldsymbol{tf}}$ (as an auxiliary step toward a proof of the decidability of CLL, classical propositional linear logic).

Emulating proofs. We characterized $R_{\rightarrow}$ as implicational IBCW. $T_{\rightarrow}$, *implicational ticket entailment* can be obtained from $R_{\rightarrow}$ by replacing the simple type of C by the simple type of B', that is, $(A \rightarrow B) \rightarrow ((B \rightarrow C) \rightarrow (A \rightarrow C))$. The three implicational logics $R_{\rightarrow}$, $E_{\rightarrow}$ and $T_{\rightarrow}$ may be compared as follows. $R_{\rightarrow}$ allows all permutations, $E_{\rightarrow}$ permits permuting implicational formulas, whereas $T_{\rightarrow}$ only allows formulas in certain positions to be permuted with re-associating the formula. This may sound a bit abstract, but we should note that it is not possible to characterize permutations by the shape of the formula in $T_{\rightarrow}$, which means that a sequent calculus cannot use multisets (possibly, with a side condition, like Kripke's LI does). These major relevance logics can be enriched with $\boldsymbol{t}$, without creating new theorems in the old language. At the same time the differences between R, E and T correspond to properties of $\boldsymbol{t}$.

Lambek invented the first sequent calculus in which the structural connective was not associative. Along similar lines, and essentially using $\boldsymbol{t}$, [9] introduced a sequent calculus for $LT^{\boldsymbol{t}}_{\rightarrow}$ (and also for the positive fragment of T with $\boldsymbol{t}$, $\circ$, $\leftarrow$, $\wedge$ and $\vee$). Relying on the insight about $\boldsymbol{t}$, [15] introduced a sequent calculus for $R^{\boldsymbol{t}}_{\rightarrow}$, which extends $LT^{\boldsymbol{t}}_{\rightarrow}$ with two rules that involve $\boldsymbol{t}$. In $T^{\circ\boldsymbol{t}}_{\rightarrow}$, $\boldsymbol{t}$ is left identity for $\circ$, but in $R^{\circ\boldsymbol{t}}_{\rightarrow}$, $\boldsymbol{t}$ is left-right identity. The two new rules in $LT^{\textcircled{t}}_{\rightarrow}$ add *exactly the difference* between a left identity and a full identity to the sequent calculus. (The rules do not depend on the presence of $\circ$, but it is easier to describe $\boldsymbol{t}$'s properties with reference to the connective $\circ$ of which $\rightarrow$ is a residual.) [16] defined a procedure to imitate proofs in the sequent calculus $LR^{\boldsymbol{t}}_{\rightarrow}$ by proofs in $LT^{\textcircled{t}}_{\rightarrow}$. With a minor extension of Kripke's result, $LR^{\boldsymbol{t}}_{\rightarrow}$ is decidable. Every theorem of $T^{\boldsymbol{t}}_{\rightarrow}$ has finitely many irredundant proofs in $LR^{\boldsymbol{t}}_{\rightarrow}$, each of which can be transformed into a finite set of proofs in $LT^{\textcircled{t}}_{\rightarrow}$. To find the theorems of $T_{\rightarrow}$ amongst the theorems of $R_{\rightarrow}$, it remains to scrutinize a finite set of proofs to see if there is a proof which does not use the rules that add right identity to $\boldsymbol{t}$.

Subsuming special contractions. We already mentioned that modalities had been added to lattice-R by Meyer. In [36], Kripke introduced rules for $\Diamond$ (possibility), which in the context of $S4$, allowed for a proof of the formulas that are customarily used in the definition of one modal connective in terms of the other and negation (e.g., $\Diamond A := \neg\Box\neg A$). The same rules may be added to lattice-R and its fragments. Furthermore, the structural rules, especially, the thinning and contraction rules may be varied, including restricting the applicability of these rules to modalized formulas. A whole range of logics was investigated in [17], and all the logics in that paper were proved decidable. The logics that do not contain modalized structural rules, really fall into the first kind of extension considered in this section. That is, the main issue is to appropriately formulate the connective rules, as well as their version that permit some contraction, and then, to prove the cut theorem and Curry's lemma.

The logics that contain modalized structural rules, however, require yet another idea. Such logics can be paired with logics in which the same structural rules are not restricted. This matching may lead to new theorems, but cannot eliminate theorems. Thus, the task is again to sift out from the set of theorems those that are not theorems of the subsumed logic. Let us assume that logic LX is subsumed by logic LY. If A is found to be a theorem of LY, then (in all interesting cases) A has more than one subformula. The Curry–Kripke method yields a finite set of irredundant proofs in LY for A. From these proofs, we can define a *heap number* for the formula by totaling up contractions on ancestors of A and selecting the largest number. Then, the number of modalized contractions is bounded by the heap number in a proof-search in LX. (For example, a modalized formula such as $\Box B$ or $\Diamond B$ has at least two subformulas, and B is not modalized if the number is two.) These ideas were used in [12] to prove the decidability of MELL (the multiplicative–exponential fragment of LL). In [13], I proved a series of semi-lattice based logics decidable, including NLL, the normal fragment of linear logic, which was introduced by Kopylov in [34].

Of course, there may be further ways to extend the Curry–Kripke technique to prove even further (propositional) logics decidable. However, there are also known obstacles. There are sequent calculi for the positive fragments of R, E and T, but these logics (i.e., R_+, E_+ and T_+) are known to be undecidable. Informally speaking, the introduction of two kinds of structural connectives with the resulting interaction of two kinds of contractions may be a reason behind the complex nature of these logics.

7 Conclusions

We traced the main changes in sequent calculus based proofs of decidability for various (propositional) logics. Curry introduced most of the new ideas in proof theory in the second half of the 20th century. These range from a conceptually different proof of the cut theorem to replacing sequences with sets or multisets.

For proofs of decidability, Curry's idea of modifying a sequent calculus so that the contraction rules become admissible rules is paramount. Turning the proof-search tree upside down (or alternatively, building the proof-search tree in a bottom-up fashion) is due to Curry. The next crucial step forward is Kripke's lemma together with a new notion of irredundancy, which avoid the need to calculate the number of possible sequents concretely, yet guarantee the finiteness of the proof-search tree. Another innovation introduced by Kripke that is crucial for decidability results in relevance and some other substructural logics is the careful handling of contraction in connective rules—no contractions are obligatory, but some are permitted.

Acknowledgments

I am grateful to Yale Weiss, who is the assistant director of the *Saul Kripke Center* at the City University of New York (CUNY), for kindly providing me with access to some of Saul Kripke's correspondence that is preserved in the Kripke Archives.

References

[1] Wilhelm Ackermann *Begründung einer strengen Implikation*, The Journal of Symbolic Logic 21, 1956, pp. 113–128.

[2] Kazimierz Ajdukiewicz *Die syntaktische Konnexität*, Studia Philosophica 1, 1935, pp. 1–27.

[3] Alan R. Anderson *Completeness theorems for the systems of E of entailment and EQ of entailment with quantification*, Zeitschrift für mathematische Logik und Grundlagen der Mathematik 6, 1960, pp. 201–216.

[4] — *Entailment shorn of modality* (abstract), The Journal of Symbolic Logic 25, 1960, page 388.

[5] Alan R. Anderson, and Nuel D. Belnap Jr. *Modalities in Ackermann's "Rigorous implication"*, The Journal of Symbolic Logic 24, 1959, pp. 107–111.

[6] — Entailment, *The Logic of Relevance and Necessity*, 1, Princeton University Press, Princeton NJ 1975.

[7] — *A decision procedure for the system $E_{\overline{I}}$ of entailment with negation*, Technical Report 11, Contract No. SAR/609 (16), Office of Naval Research, New Haven CT 1961.

[8] — *A decision procedure for the system $E_{\overline{I}}$ of entailment with negation*, Zeitschrift für mathematische Logik und Grundlagen der Mathematik 11, 1965, pp. 277–289.

[9] Katalin Bimbó *Relevance logics*, in: Dov Gabbay, Paul Thagard, and John Woods (eds.). Handbook of the Philosophy of Science 5, *Philosophy of Logic* (edited by Dale Jacquette), Elsevier / North-Holland, Amsterdam, 2007, pp. 723–789.

[10] — *Dual gaggle semantics for entailment*, Notre Dame Journal of Formal Logic 50, 2009, pp. 23–41.

[11] — Proof Theory. Sequent Calculi and Related Formalisms, CRC Press, Boca Raton FL 2015.

[12] — *The decidability of the intensional fragment of classical linear logic*, Theoretical Computer Science 597, 2015, pp. 1–17.

[13] — *On the decidability of certain semi-lattice based modal logics*, in: Renate A. Schmidt, and Cláudia Nalon (eds.) Automated Reasoning with Analytic Tableaux and Related Methods (Proceedings of the 26th International Conference, TABLEAUX 2017, Brasília, Brazil, September 25–28, 2017), Springer International Publishing, Cham 2017 [*Lecture Notes in Artificial Intelligence 10501*], pp. 44–61.

[14] Katalin Bimbó, and J. Michael Dunn Generalized Galois Logics: Relational Semantics of Nonclassical Logical Calculi, CSLI Publications, Stanford CA 2008 [*CSLI Lecture Notes 188*].

[15] — *New consecution calculi for* $R^{t}_{\rightarrow}$, Notre Dame Journal of Formal Logic 53 (4), 2012, pp. 491–509.

[16] — *On the decidability of implicational ticket entailment*, The Journal of Symbolic Logic 78 (1), 2013, pp. 214–236.

[17] — *Modalities in lattice-R*, 2015 [manuscript, 39 pages].

[18] Wayne D. Blizard *Multiset theory*, Notre Dame Journal of Formal Logic 30 (1), 1989, pp. 36–66.

[19] Alonzo Church *A formulation of the simple theory of types*, The Journal of Symbolic Logic 5, 1940, pp. 56–68.

[20] — *The weak theory of implication*, in: A. Menne, A. Wilhelmy, and H. Angstl (eds.) Kontrolliertes Denken, *Untersuchungen zum Logikkalkül und zur Logik der Einzelwissenschaften*, Kommissions-Verlag Karl Alber, Freiburg and Munich 1951, pp. 22–37.

[21] Haskell B. Curry *The combinatory foundations of mathematical logic*, The Journal of Symbolic Logic 7, 1942, pp. 49–64.

[22] — A Theory of Formal Deducibility, University of Notre Dame Press, Notre Dame IN 1950 [*Notre Dame Mathematical Lectures 6*].

[23] — Foundations of Mathematical Logic, McGraw-Hill Book Company, New York 1963. (Revised reprint by Dover Publications, New York 1977).

[24] J. Michael Dunn *A 'Gentzen system' for positive relevant implication* (abstract), The Journal of Symbolic Logic 38, 1973, pp. 356–357.

[25] — *Relevance logic and entailment*, in: Dov Gabbay, and Franz Guenthner (eds.), Handbook of Philosophical Logic 3, D. Reidel Publishing Company, Dordrecht 1986 [first edition], pp. 117–224.

[26] J. Michael Dunn, and Greg Restall *Relevance logic*, in: Dov Gabbay, and Franz Guenthner (eds.), Handbook of Philosophical Logic, vol. 6, Kluwer, Amsterdam 2002 [second edition], pp. 1–128.

[27] Gerhard Gentzen *Untersuchungen über das logische Schließen*, Mathematische Zeit-

schrift 39, 1935, pp. 176–210.

[28] — *Untersuchungen über das logische Schließen* II, Mathematische Zeitschrift 39, 1935, pp. 405–431.

[29] — *Investigations into logical deduction,* American Philosophical Quarterly 1, 1964, pp. 288–306.

[30] — *Investigations into logical deduction* II, American Philosophical Quarterly 2, 1965, pp. 204–218.

[31] Kurt Gödel *Zum intuitionistischen Aussagenkalkül,* in: S. Feferman (ed.) KURT GÖDEL Collected Works 1, Oxford University Press and Clarendon Press, New York & Oxford 1986, pp. 222–225.

[32] Arend Heyting *Die formalen Regeln der intuitionistischen Logik,* Sitzungsberichte der Preußischen Akademie der Wissenschaften (Phys.-Math. Kl.), 1930, pp. 42–56.

[33] Stephen C. Kleene Introduction to Metamathematics, D. Van Nostrand Company Inc., Princeton NJ 1952.

[34] Alexei P. Kopylov *Decidability of linear affine logic,* Information and Computation 164 (1), 2002, pp. 173–198.

[35] Saul A. Kripke *The problem of entailment* (abstract), The Journal of Symbolic Logic 24, 1959, p. 324.

[36] — *Semantical analysis of modal logic I. Normal modal propositional calculi,* Zeitschrift für mathematische Logik und Grundlagen der Mathematik 9, 1963, pp. 67–96.

[37] Joachim Lambek *The mathematics of sentence structure,* American Mathematical Monthly 65 (3), 1958, pp. 154–170.

[38] — *On the calculus of syntactic types,* in: R. Jacobson (ed.), Structure of Language and its Mathematical Aspects, American Mathematical Society, Providence RI 1961, pp. 166–178.

[39] Robert K. Meyer Topics in Modal and Many-valued Logic, PhD Dissertation, University of Pittsburgh, Ann Arbor (UMI), 1966.

[40] Sara Negri, and Jan von Plato Structural Proof Theory, Cambridge University Press, Cambridge UK 2001.

[41] — Proof Analysis, *A Contribution to Hilbert's Last Problem,* Cambridge University Press, Cambridge UK, 2011.

[42] Jacques Riche, and Robert K. Meyer *Kripke, Belnap, Urquhart and relevant decidability & complexity,* in: G. Gottlob, E. Grandjean, and K. Seyr (eds.), Computer Science Logic (Brno, 1998), Springer-Verlag, Berlin & Heidelberg 1999 [*Lecture Notes in Computer Science 1584*], pp. 224–240.

[43] Manfred E. Szabo (ed.) The Collected Papers of Gerhard Gentzen, North-Holland Publishing Company, Amsterdam 1969 [*Studies in Logic and the Foundations of Mathematics*].

NON-DISTRIBUTIVE LOGICS: FROM SEMANTICS TO MEANING

WILLEM CONRADIE
School of Mathematics, University of the Witwatersrand
willem.conradie@wits.ac.za

ALESSANDRA PALMIGIANO
School of Business and Economics, Vrije Universiteit, Amsterdam
Department of Mathematics and Applied Mathematics, University of Johannesburg
alessandra.palmigiano@vu.nl

CLAUDETTE ROBINSON
Department of Mathematics and Applied Mathematics, University of Johannesburg
claudetter@uj.ac.za

NACHOEM WIJNBERG
Amsterdam Business School, University of Amsterdam
College of Business and Economics, University of Johannesburg
n.m.wijnberg@uva.nl

ABSTRACT We discuss an ongoing line of research in the relational (non topological) semantics of non-distributive logics. The developments we consider are technically rooted in dual characterization results and insights from unified correspondence theory. However, they also have broader, conceptual ramifications for the intuitive meaning of non-distributive logics, which we explore.

KEYWORDS Non-distributive modal logic, Polarity-based semantics, Graph-based semantics, Formal Concept Analysis, Rough concepts, Categorization theory.

1 Introduction

The term 'non-distributive logics' (cf. [10]) refers to the wide family of non-classical propositional logics in which the distributive laws $\alpha \wedge (\beta \vee \gamma) \vdash (\alpha \wedge \beta) \vee (\alpha \wedge \gamma)$ and $(\alpha \vee \beta) \wedge (\alpha \vee \gamma) \vdash \alpha \vee (\beta \wedge \gamma)$ do not need to hold. Since the rise of very well known instances such as quantum logic [52], interest in non-distributive logics has been building steadily over the years. This interest has been motivated by insights from a range of fields in logic and neighbouring disciplines. Techniques and ideas have come from areas in pure mathematics such as lattice theory, duality and representation (cf. [68, 28, 48, 1, 61, 47, 43, 34], and more recently [29, 36, 26, 55, 56, 25, 44, 45, 46]), and areas in mathematical logic such as algebraic proof theory (cf. [5, 32, 11]), but also from the philosophical and formal foundations of quantum physics [8, 38, 10], philosophical logic [63], theoretical computer science, and formal linguistics [50, 37, 53].

The present paper discusses an ongoing line of research in the relational (non topological) semantics of non-distributive logics. The developments we discuss are technically rooted in dual characterization results and insights from unified correspondence theory [18, 20, 21]. These developments have also broader, *conceptual* ramifications concerning the *intuitive meaning* of non-distributive logics. Specifically, the slogan from the title, 'from semantics to meaning', intends to convey the idea that, not dissimilarly from the conceptual contribution of Kripke frames to the intuitive understanding of various modal logics, the relational semantics of non-distributive logics can help to illuminate what these logics are about also at a more fundamental and conceptual level. For instance, a natural question is whether the relational semantics of (some) non-distributive logics can provide an intuitive explanation of why, or under which circumstances, the failure of distributivity is a reasonable and desirable feature; i.e. whether a given relational semantics supports one or more intuitive interpretations under which the failure of distributivity is essential to 'correct reasoning patterns' in certain specific contexts. Perhaps even more interestingly, whether relational semantics can be used to unambiguously identify those contexts. Such an intuitive explanation also requires a different interpretation of the connectives $\vee$ and $\wedge$ which coherently fits with the interpretation of the other logical connectives, and which coherently extends to the meaning of axioms in various signatures.

The starting point and background of the present paper is the theory of those non-distributive logics which arise as the logics canonically associated with varieties of normal lattice expansions (cf. Section 2), i.e. algebras based on general lattices with additional connectives of any finite arity, which satisfy certain finite-distributivity properties coordinatewise. Throughout the paper, these logics will be referred to as *normal LE-logics*, or just *LE-logics*. By their definition, LE-logics are presented via their algebraic semantics; however, relational semantics for LE-logics in any signature can be introduced by a process of *dual characterization* which pivots on well-known adjunctions and representation theorems

[7, 61] for *complete* lattices. In Section 4, we illustrate the dual characterization methodology in detail, and apply it to obtain the definitions of two relational semantic frameworks for LE-logics: the *polarity-based* frames and the *graph-based* frames. Although polarity-based and graph-based semantics are tightly connected and stem from the application of the same methodology, they give rise to two radically different intuitive interpretations of LE-logics: namely, the polarity-based semantics supports the interpretation of LE-logics as logics of *formal concepts* (and consequently, of specific LE-signatures such as those of lattice-based normal modal logics as e.g. epistemic logics of categories and concepts); the graph-based semantics supports the idea that LE-logics can be viewed as *hyper-constructivist* logics, i.e. logics in which the principle of excluded middle fails at the *meta-linguistic level* (in the sense that, at any state of a graph-based model, formulas can be satisfied, refuted or *neither*), and hence their propositional base generalizes intuitionistic logic in the same way in which intuitionistic logic generalizes classical logic. Consequently, we will argue that graph-based semantics supports the interpretation of specific LE-signatures such as those of lattice-based normal modal logics as e.g. epistemic logics of *informational entropy*. All this is discussed in Section 4.

In the present paper, we will only scratch the surface of a broad research program, and hint at the existence of an extremely rich conceptual and technical landscape. Rather than giving an exhaustive survey of the existing results in the relational semantics of non-distributive logic, we focus on: (a) highlighting the *methodological* aspects involved in the definition of the relational semantics of LE-logics; this methodology can be applied also to different semantics than the ones discussed in the present paper (and it would be very interesting to discover which possibly different interpretations of LE-logics are supported by other semantics), and (b) substantiating the claim that relational semantics of non-distributive logics can bring the same *benefits* that Kripke semantics has brought to modal logic, both from a technical viewpoint (concerning results such as the finite model property, correspondence theory, transfer results driven by Gödel-McKinsey-Tarski-style translations, Goldblatt-Thomason-style characterization theorems, semantic cut elimination) and from a conceptual viewpoint, concerning the extent to which these logics are suited to model a range of (real-life) situations and phenomena. Finally, the picture already emerging from the preliminary account presented in this paper suggests that, similarly to what happens in the case of modal logics, many important results and insights can be obtained in complete *uniformity*, also across different types of relational semantics, so that various proposals and solutions developed for specific signatures (e.g. Routley-Meyer semantics for substructural logics) can be systematically connected and extended to other signatures.

2 LE-logics

Informally, LE-logics are the logics of (varieties of) lattice expansions: for disjoint sets of connectives $\mathcal{F}$ and $\mathcal{G}$, a *lattice expansion* (abbreviated as LE) is a tuple $\mathbb{A} = (\mathbb{L}, \mathcal{F}^{\mathbb{A}}, \mathcal{G}^{\mathbb{A}})$ such that $\mathbb{L}$ is a bounded lattice, $\mathcal{F}^{\mathbb{A}} = \{f^{\mathbb{A}} \mid f \in \mathcal{F}\}$ and $\mathcal{G}^{\mathbb{A}} = \{g^{\mathbb{A}} \mid g \in \mathcal{G}\}$, such that every $f^{\mathbb{A}} \in \mathcal{F}^{\mathbb{A}}$ (resp. $g^{\mathbb{A}} \in \mathcal{G}^{\mathbb{A}}$) is an n_f-ary (resp. n_g-ary) operation on $\mathbb{A}$. We say that an LE is *normal* if every $f^{\mathbb{A}} \in \mathcal{F}^{\mathbb{A}}$ (resp. $g^{\mathbb{A}} \in \mathcal{G}^{\mathbb{A}}$) is coordinatewise either monotone or antitone, and preserves finite (hence also empty) joins (resp. meets) in each monotone coordinate and reverses finite (hence also empty) meets (resp. joins) in each antitone coordinate. Henceforth, since every LE is assumed to be normal, the adjective will be dropped. An LE as above is *complete* if $\mathbb{L}$ is a complete lattice, and every $f^{\mathbb{A}} \in \mathcal{F}^{\mathbb{A}}$ (resp. $g^{\mathbb{A}} \in \mathcal{G}^{\mathbb{A}}$) preserves arbitrary joins (resp. meets) in each monotone coordinate and reverses arbitrary meets (resp. joins) in each antitone coordinate.

For any algebraic LE-signature as above, the set of formulas of the language $\mathcal{L}_{\mathrm{LE}} := \mathcal{L}_{\mathrm{LE}}(\mathcal{F}, \mathcal{G})$ over a denumerable set Prop of proposition letters is defined recursively as follows:

$$\varphi ::= p \mid \bot \mid \top \mid \varphi \wedge \varphi \mid \varphi \vee \varphi \mid f(\overline{\varphi}) \mid g(\overline{\varphi})$$

where $p \in \mathsf{Prop}$, $f \in \mathcal{F}$ and $g \in \mathcal{G}$.

Example 2.1. *The language $\mathcal{L}_{\mathrm{Bi}}$ of* bi-intuitionistic logic *[62] is obtained by instantiating $\mathcal{F} := \{\succ\}$ and $\mathcal{G} := \{\rightarrow\}$. Both connectives are binary, and are antitone in the first coordinate and monotone in the second one. The language $\mathcal{L}_{\mathrm{IL}}$ of* intuitionistic logic *is the $\{\succ\}$-free fragment of $\mathcal{L}_{\mathrm{Bi}}$. The language $\mathcal{L}_{\mathrm{DML}}$ of* distributive modal logic *(cf. [35], [19]) is obtained by letting $\mathcal{F} := \{\Diamond, \lhd\}$ and and $\mathcal{G} := \{\Box, \rhd\}$; all connectives are unary, and $\Diamond$ and $\Box$ are monotone while $\lhd$ and $\rhd$ negative. The language $\mathcal{L}_{\mathrm{PML}}$ of* positive modal logic *[27] is the $\{\lhd, \rhd\}$-free fragment of $\mathcal{L}_{\mathrm{DML}}$. The language $\mathcal{L}_{\mathrm{BiML}}$ of* bi-intuitionistic modal logic *[69] is obtained by instantiating $\mathcal{F} := \{\succ, \Diamond\}$ and $\mathcal{G} := \{\rightarrow, \Box\}$. The language $\mathcal{L}_{\mathrm{IML}}$ of* intuitionistic modal logic *[31] is the $\{\succ\}$-free fragment of $\mathcal{L}_{\mathrm{BiML}}$. The language $\mathcal{L}_{\mathrm{SDM}}$ of* semi-De Morgan logic *(cf. [65]) coincides with the language of Orthologic [38] and is the $\{\Diamond, \Box, \lhd\}$-free fragment of $\mathcal{L}_{\mathrm{DML}}$. The language $\mathcal{L}_{\mathrm{FL}}$ of the* Full Lambek calculus *[51, 32] is obtained by instantiating $\mathcal{F} := \{e, \circ\}$ with e nullary and $\circ$ binary and monotone in both coordinates, and $\mathcal{G} := \{/_\circ, \backslash_\circ\}$ with $/_\circ$ (resp. $\backslash_\circ$) binary and monotone in its first (resp. second) coordinate and antitone in its second (resp. first) one. The language $\mathcal{L}_{\mathrm{LG}}$ of the* Lambek-Grishin calculus *(cf. [54]) is obtained by letting $\mathcal{F} := \{e, \circ, /_\star, \backslash_\star\}$ and $\mathcal{G} := \{\partial, \star, /_\circ, \backslash_\circ\}$. Here ∂ is nullary, $\star$ is binary and monotone in both coordinates, and $/_\star$ (resp. $\backslash_\star$) binary and monotone in its first (resp. second) coordinate and antitone in its second (resp. first) one. The language $\mathcal{L}_{\mathrm{MALL}}$ of the* multiplicative-additive fragment of linear logic *(cf. [32]) is the $\{\star, /_\star, \backslash_\star, /_\circ\}$-free fragment of $\mathcal{L}_{\mathrm{LG}}$.*

In what follows, we will mainly focus on LE-signatures the 'expansion' part of which consists of unary connectives, and take them as our running examples throughout the remainder of the paper (see [41, 20] for a treatment of arbitrary signatures). The basic framework is given by the basic normal *non-distributive modal logic* $\mathbf{L} := \mathbf{L}_{\mathcal{L}_{\mathrm{DML}}}$, defined as the smallest set of sequents $\varphi \vdash \psi$ in the language $\mathcal{L}_{\mathrm{DML}}$ of distributive modal logic, containing the following axioms:

- Sequents for propositional connectives:

$$p \vdash p, \qquad \bot \vdash p, \qquad p \vdash \top,$$
$$p \vdash p \vee q, \qquad q \vdash p \vee q, \qquad p \wedge q \vdash p, \qquad p \wedge q \vdash q,$$

- Sequents for modal operators:

$$\top \vdash \Box\top \qquad \Box p \wedge \Box q \vdash \Box(p \wedge q) \qquad \Diamond\bot \vdash \bot \qquad \Diamond(p \vee q) \vdash \Diamond p \vee \Diamond q$$
$$\top \vdash \rhd\bot \qquad \rhd p \wedge \rhd q \vdash \rhd(p \vee q) \qquad \lhd\top \vdash \bot \qquad \lhd(p \wedge q) \vdash \lhd p \vee \lhd q$$

and closed under the following inference rules:

$$\frac{\varphi \vdash \chi \quad \chi \vdash \psi}{\varphi \vdash \psi} \qquad \frac{\varphi \vdash \psi}{\varphi(\chi/p) \vdash \psi(\chi/p)} \qquad \frac{\chi \vdash \varphi \quad \chi \vdash \psi}{\chi \vdash \varphi \wedge \psi} \qquad \frac{\varphi \vdash \chi \quad \psi \vdash \chi}{\varphi \vee \psi \vdash \chi}$$

$$\frac{\varphi \vdash \psi}{\Box\varphi \vdash \Box\psi} \qquad \frac{\varphi \vdash \psi}{\Diamond\varphi \vdash \Diamond\psi} \qquad \frac{\varphi \vdash \psi}{\rhd\psi \vdash \rhd\varphi} \qquad \frac{\varphi \vdash \psi}{\lhd\psi \vdash \lhd\varphi}$$

3 Polarity-based and graph-based semantics of LE-logics

In the present section, we discuss two types of relational semantics for LE-logics, starting with the common methodology used to define them, both for specific LE-signatures [36, 12, 14] and for arbitrary ones [20, 41, 24]. This methodology uses adjunctions involving complete lattices to define the interpretation of LE-formulas on relational structures by 'translating' their interpretation on algebras by means of a dual characterization process on which we expand and exemplify in the next subsection (cf. also discussions in [18, 20]). The resulting semantic environments are those we discuss below and in the appendix.

3.1 Polarities, reflexive graphs and complete lattices

Polarities and representation of complete lattices. A *formal context* or *polarity* [33] is a structure $\mathbb{P} = (A, X, I)$ such that A and X are sets and $I \subseteq A \times X$ is a binary relation.

For every polarity $\mathbb{P}$, maps $(\cdot)^{\uparrow} : \mathscr{P}(A) \to \mathscr{P}(X)$ and $(\cdot)^{\downarrow} : \mathscr{P}(X) \to \mathscr{P}(A)$ can be defined as follows: $B^{\uparrow} := \{x \in X \mid \forall a \in A(a \in B \to aIx)\}$ and $Y^{\downarrow} := \{a \in A \mid \forall x \in X(x \in Y \to aIx)\}$. The maps $(\cdot)^{\uparrow}$ and $(\cdot)^{\downarrow}$ form a *Galois connection* between $(\mathscr{P}(A), \subseteq)$ and $(\mathscr{P}(X), \subseteq)$, i.e. $Y \subseteq B^{\uparrow}$ iff $B \subseteq Y^{\downarrow}$ for all $B \in \mathscr{P}(A)$ and $Y \in \mathscr{P}(X)$.

A *formal concept* of $\mathbb{P}$ is a pair $c = ([\![c]\!], (\![c]\!))$ such that $[\![c]\!] \subseteq A$, $(\![c]\!) \subseteq X$, and $[\![c]\!]^{\uparrow} = (\![c]\!)$ and $(\![c]\!)^{\downarrow} = [\![c]\!]$. It follows immediately from this definition that if $([\![c]\!], (\![c]\!))$ is a formal concept, then $[\![c]\!]^{\uparrow\downarrow} = [\![c]\!]$ and $(\![c]\!)^{\downarrow\uparrow} = (\![c]\!)$, that is, the sets $[\![c]\!]$ and $(\![c]\!)$ are *Galois-stable*. The set $\mathbb{L}(\mathbb{P})$ of the formal concepts of $\mathbb{P}$ can be partially ordered as follows: for any $c, d \in \mathbb{L}(\mathbb{P})$,

$$c \leq d \quad \text{iff} \quad [\![c]\!] \subseteq [\![d]\!] \quad \text{iff} \quad (\![d]\!) \subseteq (\![c]\!).$$

With this order, $\mathbb{L}(\mathbb{P})$ is a complete lattice such that, for any $\mathscr{X} \subseteq \mathbb{L}(\mathbb{P})$,

$$\begin{aligned} \bigwedge \mathscr{X} &= \left(\bigcap\{[\![c]\!] \mid c \in \mathscr{X}\}, \left(\bigcap\{[\![c]\!] \mid c \in \mathscr{X}\}\right)^{\uparrow}\right) \\ \bigvee \mathscr{X} &= \left(\left(\bigcap\{(\![c]\!) \mid c \in \mathscr{X}\}\right)^{\downarrow}, \bigcap\{(\![c]\!) \mid c \in \mathscr{X}\}\right). \end{aligned}$$

This complete lattice is referred to as the *concept lattice* $\mathbb{P}^{+}$ of $\mathbb{P}$. Moreover,

Proposition 3.1. *For any polarity $\mathbb{P} = (A, X, I)$, the complete lattice $\mathbb{P}^{+}$ is completely join-generated by the set $\{\mathbf{a} := (a^{\uparrow\downarrow}, a^{\uparrow}) \mid a \in A\}$ and is completely meet-generated by the set $\{\mathbf{x} := (x^{\downarrow}, x^{\downarrow\uparrow}) \mid x \in X\}$.*

Proof. Every formal concept is both of the form $c = (Y^{\downarrow}, Y^{\downarrow\uparrow})$ for some $Y \subseteq X$ and of the form $c = (B^{\uparrow\downarrow}, B^{\uparrow})$ for some $B \subseteq A$. Since $Y = \bigcup\{x \mid x \in Y\}$ and $B = \bigcup\{a \mid a \in B\}$, and $(\cdot)^{\downarrow} : \mathscr{P}(X) \to \mathscr{P}(A)$ and $(\cdot)^{\uparrow} : \mathscr{P}(A) \to \mathscr{P}(X)$, being Galois-adjoint, are completely join-reversing, $[\![c]\!] = Y^{\downarrow} = \bigcap\{[\![\mathbf{x}]\!] \mid x \in Y\} = [\![\bigwedge\{\mathbf{x} \mid x \in Y\}]\!]$ and $[\![c]\!] = B^{\uparrow\downarrow} = (\bigcap\{(\![\mathbf{a}]\!) \mid a \in B\})^{\downarrow} = [\![\bigvee\{\mathbf{a} \mid a \in B\}]\!]$, as required. □

The following theorem is a converse to the proposition above.[1]

Theorem 3.2 (Birkhoff's representation theorem). *Any complete lattice $\mathbb{L}$ is isomorphic to the concept lattice $\mathbb{P}^{+}$ of some polarity $\mathbb{P}$.*

Proof. Let $\mathbb{P} := (L, L, \leq)$, where L is the domain of $\mathbb{L}$ and $\leq$ is the lattice order. Then it is easy to see that the formal concepts of $\mathbb{P}$ are of the form $((\bigwedge X){\downarrow}, (\bigwedge X){\uparrow})$ for any $X \subseteq L$, and since $\mathbb{L}$ is complete, the assignments $a \mapsto (a{\downarrow}, a{\uparrow})$ and $((\bigwedge X){\downarrow}, (\bigwedge X){\uparrow}) \mapsto \bigwedge X$ define a pair of order isomorphisms, inverse to each other, between $\mathbb{L}$ and $\mathbb{P}^{+}$. □

[1] For every partial order $(X, \leq)$ and every $x \in X$, we let $x{\uparrow} := \{y \in X \mid x \leq y\}$ and $x{\downarrow} := \{y \in X \mid y \leq x\}$.

Reflexive graphs and representation of complete lattices. A *reflexive graph* is a structure $\mathbb{X} = (Z, E)$ such that Z is a nonempty set, and $E \subseteq Z \times Z$ is a reflexive relation, i.e. $\Delta \subseteq E$, where $\Delta := \{(z,z) \mid z \in Z\}$. From now on, we will assume that all graphs we consider are reflexive even when we drop the adjective. Any graph $\mathbb{X} = (Z, E)$ defines the polarity $\mathbb{P}_{\mathbb{X}} = (Z_A, Z_X, I_{E^c})$ where $Z_A = Z = Z_X$ and $I_{E^c} \subseteq Z_A \times Z_X$ is defined as $aI_{E^c}x$ iff aE^cx iff $(a,x) \notin E$.

The complete lattice $\mathbb{X}^+$ associated with a graph $\mathbb{X}$ is defined as the concept lattice of $\mathbb{P}_{\mathbb{X}}$. Conversely, for any lattice $\mathbb{L}$, let $\mathsf{Flt}(\mathbb{L})$ and $\mathsf{Idl}(\mathbb{L})$ denote the set of filters and ideals of $\mathbb{L}$, respectively. The graph associated with $\mathbb{L}$ is $\mathbb{X}_{\mathbb{L}} := (Z, E)$ where Z is the set of tuples $(F, J) \in \mathsf{Flt}(\mathbb{L}) \times \mathsf{Idl}(\mathbb{L})$ such that $F \cap J = \varnothing$. For $z \in Z$, we let F_z denote the filter part of z and J_z the ideal part of z. Clearly, filter parts and ideal parts of states of $\mathbb{X}_{\mathbb{L}}$ must be proper. The (reflexive) relation E on Z is defined by zEz' if and only if $F_z \cap J_{z'} = \varnothing$.

3.2 From algebraic to relational semantics

In the present subsection, we discuss how the adjunctions outlined in the previous subsection can be used to define an interpretation of the propositional language of general lattices on polarities and reflexive graphs, starting from the standard interpretation of this logic on complete lattices. The same method will be applied to define polarity-based and graph-based semantics for any LE-language $\mathscr{L}_{\mathrm{LE}} = \mathscr{L}_{\mathrm{LE}}(\mathscr{F}, \mathscr{G})$ from complete $\mathscr{L}_{\mathrm{LE}}$-algebras, via the adjunctions above, suitably expanded to account for the interpretation of the additional connectives, in a uniform and modular way.

This method stems from the observation that interpretations on complex algebras and satisfaction relations on frames *correspond* to one another along the adjunctions outlined above.

Let L be the propositional language of the logic of general lattices. In what follows we will abuse notation and identify the logic with its language. Let us briefly recall how the correspondence between interpretations on complex algebras and satisfaction relations on frames works in the Boolean and distributive settings.

In the Boolean and distributive settings, for any partially ordered set $\mathbb{F} = (W, \leq)$ (in the Boolean case, $\leq$ coincides with the identity Δ_W), and any satisfaction relation $\Vdash \subseteq W \times \mathrm{L}$ between elements of $\mathbb{F}$ and formulas, an interpretation $\overline{v} : \mathrm{L} \to \mathbb{F}^+$ can be defined[2], which is a lattice homomorphism, and is obtained as the unique homomorphic extension of the equivalent functional representation of the relation $\Vdash$ as a map $v : \mathsf{Prop} \to \mathbb{F}^+$, defined as $v(p) = \Vdash^{-1}[p]$[3]. In this way, interpretations on complete lattices can be derived from

[2] In the Boolean setting, $\mathbb{F}^+$ is the powerset algebra $\mathscr{P}(W)$; in the distributive setting, $\mathbb{F}^+$ is the algebra $\mathscr{P}^{\uparrow}(W)$ of the upward-closed subsets of $\mathbb{F}$.

[3] Notice that in order for this equivalent functional representation to be well defined, we need to assume that the relation $\Vdash$ is $\mathbb{F}^+$-*compatible*, i.e. that $\Vdash^{-1}[p] \in \mathbb{F}^+$ for every $p \in \mathsf{Prop}$. In the Boolean case, every

satisfaction relations, so that for every $w \in W$ and every L-formula φ,

$$w \Vdash \varphi \quad \text{iff} \quad \mathbf{w} \leq \overline{v}(\varphi), \tag{1}$$

where, on the right-hand side, $\mathbf{w} \in J^{\infty}(\mathbb{F}^+)$ is the completely join-irreducible element[4] of $\mathbb{F}^+$ arising from $w \in W$. Conversely, for any such $\mathbb{F}$, any lattice homomorphism $\overline{v} : \mathrm{L} \to \mathbb{F}^+$ gives rise to a satisfaction relation $\Vdash \subseteq W \times \mathrm{L}$ defined as in (1). Instantiating condition (1) according to the syntactic shape of each formula in L, we obtain the familiar satisfaction conditions of L-formulas in the distributive setting; for instance, the satisfaction clause of $\vee$-formulas can be obtained as follows:

$w \Vdash \varphi \vee \psi$	iff	$\mathbf{w} \leq \overline{v}(\varphi \vee \psi)$	definition of $\Vdash$ as in (1)
	iff	$\mathbf{w} \leq \overline{v}(\varphi) \vee \overline{v}(\psi)$	$\overline{v}$ is a homomorphism
	iff	$\mathbf{w} \leq \overline{v}(\varphi)$ or $\mathbf{w} \leq \overline{v}(\psi)$	$\mathbf{w} \in J^{\infty}(\mathbb{F}^+)$ and $\mathbb{F}^+$ distributive
	iff	$w \Vdash \varphi$ or $w \Vdash \psi$	(1) on φ and ψ by induction hypothesis.

To define an interpretation of L on polarities and reflexive graphs, we are going to apply the homomorphism-to-relation direction of the argument illustrated above. That is, for an arbitrary polarity $\mathbb{P} = (A, X, I)$ any homomorphic assignment $\overline{v} : \mathrm{L} \to \mathbb{P}^+$ will give rise to *pairs* of relations $(\Vdash, \succ)$ such that $\Vdash \subseteq A \times \mathrm{L}$ and $\succ \subseteq X \times \mathrm{L}$, so that for every $a \in A$ and $x \in X$, and every L-formula φ,

$$a \Vdash \varphi \quad \text{iff} \quad \mathbf{a} \leq \overline{v}(\varphi), \tag{2}$$

$$x \succ \varphi \quad \text{iff} \quad \overline{v}(\varphi) \leq \mathbf{x}, \tag{3}$$

where, on the right-hand side of the equivalences above, $\mathbf{a} = (a^{\uparrow\downarrow}, a^{\uparrow}) \in \mathbb{P}^+$ and $\mathbf{x} = (x^{\downarrow}, x^{\downarrow\uparrow}) \in \mathbb{P}^+$. For instance, spelling out conditions (2) and (3) for $\vee$-formulas, we obtain the following clauses:

$x \succ \varphi \vee \psi$	iff	$\overline{v}(\varphi \vee \psi) \leq \mathbf{x}$	definition of $\succ$ as in (3)
	iff	$\overline{v}(\varphi) \vee \overline{v}(\psi) \leq \mathbf{x}$	$\overline{v}$ is a homomorphism
	iff	$\overline{v}(\varphi) \leq \mathbf{x}$ and $\overline{v}(\psi) \leq \mathbf{x}$	definition of $\vee$
	iff	$x \succ \varphi$ and $x \succ \psi$	(3) on φ and ψ by induction hypothesis.

relation from W to L is clearly $\mathbb{F}^+$-compatible, but already in the distributive case this is not so: indeed $\Vdash^{-1}[p]$ needs to be an upward- or downward-closed subset of $\mathbb{F}$. This gives rise to the persistence condition, e.g. in the relational semantics of intuitionistic logic.

[4]In the Boolean setting, $\mathbf{w} := \{w\}$; in the distributive setting, $\mathbf{w} := w\uparrow = \{w' \in W \mid w \leq w'\}$.

	$a \Vdash \varphi \vee \psi$	
iff	$\mathbf{a} \leq \overline{v}(\varphi \vee \psi)$	definition of $\Vdash$ as in (2)
iff	$\mathbf{a} \leq \overline{v}(\varphi) \vee \overline{v}(\psi)$	$\overline{v}$ homomorphism
iff	$[\![\mathbf{a}]\!] \subseteq [\![\overline{v}(\varphi) \vee \overline{v}(\psi)]\!]$	definition of order in $\mathbb{P}^+$
iff	$a^{\uparrow\downarrow} \subseteq (([\overline{v}(\varphi)]) \cap ([\overline{v}(\psi)]))^{\downarrow}$	definition of **a** and $\vee$ in $\mathbb{P}^+$
iff	$a \in (([\overline{v}(\varphi)]) \cap ([\overline{v}(\psi)]))^{\downarrow}$	definition of Galois-closure
iff	for all $x \in X$, if $x \in ([\overline{v}(\varphi)]) \cap ([\overline{v}(\psi)])$ then aIx	definition of $(\cdot)^{\downarrow}$
iff	for all $x \in X$, if $x \in ([\overline{v}(\varphi)])$ and $x \in ([\overline{v}(\psi)])$ then aIx	definition of $\cap$
iff	for all $x \in X$, if $x^{\downarrow\uparrow} \subseteq ([\overline{v}(\varphi)])$ and $x^{\downarrow\uparrow} \subseteq ([\overline{v}(\psi)])$ then aIx	definition of Galois-closure
iff	for all $x \in X$, if $\overline{v}(\varphi) \leq \mathbf{x}$ and $\overline{v}(\psi) \leq \mathbf{x}$ then aIx	definition of order in $\mathbb{P}^+$
iff	for all $x \in X$, if $x \succ \varphi$ and $x \succ \psi$ then aIx.	(3) on φ and ψ by ind. hyp.

Notice that, unlike the argument in the distributive setting, we did not need to appeal to join- or (meet-)irreducibility. Reasoning in an analogous way for the remaining connectives, we obtain the following recursive definition of $\Vdash$ and $\succ$ on polarities for all L-formulas:

$a \Vdash \bot$		aIx for all $x \in X$	$x \succ \bot$		always
$a \Vdash \top$		always	$x \succ \top$		aIx for all $a \in A$
$a \Vdash p$	iff	$a \in [\![\overline{v}(p)]\!]$	$x \succ p$	iff	$x \in ([\overline{v}(p)])$
$a \Vdash \varphi \wedge \psi$	iff	$a \Vdash \varphi$ and $a \Vdash \psi$			
$x \succ \varphi \wedge \psi$	iff	for all $a \in A$, if $a \Vdash \varphi$ and $a \Vdash \psi$ then aIx			
$a \Vdash \varphi \vee \psi$	iff	for all $x \in X$, if $x \succ \varphi$ and $x \succ \psi$ then aIx			
$x \succ \varphi \vee \psi$	iff	$x \succ \varphi$ and $x \succ \psi$.			

Likewise, for an arbitrary reflexive graph $\mathbb{X} = (Z, E)$, any homomorphic assignment $\overline{v} : \mathrm{L} \to \mathbb{X}^+$ will give rise to *pairs* of relations $(\Vdash, \succ)$ such that $\Vdash \subseteq Z \times \mathrm{L}$ and $\succ \subseteq Z \times \mathrm{L}$, so that for every $z \in Z$ and every L-formula φ,

$$z \Vdash \varphi \quad \text{iff} \quad \mathbf{z}_s \leq \overline{v}(\varphi), \tag{4}$$

$$z \succ \varphi \quad \text{iff} \quad \overline{v}(\varphi) \leq \mathbf{z}_r, \tag{5}$$

where, on the right-hand side of the equivalences above, $\mathbf{z}_s = (z^{\uparrow\downarrow}, z^{\uparrow}) \in \mathbb{X}^+$ and $\mathbf{z}_r = (z^{\downarrow}, z^{\downarrow\uparrow}) \in \mathbb{X}^+$. For instance, spelling out conditions (4) and (5) for $\vee$-formulas, we obtain the following clauses:

$z \succ \varphi \vee \psi$	iff	$\overline{v}(\varphi \vee \psi) \leq \mathbf{z}_r$	definition of $\succ$ as in (5)
	iff	$\overline{v}(\varphi) \vee \overline{v}(\psi) \leq \mathbf{z}_r$	$\overline{v}$ homomorphism
	iff	$\overline{v}(\varphi) \leq \mathbf{z}_r$ and $\overline{v}(\psi) \leq \mathbf{z}_r$	definition of $\vee$
	iff	$z \succ \varphi$ and $z \succ \psi$.	(5) on φ and ψ by induction hypothesis.

	$z \Vdash \varphi \vee \psi$	
iff	$\mathbf{z}_s \leq \overline{v}(\varphi \vee \psi)$	definition of $\Vdash$ as in (4)
iff	$\mathbf{z}_s \leq \overline{v}(\varphi) \vee \overline{v}(\psi)$	$\overline{v}$ is a homomorphism
iff	$[\![\mathbf{z}_s]\!] \subseteq [\![\overline{v}(\varphi) \vee \overline{v}(\psi)]\!]$	definition of order in $\mathbb{X}^+$
iff	$z^{\uparrow\downarrow} \subseteq ((\![\overline{v}(\varphi)]\!) \cap (\![\overline{v}(\psi)]\!))^{\downarrow}$	definition of $\mathbf{z}_s$ and $\vee$ in $\mathbb{X}^+$
iff	$z \in ((\![\overline{v}(\varphi)]\!) \cap (\![\overline{v}(\psi)]\!))^{\downarrow}$	definition of Galois-closure
iff	for all $z' \in Z$, if $z' \in (\![\overline{v}(\varphi)]\!) \cap (\![\overline{v}(\psi)]\!)$ then $zE^c z'$	definition of $(\cdot)^{\downarrow}$
iff	for all $z' \in Z$, if $z' \in (\![\overline{v}(\varphi)]\!)$ and $z' \in (\![\overline{v}(\psi)]\!)$ then $zE^c z'$	definition of $\cap$
iff	for all $z' \in Z$, if $z^{\downarrow\uparrow} \subseteq (\![\overline{v}(\varphi)]\!)$ and $x^{\downarrow\uparrow} \subseteq (\![\overline{v}(\psi)]\!)$ then $zE^c z'$	definition of Galois-closure
iff	for all $z' \in Z$, if $\overline{v}(\varphi) \leq \mathbf{x}$ and $\overline{v}(\psi) \leq \mathbf{z}'_r$ then $zE^c z'$	definition of order in $\mathbb{P}^+$
iff	for all $z' \in Z$, if $z' \succ \varphi$ and $z' \succ \psi$ then $zE^c z'$	(5) on φ and ψ by ind. hyp.
iff	for all $z' \in Z$, if zEz' then $z' \not\succ \varphi$ or $z' \not\succ \psi$.	contraposition.

Reasoning in an analogous way, we obtain the following recursive definition of $\Vdash$ and $\succ$ on graphs for all L-formulas:

$z \succ \bot$		always	$z \Vdash \bot$		never
$z \Vdash \top$		always	$z \succ \top$		never
$z \Vdash p$	iff	$z \in V(p)$	$z \succ p$	iff	for all z', if $z'Ez$ then $z' \not\Vdash p$
$z \succ \varphi \vee \psi$	iff	$z \succ \varphi$ and $z \succ \psi$			
$z \Vdash \varphi \vee \psi$	iff	for all z', if zEz' then $z' \not\succ \varphi$ or $z' \not\succ \psi$			
$z \Vdash \varphi \wedge \psi$	iff	$z \Vdash \varphi$ and $z \Vdash \psi$			
$z \succ \varphi \wedge \psi$	iff	for all z', if $z'Ez$ then $z' \not\Vdash \varphi$ or $z' \not\Vdash \psi$			

In what follows, when the assignment $\overline{v}$ is clear from the context, we will sometimes write $[\![\varphi]\!]$ and $(\![\varphi]\!)$ for $[\![\overline{v}(\varphi)]\!]$ and $(\![\overline{v}(\varphi)]\!)$, respectively.

3.3 Relational interpretation of additional connectives

In the present subsection, we apply the method discussed in the previous subsection to define polarity-based and graph-based semantics for any LE-language $\mathcal{L}_{\mathrm{LE}} = \mathcal{L}_{\mathrm{LE}}(\mathcal{F}, \mathcal{G})$. Starting from complete $\mathcal{L}_{\mathrm{LE}}$-algebras, we will translate homomorphic assignments of $\mathcal{L}_{\mathrm{LE}}$-formulas into relations $\Vdash$ and $\succ$ via suitable expansions of the adjunctions between complete lattices and polarities and (complete) lattices and reflexive graphs.

Let us first recall how the usual satisfaction relation clauses can be retrieved from the algebraic interpretation in the Boolean and distributive case for a unary diamond $\Diamond$. Let ML be the propositional language of the logic of general lattices expanded with a unary and positive f-type connective $\Diamond$. In what follows, we will abuse notation and identify the minimal normal ML-logic with its language. Let $\mathbb{W} = (W, \leq)$ be a partially ordered set (in the Boolean case, $\leq$ coincides with the identity Δ_W), and let us expand $\mathbb{W}^+$ with a completely join-preserving unary operation $\Diamond^{\mathbb{W}^+}$ so as to obtain a (Boolean or distributive) modal algebra $\mathbb{A} = (\mathbb{W}^+, \Diamond^{\mathbb{W}^+})$. Let $\overline{v} : \mathrm{ML} \to \mathbb{A}$ be a homomorphic assignment, hence $\overline{v}(\Diamond\varphi) = \Diamond^{\mathbb{W}^+}\overline{v}(\varphi)$.

As done in the previous subsection, the recursive definition of the relation $\Vdash\, \subseteq W \times \mathrm{ML}$ corresponding to the assignment $\overline{v}$ is obtained by spelling out equation (1).

For the case of $\Diamond$-formulas, since $\mathbb{W}^+$ is a perfect[5] distributive lattice and $\overline{v}(\Diamond\psi) \in \mathbb{W}^+$, we get $\overline{v}(\psi) = \bigvee\{\mathbf{w}' \in J^\infty(\mathbb{W}^+) \mid \mathbf{w}' \leq \overline{v}(\psi)\} = \bigvee\{\mathbf{w}' \in J^\infty(\mathbb{W}^+) \mid w' \Vdash \psi\}$. Since by assumption $\overline{v}$ is a homomorphism, $\overline{v}(\Diamond\psi) = \Diamond^{\mathbb{W}^+}\overline{v}(\psi) = \Diamond^{\mathbb{W}^+}(\bigvee\{\mathbf{w}' \in J^\infty(\mathbb{W}^+) \mid w' \Vdash \psi\})$, and since $\Diamond^{\mathbb{W}^+}$ is completely join-preserving, we get:

$$\overline{v}(\Diamond\psi) = \bigvee\{\Diamond^{\mathbb{W}^+}\mathbf{w}' \mid w' \Vdash \psi\}. \tag{6}$$

Hence, for any $w \in W$,

$w \Vdash \Diamond\psi$	iff	$\mathbf{w} \leq \overline{v}(\Diamond\psi)$	
	iff	$\mathbf{w} \leq \bigvee\{\Diamond^{\mathbb{W}^+}\mathbf{w}' \mid w' \Vdash \psi\}$	(6)
	iff	$\mathbf{w} \leq \Diamond^{\mathbb{W}^+}\mathbf{w}'$ for some $w' \in W$ s.t. $w' \Vdash \psi$	($\mathbf{w}$ completely join-prime)
	iff	$wR_\Diamond w'$ for some $w' \in W$ s.t. $w' \Vdash \psi$.	

So we have done two things at the same time: Firstly, we have *defined* the accessibility relation $R_\Diamond \subseteq W \times W$ corresponding to the interpretation of $\Diamond$ as $\Diamond^{\mathbb{W}^+}$ as follows:

$$wR_\Diamond w' \quad \text{iff} \quad \mathbf{w} \leq \Diamond^{\mathbb{W}^+}\mathbf{w}'.$$

Secondly, we have derived the corresponding defining clause for $\Diamond$-formulas. The same can be done in the general lattice case, starting e.g. from a polarity $\mathbb{P} = (A, X, I)$ and expanding $\mathbb{P}^+$ with a completely join preserving unary operation $\Diamond^{\mathbb{P}^+}$ so as to obtain a lattice-based complete modal algebra $\mathbb{A} = (\mathbb{P}^+, \Diamond^{\mathbb{P}^+})$. Analogously to the way we argued above and appealing to Proposition 3.1, we can write $\overline{v}(\psi) = \bigvee\{\mathbf{a} \in \mathbb{P}^+ \mid \mathbf{a} \leq \overline{v}(\psi)\} = \bigvee\{\mathbf{a} \in \mathbb{P}^+ \mid a \Vdash \psi\}$. Since by assumption $\overline{v}$ is a homomorphism, $\overline{v}(\Diamond\psi) = \Diamond^{\mathbb{P}^+}\overline{v}(\psi) = \Diamond^{\mathbb{P}^+}(\bigvee\{\mathbf{a} \in \mathbb{P}^+ \mid a \Vdash \psi\})$, and since $\Diamond^{\mathbb{P}^+}$ is completely join-preserving, we get:

$$\overline{v}(\Diamond\psi) = \bigvee\{\Diamond^{\mathbb{P}^+}\mathbf{a} \mid a \Vdash \psi\}. \tag{7}$$

However, the chain of equivalences above breaks down in the third step, since the elements $\mathbf{a}$ are not in general completely join-*prime* anymore, but only complete join-generators (cf. Proposition 3.1). However, we can obtain a reduction also in this case, by crucially making use of the completely meet-generating elements $\mathbf{x}$:

$x \succ \Diamond\psi$	iff	$\overline{v}(\Diamond\psi) \leq \mathbf{x}$	definition of $\succ$ as in (3)
	iff	$\bigvee\{\Diamond^{\mathbb{P}^+}\mathbf{a} \mid a \Vdash \psi\} \leq \mathbf{x}$	(7)
	iff	for all $a \in A$, if $a \Vdash \psi$ then $\Diamond^{\mathbb{P}^+}\mathbf{a} \leq \mathbf{x}$	definition of $\bigvee$
	iff	for all $a \in A$, if $a \Vdash \psi$ then $xR_\Diamond a$,	(8)

[5] A complete lattice $\mathbb{A}$ is *perfect* if it is both completely join-generated by the set $J^\infty(\mathbb{A})$ of its completely join-irreducible elements, and completely meet-generated by the set $M^\infty(\mathbb{A})$ of its completely meet-irreducible elements.

where we have *defined* the accessibility relation $R_\Diamond \subseteq X \times A$ corresponding to the interpretation of $\Diamond$ as $\Diamond^{\mathbb{P}^+}$ as follows:

$$xR_\Diamond a \quad \text{iff} \quad \Diamond^{\mathbb{P}^+}\mathbf{a} \leq \mathbf{x}. \tag{8}$$

Now, using the fact that the set of elements $\mathbf{x}$ for $x \in X$ are meet-generators of $\mathbb{P}^+$, we can write:

	$a \Vdash \Diamond\psi$	
iff	$\mathbf{a} \leq \overline{v}(\Diamond\psi)$	definition of $\Vdash$ as in (2)
iff	$\mathbf{a} \leq \bigwedge\{\mathbf{x} \mid \overline{v}(\Diamond\psi) \leq \mathbf{x}\}$	$\mathbf{x}$ for $x \in X$ meet-generators of $\mathbb{P}^+$
iff	for all $x \in X$, if $\overline{v}(\Diamond\psi) \leq \mathbf{x}$ then $\mathbf{a} \leq \mathbf{x}$	definition of $\bigwedge$
iff	for all $x \in X$, if $x \succ \Diamond\psi$ then aIx	definition of $\succ$ as in (3).

Similar arguments yield the recursive definition of $\Vdash$ and $\succ$ on polarity-based relational structures for formulas in any $\mathscr{L}_{\mathrm{LE}}$-signature, for instance:

$a \Vdash \Box\varphi$	iff	for all $x \in X$, if $x \succ \varphi$ then $aR_\Box x$
$x \succ \Box\varphi$	iff	for all $a \in A$, if $a \Vdash \Box\varphi$ then aIx
$a \Vdash \lhd\varphi$	iff	for all $x \in X$, if $x \succ \lhd\varphi$ then aIx
$x \succ \lhd\varphi$	iff	for all $x' \in X$, if $x' \succ \varphi$ then $xR_\lhd x'$
$a \Vdash \rhd\varphi$	iff	for all $a' \in A$, if $a' \Vdash \varphi$ then $aR_\rhd a'$
$x \succ \rhd\varphi$	iff	for all $a \in A$, if $a \Vdash \rhd\varphi$ then aIx

where the relations $R_\Box \subseteq A \times X$, $R_\lhd \subseteq X \times X$ and $R_\rhd \subseteq A \times A$ are defined as follows:

$$aR_\Box x \text{ iff } \mathbf{a} \leq \Box^{\mathbb{P}^+}\mathbf{x} \qquad xR_\lhd x' \text{ iff } \lhd^{\mathbb{P}^+}\mathbf{x}' \leq \mathbf{x} \qquad aR_\rhd a' \text{ iff } \mathbf{a} \leq \rhd^{\mathbb{P}^+}\mathbf{a}'. \tag{9}$$

More generally, for any connective $f \in \mathscr{F}$ of arity n_f and any connective $g \in \mathscr{G}$ of arity n_g, any interpretation of f and g on $\mathbb{P}^+$ will yield relations $R_f \subseteq X \times \overline{A}^{(n_f)}$ and $R_g \subseteq A \times \overline{X}^{(n_g)}$, where $\overline{A}^{(n_f)}$ denotes the n_f-fold cartesian product of A and X such that for each $1 \leq i \leq n_f$ if f is monotone in its ith coordinate then the ith coordinate of $\overline{A}^{(n_f)}$ is A, whereas if f is antitone in its ith coordinate then the ith coordinate of $\overline{A}^{(n_f)}$ is X, and $\overline{X}^{(n_g)}$ is defined in a similar way w.r.t. g. The relations R_f and R_g are defined as follows:

$$R_f(x, \overline{a}^{(n_f)}) \text{ iff } f^{\mathbb{P}^+}(\overline{\mathbf{a}}^{(n_f)}) \leq \mathbf{x} \qquad R_g(a, \overline{x}^{(n_g)}) \text{ iff } \mathbf{a} \leq g^{\mathbb{P}^+}(\overline{\mathbf{x}}^{(n_g)}). \tag{10}$$

The corresponding clauses for $\Vdash$ and $\succ$ are then the following ones:

$a \Vdash g(\overline{\varphi})$	iff	for all $\overline{x}^{(n_g)} \in \overline{X}^{(n_g)}$, if $\overline{x}^{(n_g)} \succ^{(n_g)} \overline{\varphi}$ then $R_g(a, \overline{x}^{(n_g)})$
$x \succ g(\overline{\varphi})$	iff	for all $a \in A$, if $a \Vdash g(\overline{\varphi})$ then aIx
$x \succ f(\overline{\varphi})$	iff	for all $\overline{a}^{(n_f)} \in \overline{A}^{(n_f)}$, if $\overline{a}^{(n_f)} \Vdash^{(n_f)} \overline{\varphi}$ then $R_f(x, \overline{a}^{(n_f)})$
$a \Vdash f(\overline{\varphi})$	iff	for all $x \in X$, if $x \succ g(\overline{\varphi})$ then aIx.

where, if $\overline{a}^{(n_f)} \in \overline{A}^{(n_f)}$, the notation $\overline{a}^{(n_f)} \Vdash^{(n_f)} \overline{\varphi}$ refers to the conjunction over $1 \leq i \leq n_f$ of statements of the form $a_i \Vdash \varphi_i$, if f is positive in its ith coordinate, or $x_i \succ \varphi_i$, if f is negative in its ith coordinate, whereas if $\overline{x}^{(n_g)} \in \overline{X}^{(n_g)}$, the notation $\overline{x}^{(n_g)} \succ^{(n_g)} \overline{\varphi}$ refers to the conjunction over $1 \leq i \leq n_g$ of statements of the form $x_i \succ \varphi_i$, if g is positive in its ith coordinate, or $a_i \Vdash \varphi_i$, if g is negative in its ith coordinate.

As an example, we instantiate the above general clauses for a binary implication-like connective $\rightarrow$ in $\mathcal{G}$ which is antitone in the first coordinate and monotone in the second. The relation corresponding to this connective is then $R_{\rightarrow} \subseteq A \times A \times X$ such that $R_{\rightarrow}(a,b,x)$ iff $\mathbf{a} \leq \mathbf{b} \rightarrow^{\mathbb{P}^+} \mathbf{x}$, and the clauses for $\Vdash$ and $\succ$ become:

$$
\begin{array}{lll}
a \Vdash \varphi \rightarrow \psi & \text{iff} & \text{for all } b \in A \text{ and } x \in X, \text{ if } b \Vdash \varphi \text{ and } x \succ \psi, \text{ then } R_{\rightarrow}(a,b,x) \text{ and} \\
x \succ \varphi \rightarrow \psi & \text{iff} & \text{for all } a \in A, \text{ if } a \Vdash \varphi \rightarrow \psi \text{ then } aIx.
\end{array}
$$

Likewise, starting from a reflexive graph $\mathbb{X} = (Z,E)$ and a homomorphic assignment $\overline{v}: \mathcal{L}_{\mathrm{LE}} \rightarrow \mathbb{X}^+$ we will extend the relation $\Vdash$ and $\succ$ to the whole of $\mathcal{L}_{\mathrm{LE}}$ via suitable relations associated with each logical connective. We do not expand further on the derivations (some of which are more extensively reported in Section A.1), but limit ourselves to reporting the definition for the unary modalities and the general $f \in \mathcal{F}$ and $g \in \mathcal{G}$.

$$
\begin{array}{lll|lll}
z \succ \Diamond\varphi & \text{iff} & \text{for all } z', \text{ if } zR_{\Diamond}z' \text{ then } z' \nVdash \varphi & z \Vdash \Diamond\varphi & \text{iff} & \text{for all } z', \text{ if } zEz' \text{ then } z' \not\succ \Diamond\varphi \\
z \Vdash \Box\psi & \text{iff} & \text{for all } z', \text{ if } zR_{\Box}z' \text{ then } z' \not\succ \psi & z \succ \Box\psi & \text{iff} & \text{for all } z', \text{ if } z'Ez \text{ then } z' \nVdash \Box\psi \\
z \succ \lhd\psi & \text{iff} & \text{for all } z', \text{ if } zR_{\lhd}z' \text{ then } z' \not\succ \psi & z \Vdash \lhd\psi & \text{iff} & \text{for all } z', \text{ if } zEz' \text{ then } z' \not\succ \psi \\
z \Vdash \rhd\psi & \text{iff} & \text{for all } z', \text{ if } zR_{\rhd}z' \text{ then } z' \nVdash \psi & z \succ \rhd\psi & \text{iff} & \text{for all } z', \text{ if } z'Ez \text{ then } z' \nVdash \rhd\psi
\end{array}
$$

where the relations $R_{\Box}$, $R_{\Diamond}$, $R_{\lhd}$ and $R_{\rhd}$ are defined as follows:

$$
\begin{array}{ll}
zR_{\Diamond}z' \text{ iff } \Diamond^{\mathbb{X}^+}\mathbf{z'}_s \not\leq \mathbf{z}_r & zR_{\Box}z' \text{ iff } \mathbf{z}_s \not\leq \Box^{\mathbb{X}^+}\mathbf{z}'_r \\
zR_{\lhd}z' \text{ iff } \lhd^{\mathbb{X}^+}\mathbf{z'}_r \not\leq \mathbf{z}_r & zR_{\rhd}z' \text{ iff } \mathbf{z}_s \not\leq \rhd^{\mathbb{X}^+}\mathbf{z'}_s.
\end{array}
\tag{11}
$$

As in the case of polarities, we can also generalize these definitions to arbitrary connectives $f \in \mathcal{F}$ and $g \in \mathcal{G}$ with arity n_f and n_g, respectively. Any interpretation of f and g on $\mathbb{X}^+$ will yield relations $R_f \subseteq Z^{n_f+1}$ and $R_g \subseteq Z^{n_g+1}$ defined as follows:

$$
R_f(z,\overline{z}) \text{ iff } f^{\mathbb{X}^+}(\overline{\mathbf{z}}_s^{(n_f)}) \not\leq \mathbf{z}_r \qquad R_g(z,\overline{z}^{(n_g)}) \text{ iff } \mathbf{z}_s \not\leq g^{\mathbb{X}^+}(\overline{\mathbf{z}}_r^{(n_g)}),
\tag{12}
$$

where the i-th component of $\overline{\mathbf{z}}_s^{(n_f)}$ is $\mathbf{z}_s$ (resp. $\mathbf{z}_r$) if f is monotone (resp. antitone) in its i-th argument, and the i-th component of $\overline{\mathbf{z}}_r^{(n_g)}$ is $\mathbf{z}_r$ (resp. $\mathbf{z}_s$) if g is monotone (resp. antitone) in its i-th argument.

Given an n_g-tuple of formulas $\overline{\varphi} = (\varphi_1,\ldots,\varphi_{n_g})$ and $\overline{z} = (z_1,\ldots,z_{n_g}) \in Z^{n_g}$, we write $\overline{z} \succ^{(n_g)} \overline{\varphi}$ to indicate that $z_i \succ \varphi_i$ for all $1 \leq i \leq n_g$ for which g is monotone in the i-th coordinate and that $z_j \Vdash \varphi_j$ for all $1 \leq j \leq n_g$ for which g is antitone in the j-th coordinate.

Similarly, given an n_f-tuple of formulas $\overline{\varphi} = (\varphi_1, \ldots, \varphi_{n_f})$ and $\overline{z} = (z_1, \ldots, z_{n_f}) \in Z^{n_f}$, we write $\overline{z} \Vdash^{(n_f)} \overline{\varphi}$ to indicate that $z_i \Vdash \varphi_i$ for all $1 \leq i \leq n_f$ for which f is monotone in the i-th coordinate and that $z_j \succ \varphi_j$ for all $1 \leq j \leq n_f$ for which f is antitone in the j-th coordinate. Using this notation, the corresponding clauses for $\Vdash$ and $\succ$ are:

$$
\begin{array}{rll}
z \Vdash g(\overline{\varphi}) & \text{iff} & \text{for all } \overline{z} \in Z^{n_g}, \text{ if } R_g(z, \overline{z}) \text{ then it is not the case that } \overline{z} \succ^{(n_g)} \overline{\varphi} \\
z \succ g(\overline{\varphi}) & \text{iff} & \text{for all } z' \in Z, \text{ if } z'Ez \text{ then } z' \not\Vdash g(\overline{\varphi}) \\
z \succ f(\overline{\varphi}) & \text{iff} & \text{for all } \overline{z} \in Z^{n_f}, \text{ if } R_f(z, \overline{z}) \text{ then it is not the case that } \overline{z} \Vdash^{(n_f)} \overline{\varphi} \\
z \Vdash f(\overline{\varphi}) & \text{iff} & \text{for all } z' \in Z, \text{ if } zEz' \text{ then } z' \not\succ f(\overline{\varphi}).
\end{array}
$$

3.4 Projecting onto the classical setting

We finish this section by specifying how the polarity-based and graph-based semantics discussed in the previous subsections project onto the Kripke semantics of classical normal modal logic. For ease of presentation we address this issue in the signature of $\mathcal{L}_{\mathrm{PML}}$.

Algebraically, the polarity-based and graph-based structures $\mathbb{F}$ that can be recognized as "classical" are exactly those the complex algebra $\mathbb{F}^+$ of which is (isomorphic to) a powerset algebra (possibly endowed with extra operations). This is the case of structures based on graphs $\mathbb{X} = (Z, E)$ such that $E = \Delta$, or based on polarities $\mathbb{P} = (A, X, I)$ such that $A = X = Z$ for some set Z, and aIx iff $a \neq x$ for all $a, x \in Z$. In these cases, the powerset algebra $\mathscr{P}(Z)$ can be represented as a concept lattice each element of which is of the form (Y, Y^c) for any $Y \subseteq Z$. Therefore, for any formula φ interpreted in these structures, $(\![\varphi]\!) = \{z \in Z \mid z \succ \varphi\} = [\![\varphi]\!]^c$, and hence $z \succ \varphi$ iff $z \not\Vdash \varphi$. This of course provides a more direct way to reduce $\succ$ to $\Vdash$ than the one defined in terms of the operations $(\cdot)^\uparrow$ and $(\cdot)^\downarrow$, which allows to formulate the well known recursive definition of satisfaction in the classical setting purely in terms of $\Vdash$. In particular, the satisfaction clause for $\vee$-formulas on the "classical" polarity-based models described above can be rewritten as follows: for any $a \in Z$ and all formulas φ and ψ,

$$
\begin{array}{ll}
 & a \Vdash \varphi \vee \psi \\
\text{iff} & \text{for all } x \in X, \text{ if } x \succ \varphi \text{ and } x \succ \psi \text{ then } aIx \\
\text{iff} & \text{for all } x \in X, \text{ if } x \not\Vdash \varphi \text{ and } x \not\Vdash \psi \text{ then } a \neq x \\
\text{iff} & \text{for all } x \in X, \text{ if } a = x \text{ then } x \Vdash \varphi \text{ or } x \Vdash \psi \\
\text{iff} & a \Vdash \varphi \text{ or } a \Vdash \psi.
\end{array}
$$

Likewise, the satisfaction clause for $\bot$ can be rewritten as

$$a \Vdash \bot \quad \text{iff} \quad a \not\succ \bot \quad \text{which is never the case.}$$

As for the interpretation of $\Diamond$-formulas, if $J_{R_\Diamond^c} \subseteq X \times A$ is the corresponding relation on the given "classical" polarity-based models as described above, applying the following clause

$$x \succ \Diamond\psi \quad \text{iff} \quad \text{for all } a \in A, \text{ if } a \Vdash \psi \text{ then } xJ_{R^c_\Diamond}a$$

yields

$$\begin{array}{rll} x \Vdash \Diamond\psi & \text{iff} & x \not\succ \Diamond\psi \\ & \text{iff} & \text{for some } a \in A,\ a \Vdash \psi \text{ and } (x,a) \notin J_{R^c_\Diamond} \\ & \text{iff} & \text{for some } a \in A,\ a \Vdash \psi \text{ and } xR_\Diamond a. \end{array}$$

where $xJ_{R^c_\Diamond}a$ iff $(x,a) \notin R_\Diamond$ for every $x, a \in Z$. As for the interpretation of $\Box$-formulas, stipulating, likewise, that $aI_{R^c_\Box}x$ iff $(a,x) \notin R_\Box$ for every $x, a \in Z$, the satisfaction clause for $\Box$-formulas can be rewritten as follows:

$$\begin{array}{rll} a \Vdash \Box\varphi & \text{iff} & \text{for all } x \in X, \text{ if } x \succ \varphi \text{ then } aI_{R^c_\Box}x \\ & \text{iff} & \text{for all } x \in X, \text{ if } x \nVdash \varphi \text{ then } (a,x) \notin R_\Box \\ & \text{iff} & \text{for all } x \in X, \text{ if } aR_\Box x \text{ then } x \Vdash \varphi. \end{array}$$

Similar computations show that also in the setting of the graph-based semantics, the satisfaction and refutation clauses project to the well known ones. For instance, when zEz' iff $z = z'$ and $z \succ \varphi$ iff $z \nVdash \varphi$,

$$\begin{array}{rll} z \Vdash \varphi \vee \psi & \text{iff} & \text{for all } z' \in Z, \text{ if } zEz' \text{ then } z' \not\succ \varphi \text{ or } z' \not\succ \psi \\ & \text{iff} & \text{for all } z' \in Z, \text{ if } z = z' \text{ then } z' \Vdash \varphi \text{ or } z' \Vdash \psi \\ & \text{iff} & z \Vdash \varphi \text{ or } z \Vdash \psi. \end{array}$$

Analogously, the following clause

$$z \succ \Diamond\varphi \quad \text{iff} \quad \text{for all } z', \text{ if } zR_\Diamond z' \text{ then } z' \nVdash \varphi$$

yields

$$\begin{array}{rll} z \Vdash \Diamond\varphi & \text{iff} & z \not\succ \Diamond\varphi \\ & \text{iff} & \text{for some } z', \text{ both } zR_\Diamond z' \text{ and } z' \Vdash \varphi. \end{array}$$

The remaining computations are omitted.

In conclusion, in the present section we have discussed a uniform methodology for defining relational semantics for normal LE-logics in any signature on the basis of their standard interpretation on complete LE-algebras of compatible signature. We have concretely illustrated how this methodology works in the case of two different semantic environments, and we have discussed how these environments project onto the Kripke semantics of classical normal modal logic. However, in order for these environments to 'make sense' in a more fundamental way, we need to couple them with extra-mathematical interpretations which simultaneously account for the meaning of *all* connectives, and which coherently extend to the meaning of logical axioms and of their first-order correspondents. In the next section, we discuss two such possible interpretations, and the views on LE-logics elicited by each of them.

4 From semantics to meaning

Any extra-mathematical interpretation of LE-logics must account for the failure of distributivity. Although, as discussed in the previous section, polarity-based and graph-based semantic structures arise from the standard interpretation in LE-algebras via the same dual-characterization methodology, they give rise to two radically different views on what LE-logics are and *mean*. The key difference lies in a dichotomy between two interpretive strategies, each of which identifies different sources of non-distributivity. The first such strategy, supported by the polarity-based semantics, drops the interpretation of $\wedge$ and $\vee$ as conjunction and disjunction in natural language and stipulates that LE-formulas do not denote sentences describing states of affairs, but rather, denote objects with a different ontology, such as categories, concepts, questions and theories, to which a truth value might not necessarily be applicable. The second interpretive strategy is supported by the graph-based semantics and retains the sentential denotation of formulas in a context which can be thought of as 'hyper-constructivist'.

4.1 LE-logics as logics of formal concepts, categories, theories, interrogative agendas, and more

Polarities as abstract databases. The idea that lattices are the proper mathematical environment for discussing "especially systems which are in any sense hierarchies" goes back to Birkhoff [6]. Based on this idea, Wille [33] and his collaborators developed Formal Concept Analysis (FCA) as a theory in information science aimed at the formal representation and analysis of conceptual structures. FCA has been applied to a wide range of fields ranging from psychology, sociology, and linguistics to biology and chemistry.

Building on philosophical insights developed by the school of Port-Royal [2], Wille specified concepts in terms of their *extension*, i.e. the set of objects which exemplify the given concept, and their *intension*, i.e. the set of attributes shared by the objects in the extension of the given concept, and identified Birkhoff's polarities $\mathbb{P} = (A, X, I)$ (cf. Section 3.1), also referred to as *formal contexts*, as the appropriate mathematical representation of these ideas. Indeed, a polarity $\mathbb{P}$ can be understood as an abstract representation of a *database*, recording information about a given set A of *objects* (relevant to a given context or situation), and a set X of relevant attributes or *features*. In this representation, the (incidence) relation $I \subseteq A \times X$ records the fact that object $a \in A$ has feature $x \in X$ as aIx. The Galois-adjoint pair of maps $(\cdot)^{\uparrow} : \mathscr{P}(A) \to \mathscr{P}(X)$ and $(\cdot)^{\downarrow} : \mathscr{P}(X) \to \mathscr{P}(A)$ can be understood as *concept-generating maps*: namely, as maps taking any set B of objects to the intension $B^{\uparrow}$ which uniquely determines the formal concept $(B^{\uparrow\downarrow}, B^{\uparrow})$ generated by B, and any set Y of attributes to the extension $Y^{\downarrow}$ which uniquely determines the formal concept $(Y^{\downarrow}, Y^{\downarrow\uparrow})$ generated by Y. Hence, the philosophical and cognitive insight that concepts do not occur in isolation, but rather arise within a hierarchy of other concepts, finds a very natural representation in the construction of the complete lattice $\mathbb{P}^+$ and its natural order as the *sub-concept* relation. Indeed, a subconcept of a given concept, understood as a more restrictive concept, will have a smaller extension (i.e. fewer examples) and a larger intension (i.e. a larger set of requirements that objects need to satisfy in order to count as examples of the given sub-concept). This interpretation accounts for the failure of distributivity, as we will concretely illustrate below.

Propositional lattice logic as the basic logic of formal concepts. Imposing the FCA interpretation of polarities discussed above on the polarity-based semantics of the logic L (Section 3.2) yields an interpretation of L-formulas as terms (i.e. names) denoting formal concepts. Starting from assignments to proposition variables, any L-formula φ is then interpreted on a given polarity $\mathbb{P} = (A, X, I)$ as a formal concept $([\![\varphi]\!], (\![\varphi]\!)) \in \mathbb{P}^+$; specifically, for each object $a \in A$ and feature $x \in X$, the relations $a \Vdash \varphi$ and $x \succ \varphi$ can be respectively understood as 'object a is a member of (or exemplifies) concept φ' and 'feature x *describes* concept φ', in the sense that x is a required attribute of every example/member of φ. Accordingly, this reading suggests that $\varphi \wedge \psi$ can be understood as 'the greatest (i.e. least restrictive) common subconcept of concept φ and concept ψ', i.e. the concept the extension of which is the intersection of the extensions of φ and ψ. Similarly, $\varphi \vee \psi$ is 'the least (i.e. most restrictive) common superconcept of concept φ and concept ψ', i.e. the concept the intension of which is the intersection of the intensions of φ and ψ; the constant $\top$ can be understood as the most generic (or comprehensive) concept (i.e. the one that, when interpreted in any given polarity $\mathbb{P}$ as above, allows all objects $a \in A$ as examples) while $\perp$ as the most restrictive (i.e. the one that, when interpreted in any given polarity $\mathbb{P}$ as above, requires its examples to have all attributes $x \in X$). Finally, $\varphi \vdash \psi$ can be understood as the statement that 'concept φ is a sub-concept of concept ψ'.

As mentioned above, this interpretation accounts for the failure of distributivity. Indeed, objects in the extension of concept $\varphi \vee \psi$ are only required to have all attributes common to the intentions of concepts φ and ψ; this weaker requirement potentially allows objects in $[\![\varphi \vee \psi]\!]$ which belong to neither $[\![\varphi]\!]$ nor to $[\![\psi]\!]$. To illustrate this point concretely, consider the 'database' of theatrical plays $\mathbb{P} = (A, X, I)$ the set of objects of which is $A := \{a, b, c\}$, where a is *A Midsummer Night's Dream*, b is *King Lear*, and c is *Julius Caesar*, while its set of features is $X := \{x, y, z\}$, where x is *'no happy end'*, y is *'some characters are real historical figures'*, and z is *'two characters fall in love with each other'*. The following picture represents $\mathbb{P}$ and its associated concept lattice.

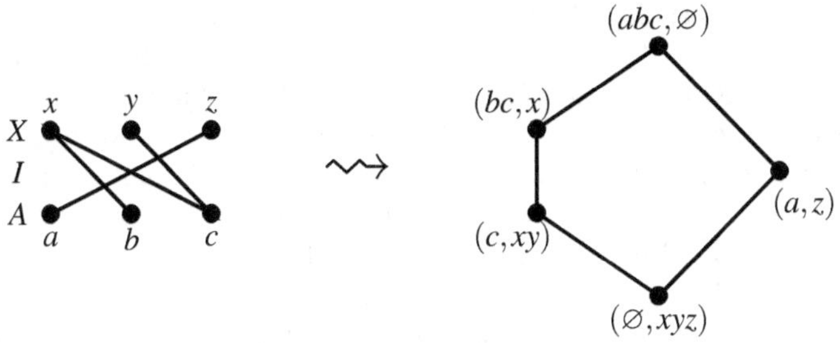

Consider the atomic concept-variables r, d, h where r stands for 'romantic comedy', d for 'drama' and h for 'historical drama'. Consider the assignment into $\mathbb{P}^+$ which maps r to (a,z), d to (bc,x) and h to (c,yz). Notice that $b \in [\![h \vee r]\!] = [\![\top]\!]$ even though $b \notin [\![h]\!] \cup [\![r]\!]$. Accordingly, under this assignment, $h \vee r$ is interpreted as the top element of the concept lattice. The concept $d \wedge r$ is interpreted as the bottom element and, as historical drama is a subgenre of drama, $d \wedge h$ coincides with h. Hence $(d \wedge h) \vee (d \wedge r)$ coincides with h while $d \wedge (h \vee r)$ coincides with d. Thus, distribution of $\wedge$ over $\vee$ fails.

Dually, objects in the extension of concept $\varphi \wedge \psi$ might have more attributes in common than those in $(\![\varphi]\!) \cup (\![\psi]\!)$; for instance, in the example above, $r \wedge d$ is the concept $(\varnothing, xyz)$ which requires feature y which is neither required by r nor by d; hence $h \vee (r \wedge d)$ coincides with h, while $(h \vee d) \wedge (h \vee r)$ coincides with d, witnessing also the failure of the distribution of $\vee$ over $\wedge$.

Lattice-based normal modal logic as an epistemic logic of formal concepts. So far, we have discussed how the polarity-based semantics of propositional lattice logic L allows for an interpretation of L-formulas as names of formal concepts, and for a coherent interpretation of the meaning of all propositional lattice connectives so that the failure of distributivity becomes essential to capturing 'correct reasoning' in the context of conceptual hierarchies. Next, based on [16, 17], we discuss how this interpretation can be extended also to the modal connectives. For the sake of simplicity, let us consider the minimal normal LE-logic $\mathrm{L}_\Box$, in the language specified according to the notation of Section 2 by instantiating $\mathcal{F} := \varnothing$ and $\mathcal{G} := \{\Box\}$, with $\Box$ unary and monotone. In what follows, we will abuse notation and identify $\mathrm{L}_\Box$ with its language. As discussed in Section 3.3, this logic can be interpreted on relational structures $\mathbb{F} = (\mathbb{P}, R_\Box)$ such that $\mathbb{P} = (A, X, I)$ is a polarity, and $R_\Box \subseteq A \times X$ is a (compatible, see Definition A.1) relation such that, for any assignment $v : \mathsf{Prop} \to \mathbb{P}^+$, corresponding relations $\Vdash \subseteq A \times \mathrm{L}_\Box$ and $\succ \subseteq X \times \mathrm{L}_\Box$ can be defined. In the case of $\Box$-formulas, this yields

$$
\begin{array}{lll}
a \Vdash \Box\varphi & \text{iff} & \text{for all } x \in X, \text{ if } x \succ \varphi \text{ then } aR_\Box x \\
x \succ \Box\varphi & \text{iff} & \text{for all } a \in A, \text{ if } a \Vdash \Box\varphi \text{ then } aIx.
\end{array}
$$

Building on the understanding of polarities as abstract representation of databases, the relational structures $\mathbb{F} = (\mathbb{P}, R_\Box)$ can be understood as (abstract representations of) enriched databases which, together with objective information about objects and their features encoded in the incidence relation I of $\mathbb{P}$, encode also *subjective* information regarding whether given objects have given attributes *according to a given agent*; this understanding allows us to read $aR_\Box x$ as 'object a has attribute x *according to agent i*'. Of course, this interpretation can be further specialized so as to represent agents' knowledge ($aR_\Box x$ iff 'agent i knows that object a has attribute x'), beliefs ($aR_\Box x$ iff 'agent i believes that object a has attribute x'), perceptions ($aR_\Box x$ iff 'agent i sees that object a has attribute x'), evidential reasoning ($aR_\Box x$ iff 'agent i has evidence that object a has attribute x'), and so on. Each of these epistemic interpretations will give rise to a different epistemic reading of $\Box\varphi$ as 'concept φ according to the given agent i': namely, 'concept φ as is known/believed/perceived/experienced by agent i'. Also in the case of $\mathrm{L}_\Box$-formulas, for every object a and attribute x, the symbols $a \Vdash \Box\varphi$ and $x \succ \Box\varphi$ can be understood as 'object a is a member/example of $\Box\varphi$' and 'attribute x describes $\Box\varphi$', respectively. Interestingly, the condition that

$$a \Vdash \Box\varphi \quad \text{iff} \quad \text{for all } x \in X, \text{ if } x \succ \varphi \text{ then } aR_\Box x$$

can then be informally understood as saying that any object a is a member/example of concept φ according to agent i if and only if agent i attributes to a all the defining features of concept φ. This reading is indeed coherent with our informal understanding of which objects should count as members of 'concept φ according to agent i'.

Finally, one would also expect that the different variants of epistemic interpretations would satisfy different axioms; for instance, if $\Box\varphi$ is interpreted as 'concept φ as *known* by agent i', one would ask whether there are some $\mathrm{L}_\Box$-axioms which would encode the counterparts, in the lattice-based setting, of well known classical epistemic principles such as the *factivity* condition which distinguishes knowledge from belief, and what these conditions would look like in the context of polarity-based relational structures. As is well known, in the setting of classical normal modal logic, factivity is formalized as the modal axiom $\Box\varphi \to \varphi$ (which reads 'if agent i knows that φ, then φ is indeed the case'). Moreover, this is a Sahlqvist formula and corresponds on Kripke frames (W,R) to R being reflexive or, equivalently, to $\Delta \subseteq R$. Since $\mathrm{L}_\Box$ is not sentential, the closest approximation to the classical formula $\Box\varphi \to \varphi$ is the $\mathrm{L}_\Box$-sequent $\Box\varphi \vdash \varphi$, which turns out (cf. [15, Proposition 4.3]) to correspond on polarity-based structures $\mathbb{F}$ as above to the first-order condition $R_\Box \subseteq I$. That is, for every object a and feature x, if $aR_\Box x$ (i.e. if a is endowed with x according to agent i) then aIx (i.e. object a indeed has feature x). This condition is arguably an appropriate rendering of factivity in the setting of polarity-based relational structures, which suggests that more modal epistemic principles might retain their intended interpretation even under a substantial generalization step such as the one from the classical (i.e. Boolean) to the lattice-based setting. Indeed, this is also the case for *positive introspection*, which in the language of classical modal logic is formalized as $\Box\varphi \to \Box\Box\varphi$ (which reads 'if agent i knows that φ, then agent i knows that she knows φ'). As is well known, this axiom is a Sahlqvist formula the first-order correspondent of which on Kripke frames (W,R) is the condition $R \circ R \subseteq R$, concisely expressing that the relation R is transitive. Again, the $\mathrm{L}_\Box$-sequent $\Box\Box\varphi \vdash \Box\varphi$ turns out (cf. [15, Proposition 4.3]) to correspond on polarity-based structures $\mathbb{F}$ to the first-order condition[6] that reads: for every object a and feature x, if agent i thinks that a has feature x, then (agent i must recognize a as an example of what i understands x-*objects* to be, i.e. as a member of i's understanding of the formal concept generated by feature x, and hence) agent i must attribute to a also all the features that, according to i, are shared by all x-objects. As in the case of factivity, one can argue that this condition is an appropriate rendering of the principle of positive introspection in the setting of polarity-based relational structures, since it is clearly an internal coherence requirement which seeks to justify any given attribution of a feature to an object by linking it to the wider context of those (other) features that are consequences of the given attribution. Lastly, the notion of *omniscience*, stipulating that the agent knows everything that is the case, is classically captured by the axiom $p \to \Box p$ corresponding on Kripke frames to the first-order condition $R \subseteq \Delta$. On polarity-based structures $\mathbb{F}$, the $\mathrm{L}_\Box$-sequent $\varphi \vdash \Box\varphi$ (cf. [15, Proposition 4.3]) corresponds to the first-order condition $I \subseteq R_\Box$, indicating that, whenever an object has a feature, the agent knows this.

Lattice-based normal modal logic as the logic of rough concepts. As discussed above, the interpretation of $\mathrm{L}_\Box$ as an epistemic logic of formal concepts, facilitated by the polarity-based semantics, extends coherently from the meaning of the defining clauses of $\Vdash$ and $\succ$ relative to $\Box$-formulas, all the way to the preservation of the meaning of well known epistemic principles. However, the epistemic interpretation is not the only possible one; in what follows, we give pointers to another family of possible interpretations, proposed in [15], where polarity-based relational structures for the languages $\mathscr{L}_{\mathrm{PML}}$ and $\mathscr{L}_{\mathrm{DML}}$ (cf. Example 2.1) are used to generalize Rough Set Theory (RST) [60] to the setting of *rough concepts*. The basic models in RST are pairs (X,R), called *approximation spaces*, with X a non-empty set and R an equivalence relation on X. The set X is to be thought of as the *domain of discourse* and R as an *indiscernibility relation*. The equivalence classes of R establish the granularity of the discourse by setting the limits to the distinctions that can be drawn. This granularity is captured algebraically by the upper and lower approximation operators arising from approximation spaces, which, when applied to any given subset $T \subseteq X$, encode the available information about T as follows. The lower approximation of $\underline{T}$ of T consists of those elements whose R-equivalence classes are contained in T, while

[6] With the aid of the notation $;_I$ for relational composition modulo the polarity relation I (cf. [15, Section 3.4] for the full definition) this condition can be succinctly captured as $R_\Box \subseteq R_\Box \,;_I R_\Box$.

the upper approximation of T consists of those elements whose R-equivalence classes have non-empty intersection with T. In other words,

$$\underline{T} := \bigcup\{R[z] \mid z \in T \text{ and } R[z] \subseteq T\} \quad \text{and} \quad \overline{T} := \bigcup\{R[z] \mid z \in T\}.$$

The lower approximation $\underline{T}$ can be thought of as the set of all objects that are *definitely* in T, while the upper approximation $\overline{T}$ consists of those objects that are *possibly* in T.

As the reader would have remarked, an approximation space is nothing but a frame for the modal logic S5, and the lower and upper approximation of $T \subseteq S$ are obtained by applying the interior and closure operators given by the S5 box and diamond operators associated with the indiscernibility relation R, respectively. This connection with modal logic has indeed not gone unnoticed in the literature and has been elaborated in e.g. [58], [4] and [59].

In [15], *conceptual approximation spaces* were defined as polarity-based $\mathcal{L}_{\mathrm{PML}}$-structures $\mathbb{F} = (\mathbb{P}, R_\Box, R_\Diamond)$ such that $\mathbb{P} = (A, X, I)$ is a polarity, and $R_\Box \subseteq A \times X$ and $R_\Diamond \subseteq X \times A$ are (compatible, see Definition A.1) relations verifying the first-order conditions corresponding to the following modal axioms: $\Box\varphi \vdash \Diamond\varphi$ (*seriality*)[7]; $\Box\varphi \vdash \varphi$ and $\varphi \vdash \Diamond\varphi$ (*reflexivity*); $\Box\varphi \vdash \Box\Box\varphi$ and $\Diamond\Diamond\varphi \vdash \Diamond\varphi$ (*transitivity*); $\varphi \vdash \Box\Diamond\varphi$ and $\Diamond\Box\varphi \vdash \varphi$ (*symmetry*).

Taken together, these conditions guarantee that $\mathbb{F}^+ := (\mathbb{P}, [R_\Box], \langle R_\Diamond \rangle)$ is a complete lattice-based algebra such that $[R_\Box]$ and $\langle R_\Diamond \rangle$ are an interior and a closure operator, respectively; moreover, $\langle R_\Diamond \rangle$ is the left adjoint of $[R_\Box]$ (i.e. $aR_\Box x$ iff $xR_\Diamond a$ for every $a \in A$ and $x \in X$).

Under the usual interpretation of $\mathbb{P} = (A, X, I)$ as a database, one possible way to understand $aR_\Box x$ or equivalently $xR_\Diamond a$ is 'there is *evidence* that object a has feature x', or 'object a *demonstrably* has feature x' (cf. [15, Section 5.1]). This intuitive understanding makes it plausible to assume that $R_\Box \subseteq I$. Recall that $\Vdash$ for $\Box$-formulas and of $\succ$ for $\Diamond$-formulas (cf. discussion in Section 3.3) were defined as follows:

$$\begin{array}{lll} a \Vdash \Box\varphi & \text{iff} & \text{for all } x \in X, \text{ if } x \succ \varphi \text{ then } aR_\Box x \\ x \succ \Diamond\varphi & \text{iff} & \text{for all } a \in A, \text{ if } a \Vdash \varphi \text{ then } xR_\Diamond a. \end{array}$$

Under the interpretation discussed above, these clauses can be understood as saying that $\Box\varphi$ is the concept the examples/members of which are exactly those objects that *demonstrably* have all the features shared by φ-objects, and that $\Diamond\varphi$ is the concept described by the features which all φ-objects *demonstrably* have. Hence, $\Box\varphi$ can be understood as the (sub)concept of the *certified members* of φ, while $\Diamond\varphi$ as the (super) concept of the *potential members* of φ.

[7]In the presence of reflexivity, seriality becomes redundant; however, for the sake of making the generalization more modular, in [15] the basic framework of conceptual approximation spaces only requires seriality.

Thus, under the interpretation of $R_\Box$ and $R_\Diamond$ proposed above, the polarity-based semantics of $\mathcal{L}_{\mathrm{PML}}$ supports the understanding of $\Box\varphi$ and $\Diamond\varphi$ as the lower and upper approximations of concept φ, respectively. Notice that, while in approximation spaces the relation R relates indiscernible states, and thus directly encodes the extent of our *ignorance*, in the setting of conceptual approximation spaces, $R_\Box$ (or equivalently $R_\Diamond$) directly encode the (possibly partial) extent of our *knowledge* or information.

From concepts to other ontologies. In [15], other more specific interpretations are proposed concerning situations which span from the analysis of text databases to medical diagnoses and the analysis of markets. Accordingly, in each of these situations, $\Box\varphi$ and $\Diamond\varphi$ can be given more specific interpretations as lower and upper approximations of concepts or categories or relevant clusters.

For instance (cf. [15, Section 5.4,] modified), text databases can be modelled as polarity-based $\mathcal{L}_{\mathrm{PML}}$-structures $\mathbb{F} = (\mathbb{P}, R_\Box, R_\Diamond)$ such that $\mathbb{P} = (A, X, I)$ with A being a set of documents, X a set of words, and aIx being understood as 'document a has word x as a keyword'. Formal concepts arising from such an $\mathbb{F}$ can be understood as *themes* or *topics*, intensionally described by Galois-stable sets of (key)words. In this situation, one of the many possible interpretations of $aR_\Box x$ or equivalently $xR_\Diamond a$ is 'document a has word x as its *first or second* keyword', which again makes it plausible to assume that $R_\Box \subseteq I$.

As another example (cf. [15, Section 5.5] modified), let $\mathbb{P} = (A, X, I)$ represent a hospital, where A is the set of patients, X is the set of symptoms, and aIx iff "patient a has symptom x". Concepts arising from this representation are *syndromes*, intensionally described by Galois-stable sets of symptoms. In this situation, let $aR_\Box x$, or equivalently $xR_\Diamond a$, iff 'a has been *tested* for symptom x with positive outcome'.

As a third example (cf. [15, Section 5.8] modified), let $\mathbb{P} = (A, X, I)$ where A is a set of consumers, X is a set of market-products, and aIx iff 'consumer a buys product x'. Concepts arising from this representation are *consumer segments*, intensionally described by Galois-stable sets of market-products. In this situation, let $aR_\Box x$, or equivalently $xR_\Diamond a$, iff 'a buys x from a certain producer i'. Then $\Box\varphi$ denotes the market share of producer i in consumer segment φ.

As a fourth example, let $\mathbb{P} = (A, X, I)$ where A is a set of empirical hypotheses, X is a set of variables, and aIx iff 'hypothesis a is formulated in terms of variable x'. Concepts arising from this representation are empirical *theories*, extensionally described by Galois-stable sets of hypotheses and intensionally described by Galois-stable sets of variables. In this situation, let $aR_\Box x$, or equivalently $xR_\Diamond a$, iff 'x is a *dependent* variable for hypothesis a'. Then $\Box\mathbf{x}$ contains all hypotheses that compete with each other to predict variable x.

Finally, let $\mathbb{P} = (A, X, I)$ represent a decision-making situation in which A is the set of decision-makers, X is the set of issues, and aIx iff 'agent a finds issue x relevant'. Concepts arising from this representation are *interrogative agendas*, extensionally described by Galois-stable *coalitions* and intensionally described by Galois-stable sets of issues. In this situation, let $aR_\Box x$, or equivalently $xR_\Diamond a$, iff 'agent a regards x as a positive issue'. For example, if A is the set of the members of a hiring committee and X is the set of the features of potential applicants, agent a could regard "the candidate obtained their PhD recently" as a desirable characteristic, i.e. a positive issue, while other agents might prefer a more experienced candidate and therefore not regard this as positive. This would mean that $R_\Box \subseteq I$ and that, extensionally, $\Box\varphi$ would be the coalition of all agents who are positive towards all issues on interrogative agenda φ.

4.2 LE-logics as logics of informational entropy

Reflexive graphs as generalized intuitionistic frames. As discussed above, polarity-based semantics supports an interpretation of non-distributive logics as *logics of formal concepts* (for specific signatures: epistemic logic of concepts, logic of rough concepts etc). Under this interpretation, formulas do not denote states of affairs but rather are names for formal concepts in the sense of FCA. This interpretation can be further specialized to entities such as categories, theories, and interrogative agendas.

Graph-based semantics suggests quite another interpretation of nondistributive logics, in which formulas do denote states of affairs; below we will argue that, under this interpretation, non-distributive lattice logic can be understood as a hyper-constructivist logic which generalizes intuitionistic logic just like intuitionistic logic generalizes classical logic.

As discussed in Section 3.2, for any reflexive graph $\mathbb{X} = (Z, E)$, homomorphic assignments $\overline{v} : \mathrm{L} \to \mathbb{X}^+$ map L-formulas φ to tuples $([\![\varphi]\!], (\![\varphi]\!))$ which, as discussed in Section 3.4, reduce to $([\![\varphi]\!], [\![\varphi]\!]^c)$ when $E := \Delta$. Hence, $[\![\varphi]\!] = \{z \in Z \mid z \Vdash \varphi\}$ and $(\![\varphi]\!) = \{z \in Z \mid z \succ \varphi\}$ can be respectively understood as the *satisfaction* and *refutation set* of φ under $\overline{v}$. Since $([\![\varphi]\!], (\![\varphi]\!))$ is a Galois-stable pair, also when $E := \Delta$, the satisfaction and refutation set of a given formula completely determine one another via the identities $(\![\varphi]\!) = [\![\varphi]\!]^\uparrow$ and $[\![\varphi]\!] = (\![\varphi]\!)^\downarrow$. However, as we will see, in contrast with the classical and intuitionistic setting, at given state z, the truth value of a given formula φ can be *undefined* (i.e. $z \nVdash \varphi$ and $z \not\succ \varphi$). This *potential indeterminacy* of formulas at states of graph-based models is the main characterizing feature of this semantic setting, and can be understood as witnessing the failure of the principle of excluded middle not just anymore at the level of the object language (as is the case of intuitionistic logic) but at the more fundamental, *meta-linguistic* level of the satisfaction and refutation of formulas. This property of the graph-based semantics of non-distributive logics justifies our view of non-distributive logics as 'hyper-constructivist' logics.

In order to discuss this generalization, it will be useful to first recall how the state-based semantics of intuitionistic logic generalizes the state-based semantic of classical propositional logic. As discussed in Section 3.2, a 'relational structure' for classical propositional logic is a structure $\mathbb{F} = (S, \Delta)$ where S a nonempty set and $\Delta \subseteq S \times S$ is the identity relation. As is well known, given an assignment $v : \mathsf{Prop} \to \mathbb{F}^+$, any formula is either true or false at each state (but at no state can a formula be *both* true *and* false), it is true precisely when its negation is false, and its truth value only depends on the values of its occurring propositional variables at the given state. A relational structure for intuitionistic propositional logic is a structure $\mathbb{F} = (S, \leq)$ where S a nonempty set and $\leq$ is a reflexive and transitive relation. In this case, for a given assignment $v : \mathsf{Prop} \to \mathbb{F}^+$, any formula is again either true or false at each state (but never both true and false); however, the fact that a given formula is refuted at a given state does not imply that the negation of that formula is satisfied at that state; indeed, the given state might refute both the formula and its negation, yielding the well known failure of the classical principle of excluded middle. More generally, the *satisfaction* of a formula at a given state might depend on the truth values of its occurring propositional variables at the *successors* of the given state. The characterizing properties of intuitionistic satisfaction, those that mark its being different from classical satisfaction, are all grounded on the fact that (homomorphic) assignments of proposition variables are *persistent*, in the sense of being upward-closed with respect to $\leq$, i.e. the fact that $v(\varphi) \in \mathbb{F}^+ \cong \mathscr{P}^{\uparrow}(S)$ for any formula φ, and hence if φ is true at a given state s, it will remain true along any forward-looking $\leq$-branch stemming from s (and dually, if φ is false at a given state s, it is so at any node of any backward-looking $\leq$-branch stemming from s).

Persistence supports our understanding of intuitionistic truth as an inherently *procedural* truth: under the *higher standard* required by having to declare true at a given state only those statements that e.g. are backed by evidence in support of their truth at that state, or for which a procedure has been completed at that state which effectively establishes their truth, the failure of the principle of excluded middle captures an essential aspect of 'correct reasoning' in this context, since, e.g. at a given state, there might not be enough evidence in support of a given statement or in support of its negation; however, it is also reasonable to require that, once proven at a given state, a statement cannot be unproven, therefore its (proven) truth persists at the successors of the given state. As mentioned above, when moving from the Boolean to the intuitionistic setting, the meaning of the implication becomes "intensional", in the specific sense that the satisfaction of $p \to q$ at any given state s of an intuitionistic frame $(S, \leq)$ does not just depend on the value of p and q at s, but depends also on the values of p and q at the $\leq$-successors of s. This change in the interpretation of $\to$-formulas can be explained from a technical ground as the consequence of translating the interpretation of $\to$-formulas from $\mathbb{F}^+$ to $\mathbb{F} = (S, \leq)$, as discussed in Section 3.2, when $\mathbb{F}^+$ is the perfect Heyting algebra $\mathscr{P}^{\uparrow}(S)$ rather than the perfect Boolean algebra $\mathscr{P}(S)$; however, this interpretation is also coherent with our understanding of intuitionistic truth as a procedural truth, since, as is well known, if a procedure is available for establishing that $\varphi \to \psi$ is true at a given state s, then at any given evolution of s (including s itself) it will be possible to use this same procedure to transform a proof of φ (whenever available) into a proof of ψ, and conversely.

Having summarised the salient features of the state-based semantics of intuitionistic logic, the following proposition establishes how it can be seen as a special instance of graph-based semantics.

Proposition 4.1. *For any reflexive graph $\mathbb{X} = (Z, E)$,*

1. *if E is transitive (i.e. E is a preorder), then $\mathbb{X}^+$ is a complete and completely distributive lattice (hence a perfect Heyting algebra).*

2. *if E is antisymmetric (i.e. zEz' and $z'Ez$ imply $z = z'$ for any $z, z' \in Z$) and $\mathbb{X}^+$ is completely distributive, then E is transitive.*[8]

Proof. As to item 1, it is enough to show that the Galois-stable sets of $\mathbb{P}^+_{\mathbb{X}} \subseteq \mathscr{P}(Z_A)$ (resp. $\mathbb{P}^+_{\mathbb{X}} \subseteq \mathscr{P}(Z_X)$) are exactly the upward-closed (resp. the downward-closed) subsets. To this end, it is enough to show that $Y^{\uparrow\downarrow} = Y{\uparrow} := \{z \in Z \mid xEz \text{ for some } x \in Y\}$ for any $Y \subseteq Z_A$ (or dually, that $Y^{\downarrow\uparrow} = Y{\downarrow} := \{z \in Z \mid zEx \text{ for some } x \in Y\}$ for any $Y \subseteq Z_X$). Indeed, $Y^{\uparrow} = \{x \in Z_X \mid \text{for all } a, \text{ if } a \in Y \text{ then } aE^c x\} = Y{\uparrow}^c$. Then $Y^{\uparrow\downarrow} = Y{\uparrow}^c{\downarrow}^c$. Given that $Y{\uparrow}$ is by definition an upward-closed set, $Y{\uparrow}^c$ is a downward-closed set, hence $Y{\uparrow}^c{\downarrow} = Y{\uparrow}^c$; hence, $Y{\uparrow}^c{\downarrow}^c = Y{\uparrow}^{cc} = Y{\uparrow}$.

As to item 2, for every $z \in Z$, we let $\underline{\mathbf{z}} = \mathbf{z}_s = (z^{\uparrow\downarrow}, z^{\uparrow})$, and $\overline{\mathbf{z}} = \mathbf{z}_r = (z^{\downarrow}, z^{\downarrow\uparrow})$. To show that E is transitive, it is enough to show that, for any $u, z \in Z$,

$$zEu \quad \text{iff} \quad \underline{\mathbf{u}} \leq \underline{\mathbf{z}}. \tag{13}$$

By definition, $\underline{\mathbf{u}} \leq \underline{\mathbf{z}}$ iff $z^{\uparrow} \subseteq u^{\uparrow}$ iff $\{z' \in Z \mid uEz'\} = (u^{\uparrow})^c \subseteq (z^{\uparrow})^c = \{z' \in Z \mid zEz'\}$. Since uEu by the reflexivity of E, this inclusion implies that zEu, as required. Conversely, let us assume that zEu. Then $z \in z^{\uparrow\downarrow}$ and $z \notin u^{\downarrow}$, hence $\underline{\mathbf{z}} \nleq \overline{\mathbf{u}}$. Notice that when $z = u$ the same argument yields $\underline{\mathbf{u}} \nleq \overline{\mathbf{u}}$. To finish the proof, let

$$\kappa(\underline{\mathbf{u}}) := \bigvee\{\underline{\mathbf{w}} \mid w \in Z \text{ and } \underline{\mathbf{u}} \nleq \underline{\mathbf{w}}\} = \bigvee\{\underline{\mathbf{w}} \mid w \in Z \text{ and } u \notin w^{\uparrow\downarrow}\} = \bigvee\{\underline{\mathbf{w}} \mid w \in Z \text{ and } w^{\uparrow} \nsubseteq u^{\uparrow}\}.$$

It is enough to show that: (a) $\underline{\mathbf{u}} \leq \underline{\mathbf{z}}$ iff $\underline{\mathbf{z}} \nleq \kappa(\underline{\mathbf{u}})$, and (b) $\overline{\mathbf{u}} = \kappa(\underline{\mathbf{u}})$.

Before addressing (a), let us preliminarily show that $\underline{\mathbf{z}}$ is completely join-prime for every $z \in Z$. Because by assumption $\mathbb{X}^+$ is completely distributive, it is enough to show that $\underline{\mathbf{z}}$ is completely join-irreducible, and for this latter claim, it is enough to show that $\underline{\mathbf{z}} \wedge \overline{\mathbf{z}}$ is the greatest element of $\mathbb{X}^+$ that is strictly less than $\underline{\mathbf{z}}$. Since, as discussed above, $\underline{\mathbf{z}} \nleq \overline{\mathbf{z}}$, it follows that $\underline{\mathbf{z}} \wedge \overline{\mathbf{z}} < \underline{\mathbf{z}}$. Let us show that $\underline{\mathbf{z}} \wedge \overline{\mathbf{z}}$ is the greatest such element by showing

[8] The antisymmetry assumption is necessary: in personal communication, Andrew Craig observed that $\mathbb{X} = (Z, E)$ with $Z := \{u, v, z\}$ and $E := Z \times Z \setminus \{(u, z)\}$ is an example of a reflexive, non antisymmetric and non-transitive graph such that $\mathbb{X}^+$ is the 3-element chain (hence is distributive). We thank Apostolos Tzimoulis for suggesting the structure of the proof of item 2 of Proposition 4.1.

that $[\![\underline{\mathbf{z}}\wedge\overline{\mathbf{z}}]\!] = z^{\uparrow\downarrow}\setminus\{z\}$. By definition, $[\![\underline{\mathbf{z}}\wedge\overline{\mathbf{z}}]\!] = [\![\underline{\mathbf{z}}]\!]\cap[\![\overline{\mathbf{z}}]\!] = z^{\uparrow\downarrow}\cap z^{\downarrow}$. Hence, we are left to show that $z^{\uparrow\downarrow}\cap z^{\downarrow} = z^{\uparrow\downarrow}\setminus\{z\}$. By the reflexivity of E, if $w\in z^{\uparrow\downarrow}\cap z^{\downarrow}$ then $w\neq z$, and hence $w\in z^{\uparrow\downarrow}\setminus\{z\}$. Conversely, if $w\in z^{\uparrow\downarrow}\setminus\{z\}$, the antisymmetry of E implies that either $wE^c z$ (i.e. $w\in z^{\downarrow}$, which is what we require) or $zE^c w$ (i.e. $w\in z^{\uparrow}$). This latter condition is excluded, again by the reflexivity of E and the assumption that $w\in z^{\uparrow\downarrow}$.

As to (a), the complete join-primeness of $\underline{\mathbf{z}}$ implies that $\underline{\mathbf{z}}\leq\kappa(\underline{\mathbf{u}}) = \bigvee\{\underline{\mathbf{w}}\mid w\in Z \text{ and } \underline{\mathbf{u}}\not\leq\underline{\mathbf{w}}\}$ iff $\underline{\mathbf{z}}\in\{\underline{\mathbf{w}}\mid w\in Z \text{ and } \underline{\mathbf{u}}\not\leq\underline{\mathbf{w}}\}$ iff $\underline{\mathbf{u}}\not\leq\underline{\mathbf{z}}$, as required.

Notice that (a) instantiated to $z=u$ yields $\underline{\mathbf{u}}\not\leq\kappa(\underline{\mathbf{u}})$. Notice also that (a) can be strengthened to (a'): $\underline{\mathbf{u}}\leq\mathbf{c}$ iff $\mathbf{c}\not\leq\kappa(\underline{\mathbf{u}})$ for all $u\in Z$ and $\mathbf{c}\in\mathbb{X}^+$. Indeed, by Proposition 3.1, $\bigvee\{\underline{\mathbf{z}}\mid\underline{\mathbf{z}}\leq\mathbf{c}\} = \mathbf{c}\leq\kappa(\underline{\mathbf{u}})$ iff $\underline{\mathbf{z}}\leq\kappa(\underline{\mathbf{u}})$ for every $z\in Z$ such that $\underline{\mathbf{z}}\leq\mathbf{c}$, iff, by (a), $\underline{\mathbf{u}}\not\leq\underline{\mathbf{z}}$ for every $z\in Z$ such that $\underline{\mathbf{z}}\leq\mathbf{c}$. By the complete join-primeness of $\underline{\mathbf{u}}$, the last condition is equivalent to $\underline{\mathbf{u}}\not\leq\mathbf{c}$.

These remarks imply that $\kappa(\underline{\mathbf{u}})$ is completely meet-irreducible. Indeed, let $Y\subseteq Z$ such that $\bigwedge\{\overline{\mathbf{z}}\mid z\in Y\} = \kappa(\underline{\mathbf{u}})$. If $\overline{\mathbf{z}}>\kappa(\underline{\mathbf{u}})$ for every $z\in Y$, then $\overline{\mathbf{z}}\not\leq\kappa(\underline{\mathbf{u}})$, i.e. by (a'), $\underline{\mathbf{u}}\leq\overline{\mathbf{z}}$ for every $z\in Y$, and hence $\underline{\mathbf{u}}\leq\kappa(\underline{\mathbf{u}})$, contradicting $\underline{\mathbf{u}}\not\leq\kappa(\underline{\mathbf{u}})$.

As to (b), by Proposition 3.1, $\overline{\mathbf{u}} = \bigvee\{\underline{\mathbf{w}}\mid w\in Z \text{ and } \underline{\mathbf{w}}\leq\overline{\mathbf{u}}\} = \bigvee\{\underline{\mathbf{w}}\mid w\in Z \text{ and } w\in u^{\downarrow}\}$. Hence, to show that $\overline{\mathbf{u}}\leq\kappa(\underline{\mathbf{u}})$, it is enough to show that, if $w\in u^{\downarrow}$, then $u\notin w^{\uparrow\downarrow} = \{z\in Z\mid\forall z'(z'\in w^{\uparrow}\Rightarrow zE^c z')\}$. By definition, $w\in u^{\downarrow}$ iff $wE^c u$ iff $u\in w^{\uparrow}$. Moreover, by reflexivity, uEu. Hence, $u\notin w^{\uparrow\downarrow}$, as required. Conversely, to show that $\kappa(\underline{\mathbf{u}})\leq\overline{\mathbf{u}}$, notice that Proposition 3.1 and $\kappa(\underline{\mathbf{u}})$ being completely meet-irreducible imply that $\kappa(\underline{\mathbf{u}}) = \overline{\mathbf{w}}$ for some $w\in Z$. Hence, from $\underline{\mathbf{u}}\not\leq\kappa(\underline{\mathbf{u}}) = \overline{\mathbf{w}}$ it follows that $u\notin w^{\downarrow}$ (i.e. uEw), while from $\overline{\mathbf{u}}\leq\kappa(\underline{\mathbf{u}}) = \overline{\mathbf{w}}$ it follows that $w\in u^{\downarrow\uparrow}$, i.e. $u^{\downarrow}\subseteq w^{\downarrow}$. Suppose for contradiction that $u\neq w$. Then, by antisymmetry, uEw implies that $wE^c u$, i.e. $w\in u^{\downarrow}\subseteq w^{\downarrow}$. Hence, $wE^c w$, contradicting the reflexivity of E. □

The proposition above shows that the Galois-stability of satisfaction (resp. refutation) sets projects onto their being $\leq$-upward (resp. $\leq$-downward) closed in the intuitionistic setting. This establishes a tight link between, on the one hand, the transitivity of intuitionistic frames and, on the other, a package of three characterizing properties of intuitionistic logic, namely *persistence* of the satisfaction of intuitionistic formulas, *distributivity* of intuitionistic $\wedge$ and $\vee$, and the relation of refutation of intuitionistic formulas at states coinciding with the one of *non-satisfaction*. In the absence of transitivity, all three properties in this package are expected to fail. Indeed, let us illustrate this fact by considering the reflexive (and antisymmetric but not transitive) graph $\mathbb{X} = (Z,E)$ shown in the left-hand side of the picture below. The polarity drawn in the centre of the picture is the polarity $\mathbb{P}_{\mathbb{X}}$ associated with $\mathbb{X}$, and the (nondistributive) lattice on the right is $\mathbb{X}^+ = \mathbb{P}_{\mathbb{X}}^+$.

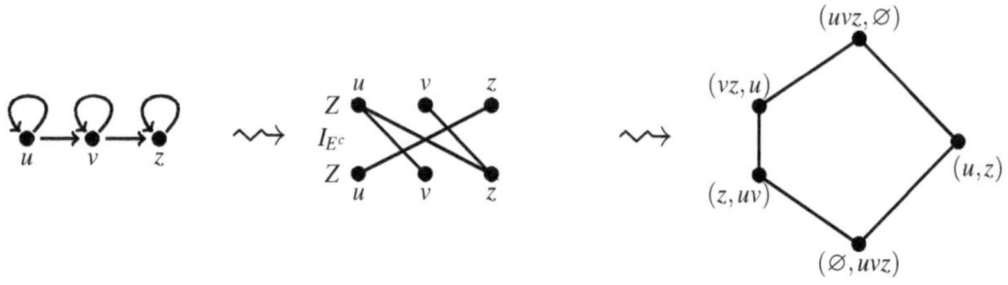

Let us consider a (homomorphic) assignment $\overline{v} : \mathrm{L} \to \mathbb{X}^+$ such that $\overline{v}(p) = (z, uv)$ and $\overline{v}(q) = (u, z)$. The following picture shows how this assignment translates into the interpretation of p and q on $\mathbb{X}$. We follow the convention that for every $y \in Z$ and formula φ, if $y \Vdash \varphi$ then φ appears above y, and if $y \succ \varphi$ then φ appears below y.

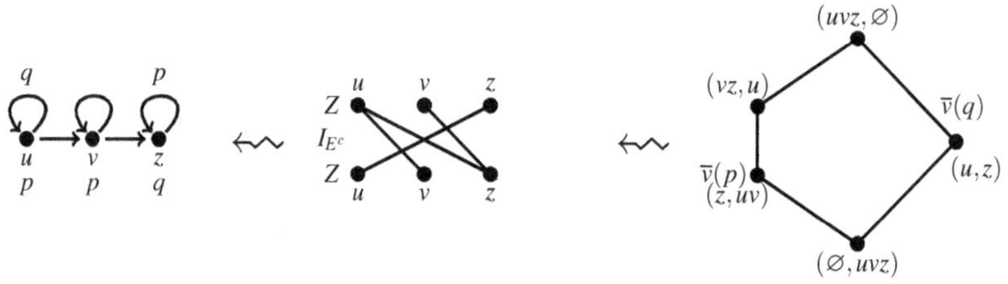

The left-hand side of the picture above shows that indeed, the formula q which is true at u becomes first indeterminate at v and then false at z. Hence, non-distributive satisfaction does not need to be persistent, and non-distributive refutation is *different* from non-satisfaction, since $v \nVdash q$ and $v \not\succ q$. Together with the failure of distributivity (which in the example above is yielded by the fact that $\mathbb{X}^+$ is nondistributive), these are the key differences setting apart non-distributive evaluation from intuitionistic evaluation. Their role in our conceptual interpretation of this semantics will be discussed below. So far, we have discussed these differences in terms of properties holding in intuitionistic models but failing in graph-based models. The following proposition identifies a property holding in graph-based models. We will refer to it as *weak persistence*.

Proposition 4.2. *For every (reflexive) graph $\mathbb{X} = (Z, E)$ any* L*-formula φ, any $z, z' \in Z$ and any homomorphic assignment $\overline{v} : \mathrm{L} \to \mathbb{X}^+$, if zEz' and $z \Vdash \varphi$, then $z' \not\succ \varphi$.*

Proof. By definition, $\overline{v}(\varphi) = ([\![\varphi]\!], (\![\varphi]\!)) \in \mathbb{X}^+$, and hence $z' \succ \varphi$ iff $z' \in (\![\varphi]\!) = [\![\varphi]\!]^{\downarrow} = \{z' \in Z \mid \text{for all } z, \text{ if } z \in [\![\varphi]\!] \text{ then } zE^c z'\}$. The assumptions imply that some $z \in Z$ exists such that $z \in [\![\varphi]\!]$ and zEz'. □

From this proposition it follows that weak persistence projects onto persistence in the intuitionistic setting: indeed, as shown in the proof of item 1 of Proposition 4.1, if E is transitive, then $[\![\varphi]\!] = ([\varphi])^c$, i.e. $z' \not\succ \varphi$ iff $z' \Vdash \varphi$. Hence, in every reflexive and transitive graph-based model, if $z \Vdash \varphi$ and zEz' then $z' \Vdash \varphi$.

However, as its name suggests, in the wider context of reflexive but not necessarily transitive graph-based models, weak persistence is strictly weaker than persistence, and hence this notion captures the difference between intuitionistic and non-distributive evaluation from yet another angle. Namely, in intuitionistic frames, any given state is bound to *accept* any formula supported by any of its predecessors, including itself, while in graph-based frames, any given state is only bound to *not reject* any formula supported by any of its predecessors, including itself; as discussed above, because of potential indeterminacy, not rejecting does not necessarily imply accepting. Hence, together with reflexivity, weak persistence guarantees that, at any given state, any given formula cannot be both accepted and rejected.

As discussed above, intuitionistic truth can be construed as a procedural truth, so that $z \Vdash \varphi$ can be read e.g. as 'at z, some finite procedure has effectively established that φ is the case'. We also observed that, in intuitionistic frames, the refutation relation $\succ$ coincides with the relation $\nVdash$ of *non-satisfaction*; that is, under the previous reading of the intuitionistic satisfaction of a proposition, $z \succ \varphi$ iff 'at z, no finite procedure has effectively established (yet) that φ is the case'. Non-distributive truth can be then construed along these same lines, but requiring an even *higher standard*[9] than intuitionistic truth: for instance, to conclude that φ is refuted at z, the *absence* of a procedure effectively establishing φ is not enough; one needs *there to be* a procedure effectively *disproving* φ.

Above, we remarked that, when moving from classical to intuitionistic logic, the interpretation of $\rightarrow$ changes and becomes intensional, in the sense that the truth value of $p \rightarrow q$ at a given state s does not just depend on the value of p and q at that state, but also on the truth values of p and q at the $\leq$-successors of s. Similarly, in moving from the intuitionistic to the non-distributive setting, conjunction and disjunction become "intensional" in the same sense, i.e. that the satisfaction of $p \vee q$ (and dually, the refutation of $p \wedge q$) at a given state s does not just depend on the value of p and q at that state, but also on the truth values of p and q at the E-successors of that state. Indeed, under the assignment above, $\overline{v}(p \vee q) = (uvz, \varnothing)$. The following picture shows how this assignment translates as

[9]For the sake of the telescopic progression of this presentation, we have discussed a possible interpretation of non-distributive evaluation which allows it to be described as being *stricter* than the intuitionistic one. However, other readings might suggest that it is just different: for instance, one can read $z \Vdash \varphi$ as 'z provides an argument/evidence supporting φ', and $z \succ \varphi$ as 'z provides an argument/evidence against φ'. The essential difference lies in the fact that, in the non-distributive setting, $\Vdash$ and $\succ$ determine each other in a way that is defined in terms of E, and, in contrast to the intuitionistic setting, cannot be simply reduced to identifying one with the complement of the other.

satisfaction/refutation of $p \vee q$ at states of $\mathbb{X}$.

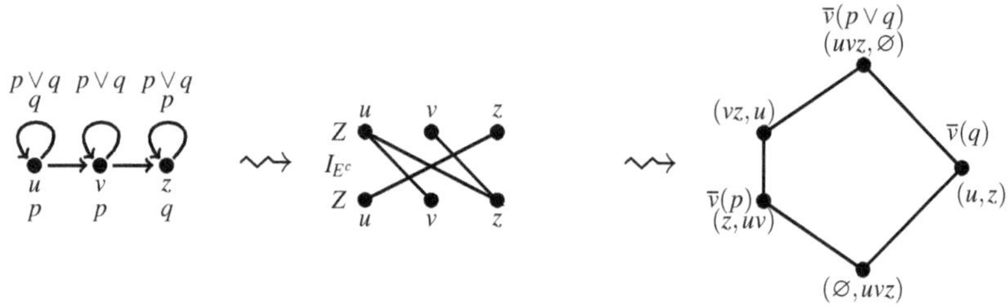

In the example above, $p \vee q$ is true at state v although neither p nor q are. More specifically, p is false at v and q is neither true nor false. Indeed, the defining clauses of $\succ$ and $\Vdash$ for $\vee$-formulas (cf. discussion in Section 3.2), repeated below for the reader's convenience,

$z \succ \varphi \vee \psi$	iff	$z \succ \varphi$ and $z \succ \psi$	$z \Vdash \varphi \vee \psi$	iff	for all z', if zEz' then $z' \not\succ \varphi \vee \psi$
$z \Vdash \varphi \wedge \psi$	iff	$z \Vdash \varphi$ and $z \Vdash \psi$	$z \succ \varphi \wedge \psi$	iff	for all z', if zEz' then $z' \nVdash \varphi \wedge \psi$

can be understood as saying that a $\vee$-formula (resp. $\wedge$-formula) is refuted (resp. satisfied) at a state iff both its immediate subformulas are refuted (resp. satisfied) at that state (which is verbatim the intuitionistic and even the *classical* refutation condition for $\vee$-formulas and satisfaction condition for $\wedge$-formulas), and that a $\vee$-formula (resp. $\wedge$-formula) is satisfied (resp. refuted) at a given state iff no successor of that state (including itself) refutes (resp. satisfies) it, i.e. no successor of that state refutes (resp. satisfies) both its immediate subformulas. The case of disjunction in the example above is then clear-cut: there is no state which refutes both p and q, and hence the satisfaction condition of $p \vee q$ is vacuously verified by all states of the graph above. However, to mark the difference with the distributive interpretation of disjunction, $p \vee q$ is still true at v even if: (a) the falsehood of p is positively established at v; (b) at v, the truth of q *can* be excluded, given that v does not access any state at which q is true, and (c) at v, the falsehood of q *cannot* be excluded, given that v accesses a state at which q is false.

Below, we discuss some possible conceptual interpretations of the graph-based semantics which will account for the intensional meaning of $\vee$ in a hyper-constructivist context in which effective procedures are needed both for accepting and for rejecting a proposition.

Reflexive graphs and the inherent bounds of knowability. In [14], it is argued that, based on their graph-based semantics, LE-logics can be taken as the logics of *informational entropy*, where informational entropy is understood as an inherent boundary to knowability due e.g. to perceptual, theoretical, technological, evidential or linguistic limits. The starting point of this interpretation is to regard reflexive graphs $\mathbb{X} = (Z, E)$ in analogy with approximation spaces (cf. discussion in penultimate paragraph of Section 4.1), and hence to regard E as an *indiscernibility* relation[10]. However, while indiscernibility is modelled as an equivalence relation in the best known settings in rough set theory and epistemic logic (e.g. [60, 30]), transitivity fails in concrete instances in which e.g. states are indiscernible when their distance is closer than a certain threshold, and, as it has been argued in the psychological literature (cf. [67, 57]), symmetry fails in situations in which indiscernibility is defined in terms of a relation of similarity where e.g. z being similar to z' does not necessarily imply that z' is similar to z. Hence, under this broader understanding, *reflexivity*, i.e. $\Delta \subseteq E$, is the minimal requirement of an indiscernibility relation[11]. The limit case in which $E := \Delta$ is the (classical) case in which there are no boundaries to perfect knowability. So, the generalization from the classical to the non-distributive propositional environment is captured by the graph-based semantics as the logical internalization of the *shift* from Δ to E.

As discussed above, rather than generating modal operators (e.g. upper and lower approximations, as in rough set theory), E is used to generate the complete lattice $\mathbb{X}^+$ as the concept lattice of the polarity $\mathbb{P}_{\mathbb{X}} = (Z_A, Z_X, I_{E^c})$. The elements of this lattice are the closures of subsets $B \subseteq Z$ (representing states of affairs) under all that is discernible from everything from which everything in B is discernible (or, equivalently, closing under everything from which all that is discernible from everything in B is discernible). Either of these closures represent the theoretical horizon to the knowability of B, given the inherent boundary encoded into E.

Like the persistence property of intuitionistic satisfaction we discussed above, the requirement of *Galois-stability* in the non-distributive setting is not only mathematically justified by the need to define a compositional semantics for L, but can also be understood at a more fundamental level: if E encodes an inherent limit to knowability, this limit should be incorporated in the meaning of formulas which are both satisfied and refuted 'up to E', i.e. the semantic representation of each formula should not be given in terms of arbitrary subsets of the domain of the graph, but only in terms of those subsets which are preserved (i.e. faithfully translated) in the shift from Δ to E; these are exactly the Galois-stable sets. In particular, the closure $a^{\uparrow\downarrow}$ of any $a \in Z$ arises by first considering the set $a^{\uparrow}$ of all the states from which a is not indiscernible, and then the set of all the states that can be told apart from every state in $a^{\uparrow}$. Then a is clearly an element of $a^{\uparrow\downarrow}$, but other states b in Z might be as well, so this is as far as we can go: $a^{\uparrow\downarrow}$ represents the *horizon*, defined in terms of E, to the possibility of completely 'knowing' a. This horizon might be epistemic, cognitive, technological, or evidential. Under this understanding of E, the following defining clauses of $\succ$ and $\Vdash$ for $\vee$-formulas (cf. discussion in Section 3.2)

$$z \succ \varphi \vee \psi \quad \text{iff} \quad z \succ \varphi \text{ and } z \succ \psi \qquad z \Vdash \varphi \vee \psi \quad \text{iff} \quad \text{for all } z', \text{ if } zEz' \text{ then } z' \not\succ \varphi \text{ or } z' \not\succ \psi$$

[10]That is, zEz' reads as *z is indiscernible from z'*. In what follows, we propose that zEz' can be sometimes interpreted as *z is consistent with z'*, or *z does not exclude z'*.

[11]These objections to modelling indiscernibility as an equivalence relation have also been addressed in rough set theory and epistemic logic, see e.g. [70, 3]

reflect a kind of *conservative* or *cautious* interpretation of the meaning of $\vee$: indeed, if states of a reflexive graphs represent e.g. epistemic situations, or hypothetical scenarios or alternative theories, then a given scenario s refutes a $\vee$-formula iff that scenario (contains enough evidence or explanatory power so that it) can refute both disjuncts; however, if no scenario that can be accessed from s (contains enough evidence so that it) can refute *both disjuncts*, then s must accept the given $\vee$-formula.

Concrete examples in which this interpretation of $\vee$ is arguably closer to reality than the classical or intuitionistic one arise e.g. in *legal* domains. Consider for instance the Rashomon-type story of a judge who is to establish whether a (female) defendant is to be declared not guilty of the physical injuries sustained by a (male) friend of hers. To simplify matters, the defendant is *not guilty* if and only if $p \vee q$, where p stands for 'the defendant has not willingly caused harm to her friend' and q stands for 'the defendant acted in self-defence'. In the trial, three witnesses testify as follows: Witness A: "I saw the defendant grabbing a tennis racket and hitting her friend. She looked terrified and didn't utter a sound." Witness B: "I saw the defendant grabbing a tennis racket and hitting her friend. She looked frightened, but not necessarily by her friend." Witness C: "I heard the defendant scream that there was a poisonous spider on her friend's shoulder, so she killed the spider." The following graph summarizes the information gathered by the judge:

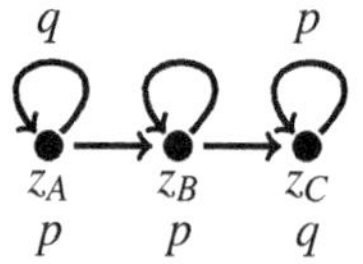

The states of the graph above represent the different testimonies of the three witnesses; the arrows represent the accessibility relation among states, where e.g. $z_A E z_B$ means 'A's testimony *is consistent with* B's testimony', and $z_B E z_C$ means 'B's testimony *is consistent with* C's' (notice, however, that z_B is arguably not consistent with z_A, since the evidence brought by z_B challenges the *conclusions* that can be drawn on the base of the evidence brought by z_A, and by the same argument, z_C is not consistent with z_B). Finally, z_A and z_C are mutually inconsistent, since they challenge each other's evidence, and each testimony is consistent with itself. Witness A's testimony offers evidence against p (A saw the defendant hitting her friend with a tennis racket) and evidence supporting q (she was frightened); Witness B's testimony offers evidence against p (B too saw the defendant hitting her friend with a tennis racket) while not offering any evidence in support of or against q (according to B, she might or might not have been frightened by her friend); Witness C's testimony offers evidence against q and supporting p. The judge is not in a position to establish whether the witnesses are lying or not, or which witness is right; however, the judge's task is to decide on $p \vee q$ on the basis of the available pieces of evidence and how these pieces fit with each other. There is no witness that provides enough evidence to refute both p and q, hence, no matter how different their versions are, all testimonies lead to the acceptance of a not guilty verdict.

LE-logics as epistemic logics of information entropy. In Section 4.1, the first proposal we discussed for an informal understanding of the modal operators on polarity-based semantics was *epistemic*, and was based on the possibility of taking the incidence relation I of any polarity as *objectively* encoding which objects have which features, while, in contrast, the relations $R_\Box$ and $R_\Diamond$ were understood as representing the *subjective* perspective of e.g. an agent on the same issues. Likewise, as discussed in [14], in the graph-based setting, additional relations on graphs-based frames can be regarded as encoding *subjective indiscernibility*, i.e. $zR_\Box y$ can be understood as 'z is indiscernible from y *according to a given agent*', as opposed to the objective, or inherent indiscernibility encoded by E. Under this interpretation, the following defining clauses of $\succ$ and $\Vdash$ for $\Box$-formulas and $\Diamond$-formulas (cf. discussion in Section 3.3)

$z \succ \Diamond\varphi$	iff	for all z', if $zR_\Diamond z'$ then $z' \nVdash \varphi$	$z \Vdash \Diamond\varphi$	iff	for all z', if zEz' then $z' \not\succ \Diamond\varphi$
$z \Vdash \Box\psi$	iff	for all z', if $zR_\Box z'$ then $z' \not\succ \psi$	$z \succ \Box\psi$	iff	for all z', if $z'Ez$ then $z' \nVdash \Box\psi$

can be understood as saying that $\Box\varphi$ is satisfied at a given state if no state that the agent considers indistinguishable from the given one refutes φ, and $\Diamond\varphi$ is refuted at a given state if no state that the agent considers indistinguishable from the given one accepts φ. Hence, under the interpretation indicated above, these semantic clauses support the usual reading of $\Box\varphi$ as 'the agent knows/believes φ' and $\Diamond\varphi$ as 'the agent considers φ plausible'.

In Section 4.1, we argued that the epistemic interpretation of $\Box$ in the polarity-based semantics carries as far as the preservation of the meaning of well known epistemic principles. Let us now finish this section by discussing that, also in the setting of graph-based semantics, the epistemic interpretation of modal operators coherently preserves the meaning of the same epistemic principles. In [14, Proposition 4], the first-order correspondents of well known modal axioms from epistemic logic have been computed, which turn out to be the parametrized 'E-counterparts' of the first-order correspondents on Kripke frames. That is, the factivity axiom $\Box\varphi \vdash \varphi$, which on classical Kripke frames corresponds to reflexivity ($\Delta \subseteq R_\Box$), corresponds to the first-order condition $E \subseteq R_\Box$ on graph-based frames, which in [14] is referred to as *E-reflexivity*; the omniscience axiom $\varphi \vdash \Box\varphi$, which on classical Kripke frames corresponds to the first-order condition $R_\Box \subseteq \Delta$, corresponds to its E-counterpart $R_\Box \subseteq E$ on graph-based frames; the positive introspection axiom $\Box\varphi \vdash \Box\Box\varphi$, which on classical Kripke frames corresponds to transitivity ($R_\Box \circ R_\Box \subseteq R_\Box$), corresponds to the first-order condition $R_\Box \bullet_E R_\Box \subseteq R_\Box$ on graph-based frames [12] which in [14] is referred to as *E-transitivity*. So it seems possible to capture the shift from Δ to E also at the

[12]For any graph $\mathbb{X} = (Z, E)$ and relations $R, S \subseteq Z \times Z$, the relation $R \bullet_E S \subseteq Z \times Z$ is defined as follows: for

level of the first-order correspondents, by establishing a sort of systematic replacement of the role of Δ with E.[13] This similarity in shape is also what guarantees that the intended meaning of the epistemic principles is preserved in the translation from Δ to E. Indeed, as discussed in [14], the E-reflexivity condition $E \subseteq R_\Box$ requires that if the agent is able to distinguish two given states, then these states are not inherently indistinguishable. That is, the agent's assessments are correct, which preserves the meaning of *factivity* modulo informational entropy. Similarly, the condition $R_\Box \subseteq E$ requires the agent to be able to distinguish any two states that are not inherently indistinguishable, which is indeed what an *omniscient* agent should be able to do, where we understand an omniscient agent to be one who knows everything that can possibly be known within the inherent limits of the situation. Finally, the condition $R_\Box \bullet_E R_\Box \subseteq R_\Box$, i.e. $\forall a \forall x [\exists y(aR_\Box y \ \&\ \forall b(bEy \Rightarrow bR_\Box x)) \Rightarrow aR_\Box x]$ requires that if the agent cannot distinguish y from a and x from any state from which y is inherently indistinguishable, then she cannot distinguish x from a. Equivalently, if the agent can distinguish x from a, then (x does not belong to the set Y of states indistinguishable from a according to the agent, and hence) every state y which belongs to Y must be inherently indistinguishable from some state b from which she can distinguish x. That is, if $x \notin Y$ as above, then for every $y \in Y$ the agent must be able to find an 'E-proxy' b of y from which she can tell x apart. Hence, the first-order correspondent of positive introspection on graphs requires that, if the agent can distinguish x from a, there is a witness b (a 'justification') for x to be distinct from any state that the agent cannot distinguish from a. Similarly to the first-order condition corresponding to positive introspection on polarities (cf. page 19), this requirement for a justification is indeed what positive introspection is about.

5 Conclusions and further directions

The present paper discussed ongoing research on the mathematical theory and conceptual understanding of relational (non-topological) semantics for non-distributive logics, with a special focus on logics algebraically captured by varieties of normal lattice expansions in arbitrary signatures. We showed how, starting from well known adjunctions between (complete) lattices and, respectively, polarities and reflexive graphs, one can derive relational semantics for the language of lattice logic on both these types of structures. We illustrated how these basic semantic frameworks can be systematically expanded to accommodate expansions of the basic lattice language with arbitrary normal connectives. The main point

any $a, x \in Z$,

$$a(R \bullet_E S)x \quad \text{iff} \quad \exists y(aRy \ \&\ \forall b(bEy \Rightarrow bSx)).$$

If $E = \Delta$, then $\{b \mid bEy\} = \{y\}$ for every $y \in Z$, hence $(R \bullet_E S)$ reduces to $R \circ S$.

[13] Mutatis mutandis, the same phenomenon is observable in the polarity-based setting. More about this in the conclusions.

we wished to convey with this discussion is methodological: starting from an adjunction between a class of relational structures and lattices, one can suitably enrich these relational structures to systematically define the relational interpretation of the additional connectives of normal LE-logics. This methodology allows one to treat LE-logics uniformly, so that various proposals and solutions which were developed for specific signatures (e.g. the Routley-Meyer semantics for substructural logics) can be systematically connected and extended to other signatures.

Having obtained these relational semantics, one can try and extract from them insights into the nature of non-distributive logics, much in the same way in which classical Kripke semantics provides insight into the nature of modal logics by enabling one to reason about possibility and necessity in terms of possible worlds. As we saw, the failure of distributivity is born out by both the polarity-based and graph-based semantics, which offer different perspectives on this phenomenon. The polarity-based semantics builds on insights from formal concept analysis, seeing polarities as abstract databases of objects and features, and gives rise to an interpretation of LE-logics where formulas act as names for formal concepts. Under this interpretation, the logical connectives $\wedge$ and $\vee$ come to stand, not for counterparts of the natural language conjunction ('and') and disjunction ('or'), but rather for operations returning, respectively, the greatest common sub-concept and least common super-concept of the concepts named by their arguments. This interpretation makes it possible to intuitively understand the (satisfaction) relation $\Vdash$ as 'being a member' in a category, or 'exemplifying' a concept, and the (refutation) relation $\succ$ as 'being a defining feature of' the given category or concept.

On the other hand, graph-based semantics evaluates formulas at states as in Kripke frames, and gives rise to an *intensional* interpretation of $\wedge$ and $\vee$, where successor states need to be taken into account in the evaluation of these connectives. The graph-based semantics supports a reading of $\wedge$ and $\vee$ which can be described as a conservative or cautious version of logical conjunction and disjunction, in a type of *hyper-constructivist* logic in which the failure of the law of excluded middle happens at the *meta-language* level, in the sense that, at any given state, any given formula can be satisfied, refuted or *neither*.

Informed by such general insights into the nature(s) of non-distributive logics, these semantic paradigms can be used to model and reason about more specific phenomena. Regarding polarities as abstract databases enriched with relations modelling other (e.g. agent specific) types of information, we saw how non-distributive logics in specific signatures emerge as modal and epistemic logics of formal concepts. Specializing and varying this interpretation allows us to reason about themes and topics in text databases, syndromes in hospitals, consumer segments within markets, competing scientific theories and interrogative agendas.

The graph-based semantics is suitable for capturing situations where relations of similarity, proximity, indistinguishability or other types of relative informational entropy between states play a role. Such relations encode certain bounds to knowability, inherent in the situation under consideration. These could be bounds on human perception, like the inability to distinguish between the colours of light frequencies which are less the two hertz apart, or bounds on the expressiveness of a natural language imposed by relations of synonymy. In this paper we developed an example where the bounds were evidential, and were encoded by the relation of consistency between a piece of testimony and its conclusions and another piece of testimony. Here the 'cautious' non-distributive disjunction functioned naturally in formulas expressing the not guilty status of the defendant.

In the present paper, we have only scratched the surface of a wide-ranging and presently ongoing research programme. Many developments are currently being investigated and will be investigated which build on the insights, techniques and methodologies presented in this paper. In the following paragraphs we outline some of the main ones.

Many-valued semantics. The interpretation of *polarity-based semantics* discussed in the present paper generalizes very naturally to many-valued versions. Objects in databases may posses features to a certain extent, rather than absolutely: instead of a word being a keyword of a document, we might want to know with what frequency it appears in the document and, accordingly, a document may be about a topic to a certain extent; members of a selection committee naturally find some issues more important than others and so the emergent interrogative agendas emphasise some issues more than others. Similarly, also the interpretation of *graph-based semantics* generalizes to many-valued versions: testimonies may be more or less compatible with one another and may support conclusions or claims to varying extents; there are different proportions of overlap between the variables featuring in databases constructed to test scientific theories, and they may offer support hypothesis to a variable extent. In line with these ideas, *many valued* versions of polarity-based and graph-based semantics of LE-logics have been proposed (see e.g. [15, 22, 13, 23]) and deserve further study and development.

Parametricity phenomena. We have illustrated how the semantic frameworks we have introduced generalize and project onto the standard relational semantics of classical and intuitionistic (modal) logic. Moreover, we saw how the level in this hierarchy is parametric in—and completely determined by—the properties of the polarity relation I or the graph relation E. We have also observed how this parametricity phenomenon extends to the interpretation of modal operators, and particularly in how the (suitably expressed) frame-properties corresponding to well-known modal axioms generalize by simply substituting the appropriate relation in containments and compositions. We conjecture that, in a way that will have to be made suitably precise, this holds for all Sahlqvist and inductive formulas, even when we consider many-valued versions of the semantics. This conjecture receives support from the fact that, in many-valued modal logic, all Sahlqvist formulas have verbatim the same first frame-correspondents, albeit with possibly different many-valued meaning [9]. This suggests that other constructions and metatheorems, including truth preserving model-constructions and characterizations of expressivity (see [40, 39, 24]), might also lend themselves to a unified, parametric treatment.

Vectors space semantics. Throughout the paper, we have particularly stressed the methodological aspects of our approach, which allow for the systematic introduction of relational semantics for non-distributive logics using some forms of representation results for lattices (or subclasses thereof) obtained via dualities or adjunction results. The polarity-based and graph-based semantics discussed in the present paper are by no means the only structures which lend themselves to the application of this methodology, which can be extended also *beyond* non-distributive logics. Another instance of this methodology being applied is [42], where a relational semantic framework based on vector spaces has been introduced for the basic (modal) Lambek calculus. Vector-based models are widely used in various areas of computer science such as computational linguistics [66], information retrieval [64] and machine learning [49], as outcomes of statistical analyses of large databases. The possibility of regarding these structures as relational models of (non-distributive) propositional logics offers a concrete route for addressing the systematic integration of symbolic and sub-symbolic methods in AI.

A Appendix

A.1 Relational interpretation of additional connectives

In this section we provide more detail on how the defining clauses of the relations $\Vdash$ and $\succ$ on models based on polarities and reflexive graphs can be retrieved from the algebraic interpretation in complete $\mathscr{L}_{\mathrm{LE}}$-algebras for $\Box$, $\lhd$, $\rhd$ and, in general, for any connective $f \in \mathscr{F}$ of arity n_f and any connective $g \in \mathscr{G}$ of arity n_g.

We start with the case of polarities. Appealing to Proposition 3.1, we can write $\overline{v}(\psi) = \bigwedge\{\mathbf{x} \in \mathbb{P}^+ \mid \overline{v}(\psi) \leq \mathbf{x}\} = \bigwedge\{\mathbf{x} \in \mathbb{P}^+ \mid x \succ \psi\}$. Since by assumption $\overline{v}$ is a homomorphism, $\overline{v}(\Box\psi) = \Box^{\mathbb{P}^+}\overline{v}(\psi) = \Box^{\mathbb{P}^+}(\bigwedge\{\mathbf{x} \in \mathbb{P}^+ \mid x \succ \psi\})$, and since $\Box^{\mathbb{P}^+}$ is completely meet-preserving, we have:

$$\overline{v}(\Box\psi) = \bigwedge\{\Box^{\mathbb{P}^+}\mathbf{x} \mid x \succ \psi\}. \tag{14}$$

Hence, for any $a \in A$, we have

$a \Vdash \Box\psi$	iff	$\mathbf{a} \leq \overline{v}(\Box\psi)$	definition of $\Vdash$ as in (2)
	iff	$\mathbf{a} \leq \bigwedge\{\Box^{\mathbb{P}^+}\mathbf{x} \mid x \succ \psi\}$	(14)
	iff	for all $x \in X$, if $x \succ \psi$ then $\mathbf{a} \leq \Box^{\mathbb{P}^+}\mathbf{x}$	definition of $\bigwedge$
	iff	for all $x \in X$, if $x \succ \psi$ then $aR_\Box x$,	(15)

where we have *defined* the accessibility relation $R_\Box \subseteq A \times X$ corresponding to the interpretation of $\Box$ as $\Box^{\mathbb{P}^+}$ as follows:

$$aR_\Box x \quad \text{iff} \quad \mathbf{a} \leq \Box^{\mathbb{P}^+}\mathbf{x}. \tag{15}$$

Now, using that the set of elements $\mathbf{a}$ for $a \in A$ are join-generators of $\mathbb{P}^+$, we obtain $\succ$-clause for $\Box$:

$x \succ \Box\psi$	iff	$\overline{v}(\Box\psi) \leq \mathbf{x}$	definition of $\succ$ as in (3)
	iff	$\bigvee\{\mathbf{a} \mid \mathbf{a} \leq \overline{v}(\Box\psi)\} \leq \mathbf{x}$	$\mathbf{a}$ for $a \in A$ join-generators of $\mathbb{P}^+$
	iff	for all $a \in A$, if $\mathbf{a} \leq \overline{v}(\Box\psi)$ then $\mathbf{a} \leq \mathbf{x}$	definition of $\bigvee$
	iff	for all $a \in A$, if $a \Vdash \Box\psi$ then aIx	definition of $\Vdash$ as in (2).

Next, we consider $\lhd$. Since $\overline{v}$ is a homomorphism, $\overline{v}(\lhd\psi) = \lhd^{\mathbb{P}^+}\overline{v}(\psi) = \lhd^{\mathbb{P}^+}(\bigwedge\{\mathbf{x} \in \mathbb{P}^+ \mid x \succ \psi\})$, and since $\lhd^{\mathbb{P}^+}$ is completely meet-reversing, we have:

$$\overline{v}(\lhd\psi) = \bigvee\{\lhd^{\mathbb{P}^+}\mathbf{x} \mid x \succ \psi\}. \tag{16}$$

Hence, for any $x \in X$, we have

$x \succ \lhd\psi$	iff	$\overline{v}(\lhd\psi) \leq \mathbf{x}$	definition of $\succ$ as in (3)
	iff	$\bigvee\{\lhd^{\mathbb{P}^+}\mathbf{x}' \mid x' \succ \psi\} \leq \mathbf{x}$	(16)
	iff	for all $x' \in X$, if $x' \succ \psi$ then $\lhd^{\mathbb{P}^+}\mathbf{x}' \leq \mathbf{x}$	definition of $\bigvee$
	iff	for all $x' \in X$, if $x' \succ \psi$ then $xR_\lhd x'$,	(17)

where we have *defined* the accessibility relation $R_\lhd \subseteq X \times X$ corresponding to the interpretation of $\lhd$ as $\lhd^{\mathbb{P}^+}$ as follows:

$$xR_\lhd x' \quad \text{iff} \quad \lhd^{\mathbb{P}^+}\mathbf{x}' \leq \mathbf{x}. \tag{17}$$

As the set of elements $\mathbf{x}$ for $x \in X$ are meet-generators of $\mathbb{P}^+$, we obtain the following equivalences:

	$a \Vdash \lhd\psi$	
iff	$\mathbf{a} \leq \overline{v}(\lhd\psi)$	definition of $\Vdash$ as in (2)
iff	$\mathbf{a} \leq \bigwedge\{\mathbf{x} \mid \overline{v}(\lhd\psi) \leq \mathbf{x}\}$	$\mathbf{x}$ for $x \in X$ meet-generators of $\mathbb{P}^+$
iff	for all $x \in X$, if $\overline{v}(\lhd\psi) \leq \mathbf{x}$ then $\mathbf{a} \leq \mathbf{x}$	definition of $\bigwedge$
iff	for all $x \in X$, if $x \succ \lhd\psi$ then aIx	definition of $\succ$ as in (3).

Next, we consider $\rhd$. Since $\overline{v}$ is a homomorphism, $\overline{v}(\rhd\psi) = \rhd^{\mathbb{P}^+}\overline{v}(\psi) = \rhd^{\mathbb{P}^+}(\bigvee\{\mathbf{a} \in \mathbb{P}^+ \mid a \Vdash \psi\})$, and since $\rhd^{\mathbb{P}^+}$ is completely join-reversing, we have:

$$\overline{v}(\rhd\psi) = \bigwedge\{\rhd^{\mathbb{P}^+}\mathbf{a} \mid a \Vdash \psi\}. \tag{18}$$

Hence, for any $a \in A$, we obtain the following:

$a \Vdash \rhd\psi$	iff	$\mathbf{a} \leq \overline{v}(\rhd\psi)$	definition of $\Vdash$ as in (2)
	iff	$\mathbf{a} \leq \bigwedge\{\rhd^{\mathbb{P}^+}\mathbf{a}' \mid a' \Vdash \psi\}$	(18)
	iff	for all $a' \in A$, if $a' \Vdash \psi$ then $\mathbf{a} \leq \rhd^{\mathbb{P}^+}\mathbf{a}'$	definition of $\bigwedge$
	iff	for all $a' \in A$, if $a' \Vdash \psi$ then $aR_{\rhd}a'$,	(19)

where we have *defined* the accessibility relation $R_{\rhd} \subseteq A \times A$ corresponding to the interpretation of $\rhd$ as $\rhd^{\mathbb{P}^+}$ as follows:

$$aR_{\rhd}a' \quad \text{iff} \quad \mathbf{a} \leq \rhd^{\mathbb{P}^+}\mathbf{a}'. \tag{19}$$

As the set of elements $\mathbf{a}$ for $a \in A$ are join-generators of $\mathbb{P}^+$, we obtain the following equivalences:

	$x \succ \rhd\psi$	
iff	$\overline{v}(\rhd\psi) \leq \mathbf{x}$	definition of $\succ$ as in (3)
iff	$\bigvee\{\mathbf{a} \mid \mathbf{a} \leq \overline{v}(\rhd\psi)\} \leq \mathbf{x}$	$\mathbf{a}$ for $a \in A$ join-generators of $\mathbb{P}^+$
iff	for all $a \in A$, if $\mathbf{a} \leq \overline{v}(\rhd\psi)$ then $\mathbf{a} \leq \mathbf{x}$	definition of $\bigvee$
iff	for all $a \in A$, if $a \Vdash \rhd\psi$ then aIx	definition of $\Vdash$ as in (2).

More generally, if f is monotone in its ith coordinate then, by appealing to Proposition 3.1, we write $\overline{v}(\varphi_i) = \bigvee\{\mathbf{a} \in \mathbb{P}^+ \mid \mathbf{a} \leq \overline{v}(\varphi_i)\} = \bigvee\{\mathbf{a} \in \mathbb{P}^+ \mid a \Vdash \varphi_i\}$, and if f is antitone in its i-th coordinate then we write $\overline{v}(\varphi_i) = \bigwedge\{\mathbf{x} \in \mathbb{P}^+ \mid \overline{v}(\varphi_i) \leq \mathbf{x}\} = \bigwedge\{\mathbf{x} \in \mathbb{P}^+ \mid x \succ \varphi_i\}$. Since by assumption $\overline{v}$ is a homomorphism, $\overline{v}(f(\overline{\varphi})) = f^{\mathbb{P}^+}(\overline{\overline{v}(\varphi)})$, and so, since $f^{\mathbb{P}^+}$ preserves arbitrary joins in each monotone coordinate and reverses arbitrary meets in each antitone coordinate, we get

$$\overline{v}(f(\overline{\varphi})) = \bigvee\{f^{\mathbb{P}^+}\left(\overline{\mathbf{a}}^{(n_f)}\right) \mid \overline{a}^{(n_f)} \Vdash^{(n_f)} \overline{\varphi}\}, \tag{20}$$

where $\overline{\mathbf{a}}^{(n_f)}$ is $\mathbf{a}$ and $\Vdash^{(n_f)}$ is $\Vdash$ if f is monotone in its i-th coordinate, whereas $\overline{\mathbf{a}}^{(n_f)}$ is $\mathbf{x}$ and $\Vdash^{(n_f)}$ is $\succ$ if f is antitone in its i-th coordinate. Hence, for any $x \in X$, we obtain the following:

$x \succ f(\overline{\varphi})$	iff	$\overline{v}(f(\overline{\varphi})) \leq \mathbf{x}$	definition of $\succ$ as in (3)
	iff	$\bigvee\{f^{\mathbb{P}^+}\left(\overline{\mathbf{a}}^{(n_f)}\right) \mid \overline{a}^{(n_f)} \Vdash^{(n_f)} \varphi\} \leq \mathbf{x}$	(20)
	iff	for all $\overline{a}^{(n_f)} \in \overline{A}^{(n_f)}$, if $\overline{a}^{(n_f)} \Vdash \overline{\varphi}$ then $f^{\mathbb{P}^+}(\overline{\mathbf{a}}^{(n_f)}) \leq \mathbf{x}$	definition of $\bigvee$
	iff	for all $\overline{a}^{(n_f)} \in \overline{A}^{(n_f)}$, if $\overline{a}^{(n_f)} \Vdash^{(n_f)} \overline{\varphi}$ then $R_f(x, \overline{a}^{(n_f)})$	(21).

Here we have *defined* the accessibility relation $R_f \subseteq X \times \overline{A}^{(n_f)}$, where $\overline{A}^{(n_f)}$ denotes the n_f-fold cartesian product of A and X such that for each $1 \leq i \leq n_f$ if f is monotone in its i-th coordinate then the i-th coordinate of $\overline{A}^{(n_f)}$ is A, whereas if f is antitone in its i-th coordinate then the i-th coordinate of $\overline{A}^{(n_f)}$ is X, corresponding to the interpretation of f as $f^{\mathbb{P}^+}$ as follows:

$$R_f(x, \overline{a}^{(n_f)}) \text{ iff } f^{\mathbb{P}^+}(\overline{\mathbf{a}}^{(n_f)}) \leq \mathbf{x}. \tag{21}$$

Now, using the fact that the set of elements $\mathbf{x}$ for $x \in X$ are meet-generators of $\mathbb{P}^+$, we can write:

	$a \Vdash f(\overline{\varphi})$	
iff	$\mathbf{a} \leq \overline{v}(f(\overline{\varphi}))$	definition of $\Vdash$ as in (2)
iff	$\mathbf{a} \leq \bigwedge\{\mathbf{x} \mid \overline{v}(f(\overline{\varphi})) \leq \mathbf{x}\}$	$\mathbf{x}$ for $x \in X$ meet-generators of $\mathbb{P}^+$
iff	for all $x \in X$, if $\overline{v}(f(\overline{\varphi})) \leq \mathbf{x}$ then $\mathbf{a} \leq \mathbf{x}$	definition of $\bigwedge$
iff	for all $x \in X$, if $x \succ f(\overline{\varphi})$ then aIx	definition of $\succ$ as in (3).

If g is monotone in its i-th coordinate then, by Proposition 3.1, we can write $\overline{v}(\varphi_i) = \bigwedge\{\mathbf{x} \in \mathbb{P}^+ \mid \overline{v}(\varphi_i) \leq \mathbf{x}\} = \bigwedge\{\mathbf{x} \in \mathbb{P}^+ \mid x \succ \varphi_i\}$, and if f is antitone in its i-th coordinate then we write $\overline{v}(\varphi_i) = \bigvee\{\mathbf{a} \in \mathbb{P}^+ \mid \mathbf{a} \leq \overline{v}(\varphi_i)\} = \bigvee\{\mathbf{a} \in \mathbb{P}^+ \mid a \Vdash \varphi_i\}$. Now, since $\overline{v}$ is a homomorphism, $\overline{v}(g(\overline{\varphi})) = g^{\mathbb{P}^+}(\overline{\overline{v}(\varphi)})$, and since $g^{\mathbb{P}^+}$ preserves arbitrary meets in each monotone coordinate and reverses arbitrary joins in each antitone coordinate, we get:

$$\overline{v}(g(\overline{\varphi})) = \bigwedge\{g^{\mathbb{P}^+}\left(\overline{\mathbf{x}}^{(n_g)}\right) \mid \overline{x}^{(n_g)} \succ^{(n_g)} \overline{\varphi}\}, \tag{22}$$

where $\overline{\mathbf{x}}^{(n_g)}$ is $\mathbf{x}$ and $\succ^{(n_g)}$ is $\succ$ if g is monotone in its i-th coordinate, whereas $\overline{\mathbf{x}}^{(n_g)}$ is $\mathbf{a}$ and $\succ^{(n_g)}$ is $\Vdash$ if g is antitone in its i-th coordinate. Thus, for any $a \in A$, we obtain the following equivalences:

$a \Vdash g(\overline{\varphi})$	iff	$\mathbf{a} \leq \overline{v}(f(\overline{\varphi}))$	definition of $\Vdash$ as in (2)
	iff	$\mathbf{a} \leq \bigwedge\{g^{\mathbb{P}^+}\left(\overline{\mathbf{x}}^{(n_g)}\right) \mid \overline{x}^{(n_g)} \succ^{(n_g)} \varphi\}$	(22)
	iff	for all $\overline{x}^{(n_g)} \in \overline{X}^{(n_g)}$, if $\overline{x}^{(n_g)} \succ \overline{\varphi}$ then $\mathbf{a} \leq g^{\mathbb{P}^+}(\overline{\mathbf{x}}^{(n_g)})$	definition of $\bigwedge$
	iff	for all $\overline{x}^{(n_g)} \in \overline{X}^{(n_g)}$, if $\overline{x}^{(n_g)} \succ^{(n_g)} \overline{\varphi}$ then $R_g(a, \overline{x}^{(n_g)})$	(23).

Here we have *defined* the accessibility relation $R_g \subseteq A \times \overline{X}^{(n_g)}$, where $\overline{X}^{(n_g)}$ denotes the n_g-fold cartesian product of A and X such that for each $1 \leq i \leq n_g$ if g is monotone in its i-th coordinate then the i-th coordinate of $\overline{X}^{(n_g)}$ is X, whereas if g is antitone in its i-th coordinate then the i-th coordinate of $\overline{X}^{(n_g)}$ is A, corresponding to the interpretation of g as $g^{\mathbb{P}^+}$ as follows:

$$R_g(a, \overline{x}^{(n_g)}) \text{ iff } \mathbf{a} \leq g^{\mathbb{P}^+}(\overline{\mathbf{x}}^{(n_g)}). \tag{23}$$

Since the set of elements $\mathbf{a}$ for $a \in A$ are join-generators of $\mathbb{P}^+$, we can write:

	$x \succ g(\overline{\varphi})$	
iff	$\overline{v}(g(\overline{\varphi})) \leq \mathbf{x}$	definition of $\succ$ as in (3)
iff	$\bigvee\{\mathbf{a} \mid \mathbf{a} \leq \overline{v}(g(\overline{\varphi}))\} \leq \mathbf{x}$	$\mathbf{a}$ for $a \in A$ join-generators of $\mathbb{P}^+$
iff	for all $a \in A$, if $\mathbf{a} \leq \overline{v}(g(\overline{\varphi}))$ then $\mathbf{a} \leq \mathbf{x}$	definition of $\bigvee$
iff	for all $a \in A$, if $a \Vdash g(\overline{\varphi})$ then aIx	definition of $\Vdash$ as in (2).

Finally, we show how the defining clauses of the relations $\Vdash$ and $\succ$ on models based on graphs can be retrieved from the algebraic interpretation in complete $\mathcal{L}_{\mathrm{LE}}$-algebras for $\Diamond$ and $\Box$. The clauses for $\lhd$, $\rhd$, f and g follow by similar arguments.

We first treat $\Diamond$. Analogously to the way we argued above, by appealing to Proposition 3.1, we can write $\overline{v}(\psi) = \bigvee\{\mathbf{z}_s \in \mathbb{X}^+ \mid \mathbf{z}_s \leq \overline{v}(\psi)\} = \bigvee\{\mathbf{z}_s \in \mathbb{X}^+ \mid z \Vdash \psi\}$. Since by assumption $\overline{v}$ is a homomorphism, $\overline{v}(\Diamond\psi) = \Diamond^{\mathbb{X}^+}\overline{v}(\psi) = \Diamond^{\mathbb{X}^+}(\bigvee\{\mathbf{z}_s \in \mathbb{X}^+ \mid z \Vdash \psi\})$, and since $\Diamond^{\mathbb{X}^+}$ is completely join-preserving, we obtain:

$$\overline{v}(\Diamond\psi) = \bigvee\{\Diamond^{\mathbb{X}^+}\mathbf{z}_s \mid z \Vdash \psi\}. \tag{24}$$

$x \succ \Diamond\psi$	iff	$\overline{v}(\Diamond\psi) \leq \mathbf{z}_r$	definition of $\succ$ as in (5)
	iff	$\bigvee\{\Diamond^{\mathbb{X}^+}\mathbf{z}_s \mid z \Vdash \psi\} \leq \mathbf{z}_r$	(24)
	iff	for all $z' \in Z$, if $z' \Vdash \psi$ then $\Diamond^{\mathbb{X}^+}\mathbf{z}_s \leq \mathbf{z}_r$	definition of $\bigvee$
	iff	for all $z' \in Z$, if $\Diamond^{\mathbb{X}^+}\mathbf{z}_s \not\leq \mathbf{z}_r$ then $z' \not\Vdash \psi$	contraposition
	iff	for all $z' \in Z$, if $zR_\Diamond z'$ then $z' \not\Vdash \psi$,	(25)

where we have *defined* the accessibility relation $R_\Diamond$ corresponding to the interpretation of $\Diamond$ as $\Diamond^{\mathbb{X}^+}$ as follows:

$$zR_\Diamond z' \quad \text{iff} \quad \Diamond^{\mathbb{X}^+}\mathbf{z}_z \not\leq \mathbf{z}_r. \tag{25}$$

	$z \Vdash \Diamond\psi$	
iff	$\mathbf{z}_s \leq \overline{v}(\Diamond\psi)$	definition of $\Vdash$ as in (4)
iff	$\mathbf{z}_s \leq \bigwedge\{\mathbf{z}_r \mid \overline{v}(\Diamond\psi) \leq \mathbf{z}_r\}$	$\mathbf{z}_r$ for $z \in Z$ meet-generators of $\mathbb{X}^+$
iff	for all $z' \in Z$, if $\overline{v}(\Diamond\psi) \leq \mathbf{z}_r$ then $\mathbf{z}_s \leq \mathbf{z}_r$	definition of $\bigwedge$
iff	for all $z' \in Z$, if $z' \succ \Diamond\psi$ then $zI_{E^c}z'$	definition of $\succ$ as in (5).
iff	for all $z' \in Z$, if $z' \succ \Diamond\psi$ then $(z,z') \notin E$	definition of I_{E^c}
iff	for all $z' \in Z$, if zEz' then $z' \not\succ \Diamond\psi$	contraposition.

Since by assumption $\overline{v}$ is a homomorphism, $\overline{v}(\Box\psi) = \Box^{\mathbb{X}^+}\overline{v}(\psi) = \Box^{\mathbb{X}^+}(\bigwedge\{\mathbf{z}_r \in \mathbb{X}^+ \mid z \succ \psi\})$, and since $\Box^{\mathbb{X}^+}$ is completely meet-preserving, we have:

$$\overline{v}(\Box\psi) = \bigwedge\{\Box^{\mathbb{X}^+}\mathbf{z}_r \mid z \succ \psi\}. \tag{26}$$

Hence, for any $z \in Z$, we have

$z \Vdash \Box\psi$	iff	$\mathbf{z}_s \leq \overline{v}(\Box\psi)$	definition of $\Vdash$ as in (4)
	iff	$\mathbf{z}_s \leq \bigwedge\{\Box^{\mathbb{X}^+}\mathbf{z}_r \mid z \succ \psi\}$	(26)
	iff	for all $z' \in Z$, if $z' \succ \psi$ then $\mathbf{z_s} \leq \Box^{\mathbb{X}^+}\mathbf{z}_r$	definition of $\bigwedge$
	iff	for all $z' \in Z$, if $\mathbf{z_s} \nleq \Box^{\mathbb{X}^+}\mathbf{z}_r$ then $z' \not\succ \psi$	contraposition
	iff	for all $z' \in Z$, if $zR_\Box z'$ then $z' \not\succ \psi$	(27)

where we have *defined* the accessibility relation $R_\Box$ corresponding to the interpretation of $\Box$ as $\Box^{\mathbb{X}^+}$ as follows:

$$zR_\Box z' \quad \text{iff} \quad \mathbf{z}_s \nleq \Box^{\mathbb{X}^+}\mathbf{z}_r. \tag{27}$$

	$z \succ \Box\psi$	
iff	$\overline{v}(\Box\psi) \leq \mathbf{z}_r$	definition of $\succ$ as in (5)
iff	$\bigvee\{\mathbf{z}_s \mid \mathbf{z}_s \leq \overline{v}(\Box\psi)\} \leq \mathbf{z}_r$	$\mathbf{z}_s$ for $z \in Z$ join-generators of $\mathbb{X}^+$
iff	for all $z' \in Z$, if $\mathbf{z}_s \leq \overline{v}(\Box\psi)$ then $\mathbf{z}_s \leq \mathbf{z}_r$	definition of $\bigvee$
iff	for all $z' \in Z$, if $z' \Vdash \Box\psi$ then $z' I_{E^c} z$	definition of $\Vdash$ as in (4).
iff	for all $z' \in Z$, if $z' \Vdash \Box\psi$ then $(z', z) \notin E$	definition of I_{E^c}
iff	for all $z' \in Z$, if $z'Ez$ then $z' \nVdash \Box\psi$	contraposition.

A.2 Polarity-based frames

In this subsection we give the definition of a polarity-based frame, as well as a suitable expanded adjunction based on the adjunction between complete lattices and polarities described in Section 3.1. We will focus mainly on the treatment of the additional connectives.

Before we give the definition of a polarity-based frame, we introduce some notation. For any sets A, B and any relation $S \subseteq A \times B$, we let, for any $A' \subseteq A$ and $B' \subseteq B$,

$$S^{(1)}[A'] := \{b \in B \mid \forall a(a \in A' \Rightarrow aSb)\} \text{ and } S^{(0)}[B'] := \{a \in A \mid \forall b(b \in B' \Rightarrow aSb)\}.$$

For all sets $A, B_1, \ldots B_n$, and any relation $S \subseteq A \times B_1 \times \cdots \times B_n$, for any $\overline{C} := (C_1, \ldots, C_n)$ where $C_i \subseteq B_i$ and $1 \leq i \leq n$ we let, for all A',

$$\overline{C}^i := (C_1, \ldots, C_{i-1}, C_{i+1}, \ldots, C_n)$$

$$\overline{C}^i_{A'} := (C_1 \ldots, C_{i-1}, A', C_{i+1}, \ldots, C_n)$$

When $C_i := \{c_i\}$ and $A' := \{a'\}$, we will write $\overline{c}$ for $\overline{\{c\}}$, and $\overline{c}^i$ for $\overline{\{c\}}^i$, and $\overline{c}^i_{a'}$ for $\overline{\{c\}}^i_{\{a'\}}$. We also let:

1. $S^{(0)}[\overline{C}] := \{a \in A \mid \forall \overline{b}(\overline{b} \in \overline{C} \Rightarrow aS\overline{b})\}$.
2. $S_i \subseteq B_i \times B_1 \times \cdots \times B_{i-1} \times A \times B_{i+1} \times \cdots \times B_n$ be defined by

$$(b_i, \overline{c}^i_a) \in S_i \text{ iff } (a, \overline{c}) \in S.$$

3. $S^{(i)}[A', \overline{C}^i] := S_i^{(0)}[\overline{C}^i_{A'}]$.

Definition A.1. *A* polarity-based frame *for $\mathcal{L}_{LE}$ is a tuple $\mathbb{F} = (\mathbb{P}, \mathscr{R}_{\mathcal{F}}, \mathscr{R}_{\mathcal{G}})$, where $\mathbb{P} = (A, X, I)$ is a polarity, $\mathscr{R}_{\mathcal{F}} = \{R_f \mid f \in \mathcal{F}\}$, and $\mathscr{R}_{\mathcal{G}} = \{R_g \mid g \in \mathcal{G}\}$, such that for each $f \in \mathcal{F}$ and $g \in \mathcal{G}$, the symbols R_f and R_g respectively denote (n_f+1)-ary and (n_g+1)-ary relations on $\mathbb{P}$,*

$$R_f \subseteq X \times \overline{A}^{(n_f)} \text{ and } R_g \subseteq A \times \overline{X}^{(n_g)},$$

where $\overline{A}^{(n_f)}$ denotes the n_f-fold cartesian product of A and X such that for each $1 \leq i \leq n_f$ if f is monotone in its i-th coordinate then the ith coordinate of $\overline{A}^{(n_f)}$ is A, whereas if f is antitone in its i-th coordinate then the ith coordinate of $\overline{A}^{(n_f)}$ is X, and $\overline{X}^{(n_g)}$ denotes the n_g-fold cartesian product of A and X such that for each $1 \leq i \leq n_g$ if g is monotone in its i-th coordinate then the i-th coordinate of $\overline{X}^{(n_g)}$ is X, whereas if g is antitone in its i-th coordinate then the i-th coordinate of $\overline{X}^{(n_g)}$ is A. In addition, all relations R_f and R_g are required to be compatible, *i.e. the following sets are assumed to be Galois-stable for all $a \in A$, $x \in X$, $\overline{a} \in A^{(n_f)}$, and $\overline{x} \in X^{(n_g)}$:*

$$R_f^{(0)}[\overline{a}] \text{ and } R_f^{(i)}[x, \overline{a}^i] \qquad R_g^{(0)}[\overline{x}] \text{ and } R_g^{(i)}[a, \overline{x}^i].$$

Definition A.2. *For a complete lattice expansion $\mathbb{A} = (\mathbb{L}, \mathcal{F}^{\mathbb{L}}, \mathcal{G}^{\mathbb{L}})$, the $\mathcal{L}_{LE}$-frame associated with $\mathbb{A}$ is the structure $\mathbb{F}_{\mathbb{A}} = (\mathbb{P}_{\mathbb{L}}, R_{\mathcal{F}}, \mathscr{R}_{\mathcal{G}})$, where $\mathbb{P}_{\mathbb{L}}$ is the polarity $(L, L, \leq)$, $\mathscr{R}_{\mathcal{F}} = \{R_f \mid f \in \mathcal{F}\}$, and $\mathscr{R}_{\mathcal{G}} = \{R_g \mid g \in \mathcal{G}\}$, such that for each $f \in \mathcal{F}$ and $g \in \mathcal{G}$, the relations R_f and R_g are defined as follows:*

$$R_f(x, \overline{a}^{(n_f)}) \text{ iff } f^{\mathbb{P}^+_{\mathbb{L}}}(\overline{\mathbf{a}}^{(n_f)}) \leq \mathbf{x} \qquad R_g(a, \overline{x}^{(n_g)}) \text{ iff } \mathbf{a} \leq g^{\mathbb{P}^+_{\mathbb{L}}}(\overline{\mathbf{x}}^{(n_g)}).$$

Proposition A.3. *For a complete lattice expansion $\mathbb{A} = (\mathbb{L}, \mathcal{F}^{\mathbb{L}}, \mathcal{G}^{\mathbb{L}})$, the associated $\mathcal{L}_{LE}$-frame $\mathbb{F}_{\mathbb{A}}$ is a polarity-based frame.*

The *complex algebra* of a polarity-based frame $\mathbb{F} = (\mathbb{P}, \mathscr{R}_{\mathcal{F}}, \mathscr{R}_{\mathcal{G}})$ for $\mathcal{L}_{LE}$ is the algebra

$$\mathbb{F}^+ = (\mathbb{L}, \{f_{R_f} \mid f \in \mathcal{F}\}, \{g_{R_g} \mid g \in \mathcal{G}\}),$$

where $\mathbb{L} := \mathbb{P}^+$, and for all $f \in \mathcal{F}$ and all $g \in \mathcal{G}$, we let

1. $f_{R_f} : \mathbb{L}^{n_f} \to \mathbb{L}$ be defined by the assignment $f_{R_f}(\overline{c}) = \left(\left(R_f^{(0)}[\overline{[\![c]\!]}^{(n_f)}]\right)^{\downarrow}, R_f^{(0)}[\overline{[\![c]\!]}^{(n_f)}]\right)$;
2. $g_{R_g} : \mathbb{L}^{n_g} \to \mathbb{L}$ be defined by the assignment $g_{R_g}(\overline{c}) = \left(R_g^{(0)}[\overline{(\![c]\!)}^{(n_g)}], \left(R_g^{(0)}[\overline{(\![c]\!)}^{(n_g)}]\right)^{\uparrow}\right)$.

Here $\overline{[\![c]\!]}^{(n_f)}$ denotes the n_f-tuple of $[\![c]\!]$ and $(\![c]\!)$ such that for each $1 \leq i \leq n_f$ if f is monotone in its i-th coordinate then the i-th coordinate of $\overline{[\![c]\!]}^{(n_f)}$ is $[\![c]\!]$, whereas if f is antitone in its i-th coordinate then the ith coordinate of $\overline{[\![c]\!]}^{(n_f)}$ is $(\![c]\!)$, and $\overline{(\![c]\!)}^{(n_g)}$ denotes the n_g-tuple of $[\![c]\!]$ and $(\![c]\!)$ such that for each $1 \leq i \leq n_g$ if g is monotone in its i-th coordinate then the i-th coordinate of $\overline{[\![c]\!]}^{(n_f)}$ is $(\![c]\!)$, whereas if g is antitone in its i-th coordinate then the i-th coordinate of $\overline{(\![c]\!)}^{(n_g)}$ is $[\![c]\!]$.

Proposition A.4 (cf. [41] Proposition 21). *If $\mathbb{F} = (\mathbb{P}, \mathscr{R}_{\mathscr{F}}, \mathscr{R}_{\mathscr{G}})$ is a polarity-based frame for $\mathscr{L}_{\mathrm{LE}}$, then $\mathbb{F}^+ = (\mathbb{L}, \{f_{R_f} \mid f \in \mathscr{F}\}, \{g_{R_g} \mid g \in \mathscr{G}\})$ is a complete lattice expansion.*

A.3 Graph-based frames

Here we define a graph-based frame and give a suitable expanded adjunction based on the adjunction between lattices and graphs described in Section 3.1.

For any sets A, B and any relation $S \subseteq A \times B$, we let, for any $A' \subseteq A$ and $B' \subseteq B$,

$$A^{[1]}[A'] := \{b \mid \forall a(a \in A' \Rightarrow aS^c b)\} \qquad S^{[0]}[B'] := \{a \mid \forall b(b \in B' \Rightarrow aS^c b)\}^{14}.$$

Hence, $S^{[1]}[A'] = (S^c)^{(1)}[A']$ and $S^{[0]}[B'] = (S^c)^{(0)}[B']$. More generally, for all sets $A, B_1, \ldots, B_n$, and any relation $S \subseteq A \times B_1 \times \cdots \times B_n$, for any $\overline{C} := (C_1, \ldots, C_n)$ where $C_i \subseteq B_i$ and $1 \leq i \leq n$ we let, for all A',

$$S^{[0]}[\overline{C}] := \{a \in A \mid \forall \overline{b}(\overline{b} \in \overline{C} \Rightarrow aS^c \overline{b})\} \qquad S^{[i]}[A', \overline{C}^i] := S_i^{[0]}[\overline{C}^i_{A'}].$$

Definition A.5. *A* graph-based frame *for $\mathscr{L}_{\mathrm{LE}}$ is a tuple $\mathbb{F} = (\mathbb{X}, \mathscr{R}_{\mathscr{F}}, \mathscr{R}_{\mathscr{G}})$, where $\mathbb{X} = (Z, E)$ is a reflexive graph, $\mathscr{R}_{\mathscr{F}} = \{R_f \mid f \in \mathscr{F}\}$, and $\mathscr{R}_{\mathscr{G}} = \{R_g \mid g \in \mathscr{G}\}$, such that for each $f \in \mathscr{F}$ and $g \in \mathscr{G}$, the symbols R_f and R_g respectively denote $(n_f + 1)$-ary and $(n_g + 1)$-ary relations on $\mathbb{P}_{\mathbb{X}}$ satisfying the following* compatibility *conditions: for all $z' \in Z$, $\overline{z}'' \in Z^{n_f}$ and $\overline{z}''' \in Z^{n_g}$,*

$$(R_g^{[0]}[\overline{z}'''])^{[10]} \subseteq R_g^{[0]}[\overline{z}'''] \qquad (R_f^{[0]}[\overline{z}''])^{[01]} \subseteq R_f^{[0]}[\overline{z}''],$$

and for every $1 \leq i \leq n_g$ and every $1 \leq j \leq n_f$ such that g and f are positive in their ith (resp. jth) coordinate,

$$(R_g^{[i]}[z', \overline{z}^i])^{[01]} \subseteq R_g^{[i]}[z', \overline{z}^i] \qquad (R_f^{[j]}[z', \overline{z}^j])^{[10]} \subseteq R_f^{[i]}[z', \overline{z}^i],$$

[14]We will sometimes abbreviate $E^{[0]}[X]$ and $E^{[1]}[Y]$ as $X^{[0]}$ and $Y^{[1]}$, respectively, for each $X, Y \subseteq Z$. If $X = \{x\}$ and $Y = \{y\}$, we write $x^{[0]}$ and $y^{[1]}$ for $\{x\}^{[0]}$ and $\{y\}^{[1]}$, and write $X^{[01]}$ and $Y^{[10]}$ for $(X^{[0]})^{[1]}$ and $(Y^{[1]})^{[0]}$, respectively. Notice that $X^{[0]} = X^{\downarrow}$ and $Y^{[1]} = Y^{\uparrow}$, where the maps $(\cdot)^{\downarrow}$ and $(\cdot)^{\uparrow}$ are those associated with the polarity $\mathbb{P}_{\mathbb{X}}$.

and for every $1 \leq i \leq n_g$ *and every* $1 \leq j \leq n_f$ *such that* g *and* f *are negative in their ith (resp. jth) coordinate,*

$$(R_g^{[i]}[z', \overline{z}^i])^{[10]} \subseteq R_g^{[i]}[z', \overline{z}^i] \qquad (R_f^{[j]}[z', \overline{z}^j])^{[01]} \subseteq R_f^{[i]}[z', \overline{z}^i].$$

Remark A.6. *Let* $\mathbb{F}$ *be a graph based frame as above. By Proposition 4.1, when* $E := \leq$ *is reflexive and transitive, the closure operator* $(-)^{[10]}$ *coincides with the upward closure* $(-)\uparrow$ *of* E *(and order-dually, the closure operator* $(-)^{[01]}$ *coincides with the downward closure* $(-)\downarrow$ *of* E*). In this case, the compatibility conditions on the additional relations of* $\mathbb{F}$ *are equivalent to the well-known interaction conditions between accessibility relations and order on Kripke frames for distributive modal logic. As an illustration, let us show that the compatibility conditions on* $R_\Box$ *project to* $(\leq \circ R_\Box \circ \leq) \subseteq R_\Box$.

By [14, Lemma 4], the conditions $(R_\Box^{[1]}[b])^{[01]} \subseteq R_\Box^{[1]}[b]$ *and* $(R_\Box^{[0]}[y])^{[10]} \subseteq R_\Box^{[0]}[y]$ *for all* $b, y \in Z$ *are respectively equivalent to* $R_\Box^{[0]}[Y^{[01]}] = R_\Box^{[0]}[Y]$ *and* $(R_\Box^{[0]}[Y])^{[10]} \subseteq R_\Box^{[0]}[Y]$ *for every* $Y \subseteq Z$*, which, as discussed above, can be equivalently rewritten as* $R_\Box^{[0]}[Y\downarrow] = R_\Box^{[0]}[Y]$ *and* $(R_\Box^{[0]}[Y])\uparrow \subseteq R_\Box^{[0]}[Y]$ *for every* $Y \subseteq Z$.

The former condition is clearly equivalent to $(R_\Box^{[0]}[Y\downarrow])^c = (R_\Box^{[0]}[Y])^c$*, while the latter requires* $R_\Box^{[0]}[Y]$ *to be an up-set, and is hence equivalent to the one requiring* $(R_\Box^{[0]}[Y])^c$ *to be a down-set for every* $Y \subseteq Z$*, i.e.* $(R_\Box^{[0]}[Y])^c\downarrow \subseteq (R_\Box^{[0]}[Y])^c$*. Note that, for every* $Y' \subseteq Z$,

$$\begin{aligned}
(R_\Box^{[0]}[Y'])^c : &= (\{b \in Z \mid \forall y (y \in Y' \Rightarrow (b,y) \notin R_\Box)\})^c \\
&= (\{b \in Z \mid \forall y (b R_\Box y \Rightarrow y \notin Y')\})^c \\
&= \{b \in Z \mid \exists y (b R_\Box y \ \& \ y \in Y')\} \\
&= R_\Box^{-1}[Y'].
\end{aligned}$$

Hence, $(R_\Box^{[0]}[Y\downarrow])^c = (R_\Box^{[0]}[Y])^c$ *and* $(R_\Box^{[0]}[Y])^c\downarrow \subseteq (R_\Box^{[0]}[Y])^c$ *can be equivalently rewritten as* $R_\Box^{-1}[Y\downarrow] = R_\Box^{-1}[Y]$ *and* $(R_\Box^{-1}[Y])\downarrow \subseteq R_\Box^{-1}[Y]$*, respectively. These conditions are equivalent to* $(R_\Box \circ \leq)^{-1}[Y] = (\geq \circ R_\Box^{-1})[Y] = R_\Box^{-1}[Y]$ *and* $(\leq \circ R_\Box)^{-1}[Y] = (R_\Box^{-1} \circ \geq)[Y] \subseteq R_\Box^{-1}[Y]$*, which are equivalent to* $(R_\Box \circ \leq)[Y] = R_\Box[Y]$ *and* $(\leq \circ R_\Box)[Y] \subseteq R_\Box[Y]$*, which are equivalent to* $(R_\Box \circ \leq) = R_\Box$ *and* $(\leq \circ R_\Box) \subseteq R_\Box$*, as required.*

Definition A.7. *The associated* $\mathcal{L}_{\mathrm{LE}}$*-frame of any lattice expansion* $\mathbb{A} = (\mathbb{L}, \mathcal{F}^{\mathbb{L}}, \mathcal{G}^{\mathbb{L}})$ *is defined by* $\mathbb{F}_{\mathbb{A}} = (\mathbb{X}_{\mathbb{L}}, R_{\mathcal{F}}, \mathcal{R}_{\mathcal{G}})$*, where* $\mathbb{X}_{\mathbb{L}}$ *is defined as in Section 3.1,* $\mathcal{R}_{\mathcal{F}} = \{R_f \mid f \in \mathcal{F}\}$*, and* $\mathcal{R}_{\mathcal{G}} = \{R_g \mid g \in \mathcal{G}\}$*, such that for each* $f \in \mathcal{F}$ *and* $g \in \mathcal{G}$*, the relations* R_f *and* R_g *are defined as follows:*

$$R_f(z, \overline{z}) \ \textit{iff} \ f^{\mathbb{X}^+}(\overline{\mathbf{z}_s}^{(n_f)}) \nleq \mathbf{z}_r \qquad R_g(z, \overline{z}) \ \textit{iff} \ \mathbf{z}_s \nleq g^{\mathbb{X}^+}(\overline{\mathbf{z}_r}^{(n_g)}),$$

where the i-th component of $\overline{\mathbf{z}_s}^{(n_f)}$ is $\mathbf{z}_s$ is $(\mathbf{z}_r)$ if f is monotone (antitone) in its i-th argument, and the i-th component of $\overline{\mathbf{z}_r}^{(n_g)}$ is $\mathbf{z}_r$ $(\mathbf{z}_s)$ if g is monotone (antitone) in its i-th argument.

Proposition A.8. *The $\mathscr{L}_{LE}$-frame $\mathbb{F}_{\mathbb{A}}$ associated with any lattice expansion $\mathbb{A} = (\mathbb{L}, \mathscr{F}^{\mathbb{L}}, \mathscr{G}^{\mathbb{L}})$ is a graph-based frame.*

The *complex algebra* of a graph-based frame $\mathbb{F} = (\mathbb{X}, \mathscr{R}_{\mathscr{F}}, \mathscr{R}_{\mathscr{G}})$ is the algebra $\mathbb{F}^+ = (\mathbb{L}, \{f_{R_f} \mid f \in \mathscr{F}\}, \{g_{R_g} \mid g \in \mathscr{G}\})$, where $\mathbb{L} := \mathbb{P}^+_{\mathbb{X}}$, and for all $f \in \mathscr{F}$ and all $g \in \mathscr{G}$, we let

1. $f_{R_f} : \mathbb{L}^{n_f} \to \mathbb{L}$ be defined by the assignment $f_{R_f}(\overline{c}) = \left(\left(R_f^{[0]}[\overline{[\![c]\!]}^{(n_f)}] \right)^{[0]}, R_f^{[0]}[\overline{[\![c]\!]}^{(n_f)}] \right)$;
2. $g_{R_g} : \mathbb{L}^{n_g} \to \mathbb{L}$ be defined by the assignment $g_{R_g}(\overline{c}) = \left(R_g^{[0]}[\overline{(\![c]\!)}^{(n_g)}], \left(R_g^{[0]}[\overline{(\![c]\!)}^{(n_g)}] \right)^{[1]} \right)$.

As before $\overline{[\![c]\!]}^{(n_f)}$ denotes the n_f-tuple of $[\![c]\!]$ and $(\![c]\!)$ such that for each $1 \leq i \leq n_f$ if f is monotone in its i-th coordinate then the i-th coordinate of $\overline{[\![c]\!]}^{(n_f)}$ is $[\![c]\!]$, whereas if f is antitone in its i-th coordinate then the i-th coordinate of $\overline{[\![c]\!]}^{(n_f)}$ is $(\![c]\!)$, and similarly for $\overline{(\![c]\!)}^{(n_g)}$.

Proposition A.9. *If $\mathbb{F} = (\mathbb{X}, \mathscr{R}_{\mathscr{F}}, \mathscr{R}_{\mathscr{G}})$ is a graph-based frame for $\mathscr{L}_{LE}$, then $\mathbb{F}^+ = (\mathbb{P}^+_{\mathbb{X}}, \{f_{R_f} \mid f \in \mathscr{F}\}, \{g_{R_g} \mid g \in \mathscr{G}\})$ is a complete lattice expansion.*

References

[1] Gerhard Allwein, and Chrysafis Hartonas *Duality for bounded lattices*, Indiana University Logic Group Preprint Series, IULG-93–25, 1993.

[2] Antoine Arnauld, and Pierre Nicole Logic or the Art of Thinking (translated and edited by Jill Vance Buroker), Cambridge University Press, Cambridge 1996 [*Cambridge Texts in the History of Philosophy*].

[3] Alexandru Baltag, Lawrence S. Moss, and Sławomir Solecki *The logic of public announcements, common knowledge, and private suspicions*, in: Horacio Arló-Costa, Vincent Hendricks, and Johan van Benthem (eds.) Readings in Formal Epistemology, *Sourcebook*, Springer, Cham, etc. 2016 [*Springer Graduate Texts in Philosophy 1*], pp. 773–812.

[4] Mohua Banerjee, and Mihir K. Chakraborty *Rough sets through algebraic logic*, Fundamenta Informaticae 28 (3, 4), 1996, pp. 211–221.

[5] Francesco Belardinelli, Peter Jipsen, and Hiroakira Ono *Algebraic aspects of cut elimination*, Studia Logica 77 (2), 2004, pp. 209–240.

[6] Garrett Birkhoff *Lattices and their applications*, Bulletin of the American Mathematical Society 44 (12), 1938, pp. 793–800.

[7] — Lattice Theory, American Mathematical Society, New York 1940 [*AMS Colloquium Publications 25*].

[8] Garrett Birkhoff, and John von Neumann *The logic of quantum mechanics*, Annals of Mathematics, 1936, pp. 823–843.

[9] Cecelia Britz Correspondence Theory in Many-valued Modal Logics, Master's thesis, University of Johannesburg, South Africa, 2016.

[10] Maria Luisa Dalla Chiara *A general approach to non-distributive logics*, Studia Logica 35 (2), 1976, pp. 139–162.

[11] Agata Ciabattoni, Nikolaos Galatos, and Kazushige Terui *Algebraic proof theory for substructural logics: cut-elimination and completions*, Annals of Pure and Applied Logic 163 (3), 2012, pp. 266–290.

[12] Willem Conradie, and Andrew Craig *Relational semantics via TiRS graphs*, in: Topology, Algebra, and Categories in Logic (Proceedings TACL, Salerno, 15–26 June 2015) [extended abstract].

[13] Willem Conradie, Andrew Craig, Alessandra Palmigiano, and Nachoem M. Wijnberg *Modelling competing theories*, in: Proceedings of the 11th Conference of the European Society for Fuzzy Logic and Technology (EUSFLAT 2019), Atlantis Press, Amsterdam, etc. 2019 [*Atlantis Studies in Uncertainty Modelling*, pp. 721–739.

[14] — *Modelling informational entropy*, in: Rosalie Iemhoff, Michael Moortgat, and Ruy de Queiroz (eds.) Logic, Language, Information, and Computation (Poceedings of the 26th International Workshop, WoLLIC 2019, Utrecht), Springer, Berlin 2019 [*Lecture Notes in Computer Science 11541*], pp. 140–160.

[15] Willem Conradie, Sabine Frittella, Krishna Manoorkar, Sajad Nazari, Alessandra Palmigiano, Apostolos Tzimoulis, and Nachoem M. Wijnberg *Rough concepts*, Information Sciences [accepted, 2020], arXiv:1907.00359.

[16] Willem Conradie, Sabine Frittella, Alessandra Palmigiano, Michele Piazzai, Apostolos Tzimoulis, and Nachoem M. Wijnberg, *Categories: How I learned to stop worrying and love two sorts*, in: Jouko Väänänen, Åsa Hirvonen, and Ruy de Queiroz (eds.) Logic, Language, Information, and Computation (Proceedings of the 23rd International Workshop, WoLLIC 2016), Springer, Berlin 2016 [*Lecture Notes in Computer Science 9803*], pp. 145–164.

[17] — *Toward an epistemic-logical theory of categorization*, in: Jérôme Lang (ed.) Proceedings of the Sixteenth Conference on Theoretical Aspects of Rationality and Knowledge (TARK 2017, Liverpool, 24–26 July 2017) [*Electronic Proceedings in Theoretical Computer Science 251*], pp. 167–186.

[18] Willem Conradie, Silvio Ghilardi, and Alessandra Palmigiano *Unified correspondence*, in: Alexandru Baltag, and Sonja Smets (eds.), Johan van Benthem on Logic and Information Dynamics, Springer International Publishing, New York 2014 [*Outstanding Contributions to Logic 5*], pp. 933–975.

[19] Willem Conradie, and Alessandra Palmigiano *Algorithmic correspondence and canonicity for distributive modal logic*, Annals of Pure and Applied Logic 163 (3), 2012, pp. 338–376.

[20] — *Algorithmic correspondence and canonicity for non-distributive logics*, Annals of Pure and Applied Logic 170 (9), 2019, pp. 923–974.

[21] — *Constructive canonicity of inductive inequalities*, in: Logical Methods in Computer Sci-

ence [accepted, 2020] (see also arXiv:1603.08341).

[22] Willem Conradie, Alessandra Palmigiano, Claudette Robinson, Apostolos Tzimoulis, and Nachoem M. Wijnberg *Modelling socio-political competition*, Fuzzy Sets and Systems, 2020 [in press].

[23] — *The logic of vague categories*, arXiv:1908.04816, 2019.

[24] Willem Conradie, Alessandra Palmigiano, and Apostolos Tzimoulis *Goldblatt-Thomason for LE-logics*, arXiv:1809.08225, 2018.

[25] Andrew Craig, Maria Gouveia, and Miroslav Haviar *TiRS graphs and TiRS frames: a new setting for duals of canonical extensions*, Algebra Universalis 74 (1–2), 2015, pp. 123–138.

[26] Andrew Craig, Miroslav Haviar, and Hilary A. Priestley *A fresh perspective on canonical extensions for bounded lattices*, Applied Categorical Structures 21 (6), 2013, pp. 725–749.

[27] Jon M. Dunn *Positive modal logic*, Studia Logica 55 (2), 1995, pp. 301–317.

[28] — *Gaggle theory: an abstraction of Galois connections and residuation, with applications to negation, implication, and various logical operators*, in: J. van Eijck (ed.) Logics in AI (Proceedings of the European Workshop JELIA '90 Amsterdam, September 10–14, 1990), Springer, Berlin, etc. 1991 [*Lecture Notes in Artificial Intelligence 478*], pp. 31–51.

[29] Jon M. Dunn, Mai Gehrke, and Alessandra Palmigiano *Canonical extensions and relational completeness of some substructural logics*, The Journal of Symbolic Logic 70 (3), 2005, pp. 713–740.

[30] Ronald Fagin, Joseph Y. Halpern, Yoram Moses, and Moshe Y. Vardi Reasoning about Knowledge, The MIT Press, Cambridge MA 2004.

[31] Gisèle Fischer Servi *On modal logic with an intuitionistic base*, Studia Logica 36 (3), 1977, pp. 141–149.

[32] Nikolaos Galatos, Peter Jipsen, Tomasz Kowalski, and Hiroakira Ono Residuated Lattices, *An Algebraic Glimpse at Substructural Logics*, Elsevier, Amsterdam 2007.

[33] Bernhard Ganter, and Rudolf Wille Formal Concept Analysis, *Mathematical Foundations*, Springer, Berlin 2012.

[34] Mai Gehrke, and John Harding *Bounded lattice expansions*, Journal of Algebra 238, 2012, pp. 345–371.

[35] Mai Gehrke, Hideo Nagahashi, and Yde Venema *A Sahlqvist theorem for distributive modal logic*, Annals of Pure and Applied Logic 131, 2005, pp. 65–102.

[36] Mai Gehrke *Generalized Kripke frames*, Studia Logica 84 (2), 2006, pp. 241–275.

[37] Jean-Yves Girard *Linear logic*, Theoretical Computer Science 50 (1), 1987, pp. 1–101.

[38] Robert I. Goldblatt *Semantic analysis of orthologic*, Journal of Philosophical Logic 3 (1–2), 1974, pp. 19–35.

[39] — *Morphisms and duality for polarities and lattices with operators*, arXiv:1902.09783, 2019.

[40] — *Canonical extensions and ultraproducts of polarities*, Algebra Universalis 78 [paper no 80], 2018 [DOI 10.1007/s00012-018-0562-4].

[41] Giuseppe Greco, Peter Jipsen, Fey Liang, Alessandra Palmigiano, and A. Tzimoulis *Algebraic proof theory for LE-logics*, arXiv:1808.04642, 2018.

[42] Giuseppe Greco, Fey Liang, Michael Moortgat, Alessandra Palmigiano, and Apostolos Tzimoulis *Vector spaces as Kripke frames*, arXiv:1908.05528, 2019.

[43] Chrysafis Hartonas *Duality for lattice-ordered algebras and for normal algebraizable logics*, Studia Logica 58 (3), 1997, pp. 403–450.

[44] — *Order-dual relational semantics for non-distributive propositional logics*, Logic Journal of the IGPL 25 (2), 2017, pp. 145–182.

[45] — *Order-dual relational semantics for non-distributive propositional logics: a general framework*. Journal of Philosophical Logic 47 (1), 2018, pp. 67–94.

[46] — *Stone duality for lattice expansions*, Logic Journal of the IGPL 26 (5), 2018, pp. 475–504.

[47] Chrysafis Hartonas, and Jon M. Dunn *Stone duality for lattices*, Algebra Universalis, 37 (3), 1997, pp. 391–401.

[48] Gerd Hartung *A topological representation of lattices*, Algebra Universalis 29 (2), 1992, pp. 273–299.

[49] Mikael Kågebäck, Olof Mogren, Nina Tahmasebi, and Devdatt Dubhashi *Extractive summarization using continuous vector space models*, in: Alexandre Allauzen, Raffaella Bernardi, Edward Grefenstette, Hugo Larochelle, Christopher Manning, and Scott Wen-tau Yih (eds.) Proceedings of the 2nd Workshop on Continuous Vector Space Models and their Compositionality (CVSC, Gothenburg, April 26–30, 2014), EACL 2014, pp. 31–39.

[50] Joachim Lambek *The mathematics of sentence structure*, The American Mathematical Monthly 65 (3), 1958, pp. 154–170.

[51] — *On the calculus of syntactic types*, in: Roman Jakobson (ed.) Structure of Language and its Mathematical Aspects (Proceedings of the Twelfth Symposium in Applied Mathematics, held in New York City, April 14–15, 1960), AMS, Providence RI 1961 [*Proceedings of Symposia in Applied Mathematics 12*], pp. 166–178.

[52] George W. Mackey *Quantum mechanics and Hilbert space*, The American Mathematical Monthly 64 (8P2), 1957, pp. 45–57.

[53] Michael Moortgat *Categorial type logics*, in: Johan van Benthem, and Alice ter Meulen (eds.) Handbook of Logic and Language, Elsevier, Amsterdam 1997, pp. 93–177.

[54] — *Symmetric categorial grammar*, Journal of Philosophical Logic 38 (6), 2009, pp. 681–710.

[55] M. Andrew Moshier, and Peter Jipsen *Topological duality and lattice expansions, I: A topological construction of canonical extensions*, Algebra Universalis 71 (2), 2014, pp. 109–126.

[56] — *Topological duality and lattice expansions, II: Lattice expansions with quasioperators*, Algebra Universalis 71 (3), 2014, pp. 221–234.

[57] Robert M. Nosofsky *Stimulus bias, asymmetric similarity, and classification*, Cognitive Psychology 23 (1), 1991, pp. 94–140.

[58] Ewa Orłowska *Rough set semantics for non-classical logics*, in: Wojciech P. Ziarko (ed.) Rough Sets, Fuzzy Sets and Knowledge Discovery, Springer, Berlin 1994, pp. 143–148.

[59] Ewa Orłowska (ed.) Incomplete Information, *Rough Set Analysis*, Physica-Verlag, Heidelberg 2013 [*Studies in Fuzziness and Soft Computing 13*].

[60] Zdzislaw Pawlak *Rough set theory and its applications to data analysis*, Cybernetics & Systems 29 (7), 1998, pp. 661–688.

[61] Miroslav Ploščica *A natural representation of bounded lattices*, Tatra Mountains Mathematical Publications 5, 1995, pp. 75–88.

[62] Cecylia Rauszer *A formalization of the propositional calculus of H-B logic*, Studia Logica, 33 (1), 1974, pp. 23–34.

[63] Richard Routley, and Robert K. Meyer *The semantics of entailment*, in: Hugues Leblanc (ed.) Truth, Syntax and Modality, North-Holland Publishing Company, Amsterdam 1973 [*Studies in Logic and the Foundations of Mathematics 68*], pp. 199–243.

[64] Gerard Salton, Anita Wong, and Chung-Shu Yang *A vector space model for automatic indexing*, Communications of the ACM 18 (11), 1975, pp. 613–620.

[65] Hanamantagouda P. Sankappanavar *Semi-De Morgan algebras*, The Journal of Symbolic Logic 52 (3), 1987, pp. 712–724.

[66] Peter D. Turney, and Patrick Pantel *From frequency to meaning: Vector space models of semantics*, Journal of Artificial Intelligence Research 37, 2010, pp. 141–188.

[67] Amos Tversky *Features of similarity*, Psychological Review 84 (4), 1977, pp. 327.

[68] Alasdair Urquhart *A topological representation theory for lattices*, Algebra Universalis 8 (1), 1978, pp. 45–58.

[69] Frank Wolter *On logics with coimplication*, Journal of Philosophical Logic 27 (4), 1998, pp. 353–387.

[70] Yiyu Yao, and Pawan J. Lingras *Interpretations of belief functions in the theory of rough sets*, Information Sciences 104 (1–2), 1998, pp. 81–106.

Bounded functional interpretation with an abstract type

Patrícia Engrácia
Direção-Geral de Estatísticas da Educação e Ciência
p.engracia@gmail.com

Fernando Ferreira
Faculdade de Ciências da Universidade de Lisboa
fjferreira@fc.ul.pt

ABSTRACT We explain what is the bounded functional interpretation and discuss its main theorem. We work in classical logic and include an abstract type. The type can stand for a metric space, a normed space, a ring, etc. The inclusion of an abstract type is important for applications in proof mining. Some examples are given.

1 Logical roots

Functional interpretations for classical logic are generalizations and extensions of Herbrand's theorem:

Theorem 1 (Herbrand's theorem). *Let $\mathcal{L}$ be a language of first-order logic without equality with at least a constant symbol. Suppose that $\exists x A(x)$ is an existential sentence of $\mathcal{L}$ (i.e., $A(x)$ is quantifier-free). If $\exists x A(x)$ is provable in first-order classical logic, then there are finitely many closed terms $t_1,\ldots,t_n$ such that the quantifier-free sentence*

$$A(t_1) \vee \cdots \vee A(t_n)$$

is also provable in first-order classical logic.

A sabbatical leave from Faculdade de Ciências da Universidade de Lisboa freed Fernando Ferreira from teaching duties during the academic year of 2019/2020. The work was partially supported by Fundação para a Ciência e a Tecnologia under the grant UIDB/04561/2020 and Centro de Matemática, Aplicações Fundamentais e Investigação Operacional of Universidade de Lisboa.

This is Jacques Herbrand's *theorème fondamental* of his 1930 doctoral dissertation [24]. The proof of the theorem has an involved story because Herbrand's argument had a serious flaw. The interested reader can consult [23] for this story. Nowadays, there are many proofs of the theorem. It can be proved using a very simple compactness argument (together with the completeness theorem of first-order logic). It can also be proved using Gentzen's *Hauptsatz*. In the review [33] of [38], Georg Kreisel mentioned with nonchalance that Herbrand's theorem can be proved using a functional interpretation. This is illuminating and, decades later, the deed was carried out: see [19] and, especially, [15].

The theorem can be easily generalized in various ways. Instead of a single variable x, we may have a tuple of variables. Instead of provability without assumptions, we may have universal assumptions. As a consequence, the theorem also holds for theories with universal axiomatizations (do notice that these theories may contain equality, since equality has a universal axiomatization). Instead of an existential sentence, we may have a $\forall\exists$ sentence (adapting the end result accordingly). An important case is when the theorem is applied to the (universal) theory of first-order primitive recursive arithmetic:

Theorem 2. *Let* PRA *be the theory of first-order primitive recursive arithmetic. Consider* $\forall x \exists y A(x, y)$ *a sentence of the language of* PRA *with* $A(x, y)$ *a quantifier-free formula. Suppose that* $\mathsf{PRA} \vdash \forall x \exists y A(x, y)$*. Then there is a primitive recursive term* $t(x)$ *such that* $\mathsf{PRA} \vdash \forall x A(x, t(x))$*.*

The Herbrand disjunctions disappear because the theory PRA allows definitions by cases. The *dialectica* interpretation of Kurt Gödel [21], adapted to classical logic, is a generalization of this result.

In 1980, Jeffrey Paris [35] published a proof of the following result:

Theorem 3. *Suppose that the theory* $\mathsf{I}\Delta_0 + \mathsf{B}\Sigma_0$ *proves the sentence* $\forall x \exists y A(x, y)$*, where* A *is a bounded formula. Then there is a term* $t(x)$ *of the language of* $\mathsf{I}\Delta_0$ *such that* $\mathsf{I}\Delta_0 \vdash \forall x \exists y \leq t(x)\, A(x, y)$*.*

In the above, $\mathsf{I}\Delta_0$ is the subsystem of Peano arithmetic with induction restricted to the bounded formulas (a good reference for first-order arithmetic, including bounded arithmetic, is [6]). The scheme $\mathsf{B}\Sigma_0$ is the scheme of bounded collection:

$$\forall x \leq t\, \exists y\, A(x, y) \rightarrow \exists w\, \forall x \leq t\, \exists y \leq w\, A(x, y),$$

where A is a bounded formula and t is a term in which x does not occur. Note that it is known that $\mathsf{B}\Sigma_0$ is not a consequence of $\mathsf{I}\Delta_0$ (see [34] or [36]). Paris proved the theorem using a compactness argument and bounded induction.

The result, however, has nothing to do with induction (nor recursion). This was first observed by Samuel Buss, who generalized in [5] the above theorem to theories with no induction. It is a quite general result. Buss gave two proofs, one proof-theoretic, using a cut-elimination argument, and the other model-theoretic, using resplendent models. However, the result of Buss can also be proved model-theoretically in a much simpler way, as shown in [10]. The proof given in this latter paper is distilled – as it is said explicitly in the paper – from a proof in [9] showing the conservativity of weak König's lemma over a certain theory of bounded arithmetic. The point of the present discussion is that $\mathsf{B}\Sigma_0$ has analogues in higher types, namely weak König's lemma. We can formulate (the contrapositive of) weak König's lemma in the following way:

$$\forall X \exists y\, A(X, y) \to \exists w \forall X \exists y \leq w\, A(X, y),$$

where A is a bounded formula (of the language of second-order arithmetic). There are evident formal similarities between $\mathsf{B}\Sigma_0$ and the above display of weak König's lemma (as we will see, the second-order quantification $\forall X$ is considered a bounded quantification).

The bounded functional interpretation introduced by the second author and Paulo Oliva in [17], when adapted to the classical setting as in [11], is a generalization of the above result of Paris. Do notice that all the papers [35], [5], [9] and [10] appeared before the bounded functional interpretation was discovered.

Functional interpretations effect a trade-off between quantifications and higher order functionals. We must go beyond the language of first-order logic, to languages with finite-types, in order to effect the trade-off. The perspective allowed by going to the finite-type languages widens the view immensely and uncovers hidden phenomena. Let us briefly discuss this. By its own nature, some so-called *characteristic principles* emerge from the trade-off. These principles are appropriately conservative over the base theories. The characteristic principles for the *dialectica* interpretation (adapted to the classical logic) take the form of the scheme of quantifier-free choice:

$$\forall x \exists y A(x, y) \to \exists f \forall x A(x, f(x)),$$

where A is a quantifier-free formula, x and y are of certain given types and f is variable of the type of a function from the elements of type of x to the elements of type of y. We do not see any of the choice principles in the first-order language. The bounded functional interpretation (for classical logic) has, on the other hand, three forms of characteristic principles. One of them is also a form of choice. The other two are the majorizability principles (we sidestep them at this moment) and the so-called bounded collection principles. There are instances of the bounded collection principles that can be formulated in the original first-order base language:

they are, precisely, the instances of $\mathsf{B}\Sigma_0$. We can see already a trace of the bounded collection principles in the base first-order language, but there are more of these principles in higher types. Weak König's lemma is also a characterictic principle of the bounded functional interpretation, but it is only detected at the level of second-order languages. It goes without saying that weak König's lemma is mathematically very interesting. When we add an abstract type to the bounded functional interpretation, the bounded collection principles can take the form of other interesting mathematical principles. We discuss some of these principles in Subsection 7.1 and Subsection 7.2. These mathematical principles have implications for the applied program of proof mining.

2 A brief on the base arithmetical setting

We are not going to describe the language of finite-type arithmetic, nor the axioms of finite-type Peano arithmetic, in any detail. In this section, we just establish some notation, make some observations and direct the reader to [1] for a proper full treatment. The language of finite-type arithmetic is a predicate language with infinitely many sorts (domains of variation), one sort for each finite arithmetical type. There is the ground type N of the natural numbers. Given types σ and τ, there is the arrow type $\sigma \to \tau$. These are all the (arithmetical) types there are. The standard full set-theoric interpretation of these domains of variation is given by the family of sets $S^\omega = \langle S_\sigma \rangle_{\sigma \text{ a type}}$, where $S_{\mathrm{N}} = \mathbb{N}$ and $S_{\sigma\to\tau} = {}^{S_\sigma}S_\tau$ (the set of all functions from S_σ to S_τ).

The language has denumerably many variables of each type, usually written with type superscripts u^{N}, x^σ, y^σ, z^τ, $w^{\sigma\to\tau}$, etc., when it is necessary to indicate the type of the variable. It includes the basic arithmetical zero constant 0 of type N and the successor constant S of type $\mathrm{N} \to \mathrm{N}$. The language also has constants called recursors, one recursor R_σ associated to each type σ (the recursor itself is not of type σ). We do not describe the recursors, but add two notes: (1) the most basic recursor, the recursor R_{N}, allows the definition of the primitive recursive functions; (2) with the aid of the other recursors, we are able "to interpret" the induction scheme or, to put the matter in the terms of the previous section, we are able to obtain the induction scheme from a universal theory as a consequence of the characteristic principles (see the end of Section 6). With the variables and the constants, we can build the terms of the language by means of the primitive operation of application: variables and constants are terms of their corresponding type and, given t a term of type $\sigma \to \tau$ and q a term of type σ, then tq (the result of applying t to q) is a term of type τ.

In finite-type arithmetic, we need to have a way of associating to each term t of type τ and variable x of type ρ (possibly, but not necessarily, occurring in t), a term of type $\rho \to \tau$ which, intuitively, denotes the function from elements of type ρ to elements of type τ given by the rule $x \rightsquigarrow t$. This new term of type $\rho \to \tau$ is denoted in lambda notation by $\lambda x.t$. A very elegant way of building these lambda terms (combinatorial completeness) is by using special constants called the combinators.

The formulation of equality in finite-type theories has been subject to many proposals (see [2] for a good discussion on these matters). For the purpose of the bounded functional interpretation, we may settle for a neutral theory of equality in which we have a binary relation symbol of equality Eq_σ at each type σ. These are the only relation symbols of our pure arithmetical language. As mentioned, the axioms of equality are universal: $\mathrm{Eq}_\sigma(x, x)$ and $\mathrm{Eq}_\sigma(x, y) \wedge A \to A'$, for A quantifier-free, x, y variables of type σ and A' a formula obtained from A by replacing some (maybe all) occurrences of the variable x by the variable y. At the ground type, we use the standard notation $x =_{\mathrm{N}} y$, or simply $x = y$, instead of $\mathrm{Eq}_{\mathrm{N}}(x, y)$. The usual universal axioms concerning the combinators and the recursors are stated with these equality relation symbols.

In set-theoretic mathematics, the definition of equality is sameness of extensions. In type $\mathrm{N} \to \mathrm{N}$, the set theoretic definition takes the form: $f =_{\mathrm{N}\to\mathrm{N}} g$ if, and only if, $\forall n^{\mathrm{N}}(fn =_{\mathrm{N}} gn)$. However, by itself, this definition does not entail the indiscernibility of identicals. We do not have

$$\forall f^{\mathrm{N}\to\mathrm{N}} \forall g^{\mathrm{N}\to\mathrm{N}} (\forall n^{\mathrm{N}}(fn =_{\mathrm{N}} gn) \to \forall \Phi^{(\mathrm{N}\to\mathrm{N})\to\mathrm{N}}(\Phi f =_{\mathrm{N}} \Phi g)),$$

unless it is postulated. This postulate is, of course, a form of the axiom of extensionality. It is important to remark that there is a serious problem with this principle of extensionality vis-à-vis obtaining a Herbrand like theorem: it is not a universal statement. In fact, the principle is refuted by the bounded functional interpretation, as was shown in [12]. In proof mining studies – as we presently understand them – we simply cannot allow the principle of extensionality.

3 A brief on majorizability

It is a distinctive mark of the bounded functional interpretation that its estimates are given by bounds, not by precise witnesses. We introduce, simultaneously by recursion on the type σ, the notion of (strong) majorizability $\leq^*_\sigma$ and the domain of the strongly majorizable functionals M_σ. At the ground type N, M_{N} is $\mathbb{N}$ and $\leq^*_{\mathrm{N}}$ is the usual less than or equal binary relation $\leq$ between natural numbers. Given types σ and τ, we define the binary relation $\leq^*_{\sigma\to\tau}$ between elements x and y of

${}^{M_\sigma}M_\tau$ as follows:

$$x \leq^*_{\sigma\to\tau} y \text{ is, by definition, } \forall u, v \in M_\sigma\, (u \leq^*_\sigma v \to xu \leq^*_\tau yv \wedge yu \leq^*_\tau yv).$$

The notion of (strong) majorizability is transitive but it is not reflexive. We say that a functional $x \in {}^{M_\sigma}M_\tau$ is majorizable if there is $y \in {}^{M_\sigma}M_\tau$ such that $x \leq^*_{\sigma\to\tau} y$. In this situation, it is automatic that $y \leq^*_{\sigma\to\tau} y$ (so-called monotone functionals). Note that monotone functionals are majorizable. Finally, the domain $M_{\sigma\to\tau}$ is, by definition, $\{x \in {}^{M_\sigma}M_\tau : x \text{ is majorizable}\}$.

These notions are due to Marc Bezem [3], building upon (and modifying) a previous notion of majorizability defined by William Howard in [27]. The majorizability structure M^ω is the family of domains $\langle M_\sigma \rangle_{\sigma \text{ a type}}$. Clearly, M^ω is closed with respect to application. All the constants of the language of finite-type Peano arithmetic, when interpreted in the natural way, are members M^ω. Therefore, M^ω is a structure for finite-type arithmetic. For details, clarifications and other properties of majorizability, we direct the reader to [30]. It is clear that $M_{\mathrm{N}} = S_{\mathrm{N}}$, $M_{\mathrm{N}\to\mathrm{N}} = S_{\mathrm{N}\to\mathrm{N}}$, but $M_{(\mathrm{N}\to\mathrm{N})\to\mathrm{N}}$ is a proper subset of $S_{(\mathrm{N}\to\mathrm{N})\to\mathrm{N}}$.

For reasons that will be apparent in a short while, we extend the language of finite-type arithmetic with a primitive binary relation symbol $\trianglelefteq_\sigma$, for each type σ. The new symbols stand for ersatz relations of the corresponding relations $\leq^*_\sigma$, sharing between them some properties, but not all. At the ground type, $\trianglelefteq_{\mathrm{N}}$ and $\leq^*_{\mathrm{N}}$ are the same. They would be the same at the other types if we had the equivalences:

$$x \trianglelefteq_{\sigma\to\tau} y \leftrightarrow \forall u^\sigma, v^\sigma (u \trianglelefteq_\sigma v \to xu \trianglelefteq_\tau yv \wedge yu \trianglelefteq_\tau yv).$$

We can postulate the implication from left to right but, unfortunately, we cannot postulate the implication from right to left. The reason for this obstruction was already discussed in the previous section: the latter implication is not a universal statement. When the bounded functional interpretation appeared in [17], the left to right implication was accepted as an axiom and the right to left implication was turned into a rule. This is a viable approach, but the use of rules makes the system a bit opaque. This is certainly a difficulty for the common mathematician, since rules – as opposed to axioms – never show up in ordinary mathematics.

If the aim of our investigation is mainly proof mining, there is a simple way out of the conundrum and avoid rules. In order to describe the solution, we introduce bounded quantifiers as a syntactic device. For each type σ, we have bounded quantifiers $\forall x \trianglelefteq_\sigma t$, where t is a term of type σ in which the variable x does not occur. Bounded quantifications of the form $\forall x \trianglelefteq_\sigma t\,(\ldots)$ are axiomatically equivalent to $\forall x^\sigma (x \trianglelefteq_\sigma t \to (\ldots))$. A bounded formula is a formula in which all occurring quantifiers are bounded. Given a formula in the extended language with the new binary relation symbols, its flattening is the formula of the original language obtained from the given formula by replacing every symbol $\trianglelefteq_\sigma$ by the corresponding relation symbol $\leq^*_\sigma$ (bounded quantifiers are first replaced by their axiomatic equivalents, as preparation). We are now ready to describe which truths do we accept regarding the new symbols. We accept all the universal closures of bounded formulas whose flattenings are true in the majorizability structure M^ω. Here are some examples of formulas whose universal closures we do accept:

$x \trianglelefteq_\sigma y \to y \trianglelefteq_\sigma y$

$x \trianglelefteq_\sigma y \wedge y \trianglelefteq_\sigma z \to x \trianglelefteq_\sigma z$

$x \trianglelefteq_{\mathrm{N}\to\mathrm{N}} \tilde{x}$, where $\tilde{x}(n) = \max_{k\leq n} xk$

$z \trianglelefteq_\sigma x \wedge w \trianglelefteq_\sigma y \to \max_\sigma(z, w) \trianglelefteq_\sigma \max_\sigma(x, y)$, where $\max_{\mathrm{N}}$ gives the maximum of the given natural numbers and $\max_{\sigma\to\tau}(x, y) = \lambda u^\sigma. \max_\tau(xu, yv)$

$x \trianglelefteq_{\mathrm{N}\to\mathrm{N}} y \to x \leq^*_{\mathrm{N}\to\mathrm{N}} y$

On the other hand, we do not accept $x \leq^*_{\mathrm{N}\to\mathrm{N}} y \to x \trianglelefteq_{\mathrm{N}\to\mathrm{N}} y$, nor do we accept $\forall u^\sigma, v^\sigma (u \trianglelefteq_\sigma v \to xu \trianglelefteq_\tau yv \wedge yu \trianglelefteq_\tau yv) \to x \trianglelefteq_{\sigma\to\tau} y$. These formulas, even though their flattenings are true in M^ω, are nevertheless existential.

Real numbers are represented by functionals of type $\mathrm{N} \to \mathrm{N}$. Ulrich Kohlenbach uses in [30] a representation based on Cauchy sequences. Engrácia uses the signed digit representation in [7]. Be that as it may, it is important to note that the (defined) relations of $=_\mathbb{R}$ and $\leq_\mathbb{R}$ between (representations of) real numbers are given by Π^0_1-formulas, and strict equality $<_\mathbb{R}$ is given by a Σ^0_1-formula. Suppose that $x \leq_\mathbb{R} y$ is given by the formula $\forall n^{\mathrm{N}}(p(x, y, n) =_{\mathrm{N}} 0)$. We define $x \trianglelefteq_\mathbb{R} y$ by $\lambda n.p(x, y, n) \trianglelefteq_{\mathrm{N}\to\mathrm{N}} \lambda n.0$. We have

$x <_\mathbb{R} y \to x \trianglelefteq_\mathbb{R} y$

$x \trianglelefteq_\mathbb{R} y \to x \leq_\mathbb{R} y$

because the above two formulas are universal and their flattenings are obviously true. The converse statements are not accepted because they are given by existential formulas. The following equivalences are clear (and useful):

$$\begin{array}{lcl} x =_\mathbb{R} y & \leftrightarrow & \forall n(|x - y| \trianglelefteq_\mathbb{R} \frac{1}{n+1}) \\ x \leq_\mathbb{R} y & \leftrightarrow & \forall n(x \trianglelefteq_\mathbb{R} y + \frac{1}{n+1}) \\ x <_\mathbb{R} y & \leftrightarrow & \exists n(x + \frac{1}{n+1} \trianglelefteq_\mathbb{R} y) \end{array}$$

4 Adding an abstract type

In the seminal paper [28], Kohlenbach introduced abstract types in proof mining studies (the sequel with Philipp Gerhardy [20] extends the analysis further). The consideration of abstract types enlarges our purview considerably since now we can consider spaces whose elements are not presented via representations. The elements of the space are given abstractly, as in ordinary mathematics, and not be means of codes. In particular, the spaces under study need not be countable or countably based (as is separable metric spaces). The first author of this paper introduced abstract types within the framework of the bounded functional interpretation in her doctoral dissertation [7]. This was done for normed spaces. In the following, we strive for generality with the aim of laying a common framework for many particular cases.

We add a new abstract ground type X to the ground type N of the natural numbers. The extended types are obtained from the ground types N and X using the arrow. We have the extended types X, N $\to$ X, X $\to$ X, (X $\to$ N) $\to$ X, etc. Types now go beyond the mere arithmetical. Given a nonempty set X, the set-theoretic domains of variation associated with the new types are obtained as in the arithmetical case, with the further ground clause $S_{\mathrm{X}} = X$. For instance, $S_{\mathrm{N}\to\mathrm{X}}$ is the space of all sequences of elements of X. We denote the set-theoretic interpretation by $S^{\omega,X}$.

As before, we have denumerably many variables associated with each type. The combinators are extended to the new types, as well as the recursors (see [30]). We have equality symbols Eq_σ for the extended types σ, as well. This is the extended language of finite-type arithmetic. We formulate the (universal) equality axioms for these symbols and, by means of them, state the usual axioms concerning the extended combinators and recursors. The logical part of the system is settled at this point. Before discussing the mathematical side of the systems (the ground type X can stand for a ring, a metric space, etc.), we now turn to the majorizability issues needed to set up the bounded functional interpretation.

Given an non-empty set X, in order to speak of majorizabilty within $S^{\omega,X}$, we need to endow X with a binary relation $\leq^*_X$ between the elements of X and the natural numbers. The relation $\leq^*_X$ must satisfy $x \leq^*_X n \wedge n \leq m \to x \leq^*_X m$, for all $x \in X$ and n, m natural numbers. To round things up, we also require that $\forall x \in X \exists n \in \mathbb{N}\, (x \leq^*_X n)$. We follow [20] in associating, to each extended type σ, the arithmetical type $\hat{\sigma}$ that is obtained from σ by replacing every occurrence of the type symbol X by the type symbol N. So, $\hat{\mathrm{X}} = \mathrm{N}$, $\widehat{\mathrm{N} \to \mathrm{X}} = \widehat{\mathrm{X} \to \mathrm{N}} = \widehat{\mathrm{X} \to \mathrm{X}} = \mathrm{N} \to \mathrm{N}$ and $\widehat{\sigma \to \tau} = \hat{\sigma} \to \hat{\tau}$. Of course, if σ is an arithmetical type, $\hat{\sigma} = \sigma$.

As in [20], we extend the notion of majorizability $\leq^*_\sigma$ and the domain of the strongly majorizable functionals M_σ to the new types. At the new ground type X, M_{X} is X and $\leq^*_{\mathrm{X}}$ is the given relation. Given extended types σ and τ, we define the binary relation $\leq^*_{\sigma\to\tau}$ between elements x of ${}^{M_\sigma}M_\tau$ and y of ${}^{M_{\hat\sigma}}M_{\hat\tau}$ as follows:

$$x \leq^*_{\sigma\to\tau} y \text{ is, by definition, } y \leq^*_{\hat\sigma\to\hat\tau} y \wedge \forall u \in M_\sigma \forall v \in M_{\hat\sigma}\, (u \leq^*_\sigma v \to xu \leq^*_\tau yv).$$

We say that a functional $x \in {}^{M_\sigma}M_\tau$ is majorizable if there is $y \in {}^{M_{\hat\sigma}}M_{\hat\tau}$ such that $x \leq^*_{\sigma\to\tau} y$. The domain $M_{\sigma\to\tau}$ is, by definition, $\{x \in {}^{M_\sigma}M_\tau : x \text{ is majorizable}\}$. The extended majorizability structure $M^{\omega,X}_{\leq^*}$ is the family $\langle M_\sigma\rangle_{\sigma \text{ an extended type}}$. This family is naturally a structure for the extended language of finite-type arithmetic and a model of its axioms.

In analogy with the arithmetic case, we add to the syntax primitive binary relation symbols $\trianglelefteq_\sigma$ infixing between terms of type σ and arithmetical terms of type $\hat\sigma$. Bounded quantifiers are introduced in the usual manner, and there is the attendant notion of bounded formula. The flattening of a formula also makes sense. We accept as axioms all the universal closures of bounded formulas whose flattenings are true in every extended majorizability structure $M^{\omega,X}_{\leq^*}$.

We have been discussing a typed language with an abstract type and a majorizability notion, and axioms for this language. On top of this language, we can add (mathematical) constants to the language and formulate axioms with mathematical content. For each constant c of type σ that we add, there must be a companion closed term t_c of arithmetical type $\hat\sigma$. This is called a *majorizability language*. We postulate $c \trianglelefteq_\sigma t_c$. Adapting (and complementing) a notion first introduced in [17], a majorizability theory is a theory formulated in a majorizability language constituted only by universal closures of bounded formulas.

5 Examples

In this section, we give some examples of majorizability theories. They will be further worked out in Section 7 after the main theorem of the bounded functional interpretation is stated and proved. The first four examples are of well-known mathematical structures, complemented with a majorizability relation and with the companions of the mathematical constants. The axioms of the theories are true in the intended structures (after flattening). In the last example, we introduce the general setting that will be needed for Section 6.

5.1 Commutative rings

We have the language of ring theory, with the constants 0^{R}, 1^{R} and $+$ and $\cdot$ of type $\mathrm{R} \to (\mathrm{R} \to \mathrm{R})$ and $-$ (for the symmetric) of type $\mathrm{R} \to \mathrm{R}$. Here, R is the ground abstract type. Given a ring R, the majorizability relation $x \leq^*_{\mathrm{R}} n$ at the ground type is the trivial one, holding universally. So, we have the universal axiom $\forall x \forall n (x \trianglelefteq_{\mathrm{R}} n)$. With this majorizability relation, quantifications over the ground type (over the ring) are bounded quantifications. In a sense, the elements of the ring are computationally inert. The constants 0^{R} and 1^{R} have 0^{N} as their common companion. The common companion of the constants $+$ and $\cdot$ is $\lambda n, m.0^{\mathrm{N}}$. The companion of the constant $-$ is $\lambda n.0^{\mathrm{N}}$. The mathematical axioms of our theory are the axioms of nontrivial commutative rings with identity. E.g., $\forall x^{\mathrm{R}}, y^{\mathrm{R}}(x + y = y + x)$. Here, $x + y = y + z$ abbreviates $\mathrm{Eq}_{\mathrm{X}}(x + y, y + x)$.

An element e of a ring R is called nilpotent if there is $n \in \mathbb{N}$ such that $e^n = 0$. It is a classical theorem of commutative ring theory that e is nilpotent if, and only if, e lies in every prime ideal of R. The substantial direction of this result is the right-to-left direction:

$$\forall e \in R\, [\forall n (e^n \neq 0) \to \exists X \subseteq R\, (X \text{ is a prime ideal and } e \notin X)].$$

The textbook proof of this result uses Zorn's lemma. We can state the above in our language:

$$\forall e^{\mathrm{R}}\, [\forall n^{\mathrm{N}} (e^n \neq 0^{\mathrm{R}}) \to \exists X \trianglelefteq_{\mathrm{R}\to\mathrm{N}} 1^{\mathrm{N}\to\mathrm{N}}\, (X \text{ is a prime ideal and } Xe \neq 0^{\mathrm{N}})]. \quad (1)$$

The application $(e, n) \rightsquigarrow e^n$ is defined by recursion, $1^{\mathrm{N}\to\mathrm{N}}$ is $\lambda k^{\mathrm{N}}.1^{\mathrm{N}}$ and X stands for the characteristic function of the prime ideal. Note that the statement "X is a prime ideal" is a bounded statement. The quantification over subsets of the ring is also bounded. In fact, the only unbounded quantifier in the above formula is the numerical quantifier $\forall n^{\mathrm{N}}$.

The classical theorem is not amenable to a proof mining treatment using the bounded functional interpretation. We explain why this is so in Subsection 7.1. That notwithstanding, we present in that section a modification of the theorem that can be proof mined.

The next three examples were considered first in [28] and [20] within the framework of the monotone functional interpretation of Ulrich Kohlenbach. Here, we adapt (and, occasionally, improve/simplify) them for the framework of the bounded functional interpretation.

5.2 Bounded metric spaces

The language of metric spaces has only one constant d, whose type is $\mathrm{X} \to \mathrm{X} \to (\mathrm{N} \to \mathrm{N})$ (the distance function takes values in elements of type $\mathrm{N} \to \mathrm{N}$ because the real numbers are represented by elements of this type). Fix $b \in \mathbb{N}$. We are interested in metric spaces with diameter bounded by b. We have the universal axiom $\forall x^{X}, y^{X}(d(x,y) \leq_{\mathbb{R}} b_{\mathbb{R}})$, where $b_{\mathbb{R}}$ is a canonical representative of the real number associated with b. The majorizability relation at the ground type X is, again, the trivial one. Therefore, we accept the axiom $\forall x \forall n\,(x \trianglelefteq_{X} n)$.

In order to describe the companion of d, we must address a subtle technical issue. It concerns the intercourse between the relations $\leq^{*}_{\mathrm{N}\to\mathrm{N}}$ and $\leq_{\mathbb{R}}$ when infixing between representatives of real numbers. In an interpretion of the distance function, the value of $d(x,y)$ must be a function from $\mathbb{N}$ to $\mathbb{N}$ that represents the real number that actually gives the distance between x and y. Which representative, we may ask? For the signed digit representation of the real numbers, any representative will do. For the Cauchy representation this is not the case, and we must be careful in chosing the representatives (see [30]). If we adopt the signed digit representation, the companion of d can be taken to be $\lambda n, m, k.\,(2b+3)$. The reason why this companion works relies on the nice fact that the signed digit representations of the real numbers α of the closed bounded interval $[-b,b]$ are such that $\alpha(k) \leq 2b+3$, for all natural numbers k.

The mathematical axioms are:

$$\forall x^{X}(d(x,x) =_{\mathbb{R}} 0_{\mathbb{R}})$$

$$\forall x^{X}, y^{X}(d(x,y) =_{\mathbb{R}} d(y,x))$$

$$\forall x^{X}, y^{X}, z^{X}(d(x,z) \leq_{\mathbb{R}} d(x,y) + d(y,z))$$

The above axioms are universal closures of bounded formulas (take notice of the presence of a hidden universal numerical quantifier in each axiom).

For proof mining reasons, the mathematical work in metric spaces is done with an equality at ground type different from the logical equality Eq_{X}. The mathematical equality $x =_{X} y$ is defined by the universal formula $d(x,y) =_{\mathbb{R}} 0_{\mathbb{R}}$. Note that whereas the logical equality Eq_{X} is quantifier-free, the mathematical equality $=_{X}$ is universal. We have $\mathrm{Eq}_{X}(x,y) \to x =_{X} y$, but not the converse (the converse is an existential formula). The equality $=_{X}$ is mathematically informative (it enforces a kind of arquimedian property) and provides a connection between the abstract metric space X and the natural numbers. It must clearly be said that this connection is fulcral for applications. The mathematical equality $=_{X}$ is reflexive, symmetric and transitive (thanks to the triangle inequality), but does not enjoy the property of indiscernibility of identicals. However, some indiscernibility is provable, and this is important. For instance, the equality $=_{X}$ is congruent with respect to the distance function: $x =_{X} y \wedge z =_{X} w \to d(x,z) =_{X} d(y,w)$.

5.3 Metric spaces

When the metric space is unbounded, we should consider a pointed metric space, i.e., a metric space with a distinguished point a. In syntactic terms, we have a constant a of type X. The majorizability relation $x \leq^*_X n$ at the ground type is now defined by $d(a,x) \leq_{\mathbb{R}} n_{\mathbb{R}}$. We accept the universal axiom: $\forall x^{\mathrm{X}} \forall n^{\mathrm{N}} (x \trianglelefteq_{\mathrm{X}} n \leftrightarrow d(a,x) \trianglelefteq_{\mathbb{R}} n_{\mathbb{R}})$. The companion of a is 0^{N}. If we use the signed digit representation of the real numbers, a companion to d is $\lambda n, m, k.(2(n+m)+3)$.

Suppose that we are given a condition $A(x)$ in a pointed metric space (X,d), with point a. Assume that there is at least one element falling under the condition. In mathematics, we can consider the distance between a and the class of elements satisfying the condition $A(x)$: it is $\inf_{x \text{ such that } A(x)} d(a,x)$. In the theories of this paper, this infimum need not exist (for lack of comprehension). Consider, nevertheless, the following weakening:

$$\forall k \in \mathbb{N}\, \exists x \in X \left(A(x) \wedge \forall y \in X \left(A(y) \to d(a,x) \leq d(a,y) + \frac{1}{k+1} \right) \right). \tag{2}$$

This result can be mined within our framework. Moreover, a simple case of it has been used in many proof mining studies. We look into this matter in Subsection 7.3.

5.4 Normed spaces

We have the language of vector spaces, with a constant 0^{X} for the zero vector, a constant $+$ of type $\mathrm{X} \to (\mathrm{X} \to \mathrm{X})$ for the vector sum, a constant $-$ of type $\mathrm{X} \to \mathrm{X}$ for the symmetric operation, and a constant $\cdot$ of type $(\mathrm{N} \to \mathrm{N}) \to (\mathrm{X} \to \mathrm{X})$ for scalar multiplication. The scalars are the reals and, as observed, they are represented by elements of type $\mathrm{N} \to \mathrm{N}$. It is simple to arrange things so that the scalar multiplication applies to every functional of type $\mathrm{N} \to \mathrm{N}$, not just to those that represent reals. In fact, it is easy to arrange things so that quantifying over the reals gets reduced to quantifying over functionals of type $\mathrm{N} \to \mathrm{N}$. See [30] or [7] for details on how to do this for the Cauchy representation and the signed digit representation, respectively. Of course, the language has also a constant of type $\mathrm{X} \to (\mathrm{N} \to \mathrm{N})$ for the norm.

Given a normed space X, the majorizability relation $x \leq^*_X n$ at the ground type is defined by $\|x\| \leq_{\mathbb{R}} n_{\mathbb{R}}$. So, we accept the axiom: $\forall x^{\mathrm{X}} \forall n^{\mathrm{N}} (x \trianglelefteq_{\mathrm{X}} n \leftrightarrow \|x\| \trianglelefteq_{\mathbb{R}} n_{\mathbb{R}})$. Clearly, we can take 0^{N} as companion for the zero vector, $\lambda n, m.(n+m)$ as companion of the sum and $\lambda n.n$ as companion of the symmetric operation. Using the signed digit representation, we can take $\lambda f, n.n(1+f0)$ as companion of the scalar multiplication (for details, see [7]) and $\lambda n.(2n+3)$ as companion of the norm.

The axioms of vector space are formulated with the logical equality. For instance, we have $\mathrm{Eq}_{\mathrm{X}}(x+y, y+x)$, $\mathrm{Eq}_{\mathrm{X}}(x+(-x), 0^{\mathrm{X}})$ and $\mathrm{Eq}_{\mathrm{X}}(\alpha(x+y), \alpha x+\alpha y)$, for all x, y of type X and α of type $\mathrm{N} \to \mathrm{N}$. The axioms of the norm are:

$$\|0^{\mathrm{X}}\| =_{\mathbb{R}} 0_{\mathbb{R}}$$

$$\forall x^{\mathrm{X}} \forall \alpha^{\mathrm{N}\to\mathrm{N}}(\|\alpha x\| =_{\mathbb{R}} |\alpha|_{\mathbb{R}}\|x\|)$$

$$\forall x^{\mathrm{X}}, y^{\mathrm{X}}(\|x+y\| \leq_{\mathbb{R}} \|x\| + \|y\|)$$

Clearly, $d(x, y) = \|x-y\|$ is a distance function. Hence, a normed space is a pointed metric space with the zero vector as the distinguished point. As discussed in Subsection 5.2, in proof mining studies we use the mathematical equality $x =_{\mathrm{X}} y$ defined by $\|x-y\| =_{\mathbb{R}} 0_{\mathbb{R}}$. If the reader consults [30], he will see that Kohlenbach has a delicate set of axioms for normed spaces because he does not have logical equality and must formulate the axioms of vector space with $=_{\mathrm{X}}$. The delicate axiomatics is needed to ensure that the equality $=_{\mathrm{X}}$ is congruent with respect to the vector space notions. For instance: $\alpha =_{\mathbb{R}} \beta \wedge x =_{\mathrm{X}} y \to \alpha x =_{\mathrm{X}} \beta y$. This is clear with our axioms:

$$\|\alpha x - \beta y\| =_{\mathbb{R}} \|(\alpha x - \alpha y) + (\alpha y - \beta y)\| \leq_{\mathbb{R}} \|\alpha x - \alpha y\| + \|\alpha y - \beta y\| =_{\mathbb{R}}$$

$$\|\alpha(x-y)\| + \|(\alpha-\beta)y\| =_{\mathbb{R}} |\alpha|_{\mathbb{R}}\|x-y\| + |\alpha-\beta|_{\mathbb{R}}\|y\| =_{\mathbb{R}} 0_{\mathbb{R}},$$

where the first two equalities are explained because we have the logical equalities $\mathrm{Eq}(\alpha x - \beta y, (\alpha x - \alpha y) + (\alpha y - \beta y))$, $\mathrm{Eq}(\alpha x - \alpha y, \alpha(x-y))$ and $\mathrm{Eq}(\alpha y - \beta y, (\alpha-\beta)y)$.

Real inner product spaces can also be axiomatized adequately. In our framework, they can be directly axiomatized. Alternatively, they can be axiomatized using the axioms of normed spaces together with the universal sentence that postulates the parallelogram law.

Many other spaces have been considered in proof mining studies in nonlinear analysis (and related areas) within the framework of the so-called monotone functional interpretation (see [30] for this interpretation). The reader can find a recent list of these spaces in [32]. They can also be formalized within our framework of a majorizability language.

5.5 Semantic theories

Let X be a nonempty set and $\leq_X^*$ be a majorizability relation on X. With this data, it makes sense to consider the extended majorizability structure $M_{\leq^*}^{\omega,X}$ with the (extended) combinators and the (extended) recursors interpreted in the usual way. This is called a standard structure. Now, given a majorizability language, we may consider the interpretations of this language in standard structures for which $c \leq_\sigma^* t_c$, for every (mathematical) constant c^σ of the language. These structures are the standard interpretations of the majorizability language.

Let be given a majorizability theory Γ in a given majorizability language. We may consider all the standard interpretations of this language in which the flattenings of the sentences of Γ are true. We refer to this class of structures by $\mathscr{C}_\Gamma$. In the example of Subsection 5.1, this class is constituted exactly by all the interpretations built upon nontrivial commutative rings. In Subsection 5.2, any metric space (X, d) with diameter less than or equal to b determines an interpretation on $M_{\leq^*}^{\omega,X}$ (modulo the representation of the reals that give distances).

For proof mining purposes, we can take a much broader view. Given a majorizability language, we can consider any class $\mathscr{C}$ of standard interpretations of this language. These could be our intended structures, or the class $\mathscr{C}$ could just have one single structure. As discussed in the previous paragraph, every majorizability theory determines one such class, namely $\mathscr{C}_\Gamma$. The class $\mathscr{C}$ can be empty, of course. This is the case for $\mathscr{C}_\Gamma$, when Γ is contradictory. Anyways, given a class $\mathscr{C}$ of standard interpretations for a majorizabilty language, let $\mathrm{Th}_{\mathrm{bd}}(\mathscr{C})$ be the theory constituted by all closures of bounded formulas whose flattenings are true in each member of $\mathscr{C}$. Note that $\Gamma \subseteq \mathrm{Th}_{\mathrm{bd}}(\mathscr{C}_\Gamma)$.

6 The main theorem

In the introductory section, we mentioned the so-called characteristic principles of the bounded functional interpretation. In order to formulate the main theorem of this section, we need to describe these principles (now, extended to the language with an abstract type). To simplify, we formulate them with single variables but the principles also include the tuple case.

Monotone bounded choice $\mathsf{mAC}^\omega_{\mathsf{bd,X}}$:

$$\tilde{\forall} a \tilde{\exists} b A(a, b) \to \tilde{\exists} f \tilde{\forall} a \tilde{\exists} b \trianglelefteq fa\, A(a, b),$$

where A is a bounded formula. The tilde in the quantifiers $\tilde{\forall}$ and $\tilde{\exists}$ relativize the quantifications to self-majorizable functionals, where a functional a is self-majorizable if $a \trianglelefteq a$. For instance, $\tilde{\forall} a\, (\ldots)$ is $\forall a\, (a \trianglelefteq a \to (\ldots))$. These so-called monotone quantifiers only apply to variables of arithmetical type and are crucial in defining the bounded functional interpretation below.

Bounded collection $\mathsf{bC}^\omega_{\mathsf{bd,X}}$:

$$\forall x \trianglelefteq a \exists y A(x, y) \to \exists b \forall x \trianglelefteq a \exists y \trianglelefteq b\, A(x, y),$$

where A is a bounded formula. Note that a and b are of arithmetical type.

Majorizability principles $\mathsf{MAJ}^{\omega}_{\mathrm{X}}$:

$$\forall x \exists a\, (x \trianglelefteq a)$$

Note that a is of arithmetical type.

Main Theorem. *Let be given a majorizability language and $\mathscr{C}$ a class of standard interpretations for this language. Suppose that*

$$\mathrm{Th}_{\mathsf{bd}}(\mathscr{C}) + \mathsf{mAC}^{\omega}_{\mathsf{bd},\mathrm{X}} + \mathsf{bC}^{\omega}_{\mathsf{bd},\mathrm{X}} + \mathsf{MAJ}^{\omega}_{\mathrm{X}} \vdash \forall x^{\sigma} \exists y^{\tau} A(x, y),$$

where A is a bounded formula (whose free variables are among x and y). Then there is a closed monotone arithmetical term t of type $\hat{\sigma} \to \hat{\tau}$ such that, for every structure $M^{\omega,X}_{\leq^*}$ in $\mathscr{C}$,

$$M^{\omega,X}_{\leq^*} \models \forall a^{\hat{\sigma}} \forall x \leq^*_{\sigma} a \exists y \leq^*_{\tau} ta\, A^*(x, y),$$

where A^* is the flattening of A.

In the above, a closed arithmetical term is called monotone if it denotes a monotone functional. The interest of the theorem is twofold. Firstly, the characteristic principles can be mathematically useful in proof mining because they can help in finding a proof of $\forall x \exists y A(x, y)$. See the examples in subsections 7.1 and 7.4. Secondly, once we have such a proof, the theorem guarantees that there is a bound t as above. Ultimately, it is a bound like this that we are after in a proof mining application. We cannot leave these matters without drawing attention to the fact that the flattenings of the characteristic principles are not true in standard interpretations. *Au contraire*, the flattenings of $\mathsf{bC}^{\omega}_{\mathsf{bd},\mathrm{X}}$ even lead to contradiction (see [11]).

The proof of the theorem above is via a functional interpretation, being a consequence of the Soundness Theorem below. It is an interesting proof in itself for four main reasons. Firstly, the technique of functional interpretations proved to be very useful in very many circumstances. The paper [21] of Kurt Gödel, which introduces functional interpretations, is truly seminal. Secondly, the proof gives an effective way of obtaining the bounding term t from a concrete proof of $\forall x \exists y A(x, y)$. Thirdly, even though there are affinities between the above theorem and the metatheorems of the monotone functional interpretation [28], [29], [20], the results are quite different. For instance, the Soundness Theorem upon which the above result rests pertains to a new systematic interpretation which yields novel conservation results. At last, the proof is also interesting in a somewhat neglected way. The transformation of formulas effected by the bounded functional interpretation provides "uniform quantitative versions" of mathematical statements. See the introduction of [16] for a discussion of these matters and subsections 7.1 and 7.3 for concrete mathematical examples.

The definition below presents the bounded functional interpretation A^{U} of an arbitrary formula A of a given majorizability language (the flattening of A^{U} is the quantitative version of A, as given by the bounded functional interpretation). The primitive logical symbols used are negation, disjunction and universal quantification (including bounded universal quantification). This choice is acceptable because we are in classical logic.

Definition. *Let be given a majorizability language. To each formula A of the language, we assign formulas A^{U} and A_{U} so that A^{U} is of the form $\tilde{\forall} b \tilde{\exists} c\, A_{\mathrm{U}}(b, c)$, with b and c tuples of variables of arithmetical type and $A_{\mathrm{U}}(b, c)$ a bounded formula, according to the following clauses:*

1. *A^{U} and A_{U} are A, for bounded formulas A.*

For the remaining cases, if we have already interpretations for A and B given by $\tilde{\forall} b \tilde{\exists} c\, A_{\mathrm{U}}(b, c)$ and $\tilde{\forall} d \tilde{\exists} e\, B_{\mathrm{U}}(d, e)$ (respectively) then we define:

2. *$(A \vee B)^{\mathrm{U}}$ is $\tilde{\forall} b, d \tilde{\exists} c, e\, [A_{\mathrm{U}}(b, c) \vee B_{\mathrm{U}}(d, e)]$,*
3. *$(\forall x A(x))^{\mathrm{U}}$ is $\tilde{\forall} a \tilde{\forall} b \tilde{\exists} c\, [\forall x \trianglelefteq a\, A_{\mathrm{U}}(x, b, c)]$,*
4. *$(\neg A)^{\mathrm{U}}$ is $\tilde{\forall} f \tilde{\exists} b\, [\tilde{\exists} b' \trianglelefteq b \neg A_{\mathrm{U}}(b', fb')]$,*
5. *$(\forall x \trianglelefteq t A(x))^{\mathrm{U}}$ is $\tilde{\forall} b \tilde{\exists} c\, [\forall x \trianglelefteq t\, A_{\mathrm{U}}(x, b, c)]$.*

The formulas between square parentheses are the corresponding lower U *formulas.*

Note that A and A^{U} have exactly the same free variables. In clause 1, the tuples of universal and existential variables are empty. In clause 4, b' is of arithmetical type, namely the type of b. Also, this clause has the aspect that it has in order to enforce in every standard interpretation the following crucial monotonicity condition: $A_{\mathrm{U}}^*(b, c) \wedge c \leq^* d \rightarrow A_{\mathrm{U}}^*(b, d)$, where A_{U}^* is the flattening of A_{U}.

Soundness Theorem. *Let be given a majorizability language and $\mathscr{C}$ a class of standard interpretations for this language. Suppose that*

$$\mathrm{Th}_{\mathsf{bd}}(\mathscr{C}) + \mathsf{mAC}^{\omega}_{\mathsf{bd},\mathrm{X}} + \mathsf{bC}^{\omega}_{\mathsf{bd},\mathrm{X}} + \mathsf{MAJ}^{\omega}_{\mathrm{X}} \vdash A(x),$$

where $A(x)$ is an arbitrary formula of the majorizability language with free variables x (x can be a tuple of variables). Then there are closed monotone arithmetical terms t such that,

$$\mathrm{Th}_{\mathsf{bd}}(\mathscr{C}) \vdash \tilde{\forall} a \tilde{\forall} b \forall x \trianglelefteq a\, A_{\mathrm{U}}(x, b, tab),$$

where $A^{\mathrm{U}}(x)$ *is* $\tilde{\forall} b \,\tilde{\exists} c\, A_{\mathrm{U}}(x,b,c)$. *As a consequence, for every structure* $M^{\omega,X}_{\leq^*}$ *in* $\mathscr{C}$,

$$M^{\omega,X}_{\leq^*} \models \forall a \forall b\, (b \leq^* b \to \forall x \leq^* a\, A^*_{\mathrm{U}}(x, b, tab)).$$

Remark. *The interpretation of a formula of the form* $\exists y A(x,y)$, *with* A *a bounded formula, is equivalent to* $\exists c\, \exists y \trianglelefteq c\, A(x,y)$. *Therefore, the main theorem is a simple consequence of the above theorem.*

Proof. We can argue like the proof of the soundness theorem in [11]. It is a proof by induction on the length of the derivation of $A(x)$, relying on the complete axiomatization of classical logic as described by Joseph Shoenfield in his textbook [38]. There is hardly anything to add or change to the proof of [11]. The sentences in $\mathrm{Th}_{\mathrm{bd}}(\mathscr{C})$ are of the form $\forall z B(z)$, with $B(z)$ a bounded formula. By definition of $\mathrm{Th}_{\mathrm{bd}}(\mathscr{C})$, the flattening of $\forall z B(z)$ is true in every standard interpretation of $\mathscr{C}$. So, the same happens to $\tilde{\forall} a \forall z \trianglelefteq a B(z)$. Therefore, this sentence is an element of the set $\mathrm{Th}_{\mathrm{bd}}(\mathscr{C})$ and, in particular, it is a consequence of $\mathrm{Th}_{\mathrm{bd}}(\mathscr{C})$. This is what we want. The logical axioms and transitions follow the proof in [11]. At this point, we should perhaps remark that, as opposed to the *dialectica* interpretation, the contraction principle (here in the form of the inference of A from $A \vee A$) poses no problem in the setting of the bounded functional interpretation (see the discussion in [11]). This is why we can use the neutral theory of equality. The only other thing to note concerning logic is that, when considering the substitution axioms $\forall x A(x) \to A(t)$, we must make sure that t is majorizable. Given that mathematical constants have companions, it is clear that all closed terms of our language denote majorizable functionals. As in [11], the characteristic principles trivialize under the interpretation. Finally, note that all the properties needed for the proof regarding the new majorizability signs $\trianglelefteq$ are available, given the way that $\mathrm{Th}_{\mathrm{bd}}(\mathscr{C})$ is defined. □

The reader may find odd that we did not include in our theories the full induction scheme (i.e., induction for every formula of the language). Normally, it is part of a soundness theorem to see that the instances of the induction scheme are interpretable. They indeed are, and the usual proof (which uses the rule of induction instead of the axiom) works in our setting as well. However, we choose to present the matters in a different way:

Lemma. *The theory* $\mathrm{Th}_{\mathrm{bd}}(\mathscr{C}) + \mathsf{mAC}^{\omega}_{\mathsf{bd,X}} + \mathsf{bC}^{\omega}_{\mathsf{bd,X}} + \mathsf{MAJ}^{\omega}_{\mathrm{X}}$ *proves the induction scheme.*

In order to prove this lemma, we need the so-called characterization theorem:

Characterization Theorem. *Let be given a majorizability language and A a formula of that language. The theory* $\mathrm{Th_{bd}}(\mathscr{C}) + \mathsf{mAC}^{\omega}_{\mathsf{bd,X}} + \mathsf{bC}^{\omega}_{\mathsf{bd,X}} + \mathsf{MAJ}^{\omega}_{\mathrm{X}}$ *proves the equivalence between A and A^{U}.*

This theorem is proved in the same way as the corresponding theorem in [11]. Now, we have all the ingredients to prove the lemma. Let A be an instance of the induction scheme, and suppose that A^{U} is $\tilde{\forall} b \tilde{\exists} c\, A_U(b,c)$. By the ordinary proof of the interpretability of A in a soundness theorem, there is a closed monotone term t such that $\mathrm{Th_{bd}}(\mathscr{C}) \vdash \tilde{\forall} b\, A_U(b, tb)$. It should be remarked that the provability of $\tilde{\forall} b\, A_U(b, tb)$ needs induction for bounded formulas. However, this kind of induction can be formulated by the universal closure of the bounded formulas:

$$B(0) \wedge \forall k < n\, (B(k) \to B(k+1)) \to B(n),$$

with B a bounded formula. Therefore, this formulation is in $\mathrm{Th_{bd}}(\mathscr{C})$. Once we have $\tilde{\forall} b\, A_U(b, tb)$, *a fortiori* we have A^{U} in $\mathrm{Th_{bd}}(\mathscr{C})$. By the characterization theorem, we are done.

7 Retaking the examples

With the theorems of the previous section, we can discuss further the examples of Section 5 and point to a few subtleties and some nice features of the bounded functional interpretation.

7.1 Commutative rings

Let Γ be the theory described in Subsection 5.1. The classical theorem of commutative algebra, as expressed by (1), is not provable in $\mathrm{Th_{bd}}(\mathscr{C}_\Gamma) + \mathsf{mAC}^{\omega}_{\mathsf{bd,X}} + \mathsf{bC}^{\omega}_{\mathsf{bd,X}} + \mathsf{MAJ}^{\omega}_{\mathrm{X}}$. In effect, by bounded collection, (1) implies

$$\exists n^{\mathrm{N}} [\forall e^{\mathrm{R}}\, (e^n \neq 0^{\mathrm{R}} \to \exists X \trianglelefteq_{\mathrm{R}\to\mathrm{N}} 1\, (X \text{ is a prime ideal and } Xe \neq 0^{\mathrm{N}}))].$$

Therefore, by the main theorem, there would be an absolute constant $n_0 \in \mathbb{N}$ such that, for every commutative ring R and every element $e \in R$,

$$e^{n_0} \neq 0 \to \exists X \subseteq R\, (X \text{ is a prime ideal and } e \notin X).$$

This is, of course, false. It is in this sense that we claimed that (1) is not amenable to a proof mining treatment. However, a weakening of (1) is amenable to such a treatment.

Definition. *Let R be a nontrivial commutative ring and f, g functions from $\mathbb{N}$ to R. An (f, g)-prime ideal of R is a (proper) ideal X of R such that*

$$\forall k\, (f(k)g(k) \in X \to f(k) \in X \vee g(k) \in X).$$

It is an obvious consequence of the classical theorem on nilpotency that

$$\forall e\, [\forall n(e^n \neq 0) \to \forall f, g\, \exists X(X \text{ is an } (f, g)\text{-prime ideal of } R \text{ and } e \notin X)]. \quad (3)$$

However, as opposed to the classical theorem, the above weakening is provable in $\mathrm{Th}_{\mathsf{bd}}(\mathscr{C}_\Gamma) + \mathsf{mAC}^{\omega}_{\mathsf{bd,X}} + \mathsf{bC}^{\omega}_{\mathsf{bd,X}} + \mathsf{MAJ}^{\omega}_{\mathsf{X}}$. This is shown in [14]. The proof does not use Zorn's lemma but instead uses a principle of the form

$$\forall k \exists X \forall n \leq k\, B(n, X) \to \exists X \forall n\, B(n, X),$$

where X is of type $\mathrm{R} \to \mathrm{N}$ and stands for a characteristic function, and $B(n, X)$ is a bounded formula. This is the contrapositive of a bounded collection principle. It has mathematical content in the same sense that weak König's lemma has. It is nevertheless a principle which is set-theoretically false.

If we compute (the flattening of) the bounded functional interpretation of (3), we get

$$\forall \ell \exists n\, [\forall e\, (e^n \neq 0 \to \forall f, g \exists X(X \text{ is an } (f, g)\text{-prime ideal of } R \text{ up to } \ell \text{ and } e \notin X))],$$

where an ideal X is (f, g)-prime up to ℓ if

$$\forall k \leq \ell\, (f(k)g(k) \in X \to f(k) \in X \vee g(k) \in X).$$

This is the quantitative version of (3), as given by the bounded functional interpretation. The main theorem of the previous section guarantees the existence of a number theoretic function, given by a term of our language, that bounds n in terms of ℓ. In fact, n is bounded by $2^{\ell+1}$. See [14] for more discussions, details and an application.

7.2 Bounded metric spaces

For the logician (and for proof mining), it is important to distinguish between two characterizations of compactness. In the case of the compact unit interval $[0, 1]$, there is, on the one hand, the notion of Heine/Borel compactness (every denumerable open cover has a finite subcover) and, on the other hand, the notion of sequential compactness (every sequence has a convergent subsequence). The former notion is equivalent to weak König's lemma, whereas the latter is equivalent to arithmetical comprehension (for these facts, see [39]). The former characterization is amenable to a simple proof mining treatment whereas the latter is not.

A bounded metric space (X, d) need not be compact, but the bounded collection principle entails that if the space X is covered by a denumerable family $(\Omega_n)_{n\in\mathbb{N}}$ of open sets, then it has a finite subcover. That is,

$$\forall x \in X \exists n \in \mathbb{N} (x \in \Omega_n) \to \exists \ell \in \mathbb{N} \forall x \in X \exists n \leq \ell (x \in \Omega_n).$$

Note that (a) the quantification over elements of X is bounded because the metric space is bounded; (b) we are assuming that the openness of the sets Ω_n translates syntactically into a condition that allows the application of the bounded collection principle. It must be said that this assumption does not hold in full generality. In Reverse Mathematics this proviso is not necessary because open sets are – *by definition* – given by codes, and membership in the open sets is translated into Σ^0_1 conditions (see [39]). We thank Ulrich Kohlenbach and an anonymous referee for urging us to make this clarification.

The above Heine/Borel principles are not a mere curiosity since they do play a role in proof mining. See the first two sections of [16] and Subsection 7.4 below.

7.3 Metric spaces

The statement (2) has a simple proof by *reductio ad absurdum*. Suppose that there is a natural number k_0 such that

$$\forall x \in X \left(A(x) \to \exists y \in X \left(A(y) \wedge d(a, x) > d(a, y) + \frac{1}{k_0 + 1} \right) \right).$$

By induction on n, there is a sequence $b_0, b_1, \ldots, b_n$ of points of X such that $A(b_0)$, $A(b_1)$,..., $A(b_n)$ and, for all $i < n$, $d(a, b_i) > d(a, b_{i+1}) + \frac{1}{k_0+1}$. We get

$$d(a, b_0) > d(a, b_n) + \frac{n}{k_0 + 1} \geq \frac{n}{k_0 + 1}.$$

This is a contradiction for n sufficiently large. Note that the argument takes place in the closed ball centered at a with radius $d(a, b_0)$.

This argument is formalizable in the theory Γ of metric spaces described in Subsection 5.3 (with induction). The theory $\mathrm{Th}_{\mathrm{bd}}(\mathscr{C}_\Gamma) + \mathsf{mAC}^\omega_{\mathsf{bd},X} + \mathsf{bC}^\omega_{\mathsf{bd},X} + \mathsf{MAJ}^\omega_X$ proves that, given b, if $A(b)$ and $b \trianglelefteq_X \ell$ then

$$\forall k^{\mathrm{N}} \exists x \trianglelefteq_X \ell \left(A(x) \wedge \forall y \trianglelefteq_X \ell \left(A(y) \to d(a, x) \trianglelefteq_{\mathbb{R}} d(a, y) + \frac{1}{k+1} \right) \right).$$

The soundness theorem of the previous section applies, of course. The structure of the bounding term (namely, the recursors used), whose existence is guaranteed by the soundness theorem, depends on the complexity of the formula $A(x)$. For simplicity, let us consider the case in which the formula $A(x)$ is of the form $\forall n^{\mathrm{N}} B(n, x)$, with $B(n, x)$ a bounded formula. We can suppose, without loss of generality, that $n \leq m \wedge B(m, x) \rightarrow B(n, x)$.

Therefore, under the assumption that $\forall n^{\mathrm{N}} B(n, b)$ and $b \trianglelefteq \ell$, we have

$$\forall k\, \exists x \trianglelefteq \ell \left(\forall n B(n, x) \wedge \forall y \trianglelefteq \ell \left(\forall n B(n, x) \rightarrow d(a, x) \trianglelefteq_{\mathbb{R}} d(a, y) + \frac{1}{k+1} \right) \right).$$

The bounded functional interpretation of the conclusion above is equivalent to

$$\forall k \tilde{\forall} f \exists n \left[\exists x \trianglelefteq \ell \left(B(fn, x) \wedge \forall y \trianglelefteq \ell \left(B(n, y) \rightarrow d(a, x) \trianglelefteq_{\mathbb{R}} d(a, y) + \frac{1}{k+1} \right) \right) \right].$$

The flattening of this formula is the quantitative version of (2), as given by the bounded functional interpretation, for the particular case at hand. It can be shown that, if there is b such that $d(a, b) \leq \ell$ and $\forall m B^*(m, b)$, then for all natural numbers k and for every monotone function $f : \mathbb{N} \rightarrow \mathbb{N}$, there is a natural number n with $n \leq f^{(k+1)\ell}(0)$ and $x \in X$ such that $d(a, x) \leq \ell$ and

$$B^*(f(n), x) \wedge \forall y \in X \left(B^*(n, y) \rightarrow d(a, x) \leq d(a, y) + \frac{1}{k+1} \right)$$

holds in M^ω. Here, f^r denotes de r-fold composition of f with itself and B^* is the flattening of B. The existence of a bound depending only on f, k and ℓ can be explained by the results of the previous section, and the computation of the concrete bound is a proof mining exercise. For details, see [37] or [16].

7.4 Normed spaces

Felix Browder proved in [4] an important theorem concerning the convergence of certain sequences (let us call them Browder sequences) to fixed points of nonexpansive mappings in Hilbert spaces. A mining analysis of the proof of this theorem amounts to computing the rate of metastability of the Browder sequences (see [30] for an explanation of metastability). This was done by Kohlenbach in [31] using the monotone functional interpretation. Recently, the analysis was redone in [16] using, instead, the bounded functional interpretation. The new proof mining analysis follows closely Kohlenbach's treatment, obtains essentially the same bounds, but brings two novelties: (1) it relies on Heine/Borel compactness as discussed in Subsection 7.2; and (2) it uses the distance analysis of Subsection 7.3. Heine/Borel compactness explains why a certain sequential weak compactness argument of Browder does not pose difficulties for a proof mining analysis (it is replaced by a Heine/Borel argument, as shown in section 2 of [16]). The bounded functional interpretation of (2), for the particular case at hand, greatly simplifies the corresponding analysis of [31].

Let us make one last observation concerning the theorem of Browder. If the proof of Browder were intuitionistic, a functional interpretation (be it bounded or monotone) for intuitionistic logic would ensure the existence of a computable rate of Cauchyness for the Browder sequences, not just a mere rate of metastability. It is known that, in general, there is no such computable rate of Cauchyness. So, Browder's proof is classical. The only place in the proof that is irremovably classical is precisely the argument by *reductio* used in Subsection 7.3.

The bounded functional interpretation has been playing an important background role in certain very recent proof mining analyses. At the time of this writing, this work – except for [16] – has not yet been published. Pedro Pinto, Bruno Dinis and Laurenţiu Leuştean are the actors in this recent trend.

8 Final short annotations

The soundness theorem of Section 6 is a conservation result. It can be refined in various ways and be viewed as a generalization of Harvey Friedman's well-known conservation result of the theory WKL_0 over RCA_0 (cf. [11]). It can be applied to other theories, as we will comment below. It can also be adapted to suitable recursively axiomatizable systems. In fact, this is how the systems were first presented.

In this paper, we opted to work with the semantic (non recursively axiomatizable) theories $\mathrm{Th}_{\mathrm{bd}}(\mathscr{C})$ because this option is more fit for proof mining and, additionally, it allowed us to avoid mentioning "intensional" aspects of the new primitive symbols $\trianglelefteq_\sigma$. Due to this option, a part of the soundness theorem may look a bit silly. Given its form, the sentence $\tilde{\forall} a \tilde{\forall} b \forall x \trianglelefteq a A_{\mathrm{U}}(x, a, tab)$ has essentially a one line proof in $\mathrm{Th}_{\mathrm{bd}}(\mathscr{C})$. The one-liner is not silly, though. It is a substantial fact that the characteristic principles do not lead us outside of the truth in M^ω for (flattenings of) sentences of that form. It is also not silly in a more pragmatic manner. How do we come to know, in concrete cases, that $A(x)$ is provable in $\mathrm{Th}_{\mathrm{bd}}(\mathscr{C})$ plus the characteristic principles? Surely, by using the characteristic principles and, well, by relying on some known truths coming from $\mathrm{Th}_{\mathrm{bd}}(\mathscr{C})$. For all we know, these can come from proofs in ZFC, say. The soundness theorem then guarantees the existence of the term t and, therefore, the *discovery* of a new member of $\mathrm{Th}_{\mathrm{bd}}(\mathscr{C})$. This is essentially what a proof mining study amounts to.

From an entirely different point of view, we may try to find base recursive axiomatizations of systems for which the soundness theorem holds, but the present paper is not the proper place to discuss this issue in any detail. Nevertheless, we do want to make two observations on this regard. As mentioned in Section 3, the presentation of the bounded functional interpretation in [17] relied on a rule. The rule, adapted to the situation of an abstract type, is the following:

$$\frac{A \wedge u \trianglelefteq_\sigma v \wedge v' \trianglelefteq_{\hat{\sigma}} v \to su \trianglelefteq_\tau tv \wedge tv' \trianglelefteq_{\hat{\tau}} tv}{A \to s \trianglelefteq_{\sigma \to \tau} t}$$

where A is a bounded formula, s is a term of type $\sigma \to \tau$ and t is a term of type $\hat{\sigma} \to \hat{\tau}$, and where u, v and v' are new variables that do not occur in A, s or t. Note, by the way, that $\mathrm{Th_{bd}}(\mathscr{C})$ is closed under this rule. As a matter of fact, $\mathrm{Th_{bd}}(\mathscr{C})$ is also closed under the (quantifier-free) extensionality rule, the preferred way in [30] for dealing with equality.

The second observation is as follows: whereas the semantic theories $\mathrm{Th_{bd}}(\mathscr{C})$ do not have characteristic functions for bounded predicates, they do have something close enough:

$$\forall a^{\hat{\sigma}} \exists f \trianglelefteq_{\sigma \to \mathrm{N}} 1\, \forall x \trianglelefteq_\sigma a\, (f(x) =_{\mathrm{N}} 0 \leftrightarrow A(x)),$$

for $A(x)$ a bounded formula. Obviously, in the presence of (the contrapositive of) bounded collection, we obtain the said characteristic functions.

The framework of the bounded functional interpretation is very stable and can be altered without detriment in a number of ways. For instance, instead of one abstract type we can surely have more than one abstract type simultaneously (as in [30]). The arithmetical terms of the theories that we have been discussing in this paper are the primitive recursive functionals in the sense of Gödel but, if we only allow the basic recursor R_{N}, we get the primitive recursive functionals in the sense of Kleene (see [30] for the terminology). The amount of induction proved in these circumstances (with the aid of monotone bounded choice) is Σ^0_1-induction, with parameters of arbitrary type and with a bounded formula for the matrix of the Σ^0_1-formula. We can cook up many variants, even down to feasible theories as in [18]. However, the deepest variants come from the work of Clifford Spector in [40], where bar recursive functionals of finite type were introduced. These functionals are not well-defined in the set-theoretic structure S^ω, but they are so in M^ω (a result due to Bezem in [3]). In the presence of these recursors, the basic theories studied in this paper, together with the characteristic principles, prove full numerical comprehension. Nay, they prove the principle of dependent choices. In the framework of the *dialectica* interpretation this was shown by Howard in [26] (see [13] for a recent exposition of this result), and extended to abstract types in [28], [20] for the monotone variant. In the framework of the bounded functional interpretation, the deduction of the principle of dependent choices was checked in detail in [8] without the abstract type. The situation with the abstract type has not been checked in detail but should pose no new difficulties.

References

[1] Jeremy Avigad, and Solomon Feferman *Gödel's functional ("Dialectica") interpretation*, in: Samuel Buss (ed.) Handbook of Proof Theory, North Holland, Amsterdam 1998, pp. 337–405.

[2] Benno van den Berg *A note on equality in finite-type arithmetic*, Mathematical Logic Quarterly 63, 2017, pp. 283–288.

[3] Marc Bezem *Strongly majorizable functionals of finite type: a model for bar recursion containing discontinuous functionals*, The Journal of Symbolic Logic 24, 1985, pp. 652–660.

[4] Felix Browder *Convergence of approximants to fixed points of nonexpansive nonlinear mappings in Banach spaces*, Archive for Rational Mechanics and Analysis 24, 1967, pp. 82–90.

[5] Samuel Buss *A conservation result concerning bounded theories and the collection scheme*, Proceedings of the American Mathematical Society 100, 1987, pp. 109–116.

[6] —— *First-order proof theory of arithmetic*, in: Samuel Buss (ed.) Handbook of Proof Theory, North Holland Publishing Company, Amsterdam 1998, pp. 79–147.

[7] Patrícia Engrácia Proof-theoretic Studies on the Bounded Functional Interpretation, PhD Dissertation, University of Lisbon, 2009.

[8] —— *The bounded functional interpretation of bar induction*, Annals of Pure and Applied Logic 163, 2012, pp. 1183–1195.

[9] Fernando Ferreira *A feasible theory for analysis*, The Journal of Symbolic Logic 59, 1994, pp. 1001–1011.

[10] —— *A note on a result of Buss concerning bounded theories and the collection scheme*, Portugaliae Mathematica 52, 1995, pp. 331–336.

[11] —— *Injecting uniformities into Peano arithmetic*, Annals of Pure and Applied Logic 157, 2009, pp. 122–129.

[12] —— *Proof interpretations and majorizablity*, in: Françoise Delon *et al.* (eds.) Logic Colloquium 2007, Cambridge University Press, Cambridge 2010, pp. 32–81.

[13] —— *Spector's proof of the consistency of analysis*, in: Reinhard Kahle, and Michael Rathjen (eds.) Gentzen's Centenary, *Tthe Quest for Consistency*, Springer, Cham etc. 2015, pp. 279–300.

[14] —— *Bounds for indexes of nilpotency in commutative ring theory: a proof-mining approach*, manuscript, 2019.

[15] Fernando Ferreira, and Gilda Ferreira *A herbrandized functional interpretation of classical first-order logic*, Archive for Mathematical Logic 56, 2017, pp. 523–539.

[16] Fernando Ferreira, Laurenţiu Leuştean, and Pedro Pinto *On the removal of weak compactness arguments in proof mining*, Advances in Mathematics 354, 2019.

[17] Fernando Ferreira, and Paulo Oliva *Bounded functional interpretation*, Annals of Pure and Applied Logic 135, 2005, pp. 73–112.

[18] —— *Bounded functional interpretation and feasible analysis*, Annals of Pure and Applied Logic 145, 2007, pp. 115–129.

[19] Philipp Gerhardy, and Ulrich Kohlenbach *Extracting Herbrand disjunctions by functional interpretation*, Archive for Mathematical Logic 44, 2005, pp. 633–644.

[20] —— *General logical metatheorems for functional analysis*, Transactions of the American Mathematical Society 360, 2008, pp. 2615–2660.

[21] Kurt Gödel *Über eine bisher noch nicht benützte Erweiterung des finiten Standpunktes*, Dialectica 12, 1958, pp. 280–287. (English translation in [22], pp. 240–251.)

[22] —— Collected Works 2 (edited by Solomon Feferman *et al.*), Oxford University Press, Oxford 1990.

[23] Warren Goldfarb *Herbrand's error and Gödel's correction*, Modern Logic 3, 1993, pp. 103–118.

[24] Jacques Herbrand Recherches sur la théorie de la démonstration, PhD Dissertation, University of Paris (Sorbonne), 1930. (English translation in [25], pp. 44–202.)

[25] —— Logical Writings (edited by Warren Goldfarb), D. Reidel Publishing Company, Dordrecht 1971.

[26] William Howard *Functional interpretation of bar induction by bar recursion*, Compositio Mathematica 20, 1968, pp. 107–124.

[27] —— *Hereditarily majorizable functionals of finite type*, in: Anne S. Troelstra (ed.) Metamathematical Investigation of Intuitionistic Arithmetic and Analysis, Springer-Verlag, Berlin etc. 1973, pp. 454–461 [*Lecture Notes in Mathematics 344*].

[28] Ulrich Kohlenbach *Some logical metatheorems with applications in functional analysis*, Transactions of the American Mathematical Society 357, 2005, pp. 89–128.

[29] —— *A logical uniform boundedness principle for abstract metric and hyperbolic spaces*, Electronic Notes in Theoretical Computer Science 165, 2006, pp. 81–93.

[30] —— Applied Proof Theory, *Proof Interpretations and their Use in Mathematics*, Springer-Verlag, Berlin 2008 [*Springer Monographs in Mathematics*].

[31] —— *On quantitative versions of theorems due to F. E. Browder and R. Wittmann*, Advances in Mathematics 226, 2011, pp. 2764–2795.

[32] —— *Recent progress in proof mining in nonlinear analysis*, IFCoLog Journal of Logics and its Applications 10, 2017, pp. 3357–3406.

[33] Georg Kreisel Review of [38], Mathematical Reviews: MR0225631 #37, 1224.

[34] Charles Parsons *On a number theoretic choice schema and its relation to induction*, in: John Myhill, Akiko Kino, and Richard E. Vesley (eds.) Intuitionism and Proof Theory, North-Holland, Amsterdam 1970, pp. 459–473.

[35] Jeffrey Paris *Some conservation results for fragments of arithmetic*, in: Chantal Ber-

line, Kenneth McAloon, and Jean-Pierre Ressayre (eds.) Model Theory and Arithmetic, Springer-Verlag, New York 1980, pp. 251–261 [*Lecture Notes in Mathematics 890*].

[36] Jeffrey Paris, and Laurence Kirby *Σ_n-collection schemas in arithmetic*, in: Angus Macintyre, Leszek Pacholski, and Jeffrey Paris (eds.) Logic Colloquium 77, North-Holland, Amsterdam 1978, pp. 199–209 [*Studies in Logic and the Foundations of Mathematics 96*].

[37] Pedro Pinto Proof Mining with the Bounded Functional Interpretation, PhD Dissertation, University of Lisbon, 2019.

[38] Joseph R. Shoenfield Mathematical Logic, Addison-Wesley Publishing Company, Reading MA, etc. 1967 [*Addison-Wesley Series in Logic*]. (Republished in 2001 by AK Peters.)

[39] Stephen Simpson Subsystems of Second Order Arithmetic, Springer-Verlag, Berlin 1999 [*Perspectives in Mathematical Logic*].

[40] Clifford Spector *Provably recursive functionals of analysis: a consistency proof of analysis by an extension of principles in current intuitionistic mathematics*, in: Jacob C. E. Dekker (ed.) Recursive Function Theory, American Mathematical Society, Providence RI 1962, pp. 1–27 [*Proceedings of Symposia in Pure Mathematics 5*].

Twins in Logic – Identical and Otherwise

Lloyd Humberstone
Monash University
lloyd.humberstone@monash.edu

ABSTRACT Connectives are twins in a logic, according to a metaphor of Łukasiewicz, when they behave 'in the same way' according to that logic. There are, however, looser and stricter ways of understanding that phrase, informally contrasted in §1 and then precisely defined and illustrated in §3, after a glance at related work by Michael Byrd and Evgeni Zolin in §2. §4 returns to the motivating case of the Ł-modal system, with its 'twin possibility operators.' One rather detailed discussion arising from §3 is deferred to an Appendix, so as not to interrupt the flow.

KEYWORDS Connectives; Łukasiewicz.

1 Introduction

Yes, the reader is right to be reminded by the title – its first three words, at least – of Łukasiewicz and the intriguing 'twin possibility' operators of his Ł-modal logic, to which we shall come in due course. First, though, as background for the present exploration of what exactly those three words might be taken to mean, we look at an example from the literature in which Łukasiewicz [42] is recalled not quite accurately. The details of the example – such as the proof (or even the correctness) of Proposition 1.1 below – are not important for later sections. The example serves only to make intelligible the remark embodying the inaccuracy just alluded to. (And in any case, the Łukasiewicz example serves chiefly as a lively prompt to investigate the 'twins' issue, which we illustrate with other examples in the following two sections before returning to Łukasiewicz to tidy up the discussion of that example.)

Humberstone [28] discusses the modal logic **S5**, formulated in some language with a functionally complete stock of Boolean connectives, of which one to be taken as primitive is the material conditional, to be written as $\supset$ just for this section (later

For their many corrections and suggestions, I am very grateful to Sam Butchart, Rohan French, and Dave Ripley.

we use $\rightarrow$ in this capacity), and the modal primitive $\Box$. The best known distinctively modal relative of the material conditional is the strict conditional $A \prec B$, defined as $\Box(A \supset B)$, but [28] takes an interest, instead, in two other implicational connectives with a distinctly modal flavour, written here as $\Rightarrow$ and $\rightarrow$, defined thus, with $\equiv$ and $\vee$ being the usual material biconditional and (inclusive) disjunction connectives:

$$A \Rightarrow B = \Box(A \equiv B) \vee B \qquad\qquad A \rightarrow B = \Box A \supset B.$$

For the moment, restrict attention to the implicational fragment of classical (non-modal) propositional logic – or CL for short – whose formulas we call pure $\supset$-formulas. We denote by $C[\Rightarrow]$ and $C[\rightarrow]$ the results, respectively, of replacing all occurrences of $\supset$ in a such a formula with $\Rightarrow$ and with $\rightarrow$, and for uniformity we write C itself as $C[\supset]$. By $\vdash_{\mathsf{CL}}$ is meant the consequence relation most commonly associated with CL, sometimes called tautological (or truth-functional) consequence, though for the moment we consider only the consequences of the empty set, as usual writing "$\vdash_{\mathsf{CL}} A$" for "$\varnothing \vdash_{\mathsf{CL}} A$". Similarly in Proposition 1.1 $\vdash_{\mathsf{S5}}$ indicates provability in S5; a simplified version of Proposition 1.5 from [28] reads as follows:

Proposition 1.1. *For any pure $\supset$-formula $A[\supset]$, the following are equivalent: (i) $\vdash_{\mathsf{CL}} A[\supset]$; (ii) $\vdash_{\mathsf{S5}} A[\Rightarrow]$; (iii) $\vdash_{\mathsf{S5}} A[\rightarrow]$.*

What has been simplified away here is restored in a footnote for those interested.[1]

An earlier publication (namely [25]) concentrated on the equivalence of *(ii)* and *(iii)* (or their more general versions in note 1) to throw light on the relation between the two implicational connectives in Matthew Spinks' BCSK logic,[2] in which they are far from interchangeable, the equivalence of *(ii)* and *(iii)* (in which each occurs without the other) notwithstanding. In view of this, a comment from [25] (p. 6, with some grammatical garbling corrected here) is recalled in [28] (p. 439), likening this situation to

[1]Proposition 1.5 of [28] is actually the following considerably stronger claim concerning the classical consequence relation $\vdash_{\mathsf{CL}}$ and the global consequence relation $\vdash^{glo}_{\mathsf{S5}}$ associated with S5 (which we can think of syntactically as saying that the formula on the right can be obtained from those on the left together with the theorems of S5 and the rules of Modus Ponens and Necessitation): The following are equivalent, for all pure $\supset$-formulas $C_1, \ldots, C_n, A$: *(i)* $C_1[\supset], \ldots, C_n[\supset] \vdash_{\mathsf{CL}} A[\supset]$; *(ii)* $C_1[\Rightarrow], \ldots, C_n[\Rightarrow] \vdash^{glo}_{\mathsf{S5}} A[\Rightarrow]$; *(iii)* $C_1[\rightarrow], \ldots, C_n[\rightarrow] \vdash^{glo}_{\mathsf{S5}} A[\rightarrow]$. Proposition 1.1 in the text is the $n = 0$ case of this more general claim. We do not need the "*glo*" superscript for that, because the global and local consequences of $\varnothing$ for S5 (or any normal modal logic) coincide. (A corresponding syntactic formulation of the local version would drop the reference to necessitation figuring in the gloss given for the global relation.)

[2]See for instance p. 6 of Veroff and Spinks [70] where the two connectives appear as notated here.

> that which prompted Łukasiewicz in [42] to speak of two connectives as being like a pair of identical twins on the grounds that any formula in which only one of them made an appearance (perhaps alongside further connectives) did not have its (un)provability affected when it was replaced (throughout) by the other, even though the provability status of a formula containing both was prone to be affected by interchanging them. The intended analogy is with those identical human twins said to be indistinguishable except when appearing together.

The reference to identical twins in [25] even found its way into the title of [28].

There is only one problem with all this: Łukasiewicz makes no mention of *identical twins* – at least not in [42], the only reference cited in [25] and [28]. (We return to this point at the end of Section 4.) He speaks there only of *twins*. Now, is this inaccuracy in paraphrase just an excusable flourish in getting the gist of Łukasiewicz's point across? Perhaps so; but perhaps not: the choice between thinking of twins in general and thinking of identical twins in particular is naturally connected with two different ways of taking the idea that the connectives in question can only be told apart when both are present. On the human side of this simile, with identical twins, especially those making a point of exploiting this effect, one may not only fail to realise that there are two of them if they are seen separately, but even on encountering them together, be able to tell them apart only in the weak sense of seeing that they are different people. On the other hand, non-identical twins can still be sufficiently similar in appearance that when met singly, they may be taken to be the same person, while if one encounters them together, one can see not only *that* they are different – as in the identical twins case – but *how* they are different: one, for example, is now seen to be slightly taller than the other.[3] Might a similar contrast be operative in the logical case? And if so, on which side of the contrast would Łukasiewicz's own example (of the twin possibility operators in [42]) fall? We will look at that example more closely in Section 4, sampling some more recent literature bearing on the issue and honing the concepts needed, in Sections 2 and 3, respectively.

[3]That is: one can tell them apart in the sense of being aware of some differentiating respect which is intrinsic to the pair. In the identical twins case these last words are intended to exclude such a response as this: yes, I can tell them apart – Tweedledum is the one on the left and Tweedledee is the one on the right.

2 Concepts and Results from Byrd and Zolin

The authors whose names feature in the present section title will supply us with useful materials for approaching the subject of twins in logic.[4] Each of them distinguishes an internal from an external perspective. Thus Zolin [72], p. 861, writes as follows, understanding by a *modality* any formula $A(p)$ in the language of monomodal logic in which only the only sentence letter to appear is p.[5]

> According to the first, or *internal*, approach, modalities are identified if they are equivalent in L, i.e., if the equivalence of formulas they are induced by is a theorem of L. (...) The second, or *external*, approach prescribes not to distinguish between modalities having an identical "behaviour" over L.

Zolin uses ∇, Δ as variables over arbitrary modalities,[6] and explains the distinctive "$L(\nabla)$" notation he makes use of in the passage to be quoted as soon as we get this announcement out of the way:

[4]Both of them put in an appearance on p. 483 of Humberstone [30], which the present discussion complements.

[5]All propositional languages considered here have the same set of sentence letters, $p_1, p_2, \ldots, p_n \ldots$, the first three of which we write for convenience as p, q, r. Zolin takes $\Box$ as primitive, with $\Diamond$ defined from it (and $\neg$) in the usual way. What is often meant by a modality given such a choice of primitives, namely a (possibly empty) string of occurrence of $\Box$ and $\neg$, Zolin calls a *linear* modality.

[6]Except when discussing the specific noncontingency modality $\Delta p = \Box p \vee \Box\neg p$. (Historically, the literature on noncontingency-based logics uses Δ and ∇ for noncontingency and contingency, respectively; Zolin himself has, incidentally, been a prominent contributor this literature, the striking paper [73] being only one example.) For present purposes the notational difficulties become more acute still – and Convention 2.1 will reduce them somewhat – since Łukasiewicz [42] uses Δ as a possibility operator, rather than $\Diamond$, and then when he wants to introduce his 'twin possibility' operator, he writes the latter as ∇. Similarly Łukasiewicz writes Γ in place of $\Box$. It is for this reason that when we want to use (specifically) Greek capital letters for sets of formulas we reach for Θ in the following section. Notationally, matters would be even more complicated had the suggestion of Schock [56], p. 12, note 1, caught on to any extent: that one use upward and downward pointing triangles in place of $\Box$ and $\Diamond$ (for continuity with "$\wedge$" and "$\vee$" and their occasional enlarged use as universal and existential quantifiers). In saying this, differences in respect of stroke modulation – e.g., between a capital delta and an upward pointing triangle of similar size – are ignored, since this is often a typesetter's decision: for instance the former, from Łukasiewicz [42] appear as the latter in [43] (and Γ turns into a rotated version of what would now be taken as a negation symbol). By Chapter 7 of Łukasiewicz [44] Γ and Δ have become L and M, with the latter (more or less inverted) W in place of ∇. (Simons [59] has the idea of writing L – we shall use this later in the form "L" – as the inverted form of the earlier Γ for the dual of Łukasiewicz's ∇.) L and M in this usage gained considerable currency especially through the books of Hughes and Cresswell from 1968 onward. But a reader coming across Prior's discussion of Łukasiewicz in [52] is at some risk of misinterpreting the the symbol "M", especially on p. 204, which Prior uses as a contingency operator rather than a possibility operator.

Convention 2.1. *From now, in view of the potential for confusion mentioned note 6 from the triple use of "Δ" and "∇" mentioned there, whenever Zolin's use of these symbols for arbitrary modalities is being echoed, even in direct quotation, they will be written with a dot beneath them: $\underset{\cdot}{\Delta}$ and $\underset{\cdot}{\nabla}$. (This still leaves the double use of the undotted versions, for noncontingency and contingency on the one had, and for Łukasiewicz's two possibility operators on the other, but no use is made of them here in the first of these two roles.)*

We can now proceed with the passage from Zolin ([72], p. 862) explaining the (as we now call it) "$L(\underset{\cdot}{\nabla})$" notation he makes use of in characterizing several further concepts:

> [W]e define a logic $L(\underset{\cdot}{\nabla})$ of a modality $\underset{\cdot}{\nabla}$ over a logic L as the set of all formulas whose $\underset{\cdot}{\nabla}$-translations are theorems of L.

The $\underset{\cdot}{\nabla}$-translation of a formula A, $\tau_{\underset{\cdot}{\nabla}}(A)$, is defined inductively on the complexity of A by setting it to be the identity map except for the case in which A is $\Box B$ for some formula B, in which case $\tau_{\underset{\cdot}{\nabla}}(A)$ is $\underset{\cdot}{\nabla}(\tau_{\underset{\cdot}{\nabla}}(B))$. Thus, $L(\underset{\cdot}{\nabla})$ is the logic that says about $\Box$ what L says about $\underset{\cdot}{\nabla}$, and L is itself $L(\Box)$. Zolin's paper includes many examples of this relationship, which in the terminology of Humberstone [33] is put by saying that $\tau_{\underset{\cdot}{\nabla}}$ is a $\Box$-definitional translation faithfully embedding the logic $L(\underset{\cdot}{\nabla})$ into L.[7] Cases mentioned in Zolin [72] include (*i*) and (*ii*) of the following; all of these concern normal modal logics:

Examples 2.2. (i) $\mathsf{K}(\boxdot) = \mathsf{KT}$, *where* $\boxdot$ *is the modality induced by the formula* $\Box p \wedge p$.
(*ii*) $\mathsf{K4}(\boxdot) = \mathsf{KD4}(\boxdot) = \mathsf{S4}$.
(*iii*) $\mathsf{S4.2}(\Diamond\Box) = \mathsf{KD45}$. ◀

Examples 2.2(*i*) and (*ii*) are sufficiently well known and easily established as to be describable as folklore. A generalization subsuming them, given (as Lemma 5.9) in Zolin [72] says that whenever for normal L, $L \subseteq L(\boxdot)$, then $L(\boxdot)$ is the normal extension of L by $\Box p \to p$. The result for (*iii*) is due independently to E. E. Dawson and W. Lenzen, for further details on which, see Example 4.4.27 of [33] and §4.2 in French [16]. More discussion concerning such embeddings (including proofs of their fidelity), whether or not expressed using Zolin's $L(\cdot)$ notation, can be found in French [14], [16], Humberstone [29], [33] (passim), and – of special significance for $\tau_{\boxdot}$ – Jeřábek [36] (settling a question from French and Humberstone [15]).

[7] The $\tau_{\underset{\cdot}{\nabla}}$-style notation is common in the works cited below; Zolin actually uses "tr" rather than "τ".

Finally, giving the equivalence relations of focal of interest for the internal and external perspectives, respectively, Zolin defines (p. 863) modalities $\underset{\cdot}{\nabla}$ and $\underset{\cdot}{\Delta}$ to be *equivalent in* L if $\vdash_L \underset{\cdot}{\nabla}(p) \leftrightarrow \underset{\cdot}{\Delta}(p)$, and to be *analogous over* L when $L(\underset{\cdot}{\nabla}) = L(\underset{\cdot}{\Delta})$. With respect to any congruential modal logic (i.e., an L for which $\vdash_L A \leftrightarrow B$ implies $\vdash_L \Box A \leftrightarrow \Box B$) any equivalent modalities are analogous, though not in general conversely. An interesting case in which the converse does hold is that of S5. In 1946, Carnap [7] showed that this logic provides sixteen non-equivalent modalities. Zolin shows that no two of them are even analogous (over S5), in the course of proving Theorem 4.21 of [72].[8] What bears more obviously on the Łukasiewicz 'twins' theme, though, is the other outcome: modalities which are analogous over L but not equivalent in L. Zolin provides numerous examples of this phenomenon, citing from the literature a proof that $\Box$ and $\Box\Box$ are analogous over – though evidently not equivalent in – K, and giving, himself, a proof that the same is so for KTB. Chellas [9], Exercise 7.8 on p. 211, had implicitly made a similar observation concerning the modalities $\Box$ and $\Diamond$ in E, the smallest congruential modal logic. In fact the points about $\Box$ and $\Box\Box$ hold, as Zolin notes, for any $\Box^m$ and $\Box^n$ with $m, n \geq 1$ and $m \neq n$ – setting the $m = n$ aside since we want *non-equivalent* (and so certainly distinct) though analogous modalities – and he also mentions at Remark 3.2 in [72], generalizing Chellas's example, that there are only three non-analogous linear modalities over E. We turn our attention presently to Byrd [5], though readers with no concern for the background of that discussion should skim the paragraphs after Remark 2.3, absorbing only the bold italic notation, down to Proposition 2.4. (Although, coincidentally, all the letters K, T and B will appear in this font, they have no connection with the K, T, and B appearing in the label "KTB" just seen.)

[8]The theorem itself reads: "$\varepsilon(\mathsf{S5}) = \alpha(\mathsf{S5}) = 16$." Here $\varepsilon(L)$ (resp. $\alpha(L)$) denotes the number of non-equivalent (resp. non-analogous) modalities in L. Zolin remarks in the course of the proof of Theorem 4.21 pertaining to non-equivalent modalities (which incidentally does not mention Carnap): "It is interesting to observe that each of these modalities is equivalent in S5 to a boolean combination of $\bigcirc$ and $\odot$." Here $\bigcirc$ is the null modality and (quoting Zolin's own shorthand description) $\odot$ is $\boxplus \wedge \Diamond$, where $\boxplus$ is $\bigcirc \rightarrow \Box$. A quick way of seeing this is to notice that the Hasse diagram of the 16 S5 modalities, partially ordered by provable implication – as at the base of p. 605 of Humberstone [27] – depicts the 16-element Boolean algebra, so taking the equivalence class of a sentence letter, this Boolean algebra is freely generated by that element and any other element at the same level in the diagram other than that element's complement. In [27] the sentence letter chosen was q and Zolin's $\odot q$ appears there as $X(q)$. (Put for convenience in alethic terms, this amounts to: it is either necessarily true or contingently false that q.) The same modality (labelled 'Q') was noted to be capable of playing this role in Canty and Scharle [6], where it was also erroneously claimed to be the only such candidate, overlooking the remaining three options in the same 'row'. (Remaining, that is, after setting aside the equivalence class of the chosen sentence letter, as well as its complement, and $\odot q$, from the six elements there.) The mistake was pointed out in Massey [46].

Remark 2.3. A unilateral version of Zolin's (symmetric) *analogousness* relation appears as the *subconnective* relation in Humberstone [22], p. 37*f.*, where the 'logically loaded' notion of a connective is in play: a connective as an equivalence class of pairs $\langle\underset{\cdot}{\nabla}, L\rangle$, w.r.t. the equivalence relation "is analogous to" – to use Zolin's notation and terminology, so that (the equivalence class of) $\langle\Box, \mathsf{S4}\rangle$ can be thought of as "$\Box$-as-it-behaves-in-S4". In the case where a single logic L is involved rather than a cross-logical comparison, then what in Zolin's notation would be written as $L(\underset{\cdot}{\nabla}) \subseteq L(\underset{\cdot}{\Delta})$ would be read as: "$\underset{\cdot}{\nabla}$ is a subconnective of $\underset{\cdot}{\Delta}$ in L", though to avoid various complications, as when this comes up in the following section, primitive (rather than defined) n-ary connectives κ and κ'; a definition in the 'one logic' case can by-pass the abstraction to equivalence classes: see Definition 3.6(*iii*) (for purposes of which, logics will be taken as consequence relations rather than sets of formulas). When $L = L'$ – or looking ahead to the consequence relation case, $\vdash = \vdash'$ – the possibility arises of a confusion between κ's being deductively stronger than (or: at least as strong as) κ' in L, in the sense that for all $A_1, \ldots, A_n$, we have $\vdash_L \kappa(A_1, \ldots, A_n) \to \kappa'(A_1, \ldots, A_n)$ (or, in the consequence relational version, which does not presume the availability of a suitably behaving $\to$ connective: $\kappa(A_1, \ldots, A_n) \vdash \kappa'(A_1, \ldots, A_n)$, all A) and κ's being a subconnective of κ in (or 'over') L (i.e., $\langle\kappa, L\rangle$'s being a subconnective of $\langle\kappa', L\rangle$). This particular external/internal confusion – the *subconnective fallacy*, we might call it – was illustrated in [22] of binary κ, κ' with an example treated in a conference presentation by Phil Staines, who cited evidence of writers' thinking along the following lines. Since a connective representing a natural language conditional construction – for example, the indicative conditional in English – would plausibly be taken as deductively stronger than the material conditional, any argument valid for the former would be valid for the latter (replacing the one conditional by the other). This amounts to passing without further ado – and thus fallaciously – from an internal deductive strength comparison to a claim that subconnective relation holds.[9] ◀

Turning now to Byrd, we are concerned with the discussion in [5] of a claim in Hintikka [19] to the effect that knowledge and true belief have different logics. To assess this contention, Byrd has to contend with another of Hintikka's claims: that the verbs *believe* and *know* are each amenable to either a weaker or a stronger interpretation. The detailed references to Hintikka's discussion are all to be found in Byrd [5], where they are treated using Hintikka's rather clunky 'model set' analysis

[9]Staines [65] is a published record of much of the material from the conference presentation cited in [22], though not, it appears, of the examples of those committing the 'subconnective fallacy' in respect of conditionals. For further discussion, see subsection 3.24 in Humberstone [30]; a bilateralized version of the subconnective fallacy would amount to confusing Zolin-style analogousness with equivalence in a logic – cf. the criticism of E. E. Dawson before Remark 4.5.1 on p. 279 of [33].

with all of its various conditions (A.PK*, A.PKK*, A.CBB*, etc.); explaining the contrast in what are now more familiar terms, we start with the representation, for a given cogniser a, of knowledge and belief attributions to a using respectively $\boldsymbol{K}_a$ and $\boldsymbol{B}_a$, taken as normal $\Box$ operators. The strong versions are those for which the 4 schema $\Box A \to \Box\Box A$ is appropriate, and the weak versions are those for which this is not so. Byrd shows that if we stick to the strong versions of both belief and knowledge, or else to the weak versions of both, then we find that knowledge and true belief in Hintikka's treatment, "when the two notions are considered in isolation (...) have the same internal logic."[10] The phrase "in isolation" is reminiscent of Łukasiewicz's earlier discussion of twins encountered one at a time, while the "internal logic" matches Zolin's later discussion of matters in those terms.

Byrd's official formulation of the result he is interested in is couched in terms of another Hintikka-specific notion, that of *defensibility* (and indefensibility), but this can be put for present purposes in the more familiar and less problematic terminology of consistency (and inconsistency), a set of formulas being understood as consistent in a particular logic if the logic does not prove any conjunction of formulas from the set.[11] For a formula A, Byrd writes $\boldsymbol{TB}_aA$ to abbreviate $\boldsymbol{B}_aA \wedge A$, but in order to recall the '$\boxdot$' notation of the discussion above, as well as to avoid the impression (with $\boldsymbol{T}$ followed by $\boldsymbol{B}$) of a composite notation, let us instead write $\dot{\boldsymbol{B}}_aA$ for this. From this point on, we drop the subscript a, since the knower/believer is taken as fixed for present purposes.[12]

The reference to inconsistency as defined above can be understood relative to the uniformly weaker ("4-less") logic or to the uniformly stronger one, and where Byrd's formulation speaks of epistemic operators because both $\boldsymbol{K}$ and its dual – the epistemic $\Diamond$ operator written, as $\boldsymbol{P}$ – are taken as primitive, of these we take only

[10]Byrd [5], p. 183. Byrd's discussion, following Hintikka's lead in this respect, abounds with talk of the strong and weak epistemic operators, and the strong and weak doxastic operators, where strength is a matter of being subject to the 4 axiom for the operator in question. Humberstone [30] accordingly complains (p. 483) that such talk is a standing invitation to – if not already an instance of – the subconnective fallacy, as Remark 2.3 puts it; if one operator has a logic that is stronger than (satisfies a proper superset of logical principles) that of another, this of course does *not* imply that when considered together the former is (deductively) stronger than the latter, or indeed vice versa.

[11]A classic critical discussion of indefensibility as originally explained by Hintikka is provided by Johnson Wu [37], though the end of the second last paragraph of Geach [17] is also interesting.

[12]Byrd follows Hintikka in writing the epistemic and doxastic operators in italic rather than in bold italic, which is done here to prevent them looking like the italic capitals $A, B, \ldots$ used here as schematic letters. For this reason Humberstone [33] switches to $\varphi, \psi, \ldots$ for schematic letters when a clash of this kind threatens. Hintikka (with Byrd following suit) uses $p, q, \ldots$ in this capacity – highly confusing in view of their use by almost everyone else as specifically sentence letters (propositional variables) rather than (standing in for) formulas of arbitrary complexity.

$\boldsymbol{K}$ as primitive, so as to have a standard bimodal logic under consideration. Lightly paraphrasing [5], p. 183*f.*, we have:

(†) Let Σ be a set of formulas in which the only non-Boolean connective to appear is $\boldsymbol{K}$. Then Σ is inconsistent if and only if the set Σ' is inconsistent, where Σ' differs from Σ only by having $\dot{\boldsymbol{B}}$ where Σ has $\boldsymbol{K}$.

The stronger logic favoured by Hintikka in [19] is S4 for $\boldsymbol{K}$ and either K4 or KD4 for $\boldsymbol{B}$, along with the bridging axiom $\boldsymbol{K}p \to \boldsymbol{B}p$, while the weaker logic combines KT for $\boldsymbol{K}$ with either K or KD for $\boldsymbol{B}$, and the same bridging axiom as before. What is all this "either/or"? Hintikka's discussion is less than ideally clear as to what is intended. On p. 26 of [19] he writes that the formula (to put it in the present notation) $\boldsymbol{B}p \wedge \boldsymbol{B}\neg\boldsymbol{B}p$ "is easily seen to be inconsistent by means of (A.CBB*), together with [some conditions relating only to the Boolean connectives]", which amounts to saying that $\boldsymbol{B}p$ provably implies $\neg\boldsymbol{B}\neg\boldsymbol{B}p$ (or $\boldsymbol{C}\boldsymbol{B}p$, where $\boldsymbol{C}$ is the doxastic $\Diamond$ operator, taken, as with $\boldsymbol{P}$ as a defined symbol rather than a separate primitive), which is certainly not the case in K4 as the logic of $\boldsymbol{B}$, though it is when we pass to to KD4, as Hintikka does only later in the book, on p. 48, with the introduction of the condition (C.b*) – which conspicuously fails to put in any appearance in the official list of labelled conditions at the back (pp.169–173) of the book; compare, in this connection, note 3 (and the text to which it is appended) in Johnson Wu [37]. Nor does that condition figure in Byrd's own proof of the above result, though Lemmon explicitly takes D for $\boldsymbol{B}$, (i.e., $\boldsymbol{B}p \to \boldsymbol{C}p$) to be part of Hintikka's doxastic logic in his review [39] (p. 382, line 4).

Let us reformulate (†) above in the style of Proposition 1.1, for one specific case, the strongest of the logics mentioned above (with 4 for both operators and D replacing T for $\boldsymbol{B}$):

Proposition 2.4. *Let* S *be the smallest normal bimodal logic in* $\boldsymbol{K}$ *and* $\boldsymbol{B}$ *containing the formulas* $\boldsymbol{K}p \to p$, $\boldsymbol{K}p \to \boldsymbol{K}\boldsymbol{K}p$, $\boldsymbol{B}p \to \boldsymbol{C}p$, $\boldsymbol{B}p \to \boldsymbol{B}\boldsymbol{B}p$ *and* $\boldsymbol{K}p \to \boldsymbol{B}p$. *Then where* $A[\boldsymbol{K}]$ *is any formula in which* $\boldsymbol{K}$ *is the only non-Boolean connective to appear:*

$$\vdash_{\mathsf{S}} A[\boldsymbol{K}] \text{ if and only if } \vdash_{\mathsf{S}} A[\dot{\boldsymbol{B}}].$$

Although we are here considering a bimodal logic and the discussion of Zolin above pertained to monomodal logic, the simplest proof of Proposition 2.4 appeals to the embedding results reported in Example 2.2(*ii*), more specifically that concerning KD4, since the latter is the logic of the doxastic fragment. (We will not directly deal with Byrd's further 'weak' version of Proposition 2.4, with the 4 principles $\boldsymbol{K}p \to \boldsymbol{K}\boldsymbol{K}p$ and $\boldsymbol{B}p \to \boldsymbol{B}\boldsymbol{B}p$ omitted.)

Let us reformulate the relevant part of Example 2.2(*ii*) in the notation used for Proposition 1.1, which involves treating $\Box$ and $\boxdot$ on a par, since they correspond to the $\boldsymbol{K}$ and $\dot{\boldsymbol{B}}$ of the current discussion:

$$\vdash_{\mathsf{S4}} A[\Box] \text{ if and only if } \vdash_{\mathsf{KD4}} A[\boxdot],$$

for any formula $A[\Box]$. We do not need to add here that the only non-Boolean vocabulary (if any) used in its construction is $\Box$, since that is the only such vocabulary in the language of S4. To get from this to Proposition 2.4, which replaces both subscripts with "S" as specified there, we need to check that the epistemic and doxastic fragments of S are given by the epistemic and doxastic subsystems of that axiomatic specification. In other words, we should be sure that (1) adding the $\boldsymbol{B}$-involving axioms listed in Prop. 2.4 – $\boldsymbol{B}p \to \boldsymbol{C}p$, $\boldsymbol{B}p \to \boldsymbol{B}\boldsymbol{B}p$, $\boldsymbol{K}p \to \boldsymbol{B}p$ – does not produce any new $\boldsymbol{B}$-free formulas over and above those yielded by only the $\boldsymbol{B}$-free axioms $\boldsymbol{K}p \to p$, $\boldsymbol{K}p \to \boldsymbol{K}\boldsymbol{K}p$, and (2) adding the $\boldsymbol{K}$-involving axioms does not yield any new $\boldsymbol{K}$-free theorems other than those provable using the $\boldsymbol{K}$-free axioms.[13] But this is straightforward, since for (1), we can simply interpret $\boldsymbol{B}$ as $\boldsymbol{K}$ itself and all the $\boldsymbol{B}$-involving axioms are already provable from $\boldsymbol{K}p \to p$, $\boldsymbol{K}p \to \boldsymbol{K}\boldsymbol{K}p$, and thus could not have the envisaged non-conservative effects, while for (2) we similarly read $\boldsymbol{K}$ as $\dot{\boldsymbol{B}}$.

It was, as already mentioned, with a view to treating $\Box$ and $\boxdot$ 'on a par' above that we wrote the claim inset above as "$\vdash_{\mathsf{S4}} A[\Box]$ iff $\vdash_{\mathsf{KD4}} A[\boxdot]$," rather than, in something closer to the style used by Zolin, as "$\vdash_{\mathsf{S4}} A$ iff $\vdash_{\mathsf{KD4}} A[\boxdot]$". Similarly, what Zolin writes as (2.1)

$$L(\underset{\cdot}{\nabla}) = L'(\underset{\cdot}{\Delta}) \tag{2.1}$$

as the special case of $L = L'$ in the definition of the relation of being analogous over L, would in the style of Proposition 1.1 appear as (2.2), understood as prefaced with "for all formulas A":

$$\vdash_L A[\underset{\cdot}{\nabla}] \text{ if and only if } \vdash_{L'} A[\underset{\cdot}{\Delta}] \tag{2.2}$$

The "$A[\underset{\cdot}{\nabla}]$","$A[\underset{\cdot}{\Delta}]$", here represent the result of replacing every occurrence of $\Box$ with an application of $\underset{\cdot}{\nabla}$, $\underset{\cdot}{\Delta}$, respectively, and so $\Box$ is in effect here playing what we might call an *anchoring* role in the sense that it marks a position in which it awaits replacement at the hands of $\tau_{\underset{\cdot}{\nabla}}$ by $\underset{\cdot}{\nabla}$, and at the hands of $\tau_{\underset{\cdot}{\Delta}}$ by $\underset{\cdot}{\Delta}$, giving the formulas schematically indicated on the left and right sides of (2.2).[14] The fact that the anchor $\Box$ is itself a primitive connective of the object language in modal logic –

[13]If we did not have the bridging axiom $\boldsymbol{K}p \to \boldsymbol{B}p$ to contend with, we could just simply appeal to Thomason [66]. Note also that for brevity, the formulation here mentions only axioms but rules – the necessitation rules for the two modal primitives – should also be understood as included.

[14]In Proposition 2.4, $\boldsymbol{K}$ plays both the anchoring role and the $\underset{\cdot}{\nabla}$ role.

and hence to be able to appear in the replacing formulas themselves – is in a way incidental, and even, for current purposes, potentially confusing, and the anchoring role might be better played by a special purpose 'dummy connective' not in the object language of the logic under discussion, though we make only incidental reference to such devices in the main body of the present discussion. Such a connective merely marks a spot for the insertion of the modalities or (primitive or compositionally derived) connectives of the object language – we may call the formula-like expression in which dummy connectives occur *preformulas* for the object language – and does not appear in what replaces it.[15] The symmetrical treatment thus afforded to $\underset{\cdot}{\nabla}$ and $\underset{\cdot}{\Delta}$, whether or not the symbol playing the $\Box$ role in specifying what these formulas are is itself part of the object language, seems in the end not to live up to its initial promise, as we shall see in the final paragraph of the following section.

On the final page of Byrd [5] one reads the following, in which the reference to specific results from earlier in the paper is not relevant to the point to be made about the passage:

> More generally, Lemma 6 and 7 together show that if a set whose sole non-truth-functional operators are true belief operators is indefensible, then so is the set obtained from it by replacing 'TB' throughout with 'K'.

Byrd is saying that if a formula in which the only non-Boolean connective to occur is $\dot{\boldsymbol{B}}$, then so is the formula obtained from it by replacing '$\dot{\boldsymbol{B}}$' throughout with '$\boldsymbol{K}$'. But $\dot{\boldsymbol{B}}$ can't be the only non-Boolean connective to occur in a formula since any subformula $\dot{\boldsymbol{B}}A$ is just the formula $\boldsymbol{B}A \wedge A$, so $\boldsymbol{B}$ itself occurs in any formula in which $\dot{\boldsymbol{B}}$ occurs. Perhaps Byrd meant that only non-Boolean connective occurring was $\boldsymbol{B}$, and all of the latter's occurrences were in subformulas of the form $\boldsymbol{B}A \wedge A$ (i.e. $\dot{\boldsymbol{B}}A$).

The minor criticism of Byrd in the preceding paragraph arises from his taking $\dot{\boldsymbol{B}}$ (or $\boldsymbol{TB}$, as he writes it) to be defined by $\dot{\boldsymbol{B}}A =_{\mathsf{df}} \boldsymbol{B}A \wedge A$, along with one of two possible views of definition, and in particular, the metalinguistic view rather

[15]In general one would want to allow the preformulas to be constructed with the aid of dummy connectives $*_n^m$ – the nth m-place dummy connective, for all $n \geq 1, m \geq 0$ (in addition to the vocabulary of object language under discussion). Occasionally below, we write just "$*$" for $*_1^1$. Dummy connectives in essentially the present sense can be found on p. 361 of Makinson [45] (and no doubt elsewhere), where they are used for a convenient formulation of conditions on the form of rules; compare also the 'nominal symbols' of Schütte [57], p. 11, used for a convenient description of positive and negative occurrences of a formula within a formula, though not themselves part of the language from which those formulas are drawn. Though the phrase is not used and the setting is slightly different, the symbol $\Box$ itself plays such a dummy (1-ary) connective role in the discussion in Williamson [71] (esp. pp. 101–106), for purposes of explaining what principles are satisfied by this or that *bona fide* sentence operator.

than the object-linguistic view of definition. On the former view the defined symbol is introduced into the metalanguage to abbreviate reference to the formulas of the object language, and the "$=_{\mathsf{df}}$" (or ":=") is just a special case of "=" as identity: one and the same thing – in the present case a linguistic expression – is spoken of in two ways. On the object-linguistic view of definition, by contrast, a definition adds a new symbol to the object language which is intended to be interreplaceable with the material in the *definiens* according to the logic (or the non-logical theory) under discussion.[16] These are really two different approaches to definition, whose relative convenience depends on the purposes at hand, rather than two 'views' in sense of conflicting opinions. For current purposes the object-linguistic approach to definition does not seem particularly convenient, since it means we have three separate non-Boolean connectives to keep track of in the language under discussion: $\boldsymbol{K}$, $\boldsymbol{B}$, and now $\dot{\boldsymbol{B}}$ as well. Accordingly our default understanding of the defined connectives will be the metalinguistic one: they are not new symbols of the object language but functions from that language to itself derived by composing applications of the primitive connectives. However, it is often more desirable to proceed to (what in the present case constitutes) the bimodal setting by throwing away (what appears in the present case as) $\boldsymbol{B}$ by first promoting $\dot{\boldsymbol{B}}$ to the status of a primitive – even though definable – connective, and then passing to the $\boldsymbol{B}$-free fragment of the resulting logic, so as to compare directly with $\boldsymbol{K}$ without interference from $\boldsymbol{B}$: see Example 3.3(i) below on the kind of interference at issue here.

3 Twins

We turn to the distinction gestured at in the final paragraph of Section 1 between twins in general and identical twins in particular, using some of the discussion of Section 2 to illuminate the topic here. The $A[\Rightarrow]/A[\rightarrow]$ style of notation, as it appeared in Proposition 1.1 is not ideal if we want, as we shall, to consider formulas corresponding to those which in the present case are constructed using both $\Rightarrow$ and $\rightarrow$ at once. Our version of Byrd's result, Proposition 2.4 had us asserting the equi-provability in a certain logic of $A[\boldsymbol{K}]$ and $A[\dot{\boldsymbol{B}}]$, but the condition on the first of these was that $\boldsymbol{K}$ was the only non-Boolean connective to appear in $A[\dot{\boldsymbol{B}}]$, so again we ended up comparing certain monomodal formulas drawn from an originally bimodal language. (In that case we took $A[\dot{\boldsymbol{B}}]$ as defined but exploited the fact that $\boldsymbol{K}$ was primitive and so could perform the 'anchoring' role played by $\Box$ in Zolin's discussion, and envisaged for $*$, in notes 15 above and 18 below.) To address the

[16]For more on this, see the index entries in Humberstone [30] under 'defined connectives: object-linguistic *vs.* metalinguistic view'.

topic of twins along the lines suggested in Section 1 we need a notation which will apply whether only one, or instead both, of the connectives being compared appear in a given formula. This will provide a convenient way of drawing the distinction foreshadowed there between connectives behaving as twins and connectives behaving as identical twins (in a given logic).

Let us use κ, λ, sometimes decorated (κ' etc.) to stand for primitive connectives, with $ar(\kappa)$ denoting the arity of κ. Where $ar(\kappa) = ar(\kappa')$ it is convenient to have a notation for the result of interchanging occurrences of κ in a formula A with occurrences of κ', the official notation for which will be $A^{\kappa\bowtie\kappa'}$ though when the identity of κ and κ' is clear from the context, we will simply write $A^{\bowtie}$ ("A swap"). There are conceptual difficulties in trying to work with derived connectives (or Zolin's 'modalities') in settings like this, some taken up in the final paragraph of this section. For the moment we indicate how the issue raises its head in the present instance after giving an inductive definition of this $\bowtie$ notation. Here let us fix n as the arity of κ, κ', and suppress reference to those two connectives in the $\bowtie$ superscript:

- $(p_i)^{\bowtie} = p_i$
- $\left(\kappa(B_1, \ldots, B_n)\right)^{\bowtie} = \kappa'(B_1^{\bowtie}, \ldots, B_n^{\bowtie})$
- $\left(\kappa'(B_1, \ldots, B_n)\right)^{\bowtie} = \kappa(B_1^{\bowtie}, \ldots, B_n^{\bowtie})$
- $\left(\lambda(B_1, \ldots, B_{ar(\lambda)})\right)^{\bowtie} = \lambda(B_1^{\bowtie}, \ldots, B_{ar(\lambda)}^{\bowtie})$, for all primitive λ other than κ, κ'.

Remark 3.1. The need to restrict attention to primitive connectives in a treatment like this, rather than trying to exchange derived connectives such as Zolin's modalities, is clear from the inductive clauses here. Suppose we have $\boxdot$, for instance, not as a primitive connective, but instead used, as in Example 2.2(i), for the modality that maps A to $\Box A \wedge A$ – often indicated in Zolin [72] by writing the value of this function as $\boxdot(A)$ rather than $\boxdot A$. A serious problem then arises with the idea of interchanging $\Box$ and $\boxdot$ in a formula using the above explanation of $(\cdot)^{\bowtie}$: the fate of (non-primitive) $\boxdot$ is already settled by the inductive steps concerning $\Box$ and $\wedge$, so any attempt to settle it again is either redundant or inconsistent, depending on whether it agrees with or differs from the composite story already told for the primitive connectives. (There would be a similar problem in trying to add in an inductive clause for the compositionally derived connectives in defining the τ_{∇} translations in play in the preceding section.) ◀

By induction (if necessary), we see that taking $\bowtie$ as $\kappa\bowtie\kappa'$ as above, for any formula A of a language in which these are primitive connectives of the same arity,

$(A^{\bowtie})^{\bowtie}$ is the same formula as A, and if A is constructed with the aid of only one of κ, κ' (and any further connectives), $A^{\bowtie}$ is the result of replacing (all occurrences of) κ with κ' or vice versa. Thus in the notation of Proposition 1.1, if A is $A[\kappa]$, then $A^{\bowtie}$ is $A[\kappa']$ – though for Prop. 1.1 there were stipulated to be no connectives present other than κ = "$\Rightarrow$", κ' = "$\rightarrow$" (or κ = "$\supset$"); no such restriction is in force here. Again allowing for the possible appearance of primitive connectives other than a given pair κ, κ' (with $ar(\kappa) = ar(\kappa')$), let us call a formula *mixed* if it is constructed with the aid of both κ, κ' and *unmixed* if it is constructed with the aid of at most one out of κ, κ'.

Definitions 3.2. *Suppose that L is a logic on a language with (possibly inter alia) primitive n-ary connectives κ, κ' (in terms of which the 'unmixed' terminology and the "$\bowtie$" in what follows is to be understood). Then, meaning by 'formula', 'formula of the language of L':*

(i) κ, κ' are twins in L if for all unmixed formulas A, we have $\vdash_L A$ iff $\vdash_L A^{\bowtie}$;

(ii) κ, κ' are identical twins in L if for all formulas A, we have $\vdash_L A$ iff $\vdash_L A^{\bowtie}$.

Note that the 'iff's in (*i*) and (*ii*) can be replaced by 'if's (or by 'only if's) without loss. For example, in the case of (*i*) the "all unmixed formulas A" covers the case of $A = A[\kappa]$, with $A^{\bowtie}$ being $A[\kappa']$, as well as the case of $A[\kappa']$, with $(\cdot)^{\bowtie}$ returning us to $A[\kappa]$. The present notion of twinhood is a close relation of Zolin's notion of analogousness, though the the role played $\Box$ in [72] would have to be played by a dummy connective of suitable arity (see note 15 above and note 18 below).

The idea, from Section 1, of twins being indistinguishable on the basis of separate encounters with them is embodied in Definition 3.2(*i*) by considering only formulas in which one of the candidate twins appears at a time: the logic in question returns the same provability verdict on all such formulas when either candidate is replaced with the other. But the stronger kind of indistinguishability raised in the introduction – there being no difference to be registered even when both twins are present – is embodied in Definition 3.2(*ii*), with its admission of mixed formulas to those in which interchanging them still leads to identical verdicts.

Here we take no interest in an 'inter-logical' version of the twin concept because the extra effort involved would have no pay-off for cases like that for which Łukasiewicz originally introduced this idea, though of course Definition 3.2(*i*) could be loosened up to allow for κ-in-L to be a 'twin' of κ'-in-L', with two one-directional versions of the $\bowtie$ notation, and the remaining vocabulary of L, L' were suitably constrained. (This amounts to something along the lines of (2.2), though the present κ, κ', unlike $\underset{\cdot}{\nabla}, \underset{\cdot}{\Delta}$ there, are supposed to primitive connectives.) Similarly, to apply this vocabulary to describe Byrd's observations in [5], L would be taken, not as the

S of Proposition 2.4, but as the $\{\boldsymbol{K}, \dot{\boldsymbol{B}}\}$-fragment of that S, now taking $\dot{\boldsymbol{B}}$ as a primitive connective in its own right – and getting rid of $\boldsymbol{B}$ itself for reasons explained in the first of the following illustrations of the 'twin' terminology:

Examples 3.3. (*i*) For S from Prop. 2.4 as just modified, we have: $\boldsymbol{K}$ and $\dot{\boldsymbol{B}}$ are twins, though not identical twins, in S. The first claim is just a restatement of Prop. 2.4, in the current terminology. But the change of terminology forces the new understanding of what S is: its language must have as the only non-Boolean connectives, $\boldsymbol{K}$ and $\dot{\boldsymbol{B}}$, throwing $\boldsymbol{B}$ out of the language and taking $\dot{\boldsymbol{B}}$ to be taken as primitive. We need the latter as primitive in any case, as observed in Remark 3.1, but we also need to exclude $\boldsymbol{B}$ from the language (of the new S) since if it were still present, taking $\bowtie$ as $\boldsymbol{K} \bowtie \dot{\boldsymbol{B}}$, we should have for $A = (\boldsymbol{B}p \wedge p) \to \dot{\boldsymbol{B}}p$

$$\vdash_{\mathsf{S}} A, \text{ whereas } \nvdash_{\mathsf{S}} A^{\bowtie},$$

since $A^{\bowtie}$ is $(\boldsymbol{B}p \wedge p) \to \boldsymbol{K}p$. (If we call connectives other than the κ, κ' being swapped by $\bowtie$, *extraneous* connectives – relative to the twinhood question under consideration – then what we have here is what might be called *interference* with the twin status of $\boldsymbol{K}$ and $\dot{\boldsymbol{B}}$ by the extraneous connective $\boldsymbol{B}$.) The second claim, that though twins in (revised) S, $\boldsymbol{K}$ and $\dot{\boldsymbol{B}}$ are not identical twins, is just a reflection of the fact that

$$\vdash_{\mathsf{S}} \boldsymbol{K}p \to \dot{\boldsymbol{B}}p \text{ whereas } \nvdash_{\mathsf{S}} \dot{\boldsymbol{B}}p \to \boldsymbol{K}p \text{ (i.e. } \nvdash_{\mathsf{S}} (\boldsymbol{K}p \to \dot{\boldsymbol{B}}p)^{\bowtie}).$$

(Byrd adds, p. 186: "There are other more interesting contrasts between knowledge and true belief in their strong senses [i.e., with the 4 principles in force], if mixed sets are allowed." He cites – putting it in the present terminology and notation – the provability of $\boldsymbol{B}p \to \dot{\boldsymbol{B}}\boldsymbol{B}p$ alongside the unprovability of $\boldsymbol{B}p \to \boldsymbol{K}\boldsymbol{B}p$ as an example. But allowing such 'mixed cases' involves going back to the language with $\boldsymbol{B}$ present in addition to $\boldsymbol{K}$ and $\dot{\boldsymbol{B}}$, which would lead straight back to the case of the A just considered, preventing $\boldsymbol{K}$ and $\dot{\boldsymbol{B}}$ from being even non-identical twins.)

(*ii*) For this example, we assume familiarity with propositional tense logic and use A. N. Prior's notation G, H for the future and past tense $\Box$-operators, to be interpreted by accessibility relations which are each other's converses. It is G and H that are to be compared and in terms of which the "$\bowtie$" notation is to be understood (the corresponding $\Diamond$-operators being F and P, taken as defined rather than primitive). The basic system K_t of tense logic (among many others) has the property that in it these two operators are identical twins in the sense of Definition 3.2(*ii*). In the tense-logical literature this property is often called the *mirror image* property, with $A^{\bowtie}$ being the 'mirror image' of the formula A. Steadfastly resisting any temptation to extend the biological metaphor by purloining talk of 'mirror twins' from that quarter, we note that this is a special case of a more general phenomenon:

Let L be any normal bimodal logic determined by (i.e., sound and complete w.r.t.) a class $\mathbb{M}$ of models $\langle W, R, S, V\rangle$, R, S binary relations on $W \neq \varnothing$ the first listed being an accessibility relation for $\Box_1$, and the second, for $\Box_2$ (and V supplies subsets of W to the sentence letters, with the inductive definition of truth at a point proceeding as usual). Then if $\mathcal{M} = \langle W, R, S, V\rangle \in \mathbb{M}$ always implies $\langle W, S, R, V\rangle \in \mathbb{M}$, then $\Box_1$ and $\Box_2$ are identical twins in L. For the proof it is convenient to use the notation $\mathcal{M}^{\bowtie}$ for the result of interchanging the two accessibility relations; $A^{\bowtie}$ is the result of interchanging the corresponding non-Boolean primitives $\Box_1, \Box_2$ in A. One shows by induction on the complexity of A that for all $w \in W$, where $\mathcal{M} = \langle W, R, S, V\rangle$ that

$$\mathcal{M} \models_w A \text{ if and only if } \mathcal{M}^{\bowtie} \models_w A^{\bowtie},$$

and then argues that if $\vdash_L A$, we must have $\vdash_L A^{\bowtie}$ by appeal to the result inset above and the fact that L is determined by $\mathbb{M}$, which is closed under the $\bowtie$ operation on models.

In the case of K_t determined by the class of models in which $S = R^{-1}$, the desired condition of closure under the $\bowtie$ operation on models is evidently satisfied, as is also the case of S as the complement of R (relative to $W \times W$), considered in [21]. In both cases given W and either accessibility relation, the other accessibility relation is uniquely fixed (as the converse or the complement, respectively) but in general the $\bowtie$-closure condition does not require this. For example the class of models $\langle W, R, S, V\rangle$ such that for each $x \in W$ there exists $y \in W$ such that Rxy and Sxy, also satisfies the condition, though neither S nor R is uniquely fixed in terms of the other in this case. (The $\bowtie$-closure condition would most naturally be conducted in terms of frames rather than models, but one wants to avoid conveying the impression that these considerations bear only on the case of Kripke-complete normal modal logics.)

(iii) This example comes from deontic logic as a variation on the use in such papers as Anderson [3] of a sentential constant (or nullary connective) $\boldsymbol{s}$ informally read as "the sanction is applied" (or more accurately "there has been an infringement of the moral code") added to the vocabulary of a suitable alethically interpreted modal logic and governed by the axiom $\Diamond\neg\boldsymbol{s}$ to yield a deontic logic in which "It is forbidden/would be wrong that A" could be represented by $\Box(A \to \boldsymbol{s})$. Humberstone [20] suggests the addition of a second sentential constant $\boldsymbol{r}$ – the letter chosen to suggest "reward" with a view to reading $\Box(A \to \boldsymbol{r})$ as saying "It would be supererogatory that A", and examining the interrelations between these different deontic notions after trading in the previously cited axiom for $\Diamond(\neg\boldsymbol{s} \wedge \neg\boldsymbol{r})$. Evidently $\boldsymbol{s}$ and $\boldsymbol{r}$ are identical twins

(though not equivalent formulas) in the resulting logic – something referred as the 'exchange property' in [20], *q.v.* for acknowledgment that the deontic notions in play here really deserve a more explicitly agent-relative treatment. (However, the description in [20] of $\boldsymbol{s}, \boldsymbol{r}$ as propositional constants rather than sentential constants is ill-advised for reasons not germane to our present concerns. [33], p. 275, gives a brief discussion of the issue, and further references.) ◀

Remarks 3.4. (*i*) In connection with Example 3.3(*ii*), it is important to recall the metalinguistic understanding of definitions mentioned at the end of the previous section. $P\top$ for instance, is none other than the formula $\neg H \neg \top$; similarly, references to F are just our abbreviated way of referring to $\neg G \neg$. Without this understanding in place, we could not have given K_t as a logic according to which G and H are (non-equivalent) identical twins, since the provable $Gp \leftrightarrow \neg F \neg p$ would have as its mirror image (its $G \bowtie H$-swap) the unprovable $Hp \leftrightarrow \neg F \neg p$, rather than, as intended, discerning the occurrence of "G" concealed by the "F" notation, the provable formula $Hp \leftrightarrow \neg P \neg p$ ($= Hp \leftrightarrow \neg\neg H \neg\neg p$, in primitive notation). Similarly if we had been considering the equivalences corresponding to the definitions that would have related distinct primitives on the object-linguistic conception of definition – $Fp \leftrightarrow \neg G \neg p$ in the future tense case. But the present point has nothing specifically to do with biconditionals: the addition of any new primitive stipulated to stand in some non-trivial logical relation[17] to G but not H or vice versa – for example adding a one-place connective O with just the axiom $Op \vee Gp$ would stop G and H from being twins (since $Op \vee Hp$) would not be provable. In the case of duals, if one wanted to treat H and P (and G and F) as separate primitives, one would have to re-work the account so as to accommodate the more complicated relation "G is to F as H is to P" – or that *pair* $\langle G, F \rangle$ is a twin (or identical twin) of the pair $\langle H, P \rangle$. (Compare the even more general idea of (esp. 'weak') duality under a permutation introduced on p. 159 of McKinsey and Tarski [47].) However, note that we may be dealing with a case in which G and H are both primitive and only one of F, P is, in which case this move would not be available; whichever of the latter two was primitive would then be an extraneous connective spoiling – or interfering with – the twin relationship, to use the terminology introduced in Example 3.3(*i*).

(*ii*) As an alternative to Example 3.3(*i*), it would be nice to present alongside Example 3.3(*ii*) a tense-logical version, with G and H being non-identical twins. For example, we might consider the normal extension of K_t by (one of) Hamblin's discreteness axiom(s) $(p \wedge Hp) \to FHp$, together with $P\top$. A simple semantic argument shows that this logic does not contain the mirror image of that discreteness axiom;

[17] More explicitly: in some *proper coercive logical relation*, as this is explained in Definition 2.7(*ii*) in Humberstone [35].

but since the given axiom has $F\top$ as a consequence we have added $P\top$ so as to keep our past and future operators twins even though not identical twins. If this logic has further unmixed theorems whose mirror images are not theorems, the latter would have to be added, and one would want some guarantee that this could be done without bringing in their wake $((p \wedge Hp) \rightarrow FHp)^{\bowtie}$ – so the author can only conjecture that an example of the kind desired can be produced in this way. (Of course we don't always have to add the mirror images of the unmixed theorems since they are often consequences – even just against the background of K_t; $P\top$ was not thus automatically forthcoming as a theorem, reflecting the fact that a relation's being serial does not imply that its converse is, contrasting in this respect with adding the 4 of axiom $Gp \rightarrow GGp$ which yields its mirror image as a theorem, reflecting the fact that the converse of a transitive relation is transitive. A semantically more nuanced presentation of these issues can be found in [33], p.185*ff.*)

(*iii*) The proof given in the course of the discussion in Example 3.3(*ii*) – concerning models $\langle W, R, S, V\rangle$ – that G and H are identical twins in K_t is matched by a different strategy for the case of logics given proof-theoretically rather than semantically, by picking an axiomatization concerning which one shows that the mirror images of the axioms are provable and that the rules preserve the property of having a provable mirror image. (Indeed the sanction-and-reward case as presented in Example 3.3(*iii*) is such a case.) But here we pause to note that a similar semantic argument can be used whenever we are dealing with two items in the models – not just accessibility relations – which can be interchanged, keeping us inside a class of models determining the logic of interest, to show that the connectives they respectively interpret are identical twins in the logic determined. One cannot, incidentally, help but notice a resemblance between pairs of non-equivalent identical twins (in a logic) and pairs of objects which are not pairs of individuals, as this terminology is explained in Caulton and Butterfield [8]; non-equivalence of connectives should be understood for in terms of the non-synonymy – in the sense of [63], p. 116 – of compounds formed from the same components with their aid. Łukasiewicz's reaction this this phenomenon, mentioned in the final paragraph of Section 4 is likewise reminiscent of the puzzlement sometimes raised by the the idea that the positive and negative square roots of -1 are distinct – touched on again in [8] and many references there cited. ◀

In all of Examples 3.3, the two connectives compared are surrounded by a bevy of Boolean connectives left unaffected by the passage from A to $A^{\bowtie}$, and in general there the risk of interference from extraneous connectives alluded to in Example 3.3(*i*), which becomes acute when, to avoid comparing derived connectives, one gives those to be compared primitive status and the inevitability of what would have been used in the a definition (had they been non-primitive) to interfere – as that example (or, originally, in the last paragraph of the preceding section). One accordingly often has occasion to study the logic of the connectives concerned in the absence of other – even all other – connectives. Indeed this happened in the case of Proposition 1.1. However, in the case of some such pairs of connectives, there are no theorems to be found in the fragments thus purified:

Example 3.5. As a case in point, consider what in Zolin-inspired notation and terminology might be called the two 1-ary 'modalities' $\underset{\cdot}{\nabla} p = p \to \bot$ and $\underset{\cdot}{\Delta} p = ((\bot \to p) \to p) \to \bot$.[18] To avoid unwanted interference it may desirable to study the pure logic of these two connectives and compare them with each other when they are taken not only as primitives in their own right – to be written respectively as $\neg$ and $\neg'$ – rather than defined, but as the *only* primitive connectives of the language considered. For a more interesting case study in which the two connectives are not equivalent, we need to drop below not only classical but even intuitionistic logic, to – this atheorematic fragment of – Minimal logic. Thus, roughly speaking:

$$\neg A = A \to \bot \qquad \text{and} \quad \neg' A = ((\bot \to A) \to A) \to \bot,$$

with $\to$ and $\bot$ as in Minimal Logic, have the same logic when each is taken as the sole connective and the logic is given by the consequence relation, the restriction of which to the negation fragment we call $\vdash_{\mathsf{ML}\neg}$ and to the non-standardly interpreted negation fragment, $\vdash_{\mathsf{ML}\neg'}$.[19] ◀

This case gives us three logics of possible interest, $\mathsf{ML}_\neg$, the standard pure negation fragment of Minimal Logic, $\mathsf{ML}_{\neg'}$, a version of Minimal logic with the 'deviant' negation $\neg'$ as the sole primitive connective, and, naturally of greatest interest here since we are concerned with twinhood, $\mathsf{ML}_{\neg,\neg'}$ which has the two primitive connectives $\neg$ and $\neg'$. But here there is a problem: if the logic is the set of provable formulas, there is no logic to speak of in any of these three cases. Accordingly for the further treatment of this issue in Example 3.7(*i*) we pass to the associated consequence relations $\vdash_{\mathsf{ML}\neg}$ etc., as was done – though this particular ('logic without

[18] In this case, unlike Zolin's, these are 1-ary modalities (compositionally derived connectives) with no 1-ary primitive connective in the language to play the anchoring role of $\Box$ in the modal case. One could use a 1-ary dummy connective $*$ (see note 15) and adapt the mapping $\tau_{\underset{\cdot}{\nabla}}$ so that it now maps preformulas involving $*$ by acting as the identity translation on sentence letters and for the (non-dummy) connectives of the object language by themselves, coming to life with the inductive step $\tau_{\underset{\cdot}{\nabla}}(*B) = \underset{\cdot}{\nabla}(\tau_{\underset{\cdot}{\nabla}}(B))$, and correspondingly in the case of $\tau_{\underset{\cdot}{\Delta}}$, *mutatis mutandis.*

[19] "Roughly speaking" because the above equalities are not to be taken as definitions within the current language, in which $\to$ and $\bot$ are not available to do the defining; rather, we intend the two negations to be taken as primitive but with $\vdash_{\mathsf{ML}\neg}$ coinciding with $\{\neg, \neg'\}$ fragment of $\vdash_{\mathsf{ML}}$, and similarly with the further subfragments involving only one or other of $\neg, \neg'$.

theorems') problem did not arise there – in note 1 for the case of BCSK. We need this move also for another example in this section (Example 3.7(i)) as well as for the discussion of Łukasiewicz below (Section 4).

We assume familiarity with the notion of a consequence relation but rather than the cumbersome notation of note 1 in which one reads such things as the following (where $\vdash$ was actually something called $\vdash^{glo}_{\mathsf{S5}}$):

$$C_1[\Rightarrow], \ldots, C_n[\Rightarrow] \vdash A[\Rightarrow] \text{ iff } C_1[\rightarrow], \ldots, C_n[\rightarrow] \vdash A[\rightarrow],$$

it is more convenient to think of the pairs $\langle\{C_1, \ldots, C_n\}, A\rangle$ which are elements of a consequence relation as provable objects – 'sequents' – in their own right and use the 'context for a connective' notation on a sequent itself, and where $\sigma = \langle\Theta, A\rangle$ write $\vdash_L \sigma$ in place of the explicit notation $\Theta \vdash_L A$. Let us focus on the case in which Θ is $\{C_1, \ldots, C_n\}$ – though there is no general restriction to finite sets here – and σ is $\langle\Theta, A\rangle$, for which a more suggestive notation is usually employed, such as $\Theta \succ A$ or $\Theta : A$, the former notation being preferred here.[20] Then we can write the above more succinctly as

$$\vdash \sigma[\Rightarrow] \text{ iff } \vdash \sigma[\rightarrow].$$

More to the point, since we want to include the case in which both of the connectives of interest are present together (in the same sequent, whether or not they co-occur in any of the formulas making up that sequent), we can similarly denote by $\sigma^{\bowtie}$ the result of interchanging those connectives in all the formulas of σ. That is, with $(\cdot)^{\bowtie}$ for formulas as above, if σ is $\Theta \succ A$, then $\sigma^{\bowtie}$ is $\{C^{\bowtie} | C \in \Theta\} \succ A^{\bowtie}$.

Just for the record, let us give here the obvious reformulation of Definitions 3.2 to apply to the case of consequence relations. With respect to a pair of primitive connectives, of the same arity, in the language of a consequence relation $\vdash$ we call a sequent *mixed* if among its constituent formulas, both connectives appear, otherwise *unmixed*. For good measure we include a definition of the subconnective relation.

Definitions 3.6. *Suppose that $\vdash$ is a consequence relation on a language whose primitive connectives include n-ary κ, κ', in terms of which, as for Definitions 3.2, the 'unmixed' terminology and the "$\bowtie$" notation is to be understood (i.e., $\bowtie$ means $\kappa \bowtie \kappa'$). Then:*

(i) κ, κ' are twins according to $\vdash$ if for all unmixed sequents σ of that language $\vdash \sigma$ iff $\vdash \sigma^{\bowtie}$;

[20] Sometimes instead, "$\rightarrow$" or "$\Rightarrow$" is used as a sequent separator, which would be very confusing in the present discussion, since those are both already in service here as sentence connectives. As to the use of "Θ", see note 6.

(ii) κ, κ' *are identical twins according to* $\vdash$ *if for all sequents* σ *of that language* $\vdash \sigma$ *iff* $\vdash \sigma^{\bowtie}$.

(iii) κ *is a subconnective of* κ' *according to* $\vdash$ *if for all sequents* $\sigma[\kappa]$ *in which* κ *and any additional primitive connectives (of the language of* $\vdash$*) with the exception of* κ' *may occur, if* $\vdash \sigma[\kappa]$ *then* $\vdash \sigma[\kappa']$.

Note that under *(iii)* we could equally well have made the definition be: for all σ composed of formulas not constructed with the aid of κ', $\vdash \sigma$ implies $\vdash \sigma^{\bowtie}$, and also that the "iff"s in *(i)*, *(ii)* add nothing to what we would have with "only if" instead.

The following pair of examples concern twins in subclassical logics. The first concerns the case of $\mathsf{ML}_{\neg,\neg'}$, introduced after Example 3.5 above;the second is borrowed from [30].

Examples 3.7. *(i)* The connectives $\neg$ and $\neg'$ from Example 3.5 are twins according to $\mathsf{ML}_{\neg,\neg'}$, though not identical twins. A proof of the first assertion is to be found in the Appendix (see Proposition 5.2 there). For the second, note that for the mixed sequent $\sigma = \neg' p \succ \neg p$, we have $\vdash_{\mathsf{ML}_{\neg,\neg'}} \sigma$ while $\nvdash_{\mathsf{ML}_{\neg,\neg'}} \sigma^{\bowtie}$. That is, $\neg p$ is a consequence of $\neg' p$ according to the current logic, though not conversely (much as in Example 3.3*(i)*). To see this note that $((q \to p) \to p) \to q$ has $p \to q$ an intuitionistic (or 'positive logical' or indeed 'Minimal') consequence, so we can just substitute $\bot$ for q; on the other hand, since $\bot$ has no special logical powers in Minimal Logic, we could only have $p \to \bot$ as an ML-consequence of $((\bot \to p) \to p) \to \bot$ if $p \to q$ had $((q \to p) \to p) \to q$ as an ML-consequence – or, equivalently, an intuitionistic consequence, since only $\to$ is involved – which is easily checked not to be the case, e.g., using the Kripke semantics. (Alternatively, much as in Exercise 4.22.11 of [30]: observe that if this were so, then $(p \to q) \to (((q \to p) \to p) \to q)$ would be intuitionistically provable – an IL-consequence of the empty set, that is – and so, therefore, permuting antecedents, would

$$((q \to p) \to p) \to ((p \to q) \to q)$$

be. But on substituting $q \to p$ for all occurrences of p, we get a conditional that would then deliver Peirce's Law from Contraction by Modus Ponens.)

(ii) In the setting of intuitionistic logic (IL), consider the binary connectives (a) of *alternative denial* (to use Quine's phrase, though he did not take an interest in the present setting), $\downarrow$, with $A \downarrow B = \neg A \vee \neg B$ and (b) *nand* (or negated conjunction) $\bar{\wedge}$ with $A \bar{\wedge} B = \neg(A \wedge B)$. As with the two minimal negations of *(i)*, we consider the fragment (now, of IL) in which only these two connectives figure, in which, as is familiar, they do not yield intuitionistically equivalent compounds. Specifically, for $\sigma = p \downarrow q \succ p \bar{\wedge} q$ we have $\vdash_{\mathsf{IL}} \sigma$ while $\nvdash_{\mathsf{IL}} \sigma^{\bowtie}$, so these two connectives are out of the

running for being *identical* twins according to the current fragmentary subrelation of $\vdash_{\mathsf{IL}} - \vdash_{\mathsf{IL}\downarrow,\bar{\wedge}}$, let's call it – which still leaves open the possibility that they are nevertheless, *twins* according to that consequence relation, which would then match the situation described under (*i*). But no, it is shown in [30] (Example 8.24.7 and Remark 8.24.8, p. 1245, where $\downarrow$ is written as $*$), that according to $\vdash_{\mathsf{IL}\downarrow,\bar{\wedge}}$, $\downarrow$ is a proper subconnective of $\bar{\wedge}$. (That is, $\downarrow$ is a subconnective of $\bar{\wedge}$ but not conversely.) ◀

Remarks 3.8. (*i*) Concerning Example 3.7(*ii*): As is also pointed out in [30], p. 397, the situation is quite different from that illustrated by the example in the case of *classical* logic ($\vdash_{\mathsf{CL}}$, as in Section 1), where no connective can be a subconnective of a non-equivalent connective (κ, κ' being said to be equivalent according to $\vdash$ when $\kappa(A_1, \ldots, A_n) \dashv\vdash \kappa'(A_1, \ldots, A_n)$ for all $A_1, \ldots, A_n$, where $ar(\kappa) = ar(\kappa') = n$).
(*ii*) Both the observation under (*i*) here and the proof in [30] of the '(proper) subconnective' claim alluded to in Example 3.7(*ii*) rely heavily on results established by W. Rautenberg. (The relevant references can be found in [30].) ◀

In this final paragraph of the present section, which can be skipped without jeopardizing the intelligibility of the remainder of the paper, the author confesses to being puzzled as to exactly how the present account of twins (Definitions 3.2(*i*) and 3.6(*i*)) is related to Zolin's account of analogousness. Zolin had no difficulty describing derived connectives ('modalities') as analogous in a given (monomodal) logic, as for example $\Box\Box$ and $\Box\Box\Box$ – or $\Box^2$ and $\Box^3$ for short – in K, whereas the $\bowtie$-based Definition 3.2(*i*), was seen (Remark 3.1) to be highly problematic for the case in which non-primitive logical vocabulary was under discussion. We might well have expected trouble with the 'identical twins' notion introduced in part (*ii*) of 3.2(*i*) (or 3.6) since it is not clear that there is any intuitive idea (for a generalization to non-primitive connectives of the $\bowtie$ operation to capture) of "interchanging $\Box^2$ and $\Box^3$" in a formula such as, for instance, the formula we might write for brevity as $\Box^8 p$. In the interests of not privileging any particular connective of the object language in the way in which $\Box$ is privileged in Zolin's treatment we might recruit the idea of a preformula with a dummy connective of the appropriate arity, here taken for simplicity to be 1. Thus where $A = A[*]$ is such a preformula, we might enquire as the equi-provability of $A[\underset{\cdot}{\nabla}]$ and $A[\underset{\cdot}{\Delta}]$ for the derived $\underset{\cdot}{\nabla}$ and $\underset{\cdot}{\Delta}$ of interest, where $A[\underset{\cdot}{\nabla}]$ is $\tau_{\underset{\cdot}{\nabla}}(*A)$ from note 18, and correspondingly for $A[\underset{\cdot}{\Delta}]$. But, for which preformulas $A[*]$ should one demand equi-provability of $A[\underset{\cdot}{\nabla}]$ and $A[\underset{\cdot}{\Delta}]$? The account should at least coincide with that given for twinhood of primitive connectives, extending it to cover non-primitives, so since we want primitive G and H in a formulation of K_t (with no other non-Boolean primitives) to count as twins – having seen them even to constitute *identical* twins in Example 3.3(ii), though here we are not worrying about the 'identical' aspect of this case.[21] For the preformula $A[\ast] = Hp \to \ast p$ we will have $A[H]$ ($= Hp \to Hp$) provable without $A[G]$ ($= Hp \to Gp$) being provable, which we do not want to count against the twin status of G, H, and can easily guard against by requiring that G, H do not themselves occur in the preformula $A[\ast]$.[22] This makes perfect sense because G, H happen to primitive connectives, and what would be needed is to make sense of this generally, but what exactly is it for a *derived* connective to occur in a formula? (Where $\boxdot A$ is, as in Example 2.2(i) $\Box A \wedge A$, does $\boxdot$ 'occur in' the formula $(\Box p \wedge q) \wedge p$? Or in $q \wedge \Box q$?) Finally, while this discussion has supposed that we can separate the 'twins' and 'identical twins' issues for derived connectives, that distinction hung on the contrast between mixed and unmixed formulas, a distinction which is problematic when non-primitive connectives are involved: returning to the above case of $\Box^2$ and $\Box^3$, is $\Box^6$ a mixed formula or an unmixed formula? If we are comparing $\Box$ and $\boxdot$, isn't any formula constructed using $\boxdot$ automatically a mixed formula? (This was the 'minor criticism' of Byrd in Section 2, where these were written as $\boldsymbol{B}$ and $\dot{\boldsymbol{B}}$.)

4 Back to Łukasiewicz

So frequently does commentary on the Ł-modal system of Łukasiewicz go astray that it is with some trepidation that one ventures into this territory. But venture we must, given the topic under discussion. That begins in the following paragraph, after we pause to illustrate the perils that have afflicted otherwise impeccably credentialed logicians. Segerberg's [58] broaches the subject of the Ł-modal logic on p. 209, where we are told that it is the smallest modal logic extending C with (all instances of the schema) – called by Segerberg (as is the logic itself) Ł: $\Box\top \to (A \to \Box A)$. Segerberg remarks that this is itself a regular modal logic, something not immediately evident from the above description of it since one of the defining conditions ([58], p. 12) for regular modal logics is closure under the rule which takes us from $A \to B$ to $\Box A \to \Box B$, making regularity a condition not automatically passed from modal logics to their extensions.[23] The description given obviously does not suffice to pick

[21]Conceivably, the pessimistic attitude already taken to identical twinhood between derived connectives notwithstanding, something could be made of this using preformulas $A[\ast_1, \ast_2]$ with two dummy connectives of the relevant arity.

[22]This is like 'extraneous interference' *à la* Example 3.3(i), except that here the interfering material – the "H" in the preformula – is here far from extraneous. The present example is highly reminiscent of Examples 6.1(i) and (ii) urged as making trouble for a particular account of logical independence in Humberstone [35].

[23]Here of course we are thinking of modal logics as sets of formulas (extensions thus being supersets), and more specifically as such sets as contain all classical truth-functional tautologies

out the Ł-modal logic since Segerberg's Ł does not containing every formula of the form $\Box A \to A$ – or more simply put, does not contain the formula $\Box p \to p$ – as the Ł-modal logic does. (At least this will be obvious in view of the matrix-based description of the logic below. This mistake is noted at the base of p. 48 in [33], no doubt among several other places.) Erring in the other direction, over-axiomatizing rather than under-axiomatizing the logic, the generally invaluable Font and Hájek ([13], p. 168) give an axiomatization of the Ł-modal system, using Modus Ponens as its sole rule, along with any axioms sufficient for classical propositional logic, and the following special axioms involving $\Box$:

$$\Box(A \to B) \to (\Box A \to \Box B) \qquad \Box A \to A \qquad \Box B \to (A \to \Box A)$$

but the first of these axioms (or more accurately, axiom schemata) is easily seen to be redundant given the other two. To be fair to Font and Hájek, (1) this is an oversight rather than an error, since the authors do not claim to be providing an independent axiomatization, and (2) it's not (originally, at least) *their* oversight, since, as they mention, they are take this axiomatization straight from Lemmon [40].[24] Font and Hájek, incidentally, also discuss consequence relations associated with the current set-of-formulas logic, though for reasons of space we defer entering into that topic until a later occasion, save for a parenthetical comment under Remark 4.1(i). (Similarly deferred is any discussion of the reaction by others to Łukasiewicz's treatment of the twins issue.)

Finally, let us note that Gottwald [18], p. 693, misidentifies the Ł-modal logic entirely, writing:

> In contrast to the situation with three-valued systems, where there are a lot of approaches and interpretations, only a few approaches concern four-valued systems and give particular interpretations to the four truth degrees. One of these rare exceptions is Łukasiewicz [1953] [= our [42]] who, in his later years, preferred a four-valued approach via his system L_4 toward a modal reading of the truth degrees over his original three-valued one via L_3 in [Łukasiewicz, 1920].

From the fact that Gottwald calls the famous three-valued Łukasiewicz logic of the

and are closed under uniform substitution and Modus Ponens. The other condition defining regular modal logics, aside from the monotonicity rule above (confusingly called the regularity rule by Segerberg) is that every formula of the form $(\Box A \wedge \Box B) \to \Box(A \wedge B)$ should belong to the logic. The two conditions are often conveniently combined into one, requiring closure under the rule taking us from $(A_1 \wedge \ldots \wedge A_n) \to B$ to $(\Box A_1 \wedge \ldots \wedge \Box A_n) \to \Box B$ (for $n \geq 1$). See also note 6 on other notations for $\Box$ and $\Diamond$ in use – including later in the present section.

[24] The axiomatization appears at p. 214 of [40], with the labels for the various axioms explained on p. 192.

1930s L_3, and from sentence after the passage quoted, with its reference to "L_4 with its linearly ordered truth degree set," it is clear that Gottwald has isolated the wrong four-valued logic here: even restricting attention to the modal connectives, the four-valued logic in the 1930s sequence of finitely many-valued logics explored by Łukasiewicz (and Tarski) has nothing to do with the four-valued logic on offer in [42], the partial order underlying the matrix treatment of the lattice connectives $\wedge$ and $\vee$ in this case being anything but a linear order.[25]

The third schema listed above could be replaced by

$$(A \leftrightarrow B) \rightarrow (\Box A \leftrightarrow \Box B)$$

explicitly revealing $\Box$ to be an extensional connective in the Ł-modal logic (alternatively put: revealing this to be an extensional modal logic) according to one natural use of the term 'extensional' as applied in propositional logic.[26] Given the classical background for the non-modal connectives here, the second occurrence of "$\leftrightarrow$" in the above schema can be replaced (without loss) by "$\rightarrow$", and given the schema $\Box A \rightarrow A$, both occurrences can be so replaced – a formulation Łukasiewicz was especially fond of, though Łukasiewicz's own preferred definition of the extensionality of a context $C(p)$ – in the above case, with $C(p) = \Box p$ – was not that $A \leftrightarrow B$ should provably imply $C(A) \leftrightarrow C(B)$ but rather that $C(A) \wedge C(\neg A)$ should provably imply $C(B)$.[27] Of course, this extensionality is widely taken to be responsible for every-

[25]As Simons ([60], p. 120) helpfully explains, to ward off the current misconception about the Ł-modal logic, "Suffice it to say that the logic is very unlike Łukasiewicz's earlier multivalent systems and also very unlike other modal systems. It is unlike his own systems in that it is an extension of classical bivalent logic and includes all bivalent tautologies."

[26]This is the usage to be found in Humberstone [22], where the present logic is one of those discussed, as well as in [30], in §3.2. The subsection there (3.24 – mentioned above in Remark 2.3) called 'Hybrids and the subconnective relation' devotes a Digression to the Ł-modal logic, beginning on p. 470. It appears in that subsection because necessity in this logic is obtained by hybridizing – isolating, that is, the common logical properties of – the one-place constant false truth-function and the identity truth-function, as will be clarified in the paragraph which follows. The consistent extensional modal logics are called *prime* logics in Zolin [72], though they are defined rather differently: as the logics of 'prime' modalities, the 1-ary modalities induced by formulas in which the sentence letter concerned does not appear in the scope of a modal operator – such as the formulas $\neg p$, $p \rightarrow \Box\bot$. In the case of the Ł-modal logic, as Zolin ([72], p. 866) notes, we are dealing with the modal logic in which $\Box p$ is equivalent to the prime modality $p \wedge \Box\top$; this modality figures extensively in translations between the present logic and others in Font and Hájek [13].

[27]The relation between these two notions of extensionality – the second of which, as one might expect, fails for intuitionistic logic – is explored in §5 (= Appendix A) of Humberstone [32]. Łukasiewicz's preferred way of expressing the negation-involving notion of extensionality would use his (one-place) variable functor "δ" and read as follows – though here we use infix notation rather than Polish notation for the binary connectives (and $\neg$ in place of N): $(\delta p \wedge \delta\neg p) \rightarrow \delta q$, or, in

thing philosophically objectionable about the Ł-modal logic as a plausible alethic modal logic,[28] though our current concern is not with pressing such objections.[29]

Historically, what Łukasiewicz wanted was a propositional logic intended for alethic modal applications, he favoured taking a $\Diamond$-operator in its language rather than a $\Box$-operator as primitive, but as mentioned in note 6, Łukasiewicz wrote Δ for $\Diamond$ (and Γ for $\Box$), so we follow suit when quoting from or otherwise presenting Łukasiewicz's material. The passage also mentions something written as ∇ – which again (repeating from Section 1) has nothing to do with the use of this symbol as a contingency operator, or with the ∇ introduced in Convention 2.1. The semantic description Łukasiewicz provides is four-valued: we take the direct product of the usual two-element matrix with itself for the Boolean connectives ($\to$ and $\neg$ in the case [42]), and interpret Δ by using the product of the identity and the (1-ary) constant true truth-functions. That is, denoting these functions by **I** and **V** respectively, and their product (in that order) by $\mathbf{I} \cdot \mathbf{V}$, the latter maps $\langle x, y\rangle$ ($x, y \in \{\mathrm{T}, \mathrm{F}\}$) to $\langle \mathbf{I}(x), \mathbf{V}(y)\rangle$.[30] The new idea occurring to Łukasiewicz in Section 7 ("The twin possibilities") of [42] is that we might also consider the product of these two two-valued truth-functions in the reverse order – $\mathbf{V} \cdot \mathbf{I}$ – and regard this as the interpretation of another possibility-like operator, namely the above ∇. For the language with both Δ and ∇, writing 1, 2, 3, 4, for $\langle \mathrm{T}, \mathrm{T}\rangle$, $\langle \mathrm{T}, \mathrm{F}\rangle$, $\langle \mathrm{F}, \mathrm{T}\rangle$ and $\langle \mathrm{F}, \mathrm{F}\rangle$, respectively we have the matrix depicted in Figure 1, with tables for the Boolean connectives and each of Δ, ∇. (As usual the designated value(s) – just one such in the present case – are indicated by an asterisk at first occurrence.)

Remarks 4.1. (*i*) We may take the Ł-modal logic to be the set of formulas valid in this matrix, since Smiley [62] showed (and Łukasiewicz seemed to have already known or at least presumed, in [42]) that the formulas provable on the basis of his

'exported' form $\delta p \to (\delta\neg p \to \delta q)$. It is interesting to read Łukasiewicz, in [44], p. 163, commenting, concerning several principles formulated with the aid of "δ", such as $\delta(p \to q) \to (p \to \delta q)$ (or, as he writes this: $C\delta CpqCp\delta q$), that these principles are "all very important but unknown to almost all logicians." Still correct today, on both counts – but the principles in question can be formulated as metalinguistic claims about contexts, without having to have a variable functor in the object language: see Humberstone [32] and (especially) [31] for this way of proceeding.

[28]See the end of §5.5 in Simons [61].

[29]On the subject of alternative axiomatization, let us note that of Tkaczyk [67] provides an interesting further example, intimately connected with Porte's representation (below) using Ω.

[30]This – along with **F** and **N** – for the constant false and negation truth-functions – is the notation used, e.g., in [28], p. 471; Łukasiewicz writes S, V, for **I**, **V**, and has no explicit product notation; to avoid the unintended (non)contingency associations of Łukasiewicz's Δ/∇ notation, these are written in [28] as $\Diamond$ and $\blacklozenge$ respectively. And naturally, Łukasiewicz uses Polish notation – $NCpNq$ for $\neg(p \to \neg q)$ etc. See [28], p. 469 for the rationale behind interpreting possibility as a product of the identity and constant true ('Verum') truth-functions. For much further background and information on the Ł-modal logic, see Font and Hájek [13] and references there given.

$\to$	1	2	3	4	$\neg$	Δ	∇
*1	1	2	3	4	4	1	1
2	1	1	3	3	3	1	2
3	1	2	1	2	2	3	1
4	1	1	1	1	1	3	2

Figure 1: Łukasiewicz's Matrix with Both Possibility Operators

axiomatization were precisely those valid in the matrix. It is clear that the twin possibility operators Δ and ∇ are indeed twins according to this logic, since the set of formulas valid in the product of two matrices is just the intersection of the sets of formulas valid in the respective factor matrices, the order of the factor matrices therefore being irrelevant. Whether a symbol is interpreted as $\mathbf{V} \cdot \mathbf{I}$ or as $\mathbf{I} \cdot \mathbf{V}$ it still exhibits just the logical behaviour it would exhibit both when interpreted as the identity truth-function and when interpreted as the constant true truth-function. As is well known, however, when logics are thought of as consequence relations, we lose the guarantee that the consequence relation determined by the product of two matrices coincides with (though it always includes) the intersection of the consequence relations determined by the factor matrices. (With generalized or multiple conclusion consequence relations, we lose even that. See Observation 2.12.6 in Humberstone [30] and the preceding discussion, for examples, proofs and references to the literature. Slogan version in the terminology of that discussion and of the already mentioned subsection 3.24 of the same work: in the framework FMLA, products of connectives are hybrids, but in SET-FMLA and SET-SET this is not guaranteed to be so.) We return to the question, raised in Section 1, of whether Δ and ∇ are not only twins but identical twins in the (formula) logic under consideration here in due course (Prop.4.4), after some further terrain familiarization. Readers for whom the answer that question is already obvious should still be able to enjoy the ride.

(*ii*) We may take the remaining Boolean connectives as defined in terms of $\to$ and $\neg$ in any of the usual ways, which will give them tables like those familiar from discussions of the product of (the matrix operation interpreting) a connective with itself (e.g., Bolc and Borowik [4], p. 21, or Rescher [55], pp. 96–98); this is just the usual notion of taking direct products of algebras, coupled with taken a value as designated in the product matrix iff each coordinate is designated in the respective factor matrix.[31] Note that this has nothing to do with the use of phrases like 'product connective' in the fuzzy logic subgenre of many-valued logic, where the reference is

[31]This, the standard approach, is Rescher's 'Policy 1' on p. 97 of [55]; ignore what he calls 'Policy 2' which replaces "each' with "at least one".

to arithmetical multiplication. (See the index entries starting with 'Product' in Metcalfe et al. [48].) Łukasiewicz [42] also observed that in this logic ∇ could be defined in terms of $\to$ and Δ: $\nabla A = \Delta A \to A$. If we are taking both of these as primitive, as suits the present discussion better (see Convention 4.2), then to secure the interreplaceability of what would otherwise be *definiens* and *definiendum*, it suffices to add as a further axiom schema $\nabla A \leftrightarrow (\Delta A \to A)$ (or the corresponding equivalence for the dual operators). ◀

Since modal matters are often discussed with $\Box$ rather than $\Diamond$ taken as primitive (cf. the opening paragraph of this section), and Łukasiewicz endorses the familiar equivalences of $\Box$ with $\neg\Diamond\neg$ and $\Diamond$ with $\neg\Box\neg$, we should note that the columns under a table for $\Box$ defined by the former equivalence when $\Diamond$ is taken as Δ would be (reading downward) 2, 2, 4, 4, while if $\Diamond$ is taken as ∇, for the defined $\Box$ we should have, instead: 3, 4, 3, 4. The associated truth-function products are (using, as well as the pair-forming $\cdot$ above, notation from note 30) $\mathbf{I} \cdot \mathbf{F}$ and $\mathbf{F} \cdot \mathbf{I}$, respectively.

Łukasiewicz concentrates on possibility rather than necessity, introducing no special notation for the $\Box$-operator dual to ∇ in the way that his Γ is dual to Δ, and writes:

> Δ and ∇ are indistinguishable when they occur separately, but their difference appears at once when they occur in the same formula. They are like twins who cannot be distinguished when met separately, but are instantly recognised as two when seen together. Take, for instance, the formulae $\Delta\Delta p$, $\nabla\nabla p$, $\Delta\nabla p$ and $\nabla\Delta p$. $\Delta\Delta p$ is equivalent to Δp which is rejected, and likewise $\nabla\nabla p$ is equivalent to ∇p which is rejected too. But $\Delta\nabla p$ and $\nabla\Delta p$ must be asserted (...) We cannot, therefore, replace in the two last formulae Δ by ∇ or vice versa (...).[32]

The talk of accepted and rejected formulas in the quoted passage may be understood in terms of the usual notion of validity and invalidity in matrix semantics, though Łukasiewicz also seeks to parallel these two with matching syntactic characterizations using counter-axioms and rules of rejection alongside the usual axioms and rules of proof – an aspect of his methodology of no concern to us here (and famously eliminated from the streamlined early presentation in Smiley [62][33]), except to note a possible confusion that can arise because of it. The risk in using

[32] This passage is from p. 370 of [42]; I have changed the word "undistinguishable" in the opening sentence to the more idiomatically suitable "indistinguishable". It would perhaps have been clearer to add after the "vice versa" the words "unless both replacements are made at once".

[33] Smiley begins by showing that the axioms and non-rejection-involving rules yield proofs of exactly the formulas valid in the matrix of Fig. 1 – the Ł-*matrix* for short; later he goes on to show that the formulas invalid in the matrix are derivable as rejected formulas in the combined proof + rejection system. Thus the rejected formulas end up being exactly those that are not provable.

terminology such as "the Ł-modal system" is that this may be taken to refer to Łukasiewicz's modal logic, understood as the phrase 'modal logic' usually is when logics are thought of as sets of (provable) formulas, and it may be taken to refer to the pair comprising the logic in that sense together with the set of explicitly rejected formulas, which may or may not coincide with the set of formulas not provable.[34] To see the need for this distinction, let us recall the discussion in Prior ([54] p. 126) where it is remarked that not only does the Ł-modal system lack theorems of the form $\Box A$ (of the form $L\alpha$, as Prior says – or $\Gamma\alpha$ as Łukasiewicz himself would put it), but that it would become inconsistent if any such formulas were added as further axioms; here he is, saying that and a bit more, and using Łukasiewicz's notation (from [44]) of L and M in place of Γ and Δ (or $\Box$ and $\Diamond$):

> Like S1–S3 it has neither the rule to infer $L\alpha$ from α nor the law MMp, but unlike them it is not merely consistent with both without implying either but positively inconsistent with both.

This may come as a surprise to someone whose first encounter with the Ł-modal logic was as in the opening paragraph of the present section where we listed three axioms (well, strictly, three axiom schemata) deliberately including the first, redundant one not only to comment on its redundancy, but also because of its fame as the K-axiom often used as part of the characterization of normality among modal logics, and in the axiomatization of the smallest such logic, K itself. In these last two roles, one needs also closure under the rule of necessitation, conspicuously missing in the present case – and apparently according to Prior, not a closure condition that one could consistently impose. And this claim of inconsistency will seem surprising since obviously the consistent normal modal logic KT! (in the nomenclature of Chellas [9]), often called the Trivial modal logic, satisfies all these conditions (and only slightly less obviously, is the least normal modal logic to do so).

The resolution of the puzzle presented by these conflicting considerations is that Prior is not discussing the Ł-modal *logic* in the 'logic as a set of (provable) formulas' sense, but the Ł-modal *system* understood in the combined assertion + rejection sense. (KT! proves $\Diamond p \to p$, which is one of Łukasiewicz's counter-axioms. Note that this does not mean that all formulas of the form $\Diamond A \to A$ are rejected – many such formulas are provable – since Uniform Substitution moves in reverse for

[34]This contrast cross-cuts the distinction between logics – with or without the rejective component – and particular axiomatizations of them (also sometimes terminologically marked by the 'logic'/'system' distinction). When the rejective component is present, we have counter-axioms or anti-axioms (the initially rejected formulas) as well as rules involving both rejected and asserted (or provable) formulas. By 'explicitly rejected' in the text to which this footnote is appended is meant those formulas which are initially rejected or whose rejection follows from the axioms and the rules (rejection-involving or otherwise) of the proof system.

rejected formulas, from a rejected formula to the rejection of any formula *of which it is* a substitution instance.[35]) Let us denote the Ł-modal logic, without any of the rejective apparatus of the full Ł-modal system, and with its language shorn of the variable functor δ (see note 27), by Ⱡ.

Convention 4.2. *More specifically, since when asking about the potential twinhood of pairs of connectives, we presume both connectives are primitive, the language of* Ⱡ *should have both* Δ *and* ∇ *as its non-Boolean primitives, or else both of their duals,* Γ *and* L – *inspired by Simons' use of L in Simons [59] rather than Łukasiewicz's own use of the latter symbol (see note 6).*

The above-mentioned non-normality – in view of the failure of necessitation to preserve provability – of Ⱡ has naturally led to its semantic treatment using the ideas introduced in Kripke [38], at the hands of Lemmon [40], as recapitulated in Font and Hájek [13] and also Tkaczyk [67]. A *frame* in this setting with universe W is equipped with a special subset N of W as well as an accessibility relation $R \subseteq W \times W$ whose elements are called *normal* worlds, and is converted into a model by the addition of a V assigning subsets of W to the sentence letters, and a formula is valid on a frame if it is true at every point (not just the normal worlds – this is feature introduced by Lemmon) in every model on the frame. $\Box A$ is deemed to be true at $w \in W$ in such a model if $w \in N$ and for every $y \in W$ with Rxy, we have A true at y. For certain purposes it is useful to break this up into its two components and introduce a propositional constant stipulated to be true in any model at precisely the normal worlds of that model, so that $\Box A$ as just defined amounts to the conjunction of that constant with $\Box' A$ where $\Box'$ is interpreted as in the semantics for normal modal logics, as just quantifying universally over accessible points. Cresswell [11] wrote a for such a constant, as did Aanderaa [1], though he took this to be a sentence letter or propositional variable, specified as one not occurring in a formula one was using the constant to translate in various embeddings between normal and non-normal modal logics. That was also Cresswell's concern, for which reason in neither case does a supplementary notation such as $\Box'$ above need to put in an appearance, each logic being cast in a language with a single necessity operator. For reasons that will

[35]As well as $\Delta p \to p$, just mentioned in the more familiar $\Diamond$ notation, Łukasiewicz has Δp as a counter-axiom, and from either of these by the rejective form of Uniform Substitution, we can derived p as a rejected formula – a counter-axiom used by Łukasiewicz when presenting non-modal propositional logic. There is also a rejective formulation of Modus Ponens: from asserted $A \to B$ and rejected B, to rejected A. This looks a lot like Modus Tollens but rejection in the present sense is not to be confused with rejection in the sense of denial (cf. [24]), and counter-axioms are not doing the work that presenting the negation of the formula concerned as an axiom would do – as the example of those counter-axioms just cited shows.

become evident presently, however, we will write this normality constant as Ω. Thus the equivalence mentioned above would be written as

$$\Box A \leftrightarrow (\Omega \wedge \Box' A).$$

Let us put this to the back of our minds for the moment as we complete this summary of the Kripke-inspired background for a range of non-normal modal logics before we get specifically the application of current interest.

Sotirov [64] notes that for the case of Ł we can discard the accessibility relation and use the simple 'accessibility-free' clauses below in the definition of truth at a point $x \in W$ in a model $\mathcal{M} = \langle W, N, V \rangle$; We write $\Box$ as Γ here so that it matches Δ, as we write the corresponding dual operator so that we can compare it with Łukasiewicz's ∇, and give the induced clause for this as (4.2):

$$\mathcal{M} \models_x \Gamma A \Leftrightarrow (x \in N \textit{ and } x \models A) \tag{4.1}$$

$$\mathcal{M} \models_x \Delta A \Leftrightarrow (x \notin N \textit{ or } x \models A) \tag{4.2}$$

Also noted in [64] is the fact that we can actually focus attention on a single frame, with one element in N and one in $W \setminus N$, which validates precisely the formulas valid on all such frames. And these formulas are exactly the Ł-theorems, since at the normal point Γ is interpreted as **I** and, at the non-normal point, as **F**.

Observe that we could equally have used a metalinguistic material conditional to write the right-hand side of (4.2) as "$x \in N \Rightarrow x \models A$," which is a metalinguistic way of bringing out what Prior had in mind in writing the following ([53], p. 189), in the course of his description of a summer logic workshop in the Oxford of 1956:

> D. P. Henry (...) told me just before Lemmon's lecture that he had defined a modal logic within the propositional calculus by defining *Mp* as *Czp*, where z is a variable not put to any other purpose. (As the answer to "p?", "Possibly" means "Yes, if —".) From this information alone it would seem that this system would boil down to Łukasiewicz's, for *CpMp* follows by substitution (q/z) in *CpCqp*, and *CMpCMNpNq* by substitution in *CCrpCCrNpCrq*.[36]

[36] As Smiley [63], note 6, does in quoting this same passage, I have changed "L" to "Ł." The parenthetical sentence is a bit slick, adding a "Yes" to the "If —". By itself "If A then B" has no tendency to convey that it is possible that B unless it is presumed that it is possible that A. This is a point associated with Chisholm ([10], p. 5f.) in connection with the supposed equivalence, in some presentations of compatibilism, between "X could have done otherwise" and "If X had chosen to do otherwise, then X would have done otherwise," an equivalence undermined by cases in which the conditional is true but X could not have chosen to do otherwise. Note, incidentally, that the conditionals here are subjunctive, and with indicative conditionals there would be trouble with Prior's "Yes, if" line, on a straight material implication account (as the use of Polish notation's "C" suggests here): even if it is possible that A, from the additional information that it is (contingently)

Now this "z" in Henry's suggestion is an early appearance of the "a" of Aanderaa and Cresswell, which was written as Ω above because that is how Jean Porte wrote it, and it was Porte who first noted the light it shed on Ł.[37] As already remarked, while the Kripke-semantical route to this logic helps locate it relative to other (regular) non-normal modal logics, so little of the semantic apparatus ends up being exploited that we might as well simply start (as indeed Porte might have[38]) with the binary division into the normal points, at which Γ ends up expressing the truth-function **I** (and Δ, **I** also), and the non-normal points, at which it expresses **F** (and Δ, **V**). Since validity requires us to take both into account we are successfully hybridizing these truth-functions with the above treatment. And we could just as well have begun with the truth-table on the left of Figure 2, focusing on the Γ case for $\Omega \wedge p$, treating Ω as though it were just another sentence letter.

Ω	$\wedge$	p
T	T	T
T	F	F
F	F	T
F	F	F

Ω	$\to$	p
T	T	T
T	F	F
F	T	T
F	T	F

Figure 2: Porte's perspective on Ł-modal necessity and possibility

Concentrating on the first table in Figure 2, we see just what we expect we for the conjunction of two formulas that can assume together all combinations of the values T and F. Focus first on the cases in which Ω gets the value T (the top two lines), thinking of this as a single operator $\Omega \wedge \_$ we apply to p, putting p into the blank indicated by underlining. What 1-ary (bivalent) truth-function of p's truth-value does this operator represent? In these two lines, we see that the whole formula has the same truth-valued as p itself, so the truth-function in question is

false that A it would follow that (material conditionally) *if A then B*, for any B, with, again, nothing to make one think on this basis that it is possible that B.

[37] An interesting coincidence, especially as all four suggestions were independent of one another: Aanderaa, Cresswell, Henry and Porte all select the first or last letters from the alphabets they are drawn from in this capacity. In the course of their re-discovery of this approach, Font and H'ajek [13] use L to play this role. For reasons of space the details are omitted here of the differences among these authors as to whether the symbol chosen should be regarded as a propositional variable (sentence letter) or as a nullary connective (sentential constant), as is any discussion of the comparative merits of these alternatives.

[38] ... but in fact did not, at least not in *quite* the form in which Lemmon presents the material, crediting work in the 1940s by Marcel Boll: [49], p. 918 base; for more detail, including on the relations between the Boll–Reinhart logics and Lemmon's E-systems see Porte [50].

the identity truth function, **I**. Similarly, looking at the bottom two lines, in which Ω has the value F, the whole conjunction has the value F regardless of p's truth-value, so in these cases $\Omega \wedge \_$ is delivering the constant false truth-function, **F**. So, if we count a formula involving Ω as valid when regardless of whether Ω is assigned T or F, the formula will have to have the value T when $\Omega \wedge \_$ expresses the identity truth function and also when $\Omega \wedge \_$ expresses the constant false function. But that is exactly what validity in the Ł-matrix demands of $\Box\_$ (writing in the blank just for the sake of a parallel notation here). Likewise in the case of Δ, the first two lines of the table on the right of Figure 2 present us with **I**, and in this case the bottom two lines, with **V**.

To see what light this throws on the 'twins' issue, recall from Remark 4.1(*ii*) that ∇A is Ł-equivalent to $\Delta A \to A$, which means that ∇A amounts to $(\Omega \to A) \to A$, or more briefly put, $\Omega \vee A$, or perhaps more usefully for present purpose, $\neg\Omega \to A$: more usefully because we can see that in passing to Γ's twin, L, what we have done is moved from Porte's constant to its negation, keeping us well inside the garden of modal operators (in the broadest sense) "hidden in classical logic", to quote from the title of Sotirov [64], where Vakarelov is credited alongside – indeed ahead of[39] – Porte for his horticultural endeavours. Figure 3 gives a fuller picture of the 1-ary operators O derived by applying the 16 binary truth-functions to a Porte-style Ω and another formula (represented by the sentence letter p). Each binary truth-functional connective # – for which we use infix notation here – yields an ordered pair $\langle f, g\rangle$, of 1-ary truth-functions satisfying the condition that for any Boolean (bivalent) valuation v, when $v(\Omega) = T$, $v(\Omega \# p)$ is $f(v(p))$ and when $v(\Omega) = F$, $v(\Omega \# p)$ is $g(v(p))$. Thus each such pair renders O as a product connective whose interpretation is $f \cdot g$. To save space we use overlining for negation, and write (connectives for) the constant true and constant false binary truth-functions as Ⓣ and Ⓕ, and the projections to the first and second coordinate as ①, ②, respectively.

Thus in particular, the top entry on the left with Op as $\Omega \wedge p$ gives us Łukasiewi-

[39]This claim of priority is not clearly correct: see note 40 of Humberstone [34]. The relevant Vakarelov references are can be found in Sotirov [64]; see also p. 171*f.* in Font and Hájek [13]. Sotirov also discusses Vakarelov's use of the $\langle \mathbf{V}, \mathbf{N}\rangle$ entry in Figure 3 ($p \to \Omega$) for handling a subclassical negation operator, so the for the record we include also Porte [51] in our bibliography. The fact that Ł can be treated in this way was already known to Prior in 1956, as we see from the earlier quotation from [53] concerning D. P. Henry. (See also p. 172 of Font and Hájek [13], for discussion of a similar idea in work by H. B. Curry.) The possibility of such a presentation is implicit in Aanderaa [1], disguised in his faithful embedding of Ł in the 'trivial modal logic' KT!, so it is not actually mentioned that we can throw away the 'modal' part of this description. Just to make the relevant observation even harder to extract, Aanderaa introduces KT! as TM at the start of [1], but when the embedding result is stated in its final line it is referred to as LT. (At least, that is the best sense I can make of Aanderaa's discussion.)

Ω compound	Hybridized pair	Ω compound	Hybridized pair
$\Omega \wedge p$	$\langle \mathbf{I}, \mathbf{F} \rangle$	$\Omega \vee p$	$\langle \mathbf{V}, \mathbf{I} \rangle$
$\Omega \wedge \overline{p}$	$\langle \mathbf{N}, \mathbf{F} \rangle$	$\Omega \vee \overline{p}$	$\langle \mathbf{V}, \mathbf{N} \rangle$
$\overline{\Omega} \wedge p$	$\langle \mathbf{F}, \mathbf{I} \rangle$	$\overline{\Omega} \vee p$	$\langle \mathbf{I}, \mathbf{V} \rangle$
$\overline{\Omega} \wedge \overline{p}$	$\langle \mathbf{F}, \mathbf{N} \rangle$	$\overline{\Omega} \vee \overline{p}$	$\langle \mathbf{N}, \mathbf{V} \rangle$
$\Omega \leftrightarrow p$	$\langle \mathbf{I}, \mathbf{N} \rangle$	Ω ① p	$\langle \mathbf{V}, \mathbf{F} \rangle$
$\overline{\Omega} \leftrightarrow p$	$\langle \mathbf{N}, \mathbf{I} \rangle$	$\overline{\Omega}$ ① p	$\langle \mathbf{F}, \mathbf{V} \rangle$
Ω Ⓣ p	$\langle \mathbf{V}, \mathbf{V} \rangle$	Ω ② p	$\langle \mathbf{I}, \mathbf{I} \rangle$
Ω Ⓕ p	$\langle \mathbf{F}, \mathbf{F} \rangle$	Ω ② $\overline{p}$	$\langle \mathbf{N}, \mathbf{N} \rangle$

Figure 3: Porte–Vakarelov Constant-induced Operators

cz's initially introduced $\Box$ connective Γ with interpretation $\mathbf{I} \cdot \mathbf{F}$, while the third entry down on the right represents his first stab at $\Diamond$ with Δ as $\mathbf{I} \cdot \mathbf{V}$. The twin possibility, ∇, appears at the top of that column, with interpretation $\mathbf{V} \cdot \mathbf{I}$, as mentioned in the vicinity of Figure 1, though now we see how the interaction with Ω gives rise to these variations. The dual operator, L, is the third entry in the first column (interpretation: $\mathbf{F} \cdot \mathbf{I}$).

Attending to $\neg\Omega$ no less than to Ω throws light on some of the Δ/∇ interactions commented on in the earlier quotation in this section from Łukasiewicz [42], in particular the remark that unlike the invalid $\Delta\Delta p$ and $\nabla\nabla p$, $\Delta\nabla p$ and $\nabla\Delta p$ were valid. (Invalid and valid in the Ł-matrix of Figure 1, that is; of course what Lukasiewicz actually said was "rejected" and "asserted".) $\Delta\nabla p$ becomes, when expressed in our current terms, $\Omega \to (\neg\Omega \to p)$ (or $\Omega \to (\Omega \vee p)$), while its Δ/∇ ⋈-switch just interchanges Ω with its negation and so again is a truth-functional tautology (treating Ω as though it were formula of CL). For our current agenda, however, more significant is the fact that since Ω was a way of coding 'normality' in the $\langle W, N, V \rangle$ models in play for (4.1) and(4.2) above, its negation serves as a marker for non-normality, so we can now treat L and ∇ in such models by corresponding clauses in the definition of truth:

$$\models_x \mathrm{L}A \;\Leftrightarrow\; (x \notin N \textit{ and } x \models A) \tag{4.3}$$

$$\models_x \nabla A \;\Leftrightarrow\; (x \in N \textit{ or } x \models A) \tag{4.4}$$

Remark 4.3. Porte [49] closes with the following observation, credited to D. Lacombe, in which Porte's use of "N" (for necessity) has been replaced by "Γ":

> It is possible to generalize the Ω-system by introducing (in the propositional calculus) any number of constants similar to Ω: $\Omega_1, \ldots, \Omega_n$. We will eventu-

ally get a characteristic matrix with 2^{n+1} elements, and we can define n 1-ary connectives similar to Γ.

What is missing here is the observation that already without adding any further Ω-style constants, we already have a second endogenous 1-ary connective 'similar to' Γ – namely the dual of ∇ (rather than of Δ) – and that, correspondingly, passing to the general case of $\Omega_1, \ldots, \Omega_n$ gives $2n$ such connectives rather than n, since in each case we can use either Ω_i or its negation to conjoin with a formula A to produce an Ł-style necessitation of A. ◀

Recall also that since the right-hand sides of these clauses do not direct us away from the point x of evaluation, we can (following Sotirov [64]) conduct the discussion entirely in terms of the characteristic two-element frame mentioned above – two elements so that the case of normal and nonnormal points are both covered. So we have a simple proof in the style of Example 3.3(*ii*) for the expected strengthening of Remark 4.1, inserting "identical":

Proposition 4.4. Γ *and* L *are identical twins in* Ł. *Alternatively, if* Δ *and* ∇ *are taken as the non-Boolean primitives: these two connectives are identical twins in* Ł.

Proof. For a model $\mathcal{M} = \langle W, N, V\rangle$ let $\mathcal{M}^{\bowtie}$ be the model $\mathcal{M} = \langle W, W \smallsetminus N, V\rangle$. Then induction on the complexity of arbitrary A (mixed or otherwise) shows that for w in W, $\mathcal{M} \models_w A$ if and only if $\mathcal{M}^{\bowtie} \models_w A^{\bowtie}$, from which we conclude that A and $A^{\bowtie}$ are equi-provable in Ł. □

This is not meant to be a new result, except insofar as it clarifies the issue of identical as opposed to 'merely fraternal' twinhood in its formulation, and places it in the context of other such results. It is not even a new proof, but a model-theoretic formulation of Łukasiewicz's matrix-theoretic proof in [42] (p. 372), which consists in observing that the function mapping 2 to 3 and 3 to 2 is a matrix isomorphism of the Ł-matrix of Figure 1: that is, an isomorphism of the algebras which preserves and reflects designation. This is just the matrix corresponding to the two-element characteristic frame in the familiar way, with 2 corresponding to (the one-element subset) N and 3 to $W \smallsetminus N$. Remark 3.4(*iii*) applies again here: one could equally well argue the case by an induction on the length of proofs in a suitable axiomatization, whether that for which Smiley [62] showed the Ł-matrix to be characteristic, or any of the alternatives suggested at the start of the present section. In view of Convention 4.2, one should bear in mind that for the latter □-based axiomatizations, in which we envisage □ re-written as Γ, the comment at the end of Remark 4.1(*ii*) about the "corresponding equivalence for the dual operators," which needs to be counted as one of the axioms, the definition-replacing equivalence in question is the perhaps surprising-looking: $\mathrm{L}A \leftrightarrow (A \wedge \neg\Gamma A)$, whose $(\Gamma \bowtie \mathrm{L})$-switch therefore needs checking in the basis part of that induction, and which incidentally reveals rather starkly the fact that the twin necessity Γ and L notions are mutually exclusive – as indeed is evident from the Porte–Vakarelov representations of ΓA and $\mathrm{L}A$ as having respectively Ω and $\neg\Omega$ as conjuncts. This last consideration shows that the situation is more extreme than that: not only are ΓA and $\mathrm{L}A$ incompatible according to Ł for all A – that is, the negation of their conjunction is Ł-provable – but so are ΓA and $\mathrm{L}B$ for any A, B. This is really just a re-phrasing of something already familiar as exhibiting the Halldén incompleteness of Ł: $\Delta A \vee \nabla B$ is always provable – so choose A, B to be variable-disjoint (e.g., to be p, q).[40]

In Section 1, it was noted out that Łukasiewicz did not – the present author's loose summaries of the status of Δ and ∇ as cited there – ever refer to these operators in [42] as being like identical twins. This provoked the present current enquiry as to whether one might, picking up on aspects of Łukasiewicz's discussion, distinguish a logical analogue of twins in general from a logical analogue of identical twins in particular. We have indeed found such a distinction to be sustainable, and of Łukasiewicz's Δ and ∇ to stand in the narrower relation of identical twinhood in his favoured logic – something we have now seen Łukasiewicz to be aware of, even though that exact terminology is not employed in [42]. It is, accordingly, gratifying to see that the passage, quoted at greater length above, in [42] (p. 37), reading:

> They are like twins who cannot be distinguished when met separately, but are instantly recognized as two when seen together

is subtly altered for its subsequent appearance on p. 173 of [44]:

> They are like identical twins who cannot be distinguished when met separately, but are instantly recognised as two when seen together.

Whether the alteration was intended to be anything more than stylistic is hard to say, but, as we have seen, the upshot can retrospectively be construed as providing a more precise description of the situation. On the other hand, in retrospect, we have also found (in effect) that Łukasiewicz's reaction – mentioned at the end of Remark 3.4(*iii*) – to the presence of twin connectives which yield non-equivalent formulas as

[40] Tkaczyk [67], p. 226, mentions this as though it were a new discovery; the observation can be found in Anderson [2]. The explanation Tkaczyk then proceeds to give – basically, that any hybrid of distinct truth-functional connectives engenders a Halldén-unreasonable disjunction with the disjuncts corresponding to the different truth-functions hybridized – was given in Humberstone [22], p. 33*f.* However, Tkaczyk does go on after that to formulate the observation in terms of the Kripke–Lemmon semantic treatment of Ł.

representing a "logical paradox" is certainly an overreaction,[41] the phenomenon in question being thoroughly commonplace – as, for instance, tense logics (like K_t, in Example 3.3(*ii*)) with the mirror image property.[42]

5 Appendix: Postscript to Section 3 on the Two Minimal Negations

Reminder: the wording "the two" in the title here is a reference to the $\neg$ and $\neg'$ introduced in Example 3.5 and last seen in Example 3.7(*i*); it is not intended to suggest that these are the only two negation-like connectives available in $\vdash_{\mathsf{ML}}$.[43] We recall from the discussion after Remarks 3.4 the two consequence relations $\vdash_{\mathsf{ML}\neg}$ and $\vdash_{\mathsf{ML}\neg'}$ and their common (conservative) extension $\vdash_{\mathsf{ML}\neg,\neg'}$. Our task here is to show that $\neg$ and $\neg'$ are twins according to this third consequence relation, which amounts to showing that the first two consequence relations coincide, modulo the notational shift between $\neg$ and $\neg'$. We begin by noting that in the present expressively impoverished setting, there is little opportunity for premisses to combine to yield conclusions:

Lemma 5.1. *(i) If $\Theta \vdash_{\mathsf{ML}\neg} C$ then there is $\Theta_0 \subseteq \Theta$ with $|\Theta_0| \leq 2$ with $\Theta_0 \vdash_{\mathsf{ML}\neg} C$, and (ii) if $\Theta \vdash_{\mathsf{ML}\neg'} C$ then there is $\Theta_0 \subseteq \Theta$ with $|\Theta_0| \leq 2$ with $\Theta_0 \vdash_{\mathsf{ML}\neg'} C$, and further, Θ_0 can be chosen in such a way that there are at most two sentence letters occurring in the sequent $\Theta_0 \succ C$.*

Proof. Note that every formula of the language of $\vdash_{\mathsf{ML}\neg}$ or $\vdash_{\mathsf{ML}\neg'}$ is the result of

[41] From just before the previous quotation on p. 370 of [42]: "We encounter here a logical paradox: although Δ and ∇ can be defined by the same matrix, they are not identical." Here the talk of identity is an allusion to the matrix isomorphism mentioned in our previous paragraph.

[42] At the semantic level, the situation with K_t is that, adapting the mirror image operation on models to one (similarly notated) on their underlying frames, it suffices for the mirror image property that the class of frames determining our tense logic is closed under the operation taking us from $\mathcal{F} = \langle W, R, S\rangle$ to $\mathcal{F}^{\bowtie} = \mathcal{F} = \langle W, S, R\rangle$ (from $\langle W, R\rangle$ to $\langle W, R^{-1}\rangle$, on the more familiar way of presenting such frames). Since a matrix corresponds to a single frame, the closer tense-logical analogue would be to have a single point-generated frame $\mathcal{F}$ which is isomorphic to $\mathcal{F}^{\bowtie}$. In the terminology of note 78 of Humberstone [33], we want a frame which is symmetrical but not symmetric – the latter because we want G and H to be identical twins without being equivalent. The logic determined by the two-element frame in which one element bears R (with converse S) to the other, or the reflexive closure of this frame, would do nicely, especially as the induced matrix has four values, as with the Ł-matrix.

[43] Numerous 1-ary connectives of interest are ML-definable which are equivalent to $\neg$ in IL though not in ML; one such, definable in ML as $A \leftrightarrow \bot$ is given some attention in [30], Observation 8.33.10, which shows that neither this nor the standard ML negation is a subconnective of the other (according to $\vdash_{\mathsf{ML}}$).

applying $\neg$ or $\neg'$, respectively, to some sentence letter p_i, and we refer to this sentence letter as the core of the formula in question. We begin with the case of $\vdash_{\mathsf{ML}\neg}$, in which $\neg(\cdot)$ is interpreted as $\cdot \to \bot$, with $\bot$ as in ML, and with the reminder that any formulas A_0, A_1 differing at most in the replacement of a subformula $\neg\neg\neg D$ with $\neg D$ are equivalent ($A_0 \dashv\vdash_{\mathsf{ML}\neg} A_1$). With this in mind, it is evident that $\Theta \vdash_{\mathsf{ML}\neg} C$ if and only if one of the following three situations obtains:

(1) "Shared core across $\vdash$, equivalence case" – C is in the same equivalence class as some formula in Θ, where formulas are equivalent if they are identical or if they have the form $\neg^m p_i$ and $\neg^n p_i$ where m, n differ by an even number and neither m nor n is 0;

(2) "Shared core across $\vdash$, one-way consequence case" – some formula in Θ properly implies C, which happens when the formula in Θ is a sentence letter and C prefixes an even number of negations to that sentence letter;

(3) "Shared core on the left case" – there are formulas $A, B \in \Theta$ with $A = \neg^m p_i$, $B = \neg^n p_i$, m, n differ by an odd number, and C is of the form $\neg D$.

Thus when $\Theta \vdash_{\mathsf{ML}\neg} C$ if we are in cases (1) or (2), there is a one-element subset Θ_0 of Θ for which $\Theta_0 \vdash_{\mathsf{ML}\neg} C$, while in case (3) there is a two-element subset Θ_0 for which this holds. (The situation with intuitionistic logic here would be identical except that in this last case there is no condition on the form of C.) Further in the sequent $\Theta_0 \succ C$ chosen on the basis of the formulas mentioned under (1)–(3), i.e., with no additional weakening in cases (1)–(2), we have only one sentence letter occurring (as the 'shared core') or else only two (one as the common core of the two formulas on the left, and one as the core of the formula on the right).

Turning to the case of $\neg'$, for which we recall that $\neg(\cdot)$ is interpreted as $((\bot \to \cdot) \to \cdot) \to \bot$, we find that exactly the same reasoning applies: we have the law of Triple Negation (for which it may help to note that, using both negations at once and taking $\vdash$ as $\vdash_{\mathsf{ML}\neg,\neg'}$, we have

$$\neg'\neg' A \dashv\vdash \neg\neg' A \dashv\vdash \neg'\neg A$$

for all A) and also precisely the same cases (1)–(3), with $\neg$ replaced by $\neg'$ as the alternative possibilities when $\Theta \vdash_{\mathsf{ML}\neg'} C$. □

Next we adopt the perspective taken in Example 3.7(i), where we recall that $\bot$ in ML behaves as though it were nothing but a further sentence letter. This simplifies the calculation of all consequence relationships among the formulas arising from Lemma 5.1. First, since whenever $\Theta \vdash C$, where $\vdash$ is either $\vdash_{\mathsf{ML}\neg}$ or $\vdash_{\mathsf{ML}\neg'}$, $A, B \in \Theta$

with $A, B \vdash C$ with $A = B$ (the 1-element Θ_0 case), in terms of the proof of Lemma 5.1 or else $A \neq B$ and neither $A \vdash C$ or $B \vdash C$ (the irredundant 2-element Θ_0 case). Since only two sentence letters are involved, we can assume without loss of generality that they are p, q, so that the formulas A, B, C are all the result of negating these sentence letters zero or more times. Reducing all triple to single negations – and here we write $\neg$, though the same applies if we are negating with $\neg'$ instead – A, B, C can be taken to be drawn from the following list: $p, \neg p, \neg\neg p, q, \neg q, \neg\neg q$. Now we write out the $\neg$ in terms of $\rightarrow$ and the first unused sentence letter (for $\bot$), which is r, since the ML logical relations among our six formulas are given by the IL logical relations among these translations. (We address the different translations we get for $\neg'$ in place of $\neg$ below.) This gives us the following six formulas, to contend with:

$$p, p \rightarrow r, (p \rightarrow r) \rightarrow r, q, q \rightarrow r, (q \rightarrow r) \rightarrow r. \qquad (\dagger)$$

Let us now recall C. A. Meredith's faithful embedding of the implicational fragment of classical logic – let us call it $\vdash_{\mathsf{CL}\rightarrow}$ – into the corresponding fragment of intuitionistic logic $\vdash_{\mathsf{IL}\rightarrow}$, with "$\rightarrow$" serving as the material implication connective, by contrast with the notational choices made for Proposition 1.1 and the discussion leading up to it, where "$\supset$" served in that capacity, the latter being used in this Postscript as a derived connective of the language of $\vdash_{\mathsf{IL}\rightarrow}$:

$$A \supset B \;=\; ((B \rightarrow A) \rightarrow A) \rightarrow B.$$

Note that this permits us to abbreviate the $\bot, \rightarrow$ definition of $\neg'$ to

$$\neg' A \;=\; A \supset \bot.$$

With the aid of $\supset$, so defined, Meredith's translation for taking us from classical to intuitionistic implication, call it τ_{Mer}, is defined inductively as follows:

- $\tau_{\mathsf{Mer}}(p_i) \;=\; p_i \; (i = 1, 2 \ldots)$;
- $\tau_{\mathsf{Mer}}(A \rightarrow B) \;=\; \tau_{\mathsf{Mer}}(A) \supset \tau_{\mathsf{Mer}}(B)$.

Lifting τ_{Mer} from formulas to sequents in the obvious way, then, one has for all (pure implicational) sequents σ:

$$\vdash_{\mathsf{CL}_\rightarrow} \sigma \text{ if and only if } \vdash_{\mathsf{IL}_\rightarrow} \tau_{\mathsf{Mer}}(\sigma). \qquad (\dagger\dagger)$$

(See [30], pp. 1081–1088 for further information and references.) We are now in a position to wrap things up; although our two negations are not identical twins according to the present consequence relation (see Example 3.7(i)), we do have:

Proposition 5.2. *$\neg$ and $\neg'$ are twins according to $\vdash_{\mathsf{ML}\neg,\neg'}$.*

Proof. It suffices to show that for any sequents σ, σ' constructed using only $\neg, \neg'$ respectively, $\vdash_{\mathsf{ML}\neg} \sigma[\neg]$ if and only if $\vdash_{\mathsf{ML}\neg'} \sigma[\neg']$. By Lemma 5.2 and the subsequent discussion it suffices to show that we have this equivalence all for $\sigma[\neg]$ constructed from the formulas $p, \neg p, \neg\neg p, q, \neg q, \neg\neg q$ and $\sigma[\neg']$ constructed correspondingly from $p, \neg' p, \neg'\neg' p, q, \neg' q, \neg'\neg' q$, for which it in turn suffices to show that for all sequents $\sigma[\rightarrow]$, constructed from the six formulas listed as (†), and the sequents $\sigma[\supset]$ constructed corresponding from the formulas replacing $\rightarrow$ by $\supset$ in those six formulas, we have

$$\vdash_{\mathsf{IL}} \sigma[\rightarrow] \text{ if and only if } \vdash_{\mathsf{IL}} \sigma[\supset].$$

(We do not need to add $\rightarrow$ to the "IL" subscript here because the sequents in play are all pure implicational sequents.) Since $\sigma[\supset]$ is $\tau_{\mathsf{Mer}}(\sigma[\rightarrow])$, this is equivalent by (††) to the claim that for the sequents in question

$$\vdash_{\mathsf{IL}} \sigma[\rightarrow] \text{ if and only if } \vdash_{\mathsf{CL}} \sigma[\rightarrow],$$

where we could have equally well written "$\vdash_{\mathsf{CL}} \sigma[\supset]$" on the right, since $\rightarrow$ and $\supset$ are classically equivalent, and indeed we might just as well omit the "$[\rightarrow]$ given that the formulation given uses the same connective on both sides. Since $\vdash_{\mathsf{IL}} \subseteq \vdash_{\mathsf{CL}}$ we have he "only if" direction automatically, so for the "if" direction we need only check that the additional strength of $\vdash_{\mathsf{CL}}$, or more particularly $\vdash_{\mathsf{CL}\rightarrow}$, does not show up when attention is restricted to the consequence relationships among the formulas (†). But this is easily done. Setting aside the $\vdash$-statements among these six formulas whose correctness is ensured simply because of the definition of a consequence relation or because of weakening an earlier such statement (adding more formulas to the left, that is), we have for $\vdash = \vdash_{\mathsf{CL}}$ just the following cases:

$p \vdash (p \rightarrow r) \rightarrow r$	$q \vdash (q \rightarrow r) \rightarrow r$
$p, p \rightarrow r \vdash q \rightarrow r$	$q, q \rightarrow r \vdash p \rightarrow r$
$p, p \rightarrow r \vdash (q \rightarrow r) \rightarrow r$	$q, q \rightarrow r \vdash (p \rightarrow r) \rightarrow r$
$(p \rightarrow r) \rightarrow r, p \rightarrow r \vdash q \rightarrow r$	$(q \rightarrow r) \rightarrow r, q \rightarrow r \vdash p \rightarrow r$
$(p \rightarrow r) \rightarrow r, p \rightarrow r \vdash (q \rightarrow r) \rightarrow r$	$(q \rightarrow r) \rightarrow r, q \rightarrow r \vdash (p \rightarrow r) \rightarrow r$

(In fact there is still some redundancy here, in that the fourth entry in the left (resp. right) column follows for arbitrary $\vdash$ the first and second entries in the left (resp. right) column.) And all of these relationships hold for $\vdash = \vdash_{\mathsf{IL}}$ no less than for $\vdash = \vdash_{\mathsf{CL}}$. □

The above proof may seem unnecessarily indirect in getting rid of negation (whether $\neg$ or $\neg'$) and then of $\bot$ so as to end up in the pure implicational fragment of IL: why not work in the $\{\to, \bot\}$ fragment of IL, for instance? The answer is that Meredith's embedding does not work with $\bot$ (or $\neg$) present: the subscripts on the turnstiles in (††) if thus enlarged turn something true into something false. Tokarz and Wójcicki [68] show that there is no definitional translation at all that embeds $\vdash_{\mathsf{CL}\to,\bot}$ faithfully in $\vdash_{\mathsf{IL}\to,\bot}$: it is not just that for the particular case of τ_{Mer} that we have a failure of the envisaged variation on (††). And indeed this remains the case even if we drop the word "faithfully" – which amounts to dropping the "if" direction of (††) – as is shown in pp. 467–469 of Humberstone [26], where the discussion is couched in terms of $\to$ and $\neg$ rather than $\to$ and $\bot$ (an inconsequential difference for present purposes). The "definitional", however, cannot be dropped: the various 'negative translations' Troelstra and van Dalen ([69], §2.3) all fail to translate the (non-logical) atomic formulas by themselves or fail to be compositional, or both.

References

[1] Stal Aanderaa *Relation between different systems of modal logic*, Notices of the American Math. Soc. 13 (3), 1966, p. 391.

[2] Alan Ross Anderson *On the interpretation of a modal system of Łukasiewicz*, Journal of Computing Systems 1 (4), 1954, pp. 209–210.

[3] — *The formal analysis of normative systems*, in: N. Rescher (ed.), The Logic of Decision and Action, University of Pittsburgh Press, Pittsburgh PA, 1967, pp. 147–213.

[4] Leonard Bolc, and Piotr Borowik Many-Valued Logics 1, *Theoretical Foundations*, Springer-Verlag, Berlin 1992

[5] Michael Byrd *Knowledge and true belief in Hintikka's epistemic logic*, Journal of Philosophical Logic 2 (2), 1973, pp. 181–192.

[6] John Thomas Canty, and Thomas Scharle *Note on the singularies of S5*, Notre Dame Journal of Formal Logic 7 (1), 1966, p. 108.

[7] Rudolf Carnap *Modalities and quantification*, Journal of Symbolic Logic 11 (2), 1946, 33–66.

[8] Adam Caulton, and Jeremy Butterfield *On kinds of indiscernibility in logic and metaphysics*, British Journal for the Philosophy of Science 63 (1), 2012, pp. 27–84.

[9] Brian Chellas Modal Logic, *An Introduction*, Cambridge University Press, Cambridge UK 1980.

[10] Roderick Chisholm *Human freedom and the self*, The Lindley Lecture, University of Kansas, April 23, 1964; Department of Philosophy, University of Kansas, 1964.

[11] Maxwell John Cresswell *The interpretation of some Lewis systems of modal logic*, Australasian Journal of Philosophy 45 (2), 1967, pp. 198–206.

[12] — *Necessity and contingency*, Studia Logica 47 (2), 1988, pp. 145–149.

[13] Josep Maria Font, and Petr Hájek *On Łukasiewicz's four-valued modal logic*, Studia Logica 70 (2), 2002, pp. 157–182.

[14] Rohan French *A simplified embedding of E into monomodal K*, Logic Journal of the IGPL 17 (4), 2009, 421–28.

[15] Rohan French, and Lloyd Humberstone *Partial confirmation of a conjecture on the boxdot translation in modal logic*', Australasian Journal of Logic 7, 2009, pp. 56–61.

[16] — Translational Embeddings in Modal Logic, PhD Dissertation, Monash University 2010.

[17] Peter Thomas Geach, Review of Hintikka [19], Philosophical Books 4 (2), 1963, pp. 7–8.

[18] Siegfried Gottwald *Many-Valued Logics*, in: D. Jacquette (ed.) Handbook of the Philosophy of Science, *Philosophy of Logic* (series editors: D. M. Gabbay, P. Thagard and J. Woods), Elsevier, Amsterdam 2007, pp. 675–722.

[19] Jaakko Hintikka Knowledge and Belief, *An Introduction to the Logic of the Two Notions*, Cornell University Press, Ithaca, NY 1962.

[20] Lloyd Humberstone *Logic for saints and heroes*, Ratio 16 (1), 1974, pp. 103–114.

[21] — *Inaccessible worlds*, Notre Dame Journal of Formal Logic 24 (3), 1983, pp. 346–352.

[22] — *Extensionality in sentence position*, Journal of Philosophical Logic 15 (1), 1986, pp. 27–54; also *ibid.* 17 (3), 1988, pp. 221–223, *The lattice of extensional connectives: a correction.*

[23] — *Singulary extensional connectives: a closer look*, Journal of Philosophical Logic 26 (3), 1997, pp. 341–356.

[24] — *The revival of rejective negation*, Journal of Philosophical Logic 29 (4), 2000, pp. 331–381.

[25] — *An intriguing logic with two implicational connectives*, Notre Dame Journal of Formal Logic 41 (1), 2000, pp. 1–40.

[26] — *Contra-classical logics*, Australasian Journal of Philosophy 78 (4), 2000, 437–474.

[27] — *Modality*, in: F. C. Jackson and M. Smith (eds.) The Oxford Handbook of Contemporary Philosophy, Oxford University Press, Oxford and New York, 2005, pp. 534–614 (= Chapter 20).

[28] — *Identical twins, deduction theorems, and pattern functions: exploring the implicative BCSK fragment of S5*, Journal of Philosophical Logic 35 (5), 2006, pp. 435–487.

[29] — *Weaker-to-stronger translational embeddings in modal logic*, in: G. Governatori, I. Hodkinson, and Y. Venema (eds.), Advances in Modal Logic 6, College Publications, London 2006, pp. 279–297.

[30] — The Connectives, MIT Press, Cambridge MA 2011.

[31] — *Replacement in logic*, Journal of Philosophical Logic 42 (1), 2013, pp. 49–89.

[32] — *Aggregation and idempotence*, Review of Symbolic Logic 6 (4), 2013, pp. 680–708.

[33] — Philosophical Applications of Modal Logic, College Publications, London 2016.

[34] — *Semantics without toil? Brady and Rush meet Halldén*, Organon F 26 (3), 2019, pp. 340–404.

[35] — *Explicating logical independence*, Journal of Philosophical Logic 49 (1), 2020, pp. 135–

218.

[36] Emil Jeřábek *Cluster expansion and the boxdot conjecture*, Mathematical Logic Quarterly 62 (6), 2016, pp. 608–614.

[37] Kathleen Johnson Wu *Hintikka and defensibility*, Ajatus 32, 1970, pp. 25–31.

[38] Saul Kripke *Semantical analysis of modal logic II: Non-normal modal propositional calculi*, in: J. W. Addison, L. Henkin, and A. Tarski (eds.) The Theory of Models, North-Holland Publishing company, Amsterdam 1965, pp. 206–220.

[39] Edward John Lemmon *Review of Hintikka* [19], Philosophical Review 74 (3), 1965, pp. 381–384.

[40] — *Algebraic semantics for modal logics: II*, The Journal of Symbolic Logic 31 (2), 1966, 191–218.

[41] Clarence Irving Lewis, and Cooper Harold Langford Symbolic Logic, Dover Publications, New York 1959. (Originally published in 1932, without Appendix III.)

[42] Jan Łukasiewicz *A system of modal logic*, in: L. Borkowski (ed.) JAN ŁUKASIEWICZ Selected Works, North-Holland Publishing Company, Amsterdam 1970, pp. 352–390 (first appeared in Journal of Computing Systems 1 (3), 1953, 111–149).

[43] — *A system of modal logic*, Proceeding of the Eleventh International Congress of Philosophy, Brussels 1953, *14*, ('Additional Volume and Contributions to the Symposium on Logic'), North-Holland Publishing Company, Amsterdam 1953, pp. 82–87.

[44] — Aristotle's Syllogistic, Oxford University Press, Oxford 1957 (second enlarged edition).

[45] David Makinson *Intelim rules for classical connectives*, in: S. O. Hansson (ed.) David Makinson on Classical Methods for Non-Classical Problems, Springer, Dordrecht 2014, pp. 359–382.

[46] Gerald Massey *Normal form generation of S5 functions via truth functions*, Notre Dame Journal of Formal Logic 9 (1), 1968, pp. 81–85.

[47] John C. C. McKinsey, and Alfred Tarski *On closed elements in closure algebras*, Annals of Mathematics 47 (1), 1946, pp. 122–162.

[48] George Metcalfe, Nicola Olivetti, and Dov Gabbay Proof Theory for Fuzzy Logics, Springer International, 2009.

[49] Jean Porte *The Ω-system and the Ł-system of modal logic*, Notre Dame Journal of Formal Logic 20 (4), 1979, pp. 915–920.

[50] — *Boll-Reinhart modal logic*, Logique et Analyse 25 (98), 1982, pp. 181–190.

[51] — *Łukasiewicz's Ł-modal system and classical refutability*, Logique et Analyse 27 (105), 1984, pp. 87–92.

[52] Arthur Norman Prior *The interpretation of two systems of modal logic*, Journal of Computing Systems 1 (4), 1954, pp. 201–208.

[53] — *Logicians at play; or Syll, Simp and Hilbert*, Australasian Journal of Philosophy 34 (3), 1956, pp. 182–192.

[54] — Time and Modality, Oxford University Press, Oxford 1957.

[55] Nicholas Rescher Many-Valued Logic, McGraw-Hill, New York 1969.

[56] Rolf Schock Quasi-connectives Definable in Concept Theory, C. W. K. Gleerup, Lund 1971.

[57] Kurt Schütte Proof Theory, Springer-Verlag, Berlin 1977. (Translated by J. N. Crossley; German original: 1960.)

[58] Krister Segerberg An Essay in Classical Modal Logic, Filosofiska Studier, Uppsala 1971.

[59] Peter Simons *Łukasiewicz, Meinong and many-valued logic*, in: K. Szaniawski (ed.), The Vienna Circle and the Lvov-Warsaw School, Kluwer, Dordrecht 1989, pp. 249–291. (Reprinted, with a re-punctuated title, as Chapter 8 of P. Simons, Philosophy and Logic in Central Europe from Bolzano to Tarski, Springer, Dordrecht 1992.)

[60] — *Łukasiewicz and the several senses of possibility*, European Review 23 (1), 2015, pp. 114–124.

[61] — *Jan Łukasiewicz*, in: Edward N. Zalta (ed.) Stanford Encyclopedia of Philosophy (Spring 2017), <https://plato.stanford.edu/archives/spr2017/entries/lukasiewicz/>

[62] Timothy John Smiley *On Łukasiewicz's Ł-modal system*, Notre Dame Journal of Formal Logic 2 (3), 1961, pp. 149–153.

[63] — *Relative necessity*, Journal of Symbolic Logic 28 (2), 1963, pp. 113–134.

[64] Vladimir Sotirov *Non-classical operations hidden in classical logic*, Journal of Applied Non-Classical Logics 18 (2–3), 2008, pp. 309–324.

[65] Phillip Staines *Some formal aspects of the argument-symbolisation relation*, Australian Logic Teachers' Journal 5, 1981, pp. 1–15.

[66] Steven Karl Thomason *Independent propositional modal logics*, Studia Logica 39 (2–3), 1980, pp. 143–144.

[67] Marcin Tkaczyk *On axiomatization of Łukasiewicz's four-valued modal logic*, Logic and Logical Philosophy 20 (3), 2011, pp. 215–232.

[68] Marek Tokarz, and Ryszard Wójcicki *The problem of reconstructability of propositional calculi*, Studia Logica 28 (1), 1971, pp. 119–127.

[69] Anne S. Troelstra, and Dirk van Dalen Constructivism in Mathematics, *An Introduction*, 1, North-Holland Publishing Company, Amsterdam 1988.

[70] Robert Veroff, and Matthew Spinks *Axiomatizing the skew Boolean propositional calculus*, Journal of Automated Reasoning 37 (1–2), 2006, pp. 3–20.

[71] Timothy Williamson *Iterated attitudes*, in: Timothy J. Smiley (ed.) Philosophical Logic (Proceedings of the British Academy 95), Oxford University Press, Oxford 1998, pp. 85–133.

[72] Evgeni E. Zolin *Embeddings of propositional monomodal logics*, Logic Journal of the IGPL 8 (6), 2000, pp. 861–882.

[73] — *Infinitary expressibility of necessity in terms of contingency*, in: K. Striegnitz (ed.), Proceedings of the Sixth ESSLLI Student Session, 2001, pp. 325–334.

ALGEBRAS OF LOGIC VS. ALGEBRAS

AFRODITA IORGULESCU
Academy of Economic Studies, Bucharest
afrodita.iorgulescu@ase.ro

ABSTRACT Inspired from twenty two of the M algebras (RM algebras, ..., BE algebras, ..., BCK algebras) in the *'world' of algebras of logic*, we introduce twenty two corresponding new commutative unital magmas with additional operations (m-RM algebras, ..., m-BE algebras , ... m-BCK algebras), not all distinct, in the *'world' of algebras* and we establish connections between the new algebras and the MV algebras, the ortholattices and the Boolean algebras. Comming back to M algebras, inspired from the commutative unital magmas with additional operations, we obtain new results; we introduce the implicative-ortholattice, the term equivalent definition of the ortholattice, and we establish its connection with the implicative-Boolean algebra. We establish two general theorems connecting the two 'worlds', in the involutive case.

In Sections 2 - 4, we overview in an unifying way the *'world' of M algebras*, with new results, we introduce and study the implicative-ortholattices and we establish the connections between BE algebras, BCK algebras, Wajsberg algebras, implicative-ortholattices and implicative-Boolean algebras.

In Sections 5 - 8, we develop in an analogous unifying way the *'world' of commutative unital magmas* with additional operations, by introducing 22 new algebras (not all distinct) and establishing the connections between m-BE algebras, m-BCK algebras, MV algebras, ortholattices and Boolean algebras.

In Section 9, we establish two general theorems connecting the two 'worlds', in the involutive case, by the mutually inverse transformations Φ and Ψ.

KEYWORDS M algebra, BE algebra, BCK algebra, Wajsberg algebra, implicative-ortholattice, implicative-Boolean algebra, commutative implicative-group, commutative implicative-goop; commutative unital magma, m-BE algebra, m-BCK algebra, MV algebra, ortholattice, Boolean algebra, commutative group, commutative goop

In memoriam
Sergiu Rudeanu

1 Introduction

This long paper deals with the two parallel "worlds" of *algebras of logic* (M algebras with additional operations) and of *algebras* (commutative unital magmas with additional operations) – in the commutative case – and with some connections between them, in the involutive case.

The *"world" of commutative algebras of logic* is very large. We are dealing here with the M algebras with additional operations, namely with:
- algebras $(A, \to, 0, 1)$ or $(A, \to, ^-, 1)$, with $1^- = 0$, where 1 *is the last element* (Hierarchies 2 and 2') and
- algebras $(A, \to, 1)$ or $(A, \to, ^{-1}, 1)$, with $1^{-1} = 1$, where 1 *is not the last element* (Hierarchies 1 and 1').

Here are some algebras belonging to this 'world'.

Hilbert algebras were introduced in 1950, in a dual form, by Henkin [24], under the name "implicative model", as a model of positive implicative propositional calculus – an important fragment of classical propositional calculus introduced by Hilbert [25], [26].

BCK algebras and *BCI algebras* were introduced in 1966 by K. Iséki [44], as algebraic models of BCK-logic and of BCI-logic, respectively. Hilbert algebras are particular cases of BCK algebras [32].

Hundred of papers were written on BCK and BCI algebras, and the books [60] and [32] on BCK algebras and the book [30] on BCI algebras.
Most of the commutative algebras of logic (such as Wajsberg algebras, implicative-Boolean algebras, Hilbert algebras, R_0 algebras, weak-R_0 algebras etc.) are particular cases of BCK algebras (more precisely, of reversed left-BCK algebras) (see [32]; see also [63], [57]); and the BCK algebras are particular cases of BCI algebras. The commutative groups were connected with particular cases of BCI algebras [59].

Several generalizations of BCI or of BCK algebras were introduced in time, namely:

BCH algebras were introduced in 1983 by Q.P. Hu and X. Li [28]. There are many papers on BCH algebras since then, but the exact connection between BCH algebras and BCI algebras was presented in [37].

BCC algebras, also called *BIK$^+$ algebras*, were introduced in 1984 by Y. Komori [53], [54] (see [73]).

BZ algebras, also called *weak-BCC algebras*, were introduced in 1995 by X.H. Zhang and R. Ye [74].

BH algebras were introduced in 1998 by Y.B. Jun, E.H. Roh and H.S. Kim [46], as a generalization of BCH and BZ algebras.

BE algebras were introduced in 2006 by H.S. Kim and Y.H. Kim [48].

CI algebras were introduced in 2010 by B.L. Meng [58], as a generalization of BE algebras.

Pre-BCK algebras were introduced in 2010 by D. Buşneag and S. Rudeanu [7].

Another notion of *pre-BCK algebra* was introduced and studied in 2003 by Matthew Spinks in his PhD Thesis [65]. Spinks's pre-BCK algebras are more general than Buşneag and Rudeanu's pre-BCK algebras: more precisely, Spinks's *pre-BCK-algebras* are just the *pre-BBBCC algebras*, introduced and studied in [37].

Jänis Cirulis introduced in 2006 the class of *weak BCK-algebras*. They were studied in [9] and as algebras with subtraction in [10], [11]. We noticed, in november 2013, that Cirulis's *weak BCK-algebras* are in fact the *aRML** algebras [37] satisfying property (D).*

In the papers [37] from 2013, 2016, starting with a list of properties of BCK algebras and with the BCH, BCC, BZ, BE, pre-BCK algebras, old generalizations of BCI algebras or of BCK algebras, we have introduced new generalizations of BCI or of BCK algebras (*RM, pre-BZ, RME (= CI), RME**, pre-BCI, aRM (= BH), BCH** algebras* and *RML, pre-BCC, aRML, BE**, aBE, aBE** algebras*, respectively, and many others) and, consequently, new generalizations of Hilbert algebras. Namely, we have found **thirty one** new distinct generalizations of BCI or of BCK algebras and **twenty** new distinct generalizations of Hilbert algebras. We have presented the hierarchies existing between all these algebras, old or new ones. We have presented proper examples for each old or new algebra.

In the book [42] from 2018, we have introduced the *M, ME, ML, MEL algebras* (in the non-commutative case), thus in the top of the hierarchy there are the *M algebras*, i.e. the algebras $(A, \to, 1)$ verifying the property (M): $1 \to x = x$.

We have also introduced the *implicative-group*, a term-equivalent definition of the group, as a particular BCI algebra, in the preprints [34], [35] and in the papers [36], [38] – in the non-commutative case. We have developed the notion in [42].

Wajsberg algebras were introduced in 1984, by Font, Rodriguez and Torrens [19], but they were also considered earlier by Komori in [50], [52], under the name of CN algebras; they are a model of $\aleph_0$-valued Łukasiewicz logic, studied by Wajsberg in 1935 [70]. They are termwise equivalent to MV algebras [19].

Implicative-Boolean algebras, a term equivalent definition of Boolean algebras, were introduced in 2009 [33] and presented also in [20], motivated by the axioms system of the classical propositional logic:

(G1) $\varphi \to (\psi \to \varphi)$,
(G2) $(\varphi \to (\psi \to \chi)) \to ((\varphi \to \psi) \to (\varphi \to \chi))$,
(G3) $(\neg\psi \to \neg\varphi) \to (\varphi \to \psi)$.

The *"world" of comutative algebras* is large also. We are dealing here with the commutative unital magmas with additional operations, namely with:
- algebras $(A, \odot, ^-, 1)$ with $1^- = 0$, where 1 *is the last element* (Hierarchies m-2 and m-2') and
- algebras $(A, \cdot, ^{-1}, 1)$ with $1^{-1} = 1$, where 1 *is not the last element* (Hierarchies m-1 and m-1').

Here are some algebras belonging to this 'world': the commutative monoids, the commutative groups, the (bounded) lattices, the residuated lattices, the BL algebras, the MTL algebras, the NM algebras, the MV algebras, the ortholattices, the Boolean algebras etc.

The earliest study of groups as such probably goes back to the work of Lagrange in the late 18th century. However, this work was somewhat isolated, and 1846 publications of Augustin Louis Cauchy and Galois are more commonly referred to as the beginning of group theory. *Lattice-ordered groups or l-groups* have been studied in different contexts, cf. [1]. The theory blossomed under the leadership of Paul Conrad in the 1960s, cf. [1].

Residuated lattices, the algebraic counterpart of logics without contraction rule, were introduced in 1924 by Krull [56].

Boolean algebras were introduced in 1847 by George Boole [3], [4].

MV algebras were introduced in 1958, by C. C. Chang [12], as a model of $\aleph_0$-valued Łukasiewicz logic.

Divisible residuated lattices (or "divisible integral, residuated, commutative l-monoids" [27]) were introduced in 1965, in a dual, more general form, by Swamy [66].

The ortholattice is an important example of *sharp* structure (which satisfies the noncontradiction principle) from sharp quantum theory [14] (Birkhoff, 1967; Kalmbach, 1983).

BL algebras were introduced in 1996, by Petr Hájek [21], [23], [22], as a common generalization of *MV algebras*, *Product algebras* and *Gödel algebras*, in connection with continuous t-norms on the real unit interval [0,1].

MTL algebras were introduced in 2001 [17], as algebraic model for the monoidal t-norm logic, a generalization of Hájek's Basic Logic.

WNM, *IMTL* and *NM algebras* are particular classes of MTL algebras [17], introduced in 2001 too.

Note that the *algebraists* work usually with the commutative additive groups and with the positive (right) cone of a partially-ordered commutative group $(G, \leq, +, -, 0)$, where there are essentially a sum $\oplus = +$ and an element 0. Sometimes, the negative (left) cone is needed also, where there are essentially a product $\odot = +$ and an element $1 = 0$. Hence, the algebraists usually work with the commutative *right-unital magmas*.

By contrary, note that the *logicians* work with the logic of *truth*, where the *truth* is represented by 1, and there is essentially one implication (two implications, in the non-commutative case); we could name this logic "left-logic". One can imagine also a "right-logic", as a logic of *false*, where the *false* is represented by 0. Hence, the logicians usually work with the *left-algebras of logic* (or the *algebras of left-logic*).

Summarizing, for algebraists, the appropriate algebras are the unital magmas, not the algebras of logic, and among the unital magmas, the appropriate algebras are the right-algebras. For logicians, by contrary, the appropriate algebras are the algebras of logic, not the unital magmas, and among the algebras of logic, the appropriate algebras are the left-algebras. This explains why, for examples, the MV algebras were initially introduced as right-unital magmas, while the Wajsberg algebras were initially introduced as left-algebras of logic.

Consequently, in this paper dealing with the two 'worlds', *since we are comming from algebras of logic side, we shall put on first line the algebras of logic (Sections 2-4) and the left-algebras*; therefore, the unital magmas and the group (Sections 5-8) will be defined multiplicatively.

The whole paper is organized in nine sections, as follows:

In Sections 2–4, we are dealing with commutative algebras of logic, namely with the M algebras (with additional operations). In Section 2, we make an overview in an unifying way of the most important 22 M algebras introduced in [37] and [42], with many clarifications and new results proved in the preprint [40]. In Section 3, we analyse in some detail the M algebras *with last element,* 1, with some additional operations, of the form $(A, \rightarrow, 0, 1)$ or $(A, \rightarrow, ^-, 1)$ with $1^- = 0$, which determine the Hierarchies 2, 2'. We present new results, some of them proved in the preprint [41]. We introduce the new notion of *implicative-ortholattice* as a term-equivalent definition of the ortholattice recalled and studied in Section 6. We establish connections mainly between BE algebras, BCK algebras, Wajsberg algebras, implicative-ortholattices and implicative-Boolean algebras. In Section 4, we analyse in some detail the M algebras *without last element*, with some additional operations, of the form $(A, \rightarrow, 1)$ or $(A, \rightarrow, ^{-1}, 1)$ with $1^{-1} = 1$, which determine the Hierarchies 1, 1'. We present new results and the connection with the implicative-group and the implicative-goop.

In Sections 5–8, we are dealing with commutative algebras, namely with the commutative unital magmas (with additional operations). In Section 5, we recall the definitions of the commutative unital magmas, commutative monoids and (bounded) lattices in an unifying way. In Section 6, we introduce 12 new algebras, the analogous of those 12 M algebras *with last element*, 1, from Section 2, as particular cases of commutative unital magmas with some additional operation of the form $(A, \odot, ^-, 1)$ with $1^- = 0$, which determine the Hierarchies m-2, m-2'. We establish connections mainly between m-BE algebras, m-BCK algebras, MV algebras, ortholattices and Boolean algebras. In Section 7, we introduce 10 new algebras, the analogous of those 10 M algebras *without last element*, as particular cases of commutative unital magmas with some additional operation of the form $(A, \cdot, ^{-1}, 1)$ with $1^{-1} = 1$, which determine the Hierarchies m-1, m-1'. We make the connections with the commutative group, moon [42], goop [42]. In Section 8, we establish the global hierarchies of the new algebras, containing the Hierarchies m-1 and m-2, m-1' and m-2'.

In Section 9, we establish two general theorems connecting the two 'worlds' in the involutive case, one for algebras *with last element*, the other for algebras *without last element*. These theorems can be used to prove the *definitionally equivalence* (d.e.) (hence their categories are isomorphic) between the analogous involutive algebras from the two 'worlds' simply by choosing appropriate definitions of the algebras. As example, the d.e. between implicative-ortholattices and ortholattices is proved.

The *unifying way* of the presentation consists in fixing unique names for the defining properties, making lists of these properties and then using them for defining the different algebras and for obtaining results.

2 Algebras of logic: the M algebras

2.1 The list A of basic properties and the M algebras

Let $\mathcal{A}^L = (A^L, \rightarrow = \rightarrow^L, 1)$ ('L' comes from 'left') be an algebra of type $(2, 0)$ through this paper, where an *internal* binary relation $\leq$ can be defined by: for all $x, y \in A^L$,

$$(dfrelR) \qquad x \leq y \overset{def.}{\Longleftrightarrow} x \rightarrow y = 1.$$

Equivalently, let $\mathcal{A}^L = (A^L, \leq, \rightarrow, 1)$ be a structure, where $\leq$ is an *internal* binary relation on A^L, $\rightarrow$ is a binary operation on A^L and $1 \in A^L$, all connected by the equivalence:

$$(EqrelR) \qquad x \leq y \Longleftrightarrow x \rightarrow y = 1.$$

Consider the following list **A** of basic properties that can be satisfied by $\mathcal{A}^L$ (in fact, the properties in the list are the most important properties satisfied by a BCK algebra), where almost each property is presented in two equivalent forms, determined by the corresponding two equivalent above definitions of $\mathcal{A}^L$:

(An) (Antisymmetry) $(x \to y = 1$ and $y \to x = 1) \implies x = y$,
(An') (Antisymmetry) $(x \le y$ and $y \le x) \implies x = y$;

(B) $(y \to z) \to [(x \to y) \to (x \to z)] = 1$,
(B') $y \to z \le (x \to y) \to (x \to z)$,

(BB) $(y \to z) \to [(z \to x) \to (y \to x)] = 1$,
(BB') $y \to z \le (z \to x) \to (y \to x)$;

(*) $y \to z = 1 \implies (x \to y) \to (x \to z) = 1$,
(*') $y \le z \implies x \to y \le x \to z$;

(**) $y \to z = 1 \implies (z \to x) \to (y \to x) = 1$,
(**') $y \le z \implies z \to x \le y \to x$;

(C) $[x \to (y \to z)] \to [y \to (x \to z)] = 1$,
(C') $x \to (y \to z) \le y \to (x \to z)$;

(D) $y \to [(y \to x) \to x] = 1$,
(D') $y \le (y \to x) \to x$;

(Ex) (Exchange) $x \to (y \to z) = y \to (x \to z)$;

(F) (First element) $0 \to x = 1$,
(F') (First element) $0 \le x$;

(K) $x \to (y \to x) = 1$,
(K') $x \le y \to x$;

(L) (Last element) $x \to 1 = 1$,
(L') (Last element) $x \le 1$;

(M) $1 \to x = x$;

(N) $1 \to x = 1 \implies x = 1$,
(N') $1 \le x \implies x = 1$;

(Re) (Reflexivity) $x \to x = 1$ (we prefered the notation (Re) instead of the original (I)),
(Re') (Reflexivity) $x \leq x$;

(S) $x = y \implies x \to y = 1$,
(S') $x = y \implies x \leq y$;

(Tr) (Transitivity) $(x \to y = 1 \text{ and } y \to z = 1) \implies x \to z = 1$,
(Tr') (Transitivity) $(x \leq y \text{ and } y \leq z) \implies x \leq z$;

(#) $x \to (y \to z) = 1 \Longrightarrow y \to (x \to z) = 1$,
(#') $x \leq y \to z \Longrightarrow y \leq x \to z$;

(H) $((y \to x) \to x) \to x = y \to x$.

Remark 2.1. The basic properties (M) and (Ex) are very special, because they have a unique form. The property (H) also has a unique form, but it is of secondary importance.

Dually, let $\mathcal{A}^R = (A^R, \to^R, 0)$ ('R' comes from 'right') be an algebra of type $(2,0)$, where an *internal* binary relation $\geq$ can be defined by: for all $x, y \in A^R$,

$$(dfrelcoR) \qquad x \geq y \overset{def.}{\Longleftrightarrow} x \to^R y = 0.$$

Equivalently, let $\mathcal{A}^R = (A^R, \geq, \to^R, 0)$ be a structure, where $\geq$ is an *internal* binary relation on A^R, $\to^R$ is a binary operation on A^R and $0 \in A^R$, all connected by the equivalence:

$$(EqrelcoR) \qquad x \geq y \Longleftrightarrow x \to^R y = 0.$$

The list of dual properties, (An^R), ... , (H^R), is omitted.

2.1.1 The M algebras of logic

Recall from [37], [42] the following definitions as algebras (the dual ones are omitted) (the equivalent definitions as structures are immediate):

• Algebras **without last element** (Hierarchies 1 and 1'):
  - without exchange property:
An algebra $(A^L, \to, 1)$ is a:
(1) *left-M algebra*, if it verifies (M);

(2) *left-RM algebra*, if it verifies (M), (Re);
(3) *left-pre-BZ algebra*, if it verifies (M), (Re), (B);
(4) *left-aRM = left-BH algebra*, if it verifies (M), (Re), (An);
(5) *left-BZ algebra*, if it verifies (M), (Re), (An), (B).

- with exchange property:

An algebra $(A^L, \rightarrow, 1)$ is a:
(E1) *left-ME algebra*, if it verifies (M), (Ex);
(E2) *left-RME = left-CI algebra*, if it verifies (M), (Ex), (Re);
(E3) *left-pre-BCI algebra*, if it verifies (M), (Ex), (Re), (B);
(E4) *left-BCH algebra*, if it verifies (Ex), (Re), (An);
(E5) *left-BCI algebras*, if it verifies (BB), (D), (Re), (N), (An) or, equivalently, (BB), (D), (Re), (An), or, equivalently, (BB), (M), (An)
and, additionally,
(E6) *left-RME** algebra*, if it is a left-RME=CI algebra verifying (**);
(E7) *left-BCH** algebra*, if it is a left-BCH algebra verifying (**).

• Algebras **with last element** (Hierarchies 2 and 2'):

- without exchange property:

An algebra $(A^L, \rightarrow, 1)$ is a:
(L1) *left-ML algebra*, if it verifies (M), (L);
(L2) *left-RML algebra*, if it verifies (M), (L), (Re);
(L3) *left-pre-BCC algebra*, if it verifies (M), (L), (Re), (B);
(L4) *left-aRML algebra*, if it verifies (M), (L), (Re), (An);
(L5) *left-BCC algebra*, if it verifies (M), (L), (Re), (B), (An).

- with exchange property:

An algebra $(A^L, \rightarrow, 1)$ is a:
(LE1) *left-MEL algebra*, if it verifies (M), (Ex), (L);
(LE2) *left-BE algebra*, if it verifies (M), (Ex), (L), (Re);
(LE3) *left-pre-BCK algebra*, if it verifies (M), (Ex), (L), (Re), (B);
(LE4) *left-aBE algebra*, if it verifies (M), (Ex), (L), (Re), (An);
(LE5) *left-BCK algebra*, if it verifies (BB), (D), (Re), (L), (An) or, equivalently, (BB), (M), (L), (An) or, equivalently, [5], [37] (B), (C), (K), (An)
and, additionally,
(LE6) *left-BE** algebra*, if it is a left-BE algebra verifying (**),
(LE7) *left-aBE** algebra*, if it is a left-aBE algebra verifying (**).

Many other subclasses of M algebras (structures) of logic were defined in [37], with examples; their connections are presented there also.

Remark 2.2. The *M algebras* and the *ME algebras* are special algebras, because are defined only using the special basic properties (M) and (Ex). By their analogy with the *commutative unital magmas* and the *monoids*, respectively (Section 5), it came the ideea to call ME algebras also as *residoids*.

Denote by **M**, ..., **aBE**** the classes of all left-M algebras, ..., left-aBE** algebras, respectively.

Proposition 2.3. [37] *Let* $(A^L, \to, 1)$ *be an algebra of type* $(2,0)$*. Then the following are true:*
(A0) (Re) $\Longrightarrow$ *(S);*
(A00) (M) $\Longrightarrow$ *(N);*
(A1) (L) + (An) $\Longrightarrow$ *(N);*
(A2) (K) + (An) $\Longrightarrow$ *(N);*
(A3) (C) + (An) $\Longrightarrow$ *(Ex);* *(A3') (Ex) + (Re)* $\Longrightarrow$ *(C);*
(A4) (Re) + (Ex) $\Longrightarrow$ *(D);* *(A4') (D) + (Re) + (An)* $\Longrightarrow$ *(N);*
(A5) (Re) + (Ex) + (An) $\Longrightarrow$ *(M);*
(A6) (Re) + (K) $\Longrightarrow$ *(L);*
(A7) (N) + (K) $\Longrightarrow$ *(L);* *(A7') (M) + (K)* $\Longrightarrow$ *(L);*
(A8) (Re) + (L) + (Ex) $\Longrightarrow$ *(K);*
(A9) (M) + (L) + (B) $\Longrightarrow$ *(K);* *(A9') (M) + (L) + (**)* $\Longrightarrow$ *(K);*
(A10) (Ex) $\Longrightarrow$ *((B)* $\Leftrightarrow$ *(BB));*
(A10') (Ex) + (B) $\Longrightarrow$ *(BB);* *(A10") (Ex) + (BB)* $\Longrightarrow$ *(B);*
(A11) (Re) + (Ex) + (*) $\Longrightarrow$ *(BB);*
(A12) (N) + (B) $\Longrightarrow$ *(*);* *(A12') (M) + (B)* $\Longrightarrow$ *(*);*
(A13) (N) + (*) $\Longrightarrow$ *(Tr);* *(A13') (M) + (*)* $\Longrightarrow$ *(Tr);*
(A14) (N) + (B) $\Longrightarrow$ *(Tr);* *(A14') (M) + (B)* $\Longrightarrow$ *(Tr);*
(A15) (N) + (BB) $\Longrightarrow$ *(**);* *(A15') (M) + (BB)* $\Longrightarrow$ *(**);*
(A16) (N) + (**) $\Longrightarrow$ *(Tr);* *(A16') (M) + (**)* $\Longrightarrow$ *(Tr);*
(A17) (N) + (BB) $\Longrightarrow$ *(Tr);* *(A17') (M) + (BB)* $\Longrightarrow$ *(Tr);*
(A18) (M) + (BB) $\Longrightarrow$ *(Re);* *(A18') (M) + (BB)* $\Longrightarrow$ *(D);*
(A19) (M) + (B) $\Longrightarrow$ *(Re);*
(A20) (BB) + (D) + (N) $\Longrightarrow$ *(C);* *(A20') (M) + (BB)* $\Longrightarrow$ *(C);*
(A21) (BB) + (D) + (N) + (An) $\Longrightarrow$ *(Ex);*
(A21') (BB) + (D) + (L) + (An) $\Longrightarrow$ *(Ex);*
(A21") (M) + (BB) + (An) $\Longrightarrow$ *(Ex);*
(A22) (K) + (Ex) + (M) $\Longrightarrow$ *(Re);*
(A23) (C) + (K) + (An) $\Longrightarrow$ *(Re);*
(A24) (Re) + (Ex) + (Tr) $\Longrightarrow$ *(**).*

Theorem 2.4. *([37], Theorem 1) (Generalization of ([7], Lemma 1.2 and Proposition 1.3))*

If an algebra $(A^L, \rightarrow, 1)$ verifies the properties (Re), (M), (Ex), then:

$$(B) \iff (BB) \iff (*).$$

Theorem 2.5. *([37], Theorem 2)*

If an algebra $(A^L, \rightarrow, 1)$ verifies the properties (Re), (M), (Ex), then:

$$(**) \iff (Tr).$$

Theorem 2.6. *([37], Theorem 3)*

If an algebra $(A^L, \rightarrow, 1)$ verifies the properties (M), (B), (An), then:

$$(Ex) \iff (BB).$$

Proposition 2.7. [37] *An aBE** algebra verifying one of the equivalent properties (BB), (B), (*) is a BCK algebra.*

2.1.2 The global hierarchy of above M algebras

We have introduced in [37] the following definitions (the dual ones are omitted).

Definitions 2.8. Let $\mathcal{A}^L = (A^L, \rightarrow, 1)$ be an algebra of type $(2, 0)$ and let $\leq$ be the associated binary relation defined by (dfrelR).

1) We shall say that $\mathcal{A}^L$ is *reflexive*, if $\leq$ is reflexive (i.e. it satisfies the property (Re) or (Re')).

2) We shall say that $\mathcal{A}^L$ is *antisymmetric*, if $\leq$ is antisymmetric (i.e. it satisfies the property (An) or (An')).

3) We shall say that $\mathcal{A}^L$ is *transitive*, if $\leq$ is transitive (i.e. it satisfies the property (Tr) or (Tr')).

4) We shall say that $\mathcal{A}^L$ is *pre-ordered*, if $\leq$ is a pre-order relation (i.e. it is reflexive and transitive).

5) We shall say that $\mathcal{A}^L$ is *ordered*, if $\leq$ is a partial-order relation (i.e. it is reflexive, antisymmetric and transitive).

6) We shall say that $\mathcal{A}^L$ is a *lattice*, if $\leq$ is a lattice-order relation (i.e. it is a partial-order such that there exists sup(x,y) and inf(x,y) for each $x, y \in A^L$); we shall use the notation $x \wedge y$ for inf(x,y) and $x \vee y$ for sup(x,y), with $x \leq y \Leftrightarrow x \wedge y = x \Leftrightarrow x \vee y = y$ (by the equivalence between an Ore lattice $(A^L, \leq)$ and a Dedekind lattice $(A^L, \wedge, \vee)$, equivalence recalled in Section 5).

Remark 2.9. [37] In the diagram of a hierarchy of classes of algebras, we shall represent:

- *reflexive* algebras by $\bigcirc$
- *antisymmetric* algebras by $\circ$
- *transitive* algebras by $\bullet$
- *reflexive* and *antisymmetric* algebras by $\circledcirc$
- *reflexive* and *transitive* algebras by $\odot$
- *ordered* algebras by ●

Then, the above subclasses of M algebras are connected as in the following Figures 1 and 2 [37], [42]. Note that:
- the subclasses of M algebras *without last element*, namely the M, RM, pre-BZ, BH, BZ algebras and the ME, RME=CI, RME**, pre-BCI, BCH, BCH**, BCI algebras, determine the *Hierarchies 1* and *1'*,
- the subclasses of M algebras *with last element* 1, namely the ML, RML, pre-BCC, aRML, BCC algebras and the MEL, BE, BE**, pre-BCK, aBE, aBE**, BCK algebras, determine the *Hierarchies 2* and *2'*.

2.1.3 The p-semisimple property of algebras in Hierarchies 1 and 1'

Recall also from [42], Definition 2.1.3, the following definition:

Definition 2.10. The algebra $(A^L, \rightarrow, 1)$ verifying (Re) is said to be *p-semisimple*, if the following holds: for all $x, y \in A^L$,
(p-s) $x \leq y \Longrightarrow x = y$ (or, equivalently, by (Re), $x \leq y \Longleftrightarrow x = y$).

Note that the p-semisimple algebras belonging to the Hierarchies 2 and 2' (i.e. verifying the property (L)) are **trivial**, because $A^L = \{1\}$ in these cases (see [42], Theorem 13.1.23).

2.2 The list B of particular properties. Some other M algebras

Consider the following list **B** of particular properties that can be satisfied by $\mathcal{A}^L$ [40]:

(impl) (implicativity) $(x \rightarrow y) \rightarrow x = x$;

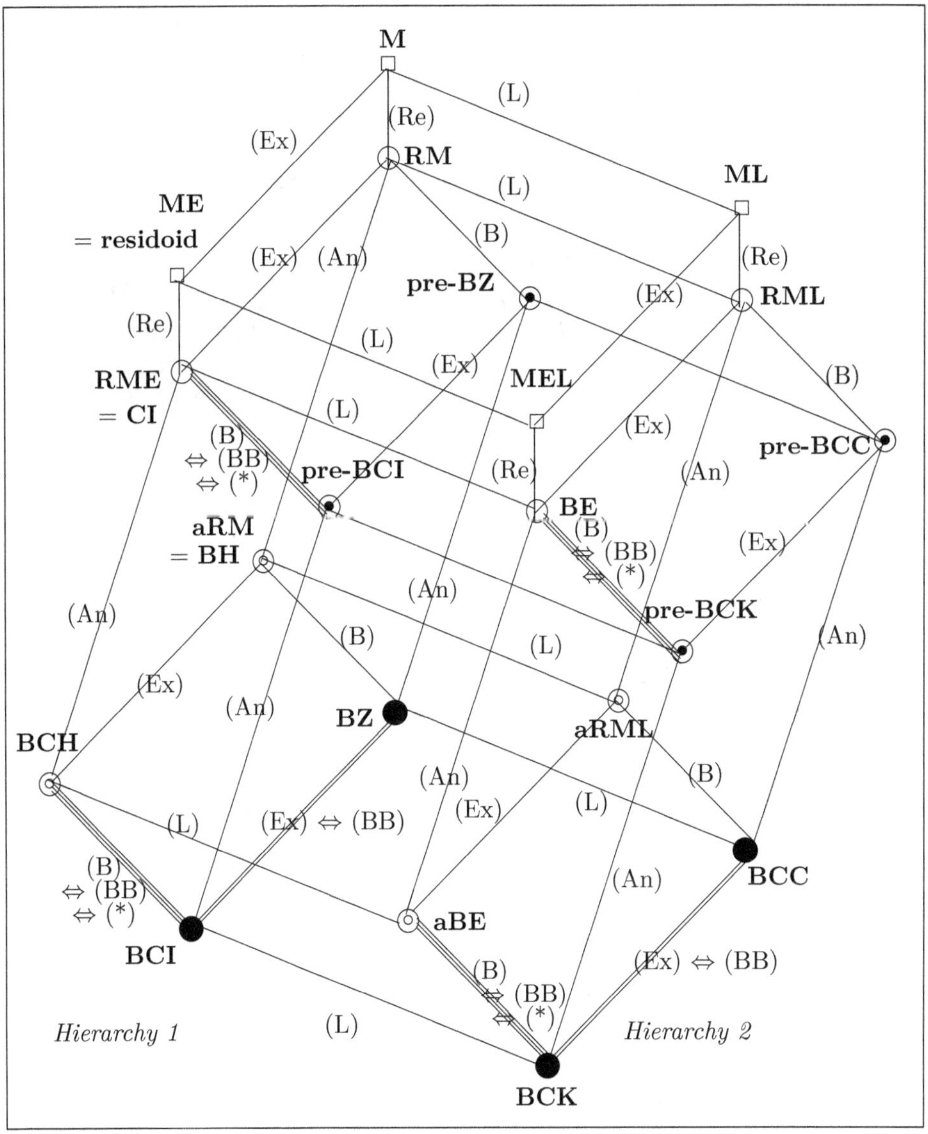

Figure 1: The global hierarchy of M algebras

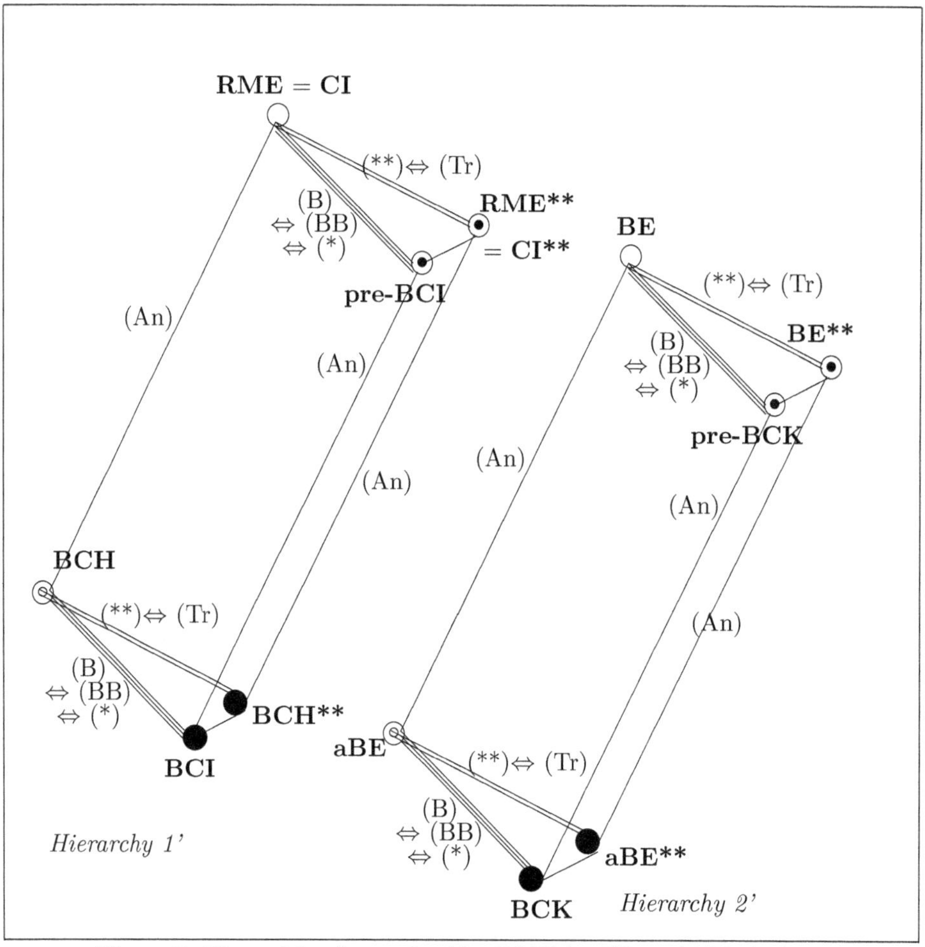

Figure 2: Hierarchies determined by RME (= CI) and BE algebras

(pi) (general positive-implicativity) $x \to (x \to y) = x \to y$,
(pimpl) (positive-implicativity) $x \to (y \to z) = (x \to y) \to (x \to z)$,
(pimpl-1) $[x \to (y \to z)] \to [(x \to y) \to (x \to z)] = 1$,
(pimpl-2) $[(x \to y) \to (x \to z)] \to [x \to (y \to z)] = 1$;

($\vee$-comm) ($\vee$-commutativity) $(x \to y) \to y = (y \to x) \to x$,
($\vee$-comm-1) $[(x \to y) \to y] \to x = (x \to (y \to x)) \to (y \to x)$.

Definitions 2.11. (See [45] for BCK algebras)

Let $\mathcal{A} = (A, \rightarrow, 1)$ be an M algebra. We say that $\mathcal{A}$ is
- *positive-implicative*, if property (pimpl) is satisfied;
- $\vee$-*commutative*, if property ($\vee$-comm) is satisfied;
- *implicative*, if property (impl) is satisfied.

Theorem 2.12. *([45], Theorem 8)*

A BCK algebra is positive-implicative if and only if the property (pi) holds (or, in a BCK algebra the properties (pimpl) and (pi) are equivalent).

Following ([45], Remark 1), $\vee$-commutative BCK algebras were introduced by S. Tanaka [67]; moreover, H. Yutani proved that [68]: *the class of* $\vee$-*commutative BCK algebras is a variety* and found the following short equivalent system of axioms [69]: ($\vee$-comm), (M), (Re), (Ex). Thus, the $\vee$-commutative BCK algebras coincide with the $\vee$-commutative RME=CI algebras.

Proposition 2.13. [40] *We have:*
(B12) ($\vee$-comm) + (M) $\Longrightarrow$ (An);
(B29) ($\vee$-comm) + (L) + (M) $\Longrightarrow$ (Re);
(B30) ($\vee$-comm) + (Re) + (M) + (Ex) $\Longrightarrow$ (L);
(B31) ($\vee$-comm) + (Re) + (M) + (Ex) $\Longrightarrow$ (*);
(B32) ($\vee$-comm) + (M) + (BB) $\Longrightarrow$ (L);
(B33) ($\vee$-comm) + (K) + (BB) + (Tr) + (M) $\Longrightarrow$ (#);
(B34) ($\vee$-comm) $\Longrightarrow$ ($\vee$-comm-1).

Proposition 2.14. *We have:*
(B31') ($\vee$-comm) + (Re) + (Ex) + (L) $\Longrightarrow$ (BB).

Proof. $(z \rightarrow x) \rightarrow (y \rightarrow x) \overset{(Ex)}{=} y \rightarrow [(z \rightarrow x) \rightarrow x]$
$\overset{(\vee-comm)}{=} y \rightarrow [(x \rightarrow z) \rightarrow z] \overset{(Ex)}{=} (x \rightarrow z) \rightarrow (y \rightarrow z)$.
Then, $(y \rightarrow z) \rightarrow [(z \rightarrow x) \rightarrow (y \rightarrow x)] = (y \rightarrow z) \rightarrow [(x \rightarrow z) \rightarrow (y \rightarrow z)]$
$\overset{(Ex)}{=} (x \rightarrow z) \rightarrow [(y \rightarrow z) \rightarrow (y \rightarrow z)] \overset{(Re)}{=} (x \rightarrow z) \rightarrow 1 \overset{(L)}{=} 1$, hence (BB) holds. □

Then, we obtain the following result [40]:

Theorem 2.15. $\vee$-*commutative RME = CI, BE, BCI, BCK algebras coincide.*

Proposition 2.16. [40] *We have:*
(BIM1) (impl) $\Longrightarrow$ (pi);
(BIM2) ($\vee$-comm) + (pi) + (Re) + (K) + (M) $\Longrightarrow$ (impl);
(BIM2') ($\vee$-comm) + (L) + (K) + (M) $\Longrightarrow$ ((pi) $\Leftrightarrow$ (impl));
(BIM3) (impl) + (Ex) + (B) + (An) $\Longrightarrow$ ($\vee$-comm);

(BIM4) (impl) + (Re) $\Longrightarrow$ *(M);*
(BIM5) (impl) + (L) $\Longrightarrow$ *(M);*
(BIM5') (impl) + (M) $\Longrightarrow$ *(L);*
(BIM5") (impl) $\Longrightarrow$ *((L)* $\Leftrightarrow$ *(M));*
(BIM6) (impl) + (K) $\Longrightarrow$ *(Re);*
(BIM7) (L) + (K) + (M) + (Ex) + (B) + (An) $\Longrightarrow$
((impl) $\Leftrightarrow$ *((*$\vee$*-comm) + (pi))).*

Proposition 2.17. [72]
Let $(A, \to, 1)$ *be an algebra verifying (Re), (D), (**) and (impl). Then, for all* $x, y \in A$,

$$y \leq x \Longrightarrow (x \to y) \to y \leq x.$$

Definitions 2.18. [37]
(i) A *left-RML** algebra* is a left-RML algebra verifying (**).
(ii) An *left-aRML** algebra* is a left-aRML algebra verifying (**).

Theorem 2.19. [72]
Let $(A, \to, 1)$ *be an implicative RML** algebra satisfying (D). Then, we have:*
(wComm) $(x \to y) \to y \leq (y \to x) \to x$.

Corollary 2.20. *Every implicative aRML** algebra satisfying (D) (that is, implicative weak BCK-algebra) is* $\vee$*-commutative.*

Since any aBE** algebra is an aRML** algebra verifying (Ex), hence (D), by (A4), it follows that we have:

Corollary 2.21. *Every implicative aBE** algebra is* $\vee$*-commutative.*

Definition 2.22. (see [15])
A *Hilbert algebra* is an algebra $\mathcal{A} = (A, \to, 1)$ of type $(2, 0)$, satisfying, for all $x, y, z \in A$:
(K) $x \to (y \to x) = 1$,
(pimpl-1) $(x \to (y \to z)) \to ((x \to y) \to (x \to z)) = 1$,
(An) $(x \to y = 1$ and $y \to x = 1)$ imply $x = y$.

Theorem 2.23. *([32], Remarks 2.1.32 (1)) (see [40] for a direct proof)*
Hilbert algebras are categorically equivalent to positive-implicative BCK algebras.

Theorem 2.24. *([45], Theorem 9)*
In a $\vee$*-commutative BCK algebra, the properties (pi) and (impl) are equivalent.*

Theorem 2.25. *([45], Theorem 10)*
Any implicative BCK algebra is $\vee$-commutative and positive-implicative.

Theorem 2.26. [40]
Any $\vee$-commutative and positive-implicative BCK algebra is implicative.

Corollary 2.27. [40] *In a BCK algebra, we have:*

$$(impl) \iff ((\vee - comm) + (pi) (\Leftrightarrow (pimpl)))$$

Corollary 2.28. [40] *Any $\vee$-commutative Hilbert algebra is implicative.*

2.3 The (involutive) negations

• In the above algebras $(A^L, \rightarrow, 1)$ where 1 is not the last element (*without last element*) (from Hierarchies 1 and 1'), we can define a *negation* (an *inverse*) as follows:

$$(dfneg1) \qquad x^{-1} \stackrel{def.}{=} x \rightarrow 1$$

and we have (N1) $(1^{-1} = 1)$, by (M).

Alternatively, in an algebra $(A^L, \rightarrow, ^{-1}, 1)$ *without last element* such that $1^{-1} = 1$, the *negation* (*inverse*) $^{-1}$ is connected with $\rightarrow$ and 1 by the equality:

$$(Neg_1) \qquad x^{-1} = x \rightarrow 1.$$

Remark 2.29. (see [42], Remark 5.1.6)

Just as the *equivalence* $x \leq y \Leftrightarrow x \rightarrow y = 1$ was used:
- either as the *definition* (dfrelR) of the binary relation $\leq$ in the algebra $(A^L, \rightarrow, 1)$,
- or as the *connection* (EqrelR) between the binary relation $\leq$ and $\rightarrow, 1$ in the structure $(A^L, \leq, \rightarrow, 1)$,

the same, the *equality* $x^{-1} = x \rightarrow 1$ was used:
- either as the *definition* (dfneg1) of the negation $^{-1}$ in the algebra $(A^L, \rightarrow, 1)$ and in this case $1^{-1} = 1$,
- or as the *connection* (Neg_1) between the negation $^{-1}$ and $\rightarrow, 1$ in the algebra $(A^L, \rightarrow, ^{-1}, 1)$ such that $1^{-1} = 1$.

If the negation $^{-1}$ verifies:
(DN) (Double Negation) $(x^{-1})^{-1} = x$,
we say that it is *involutive* and that the corresponding algebra is *involutive*.

• In the above algebras $(A^L, \rightarrow, 1)$ *with last element* 1 (from Hierarchies 2 and 2'), the above definition of the negation $^{-1}$ does not work, because it produces $x^{-1} = 1$, for all x, by (L). In this case, we add a first element, 0, verifying: (F) (First element) $0 \to x = 1$.

The algebra $(A^L, \to, 1)$ is said to be *bounded*, if besides the last element 1 (verifying (L)), there exists a first element also, 0 (verifying (F)); it is denoted by $(A^L, \to, 0, 1)$.

Denote by $\mathbf{ML}^b$, ..., $\mathbf{BCK}^b$ the corresponding classes of bounded algebras (where 'b' means 'bounded').

In such algebra $(A^L, \to, 0, 1)$, a *negation* $^-$ can be defined as follows:

$$(dfneg) \qquad x^- \overset{def.}{=} x \to 0$$

and we have (Neg1-0) $1^- = 0$, by (M), and (Neg0-1) $0^- = 1$, by (Re).

Alternatively, in an algebra $(A^L, \to, ^-, 1)$ *with last element* 1 such that $0 \overset{def.}{=} 1^-$, the *negation* $^-$ is connected with $\to$ and 0 by the equality:

$$(Neg) \qquad x^- = x \to 0.$$

Remark 2.30. (see [42], Remark 5.1.6)

Just as the *equivalence* $x \leq y \Leftrightarrow x \to y = 1$ was used:
- either as the *definition* (dfrelR) of the binary relation $\leq$ in the algebra $(A^L, \to, 1)$,
- or as the *connection* (EqrelR) between the binary relation $\leq$ and $\to, 1$ in the structure $(A^L, \leq, \to, 1)$,

the same, the *equality* $x^- = x \to 0$ was used:
- either as the *definition* (dfneg) of the negation $^-$ in the algebra $(A^L, \to, 0, 1)$ and in this case $1^- = 0$ and $0^- = 1$,
- or as the *connection* (Neg) between the negation $^-$ and $\to, 0$ in the algebra $(A^L, \to, ^-, 1)$ such that $0 = 1^-$.

If the negation $^-$ verifies:
(DN) (Double Negation) $(x^-)^- = x$,
we say that it is *involutive* and that the algebra is *involutive*.

The subclasses of M algebras *with last element* 1 *and first element* 0, i.e. the bounded ML, RML, pre-BCC, aRML, BCC algebras and the bounded MEL, BE, BE**, pre-BCK, aBE, aBE**, BCK algebras determine the *Hierarchies* $\mathcal{2}^b$ and 2^b.

Remark 2.31. Remark that the special algebras, the *M algebras* and the *ME algebras = residoids*, can be considered also as belonging to the Hierarchy 2^b.

The subclass of involutive algebras of the class $\mathbf{X}$ will be denoted by $\mathbf{X}_{(DN)}$.

The algebras without last element (Hierarchy 1) and the bounded algebras (Hierarchy 2^b) are presented in the following Figure 3. The Hierarchy $2'^b$ is obvious and is omitted.

3 Algebras $(A, \to, 0, 1)$ or $(A, \to, ^-, 1)$ with $1^- = 0$ (Hierarchies 2^b and $2'^b$)

Let $\mathcal{A}^L = (A^L, \to, 0, 1)$ be a bounded M algebra *not verifying (Ex)*. Let us introduce in this case the weaker condition:
(Ex0) $x \to (y \to 0) = y \to (x \to 0)$.

We shall say that $\mathcal{A}^L$ is *strong-0*, if it verifies (Ex0).

Note that (Ex) implies (Ex0).

3.1 The list C of the properties of the negation $^-$ and some results

Consider the following list **C** of properties that can be satisfied by an (involutive) negation $^-$ defined on A^L:
(dfneg) $x^- \overset{def.}{=} x \to 0$,
(Neg) $x^- = x \to 0$;

(DN) (Double negation) $(x^-)^- = x$;
(Neg1-0) $1^- = 0$,
(Neg0-1) $0^- = 1$;

(Neg1) $(x \to y) \to (y^- \to x^-) = 1$,
(Neg1') $x \to y \leq y^- \to x^-$;
(Neg2) (contraposition) $x \to y = 1 \Longrightarrow y^- \to x^- = 1$,
(Neg2') (contraposition) $x \leq y \Longrightarrow y^- \leq x^-$;
(Neg3) $y \to x^- = x \to y^-$;
(Neg4) $x \to (x^-)^- = 1$,
(Neg4') $x \leq (x^-)^-$;
(Neg5) $(x \to y)^- \to x = 1$,
(Neg5') $(x \to y)^- \leq x$;
(Neg6) $x \to x^- = x^-$;

(DN1) $(y^- \to x^-) \to (x \to y) = 1$,
(DN1') $y^- \to x^- \leq x \to y$;
(DN2) $x \to y = y^- \to x^-$;

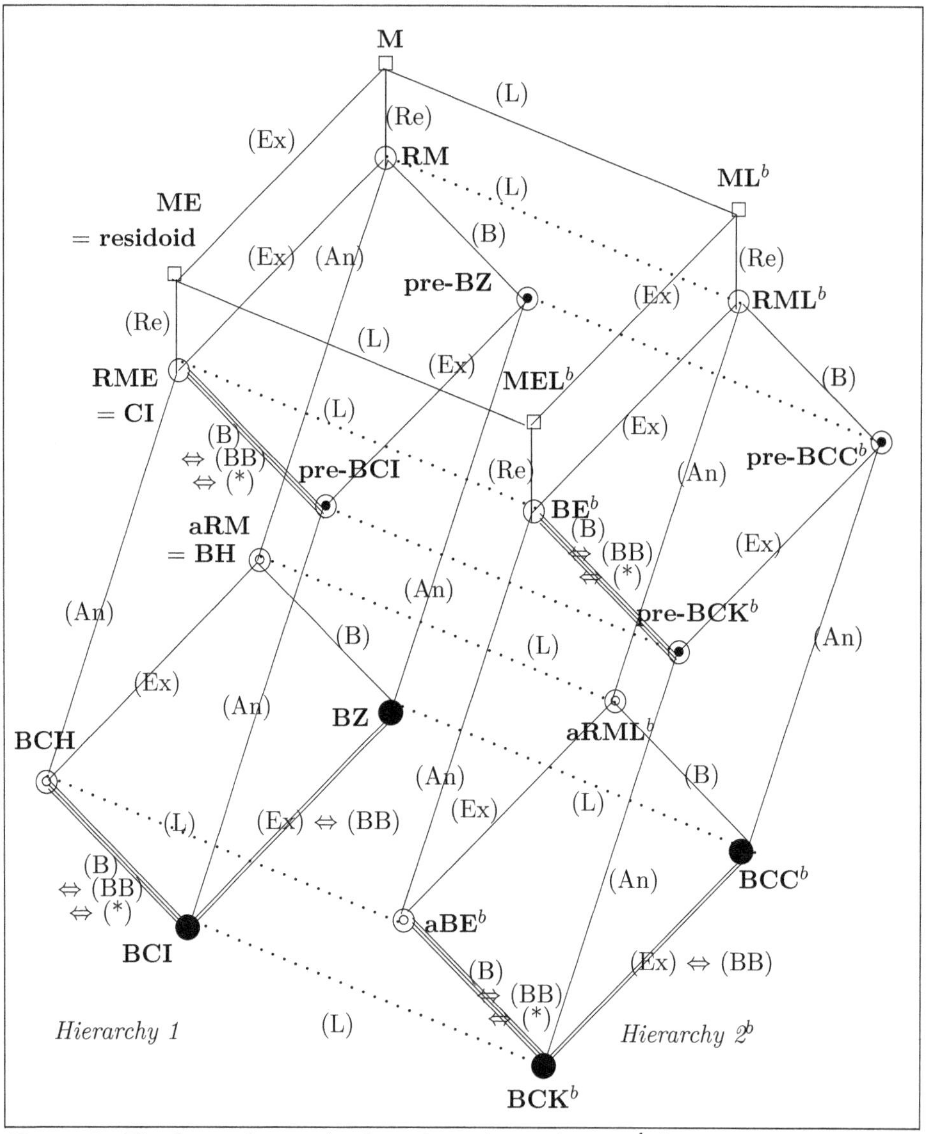

Figure 3: The hierarchies 1 (without last element) and 2^b (with last element, and first element) of M algebras of logic

(DN3) $y^- \to x = x^- \to y$;
(DN4) $x \to y = 1 \Longleftrightarrow y^- \to x^- = 1$,
(DN4') $x \leq y \Longleftrightarrow y^- \leq x^-$;
(DN5) $x^- \to (x \to y) = 1$,
(DN5') $x^- \leq x \to y$;
(DN6) $x^- \to x = x$;

(WRe) $x \wedge x^- = 0$,
(VRe) $x \vee x^- = 1$;

(@) $(y^- \to x) \to y = x \to y$.

Proposition 3.1. *We have:*
(CE1) (dfneg) + (Ex0) $\Longrightarrow$ (Neg3),
(CE1') (dfneg) + (Neg3) $\Longrightarrow$ (Ex0),
(CE1") (dfneg) $\Longrightarrow$ ((Ex0) $\Longleftrightarrow$ (Neg3));
(CE2) (Neg) + (Ex0) $\Longrightarrow$ (Neg3),
(CE2') (Neg) + (Neg3) $\Longrightarrow$ (Ex0),
(CE2") (Neg) $\Longrightarrow$ ((Ex0) $\Longleftrightarrow$ (Neg3)).

Proof. (CE1): $x \to y^- \stackrel{(dfneg)}{=} x \to (y \to 0) \stackrel{(Ex0)}{=} y \to (x \to 0) \stackrel{(dfneg)}{=} y \to x^-$; thus, (Neg3) holds.

(CE1'): $x \to (y \to 0) \stackrel{(dfneg)}{=} x \to y^- \stackrel{(Neg3)}{=} y \to x^- \stackrel{(dfneg)}{=} y \to (x \to 0)$; thus, (Ex0) holds.

(CE1"): By (CE1) and (CE1').

(CE2), (CE2'), (CE2"): similarly. □

Proposition 3.2. *We have* [41]*:*
(C0) (DN) + (Neg1-0) $\Longrightarrow$ (Neg0-1), *(C0') (DN) + (Neg0-1) $\Longrightarrow$ (Neg1-0),*
(C0") (DN) $\Longrightarrow$ ((Neg1-0) $\Leftrightarrow$ (Neg0-1));

(C1) (DN) + (Neg1) $\Longrightarrow$ (DN1), *(C1') (DN) + (DN1) $\Longrightarrow$ (Neg1),*
(C1") (DN) $\Longrightarrow$ ((Neg1) $\Leftrightarrow$ (DN1));

(C2) (DN) + (DN2) $\Longrightarrow$ (DN3), *(C2') (DN) + (DN3) $\Longrightarrow$ (DN2) ;*
(C2") (DN) $\Longrightarrow$ ((DN2) $\Leftrightarrow$ (DN3));

(C3) (DN) + (Neg3) $\Longrightarrow$ (DN3), *(C3') (DN) + (DN3) $\Longrightarrow$ (Neg3),*

(C3") (DN) $\Longrightarrow$ *((Neg3)* $\Leftrightarrow$ *(DN3));*

(C4) (DN2) $\Longrightarrow$ *(DN4);*

(C5) (DN) + (DN2) + (Neg5) $\Longrightarrow$ *(DN5),*
(C5') (DN) + (DN2) + (DN5) $\Longrightarrow$ *(Neg5),*
(C5") (DN) + (DN2) $\Longrightarrow$ *((Neg5)* $\Leftrightarrow$ *(DN5));*

(C6) (DN) + (Neg6) $\Longrightarrow$ *(DN6),* *(C6') (DN) + (DN6)* $\Longrightarrow$ *(Neg6),*
(C6") (DN) $\Longrightarrow$ *((Neg6)* $\Leftrightarrow$ *(DN6)).*

Lemma 3.3. *Let* $\mathcal{A}^L = (A^L, \rightarrow, ^-, 1)$ *be an algebra of type* $(2,1,0)$ *verifying (M) and (Neg3). Define* $0 \stackrel{def.}{=} 1^-$. *Then, (Neg) holds.*

Proof. $x \rightarrow 0 \stackrel{def.}{=} x \rightarrow 1^- \stackrel{(Neg3)}{=} 1 \rightarrow x^- \stackrel{(M)}{=} x^-$; thus, (Neg) holds. □

The following two theorems allows us to say that, roughly speaking, the M algebras $(A^L, \rightarrow, 0, 1)$ and $(A^L, \rightarrow, ^-, 1)$ are definitionally equivalent.

Theorem 3.4.
(1) Let $\mathcal{A}^L = (A^L, \rightarrow, 0, 1)$ *be an algebra of type* $(2,0,0)$ *verifying (M) and (Ex0). Define a negation* $^-$ *by (dfneg).*
Then, $\alpha(\mathcal{A}^L) = (A^L, \rightarrow, ^-, 1)$ *is an algebra of type* $(2,1,0)$ *verifying (M) and (Neg3).*
(1') Let $\mathcal{A}^L = (A^L, \rightarrow, ^-, 1)$ *be an algebra of type* $(2,1,0)$ *verifying (M) and (Neg3). Define* $0 \stackrel{def.}{=} 1^-$.
Then, $\beta(\mathcal{A}^L) = (A^L, \rightarrow, 0, 1)$ *is an algebra of type* $(2,0,0)$ *verifying (M) and (Ex0).*
(2) The maps α *and* β *are inverse to each other.*

Proof. (1): By (CE1), (dfneg) + (Ex0) imply (Neg3); thus, (Neg3) holds.
(1'): By Lemma 3.3, (Neg) holds. Then, by (CE2'), (Neg) + (Neg3) imply (Ex0); thus, (Ex0) holds.
(2): Let

$$(A^L, \rightarrow, 0, 1) \quad \stackrel{\alpha}{\longrightarrow} \quad (A^L, \rightarrow, ^-, 1) \quad \stackrel{\beta}{\longrightarrow} \quad (A^L, \rightarrow, \mathbf{0}, 1).$$

Then, we have: $\mathbf{0} \stackrel{def.}{=} 1^- \stackrel{(dfneg)}{=} 1 \rightarrow 0 \stackrel{(M)}{=} 0$.
Conversely, let

$$(A^L, \rightarrow, ^-, 1) \quad \stackrel{\beta}{\longrightarrow} \quad (A^L, \rightarrow, 0, 1) \quad \stackrel{\alpha}{\longrightarrow} \quad (A^L, \rightarrow, ', 1).$$

Then, for all $x \in A^L$, we have: $x' \stackrel{(dfneg)}{=} x \rightarrow 0 \stackrel{def.}{=} x \rightarrow 1^- \stackrel{(Neg3)}{=} 1 \rightarrow x^- \stackrel{(M)}{=} x^-$. □

Theorem 3.5.

(1) Let $\mathcal{A}^L = (A^L, \rightarrow, 0, 1)$ *be an algebra of type* $(2,0,0)$ *verifying (M), (L), (F) and (Ex0). Define a negation* $^-$ *by (dfneg) such that (DN) holds.*
Then, $\alpha(\mathcal{A}^L) = (A^L, \rightarrow, ^-, 1)$ *is an algebra of type* $(2,1,0)$ *verifying (M), (L), (DN) and (Neg3).*

(1') Let $\mathcal{A}^L = (A^L, \rightarrow, ^-, 1)$ *be an algebra of type* $(2,1,0)$ *verifying (M), (L), (DN) and (Neg3). Define* $0 \stackrel{def.}{=} 1^-$.
Then, $\beta(\mathcal{A}^L) = (A^L, \rightarrow, 0, 1)$ *is an algebra of type* $(2,0,0)$ *verifying (M), (L), (F) (and (DN)) and (Ex0).*

(2) The maps α *and* β *are inverse to each other.*

Proof. By Theorem 3.4, it remains to prove that (F) holds, in the case (1'). Indeed, by (C3), (DN) + (Neg3) imply (DN3); then, $0 \rightarrow x = 1^- \rightarrow x \stackrel{(DN3)}{=} x^- \rightarrow 1 \stackrel{(L)}{=} 1$; thus, (F) holds. □

Proposition 3.6. *We have:*
(CB0) (pimpl) + (Ex) + (DN6) $\Longrightarrow$ *(@),*
(CB1) ($\vee$*-comm) + (Neg) + (F) + (M)* $\Longrightarrow$ *(DN)* [41],
(CB2) (impl) + (Neg) $\Longrightarrow$ *(DN6)* [41],
(CB5) (DN) + (DN2) + (DN5) + (DN6) + (K) + (*) + (An) $\Longrightarrow$ *(impl)* [41].

Proof.

(CB0): In (pimpl) $(x \rightarrow (y \rightarrow z) = (x \rightarrow y) \rightarrow (x \rightarrow z))$ take $x = x^-$ and $z = x$; we obtain:
$y \rightarrow x \stackrel{(DN6)}{=} y \rightarrow (x^- \rightarrow x) \stackrel{(Ex)}{=} x^- \rightarrow (y \rightarrow x) \stackrel{(pimpl)}{=} (x^- \rightarrow y) \rightarrow (x^- \rightarrow x) \stackrel{(DN6)}{=} (x^- \rightarrow y) \rightarrow x$; thus, (@) holds. □

Proposition 3.7. *We have:*
(CC1) (@) + (DN2) + (DN3) $\Longrightarrow$ *(H),*
(CC2) (@) + (H) + (DN3) $\Longrightarrow$ *(*$\vee$*-comm).*

Proof.

(CC1): $((x \rightarrow y) \rightarrow y) \rightarrow y \stackrel{(DN2)}{=} ((y^- \rightarrow x^-) \rightarrow y) \rightarrow y \stackrel{(@)}{=} (x^- \rightarrow y) \rightarrow y \stackrel{(DN3)}{=} (y^- \rightarrow x) \rightarrow y \stackrel{(@)}{=} x \rightarrow y$; thus, (H) holds.

(CC2): We must prove that $(x \rightarrow y) \rightarrow y = (y \rightarrow x) \rightarrow x$.
Indeed, $(x \rightarrow y) \rightarrow y \stackrel{(@)}{=} ((y^- \rightarrow x) \rightarrow y) \rightarrow y \stackrel{(DN3)}{=} ((x^- \rightarrow y) \rightarrow y) \rightarrow y \stackrel{(H)}{=} x^- \rightarrow y$ and
$(y \rightarrow x) \rightarrow x \stackrel{(@)}{=} ((x^- \rightarrow y) \rightarrow x) \rightarrow x \stackrel{(DN3)}{=} ((y^- \rightarrow x) \rightarrow x) \rightarrow x \stackrel{(H)}{=} y^- \rightarrow x \stackrel{(DN3)}{=}$

$x^- \to y$; thus, ($\vee$-comm) holds. □

We add now the following results:

Proposition 3.8. *(The ideea came from the analogous result (mCDN3) from Section 6)*

We have:
(i) (DN) + (Ex0) + (dfneg) $\Longrightarrow$ *((*)* $\Leftrightarrow$ *(**)),*
(ii) (DN) + (Neg3) + (Neg) $\Longrightarrow$ *((*)* $\Leftrightarrow$ *(**)).*

Proof. (i): First note that (Ex0) + (dfneg) imply (Neg3), by (CE1); then, (DN) + (Neg3) imply (DN2), by (C3), (C2"); then, (DN2) implies (DN4), by (C4). If $x \le y$, then by (DN4'), $y^- \le x^-$; then, by (*'), $z^- \to y^- \le z^- \to x^-$, hence $y \to z \le x \to z$, by (DN2); thus, (**') holds. Conversely, if $x \le y$, then $y^- \le x^-$, by (DN4'); then, by (**'), $x^- \to z^- \le y^- \to z^-$, hence $z \to x \le z \to y$, by (DN2); thus, (*') holds.

(ii): Similarly. □

Theorem 3.9. *(This Theorem is inspired by Theorem 6.12)*

If an algebra $(A, \to, 0, 1)$ *verifies the properties (M), (Re), (Ex), (dfneg) and (DN), then:*

$$(B) \Leftrightarrow (BB) \Leftrightarrow (*) \Leftrightarrow (**) \Leftrightarrow (Tr).$$

Proof. By Theorems 2.4, 2.5 and Proposition 3.8 (i), since (Ex) implies (Ex0). □

Open problem 3.10. Does Theorem 3.9 remain valid for (Ex0) instead of (Ex)?

By Propositions 3.8 and 2.7, we obtain the following result (inspired from Theorem 6.13):

Proposition 3.11. *The involutive aBE** algebras coincide with the involutive BCK algebras.*

By Corollary 2.21 and (CB1), we obtain:

Corollary 3.12. *Every bounded implicative aBE** algebra satisfies (DN) (i.e. it is involutive).*

By above Corollary 3.12 and Proposition 3.11, we obtain immediately that:

Corollary 3.13. *The bounded implicative aBE** algebras coincide with the bounded implicative BCK algebras.*

3.2 Wajsberg algebras and i-Boolean algebras. Connections

3.2.1 Wajsberg algebras. Connections

Recall from [19] the definition of Wajsberg algebras:

Definition 3.14. (Definition 1) (the dual case is omitted)
An algebra $(A, \rightarrow, ^{-}, 1)$ of type $(2, 1, 0)$ is a *left-Wajsberg algebra*, if the properties (M), (BB), (DN1), ($\vee$-comm) hold.

Denote by $\mathbf{W}$ the class of all left-Wajsberg algebras.

Recall that the bounded $\vee$-commutative BCK algebras are categorically equivalent to the MV algebras [61] and that the MV algebras are categorically equivalent (in fact are d.e.) to the Wajsberg algebras [19]. Hence we have:

Theorem 3.15. *The bounded $\vee$-commutative BCK algebras are categorically equivalent to the Wajsberg algebras.*

The ideea of the following two results came from Remark 6.16 concerning the definition of MV algebras.

Theorem 3.16. *The bounded $\vee$-commutative MEL, BE, aBE, BCK algebras coincide.*

Proof. Let $\mathcal{A}^L = (A^L, \rightarrow, 0, 1)$ be a bounded $\vee$-commutative MEL algebra, i.e. (M), (Ex), (L), (F), ($\vee$-comm) hold. Then, by (B29), ($\vee$-comm) + (L) + (M) $\Longrightarrow$ (Re), hence $\mathcal{A}$ is a bounded $\vee$-commutative BE algebra; by (B12), ($\vee$-comm) + (M) $\Longrightarrow$ (An), hence $\mathcal{A}$ is a bounded $\vee$-commutative aBE algebra; finally, by (B31'), ($\vee$-comm) + (Re) + (L) + (Ex) $\Longrightarrow$ (BB), hence $\mathcal{A}$ is a bounded $\vee$-commutative BCK algebra. The converse is obvious. □

Theorem 3.17. *The bounded $\vee$-commutative MEL algebras are categorically equivalent to the Wajsberg algebras.*

Proof. By Theorems 3.15 and 3.16. □

It follows, by above Theorem, a second equivalent definition of Wajsberg algebras:

Definition 3.18. (Definition 2)
A *Wajsberg algebra* is an algebra $(A, \rightarrow, 0, 1)$ of type $(2, 0, 0)$ verifying: (M), (Ex), (L), (F), ($\vee$-comm).

3.2.2 Implicative-Boolean algebras. Connections

There are many term-equivalent definitions of Boolean algebras. We shall use here the term-equivalent definition introduced in 2009 [33] and presented also in [20], motivated by the axioms system of the classical propositional logic (of truth).

Definition 3.19. (The dual case is omitted)

A *left-implicative-Boolean algebra*, or a *left-i-Boolean algebra* for short, is an algebra $\mathcal{B}^L = (B^L, \rightarrow, ^-, 1)$ of type $(2, 1, 0)$ verifying: (K), (pimpl-1), (DN1), (An).

Denote by **i-Boole** the class of all left-i-Boolean algebras.

We recall their term-equivalence (d.e.) with the Boolean algebras from [20]:

Theorem 3.20. [20]

(1) Let $\mathcal{B}^L = (B^L, \rightarrow, ^-, 1)$ be a left-i-Boolean algebra.

Define $\Phi(\mathcal{B}^L) \stackrel{def.}{=} (B^L, \wedge, \vee, ^-, 0, 1)$ as follows: for every $x, y \in B^L$,
$x \wedge y \stackrel{def.}{=} (x \rightarrow y^-)^-$, $x \vee y \stackrel{def.}{=} (x^- \wedge y^-)^- = x^- \rightarrow y$, $0 \stackrel{def.}{=} 1^-$.

Then $\Phi(\mathcal{B}^L)$ is a left-Boolean algebra.

(1') Conversely, let $\mathcal{B}^L = (B^L, \wedge, \vee, ^-, 0, 1)$ be a left-Boolean algebra.

Define $\Psi(\mathcal{B}^L) \stackrel{def.}{=} (B^L, \rightarrow, ^-, 1)$ as follows: for every $x, y \in B^L$,
$x \rightarrow y \stackrel{def.}{=} (x \wedge y^-)^- = x^- \vee y$.

Then $\Psi(\mathcal{B}^L)$ is a left-i-Boolean algebra.

(2) The mappings Φ and Ψ are mutually inverse.

Theorem 3.21. *(See* [45]*, Theorem 12)*

The bounded implicative BCK algebras are categorically equivalent to the Boolean algebras.

By this Theorem and the above Theorem 3.20, it follows that:

Corollary 3.22. *The bounded implicative BCK algebras are categorically equivalent to the i-Boolean algebras.*

By this Corollary 3.22 and the Corollaries 2.21, 3.12, we obtain:

Theorem 3.23. *The bounded implicative aBE** algebras are categorically equivalent to the i-Boolean algebras.*

Theorem 3.24. *The bounded positive-implicative BCK algebras with involutive negation are categorically equivalent to the i-Boolean algebras.*

It is known that any bounded Hilbert algebra with involutive negation is a Boolean algebra [6]. Moreover, we have the following result.

Corollary 3.25. *The bounded Hilbert algebras with involutive negation are categorically equivalent to the i-Boolean algebras.*

3.2.3 The connection between Wajsberg algebras and i-Boolean algebras

We now make the connection between Wajsberg algebras and i-Boolean algebras:

Theorem 3.26. *The Wajsberg algebras verifying the property (DN6) are d.e. to the i-Boolean algebras.*

Proof. $\Longrightarrow$: Let $\mathcal{A} = (A, \rightarrow, ^-, 1)$ be a left-Wajsberg algebra verifying (DN6). By Theorem 3.15, $\mathcal{A}$ is a bounded $\vee$-commutative BCK algebra. Then, by (CB5), (DN) + (DN2) + (DN5) + (K) + (*) + (An) + (DN6) $\Longrightarrow$ (impl), hence $\mathcal{A}$ is a bounded implicative left-BCK algebra. Then, by Corollary 3.22, $\mathcal{A}$ is a left-i-Boolean algebra.

$\Longleftarrow$: Let $\mathcal{A} = (A, \rightarrow, ^-, 1)$ be a left-i-Boolean algebra, i.e. (K), (pimpl-1), (DN1), (An) hold. By Corollary 3.22, $(A, \rightarrow, 0, 1)$ is a bounded implicative left-BCK algebra, i.e. (BB), (M), (Neg), (impl) hold, where $0 = 1^-$. By Theorem 2.25, ($\vee$-comm) holds. By (CB2), (impl) + (Neg) imply (DN6). Thus, $\mathcal{A}$ satisfy the properties (M), (BB), ($\vee$-comm), (DN1) and (DN6), i.e. it is a left-Wajsberg algebra (Definition 1) verifying (DN6). □

3.3 Implicative-ortholattices

We introduce here the notion of implicative-ortholattice as the term-equivalent definition of the ortholattice recalled and studied in Section 6.

Definitions 3.27.

(i) A *left-implicative-ortholattice*, or a *left-i-ortholattice* for short, is an algebra $\mathcal{A}^L = (A^L, \rightarrow = \rightarrow^L, ^- = ^{-^L}, 1)$ of type $(2, 1, 0)$ such that $x^- = x \rightarrow 0$ (i.e. (Neg) holds), where $0 \overset{def.}{=} 1^-$ (i.e. (Neg1-0) holds), and (M), (Ex), (Re), (L), (impl), (DN) hold.

(i') Dually, a *right-implicative-ortholattice*, or a *right-i-ortholattice* for short, is an algebra $\mathcal{A}^R = (A^R, \rightarrow^R, ^- = ^{-^R}, 0)$ of type $(2, 1, 0)$ such that $x^- = x \rightarrow^R 1$ (i.e. (Neg^R) holds), where $1 \overset{def.}{=} 0^-$ (i.e. (Neg0-1) holds), and (M^R), (Ex^R), (Re^R), (F^R), (impl^R), (DN) hold.

Remark 3.28. The name 'implicative-ortholattice' means an ortholattice defined equivalently with implication. Note that an implicative-ortholattice is *implicative* indeed.

Note that we have immediately the following equivalent definition:

Definition 3.29. (Definition 2) (the dual case is omitted)

A *left-implicative-ortholattice*, or a *left-i-ortholattice* for short, is an algebra $\mathcal{A}^L = (A^L, \to, 0, 1)$ of type $(2,0,0)$ such that (M), (Ex), (Re), (F), (L), (impl) and (DN) hold, where $x^- \stackrel{def.}{=} x \to 0$ by (dfneg), i.e. it is an *implicative involutive left-BE algebra*.

Denote by **i-OL** the class of all left-i-ortholattices.

Proposition 3.30. *Let $\mathcal{A}^L = (A^L, \to, ^-, 1)$ be a left-i-ortholattice. Then, the following properties hold: (pi), (Neg0-1), (Neg3), (DN3), (DN2), (DN6), (Neg6).*

Proof. By (BIM1), (pi) holds. By (C0), (Neg0-1) holds. (Ex) implies (Ex0), then by (CE2), (Neg) + (Ex0) imply (Neg3), hence (Neg3) holds. By (C3), (DN3) holds. By (C2'), (DN2) holds. By (CB2), (DN6) holds. By (C6'), (Neg6) holds. □

Proposition 3.31. *Let $\mathcal{A}^L = (A^L, \to, ^-, 1)$ be a left-i-ortholattice. Define, for all $x, y \in A^L$,*

$$x \leq^O y \stackrel{def.}{\Longleftrightarrow} (x \to y^-)^- = x \ (\stackrel{Theorem\ 9.2}{\Longleftrightarrow} x \wedge y = x),$$

where "O" comes from Ore (lattice). Then, we have:

(i) $(A, \leq^O, 0, 1)$ is a bounded po-set.

(ii) $x \leq^O y \Longrightarrow x \leq y$ (i.e. $x \to y = 1$, by (dfrelR)).

*(iii) The property (O-**') (if $x \leq^O y$ then $y \to z \leq^O x \to z$) holds.*

(iv) The property (O-K') ($x \leq^O y \to x$) holds.

Proof. (i): Reflexivity, i.e. for all $x \in A$, $x \leq^O x$: indeed, $x \leq^O x \stackrel{def.}{\Longleftrightarrow} (x \to x^-)^- = x$ and $(x \to x^-)^- \stackrel{(DN6)}{=} (x^-)^- \stackrel{(DN)}{=} x$.

Antisymmetry, i.e. for all $x, y \in A$, $x \leq^O y$ and $y \leq^O x$ imply $x = y$: indeed, by definition, $(x \to y^-)^- = x$ and $(y \to x^-)^- = y$, hence $x^- \stackrel{(DN)}{=} x \to y^- \stackrel{(Neg3)}{=} y \to x^- \stackrel{(DN)}{=} y^-$; hence, $x = y$, by (DN) again.

Transitivity, i.e. for all $x, y, z \in A$, $x \leq^O y$ and $y \leq^O z$ imply $x \leq^O z$: indeed, by definition, we have $(x \to y^-)^- = x$ and $(y \to z^-)^- = y$, hence $x \to y^- = x^-$ and $y \to z^- = y^-$, by (DN); then, $(x \to z^-)^- \stackrel{(Neg3)}{=} (z \to x^-)^-$
$= (z \to (x \to y^-))^- \stackrel{(Ex)}{=} (x \to (z \to y^-))^- \stackrel{(Neg3)}{=} (x \to (y \to z^-))^-$
$= (x \to y^-)^- = x$, i.e. $x \leq^O z$.

Thus, $\leq^O$ is an order relation.

$0 \leq^O x \stackrel{def.}{\Longleftrightarrow} (0 \to x^-)^- = 0$, that is true by (F) and (Neg0-1).

$x \leq^O 1 \stackrel{def.}{\Longleftrightarrow} (x \to 1^-)^- = x \Longleftrightarrow (x^-)^- \to 0 = x$, that is true by (Neg1-0), (Neg), (DN).

(ii): Suppose $x \leq^O y$, i.e. $(x \to y^-)^- = x$. Then, $x \to y = (x \to y^-)^- \to y \stackrel{(DN3)}{=}$ $y^- \to (x \to y^-) \stackrel{(Ex)}{=} x \to (y^- \to y^-) \stackrel{(Re)}{=} x \to 1 \stackrel{(L)}{=} 1$.

(iii): Suppose $x \leq^O y$, i.e. $(x \to y^-)^- = x$; we must prove that $y \to a \leq^O x \to a$, i.e. $(y \to a) \to (x \to a)^- = (y \to a)^-$, by (DN).
Indeed, $x \to a = (x \to y^-)^- \to a \stackrel{(DN3)}{=} a^- \to (x \to y^-)$
$\stackrel{(Ex)}{=} x \to (a^- \to y^-) \stackrel{(DN2)}{=} x \to (y \to a)$.
Then, $(y \to a) \to (x \to a)^- \stackrel{(Neg3)}{=} (x \to a) \to (y \to a)^-$
$= [x \to (y \to a)] \to (y \to a)^- \stackrel{(DN6)}{=} [x \to ((y \to a)^- \to (y \to a))] \to (y \to a)^-$
$\stackrel{(Ex)}{=} [(y \to a)^- \to (x \to (y \to a))] \to (y \to a)^- \stackrel{(impl)}{=} (y \to a)^-$.

(iv): We must prove that $x \leq^O y \to x$, i.e. $x \to (y \to x)^- = x^-$, by (DN). Indeed, $x \to (y \to x)^- \stackrel{(Neg3)}{=} (y \to x) \to x^- \stackrel{(DN2)}{=} (x^- \to y^-) \to x^- \stackrel{(impl)}{=} x^-$. □

Note that $\leq$ is reflexive, by (Re), but it is not antisymmetric or transitive, as the following examples show.

Examples 3.32. (See Examples 6.29)

Examples of left-i-ortholattices with six elements:

(1) The algebra $\mathcal{IO}_1 = (A_6 = \{0, a, b, c, d, 1\}, \to_1, ^{-1}, 1)$ with

$\to_1$	0	a	b	c	d	1
0	1	1	1	1	1	1
a	d	1	1	1	d	1
b	c	1	1	c	1	1
c	b	1	b	1	1	1
d	a	a	1	1	1	1
1	0	a	b	c	d	1

and

x	x^{-1}
0	1
a	d
b	c
c	b
d	a
1	0

is an i-ortholattice with the bounded po-set $(A_6, \leq^O, 0, 1)$ represented by the Hasse diagram from Figure 4. It does not verify (BB) for (a, d, b), (B) for (a, b, d), (**) for (a, d, b), (*) for (a, b, d), (An) for (a, b), (Tr) for (a, b, d), ($\vee$-comm) for (a, b).

(2) The algebra $\mathcal{IO}_2 = (A_6 = \{0, a, b, c, d, 1\}, \to_2, ^{-2}, 1)$ with

$\rightarrow_2$	0	a	b	c	d	1
0	1	1	1	1	1	1
a	b	1	b	1	1	1
b	a	a	1	1	1	1
c	d	1	1	1	d	1
d	c	1	1	c	1	1
1	0	a	b	c	d	1

and

x	x^{-2}
0	1
a	b
b	a
c	d
d	c
1	0

is an i-ortholattice with the bounded po-set $(A_6, \leq^O, 0, 1)$ represented by the same Hasse diagram from Figure 4. It does not verify (BB) for (a, b, c), (B) for (a, c, b), (**) for (a, b, c), (*) for (a, c, b), (An) for (b, c), (Tr) for (a, c, b), ($\vee$-comm) for (b, c).

Note that the i-ortholattices $\mathcal{IO}_1$ and $\mathcal{IO}_2$ correspond to the ortholattices $\mathcal{O}_1$ and $\mathcal{O}_2$, respectively, from Examples 6.29, by Theorem 9.2.

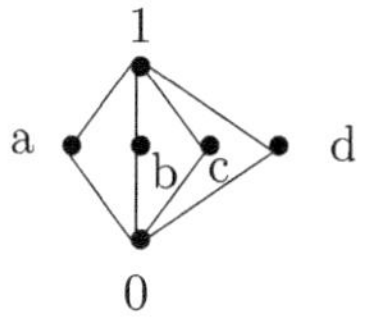

Figure 4: An i-ortholattice with 6 elements

Remarks 3.33.
(1) $(A, \leq^O, 0, 1)$ is in fact a bounded *Ore lattice* (see the ortholattices in Section 6).
(2) The converse of Proposition 3.31 (ii) does not hold. For example, in the i-ortholattice $\mathcal{IO}_1$ from Examples 3.32, we have $a \leq b$ (i.e. $a \rightarrow b = 1$), but $a \not\leq^O b$, since $(a \rightarrow b^-)^- = (a \rightarrow c)^- = 1^- = 0 \neq a$.

Proposition 3.34. *Let $\mathcal{A}^L = (A^L, \rightarrow, ^-, 1)$ be a left-i-ortholattice.*
If (@) $((y^- \rightarrow x) \rightarrow y = x \rightarrow y$, i.e. $(x \vee y) \wedge x^- = y \wedge x^-$, by Φ) holds, then we have:

(H) $((x \rightarrow y) \rightarrow y) \rightarrow y = x \rightarrow y,$
($\vee$-comm) $(x \rightarrow y) \rightarrow y = (y \rightarrow x) \rightarrow x,$
(EqOR) $x \leq^O y \Longleftrightarrow x \leq y.$

Proof. By Proposition 3.30, (DN2) and (DN3) hold. If (@) holds, then by (CC1), (@) + (DN2) + (DN3) imply (H), thus (H) holds.

By (CC2), (@) + (H) + (DN3) imply ($\vee$-comm); thus, ($\vee$-comm) holds.

By Proposition 3.31 (ii), $x \leq^O y$ implies $x \leq y$. Conversely, if $x \leq y$, i.e. $x \rightarrow y = 1$, then: $x \rightarrow y^- \overset{(@),(DN)}{=} (y \rightarrow x) \rightarrow y^- \overset{(DN2)}{=} (x^- \rightarrow y^-) \rightarrow y^- \overset{(\vee-comm)}{=}$

$(y^- \to x^-) \to x^- \overset{(DN2)}{=} (x \to y) \to x^- = 1 \to x^- \overset{(M)}{=} x^-$, hence $(x \to y^-)^- = x$, by (DN), i.e. $x \leq^O y$. Thus, (EqOR) holds. □

3.3.1 The connection between i-ortholattices and i-Boolean algebras

Theorem 3.35. *The i-ortholattices satisfying (@) are d.e. to the i-Boolean algebras.*

Proof. (1): Let $\mathcal{A} = (A, \to, ^-, 1)$ be an i-ortholattice satisfying the property (@). Then, by Proposition 3.34, ($\vee$-comm) holds. By (B31), ($\vee$-comm) + (Re) + (M) + (Ex) $\Longrightarrow$ (*). By Theorem 2.4, (BB) holds. By (B12), ($\vee$-comm) + (M) imply (An). Thus, (BB), (M), (L), (An) and (impl) hold, hence $(A, \to, 1)$ is an implicative BCK algebra (Definition 2), hence $(A, \to, 0, 1)$ is a bounded implicative BCK algebra; hence, $\mathcal{A}^L$ is an i-Boolean algebra (by Corollary 3.22).

(1'): Conversely, let $\mathcal{A} = (A, \to, ^-, 1)$ be an i-Boolean algebra. Then, $(A, \to, 0, 1)$ is a bounded, implicative BCK algebra, where $0 = 1^-$; hence, the properties (M), (Ex), (Re), (F), (L), (impl), (DN) hold; hence, (pimpl), (DN6) hold too. We prove that (@) holds. Indeed, by (CB0), (pimpl) + (Ex) + (DN6) imply (@). Thus, $\mathcal{A}$ is an i-ortholattice verifying (@). □

Theorem 3.35 gives us an equivalent definition of i-Boolean algebras:

Definition 3.36. (Definition 2) (The dual case is omitted)

A *left-i-Boolean algebra* is an algebra $\mathcal{A}^L = (A^L, \to, ^-, 1)$ of type $(2, 1, 0)$ such that $0 \overset{def.}{=} 1^-$ (i.e. (Neg1-0) holds), $x^- = x \to 0$ (i.e. (Neg) holds) and (M), (Ex), (Re), (L), (impl), (DN) and (@) hold.

3.4 Final remarks in the case $1^- = 0$

Resuming, we have the connections from Figure 5.

Remark 3.37.

(1) The Wajsberg algebras and the i-ortholattices are incomparable, since any Wajsberg algebra does not verify (impl) and any i-ortholattice does not verify ($\vee$-comm).

(2) The i-ortholattices are d.e. with the ortholattices, as we shall prove in Section 9, Theorem 9.2.

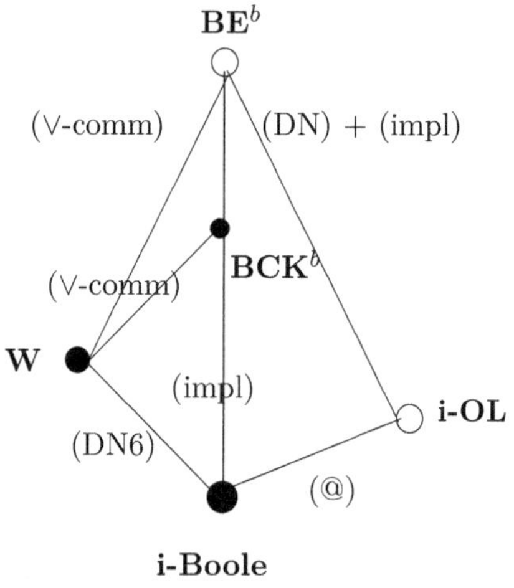

Figure 5: Connections between bounded BE algebras, bounded BCK algebras, Wajsberg algebras, i-ortholattices and i-Boolean algebras

4 Algebras $(A, \to, 1)$ or $(A, \to, ^{-1}, 1)$ with $1^{-1} = 1$ (Hierarchies 1 and 1')

Let $\mathcal{A}^L = (A, \to, 1)$ be an M algebra *without last element and not verifying (Ex).*

Let us introduce the weaker condition [42]:
(Ex1) $x \to (y \to 1) = y \to (x \to 1)$.

We shall say that $\mathcal{A}^L$ is *strong*, if it verifies (Ex1) [42].
Note that (Ex) implies (Ex1).

4.1 The list $\mathbf{C}_1$ of the properties of the negation $^{-1}$ and some results

Consider the following list $\mathbf{C}_1$ of properties that can be satisfied by the negation (inverse) $^{-1}$ defined on A^L:
(dfneg1) $x^{-1} \overset{def.}{=} x \to 1$,
(Neg$_1$) $x^{-1} = x \to 1$;

(DN) (Double Negation) $(x^{-1})^{-1} = x$;
(N1) $1^{-1} = 1$,

(Neg1) $(x \to y) \to (y^{-1} \to x^{-1}) = 1$,
(Neg1') $x \to y \leq y^{-1} \to x^{-1}$;
(Neg2) (contraposition) $x \to y = 1 \Longrightarrow y^{-1} \to x^{-1} = 1$,
(Neg2') (contraposition) $x \leq y \Longrightarrow y^{-1} \leq x^{-1}$;
(Neg3) $x \to y^{-1} = y \to x^{-1}$;
(Neg4) $x \to (x^{-1})^{-1} = 1$,
(Neg4') $x \leq (x^{-1})^{-1}$;
(Neg5) $(x \to y)^{-1} \to x = 1$,
(Neg5') $(x \to y)^{-1} \leq x$;
(Neg6) $x \to x^{-1} = x^{-1}$;

(DN1) $(y^{-1} \to x^{-1}) \to (x \to y) = 1$,
(DN1') $y^{-1} \to x^{-1} \leq x \to y$;
(DN2) $x \to y = y^{-1} \to x^{-1}$;
(DN3) $x^{-1} \to y = y^{-1} \to x$;
(DN4) $x \to y = 1 \Longleftrightarrow y^{-1} \to x^{-1} = 1$,
(DN4') $x \leq y \Longleftrightarrow y^{-1} \leq x^{-1}$;
(DN5) $x^{-1} \to (x \to y) = 1$,
(DN5') $x^{-1} \leq x \to y$;
(DN6) $x^{-1} \to x = x$.

Proposition 4.1. *We have:*
(1CE1) (dfneg$_1$) + (Ex1) $\Longrightarrow$ (Neg3),
(1CE1') (dfneg$_1$) + (Neg3) $\Longrightarrow$ (Ex1),
(1CE1") (dfneg$_1$) $\Longrightarrow$ ((Ex1) $\Leftrightarrow$ (Neg3));
(1CE2) (Neg$_1$) + (Ex1) $\Longrightarrow$ (Neg3),
(1CE2') (Neg$_1$) + (Neg3) $\Longrightarrow$ (Ex1),
(1CE2") (Neg$_1$) $\Longrightarrow$ ((Ex1) $\Leftrightarrow$ (Neg3)).

Proof. (1CE1): $x \to y^- \overset{(dfneg_1)}{=} x \to (y \to 0) \overset{(Ex1)}{=} y \to (x \to 0) \overset{(dfneg_1)}{=} y \to x^-$; thus, (Neg3) holds.

(1CE1'): $x \to (y \to 0) \overset{(dfneg_1)}{=} x \to y^- \overset{(Neg3)}{=} y \to x^- \overset{(dfneg_1)}{=} y \to (x \to 0)$; thus, (Ex1) holds.

(1CE1"): By (1CE1) and (1CE1').

(1CE2), (1CE2'), (1CE2"): similarly. □

Proposition 4.2. *(See Proposition 3.2)*

We have:
(1C1) (DN) + (Neg1) $\Longrightarrow$ (DN1), *(1C1') (DN) + (DN1) $\Longrightarrow$ (Neg1),*

(1C1") (DN) $\Longrightarrow$ ((Neg1) $\Leftrightarrow$ (DN1));

(1C2) (DN) + (DN2) $\Longrightarrow$ (DN3), *(1C2') (DN) + (DN3) $\Longrightarrow$ (DN2) ;*
(1C2") (DN) $\Longrightarrow$ ((DN2) $\Leftrightarrow$ (DN3));

(1C3) (DN) + (Neg3) $\Longrightarrow$ (DN3), *(1C3') (DN) + (DN3) $\Longrightarrow$ (Neg3),*
(1C3") (DN) $\Longrightarrow$ ((Neg3) $\Leftrightarrow$ (DN3));

(1C4) (DN2) $\Longrightarrow$ (DN4);

(1C5) (DN) + (DN2) + (Neg5) $\Longrightarrow$ (DN5),
(1C5') (DN) + (DN2) + (DN5) $\Longrightarrow$ (Neg5),
(1C5") (DN) + (DN2) $\Longrightarrow$ ((Neg5) $\Leftrightarrow$ (DN5));

(1C6) (DN) + (Neg6) $\Longrightarrow$ (DN6), *(1C6') (DN) + (DN6) $\Longrightarrow$ (Neg6),*
(1C6") (DN) $\Longrightarrow$ ((Neg6) $\Leftrightarrow$ (DN6)).

Proof. The same proof as for Proposition 3.2. $\square$

Lemma 4.3. *Let $\mathcal{A}^L = (A^L, \to, ^{-1}, 1)$ be an algebra of type $(2,1,0)$ verifying (M) and (Neg3) and such that $1^{-1} = 1$. Then, (Neg$_1$) holds.*

Proof. $x \to 1 = x \to 1^{-1} \stackrel{(Neg3)}{=} 1 \to x^{-1} \stackrel{(M)}{=} x^{-1}$; thus, (Neg$_1$) holds. $\square$

The following theorem allows us to say that, roughly speaking, the M algebras $(A^L, \to, 1)$ and $(A^L, \to, ^{-1}, 1)$ are definitionally equivalent.

Theorem 4.4.
(1) Let $\mathcal{A}^L = (A^L, \to, 1)$ be an algebra of type $(2,0)$ verifying (M) and (Ex1). Define a negation $^{-1}$ by (dfneg$_1$).
Then, $\alpha_1(\mathcal{A}^L) = (A^L, \to, ^{-1}, 1)$ is an algebra of type $(2,1,0)$ verifying (M) and (Neg3).
(1') Let $\mathcal{A}^L = (A^L, \to, ^{-1}, 1)$ be an algebra of type $(2,1,0)$ verifying (M) and (Neg3) and such that $1^{-1} = 1$.
Then, $\beta_1(\mathcal{A}^L) = (A^L, \to, 1)$ is an algebra of type $(2,0)$ verifying (M) and (Ex1).
(2) The maps α_1 and β_1 are inverse to each other.

Proof. (1): By (1CE1), (dfneg$_1$) + (Ex1) imply (Neg3); thus, (Neg3) holds.
(1'): By Lemma 4.3, (Neg$_1$) holds. Then, by (1CE2'), (Neg$_1$) + (Neg3) imply (Ex1); thus, (Ex1) holds.

(2): Let

$$(A^L, \to, 1) \quad \xrightarrow{\alpha_1} \quad (A^L, \to, ^{-1}, 1) \quad \xrightarrow{\beta_1} \quad (A^L, \to, 1).$$

Nothing to prove.

Conversely, let

$$(A^L, \to, ^{-1}, 1) \quad \xrightarrow{\beta_1} \quad (A^L, \to, 1) \quad \xrightarrow{\alpha_1} \quad (A^L, \to, ', 1).$$

Then, for all $x \in A^L$, we have: $x' \overset{(dfneg_1)}{=} x \to 1 = x \to 1^{-1} \overset{(Neg3)}{=} 1 \to x^{-1} \overset{(M)}{=} x^{-1}$.
□

Proposition 4.5. *(See Proposition 3.8) We have:*
(i) (DN) + (Ex1) + (dfneg$_1$) $\Longrightarrow$ ((*) $\Leftrightarrow$ (**)),
(ii) (DN) + (Ex1) + (Neg$_1$) $\Longrightarrow$ ((*) $\Leftrightarrow$ (**)).

Proof. (i): First note that (Ex1) + (dfneg$_1$) imply (Neg3), by (1CE1); then, (DN) + (Neg3) imply (DN2), by (1C3) and (1C2"); then, (DN2) implies (DN4), by (1C4).
If $x \leq y$, then by (DN4'), $y^{-1} \leq x^{-1}$; then, by (*'), $z^{-1} \to y^{-1} \leq z^{-1} \to x^{-1}$, hence $y \to z \leq x \to z$, by (DN2); thus, (**') holds. Conversely, if $x \leq y$, then $y^{-1} \leq x^{-1}$, by (DN4'); then, by (**'), $x^{-1} \to z^{-1} \leq y^{-1} \to z^{-1}$, hence $z \to x \leq z \to y$, by (DN2); thus, (*') holds.

(ii): Similarly. □

4.2 The p-semisimple algebras

We have seen in Section 2 that:

Definition 4.6. The algebra $(A^L, \to, 1)$ verifying (Re) is said to be *p-semisimple*, if for all $x, y \in A^L$,
(p-s) $x \leq y \Longrightarrow x = y$ (or, equivalently, by (Re), $x \leq y \Longleftrightarrow x = y$).

The subclass of p-semisimple algebras of the class $\mathbf{X}$ will be denoted by $\mathbf{X}_{(p-s)}$.

Consider the following additional properties of the list $\mathbf{A}$:
(Eq=) $x = y \Longleftrightarrow x \to y = 1$;
(BB=) $y \to z = (z \to x) \to (y \to x)$,
(B=) $y \to z = (x \to y) \to (x \to z)$,
(D=) $x = (x \to y) \to y$,
(D=1) $x = (x \to 1) \to 1$,
(DB=) $x = (y \to 1) \to (y \to x)$.

Remark 4.7. *The equivalence (Eq=) was introduced in [42], Chapter 8 (in the non-commutative case).*

Proposition 4.8. *(See [42], Proposition 8.1.2)*
Let $\mathcal{A}^L = (A^L, \to, 1)$ be an algebra verifying (Re). We have:
(0) (dfrelR) + (p-s) $\Longrightarrow$ (Eq=),
(0') (dfrelR) + (Eq=) $\Longrightarrow$ (p-s),
(0") (dfrelR) $\Longrightarrow$ ((p-s) $\Leftrightarrow$ (Eq=));
(1) (p-s) $\Longrightarrow$ (An'),
(1') (p-s) $\Longrightarrow$ ('),*
*(1") (p-s) $\Longrightarrow$ (**'),*
(1"') (p-s) $\Longrightarrow$ (Tr');
(2) (p-s) + (BB') $\Longrightarrow$ (BB=),
(2') (p-s) + (B') $\Longrightarrow$ (B=),
(2") (p-s) + (D') $\Longrightarrow$ (D=);
(3) (Ex1) + (Re) + (Eq=) + (dfneg1) $\Longrightarrow$ (DN);
(4) (Eq=) $\Longrightarrow$ (Re).

Proof. (0), (0'), (0"): Obviously. (1), (1'), (1"), (1"'): Obviously.
(2), (2'), (2"): Immediately.
(3): $x \to (x^{-1})^{-1} \overset{(dfneg1)}{=} x \to (x^{-1} \to 1) \overset{(Ex1)}{=} x^{-1} \to (x \to 1) \overset{(dfneg1)}{=} x^{-1} \to x^{-1} \overset{(Re)}{=} 1$; then, by (Eq=), we have: $(x^{-1})^{-1} = x$, i.e. (DN) holds.
(4): (Re) $\overset{def.}{\Longleftrightarrow} x \to x = 1 \overset{(Eq=)}{\Longleftrightarrow} x = x$, that is true. □

Proposition 4.9. *(See [42], Proposition 8.1.2)*
We have:
(1A1) (Ex) + (Re) + (Eq=) $\Longrightarrow$ (D=),
(1A1') (Ex) + (Re) + (Eq=) $\Longrightarrow$ (BB=); (Ex) + (D=) $\Longrightarrow$ (BB=);
(1A2) (Ex) + (Eq=) + (BB=) $\Longrightarrow$ (B=),
(1A2') (Ex) + (Eq=) + (B=) $\Longrightarrow$ (BB=),
(1A2") (Ex) + (Eq=) $\Longrightarrow$ ((B=) $\Leftrightarrow$ (BB=));
(1A3) (D=) + (Eq=) $\Longrightarrow$ (M);
(1A4) (M) + (BB=) $\Longrightarrow$ (Re),
(1A5) (M) + (BB=) $\Longrightarrow$ (D=),
(1A6) (Re) + (D=) $\Longrightarrow$ (Eq=),
(1A7) (M) + (BB=) $\Longrightarrow$ (Ex).

Proof.
(1A1): $y \to [(y \to x) \to x] \overset{(Ex)}{=} (y \to x) \to (y \to x) \overset{(Re)}{=} 1$; then, by (Eq=), it follows that $y = (y \to x) \to x$, i.e. (D=) holds.

(1A1'): By (1A1), (Ex) + (Re) + (Eq=) $\Longrightarrow$ (D=), i.e. $z = (z \to x) \to x$; then, $y \to z = y \to [(z \to x) \to x] \overset{(Ex)}{=} (z \to x) \to (y \to x)$; thus, (BB=) holds.

(1A2): By (BB=), $y \to z = (z \to x) \to (y \to x)$; then, by (Eq=), $(y \to z) \to [(z \to x) \to (y \to x)] = 1$; then, by (Ex), $(z \to x) \to [(y \to z) \to (y \to x)] = 1$; then, by (Eq=) again, $z \to x = (y \to z) \to (y \to x)$, i.e. (B=) holds.

(1A2'): By (B=), $z \to x = (y \to z) \to (y \to x)$; then, by (Eq=), $(z \to x) \to [(y \to z) \to (y \to x)] = 1$; then, by (Ex), $(y \to z) \to [(z \to x) \to (y \to x)] = 1$; then, by (Eq=) again, $y \to z = (z \to x) \to (y \to x)$, i.e. (BB=) holds.

(1A2"): By (1A1) and (1A2').

(1A3): $1 \to x = x \overset{(Eq=)}{\Longleftrightarrow} (1 \to x) \to x = 1 \overset{(D=)}{\Longleftrightarrow} 1 = 1$ that is true.

(1A4): $x \to x \overset{(M)}{=} (1 \to x) \to (1 \to x) \overset{(BB=}{=} 1 \to 1 \overset{(M}{=} 1$, i.e. (Re) holds.

(1A5): $(x \to y) \to y \overset{(M)}{=} (x \to y) \to (1 \to y) \overset{(BB=)}{=} 1 \to x \overset{(M)}{=} x$, i.e. (D=) holds.

(1A6): If $x = y$, then $x \to y = y \to y \overset{(Re)}{=} 1$. Conversely, if $x \to y = 1$, then $x \overset{(D=)}{=} (x \to y) \to y = 1 \to y \overset{(Re)}{=} (y \to y) \to y \overset{(D=)}{=} y$. Thus, (Eq=) holds.

(1A7): By (1A4), (M) + (BB=) $\Longrightarrow$ (Re), thus (Re) holds.
By (1A5), (M) + (BB=) $\Longrightarrow$ (D=), thus (D=) holds.
By (1A6), (Re) + (D=) $\Longrightarrow$ (Eq=), thus (Eq=) holds.
By (BB=), $Y \to Z = (Z \to X) \to (Y \to X)$;
take $X = x \to z$, $Y = y$ and $Z = (y \to z) \to z \overset{(D=)}{=} y$; we obtain:
$1 \overset{(Re)}{=} y \to y \overset{(D=)}{=} y \to [(y \to z) \to z]$
$\overset{(BB=)}{=} [((y \to z) \to z) \to (x \to z)] \to [y \to (x \to z)]$
$\overset{(BB=)}{=} [x \to (y \to z)] \to [y \to (x \to z)]$;
then, by (Eq=), $x \to (y \to z) = y \to (x \to z)$, i.e. (Ex) holds. □

Proposition 4.10. *(Following the numbering from Proposition 4.9)*

We have:
(1AB0) (dfneg1) + (D=1) $\Longrightarrow$ (DN), (Neg_1) + (DN) $\Longrightarrow$ (D=1),
(1AB00) (D=) $\Longrightarrow$ (D=1);
(1AB1) (Ex) + (Re) + (Eq=) $\Longrightarrow$ (DB=),
(1AB1') (Ex) + (Re) + (Eq=) $\Longrightarrow$ (B=);
(1AB5) (B=) + (DB=) $\Longrightarrow$ (M),
(1AB5') (B=) + (M) $\Longrightarrow$ (DB=),
(1AB5") (B=) $\Longrightarrow$ ((DB=) $\Leftrightarrow$ (M));
(1AB6) (Re) + (DB=) $\Longrightarrow$ (Eq=),
(1AB6') (Re) + (DB=) $\Longrightarrow$ (D=1);
(1AB7) (M) + (B=) + (Eq=) $\Longrightarrow$ (D=),

(1AB7') (M) + (B=) + (Eq=) $\Longrightarrow$ (Ex),
(1AB7") (M) + (B=) + (Eq=) $\Longrightarrow$ (BB=).

Proof.

(1AB0): Immediately.

(1AB00): Take $y = 1$ in (D=).

(1AB1): $y \to [(x \to 1) \to (x \to y)] \overset{(Ex)}{=} (x \to 1) \to (y \to (x \to y))$
$\overset{(Ex)}{=} (x \to 1) \to [x \to (y \to y)] \overset{(Re)}{=} (x \to 1) \to (x \to 1) \overset{(Re)}{=} 1$;
hence, by (Eq=), $y = (x \to 1) \to (x \to y)$, i.e. (DB=) holds.

(1AB1'): By (1A1), (Ex) + (Re) + (Eq=) imply (D=);
by (1AB1), (Ex) + (Re) + (Eq=) imply (DB=).
By (DB=), $z = (x \to 1) \to (x \to z)$; then,
$y \to z = y \to [(x \to 1) \to (x \to z)]$
$\overset{(Ex)}{=} (x \to 1) \to [y \to (x \to z)]$
$\overset{(Re)}{=} (x \to (y \to y)) \to [y \to (x \to z)]$
$\overset{(Ex)}{=} [y \to (x \to y)] \to [y \to (x \to z)]$.
It remains to prove that

$$[y \to (x \to y)] \to [y \to (x \to z)] = (x \to y) \to (x \to z),$$

i.e., by (Eq=), that

$$A \overset{notation}{=} [(x \to y) \to (x \to z)] \to [(y \to (x \to y)) \to (y \to (x \to z))] = 1 \qquad (1)$$

Indeed, $A \overset{(Ex)}{=} [y \to (x \to y)] \to [((x \to y) \to (x \to z)) \to (y \to (x \to z))]$
$\overset{(Ex)}{=} [y \to (x \to y)] \to [y \to (((x \to y) \to (x \to z)) \to (x \to z))]$
$\overset{(D=)}{=} [y \to (x \to y)] \to [y \to (x \to y)] \overset{(Re)}{=} 1$, hence (5) holds.
Thus, $y \to z = (x \to y) \to (x \to z)$, i.e. (B=) holds.

(1AB5): $1 \to x \overset{(B=)}{=} (y \to 1) \to (y \to x) \overset{(DB=)}{=} x$, i.e. (M) holds.

(1AB5'): $(x \to 1) \to (x \to y) \overset{(B=)}{=} 1 \to y \overset{(M)}{=} y$, i.e. (DB=) holds.

(1AB5"): By (1AB5) and (1AB5').

(1AB6): If $x = y$, then $x \to y = y \to y \overset{(Re)}{=} 1$. Conversely, if $x \to y = 1$, then $y \overset{(DB=)}{=} (x \to 1) \to (x \to y) = (x \to 1) \to 1 \overset{(Re)}{=} (x \to 1) \to (x \to x) \overset{(DB=)}{=} x$. Thus, (Eq=) holds.

(1AB6'): $x \overset{(DB=)}{=} (y \to 1) \to (y \to x)$ and for $y = x$, we obtain:
$x = (x \to 1) \to (x \to x) \overset{(Re)}{=} (x \to 1) \to 1$; thus, (D=1) holds.

(1AB7): By Proposition 4.8 (4), (Eq=) implies (Re); by (1AB5'), (M) + (B=) $\Longrightarrow$ (DB=); by (1AB6'), (DB=) + (Re) $\Longrightarrow$ (D=1).
First, we prove

$$(z \to x) \to (y \to x) = [(y \to z) \to (z \to x)] \to (z \to x). \tag{2}$$

Indeed, if $T \overset{notation}{=} (z \to x) \to (y \to x) \overset{(B=)}{=} [X \to (z \to x)] \to [X \to (y \to x)]$, then, for $X = y \to z$, we obtain:
$T = [(y \to z) \to (z \to x)] \to [(y \to z) \to (y \to x)]$
$\overset{(B=)}{=} [(y \to z) \to (z \to x)] \to (z \to x)$, i.e. (2) holds.
Second, we prove

$$x \to y = (y \to x) \to 1. \tag{3}$$

Indeed, $x \to y \overset{(B=)}{=} (X \to x) \to (X \to y)$ and for $X = y$, we obtain:
$x \to y = (y \to x) \to (y \to y) \overset{(Re)}{=} (y \to x) \to 1$, i.e. (3) holds.
Now we prove

$$x = [[(x \to y) \to y] \to (y \to 1)] \to (y \to 1). \tag{4}$$

Indeed, $x \overset{(DB=)}{=} (y \to 1) \to (y \to x) \overset{(3)}{=} (y \to 1) \to ((x \to y) \to 1)$
$\overset{(2)}{=} [[(x \to y) \to y] \to (y \to 1)] \to (y \to 1)$, i.e. (4) holds.
Finally, suppose that $(x \to y) \to y = a$; then, $x \overset{(4)}{=} [a \to (y \to 1)] \to (y \to 1)$ and for $y = 1$, we obtain:
$x = [a \to (1 \to 1)] \to (1 \to 1) \overset{(Re)}{=} [a \to 1] \to 1 \overset{(D=1)}{=} a$, hence $(x \to y) \to y = x$, i.e. (D=) holds.

(1AB7'): By (1AB7), (M) + (B=) + (Eq=) $\Longrightarrow$ (D=). Then,
$y \to (x \to z) \overset{(B=)}{=} (X \to y) \to (X \to (x \to z))$ and for $X = x \to y$, we obtain:
$y \to (x \to z) = ((x \to y) \to y) \to ((x \to y) \to (x \to z))$
$\overset{(D=)}{=} x \to ((x \to y) \to (x \to z)) \overset{(B=)}{=} x \to (y \to z)$, i.e. (Ex) holds.

(1AB7"): By (1AB7), (M) + (B=) + (Eq=) $\Longrightarrow$ (D=). Then,
$(z \to x) \to (y \to x) \overset{(2)}{=} [(y \to z) \to (z \to x)] \to (z \to x) \overset{(D=)}{=} y \to z$, i.e. (BB=) holds. □

Theorem 4.11. *(See Proposition 4.14) If (Ex), (Eq=) hold, then:*

$$(B =) \Longleftrightarrow (BB =).$$

Proof. By (1A2"). □

Proposition 4.12. *We have:*
(1CN1) (Ex) + (Re) + (Eq=) + (dfneg1) $\Longrightarrow$ (DN),
(1CN2) (DN) + (Ex) + (DN3) + (Re) + (dfneg1) $\Longrightarrow$ (Eq=).

Proof.
(1CN1): By (1A1), (Ex) + (Re) + (Eq=) imply (D=); then, (D=) implies (D=1), by (1AB00), and (D=1) + (dfneg1) imply (DN), by (1AB0); thus, (DN) holds.

(1CN2): If $x = y$, then $x \to y = x \to x \stackrel{(Re)}{=} 1$. Conversely, if $x \to y = 1$, then: $x \stackrel{(DN)}{=} (x^{-1})^{-1} \stackrel{(dfneg1)}{=} x^{-1} \to 1 = x^{-1} \to (x \to y) \stackrel{(Ex)}{=} x \to [x^{-1} \to y] \stackrel{(DN3)}{=} x \to [y^{-1} \to x] \stackrel{(Ex)}{=} y^{-1} \to (x \to x) \stackrel{(Re)}{=} y^{-1} \to 1 = (y^{-1})^{-1} \stackrel{(DN)}{=} y$. Thus, (Eq=) holds. □

In [42], the notion of *implicative-group* was introduced, in the non-commutative case, as a term equivalent definition of the group; there are five equivalent definitions of implicative-groups:

Definition 4.13. [42] A *commutative implicative-group*, or a *commutative i-group* for short, is an algebra $\mathcal{G} = (G, \to, 1)$ of type $(2, 0)$ verifying one of the following:
- (M), (BB=) (Definition 1) ([42] Definition 8.1.1),
- (M), (B=), (Eq=) (Definition 2) (note that [42] Definition 8.1.10 is incomplete),
- (BB=), (D=), (Eq=) (Definition 3),
- (M), (BB=), (Eq=) (Definition 4),
- (Ex), (Eq=) (Definition 5) ([42] Definition 8.1.14).

We prove the following results, their ideea coming from Remarks 7.1 (ii).

Proposition 4.14. *The p-semisimple left-RME = left-CI algebra coincides with the commutative i-group.*

Proof.
(1): Let $\mathcal{G} = (G, \to, 1)$ be a commutative i-group, i.e., by Definition 5, (Ex), (Eq=) hold. We must prove that (M), (Re) and (p-s) hold. Indeed, by Proposition 4.8 (4), (Eq=) implies (Re); thus, (Re) holds; also, by Proposition 4.8 (0'), (dfrelR) + (Eq=) imply (p-s), thus (p-s) holds; then, by (1A1), (Ex) + (Re) + (Eq=) imply (D=) and by (1A3), (D=) + (Eq=) imply (M), thus (M) holds. Thus, $\mathcal{G}$ is a p-semisimple left-CI algebra.

(2): Let $\mathcal{A}^L = (A^L, \to, 1)$ be a p-semisimple left-CI algebra, i.e. (M), (Ex), (Re), (p-s) hold. We prove that (Eq=) holds. Indeed, by Proposition 4.8, (0), (dfrelR) + (p-s) imply (Eq=). Thus, $\mathcal{A}^L$ is a commutative i-group, by Definition 5. □

Theorem 4.15. *We have:*
(i) Any p-semisimple left-CI algebra is involutive.
(ii) Any involutive left-CI algebra is p-semisimple.
(iii) The p-semisimple left-CI algebras coincide with the involutive left-CI algebras.

Proof. (i): Let $\mathcal{A}^L = (A^L, \to, 1)$ be a p-semisimple left-CI algebra, i.e. (M), (Ex), (Re), (p-s) hold. Then, by Proposition 4.8 (0), (dfrelR) + (p-s) imply (Eq=); by (1CN1), (Ex) + (Re) + (Eq=) + (dfneg1) $\Longrightarrow$ (DN), thus (DN) holds, where $x^{-1} \overset{def.}{=} x \to 1$.

(ii): Let $\mathcal{A}^L = (A^L, \to, 1)$ be an involutive left-CI algebra, i.e. (M), (Ex), (Re), (Neg$_1$), (DN) hold, where $x^{-1} \overset{def.}{=} x \to 1$, by (dfneg1). Then, (Neg$_1$) + (Ex) imply (Neg3) and (DN) + (Neg3) imply (DN3), by (1C3). Now, by (1CN2), (DN) + (Ex) + (DN3) + (Re) + (dfneg1) imply (Eq=) and by Proposition 4.8 (0'), (dfrelR) + (Eq=) imply (p-s); thus, (p-s) holds.

(iii): By above (i), (ii). □

We write:

$\mathbf{CI}_{(p-s)} = \mathbf{CI}_{(DN)} =$ **commutative i-group**.

Corollary 4.16. *The p-semisimple CI**, pre-BCI, BCH, BCH**, BCI algebras coincide with the involutive corresponding ones.*

Proof. All these algebras are p-semisimple CI algebras, by definition; then, apply Theorem 4.15 (iii). □

Corollary 4.17.
*(i) The involutive CI algebras coincide with the involutive CI** algebras.*
*(ii) The involutive BCH algebras coincide with the involutive BCH** algebras.*

Proof. (i): Let $\mathcal{A}^L = (A^L, \to, 1)$ be an involutive left-CI algebra, i.e. (M), (Ex), (Re), (Neg$_1$), (DN) hold, where $x^{-1} = x \to 1$, by (dfneg1). By Theorem 4.15 (ii), (p-s) holds and by Proposition 4.8 (1"), (**) holds; thus, $\mathcal{A}$ is a left-CI** algebra. Since, conversely, any involutive left-CI** algebra is an involutive left-CI algebra, they coincide.

(ii): Let $\mathcal{A}^L = (A^L, \to, 1)$ be an involutive left-BCH algebra; then, $\mathcal{A}^L$ is an involutive left-CI algebra, by definition; then, by above (i), (p-s) and (**) hold; thus, $\mathcal{A}$ is a left-BCH** algebra. Since, conversely, any involutive left-BCH** algebra is an involutive left-BCH algebra, they coincide. □

Moreover, we have:

Proposition 4.18. *Let $\mathcal{A}^L = (A^L, \to, 1)$ be a p-semisimple left-RME = left-CI algebra. Then, $\mathcal{A}^L$ is a (p-semisimple) left-BCI algebra.*

Proof. Let $\mathcal{A}^L = (A^L, \to, 1)$ be a p-semisimple left-RME= left-CI algebra, i.e. (M), (Ex), (Re) and (p-s) hold. We prove that (An) and (BB=) hold. Indeed, by Proposition 4.8 (1), (p-s) implies (An'), thus (An) holds. By Proposition 4.8 (0), (dfrelR) + (p-s) imply (Eq=) and by (1A1'), (Ex) + (Re) + (Eq=) imply (BB=), thus (BB=) holds. Thus, $\mathcal{A}^L$ is a p-semisimple left-BCI algebra, by Definition 3. □

Since conversely, any p-semisimple left-BCI algebra is a p-semisimple left-CI algebra, then, by Propositions 4.14, 4.18 and Theorem 4.15, we obtain:

Corollary 4.19. *(See [42], Remark 13.1.16) (See the inspireing Corollary 7.18)*
*The p-semisimple (and involutive) left-CI, left-CI**, left-pre-BCI, left-BCH, left-BCH** and left-BCI algebras coincide (with the commutative i-groups).*

We write:
$\mathbf{CI}_{(p-s)} = \mathbf{CI}_{(DN)} = \mathbf{CI^{**}}_{(p-s)} = \mathbf{CI^{**}}_{(DN)} = \mathbf{pre\text{-}BCI}_{(p-s)} = \mathbf{pre\text{-}BCI}_{(DN)} = \mathbf{BCH}_{(p-s)} = \mathbf{BCH}_{(DN)} = \mathbf{BCH^{**}}_{(p-s)} = \mathbf{BCH^{**}}_{(DN)} = \mathbf{BCI}_{(p-s)} = \mathbf{BCI}_{(DN)}$ = **commutative i-group.**

By Corollary 4.19, it follows that the left-CI, left-CI**, left-pre-BCI, left-BCH, left-BCH**, left-BCI algebras are all generalizations *with the exchange property (Ex)* of the commutative i-group, while the left-RM, left-pre-BZ, left-BH, left-BZ algebras are all generalizations *without the exchange property (Ex)* of the commutative i-group.

Proposition 4.20.
(i) The p-semisimple RM algebras coincide with the p-semisimple aRM=BH algebras.
(ii) The p-semisimple pre-BZ algebras coincide with the p-semisimple BZ algebras.

Proof. By Proposition 4.8 (1), (p-s) implies (An), then use the definitions. □
We write:
$\mathbf{RM}_{(p-s)} = \mathbf{BH}_{(p-s)}$ and $\mathbf{pre\text{-}BZ}_{(p-s)} = \mathbf{BZ}_{(p-s)}$.

In [42] also, the notions of *implicative-hub* ([42] Definition 11.1.1), *implicative-moon* ([42] Definition 11.1.10) and *implicative-goop* ([42] Definition 11.1.35) were introduced and studied, in the non-commutative case, as generalizations *without the exchange property* of the i-group. In the commutative case, we have the following definitions:

Definitions 4.21.

(i) A *commutative implicative-hub*, or a *commutative i-hub* for short, is an algebra $\mathcal{G} = (G, \rightarrow, 1)$ of type $(2, 0)$ verifying (M), (Eq=).

(ii) A *commutative implicative-moon*, or a *commutative i-moon* for short, is an algebra $\mathcal{G} = (G, \rightarrow, ^{-1}, 1)$ of type $(2, 1, 0)$ verifying (M), (Eq=) and (Neg_1).

(iii) A *commutative implicative-goop*, or a *commutative i-goop* for short, is a strong commutative i-moon (i.e. verifying (Ex1)).

Proposition 4.22. *The commutative i-hubs are definitionally equivalent (d.e.) to the commutative i-moons.*

Proof. Immediately. □

Proposition 4.23. *The p-semisimple left-BH algebras are d.e. (coincide) with the commutative i-hubs.*

Proof. (1): Let $\mathcal{A}^L = (A^L, \rightarrow, 1)$ be a p-semisimple left-BH algebra, i.e. (M), (Re), (An) and (p-s) hold. By Proposition 4.8 (0), (dfrelR) + (p-s) imply (Eq=); thus, $\mathcal{A}^L$ is a commutative i-hub.

(1'): Conversely, let $\mathcal{A}^L = (A^L, \rightarrow, 1)$ be a commutative i-hub, i.e. (M) and (Eq=) hold. By Proposition 4.8 (4), (Eq=) implies (Re), thus (Re) holds; by Proposition 4.8 (0'), (dfrelR) + (Eq=) imply (p-s) and by (1), (p-s) implies (An'), thus (An) holds. Thus, $\mathcal{A}^L$ is a p-semisimple left-BH algebra. □

Corollary 4.24. *The strong p-semisimple left-BH algebras are d.e. (coincide) with the commutative i-goops.*

Proof. Since the p-semisimple left-BH algebras are d.e. to the commutative i-moons, by Propositions 4.23 and 4.22. □

Proposition 4.25. *Any strong p-semisimple left-BH algebra (i.e. any commutative i-goop) is involutive.*

Proof. Let $\mathcal{A}^L = (A^L, \rightarrow, 1)$ be a strong p-semisimple left-BH algebra, i.e. (M), (Re), (An), (p-s), (Ex1) hold. By Proposition 4.8 (0), (3), (dfrelR) + (p-s) imply (Eq=) and (Ex1) + (Re) + (Eq=) imply (DN), thus (DN) holds, where $x^{-1} \stackrel{def.}{=} x \rightarrow 1$. □

The converse of Proposition 4.25 does not hold: there are involutive p-semisimple left-BH algebras (i.e. involutive commutative i-moons) that are not strong; see the following example (2).

Examples 4.26. (See Examples 7.22)

(1) $\mathcal{A}_4^1 = (A_4 = \{a, b, c, 1\}, \to_1, ^{-1}, 1)$, with

$\to_1$	a	b	c	1
a	1	a	a	a
b	a	1	a	a
c	b	c	1	a
1	a	b	c	1

and

x	x^{-1}
a	a
b	a
c	a
1	1

, is a **commutative i-moon that is not involutive**, because (DN) is not verified for $x = b$.

Equivalently, $(A_4 = \{a, b, c, 1\}, \to_1, 1)$, with the above table of $\to_1$, is a **p-semisimple left-BH algebra that is not involutive**, since (M), (Re), (An), (p-s) hold, while (Ex) does not hold for (a, b, a), (Ex1) for (a, b), (B) for (a, b, c), (BB) for $(a, b, 1)$, (DN) for $x = b$, where $x^{-1} \stackrel{def.}{=} x \to 1$.

(2) $\mathcal{A}_4^2 = (A_4 = \{a, b, c, 1\}, \to_2, ^{-1}, 1)$, with

$\to_2$	a	b	c	1
a	1	a	b	c
b	a	1	b	b
c	a	a	1	a
1	a	b	c	1

and

x	x^{-1}
a	c
b	b
c	a
1	1

, is an **involutive commutative i-moon that is not strong** (see [42], Examples 11.1.40), because (Ex1) is not verified for $(x, y) = (a, b)$: $a = a \to b = a \to (b \to 1) \neq b \to (a \to 1) = b \to c = b$.

Equivalently, $(A_4 = \{a, b, c, 1\}, \to_2, 1)$, with the above table of $\to_2$, is an **involutive p-semisimple left-BH algebra that is not strong**, since (M), (Re), (An), (p-s), (DN) hold, where $x^{-1} \stackrel{def.}{=} x \to 1$, while (Ex) does not hold for (a, b, a), (Ex1) for (a, b), (B) for (a, b, a), (BB) for (a, a, b). Note that Theorem 9.3 cannot be applied, because (Ex1) is not verified.

(3) $\mathcal{A}_4^3 = (A_4 = \{a, b, c, 1\}, \to_3, ^{-1}, 1)$, with

$\to_3$	a	b	c	1
a	1	b	b	b
b	b	1	a	a
c	a	b	1	c
1	a	b	c	1

and

x	x^{-1}
a	b
b	a
c	c
1	1

, is a **commutative i-goop**.

Equivalently, $(A_4 = \{a, b, c, 1\}, \to_3, 1)$, with the above table of $\to_3$, is a **strong p-semisimple left-BH algebra**, since (M), (Re), (An), (p-s), (Ex1) (hence (DN)) hold (where $x^{-1} \stackrel{def.}{=} x \to 1$), while (Ex) does not hold for (a, b, a), (B) for (a, b, a), (BB) for (a, a, b). Note that Theorem 9.3 can now be applied, because (Ex1) is verified, and we obtain the algebra from Examples 7.22 (3m).

Thus, the connections between the p-semisimple left-BH algebras (commutative i-moons), the involutive p-semisimple left-BH algebras (involutivecommutative i-moons) and the strong p-semisimple left-BH algebras (commutative i-goops) are presented in the following Figure 6 (see [42], Figure 11.4, in the non-commutative case).

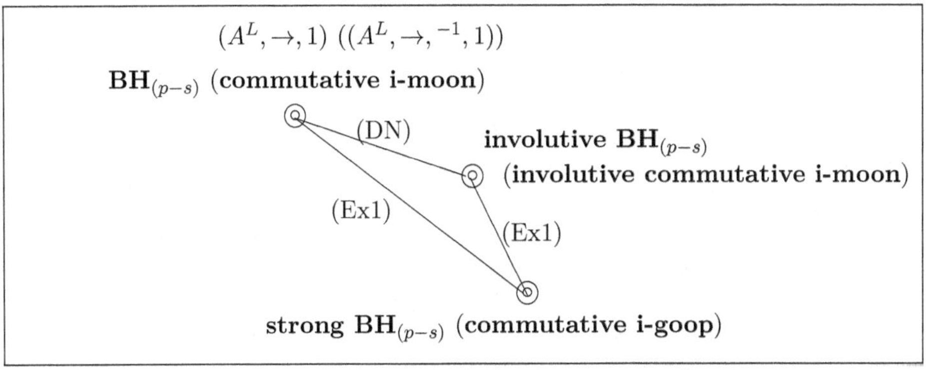

Figure 6: The hierarchy between commutative i-moons, involutive commutative i-moons and commutative i-goops

Theorem 4.27. *(See [42], Remark 13.1.16)*

The p-semisimple BZ algebras (pre-BZ algebras) and the p-semisimple BCI algebras coincide (with the commutative i-groups).

Proof. By Proposition 4.20 (ii), the p-semisimple pre-BZ, BZ algebras coincide.

We prove that any p-semisimple BZ algebra is a (p-semisimple) BCI algebra. Indeed, let $\mathcal{A}^L = (A^L, \rightarrow, 1)$ be a p-semisimple left-BZ algebra, i.e. (M), (Re), (B), (An), (p-s) hold. By Proposition 4.8 (0) and (2'), (Eq=) and (B=) hold; then, by (1AB7"), (M) + (B=) + (Eq=) imply (BB=), hence $\mathcal{A}^L$ is a p-semisimple left-BCI algebra, i.e. a commutative i-group, by Corollary 4.19. Since, obviously, any p-semisimple left-BCI algebra is a (p-semisimple) left-BZ algebra, the proof is complete. □

We write:

pre-BZ$_{(p-s)}$ = **BZ**$_{(p-s)}$ = **BCI**$_{(p-s)}$ = **commutative i-group**.

Note that we have no an example of involutive BZ algebra that is not p-semisimple (commutative i-group).

5 Preliminaries on algebras

Recall the following notions from the 'world' of *commutative algebras*, needed in this paper.

5.1 Commutative unital magmas and monoids

Definition 5.1.

A *commutative unital magma* is an algebra $(A, \otimes, e)$ of type $(2, 0)$ verifying the axioms: for all $x, y \in A$,

(U) $e \otimes x = x (= x \otimes e)$, i.e. e is the *unit* element of A,

(comm) $x \otimes y = y \otimes x$ (commutativity).

Note that a commutative unital magma can be denoted:
- multiplicatively, by $(A^L, \odot, 1)$ or $(A^L, \cdot, 1)$ or $(A^L, \wedge, 1)$, in which case the unit 1 is called the *identity* element and the unital magma is called *left*, or
- additively, by $(A^R, \oplus, 0)$ or $(A^R, +, 0)$ or $(A^R, \vee, 0)$, in which case the unit 0 is called the *neutral* element and the unital magma is called *right*.

Denote by **cUm** the class of all commutative left-unital magmas, where 'm' comes from 'magma'. Dually, denote by $\mathbf{cUm}^R$ the class of all commutative right-unital magmas.

Consequently, we define the following "dual" notions:

Definitions 5.2.

(i) A *commutative left-monoid* is an algebra $(A^L, \odot, 1)$ of type $(2, 0)$ verifying the axioms: for all $x, y, z \in A^L$,

(PU) (unit element of product) $1 \odot x = x \ (= x \odot 1)$,

(Pcomm) (commutativity of product) $x \odot y = y \odot x$,

(Pass) (associativity of product) $x \odot (y \odot z) = (x \odot y) \odot z$,

i.e. it is a commutative left-unital magma verifying (Pass).

(i') A *commutative right-monoid* is an algebra $(A^R, \oplus, 0)$ of type $(2, 0)$ verifying the axioms: for all $x, y, z \in A^R$,

(SU) (unit element of sum) $0 \oplus x = x \ (= x \oplus 0)$,

(Scomm) (commutativity of sum) $x \oplus y = y \oplus x$,

(Sass) (associativity of sum) $x \oplus (y \oplus z) = (x \oplus y) \oplus z$,

i.e. it is a commutative right-unital magma verifying (Sass).

Denote by **monoid** the class of all commutative left-monoids and, dually, denote by $\mathbf{monoid}^R$ the class of all commutative right-monoids.

Remark 5.3. We can define, alternatively, a commutative left-unital magma and a commutative left-monoid as algebras $(A^L, \cdot, ^{-1}, 1)$ of type $(2, 1, 0)$ with $1^{-1} = 1$

(Hierarchy m-1) or as algebras $(A^L, \odot, ^-, 1)$ of type $(2, 1, 0)$ with $1^- = 0$ (Hierarchy m-2). These alternative definitions are useful in this paper.

5.2 Ore and Dedekind lattices. Their equivalence

Definition 5.4. (The dual one is omitted)

A poset (po-set) (partially ordered set) $\mathcal{A} = (A, \leq^O)$ is called an *Ore lattice*, if, for each two elements $x, y \in A$, there exist $\inf(x, y)$ and $\sup(x, y)$.

Moreover, if there exist $0, 1 \in A$ such that $0 \leq^O x \leq^O 1$ for all $x \in A$, then $\mathcal{A}$ is said to be a *bounded* Ore lattice (with last (top) element 1 and first (bottom) element 0) and is denoted by $\mathcal{A} = (A, \leq^O, 0, 1)$.

An Ore lattice can be represented by the Hasse diagram.

In an Ore lattice, the following are equivalent: for all $x, y \in A$,
(i) $x \leq^O y$, (ii) $\inf(x, y) = x$, (iii) $\sup(x, y) = y$.

Definition 5.5. An algebra $\mathcal{A} = (A, \wedge, \vee)$ or, dually, $\mathcal{A} = (A, \vee, \wedge)$, of type $(2, 2)$, is a *Dedekind lattice*, if the following properties hold: for all $x, y, z \in A$,

(m-Wid) (idempotency of $\wedge$) $\quad x \wedge x = x$,
(m-Wcomm) (commutativity of $\wedge$) $\quad x \wedge y = y \wedge x$,
(m-Wass) (associativity of $\wedge$) $\quad x \wedge (y \wedge z) = (x \wedge y) \wedge z$,
(m-Wabs) (absorption of wedge over vee) $\quad x \wedge (x \vee y) = x$;
(m-Vid) (idempotency of $\vee$) $\quad x \vee x = x$,
(m-Vcomm) (commutativity of $\vee$) $\quad x \vee y = y \vee x$,
(m-Vass) (associativity of $\vee$) $\quad (x \vee y) \vee z = x \vee (y \vee z)$,
(m-Vabs) (absorption of vee over wedge) $\quad x \vee (x \wedge y) = x$,

where "W" comes from "wedge" and "V" comes from "vee".

Moreover, if there exist $0, 1 \in A$ such that: for all $x \in A$,
(m-WU) $1 \wedge x = x$ and
(m-VU) $0 \vee x = x$,
then $\mathcal{A}$ is said to be a *bounded* Dedekind lattice (with last element 1 and first element 0) and is denoted by $\mathcal{A} = (A, \wedge, \vee, 0, 1)$ or, dually, by $\mathcal{A} = (A, \vee, \wedge, 0, 1)$.

Note that a *bounded Dedekind lattice* is (a pair of) both a *commutative left-monoid* (for $\odot = \wedge$, since (m-WU), (m-Wcomm), (m-Wass) hold) and a *commutative right-monoid* (for $\oplus = \vee$, since (m-VU), (m-Vcomm), (m-Vass) hold), plus additional properties.

In a Dedekind lattice, we have the equivalence: for all $x, y \in A$,
(EqWV) $x \wedge y = x \Leftrightarrow x \vee y = y$.

Recall finally that the two kinds of lattices are *definitionally equivalent* (d.e.):

Theorem 5.6.

(1) Let $\mathcal{A} = (A, \leq^O)$ *be an Ore lattice. Define* $\delta(\mathcal{A}) \stackrel{def.}{=} (A, \wedge, \vee)$*, where for all* $x, y \in A$*:*
$x \wedge y \stackrel{def.}{=} \inf(x, y)$ *and* $x \vee y \stackrel{def.}{=} \sup(x, y)$.

Then, $\delta(\mathcal{A})$ *is a Dedekind lattice.*

(1') Let $\mathcal{A} = (A, \wedge, \vee)$ *be a Dedekind lattice. Define* $\omega(\mathcal{A}) \stackrel{def.}{=} (A, \leq^O)$*, where for all* $x, y \in A$*:*
(m-dfOW) $x \leq^O y \stackrel{def.}{\Longleftrightarrow} x \wedge y = x$ *or, equivalently,*
(m-dfOV) $x \leq^O y \stackrel{def.}{\Longleftrightarrow} x \vee y = y$.

Then, $\omega(\mathcal{A})$ *is an Ore lattice.*

(2) The two maps, δ *and* ω*, are mutually inverse.*

6 Algebras $(A, \odot, ^-, 1)$ with $1^- = 0$ (Hierarchies m-2, m-2')

Some of the results from this section were presented in [43].

6.1 The list m-A of basic properties. New algebras

Let $\mathcal{A}^L = (A^L, \odot, ^- = ^{-^L}, 1)$ be an algebra of type $(2, 1, 0)$ and define $0 \stackrel{def.}{=} 1^-$. Define an *internal* binary relation $\leq_m$ on A^L by: for all $x, y \in A^L$,

$$(m - dfrelP) \qquad x \leq_m y \stackrel{def.}{\Longleftrightarrow} x \odot y^- = 0.$$

Equivalently, consider a structure $\mathcal{A}^L = (A^L, \leq_m, \odot, ^-, 1)$ such that, if $0 \stackrel{def.}{=} 1^-$, then the *internal* binary relation $\leq_m$, the binary operation $\odot$, the unary operation $^-$ and the element 0 are connected by the equivalence:

$$(m - EqrelP) \qquad x \leq_m y \Longleftrightarrow x \odot y^- = 0.$$

Consider the following list **m-A** of basic properties that can be satisfied by $\mathcal{A}^L$ (in fact, the properties in the list are the most important properties satisfied by a left-m-BCK algebra, a new algebra that will be introduced in this section), where almost all properties can be presented in two equivalent forms, determined by the corresponding two equivalent above definitions of $\mathcal{A}^L$:

(PU) $1 \odot x = x = x \odot 1$ (unit element of product, the *identity*);
(Pcomm) $x \odot y = y \odot x$ (commutativity of product),
(Pass) $x \odot (y \odot z) = (x \odot y) \odot z$ (associativity of product);

(Neg1-0) $1^- = 0$;
(Neg0-1) $0^- = 1$;

(m-An) $(x \odot y^- = 0$ and $y \odot x^- = 0) \Longrightarrow x = y$ (antisymmetry),
(m-An') $(x \leq_m y$ and $y \leq_m x) \Longrightarrow x = y$ (antisymmetry);

(m-B) $[(x \odot y^-)^- \odot (x \odot z)] \odot (y \odot z)^- = 0$,
(m-B') $(x \odot y^-)^- \odot (x \odot z) \leq_m y \odot z$;

(m-BB) $[(z \odot x)^- \odot (y \odot x)] \odot (y \odot z^-)^- = 0$,
(m-BB') $(z \odot x)^- \odot (y \odot x) \leq_m y \odot z^-$;

(m-*) $x \odot y^- = 0 \Longrightarrow (z \odot y^-) \odot (z \odot x^-)^- = 0$,
(m-*') $x \leq_m y \Longrightarrow z \odot y^- \leq_m z \odot x^-$;

(m-**) $x \odot y^- = 0 \Longrightarrow (x \odot z) \odot (y \odot z)^- = 0$,
(m-**') $x \leq_m y \Longrightarrow x \odot z \leq_m y \odot z$;

(m-C) $[y \odot (x \odot z)] \odot [x \odot (y \odot z)]^- = 0$,
(m-C') $y \odot (x \odot z) \leq_m x \odot (y \odot z)$;

(m-D) $[(y^- \odot x)^- \odot x] \odot y^- = 0$,
(m-D') $(y^- \odot x)^- \odot x \leq_m y$;

(m-F) $\quad 0 \odot x^- = 0$ (first element),
(m-F') $\quad 0 \leq_m x$ (first element);

(m-K) $\quad x \odot [(y \odot x^-)^-]^- = 0$,
(m-K') $\quad x \leq_m (y \odot x^-)^-$;

(m-L) $\quad x \odot 0 = 0$ (last element),
(m-L') $\quad x \leq_m 1$ (last element);

(m-N) $\quad 1 \odot x^- = 0 \Longrightarrow x = 1$,
(m-N') $\quad 1 \leq_m x \Longrightarrow x = 1$;

(m-V) $\quad x \odot 1 = 0 \Longrightarrow x = 0$,
(m-V') $\quad x \leq_m 0 \Longrightarrow x = 0$;

(m-Re) $\quad x \odot x^- = 0$ (reflexivity),
(m-Re') $\quad x \leq_m x$ (reflexivity);

(m-Tr) $\quad (x \odot y^- = 0$ and $y \odot z^- = 0) \Longrightarrow x \odot z^- = 0$ (transitivity),
(m-Tr') $\quad (x \leq_m y$ and $y \leq_m z) \Longrightarrow x \leq_m z$ (transitivity);

(m-P- -) $\quad x \odot y^- = 0$ and $a \odot b^- = 0 \Longrightarrow (x \odot a) \odot (y \odot b)^- = 0$,
(m-P- -') $\quad x \leq_m y$ and $a \leq_m b \Longrightarrow x \odot a \leq_m y \odot b$;

(m-Pleq) $\quad (x \odot y) \odot x^- = 0$ and $(x \odot y) \odot y^- = 0$,
(m-Pleq') $\quad x \odot y \leq_m x, y$.

Dually,
let $\mathcal{A}^R = (A^R, \oplus, ^- = ^{-^R}, 0)$ be an algebra of type $(2, 1, 0)$ and define $1 \stackrel{def.}{=} 0^- \neq 0$. Define an *internal* binary relation $\geq_m$ on A^R by: for all $x, y \in A^R$,

$$(m - dfrelS) \qquad x \geq_m y \stackrel{def.}{\Longleftrightarrow} x \oplus y^- = 1.$$

Equivalently, consider a structure $\mathcal{A}^R = (A^R, \geq_m, \oplus, ^-, 0)$ such that, if $1 \stackrel{def.}{=} 0^- \neq 0$, then the *internal* binary relation $\geq_m$, the binary operation $\oplus$, the unary operation $^-$ and the element 1 are connected by the equivalence:

$$(m - EqrelS) \qquad x \geq_m y \Longleftrightarrow x \oplus y^- = 1.$$

The list of dual properties is omitted.

Remark 6.1. If $\odot = \wedge$ (Wedge) and/or $\oplus = \vee$ (Vee), then the property (m-An) becomes (m-WAn) and/or the dual property (m-An^R) becomes (m-VAn) and so on.

6.1.1 New algebras

Let $\mathcal{A}^L = (A^L, \odot, ^- = ^{-^L}, 1)$ be an algebra of type $(2,1,0)$ through this section. Define $0 \overset{def.}{=} 1^-$ (hence (Neg1-0) holds) and suppose that $0^- = 1$ (hence (Neg0-1) holds too).

We now introduce the following notions, the analogous of those bounded M algebras of logic from Section 2 (Hierarchies 2^b and $2'^b$) (the dual ones are omitted).

- Algebras with last element, without associativity:
$\mathcal{A}^L$ is a:

(mL1) *left-m-ML algebra*, if it verifies (PU), (Pcomm), (m-L);

(mL2) *left-m-RML algebra*, if it verifies (PU), (Pcomm), (m-L), (m-Re);

(mL3) *left-pre-m-BCC algebra*, if it verifies (PU), (Pcomm), (m-L), (m-Re), (m-B);

(mL4) *left-m-aRML algebra*, if it verifies (PU), (Pcomm), (m-L), (m-Re), (m-An);

(mL5) *left-m-BCC algebra*, if it verifies (PU), (Pcomm), (m-L), (m-Re), (m-B), (m-An).

- Algebras with last element, with associativity:
$\mathcal{A}^L$ is a:

(mLE1) *left-m-MEL algebra*, if it verifies (PU), (Pcomm), (Pass), (m-L), i.e. if it is a left-monoid verifying the additional axiom (m-L) of last element;

(mLE2) *left-m-BE algebra*, if it verifies (PU), (Pcomm), (Pass), (m-L), (m-Re), i.e. if it is a left-m-MEL algebra verifying the additional axiom (m-Re) of reflexivity;

(mLE3) *left-pre-m-BCK algebra*, if it verifies (PU), (Pcomm), (Pass), (m-L), (m-Re) and (m-BB), i.e. if it is a left-m-BE algebra verifying the additional axiom (m-BB);

(mLE4) *left-m-aBE algebra*, if it verifies (PU), (Pcomm), (Pass), (m-L), (m-Re), (m-An), i.e. if it is a left-m-BE algebra verifying the additional axiom (m-An) of antisymmetry;

(mLE5) *left-m-BCK algebra*, if it verifies (PU), (Pcomm), (Pass), (m-L), (m-Re), (m-An) and (m-BB), i.e. it is a left-m-aBE algebra verifying the additional axiom (m-BB);

and, additionally,

(mLE6') *left-m-BE* algebra*, if it verifies (PU), (Pcomm), (Pass), (m-L), (m-Re) and (m-*), i.e. if it is a left-m-BE algebra verifying the additional axiom (m-*).

(mLE7') *left-m-aBE* algebra*, if it verifies (PU), (Pcomm), (Pass), (m-L), (m-Re), (m-An) and (m-*), i.e. if it is a left-m-aBE algebra verifying the additional axiom (m-*).

Denote by **m-ML**, ... **m-aBE***, the classes of left-m-ML, ..., left-m-aBE* algebras, respectively.

Note that sometimes we shall simply write *m-ML algebra* instead of *left-m-ML algebra* and so on.

6.1.2 Connections between the properties

Proposition 6.2. *We have (the dual case is omitted):*
(mA0) (m-L) + (Pcomm) $\Longrightarrow$ (m-F);
(mA1) (PU) + (Neg0-1) $\Longrightarrow$ (m-V'),
(mA2) (m-C) + (m-An) + (Pcomm) $\Longrightarrow$ (Pass),
(mA3) (PU) + (m-BB) + (Pcomm) + (Neg1-0) + (Neg0-1) $\Longrightarrow$ (m-Re),
*(mA4) (PU) + (m-BB) + (Pcomm) + (Neg0-1) $\Longrightarrow$ (m-**),*
(mA5) (PU) + (m-B) + (Pcomm) + (Neg0-1) $\Longrightarrow$ (m-);*
*(mA6) (PU) + (m-**) + (Neg0-1) $\Longrightarrow$ (m-Tr),*
(mA7) (PU) + (m-) + (Neg0-1) $\Longrightarrow$ (m-Tr);*
(mA8) (m-Re) + (Pass) + (Pcomm) $\Longrightarrow$ (m-D),
(mA9) (m-Tr) + (Pass) + (Pcomm) + (m-Re) $\Longrightarrow$ (m-),*
*(mA10) (m-Re) + (Pass) + (Pcomm) + (m-**) $\Longrightarrow$ (m-B);*
(mA11) (Pass) + (Pcomm) $\Longrightarrow$ ((m-B) $\Leftrightarrow$ (m-BB));
*(mA12) (m-**) + (Pcomm) + (m-Tr) $\Longrightarrow$ (m-P- -),*
*(mA13) (m-**) + (PU) + (Pcomm) + (m-L) $\Longrightarrow$ (m-Pleq);*
(mA14) (Pcomm) + (Pass) + (m-L) + (m-Re) $\Longrightarrow$ (m-Pleq).

Proof.
(mA0): $0 \odot x^- \overset{(Pcomm)}{=} x^- \odot 0 \overset{(m-L)}{=} 0$, i.e. $0 \leq_m x$; thus, (m-F) holds.

(mA1): Suppose $x \leq_m 0$, i.e. $x \odot 0^- = 0$; then, by (Neg0-1), $x \odot 1 = 0$, hence $x = 0$, by (PU); thus, (m-V') holds.

(mA2): By (m-C'), $x \odot (z \odot y) \leq_m z \odot (x \odot y)$ and also $z \odot (x \odot y) \leq_m x \odot (z \odot y)$; then, by (m-An), $x \odot (z \odot y) = z \odot (x \odot y)$, which gives $x \odot (y \odot z) = (x \odot y) \odot z$, by (Pcomm).

(mA3): In (m-BB) take $y = z = 1$; we obtain $0 = [(1 \odot x)^- \odot (1 \odot x)] \odot (1 \odot$

$1^-)^- \overset{(PU),(Neg1-0)}{=} [x^- \odot x] \odot 0^- \overset{(Neg0-1)}{=} [x^- \odot x] \odot 1 \overset{(PU)}{=} x^- \odot x \overset{(Pcomm)}{=} x \odot x^-$; thus, (m-Re) holds.

(mA4): Suppose $y \leq_m z$, i.e. $y \odot z^- = 0$; then, by (m-BB), $0 = [(z \odot x)^- \odot (y \odot x)] \odot 0^- \overset{(Neg0-1)}{=} [(z \odot x)^- \odot (y \odot x)] \odot 1 \overset{(Pcomm),(PU)}{=} (z \odot x)^- \odot (y \odot x) \overset{(Pcomm)}{=} (y \odot x) \odot (z \odot x)^-$, i.e. $y \odot x \leq_m z \odot x$; thus, (m-**) holds.

(mA5): Suppose $y \leq_m z$, i.e. $y \odot z^- = 0$; then, by (m-B), $0 = [(x \odot y^-)^- \odot (x \odot z^-)] \odot (y \odot z^-)^-$
$\overset{(Neg0-1)}{=} [(x \odot y^-)^- \odot (x \odot z^-)] \odot 1 \overset{(PU)}{=} (x \odot y^-)^- \odot (x \odot z^-) \overset{(Pcomm)}{=} (x \odot z^-) \odot (x \odot y^-)^-$,
i.e. $x \odot z^- \leq_m x \odot y^-$; thus, (m-*) holds.

(mA6): Suppose $x \leq_m y$ and $y \leq_m z$, i.e. $x \odot y^- = 0$ and $y \odot z^- = 0$.
Then, by (m-**'), $x \odot z^- \leq_m y \odot z^-$, i.e. $(x \odot z^-) \odot (y \odot z^-)^- = 0$; hence, $(x \odot z^-) \odot 0^- = 0$.
Then, by (Neg0-1), (PU), we obtain $x \odot z^- = 0$, i.e. $x \leq_m z$. Thus, (m-Tr') holds.

(mA7): Suppose $x \leq_m y$ and $y \leq_m z$, i.e. $x \odot y^- = 0$ and $y \odot z^- = 0$.
Then, by (m-*'), $x \odot z^- \leq_m x \odot y^-$, i.e. $(x \odot z^-) \odot (x \odot y^-)^- = 0$; then,
$0 = (x \odot z^-) \odot 0^- \overset{(Neg0-1)}{=} (x \odot z^-) \odot 1 \overset{(PU)}{=} x \odot z^-$; it follows that $x <_m z$; thus, (m-Tr) holds.

(mA8): $[(y^- \odot x)^- \odot x] \odot y^- \overset{(Pass)}{=} (y^- \odot x)^- \odot (x \odot y^-) \overset{(Pcomm)}{=} (x \odot y^-) \odot (x \odot y^-)^- \overset{(m-Re)}{=} 0$; thus, (m-D) holds.

(mA9): Suppose $y \leq_m z$; by (mA8), (m-Re) + (Pass) + (Pcomm) $\Longrightarrow$ (m-D), hence $(y^- \odot x)^- \odot x \leq_m y$; then, by (m-Tr), $(y^- \odot x)^- \odot x \leq_m z$, i.e. $[(y^- \odot x)^- \odot x] \odot z^- = 0$; then, by (Pass), $0 = (y^- \odot x)^- \odot (x \odot z^-) \overset{(Pcomm)}{=} (x \odot z^-) \odot (x \odot y^-)^-$; it follows that $x \odot z^- \leq_m x \odot y^-$; thus, (m-*) holds.

(mA10): By (mA8), (m-Re) + (Pass) + (Pcomm) $\Longrightarrow$ (m-D), hence $(y^- \odot x)^- \odot x \leq_m y$;
then, by (m-**), $[(y^- \odot x)^- \odot x] \odot z \leq_m y \odot z$; hence, $(y^- \odot x)^- \odot (x \odot z) \leq_m y \odot z$, by (Pass);
it follows $(x \odot y^-)^- \odot (x \odot z) \leq_m y \odot z$, by (Pcomm); thus, (m-B') holds.

(mA11): If (m-BB) holds, i.e. $[(z \odot x)^- \odot (y \odot x)] \odot (y \odot z^-)^- = 0$, then, by (Pass), we obtain:
$(z \odot x)^- \odot [(y \odot x) \odot (y \odot z^-)^-] = 0$, hence, by (Pcomm), we obtain:
$[(y \odot z^-)^- \odot (y \odot x)] \odot (z \odot x)^- = 0$; thus, (m-B) holds.

Conversely, if (m-B) holds, i.e. $[(x \odot y^-)^- \odot (x \odot z)] \odot (y \odot z)^- = 0$, then, by (Pass), we obtain:
$(x \odot y^-)^- \odot [(x \odot z) \odot (y \odot z)^-] = 0$, hence, by (Pcomm), we obtain:
$[(y \odot z)^- \odot (x \odot z)] \odot (x \odot y^-)^- = 0$; thus, (m-BB) holds.

(mA12): Suppose $x \leq_m y$ and $a \leq_m b$. By (m-**'), $a \leq_m b$ implies $a \odot x \leq_m b \odot x$ and $x \leq_m y$ implies $x \odot b \leq_m y \odot b$; hence, $x \odot a \overset{(Pcomm)}{=} a \odot x \leq_m b \odot x \overset{(Pcomm)}{=} x \odot b \leq_m y \odot b$; then, $x \odot a \leq_m y \odot b$, by (m-Tr').

(mA13): Since $x \leq_m 1$, by (m-L'), it follows, by (m-**'), that $x \odot y \leq_m 1 \odot y \overset{(PU)}{=} y$. Similarly, we prove that $x \odot y \leq_m x$, by (Pcomm).

(mA14): $x \odot y \leq_m x \overset{def.}{\Longleftrightarrow} (x \odot y) \odot x^- = 0 \overset{(Pcomm)}{\Longleftrightarrow} (y \odot x) \odot x^- = 0 \overset{(Pass)}{\Longleftrightarrow} y \odot (x \odot x^-) = 0 \overset{(m-Re)}{\Longleftrightarrow} y \odot 0 = 0$, that is true by (m-L). Thus, $x \odot y \leq_m x$. Similarly, we prove that $x \odot y \leq_m y$. □

Remark 6.3. By (mA0), all above defined algebras verify both (m-L) and (m-F), i.e. are *bounded*: for all x, $0 \leq_m x \leq_m 1$.

We can obtain now the following two important results (see the corresponding Theorems 2.4 and 2.5 from Section 2).

Theorem 6.4. *Let* $\mathcal{A}^L = (A^L, \odot, ^-, 1)$ *be an algebra of type* $(2,1,0)$ *and let* $0 \overset{def.}{=} 1^-$, *hence (Neg1-0) holds. If (Neg0-1), (PU), (Pcomm), (Pass), (m-Re) hold, then*

$$(m-B) \Longleftrightarrow (m-BB) \Longleftrightarrow (m-**).$$

Proof. By (mA11), (Pass) + (Pcomm) $\Longrightarrow$ ((m-B) $\Leftrightarrow$ (m-BB));
by (mA4), (PU) + (m-BB) + (Pcomm) + (Neg0-1) $\Longrightarrow$ (m-**);
by (mA10), (m-Re) + (Pass) + (Pcomm) + (m-**) $\Longrightarrow$ (m-B). □

Theorem 6.5. *Let* $\mathcal{A}^L = (A^L, \odot, ^-, 1)$ *be an algebra of type* $(2,1,0)$ *and let* $0 \overset{def.}{=} 1^-$, *hence (Neg1-0) holds. If (Neg0-1), (PU), (Pcomm), (Pass), (m-Re) hold, then*

$$(m-Tr) \Longleftrightarrow (m-*).$$

Proof. By (mA7) and (mA9). □

6.2 The list m-B of particular properties. Some other new algebras

Consider now the following list **m-B** of particular properties:

($\wedge_m$-comm) ($\wedge_m$-commutativity)	$(x^- \odot y)^- \odot y = (y^- \odot x)^- \odot x$,
(m-Pimpl) (m-Pimplicativity)	$[(x \odot y^-)^- \odot x^-]^- = x$,
(m-Wimpl) (m-Wimplicativity)	$[(x \wedge y^-)^- \wedge x^-]^- = x$,
(m-Pimpl-1)	$(x^- \odot y)^- \odot x = x$,
(m-DN6)	$(x^- \odot x^-)^- = x$,
(G)	$x \odot x = x$,

(m-Wdis) (distributivity of wedge over vee) $z \wedge (x \vee y) = (z \wedge x) \vee (z \wedge y)$,
and, dually,

($\vee_m$-comm) ($\vee_m$-commutativity)	$(x^- \oplus y)^- \oplus y = (y^- \oplus x)^- \oplus x$;
(m-Simpl) (m-Simplicativity)	$[(x \oplus y^-)^- \oplus x^-]^- = x$,
(m-Vimpl) (m-Vimplicativity)	$[(x \vee y^-)^- \vee x^-]^- = x$,
(m-Simpl-1)	$(x^- \oplus y)^- \oplus x = x$,
(m-DN6^R)	$(x^- \oplus x^-)^- = x$,
(G^R)	$x \oplus x = x$;

(m-Vdis) (distributivity of vee over wedge) $z \vee (x \wedge y) = (z \vee x) \wedge (z \vee y)$.

Note that (G) becomes (m-Wid), for $\odot = \wedge$, while (G^R) becomes (m-Vid), for $\oplus = \vee$, where (m-Wid) and (m-Vid) are axioms of Dedekind lattices.

Definitions 6.6.

An algebra $(A^L, \odot, ^-, 1)$ is said to be:

(i) $\wedge_m$-*commutative*, if it verifies the property ($\wedge_m$-comm).

(ii) *m-Pimplicative*, if it verifies the property (m-Pimpl).

Dually, an algebra $(A^R, \oplus, ^-, 0)$ is said to be:

(i') $\vee_m$-*commutative*, if it verifies the property ($\vee_m$-comm).

(ii') *m-Simplicative*, if it verifies the property (m-Simpl).

Proposition 6.7. *We have (the dual case is omitted):*
(mB1) ($\wedge_m$-comm) + (Neg0-1) + (PU) + (Pcomm) + (m-L) $\Longrightarrow$ (m-Re),
(mB2) ($\wedge_m$-comm) + (Neg0-1) + (PU) + (Pcomm) $\Longrightarrow$ (m-An),
(mB3) (m-Pimpl) + (Neg0-1) + (PU) $\Longrightarrow$ (m-DN6).

Proof.

(mB1): $x \odot x^- \overset{(Pcomm),(PU)}{=} (1 \odot x)^- \odot x \overset{(Neg0-1)}{=} (0^- \odot x)^- \odot x \overset{(\wedge_m-comm)}{=} (x^- \odot 0)^- \odot 0 \overset{(m-L)}{=} 0$; thus, (m-Re) holds.

(mB2): Suppose $x \leq_m y$ and $y \leq_m x$, i.e. $x \odot y^- = 0$ and $y \odot x^- = 0$; then, by (Pcomm), ($\wedge_m$-comm), (Neg0-1), we obtain: $1 \odot x = 1 \odot y$, hence $x = y$, by (PU); thus, (m-An') holds.

(mB3): Take $y = 0$ in (m-Pimpl); we obtain: $x = [(x \odot 0^-)^- \odot x^-]^- \overset{(Neg0-1)}{=} [(x \odot 1)^- \odot x^-]^- \overset{(PU)}{=} [x^- \odot x^-]^-$; thus, (m-DN6) holds. □

6.3 The list m-C of properties of the negation. Connections

Consider now the following list **m-C** of properties of the negation $^-$:

(m-Neg2') $x \leq_m y \Longrightarrow y^- \leq_m x^-$ and, dually,

(m-Neg2^{R}') $x \geq_m y \Longrightarrow y^- \geq_m x^-$;
(m-Neg4') $(x^-)^- \leq_m x$ and, dually,
(m-Neg4^{R}') $(x^-)^- \geq_m x$;
(m-Neg7) $((x^-)^-)^- = x^-$;

(DN) (Double Negation) $(x^-)^- = x$;

(m-DN4') $x \leq_m y \Longleftrightarrow y^- \leq_m x^-$ and, dually,
(m-DN4^{R}') $x \geq_m y \Longleftrightarrow y^- \geq_m x^-$;

(m-WRe) $x \wedge x^- = 0$ and, dually,
(m-VRe) $x \vee x^- = 1$;
(m-Pdiv) $x \odot (x \odot y^-)^- = x \odot y$ and, dually,
(m-Sdiv) $x \oplus (x \oplus y^-)^- = x \oplus y$;
(m-@)=(m-Wdiv) $x \wedge (x \wedge y^-)^- = x \wedge y$ and, dually,
(m-@^{R})=(m-Vdiv) $x \vee (x \vee y^-)^- = x \vee y$.

Proposition 6.8. *We have (the dual case is omitted):*
(mC1) (m-) + (PU) $\Longrightarrow$ (m-Neg2'),*
(mC2) (m-Re) + (Pcomm) $\Longrightarrow$ (m-Neg4'),
(mC2') (m-D) + (PU) $\Longrightarrow$ (m-Neg4');
(mC3) (m-Neg4') + (m-) + (PU) + (m-An) $\Longrightarrow$ (m-Neg7).*

Proof.
(mC1): Suppose $x \leq_m y$; then, by (m-*), $z \odot y^- \leq_m z \odot x^-$; then, for $z = 1$, we obtain $1 \odot y^- \leq_m 1 \odot x^-$; hence, $y^- \leq_m x^-$, by (PU); thus, (m-Neg2') holds.

(mC2): $(x^-)^- \odot x^- \overset{(Pcomm)}{=} x^- \odot (x^-)^- \overset{(m-Re)}{=} 0$, i.e. $(x^-)^- \leq_m x$; thus, (m-Neg4') holds.

(mC2'): Take $x = 1$ in (m-D); we obtain $(y^- \odot 1)^- \odot 1 \leq_m y$; hence, $(y^-)^- \leq_m y$, by (PU); thus, (m-Neg4') holds.

(mC3): By (m-Neg4'), $((y^-)^-)^- \leq_m y^-$. On the other hand, by (m-Neg4'), $(y^-)^- \leq_m y$, hence $1 \odot y^- \leq_m 1 \odot ((y^-)^-)^-$, by (m-*); then, $y^- \leq_m ((y^-)^-)^-$, by (PU). Now apply (m-An) to obtain $((y^-)^-)^- = y^-$; thus, (m-Neg7) holds. □

Definition 6.9. We shall say that an algebra $\mathcal{A}^L = (A^L, \odot, ^-, 1)$ is *involutive*, if it verifies (DN).

The subclass of involutive algebras of the class $\mathbf{X}$ will be denoted by $\mathbf{X}_{(DN)}$.

Proposition 6.10. *We have (the dual case is omitted):*
(mCBN1) ($\wedge_m$-comm)+(DN)+(Pass)+(Pcomm)+(m-Re)+(m-L) $\Longrightarrow$ (m-BB);

(mCIM1) (m-Pimpl) + (Neg1-0) + (Neg0-1) + (m-L) + (PU) $\Longrightarrow$ (DN),
(mCIM1') (m-Pimpl) + (m-Re) + (Neg0-1) + (PU) $\Longrightarrow$ (DN);
(mCIM2) (m-Pimpl) + (DN) $\Longrightarrow$ (m-Pimpl-1),
(mCIM3) (m-Pimpl-1) + (PU) + (DN) $\Longrightarrow$ (G).

Proof.

(mCBN1): We must prove that $T \overset{notation}{=} [(z \odot x)^- \odot (y \odot x)] \odot (y \odot z^-)^- = 0$.
Indeed, $(z \odot x)^- \odot (y \odot x) \overset{(Pcomm),(Pass)}{=} y \odot [(z \odot x)^- \odot x]$
$\overset{(DN),(\wedge_m-comm)}{=} y \odot [(x^- \odot z^-)^- \odot z^-]$. Then,
$T = (y \odot [(x^- \odot z^-)^- \odot z^-]) \odot (y \odot z^-)^-$
$\overset{(Pass),(Pcomm)}{=} (x^- \odot z^-)^- \odot [(y \odot z^-) \odot (y \odot z^-)^-]$
$\overset{(m-Re)}{=} (x^- \odot z^-)^- \odot 0 \overset{(m-L)}{=} 0$. Thus, (m-BB) holds.

(mCIM1): Take $y = 1$ in (m-Pimpl); we obtain: $x = [(x \odot 1^-)^- \odot x^-]^- \overset{(Neg1-0)}{=}$ $[(x \odot 0)^- \odot x^-]^- \overset{(m-L)}{=} [0^- \odot x^-]^- \overset{(Neg0-1)}{=} [1 \odot x^-]^- \overset{(PU)}{=} (x^-)^-$; thus, (DN) holds.

(mCIM1'): Take $y = x$ in (m-Pimpl); we obtain: $x = [(x \odot x^-)^- \odot x^-]^- \overset{(m-Re)}{=}$ $[0^- \odot x^-]^- \overset{(Neg0-1)}{=} [1 \odot x^-]^- \overset{(PU)}{=} (x^-)^-$.

(mCIM2): By (m-Pimpl), $[(x \odot y^-)^- \odot x^-]^- = x$; hence, by (DN), we obtain: $(x \odot y^-)^- \odot x^- = x^-$; then, by replacing x by x^- and y by y^-, we obtain, by (DN) again: $(x^- \odot y)^- \odot x = x$; thus, (m-Pimpl-1) holds.

(mCIM3): Take $y = 1$ in (m-Pimpl-1); we obtain: $x = (x^- \odot 1)^- \odot x \overset{(PU)}{=}$ $(x^-)^- \odot x \overset{(DN)}{=} x \odot x$; thus, (G) holds. □

Proposition 6.11. *We have (the dual case is omitted):*
(mCDN0) (DN) + (PU) + (Neg0-1) $\Longrightarrow$ (m-N'),
(mCDN1) (DN) $\Longrightarrow$ ((Neg1-0) $\Leftrightarrow$ (Neg0-1)),
(mCDN2) (DN) + (Pcomm) $\Longrightarrow$ (m-DN4),
(mCDN3) (DN) + (Pcomm) $\Longrightarrow$ ((m-) $\Leftrightarrow$ (m-**)),*
(mCDN4) (DN) + (Pass) + (m-Re) + (m-L) + (Pcomm) $\Longrightarrow$ (m-K);
(mCDN5) (PU) + (Pcomm) + (m-Re) + (m-) + (m-An) $\Longrightarrow$ (DN);*
*(mCDN6) (Pcomm) + (m-Re) + (m-**) + (m-F) + (m-An) $\Longrightarrow$ (DN),*
*(mCDN6') (Pcomm) + (m-Re) + (m-**) + (m-V) + (m-An) $\Longrightarrow$ (DN).*

Proof.

(mCDN0): Suppose $1 \leq_m x$, i.e. $1 \odot x^- = 0$; then, by (PU), $x^- = 0$, hence $x = 1$, by (DN) and (Neg0-1); thus, (m-N') holds.

(mCDN1): Suppose (Neg1-0) holds; then $0^- = (1^-)^- \overset{(DN)}{=} 1$, i.e. (Neg0-1) holds too. Similarly, (Neg0-1) implies (Neg1-0).

(mCDN2): $x \leq_m y \Longleftrightarrow x \odot y^- = 0 \stackrel{(DN)}{\Longleftrightarrow} (x^-)^- \odot y^- = 0 \stackrel{(Pcomm)}{\Longleftrightarrow} y^- \odot (x^-)^- = 0 \Longleftrightarrow y^- \leq_m x^-$; thus, (m-DN4') holds.

(mCDN3): By (mCDN2), (DN) + (Pcomm) $\Longrightarrow$ (m-DN4').
If $x \leq_m y$, then, by (m-DN4'), $y^- \leq_m x^-$; then, by (m-* '), $z \odot (x^-)^- \leq_m z \odot (y^-)^-$; then, by (DN), $z \odot x \leq_m z \odot y$; then, by (Pcomm), $x \odot z \leq_m y \odot z$; thus, (m-** ') holds.
Conversely, if $x \leq_m y$, then, by (m-DN4'), $y^- \leq_m x^-$; then, by (m-** '), $y^- \odot z \leq_m x^- \odot z$; then, by (Pcomm), $z \odot y^- \leq_m z \odot x^-$; thus, (m-* ') holds.

(mCDN4): $x \leq_m (y \odot x^-)^- \Longleftrightarrow x \odot ((y \odot x^-)^-)^- = 0$ and, indeed,
$x \odot ((y \odot x^-)^-)^- \stackrel{(DN)}{=} x \odot (y \odot x^-) \stackrel{(Pcomm),(Pass)}{=} y \odot (x \odot x^-) \stackrel{(m-Re)}{=} y \odot 0 \stackrel{(m-L)}{=} 0$; thus, (m-K) holds.

(mCDN5): By (mC2), (m-Re) + (Pcomm) $\Longrightarrow$ (m-Neg4'), i.e. $(x^-)^- \leq_m x$.
On the other hand, by (mC3), (m-Neg4') + (m-*) + (PU) + (m-An) $\Longrightarrow$ (m-Neg7), i.e. $((x^-)^-)^- = x^-$; then, $x \odot ((x^-)^-)^- = x \odot x^- \stackrel{(m-Re)}{=} 0$, i.e. $x \leq_m (x^-)^-$.
Now, by (m-An), we obtain $(x^-)^- = x$, i.e. (DN) holds.

(mCDN6): $(x^-)^- \leq_m x \Longleftrightarrow (x^-)^- \odot x^- = 0$, that is true by (Pcomm) and (m-Re); thus, $(x^-)^- \leq_m x$. On the other hand, by above inequality, we have $((x^-)^-)^- \leq_m x^-$; then, by (m-**'),
$x \odot ((x^-)^-)^- \leq_m x \odot x^- \stackrel{(m-Re)}{=} 0$; by (m-F'), we also have $0 \leq_m x \odot ((x^-)^-)^-$; hence, by (m-An'), we obtain $x \odot ((x^-)^-)^- = 0$; thus, $x \leq_m (x^-)^-$.
From $(x^-)^- \leq_m x$ and $x \leq_m (x^-)^-$, we obtain $(x^-)^- = x$, by (m-An') again; thus, (DN) holds.

(mCDN6'): $(x^-)^- \leq_m x \Longleftrightarrow (x^-)^- \odot x^- = 0$, that is true by (Pcomm) and (m-Re); thus, $(x^-)^- \leq_m x$. On the other hand, by above inequality, we have $((x^-)^-)^- \leq_m x^-$, hence, by (m-**'),
$x \odot ((x^-)^-)^- \leq_m x \odot x^- \stackrel{(m-Re)}{=} 0$; by (m-V'), we obtain $x \odot ((x^-)^-)^- = 0$, hence, $x \leq_m (x^-)^-$.
From $(x^-)^- \leq_m x$ and $x \leq_m (x^-)^-$, we obtain $(x^-)^- = x$, by (m-An'); thus, (DN) holds. □

Theorem 6.12. *Let $\mathcal{A}^L = (A^L, \odot, ^-, 1)$ be an algebra of type $(2,1,0)$ and let $0 \stackrel{def.}{=} 1^-$, hence (Neg1-0) holds. If (Neg0-1), (PU), (Pcomm), (Pass), (m-Re) and (DN) hold, then*

$$(m-BB) \Longleftrightarrow (m-B) \Longleftrightarrow (m-**) \Longleftrightarrow (m-*) \Longleftrightarrow (m-Tr).$$

Proof. By Theorems 6.4, 6.5 and (mCDN3). □

Theorem 6.13. *We have:*
(1) any left-m-aBE algebra is involutive,*
(2) any left-m-BCK algebra is involutive and
(3) the (involutive) left-m-aBE algebras coincide with the (involutive) left-m-BCK algebras;*
(4) any left-m-BCC algebra is involutive.

Proof.
(1): By (mCDN5), (PU) + (Pcomm) + (m-Re) + (m-*) + (m-An) $\Longrightarrow$ (DN).
(2): By (mA1), (m-V) holds; by Theorem 6.4, (m-**) holds; by (mCDN6'), (Pcomm) + (m-Re) + (m-**) + (m-V) + (m-An) $\Longrightarrow$ (DN).
(3): By (mCDN3), (DN) + (Pcomm) $\Longrightarrow$ ((m-*) $\Longleftrightarrow$ (m-**)).
(4): Let $\mathcal{A}^L = (A^L, \odot, ^-, 1)$ be a left-m-BCC algebra, i.e. (Neg1-0), (Neg0-1), (PU), (Pcomm), (m-L), (m-Re), (m-An), (m-B) hold. Then, by (mA5), (PU) + (m-B) + (Pcomm) + (Neg0-1) $\Longrightarrow$ (m-*), thus (m-*) holds; by (mCDN5), (PU) + (Pcomm) + (m-Re) + (m-*) + (m-An) $\Longrightarrow$ (DN), thus (DN) holds. □

By above Theorem 6.13, we write:
m-aBE*=m-aBE*$_{(DN)}$= **m-BCK= m-BCK**$_{(DN)}$, **m-BCC=m-BCC**$_{(DN)}$.

Note that an (involutive) m-BCK algebra satisfies all the properties in the first list **m-A** of properties and, additionally, (DN) and (m-DN4).

Proposition 6.14. *We have:*
(mCBB1) (DN)+(Pcomm)+(Pass)+(m-Re)+(m-L)+(m-Pdiv) $\Longrightarrow$ *(m-BB),*
(mCBB2) (m-Pimpl) + (PU) + (Pcomm) + (Pass) + (Neg0-1) + (m-Re) + (m-L) + (m-An) + (m-BB) $\Longrightarrow$ *(m-Pdiv).*

Proof.
(mCBB1): $[(z \odot x)^- \odot (y \odot x)] \odot (y \odot z^-)^- \overset{(Pass)}{=} [((z \odot x)^- \odot x) \odot y] \odot (y \odot z^-)^- \overset{(DN)}{=} [(((z^-)^- \odot x)^- \odot x) \odot y] \odot (y \odot z^-)^- \overset{(m-Pdiv)}{=} [(x \odot z^-) \odot y] \odot (y \odot z^-)^- \overset{(Pcomm),(Pass)}{=} x \odot [(y \odot z^-) \odot (y \odot z^-)^-] \overset{(m-Re)}{=} x \odot 0 \overset{(m-L)}{=} 0$; thus, (m-BB) holds.
(mCBB2): By (mA8), (m-Re) + (Pass) + (Pcomm) imply (m-D); by (m-D'),

$$x \odot (x \odot y^-)^- \leq_m y. \tag{5}$$

By (mA4), (PU) + (Pcomm) + (m-BB) + (Neg0-1) imply (m-**).
From (5), by (m-**'), we obtain $(x \odot (x \odot y^-)^-) \odot x \leq_m y \odot x$; hence, by (Pcomm), (Pass), we obtain

$$(x \odot x) \odot (x \odot y^-)^- \leq_m x \odot y. \tag{6}$$

By (mA0), (m-L) + (Pcomm) imply (m-F); by (mCDN6), (Pcomm) + (m-Re) + (m-**) + (m-F) + (m-An) imply (DN) (or by (mCIM1')). Then, by (mCIM2), (m-Pimpl) + (DN) imply (m-Pimpl-1);
by (mCIM3), (m-Pimpl-1) + (PU) + (DN) imply (G).
From (6) and (G), we obtain:

$$x \odot (x \odot y^-)^- \leq_m x \odot y. \tag{7}$$

On the other hand, by (mCDN4), (DN) + (Pass) + (m-Re) + (m-L) + (Pcomm) imply (m-K); by (m-K'), $y \leq_m (x \odot y^-)^-$; then, by (m-**'), we obtain:

$$x \odot y \leq_m x \odot (x \odot y^-)^-. \tag{8}$$

Now, from (7) and (8) and (m-An), we obtain $x \odot (x \odot y^-)^- = x \odot y$. Thus, (m-Pdiv) holds. □

With analogous definitions for the algebra $\mathcal{A} = (A, \odot, ^-, 1)$ and the associated binary relation $\leq_m$ as in Definitions 2.8 and with the same representations as in Remark 2.9, the new left-algebras defined in this section are connected (following their definitions and the above results) as in the *Hierarchy m-2* from Figure 10 and *Hierarchy m-2'* from Figure 11, in Section 8.

6.4 Connections of the new algebras with the MV algebras

6.4.1 The MV algebras

The our days definition of MV algebras is the following:

Definitions 6.15.

(i) A *left-MV algebra* is an algebra $\mathcal{A}^L = (A^L, \odot, ^- = ^{-^L}, 1)$ verifying: for all $x, y, z \in A^L$,

(PU) $1 \odot x = x (= x \odot 1)$,
(Pcomm) $x \odot y = y \odot x$,
(Pass) $x \odot (y \odot z) = (x \odot y) \odot z$,
(m-L) $x \odot 0 = 0$, where $0 \overset{def.}{=} 1^-$ (i.e. (Neg1-0) holds),
(DN) $(x^-)^- = x$,
($\wedge_m$-comm) $(x^- \odot y)^- \odot y = (y^- \odot x)^- \odot x$.

(i') A *right-MV algebra* is an algebra $\mathcal{A}^R = (A^R, \oplus, ^- = ^{-^R}, 0)$ verifying: for all $x, y, z \in A^R$,

(SU) $0 \oplus x = x (= x \oplus 0)$,
(Scomm) $x \oplus y = y \oplus x$,
(Sass) $x \oplus (y \oplus z) = (x \oplus y) \oplus z$,
(m-L^R) $x \oplus 1 = 1$, where $1 \stackrel{def.}{=} 0^-$ (i.e. (Neg0-1) holds),
(DN) $(x^-)^- = x$,
($\vee_m$-comm) $(x^- \oplus y)^- \oplus y = (y^- \oplus x)^- \oplus x$.

Denote by **MV** the class of all left-MV algebras and by $\mathbf{MV}^R$ the class of all right-MV algebras.

Note that, in a left-MV algebra, by (mCDN1), the property (Neg0-1) holds too, and in a right-MV algebra, by (mCDN1^R), the property (Neg1-0) holds too.

Note that the left-MV algebra verifies (DN)=(DN^L) (it is *involutive*), hence it is self-dual, i.e. the dual of $(A^L, \odot, ^-, 1)$ is $(A^L, \oplus, ^-, 0)$, where $x \oplus y = (x^- \odot y^-)^-$, and vice-versa.

Dually, the right-MV algebra verifies (DN) (it is *involutive*), hence it is self-dual, i.e. the dual of $(A^R, \oplus, ^-, 0)$ is $(A^R, \odot, ^-, 1)$, where $x \odot y = (x^- \oplus y^-)^-$, and vice-versa.

We make the following important remark, which was the motivation of this paper:

Remark 6.16.
(i) The left-MV algebra is just the $\wedge_m$-commutative involutive left-m-MEL algebra.

(i') The right-MV algebra is just the $\vee_m$-commutative involutive right-m-MEL algebra.

Theorem 6.17. *Let $\mathcal{A}^L = (A^L, \odot, ^-, 1)$ be a left-MV algebra. Define*

$$x \wedge_m y \stackrel{def.}{=} (x^- \odot y)^- \odot y, \quad x \vee_m y \stackrel{def.}{=} (x^- \wedge_m y^-)^- = (x^- \oplus y)^- \oplus y,$$

where $x \oplus y \stackrel{def.}{=} (x^- \odot y^-)^-$, and $0 \stackrel{def.}{=} 1^-$.
Then, $(A^L, \wedge = \wedge_m, \vee = \vee_m, 0, 1)$ is a bounded distributive (Dedekind) lattice.

In a left-MV algebra, we have: for all x, y, z,
(DPW) $(x \wedge y) \odot z = (x \odot z) \wedge (y \odot z)$.

Recall also that in a left-MV algebra $(A^L, \odot, ^-, 1)$ we can define two equivalent order relations:

$$(m - dfrelP) \qquad x \leq_m y \stackrel{def.}{\Longleftrightarrow} x \odot y^- = 0$$

and

$$(m - dfO_mW_m) \qquad x \leq_m^{O_m} y \stackrel{def}{\Longleftrightarrow} x \wedge_m y = x,$$

where 'O' comes from Ore and $x \wedge_m y \stackrel{def.}{=} (x^- \odot y)^- \odot y$;
we have the equivalence:

$$(m - EqO_mP) \qquad x \leq_m^{O_m} y \Longleftrightarrow x \leq_m y.$$

Theorem 6.18. *Let $\mathcal{A}^L = (A^L, \odot, ^-, 1)$ be a left-MV algebra. If (G) holds, then (G^R) holds too and*

$$x \wedge_m y = x \odot y, \quad x \vee_m y = x \oplus y.$$

Proof. $x \oplus x = (x^- \odot x^-)^- \stackrel{(G)}{=} (x^-)^- \stackrel{(DN)}{=} x$; thus, ($G^R$) holds too.
$x \wedge_m y \stackrel{def.}{=} (y^- \odot x)^- \odot x \stackrel{(G)}{=} (y^- \odot x)^- \odot (x \odot x) \stackrel{(Pass)}{=} [(y^- \odot x)^- \odot x] \odot x = (x \wedge_m y) \odot x \stackrel{(DPW)}{=} (x \odot x) \wedge_m (y \odot x) \stackrel{(G)}{=} x \wedge_m (y \odot x) \stackrel{(PU)}{=} (1 \odot x) \wedge_m (y \odot x) \stackrel{(DPW)}{=} (1 \wedge_m y) \odot x \stackrel{(m-WU)}{=} y \odot x \stackrel{(Pcomm)}{=} x \odot y$. Thus, $x \wedge_m y = x \odot y$. By a dual proof, $x \vee_m y = x \oplus y$. □

Proposition 6.19. *We have (the dual results are omitted):*
(mCDN7) (G) + (DN) + (m-Re) + (Pcomm) $\Longrightarrow$ (m-WRe),
(mCDN8) (m-WRe) + (DN) + (m-Pleq) + (m-An) $\Longrightarrow$ (G).

Proof.
(mCDN7): $x \wedge x^- \stackrel{def.}{=} ((x^-)^- \odot x)^- \odot x \stackrel{(DN)}{=} (x \odot x)^- \odot x \stackrel{(G)}{=} x^- \odot x \stackrel{(Pcomm)}{=} x \odot x^- \stackrel{(m-Re)}{=} 0$.

(mCDN8): First, we have $x \odot x \leq_m x$, by (m-Pleq'). We prove that $x \leq_m x \odot x$ also. Indeed, $x \odot (x \odot x)^- \stackrel{(DN)}{=} x \odot (x \odot (x^-)^-)^- \stackrel{def.}{=} x \wedge x^- \stackrel{(m-WRe)}{=} 0$, hence $x \leq_m x \odot x$. Now, by (m-An'), we obtain (G). □

Proposition 6.20. *Let $\mathcal{A}^L = (A^L, \odot, ^-, 1)$ be a left-MV algebra. Then, we have the equivalences:*

$$(G) \Longleftrightarrow (m - WRe), \qquad (G^R) \Longleftrightarrow (m - VRe).$$

Proof. By (mCDN1), (Neg0-1) holds and, by (mB1), (m-Re) holds. By (mA14), (m-Pleq) holds and, by (mB2), (m-An) holds. By (mCDN7), (G) + (DN) + (m-Re) + (Pcomm) $\Longrightarrow$ (m-WRe), and by (mCDN8), (m-WRe) + (DN) + (m-Pleq) + (m-An) $\Longrightarrow$ (G). Thus, (G) $\Longleftrightarrow$ (m-WRe). Dually, (G^R) $\Longleftrightarrow$ (m-VRe). □

6.4.2 Connections with the MV algebras

Theorem 6.21. *The class of $\wedge_m$-commutative (involutive) left-m-BCK algebras is d.e. to the class of left-MV algebras.*

Proof. (1): Let $\mathcal{A}^L = (A^L, \odot, ^-, 1)$ be a $\wedge_m$-commutative (involutive) left-m-BCK algebra (see Theorem 6.13). By definition, $\mathcal{A}^L$ is a $\wedge_m$-commutative involutive left-m-MEL algebra, i.e. it is a left-MV algebra.

(1'): Conversely, let $\mathcal{A}^L = (A^L, \odot, ^-, 1)$ be a left-MV algebra, i.e. a $\wedge_m$-commutative involutive left-m-MEL algebra, i.e. (PU), (Pcomm), (Pass), (m-L), (Neg1-0), (DN), ($\wedge$-comm) hold. We must prove that $\mathcal{A}^L$ is a left-m-BCK algebra, i.e. (m-Re), (m-An), (m-BB) hold. Indeed,
by (mCDN1), (DN) + (Neg1-0) imply (Neg0-1);
by(mB1), ($\wedge_m$-comm) + (Neg0-1) + (PU) + (Pcomm) + (m-L) imply (m-Re);
by (mB2), ($\wedge_m$-comm) + (Neg0-1) + (PU) + (Pcomm) imply (m-An);
by (mCBN1), ($\wedge_m$-comm) + (DN) + (Pass) + (Pcomm) + (m-Re) + (m-L) imply (m-BB). □

By Theorem 6.21 and Remark 6.16, we obtain immediately:

Corollary 6.22. *$\wedge_m$-commutative involutive left-m-MEL, left-m-BE, left-m-BE*, left-pre-m-BCK, left-m-aBE, left-m-aBE* = left-m-BCK algebras coincide (with left-MV algebras).*

6.5 Connections of the new algebras with the ortholattices

6.5.1 The ortholattices

Recall first the following definition.

Definitions 6.23. [14] A *(left-) bounded involutive poset*, or a *bounded involution poset*, is a structure $(A, \leq, ^-, 0, 1)$, where:
(i) $(A, \leq, 0, 1)$ is a bounded poset,
(ii) $^-$ is a unary operation (called *involution* or *generalizaed complement*) that satisfies the following conditions:
(DN) $(x^-)^- = x$ (Double Negation),
(Neg2') $x \leq y$ implies $y^- \leq x^-$ (contraposition).

A *bounded involutive lattice* is a bounded involutive poset that is also a lattice.

Definitions 6.24. (Definition 1) (see [62]) (see also [14])

(i) A *left-ortholattice* is an algebra $\mathcal{A}^L = (A^L, \wedge, \vee, ^- = ^{-^L}, 0, 1)$ such that:
- the reduct $(A^L, \wedge, \vee, 0, 1)$ is a bounded Dedekind lattice, i.e. the following properties hold:

(m-Wid)	$x \wedge x = x$,	(m-Vid)	$x \vee x = x$,
(m-Wcomm)	$x \wedge y = y \wedge x$,	(m-Vcomm)	$x \vee y = y \vee x$,
(m-Wass)	$x \wedge (y \wedge z) = (x \wedge y) \wedge z$,	(m-Vass)	$x \vee (y \vee z) = (x \vee y) \vee z$,
(m-Wabs)	$x \wedge (x \vee y) = x$,	(m-Vabs)	$x \vee (x \wedge y) = x$;
(m-WU)	$1 \wedge x = x$,	(m-VU)	$0 \vee x = x$;

- the unary operation $^-$ satisfies the law of Double Negation, the De Morgan laws and the complementation laws, where:
(DN) $(x^-)^- = x$;
(DeM1) $(x \vee y)^- = x^- \wedge y^-$ (De Morgan law 1),
(DeM2) $(x \wedge y)^- = x^- \vee y^-$ (De Morgan law 2);
(m-WRe) $x \wedge x^- = 0$ (noncontradiction principle),
(m-VRe) $x \vee x^- = 1$ (excluded middle principle).

(i') A *right-ortholattice* is an algebra $\mathcal{A}^R = (A^R, \vee, \wedge, ^- = ^{-^R}, 0, 1)$ such that the reduct $(A^R, \vee, \wedge, 0, 1)$ is a bounded Dedekind lattice and the unary operation $^-$ satisfies the law of Double Negation, the dual De Morgan laws and the dual complementation laws.

Denote by **OL** the class of all left-ortholattices and by $\mathbf{OL}^R$ the class of all right-ortholattices.

Note that a left-ortholattice verifies (DN) (it is *involutive*), hence it is self-dual, i.e. the dual of $(A^L, \wedge, \vee, ^{-^L}, 0, 1)$ is $(A^L, \vee, \wedge, ^{-^L}, 0, 1)$, and vice-versa.

Dually, a right-ortholattice verifies (DN) (it is *involutive*), hence it is self-dual, i.e. the dual of $(A^R, \vee, \wedge, ^{-^R}, 0, 1)$ is $(A^R, \wedge, \vee, ^{-^R}, 0, 1)$, and vice-versa.

Note [62] that it suffices to postulate only one of the De Morgan laws. For instance, if (DeM1) holds, then $(x \wedge y)^- \overset{(DN)}{=} ((x^-)^- \wedge (y^-)^-)^- \overset{(DeM1)}{=} ((x^- \vee y^-)^-)^- \overset{(DN)}{=} x^- \vee y^-$, i.e. (DeM2) holds too.

Roughly speaking, one could say that an ortholattice is a Boolean algebra without distributivity.

Proposition 6.25. *Let $\mathcal{A}^L = (A^L, \wedge, \vee, ^-, 0, 1)$ be a left-ortholattice. Define, for all $x, y \in A^L$,*
(m-dfOW) $x \leq_m^O y \overset{def.}{\Longleftrightarrow} x \wedge y = x$,

or, equivalently,
(m-dfOV) $x \leq_m^O y \stackrel{def.}{\Longleftrightarrow} x \vee y = y$.
Then, $(A^L, \leq_m^O, 0, 1)$ *is a bounded Ore lattice.*

Proof. Obviously, by Theorem 5.6, since $(A^L, \wedge, \vee, 0, 1)$ is a bounded Dedekind lattice. □

Proposition 6.26. [14] *Let* $\mathcal{A}^L = (A^L, \wedge, \vee, ^-, 0, 1)$ *be a left-ortholattice. The following are equivalent:*
(m-O-Neg2') $x \leq_m^O y \Longrightarrow y^- \leq_m^O x^-$,
(DeM1) $(x \vee y)^- = x^- \wedge y^-$,
(DeM2) $(x \wedge y)^- = x^- \vee y^-$.

Remarks 6.27. Let $\mathcal{A}^L = (A^L, \wedge, \vee, ^-, 0, 1)$ be a left-ortholattice. Note that:
(i) the Dedekind lattice operations $\wedge$ and $\vee$ are connected as follows: for al $x, y \in A^L$,

$$x \vee y = (x^- \wedge y^-)^-, \quad x \wedge y = (x^- \vee y^-)^-;$$

(ii) $(A^L, \leq_m^O, ^-, 0, 1)$ is a bounded involutive lattice.

Proposition 6.28. *Let* $\mathcal{A}^L = (A^L, \wedge, \vee, ^-, 0, 1)$ *be a left-ortholattice. Then, we have:*
(Neg1-0) $1^- = 0$,
(Neg0-1) $0^- = 1$,
(m-Wimpl) $[(x \wedge y^-)^- \wedge x^-]^- = x$,
(m-WL) $x \wedge 0 = 0$,
*(m-O-** ')* *If* $x \leq_m^O y$, *then* $x \wedge a \leq_m^O y \wedge a$.

Proof.
(Neg1-0): $1^- \stackrel{(m-WU)}{=} 1 \wedge 1^- \stackrel{(m-WRe)}{=} 0$.
(Neg0-1): By (mCDN1).
(m-Wimpl): $[(x \wedge y^-)^- \wedge x^-]^- \stackrel{(DeM2)}{=} [(x^- \vee (y^-)^-) \wedge x^-]^- \stackrel{(m-Wcomm),(DN)}{=} [x^- \wedge (x^- \vee y)]^- \stackrel{(m-Wabs)}{=} [x^-]^- \stackrel{(DN)}{=} x$.
(m-WL): $x \wedge 0 = 0 \stackrel{(m-Wcomm)}{\Longleftrightarrow} 0 \wedge x = 0 \stackrel{(m-dfOW)}{\Longleftrightarrow} 0 \leq_m^O x$, that is true by Proposition 6.25; thus, (m-WL) holds.
(m-O-**'): Suppose $x \leq_m^O y$, i.e. $x \wedge y = x$; then, $(x \wedge a) \wedge (y \wedge a) \stackrel{(m-Wcomm),(m-Wass)}{=} (x \wedge y) \wedge (a \wedge a) \stackrel{(m-Wid)}{=} x \wedge a$, i.e. $x \wedge a \leq_m^O y \wedge a$; thus, (m-O-** ') holds. □

Consider the following binary relation:
(m-dfrelW) $x \leq_m y \overset{def.}{\Longleftrightarrow} x \wedge y^- = 0$.

Note that $\leq_m$ is reflexive, by (m-WRe), but it is not antisymmetric or transitive, as the following examples show.

Examples 6.29. (See Examples 3.32)
Examples of ortholattices with six elements.
(m1) The algebra $\mathcal{O}_1 = (A_6 = \{0, a, b, c, d, 1\}, \wedge_1, \vee_1, ^{-1}, 0, 1)$ with

$\wedge_1$	0	a	b	c	d	1
0	0	0	0	0	0	0
a	0	a	0	0	0	a
b	0	0	b	0	0	b
c	0	0	0	c	0	c
d	0	0	0	0	d	d
1	0	a	b	c	d	1

,

x	x^{-1}
0	1
a	d
b	c
c	b
d	a
1	0

and $x \vee_1 y = (x^{-1} \wedge_1 y^{-1})^{-1}$ is an ortholattice with the bounded po-set $(A_6, \leq_m^O, 0, 1)$ represented by the Hasse diagram from Figure 7. It does not verify (m-WBB) for (a, a, b), (m-WB) for (a, b, a), (m-W**) for (a, b, a), (m-W*) for (a, b, c), (m-WAn) for (a, b), (m-WTr) for (a, b, d), ($\wedge_m$-comm) for (a, b).

(m2) The algebra $\mathcal{O}_2 = (A_6 = \{0, a, b, c, d, 1\}, \wedge_2 = \wedge_1, \vee_2, ^{-2}, 0, 1)$, with
$(0, a, b, c, d, 1)^{-2} = (1, b, a, d, c, 0)$ and $x \vee_2 y = (x^{-2} \wedge_2 y^{-2})^{-2}$,
is an ortholattice with the bounded po-set $(A_6, \leq_m^O, 0, 1)$ represented by the same Hasse diagram from Figure 7. It does not verify (m-WBB) for (a, a, c), (m-WB) for (a, c, a), (m-W**) for (a, c, a), (m-W*) for (a, c, d), (m-WAn) for (a, c), (m-WTr) for (a, c, b), ($\wedge_m$-comm) for (a, c).

Note that the ortholattices $\mathcal{O}_1$ and $\mathcal{O}_2$ correspond to the i-ortholattices $\mathcal{IO}_1$ and $\mathcal{IO}_2$, respectively, from Examples 3.32, by Theorem 9.2.

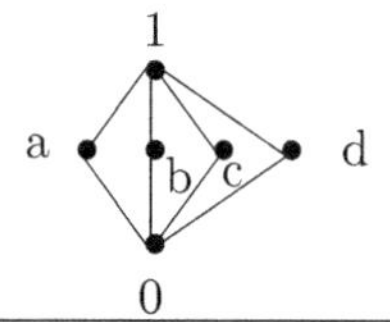

Figure 7: An ortholattice with 6 elements

Proposition 6.30. *Let $\mathcal{A}^L = (A^L, \wedge, \vee, ^-, 0, 1)$ be a left-ortholattice. Then, we have:*
(1) $x \leq_m^O y \Longrightarrow x \leq_m y$.
(2) If (m-@)=(m-Wdiv) $(x \wedge (x \wedge y^-)^- = x \wedge y)$ holds, then we have:
(m-EqOW) $x \leq_m^O y \Longleftrightarrow x \leq_m y$.

Proof.
(1): Suppose $x \leq_m^O y$, i.e. $x \wedge y = x$, by (m-dfOW); then, $x \wedge y^- = (x \wedge y) \wedge y^- \overset{(m-Wass)}{=} x \wedge (y \wedge y^-) \overset{(m-WRe)}{=} x \wedge 0 \overset{(m-WL)}{=} 0$; thus, $x \leq_m y$.
(2): If $x \leq_m y$, i.e. $x \wedge y^- = 0$, then $x \wedge y \overset{(m-Wdiv)}{=} x \wedge (x \wedge y^-)^- = x \wedge 0^- \overset{(Neg0-1)}{=} x \wedge 1 \overset{(m-Wcomm)}{=} 1 \wedge x \overset{(m-WU)}{=} x$, i.e. $x \leq_m^O y$, by (m-dfOW) again. Thus, $x \leq_m^O y \Longleftrightarrow x \leq_m y$, by above (1), i.e. (m-EqOW) holds. □

Remark 6.31. The converse of Proposition 6.30 (1) does not hold. For example, in the ortholattice $\mathcal{O}_1$ from Examples 6.29, we have $a \leq_m b$, i.e. $a \wedge_1 b^{-1} = a \wedge_1 c = 0$, but $a \not\leq_m^O b$.

Consider the following equivalent (dual) properties of distributivity:
(m-Wdis) (distributivity of wedge over vee) $z \wedge (x \vee y) = (z \wedge x) \vee (z \wedge y)$ and, dually,
(m-Vdis) (distributivity of vee over wedge) $z \vee (x \wedge y) = (z \vee x) \wedge (z \vee y)$.

Then, we have the following important result (the dual one is omitted):

Theorem 6.32. *Let $\mathcal{A}^L = (A^L, \wedge, \vee, ^-, 0, 1)$ be a left-ortholattice. The following are equivalent:*
(m-@)=(m-Wdiv) $\quad x \wedge (x \wedge y^-)^- = x \wedge y$;
(m-Wdis) $\quad z \wedge (x \vee y) = (z \wedge x) \vee (z \wedge y)$.

Proof.
(m-Wdiv) $\Longrightarrow$ (m-Wdis): By Proposition 6.30 (2), (m-EqOW) holds.
We prove first that:

$$(z \wedge x) \vee (z \wedge y) \leq_m^O z \wedge (x \vee y). \tag{9}$$

Indeed, by (m-O-Vgeq'), x, $y \leq_m^O x \vee y$; then, by (m-Wcomm), (m-O-*), $z \wedge x$, $z \wedge y \leq_m^O z \wedge (x \vee y)$, i.e. $z \wedge (x \vee y)$ is a majorant of $\{z \wedge x, z \wedge y\}$; then, $(z \wedge x) \vee (z \wedge y) \leq_m^O z \wedge (x \vee y)$, i.e. (9) holds.
Now we prove that:

$$z \wedge (x \vee y) \leq_m^O (z \wedge x) \vee (z \wedge y). \tag{10}$$

Indeed, first: $x \vee y \overset{(m-O-Vgeq')}{\leq_m^O} z^- \vee (x \vee y) \overset{(DeM1),(DN)}{=} (z \wedge x^- \wedge y^-)^-$
$\overset{(m-Wid)}{=} (z \wedge z \wedge x^- \wedge y^-)^- \overset{(m-Wcomm),(m-Wass)}{=} ((z \wedge x^-) \wedge (z \wedge y^-))^-$
$\overset{(DN),(m-Wdiv)}{=} [(z \wedge (z \wedge x)^-) \wedge (z \wedge (z \wedge y)^-)]^-$
$\overset{(m-Wcomm),(m-Wass),(m-Wid)}{=} [z \wedge (z \wedge x)^- \wedge (z \wedge y)^-]^-$.
Then, by (m-EqOW), (DN), we obtain: $0 = (x \vee y) \wedge [z \wedge (z \wedge x)^- \wedge (z \wedge y)^-]$
$\overset{(m-Wass)}{=} [(x \vee y) \wedge z] \wedge [(z \wedge x)^- \wedge (z \wedge y)^-] \overset{(DeM1)}{=} [(x \vee y) \wedge z] \wedge [(z \wedge x) \vee (z \wedge y)]^-$.
Then, by (m-EqOW) again, $z \wedge (x \vee y) \overset{(m-Wcomm)}{=} (x \vee y) \wedge z \leq_m^O (z \wedge x) \vee (z \wedge y)$, i.e. (10) holds.
Finally, by (9), (10) and (m-O-An'), it follows that (m-Wdis) holds.

(m-Wdis) $\Longrightarrow$ (m-Wdiv): First, $x \wedge (x^- \vee y) \overset{(m-Wdis)}{=} (x \wedge x^-) \vee (x \wedge y) \overset{(m-WRe)}{=} 0 \vee (x \wedge y) \overset{(m-VU)}{=} x \wedge y$. Then, $x \wedge y = x \wedge (x^- \vee y) \overset{(DeM1)}{=} x \wedge ((x^-)^- \wedge y^-)^- \overset{(DN)}{=} x \wedge (x \wedge y^-)^-$, i.e. (m-Wdiv) holds. □

6.5.2 Connections with the ortholattices

Proposition 6.33. *Let $\mathcal{A}^L = (A^L, \wedge, \vee, ^-, 0, 1)$ be a left-ortholattice. Then,*
$(A^L, \odot = \wedge, ^-, 1)$
is a m-Pimplicative, involutive left-m-BE algebra.

Proof. By Proposition 6.28, (Neg1-0), (Neg01), (m-WL) and (m-Wimpl) hold. Then:
(PU): $1 \odot x = 1 \wedge x \overset{(m-WU)}{=} x$.
(Pcomm): $x \odot y = x \wedge y \overset{(m-Wcomm)}{=} y \wedge x = y \odot x$.
(Pass): $x \odot (y \odot z) = x \wedge (y \wedge z) \overset{(m-Wass)}{=} (x \wedge y) \wedge z = (x \odot y) \odot z$.
(m-L): $x \odot 0 = x \wedge 0 \overset{(m-WL)}{=} 0$.
(m-Re): $x \odot x^- = x \wedge x^- \overset{(m-WRe)}{=} 0$.
(m-Pimpl): $[(x \odot y^-)^- \odot x^-]^- = [(x \wedge y^-)^- \wedge x^-]^- \overset{(m-Wimpl)}{=} x$.
(DN) holds by hypothesis. □

In order to establish the converse result, first note that we have:

Proposition 6.34. *Let $\mathcal{A}^L = (A^L, \odot, ^-, 1)$ be an m-Pimplicative left-m-BE algebra. Then, (DN), (m-Pimpl-1) and (G) hold.*

Proof. By (mCIM1) or (mCIM1'), (DN) holds. Then, by (mCIM2), (m-Pimpl-1) holds and by (mCIM3), (G) holds. □

Now, we can establish the converse of Proposition 6.33.

Proposition 6.35. *Let $\mathcal{A}^L = (A^L, \odot, ^-, 1)$ be an m-Pimplicative (involutive) left-m-BE algebra. Define, for all $x, y \in A^L$,*

$$x \wedge y \stackrel{def.}{=} x \odot y, \quad x \vee y \stackrel{def.}{=} x \oplus y, \quad 0 \stackrel{def.}{=} 1^-,$$

where $x \oplus y \stackrel{def.}{=} (x^- \odot y^-)^-$.

Then, $(A^L, \wedge, \vee, ^-, 0, 1)$ is a left-ortholattice.

Proof. By Proposition 6.34, (DN), (m-Pimpl-1) and (G) hold.

(1): (m-Wid): $x \wedge x = x \odot x \stackrel{(G)}{=} x$.

(m-Wcomm): $x \wedge y = x \odot y \stackrel{(Pcomm)}{=} y \odot x = y \wedge x$.

(m-Wass): $(x \wedge y) \wedge z = (x \odot y) \odot z \stackrel{(Pass)}{=} x \odot (y \odot z) = x \wedge (y \wedge z)$.

(m-WU): $1 \wedge x = 1 \odot x \stackrel{(PU)}{=} x$.

(m-Wabs): $x \wedge (x \vee y) = x \odot (x \oplus y) = x \odot (x^- \odot y^-)^- \stackrel{(m-Pimpl-1)}{=} x$.

(m-Vid): $x \vee x = x \oplus x = (x^- \odot x^-)^- \stackrel{(G)}{=} (x^-)^- \stackrel{(DN)}{=} x$.

(m-Vcomm): $x \vee y = x \oplus y = (x^- \odot y^-)^- \stackrel{(Pcomm)}{=} (y^- \odot x^-)^- = y \oplus x = y \vee x$.

(mVass): $(x \vee y) \vee z = (x \oplus y) \oplus z = (x^- \odot y^-)^- \oplus z \stackrel{(DN)}{=} [(x^- \odot y^-) \odot z^-]^- \stackrel{(Pass)}{=}$
$[x^- \odot (y^- \odot z^-)]^- \stackrel{(DN)}{=} [x^- \odot (y \oplus z)^-]^- = x \oplus (y \oplus z) = x \vee (y \vee z)$.

(m-VU): $0 \vee x = 0 \oplus x = (0^- \odot x^-)^- \stackrel{(Neg0-1)}{=} (1 \odot x^-)^- \stackrel{(PU)}{=} (x^-)^- \stackrel{(DN)}{=} x$.

(m-Vabs): $x \vee (x \wedge y) = x \oplus (x \odot y) = [x^- \odot (x \odot y)^-]^- \stackrel{(m-Pimpl)}{=} x$.

Thus, $(A^L, \wedge, \vee, 0, 1)$ is a bounded Dedekind lattice.

(2): The unary operation $^-$ satisfies (DN), by Proposition 6.34.

(DeM1): $(x \vee y)^- = (x \oplus y)^- = ((x^- \odot y^-)^-)^- \stackrel{(DN)}{=} x^- \odot y^- = x^- \wedge y^-$;

(DeM2): $(x \wedge y)^- = (x \odot y)^- \stackrel{(DN)}{=} ((x^-)^- \odot (y^-)^-)^- = x^- \oplus y^- = x^- \vee y^-$.

(m-WRe): $x \wedge x^- = x \odot x^- \stackrel{(m-Re)}{=} 0$ and

(m-VRe): $x \vee x^- = (x^- \wedge (x^-)^-)^- \stackrel{(m-WRe)}{=} 0^- \stackrel{(Neg0-1)}{=} 1$. □

The next result says that the class of m-Pimplicative (involutive) m-BE algebras is *definitionally equivalent* (d.e.) to the class of ortholattices (hence their categories are isomorphic).

Theorem 6.36.

(1) Let $\mathcal{A}^L = (A^L, \odot, ^-, 1)$ *be an m-Pimplicative (involutive) left-m-BE algebra. Define*

$$x \wedge y \overset{def.}{=} x \odot y, \quad x \vee y \overset{def.}{=} x \oplus y = (x^- \wedge y^-)^-, \quad 0 \overset{def.}{=} 1^-.$$

Then, $F(\mathcal{A}^L) = (A^L, \wedge, \vee, ^-, 0, 1)$ *is a left-ortholattice.*

(1') Conversely, let $\mathcal{A}^L = (A^L, \wedge, \vee, ^-, 0, 1)$ *be a left-ortholattice. Define*

$$x \odot y \overset{def.}{=} x \wedge y.$$

Then, $G(\mathcal{A}^L) = (A^L, \odot, ^-, 1)$ *is an m-Pimplicative (involutive) left-m-BE algebra.*

(2) The mappings F *and* G *are mutually inverse.*

Proof. By Proposition 6.33 and Proposition 6.35. □

The above Theorem 6.36 helps us to give an equivalent definition of ortholattices, as follows.

Definitions 6.37. (Definition 2)

(i) A *left-ortholattice* is an algebra $\mathcal{A}^L = (A^L, \wedge, ^-, 1)$ of type $(2, 1, 0)$ such that (Neg1-0), (Neg0-1), (m-WU), (m-Wcomm), (m-Wass), (m-WL), (m-WRe), (m-Wimpl) (hence (DN)) hold.

(i') A *right-ortholattice* is an algebra $\mathcal{A}^R = (A^R, \vee, ^-, 0)$ of type $(2, 1, 0)$ such that the dual conditions, (Neg0-1), (Neg1-0), (m-VU), (m-Vcomm), (m-Vass), (m-VL), (m-VRe), (m-Vimpl) (hence (DN)) hold.

6.6 Connections of the new algebras with the Boolean algebras

6.6.1 Boolean algebras

Boolean algebras are only commutative. The most used definition is as a *complemented, distributive, bounded (Dedekind) lattice*, namely:

Definitions 6.38. (Definition 1)

(i) A *left-Boolean algebra* is an algebra $\mathcal{A}^L = (A^L, \wedge = \wedge_m, \vee = \vee_m, ^- = ^{-^L}, 0, 1)$ verifying: for all $x, y, z \in A^L$,

(m-Wid)	$x \wedge x = x$,	(m-Vid)	$x \vee x = x$,
(m-Wcomm)	$x \wedge y = y \wedge x$,	(m-Vcomm)	$x \vee y = y \vee x$,
(m-Wass)	$x \wedge (y \wedge z)$ $= (x \wedge y) \wedge z$,	(m-Vass)	$x \vee (y \vee z)$ $= (x \vee y) \vee z$,
(m-Wabs)	$x \wedge (x \vee y) = x$,	(m-Vabs)	$x \vee (x \wedge y) = x$;
(m-WU)	$1 \wedge x = x$,	(m-VU)	$0 \vee x = x$,
(m-Wdis)	$x \wedge (y \vee z)$ $= (x \wedge y) \vee (x \wedge z)$,	(m-Vdis)	$x \vee (y \wedge z)$ $= (x \vee y) \wedge (x \vee z)$,
(m-WRe)	$x \wedge x^- = 0$,	(m-VRe)	$x \vee x^- = 1$.

(i') Dually, a *right-Boolean algebra* is an algebra $\mathcal{A}^R = (A^R, \vee = \vee_m, \wedge = \wedge_m, ^- = ^{-^R}, 0, 1)$ verifying: (m-Vid), (m-Vcomm), (m-Vass), (m-Vabs), (m-VU), (m-Vdis), (m-VRe) and, dually, (m-Wid), (m-Wcomm), (m-Wass), (m-Wabs), (m-WU), (m-Wdis), (m-WRe).

Denote by **Boole** the class of all left-Boolean algebras and by **Boole**^{R} the class of all right-Boolean algebras.

Note that a left-Boolean algebra verifies (DN) (it is *involutive*), hence it is self-dual, i.e. the dual of $(A^L, \wedge, \vee, ^{-^L}, 0, 1)$ is $(A^L, \vee, \wedge, ^{-^L}, 0, 1)$, and vice-versa.

Dually, a right-Boolean algebra verifies (DN) (it is *involutive*), hence it is self-dual, i.e. the dual of $(A^R, \vee, \wedge, ^{-^R}, 0, 1)$ is $(A^R, \wedge, \vee, ^{-^R}, 0, 1)$, and vice-versa.

Recall that in a left-Boolean algebra $(A^L, \wedge, \vee, ^-, 0, 1)$ we can define three equivalent order relations:

$$(m - dfrelW) \qquad x \leq_m y \overset{def.}{\Longleftrightarrow} x \wedge y^- = 0$$

and

$$(m - dfOW) \qquad x \leq_m^O y \overset{def.}{\Longleftrightarrow} x \wedge y = x,$$

$$(m - dfO_mW_m) \qquad x \leq_m^{O_m} y \overset{def.}{\Longleftrightarrow} x \wedge_m y = x,$$

where $x \wedge_m y = (x^- \wedge y)^- \wedge y$.
We have $\wedge_m = \wedge$, hence $x \leq_m^O y \Longleftrightarrow x \leq_m^{O_m} y$, and we have the equivalence:

$$(m - EqOW) \qquad x \leq_m^O y \; (\Longleftrightarrow x \leq_m^{O_m} y\;) \Longleftrightarrow x \leq_m y.$$

6.6.2 Connections with the Boolean algebras

The following result says that the class of m-Pimplicative (involutive) m-BCK algebras and the class of Boolean algebras are *definitionally equivalent* (d.e.) (hence their categories are isomorphic).

Theorem 6.39.

(1) Let $\mathcal{A}^L = (A^L, \odot, ^-, 1)$ *be an m-Pimplicative (involutive) left-m-BCK algebra. Define*

$$x \wedge y \overset{def.}{=} x \odot y, \quad x \vee y \overset{def.}{=} x \oplus y = (x^- \wedge y^-)^-, \quad 0 \overset{def.}{=} 1^-.$$

Then, $F(\mathcal{A}^L) = (A^L, \wedge, \vee, ^-, 0, 1)$ *is a left-Boolean algebra.*

(1') Conversely, let $\mathcal{A}^L = (A^L, \wedge, \vee, ^-, 0, 1)$ *be a left-Boolean algebra. Define*

$$x \odot y \overset{def.}{=} x \wedge y.$$

Then, $G(\mathcal{A}^L) = (A^L, \odot, ^-, 1)$ *is an m-Pimplicative (involutive) left-m-BCK algebra.*

(2) The mappings F *and* G *are mutually inverse.*

Proof

(1): Let $\mathcal{A}^L = (A^L, \odot, ^-, 1)$ be an m-Pimplicative (involutive) left-m-BCK algebra. Then, $\mathcal{A}^L$ is an m-Pimplicative left-m-BE algebra, hence $(A^L, \wedge, \vee, ^-, 0, 1)$ is a left-ortholattice, by Proposition 6.35. By (mCBB2), (m-Pimpl) + (PU) + (Pcomm) + (Pass) + (Neg0-1) + (m-Re) + (m-L) + (m-An) + (m-BB) $\Longrightarrow$
(m-Pdiv), thus (m-@)=(m-Wdiv) holds. Then, (m-Wdis) holds, by Theorem 6.32. Thus, $F(\mathcal{A}^L)$ is a left-Boolean algebra.

(1'): Let $\mathcal{A}^L = (A^L, \wedge, \vee, ^-, 0, 1)$ be a left-Boolean algebra. We prove that $(A^L, \odot, ^-, 1)$ is an m-Pimplicative (involutive) left-m-BCK algebra. Indeed,
(PU): $1 \odot x = 1 \wedge x \overset{(m-WU)}{=} x$.
(Pcomm): $x \odot y = x \wedge y \overset{(m-Wcomm)}{=} y \wedge x = y \odot x$.
(Pass): $x \odot (y \odot z) = x \wedge (y \wedge z) \overset{(m-Wass)}{=} (x \wedge y) \wedge z = (x \odot y) \odot z$.
(m-L): $x \odot 0 = x \wedge 0 \overset{(m-WL)}{=} 0$.
(m-Re): $x \odot x^- = x \wedge x^- \overset{(m-WRe)}{=} 0$. Hence, $(A^L, \odot, ^-, 1)$ is until now a left-m-BE algebra.
(DN) holds and $\leq_m$ is an order relation, hence (m-An) holds.
(m-BB): by (m-Wdis), (m-@)=(m-Wdiv) holds, hence (m-Pdiv) holds; then, by (mCBB1), (DN) + (Pcomm) + (Pass) + (m-Re) + (m-L) + (m-Pdiv) $\Longrightarrow$
(m-BB). Hence, $(A^L, \odot, ^-, 1)$ is until now an (involutive) left-m-BCK algebra.
(m-Pimpl): $[(x \odot y^-)^- \odot x^-]^- = [(x \wedge y^-)^- \wedge x^-]^- = x \vee (x \wedge y^-) \overset{(m-Wabs)}{=} x$. Hence, $G(\mathcal{A}^L)$ is an m-Pimplicative (involutive) left-m-BCK algebra.

(2): Routine. □

Recall the well-known result:

Theorem 6.40. *The class of MV algebras of Gödel type (i.e. satisfying the property (G)) is d.e. to the class of Boolean algebras.*

Theorem 6.41. *The class of ortholattices satisfying one of the equivalent conditions (m-@)=(m-Wdiv), (m-Wdis), (m-Vdis) is d.e. to the class of Boolean algebras.*

Proof. Obviously. □

By Theorem 6.41 and Definition 2 of ortholattices, we obtain an equivalent definition of Boolean algebras:

Definitions 6.42. (Definition 2)

(i) A *left-Boolean algebra* is an algebra $\mathcal{A}^L = (A^L, \wedge, ^-, 1)$ of type $(2,1,0)$ such that the following properties hold: for all $x, y, z \in A^L$,

(Neg0-1)	$0^- = 1$, where $0 \overset{def.}{=} 1^-$,
(m-WU)	$1 \wedge x = x$,
(m-Wcomm)	$x \wedge y = y \wedge x$,
(m-Wass)	$x \wedge (y \wedge z) = (x \wedge y) \wedge z$,
(m-WL)	$x \wedge 0 = 0$,
(m-WRe)	$x \wedge x^- = 0$,
(m-Wimpl)	$[(x \wedge y^-)^- \wedge x^-]^- = x$,
(m-@)= (m-Wdiv)	$x \wedge (x \wedge y^-)^- = x \wedge y$.

(i') A *right-Boolean algebra* is an algebra $\mathcal{A}^R = (A^R, \vee, ^-, 0)$ of type $(2,1,0)$ such that the following properties hold: for all $x, y, z \in A^R$,

(Neg1-0)	$1^- = 0$, where $1 \overset{def.}{=} 0^-$,
(m-VU)	$0 \vee x = x$,
(m-Vcomm)	$x \vee y = y \vee x$,
(m-Vass)	$x \vee (y \vee z) = (x \vee y) \vee z$,
(m-VF)	$x \vee 1 = 1$,
(m-VRe)	$x \vee x^- = 1$,
(m-Vimpl)	$[(x \vee y^-)^- \vee x^-]^- = x$,
(m-@^{R})= (m-Vdiv)	$x \vee (x \vee y^-)^- = x \vee y$.

6.7 Final remarks in the case $1^- = 0$

Resuming, we have the connections from the following Figure 8.

Remark 6.43. The MV algebras and the ortholattices are incomparable: any MV algebra does not verify (m-Pimpl) and any ortholattice does not verify ($\wedge_m$-comm).

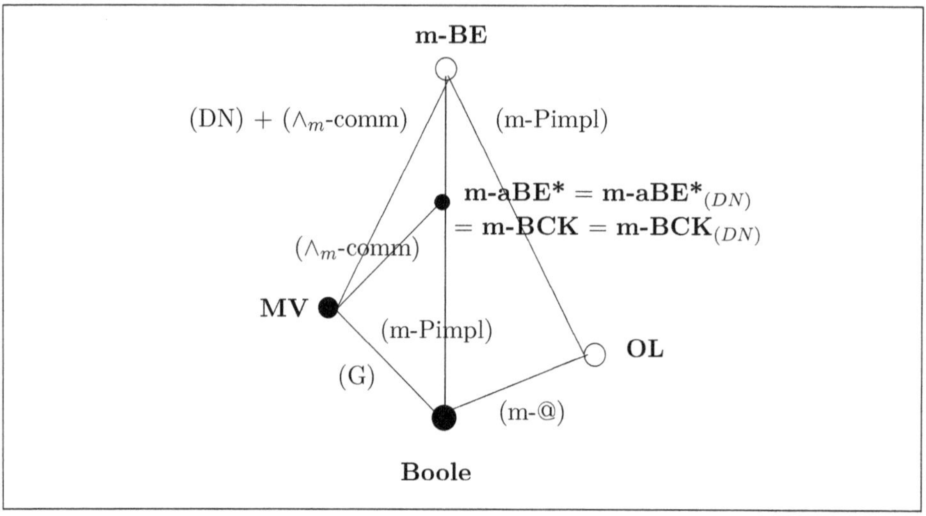

Figure 8: Connections between m-BE algebras, m-BCK algebras, MV algebras, ortholattices and Boolean algebras

It is an open problem to define the non-associative MV algebras, the non-associative ortholattices and the non-associative Boolean algebras and to connect them with the non-associative algebras defined above (the m-ML, ..., m-BCC algebras).

7 Algebras $(A, \cdot, ^{-1}, 1)$ with $1^{-1} = 1$ (Hierarchies m-1, m-1')

In this section, the product operation will be denoted by $\cdot$ instead of $\odot$, just as the negation operation is denoted by $^{-1}$ (and is called the *inverse*) instead of $^{-}$.

7.1 The list m$_1$-A of basic properties. New algebras

Let $\mathcal{A}^L = (A^L, \cdot, ^{-1}, 1)$ be an algebra of type $(2, 1, 0)$ such that $1^{-1} = 1$. Define an *internal* binary relation $\leq_{m_1}$ on A^L by: for all $x, y \in A^L$,

$$(m - dfrelP_1) \qquad x \leq_{m_1} y \overset{def.}{\Longleftrightarrow} x \cdot y^{-1} = 1.$$

Equivalently, consider a structure $\mathcal{A}^L = (A^L, \leq_{m_1}, \cdot, ^{-1}, 1)$ such that $1^{-1} = 1$ and the *internal* binary relation $\leq_{m_1}$, the binary operation $\cdot$, the unary operation

$^{-1}$ and the element 1 are connected by the equivalence:

$$(m - EqrelP_1) \qquad x \leq_{m_1} y \Longleftrightarrow x \cdot y^{-1} = 1.$$

Consider the following list $\mathbf{m}_1$-$\mathbf{A}$ of basic properties determined by the corresponding two above equivalent definitions of $\mathcal{A}^L$:

(PU) $1 \cdot x = x = x \cdot 1$ (unit element of product, the *identity*);
(Pcomm) $x \cdot y = y \cdot x$ (commutativity of product),
(Pass) $x \cdot (y \cdot z) = (x \cdot y) \cdot z$ (associativity of product);

(N1) $1^{-1} = 1$;

(m$_1$-An) $(x \cdot y^{-1} = 1$ and $y \cdot x^{-1} = 1) \Longrightarrow x = y$ (antisymmetry),
(m$_1$-An') $(x \leq_{m_1} y$ and $y \leq_{m_1} x) \Longrightarrow x = y$ (antisymmetry);

(m$_1$-B) $[(x \cdot y^{-1})^{-1} \cdot (x \cdot z)] \cdot (y \cdot z)^{-1} = 1$,
(m$_1$-B') $(x \cdot y^{-1})^{-1} \cdot (x \cdot z) \leq_{m_1} y \cdot z$;
(m$_1$-B=) $(x \cdot y^{-1})^{-1} \cdot (x \cdot z) = y \cdot z$;

(m$_1$-BB) $[(z \cdot x)^{-1} \cdot (y \cdot x)] \cdot (y \cdot z^{-1})^{-1} = 1$,
(m$_1$-BB') $(z \cdot x)^{-1} \cdot (y \cdot x) \leq_{m_1} y \cdot z^{-1}$;
(m$_1$-BB=) $(z \cdot x)^{-1} \cdot (y \cdot x) = y \cdot z^{-1}$;

(m$_1$-*) $x \cdot y^{-1} = 1 \Longrightarrow (z \cdot y^{-1}) \cdot (z \cdot x^{-1})^{-1} = 1$,
(m$_1$-*') $x \leq_{m_1} y \Longrightarrow z \cdot y^{-1} \leq_{m_1} z \cdot x^{-1}$;

(m$_1$-**) $x \cdot y^{-1} = 1 \Longrightarrow (x \cdot z) \cdot (y \cdot z)^{-1} = 1$,
(m$_1$-**') $x \leq_{m_1} y \Longrightarrow x \cdot z \leq_{m_1} y \cdot z$;

(m$_1$-D) $[(y^{-1} \cdot x)^{-1} \cdot x] \cdot y^{-1} = 1$,
(m$_1$-D') $(y^{-1} \cdot x)^{-1} \cdot x \leq_{m_1} y$;
(m$_1$-D=) $(y^{-1} \cdot x)^{-1} \cdot x = y$;

(m$_1$-Re) $x \cdot x^{-1} = 1$ (reflexivity),
(m$_1$-Re') $x \leq_{m_1} x$ (reflexivity);

(m$_1$-Tr) $(x \cdot y^{-1} = 1$ and $y \cdot z^{-1} = 1) \Longrightarrow x \cdot z^{-1} = 1$ (transitivity),
(m$_1$-Tr') $(x \leq_{m_1} y$ and $y \leq_{m_1} z) \Longrightarrow x \leq_{m_1} z$ (transitivity);

(m-p-s) $x \leq_{m_1} y \Longleftrightarrow x = y$ (m-p-semisimple);
(EqP=) $x = y \Longleftrightarrow x \cdot y^{-1} = 1$.

Dually,
let $\mathcal{A}^R = (A^R, +, -, 0)$ be an algebra of type $(2, 1, 0)$ such that $-0 = 0$. Define an *internal* binary relation $\geq_{m_0}$ on A^R by: for all $x, y \in A^R$,

$$(m - dfrelS_0) \qquad x \geq_{m_0} y \overset{def.}{\Longleftrightarrow} x + (-y) = 0.$$

Equivalently, consider a structure $\mathcal{A}^R = (A^R, \geq_{m_0}, +, -, 0)$ such that $-0 = 0$ and the *internal* binary relation $\geq_{m_0}$, the binary operation $+$, the unary operation $-$ and the element 0 are connected by the equivalence:

$$(m - EqrelS_0) \qquad x \geq_{m_0} y \Longleftrightarrow x + (-y) = 0.$$

The list of basic dual properties of the dual algebra $\mathcal{A}^R = (A^R, +, -, 0)$ (structure $\mathcal{A}^R = (A^R, \geq_{m_0}, +, -, 0)$) is omitted.

7.1.1 New algebras

Let $\mathcal{A}^L = (A^L, \cdot, ^{-1}, 1)$ be an algebra of type $(2, 1, 0)$ through this section such that (N1) holds $(1^{-1} = 1)$.

We now introduce the following notions, the analogous of those *M algebras without last element* from Section 2 (Hierarchies 1 and 1') (the dual ones are omitted).

- Algebras without last element, without associativity:
$\mathcal{A}^L$ is a:

(m2) *left-m-RM algebra*, if it verifies (PU), (Pcomm), (m_1-Re), i.e. it is a *commutative left-unital magma* verifying (m_1-Re),

(m3) *left-pre-m-BZ algebra*, if it verifies (PU), (Pcomm), (m_1-Re), (m_1-B),

(m4) *left-m-aRM = left-m-BH algebra*, if it verifies (PU), (Pcomm), (m_1-Re), (m_1-An),

(m5) *left-m-BZ algebra*, if it verifies (PU), (Pcomm), (m_1-Re), (m_1-An), (m_1-B);

- Algebras without last element, with associativity:
$\mathcal{A}^L$ is a:

(mE2) *left-m-RME =left-m-CI algebra*, if it verifies (PU), (Pcomm), (Pass), (m_1-Re), i.e. it is a *commutative left-monoid* verifying (m_1-Re),

(mE3) *left-pre-m-BCI algebra*, if it verifies (PU), (Pcomm), (Pass), (m_1-Re), (m_1-B), (mE4) *left-m-BCH algebra,* if it verifies (PU), (Pcomm), (Pass), (m_1-Re), (m_1-An),

(mE5) *left-m-BCI algebra,* if it verifies (PU), (Pcomm), (Pass), (m_1-Re), (m_1-An), (m_1-B)

and, additionally:

(mE6) *left-m-RME* = left-m-CI* algebra,* if it is a left-m-RME = left-m-CI algebra verifying (m_1-*),

(mE7) *left-m-BCH* algebra,* if it is a left-m-BCH algebra verifying (m_1-*).

Denote by **m-RM**, ... **m-BCH***, the classes of left-m-RM, ..., left-m-BCH* algebras, respectively.

Remarks 7.1.

(i) Note that the left-m-RM algebra is just the commutative (multiplicative) *moon*, as defined in [42], Definition 10.1.5.

(ii) Note that the left-m-CI algebra is just the commutative (multiplicative) *group* (recalled and studied in [42], Definition 7.1.1).

7.1.2 Connections between the properties

Proposition 7.2. *(See Proposition 6.2)*

We have (the dual case is omitted):

*(1mA4) (PU) + (Pcomm) + (N1) + (m_1-BB) $\Longrightarrow$ (m_1-**),*

(1mA5) (PU) + (Pcomm) + (N1) + (m_1-B) $\Longrightarrow$ (m_1-);*

*(1mA6) (PU) + (N1) + (m_1-**) $\Longrightarrow$ (m_1-Tr),*

(1mA7) (PU) + (N1) + (m_1-) $\Longrightarrow$ (m_1-Tr).*

Proof.

(1mA4): Suppose $y \leq_{m_1} z$, i.e. $y \cdot z^{-1} = 1$; then, by (m_1-BB),

$1 = [(z \cdot x)^{-1} \cdot (y \cdot x)] \cdot 1^{-1} \overset{(N1)}{=} [(z \cdot x)^{-1} \cdot (y \cdot x)] \cdot 1 \overset{(Pcomm),(PU)}{=} (z \cdot x)^{-1} \cdot (y \cdot x)$ $\overset{(Pcomm)}{=} (y \cdot x) \cdot (z \cdot x)^{-1}$, i.e. $y \cdot x \leq_{m_1} z \cdot x$; thus, ($m_1$-**') holds.

(1mA5): Suppose $y \leq_{m_1} z$, i.e. $y \cdot z^{-1} = 1$; then, by (m_1-B),

$1 = [(x \cdot y^{-1})^{-1} \cdot (x \cdot z^{-1})] \cdot (y \cdot z^{-1})^{-1} \overset{(N1)}{=} [(x \cdot y^{-1})^{-1} \cdot (x \cdot z^{-1})] \cdot 1$ $\overset{(Pcomm),(PU)}{=} (x \cdot y^{-1})^{-1} \cdot (x \cdot z^{-1}) \overset{(Pcomm)}{=} (x \cdot z^{-1}) \cdot (x \cdot y^{-1})^{-1}$,

i.e. $x \cdot z^{-1} \leq_{m_1} x \cdot y^{-1}$; thus, ($m_1$-*') holds.

(1mA6): Suppose $x \leq_{m_1} y$ and $y \leq_{m_1} z$, i.e. $x \cdot y^{-1} = 1$ and $y \cdot z^{-1} = 1$; then, by (m_1-**'), $x \cdot z^{-1} \leq_{m_1} y \cdot z^{-1}$, i.e. $(x \cdot z^{-1}) \cdot (y \cdot z^{-1})^{-1} = 1$; hence, $(x \cdot z^{-1}) \cdot 1^{-1} = 1$; then, by (N1), (PU), we obtain $x \cdot z^{-1} = 1$, i.e. $x \leq_{m_1} z$; thus, (m_1-Tr') holds.

(1mA7): Suppose $x \leq_{m_1} y$ and $y \leq_{m_1} z$, i.e. $x \cdot y^{-1} = 1$ and $y \cdot z^{-1} = 1$; then, by (m_1-*'), $x \cdot z^{-1} \leq_{m_1} x \cdot y^{-1}$, i.e. $(x \cdot z^{-1}) \cdot (x \cdot y^{-1})^{-1} = 1$; then, $1 = (x \cdot z^{-1}) \cdot 1^{-1} \stackrel{(N1)}{=} (x \cdot z^{-1}) \cdot 1 \stackrel{(Pcomm),(PU)}{=} x \cdot z^{-1}$, i.e. $x \leq_{m_1} z$; thus, (m_1-Tr') holds. □

7.2 The list m_1-C of properties of the inverse. Connections

Consider now the following list **m1-C** of properties of the inverse $^{-1}$ (the dual one is omitted):

(m_1-Neg4') $(x^{-1})^{-1} \leq_{m_1} x$,
(m_1-Neg7) $((x^{-1})^{-1})^{-1} = x^{-1}$;
(m_1-Neg2') $x \leq_{m_1} y \Longrightarrow y^{-1} \leq_{m_1} x^{-1}$;
(DN) (Double Negation) $(x^{-1})^{-1} = x$;
(m_1-DN4') $x \leq_{m_1} y \Longleftrightarrow y^{-1} \leq_{m_1} x^{-1}$.

Proposition 7.3. *(See Proposition 6.8)*
We have:
(m1C1) (m_1-) + (PU) $\Longrightarrow$ (m_1-Neg2'),*
(m1C2) (m_1-Re) + (Pcomm) $\Longrightarrow$ (m_1-Neg4'),
(m1C2') (m_1-D) + (PU) + (Pcomm) $\Longrightarrow$ (m_1-Neg4');
(m1C3) (m_1-Neg4') + (m_1-) + (PU) + (m_1-An) $\Longrightarrow$ (m_1-Neg7).*

Proof.
(m1C1): Suppose $x \leq_{m_1} y$; then, by (m_1-*), $z \cdot y^{-1} \leq_{m_1} z \cdot x^{-1}$; then, for $z = 1$, we obtain $1 \cdot y^{-1} \leq_{m_1} 1 \cdot x^{-1}$, hence, $y^{-1} \leq_{m_1} x^{-1}$, by (PU); thus, (m_1-Neg2') holds.

(m1C2): $(x^{-1})^{-1} \cdot x^{-1} \stackrel{(Pcomm)}{=} x^{-1} \cdot (x^{-1})^{-1} \stackrel{(m_1-Re)}{=} 1$, i.e. $(x^{-1})^{-1} \leq_{m_1} x$; thus, ($m_1$-Neg4') holds.

(m1C2'): Take $x = 1$ in (m_1-D); we obtain $(y^{-1} \cdot 1)^{-1} \cdot 1 \leq_{m_1} y$; hence, $(y^{-1})^{-1} \leq_{m_1} y$, by (Pcomm), (PU); thus, (m_1-Neg4') holds.

(m1C3): By (m_1-Neg4'), $((y^{-1})^{-1})^{-1} \leq_{m_1} y^{-1}$. On the other hand, by (m_1-Neg4'), $(y^{-1})^{-1} \leq_{m_1} y$, hence $1 \cdot y^{-1} \leq_{m_1} 1 \cdot ((y^{-1})^{-1})^{-1}$, by ($m_1$-*); then, $y^{-1} \leq_{m_1} ((y^{-1})^{-1})^{-1}$, by (PU). Now apply ($m_1$-An) to obtain $((y^{-1})^{-1})^{-1} = y^{-1}$; thus, ($m_1$-Neg7) holds. □

Definition 7.4. We shall say that an algebra $\mathcal{A}^L = (A^L, \cdot, ^{-1}, 1)$ is *involutive*, if it verifies (DN).

The subclass of involutive algebras of the class $\mathbf{X}$ will be denoted by $\mathbf{X}_{(DN)}$.

Proposition 7.5. *(See Proposition 6.11)*
We have:
(m1CDN2) (DN) + (Pcomm) $\Longrightarrow$ *(m_1-DN4),*
(m1CDN3) (DN) + (Pcomm) $\Longrightarrow$ *((m_1-*)* $\Leftrightarrow$ *(m_1-**)),*
(m1CDN5) (PU) + (Pcomm) + (m_1-Re) + (m_1-) + (m_1-An)* $\Longrightarrow$ *(DN).*

Proof.
(m1CDN2): $x \leq_{m_1} y \overset{(m-dfrelP_1)}{\Longleftrightarrow} x \cdot y^{-1} = 1 \overset{(DN)}{\Longleftrightarrow} (x^{-1})^{-1} \cdot y^{-1} = 1 \overset{(Pcomm)}{\Longleftrightarrow}$ $y^{-1} \cdot (x^{-1})^{-1} = 1 \overset{(m-dfrelP_1)}{\Longleftrightarrow} y^{-1} \leq_{m_1} x^{-1}$; thus, ($m_1$-DN4') holds.

(m1CDN3): By (m1CDN2), (DN) + (Pcomm) $\Longrightarrow$ (m_1-DN4').
Suppose that (m_1-*') holds; then, if $x \leq_{m_1} y$, then, by (m_1-DN4'), $y^{-1} \leq_{m_1} x^{-1}$, hence, by (m_1-*'), $z \cdot (x^{-1})^{-1} \leq_{m_1} z \cdot (y^{-1})^{-1}$; then, by (DN), $z \cdot x \leq_{m_1} z \cdot y$; then, by (Pcomm), $x \cdot z \leq_{m_1} y \cdot z$; thus, ($m_1$-**') holds.
Conversely, suppose that (m_1-**') holds; then, if $x \leq_{m_1} y$, then, by (m_1-DN4'), $y^{-1} \leq_{m_1} x^{-1}$; then, by (m_1-** '), $y^{-1} \cdot z \leq_{m_1} x^{-1} \cdot z$; then, by (Pcomm), $z \cdot y^{-1} \leq_{m_1} z \cdot x^{-1}$; thus, ($m_1$-*') holds.

(m1CDN5): By (m1C2), (m_1-Re) + (Pcomm) $\Longrightarrow$ (m_1-Neg4'),
i.e. $(x^{-1})^{-1} \leq_{m_1} x$.
On the other hand, by (m1C3), (m_1-Neg4') + (m_1-*) + (PU) + (m_1-An) $\Longrightarrow$ (m_1-Neg7), i.e. $((x^{-1})^{-1})^{-1} = x^{-1}$; then, $x \cdot ((x^{-1})^{-1})^{-1} = x \cdot x^{-1} \overset{(m_1-Re)}{=} 1$, i.e. $x \leq_{m_1} (x^{-1})^{-1}$.
Now, by (m_1-An), we obtain $(x^{-1})^{-1} = x$, i.e. (DN) holds. □

Theorem 7.6. *Any left-m-BZ algebra is involutive.*

Proof. Let $\mathcal{A}^L = (A^L, \cdot, ^{-1}, 1)$ be a left-m-BZ algebra, i.e. (N1), (PU), (Pcomm), (m_1-Re), (m_1-An), (m_1-B) hold. Then, by (1mA5), (PU) + (m_1-B) + (Pcomm) + (N1) $\Longrightarrow$ (m_1-*), hence (m_1-*) holds too. By (m1CDN5), (PU) + (Pcomm) + (m_1-Re) + (m_1-*) + (m_1-An) $\Longrightarrow$ (DN), thus, (DN) holds. □

By above theorem, we shall write: **m-BZ** = $\mathbf{m\text{-}BZ}_{(DN)}$.

7.3 The m-p-semisimple property

Let us introduce the following definition, by analogy with the *p-semisimple* property from the algebras of logic, recalled in Section 4, Definition 4.6.

Definition 7.7. Let $\mathcal{A}^L = (A^L, \cdot, ^{-1}, 1)$ be an algebra (or, equivalently, let $\mathcal{A}^L = (A^L, \leq_{m_1}, \cdot, ^{-1}, 1)$ be a structure) verifying (m_1-Re).
$\mathcal{A}^L$ is said to be *m-p-semisimple*, if for all $x, y \in A^L$,
(m-p-s) $x \leq_{m_1} y \Longrightarrow x = y$ (or, equivalently, by (m_1-Re), $x \leq_{m_1} y \Longleftrightarrow x = y$).

The subclass of m-p-semisimple algebras of the class **X** will be denoted by $\mathbf{X}_{(m-p-s)}$.

Remark 7.8. Note that the m-p-semisimple algebras (structures) verifying the property (m_1-L) ((m_1-L'), respectively) are trivial, because in this case $A^L = \{1\}$. Therefore, the m-p-semisimple property is studied only for algebras (structures) *without last element.*

The following equivalence was introduced in [42], Definition 10.1.24 (in the non-commutative case) to define a *sharp* moon:
(EqP=) $x = y \Longleftrightarrow x \cdot y^{-1} = 1$.

Proposition 7.9. *(See Proposition 4.8)*
Let $(A^L, \cdot, ^{-1}, 1)$ *be an algebra verifying* (m_1-Re). *We have:*
(0) (m-dfrelP$_1$*) + (m-p-s)* $\Longrightarrow$ *(EqP=),*
(0') (m-dfrelP$_1$*) + (EqP=)* $\Longrightarrow$ *(m-p-s),*
(0") (m-dfrelP$_1$*)* $\Longrightarrow$ *((m-p-s)* $\Leftrightarrow$ *(EqP=));*
(1) (m-p-s) $\Longrightarrow$ *(*m_1*-An),*
(1') (m-p-s) $\Longrightarrow$ *(*m_1*-*'),*
(1") (m-p-s) $\Longrightarrow$ *(*m_1*-**'),*
(1"') (m-p-s) $\Longrightarrow$ *(*m_1*-Tr');*
(2) (m-p-s) + (m_1*-BB')* $\Longrightarrow$ *(*m_1*-BB=),*
(2') (m-p-s) + (m_1*-B')* $\Longrightarrow$ *(*m_1*-B=),*
(2") (m-p-s) + (m_1*-D')* $\Longrightarrow$ *(*m_1*-D=);*
(3) (EqP=) + (Pcomm) + (m_1*-Re)* $\Longrightarrow$ *(DN);*
(4) (EqP=) $\Longrightarrow$ *(*m_1*-Re);*
(5) (Pcomm) + (DN) + (m_1*-B=)* $\Longrightarrow$ *(*m_1*-BB=),*
(5') (Pcomm) + (DN) + (m_1*-BB=)* $\Longrightarrow$ *(*m_1*-B=),*
(5") (Pcomm) + (DN) $\Longrightarrow$ *((*m_1*-B=)* $\Leftrightarrow$ *(*m_1*-BB=)).*

Proof. (0), (0'), (0"): Obviously.
(1), (1'), (1"), (1"'): Obviously.
(2), (2'), (2"): Immediately.
(3): $(x^{-1})^{-1} = x \overset{(EqP=)}{\Longleftrightarrow} (x^{-1})^{-1} \cdot x^{-1} = 1$ that is true by (Pcomm) and (m_1-Re); thus, (DN) holds.
(4): (m_1-Re) $\overset{def.}{\Longleftrightarrow} x \cdot x^{-1} = 1 \overset{(EqP=)}{\Longleftrightarrow} x = x$, that is true.
(5): $(z \cdot x)^{-1} \cdot (y \cdot x) \overset{(Pcomm)}{=} (x \cdot z)^{-1} \cdot (x \cdot y)$
$\overset{(DN)}{=} (x \cdot (z^{-1})^{-1})^{-1} \cdot (x \cdot y) \overset{(m_1-B=)}{=} z^{-1} \cdot y \overset{(Pcomm)}{=} y \cdot z^{-1}$; thus, ($m_1$-BB=) holds.

(5'): $(x \cdot y^{-1})^{-1} \cdot (x \cdot z) \overset{(Pcomm)}{=} (y^{-1} \cdot x)^{-1} \cdot (z \cdot x)$
$\overset{(m_1-BB=)}{=} z \cdot (y^{-1})^{-1} \overset{(DN)}{=} z \cdot y \overset{(Pcomm)}{=} y \cdot z$; thus, ($m_1$-B=) holds.

(5"): By (5) and (5'). □

Remark 7.10. By above (0"), being *m-p-semisimple* is equivalent to being *sharp*.

Proposition 7.11. *(See Proposition 4.9)*
Let $(A^L, \cdot, ^{-1}, 1)$ *be an algebra. We have:*
(m1A0) (Pass) + (Pcomm) + (PU) + (m_1-Re) $\Longrightarrow$ (EqP=),
(m1A1) (Pass) + (Pcomm) + (m_1-Re) + (EqP=) $\Longrightarrow$ (m_1-D=);
(m1A1') (Pass) + (Pcomm) + (m_1-Re) + (EqP=) $\Longrightarrow$ (m_1-BB=);
(m1A2) (Pass) + (Pcomm) + (EqP=) + (m_1-BB=) $\Longrightarrow$ (m_1-B=),
(m1A2') (Pass) + (Pcomm) + (EqP=) + (m_1-B=) $\Longrightarrow$ (m_1-BB=),
(m1A2") (Pass) + (Pcomm) + (EqP=) $\Longrightarrow$ ((m_1-B=) $\Leftrightarrow$ (m_1-BB=));
(m1A3) (m_1-D=) + (EqP=) + (Pcomm) + (N1) $\Longrightarrow$ (PU),
(m1A4) (PU) + (Pcomm) + (N1) + (m_1-BB=) $\Longrightarrow$ (m_1-Re),
(m1A5) (PU) + (DN) + (m_1-BB=) $\Longrightarrow$ (m_1-D=),
(m1A6) (PU) + (Pcomm) + (N1) + (m_1-Re) + (m_1-D=) $\Longrightarrow$ (EqP=),
(m1A7) (PU) + (Pcomm) + (N1) + (DN) + (m_1-BB=) $\Longrightarrow$ (Pass).

Proof.

(m1A0): If $x = y$, then $x \cdot y^{-1} = y \cdot y^{-1} \overset{(m_1-Re)}{=} 1$. Conversely, if $x \cdot y^{-1} = 1$, then: $y \overset{(PU)}{=} 1 \cdot y = (x \cdot y^{-1}) \cdot y \overset{(Pass)}{=} x \cdot (y^{-1} \cdot y) \overset{(Pcomm)}{=} (y \cdot y^{-1}) \cdot x \overset{(m_1-Re)}{=} 1 \cdot x \overset{(PU)}{=} x$. Thus, (EqP=) holds.

(m1A1): $[(y^{-1} \cdot x)^{-1} \cdot x] \cdot y^{-1} \overset{(Pass)}{=} (y^{-1} \cdot x)^{-1} \cdot (x \cdot y^{-1}) \overset{(Pcomm)}{=} (x \cdot y^{-1}) \cdot (x \cdot y^{-1})^{-1} \overset{(m_1-Re)}{=} 1$, hence $(y^{-1} \cdot x)^{-1} \cdot x = y$, by (EqP=); thus, ($m_1$-D=) holds.

(m1A1'): By (m1A1), (Pass) + (Pcomm) + (m_1-Re) + (EqP=) imply (m_1-D=). Then,
$[(z \cdot x)^{-1} \cdot (y \cdot x)] \cdot (y \cdot z^{-1})^{-1}$
$\overset{(Pass)}{=} (z \cdot x)^{-1} \cdot [(y \cdot x) \cdot (y \cdot z^{-1})^{-1}]$
$\overset{(Pcomm)}{=} (z \cdot x)^{-1} \cdot [(z^{-1} \cdot y)^{-1} \cdot (y \cdot x)]$
$\overset{(Pass)}{=} (z \cdot x)^{-1} \cdot [((z^{-1} \cdot y)^{-1} \cdot y) \cdot x]$
$\overset{(m_1-D=)}{=} (z \cdot x)^{-1} \cdot (z \cdot x) \overset{(Pcomm)}{=} (z \cdot x) \cdot (z \cdot x)^{-1} \overset{(m_1-Re)}{=} 1.$
Then, by (EqP=), $(z \cdot x)^{-1} \cdot (y \cdot x) = y \cdot z^{-1}$, i.e. ($m_1$-BB=) holds.

(m1A2): $[(x \cdot y^{-1})^{-1} \cdot (x \cdot z)] \cdot (y \cdot z)^{-1} \overset{(Pass)}{=} (x \cdot y^{-1})^{-1} \cdot [(x \cdot z) \cdot (y \cdot z)^{-1}] \overset{(Pcomm)}{=} [(y \cdot z)^{-1} \cdot (x \cdot z)] \cdot (x \cdot y^{-1})^{-1} \overset{(m_1-BB=),(EqP=)}{=} 1$, hence $(x \cdot y^{-1})^{-1} \cdot (x \cdot z) = y \cdot z$, by (EqP=) again, i.e. (m_1-B=) holds.

(m1A2'): $[(z \cdot x)^{-1} \cdot (y \cdot x)] \cdot (y \cdot z^{-1})^{-1} \overset{(Pass)}{=} (z \cdot x)^{-1} \cdot [(y \cdot x) \cdot (y \cdot z^{-1})^{-1}]$ $\overset{(Pcomm)}{=} [(y \cdot z^{-1})^{-1} \cdot (y \cdot x)] \cdot (z \cdot x)^{-1} \overset{(m_1-B=),(EqP=)}{=} 1$, hence $(z \cdot x)^{-1} \cdot (y \cdot x) = y \cdot z^{-1}$, by (EqP=) again, i.e. (m_1-BB=) holds.

(m1A2"): By (m1A2) and (m1A2').

(m1A3): $x = 1 \cdot x \overset{(EqP=)}{\Longleftrightarrow} x \cdot (1 \cdot x)^{-1} = 1 \overset{(Pcomm),(N1)}{\Longleftrightarrow} (1^{-1} \cdot x)^{-1} \cdot x = 1 \overset{(m_1-D=)}{\Longleftrightarrow}$ $1 = 1$, that is true; thus, (PU) holds.

(m1A4): $x \cdot x^{-1} \overset{(PU)}{=} (1 \cdot x) \cdot (1 \cdot x)^{-1} \overset{(Pcomm)}{=} (1 \cdot x)^{-1} \cdot (1 \cdot x) \overset{(m_1-BB=)}{=} 1 \cdot 1^{-1} \overset{(N1)}{=}$ $1 \cdot 1 \overset{(PU)}{=} 1$; thus, ($m_1$-Re) holds.

(m1A5): $(y^{-1} \cdot x)^{-1} \cdot x \overset{(PU)}{=} (y^{-1} \cdot x)^{-1} \cdot (1 \cdot x) \overset{(m_1-BB=)}{=} 1 \cdot (y^{-1})^{-1} \overset{(DN),(PU)}{=} y$; thus, ($m_1$-D=) holds.

(m1A6): If $x = y$, then $x \cdot y^{-1} = x \cdot x^{-1} \overset{(m_1-Re)}{=} 1$. Conversely, if $x \cdot y^{-1} = 1$, then $y \overset{(m_1-D=)}{=} (y^{-1} \cdot x)^{-1} \cdot x \overset{(Pcomm)}{=} (x \cdot y^{-1})^{-1} \cdot x = 1^{-1} \cdot x \overset{(N1)}{=} 1 \cdot x \overset{(PU)}{=} x$. Thus, (EqP=) holds.

(m1A7): By (m1A4), (PU) + (Pcomm) + (N1) + (m_1-BB=) imply (m_1-Re), by (m1A5), (PU) + (DN) + (m_1-BB=) imply (m_1-D=), and by (m1A6), (PU) + (Pcomm) + (N1) + (m_1-Re) + (m_1-D=) imply (EqP=); thus, (m_1-Re), (m_1-D=) and (EqP=) hold. Then,

$1 \overset{(m_1-Re)}{=} y \cdot y^{-1} \overset{(m_1-D=)}{=} [(y^{-1} \cdot x)^{-1} \cdot x] \cdot y^{-1}$
$\overset{(DN)}{=} ([(y^{-1} \cdot x)^{-1} \cdot x]^{-1})^{-1} \cdot y^{-1}$
$\overset{(Pcomm)}{=} y^{-1} \cdot ([(y^{-1} \cdot x)^{-1} \cdot x]^{-1})^{-1}$
$\overset{(m_1-BB=)}{=} ([(y^{-1} \cdot x)^{-1} \cdot x]^{-1} \cdot (z \cdot x))^{-1} \cdot (y^{-1} \cdot (z \cdot x))$
$\overset{(m_1-BB=)}{=} (z \cdot ((y^{-1} \cdot x)^{-1})^{-1})^{-1} \cdot (y^{-1} \cdot (z \cdot x))$
$\overset{(DN)}{=} (z \cdot (y^{-1} \cdot x))^{-1} \cdot (y^{-1} \cdot (z \cdot x))$
$\overset{(Pcomm)}{=} (y^{-1} \cdot (z \cdot x)) \cdot (z \cdot (y^{-1} \cdot x))^{-1}$.

Then, by (EqP=), $y^{-1} \cdot (z \cdot x) = z \cdot (y^{-1} \cdot x)$, hence, by (Pcomm), $y^{-1} \cdot (x \cdot z) = (y^{-1} \cdot x) \cdot z$, hence $y \cdot (x \cdot z) = (y \cdot x) \cdot z$, by (DN); thus, (Pass) holds. □

Proposition 7.12. *(See Proposition 4.10)*

We have:

(m1AB0) (m_1-D=) + (PU) $\Longrightarrow$ (DN);
(m1AB1) (Pass) + (Pcomm) + (m_1-Re) + (EqP=) $\Longrightarrow$ (m_1-B=);
(m1AB7) (PU) + (Pcomm) + (m_1-B=) $\Longrightarrow$ (m_1-D=),
(m1AB7') (PU) + (Pcomm) + (m_1-B=) $\Longrightarrow$ (Pass),
(m1AB7") (PU) + (Pcomm) + (m_1-B=) $\Longrightarrow$ (m_1-BB=).

Proof.

(m1AB0): In (m_1-D=) $((y^{-1}\cdot x)^{-1}\cdot x = y)$ take $x=1$; we obtain, by (PU), $(y^{-1})^{-1} = y$, i.e. (DN) holds.

(m1AB1): By (m1A1), (Pass) + (Pcomm) + (m_1-Re) + (EqP=) imply (m_1-D=), hence $(y^{-1}\cdot x)^{-1}\cdot x = y$; then, $[(y^{-1}\cdot x)^{-1}\cdot x]\cdot z = y\cdot z$, hence $(y^{-1}\cdot x)^{-1}\cdot(x\cdot z) = y\cdot z$, by (Pass); it follows, by (Pcomm), that $(x\cdot y^{-1})^{-1}\cdot(x\cdot z) = y\cdot z$, i.e. ($m_1$-B=) holds.

(m1AB7): By (m_1-B=), $(x\cdot y^{-1})^{-1}\cdot(x\cdot z) = y\cdot z$; then, for $z=1$, we obtain $(x\cdot y^{-1})^{-1}\cdot(x\cdot 1) = y\cdot 1$, hence $(y^{-1}\cdot x)^{-1}\cdot x = y$, by (PU), (Pcomm); thus, (m_1-D=) holds.

(m1AB7'): By (m1AB7), (PU) + (Pcomm) + (m_1-B=) imply (m_1-D=), and by (m1AB0), (m_1-D=) + (PU) imply (DN). Then,

$(y\cdot z)\cdot x \overset{(Pcomm)}{=} x\cdot(y\cdot z) \overset{(m_1-B=)}{=} (X\cdot x^{-1})^{-1}\cdot(X\cdot(y\cdot z))$ and for $X = (y\cdot x^{-1})^{-1}$, we obtain:

$$\begin{aligned}
(y\cdot z)\cdot x &= ((y\cdot x^{-1})^{-1}\cdot x^{-1})^{-1}\cdot((y\cdot x^{-1})^{-1}\cdot(y\cdot z))\\
&\overset{(m_1-B=)}{=} ((y\cdot x^{-1})^{-1}\cdot x^{-1})^{-1}\cdot(x\cdot z)\\
&\overset{(DN)}{=} (((y^{-1})^{-1}\cdot x^{-1})^{-1})\cdot x^{-1})^{-1}\cdot(x\cdot z)\\
&\overset{(m_1-D=)}{=} (y^{-1})^{-1}\cdot(x\cdot z) \overset{(DN)}{=} y\cdot(x\cdot z) \overset{(Pcomm)}{=} y\cdot(z\cdot x),
\end{aligned}$$
i.e. (Pass) holds.

(m1AB7"): By (m1AB7), (PU) + (Pcomm) + (m_1-B=) imply (m_1-D=), and by (m1AB0), (m_1-D=) + (PU) imply (DN); by Proposition 7.9 (5), (Pcomm) + (DN) + (m_1-B=) imply (m_1-BB=); thus, (m_1-BB=) holds. □

Theorem 7.13. *(See Theorem 6.4) (See Proposition 4.9)*

Let $\mathcal{A}^L = (A^L, \cdot, ^{-1}, 1)$ *be an algebra of type* $(2,1,0)$. *If (PU), (Pcomm), (Pass), (m_1-Re) hold, then*

$$(m_1 - B =) \Longleftrightarrow (m_1 - BB =).$$

Proof. By (m1A0), (Pass) + (Pcomm) + (PU) + (m_1-Re) imply (EqP=), hence (EqP=) holds; then, by (m1A2"), (Pass) + (Pcomm) + (EqP=) $\Longrightarrow$ ((m_1-B=) $\Leftrightarrow$ (m_1-BB=)). □

By above Propositions 7.9 and 7.11 we obtain:

Theorem 7.14. *Any m-CI algebra (i.e. any commutative group) is m-p-semisimple (= sharp) and involutive.*

Proof. Let $\mathcal{A}^L = (A^L, \cdot, ^{-1}, 1)$ be a left-m-CI algebra, i.e. (N1), (PU), (Pcomm), (Pass), (m_1-Re) hold. Then, by (m1A0), (Pass) + (Pcomm) + (PU) + (m_1-Re) $\Longrightarrow$ (EqP=), hence (EqP=) holds, hence (m-p-s) holds, by Proposition 7.9 (0'); by Proposition 7.9 (3), (EqP=) + (Pcomm) + (m_1-Re) $\Longrightarrow$ (DN), hence (DN) holds too. □

We write this as follows:
m-CI = **m-CI**$_{(m-p-s)}$ = **m-CI**$_{(DN)}$ = **commutative group.**

Corollary 7.15. *The pre-m-BCI, m-BCH, m-BCI algebras and m-CI*, m-BCH* algebras are m-p-semisimple and involutive.*

Proof. All these algebras are m-CI algebras (=commutative groups), by definition; then apply Theorem 7.14. □

Corollary 7.16. *(i) The m-CI algebras coincide with the m-CI* algebras.*
(ii) The m-BCH algebras coincide with the m-BCH algebras.*

Proof. (i): Let $\mathcal{A}^L$ be a left-m-CI algebra; by Theorem 7.14, (m-p-s) holds; then, by Proposition 7.9 (1'), (m_1-*) holds; hence, $\mathcal{A}^L$ is a left-m-CI* algebra. Since, conversely, any left-m-CI* algebra is a left-m-CI algebra, it follows that they coincide.
(ii): Let $\mathcal{A}^L$ be a left-m-BCH algebra; by Corollary 7.15, (m-p-s) holds; then, by Proposition 7.9 (1'), (m_1-*) holds; hence, $\mathcal{A}^L$ is a left-m-BCH* algebra. Since, conversely, any left-m-BCH* algebra is a left-m-BCH algebra, it follows that they coincide. □

Moreover, we have the following expected result.

Proposition 7.17. *Any m-CI algebra (= commutative group) is an m-BCI algebra.*

Proof. By Corollary 7.15, any m-BCI algebra is m-p-semisimple; by Proposition 7.9 (2'), (m-p-s) + (m_1-B') imply (m_1-B=).
Let now $\mathcal{A}^L$ be an m-CI algebra, i.e. (N1), (PU), (Pcomm), (Pass), (m_1-Re) hold. We must prove that (m_1-An) and (m_1-B=) hold. Indeed, by Theorem 7.14, (m-p-s) holds; by Proposition 7.9 (1), (m-p-s) implies (m_1-An'), hence (m_1-An) holds. By Proposition 7.9 (0), (m-dfrelP$_1$) + (m-p-s) $\Longrightarrow$ (EqP=), hence (EqP=) holds; by (m1AB1), (Pass) + (Pcomm) + (m_1-Re) + (EqP=) $\Longrightarrow$ (m_1-B=); thus, (m_1-B=) holds too. Hence, $\mathcal{A}^L$ is an (m-p-semisimple) m-BCI algebra. □

Since, conversely, any m-BCI algebra is an m-CI algebra, by definition, we obtain:

Corollary 7.18. *The (m-p-semisimple and involutive) left-m-CI, left-m-CI*, left-pre-m-BCI, left-m-BCH, left-m-BCH* and left-m-BCI algebras coincide (with the commutative groups).*

We write this as follows, for short:
m-CI = **m-CI*** = **pre-m-BCI** = **m-BCH** = **m-BCH*** = **m-BCI** = **commutative group.**

By Corollary 7.18, it follows that the *algebras without associativity* (or *non-associative algebras*) defined in this section (the left-m-RM, left-pre-m-BZ, left-m-BH, left-m-BZ algebras) are all (non-associative) generalizations of the commutative group.

Proposition 7.19. *(See Proposition 4.20)*

(i) The m-p-semisimple m-RM algebras (= commutative moons) coincide with the m-p-semisimple m-BH algebras.

(ii) The m-p-semisimple pre-m-BZ algebras coincide with the m-p-semisimple m-BZ algebras.

Proof. By Proposition 7.9 (1). □

Proposition 7.20. *(See Proposition 4.25)*

Any m-p-semisimple non-associative algebra is involutive.

Proof. By Proposition 7.9 (0) and (3). □

The converse of Proposition 7.20 does not hold: there are involutive non-associative algebras that are not m-p-semisimple (see the Example (2m) below).

Remark 7.21. Note that the m-p-semisimple (hence involutive) m-BH algebra is just the commutative *goop* (defined in [42] Definition 10.1.25, see also Theorem 10.1.28).

Examples 7.22. (See Examples 4.26)

(1m) The algebra $\mathcal{A}_1 = (\{a, b, c, 1\}, \cdot_1, ^{-1_1}, 1)$, with

$\cdot_1$	a	b	c	1
a	1	1	1	a
b	1	a	a	b
c	1	a	a	c
1	a	b	c	1

and

x	x^{-1_1}
a	a
b	a
c	a
1	1

, is a **m-RM algebra (i.e. a commutative moon) that is not involutive.** (Pass) is not verified for (a, a, b); (m_1-B) is not verified for (a, b, b); (m_1-BB) is not verified for (b, a, b); (m_1-An) for (a, b).

(2m) The algebra $\mathcal{A}_2 = (\{a, b, c, 1\}, \cdot_2, ^{-1_2}, 1)$, with

$\cdot_2$	a	b	c	1
a	a	1	1	a
b	1	a	1	b
c	1	1	1	c
1	a	b	c	1

and

x	x^{-1_2}
a	b
b	a
c	c
1	1

, is an **involutive m-RM algebra (i.e. an involutive commutative moon) that is not m-p-semisimple**. (Pass) is not verified for (a, a, b); (m_1-B) is not verified for (a, b, b); (m_1-BB) is not verified for (a, a, c);

(m_1-An) for (a,b); (EqP=) $\Longleftrightarrow$ (m-p-s) for (b,c): $b \neq c$ but $b \cdot_2 c^{-1} = b \cdot_2 c = 1$. Note that we can apply Theorem 9.3 and we obtain the involutive RM algebra $(\{a,b,c,1\}, \rightarrow, 1)$, with

$\rightarrow$	a	b	c	1
a	1	b	1	b
b	b	1	1	a
c	1	1	1	c
1	a	b	c	1

and $x^{-1_2} \overset{def.}{=} x \rightarrow 1$, since (Re), (M), (DN), and (Ex1) hold; it does not verify (Ex) for (a,b,a), (B) for (a,b,a), (BB) for (a,a,b), (*) for (a,b,c), (**) for (a,b,c), (Tr) for (a,c,b), (An) for (a,c).

(3m) The algebra $\mathcal{A}_3 = (\{a,b,c,1\}, \cdot_3, ^{-1_3}, 1)$, with

$\cdot_3$	a	b	c	1
a	a	1	a	a
b	1	a	b	b
c	a	b	1	c
1	a	b	c	1

and

x	x^{-1_3}
a	b
b	a
c	c
1	1

, is a **m-p-semisimple (involutive) m-BH algebra (i.e. a commutative goop)**. (Pass) is not verified for (a,a,b); (m_1-B) is not verified for (a,b,b); (m_1-BB) is not verified for (a,a,c). Note that we can apply Theorem 9.3 and we obtain the algebra from Examples 4.26 (3).

The connections between the left-m-RM algebras (commutative moons), the involutive left-m-RM algebras (involutive moons) and the m-p-semisimple left-m-BH algebras (commutative goops) are presented in the following Figure 9 (see [42], Figure 10.4, in the non-commutative case).

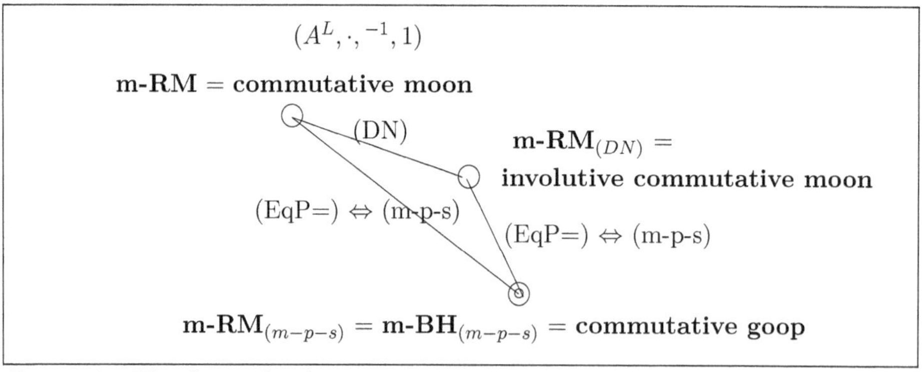

Figure 9: The hierarchy between commutative moons, involutive commutative moons and commutative goops

Theorem 7.23. *(See Theorem 4.27)*

The m-p-semisimple (involutive) m-BZ algebras coincide with the m-BCI algebras (i.e. with the commutative groups).

Proof. Let $\mathcal{A}^L = (A^L, \cdot, ^{-1}, 1)$ be a m-p-semisimple (involutive) left-m-BZ algebra, i.e. (PU), (Pcomm), (m_1-Re), (m_1-An), (m_1-B),((DN)), (m-p-s) hold. We prove that (Pass) holds. Indeed, by Proposition 7.9 (2'), (m-p-s) + (m_1-B) imply (m_1-B=). Now, by (m1AB7'), (PU) + (Pcomm) + (m_1-B=) imply (Pass). Thus, $\mathcal{A}^L$ is a (m-p-semisimple) left-m-BCI algebra. Since the converse implication is obvious, the proof is complete. □

Note that we have no an example of involutive m-BZ algebra that is not m-p-semisimple (i.e. a commutative group).

All left-algebras defined in this section are connected (following their definitions and the above results) as in the *Hierarchy m-1* from Figure 10 and *Hierarchy m-1'* from Figure 11, in Section 8.

8 Final connections between the algebras defined in Sections 6, 7

With analogous definitions as in Definitions 2.8 and with the same representations as in Remark 2.9, the final connections between the subclasses of commutative unital magmas defined in Sections 6, 7 are presented in the next Figures 10 and 11.

Other algebras can be defined following the other algebras defined in [37].

9 Connections between the two 'worlds' in the involutive case

The *'world' of M algebras* (of logic) (i.e. the 'world' of $\rightarrow, ^{-}, 1$ (or $\rightarrow, ^{-1}, 1$)) (from Sections 2 - 4) and the *'world' of commutative unital magmas* (i.e. the 'world' of $\odot/\wedge, ^{-}, 1$ (or $\cdot, ^{-1}, 1$)) (from Sections 5 - 8) are connected, *in the involutive case (when (DN) holds)*, by two mutually inverse transformations, Φ and Ψ (or Φ_1 and Ψ_1, respectively), as we shall see in the following two subsections.

9.1 Connections in the case $1^{-} = 0$

The M algebras from the Hierarchies 2, 2'^{b} (from Sections 2, 3) and the corresponding commutative unital magmas from the Hierarchies m-2, m-2' (from Sections 6, 8) are

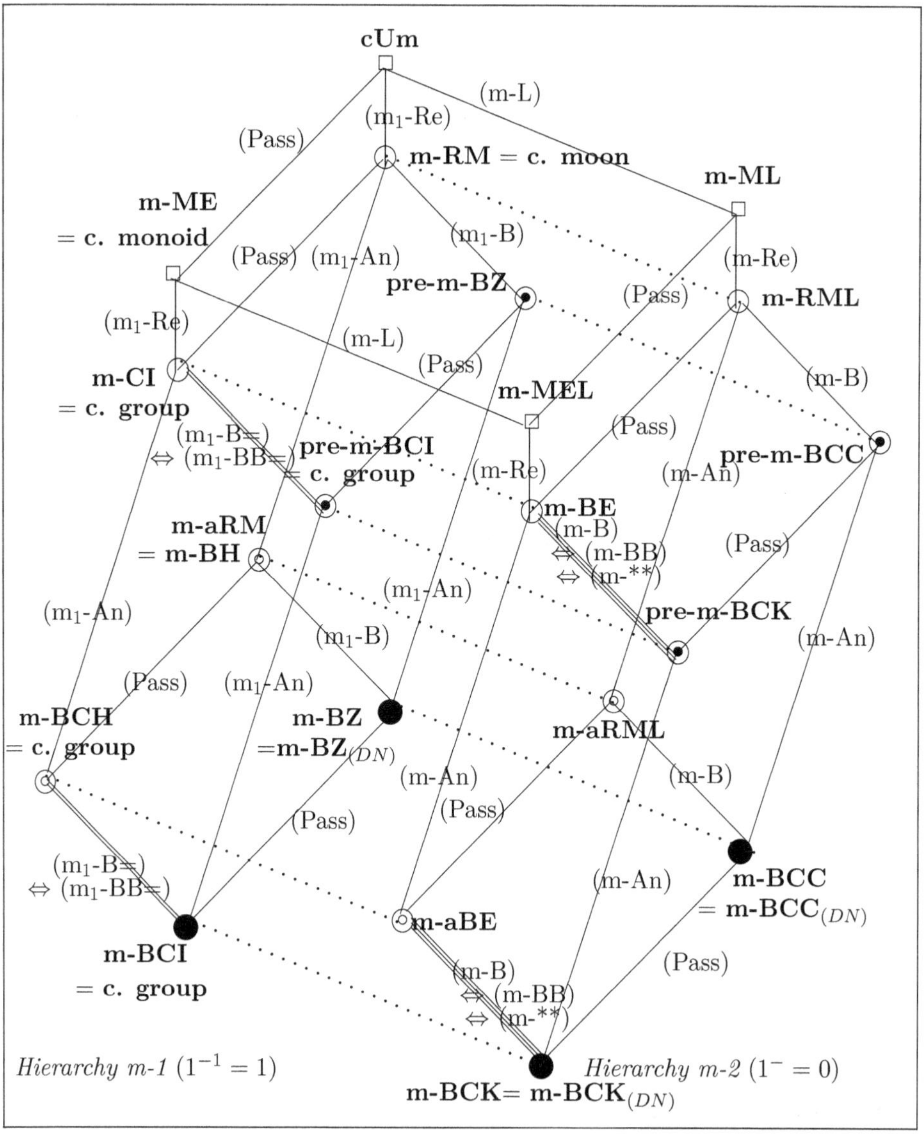

Figure 10: The hierarchies m-1 and m-2 of commutative unital magmas (**cUm**)

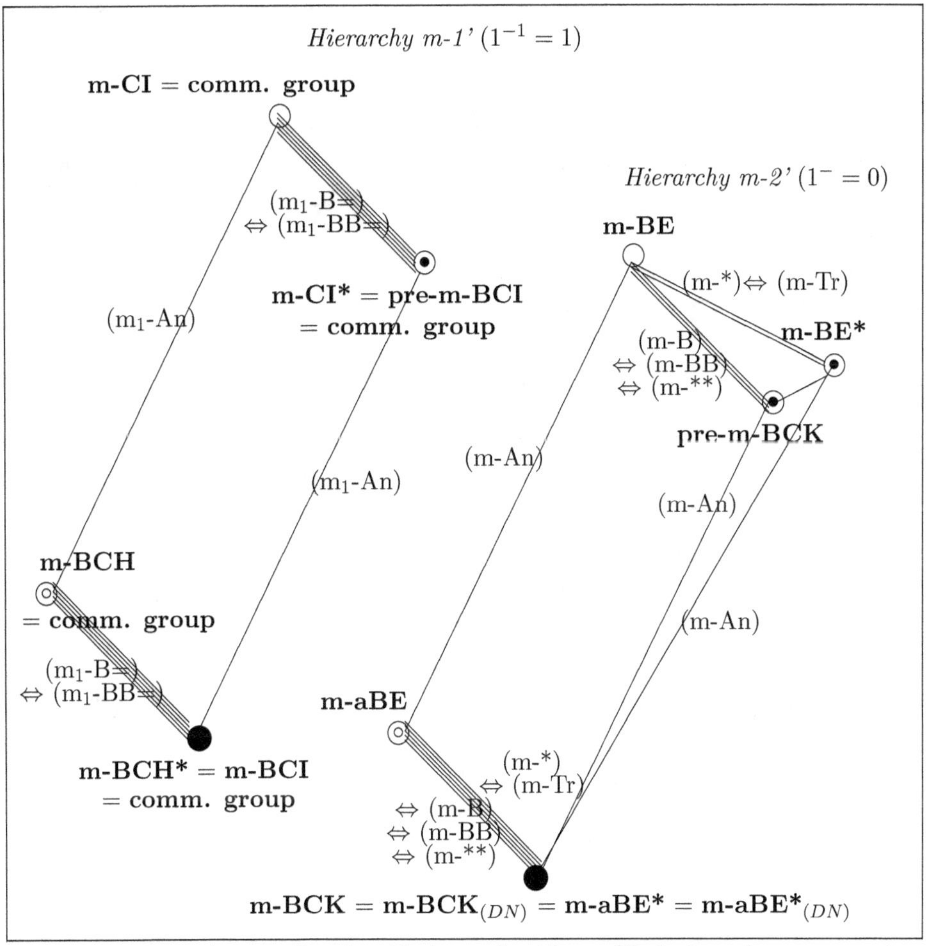

Figure 11: Hierarchies m-1' and m-2', determined by m-CI algebras (= commutative groups) and m-BE algebras, respectively

connected, *in the involutive case*, by two mutually inverse transformations, Φ and Ψ, as we shall see in next Theorem 9.1.

Recall the properties (Ex0): $x \to (y \to 0) = y \to (x \to 0)$ and (Neg3): $x \to y^- = y \to x^-$ and that (Ex) implies (Ex0). We then have the following general result:

Theorem 9.1.

(1) Let $\mathcal{A}^L = (A^L, \to, 0, 1)$ *be an algebra of type* $(2,0,0)$ *verifying (M), (Ex0) and (DN), where a negation* $^-$ *is defined by (dfneg)* $(x^- \stackrel{def.}{=} x \to 0)$. *(Equivalently, let* $\mathcal{A}^L = (A^L, \to, ^-, 1)$ *be an algebra of type* $(2,1,0)$ *verifying (M), (Neg3) and (DN), such that* $0 \stackrel{def.}{=} 1^-$*).*

Define an internal binary relation $\leq$ *by (dfrelR)* $(x \leq y \stackrel{def.}{\Longleftrightarrow} x \to y = 1)$.

Define $\Phi(\mathcal{A}^L) \stackrel{def.}{=} (A^L, \odot, ^-, 1)$ *by: for all* $x, y \in A^L$,

$$x \odot y \stackrel{def.}{=} (x \to y^-)^-.$$

Then, $\Phi(\mathcal{A}^L)$ *satisfies (PU), (Pcomm) and (DN).*

(1') Conversely, let $\mathcal{A}^L_m = (A^L, \odot, ^-, 1)$ *be an algebra of type* $(2,1,0)$. *Define* $0 \stackrel{def.}{=} 1^-$ *and suppose that (PU), (Pcomm) and (DN) hold.*

Define an internal binary relation $\leq_m$ *by (m-dfrelP)* $(x \leq_m y \stackrel{def}{\Longleftrightarrow} x \odot y^- = 0)$.

Define $\Psi(\mathcal{A}^L_m) \stackrel{def.}{=} (A, \to, 0, 1)$ *(or, equivalently,* $\Psi(\mathcal{A}^L_m) \stackrel{def.}{=} (A, \to, ^-, 1)$*) by: for all* $x, y \in A^L$,

$$x \to y \stackrel{def.}{=} (x \odot y^-)^-.$$

Then, $\Psi(\mathcal{A}^L_m)$ *satisfies (M), (Ex0) and (DN) (or (M), (Neg3), (DN), respectively).*

(2) The mappings Φ *and* Ψ *are mutually inverse.*

(3) $x \leq y \Longleftrightarrow x \leq_m y$, *for all* $x, y \in A^L$.

(4) The following properties are equivalent:

(Ex) $\Longleftrightarrow$ *(Pass), (L)* $\Longleftrightarrow$ *(m-L), (Re)* $\Longleftrightarrow$ *(m-Re),*
(An) $\Longleftrightarrow$ *(m-An), (DN4)* $\Longleftrightarrow$ *(m-DN4),*
(**) $\Longleftrightarrow$ *(m-**), (*)* $\Longleftrightarrow$ *(m-*), (BB)* $\Longleftrightarrow$ *(m-BB), (B)* $\Longleftrightarrow$ *(m-B),*
(N) $\Longleftrightarrow$ *(m-N) and (V)* $\Longleftrightarrow$ *(m-V),*
(C) $\Longleftrightarrow$ *(m-C), (Tr)* $\Longleftrightarrow$ *(m-Tr), (D)* $\Longleftrightarrow$ *(m-D),*
($\vee$*-comm)* $\Longleftrightarrow$ *(*$\wedge_m$*-comm), (impl)* $\Longleftrightarrow$ *(m-Pimpl),*
(DN6) $\Longleftrightarrow$ *(G), (@)* $\Longleftrightarrow$ *(m-@)=(m-Wdiv).*

Proof. First note that the equivalence of the conditions from the hypothesis was proved in Theorems 3.4 and 3.5 from Section 3.

(1): Let $\mathcal{A}^L = (A^L, \rightarrow, 0, 1)$ be an algebra such that (M), (Ex0) and (DN) hold, where $x^- \stackrel{def.}{=} x \rightarrow 0$, by (dfneg). Then, (Neg1-0) holds, by (M): $1^- = 1 \rightarrow 0 = 0$; (Neg0-1) holds too, by (DN): $0^- = (1^-)^- = 1$; (Neg3) holds, by (Ex0): $x \rightarrow y^- = x \rightarrow (y \rightarrow 0) = y \rightarrow (x \rightarrow 0) = y \rightarrow x^-$. By (C3), (DN) + (Neg3) imply (DN3); by (C2'), (DN) + (DN3) imply (DN2); by (C4), (DN2) implies (DN4), hence (DN4') holds too.

Equivalently, let $\mathcal{A}^L = (A^L, \rightarrow, ^-, 1)$ be an algebra such (M), (Neg3) and (DN) hold, where $0 \stackrel{def.}{=} 1^-$, hence (Neg1-0) holds; then, (Neg0-1) holds too, by (DN), and (DN4') holds, by (C3), (C2'), (C4).

Define $x \odot y \stackrel{def.}{=} (x \rightarrow y^-)^-$. We must prove that (Pcomm) and (PU) hold. Indeed,

$x \odot y \stackrel{def.}{=} (x \rightarrow y^-)^- \stackrel{(Neg3)}{=} (y \rightarrow x^-)^- = y \odot x$; thus, (Pcomm) holds.

$1 \odot x \stackrel{def.}{=} (1 \rightarrow x^-)^- \stackrel{(M)}{=} (x^-)^- \stackrel{(DN)}{=} x$; thus, (PU) holds.

(1'): Let $\mathcal{A}^L_m = (A^L, \odot, ^-, 1)$ be an algebra such that (Pcomm), (PU) and (DN) hold; let $0 \stackrel{def.}{=} 1^-$, hence (Neg1-0) holds; then, (Neg0-1) holds too, by (DN). By (mCDN2), (Pcomm) + (DN) imply (m-DN4'), hence (m-DN4') holds too.

Define $x \rightarrow y \stackrel{def.}{=} (x \odot y^-)^-$. We must prove that (M) and (Ex0) hold. Indeed,

$1 \rightarrow x \stackrel{def.}{=} (1 \odot x^-)^- \stackrel{(PU)}{=} (x^-)^- \stackrel{(DN)}{=} x$; thus, (M) holds.

$x \rightarrow (y \rightarrow 0) \stackrel{def.}{=} (x \odot ((y \odot 0^-)^-)^-)^- \stackrel{(DN),(Neg0-1)}{=} (x \odot (y \odot 1))^-$
$\stackrel{(PU)}{=} (x \odot y)^- \stackrel{(Pcomm)}{=} (y \odot x)^- \stackrel{(PU)}{=} (y \odot (x \odot 1))^-$
$\stackrel{(Neg0-1),(DN)}{=} (y \odot ((x \odot 0^-)^-)^-)^- \stackrel{def.}{=} y \rightarrow (x \rightarrow 0)$; thus, (Ex0) holds.

(2): Let

$$(A^L, \rightarrow, 0, 1) \quad \xrightarrow{\Phi} \quad (A^L, \odot, ^-, 1) \quad \xrightarrow{\Psi} \quad (A^L, \Rightarrow, \mathbf{0}, 1).$$

Then, for all $x, y \in A^L$, we have:

$x \Rightarrow y = (x \odot y^-)^- = ((x \rightarrow (y^-)^-)^-)^- \stackrel{(DN)}{=} x \rightarrow y$ and

$\mathbf{0} = 1^- = 1 \rightarrow 0 \stackrel{(M)}{=} 0$.

Conversely, let

$$(A^L, \odot, ^-, 1) \quad \xrightarrow{\Psi} \quad (A^L, \rightarrow, 0, 1) \quad \xrightarrow{\Phi} \quad (A^L, \otimes, ', 1).$$

Then, for all $x, y \in A^L$, we have:

$x' = x \rightarrow 0 = (x \odot 0^-)^- \stackrel{(Neg0-1)}{=} (x \odot 1)^- \stackrel{(PU)}{=} x^-$ and

$x \otimes y = (x \rightarrow y')' = (x \rightarrow y^-)^- = (x \odot (y^-)^-)^- \stackrel{(DN)}{=} x \odot y$.

(3): Suppose $x \leq y$, i.e. $x \to y = 1$, by (dfrelR); hence, $(x \odot y^-)^- = 1$, by Φ; then, $x \odot y^- = 0$, by (DN) and (Neg1-0), hence $x \leq_m y$, by (m-dfrelP).
Suppose now $x \leq_m y$, i.e. $x \odot y^- = 0$, by (m-dfrelP); then, $(x \odot y^-)^- = 1$, by (Neg0-1); hence $x \to y = 1$, by Ψ; hence $x \leq y$, by (dfrelR).

(4): (Ex) $\Longrightarrow$ (Pass): $x \odot (y \odot z) \overset{(Pcomm)}{=} x \odot (z \odot y)$
$= (x \to ((z \to y^-)^-)^-)^- \overset{(DN)}{=} (x \to (z \to y^-))^- \overset{(Ex)}{=} (z \to (x \to y^-))^-$
$\overset{(DN)}{=} (z \to ((x \to y^-)^-)^-)^- = z \odot (x \odot y) \overset{(Pcomm)}{=} (x \odot y) \odot z.$
(Pass) $\Longrightarrow$ (Ex): $x \to (y \to z) = (x \odot ((y \odot z^-)^-)^-)^- \overset{(DN)}{=} (x \odot (y \odot z^-))^-$
$\overset{(Pass)}{=} ((x \odot y) \odot z^-)^- \overset{(Pcomm)}{=} ((y \odot x) \odot z^-)^- \overset{(Pass)}{=} (y \odot (x \odot z^-)^-$
$\overset{(DN)}{=} (y \odot ((x \odot z^-)^-)^-)^- = y \to (x \to z).$

(L) $\Longleftrightarrow$ (m-L): (L') $\Longleftrightarrow x \leq 1 \overset{(3)}{\Longleftrightarrow} x \leq_m 1 \Longleftrightarrow$ (m-L').

(Re) $\Longleftrightarrow$ (m-Re): (Re') $\Longleftrightarrow x \leq x \overset{(3)}{\Longleftrightarrow} x \leq_m x \Longleftrightarrow$ (m-Re').

(An) $\Longleftrightarrow$ (m-An): Suppose that (An') holds and take $x \leq_m y$ and $y \leq_m x$; then, $x \leq y$ and $y \leq x$, by above (3); then, $x = y$, by (An'); thus, (m-An') holds too.
Conversely, suppose that (m-An') holds and take $x \leq y$ and $y \leq x$; then, $x \leq_m y$ and $y \leq_m x$, by above (3); then, $x = y$, by (m-An'); thus, (An') holds too.

(DN4) $\Longleftrightarrow$ (m-DN4): (DN4'), i.e. $x \leq y \Longleftrightarrow y^- \leq x^-$, is equivalent to (m-DN4'), i.e. $x \leq_m y \Longleftrightarrow y^- \leq_m x^-$, by above (3).

(**) $\Longleftrightarrow$ (m-**): Suppose that (**') holds and take $x \leq_m y$; then, $x \leq y$, by above (3); then, $y \to z^- \leq x \to z^-$, by (**'); hence, $(y \odot (z^-)^-)^- \leq_m (x \odot (z^-)^-)^-$, by (3) again; then, $(y \odot z)^- \leq_m (x \odot z)^-$, by (DN); then $x \odot z \leq_m y \odot z$, by (m-DN4'); thus, (m-**') holds too.
Conversely, suppose that (m-**') holds and take $x \leq y$; then, $x \leq_m y$, by above (3); then, $x \odot z^- \leq_m y \odot z^-$, by (m-**'); hence, $x \odot z^- \leq y \odot z^-$, by (3) again, hence $(x \to (z^-)^-)^- \leq (y \to (z^-)^-)^-$, by Ψ; then, $(x \to z)^- \leq (y \to z)^-$, by (DN); then $y \to z \leq x \to z$, by (DN4'); thus, (**') holds too.

(*) $\Longleftrightarrow$ (m-*): Suppose that (*') holds and take $x \leq_m y$; then, $x \leq y$, by above (3); then, $z \to x \leq z \to y$, by (*'); then $(z \to y)^- \leq (z \to x)^-$, by (DN4') and $(z \to (y^-)^-)^- \leq (z \to (x^-)^-)^-$, by (DN); hence, $z \odot y^- \leq_m z \odot x^-$, by (3) again and by Φ; thus, (m-*') holds too.
Conversely, suppose that (m-*') holds and take $x \leq y$; then, $x \leq_m y$, by above (3); then, $z \odot y^- \leq_m z \odot x^-$, by (m-*'); hence, $(z \to (y^-)^-)^- \leq (z \to (x^-)^-)^-$, by (3) again and by Ψ; then, $(z \to y)^- \leq (z \to x)^-$, by (DN) and $z \to x \leq z \to y$, by (DN4'); thus, (*') holds too.

(BB) $\Longleftrightarrow$ (m-BB): Suppose that (BB) holds; then,
$[(z \odot x)^- \odot (y \odot x)] \odot (y \odot z^-)^- \overset{(Pcomm)}{=} (y \odot z^-)^- \odot [(z \odot x)^- \odot (y \odot x)]$

$\overset{(DN)}{=} (y \odot z^-)^- \odot [[(z \odot (x^-)^-)^- \odot ((y \odot (x^-)^-)^-)^-]^-]^-$
$= (y \to z) \odot [[(z \to x^-) \odot (y \to x^-)^-]^-]^-$
$\overset{(DN)}{=} ((y \to z) \to [(z \to x^-) \to (y \to x^-)])^- \overset{(BB)}{=} 1^- \overset{(Neg1-0)}{=} 0$; thus, (m-BB) holds too.
Conversely, suppose that (m-BB) holds; then,
$(y \to z) \to [(z \to x) \to (y \to x)] = ((y \odot z^-)^- \odot [[(z \odot x^-)^- \odot ((y \odot x^-)^-)^-]^-]^-)^-$
$\overset{(DN)}{=} ((y \odot z^-)^- \odot [(z \odot x^-)^- \odot (y \odot x^-)])^-$
$\overset{(Pcomm)}{=} ([(z \odot x^-)^- \odot (y \odot x^-)] \odot (y \odot z^-)^-)^- \overset{(m-BB)}{=} 0^- \overset{(Neg0-1)}{=} 1$; thus, (BB) holds.

(B) $\Longleftrightarrow$ (m-B): Suppose that (B) holds; then,
$[(x \odot y^-)^- \odot (x \odot z)] \odot (y \odot z)^- \overset{(Pcomm)}{=} (y \odot z)^- \odot [(x \odot y^-)^- \odot (x \odot z)]$
$\overset{(DN)}{=} (y \odot (z^-)^-)^- \odot [[(x \odot y^-)^- \odot (x \odot (z^-)^-)]^-]^-$
$= (y \to z^-) \odot [(x \to y) \to (x \to z^-)]^-$
$\overset{(DN)}{=} ((y \to z^-) \to [(x \to y) \to (x \to z^-)])^- \overset{(B)}{=} 1^- \overset{(Neg1-0)}{=} 0$; thus, (m-B) holds too.
Conversely, suppose that (m-B) holds; then,
$(y \to z) \to [(x \to y) \to (x \to z)] = ((y \odot z^-)^- \odot [[(x \odot y^-)^- \odot ((x \odot z^-)^-)^-]^-]^-)^-$
$\overset{(DN)}{=} ((y \odot z^-)^- \odot [(x \odot y^-)^- \odot (x \odot z^-)])^-$
$\overset{(Pcomm)}{=} ([(x \odot y^-)^- \odot (x \odot z^-)] \odot (y \odot z^-)^-)^- \overset{(m-B)}{=} 0^- \overset{(Neg0-1)}{=} 1$; thus, (B) holds.

(N) $\Longleftrightarrow$ (m-N): $(1 \le x \Longrightarrow x = 1) \overset{(3)}{\Longleftrightarrow} (1 \le_m x \Longrightarrow x = 1) \Longleftrightarrow$ (m-N');
(V) $\Longleftrightarrow$ (m-V): $(x \le 0 \Longrightarrow x = 0) \overset{(3)}{\Longleftrightarrow} (x \le_m 0 \Longrightarrow x = 0) \Longleftrightarrow$ (m-V').

(C) $\Longleftrightarrow$ (m-C): Suppose that (C') holds; then,
$y \odot (x \odot z) \le_m x \odot (y \odot z) \overset{(DN)}{\Longleftrightarrow} y \odot (x \odot (z^-)^-) \le_m x \odot (y \odot (z^-)^-)$
$\overset{(m-DN4')}{\Longleftrightarrow} (x \odot (y \odot (z^-)^-))^- \le_m (y \odot (x \odot (z^-)^-))^-$
$\overset{(DN)}{\Longleftrightarrow} (x \odot ((y \odot (z^-)^-)^-)^-)^- \le_m (y \odot ((x \odot (z^-)^-)^-)^-)^-$
$\overset{(3)}{\Longleftrightarrow} x \to (y \to z^-) \le y \to (x \to z^-)$ that is true by (C'); thus, (m-C') holds too.
Conversely, suppose that (m-C') holds; then,
$x \to (y \to z) \le y \to (x \to z) \overset{(3)}{\Longleftrightarrow} (x \odot ((y \odot z^-)^-)^-)^- \le_m (y \odot ((x \odot z^-)^-)^-)^-$
$\overset{(DN)}{\Longleftrightarrow} (x \odot (y \odot z^-))^- \le_m (y \odot (x \odot z^-))^-$
$\overset{(DN4')}{\Longleftrightarrow} y \odot (x \odot z^-) \le_m x \odot (y \odot z^-)$ that is true by (m-C'); thus, (C') holds too.

(Tr) $\Longleftrightarrow$ (m-Tr): (Tr') $\Longleftrightarrow (x \le y$ and $y \le z \Longrightarrow x \le z)$
$\overset{(3)}{\Longleftrightarrow} (x \le_m y$ and $y \le_m z \Longrightarrow x \le_m z) \Longleftrightarrow$ (m-Tr').

(D) $\Longleftrightarrow$ (m-D): Suppose that (D') holds; then,

$(y^- \odot x)^- \odot x \leq_m y \overset{(m-DN4')}{\Longleftrightarrow} y^- \leq_m [(y^- \odot x)^- \odot x]^-$
$\overset{(DN)}{\Longleftrightarrow} y^- \leq_m [(y^- \odot (x^-)^-)^- \odot (x^-)^-]^-$
$\overset{(3)}{\Longleftrightarrow} y^- \leq (y^- \to x^-) \to x^-$ that is true by (D'); thus, (m-D') holds too.
Conversely, suppose that (m-D') holds; then,
$y \leq (y \to x) \to x \overset{(DN4')}{\Longleftrightarrow} [(y \to x) \to x]^- \leq y^-$
$\overset{(3)}{\Longleftrightarrow} [[(y \odot x^-)^- \odot x^-]^-]^- \leq_m y^- \overset{(DN)}{\Longleftrightarrow} (y \odot x^-)^- \odot x^- \leq_m y^-$
$\overset{(DN)}{\Longleftrightarrow} ((y^-)^- \odot x^-)^- \odot x^- \leq_m y^-$ that is true by (m-D'); thus, (D') holds too.

($\vee$-comm) $\Longleftrightarrow$ ($\wedge_m$-comm): Suppose that ($\vee$-comm) holds; then,
$(x^- \odot y)^- \odot y = (y^- \odot x)^- \odot x$
$\overset{(DN)}{\Longleftrightarrow} [[(x^- \odot (y^-)^-)^- \odot (y^-)^-]^-]^- = [[(y^- \odot (x^-)^-)^- \odot (x^-)^-]^-]^-$
$\Longleftrightarrow [(x^- \to y^-) \to y^-]^- = [(y^- \to x^-) \to x^-]^-$
$\Longleftrightarrow (x^- \to y^-) \to y^- = (y^- \to x^-) \to x^-$ that is true by ($\vee$-comm); thus, ($\wedge_m$-comm) holds too.
Conversely, suppose that ($\wedge_m$-comm) holds; then,
$(x \to y) \to y = (y \to x) \to x \Longleftrightarrow ((x \odot y^-)^- \odot y^-)^- = ((y \odot x^-)^- \odot x^-)^-$
$\overset{(DN)}{\Longleftrightarrow} (((x^-)^- \odot y^-)^- \odot y^-)^- = (((y^-)^- \odot x^-)^- \odot x^-)^-$
$\overset{(DN)}{\Longleftrightarrow} ((x^-)^- \odot y^-)^- \odot y^- = ((y^-)^- \odot x^-)^- \odot x^-$
that is true by ($\wedge_m$-comm); thus, ($\vee$-comm) holds too.

(impl) $\Longleftrightarrow$ (m-Pimpl): (impl) $\Longleftrightarrow (x \to y) \to x = x$
$\Longleftrightarrow [(x \odot y^-)^- \odot x^-]^- = x \Longleftrightarrow$ (m-Pimpl).

(DN6) $\Longleftrightarrow$ (G): Suppose that (DN6) holds; then, $x \odot x = (x \to x^-)^- \overset{(DN)}{=} ((x^-)^- \to x^-)^- \overset{(DN6)}{=} (x^-)^- \overset{(DN)}{=} x$; thus, (G) holds.
Conversely, suppose that (G) holds; then, $x^- \to x = (x^- \odot x^-)^- \overset{(G)}{=} (x^-)^- = x$; thus, (DN6) holds.

(@) $\Longleftrightarrow$ (m-@): Suppose that (@) holds; then,
$(y \odot x^-)^- \odot y = x \odot y \overset{(DN)}{\Longleftrightarrow} [[((y^-)^- \odot x^-)^- \odot (y^-)^-]^-]^- = [[x \odot (y^-)^-]^-]^-$
$\Longleftrightarrow [((y^-)^- \to x) \to y^-]^- = [x \to y^-]^-$ that is true by (@); thus, (m-@) holds.
Conversely, suppose that (m-@) holds; then,
$(y^- \to x) \to y = x \to y \Longleftrightarrow ((y^- \odot x^-)^- \odot y^-)^- = (x \odot y^-)^-$ that is true by (m-@); thus, (@) holds too. □

By this Theorem 9.1, choosing appropriate definitions, one can prove all the definitionally equivalences (in the involutive case) between the corresponding algebras from the two 'worlds', for examples between Wajsberg algebras and MV algebras, between i-Boolean algebras and Boolean algebras etc.. For example, we shall prove that the i-ortholattices (Definition 2) (introduced and studied in Section 3) are d.e. to the ortholattices (Definition 2) (studied in Section 6):

Theorem 9.2.
(o1) Let $\mathcal{A} = (A, \to, 0, 1)$ *be an i-ortholattice, where a negation* $^-$ *is defined by* $x^- \stackrel{def.}{=} x \to 0$. *Define* $\Phi(\mathcal{A}) \stackrel{def.}{=} (A, \wedge, ^-, 1)$ *as follows: for all* $x, y \in A$,

$$x \wedge y \stackrel{def.}{=} (x \to y^-)^-.$$

Then, $\Phi(\mathcal{A})$ *is an ortholattice.*

(o1') Conversely, let $\mathcal{A} = (A, \wedge, ^-, 1)$ *be an ortholattice and* $0 \stackrel{def.}{=} 1^-$. *Define* $\Psi(\mathcal{A}) \stackrel{def.}{=} (A, \to, 0, 1)$ *as follows: for all* $x, y \in A$,

$$x \to y \stackrel{def.}{=} (x \wedge y^-)^-.$$

Then, $\Psi(\mathcal{A})$ *is an i-ortholattice.*

(o2) The two mappings, Φ *and* Ψ *are mutually inverse.*

Proof. Let $\mathcal{IO} = (A, \to, 0, 1)$ be a left-i-ortholattice (Definition 2), i.e. (M), (Ex), (Re), (F), (L), (impl), (DN) hold, where $x^- = x \to 0$, by (dfneg). Since (Ex) holds, it follows that (Ex0) holds. Let also $\mathcal{O} = (A, \wedge, ^-, 1)$ be a left-ortholattice (Definition 2) and $0 \stackrel{def.}{=} 1^-$, i.e. (Neg1-0), (Neg0-1), (m-WU), (m-Wcomm), (m-Wass), (m-WL), (m-WRe), (m-Wimpl) (hence (DN)) hold.

Since $\mathcal{IO}$ verifies (M), (Ex0), (DN) and $\mathcal{O}$ verifies (m-WU), (m-Wcomm), (DN) hold, we can apply Theorem 9.1 for $\odot = \wedge$ and we have that (1), (1') and (2) hold; then, (Ex) $\Longleftrightarrow$ (m-Wass), (Re) $\Longleftrightarrow$ (m-WRe), (L) $\Longleftrightarrow$ (m-WL), (impl) $\Longleftrightarrow$ (m-Wimpl). It remains to prove that (F) holds; indeed, $0 \to x = (0 \wedge x^-)^- \stackrel{(m-Wcomm)}{=} (x^- \wedge o)^- \stackrel{(m-WL)}{=} 0^- \stackrel{(Neg0-1)}{=} 1$, i.e. (F) holds. Thus, (o1), (o1') and (o2) hold. □

9.2 Connections in the case $1^{-1} = 1$

The M algebras from the Hierarchies 1, 1' (from Sections 2, 4) and the corresponding commutative unital magmas from the Hierarchies m-1, m-1' (from Sections 7, 8) are connected, *in the involutive case*, by two mutually inverse transformations, Φ_1 and Ψ_1, as we shall see in next Theorem 9.3.

Recall the properties (Ex1): $x \to (y \to 1) = y \to (x \to 1)$ and (Neg3): $x \to y^{-1} = y \to x^{-1}$ and that (Ex) implies (Ex1). We then have the following general result:

Theorem 9.3.

(1) Let $\mathcal{A}^L = (A^L, \rightarrow, 1)$ *be an algebra of type* $(2,0)$ *verifying (M), (Ex1) and (DN), where a negation* $^{-1}$ *is defined by (dfneg1)* $(x^{-1} \stackrel{def.}{=} x \rightarrow 1)$*. (Equivalently, let* $\mathcal{A}^L = (A^L, \rightarrow, ^{-1}, 1)$ *be an algebra of type* $(2,1,0)$ *such that* $1^{-1} = 1$ *and verifying (M), (Neg3) and (DN)).*

Define an internal binary relation $\leq$ *by (dfrelR)* $(x \leq y \stackrel{def.}{\Longleftrightarrow} x \rightarrow y = 1)$*.*

Define $\Phi_1(\mathcal{A}^L) \stackrel{def.}{=} (A^L, \cdot, ^{-1}, 1)$ *by: for all* $x, y \in A^L$,

$$x \cdot y \stackrel{def.}{=} (x \rightarrow y^{-1})^{-1}.$$

Then, $\Phi_1(\mathcal{A}^L)$ *satisfies (PU), (Pcomm) and (DN).*

(1') Conversely, let $\mathcal{A}^L_{m_1} = (A^L, \cdot, ^{-1}, 1)$ *be an algebra of type* $(2,1,0)$ *such that* $1^{-1} = 1$ *and suppose that (PU), (Pcomm) and (DN) hold.*

Define an internal binary relation $\leq_{m_1}$ *by (m-dfrelP$_1$)* $(x \leq_{m_1} y \stackrel{def}{\Longleftrightarrow} x \cdot y^{-1} = 1)$*.*

Define $\Psi_1(\mathcal{A}^L_{m_1}) \stackrel{def.}{=} (A^L, \rightarrow, 1)$ *(or, equivalently,* $\Psi_1(\mathcal{A}^L_{m_1}) \stackrel{def.}{=} (A^L, \rightarrow, ^{-1}, 1)$*) by: for all* $x, y \in A^L$,

$$x \rightarrow y \stackrel{def.}{=} (x \cdot y^{-1})^{-1}.$$

Then, $\Psi_1(\mathcal{A}^L_{m_1})$ *satisfies (M), (Ex1) and (DN) (or (M), (Neg3) and (DN), respectively).*

(2) The mappings Φ_1 *and* Ψ_1 *are mutually inverse.*

(3) $x \leq y \Longleftrightarrow x \leq_{m_1} y$*, for all* $x, y \in A^L$*.*

(4) The following properties are equivalent:

(Ex) $\Longleftrightarrow$ *(Pass), (Re)* $\Longleftrightarrow$ *(m_1-Re), (An)* $\Longleftrightarrow$ *(m_1-An), (DN4)* $\Longleftrightarrow$ *(m_1-DN4),*
(**) $\Longleftrightarrow$ *(m_1-**), (*)* $\Longleftrightarrow$ *(m_1-*), (BB)* $\Longleftrightarrow$ *(m_1-BB), (BB=)* $\Longleftrightarrow$ *(m_1-BB=),*
(B) $\Longleftrightarrow$ *(m_1-B), (B=)* $\Longleftrightarrow$ *(m_1-B=), (C)* $\Longleftrightarrow$ *(m_1-C), (Tr)* $\Longleftrightarrow$ *(m_1-Tr),*
(D) $\Longleftrightarrow$ *(m_1-D), (D=)* $\Longleftrightarrow$ *(m_1-D=),*
($\vee$-comm) $\Longleftrightarrow$ *($\wedge_m$-comm), (impl)* $\Longleftrightarrow$ *(m-Pimpl), (p-s)* $\Longleftrightarrow$ *(m-p-s).*

Proof. First note that the equivalence of the conditions from the hypothesis was proved in Theorem 4.4 from Section 4.

(1): Let $\mathcal{A}^L = (A^L, \rightarrow, 1)$ be an algebra such that (M), (Ex1) and (DN) hold, where $x^{-1} \stackrel{def.}{=} x \rightarrow 1$, by (dfneg1). Then, (N1) holds, by (M): $1^{-1} = 1 \rightarrow 1 = 1$; (Neg3) holds, by (Ex1): $x \rightarrow y^{-1} = x \rightarrow (y \rightarrow 1) \stackrel{(Ex1)}{=} y \rightarrow (x \rightarrow 1) = y \rightarrow x^{-1}$. By (C3), (DN) + (Neg3) imply (DN3); by (C2'), (DN) + (DN3) imply (DN2); by (C4), (DN2) implies (DN4), hence (DN4') holds too.

Define $x \cdot y \stackrel{def.}{=} (x \rightarrow y^{-1})^{-1}$. We must prove that (Pcomm) and (PU) hold. Indeed,

$x \cdot y \overset{def.}{=} (x \to y^{-1})^{-1} \overset{(Neg3)}{=} (y \to x^{-1})^{-1} = y \cdot x$; thus, (Pcomm) holds.
$1 \cdot x \overset{def.}{=} (1 \to x^{-1})^{-1} \overset{(M)}{=} (x^{-1})^{-1} \overset{(DN)}{=} x$; thus, (PU) holds.

(1'): Let $\mathcal{A}^L_{m_1} = (A^L, \cdot, ^{-1}, 1)$ be an algebra such that (Pcomm), (PU) and (DN) hold and $1^{-1} = 1$, hence (N1) holds; then, by (m1CDN2), (Pcomm) + (DN) imply (m_1-DN4'), hence (m_1-DN4') holds too.
Define $x \to y \overset{def.}{=} (x \cdot y^{-1})^{-1}$. We must prove that (M) and (Ex1) hold. Indeed,
$1 \to x \overset{def.}{=} (1 \cdot x^{-1})^{-1} \overset{(PU)}{=} (x^{-1})^{-1} \overset{(DN)}{=} x$; thus, (M) holds.
$x \to (y \to 1) \overset{def.}{=} (x \cdot ((y \cdot 1^{-1})^{-1})^{-1})^{-1} \overset{(DN),(N1)}{=} (x \cdot (y \cdot 1))^{-1}$
$\overset{(PU)}{=} (x \cdot y)^{-1} \overset{(Pcomm)}{=} (y \cdot x)^{-1} \overset{(PU)}{=} (y \cdot (x \cdot 1))^{-1}$
$\overset{(N1),(DN)}{=} (y \cdot ((x \cdot 1^{-1})^{-1})^{-1})^{-1} \overset{def.}{=} y \to (x \to 1)$; thus, (Ex1) holds.

(2): Let

$$(A^L, \to, 1) \quad \overset{\Phi_1}{\longrightarrow} \quad (A^L, \cdot, ^{-1}, 1) \quad \overset{\Psi_1}{\longrightarrow} \quad (A^L, \Rightarrow, 1).$$

Then, for all $x, y \in A^L$, we have:
$x \Rightarrow y = (x \cdot y^{-1})^{-1} = ((x \to (y^{-1})^{-1})^{-1})^{-1} \overset{(DN)}{=} x \to y$.

Conversely, let

$$(A^L, \cdot, ^{-1}, 1) \quad \overset{\Psi_1}{\longrightarrow} \quad (A^L, \to, 1) \quad \overset{\Phi_1}{\longrightarrow} \quad (A^L, \otimes, ', 1).$$

Then, for all $x, y \in A^L$, we have:
$x' = x \to 1 = (x \cdot 1^{-1})^{-1} \overset{(N1)}{=} (x \cdot 1)^{-1} \overset{(PU)}{=} x^{-1}$ and
$x \otimes y = (x \to y')' = (x \to y^{-1})^{-1} = (x \cdot (y^{-1})^{-1})^{-1} \overset{(DN)}{=} x \cdot y$.

(3): Suppose $x \leq y$, i.e. $x \to y = 1$, by (dfrelR); hence, $(x \cdot y^{-1})^{-1} = 1$, by Φ_1; then, $x \cdot y^{-1} = 1$, by (DN) and (N1), hence $x \leq_{m_1} y$, by (m-dfrelP$_1$).
Conversely, suppose $x \leq_{m_1} y$, i.e. $x \cdot y^{-1} = 1$, by (m-dfrelP$_1$); then, $(x \cdot y^{-1})^{-1} = 1$, by (N1); hence $x \to y = 1$, by Ψ_1; hence $x \leq y$, by (dfrelR).

The proof of (4) is similar to the proof of (4) from Theorem 9.1. We prove only the following:

(BB=) $\Longleftrightarrow$ (m_1-BB=): Suppose that (BB=) holds; then,
$(z \cdot x)^{-1} \cdot (y \cdot x) = y \cdot z^{-1} \Longleftrightarrow (y \cdot z^{-1})^{-1} = [(z \cdot x)^{-1} \cdot (y \cdot x)]^{-1}$
$\overset{(DN)}{\Longleftrightarrow} (y \cdot z^{-1})^{-1} = [(z \cdot (x^{-1})^{-1})^{-1} \cdot ((y \cdot (x^{-1})^{-1})^{-1})^{-1}]^{-1}$
$\Longleftrightarrow y \to z = [(z \to x^{-1}) \cdot (y \to x^{-1})^{-1}]^{-1}$
$\Longleftrightarrow y \to z = (z \to x^{-1}) \to (y \to x^{-1})$ that is true by (BB=); thus, (m_1-BB=) holds too.
Conversely, suppose that (m_1-BB=) holds; then,

$y \to z = (z \to x) \to (y \to x) \Longleftrightarrow (y \cdot z^{-1})^{-1} = [(z \cdot x^{-1})^{-1} \cdot ((y \cdot x^{-1})^{-1})^{-1}]^{-1}$
$\overset{(DN)}{\Longleftrightarrow} (y \cdot z^{-1})^{-1} = [(z \cdot x^{-1})^{-1} \cdot (y \cdot x^{-1})]^{-1}$
$\overset{(DN)}{\Longleftrightarrow} (z \cdot x^{-1})^{-1} \cdot (y \cdot x^{-1}) = y \cdot z^{-1}$ that is true by (m_1-BB=); thus, (BB=) holds.

(B=) $\Longleftrightarrow$ (m_1-B=): Suppose that (B=) holds; then,
$(x \cdot y^{-1})^{-1} \cdot (x \cdot z) = y \cdot z \Longleftrightarrow (y \cdot z)^{-1} = [(x \cdot y^{-1})^{-1} \cdot (x \cdot z)]^{-1}$
$\overset{(DN)}{\Longleftrightarrow} (y \cdot (z^{-1})^{-1})^{-1} = [(x \cdot y^{-1})^{-1} \cdot (x \cdot (z^{-1})^{-1})^{-1})^{-1}]^{-1}$
$\Longleftrightarrow y \to z^{-1} = (x \to y) \to (x \to z^{-1})$ that is true by (B=); thus, (m_1-B=) holds.
Conversely, suppose that (m_1-B=) holds; then,
$y \to z = (x \to y) \to (x \to z) \Longleftrightarrow (y \cdot z^{-1})^{-1} = [(x \cdot y^{-1})^{-1} \cdot ((x \cdot z^{-1})^{-1})^{-1}]^{-1}$
$\overset{(DN)}{\Longleftrightarrow} (y \cdot z^{-1})^{-1} = [(x \cdot y^{-1})^{-1} \cdot (x \cdot z^{-1})]^{-1}$
$\Longleftrightarrow (x \cdot y^{-1})^{-1} \cdot (x \cdot z^{-1}) = y \cdot z^{-1}$ that is true by (m_1-B=); thus, (B=) holds.

(D=) $\Longleftrightarrow$ (m_1-D=): Suppose that (D=) holds; then,
$(y^{-1} \cdot x)^{-1} \cdot x = y \Longleftrightarrow y^{-1} = [(y^{-1} \cdot x)^{-1} \cdot x]^{-1}$
$\overset{(DN)}{\Longleftrightarrow} y^{-1} = [(y^{-1} \cdot (x^{-1})^{-1})^{-1} \cdot (x^{-1})^{-1}]^{-1}$
$\Longleftrightarrow y^{-1} = (y^{-1} \to x^{-1}) \to x^{-1}$ that is true by (D=); thus, (m_1-D=) holds too.
Conversely, suppose that (m_1-D=) holds; then,
$y = (y \to x) \to x \Longleftrightarrow [(y \to x) \to x]^{-1} = y^{-1}$
$\Longleftrightarrow [[(y \cdot x^{-1})^{-1} \cdot x^{-1}]^{-1}]^{-1} = y^{-1} \overset{(DN)}{\Longleftrightarrow} (y \cdot x^{-1})^{-1} \cdot x^{-1} = y^{-1}$
$\overset{(DN)}{\Longleftrightarrow} ((y^{-1})^{-1} \cdot x^{-1})^{-1} \cdot x^{-1} = y^{-1}$ that is true by (m_1-D=); thus, (D=) holds too.

(p-s) $\Longleftrightarrow$ (m-p-s):
(p-s) $\Longleftrightarrow (x \leq y \Longleftrightarrow x = y) \overset{(3)}{\Longleftrightarrow} (x \leq_{m_1} y \Longleftrightarrow x = y) \Longleftrightarrow$ (m-p-s). □

By this Theorem 9.3, choosing appropriate definitions, one can prove all the definitionally equivalences (in the involutive case) between the corresponding algebras from the two 'worlds', as for examples between commutative i-goops and commutative goops, between commutative i-groups and commutative groups, etc.

References

[1] Marlow Anderson, and Todd Feil Lattice-Ordered Groups, *An Introduction*, D. Reidel Publishing Company, Dordrecht, etc. 1988.

[2] Garrett Birkhoff Lattice Theory, American Mathematical Society, Providence RI 1967 (3rd edition) [*Colloquium Publications 25*].

[3] George Boole The Mathematical Analysis of Logic, Macmillan, Barclay & Macmillan, Cambridge 1847.

[4] — An Investigation into the Laws of Thought, Macmillan, London 1854 [reprint: Open Court Publ. Co., Chicago IL 1940].

[5] Martin W. Bunder *Simpler axioms for* BCK-*algebras and the connection between the axioms and the combinators* B, C *and* K, Mathematica Japonica 26, 1981, pp. 415–418.

[6] Dumitru Buşneag Contributions to the Study of Hilbert Algebras [Romanian], PhD Dissertation, University of Bucharest, 1985.

[7] Dumitru Buşneag, and Sergiu Rudeanu *A glimpse of deductive systems in algebra*, Central Eururopean Journal of Mathematics 8 (4), 2010, pp. 688–705.

[8] Roberto L. O. Cignoli, Itala M. L. D'Ottaviano, and Daniele Mundici Algebraic Foundations of Many-valued Reasoning, Kluwer Academic Publishers & Springer Science, Dordrecht 2000 [*Trends in Logic – Studia Logica Library 7*].

[9] Jänis Cirulis *Implication in sectionally pseudocomplemented posets*, Acta Scientiarum Mathematicarum (Szeged) 74, 2008, pp. 477–491.

[10] — *Subtraction-like operations in nearsemilattices*, Demonstratio Mathematica 43, 2010, pp. 725–738.

[11] — *Quasi-orthomodular posets and weak* BCK-*algebras*, Order, DOI 10.1007/s11083-013-9309-1.

[12] Chen Chung Chang *Algebraic analysis of many valued logics*, Transactions of the American Mathematical Society 88, 1958, pp. 467–490.

[13] Paul F. Conrad Lattice-ordered Groups, Tulane Lecture Notes, New Orleans, 1970.

[14] Maria Luisa Dalla Chiara, Roberto Giuntini, and Richard Greechie Reasoning in Quantum Theory, *Sharp and Unsharp Quantum Logics*, Springer, 2004 [*Trends in Logic – Studia Logica Library 22*].

[15] Antonio Diego Sur les algèbres de Hilbert, Gauthier-Villars, Paris 1966. [*Collection de Logique mathématique, Serie A, XXI*]. (A translation of the PhD Dissertation Sobre álgebras de Hilbert, Instituto de Matemática, Universidad Nacional del Sur, Bahia Blanca 1965 [*Nótas de Logica Matemática 12*].)

[16] Anatolij Dvurečenskij, and Sylvia Pulmanová New Trends in Quantum Algebras, Kluwer Academic Publishers, Dordrecht & Ister Science, Bratislava, 2000.

[17] Francesc Esteva, and Lluis Godo *Monoidal t-norm based logic: towards a logic for left-continuous t-norms*, Fuzzy Sets and Systems 124 (3), 2001, pp. 271–288.

[18] Paul Flondor, George Georgescu, and Afrodita Iorgulescu *Pseudo-t-norms and pseudo-*BL *algebras*, Soft Computing 5 (5), 2001, pp. 355–371.

[19] Josep Maria Font, Antonio J. Rodríguez, and Antoni Torrens *Wajsberg algebras*, Stochastica 8 (1), 1984, pp. 5–31.

[20] George Georgescu, and Afrodita Iorgulescu, Logică matematică [Mathematical Logic] [Romanian], Editura ASE [Academy of Economic Studies Publishing House], Bucharest 2010.

[21] Petr Hájek *Metamathematics of fuzzy logic*, Technical Report 682, Institute of Computer Science, Academy of Science of the Czech Republic, 1996.

[22] — *Basic Fuzzy Logic and* BL-*Algebras*, Soft Computing 2, 1998, pp. 124–128.

[23] — Metamathematics of Fuzzy Logic, Kluwer Academic Publishers, Dordrecht 1998 [*Trends in Logic 4*].

[24] Leon Henkin *An algebraic characterization of quantifiers*, Fundamenta Mathematicae 37 (1), 1950, pp. 63–74.

[25] David Hilbert *Die logischen Grundlagen der Mathematik*, Mathematischen Annalen 88, 1923, pp. 151–165.

[26] David Hilbert, and Paul Bernays Grundlagen der Mathematik 1, Springer Verlag, Berlin 1934.

[27] Ulrich Höhle *Commutative, residuated l-monoids.* in: Ulrich Höhle, and Erich Peter Klement (eds.), Non-Classical Logics and Their Applications to Fuzzy Subsets, Kluwer Academic Publishers, Dordrecht 1995, pp. 53–106.

[28] Qingping Hu, and Xin Li *On* BCH-*algebras*, Mathematics Seminar Notes (Kobe Univesity) 11, 1983, pp. 313–320.

[29] — *On proper* BCH-*algebras*, Mathematica Japonica 30, 1985, pp. 659–661.

[30] Yisheng Huang BCI-algebras, Science Press, Beijing 2006. (Distributed by Elsevier Science.)

[31] Yasuyuki Imai, and Kioshi Iséki *On axiom systems of propositional calculi* XIV, Proceedings of the Japan Academy 42, 1966, pp. 19–22.

[32] Afrodita Iorgulescu *Algebras of logic as* BCK *algebras*, Academy of Economic Studies Press, Bucharest 2008.

[33] — *Asupra algebrelor Booleene* [Romanian], Revista de logică, `http://egovbus.-net/rdl`, 25.01.2009, pp. 1–25.

[34] — *The implicative-group - a term equivalent definition of the group coming from algebras of logic*, Part I, Preprint nr. 11/2011, Institutul de Matematică "Simion Stoilow" al Academiei Române, Bucharest 2011.

[35] — *The implicative-group - a term equivalent definition of the group coming from algebras of logic*, Part II, Preprint nr. 12/2011, Institutul de Matematică "Simion Stoilow" al Academiei Române, Bucharest 2011.

[36] — *On l-implicative-groups and associated algebras of logic*, presented at Congmatro 2011-Braşov, Bulletin of the Transilvania University of Brașov [# Series III: Mathematics, Informatics, Physics], 5 (54) 2012, pp. 179–194. (Special Issue: Proceedings of the Seventh Congress of Romanian Mathematicians, published by Transilvania University Press, Braşov, and the Publishing House of the Romanian Academy, Bucharest.)

[37] — *New generalizations of* BCI, BCK *and Hilbert algebras*, Parts I, II (Dedicated to Dragoş Vaida), Journal of Multi-valued Logic and Soft Computing 27 (4), 2016, pp. 353–406, and 407–456. (A previous version available from December 6, 2013, at `http://arxiv.org/abs/1312.2494`.)

[38] — *New algebras and new connections between/in the algebras of logic and the monoidal algebras*, Libertas Mathematica [Arligtom TX], 35 (1), 2015, pp. 13–55. (Dedicated to Nicolae Dinculeanu and Solomon Marcus in celebration of their 90th Birthday.)

[39] — *Quasi-algebras versus regular algebras*, Part I (Dedicated to Sergiu Rudeanu), Scientific Annals of Computer Science 25 (1), 2015, pp.1–43 (doi: 10.7561/SACS.2015.1.ppp).

[40] — *Quasi-algebras vs. regular algebras*, Part II, 2015, Preprint nr. 3/2017, http://imar.ro/~increst/2017/3_2017.pdf.

[41] — *Quasi-algebras vs. regular algebras*, Part III, 2015, Preprint nr. 4/2017, http://imar.ro/~increst/2017/4_2017.pdf.

[42] — Implicative-groups vs. groups and generalizations, Matrix Rom, Bucharest 2018.

[43] — *Generalizations of MV algebras, ortholattices and Boolean algebras*, ManyVal 2019, Bucharest, Romania, November 1-3, 2019.

[44] Kiyoshi Iséki *An algebra related with a propositional calculus*, Proceedings of the Japan Academy 42, 1966, pp. 26–29.

[45] Kiyoshi Iséki, and Shôtarô Tanaka *An introduction to the theory of* BCK*-algebras*, Mathematica Japonica 23 (1), 1978, pp. 1–26.

[46] Young Bae Jun, Eun Hwan Roh, and Hee Sik Kim *On* BH*-algebras*, Scientiae Mathematicae Japonicae 1, 1998, pp. 347–354.

[47] Gudrun Kalmbach Orthomodular Lattices, Academic Press, London, New York, etc. 1983 [*London Mathematical Society Monographs 18*].

[48] Hee Sik Kim, and Young Hee Kim *On* BE*-algebras*, Scientia Mathematicae Japonicae, online e-2006 (2006), pp. 1192–1202.

[49] Jae-Doek Kim, Young-Mi Kim, and Eun-Hwan Roh *A note on* GT*-algebras*, Journal of the Korean Society of Mathematical Education (Ser. B, Pure and Applied Mathematics) 16, 2009, pp. 59–68.

[50] Yuichi Komori *The separation theorem of the $\aleph_0$-valued Łukasiewicz propositional logic*, Reports of the Faculty of Science, Shizuoka University 12, 1978, pp. 1–5.

[51] — *Super-Łukasiewicz implicational logics*, Nagoya Mathematical Journal 72, 1978, pp. 127–133.

[52] — *Super-Łukasiewicz propositional logics*, Nagoya Mathematical Journal 84, 1981, pp. 119–133.

[53] — *The variety generated by* BCC*-algebras is finitely based*, Reports of the Faculty of Science, Shizuoka University 17, 1983, pp. 13–16.

[54] — *The class of BCC-algebras is not a variety*, Mathematica Japonica 29, 1984, pp. 391–394.

[55] Tomasz Kowalski, and Francesco Paoli *Joins and subdirect products of varieties*, Algebra Universalis 65, 2011, pp. 371–391.

[56] Wolfgang Krull *Axiomatische Begründung der allgemeinen Idealtheorie*, Sitzungsberichte der physikalisch-medizinischen Societät zu Erlangen 56, 1924, pp. 47–63.

[57] Laurenţiu Leuştean Representations of Many-valued Algebras, Editura Universitară, Bucharest, 2010.

[58] Bao Long Meng CI*-algebras*, Scientiae Mathematicae Japonicae 71 (1), 2010, pp. 11–17.

[59] Jie Meng *BCI-algebras and abelian groups*, Mathematica Japonica 32, 1987, pp. 693–

696.

[60] Jie Meng, and Young Bae Jun BCK-algebras, Kyung Moon Sa Co., Seoul 1994.

[61] Daniele Mundici *MV-algebras are categorically equivalent to bounded commutative BCK-algebras*, Mathematica Japonica 31 (6), 1986, pp. 889–894.

[62] Ranganathan Padmanabhan, and Sergiu Rudeanu Axioms for Lattices and Boolean Algebras, World Scientific, Singapore, etc. 2008.

[63] Dana Piciu Algebras of Fuzzy Logic, Editura Universitaria Craiova, Craiova 2007.

[64] Marlow Sholander *Postulates for distributive lattices*, Canadian Journal of Mathematics 3 (1), 1951, pp. 28–30.

[65] Matthew Spinks Contributions to the Theory of Pre-BCK-Algebras, PhD Dissertation, Monash University, 2003.

[66] Kanduri Lakshminarasimha Swamy *Dually residuated lattice ordered semigroups*, Mathematische Annalen 159, 1965, pp. 105–114.

[67] Shôtarô Tanaka *On $\wedge$-commutative algebras*, Mathematics Seminar Notes (Kobe University) 3, (7) 1975.

[68] Hiroshi Yutani *The class of commutative BCK-algebra is equationally definable*, Mathematics Seminar Notes (Kobe University) 5, 1977, pp. 207–210.

[69] — *On a system of axioms of a commutative BCK-algebra*, Mathematics Seminar Notes (Kobe University) 5, 1977, pp. 255–256.

[70] Mordchaj Wajsberg *Beiträge zum Metaaussagenkalkül*, Monatshefte für Mathematik und Physik 42, 1935, p. 240.

[71] Andrzej Walendziak *The property of commutativity for some generalizations of BCK-algebras*, Soft Computing 23, 2029, pp. 7505–7511.

[72] — *The implicative property for some generalizations of BCK-algebras*, Journal of Multiple-Valued Logic and Soft Computing [to appear].

[73] Xiaohong Zhang *BIK^+-logic and non-commutative fuzzy logics*, Fuzzy Systems and Mathematics 21 (6), 2007, pp. 31–36.

[74] Xiaohong Zhang, and Ruifen Ye *BZ-algebras and groups*, Journal of Mathematical and Physical Sciences 29, 1995, pp. 223–233.

RELEVANT LOGICS: FROM SEMANTICS TO PROOF SYSTEMS

HIDENORI KUROKAWA
Institute of Liberal Arts and Sciences, Kanazawa University, Japan
hidenori.kurokawa@gmail.com

SARA NEGRI
Department of Mathematics, University of Genova, Italy
Department of Philosophy, University of Helsinki, Finland
sara.negri@unige.it, sara.negri@helsinki.fi

ABSTRACT Axiomatic presentations for a wide family of relevant logics as well as their Routley-Meyer semantics based on ternary accessibility relations are recalled. Following a general method, the latter are used to generate uniformly a corresponding family of sequent calculi that enjoy good structural properties and are complete with respect to the semantics.

1 Introduction

Relevant logics have a long and complicated history. They have one origin in Church's lambda-I calculus [11], another in Ackermann's work on "strenge Implikation" [1] and independently some other historical origins [13]. The fundamental idea of relevant logic is that a logical implication "$\rightarrow$" should have some connection between its antecedent and consequent (which is formally expressed by the notion of variable sharing [2]). On the other hand, relevantists claim that material, intuitionitistic, and strict implications are unsatisfactory in this respect, since, for an inference based on an implication to be called "deductive," the implication needs to ensure a tighter connection between the antecedent and the consequent than usually considered in the aforementioned implications.[1]

The first monograph on relevant logic is Anderson and Belnap's *Entailment I* [2], which presents a comprehensive survey of the contemporary literature on relevant logic. Although the book discusses quite an extensive class of relevant logics, the logics semantically studied there are mainly limited to the neighborhoods of FDE, R, E, and of a

[1]Disclaimer: the following description of the history relevant logic focusing on the proof-theoretic studies of the topic is confined to a rough historical description, which is barely sufficient to provide a context in which our work was started, and is by no means intended to be a comprehensive one.

stronger logics such as RM. It is the ternary relational semantics, introduced by Routley and Meyer [41], that has made it possible to study uniformly a much larger class of relevant logics including both relatively weak relevant logics such as B, traditional relevant logics such as R, and super-logics of those. Besides, Routley-Meyer's ternary relational semantics is usually combined with an idea of handling a de Morgan negation, called the *Routley star*.

This line of research is often called "Australian plan," whereas the one based on formulations of relevant logics based on Belnap's four-valued logic for FDE is often called "American plan." The actual relationship between the American plan and the Australian plan is quite complicated and includes overlaps of people and ideas. Moreover, relevant logics are now considered to be part of a larger class of logics called *substructural logics* [14] that include logics such as linear logic, BCK logic, etc.

However, no matter how we consider the present situation of the research on the semantics of relevant and related logics, Routley and Meyer's ternary semantics is definitely one of the most important contributions to the study of relevant logics. In particular, the ternary relational semantics as we consider it here plays a major role in logical systems which contain the distributive law of logic.[2] In this article we aim at developing a general proof theory for relevant logic based on the Routley-Meyer ternary relational semantics.

Proof-theoretic studies of relevant logics started quite early, and [2] contains some proof-theoretic discussions (the literature significantly precedes the book). So far, we have at least the following approaches in the literature of relevant logics:

1. Fitch-style natural deduction
2. Consecution calculi
3. Dunn-Mints calculi
4. Hypersequent calculi
5. Display calculi
6. Tableau systems
7. Labelled deduction systems

Let us give some comments on the above proof systems. Fitch-style natural deduction (1) is useful to express the idea of "relevance" between the antecedent and the consequent. R. Brady formulated Fitch-style natural deduction systems for an extensive class of relevant logics [7] and proved the normalization theorem for DW and related logics in [9]. The extension normalization to a wider variety of relevant logics discussed in [9] and in

[2]For comprehensive expositions of Routley-Meyer semantics, we refer to [36], [34], [25], [17]. Also, there are some substructural logics, e.g., linear logic, in which the distributive law does not hold. It is possible to handle such logics by a ternary relational semantics, but omitting the distributive law makes the semantics complicated. See [39], [33].

[7] has been announced but apparently not yet been published.[3] Consecution calculi (2) are quite close to the traditional Gentzen-style sequent calculi, and they were useful in a relatively early stage of the proof-theoretic study of relevant logics, e.g., for proving decidability of some (fragments of) logics. However, they can cover only fragments of the entire language of relevant logic, which contain both intensional and extensional conjunctions and disjunctions. In order to express both intensional and extensional contexts in a sequent style formulation, Dunn-Mints calculi (3) [15], [26] have been developed. Hypersequents (4) turn out to be useful in formulating a particularly strong system of relevant logic RM and related systems [3], [4], but apparently they are not suitable for formulating weaker systems of relevant logics. Display calculi (5) [38], [39] are sufficiently expressive to formulate practically all the systems, but they are not suitable for proof search and unsuitable for proving decidability (in display calculi, the real subformula property is lost since we freely replace logical constants and meta-symbols for them back and forth).[4] The goals of display calculi and labelled sequent calculi are directed to almost the opposite directions in the following sense. Display calculi use structural connectives to explicitly express structural rules or (more broadly) structural features of various logics, whereas in labelled sequent calculi in G3-style, we show that all the traditional structural rules are admissible.[5] Tableau systems (6) [34] and labelled deduction systems (7) [44] are somewhat similar to our systems and should be compared more closely. We discuss the issue shortly.

In this paper, we introduce labelled sequent calculi in which we have labels for possible situations (or worlds) and a ternary relational symbol for a ternary accessible relation in Routley-Meyer's semantics for relevant logics.[6] Labelled sequent calculi are introduced for handling some comprehensive classes of logics that are otherwise difficult to formulate in sequent calculi, having desirable proof theoretic properties from the viewpoint of structural proof theory.[7]

[3]It should be noted that Brady also developed "Gentzenized" systems for a variety of relevant logics. Some of these natural deduction systems and Gentzen-style systems (and their extensions to first-order predicate logic) are surveyed in [8]. Also, Brady proposes a semantics called "free semantics" based on the normalized natural deduction systems [10].

[4]Here is what Goré says concerning the issue. "Having tried to obtain decision procedures from display calculi, I must confess that I am also sometimes frustrated by this aspect of display calculi. I think that the truth probably lies somewhere in the middle. Certainly it is true that more traditional calculi give decision procedures for simple logics via the "real" subformula property. However, this usually only works for the very simplest of logics. As soon as contraction or weakening aspects surface, the problems become difficult even with a "real" subformula property as witnessed by traditional Gentzen calculi for propositional Linear Logic. In display calculi, these problems surface even at the base level (p.270, [23])."

[5]Combining the foregoing critical observations on the weakness of display calculi, one might think that we are simply dismissive of display calculi. But this is not so. For this, see the conclusion.

[6]This idea is already presented in [29], but only in outline.

[7]Labelled sequent calculi, as we use them, are introduced in [28], but there are similar works anticipating them but based on somewhat different motivations, such as [19], [24], [27]. The design of the

We cover systems of relevant logics apparently considered to be "standard" and extensions thereof (not necessarily relevant), which are presented in Ch.4 of [42]. We formulate these logics in G3-style sequent calculi [43, 31], where the following hold:

1. All the rules are invertible;
2. All the structural rules are admissible;
3. Cut is admissible.

Although our systems contain semantic information in the form of labels, our proof method of cut-elimination is a syntactic one, i.e., based on reductions of proofs.

Our work is motivated by the following reasons (all related to the issues concerning what other approaches cannot or do not do).[8] First, there are some precedent works handling relevant logics by using some sort of labels. For instance, Priest [34] and Viganò [44]. However, [34] does not address proof-theoretic issues primarily because of the presentation in an introductory textbook. Viganò, on the other hand, does not address the issue of admissibility of structural rules from the viewpoint of structural proof theory. And what is potentially interesting here is that most relevant logics we deal with in this paper are broadly classified as "substructural" logics; their characterization has been given by dropping structural rules from the traditional Gentzen-style sequent calculi, but here the structural rules are admissible. This may raise a philosophical question, "structural rules are features of logic or features of systems?" We do not take any particular position on this issue in this paper. However, the fact that we can "absorb" this feature by using labels may suggest that identifying logics is not as simple as one might have thought.

Second, Mike Dunn, e.g., in [16], proposes a kind of "thesis" concerning the relationship between logical connectives and relational semantics, which can be generally stated as follows: Every n-ary logical connective can be formulated by using $n+1$-ary relational semantics. Kripke semantics for modal logic is a binary relational semantics for unary connectives (modal operators), and Routley-Meyer semantics is a ternary relational semantics for a binary connective (relevant implication). According to this view, material implication and intuitionistic implication may be merely special cases in which we do not need a ternary relational semantics, and indeed it is possible to formulate classical and intuitionistic logic in Routley-Meyer semantics (as degenerate cases). We think it is nice to have proof systems that realize this idea.

Thirdly, this work can be considered to be an additional contribution to a general research program in which labelled sequent calculi are just special cases, i.e. the fact

systems and our proof method here are based on [31], [32] and related works.

[8]Let us quickly mention that, in addition to the reasons given below, one advantage of our framework is comprehensiveness. Labelled sequent calculi probably have a potential to cover one of the greatest varieties of relevant logics. For example, although we do not discuss relevant logics with intensional conjunction, it is possible to handle such cases in our framework. We omitted them only because our work is already huge.

that a theory axiomatized by *geometric* formulas can be presented in a cut-free sequent calculus. It turns out that the overwhelming majority of the semantic conditions in the ternary semantics for axioms of relevant logics and their extensions treated here have a formulation in geometric formulas. There is only one case whose semantic condition goes beyond geometric formula, but even this case can be handled by a method called "geometrization" in [18].

Fourthly, we have decided to use the "indexed modality" to handle the ternary relation for implication. Indexed modalities have been used in [21] for obtaining a tableau calculus for preference-based conditional logics. They have also been used in [30] to obtain labelled sequent calculi for Lewis's logic VCU and in [22] to handle Burgess's preferential conditional logic PCL, Lewis's counterfactual logic V, and their extensions. This choice is not mandatory, i.e., the ternary relation for implication could be directly handled without using the indexed modality. But via the indexed modality we can obtain a uniformity with these works on conditional logics.

The paper is organized as follows. In §2, we present Hilbert-style axiomatizations of the relevant logics that will be covered. In §3, we present the Routley-Meyer's ternary semantics and in §4 proceed with presenting labelled sequent calculi for all the relevant logics presented in §2. In §5, we prove several properties—important from the viewpoint of structural proof theory—of labelled sequent calculi for relevant logics, all of which are used later for our proof of the completeness theorem and of the cut-elimination theorem, proved, respectively, in §6 and in §7.

2 Hilbert-style systems for relevant logics

There are simply too many systems of relevant logics to present all of them in this paper. However, here we would like to formulate at least the major ones. To systematically present numerous systems, we adopt a Hilbert-style formulation. In particular, we pick up all the systems given in Chapter 4 of [42], which is one of the most comprehensive presentations of relevant logic. In this way, we identify the class of target systems that we aim to formulate by using labelled sequent calculi. Due to the existence of numerous systems of relevant logics, we give up formulating systems with Boolean negation, intensional conjunction, propositional constants t, etc., and confine ourselves to traditional relevant logics. We can describe the features of the logical systems discussed in this paper as follows:

1. Logics formulated in the language $\{\neg, \rightarrow, \wedge, \vee\}$ (we call this the language of relevant logic).
2. The best known system of relevant logic R, subsystems and supersystems thereof (in the same language).
3. Traditional relevant logics often formulated by Routley-Meyer semantics.

There are lots of substructural variants of these latter systems which lack the distributive law, but we leave them out from the scope of this work.

Following the presentation of the systems of relevant logic in Ch. 4 of [42],[9] let us first formulate the basic system B. Traditionally, its positive variant B^+ is taken as a privileged basic system (technically this is not a subsystem of B since B^+ is usually formulated with the propositional constant t). But, for greater generality, we take B to be the basic system for the whole class of relevant logics that we handle here. Also, we formulate B somewhat differently from that of [42]. Below we use i) the ordinary convention on the strength of connection among the propositional connectives and ii) $\vdash$ for expressing the derivability of a formula in B and its extensions.

1. System B:

A1. $A \to A$

A2. $A \wedge B \to A$

A3. $A \wedge B \to B$

A4. $A \to A \vee B$

A5. $B \to A \vee B$

A6. $(A \to B) \wedge (A \to C) \to (A \to B \wedge C)$

A7. $(A \to C) \wedge (B \to C) \to (A \vee B \to C)$

A8. $A \wedge (B \vee C) \to (A \wedge B) \vee (A \wedge C)$

A9. $\neg\neg A \to A$

R1. $\dfrac{\vdash A \qquad \vdash A \to B}{\vdash B}$

R2. $\dfrac{\vdash A \qquad \vdash B}{\vdash A \wedge B}$

R3. $\dfrac{\vdash A \to B}{\vdash (B \to C) \to (A \to C)}$

R4. $\dfrac{\vdash A \to B}{\vdash (C \to A) \to (C \to B)}$

R5. $\dfrac{\vdash A \to \neg B}{\vdash B \to \neg A}$

2. Additional schemata that can be added to B:

B1. $A \wedge (A \to B) \to B$

B2. $(A \to B) \wedge (B \to C) \to (A \to C)$

B3. $(A \to B) \to ((B \to C) \to (A \to C))$

B4. $(A \to B) \to ((C \to A) \to (C \to B))$

B5. $(A \to (A \to B) \to (A \to B)$ (or $(A \to (B \to C)) \to (A \wedge B \to C))$

B6. $A \to ((A \to B) \to B)$

[9]Note that the presentation in [42] is more comprehensive than the one given by Greg Restall in [37] and [34], in both of which the simplified ternary relational semantics is used and the latter of which uses tableau systems. Observe that [39] may be more comprehensive but it also covers substructural logics that are not relevant logics. Here we confine our attention only to relevant logics.

B7. $(A \to (B \to C)) \to (B \to (A \to C))$

B8. $(A \to (B \to C)) \to ((A \to B) \to (A \to C))$

B9. $(A \to B) \to ((A \to (B \to C)) \to (A \to C))$

B10. $A \to (B \to B)$

B11. $B \to (A \to B)$

B12. $A \to (B \to (C \to B))$

B13. $A \to (B \to A \wedge B)$

B14. $(A \to B) \to ((A \to C) \to (A \to B \wedge C))$
or $(A \to C) \to ((B \to C) \to (A \vee B \to C))$

B15. $(A \wedge B \to C) \to (A \to (B \to C))$

B16. $A \vee (A \to B)$

B17. $(A \to B) \vee (B \to A)$

B18. $A \to (A \to A)$

B19. $A \vee B \to ((A \to B) \to B)$

B20. $(A \wedge B \to C) \to (A \to C) \vee (B \to C)$

3. Axioms handling negation:

D1. $(A \wedge B \to C) \to (A \wedge \neg C \to \neg B)$

D2. $A \vee \neg A$

D3. $(A \to \neg A) \to \neg A$

D4. $(A \to \neg B) \to (B \to \neg A)$

D5. $B \to (A \vee \neg A)$

D6. $A \to (\neg A \to B)$

D7. $\neg(A \to B) \to (B \to A)$

D8. $(A \to \neg(B \to C)) \to (\neg B \to \neg A)$

3. Rule extensions of the systems:

DR1. $\dfrac{\vdash A}{\vdash (A \to B) \to B}$ DR2. $\dfrac{\vdash A \to B}{\vdash \neg A \vee B}$ DR3. $\dfrac{\vdash A \wedge B \to C}{\vdash A \wedge \neg C \to \neg B}$

Extensions of B by any combination of postulates B1-9, D2-4, BR1-2 are called relevant extensions. Here are some representative systems. (The numbers at the vertices of a diagram show how we can arrange relevant logics starting from B.)

1 DW = B + D4

1.1 DL = DW + { B2, D3 } = DK + D3

1.1.1 TL = DL + { B3, B4 } = TW + { B2, D3 }

1.1.1.1 T = N + B5 = TL + B5

1.1.1.1.1/2 E = T + BR1

1.1.1.1.1/2.1/2 R = RW + B5 = T + B6

1.1.1.1/2.1/2.1 RM = R + B18

1.1.1.1/2.1/2.1.1 K (= classical) = RM + B11

1.2 TW (=N) = B + {B3, B4, D4} = DW + {B3, B4}

1.2.2 RW = N (TW) + B7

2 G = B + D2

2.1 DK = G + {B2, D4}

3 DA = B + { B1, B2, D3, D4 }

4 C = B + { B1, B3, B4 }

The following is a diagram showing the relationship among relevant logics (we concentrate on relevant logics, non-relevant logics are left out).

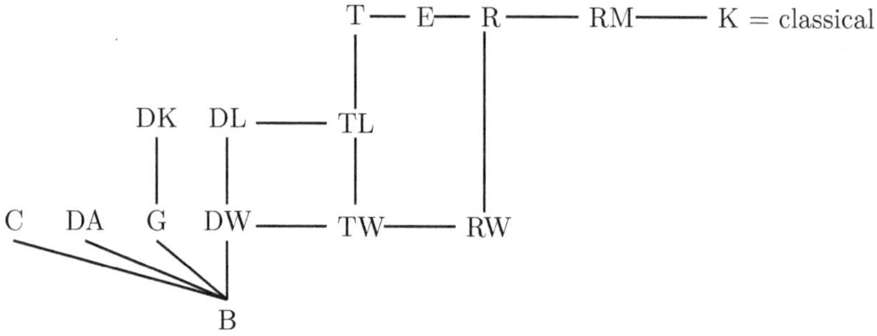

3 Routley-Meyer ternary relational semantics for relevant logics

In this section, we give a definition of the Routley-Meyer ternary relational semantics for relevant logics. For several reasons, we follow the Routley-Meyer semantics originally given in [41], although the definition is slightly more general in order to cover a wider class of logical systems. Our official source of the definition of semantics is [42].

3.1 B model structure

Let us start defining the traditional B (corresponding to the system B above) model structure in the Routley-Meyer style. The notion of model structure is more like a "frame" in modal logic, where a valuation is defined separately.

Definition 3.1. *A* B model structure $\mathfrak{M}$ *is an ordered quodruple* $\langle 0, K, R, * \rangle$, *where*

1. K *is a non-empty set (of worlds or situations)*
2. 0 *is a one-place relation on* K, *i.e. a non-empty subset of* K
3. R *is a ternary relation on* K
4. $*$ *is a one-place operation on* K.

Additionally, a designated element (picked up by using Hilbert's epsilon notation) from the subset 0 *and a preorder are defined as follows:*

d0. $T =_{df.} \epsilon x 0x$, *i.e.* T *is an arbitrary selected element of* 0.

d1. $a \leq b =_{df.} \exists x(0x \wedge aR_x b)$[10]

The following conditions are required:

p1. $a \leq a$, *i.e.,* $\exists x(0x \wedge aR_x a)$.

p2. $a \leq d$ *and* $bR_d c$ *implies* $bR_a c$.

p3. $a = a^{**}$

p4. $a \leq b$ *implies* $b^* \leq a^*$.

p5. $a \leq b$ *and* $b \leq c$ *implies* $a \leq c$.

p6. $a \leq d$ *and* $cR_b a$ *implies* $cR_b d$.

p7. $a \leq d$ *and* $dR_b c$ *implies* $aR_b c$.

Remark: 1. The situations in the nonempty subset 0 of K are called "regular" situations, and the situation T is called the "designated" (real, factual) situation.

2. Although we present the definition of T for the sake of completeness of the presentation, unlike [42], we avoid this concept in defining the notion of validity, following a suggestion in [42] on an alternative definition of validity (see below).

2. The above p4., p5., p6., p7. are stated as additional conditions in [42], but they could be stated as integrated parts of B model structure, since adding these conditions does not affect the set of provable formulas in system B.

3. When a model structure has an additional condition that 0 is confined to a singleton of an element of K (called "T"), the model structure is called a "reduced model structure." The reduced model structure is useful for giving a simpler semantic condition for relatively strong systems of relevant logic; however, it is not so for giving semantic conditions for a wider class of logics.

[10] We write $aR_x b$ for a ternary relation, which may be traditionally written as $Rxab$. For the reason for this, see 3.5.

3.2 Valuations and validity

Let us now define the notion of valuation and validity w.r.t. a given model strucrure.

Definition 3.2. *Given a (B or reduced) structure $\mathfrak{M} = \langle 0, K, R, * \rangle$, a* valuation *$v$ is a function from the set of pairs (p, a), where p is a propositional variable of the language and a is a world in K, to $\{0, 1\}$ which satisfies the following condition:*

Atomic Heredity*: for $a, b \in K$, if $a \leq b$ and $v(p, a) = 1$, then $v(p, b) = 1$.*

A valuation v is extended to the entire language inductively as follows (we write this by using $\Vdash_v$ for a fixed valuation v and make the fixed valuation explicit).

Definition 3.3. *Given a B structure $\mathfrak{M} = \langle 0, K, R, * \rangle$, and a valuation v, the associated forcing relation $\Vdash_v$ is extended to arbitrary formulas of the language as follows:*

1. *$a \Vdash_v p$ iff $v(p, a) = 1$*
2. *$a \Vdash_v A \wedge B$ iff $a \Vdash_v A$ and $a \Vdash_v B$.*
3. *$a \Vdash_v A \vee B$ iff $a \Vdash_v A$ or $a \Vdash_v B$.*
4. *$a \Vdash_v A \rightarrow B$ iff for all $b, c \in K$, if $bR_a c$ and $b \Vdash_v A$, then $c \Vdash_v B$.*
5. *$a \Vdash_v \neg A$ iff $a^* \nVdash_v A$.*

The terminology related to a valuation and an interpretation, etc. is as follows, where A is a wff. We give a definition based on the notion of "being ensured on v."[11]

Definition 3.4. 1. *A* holds *on a valuation v in $\mathfrak{M}$* at a world *$a \in K$ iff $a \Vdash_v A$*

2. *A is* ensured *on v iff A holds on v at every element of* 0*, i.e. for every $a \in 0$, $a \Vdash_v A$*
3. *A is* valid *in $\mathfrak{M}$ iff A is ensured on all valuations therein.*
4. *A is* L-valid *iff A is ensured on v, i.e., holds on v for every valuation v, at every element of* 0 *for every model structure.*

Remark: In the original version, the definition of L-valid is given by using the concept of "true on I" ("I" is an interpretation, corresponding to our $\Vdash_v$), which in turn uses the notion of "the real, factual, or base situation" T. However, as they point out[12], L-valid can be alternatively defined only by using the concept "being ensured on v (see below)," avoiding the use of both "being true on v" and T.

Definition 3.5. *A* model *is a structure $\langle \mathfrak{M}, v \rangle$, where $\mathfrak{M}$ is a model structure and v is a valuation in $\mathfrak{M}$. Truth, validity etc. can also be defined for models as before.*

[11]Truth on v is defined by A is *true* on v in $\mathfrak{M}$ iff $T \Vdash_v A$. A is valid in $\mathfrak{M}$ iff A is true on all valuation v.

[12]They say, "[u]se of both Hilbert's epsilon operator, ξ, and definition d0 ($T =_{Df} \xi x 0 x$) may be eliminated either by redefining validity in terms of holding at each element of 0 ... (p.298, [42]). Also, "[e]quivalently, A is L-valid iff A is ensured, i.e. holds at every element of 0, for every L m.s. (without a distinguished base element T)," (p.302, [42]).

3.3 Semantic implication, truth and soundness

Let us introduce some concepts that are needed to state some lemmas below.

Definition 3.6. *Let v be one of the valuations introduced in model structure $\mathfrak{M}$.*

1. A implies B on v *(or A* v-implies B*) iff for every $a \in K$, if A holds on I in $\mathfrak{M}$ at a, then B holds on I in $\mathfrak{M}$ at a.*
2. A implies B in $\mathfrak{M}$ *iff A v-implies B on every valuation in $\mathfrak{M}$.*
3. A L-implies B *iff A implies B in every model structure.*

By a routine induction on the complexity of formulas we have the following:

Lemma 3.7. *Let v be a valuation in a model structure $\mathfrak{M}$ and we consider its extension $\Vdash_v$ to the entire language of relevant logic.*

(Heredity) For every A and every $a, b \in K$, if $a \Vdash_v A$ and $a \leq b$, then $b \Vdash_v A$.

Lemma 3.8. (3′) *if A implies B on v, then $A \to B$ is ensured on v;* (4′) *if A implies B on every valuation v in $\mathfrak{M}$, then $A \to B$ is ensured on every valuation v in $\mathfrak{M}$;* (5) *if A L-implies B iff $A \to B$ is L-valid.*

As mentioned above, [42] talks about the alternative possibility of defining L-validity in terms of "being ensured on v" but they do not give a proof the lemmas based on the alternative definition. Thus let us give proofs of these lemmas here.

Proof. (3′) Suppose A implies B on v. I.e., for any $a \in K$, if $a \Vdash_v A$ then $a \Vdash_v B$.

Also, suppose $c \in 0$, $dR_c e$ and $d \Vdash_v A$. By the above, if $d \Vdash_v A$ then $d \Vdash_v B$. Hence, $d \Vdash_v B$. However, if $c \in 0$ and $dR_c e$ then $d \leq e$ (by d1). Hence, by Heredity, $e \Vdash_v B$. Thus, for any $c \in 0$, $c \Vdash_v A \to B$. That is, $A \to B$ is ensured on v.

(4′) Suppose A implies B in $\mathfrak{M}$. I.e., A implies B on v for every valuation v in $\mathfrak{M}$. Fix an arbitrary v in $\mathfrak{M}$. Then A implies B on v. Then by (3′), $A \to B$ is ensured on v. But v is arbitrary, so $A \to B$ is ensured on every valuation v in $\mathfrak{M}$.

(5) ($\Longrightarrow$) Suppse A L-implies B. I.e., A implies B for every model structure $\mathfrak{M}$. Fix an arbitrary model structure $\mathfrak{M}$. Then by (4′), we obtain $A \to B$ is ensured on every valuation v in $\mathfrak{M}$. But $\mathfrak{M}$ is arbitrary, so $A \to B$ is ensured for every L m.s. $\mathfrak{M}$. That is, $A \to B$ is L-valid.

($\Longleftarrow$) Suppose $A \to B$ is L-valid. That is, for every $c \in 0$, $c \Vdash_v A \to B$ for every L m.s. $\mathfrak{M}$. Let us further suppose $a \Vdash_v A$ by fixing an arbitrary $a \in K$ and an arbitrary valuation v in $\mathfrak{M}$. We show $a \Vdash_v B$. By the first assumption, $\forall c \in 0, \forall d, e(dR_c e\, , and\, d \Vdash_v A \Rightarrow e \Vdash_v B)$. Since $a \leq a$, $\exists g \in 0$, s.t. $aR_g a$. Then $\forall d, e(dR_g e, , and\, d \Vdash_v A \Rightarrow e \Vdash_v B)$. Thus, $aR_g a, , and\, a \Vdash_v A \Rightarrow a \Vdash_v B$. Hence, $a \Vdash_v B$.

Since $a \in K$ and v are arbitrary, this suffices to show $\forall a \in K$, $a \Vdash_v A \Rightarrow a \Vdash_v B$ on every valuation v in every L m.s. $\mathfrak{M}$. That is, A L-implies B. □

Remark: There exists another ternary relational semantics for relevant logics, called "simplified" semantics [35, 37]. Since the simplified semantics appeared after the original Routley-Meyer semantics and was presumably proposed as an improvement of the original semantics, we may need good reason to stick to the original semantics. The major changes in the simplified semantics contain the following: i) One "designated" situation is sufficient to formulate as wide a variety of logics as the original semantics; ii) $\leq$ is not defined but introduced as primitive. The point i) may look attractive, but the simplified semantics has some features unsuitable for our purpose. First, Restall's semantic condition for a general case of the simplified semantics in [37] is not quite a geometric formula (this can be fixed, but this already shows that the simplified semantics hardly fits with labelled sequents). Second, the simplified semantics at least does not appear to fit well with labelled sequents with the indexed modality (due to the behavior of the designated situation). Third, the original formulation of the simplified semantics in [37] had some error [40], and the semantic condition for the permutation axiom had to be reformulated (the reformulation works, but in a little contrived manner). So the uniformity which initially appeared to hold in the formulation in [37] is now untenable. On the contrary, the original Routley-Meyer semantics turns out to be better suited to labelled sequents in these respects. By using "regular" situations in 0 (not necessarily a singleton) we can uniformly formulate an extensive class of relevant logics without making the semantics too complicated. Hence, we have chosen to use the original Routley-Meyer semanitcs. This choice involved lots of unexpected subtle details.

3.4 Additional conditions for extensions of B

We now present the additional conditions for B model structure in order to formulate extensions of the relevant logic B. (We use $\Longrightarrow$, *and*, $\vee$ for a conditional, a conjunction and a disjunction in the meta-level with the usual convention on the strength of connections. We shall instead use $\Rightarrow$ for sequent calculi.) Note that we present each condition in the canonical form of a geometric formula (if possible).

q1. $aR_a a$ $\qquad A \wedge (A \to B) \to B$

q2. $bR_a c \Longrightarrow \exists x(bR_a x \ and \ xR_a c)$ $\qquad (A \to B) \wedge (B \to C) \to (A \to C)$

q3. $\forall x(bR_a x \ and \ cR_x d \Longrightarrow \exists y(cR_a y \ and \ yR_b d))$ $\qquad (A \to B) \to ((B \to C) \to (A \to C))$

q4. $\forall x(bR_a x \ and \ cR_x d \Longrightarrow \exists y(cR_b y \ and \ yR_a d))$ $\qquad (A \to B) \to ((C \to A) \to (C \to B))$

q5. $bRc \Longrightarrow \exists(bR_a x \ and \ bR_x c)$ $\qquad (A \to (A \to B)) \to (A \to B)$

q6. $bR_a c \Longrightarrow aR_b c$ $\qquad A \to ((A \to B) \to B)$

q7. $\forall x(bR_a x \ and \ cR_x d \Longrightarrow \exists y(cR_a y \ and \ bR_y d))$ $\qquad (A \to (B \to C)) \to (B \to (A \to C))$

q8. $\forall x(bR_a x \ and \ cR_x d \Longrightarrow \exists y \exists z(cR_a z \ and \ yR_z d \ and \ cR_b y))$
$$(A \to (B \to C)) \to ((A \to B) \to (A \to C))$$

q9. $\forall x(bR_a x \text{ and } cR_x d \Longrightarrow \exists y \exists z(cR_b z \text{ and } yR_z d \text{ and } cR_a y))$
$$(A \to B) \to ((A \to (B \to C)) \to (A \to C))$$

The logics that are within the realm of traditional relevant logic have axioms up to q9. From q10 on, we have non-relevant logics but we can characterize these logics by postulating conditions on accessibility relations. (Since $\leq$ is ultimately defined by ternary accessibility relations, we can use only R to characterize these logics. But for the sake of simplicity we use the preorder relation $\leq$ to state additional conditions.)

q10. $bR_a c \Longrightarrow b \leq c$ $\quad A \to (B \to B)$

q11. $bR_a c \Longrightarrow a \leq c$ $\quad B \to (A \to B)$

q12. $\forall x(bR_a x \text{ and } cR_x d \Longrightarrow a \leq d)$ $\quad A \to (B \to (C \to A))$

q13. $bR_a c \Longrightarrow (a \leq c \text{ and } b \leq c)$ $\quad A \to (B \to A \wedge B)$

q14. $\forall x(bR_a x \text{ and } cR_x d \Longrightarrow (cR_a d \text{ and } cR_b d))$
$$(A \to B) \to ((B \to C) \to (A \to B \wedge C))$$

q15. $\forall x(bR_a x \text{ and } cR_x d \Longrightarrow \exists y(b \leq y \text{ and } c \leq y \text{ and } yR_a d))$
$$(A \wedge B \to C) \to (A \to (B \to C))$$

q16. $a \leq b \text{ and } 0x \Longrightarrow a \leq x$ $\quad A \vee (A \to B)$

q17. $a \leq b \vee b \leq a$ $\quad (A \to B) \vee (B \to A)$

q18. $bR_a c \Longrightarrow (a \leq c \vee b \leq c)$ $\quad A \to (A \to A)$

q19. $bR_a c \Longrightarrow (aR_b c \text{ and } b \leq c)$ $\quad (A \vee B) \to ((A \to B) \to B)$

q20. $\forall x(bR_a c \text{ and } dR_a e \Longrightarrow \exists y(b \leq y \text{ and } d \leq y \text{ and } (yR_a c \vee yR_a e))$
$$(A \wedge B \to C) \to (A \to C) \vee (B \to C)$$

The following conditions are the ones that we have for handling negation.

s1. $bR_a c \Longrightarrow \exists y(b \leq y \text{ and } c^* \leq y \text{ and } yR_a b^*)$ $\quad (A \wedge B \to C) \to (A \wedge \neg C \to \neg B)$

s2. For any x such that $0x$, $x^* \leq x$ $\quad A \vee \neg A$

s3. $a^* R_a a$ $\quad (A \to \neg A) \to \neg A$

s4. $bR_a c \Longrightarrow (c^* R_a b^*)$ $\quad (A \to \neg B) \to (B \to \neg A)$

s5. $a^* \leq a$ $\quad B \to (A \vee \neg A)$

s6. $bR_a c \Longrightarrow a \leq b^*$ $\quad A \to (\neg A \to B)$

s7. $bR_a c \text{ and } dR_{a^*} e \Longrightarrow (d \leq c \vee b \leq e)$ $\quad \neg(A \to B) \to (B \to A)$

s8. $bR_a c \Longrightarrow \exists y(c^* R_a y \text{ and } (\forall d, e)(dR_{y^*} e \Longrightarrow d \leq b^*))$
$$(A \to \neg(B \to C)) \to (\neg B \to \neg A)$$

Condition s8 goes beyond geometric formulas, so we need to apply later the technique of "geometrization" [18] to formulate this in a cut-free labelled sequent calculus.

dr1. $\exists x\ (0x\ \textit{and}\ xR_a a)$ $\qquad \dfrac{\vdash A}{\vdash (A \to B) \to B}$

dr2. $0x \Longrightarrow x^* \leq x$ $\qquad \dfrac{\vdash A \to B}{\vdash \neg A \vee B}$

dr3. $a^* \leq a$ $\qquad \dfrac{\vdash A \wedge B \to C}{\vdash A \wedge \neg C \to \neg B}$

The extensions do not go beyond systems representable by axiomatic extensions.

3.5 Expanding the language by the indexed modality

In order to express some semantic concepts used in the Routley-Meyer style ternary semantics, we need to introduce a bunch of expressions in our language of labelled sequent calculi. These are the same as the expressions in the semantics $(R, \leq, *, 0)$ except ":" for "$\Vdash$". (We follow [32] in giving the usual inductive definition of a proof tree and the proof height of a proof tree, etc. and do not give definitions of these notions explicitly.)

In addition, we introduce an auxiliary device called "the indexed modality" $\Box_x$ to our language. The indexed modality is a modality associated with a situation indicated by the subscript (the variable x) written as a part of the modal operator. This index gives an accessibility relation $yR_x z$ but, unlike the traditional interpretation of the ternary relation in Routley-Meyer semantics, this is not taken to be ternary among situations but to be an assignment of a binary relation to the index.[13]

The condition of a statement holding at y on v, which has an indexed modality as the outermost logical symbol, is given as follows:

$y \Vdash_v \Box_x A$ iff $\forall z(yR_x z \Longrightarrow x \Vdash_v A)$.

By this auxiliary expression, we can formulate the condition of holding at x on v indirectly as follows:

$x \Vdash_v A \to B$ iff $\forall y(y \Vdash_v A \Longrightarrow y \Vdash_v \Box_x B)$

By the foregoing condition of $\Box_x B$ at y on v, this condition can be rewritten as:

$\forall y(y \Vdash_v A \Longrightarrow \forall z(yR_x z \Longrightarrow z \Vdash_v B)$

[13]The idea of indexed modality goes back to [20]. It is applied to labelled sequents in [30]. The interpretation of the modality in the ternary semantics for relevant logic is discussed in [5].

Since $y \Vdash_v A$ does not contain the variable z (or we can change the bound variable if there is any variable conflict), we can show that this condition is equivalent to the following, which is the one of holding at y on v directly given to $A \to B$:

$$\forall y \forall z (y R_x z \,, and\, y \Vdash_v A \Longrightarrow z \Vdash_v B)$$

Thus, by using the indexed modality, we can separate the entire condition of $A \to B$ holding at y on v into the following two parts: i) The condition for $\to$ described as an implication between A and $\Box_x B$; ii) The condition for $\Box_x B$. Based on the condition of the indexed modality and the indirect specification of the condition of a $\to$-formula, we formulate labelled sequent calculi for relevant logics.

Also, in order for the induction on the complexity of formulas to work, we need the following official definition of the weight of formulas.

Definition 3.9 (weight of a formula). *The weight $w(A)$ of a formula A is defined by the following:*

1. $w(\varphi) = 1$ *for an atomic formula or a relational atom.*
2. $w(A \circ B) = w(A) + w(B) + 1$ *for* $\circ \in \{\wedge, \vee, \to\}$
3. $w(\neg A) = w(A) + 1$
4. $w(\Box_x A) = w(A) + 1$

Note that by this definition $w(\Box_x B) < w(A \to B)$ for any formula A and the smallest weight is 1.

4 Labelled sequent calculi for the relevant logic B and its extensions

We shall now introduce labelled sequent calculi for relevant logics. In this section we first introduce a labelled sequent calculus for the basic relevant logic B with the indexed modality. We call the system for the basic relevant logic G3B, because the formulation of it is based on the traditional G3-style sequent calculi, where structural rules are admissible. Then we shall move on to formulate various extensions thereof.

4.1 Labelled sequent calculus for relevant logic B: G3B

Here is the basic system G3B:

1. Axioms[14] $\quad x : P, \Gamma \Rightarrow \Delta, x : P \quad$ (We allow x^* to occur as x.)

[14]The proof of completeness of the labelled sequent calculi given below suggests that axioms with atomic formulas other than the ones of the form $x\!:\!P$ is redundant.

2. Rules for $\rightarrow$ with the indexed modality

$$\frac{x : A \rightarrow B, \Gamma \Rightarrow \Delta, y : A \qquad y : \Box_x B, x : A \rightarrow B, \Gamma \Rightarrow \Delta}{x : A \rightarrow B, \Gamma \Rightarrow \Delta} L\rightarrow$$

$$\frac{\Gamma, y : A \Rightarrow y : \Box_x B, \Delta}{\Gamma \Rightarrow x : A \rightarrow B, \Delta} R\rightarrow \ (y \text{ fresh}^{15})$$

3. Rules for the indexed modality

$$\frac{z : A, yR_x z, y : \Box_x A, \Gamma \Rightarrow \Delta}{yR_x z, y : \Box_x A, \Gamma \Rightarrow \Delta} L\Box \qquad \frac{yR_x z, \Gamma \Rightarrow z : A, \Delta}{\Gamma \Rightarrow y : \Box_x A, \Delta} R\Box \ (z \text{ fresh})$$

4. Rules for the FDE (first-degree entailment) part of B[16]

$$\frac{x : A, x : B, \Gamma \Rightarrow \Delta}{x : A \wedge B, \Gamma \Rightarrow \Delta} L\wedge \qquad \frac{\Gamma \Rightarrow \Delta, x : A \qquad \Gamma \Rightarrow \Delta, x : B}{\Gamma \Rightarrow \Delta, x : A \wedge B} R\wedge$$

$$\frac{x : A, \Gamma \Rightarrow \Delta \qquad x : B, \Gamma \Rightarrow \Delta}{x : A \vee B, \Gamma \Rightarrow \Delta} L\vee \qquad \frac{\Gamma \Rightarrow \Delta, x : A, x : B}{\Gamma \Rightarrow \Delta, x : A \vee B} R\vee$$

$$\frac{\Gamma \Rightarrow \Delta, x^* : A}{x : \neg A, \Gamma \Rightarrow \Delta} L\neg \qquad \frac{x^* : A, \Gamma \Rightarrow \Delta}{\Gamma \Rightarrow \Delta, x : \neg A} R\neg \qquad \frac{x = x^{**}, \Gamma \Rightarrow \Delta}{\Gamma \Rightarrow \Delta} \text{Dual}$$

5. Relational rules
5.1 Relational rules for =

$$\frac{x = x, \Gamma \Rightarrow \Delta}{\Gamma \Rightarrow \Delta} \text{EqRef} \qquad \frac{y = z, x = y, x = z, \Gamma \Rightarrow \Delta}{x = y, x = z, \Gamma \Rightarrow \Delta} \text{EqETrans}^{17}$$

$$\frac{wR_x z, y = w, yR_x z, \Gamma \Rightarrow \Delta}{y = w, yR_x z, \Gamma \Rightarrow \Delta} \text{EqRel}_1 \qquad \frac{yR_w z, x = w, yR_x z, \Gamma \Rightarrow \Delta}{x = w, yR_x z, \Gamma \Rightarrow \Delta} \text{EqRel}_2$$

$$\frac{yR_x w, z = w, yR_x z, \Gamma \Rightarrow \Delta}{z = w, yR_x z, \Gamma \Rightarrow \Delta} \text{EqRel}_3 \qquad \frac{y{:}P, x = y, x{:}P, \Gamma \Rightarrow \Delta}{x = y, x{:}P, \Gamma \Rightarrow \Delta} \text{AtRepl}$$

We further propose that equality rules should be extended to handle $()^*$ as a kind of function symbol. The following rule appears to be sufficient for the admissibility of

[15] A variable is *fresh* in a rule if it does not occur in its conclusion.

[16] The rules for $\wedge$, $\vee$ are standard rules; the rules for $\neg$ is a de Morgan negation given by the Routley star.

[17] This stands for Euclidean transitivity.

replacement for an arbitrary formula (see Lemma 5.6).[18]

$$\frac{x^* = y^*, x = y, \Gamma \Rightarrow \Delta}{x = y, \Gamma \Rightarrow \Delta}\ \text{Eq}^*$$

5.2 Rules for $\leq$[19]

5.2.1 Preorder rules for $\leq$

$$\frac{a \leq a, \Gamma \Rightarrow \Delta}{\Gamma \Rightarrow \Delta}\ \text{Ref}\leq \qquad \frac{a \leq c, a \leq b, b \leq c, \Gamma \Rightarrow \Delta}{a \leq b, b \leq c, \Gamma \Rightarrow \Delta}\ \text{Trans}\leq$$

5.2.2 Rules connecting R and $\leq$

$$\frac{y \leq z, yR_x z, 0x, \Gamma \Rightarrow \Delta}{yR_x z, 0x, \Gamma \Rightarrow \Delta} \leq /R \qquad \frac{0x, yR_x z, y \leq z, \Gamma \Rightarrow \Delta}{y \leq z, \Gamma \Rightarrow \Delta}\ (x \text{ fresh})\ R/\leq$$

5.2.3 Monotonicity rules for R and $\leq$

$$\frac{bR_a c, a \leq x, bR_x c, \Gamma \Rightarrow \Delta}{a \leq x, bR_x c, \Gamma \Rightarrow \Delta}\ \text{Mon}_1 \qquad \frac{bR_a c, a \leq x, xR_a c, \Gamma \Rightarrow \Delta}{b \leq x, xR_a c, \Gamma \Rightarrow \Delta}\ \text{Mon}_2$$

$$\frac{bR_a c, x \leq c, bR_a x, \Gamma \Rightarrow \Delta}{x \leq c, bR_a x, \Gamma \Rightarrow \Delta}\ \text{Mon}_3$$

5.2.4 Atomic Heredity

$$\frac{y{:}P, x \leq y, x{:}P, \Gamma \Rightarrow \Delta}{x \leq y, x{:}P, \Gamma \Rightarrow \Delta}\ \text{AtHR}$$

6. Cut

$$\frac{\Gamma \Rightarrow \Delta, A \qquad A, \Pi \Rightarrow \Sigma}{\Gamma, \Pi \Rightarrow \Delta, \Sigma}\ \text{Cut}$$

Remark: Rules for $\rightarrow$ can be given without the indexed modality:

$$\frac{yR_x z, x : A \rightarrow B, \Gamma \Rightarrow \Delta, y : A \qquad yR_x z, z : B, x : A \rightarrow B, \Gamma \Rightarrow \Delta}{yR_x z, x : A \rightarrow B, \Gamma \Rightarrow \Delta}\ L2 \rightarrow$$

$$\frac{yR_x z, y : A, \Gamma \Rightarrow \Delta, z : B}{\Gamma \Rightarrow x : A \rightarrow B, \Delta}\ R2 \rightarrow (y,\ z \text{ fresh})$$

We remark that with the indexed modality we can uniformly treat all the logics presented above, as well others.

[18] It is easy to justify the rule, since $x = y \Rightarrow x^* = y^*$ must be correct from a semantical point of view, and the rule can be derived by this principle and cut.

[19] Although $\leq$ is ultimately definable by R in the traditional Routley-Meyer semantics, we use $\leq$ as a primitive symbol in our labelled sequent calculi for the sake of convenience.

4.2 Labelled sequent calculi for extensions of relevant logic B

Let us now formulate labelled sequent calculi for extensions of relevant logic B. For any combination of extra axioms (and semantic conditions) added to B, we can formulate a complete Hilbert-style axiomatic system for the logic. We can actually formulate a labelled sequent calculus that exactly corresponds to such a Hilbert-style axiomatic system. These calculi can be formulated by augmenting the following additional rules to the foregoing labelled sequent calculus for relevant logic B.

LB1. $\dfrac{aR_a a, \Gamma \Rightarrow \Delta}{\Gamma \Rightarrow \Delta}$

LB2. $\dfrac{bR_a x, xR_a c, bR_a c, \Gamma \Rightarrow \Delta}{bR_a c, \Gamma \Rightarrow \Delta}$ (x fresh)

LB3. $\dfrac{cR_a y, yR_b d, bR_a x, cR_x d, \Gamma \Rightarrow \Delta}{bR_a x, cR_x d, \Gamma \Rightarrow \Delta}$ (y fresh)

LB4. $\dfrac{cR_b y, yR_a d, bR_a x, cR_x d, \Gamma \Rightarrow \Delta}{bR_a x, cR_x d, \Gamma \Rightarrow \Delta}$ (y fresh)

LB5. $\dfrac{bR_a x, bR_x c, bR_a c, \Gamma \Rightarrow \Delta}{bR_a c, \Gamma \Rightarrow \Delta}$ (x fresh)

LB6. $\dfrac{aR_b c, bR_a c, \Gamma \Rightarrow \Delta}{bR_a c, \Gamma \Rightarrow \Delta}$

LB7. $\dfrac{cR_a y, bR_y d, bR_a x, cR_x d, \Gamma \Rightarrow \Delta}{bR_a x, cR_x d, \Gamma \Rightarrow \Delta}$ (y fresh)

LB8. $\dfrac{cR_a z, yR_z d, cR_b y, bR_a x, cR_x d, \Gamma \Rightarrow \Delta}{bR_a x, cR_x d, \Gamma \Rightarrow \Delta}$ (y,z fresh)

LB9. $\dfrac{cR_b z, yR_z d, cR_a y, bR_a x, cR_x d, \Gamma \Rightarrow \Delta}{bR_a x, cR_x d, \Gamma \Rightarrow \Delta}$ (y,z fresh)

Rules for any non-relevant extension of B that uses both $yR_x z$ and $\leq$

LB10. $\dfrac{b \leq c, bR_a c, \Gamma \Rightarrow \Delta}{bR_a c, \Gamma \Rightarrow \Delta}$

LB11. $\dfrac{a \leq c, bR_a c, \Gamma \Rightarrow \Delta}{bR_a c, \Gamma \Rightarrow \Delta}$

LB12. $\dfrac{a \leq d, bR_a x, cR_x d, \Gamma \Rightarrow \Delta}{bR_a x, cR_x d, \Gamma \Rightarrow \Delta}$

LB13. $\dfrac{a \leq c, b \leq c, bR_a c, \Gamma \Rightarrow \Delta}{bR_a c, \Gamma \Rightarrow \Delta}$

LB14. $\dfrac{cR_a d, cR_b d, bR_a x, cR_x d, \Gamma \Rightarrow \Delta}{bR_a x, cR_x d, \Gamma \Rightarrow \Delta}$

LB15. $\dfrac{b \leq y, c \leq y, yR_a d, bR_a x, cR_x d, \Gamma \Rightarrow \Delta}{bR_a x, cR_x d, \Gamma \Rightarrow \Delta}$ (y fresh)

LB16. $\dfrac{a \leq x, a \leq b, 0x, \Gamma \Rightarrow \Delta}{a \leq b, 0x, \Gamma \Rightarrow \Delta}$

LB17. $\dfrac{a \leq b, \Gamma \Rightarrow \Delta \qquad b \leq a, \Gamma \Rightarrow \Delta}{\Gamma \Rightarrow \Delta}$

LB18. $\dfrac{a \leq c, bR_a c, \Gamma \Rightarrow \Delta \qquad b \leq c, bR_a c, \Gamma \Rightarrow \Delta}{bR_a c, \Gamma \Rightarrow \Delta}$

LB19. $\dfrac{aR_b c, a \leq c, bR_a c, \Gamma \Rightarrow \Delta}{bR_a c, \Gamma \Rightarrow \Delta}$

LB20. $\dfrac{yR_a c, b \leq y, d \leq y, bR_a c, dR_a e, \Gamma \Rightarrow \Delta \qquad yR_a e, b \leq y, d \leq y, bR_a c, dR_a e, \Gamma \Rightarrow \Delta}{bR_a c, dR_a e, \Gamma \Rightarrow \Delta}$ (y fresh)

Let us now further move on to the extensions involving negation.

LD1. $\dfrac{b \leq y, c^* \leq y, yR_ab^*, bR_ac, \Gamma \Rightarrow \Delta}{bR_ac, \Gamma \Rightarrow \Delta}$ (y fresh)

LD2. $\dfrac{x^* \leq x, 0x, \Gamma \Rightarrow \Delta}{0x, \Gamma \Rightarrow \Delta}$

LD3. $\dfrac{a^* R_a a, \Gamma \Rightarrow \Delta}{\Gamma \Rightarrow \Delta}$

LD4. $\dfrac{c^* R_a b^*, bR_ac, \Gamma \Rightarrow \Delta}{bR_ac, \Gamma \Rightarrow \Delta}$

LD5. $\dfrac{a^* \leq a, \Gamma \Rightarrow \Delta}{\Gamma \Rightarrow \Delta}$

LD6. $\dfrac{a \leq b^*, bR_ac, \Gamma \Rightarrow \Delta}{bR_ac, \Gamma \Rightarrow \Delta}$

LD7. $\dfrac{d \leq c, bR_ac, dR_{a^*}e, \Gamma \Rightarrow \Delta \qquad b \leq e, bR_ac, dR_{a^*}e, \Gamma \Rightarrow \Delta}{bR_ac, dR_{a^*}e, \Gamma \Rightarrow \Delta}$

LD8. This case needs special care. Let us first introduce a new binary predicate $P(x, y)$ and formulate the rules of this case as follows.

LD8(1) $\dfrac{cR_ay, P(b, y), bR_ac, \Gamma \Rightarrow \Delta}{bR_ac, \Gamma \Rightarrow \Delta}$ (y fresh)

LD8(2) $\dfrac{d \leq b^*, P(b, y), dR_{y^*}e, \Gamma \Rightarrow \Delta}{P(b, y), dR_{y^*}e, \Gamma \Rightarrow \Delta}$

Remark: The semantic condition for D8 is given by a formula which is beyond the form of geometric formulas. However, by using a technique called "geometrization" [18], we can accommodate this case within the scope of the proof technique that we have used so far. We first introduce a new binary predicate $P(x, y)$ and rewrite the frame condition in the following manner:

(1) $bR_ac \Rightarrow \exists y(c^* R_a y, and P(b, y))$

(2) $\forall d, e(P(b, y), and dR_{y^*}e \Rightarrow d \leq b^*)$

Finally, let us list the conditions for the rules giving extensions of B:

LDR1. $\dfrac{0x, xR_aa, \Gamma \Rightarrow \Delta}{\Gamma \Rightarrow \Delta}$

LDR2. $\dfrac{x^* \leq x, 0x, \Gamma \Rightarrow \Delta}{0x, \Gamma \Rightarrow \Delta}$

LDR3. $\dfrac{a^* \leq a, \Gamma \Rightarrow \Delta}{\Gamma \Rightarrow \Delta}$

Note: the condition for LDR2 is identical with LD2, and that for LDR3 is with LD5.

5 Some properties used in structural proof theory of labelled sequent calculi for relevant logic

In this section we prove a series of propositions and lemmas needed to prove cut-elimination for labelled sequent calculi for relevant logics. (In the rest of this section, we show only representative or crucial cases, since there are too many cases.)

By proving a substitution lemma and then by allowing x^* to be substituted to a variable, we can make the number of rules governing a term x^* smaller, so we initially prove the substitution lemma. Also, to use induction on the height of a proof tree, we write $\vdash_n$ to mean the derivability in a labelled sequent calculus with height n.

5.1 Substitution lemma

To prove the substitution lemma, let us define a substitution for a label as follows:

Definition 5.1. *A substitution for the primitive ternary relation R and labelled formula $x : A$ is defined as follows:*

1. $yR_xz(r/l) \equiv yR_xz$ *if* $x \neq l$, $y \neq l$ *and* $z \neq l$.
2. a. $yR_xz(r/x) \equiv yR_rz$ *if* $y \neq x$ *and* $z \neq x$.
   b. $yR_xz(r/y) \equiv rR_xz$ *if* $x \neq y$ *and* $z \neq y$.
   c. $yR_xz(r/z) \equiv yR_xr$ *if* $x \neq z$ *and* $y \neq z$.
3. a. $xR_xz(r/x) \equiv rR_rz$ *if* $y = x$ *and* $z \neq x$
   b. $yR_xy(r/y) \equiv rR_xr$ *if* $y = z$ *and* $x \neq y$
   c. $yR_xx(r/x) \equiv yR_rr$ *if* $z = x$ *and* $y \neq x$
4. $xR_xx(r/x) \equiv rR_rr$ *if* $y = x$ *and* $z = x$
5. a. $x{:}A(r/w) \equiv x{:}A$ *if* $w \neq x$
   b. $x{:}A(r/x) \equiv z{:}A$

Theorem 5.2 (Substitution Lemma). *If* $\vdash_n \Gamma \Rightarrow \Delta$, *then* $\vdash_n \Gamma(r/w) \Rightarrow \Delta(r/w)$.[20]

Proof. By induction on the height n of the derivation of $\Gamma \Rightarrow \Delta$.

Case 1) $n = 0$. $\Gamma \Rightarrow \Delta$ has to be an initial sequent $x{:}P, \Gamma \Rightarrow \Delta, x{:}P$. The result of the substitution is also an initial sequent.

Case 2) $n > 0$. Consider the last rule applied in the derivation. (rules for $\wedge$, $\vee$, $\rightarrow$, and relational rules).

Subcase of L$\rightarrow$ and L$\Box$. Neither L$\rightarrow$ nor L$\Box$ has the eigenvariable condition. (Depending on which variables are identical with w, we have several cases, but we do not make them explicit since the substitution is straightforward.)

By IH, for L$\Box$, $\vdash_{n-1} z : A, yR_xz, y : \Box_xA, \Gamma(r/w) \Rightarrow \Delta(r/w)$;
for L$\rightarrow$, $\vdash_{n-1} x{:}A \rightarrow B, \Gamma(r/w) \Rightarrow y : A, \Delta(r/w)$;
$\vdash_{n-1} x{:}A \rightarrow B, y{:}\Box_xB, \Gamma(r/w) \Rightarrow \Delta(r/w)$.

Then for L$\Box$, we have the following inference:

[20] The variable w can be x, y, z or some other variable. To simplify the notation, we write (r/w) only at the last part of the antecedent and of the succedent of each sequent, but the substitution covers the entire sequent.

$$\frac{z : A, yR_x z, y : \Box_x A, \Gamma(r/w) \Rightarrow \Delta(r/w)}{yR_x z, y : \Box_x A, \Gamma(r/w) \Rightarrow \Delta(r/w)} L\Box$$

For L→, we have the following inference:

$$\frac{x{:}A \to B, \Gamma(r/w) \Rightarrow y : A, \Delta(r/w), and x{:}A \to B, y{:}\Box_x B, \Gamma(r/w) \Rightarrow \Delta(r/w)}{x{:}A \to B, \Gamma(r/w) \Rightarrow \Delta(r/w)} L \to$$

Subcase of R→, R□. We take care of subcases with or without eigenvariables.
1) The substitution is vacuous. ($w \neq x$, $w \neq y$, $w \neq z$).
2) The substitution is non-vacuous, but r is not an eigenvariable.
There is only one case for each of $R\Box$, r is substituted for x or y.
By IH, $\vdash_{n-1} yR_r z, \Gamma(r/x) \Rightarrow z{:}A, \Delta(r/x)$ or $\vdash_{n-1} rR_x z, \Gamma(r/y) \Rightarrow z{:}A, \Delta(r/y)$

In the first case, we have $\dfrac{\Gamma(r/x) \Rightarrow y{:}\Box_r A, \Delta(r/x)}{yR_r z, \Gamma(r/x) \Rightarrow z{:}A, \Delta(r/x)}$ R□

In the second case, we have $\dfrac{\Gamma(r/y) \Rightarrow r{:}\Box_x A, \Delta(r/x)}{rR_x z, \Gamma(r/y) \Rightarrow z{:}A, \Delta(r/y)}$ R□

The case of R→ is similar.

3) The substitution is non-vacuous, and r is an eigenvariable. In this case, we replace the eigenvariable r in the original deduction by a fresh variable s. By IH, $\vdash_{n-1} yR_x s, \Gamma(r/z) \Rightarrow s : A, \Delta(r/z)$ and $\vdash_{n-1} s : A, \Gamma(r/y) \Rightarrow s : \Box_x B, \Delta(r/y)$. Then apply $R\Box$ and $R \to$ as follows:

$$\frac{yR_x s, \Gamma(r/z) \Rightarrow s{:}A, \Delta(r/z)}{\Gamma(r/z) \Rightarrow y{:}\Box_x A, \Delta(r/z)} R\Box \qquad \frac{s{:}A, \Gamma(r/y) \Rightarrow s{:}\Box_x B, \Delta(r/y)}{\Gamma(r/y) \Rightarrow x{:}A \to B, \Delta(r/y)} R \to'.$$

Subcase of the relational rules. Since there is no variable condition, the argument is straightforward. Let us take some representative cases, i.e. EqRel$_1$ and AtRepl.

1) EqRel$_1$. Consider this in the form $\dfrac{vR_x z, x = v, yR_x z, \Gamma \Rightarrow \Delta}{x = v, yR_x z, \Gamma \Rightarrow \Delta}$

If we consider (r/w) and $w \neq x$, $w \neq y$, $w \neq z$ and $w \neq v$, then there is no change in the original inference. If $w = x$, then by IH $vR_r z, r = v, yR_r z, \Gamma(r/w) \Rightarrow \Delta(r/w)$, then the conclusion is $r = v, yR_r z, \Gamma(r/w) \Rightarrow \Delta(r/w)$, which is what is desired. Other cases such as $w = x$, $w = y$ or $w = x$, $w = y$, $w = z$, etc. are similar.

2) AtRepl. If $w \neq x$ and $w \neq y$, then there is no change. If $w = x$ and $w \neq y$, then by IH $y : P, r = y, r : P, \Gamma(r/w) \Rightarrow \Delta(r/w)$. The conclusion is $r = y, r : P, \Gamma(r/w) \Rightarrow \Delta(r/w)$. If $w = x$ and $w = y$, then by IH $r : P, r = r, r : P, \Gamma(r/w) \Rightarrow \Delta(r/w)$. The

conclusion is $r = r, r : P, \Gamma(r/w) \Rightarrow \Delta(r/w)$. Other cases are similar.

Subcase of the rules for extensions of relevant logic B. Let us take representative cases and some subcases thereof, since it is too tedious to check every single case.

1) LB8. We have a bunch of subcases. If $w \neq a$, $w \neq b$, $w \neq c$, $w \neq d$, $w \neq x$, $w \neq y$, $w \neq z$, then the substitution is vacuous. By IH, if $w = a$ and w is identical with none of the others, then $\vdash_{n-1} bR_r x, cR_x d, cR_r y, zR_y d, cR_b z, \Gamma(r/w) \Rightarrow \Delta(r/w)$. (In this case, if r is identical with either y or z, then we initially carry out replacing y, z by a fresh variable, then apply IH concerning r again.) Applying the rule LB8, we get the following:

$$\frac{bR_r x, cR_x d, cR_r y, zR_y d, cR_b z, \Gamma(r/w) \Rightarrow \Delta(r/w)}{bR_r x, cR_x d, \Gamma(r/w) \Rightarrow \Delta(r/w)}\ \text{LB8}$$

The lower sequent is the one obtained by substituting r to the conclusion of LB8 rule. Hence, the rule LB8 preserves the substitution. (The other cases in which w is identical with other variables and other cases of rules are similar.) □

5.2 Derivability of initial sequents for arbitrary formulas

We show that we can derive our initial sequents with respect to arbitrary formulas in our language. This is crucial to show completeness of the labelled sequent calculi with respect to Hilbert-style axiom systems, since the latter are schematically formulated.

Proposition 5.3 (Derivability of initial sequents for arbitrary formulas).
Initial sequents of the form $x : A, \Gamma \Rightarrow \Delta, x : A$ are derivable in the labelled sequent calculus G3B *for arbitrary formulas A (in the language of relevant logics and the auxiliary primitive symbol $\Box_z$) and hence all extensions thereof.*

Proof. By induction on the logical structure of A. Atomic case: $A = P$. This is identical with the initial sequents in the original system.

Inductive case: Subcase $A = B \to C$. By IH, $y : B, x : B \to C, \Gamma \Rightarrow \Delta, y : B$ and $yR_x z, y : B, x : B \to C, y : \Box_x C, z : C, \Gamma \Rightarrow \Delta, z : C$ are derivable. We then have:

$$\dfrac{\dfrac{y : B, x : B \to C, \Gamma \Rightarrow \Delta, y : B \qquad \dfrac{\dfrac{yR_x z, y : B, y : B \to C, y : \Box_x C, z : C, \Gamma \Rightarrow \Delta, z : C}{yR_x z, y : B, y : B \to C, y : \Box_x C, \Gamma \Rightarrow \Delta, z : C}\ L\Box}{y : B, y : B \to C, y : \Box_x C, \Gamma \Rightarrow \Delta, z : \Box_x C}\ R\Box}{y : B, x : B \to C, \Gamma \Rightarrow \Delta, y : \Box_x C}\ L\to}{x : B \to C, \Gamma \Rightarrow \Delta, x : B \to C}\ R\to$$

Subcase $A = \neg B$. By IH, $x^* : B, \Gamma \Rightarrow \Delta, x^* : B$ is derivable by substituting x^* to x.

$$\dfrac{\dfrac{x^* : B, \Gamma \Rightarrow \Delta, x^* : B}{\Gamma \Rightarrow \Delta, x^* : B, x : \neg B}\ R\neg}{x : \neg B, \Gamma \Rightarrow \Delta, x : \neg B}\ L\neg$$

□

5.3 Height-preserving admissibility of weakening

We prove admissibility of weakening at this stage, since it is needed to prove the lemma for the admissibility of replacement for arbitrary formulas.

Proposition 5.4. *Weakening rules are admissible in height-preserving manner in* G3B *and its extensions, i.e. if* $\vdash_n \Gamma \Rightarrow \Delta$, *then* $\vdash_n \Gamma, \Gamma' \Rightarrow \Delta, \Delta'$.

Proof. Induction on the height of a derivation of G3B and its extensions.

Base case. $n = 0$ (initial sequent cases). In this case, i.e. the case in which $\Gamma, x{:}p \Rightarrow x{:}p, \Delta$ or $\Gamma, yR_xz \Rightarrow yR_xz, \Delta$, all the following are also initial sequents. $\Gamma, \Gamma', x{:}p \Rightarrow x: p, \Delta', \Delta$ or $\Gamma, \Gamma', yR_xz \Rightarrow yR_xz, \Delta', \Delta$.

Inductive case. $n > 0$

Subcase L→. The last rule is L→.

$$\frac{x{:}A \to B, \Gamma \Rightarrow \Delta, y{:}A \qquad y{:}\Box_x, x{:}A \to B, \Gamma \Rightarrow \Delta}{x{:}A \to B, \Gamma \Rightarrow \Delta}$$

By IH, $\vdash_{n-1} x{:}A \to B, \Gamma, \Gamma' \Rightarrow \Delta', \Delta, y{:}A$ and $\vdash_{n-1} y{:}\Box_x, x{:}A \to B, \Gamma, \Gamma' \Rightarrow \Delta', \Delta$. Applying L→′, we get the following inference:

$$\frac{x{:}A \to B, \Gamma, \Gamma' \Rightarrow \Delta', \Delta, y{:}A \qquad y{:}\Box_x, x{:}A \to B, \Gamma, \Gamma' \Rightarrow \Delta', \Delta}{x{:}A \to B, \Gamma, \Gamma' \Rightarrow \Delta', \Delta} L \to$$

The lower sequent is the same as the one obtained by applying several weakenings to the lower sequent of the original inference. Hence, the claim holds in this case. Note that we have no eigenvariable condition for $L \to$.

Subcase R→. The last rule applied is $R \to$. $\quad \dfrac{y{:}A, \Gamma \Rightarrow y{:}\Box_x B, \Delta}{\Gamma \Rightarrow x{:}A \to B, \Delta}$

In this case, we have the eigenvariable condition. The weakened formulas Γ', Δ' may contain the eigenvariable y. So we need a special care. In such a case we apply the substitution lemma and replace eigenvariable y by another fresh variable, say s .

Then we get $\vdash_{n-1} s : A, \Gamma \Rightarrow s : \Box_x B, \Delta$. By IH, $\vdash_{n-1} s : A, \Gamma, \Gamma' \Rightarrow s : \Box_x B, \Delta', \Delta$ holds. Now we apply $R \to$ and get the following inference.

$$\frac{s{:}A, \Gamma, \Gamma' \Rightarrow s{:}\Box_x B, \Delta', \Delta}{\Gamma, \Gamma' \Rightarrow x{:}A \to B, \Delta', \Delta} R \to$$

The lower sequent is now the same as the result of applying several weakenings to the lower sequent of the original inference. Hence, the claim holds.

Note that by the application of the substitution lemma, there is no occurrence of eigenvariable s in Γ', Δ', so there is no problem of applying $R \to$.

Subcase L$\Box$. The last application of a rule is L$\Box$. $\dfrac{z\!:\!A, yR_x z, y\!:\!\Box_x A, \Gamma \Rightarrow \Delta}{yR_x z, y\!:\!\Box_x A, \Gamma \Rightarrow \Delta}$.

By IH, $\vdash_{n-1} z\!:\!A, yR_x z, y\!:\!\Box_x A, \Gamma, \Gamma' \Rightarrow \Delta', \Delta$. L$\Box$ has no eigenvariable condition, so we can apply L$\Box_x$ and get the following inference.

$$\dfrac{z\!:\!A, yR_x z, y\!:\!\Box_x A, \Gamma, \Gamma' \Rightarrow \Delta', \Delta}{yR_x z, y\!:\!\Box_x A, \Gamma, \Gamma' \Rightarrow \Delta', \Delta}\ L\Box.$$

The lower sequent is the sequent desired, so the claim holds.

Subcase R$\Box$. The last application of a rule is R$\Box$. $\dfrac{yR_x z, \Gamma \Rightarrow z\!:\!A, \Delta}{\Gamma \Rightarrow y\!:\!\Box_x A, \Delta}$.

In this case, we have the eigenvariable condition for z. Hence, before we apply IH, we replace z by a fresh variable, say s. Then we have $\vdash_{n-1} yR_x s, \Gamma \Rightarrow s\!:\!A, \Delta$.

By IH, $\vdash_{n-1} yR_x s, \Gamma, \Gamma' \Rightarrow s\!:\!A, \Delta', \Delta$. Applying $R\Box$, we get the following.

$$\dfrac{yR_x s, \Gamma, \Gamma' \Rightarrow s\!:\!A, \Delta', \Delta}{\Gamma, \Gamma' \Rightarrow y\!:\!\Box_x A, \Delta', \Delta}\ R\Box_x$$

The lower sequent is clearly the desired sequent, so the claim holds.

Subcase of relational rules. None of the relational rules has the eigenvariable condition. Hence, the proof is a straightforward application of IH.

Subcase of additional rules for systems extending G3B. For any additional rule that has no eigenvariable condition, the argument is essentially the same as the foregoing ones in which we have no eigenvariable condition. For any additional rule that has eigenvariable conditions, the argument follows the patterns of the foregoing cases in which we have variable changes. Let us pick a representative example, LB2 and this allows pushing up weakenings.

$$\text{LB2} \qquad \dfrac{bR_a x, xR_a c, bR_a c, \Gamma \Rightarrow \Delta}{bR_a c, \Gamma \Rightarrow \Delta}$$

By IH, we have, $bR_a x, xR_a c, bR_a c, \Gamma, \Gamma' \Rightarrow \Delta, \Delta'$. When we have no occurrence of the variable x in Γ' and Δ', we directly apply the rule LB2. Then we will obtain desired sequent $bR_a c, \Gamma, \Gamma' \Rightarrow \Delta, \Delta'$.

However, if x occurs in Γ', Δ', then directly applying the rule LB2 violates the eigenvariable condition. Hence, we first change the variable x to, say, s; hence, we obtain then, $bR_a s, sR_a c, bR_a c, \Gamma, \Gamma' \Rightarrow \Delta', \Delta$. Then the inference can go further.

$$\frac{bR_as, sR_ac, bR_ac, \Gamma, \Gamma' \Rightarrow \Delta', \Delta}{bR_ac, \Gamma, \Gamma' \Rightarrow \Delta', \Delta} LB2$$

Then the lower sequent is the desired sequent (the sequent obtained by applying weakenings to the lower sequent of the rule LB2). Therefore, the claim holds. □

5.4 Admissibility of replacement rule for arbitrary formulas

Now we show that the property of replacement and heredity represented by the rules AtRepl and AtHR can be extended to arbitrary formulas. We first show a theorem corresponding to replacement for arbitrary formulas. (Since this proof also works systems with the additional rules, the admissibility of Repl for arbitrary formulas also holds in all the extensions of G3B.)

Theorem 5.5. *The general form of Replacement Rule is admissible in a system G3B with AtRepl and Cut.*

The arguments follow [31] 6.5, where the admissibility of the general rule of identity by using the atomic rule of identity, Cut, and the lemma corresponding to the following (in full first-order logic with identity) is established. Here we have the lemma for the labelled sequent calculi for relevant logics.

Lemma 5.6 (Formulas for Eq). $x = y, x{:}A \Rightarrow y{:}A$

Since the theorem is an immediate consequence of the lemma, we concentrate on proving the lemma.

Proof of the lemma. By induction on the logical structure (weight) of A.

Base case. $A = P$. This case can be proven as follows, where the uppersequent is an instance of initial sequent in G3B:

$$\frac{y{:}P, x = y, x{:}P \Rightarrow y{:}P}{x = y, x{:}P \Rightarrow y{:}P} AtRepl$$

Inductive case. Subcase $A = \Box_z B$. A proof of this case can be given as follows (note that this does not require us to use IH):

$$\dfrac{\dfrac{\dfrac{w{:}B, xR_zw, xR_zw, x = y, x{:}\Box_z B \Rightarrow w{:}B}{xR_zw, yR_zw, x = y, x{:}\Box_z B \Rightarrow w{:}B} L\Box_z}{yR_zw, x = y, x{:}\Box_z B \Rightarrow w{:}B} EqRel_2}{x = y, x{:}\Box_z B \Rightarrow y{:}\Box_z B} R\Box_z$$

Subcase $A = B \to C$. This case can be proven as follows:

$$\dfrac{\dfrac{\dfrac{\dfrac{zR_yx, w, zR_yw, z:B, x = y, x:B \to C \Rightarrow w:C, z:A \qquad \dfrac{w:B, z:\Box_x B, zR_yw, z:B, x = y, x:B \to C \Rightarrow w:C}{z:\Box_x B, zR_yw, z:B, x = y, x:B \to C \Rightarrow w:C}\,L\Box}{zR_yx, w, zR_yw, z:B, x = y, x:B \to C \Rightarrow w:C}\,R\to}{zR_yw, z:B, x = y, x:B \to C \Rightarrow w:C}\,EqRel_2}{z:B, x = y, x:B \to C \Rightarrow z:\Box_y C}\,R\Box}{x = y, x:B \to C \Rightarrow y:B \to C}\,R\to$$

Subcase $A = \neg B$. This case can be proven as follows: By IH, $x = y, x:B \Rightarrow y:B$. By the substitution lemma, from this we can prove $y^* = x^*, y^*:B \Rightarrow x^*:B$.[21]

$$\dfrac{\dfrac{\dfrac{\dfrac{\dfrac{\dfrac{\dfrac{y^* = x^*, y^*:B \Rightarrow x^*:B}{y^* = x^*, y = x, y^*:B \Rightarrow x^*:B}\,ad.wk}{y = x, y^*:B \Rightarrow x^*:B}\,Eq^*}{y = x, x = y, x = x, y^*:B \Rightarrow x^*:B}\,ad.wk}{x = y, x = x, y^*:B \Rightarrow x^*:B}\,EqETrans}{x = y, y^*:B \Rightarrow x^*:B}\,EqRef}{y^*:B, x = y, x:\neg B \Rightarrow}\,L\neg}{x = y, x:\neg B \Rightarrow y:\neg B}\,R\neg$$

This completes the proof of the lemma. □

We can prove the theorem by using the lemma and cut (although our form of cut requires admissibility of contraction, which is proven later).

5.5 Admissibility of the heredity rule for arbitrary formulas

Here we prove the admissibility of heredity rule for arbitrary formulas based on AtHR. The proof is similar to the case of the admissibility of Repl based on AtRepl.

Theorem 5.7. *The general form of Heredity Rule is admissible in a system* G3B+ *with AtHR and Cut.*

We prove this theorem by proving the following lemma:

Lemma 5.8 (Formulas for HR). $x \leq y, x:A \Rightarrow y:A$

Proof of the lemma. By induction on the logical structure (weight) of A.

Base case. $A = P$. This case can be proven by AtHR as follows:

$$\dfrac{y:P, x \leq y, x:P \Rightarrow y:P}{x \leq y, x:P \Rightarrow y:P}\,AtHR$$

[21]Observe that, in this chapter, transitivity is formulated in the form of Euclidean transitivity.

Inductive case. Subcase $A = \Box_z B$. This case can be proven as follows:

$$\dfrac{\dfrac{\dfrac{w:B, xR_zw, yR_zw, x \leq y, x:\Box_z B \Rightarrow w:B}{xR_zw, yR_zw, x \leq y, x:\Box_z B \Rightarrow w:B}\,L\Box}{yR_zw, x \leq y, x:\Box_z B \Rightarrow w:B}\,Mon_2}{x \leq y, x:\Box_z B \Rightarrow x:\Box_z B}\,R\Box$$

Subcase $A = B \rightarrow C$. This case can be proven as follows:

$$\dfrac{\dfrac{\dfrac{\dfrac{zR_xw, zR_yw, z:B, x \leq y, x:B \rightarrow C \Rightarrow w:C, z:B \quad \dfrac{w:C, z:\Box_x C, zR_xw, zR_yw, z:B, x \leq y, x:B \rightarrow C \Rightarrow w:C}{z:\Box_x C, zR_xw, zR_yw, z:B, x \leq y, x:B \rightarrow C \Rightarrow w:C}\,L\Box}{zR_xw, zR_yw, z:B, x \leq y, x:B \rightarrow C \Rightarrow w:C}\,L\rightarrow}{zR_yw, z:B, x \leq y, x:B \rightarrow C \Rightarrow w:C}\,Mon_1}{z:B, x \leq y, x:B \rightarrow C \Rightarrow z:\Box_z C}\,R\Box}{x \leq y, x:B \rightarrow C \Rightarrow y:B \rightarrow C}\,R\rightarrow$$

This completes the proof of the lemma. □

5.6 Height-preserving invertibility of each rule of G3B and its extensions

We show that all the rules of G3B and its extensions are height-preserving invertible.

Theorem 5.9. *All rules of* G3B *are height-preserving invertible. That is,*

1. *$L\wedge$: If $\vdash_n x:A \wedge B, \Gamma \Rightarrow \Delta$, then $\vdash_n x:A, x:B, \Gamma \Rightarrow \Delta$.*
2. *$R\wedge$: If $\vdash_n \Gamma \Rightarrow \Delta, x:A \wedge B$, then $\vdash_n \Gamma \Rightarrow \Delta, x:A$ and $\vdash_n \Gamma \Rightarrow \Delta, x:B$.*
3. *$L\vee$: If $\vdash_n x:A \vee B, \Gamma \Rightarrow \Delta$, then $\vdash_n x:A, \Gamma \Rightarrow \Delta$ and $\vdash_n, x:B, \Gamma \Rightarrow \Delta$.*
4. *$R\vee$: If $\vdash_n \Gamma \Rightarrow \Delta, x:A \vee B$, then $\vdash_n \Gamma \Rightarrow \Delta, x:A, x:B$.*
5. *$L\rightarrow$: If $\vdash_n x:A \rightarrow B, \Gamma \Rightarrow \Delta$,
then $\vdash_n x:\ A \rightarrow B, \Gamma \Rightarrow \Delta, y:A$ and $\vdash_n y:\Box_x B, x:\ A \rightarrow B, \Gamma \Rightarrow \Delta$.*
6. *$R\rightarrow$: If $\vdash_n \Gamma \Rightarrow x:A \rightarrow B, \Delta$, then $y:A, \Gamma \Rightarrow \Delta, y:\Box_x B$.*
7. *$L\Box$: If $\vdash_n yR_xz, y:\Box_x A, \Gamma \Rightarrow \Delta$, then $\vdash_n z:A, yR_xz, y:\Box_x A, \Gamma \Rightarrow \Delta$.*
8. *$R\Box$: If $\Gamma \Rightarrow \Delta, y:\Box_x A$, then $yR_xz, \Gamma \Rightarrow \Delta, z:A$.*

   Rules for $\neg$ and $*$

9. *$L\neg$: If $\vdash_n x:\neg A, \Gamma \Rightarrow \Delta$, then $\vdash_n \Gamma \Rightarrow \Delta, x^*:A$.*
10. *$R\neg$: If $\vdash_n \Gamma \Rightarrow \Delta, x:\neg A$, then $\vdash_n x^*:A, \Gamma \Rightarrow \Delta$.*
11. *Dual: If $\Gamma \Rightarrow \Delta$, then $x = x^{**}, \Gamma \Rightarrow \Delta$.*
12. *$=^*$: If $x = y, \Gamma \Rightarrow \Delta$, then $x^* = y^*, x = y, \Gamma \Rightarrow \Delta$.*

    Relational rules

13. *EqRef: If* $\vdash_n \Gamma \Rightarrow \Delta$*, then* $x = x, \Gamma \Rightarrow \Delta$.

14. *EqTrans: If* $\vdash_n x = y, x = z, \Gamma \Rightarrow \Delta$*, then* $\vdash_n y = z, x = y, x = z, \Gamma \Rightarrow \Delta$.

15. *EqRel*$_1$ *If* $\vdash_n y = w, yR_xz, \Gamma \Rightarrow \Delta$*, then* $\vdash_n wR_xz, y = w, yR_xz, \Gamma \Rightarrow \Delta$.

16. *EqRel*$_2$ *If* $\vdash_n z = w, yR_xz, \Gamma \Rightarrow \Delta$*, then* $\vdash_n yR_xz, z = w, yR_xz, \Gamma \Rightarrow \Delta$.

17. *EqRel*$_3$ *If* $\vdash_n x = w, yR_xz, \Gamma \Rightarrow \Delta$*, then* $\vdash_n yR_wz, x = w, yR_xz, \Gamma \Rightarrow \Delta$.

18. *AtRepl: If* $\vdash_n x = y, x{:}P, \Gamma \Rightarrow \Delta$*, then* $y{:}P, x = y, x{:}P, \Gamma \Rightarrow \Delta$.

    Basic Rules for $\leq$

19. $\leq /R$*: If* $\vdash_n 0x, yR_xz, \Gamma \Rightarrow \Delta$*, then* $y \leq z, 0x, yR_xz, \Gamma \Rightarrow \Delta$.

20. $R/\leq$*: If* $\vdash_n y \leq z, \Gamma \Rightarrow \Delta$*, then* $0x, yR_xz, y \leq z, \Gamma \Rightarrow \Delta$.

21. *Ref* $\leq$*: If* $\vdash_n \Gamma \Rightarrow \Delta$*, then* $x \leq x, \Gamma \Rightarrow \Delta$.

22. *Trans* $\leq$*: If* $\vdash_n x \leq y, y \leq z, \Gamma \Rightarrow \Delta$*, then* $x \leq z, x \leq y, y \leq z, \Gamma \Rightarrow \Delta$.

23. *Mon*$_1$*: If* $\vdash_n a \leq x, bR_xc, \Gamma \Rightarrow \Delta$*, then* $bR_ac, a \leq x, bR_xc, \Gamma \Rightarrow \Delta$.

24. *Mon*$_2$*: If* $\vdash_n b \leq x, xR_ac, \Gamma \Rightarrow \Delta$*, then* $bR_ac, b \leq x, xR_ac, \Gamma \Rightarrow \Delta$.

25. *Mon*$_3$*: If* $\vdash_n x \leq c, bR_ax, \Gamma \Rightarrow \Delta$*, then* $bR_ac, x \leq c, bR_ax, \Gamma \Rightarrow \Delta$.

Similar cases hold for all the additional rules: LB1-LB20, LD1-LD8, LDR1-LDR3.

Proof. Let us initially make a distinction between cases in which the height-preserving invertibility is merely a special case of weakening, which applies to the rules from 11–25 in G3B and all the additional rules, and cases in which it is not.

Since we have given a proof of the general case of the height-preserving admissibility of weakening, the claim for these cases follows from the theorem. Those that are not special cases of the height-preserving admissibility of weakening require further arguments. The proof is by induction on height of a derivation n.

Base case: $n = 0$. This is the case where the derivation consists only of an instance of an initial sequent. Let us, for instance, consider $\vdash_0 \Gamma, x{:}A \wedge B \Rightarrow \Delta$. In this case, it must be the case that both Γ and Δ contain some $y{:}P$, since otherwise this would not be an instance of an initial sequent (since $x : A \wedge B$ is not atomic). But in these cases, $\Gamma, x : A, x : B \Rightarrow \Delta$ is also an instance of an initial sequent. The argument applies to all the cases of formulas the introduction of which require operational rules. (For this, restricting the initial sequent to an atomic case is crucial).

Inductive case: $n > 0$. We need to check that the claim holds both for the subcase 1) the formula explicit in each case is non-principal and the subcase 2) the formula explicit in each case is the principal formula. In principle, in the subcase 1), we need to check all the other cases in which other rules are the last rule applied and make sure that applying IH to $n - 1$ and this application of IH does not disturb any application of those other rules. However, it suffices to show schematically that for the two forms of rules in G3B, the claim works. We give some representative cases.

For 5. Subcase 5.1) The principal formula of the last rule applied is not $x{:}A \to B$.

5.1a) The last rule applied is unary. In this case, the last inference looks as follows:

$$\frac{x{:}A \to B, \Gamma' \Rightarrow \Delta'}{x{:}A \to B, \Gamma \Rightarrow \Delta}\ r1$$

The premise satisfies the property that $\vdash_{n-1} x{:}A \to B, \Gamma' \Rightarrow \Delta'$.

By IH, $\vdash_{n-1} x{:}\ A \to B, \Gamma' \Rightarrow \Delta', y{:}A$ and $\vdash_{n-1} y{:}\Box_x B, x{:}\ A \to B, \Gamma' \Rightarrow \Delta'$.

Applying $r1$, we get the following inference.

$$\frac{x{:}\ A \to B, \Gamma' \Rightarrow \Delta', y{:}A}{x{:}\ A \to B, \Gamma \Rightarrow \Delta, y{:}A}\ r1 \qquad \frac{y{:}\Box_x B, x{:}\ A \to B, \Gamma' \Rightarrow \Delta'}{y{:}\Box_x B, x{:}\ A \to B, \Gamma \Rightarrow \Delta}\ r1$$

The lower sequents of these satisfy $\vdash_n x{:}\ A \to B, \Gamma \Rightarrow \Delta, y{:}A$ and $\vdash_n y{:}\Box_x B, x{:}\ A \to B, \Gamma \Rightarrow \Delta$. These are desired statements.

5.1b) The last rule applied is binary. This looks as follows:

$$\frac{x{:}A \to B, \Gamma' \Rightarrow \Delta' \qquad x{:}A \to B, \Gamma'' \Rightarrow \Delta''}{x{:}A \to B, \Gamma \Rightarrow \Delta}\ r2$$

The premises have the form $\vdash_{n-1} x{:}A \to B, \Gamma' \Rightarrow \Delta'$ and $\vdash_{n-1} x{:}A \to B, \Gamma'' \Rightarrow \Delta''$. By IH, $\vdash_{n-1} x{:}\ A \to B, \Gamma' \Rightarrow \Delta', y{:}A$ and $\vdash_{n-1} y{:}\Box_x B, x{:}\ A \to B, \Gamma' \Rightarrow \Delta'$, and $\vdash_{n-1} x{:}\ A \to B, \Gamma'' \Rightarrow \Delta'', y{:}A$ and $\vdash_{n-1} y{:}\Box_x B, x{:}\ A \to B, \Gamma'' \Rightarrow \Delta''$. Applying $r2$ to these, we can get the following.

$$\frac{x{:}\ A \to B, \Gamma' \Rightarrow \Delta', y{:}A \qquad x{:}\ A \to B, \Gamma'' \Rightarrow \Delta'', y{:}A}{x{:}\ A \to B, \Gamma \Rightarrow \Delta, y{:}A}\ r2$$

$$\frac{y{:}\Box_x B, x{:}\ A \to B, \Gamma' \Rightarrow \Delta' \qquad y{:}\Box_x B, x{:}\ A \to B, \Gamma'' \Rightarrow \Delta''}{y{:}\Box_x B, x{:}\ A \to B, \Gamma \Rightarrow \Delta}\ r2$$

The lower sequents of these cases satisfy the following properties.

$\vdash_n x{:}\ A \to B, \Gamma \Rightarrow \Delta, y{:}A$ and $\vdash_n y{:}\Box_x B, x{:}\ A \to B, \Gamma \Rightarrow \Delta$, which are desired.

Subcase 5.2) The principal formula of the last rule applied is $x{:}A \to B$. In this case, the last inference looks as follows:

$$\frac{x{:}A \to B, \Gamma \Rightarrow \Delta, y{:}A \qquad y{:}\Box_x B, x{:}A \to B, \Gamma \Rightarrow \Delta}{x{:}\ A \to B, \Gamma \Rightarrow \Delta}\ L\Box_x$$

We can pick up the derivations for these premises. Then we have $\vdash_{n-1} x\!:\!A \to B, \Gamma \Rightarrow \Delta, y\!:\!A$ and $\vdash_{n-1} y\!:\!\Box_x B, x\!:\!A \to B, \Gamma \Rightarrow \Delta$, which are desired.

For 6. Subcase 6.1) $x\!:\!A \to B$ is not the principal formula of the last rule.

6.1a) The last rule applied is unary. In this case, the last inference looks as follows:

$$\frac{\Gamma' \Rightarrow x\!:\!A \to B, \Delta'}{\Gamma \Rightarrow x\!:\!A \to B, \Delta}\ r1$$

The premise has the property that $\vdash_{n-1} \Gamma' \Rightarrow x\!:\!A \to B, \Delta'$. By IH, $\vdash_{n-1} y\!:\!A, \Gamma' \Rightarrow y\!:\!\Box_x B, \Delta'$. Applying the rule $r1$, we get the following inference. (We need to make sure that y is not the eigenvariable used in $r1$. If so, use the substitution lemma to replace the eigenvariable to a fresh variable.)

$$\frac{y\!:\!A, \Gamma' \Rightarrow y\!:\!\Box_x B, \Delta'}{y\!:\!A, \Gamma \Rightarrow y\!:\!\Box_x B, \Delta}\ r1$$

The lower sequent satisfies the desired form, i.e., $\vdash_n y\!:\!A, \Gamma \Rightarrow y\!:\!\Box_x B, \Delta$.

6.1b) The last rule applied is binary:

$$\frac{\Gamma' \Rightarrow x\!:\!A \to B, \Delta' \qquad \Gamma'' \Rightarrow x\!:\!A \to B, \Delta''}{\Gamma \Rightarrow x\!:\!A \to B, \Delta}\ r1$$

The premises are $\vdash_{n-1} \Gamma' \Rightarrow x\!:\!A \to B, \Delta'$ and $\vdash_{n-1} \Gamma'' \Rightarrow x\!:\!A \to B, \Delta''$.
By IH, $\vdash_{n-1} y\!:\!A, \Gamma' \Rightarrow y\!:\!\Box_x B, \Delta'$ and $\vdash_{n-1} y\!:\!A, \Gamma'' \Rightarrow y\!:\!\Box_x B, \Delta''$.
Applying the rule $r2$, we get the following inference.

$$\frac{y\!:\!A, \Gamma' \Rightarrow y\!:\!\Box_x B, \Delta' \qquad y\!:\!A, \Gamma'' \Rightarrow y\!:\!\Box_x B, \Delta''}{y\!:\!A, \Gamma \Rightarrow y\!:\!\Box_x B, \Delta}\ r2$$

The lower sequent has the desired form that $\vdash_n y\!:\!A, \Gamma \Rightarrow y\!:\!\Box_x B, \Delta$.

Subcase 6.2) $x\!:\!A \to B$ is the principal formula. The last rule applies as follows:

$$\frac{y\!:\!A, \Gamma \Rightarrow y\!:\!\Box_x A, \Delta}{\Gamma \Rightarrow x\!:\!A \to B, \Delta}\ R\!\to$$

Then take the derivation up to $\vdash_{n-1} y\!:\!A, \Gamma \Rightarrow y\!:\!\Box_x A, \Delta$, which is desired.

□

Note: The height-preserving invertibility can be extended to all other additional rules for extensions of B, too, by applying the height-preserving weakening.

5.7 Height-preserving admissibility of contraction

In this subsection, we prove the height-preserving admissibility of contraction. The statement of this result goes as follows:

Theorem 5.10. *The rules of contraction are height-preserving admissible in G3B and its extensions. Here* $\varphi \in \{x = y, x \leq y, yR_xz, 0x\}$.

If $\vdash_n x:A, x:A, \Gamma \Rightarrow \Delta$, *then* $\vdash_n x:A, \Gamma \Rightarrow \Delta$.

If $\vdash_n \varphi, \varphi, \Gamma \Rightarrow \Delta$, *then* $\vdash_n \varphi, \Gamma \Rightarrow \Delta$.

If $\vdash_n, \Gamma \Rightarrow \Delta, x:A, x:A$, *then* $\vdash_n \Gamma \Rightarrow \Delta, x:A$.

Proof. By induction on the height of a proof n.

Base case: $n = 0$. Suppose $\vdash_0 x:A, x:A, \Gamma \Rightarrow \Delta$, $\vdash_0 \Gamma \Rightarrow \Delta, x:A, x:A$, $\vdash_0 \varphi, \varphi, \Gamma \Rightarrow \Delta$. These are cases of initial sequents. A sequent in which a contracted formula occurs must also be an initial sequent, whether or not the contracted formula $x:A$ is the essential part of the initial sequent (of the form $x:P \Rightarrow x:P$) or not, and φ has no occurrence in the essential part of the initial sequent.

Inductive case: $n > 0$.

Suppose $\vdash_n x:A, x:A, \Gamma \Rightarrow \Delta$, $\vdash_n \Gamma \Rightarrow \Delta, x:A, x:A$, and $\vdash_n \varphi, \varphi, \Gamma \Rightarrow \Delta$

Case 1) Contracted formula is not principal in the last rule applied.

Subcase 1a) The last rule applied is unary. Then the cases look as follows:

$$\frac{x:A, x:A, \Gamma' \Rightarrow \Delta'}{x:A, x:A, \Gamma \Rightarrow \Delta}\ r1 \qquad \frac{\Gamma' \Rightarrow \Delta', x:A, x:A}{\Gamma \Rightarrow \Delta, x:A, x:A}\ r1 \qquad \frac{\varphi, \varphi, \Gamma' \Rightarrow \Delta'}{\varphi, \varphi, \Gamma \Rightarrow \Delta}\ r1$$

Then $\vdash_{n-1} x:A, x:A, \Gamma' \Rightarrow \Delta'$, $\vdash_{n-1} \Gamma' \Rightarrow \Delta', x:A, x:A$, $\vdash_{n-1} \varphi, \varphi, \Gamma' \Rightarrow \Delta'$.

By IH, we obtain $\vdash_{n-1} x:A, \Gamma' \Rightarrow \Delta'$, $\vdash_{n-1} \Gamma' \Rightarrow \Delta', x:A$, $\vdash_{n-1} \varphi, \Gamma' \Rightarrow \Delta'$.

We now apply $r1$ to these. (There is no need to worry about the eigenvariable condition since IH introduces no new variables.) Then we get the following inferences.

$$\frac{x:A, \Gamma' \Rightarrow \Delta'}{x:A, \Gamma \Rightarrow \Delta}\ r1 \qquad \frac{\Gamma' \Rightarrow \Delta', x:A}{\Gamma \Rightarrow \Delta, x:A}\ r1 \qquad \frac{\varphi, \Gamma' \Rightarrow \Delta'}{\varphi, \Gamma \Rightarrow \Delta}\ r1$$

The lower sequents give the desired conclusions. (Observe that all the relational rules and additional rules can be covered by the third form in this subcase. Hence, we do not have to handle relational rules and additional rules, independently of this.)

Subcase 1b) The last rule applied is binary. Then the cases look as follows:

$$\frac{x:A, x:A, \Gamma' \Rightarrow \Delta' \qquad x:A, x:A, \Gamma'' \Rightarrow \Delta''}{x:A, x:A, \Gamma \Rightarrow \Delta}\ r2$$

$$\frac{\Gamma' \Rightarrow \Delta', x\!:\!A, x\!:\!A \qquad \Gamma'' \Rightarrow \Delta'', x\!:\!A, x\!:\!A}{\Gamma \Rightarrow \Delta, x\!:\!A, x\!:\!A}\, r2 \qquad \frac{\varphi, \varphi, \Gamma' \Rightarrow \Delta' \qquad \varphi, \varphi, \Gamma'' \Rightarrow \Delta''}{\varphi, \varphi, \Gamma \Rightarrow \Delta}\, r2$$

Then, by IH, $\vdash_{n-1} x\!:\!A, \Gamma' \Rightarrow \Delta'$ and $\vdash_{n-1} x\!:\!A, \Gamma'' \Rightarrow \Delta''$, $\vdash_{n-1} \Gamma' \Rightarrow \Delta', x\!:\!A$ and $\vdash_{n-1} \Gamma'' \Rightarrow \Delta'', x\!:\!A$, $\vdash_{n-1} \varphi, \Gamma' \Rightarrow \Delta'$ and $\vdash_{n-1} \varphi, \Gamma'' \Rightarrow \Delta''$. Then we have the following inferences for each of these cases:

$$\frac{x\!:\!A, \Gamma' \Rightarrow \Delta' \qquad \Gamma'' x\!:\!A, \Rightarrow \Delta''}{x\!:\!A, \Gamma \Rightarrow \Delta}\, r2 \qquad \frac{\Gamma' \Rightarrow \Delta', x\!:\!A \qquad \Gamma'' \Rightarrow \Delta'', x\!:\!A}{\Gamma \Rightarrow \Delta, x\!:\!A}\, r2$$

$$\frac{\varphi, \Gamma' \Rightarrow \Delta' \qquad \varphi, \Gamma'' \Rightarrow \Delta''}{\varphi, \Gamma \Rightarrow \Delta}\, r2$$

The lower sequent for each of these cases has the desired form $\vdash_n x\!:\!A, \Gamma \Rightarrow \Delta$, $\vdash_n \Gamma \Rightarrow \Delta, x\!:\!A$, $\vdash_n \varphi, \Gamma \Rightarrow \Delta$ or $\vdash_n \Gamma \Rightarrow \Delta, \varphi$.

Case 2) The contracted formula is principal in the last rule applied. Then we have the following further subcases:

Subcase 2a) the active formulas are only proper subformulas of the principal formula (rules for $\wedge$, $\vee$, $\neg$).

Subcase 2b) the principal formula also occurs in the premise (L$\rightarrow$, L$\Box$).

Subcase 2c) the active formulas are atoms, φ, and proper subformulas of the principal formula (R$\rightarrow$, R$\Box$).

2b.i) $L\rightarrow$. The assumption is obtained by the following inference:

$$\frac{x\!:\!A \rightarrow B, x\!:\!A \rightarrow B, \Gamma \Rightarrow \Delta, y\!:\!A \qquad y\!:\!\Box_x B, x\!:\!A \rightarrow B, x\!:\!A \rightarrow B, \Gamma \Rightarrow \Delta}{x\!:\!A \rightarrow B, x\!:\!A \rightarrow B, \Gamma \Rightarrow \Delta}\, L\rightarrow$$

The premise has the height $n-1$. So we can apply IH to the premise and obtain the following: $\vdash_{n-1} x\!:\!A \rightarrow B, \Gamma \Rightarrow \Delta, y\!:\!A$ and $\vdash_{n-1} y\!:\!\Box_x B, x\!:\!A \rightarrow B, \Gamma \Rightarrow \Delta$.

We apply the rule $L\rightarrow$. Then we obtain the desired $\vdash_n x\!:\!A \rightarrow B, \Gamma \Rightarrow \Delta$.

2b.ii) $L\Box$. The assumption is obtained by the following inference:

$$\frac{z\!:\!A, yR_x z, y\!:\!\Box_x A, y\!:\!\Box_x A, \Gamma \Rightarrow \Delta}{y\!:\!\Box_x A, y\!:\!\Box_x A, \Gamma \Rightarrow \Delta}\, L\Box$$

The premise of this inference has the height $n-1$. So we can apply IH to this.

Hence, $\vdash_{n-1} z\!:\!A, yR_x z, y\!:\!\Box_x A, \Gamma \Rightarrow \Delta$. We apply the rule $L\Box$ to this and obtain the following: $\vdash_n y\!:\!\Box_x A, \Gamma \Rightarrow \Delta$. But this is what has been desired.

2c.i) $R\rightarrow$. The assumption is obtained by the following inference:

$$\frac{y\!:\!A, \Gamma \Rightarrow \Delta, x\!:\!A \to B, y\!:\!\Box_x B}{\Gamma \Rightarrow \Delta, x\!:\!A \to B, x\!:\!A \to B}\ R\to$$

The premise of this inference has the height $n-1$. We first apply the h-p invertibility to the premise. Then we obtain $\vdash_{n-1} y\!:\!A, y\!:\!A, \Gamma \Rightarrow \Delta, y\!:\!\Box_x B, y\!:\!\Box_x B$. This has the height $n-1$, so by IH $\vdash_{n-1} y\!:\!A, \Gamma \Rightarrow \Delta, y\!:\!\Box_x B$. Then, applying $R\to$, we get the desired $\vdash_n \Gamma \Rightarrow \Delta, x\!:\!A \to B$.

2c.ii) $R\Box$. The assumption for this is obtained by the following inference:

$$\frac{yR_x z, \Gamma \Rightarrow \Delta, y\!:\!\Box_x A, z\!:\!A}{\Gamma \Rightarrow \Delta, y\!:\!\Box_x A, y\!:\!\Box_x A}\ R\Box .$$

We first apply the h-p invertibility for the premise. Then we get $\vdash_{n-1} yR_x z, yR_x z, \Gamma \Rightarrow z:A, z:A, \Delta$. Since the derivation of this sequent has the height $n-1$, by IH $\vdash_{n-1} yR_x z, \Gamma \Rightarrow \Delta, z\!:\!A$. Finally, we apply $R\Box$ to this sequent and obtain the desired $\vdash_n \Gamma \Rightarrow \Delta, y\!:\!\Box_x A$. □

6 Soundness and completeness for labelled sequent calculi with cut

To state soundness and completeness of our labelled sequent calculi, we introduce a method of formulating semantic notions in sequents.[22] For doing that, we use a notion of interpretation from a set of labels W to the underlying set K of a model structure; this is given by a map $[\![\cdot]\!]\colon W \to K$ in $\mathfrak{M}$. The notion of sequent-validity with respect to a valuation and an interpretation can be defined as follows:

Definition 6.1. *A sequent $\Gamma \Rightarrow \Delta$ is sequent-valid w.r.t a valuation v and an interpretation $[\![\cdot]\!]$ if, for all labelled formulas $x : A$ and all relational atoms $yR_x z$, $0x$, $x \le y$, $x = y$ in Γ, whenever $[\![x]\!] \Vdash_v A$ and $[\![y]\!] R_{[\![x]\!]} [\![z]\!]$, $0[\![x]\!]$, $[\![x]\!] \le [\![y]\!]$, $[\![x]\!] = [\![y]\!]$ in $\mathfrak{M}$, then for some $y : B$ in Δ, $[\![y]\!] \Vdash_v B$. A sequent $\Gamma \Rightarrow \Delta$ is sequent-valid if $\Gamma \Rightarrow \Delta$ is sequent-valid for every valuation v and every interpretation $[\![\cdot]\!]$.*

Theorem 6.2 (Soundness). *If $\Gamma \Rightarrow \Delta$ is provable in a labelled sequent calculus for a logic L, then $\Gamma \Rightarrow \Delta$ is sequent-valid for every model structure $\mathfrak{M}$ for the L.*

We prove this by induction on the derivation of $\Gamma \Rightarrow \Delta$. The initial sequent is obvious, since $x\!:\!P \in \Gamma$ and $x\!:\!P \in \Delta$. All the logical rules preserves this notion of validity due to the truth conditions for the connectives (we handle the relevant implication through the indexed modality, but this is also straightforward). In the case of relational rules for R,

[22] We basically follow [32], pp.213 and apply the method there to the ternary semantics for relevant logics. The presentation here is not completely self-contained, so we refer to [32] for more details.

$\leq$, $=$, all the rules are designed so that the conclusion of a rule can be drawn from both the premise of the rule and the semantic conditions stated in a sequent format by means of structural rules including cut (which are proven to be admissible here). The details are routine,[23] so we omit them here.

In our labelled sequent calculi, our sequent is an instrument for derivation. Unlike the case of modal logic, the connection between the notion of sequent-validity and the L-validity for a formula is only indirectly given, since the latter is based on the notion of regular situation 0. Here we adopt a hand-waving method of connecting these by combining the two steps below: 1) expressing L-validity for a formula in the labelled sequent calculi; 2) proving completeness of labelled sequent calculi via Hilbert-style systems, i.e. completeness w.r.t. formulas (instead of sequents).

Our definition of L-validity given in §3 (based on the notion of "ensured"[24]) can be formulated in our labelled sequent calculi as: $0x \Rightarrow x{:}A$. By using the foregoing notion of interpretation, it can be made clear that this sequent expresses the notion of L-validity of A, because the image of interpretation is : if $[\![x]\!] \in 0$, then $[\![x]\!] \Vdash_v A$ for any situation $[\![x]\!] \in K$, for any valuation v and for any model structure $\mathfrak{M}$.

To prove completeness of labelled sequent calculi with respect to the Hilbert-style systems, we show a useful lemma, which correspond to the aforementioned semantic lemma 3.8 on L-validity in our labelled sequent calculi. Here the notion of L-implication and L-validity (for $A \to B$) are expressed in labelled sequent calculi as follows:

1. A L-implies B iff $x{:}A \Rightarrow x{:}B$ (for arbitrary x).
2. $A \to B$ is L-valid iff $0x \Rightarrow x{:}A \to B$ (for arbitrary x).

Let us now state the completeness theorem w.r.t. Hilbert-style systems. Here let us denote "the provability in a Hilbert-style system HL" by "$\vdash_{HL}$."

Theorem 6.3. *If* $\vdash_{HL} A$, *then* $0x \Rightarrow x{:}A$ *for* G3B *and extensions thereof.*

The semantic completeness of the labelled sequent calculi is an immediate consequence of this theorem and the completeness of the Hilbert-style systems in [42].

Corollary 6.4. *If* A *is L-valid, then* $0x \Rightarrow x{:}A$ *for* G3B *and extensions thereof.*

To show the theorem, we first prove a lemma.

Lemma 6.5 (Lemma 3.8 in labelled sequent). $x{:}A \Rightarrow x{:}B$ *iff* $0y \Rightarrow y{:}A \to B$.

[23] This is so, except the case of LD8, but even in this case the verification is straightforward by replacing the auxiliary predicate $P(b, y)$ by $\forall d, e(dR_{y^*}e \Rightarrow d \leq b^*)$ on the semantic part.

[24] We do not use a designated situation to define the notion of L-validity.

Proof. $\Longrightarrow$)

$$\dfrac{\dfrac{\dfrac{\dfrac{\dfrac{x:A \Rightarrow x:B}{x:A, z \leq x, 0y, z:A \Rightarrow x:B}\;\mathit{ad.wk}}{z \leq x, 0y, z:A \Rightarrow x:B}\;\mathit{HR}}{0y, zR_y x, z:A \Rightarrow x:B}\;\leq/R}{0y, z:A \Rightarrow z:\Box_y B}\;R\Box}{0y \Rightarrow y:A \to B}\;R\to$$

$\Longleftarrow$)

$$\dfrac{\dfrac{\dfrac{\dfrac{\dfrac{\dfrac{0y \Rightarrow y:A \to B}{z:A, 0y \Rightarrow z:\Box_y B}\;\mathit{Inv.R}\to}{xR_y w, 0y, z:A \Rightarrow w:B}\;\mathit{Inv.R}\Box}{xR_y x, 0y, x:A \Rightarrow x:B}\;\mathit{subst.}}{xR_y x, 0y, x \leq x, x:A \Rightarrow x:B}\;R/\leq}{x \leq x, x:A \Rightarrow x:B}\;\mathit{ad.wk.}}{x:A \Rightarrow x:B}\;\mathit{Ref}\leq$$

□

In proving completeness, we take a rule of the form $\dfrac{\vdash_{HL} A}{\vdash_{HL} B}$, to be validity-preserving, appealing to the soundness and completeness theorem for Hilbert-style systems, and take this to mean $\dfrac{0y \Rightarrow y:A}{0x \Rightarrow x:B}$.

Proof of completeness. The proof is by induction of length of proof in the Hilbert-style system, i.e. proving that all the axioms are provable in a labelled sequent calculus and all the rules preserve provability in a labelled sequent calculus, i.e. L-validity in the sequent calculus. We first show that the rules preserves L-validity in sequents.

MP: $\dfrac{0z \Rightarrow z:A \qquad 0y \Rightarrow y:A \to B}{0z \Rightarrow z:B}$

$$\dfrac{0z \Rightarrow z:A \qquad \dfrac{0y \Rightarrow y:A \to B}{z:A \Rightarrow z:B}\;\mathit{lemma}\ 6.5}{0z \Rightarrow z:B}\;\mathit{cut}$$

Adjunction: $\dfrac{0y \Rightarrow y:A \qquad 0y \Rightarrow y:B}{0y \Rightarrow y:A \wedge B}$. The rule itself is self-evident.

Suffix : $\dfrac{0v \Rightarrow v:A \to B}{0y \Rightarrow y:(B \to C) \to (A \to C)}$

$$\dfrac{\dfrac{\dfrac{\dfrac{\dfrac{z:A \Rightarrow z:B}{zR_x w, z:A, x:B \to C \Rightarrow w:C, w:B}\;\mathit{ad.wk} \qquad \dfrac{w:C, z:\Box C, zR_x w, z:A, x:B \to C \Rightarrow w:C}{z:\Box C, zR_x w, z:A, x:B \to C \Rightarrow w:C}\;L\Box}{zR_x w, z:A, x:B \to C \Rightarrow w:C}\;L\to}{z:A, x:B \to C \Rightarrow z:\Box_x C}\;R\Box}{x:B \to C \Rightarrow x:A \to C}\;R\to}{0y \Rightarrow zy:(B \to C) \to (A \to C)}\;\mathit{lemma}\ 6.5$$

Prefix : we omit this case. The proof is similar to the case of prefix.

Weak contrapositive: $\dfrac{0v \Rightarrow v{:}A \to \neg B}{0y \Rightarrow y{:}B \to \neg A}$

$$
\dfrac{\dfrac{\dfrac{\dfrac{}{x = x^{**}, x{:}B \Rightarrow x^{**}{:}B}\ \textit{lemma for Eq}}{x{:}B \Rightarrow x^{**}{:}B}\ \textit{dual} \qquad \dfrac{\dfrac{\dfrac{0v \Rightarrow v{:}A \to \neg B}{z{:}A \Rightarrow z{:}\neg B}\ \textit{lemma 6.5}}{x^{*}{:}A \Rightarrow x^{*}{:}\neg B}\ \textit{subst}}{x^{*}{:}A, x^{**}{:}B \Rightarrow}\ L\neg}{\dfrac{x{:}B, x^{*}{:}A \Rightarrow}{\dfrac{x{:}B \Rightarrow x{:}\neg A}{0y \Rightarrow y{:}B \to \neg A}\ \textit{lemma 6.5}}\ R\neg}\ \textit{cut}
$$

Let us now move on to prove axioms of the system B. We show representative cases.

A7. $0y \Rightarrow y{:}(A \to C) \wedge (B \to C) \to (A \vee B \to C)$

$$
\dfrac{\dfrac{\overset{(1)}{\vphantom{x}}}{zR_x w, z : A, x : A \to C, x : B \to C \Rightarrow w{:}C}\ L\to \qquad \dfrac{\overset{(2)}{\vphantom{x}}}{zR_x w, z : B, x : A \to C, x : B \to C \Rightarrow w : C}\ L\to}{\dfrac{zR_x w, z : A \vee B, x : A \to C, x : B \to C \Rightarrow w{:}C}{\dfrac{z : A \vee B, x : A \to C, x : B \to C \Rightarrow z{:}\Box_x C}{\dfrac{x : A \to C, x : B \to C \Rightarrow x : A \vee B \to C}{\dfrac{x : (A \to C) \wedge (B \to C) \Rightarrow x : A \vee B \to C}{\dfrac{x{:}(A \to C) \wedge (B \to C) \Rightarrow x : A \vee B \to C}{0y \Rightarrow y{:}(A \to C) \wedge (B \to C) \to (A \vee B \to C)}\ \textit{lemma 6.5}}\ R\to}\ L\wedge}\ R\to}\ R\Box}\ L\vee
$$

(1), (2) are the following respectively:

$$
\dfrac{zR_x w, z{:}A, x{:}A \to C, x{:}B \to C \Rightarrow w{:}C, z{:}A \qquad \dfrac{w{:}C, z{:}\Box_x C, zR_x w, z{:}A, x{:}A \to B, x{:}A \to C \Rightarrow w{:}C}{z{:}\Box_x C, zR_x w, z{:}A, x{:}A \to B, x{:}A \to C \Rightarrow w{:}C}\ L\Box}{zR_x w, z : A, x : A \to B, x : A \to C :\Rightarrow w : C}\ L\to
$$

$$
\dfrac{zR_x w, z{:}B, x{:}A \to C, x{:}B \to C \Rightarrow w{:}C, z{:}B \qquad \dfrac{w{:}C, z{:}\Box_x C, zR_x w, z{:}A, x{:}A \to C, x{:}B \to C \Rightarrow w{:}C}{z{:}\Box_x C, zR_x w, z{:}A, x{:}A \to C, x{:}B \to C \Rightarrow w{:}C}\ L\Box}{zR_x w, z{:}B, x{:}A \to C, x{:}B \to C \Rightarrow w{:}C}\ L\to
$$

A8. $0y \Rightarrow y{:}A \wedge (B \vee C) \to (A \wedge B) \vee (A \wedge C)$

$$
\dfrac{\dfrac{\dfrac{(1) \qquad\qquad (2)}{x{:}A, x{:}B \vee C \Rightarrow x{:}A \wedge B, x{:}A \wedge C}\ L\vee}{\dfrac{x{:}A \wedge (B \vee C) \Rightarrow x{:}A \wedge B, x{:}A \wedge C}{x{:}A \wedge (B \vee C) \Rightarrow x{:}(A \wedge B) \vee (A \wedge C)}\ R\vee}\ L\wedge}{0y \Rightarrow y{:}A \wedge (B \vee C) \to (A \wedge B) \vee (A \wedge C)}\ \textit{lemma 6.5}
$$

Here (1) and (2) are the following respectively.

$$\dfrac{x\!:\!A, x\!:\!B \Rightarrow x\!:\!A, x\!:\!A \wedge C \qquad x\!:\!A, x\!:\!B \Rightarrow x\!:\!B, x\!:\!A \wedge C}{x\!:\!A, x\!:\!B \Rightarrow x\!:\!A \wedge B, x\!:\!A \wedge C}\ R\wedge$$

$$\dfrac{x\!:\!A, x\!:\!C \Rightarrow x\!:\!A \wedge B, x\!:\!A \qquad x\!:\!A, x\!:\!C \Rightarrow x\!:\!A \wedge B, x\!:\!C}{x\!:\!A, x\!:\!C \Rightarrow x\!:\!A \wedge B, x\!:\!A \wedge C}\ R\wedge$$

Remark: The foregoing proof shows that the distributive law can be easily shown in all of our proof systems. We take this to be an advantage of our proof systems, since the distributive law is occasionally taken to be essential in relevant logics (e.g. [6]), but let us note that it seems more challenging to formulate non-distributive relevant (or substructural) logics in our labelled sequent calculi in general.

Now we move on to additional axioms and rules.

B3. $0y \Rightarrow y\!:\!(A \to B) \to ((B \to C) \to (A \to C))$

$$\dfrac{t\!:\!C, s\!:\!\Box_z C, s\!:\!B, sR_z t, z\!:\!B \to C, \Gamma_3 \Rightarrow t\!:\!C}{s\!:\!\Box_z C, s\!:\!B, sR_z t, z\!:\!B \to C, \Gamma_3 \Rightarrow t\!:\!C}\ L\Box$$

$$\dfrac{s\!:\!B, sR_z t, z\!:\!B \to C, \Gamma_3 \Rightarrow t\!:\!C, s\!:\!B \qquad s\!:\!\Box_z C, s\!:\!B, sR_z t, z\!:\!B \to C, \Gamma_3 \Rightarrow t\!:\!C}{s\!:\!B, v\!:\!\Box_x B, vR_x s, sR_z t, z\!:\!B \to C, \Gamma_2 \Rightarrow t\!:\!C}\ L\to$$

$$\dfrac{s\!:\!B, v\!:\!\Box_x B, vR_x s, sR_z t, z\!:\!B \to C, \Gamma_2 \Rightarrow t\!:\!C}{v\!:\!\Box_x B, vR_x s, sR_z t, z\!:\!B \to C, \Gamma_2 \Rightarrow t\!:\!C}\ L\Box$$

$$\dfrac{vR_x s, \Gamma_1, v\!:\!A, x\!:\!A \to B \Rightarrow t\!:\!C, v\!:\!A \qquad v\!:\!\Box_x B, vR_x s, sR_z t, z\!:\!B \to C, \Gamma_2 \Rightarrow t\!:\!C}{vR_x s, sR_z t, vR_w t, v\!:\!A, zR_x w, z\!:\!B \to C, x\!:\!A \to B \Rightarrow t\!:\!C}\ L\to$$

$$\dfrac{\dfrac{\dfrac{\dfrac{\dfrac{\dfrac{vR_x s, sR_z t, vR_w t, v\!:\!A, zR_x w, z\!:\!B \to C, x\!:\!A \to B \Rightarrow t\!:\!C}{vR_w t, v\!:\!A, zR_x w, z\!:\!B \to C, x\!:\!A \to B \Rightarrow t\!:\!C}\ LB3}{v\!:\!A, zR_x w, z\!:\!B \to C, x\!:\!A \to B \Rightarrow v\!:\!\Box_w C}\ R\Box}{zR_x w, z\!:\!B \to C, x\!:\!A \to B \Rightarrow w\!:\!A \to C}\ R\to}{z\!:\!B \to C, x\!:\!A \to B \Rightarrow z\!:\!\Box_x(A \to C)}\ R\Box}{x\!:\!A \to B \Rightarrow x\!:\!(B \to C) \to (A \to C)}\ R\to}{0y \Rightarrow y\!:\!(A \to B) \to ((B \to C) \to (A \to C))}\ \textit{lemma } 6.5$$

Here Γ_1, Γ_2, Γ_3 stand for appropriate multi-sets of formulas.

D8. $0y \Rightarrow y : (A \to \neg(B \to C)) \to (\neg B \to \neg A)$ (This requires "geometrization.")

$$\dfrac{w^* : A, x : A \to \neg(B \to C), w^* R_x v, P(z,v), zR_x w \Rightarrow z^* : B, w^* : A \qquad (2)}{w^* : A, x : A \to \neg(B \to C), w^* R_x v, P(z,v), zR_x w \Rightarrow z^* : B}\ L\to$$

$$\dfrac{\dfrac{\dfrac{\dfrac{\dfrac{\dfrac{\dfrac{w^* : A, x : A \to \neg(B \to C), w^* R_x v, P(z,v), zR_x w \Rightarrow z^* : B}{w^* : A, x : A \to \neg(B \to C), zR_x w \Rightarrow z^* : B}\ LD8(1)}{w^* : A, z : \neg B, x : A \to \neg(B \to C), zR_x w \Rightarrow}\ L\neg}{z : \neg B, x : A \to \neg(B \to C), zR_x w \Rightarrow w : \neg A}\ R\neg}{z : \neg B, x : A \to \neg(B \to C) \Rightarrow z : \Box_x \neg A}\ R\Box}{x : A \to \neg(B \to C) \Rightarrow x : \neg B \to \neg A)}\ R\to}{y : 0 \Rightarrow y : (A \to \neg(B \to C)) \to (\neg B \to \neg A)}\ \textit{lemma } 6.5$$

$$
\begin{array}{cl}
z^* : B, w^* : \Box_x \neg(B \to C), w^* : A, x : A \to \neg(B \to C), w^* R_x v, P(z,v), z R_x w, t : B, t R_{v^*} s, t \leq z^* \Rightarrow z^* : B, s : C & \\
\hline
w^* : \Box_x \neg(B \to C), w^* : A, x : A \to \neg(B \to C), w^* R_x v, P(z,v), z R_x w, t : B, t R_{v^*} s, t \leq z^* \Rightarrow z^* : B, s : C & HR \\
\hline
w^* : \Box_x \neg(B \to C), w^* : A, x : A \to \neg(B \to C), w^* R_x v, P(z,v), z R_x w, t : B, t R_{v^*} s \Rightarrow z^* : B, s : C & LD8(2) \\
\hline
w^* : \Box_x \neg(B \to C), w^* : A, x : A \to \neg(B \to C), w^* R_x v, P(z,v), z R_x w, t : B \Rightarrow z^* : B, t : \Box_{v^*} C & R\Box \\
\hline
w^* : \Box_x \neg(B \to C), w^* : A, x : A \to \neg(B \to C), w^* R_x v, P(z,v), z R_x w \Rightarrow z^* : B, v^* : (B \to C) & R\to \\
\hline
v : \neg(B \to C), w^* : \Box_x \neg(B \to C), w^* : A, x : A \to \neg(B \to C), w^* R_x v, P(z,v), z R_x w \Rightarrow z^* : B & R\neg \\
\hline
(2)\ w^* : \Box_x \neg(B \to C), w^* : A, x : A \to \neg(B \to C), w^* R_x v, P(z,v), z R_x w \Rightarrow z^* : B & L\Box
\end{array}
$$

□

7 Cut elimination for labelled sequent calculi for relevant logics

Let us now give a proof of the cut-elimination theorem for the labelled sequent calculus for B and extensions of thereof. The theorem can be stated as follows:

Theorem 7.1 (Cut-elimination). *Cut is admissible in* G3B *and extensions thereof.*

Remark: 1. The labelled sequent calculi that we use here contain several atomic formulas in addition to propositional variables.[25] However, it turns out that we do not have to include those atomic cases as an instance of cut formula except $x : P$.[26] This implies that the subcases in the following which schematically handle unary rules and binary rules in which cut formulas are inactive or side formulas are sufficient to cover all the additional rules. Hence, our proof in the following will be sufficient to cover not only G3B but all of its extensions by the additional rules.

2. Also, we use a term "active" when an atomic formula plays an essential role in an application of a rule (without the formula, we cannot apply the rule). In this sense, all principal formulas are obviously active, but not vice versa.

Proof. By double induction on i) the weight of an uppermost cut formula in a derivation and ii) the subinduction on the height of cut, namely, the sum of the heights of derivation of the premises of a cut.

Case I. The cut formula is atomic: $x : P$.

Subcase I.1 The cut formula is active on both of the premises. We can consider four further possible subcases of this.

I.1.1 Both premises are initial sequents.

[25] We follow [42], Ch.4, in choosing a language without propositional constants. Obviously, in relevant logic, introducing a propositional constant introduces some subtlety. Our discussions are already quite complicated, so we have tried to reduce a further source of complication.

[26] This is essentially due to the fact that we do not have to consider axioms with those atomic formulas except $x : P$.

I.1.2 At least one of the premises is obtained by a relational rule.

- I.1.2.1 The left premise (L-premise) is an initial sequent, but the right premise (R-premise) is obtained by a relational rule.
- I.1.2.2 The L-premise is obtained by a relational rule, and the R-premise is an initial sequent.
- I.1.2.3 Both of the L-premise and the R-premise are obtained by relational rules.

But note that there are no actual cases of I.1.2.1 or I.1.2.1, because formulas on the succedent are inactive in each relational rule. We check the two actual subcases.

I.1.1 Both premises are initial sequents, and the cut formula is identical with the active (atomic) formula. Hence, the cut looks as follows:

$$\dfrac{x\!:\!P,\Gamma \Rightarrow \Delta, x\!:\!P \qquad x\!:\!P,\Gamma' \Rightarrow \Delta', x\!:\!P}{x\!:\!P,\Gamma,\Gamma' \Rightarrow \Delta,\Delta', x\!:\!P}\ \mathit{Cut}$$

Observe that the lower sequent can be obtained as an initial sequent with more side formulas. Hence, the cut is eliminable.

I.1.2.1 The L-premise is an initial sequent but the R-premise is obtained by a relational rule.

I.1.2.1.a The L-premise is an initial sequent, and the R-premise is obtained by AtRepl. The case of cut looks as follows:

$$\dfrac{x\!:\!P,\Gamma \Rightarrow \Delta, x\!:\!P \qquad \dfrac{y\!:\!P, x=y, x\!:\!P,\Gamma' \Rightarrow \Delta'}{x=y, x\!:\!P,\Gamma' \Rightarrow \Delta'}\ \mathit{AtRepl}}{x=y, x\!:\!P,\Gamma,\Gamma' \Rightarrow \Delta,\Delta'}\ \mathit{cut}$$

In this case, instead of the current R-premise, we can take the following $y:P, x = y, x\!:\!P,\Gamma,\Gamma' \Rightarrow \Delta,\Delta'$ by the h-p admissibility of weakening. Then we have

$$\dfrac{y\!:\!P, x=y, x\!:\!P,\Gamma,\Gamma' \Rightarrow \Delta,\Delta'}{x=y, x\!:\!P,\Gamma,\Gamma' \Rightarrow \Delta,\Delta'}\ \mathit{AtRepl}$$

We get the same conclusion without using the cut.

I.1.2.1.b The L-premise is an initial sequent, and the R-premise is obtained by AtHR. The argument is essentially the same as above.

$$\dfrac{x\!:\!P,\Gamma \Rightarrow \Delta, x\!:\!P \qquad \dfrac{y\!:\!P, x\leq y, x\!:\!P,\Gamma' \Rightarrow \Delta'}{x\leq y, x\!:\!P,\Gamma' \Rightarrow \Delta'}\ \mathit{AtHR}}{x\leq y, x\!:\!P,\Gamma,\Gamma' \Rightarrow \Delta,\Delta'}\ \mathit{cut}$$

In this case, instead of the current R-premise, we can take the following $y:P, x \leq y, x\!:\!P,\Gamma,\Gamma' \Rightarrow \Delta,\Delta'$ by the h-p admissibility of weakening. Then we have

$$\frac{y\!:\!P, x \leq y, x\!:\!P, \Gamma, \Gamma' \Rightarrow \Delta, \Delta'}{x \leq y, x\!:\!P, \Gamma, \Gamma' \Rightarrow \Delta, \Delta'}\ AtHR$$

We get the same conclusion without using the cut.

Subcase I.2 The cut formula is inactive in at least one of the premises. Since a cut formula is of the form $x\!:\!P$, it suffices to check AtRepl, AtHR. Here we have the following combinatorial possibilities.

I.2.1 In the L-premise, $x\!:\!P$ is active in initial sequent, but inactive in the R-premise.

I.2.2 In the L-premise, $x : P$ is inactive, but it is active in the R-premise as an initial sequent.

I.2.3 In the L-premise, $x\!:\!P$ is inactive, but it is active in the R-premise in either AtRepl or AtHR.

I.2.4 In the L-premise, $x : P$ is active in either AtRepl or AtHR, but inactive in the R-premise.

I.2.5 Both in the L-premise and in the R-premise, $x\!:\!P$ is inactive.

(There is no real case of I.2.4, and the case I.2.5 can be reduced to other subcases. So we check the other cases.)

I.2.1 In the L-premise, $x\!:\!P$ is active in initial sequent, but inactive in the R-premise.

I.2.1.1 The R-premise is an initial sequent where $x\!:\!P$ is not active.

$$\frac{x\!:\!P, \Gamma \Rightarrow \Delta, x\!:\!P \qquad x\!:\!P, \Gamma' \Rightarrow \Delta'}{x\!:\!P, \Gamma, \Gamma' \Rightarrow \Delta, \Delta'}\ cut$$

The conclusion is also an initial sequent by the admissibility of weakening in the R-premise. Hence, the cut is eliminable.

I.2.1.2 The R-premise is obtained by a rule (unary or binary).

I.2.1.2.a The R-premise is obtained by a unary rule.

$$\frac{x\!:\!P, \Gamma \Rightarrow \Delta, x\!:\!P \qquad \dfrac{x\!:\!P, \Gamma_1 \Rightarrow \Delta_1}{x\!:\!P, \Gamma' \Rightarrow \Delta'}\ r1}{x\!:\!P, \Gamma, \Gamma' \Rightarrow \Delta, \Delta'}\ cut$$

The conclusion follows from the R-premise by the admissibility of weakening. Hence, the cut is eliminable.

I.2.1.2.b The R-premise is obtained by a binary rule.

$$\frac{x\!:\!P, \Gamma \Rightarrow \Delta, x\!:\!P \qquad \dfrac{x\!:\!P, \Gamma_1 \Rightarrow \Delta_1 \qquad x\!:\!P, \Gamma_2 \Rightarrow \Delta_2}{x\!:\!P, \Gamma' \Rightarrow \Delta'}\ r2}{x\!:\!P, \Gamma, \Gamma' \Rightarrow \Delta, \Delta'}\ cut$$

The conclusion follows from the R-premise by the admissibility of weakening. Hence, the cut is eliminable.

I.2.2 In the L-premise, $x\!:\!P$ is inactive, but it is active in the R-premise in an initial sequent. This case is entirely symmetric to I.2.1.

I.2.3 In the L-premise, $x : P$ is inactive, but it is active in the R-premise in either AtRepl or AtHR.

I.2.3.1 The R-premise is AtRepl.

There are three further subcases of $x\!:\!P$ being inactive on the L-premise.

I.2.3.1.1 The L-premise is an initial sequent, i.e. $\Gamma \Rightarrow \Delta$ is an initial sequent without $x\!:\!P$. The cut looks as follows:

$$\dfrac{\Gamma \Rightarrow \Delta, x\!:\!P \qquad \dfrac{y\!:\!P, x = y, x\!:\!P, \Gamma' \Rightarrow \Delta'}{x = y, x\!:\!P, \Gamma' \Rightarrow \Delta'}\,\mathit{AtRepl}}{x = y, \Gamma, \Gamma' \Rightarrow \Delta, \Delta'}\,\mathit{cut}$$

The L-premise is an initial sequent even without $x : P$, since $x : P$ is inactive in the L-premise. Hence, the conclusion is also an initial sequent by the h-p admissibility of weakening. Hence, the cut is eliminable.

I.2.3.1.2 The L-Premise is obtained by a rule.

I.2.3.1.2a The L-premise is obtained by a unary rule.

$$\dfrac{\dfrac{\Gamma_1 \Rightarrow \Delta_1, x\!:\!P}{\Gamma \Rightarrow \Delta, x\!:\!P}\,r1 \qquad \dfrac{y\!:\!P, x = y, x\!:\!P, \Gamma' \Rightarrow \Delta'}{x = y, x\!:\!P, \Gamma' \Rightarrow \Delta'}\,\mathit{AtRepl}}{x = y, \Gamma, \Gamma' \Rightarrow \Delta, \Delta'}\,\mathit{cut}$$

Due to the atomic nature of $x\!:\!P$, the following reduction is sufficient in this case.

$$\dfrac{\dfrac{\Gamma_1 \Rightarrow \Delta_1, x\!:\!P \qquad \dfrac{y\!:\!P, x = y, x\!:\!P, \Gamma' \Rightarrow \Delta'}{x = y, x\!:\!P, \Gamma' \Rightarrow \Delta'}\,\mathit{AtRepl}}{x = y, \Gamma_1, \Gamma' \Rightarrow \Delta_1, \Delta'}\,\mathit{cut}}{x = y, \Gamma, \Gamma' \Rightarrow \Delta, \Delta'}\,r1$$

This cut has a smaller height than the original one, so by IH this cut is eliminable.

I.2.3.1.2b The L-premise is obtained by a binary rule.

$$\dfrac{\dfrac{\Gamma_1 \Rightarrow \Delta_1, x\!:\!P \qquad \Gamma_2 \Rightarrow \Delta_2, x\!:\!P}{\Gamma \Rightarrow \Delta, x\!:\!P}\,r2 \qquad \dfrac{y\!:\!P, x = y, x\!:\!P, \Gamma' \Rightarrow \Delta'}{x = y, x\!:\!P, \Gamma' \Rightarrow \Delta'}\,\mathit{AtRepl}}{x = y, \Gamma, \Gamma' \Rightarrow \Delta, \Delta'}\,\mathit{cut}$$

Due to the atomic nature of $x\!:\!P$, the following reduction is sufficient in this case.

$$\dfrac{\Gamma_1 \Rightarrow \Delta_1, x{:}P \quad \dfrac{y{:}P, x=y, x{:}P, \Gamma' \Rightarrow \Delta'}{x=y, x{:}P, \Gamma' \Rightarrow \Delta'}\,AtRepl}{x=y, \Gamma_1, \Gamma' \Rightarrow \Delta_1, \Delta'}\,cut \quad \dfrac{\Gamma_2 \Rightarrow \Delta_2, x{:}P \quad \dfrac{y{:}P, x=y, x{:}P, \Gamma' \Rightarrow \Delta'}{x=y, x{:}P, \Gamma' \Rightarrow \Delta'}\,AtRepl}{x=y, \Gamma_1, \Gamma' \Rightarrow \Delta_1, \Delta'}\,cut \over x=y, \Gamma, \Gamma' \Rightarrow \Delta, \Delta' \quad r2$$

This cut has a smaller height than the original one, so by IH this cut is eliminable.

I.2.3.2 The R-premise is AtHR. This case similar to 1.2.3.1.

Case II. The cut formula is not atomic. The weight of the cut formula of an uppermost cut is n, s.t. $n > 0$. The outermost logical symbols are $\wedge$, $\vee$, $\rightarrow$, $\neg$, and the indexed modality $\Box_x$. In this case, we have the following subcases.

II.1 The cut formula on the L-premise is a side formula in at least one of the L-premise and the R-premise.

- II.1.1 The cut formula on the L-premise is a side formula.
- II.1.2 The cut formula on the R-premise is a side formula.

II.2 The cut formula is principal on both the L-premise and the R-premise.

In the first two subcases, the argument depends only on the fact that in theses cases at least one of the premises of cut has the cut formulas as a side formula.

Subase II.1.1 The cut-formula on the L-premise is a side formula.

II.1.1.a The last rule on the L-premise is unary. The cut looks as follows:

$$\dfrac{\dfrac{\Gamma_1 \Rightarrow \Delta_1, x{:}A}{\Gamma \Rightarrow \Delta, x{:}A}\,r1 \qquad x{:}A, \Gamma' \Rightarrow \Delta'}{\Gamma, \Gamma' \Rightarrow \Delta, \Delta'}\,cut$$

We can reduce this case to the following.

$$\dfrac{\dfrac{\Gamma_1 \Rightarrow \Delta_1, x{:}A \qquad x{:}A, \Gamma' \Rightarrow \Delta'}{\Gamma_1, \Gamma' \Rightarrow \Delta_1, \Delta'}\,cut}{\Gamma, \Gamma' \Rightarrow \Delta, \Delta'}\,r1$$

This cut has a smaller height of cut than before. By IH, this cut can be eliminated. Unary rules may include rules that have the eigenvariable condition, and in applying the cut there may be formulas in $\Gamma' \Rightarrow \Delta'$ which violate the condition. If this occurs, by the substitution lemma we can replace the problematic variables by fresh ones.

II.1.1.b The last rule on the L-premise is binary. The cut looks as follows:

$$\dfrac{\dfrac{\Gamma_1'' \Rightarrow \Delta_1'', x{:}A \qquad \Gamma_2'' \Rightarrow \Delta_2'', x{:}A}{\Gamma \Rightarrow \Delta, x{:}A}\,r2 \qquad x{:}A, \Gamma' \Rightarrow \Delta'}{\Gamma, \Gamma' \Rightarrow \Delta, \Delta'}\,cut$$

This is reduced to the following.

$$\dfrac{\dfrac{\Gamma_1'' \Rightarrow \Delta_1'', x{:}A \qquad x{:}A, \Gamma' \Rightarrow \Delta'}{\Gamma', \Gamma_1'' \Rightarrow \Delta', \Delta_1''}\, cut \qquad \dfrac{\Gamma_2'' \Rightarrow \Delta_2'', x{:}A \qquad x{:}A, \Gamma' \Rightarrow \Delta'}{\Gamma', \Gamma_2'' \Rightarrow \Delta', \Delta_2''}\, cut}{\Gamma, \Gamma' \Rightarrow \Delta, \Delta'}\, r2$$

This cut has a strictly smaller height of cut than before. Thus, applying IH, this cut can be eliminated. (Among the additional rules, there is a binary rule that has an eigenvariable condition. In such a case, we need to appropriately replace variables to fresh ones so that the eigenvariable condition be satisfied.)

II.1.2. The cut-formula on the R-premise is a side formula. These cases are entirely symmetric to II.2a and II.2b.

II.2 Both cut formulas are principal.

II.2.1 $x{:}A = x{:}B \wedge C$, II.2.2 $x{:}A = x{:}B \vee C$. We omit these cases for brevity.

II.2.3 $x{:}A = x{:}B \to C$. The application of cut in this case looks as follows:

$$\dfrac{\dfrac{y{:}B, \Gamma \Rightarrow \Delta, y{:}\Box_x C}{\Gamma \Rightarrow \Delta, x{:}B \to C}\, R\!\to \qquad \dfrac{x{:}B \to C, \Gamma' \Rightarrow \Delta', y{:}B \qquad y{:}\Box_x C, x{:}B \to C, \Gamma' \Rightarrow \Delta'}{x{:}B \to C, \Gamma' \Rightarrow \Delta'}\, L\!\to}{\Gamma, \Gamma' \Rightarrow \Delta, \Delta'}\, cut$$

This cut can be reduced to the combination of the following four cuts:

$$\dfrac{\dfrac{\dfrac{\Gamma \Rightarrow \Delta, x{:}B \to C \qquad x{:}B \to C, \Gamma' \Rightarrow \Delta', y{:}B}{\Gamma' \Rightarrow \Delta, \Delta', y{:}B}\, cut_1 \qquad y{:}B, \Gamma \Rightarrow \Delta, y{:}\Box_x C}{\Gamma, \Gamma, \Gamma' \Rightarrow \Delta, \Delta, \Delta', y{:}\Box_x C}\, cut_2}{\Gamma, \Gamma' \Rightarrow \Delta, \Delta', y{:}\Box_x C}\, \star$$

At the step $\star$, we apply h-p admissibility of contraction.

$$\dfrac{\Gamma \Rightarrow \Delta, x{:}B \to C \qquad y{:}\Box_x C, x{:}B \to C, \Gamma' \Rightarrow \Delta'}{y{:}\Box_x C, \Gamma, \Gamma' \Rightarrow \Delta, \Delta'}\, cut_3$$

Applying another cut to the lower sequents of these, we can have the following.

$$\dfrac{\dfrac{\Gamma, \Gamma' \Rightarrow \Delta, \Delta', y{:}\Box_x C \qquad y{:}\Box_x C, \Gamma, \Gamma' \Rightarrow \Delta, \Delta'}{\Gamma, \Gamma, \Gamma', \Gamma' \Rightarrow \Delta, \Delta, , \Delta'}\, cut_4}{\Gamma, \Gamma' \Rightarrow \Delta, \Delta'}\, \star$$

At the step $\star$, we apply h-p admissibility of contraction.

Note that for cut_1 and cut_3, cut-heights are strictly reduced, and cut_2 and cut_4 have weights of cut-formula smaller than that of the original cut. Hence, by applying IH, these cuts can be eliminated.

II.2.4 $y{:}A = y{:}\Box_x B$.[27]

$$\dfrac{\dfrac{yR_xz, \Gamma \Rightarrow \Delta, z{:}B}{\Gamma \Rightarrow \Delta, y{:}\Box_x B}\,R\Box_x \qquad \dfrac{yR_xz, y{:}\Box_x B, z{:}B, \Gamma' \Rightarrow \Delta'}{y{:}\Box_x B, \Gamma' \Rightarrow \Delta'}\,L\Box_x}{\Gamma, \Gamma' \Rightarrow \Delta, \Delta'}\,cut$$

This application of cut can be reduced to the following two cuts:

$$\dfrac{\dfrac{yR_xz, \Gamma \Rightarrow \Delta, z{:}B \qquad \dfrac{\Gamma \Rightarrow \Delta, y{:}\Box_x B \qquad yR_xz, y{:}\Box_x B, z{:}B, \Gamma, \Gamma' \Rightarrow \Delta, \Delta'}{yR_xz, z{:}B, \Gamma, \Gamma' \Rightarrow \Delta, \Delta'}\,cut_1}{yR_xz, \Gamma, \Gamma, \Gamma' \Rightarrow \Delta, \Delta, \Delta'}\,cut_2}{yR_xz, \Gamma, \Gamma' \Rightarrow \Delta, \Delta'}\,\star$$

At the step $\star$, we apply h-p admissibility of contraction. Cut_1 has the weight of the cut formula strictly smaller than the original cut, cut_2 has the height strictly smaller than that of the original one. So, by IH, these cuts can be eliminated.

II.2.5 $x{:}A = x{:}\neg B$. The application of cut in this case looks as follows:

$$\dfrac{\dfrac{\Gamma, x^*{:}B \Rightarrow \Delta}{\Gamma \Rightarrow \Delta, x{:}\neg B}\,R\neg \qquad \dfrac{\Gamma' \Rightarrow x^*{:}B, \Delta'}{x{:}\neg B, \Gamma' \Rightarrow \Delta'}\,L\neg}{\Gamma, \Gamma' \Rightarrow \Delta, \Delta'}\,cut$$

This cut can be reduced to the following.

$$\dfrac{\Gamma' \Rightarrow x^*{:}B, \Delta' \qquad \Gamma, x^*{:}B \Rightarrow \Delta}{\Gamma, \Gamma' \Rightarrow \Delta, \Delta'}\,cut$$

This cut has the weight of the cut formula strictly smaller than the previous cut. Hence, by IH, this cut can be eliminated.

This completes the proof of cut-elimination for B3G and its extensions. $\Box$

Remark. Our proof of cut elimination depends on a subtle architecture of precise choices in the formulation of the rules. E.g., in case I.1.1.a above, the restriction to atomic cases of rules AtRepl and AtHR is essential to our method of cut-elimination. With an non-atomic replacement rule, we would have the following case of cut:

$$\dfrac{\dfrac{\Gamma \Rightarrow \Delta, x{:}B \qquad \Gamma \Rightarrow \Delta, x{:}C}{\Gamma \Rightarrow \Delta, x{:}B \wedge C}\,R\wedge \qquad \dfrac{y{:}B \wedge C, x = y, x{:}B \wedge C, \Gamma' \Rightarrow \Delta'}{x = y, x{:}B \wedge C, \Gamma' \Rightarrow \Delta'}\,Repl}{x = y, \Gamma, \Gamma' \Rightarrow \Delta, \Delta'}\,cut$$

[27] Since x is taken as the index variable, we use y as a label for the whole formula.

Apparently, there is no other way of reducing this case than taking the following, because the left premise simply does not have a formula $y{:}B \wedge C$.

$$\dfrac{\dfrac{\Gamma \Rightarrow \Delta, x{:}B \qquad \Gamma \Rightarrow \Delta, x{:}C}{\Gamma \Rightarrow \Delta, x{:}B \wedge C}\,R\wedge \qquad y{:}B \wedge C, x = y, x{:}B \wedge C, \Gamma' \Rightarrow \Delta'}{y{:}B \wedge C, x = y, \Gamma, \Gamma' \Rightarrow \Delta, \Delta'}\,cut$$

If we do this, there seems to be no obvious (local) way of handling $y{:}B \wedge C$.

8 Conclusion

In this paper, we presented labelled sequent calculi for an extensive class of relevant logics by using ternary relation symbols to represent the Routley-Meyer ternary relational semantics, and proved the cut-elimination theorem for these systems by using a proof method presented in [31] and [32], i.e., typical in structural proof theory. (We first show some lemmas, such as height-preserving invertibility of the rules, height-preserving admissibility of structural rules, and prove the cut-elimination theorem.)

There are still lots of systems of relevant logic or related substructural logics that have not been covered in this paper. However, partly because our work here has already become too bulky, we leave the treatment of these logics to another occasion.

Also, at least three further topics are left out from our investigation. First, we left out proving completeness of the labelled sequent calculi directly by proof-search, i.e. showing that either a given formula is provable or it has a countermodel. Our proof of the semantic completeness of our labelled sequent calculi is given only via the completeness of Hilbert-style systems. We used Hilbert-style systems since they provide a convenient medium for handling the special way of defining L-validity in relevant logics. However, a direct proof of completeness by proof-search must be possible, since labelled sequent calculi are in general suitable for proof-search and invertible rules preserve countermodels.

Second, related to the first point, once completeness by proof-search is proved, this itself naturally opens up another direction of research, i.e., giving an alternative proof-theoretic proof of decidability of (most) logics presented here. This is because finding out a bound in proof-search is one of the well-known methods of proving decidability of a logic. This method has already been applied in some venues that use labelled sequent calculi [32, 30, 22], but the case of relevant logic is particularly interesting, since some relevant logics are actually undecidable and proving decidability may involve some subtlety.

Third, syntactic translations between labelled sequent calculi (or other proof systems having devices expressing some semantic information in proof systems) and display calculi have been developed in other cases of non-classical logics, e.g., [27], [12]. On the other hand, partly due to the flexible nature of ternary relations in the Routley-Meyer semantics, one can directly establish a correspondence between ternary relations and combinators. Thus, a correspondence between ternary relations and display calculi rules tends to be discussed directly between relations in semantics and rules for structural connectives in display calculi (e.g., [39]). However, this is not quite the same as considering a translation between G3-style labelled sequent calculi for relevant logics in which structural rules are admissible (which are new) and display calculi for relevant logics, so we consider constructing a translation between our labelled sequent calculi for relevant logics and the display calculi for relevant logics to be a meaningful open problem and leave it for future investigations.

Acknowledgements

The research for this paper was supported both by the Japan Society for the Promotion of Science (JSPS), Core-to-Core Program (A. Advanced Research Networks) no. 963176 and by the Academy of Finland research project no. 1308664.

References

[1] Wilhelm Ackermann *Begründung einer strengen Implikation*, The Journal of Symbolic Logic 21 (2), 1956, pp. 113–128.

[2] Alan Anderson, and Nuel Belnap Jr. Entailment, *The Logic of Relevance and Necessity*, 1, Princeton University Press, Princeton NJ 1975.

[3] Arnon Avron *A constructive analysis of RM*, Journal of Symbolic Logic 52 (4), 1987, pp. 939–951.

[4] — *Relevance and paraconsistency – a new approach. III. Cut-free Gentzen-type systems*, Notre Dame Journal of Formal Logic 32 (1), 1991, pp. 147–160.

[5] Jc Beall, Ross Brady, J. Michael Dunn, Allen P. Hazen, Edwin Mares, Robert K. Meyer, Graham Priest, Greg Restall, David Ripley, John Slaney, and Richard Sylvan *On the ternary relation and conditionality*, Journal of Philosophical Logic 41 (3), 2012, pp. 595–612.

[6] Nuel Belap Jr. *Life in the undistributed middle*, in [14], pp. 31–41.

[7] Ross T. Brady (1984). *Natural deduction systems for some quantified relevant logics*, Logique et Analyse 27 (8), 1984, pp. 355–377.

[8] Ross T. Brady Universal Logic, Cambridge University Press, 2006.

[9] Ross T. Brady *Normalized natural deduction systems for some relevant logics I: the logic DW*, Journal of Symbolic Logic 71 (1), 2006, pp. 35 - 66.

[10] Ross T. Brady *Free semantics*, Journal of Philosophical Logic 39, 2010, pp. 511–529.

[11] Alonzo Church The Calculi of Lambda Conversion, Princeton University Press, Princeton NJ 1941.

[12] Agata Ciabattoni, Tim Lyon T., Ravantha Ramanayake *From display to labelled proofs for tense logics*, In: Sergei Artemov, and Anil Nerode (eds) Logical Foundations of Computer Science. LFCS 2018. Lecture Notes in Computer Science, vol 10703. Springer, 2018.

[13] Kosta Došen *A historical introduction to substructural logics*, in [14], pp. 1–36.

[14] Kosta Došen, and Peter Schroeder-Heister (eds.), Substructural Logics, Clarendon Press, Oxford, 1993.

[15] Michael Dunn *A "Gentzen system" for positive relevant implication*, The Journal of Symbolic Logic 38, 1973, pp. 356–357.

[16] Katalin Bimbó, and Michael Dunn Generalized Galois Logics, *Relational Semantics of Non-classical Logical Calculi*, CSLI Publications, Stanford CA 2008 [*CSLI Lectures Notes 188*].

[17] Michael Dunn, and Greg Restall. Relevance Logic, in: Dov Gabbay, and Franz Guenthner (eds.) Handbook of Philosophical Logic 6, Kluwer Academic Publishers, 2002 (second edition).

[18] Roy Dyckhoff, and Sara Negri *Geometrisation of first-order logic*, The Bulletin of Symbolic Logic 21 (2), 2015, pp. 123–163.

[19] Melvin Fitting Proof Methods for Modal and Intuitionistic Logic, Reidel Publishing Company, 1983.

[20] Olivier Gasquet, and Andreas Herzig *From classical to normal modal logics*, in: Heinrich Wansing (ed.) Proof Theory of Modal Logic, Springer, 1996, pp. 293–311.

[21] Laura Giordano, Valentina Gliozzi, Nicola Olivetti, and Camilla Schwind *Tableau calculi for preference-based conditional logics: PCL and its extensions*, ACM Transactions on Computational Logic 10 (3/21), 2008, pp. 1–45.

[22] Marianna Girlando, Sara Negri, and Giorgio Sbardolini *Uniform labelled calculi for conditional and counterfactual logics*, in: Rosalie Iemhoff, Michael Moortgat, and Ruy de Queiroz (eds.) Logic, Language, Information, and Computation (26th International Workshop, WoLLIC 2019), *LNCS 11541*, 2019, pp. 248–263.

[23] Rajeev Goré Review of 'Displaying modal logic' by Heinrich Wansing, Journal of Logic, Language and Information 9 (2), 2000, pp. 259–272.

[24] Hirohiko Kushida, and Mitsuhiro Okada *A Proof-theoretic study of the correspondence of classical logic and modal logic*, The Journal of Symbolic Logic 68 (4), 2003, pp. 1403–1414.

[25] Edwin Mares Relevant Logic, *A Philosophical Interpretation*, Cambridge University Press, 2004.

[26] Grigori E. Mints [Minc] *Teorema ob ustrannimosti sečeniĭa dlĭa relevantnyx logik* [Cut elimination theorem in relevant logics] [Russian], in: Yuriĭ V. Matiyasevič, and Anatol' O. Slisenko (eds.), Issledovaniĭa po konstruktivnoĭ matematike i matematičeskoĭ logike [Essays on constructive mathematics and mathematical logic] [Russian] 6, Nauka, Moskow 1972, pp. 90–97. (English translation in Journal of Soviet Mathematics 6, 1976, pp. 422–428.)

[27] — *Indexed systems of sequents and cut-elimination*, Journal of Philosophical Logic 26 (6), 1997, pp. 671–696.

[28] Sara Negri *Proof analysis in modal logic*, Journal of Philosophical Logic 34 (5–6), 2005, pp. 507–544.

[29] — *Proof analysis in non-classical logics*, in: Costas Dimitracopoulos, Ludomir Newelski, Dag Normann, and John Steel (eds.) Logic Colloquium '05 (Proceedings of the Annual European

Summer Meeting of the Association for Symbolic Logic) [*ASL Lecture Notes in Logic 28*], 2007, pp. 107–128.

[30] Sara Negri, and Giorgio Sbardolini *Proof analysis for Lewis counterfactuals*, The Review of Symbolic Logic 9 (1), 2016, pp. 44–75.

[31] Sara Negri, and Jan von Plato Structural Proof Theory, Cambridge University Press, 2001.

[32] — Proof Analysis, *A contribution to Hilbert's Last Problem*, Cambridge University Press, 2014.

[33] Francesco Paoli Substructural Logics, *A Primer*, Springer, 2002.

[34] Graham Priest An Introduction to Non-Classical Logic: From If to Is, Cambridge University Press, 2008 (second edition).

[35] Graham Priest, and Richard Sylvan *Simplified semantics for basic relevant logics*, Journal of Philosophical Logic 21 (2), 1992, pp. 217–232.

[36] Stephen Read Relevant Logic, *A Philosophical Examination of Inference*, Wiley-Blackwell, Oxford 1988.

[37] Greg Restall *Simplified semantics for relevant logics (and some of their rivals)*, Journal of Philosophical Logic 22 (5), 1993, pp. 481–511.

[38] — *Displaying and deciding substructural logics 1: Logics with contraposition*, Journal of Philosophical Logic 27 (2), 1998, pp. 179–216.

[39] — An Introduction to Substructural Logics, Routledge, 2000.

[40] Greg Restall, and Tony Roy *On permutation in simplified semantics*, Journal of Philosophical Logic 38 (3), 2009, pp. 333–341

[41] Richard Routley, and Robert K. Meyer *The Semantics of Entailment*, in: Hughes Leblanc (ed.) Truth, Syntax, and Modality (Proceedings Of The Temple University Conference on Alternative Semantlcs), North-Holland Publishing Company, Amsterdam 1973, pp. 199–243.

[42] Richard Routley, Val Plumwood, Robert K. Meyer, and Ross T. Brady Relevant Logics and Their Rivals 1, Ridgeview, Atascadero CA 1982.

[43] Anna Troelstra, and Helmut Schwichtenberg Basic Proof Theory, Cambridge University Press, 2000 (second edition).

[44] Luca Viganò Labelled Non-Classical Logics, Kluwer Academic Publishers, 2000

Sobociński's Nachlaß

V. Frederick Rickey
Emeritus Professsor of Mathematics
United States Military Academy
fred.rickey@me.com

1 Introduction

In the last two decades there has been a resurgence of interest in the logical systems of Stanisław Leśniewski. As is well known, when he died in 1939, his most knowledgeable student, Bolesław Sobociński (1906–1980), took charge of his papers, but they were destroyed early in World War II. Sobociński made an attempt to reconstruct some of them, but that work also perished in the War. Sobociński's reconstructive work continued after he escaped from Poland and he continued to do research on the Leśniewskian systems. There is a sizable amount of unpublished information about Leśniewski's systems in Sobociński's papers and our purpose here is to describe some of it.

When Sobociński died in 1980, I purchased his library from his wife, but she kept his papers. It was not until September of 1996 that I received an urgent email from Abraham Goetz (1926–2017), a Notre Dame Mathematics faculty member and friend of Sobociński, informing me that she had moved into a nursing home and was "extremely worried" that his papers would be discarded. She asked if I could "help in getting them to people or institutions who might be interested in them." Understanding the urgency of the situation I immediately came to South Bend, boxed up the papers and took them home with me. They consisted of about 20 boxes. They sat in my library until 2017 until two things prompted me to look at them carefully: John Derwent, a longtime Notre Dame mathematics professor, sent me a copy of Kordula Świętorzecka's paper "Bolesław Sobociński. The ace of the second generation of the LWS" [224] and Bob Clay, Sobociński's first PhD student, came to visit me. Together we looked quickly through the boxes, judging what was

---

Appendix II – documenting the students of Bolesław Sobociński – and the appended bibliography – including the relevant secondary literature concerning him and his work – have been supplied by the Editor.

most important. The paper of Świętorzecka motivated me to write my memoirs "Professor Sobociński and Logic at Notre Dame."[1]

Sadly, Sobociński's papers were in quite a mess. He was too overwhelmed with editing, teaching, and research to organize his files. Some things were in reasonable order, such as his class notes. His habit was to number the pages used in his lectures and to place the papers in a file in reverse chronological order, so page 1 was on the bottom, page 2 on top of that, etc. (I have reversed that order). In almost every one of these files, there was extraneous material that he acquired while teaching: departmental memos, book advertisements, correspondence — all manner of things. The folders were usually labeled with the title of the course — in Polish — and the semester and year. There was not a great deal of expository material in the files, just notes about what to say; as an expert on Leśniewski's logical systems, he did not need write out this detail in his notes. But the systematic development of the course — theses and proofs — was written out in detail. Most of the text was in Polish. I consider these class notes to be the most important part of Sobociński's *Nachlaß* for they are likely to prompt further research.

This paper is not intended to be a detailed inventory of what is in Sobociński's papers, but just a description of items which will be the most interesting to researchers or historians. In 2017, I deposited 6.5 linear feet of Sobociński's papers in the archive at the University of Notre Dame. The archivists have called these the "Sobociński-Rickey" papers because they include information dealing with the transfer of the editorship of the Notre Dame Journal of Formal Logic from Sobociński to Rickey. My inventory of these papers is given as an appendix to this paper and the archivists catalog of the papers is available at https://lesniewski.info. Anyone wanting to visit the archive should consult both of these lists and contact the archivists at Notre Dame: archives@nd.edu . I cannot supply more information about what is mentioned in these lists without making a trip to Notre Dame and examining the files. When the remainder of the papers are deposited in the archive a similar inventory will be posted. Naturally it will contain items not mentioned here.

The arrangement here is to list items by the cities where he worked: Warsaw, Brussels, Saint Paul, Notre Dame. Items that I am unable to place chronologically are at the end.

[1]Now published in: M. Zack, and D. Schlimm (eds.) Research in History and Philosophy of Mathematics (*Proceedings of the Canadian Society for History and Philosophy of Mathematics/ Société canadienne d'histoire et de philosophie des mathématiques*, 2019), Birkhäuser [Springer International Publishing], Basel 2020, pp. 1–24. Cf. [199].

2 Escape from Poland

When Leśniewski died right before the beginning of WW II, his papers were entrusted to Sobociński.[2] Sadly they were destroyed when Gemany invaded Poland in 1939. Sobociński reconstructed some of the work — in a thousand page manuscript, as tradition has it — but it was destroyed in 1944. One would think that nothing was saved, but a few things survive in Sobociński's Nachlaß.

As far as I am aware, Sobociński did not bring any manuscripts with him when he walked out of Poland. His Nachlaß contains a number of reprints of his papers, but it is impossible to ascertain when he acquired them, so we enumerate them here.

2.1 *LOGIKA MATEMATYCZNA*

This small unpaginated notebook on squared paper bearing the above title, is the only manuscript in Sobociński's Nachlaß that was most likely written in Poland during WWII. The evidence for this is that on page 2 there are two references:

prof. Jan Łukasiewicz— "Elementy logiki matematycznej"

doc. Bolesław Sobociński — wykłady rok 1943–1944

The second of these means "docent Bolesław Sobociński — lectures from the year 1943–1944." The first could simply mean that Łukasiewicz's 1929 book was a reference, rather than that he was one of the lecturers (but the arrangement of the document makes me doubt this) But the second definitely indicates that Sobociński was involved. This is clear evidence that he did some teaching in the underground university. In September 1946, Sobociński secretly left Poland, together with Ewa Wrześniewska, his future wife.

These notes are so carefully written that they were almost certainly recopied from rough notes taken during lectures. This may have been done after the war, but the date clearly indicates that the lectures were given in 1943–1944.

After some introductory material, there is a long section, III, on propositional logic consisting of 140 theses.[3] These compare very closely with Łukasiewicz's book, including expository remarks. Section III.a has proofs of consistency, independence, and completeness of the axiom system as in Łukasiewicz's book. Section IV deals with the theory of deduction with quantifiers (4 pages). However, Peirce's quantifier, Π, has been replaced by Leśniewski's square brackets.

[2]For biographical information about Sobociński, see Świętorzecka [224] and Rickey [199].

[3]This is where the first (1929) Polish edition of Łukasiewicz's book ends. Theses 62–65 are as in the first edition, not the replacements Słupecki inserted in the second (1958) Polish edition. There are a few corrections made with Sobociński's dull pencil as in his two Mereology Notebooks mentioned in §3.2. This is evidence that he did not acquire this notebook until he was in Belgium.

Section V deals with Protothetic (13 pages). After discussing the deductive rules of the system, he gives a four axioms base of his own, which seem not to have published before:

A1. $[pqr] : . p \equiv q . \equiv : r \equiv q . \equiv . p \equiv r$

A2. $[pq] : . p \equiv q . \equiv : [f] . f(pp) . \equiv . f(qq)$

A3. $[pq] : . p \equiv q . \equiv : [f] : f(pp) \equiv f(qq) . \equiv . p \equiv q$

A4. $[pf] :: f(p, [q].q) . \equiv : . f(p, [q].q . \equiv [q].q) . \equiv [r] : f(p, [q].q)$

Several pages of deductions follow.

Section VI deals with Ontology (49 pages). There is a nice discussion of the fundamentals, the development of numerous theses, and several examples of higher epsilons. The last section, VII, deals with the antinomies (2 pages).

Unfortunately I cannot read the colophon which includes the name of the individual who wrote these notes:

... stud. III r. ... U. W. 4/X 1945r.

3 Belgium

3.1 Three notebooks on Leśniewski's Ontology

Sobociński was a research member of the Polish Scientific Institute (L'Institut d'Études Polonaises) in Belgium, 1946–1949. When he arrived in Brussels he set about reconstructing portions of Leśniewski's work. Perhaps the most important things he prepared at this time were three notebooks dealing with Ontology. The first, "Ontology Notebook 1"⋆[4] is dated "Bruksella, 5.II.1947" and consists of deductions from 20 definitions. There are 163 pages of text. The paper is a grid of squares which are 0.5 cm on a side. Sobociński often used such paper when presenting deductions for it made it easy to keep the columns straight. This is an exceptionally important set of notes, for it is the earliest surviving set of notes in his Nachlaß dealing with Leśniewski's Ontology.

"Ontology Notebook 2"⋆ is a continuation of Notebook 1. The format is the same and the numbering of theses continues that in Notebook 1. The theses are numbered, not the pages; there are 155 pages. At the end there are several pages of scrap work.

"Ontology Notebook 3"⋆ deals with three topics:

1. Ontology: There are 39 numbered pages of text. The last page is dated 1965 and has the definition of a higher epsilon. From the handwriting and other items

[4]A star, ⋆, indicates that there is a copy of this work available at https://lesniewski.info .

in this notebook, it is clear that this is the only page from 1965. The rest of it was written in Brussels.

The text is divided into 12 sections. It consists primarily of definitions. The definitions involve higher semantical categories than the definitions in Ontology Notebooks 1 and 2. These texts are useful for they reveal the style of parentheses that Sobociński, and probably Leśniewski, used to keep track of semantical categories.

2. Notes on Carnap's Abriss der Logistik (1929). 35 unnumbered pages.

3. An analysis of Russell's Antinomie. 22 unnumbered pages. The last page is dated "Bruskella, 10.IX.1948." The deductions here follow the early deductions in Sobociński's paper "L'Analyse de l'antinomie Russellienne par Leśniewski," which was submitted on 15 December 1948 to Methodos. There is no text and there are differences in the deductions, so this will require further study.

3.2 Two notebooks on Mereology

During the years 1927–1931, Leśniewski published a long paper entitled "O podstawach matematyki" (On the foundations of mathematics, briefly OPM), in XI chapters, containing several versions of his Mereology and a final chapter on Ontology. At the time of publication Leśniewski was not comfortable with formal notation and so expressed the theses and derivations in ordinary (Polish) language. In these notebooks Sobociński has translated these deductions into symbolic form. Such a translation has never been published.

"Mereology OPM Notebook 1"⋆ contains translations of Chapters IV, VI, VII, VIII, IX, and X of OPM. Chapters I–III contain no deductions so there was no need to convert them to symbolic form. One photocopy of the original that Sobociński owned lacked Chapter V, probably because the library from which he obtained this copy lacked volume 32 of Przegląd Filozoficzny. His second copy of the original is complete and so Chapter V is in "Mereology OPM Notebook 2"⋆.

These chapters present versions of Mereology with different primitive terms and different axiom systems, so these two notebooks will be very useful to someone doing a careful study of the development of Mereology. On the other hand, the photocopies themselves are of limited interest because the Polish original has been reprinted by Jacek J. Jadacki in Pisma zebrane (2015) and an English translation is in STANISŁAW LEŚNIEWSKI: Collected Works.[5]

[5]Some caution needs to be used with the English Collected Works, for the page numbers in that work refer to the original papers, not to this edition. The page numbers in the footnotes can be particularly confusing. Jadacki's edition solves this problem even for people who do not read Polish.

3.3 Draft of a book on Mereology

This undated manuscript⋆ consists of 91 pages of well organized notes on squared paper. There are seven sections, an introductory one and then six sections for theses involving defined terms: Klass (59 theses), part (24), collection (59), outside (49), Universe (25). The manuscript breaks off after the definition of discrete. I have three reasons for believing that this is a draft manuscript for the book on Mereology that he promised North Holland: (1) There is a reasonable number of comments in the ms, something that is lacking in many later ms; (2) It is done in pencil on squared paper and the handwriting resembles other manuscripts done in Belgium: and (3) The theses are numbered decimally (this is unique to this manuscript).

3.4 Axiom Systems for PC and Protothetic

This small notebook of "Axiom Systems"⋆, with a brown paper cover, has axioms systems for various propositional calculus and a list of axioms for Protothetic. Sobociński likely prepared this as a handy record so that these things did not need to be looked up when needed. The paper has half-centimeter squares, indicating a European origin.

3.5 Publications from this period

1949a "An investigation of protothetic,"⋆ Cahiers de l'Institut d'Études Polonaises en Belgique, no. 5. Polycopié. Brussels 1949, v + 44 pp. Reviewed by Alonzo Church, JSL, 15 (1950), 64. Sobociński 1967b, in Polish Logic 1920–1931, is a new English translation from the Polish by Z. Jordan. A quick glance at the original will show that the English in the first six pages of Sobociński 1988 has been cleaned up considerably; in fact it is a new translation. The information about the fate of Collecteana Logica has not been reprinted. There is another reprint in 1998 in Leśniewski's Systems: Protothetic but the original is still of interest for it contains a x page discussion of the journal Collectanea Logica. The mimeographed offprint is quite hard to read.

1949b "L'analyse de l'antinomie russelienne par Leśniewski," Methodos, vol. 1 (1949), 94–107, 220–228, 308–316, vol. 2 (1950), 237–257. Reviewed by Prior, JSL, 18 (1953), 331–333. Sobociński 1984 is an English translation.

Both Professor and Mrs. Sobociński obtained a passport while in Belgium and they are now in the archives at Notre Dame (and pictured in Rickey [199].) Where precisely they went is unclear, but they visited Bocheński in Switzerland and also visited Paris.

4 Saint Paul, Minnesota

Sobociński's first job in the United States was teaching philosophy at the College of Saint Thomas in Minneapolis, a position that he only held for the spring semester of 1950. While there he presented two lectures on "Leśniewski's Foundations of Mathemaitics".* The manuscript is untitled, so I have added the title. It is very likely that Marian W. Heitzman[6] helped Sobociński with the English. These are quite interesting from an expository point of view.

A second set of three lectures deals with the history of logic.* They display considerable erudition on Sobociński's part.

4.1 Two lectures on "Leśniewski's Foundations of Mathematics"

There is no title on the ms, so I have added the above title. Part one is pp. 1–11; Part II is pp. 12–26 plus one two-sided page.

This is most likely a set of two lectures given at the College of Saint Thomas in the Spring of 1950.

4.2 Philosophy Seminar, March 10, 1950

This is in three parts: part I, pp. 1–10; part II, pp. 11–21; part III, 5 single spaced unnumbered pages.

The last page bears the date 5-13-50, but it has been crossed out in pencil.

These deal with the history of logic. They display considerable erudition on Sobociński's part.

4.3 Journal of Computing Systems

From 1951 to 1956 Sobociński served as director of research at the Institute of Applied Logic in Saint Paul, Minnesota. The Director of the Institute was John Goodell. Together they founded the The Journal of Computing Systems, the first number of which appeared in June 1952. There were only four issues. The Institute closed for lack of funds.

The authors who published in this short-lived journal were some of Sobociński's connections: Jan Łukasiewicz, Alan Rose, Carew A. Meredith, A. N. Prior, Alan

[6]Heitzman (1899–1964) was a faculty member at the College who knew Sobociński from Poland and likely helped Sobociński obtain a position there. He is the author of "The philosophical foundations of the Aristotelian logic and the origin of the syllogism," Proceedings of the American Catholic Philosophical Association, vol. 29 (1954), pp 131–142. Reviewed by E. J. Lemmon, The Journal of Symbolic Logic, 21, p. 389.

Ross Anderson, and Desmond Paul Henry. Of the 24 papers published in the Journal, four were by Sobociński. One of these, "On a universal decision element," appears to break new research ground for Sobociński, but on closer examination, it too deals with propositional logic. It investigates logic gates and the simplest ways to construct them from given gates.

The last, "Axiomatization of the conjunctive-negative calculus of propositions," was a reconstruction from memory of a paper originally "published" in the ill-fated journal Collectana Logica in 1939. A folder in his papers contains a typescript of the opening pages followed by pp. 18–44 of the same manuscript (the intervening pages are missing). These pages were never published (because of space limitations in the JCS?). They deal with connections with Heyting's intuitionistic calculus and Johannson's minimal calculus.

4.4 Table of Contents of the Journal of Computing Systems

We have retained the same numbering of the articles as in the journal.

4.4.1 Volume 1, number 1, June, 1952

1. John D. Goodell, "The foundations of computing machinery," pp. 1–13.
2. Tenny Lode, "The realization of a universal decision element," pp. 14–22.
3. Bolesław Sobociński, "Axiomatization of a partial system of three-value calculus of propositions," pp. 23–55.

4.4.2 Volume 1, number 2, January 1953

4. Edward C. Varnum, "Polynomial determination in a field of integers modulo p," pp. 57–70.
5. Bolesław Sobociński, "On a universal decision element," pp. 71–80.
6. Michael J. Norris, "Cofinally concentrated directed systems," pp. 81–85.
7. John D. Goodell, "The foundations of computing machinery, Part II," pp. 86–110.

4.4.3 Volume 1, number 3, January 1953

8. Jan Łukasiewicz, "A system of modal logic," pp. 111–149.
9. Norman M. Martin, "On completeness of decision element sets," pp. 150–154.
10. Carew A. Meredith, "Single axioms for the systems (C, N), (C, O) and (A, N) of the two-valued propositional calculus," pp. 155–164. Reviewed by Church, JSL, 19 (1954), 143.

11. Alan Rose, "A formalization of Sobociński's three-valued implicational propositional calculus," pp. 165–168. Reviewed by Gene Rose, JSL, 19 (1954), 144.
12. Carew A. Meredith, "A single axion of positive logic," pp. 169–170. Reviewed by Church, JSL, 19 (1954), 144.
13. Bolesław Sobociński, "Note on a model system of Feys-Von Wright," pp. 171–178.
14. W. C. Carter and A. S. Rettig, "Analytic minimization methods I: Conjunctive forms," pp. 179–195.
15. John D. Goodell, "Notes on decision element systems using various practical techniques," pp. 196–199. Reviewed by Church, JSL, 19 (1954), 143.

4.4.4 Volume 1, number 4, January 1953

16. A. N. Prior, "The interpretation of two systems of modal logic," pp. 201–208.
17. Alan Ross Anderson, "On the interpretation of a modal system of Łukasiewicz", pp. 209–210.
18. Alan Ross Anderson, "On alternative formulations of a modal system of Feys-von Wright," pp. 211–212.
19. Jan Łukasiewicz, "Arithmetic and modal logic," pp. 213–219.
20. Norman M. Martin, "Note on completeness of decision elements sets," p. 220.
21. D. P. Henry, "Expressions trivially decidable," pp. 221–224.
22. Ryóichi Takekuma, "On a nine-valued propositional calculus," pp. 225–228.
23. Bolesław Sobociński, "Axiomatization of a conjunctive-negative calculus of propositions," pp. 229–242.
24. John D. Goodell, "The relations between logical, mathematical and computing machine systems," pp. 243–254.

5 Notre Dame Journal of Formal Logic

Sometime in January 1959, Sobociński visited Fr. Paul Beichner, Dean of the Graduate School, arguing that he had a number of important papers in symbolic logic, some of which had been finished for years, and about which people had asked him. But the philosophy journals had so much material that publication would be delayed for years. The issue of an annual volume had come up several times before, and all agreed that it was important to have a publication that "comes out exclusively under our own name." Sobociński argued that it would make the university better known.

There is nothing in Sobociński's papers about the founding of the Journal. Whatever material that is in the University of Notre Dame archives is restricted.

6 Sobociński's Lecture Notes

These fall into two broad categories, his holograph notes that he used in class (which are primarily in Polish), and the notes that students took in those classes. We will consider them together, for they deal with the same topics.

This is intended to be complete list of courses that Sobociński taught while at Notre Dame. Information about the courses professors taught is available at http://archives.nd.edu/courses/.

1956 Fall: Symbolic Logic, Philosophy 111.

1957 Spring:[7]

1957 Fall: Symbolic Logic, Philosophy 111. This course dealt with Leśniewski's Ontology. John T. Kearns (1936–) was a student; his notes are reproduced as an appendix in his 1962 Yale PhD Dissertation, Leśniewski, Language, and Logic. His paper of the same name, NDJFL, 8 (1967), 61–93 does not have this appendix. [Email from John Kearns, August 17, 2018].[8]

1957 Fall: Modal Logic, Philosophy 251.[9]

In a small pamphlet on *Philosophy at Notre Dame* this course is described as "A discussion of modal logic in the context of contemporary symbolic logic centering on Aristotelian modalities and the notion of strict implication. This course is open only to students who have taken introductory courses in symbolic logic. Others must receive permission of the Head of the Department of Philosophy or Mathematics. Offered in the Fall semester. Three credits."

1958 Spring: Symbolic Logic, Philosophy 111.

From the same pamphlet: "This course will consist of a presentation and discussion of the elementary properties of relations, the fundamental notions and principle theorems in this area." The same restrictions as in Modal logic are included.

[7]The Notre Dame Archives has no Schedule of Courses for this academic year, so what Sobociński taught is unknown. What he taught in the fall is known from a letter from Fr. Herman Reith, Head of the Philosophy Department, June 12 1956.

[8]The computer generated "Preliminary Class List" for this class survives in Sobociński's Nachlass. There were 14 students, mostly sophomores. Besides Kearns, a senior philosophy major, they included Joseph A. Buckley, a graduate student in philosophy, both of whom went on to have careers as philosophy professors.

[9]These two courses are listed in the Fall 1957 Revised Schedule of Courses: "Math. Logic" (Math 291) and "Found. of set Theo." (Math 411) are also listed. No instructor is named for any of these courses.

1958 Spring: Calculus of Relations, Philosophy 242.[10] From the same pamphlet: "This course will consist of a presentation and discussion of the elementary properties of relations, the fundamental notions and principle theorems in this area." The same restrictions as in Modal logic are included.

1958 Fall: Modal Logic, Philosophy 251.

1958 Fall: Mereology Seminar. Philosophy 255. Bob Clay has notes from this course in a brown binder. The label on the front reads "Sobociński's Courses: Protothetic, Ontology, Relations, Mereology." For convenience I have split them into three folders: Clay 1: Protothetic; Clay 2: Ontology; Clay 3: Mereology.

The Protothetic section consists of 36 pages of notes (Rickey's numbering).

The Ontology section is labeled "Calculus of Name" and consists of 76 pages of notes, again in Rickey's numbering. They begin with the long axiom. There are 146 definitions and 520 theses. There are 8 pages of preliminary material, including a discussion of Leśniewski's solution of the Russell antinomie.

There are no notes on relations.

The Mereology section has 96 pages of notes. There are 16 definitions and 308 theses.

1959 Fall: Logic of Names, Philosophy 261. The folder bears the title "Advanced Calculus of Names" and the date "I s. 1959/60," i.e., first (fall) semester 1959/1960. There are 79 pages of notes, starting from the short axiom of Ontology. There are numerous miscellaneous notes, some appear to deal with the Schröder-Bernstein Theorem. The backs of some pages come from drafts of his papers (which was not uncommon).

Bob Clay has notes from this course.

1960 Fall: Symbolic Logic. Philosophy 111. Course taken by Ed Siegfried. As a student, Rickey copied them into one notebook, 96 pages. It begins with propositional logic, has some quantification theory, and then a section on Aristotelian Logic.

1961 Spring: Calculus of Names. Philosophy 258. Sobociński's folder of classnotes is entitled "Foundations of the Calculus of Names" and is dated "II

[10]The Spring 1958 Revised Schedule of Courses lists Sobociński as the "instructor in these two courses." A "Seminar in Logic" (Math 410) is announced as a graduate course in the mathematics department, but with no instructor listed. "Found of Set Theory" (Math 412) is another graduate course with Skolem as instructor.

Sem, 1960/61." There are 69 pages of notes and some miscellaneous papers. The deductions start from the short axiom of ontology.

1961 Fall: (Abstract) Calculus of Relations, Philosophy 265. F. Thomas Farrell, now a well known topologist, took this course and Rickey made a copy of his notes (by hand, as this was before photocopies were widely available). Two notebooks, 145 pages.

Sobociński has a file labeled "Teorja Stosunków II" (Theory of Relations II). I have not tried to match it up with Farrell's notes, but this appears to be working notes, before a good copy was prepared.

Seemingly related to this is a folder labelled "Wybrane twierdzenie z Ontologji" (Chosen theses of Ontology). The first page is entitled "Ogólna zasada abstrakiji" (general principle of abstraction) These notes start with a reference to Principia Mathematica *72·66. Perhaps this material was prepared to use in the Calculus of Relations course, but it is not included in Farrell's notes. Further study will be needed to determine precisely what Sobociński was doing in these notes.

1962 Spring: Symbolic Logic II, Philosophy 112. This rather tattered folder bears the title "Elementarna ontologja," i.e., Elementary Ontology. The file contains 24 numbered pages of definitions, theses, and proofs. There is one page containing ten questions in English which were undoubtedly for the final oral exam.[11]

This is the first course taught by Sobociński that Rickey attended. He was then an undergraduate in his Junior (third) year. He took 232 pages of notes in three notebooks.

1962 Fall: Mereology, Philosophy 267. Rickey has 271 pages of notes in 3 notebooks.

1963 Spring: Foundations of Logic, Philosophy 268. Sobociński's file, "Foundations of Logic. 1962/63, II semester" is quite thin.

Rickey has 378 pages of notes in four notebooks. In addition he has two notebooks (in a different style), 209 pages for the same course.

[11]There are several other items in the folder: A sheet listing Sobociński's classes for the Spring Semester of 1962. Besides Phil 112 it includes Phil 266 and Phil 270. On the back the eight students in the class have written their names: Bill Todd, Tom Schwartzbauer, Fred Rickey, Vladimir Drobot, Leonard Hauer, C. Edward Emmer, Thomas A. McCartly, and Frederick N. Sringsteel.

There is a mimeographed sheet announcing a Department of Mathematics Colloquium by Hugo Steinhaus, of Wroclaw University, who was a short term visitor. His topic for Friday April 21, 1961, was "An alternative for the axiom of choice."

Three pages of notes related to the Schöder-Bernstein Theorem, including a diagram (a rarity in Sobociński's notes).

1963 Fall: Metalogic and Metamath, Philosophy 271. This is the first time Sobociński taught this course. Rickey's notes consist of 378 pages of notes in 4 notebooks. A table of contents is on cover 2 of the first notebook.

NB: When I put these in the box of classnotes I could only find notebooks 2 and 3. The third has a bluebook in it.

1964 Spring: Calculus of Names, Philosophy 273. Advanced ontology. Rickey's notes consist of 284 holograph pages in 3 notebooks.

1964 Fall: Logic Classification of Concepts, Philosophy 277. Sobociński's folder for this course is marked "Klasyfikacja logiczna pojęć" and contains a two page outline (which is unusual) and 121 pages of notes.

Rickey has 276 pages of notes in three notebooks.

1965 Spring: Advanced Calculus of Propositions, Philosophy 278. Rickey's notes are in 3 notebooks, 296 pages.

1965 Fall: Foundations of Mathematics, Philosophy 294.

1966 Spring: Combinatory Logic, Philosophy 295. The folder with Sobociński's notes is about half an inch thick, but the pages are not numbered so it will take some effort to sort this out, especially as I cannot find my notes for the course.

1966 Summer: Set Theory, Mathematics 230. Sobociński's folder for this class notes indicates that he taught this course in the summers of 1959, 1960, 1961, ... , 1967, for a total of nine times. Quite a few of the pages are marked "1960", indicating that he added material after the first year. The course was designed for high school teachers that were enrolled in a NSF funded Master's program.

The only set of notes that I have for the course were taken by Sister Mary Treanor. She has 93 pages of notes. Ramesh K. Miglani was the Teaching Assistant. He received his PhD in algebra in 1971.

1966 Fall: Cardinal and Ordinal Numbers, Philosophy 389. Rickey took 164 pages of notes on loose sheets of paper. This course is partially based on Ontology. These is another set of notes from this course taken by Sister Mary Treanor, 136 pages.

1966 Fall: Logic Seminar. Rickey has a file of notes from the seminar for 1966–1967. Dr. Clay discussed a system strictly weaker than Mereology (7 pages). He also spoke about partially ordered sets and latices (6 pages) and the Relation of

Boolean Algebra to Mereology (4 pages). Jack Canty spoke on Leśniewski's Ontology and Gödel's Incompleteness Theorem, an abstract of his thesis (2 pages).

1967 Spring: Theory of Models, Philosophy 290. Father George R. Edmonstone, S.J. attended this course and took notes. Sister Mary Treanor photocopied some of his notes (pp. 36–159). This is followed by 99 pages of notes plus 24 unnumbered pages by Treanor that appear to be from the beginning of the course. Next are 28 pages of typed notes prepared by Charles Quinn. The first page is dated February 10, 1967. These notes are in a binder which I will call "Treanor Binder I."

1967 Spring: Logic Seminar. Rickey has a file of notes from the seminar for 1966-1967. Sobociński spoke about Boolean rings (6 pages) and Boolean Algebras (8 pages). He also suggested a problem in modal logic. Thomas Scharle discussed connections between free logic and elementary ontology (5 pages). Theodore Sullivan, a PhD student of Clay, spoke about Affine Geometry (9 pages). Clay presented sole axioms for partially ordered sets (6 pages).

1967 Fall: Theory of Relations, Philosophy 277. 91 pages of notes. Carefully written by an unknown person.

1967 Fall. Seminar in Symbolic Logic. Rickey has a file of notes from the seminar for 1967–1968. Sobociński discussed a "Completely formalized systems of ancestral relations"⋆. Rickey took 20 pages of notes.

1968 Spring: Boolean Algebras, Philosophy 384. These notes were taken by Sister Mary Treanor. She did not attend the first few lectures so there are only 110 pages of notes.

1968 Summer: Logic (taught in the Mathematics department).

1968 Fall: Deductive Theories, Philosophy 386.

1968 Spring: Logic Seminar. Rickey has a file of notes from the seminar for 1967–1968.

1969 Spring: Advanced Logic, Philosophy 388.
Luis M. Laita took notes.

1969 Spring: Logic Seminar, 482[12]

[12] Also in the Spring of 1969, B. Pahi taught Indian Philosophy (Phil 83) and Symbolic Logic (Phil 132). Canty taught Systems of Natural Deduction (Phil 288).

1969 Spring: Metalogical Set Theory, Philosophy 389. 72 pages of notes in Treanor Binder I.

1969 Fall: Ordinal Numbers, Philosophy 387. Sobociński's folder for the course is labeled "Liczby porządkowe. I semester 1969/70." There are 65 pages of notes.

Luis M. Laita (1935–2017) attended the classes and seminars in 1969–1970.

1970 Spring: Advanced Logic, Philosophy 388.

1970 Spring: Seminar in Symbolic Logic, Philosophy 482.

1970 Summer: Introduction to Symbolic Logic, Mathematics 243. Sobociński's slim folder is marked "Elementarny rachunek zdań." The class list in the folder lists 18 students; they were taught in two sections with Paul Welsh (1943–2016) as the TA. He received a PhD in 1971 under the direction of Robert Clay. The title was Primitivity in Mereology. These notes are rather disorganized. There are several lists of topics to be covered, then a bunch off numbered pages with a few missing. There are two pages of deductions of theses in propositional logic (112 to 142), but the development is not the same as in Łukasiewicz's little book.

The folder contains a letter from William Frascella (24 August 1968) indicating that he is enclosing Paul J. Cohen's paper on the independence of the Axiom of Choice. Also mentioned is Witte's false claim that he has proved the independence of set theory.

1970 Fall: Theory of Models, Philosophy 693.

1970 Fall: Logic Seminar, Philosophy 697.[13]

1971 Spring: Mereology, Philosophy 694.

1971 Spring: Seminar in Symbolic Logic, Philosophy 698N.[14]

[13]See The Fall 1970 Course Schedule. Introductory Logic (Phil 211) was taught by Fr. T. Brennan. Bishwambhar Pahi taught Indian Philosophy (Phil 259), Informal Set Theory (Phil 391), and Symbolic Logic (Phil 431). Fr. I. Thomas taught History of Medieval Logic (Phil 523). In the Mathematics Department there were two courses: "Topic Math Logic" (Math 711), and "Seminar Math Logic" (Math 671), but no instructors were listed. In the revised schedule Chapin taught Mathematical Logic (Math 412). No seminar was listed.

[14]See The Spring 1971 Course Schedule. Pahi taught Indian Philosophy (Phil 259), and Introductory Symbolic Logic (Phil 311). Chapin taught Modern Logic (Math 110) Clay taught Algebraic Structures (Math 222) and Modern Logic (Math 614).

1971 Summer: Introduction to Symbolic Logic, Mathematics 558. A course with this number, but with the title "Introduction to Mathematical Logic," is included in the "List of Courses for Teacher Training Institute." There were two sections and the Problem Seminar was conducted by Huei-shyong Lue. Since he received his PhD in 1974 in Differential Geometry, he could have been the TA in a later year.

1971 Fall: Modal Logic, Philosophy 695.

1971 Fall: Seminar in Symbolic Logic, Philosophy 697.[15]

1972 Spring: Algebraic Logic, Philosophy 696. Spiral bound document on Algebraic Logic (in Spanish) together with other notes prepared by Luis María Laita who worked under Sobociński's supervision for six years before transferring to the Department of History. His 1976 Notre Dame PhD Dissertation, directed by Michael Crowe, is entitled A Study of the Genesis of Boolean Logic. He was first person to receive a doctorate from the University of Notre Dame with training in the history and philosophy of science.

1972 Fall: Logical Foundations of Mathematics, Philosophy 696. A mimeographed list of "Graduate Courses, Fall '72" describes this course as follows: "In this course the following problems will be analyzed and discussed:

1) The connections existing between mathematics and a strict description of reality. Characterization of the mathematical theories.

2) The deductive character of mathematics will be analyzed, and it will be shown that every mathematical theory sufficiently developed can be and should be axiomatized. Using some examples the different methods of aziomatization of the mathematical and other deductive theories will be presented."

1972 Fall: Seminar in Symbolic Logic, Philosophy 697. The same memo describes this as a "Weekly seminar involving the preparation and presentation of research themes."

[15]See The Fall 1971 Course Schedule. Fr. T. Brennan taught Introductory Logic (Phil 211). There were three sections of Introduction to Symbolic Logic (Phil 213); K. Goodpaster taught one and Jack Boudreaux taught two. Pahi taught two sections of Indian Philosophy (Phil 359) and one of Intermediate Symbolic Logic (Phil 431). Vučković taught Recursive Functions (Math 625). Rickey, who was a visitor in 1971–1972, taught Principles of Math (Math 103) and Finite Mathematics (Math 104). Chapin taught Honors Calculus III (Math 265) and Complex Analysis (Math 465).

1972 Spring: Seminar in Symbolic Logic, Philosophy 698.[16]

1973 Spring: Proof Theory, Philosophy 694.

1973 Spring: Seminar in Symbolic Logic, Philosophy 698.[17]

1973 Fall: Introduction to Metamathematics, Philosophy 691.

1973 Fall: Seminar in Symbolic Logic, Philosophy 697.[18]

1974 Spring: Philosophy 698.[19]

1974 Spring: Foundations of Propositional Calculi, Philosophy 692.

1974 Fall: Philosophical Foundations of Logic, Philosophy 693. In a February 8, 1974 letter to Cornelius F. Delaney, Chairman of the Department of Philosophy, Sobociński proposed teaching the following course in the fall semester of 1974: "Philosophical and formal foundations of logic. The philosophical origin of logic in antiquity. Its influence on the further development of formal logic. The traditional formal and material logics. The platonic and nominalistic approaches to logic. Mixture of formal logic with he semantical and philosophical problems in contemporary philosophical logic. Symbolic logic and the problem of the antinomies. The theory of types of Russell and its further developments. The pre-logical assumptions in the logical systems. Logic and the problem of existence. The philosophically neutral systems of logic and their structure. The free logics and their difficulties.

[16]See The Spring 1972 Course Schedule. There were four sections of "Int to symb logic" (Phil 213); one taught by K. Goodpaster, one by T. Tumulty, and two by John Robinson. There was no instructor listed for Indian Philosophy (Phil 359). Pahi taught Intermediate Symbolic Logic (Phil 432) and "Theory Propos Calc" (Phil 593). Chapin taught "Elementary Sanskrit II" (Classics 501), Calculus IV (Math 226), and Honors Calculus IV (Math 266). Vučković taught Calc II Enriched (Math 136) and Recursive Functions (Math 626). Rickey taught Elements of Calculus I (Math 105) and Modern Logic (Math 110).

[17]See The Spring 1973 Course Schedule. During this term Introduction to Symbolic Logic was taught by P. Tumulty (Phil 213), B. Pahi taught "Indian Philosophy" (Phil 359), "Intermediate Symbolic Logic" (Phil 432), and "Deontic Logic," (Phil 692), and Guido Küng taught "Phenomenology" (Phil 444). In the mathematics department "Basic Modern Logic" (Math 613) was taught by E. William Chapin, PhD Princeton, 1969. His advisors were Alonzo Church and Simon Kochen.

[18]See The Fall 1973 Course Schedule. During this term B. Pahi taught two sections of Introduction to Symbolic Logic (Phil 213) as well as Intermediate Symbolic Logic (Phil 431). Vladeta Vučković taught Recursive Functions (Math 625). Chapin taught "Applied Math" (Math 589).

[19]See The Spring 1974 Course Schedule. C. Davis taught two sections of Introduction to Symbolic Logic (Phil 213) as well as "Philos Logic" (Phil 432). Charles C. Davis, Jr., received his PhD under Sobociński's direction in 1973. In the mathematics department, Chapin taught Calculus III (Math 225), Vučković taught "Calc IV Enriched" (Math 236) and recursive functions (Math 626).

Fall semester. Open to students interested in logic, its philosophy and applications. No prerequisite mathematical courses are required."

1974 Fall. Seminar in Symbolic Logic.

1975 Spring: Theory of Definitions, Philosophy 694. In a February 8, 1974 letter to Cornelius F. Delaney Sobociński proposed teaching the following course in the spring semester of 1975: "The theory of definitions. The traditional forms of definitions. The real and nominal definitions. The formal definitions. The primitive and defined notions in the field of formal systems. The predicative and unpredicative definitions. The creative definitions. The rules of definitions. Recursive definitions and their applications. Elimination of recursive definitions by the method of Frege. Syntactical and semantical definability. An introduction to the theory of classifications of the logical notions. Three credits. Spring Semester. Open to students with an interest in logic. No prerequisite mathematical courses are required."

1975 Spring. Seminar in Symbolic Logic. Philosophy 698.[20]

1975 Fall: Sobociński is not listed in the The Fall 1975 Course Schedule. So his last year of teaching was 1974–1975.

7 Correspondance

With some exceptions Sobociński did not keep copies of letters he sent, either research or professional. There are letters from Lejewski, Prior, and Hiż, but few others. He did not keep copies of the letters he sent them. To carefully study the correspondence, it will be necessary to visit their archives.

Sobociński in Minneapolis. The letters in this file primarily concern his futile search for a job. There is a two page typed "Life Abstract" that notes that he knew French, German, Russian, and Polish. From 1934–37 he was "Associate Professor at the Chair of Philosophy of Mathematics at Warsaw University." Clearly he has Americanized the title. From 1937–45 he was "Assistant Professor of Mathematics and Logics at Warsaw University. During the war conducted Underground University courses in Mathematics and Logics." In 1945–46 he was "Professor of Theory

[20]See The Spring 1975 Course Schedule. C. Davis Introduction to Symbolic Logic (Phil 213) as well as "Intermediate Symbolic Logic" (Phil 431) and Seminar of Natural Language (Phil 593). In the mathematics department, Chapin, C. Kalicki and T. Naps taught Modern Logic (Math 110) R. Clay taught Many-Valued Logic (Math 226), Vučković taught "Honors Calculus II" (Math 166) "Honors Calculus II" (Math 166). Chapin taught two sections of Calculus IV (Math 226)

of Deductive Sciences at Lodz university, Poland." [He did receive this position, but left Poland before taking up duties there.] At the Institute of Applied Logics he was "*Consulting* on application of Logical methods to the speeding of the arithmetical operations. Methods of automatic programming and techniques of handling informations, adaptation of theory of quantifiers, protothetic and the theory of recursive publications." There is a list of 19 publications, including three that never appeared: "17. Relations between the conjunctive-negative propositional calculus, the Minimalcalculus and the system of Heyting. The Journal of Symbolic Logic. 18. The Mereology. Studies in Logic, North Holand Publishing Company, 1955 USA. 19. The Propositional Calculus, textbook. 1955 USA". There is also a three page single spaced typescript of subscribers to the Journal of Computer Systems, dated 4-1-53.

Miscellaneous editorial correpondance, 1964–1969. No logic. Authors include E. J. Ashworth, E. John Lemmon, Karl Menger, Rolf Schock. Another folder dealing with NDJFL, V, 2 (1964). Another folder for V, 3 & 4 with wide ranging correspondence dealing with the Journal and miscellaneous matters. Four photographs of individuals that I cannot identify.

A file of 11 letters in Polish and 7 typescripts in English from Juliusz Reichbach.

Letter and two offprints from Günter Frei. He asks for a letter of recommendation for full professsor at the Université Laval in Canada.

Two letters in Polish to Eva Sobociński, both from 1971, probably from the same person. One typed, from New York; the other handwritten, from Quebec.

8 Miscellaneous Notes

There is a file labeled "Logika tradycyjna" (Traditional Logic) that lists syllogisms and provides proofs. There is an unrelated memo in the file from Jane Dunkel, secretary in the Mathematics Department, dated March 8, 1971. This does not seem to go with a course from that time.

"Research Proposal Submitted to [the] Office of Naval Research, Department of the Navy," 25 pages. The Principal Investigators were Kurt Mahler (Manchester), Thoralf Skolem (Oslo) and Bolesław Sobociński (Notre Dame). Associated Investigators were J. Richard Büchi (Illinois), William Craig (Pennsylvania State) and R. P. Reader (Punjab U.). Büchi and Craig were faculty in mathematics at the time. The research topics were number theory and foundations of mathematics. The document is undated but the proposal is for 1957–58. A "critical edition with remarks and commentaries" of Leśniewski's papers is planned, with Sobociński being assisted by Rose Rand. A vita for Sobociński with 22 publications is included.

A 15 page section entitled "Sobociński's Section" appears to have been prepared for the proposal, but not included. For the past three years Sobociński has worked in Leśniewski's systems, modal logic, and the theory of relations. Sobociński has established a general method for constructing a protothetical system based on any finite-valued propositional calculus. In Ontology he has proved that every valid formula in the functional calculus of degree ω is true in Ontology, but not vice-versa. He has established a theory of atomistic Mereology. His work on modal logics and on the theory of relations is too technical to describe here. A vita and 66 item list of publications is included (as this goes through 1971, it must have been prepared later). This proposal was not funded.

An unlabeled folder that contains an incomplete (pages 6–23 only) typescript of Sobociński's paper "Concerning the postulate-system of subtractive abelian groups," NDJFL, XVI (1975), 429–444. The folder also contains 10 pages of deductions in Protothetic.

A folder containing an early draft (53 pages) of James George Kowalski's 1975 dissertation under Sobociński. It is published as "Leśniewski's ontology extended with the axiom of choice," NDJFL, XVIII (1977), 1-78.

"Many-Link Functors of Leśniewski"*. 14 page typescript. There is also a handwritten draft. Unpublished.

The Schröeder-Bernstein Theorem.* 15 pages. There are several versions of the proof. Sobociński also presented this in class on two occasions. Lejewski thinks that this is original work by Sobociński.

8.1 Partial draft of many-link functors

This is a handwritten draft, carefully written in English, of the typescript in subsection 1. Sobociński's handwriting in English is quite easy to read; this is not so for his Polish handwriting. The pages are numbered 2 through 9 inclusive. This draft is incomplete; it only goes to page 7 of the typescript. No additional pages have been identified in this file.

Page 1 was likely never prepared. Besides a title it likely contained the bibliography (a feature of Leśniewski's publications that has been discarded in reprints). This argues for an early date, before Sobociński had absorbed more modern publication standards.

There are many small changes in the typescript. They deserve a closer inspection.

8.2 Many-Link Functors of Leśniewski

This is a 14 page transcript. There are 3 copies in the folder. This was never published.

The opening paragraph gives several references, but no bibliography is attached. The reference "Sobociński$_1$, the thesis D and note 26" is the easiest to identify. This refers to his 1934 paper "O kolejnych uproszczeniach aksjomatyki "ontologji" Prof. St. Leśniewskiego." There is an English translation, 1967a, in McCall's Polish Logic: 1920–1939. The title of the translation is "Successive simplifications of the axiom-system of Leśniewski's ontology."

Because the numbering of the footnotes was changed in the translation, we are interested in note 2, page 197. The reference to "thesis D" in the typescript is a typo; both the original and the translation have "*Df* Ω."

This footnote has enabled me to give plausible references to the four papers which are mentioned.

It refers to "Leśniewski [1], p. 66, footnote 1," i.e., to Leśniewski 1929. In the 1992 English translation in Leśniewski's Collected Works, this is footnote 69 on p. 475. Leśniewski does not use the term "many-link." Sobociński's footnote gives several references to the terminological notations in Leśniewski 1929; individuals well-versed in the TEs can easily locate the references.

Leśniewski's paper on the foundations of ontology is likely another of Sobociński's references. While multi-link functors are not mentioned in the paper the TEs allow for their introduction in ontology.

I don't find any references to many-link functors in Leśniewski's 1931 paper on definitions in the propositional calculus.

8.3 Folder labelled "*Tw. Schroedera-Bernshtaina* [?] etc."

This has two versions of the proof.

8.3.1 The Rough Version

There are 15 pages. Rickey numbered the page beginning with A1 as page 0; it is questionable if this is correct. The next pages are numbered 1–14. I labeled one last sheet page 15. Between pages 3 and 4 there was a small card; I labeled it 3a and 3a verso. Between pages 11 and 12 there was a page that I labeled 11a and 11a verso. Note that page 11a is on "The Institute of Applied Logic" stationary. The same is true of pages 2–14, but that is on the verso of the photocopies.

8.3.2 The More Polished Version

This has 10 pages, numbered I.–X. at the center top by Sobociński. This material should be compared this material with classnotes. Lejewski thinks that this is original work by Sobociński [private correspondence]. Guessing by the manila folder, this work was done at ND.

8.4 More items

A file with a 1975 letter from Stanisław J. Surma sending Sobociński a photocopy of Leśniewski's 1916 paper (in Polish) "Foundations of the general theory of sets: I." There is also a 35 page unsigned translation of the paper that appears to be the same one in Leśniewski's Works by D. I. Barnett.

A file with a 1979 letter from Georges Kalinowski, who is sending his French translation of Leśniewski's 1931 chapter on Ontology (Chapter XI of OPM). This translation, *Sur les propositions "singulières" du type "A ε b"*, was published in Stanisław Leśniewski Sur les fondements de la mathématique. Fragments (discussions préalables, méréologie, ontologie), Hermes, Paris 1989, pp. 101–114. The file also contains Rose Rand's 63 page unpublished English translation of the same work.

A file of deductions that I do not recognize is included. Perhaps they deal with geometry.

A file of deductions in Mereology. Carefully written. Two series of 13 pages each. This is no later earlier than 1964.

A folder labeled "Documents about Sobociński" that Rickey gathered.

A document labeled "Publications and Research, September 1, 1959 — September 1, 1960" lists the following publications by Sobociński:

1. "The Investigations on the Heyting Propositional Calculus."
2. "A Logical Remark Concerning Certain Metaphysical assumptions."
3. "A Classical Fromalization of a model system of Feys-von Wright."
4. "A Contribution to the Axiom Systems of the Modal Logic of Lewis."
5. "Two Polish Logicians."
6. "On the Computable Systems of Protothetic."
7. "Note on the so-called Pseudo-Definitions."
8. "*Mereology.* A book"
9. "*Propositional Calculus.* A textbook"
10. "A critical English edition of all Leśniewski's writings."

These are not marked "in preparation" and, to my knowledge, none of them ever appeared. He also mentions "Besides, about one hundred reviews in Polish, German, French and English." Not many of these have been located; probably many were unsigned.

Kotarbiński's biography of Leśniewski: "Stanisław Leśniewski: A handful of memories."* This translation from Ruch Filozoficzny was distributed at the Leśniewski meeting in Kraków in 1976.

Notes of a course in Computer Science taught by Frieder Schwenkel in the Spring of 1967. The title in the Schedule of Classes was "Comp Computability," CS 124. I took 64 pages of notes.

Notes from a presentation at Sobociński's logic seminar by Rickey dealing with a simplification of the Terminological Explanations in Leśniewski's "Über Definitionen in der Sogenannten Theorie der Deduktion" (1930). Leśniewski's original plus handwritten notes that were published, in revised form, in "Creative definitions in propositional calculi," NDJFL, XVI, 273–294.

9 Appendix I

On October 16, 2017, Rickey deposited seven boxes (6.5 linear feet) of material from Sobociński's Nachlaß in the University of Notre Dame Archives. What follows here is a description of that material that Rickey prepared and that accompanied the donation. A few comments have been added later and are included in double square brackets.

The archivists chose to call these the Sobocinski-Rickey Papers and they prepared a 17 page finding aid that contains a more detailed inventory of the material. For example, they listed Sobociński's papers individually in folders CSRK 2/13 to CSRK 2/15. These numbers identify the folder that contains the material. What CSRK means is irrelevant; it is much like the call number for a book.

Since the material is now in the University of Notre Dame Archives, I am unable to supply more detailed information about the items in the finding aid.

Here is the document given to the archivists:

Sobociński's Nachlaß

Compiled by V. Frederick Rickey

Deposited in the Notre Dame Archives, October 16, 2017

Box 1: Personal Items

- Small brown cardboard box, 21 by 14 by 5 cm.

  This contains passports of the Sobocińskis from Brussels, dated 1948. There is also a brochure with a photo of the young Bocheński. There are six letters dealing with the series "Studies in Logic." Sobociński agreed to write a monograph on Mereology, but that never happened. There are less interesting items in the box.

- Two guestbooks from the Sobocińskis home at 2405 Club Drive in South Bend. Check: Were they already in that home in 1959? The first goes from 2.VI.59 to 20.I.1974. Among the first guests were Wacław Sierpiński, Ivo Thomas, and Thoralf Skolem. The second goes from 7.IX.1974 to 8.III.1980. These are vital for tracing the logicians that passed through Notre Dame.

- *In Memoriam* book from Sobociński's funeral.

- Two copies of *Index ac status causarum Beatificationis Servorum Dei et Canonizationis Beatorum*, 1953 and 1962. One is heavily annotate; the other is heavily stuffed with newspaper clippings. These are evidence that Sobociński was a very devout Catholic.

- Yearbook from the College of St. Thomas, 1950. Sobociński is pictured on page 31; the caption reads "M.A., Ph.D., Doz. Lecturer in Philosophy." Marian Heitzmann is pictured on page 25. He was "Ph.D., Doz. Lecturer in Philosophy." He was important in bringing Sobociński to St. Thomas.

- Brown binder with manuscript of *An Introduction to Philosophy for Catholic Colleges*, by M. Heitzman, 1960.

- Five notebooks in Polish dealing with calculus, mechanics, and engineering. These almost certainly belonged to Sobociński's father, Antoni Sobociński (1873–1963).

- A folder of materials, probably by Antoni Sobociński containing word lists, and some mathematics. Two more folders.

- Folder of official documents dealing with Antoni Sobociński. Includes family photographs and his Polish passport.

- A folder of newspaper clippings.

- Miscellaneous pamphlets.

Box 2: Personal Items

- Box for an Ironing Board Pad! Lots of personal items. Letters to Mrs. Sobociński. Letters from Antoni Sobociński. NB: Rickey has removed some letters from this box dealing with Sobociński's situation in St. Paul, including his job search. They will be sent to the ND Archives.

- A folder of letters primarily to Antoni Sobociński, Prof. Sobociński's father.

- Another folder of correspondence in a hard binder.
- Thin "Personal Rite" box of correspondence, primarily financial.
- Translation of Leśniewski's "Grundzüge eines neuen Systems der Grundlagen der Mathematik," Fundamenta Mathematicae, 14 (1929), 1–81 by Michael P. O'Neil. Published, probably with changes, in Leśniewski's Collected Works, Hesburgh Library General Collection B 4691. L442 E5 1991.
- Polish original of "La génesis de la Escuela Polaca de Lógica,"
- File of correspondence between Dana Scott and John Lemmon dealing with modal logic. When Anjan Shukla was a graduate student at ND he wrote Scott about this, and Scott sent the originals. This is a photocopy. Shukla is now deceased; he probably kept the originals. So I suspect this is the only extant copy.
- Papers by Mostowski, Lacombe, Leblanc, Fraïssé, and Henkin. In a purple envelope. These were presented at the University of Montreal at a Séminaire de Mathématiques supérieures, summer 1962. Rickey, Shukla, and Payne attended. I do not know if these papers were published.
- Anjan Shukla's dissertation under Sobociński, marked up for publication.
- Partial handwritten draft of Rickey's dissertation under Sobociński.
- Luis M. Laita. Spiral bound document on Algebraic Logic (in Spanish) together with other notes. He studied with Sobociński.
- Typescript of a play, *Wałęsa*, by Jerzy Tymicki.
- Brochures about Solidarity.

Box 3: Bibliography of Bolesłáw Sobociński (1906–1980)

This is 9 pages long and will not be repeated here. There is a copy in the box. The box contains offprints of all of Sobociński's papers as given in the bibliography.

Box 4: Papers Related to NDJFL

- A blue notebook purchased from the ND Bookstore. This is a numbered list of articles for NDJFL. Information includes author, address, title, date submitted, and when published for 196 papers up through 1967. It is unclear why it stops here.

- A gray cloth-bound notebook. This lists 12 of Sobociński's early publications and who he sent copies to. This notebook should have been included in his personal papers.
- 18 folders dealing with individual numbers of NDJFL. They include correspondence with the authors, some page proofs, etc. These do not deal with all issues of the journal. Some were used for several issues and some were reused for other things.
- A few page proofs and related things. Most of these are in another box.
- Two large files of correspondence related to NDJFL. Mostly corr with authors.

Box 5: More papers related to NDJFL

This is a large box of various papers related to the NDJFL.

There is no correspondence here.

Included are original manuscripts marked up for publication.

Box 6: Translations by Rose Rand

Sobociński had hoped to publish the collected works of Stanisław Leśniewski, but that never happened.

There is a box of draft translations by Rose Rand. Her English was clearly not up to the task.

The Rose Rand Papers are at the University of Pittsburgh. A Finding Aid is on-line: `https://digital.library.pitt.edu/collection/rose-rand-papers`

I have never examined those papers, so I expect there is considerable duplication.

Box 7: NDJFL — Transition from Sobociński to Rickey

The materials in this box deal with the change of editorship from Sobociński to Rickey.

The primary issues were clearing the backlog and then substantially raising the standard of the journal.

There are no restrictions on the use of this material.

Two extremely valuable notebooks

Ontology Notebook 2

This is called Ontology Notebook 2, because there is another one that precedes it; but I will give it to the archives later. [[After further study, I decided this should be Ontology Notebook 1.]]

This is a two-hole looseleaf binder with a hard brown cover. In the upper left of the outside cover is the annotation "24—" which is probably a price. On the first leaf we find:

Bruksella, 5.II.1947

Ontologja

Here Sobociński has used the Polish word for Brussels.

There are 163 pages of text, all recto, except for one verso page. The pages are 22 cm high and 15 wide. The paper has a grid of squares, which are 0.5 cm on a side.

This is a very carefully done development of Ontology. It is divided into sections, one for each definition. There are 20 sections.

This is an exceptionally important document for it is almost certainly Sobociński's first surviving development of Leśniewski's Ontology. The material here has never been published!

Ontology Notebook 3

This is a continuation of Ontology Notebook 2. [[After further study, I decided this should be Ontology Notebook 2.]] The format is the same.

It has sections 21 to 31.

Rickey will give this to the ND archives at a later time.

10 Appendix II

This note[21] documents *the didactic activities of Sobociński* – mainly *the students and close collaborators* – extracted from notes of the author and from other sources.

Among the most notable *undergraduate* students of Sobociński *in Poland* one can mention Helena Rasiowa (1917–1994)[22], and Krysztof Tatarkiewicz (1923–2011).

[21] Written by Adrian Rezuş, in 2018.

[22] Although Jan Łukasiewicz acted as the official supervisor, Sobociński was, *de facto*, in charge with the supervision of her (first) Master's thesis at the Underground Warsaw University, which she lost during a bombing of the town. She later managed to reconstruct her early work, and eventually published it as [183] (1955).

Rasiowa graduated after the war, in 1950, at the (new) Warsaw University, on a logic subject (model theory for modal and intuitionistic logic)[23], having Andrzej Mostowski (1913–1975) as an advisor, while Tatarkiewicz took his PhD degree in mathematics proper (analysis) at *Jagellonica Cracoviensis* (Uniwersytet Jagielloński, Kraków), also in 1950, under the supervision of Tadeusz Ważewski (1896–1972).[24]

As for his *undergraduate* students in the United States, they are too many to mention. Some of them pursued a graduation under his supervision, however.

Altogether, Sobociński supervised, in the United States, 15 PhD Dissertations (in mathematics and/or in philosophy) at the University of Notre Dame, Notre Dame IN, on logical and mathematical subjects. The themes of the Dissertations reflect, at least indirectly, Sobociński's vast interests in logic and foundational matters, as well as in mathematics at large or in philosophy[25]. Here is the exhaustive list, in chronological order[26]:

- Robert E. Clay, Contributions to Mereology, August 1961 [mathematics],
- Shirley Marie Wilde, A Quaternary Relation as the Primitive Notion in Several Geometries, April 1964 [mathematics],
- William J. Frascella, Block Designs on Infinite Sets, February 1966 [mathematics],
- John Thomas Canty, Leśniewski's Ontology and Gödel's Incompleteness Theorem, June 1967 [philosophy][27]
- Anjan Shukla, Decision Procedures for Lewis' System S1 and Related Modal Systems, June 1967 [mathematics],
- V. Frederick Rickey, An Axiomatic Theory of Syntax, August 1968 [mathematics],
- Richard L. Poss, Weak Forms of the Axiom of Constructibility, August 1970 [mathematics],
- Charles Francis Quinn, An Analysis of the Concept of Constructive Categoricity, February 1971 [philosophy],
- William Russell Belding, Incidence Rings of Pre-Ordered Sets, July 1972 [mathematics],
- Thomas A. Scharle, Axiomatization of Fragments of S5, August 1973 [philosophy][28],

[23]Algebraic Treatment of the Functional Calculi of Lewis and Heyting, [182] (1951), the actual source of her later work on model theory.

[24]For a complete bibliography of Krysztof Tatarkiewicz, see, e.g., [134].

[25]William J. Frascella obtained two PhD degrees (one in mathematics and one in philosophy) under the supervision of Sobociński. So Sobociński had, in fact, only 14 PhD students.

[26]The list has been established by V. Frederick Rickey, [198] (copies of Dissertations are available in the Library of the University of Notre Dame). The current (2018) list online at the *Mathematics Genealogy Project*, *sub verbo*, is incomplete, even for mathematics. For more details, see the bibliography, at each author, and the remarks following below.

[27]Canty took also his Master's degree with Sobociński, with a thesis on A Natural Deduction System for Modal Logic, in June 1963.

[28]Scharle took also his Master's degee [MA] in philosophy with Sobociński, with a thesis on A Diagram of the Functors of the Two-valued Propositional Calculus, in June 1962, published as [213]. See also his [214].

- Charles C. Davis, An Investigation Concerning the Hilbert-Sierpiński Logical Form of the Axiom of Choice, August 1973 [philosophy],
- Jack Christopher Boudreaux, A Model-theoretic Investigation of Higher-order Functional Calculi, May 1975 [philosophy],
- James George Kowalsk, Leśniewski's Ontology Extended with the Axiom of Choice, 1975 [philosophy],
- Thomas A. Sudkamp, Algebras of N-ary Relations, August 1978 [mathematics],
- William J. Frascella, Semantics and the Concept of Evidence, 1978 [philosophy].

It appears that only Clay (PhD 1961), Canty (PhD 1967), Rickey (PhD 1968), and Kowalsk (PhD 1975) – i.e., less than a third of the *doctorandi* – have followed the initial interests of their *Doktorvater*, on typical Leśniewski matters. For the rest, the subject distribution looks rather balanced:

- classical (propositional) logic: Scharle (MA 1962),
- modal logic: Canty (MA 1963), Shukla (PhD 1967), Scharle (PhD 1973),
- set theory: Poss (PhD 1970), Davis (PhD 1973),
- model theory: Quinn (PhD 1971), Boudreaux (PhD 1975),
- mathematics proper (geometry, combinatorics, algebra) : Wilde (PhD 1964), Frascella (PhD 1966), Belding (PhD 1972), Sudkamp (PhD 1978),
- epistemology: Frascella (PhD 1978).

In particular, none of his American PhD students seems to have shared interest in his early ('Polish') concerns with axiomatizability matters and multi-valued logics. Such *adhoc* statistics are not too relevant, however, since Soboncińki had many followers – not necessarily academic 'descendants' – on Leśniewski topics, as well as on axiomatizability matters, on multi-valued and modal logics, on set-theoretical issues, and on purely algebraic (resp. model-theoretical) subjects.

Adrian Rezuş – A Sobociński Bibliography

[1] *** EUDML = The European Digital Mathematics Library. Online @ https://eudml.org/. [Fundamenta Mathematicae, Mathematische Annalen, etc.]

[2] *** IPSB = Internetowy Polski Słownik Biograficzny (Instytut Historii PAN) [The Polish Internet Biographical Dictionary, edited by the Institute of History of the Polish Academy of Sciences]. Online @ http://www.ipsb.nina.gov.pl.

[3] *** NDJFL = Notre Dame Journal of Formal Logic, Online @ Project Euclid, https://projecteuclid.org/all/euclid.ndjfl

[4] *** NK = Nowa Książka (Warsaw, 1934–1939), digitalised online @ *Małopolska Biblioteka Cyfrowa* [The Smaller Polish Digital Library], Kraków, http://mbc.malopolska.pl/dlibra/publication?id=77505&tab=3

[5] *** ORGANON ToC 2008 *Tables des matières des numeros 1, 1936 – 3, 1939 & 1, 1964 – 37 (40), 2008*, in: Organon 37 (40), 2008, pp. 269–294. Online @ *Muzeum Historii Polski w Warszawie*, bazhum.muzhp.pl/czasopismo/69/?idvol=6709

[6] *** ORGANON Index 2008a *Tables des auteurs des numeros 1, 1936 – 3, 1939 & 1, 1964 – 37 (40), 2008*, in: Organon 37 (40), 2008, pp. 295–317. Online @ *Muzeum Historii Polski w Warszawie*, bazhum.muzhp.pl/czasopismo/69/?idvol=6709

[7] *** PF = Przegląd Filozoficzny (Warsaw, 1897–1939), digitalised online @ *Wielkopolska Biblioteka Cyfrowa* [The Greater Polish Digital Library], Poznań, http://www.wbc.poznan.pl/publication/105589

[8] *** SEP = Stanford Encyclopedia of Philosophy. Online @ https://plato.stanford.edu.

Bolesław Sobociński

[9] — 1932 *Z badań nad teorią dedukcji* [Investigations on the theory of deduction], Przegląd Filozoficzny 35, 1932, pp. 171–193 (= Księga Pamiątkowa Koła Filozoficznego Słuchaczy Uniwersytetu Warszawskiego, Warsaw 1932, pp. 3–25). (Online in [7]. Partial English translation [§ 1], by Adrian Rezuş, in: [187], *IV.17 Appendix*, pp. 257–268.)

[10] — 1934 *O kolejnych uproszczeniach aksjomatyki 'ontologii' Prof. S. Leśniewskiego* [On the successive simplifications of the axiom-system of Prof. S. Leśniewski's 'ontology'], Przegląd Filosoficzny 37, 1934], pp. 143–160 (= Janina Hosiasson *et al.* (eds.) Fragmenty Filozoficzne, *Księga pamiątkowa ku uczczeniu 15-lecia pracy nauczycielskiej w Uniwersytecie Warszawskim Prof. Tadeusza Kotarbińskiego* [Philosophical Fragments, A book commemorating the fifteeen years long didactic activities of Professor Tadeusz Kotarbiński at the University of Warsaw], Druk. Kasy im. Mianowskiego [Printed by the Mianowski Foundation], Warsaw 1934. (Online in [7]. English translation, by Zbignew Jordan, published as [56], i.e., in [164], pp. 188–200.)

[11] — 1935 *Aksjomatyzacja implikacyjno-koniunkcyjnej teorii dedukcji* [Axiomatization of the implicative-conjunctive theory of deduction], Przegląd Filozoficzny 38, 1935, pp. 85–95. (Online in [7].)

[12] — 1936 *Aksjomatyzacja pewnych wielowartosciowych systemow teorii dedukcji* [Axiomatization of certain many-valued systems of the theory of deduction], Roczniki Prac Naukowych Zrzeszenia Asystentow Uniwersytetu Józefa Piłsudskiego w Warszawie 1 (*Wydzial matematyczno-przyrodniczy 1*) [Annals of the Scientific Work of the Association of the Assistants of the Józef Piłsudski University in Warsaw 1 (Faculty of Mathematics and Natural Sciences 1)], 1936 pp. 399–419. (PhD Dissertation, University of Warsaw, supervised by Jan

Łukasiewicz. Review: Alfred Tarski, The Journal of Symbolic Logic 2, p. 93.)

[13] — 1939 *Z badań nad prototetyką* [Investigations on protothetic], Collectanea Logica (Warsaw) 1, 1939, pp. 171–176. (See [18], for an English translation by the author, and [57], for a new translation by Zbigniew Jordan, published in [164], pp. 201–206.) [Review: Alonzo Church, The Journal of Symbolic Logic 15 (1), 1950, p. 64.]

[14] — 1939a *Aksjomatyzacja konjunkcyjno-negacyjne teorji dedukcji* [Axiomatisation of a conjunctive-negative theory of deduction], Collectanea Logica (Warsaw) 1, 1939, pp. 179–193. [Review: Heinrich Scholz, Zentralblatt für Mathematik und ihre Grenzgebiete 23, 1940, pp. 97–98, and Jahrbuch über die Fortschritte der Mathematik 65, 1939 (published 1941), pp. 24–25.]

[15] — 1939b *O aksjomatykach prototetyki* [On the axiomatics of protothetic], Collectanea Logica (Warsaw) 1, 1939 [lost].

[16] — 1939c *O rożnych systemach pototetyki* [On different systems of protothetic], Collectanea Logica (Warsaw) 1, 1939 [lost].

[17] — 1939d *Uwagi w zwiezku z prace p. J. Słupeckiego: "Przyczynek do prototetyki"* [Remarks concerning the paper of Mr. J. Słupecki: 'Contributions to protothetic'] Collectanea Logica (Warsaw) 1, 1939 [lost].

[18] — 1949 *An investigation of protothetic*, Cahiers de l'Institut d'Études Polonaises en Belgique (Brussels) 5 [v + 44 pp.]. (This is a translation of [13] by the author. See also the new translation by Zbigniew Jordan, without the Introduction [pp. 7–27, giving an account of the ill-fated periodical Collectanea Logica], in: [164], pp. 201–206. Also reprinted, with some omissions, in: [221], pp. 69–88.) [Review: Alonzo Church, The Journal of Symbolic Logic 15, 1960, p. 64.]

[19] — 1949–1950 *L'analyse de l'antinomie russellienne par Leśniewski*, Methodos (Milan) 1, 1949, pp. 94–107, 220–228, 308–316; 2, 1950, pp. 237–257. (English traslation, by Robert E. Clay, published as [95], i.e., in: [220], pp. 11–44.) [Review: Arthur N. Prior, The Journal of Symbolic Logic 18 (4), 1953, pp. 331–333.]

[20] — 1952 *Axiomatization of a partial system of three-value calculus of propositions*, The Journal of Computing Systems 1 (1), 1952, pp. 23–55. [Review: Gene F. Rose, The Journal of Symbolic Logic 18 (3), 1953, p. 283.]

[21] — 1953 *On a universal decision element*, The Journal of Computing Systems 1 (2), 1953, pp. 71–80. [Review: Alonzo Church, The Journal of Symbolic Logic 18 (3), [September] 1953, pp. 284–285.]

[22] — 1953a *Note on a modal system of Feys-Von Wright*, The Journal of Computing Systems 1 (3), 1953, pp. 171–178. [Review: Naoto Yonemitsu, The Journal of Symbolic Logic 19 (4), 1954, p. 293.]

[23] — 1953–1954 *Z badań nad aksjomatyk prototetyki Stanisława Leśniewskiego* [Investigations on the axiomatics of Stanisław Leśniewski's protothetic], Rocznik Polskiego Towarzystwa Naukowego na Obczyinie [Yearbook of the Polish Society of Arts and Sciences Abroad] (London), 4, 1953–1954, pp. 18–20. Review: Czesław Lejewski, The Journal of Symbolic Logic 21 (3), 1956, p. 325

[24] — 1954 *Axiomatization of a conjunctive-negative calculus of propositions*, The Journal of Computing Systems 1 (4), 1954, pp. 229–242. [Review: Gene F. Rose, The Journal of Symbolic Logic 20 (3), 1955, pp. 303–304.]

[25] — 1954–1955 *A contribution to Leśniewski's mereology*, Rocznik Polskiego Towarzystwa

Naukowego na Obczyinie [Yearbook of the Polish Society of Arts and Sciences Abroad] (London), 5, 1954–1955, pp. 34–43. [Review: Arthur N. Prior, The Journal of Symbolic Logic 21 (3), 1956, pp. 325–326.]

[26] — 1954–1955a *Studies in Leśniewski mereology*, Rocznik Polskiego Towarzystwa Naukowego na Obczyinie [Yearbook of the Polish Society of Arts and Sciences Abroad] (London), 5, 1954–1955, pp. 43–50. (Reprinted in: [220], pp. 217–227.)

[27] — 1955 *Note on a problem of Paul Bernays*, The Journal of Symbolic Logic 20 (2), 1955, pp. 109–114.

[28] — 1955–1956 *On well-constructed axiom systems*, Rocznik Polskiego Towarzystwa Naukowego na Obczyinie [Yearbook of the Polish Society of Arts and Sciences Abroad] (London), 6, 1955–1956, pp. 54–65. (Polish translation, *W sprawie dobrze skonstruowanej aksjomatyki*, by Józef Andrzej Stuchliński, in: [96].) [Review: Hugues Leblanc, The Journal of Symbolic Logic 22 (4), 1957, pp. 358–359.]

[29] — 1956 *In memoriam Jan Łukasiewicz* (1878–1956), Philosophical Studies (Maynooth, Ireland) 6, pp. 3–49. (English version of [30]. Contains a 'Curriculum vitae of Jan Łukasiewicz', and a list of the publications of Łukasiewicz.) [Review: Karl Dürr, The Journal of Symbolic Logic 22 (4), 1957, pp. 385–387.]

[30] — 1956–1957 *Jan Łukasiewicz* (1878–1956) [Polish], Rocznik Polskiego Towarzystwa Naukowego na Obczyznie [Yearbook of the Polish Society of Arts and Sciences Abroad] (London), 7, 1956–1957, pp. 3–21. (Polish version of [29], containing some additional information.)

[31] — 1957 *La génesis de la Escuela Polaca de Lógica* [Spanish], Oriente Europeo (Madrid), 7 (25), 1957, pp. 83–95. [Review: José Ferrater Mora, The Journal of Symbolic Logic 25 (1), 1960, pp. 63–64.]

[32] — 1958 *Jan Salamucha (1903–1944): A biographical note*, The New Scholasticism 32 (3), [July] 1958, pp. 327–333.

[33] — 1960 *A simple formula equivalent to the axiom of choice*, Notre Dame Journal of Formal Logic 1 (3), 1960, pp. 115–117.

[34] — 1960a *A note concerning the axiom of choice*, Notre Dame Journal of Formal Logic 1 (3), p. 122.

[35] — 1960–1961 *On the single axioms of protothetic*, Notre Dame Journal of Formal Logic, I: 1 (1–2), 1960, pp. 52–73; *Errata* I: 1 (4), 1960, 176–177; *Errata* I: 2 (4), p. 259; II: 2, 1961, pp. 110–126; III: 2 (3), pp. 129–148; *Errata* III: 2 (4), 1961, p. 259.

[36] — 1961 *Three set-theoretical formulas*, Notre Dame Journal of Formal Logic 2 (1), 1961, pp. 58–64. *Errata*, 2 (4), 1961, p. 259.

[37] — 1961a *A note concerning the many-valued propositional calculi*, Notre Dame Journal of Formal Logic 2 (2), 1960, pp. 127–128.

[38] — 1961b *Certain formulas equivalent to the axiom of choice*, Notre Dame Journal of Formal Logic 2 (4), 1961, pp. 229–235.

[39] — 1961c *A theorem on Hartogs' alephs*, Notre Dame Journal of Formal Logic 2 (4), 1961, pp. 255–258.

[40] — 1962 *A contribution to the axiomatization of Lewis' system S5*, Notre Dame Journal of Formal Logic 3 (1), 1962, pp. 51–60.

[41] — 1962a *A note on the regular and irregular modal systems of Lewis*, Notre Dame Journal of Formal Logic 3 (2), 1962, pp. 109–113.

[42] 1962b *On the generalized Brouwerian axioms*, Notre Dame Journal of Formal Logic 3 (2), 1962, pp. 123–128.

[43] — 1962c *A set-theoretical formula equivalent to the axiom of choice*, Notre Dame Journal of Formal Logic 3 (3), 1962, pp. 167–169.

[44] — 1962d *Six new sets of independent axioms for distributive lattices with O and I*, Notre Dame Journal of Formal Logic 3 (3), 1962, pp. 187–192.

[45] — 1962e *An axiom-system for {K;N}-propositional calculus related to Simons' axiomatization of S3*, Notre Dame Journal of Formal Logic 3 (3), 1962, pp. 206–208.

[46] — 1962f *A remark concerning the third theorem about the existence of successors of cardinals*, Notre Dame Journal of Formal Logic 3 (4), 1962, pp. 279–283.

[47] — 1962–1963 *A note on the generalized continuum hypothesis*, Notre Dame Journal of Formal Logic, I: 3 (4), 1962, pp. 274–278; II: 4 (1), 1963, pp. 67–79; III: 4 (3), pp. 233–240.

[48] — 1963 *A note on modal systems*, Notre Dame Journal of Formal Logic 4 (2), 1963, pp. 155–157.

[49] — 1964 *A theorem of Sierpiński on triads and the axiom of choice*, Notre Dame Journal of Formal Logic 5 (1), 1964, pp. 51–58.

[50] — 1964a *Remarks about axiomatizations of certain modal systems*, Notre Dame Journal of Formal Logic 5 (1), 1964, pp. 71–80.

[51] — 1964b *A note on Prior's systems in 'The theory of deduction'*, Notre Dame Journal of Formal Logic 5 (2), 1964, pp. 139–140.

[52] — 1964c *Family K of the non-Lewis modal systems*, Notre Dame Journal of Formal Logic 5 (4), 1964, pp. 313–318.

[53] — 1964d *On the propositional system A of Vučković and its extension*, Notre Dame Journal of Formal Logic I: 5 (2), 1964, pp. 141–153; II: 5 (3), 1964, pp. 223–237.

[54] — 1964e *Modal system S4.4*, Notre Dame Journal of Formal Logic 5 (4), pp. 305–312.

[55] — 1965 *A note on certain set-theoretical formulas*, Notre Dame Journal of Formal Logic 6 (2), 1965, pp. 157–160.

[56] — 1967 *On the successive simplifications of the axiom-system of Prof. S. Leśniewski's 'ontology'*, in: [164], pp. 188–200. (English translation of [10], by Zbigniew Jordan.)

[57] — 1967a *An investigation of prothotetic*, in: [164], pp. 201–206. (A new translation of [13], by Zbigniew Jordan.)

[58] — 1970 *Note on G. J. Massey's closure-algebraic operation*, Notre Dame Journal of Formal Logic 11 (3), 1970, pp. 343–346; Errata: 14 (4), 1973, p. 584. [Review: R. A. Bull, The Journal of Symbolic Logic 36 (4), 1971, p. 691]

[59] — 1970a *Certain extensions of modal system S4*, Notre Dame Journal of Formal Logic 11 (3), 1970, pp. 347–368.

[60] — 1970b *Note on Zeman's modal system S4.04*, Notre Dame Journal of Formal Logic 11 (3), 1970, pp. 383–384.

[61] — 1971 *Lattice-theoretical and mereological forms of Hauber's law*, Notre Dame Journal of Formal Logic 12 (1), 1971, pp. 81–85.

[62] — 1971a *Atomistic mereology*, Notre Dame Journal of Formal Logic, I: 12 (1), 1971, pp. 89–103; II: 12 (2), 1971, pp. 203–213.

[63] — 1971b *A note on an axiom-system of atomistic mereology*, Notre Dame Journal of Formal Logic 12 (2), 1971, pp. 249–251.

[64] — 1971c *Concerning some extensions of S4*, Notre Dame Journal of Formal Logic 12 (3), 1971, pp. 363–370.

[65] — 1971d *A new class of modal systems*, Notre Dame Journal of Formal Logic 12 (3), 1971, pp. 371–377.

[66] — 1971e *A proper subsystem of S4.04*, Notre Dame Journal of Formal Logic 12 (3), 1971, pp. 381-384 (1971)

[67] — 1972 *Additional note on lattice-theoretical form of Hauber's law*, Notre Dame Journal of Formal Logic 13 (1), 1972, pp. 101–102.

[68] — 1972a *Certain sets of postulates for distributive lattices with the constant elements*, Notre Dame Journal of Formal Logic 13 (1), 1972, pp. 119–123.

[69] — 1972b *An abbreviation of Croisot's axiom-system for distributive lattices with I*, Notre Dame Journal of Formal Logic 13 (1), 1972, pp. 139–141.

[70] — 1972c *A new formalization of Newman algebra*, Notre Dame Journal of Formal Logic 13 (2), 1972, pp. 255–264.

[71] — 1972d *An equational axiomatization of associative Newman algebras*, Notre Dame Journal of Formal Logic 13 (2), 1972, pp. 265–269; *Errata*, 14 (4), 1973, p. 584.

[72] — 1972e *A semi-lattice theoretical characterization of associative Newman algebras*, Notre Dame Journal of Formal Logic 13 (2), 1972, pp. 283–285.

[73] — 1972f *An equational axiomatization and a semi-lattice theoretical characterization of mixed associative Newman algebras*, Notre Dame Journal of Formal Logic 13 (3), 1972, pp. 407–423.

[74] — 1972g *Solution to the problem concerning the Boolean bases for cylindric algebras*, Notre Dame Journal of Formal Logic 13 (4), 1972, pp. 529–545.

[75] — 1973 *Remark about the Boolean parts in the postulate-systems of closure, derivative and projective algebras*, Notre Dame Journal of Formal Logic 14 (1), 1973, pp. 111–117.

[76] — 1973a *A note on Newman's algebraic systems*, Notre Dame Journal of Formal Logic 14 (1), 1973, pp. 129–133.

[77] — 1973b *A new axiomatization of modal system K1.2*, Notre Dame Journal of Formal Logic 14 (3), 1973, pp. 413–414.

[78] — 1973c *Modal system S3 and the proper axioms of S4.02 and S4.04*, Notre Dame Journal of Formal Logic 14 14 (3), 1973, pp. 415–418; *Erratum*, 15 (4), 1974, p. 648.

[79] — 1973d *Note about the Boolean parts of the extended Boolean algebras*, Notre Dame Journal of Formal Logic 14 (3), 1973, pp. 419–422.

[80] — 1973e *Concerning the quantifier algebras in the sense of Pinter*, Notre Dame Journal of Formal Logic 14 (4), 1973, pp. 547–553.

[81] — 1974 *Concerning the proper axioms of S4.02*, Notre Dame Journal of Formal Logic 15 (1), 1974, pp. 169–172.

[82] — 1974a *A theorem concerning a restricted rule of substitution in the field of propositional calculi*, Notre Dame Journal of Formal Logic, I: 15 (3), 1974, 465–476; II: 15 (4), 1974, pp. 589–597.

[83] — 1975 *A new postulate-system for modular lattices*, Notre Dame Journal of Formal Logic 16 (1), 1975, pp. 81–85.

[84] — 1975a *A short postulate-system for ortholattices*, Notre Dame Journal of Formal Logic 16 (1), 1975, pp. 141–144.

[85] — 1975b *Concerning the postulate-systems of subtractive abelian groups*, Notre Dame Journal

of Formal Logic 16 (3), 1975, pp. 429–444. (Cf. [225], [166], [140], [142], [141].)

[86] — 1976 *Pledger lemma and the modal system S3°*, Notre Dame Journal of Formal Logic 17 (2), 1976, pp. 253–256.

[87] — 1976a *A short equational axiomatization of modular ortholattices*, Notre Dame Journal of Formal Logic 17 (2), 1976, pp. 311–316.

[88] — 1976b *A short equational axiomatization of orthomodular lattices*, Notre Dame Journal of Formal Logic 17 (2), 1976, pp. 317–320.

[89] — 1976c *The modular latticoids*, Notre Dame Journal of Formal Logic 17 (4), 1976, pp. 617–621.

[90] — 1976d *The axioms for latticoids and their associative extensions*, Notre Dame Journal of Formal Logic 17 (4), 1976, pp. 625–631.

[91] — 1978 *Awkward axiom-systems*, Notre Dame Journal of Formal Logic 19 (2), 1978, pp. 315–320.

[92] — 1978a *Note about Łukasiewicz's theorem concerning the system of axioms of the implicational propositional calculus*, Notre Dame Journal of Formal Logic 19 (3), 1978, pp. 457–460.

[93] — 1978b *A new axiomatization of the mixed associative Newman algebras*, Notre Dame Journal of Formal Logic 19 (3), 1978, pp. 467–474.

[94] — 1978c *Equational two axiom bases for Boolean algebras and some other lattice theories*, Notre Dame Journal of Formal Logic 20 (4), 1979, pp. 865–875.

[95] — 1984 *Leśniewski's analysis of Russell's paradox*, in [220], pp. 11–44. (A translation of [19], by Robert E. Clay.)

[96] — 2004 *W sprawie dobrze skonstruowanej aksjomatyki* [On well-constructed axiom systems] (Polish translation of [28], by Józef Andrzej Stuchliński), Filozofia Nauki (Warsaw) 12 (1), 2004, pp. 123–136.) Online @ *Muzeum Historii Polski.*

(reviews and occasional papers)

[97] Bolesław Sobocinśki R 1931 Review of: Isidora Dąmbska *La Théorie du jugement de M. Edmond Goblot*, 1930, [128], Przegląd Filozoficzny 34 (1), 1931, p. 87. Online in [7].
`http://www.wbc.poznan.pl/dlibra/publication?id=107007&tab=3`

[98] — R 1931a Review of: *L'Orientation actuelle des sciences*, 1930, [113], Przegląd Filozoficzny 34 (4), 1931, pp. 273–274. Online in [7].
`http://www.wbc.poznan.pl/dlibra/docmetadata?id=120445&from=publication`

[99] — R 1931b Review of: Julien Pacotte, *La Pensée technique*, 1931, [173], Przegląd Filozoficzny 34 (4), 1931, p. 274. Online in [7].
`http://www.wbc.poznan.pl/dlibra/docmetadata?id=120445&from=publication`

[100] — R 1932 Review of: André Lalande, *Les Illusions évolutionnistes*, 1930, [155], Przegląd Filozoficzny 35 (1–2), 1932, p. 160–161. Onlie on [7].
`http://www.wbc.poznan.pl/dlibra/docmetadata?id=120446&from=publication`

[101] — 1933 *Informator o studjach i egzaminach z nauk filozoficznych na Uniwersytecie Warszawskim* [Guidebook about studies and examinations in philosophical sciences at the University of Warsaw], Warsaw 1933.

[102] — R 1936 *Polskie wydawnictwa filozoficzne w latach 1918–1936* [Polish philosophical publications in the years 1918–1936], Nowa Książka 3 (3), 1936, pp. 113—121. Online in [4].

http://mbc.malopolska.pl/dlibra/docmetadata?id=72165&from=publication

[103] — R 1936a *Tendencje rozwojowe współczesnej filozofii polskiej (Refleksje na marginesie III Polskiego Zjazdu Filozoficznego, Kraków 24–27.IX.1936)* [Trends in the development of contemporary Polish philosophy (*Marginalia* on the Third Polish Congress of Philosophy, Kraków, September 24–27,1936)], Nowa Książka 3 (8), 1936, pp. 433–437. Online in [4]. http://mbc.malopolska.pl/dlibra/doccontent?id=72170

[104] — R 1938 Review (in French) of: Antoni Korcik, *Teorja konwersji zdań asertorycznych u Arystotelesa w świetle teorji dedukcji* [The theory conversion of the assertoric propositions in Aristotle, in the light of the theory of deduction], 1937, [150], The Journal of Symbolic logic 3 (1), [March] 1938, p. 45.

[105] — R 1950 Review (in French) of: Jacob Ridder, *Über mehrwertige Aussagenkalküle und mehrwertige engere Prädikatenkalküle*, I–III, 1948, [200], The Journal of Symbolic Logic 14 (2), 1950, p. 139 (I), and *ibid.* 14 (4), 1950, pp. 261–262 (II–III).]

[106] — R 1950a Review (in French) of: Innocent Marie Bocheński, Wstęp do teorii analogii [Introduction to a theory of analogy], 1948, [119], The Journal of Symbolic Logic 14 (2), 1950, pp. 139–140.

[107] — R 1955 Review of: J. Barkley Rosser, and Atwell R. Turquette, *Many-valued Logics*, 1952, [210], The Journal of Symbolic Logic 20 (1), 1955, pp. 45–50.]

[108] — R 1955a Review of: Jan Łukasiewicz, *Concerning an axiom system of the implicational propositional calculus*, 1950, [162], The Journal of Symbolic Logic 20 (2), 1955, pp. 173–174.

[109] — R 1956 Review of: Henryk Stonert, *Sprawozdanie z I Konferencji Logików (Warszawa, grudzień 1952)* [Report on the Ist Conference of Logicians (Warsaw, December 1952], (1955), [222], The Journal of Symbolic Logic 21 (3), 1956, p. 311.

[110] — R 1956a Review of: Kazimierz Ajdukiewicz, *Klasyfikacia rozumowah* [The classification of reasonings], 1955, [116], The Journal of Symbolic Logic 21 (3), 1956, pp. 311–312.

[111] — R 1956b Review of: Janina Kotarbińska, *Definicja* [On definition], 1955, [152], The Journal of Symbolic Logic 21 (3), 1956, p. 312.

[112] — 1995 *Lata wojny... Listy do Józefa M. Bocheńskiego* [Years of war... Letters to Józef M. Bocheński], in: [147], 1, pp. 123–128.

SECONDARY LITERATURE

[113] *** Conférences ENS (1930) L'Orientation actuelle des sciences (Conférences faites à l'École Normale Supérieure par MM. Jean-Baptiste Perrin, Paul Langevin, Georges Urbain, Louis-Edouard Lapicque, Charles Perez, Lucien Plantefol; Introduction de Léon Brunschvicg), Librairie Félix Alcan, Paris 1930. [*Bibliothèque de philosophie contemporaine*] [Review: Bolesław Sobociński, Przegląd Filozoficzny 34 (4), 1931, pp. 273–274.] ('Conférences faites pendant l'année 1929–1930 pour tous les candidats de l'Université de Paris à l'agrégation de philosophie'. SUDOC: http://www.sudoc.fr/009470484.)

[114] *** Congress Proceedings PZF 3 (1936) Księga pamiątkowa III Polskiego Zjazdu Filozoficznego (Kraków, 1936) [Memorial Book of the Third Polish Philosophical Congress (Kraków, 1936)], Przegląd Filozoficzny 39 (4), 1936, pp. 317–548. Online in [7].

[115] *** Congress Proceedings PZM 1 (1929) Księga Pamiątkowa Pierwszego Polskiego Zjazdu Matematycznego [Memorial Book of the First Polish Mathematical Congress], Lwów, 7–10.IX.1927 [Dotatek do 'Annales de la Société Polonaise de Mathématique'], Kraków

[Czonkami Drukarni Universytetu Jagiellonskiego] 1929. Online @ www.ptm.org.pl/zawartosc/i-zjazd-ptm-1927-e-book.

Kazimierz Ajdukiewicz

[116] Kazimierz Ajdukiewicz 1955 *Klasyfikacia rozumowah* [A classification of arguments], Studia Logica 2, 1955, pp. 278–300. [Review: Bolesław Sobociński, The Journal of Symbolic Logic 21 (3), 1956, pp. 311–312.]

William Russell Belding

[117] William Russell Belding 1972 Incidence Rings of Pre-Ordered Sets, PhD Dissertation, University of Notre Dame, Notre Dame IN, 1972 (under the direction of Bolesław Sobociński, cf. [118]).

[118] —1973 *Incidence rings of pre-ordered sets*, Notre Dame Journal of Formal Logic 14 (4), 1973, pp. 481–509.

Innocent Marie Bocheński

[119] Innocent Marie Bocheński 1948 Wstęp do teorii analogii [Introduction to a theory of analogy], Roczniki filozoficzne 1, 1948, pp. 64–82. [Review (in French): Bolesław Sobociński, The Journal of Symbolic Logic 14 (2), 1950, pp. 139–140.]

[120] — *The problem of universals*, in: [121], pp. 35–54.

Innocent Marie Bocheński (ed.)

[121] Innocent Marie Bocheński (ed.) 1956 The Problem of Universals, University of Notre Dame Press, Notre Dame IN 1956. (Contains three papers 'read at the Aquinas Symposium sponsored by the Department of Philosophy of the University of Notre Dame on March 9–10, 1956.', namely: Alonzo Church, *Propositions and sentences*, pp. 3–12, Nelson Goodman, *A world of individuals*, p. 15–31; Innocent Marie Bocheński, *The problem of universals*, pp. 35–54.)

Leonard Bolc
co-authors: Piotr Borowik

[122] Leonard Bolc, and Piotr Borowik 1992 Many-valued Logics 1, *Theoretical Foundations*, Springer-Verlag, Berlin, and Heidelberg 1992. (Herein: *3.12 The Three-Valued Calculus of Sobociński*, p. 77, and *4.2 The Many-Valued Calculus of Sobociński*, pp. 82–83, referring to [12].)

Anna Brożek (ed.)
co-editors: Alicja Chybinska, Mariusz Grygianiec, Marcin Tkaczyk

[123] Anna Brożek, Alicja Chybinska, Mariusz Grygianiec, and Marcin Tkaczyk (eds.) 2016 Myśli o języku, nauce i wartościach. [Seria druga.] *Profesorowi Jackowi Jadackiemu w 70. rocznicę urodzin* [Thoughts about language, science and values. Second series. Dedicated to Professor Jacek Jadacki on the 70th birthday anniversary], Wydawnictwo Naukowe Semper [Semper Scientific Publishers], Warsaw 2016.

John Thomas Canty

[124] John Thomas Canty 1967 Leśniewski's Ontology and Gödel's Incompleteness Theorem, PhD Dissertation, University of Notre Dame, Notre Dame IN, 1967 (under the direction of Bolesław Sobociński; cf. [125], [126]).

[125] — 1969 *The numerical epsilon*, Notre Dame Journal of Formal Logic 10 (1), 1969, pp. 47–63.

[126] — 1969a *Leśniewski's terminological explanations as recursive concepts*, Notre Dame Journal of Formal Logic 10 (4), 1969, pp. 337–369.

Robert E. Clay

[127] Robert E. Clay 1961 Contributions to Mereology, PhD Dissertation, University of Notre Dame, Notre Dame IN, 1961 (under the direction of Bolesław Sobociński).

Isidora [Izydora] Dąmbska

[128] Isidora Dąmbska 1930 *La théorie du jugement de M. Edmond Goblot*, Archiwum Towarzystwa Naukowego we Lwowie, (Wydział 2, Historyczno-filozoficzny), 6 (3) Lwów 1930. [Société des Sciences et des Lettres, Léopol] [Review: Bolesław Sobociński Przegląd Filozoficzny 34 (1), 1931, p. 87.]

Charles Davis

[129] Charles Davis 1973 An Investigation Concerning the Hilbert-Sierpiński Logical Form of the Axiom of Choice, PhD Dissertation, University of Notre Dame, Notre Dame IN, 1973 (under the direction of Bolesław Sobociński, cf. [130], [131]).

[130] — 1975 *An Investigation concerning the Hilbert-Sierpiński logical form of the axiom of choice*, Notre Dame Journal of Formal Logic 16, 1975, pp. 145–184.

[131] — 1976 *A note on the axiom of choice in Leśniewski's ontology*, Notre Dame Journal of Formal Logic 17, 1976, pp. 35–43.

Anton Dumitriu

[132] Anton Dumitriu 1971 Logica polivalentă [Multi-valued logic] [Romanian], Editura enciclopedică română, Bucharest 1971. (The author refers twice [p. 151, and 153] to Sobociński – and indirectly so, by quoting other people – on his contributions to modal logic [sic]. Unfortunately, he managed to get things wrong.)

[133] — 1975 Istoria logicii [History of logic] [Romanian] Editura didactică şi pedagogică, Bucharest 1975. (Second revised and expanded edtion; first edition: 1969. The author mentions Sobociński twice, with [22], on p. 950 [on the Feys-von Wright system], and as a prewar 'Polish logician' in the shadow in Łukasiewicz, on p. 961.)

Stanisław Domoradzki

[134] Stanisław Domoradzki 2014 *Krzysztof Tatarkiewicz (1923–2011)*, Antiquitates matematicae 8 (10), 2014, pp. 151–168. (A thorough bio-bibliograpy of Tatarkiewicz.)

William J. Frascella

[135] William J. Frascella 1966 Block Designs on Infinite Sets, PhD Dissertation [mathematics], Uni-

versity of Notre Dame, Notre Dame IN, 1966 (under the direction of of Bolesław Sobociński; cf. [136]).

[136] — 1967 *Combinatorial design of infinite sets*, Notre Dame Journal of Formal Logic 8 (1–2), 1967, pp. 24–47.

[137] — 1978 Semantics and the Concept of Evidence, PhD Dissertation [philosophy], University of Notre Dame, Notre Dame IN, 1978 (under the direction of of Bolesław Sobociński).

Stanisław Fudali

[138] Stanisław Fudali (ed.) 1998 Matematycy polskiego pochodzenia na obczyźnie, *Materiały z XI Ogólnopolskiej Szkoły Historii Matematyki* (Kołobrzeg, 5–9 maja 1997), Uniwersytet Szczeciński, *Materiały-Konferencje nr. 30* [Mathematicians of Polish descent abroad. Materials from the 11th National School of History of Mathematics (Kołobrzeg, 5–9 May 1997), University of Szczecin, *Documents of the Conferences 30*], The Publishing House of the Szczecin University, Szczecin 1998.

Ángel Garrido (ed.)
co-editors: Urszula Wybraniec-Skardowska

[139] Ángel Garrido, and Urszula Wybraniec-Skardowska (eds.) 2018 The Lvov-Warsaw School, Past and Present, Birkhäuser [Springer International Publishing] [June] 2018 [*Studies in Universal Logic*].

Reiner Güting

[140] Reiner Güting 1971 *Ein neues Axiomensystem für Boolesche Verbande*, Journal für die reine und angewandte Mathematik 251, 1971, pp. 212–220.

[141] — 1975 *Subtractive abelian groups*, Notre Dame Journal of Formal Logic 16 (3), 1975, pp. 425–428.

Thomas M. Hearne
co-authors: Carl G. Wagner

[142] Thomas M. Hearne, and Carl G. Wagner 1974 *Boolean subtractive algebras*, Notre Dame Journal of Formal Logic 15 (2), 1974, pp. 317–324.

Piotr Hübner
co-authors: Jan Piskurewicz, Leszek Zasztowt

[143] Piotr Hübner, Jan Piskurewicz, and Leszek Zasztowt 1992 Kasa im. Józefa Mianowskiego [The Józef Mianowski Fund], Fundacja Popierania Nauki 1881–1991, nakł. Kasy im. J. Mianowskiego, Warsaw 1992.

[144] — 2010 *History* [of the Józef Mianowski Fund, a Foundation for the Promotion of Science, established 1881] (translated by Jacek Soszyński), online @ `http://www.mianowski.waw.pl/foundation/history/?lang=en`

[145] — 2013 A History of the Józef Mianowski Fund (translated by Jacek Soszyński), Kasa im. Józefa Mianowskiego – Fundacja Popierania Nauki, Oficyna Wydawnicza ASPRA-JR, Warsaw 2013.

Jacek Juliusz Jadacki

[146] Jacek Juliusz Jadacki 1980 *Bibliografia logiki polskiej* [A bibliography of Polish logic], Studia Filozoficzne 1, 1980, pp. 161–176; *Część II. 1918–1939* [Part II. 1918–1939], 2, 1980, pp. 151–175 (cf. page 166).

Jacek Juliusz Jadacki (ed.)
co-editors: Barbara Markiewicz; *see also* [161]

[147] Jacek Juliusz Jadacki, and Barbara Markiewicz (eds.) 1995–1996 Próg istnienia [Threshold of being], (in two volumes) 1 *Zdziesiątkowane pokolenie* [The decimated generation], 2 *Dzieło nie dokończone* [A never finished work], PTF – Polskie Towarzystwo Filozoficzne [The Publishing House the Polish Society of Philosophy], Warsaw 1: 1995, 2: 1996.

Michał Kokowski

[148] Michał Kokowski 2015 *The science of science (naukoznanstwo) in Poland: The changing theoretical perspectives and political context – A historical sketch from the 1910's to 1993*, Organon 47, 2015, pp. 147–236.

Stefan Korbonski

[149] Stefan Korbonski 1978 The Polish Underground State, *A Guide to the Underground, 1939–1945* (translated by Marta Erdman), *East European Quarterly*, Boulder, Colorado (distributed by Columbia University Press, New York) 1978. [*East European Monographs 39*] (Paperback reprint: Hippocrene Books, [June] 1981.)

Antoni Korcik

[150] Antoni Korcik 1937 Teorja konwersji zdań asertorycznych u Arystotelesa w świetle teorji dedukcji (*Studium historyczno-krytyczne*) [The theory conversion of the assertoric propositions in Aristotle, in the light of the theory of deduction (A historico-critical study)], Wydawnictwo studjów teologicznych, Vilnius 1937. [Review (in French): Bolesław Sobociński, The Journal of Symbolic logic 3 (1), [March] 1938, p. 45.] (See also [151].)

[151] — 1948 Teoria sylogizmu zdań asertorycznych u Arystotelesa na tle logiki tradycyjnej (*Studium historyczno-krytyczne*) [The theory of the syllogism of assertoric propositions in Aristotle against the background of traditional logic (A historical-critical study)], Nakładem Towarzystwa Naukowego, Lublin 1948. [*Rozprawy Wydziału historyczno-filologicznego. Sekcja filozoficzna. Towarzystwo naukowe Katolickiego Uniwersytetu lubelskiego 2*]

Janina Kotarbińska

[152] Janina Kotarbińska 1955 *Definicja* [On definition], Studia Logica 2, 1955, pp. 301–327. [Review: Bolesław Sobociński, The Journal of Symbolic Logic 21 (3), 1956, p. 312.]

André Lalande

[153] André Lalande 1898 L'Idée directrice de la dissolution, opposée à celle de l'évolution dans la méthode des sciences physiques et morales, Librairie Félix Alcan, Paris 1898. (Thèse de doctorat ès Lettres, Paris 1899. 'Une édition commerciale... a paru en 1899 sous le titre : *La Dissolution, opposée à l'évolution dans les sciences physiques et morales*, [154]. 'Une nouvelle édition, revue

et abrégée a paru en 1930' [155]. SUDOC: http://www.sudoc.fr/004663020.)

[154] — 1899 La Dissolution, opposée à l'évolution dans les sciences physiques et morales, Librairie Félix Alcan, Paris 1899. ('Une édition commerciale de la thèse [153]'. SUDOC: http://www.sudoc.fr/004663020.)

[155] — 1930 Les Illusions évolutionnistes, Librairie Félix Alcan, Paris 1930. [*Bibliothèque de philosophie contemporaine*] [Review: Bolesław Sobociński, Przegląd Filozoficzny 35 (1–2), 1932, pp. 160–161.] (SUDOC: http://www.sudoc.fr/012137537)

Czesław Lejewski

[156] Czesław Lejewski 1984 *Ś.P. Bolesław Sobociński* [In the memory of Bolesław Sobociński] [Polish], Znak (Kraków) 36 (351–352) [2–3], 1984, pp. 400–403. (Here, *Ś.P.* = *świętej pamięci* means 'deceased', lit. 'holy memory'. [VFR]. There is a draft [MS] English translation of this item, by V. Frederick Rickey.)

[157] — 1987 *Powojenna działalność logików na Zachodzie* [Postwar activity of logicians in the West], in: [223], pp. 169–172.

George I. Lerski

[158] George I. Lerski 1996 Historical Dictionary of Poland, 966–1945 (With special editing and emendations by Piotr Wrobel and Richard J. Kozicki), Greenwood Press [Greenwood Publishing Group], Westport, Connecticut & London 1996.

Stanisław Leśniewski

[159] Stanisław Leśniewski 1988 S. Leśniewski's Lecture Notes in Logic (edited by Jan T. J. Srzednicki, and Zbigniew Stachniak), Kluwer Academic Publishers, Dordrecht, Boston, and London 1988. [Review: Peter Simons, History and Philosophy of Logic 11, 1990, pp. 107—110.]

[160] — 1992 Collected Works (edited by Stanisław J. Surma, Jan T. Srzednicki, and Dene I. Barnett, with an annotated bibliography by V. Frederick Rickey), Kluwer Academic Publishers, Dordrecht, Boston 1992. (In 2 volumes, with continuous page-numbering, 1: pp. i–xvi (Introduction by the editors) and pp. 1–382, 2: pp. 383–794).

[161] — 2015 Pisma zebrane / Sobrannye sočineniya / Gesammelte Schriften (edited with an Introduction and a Postscript [*Stanisław Leśniewski's Life and Work*] by Jacek [Juliusz] Jadacki), Wydawnictwo Naukowe Semper [Semper Scientific Publishers], Warsaw 2015. (In 2 volumes, with continuous page-numbering, 1: 468 + 2: 408, i.e., 876 pp. (Contains Russian, German and Polish originals.)

Jan Łukasiewicz

[162] Jan Łukasiewicz 1950 *W sprawie aksjomatiky implikacyjnego rachunku zdań* [Concerning an axiom system of the implicational propositional calculus], VI Zjazd Matematyków Polskich (Warszawa, 20–23 IX 1948) [The Sixth Congress of the Polish Mathematicians, 20–23 IX 1948], Annales de la Société Polonaise de Mathématique (Kraków) [Supplement] 22, 1950, pp. 87–92. [Review: Bolesław Sobociński, The Journal of Symbolic Logic 20 (2), 1955, pp. 173–174.]

Marcin Łyczak
co-authors: Marek Porwolik, Kordula Świętorzecka

[163] Marcin Łyczak, Marek Porwolik, and Kordula Świętorzecka 2016 *The Universe in Leśniewski's mereology: some comments on Sobociński's reflections*, Axioms 5 (3), 23, 2016, 13 pp.

Storrs McCall (ed.)

[164] Storrs McCall (ed.) 1967 Polish Logic 1920–1931, Clarendon Press, Oxford 1967.

W. Moore McLean

[165] W. Moore McLean 1938 *Michalski, Stanisław: authors-professors and educators*, in: Notable Personalities of Polish Ancestry – Znamienite osobistości pochodzenia Polskiego, The Unique Press, Detroit IL 1938. (Electronic reproduction [s.l.]: HathiTrust Digital Library, 2010. MiAaHDL.)

Carew Arthur Meredith
co-authors: Arthur N. Prior

[166] Carew Arthur Meredith, and Arthur N. Prior 1968 *Equational logic*, Notre Dame Journal of Formal Logic 9, 1968, pp. 212–226. (Herein: Carew A. Meredith "Letter of December 13, 1957" to Arthur N. Prior. See also Arthur N. Prior, *Corrigendum to C. A. Meredith's and my paper 'Equational logic'*, Notre Dame Journal of Formal Logic 10 (4), 1969, p. 452.)

Grigore C. Moisil

[167] Grigore C. Moisil 1935 *Recherches sur l'algèbre de la logique*, Annales Scientifiques de l'Université de Jassy (Iaşi, Romania) 22, 1935, pp. 1–117. (Not included in the collection [168].)

[168] — 1972 Éssais sur les logiques non-chryssippiennes, Editura Academiei R.S.R., Bucharest 1972. (This is a collection of papers, mainly reprints and/or French versions of Romanian originals, dated 1939–1971. For Moisil, 'non-chrysippean' logic means, in general, non-classical, and, more specifically, multi-valued logic, although intuitionistic and modal logics would also fit, now and then, the description. In his – otherwise copious – bibliography, he refers explicitly only to Sobociński's work on many-valued logics, viz., to [12], [20], and [37].)

Charles W. Morris

[169] Charles W. Morris 1939 Review of: *Organon. – International Review. Published by the Mianowski Institute for the Promotion of Science and Letters. Warsaw, Poland. Editor Stanisław Michalski. Volume I, 1936; Volume II, 1938*, in: Isis (*A Journal of the History of Science Society*) 30 (2) [May] 1939, pp. 297–298.

Wojciech Jerzy Muszyński

[170] Wojciech Jerzy Muszyński 2011 Duch młodych. Organizacja Polska i Obóz Narodowo-Radykalny w latach 1934–1944, *Od studenckiej rewolty do konspiracji niepodległościowej* [The spirit of the young: the Polish National Radical Camp in the years 1934–1944. From student revolts to conspiracy for Polish independence], Instytut Pamięci Narodowej, Warsaw 2011. [Review: Jan Peczkis, *A long-overdue, magisterial analysis of the ONR (Oboz Narodowo-Radykalny / Polish National Radical Camp)*,

@ http://www.glaukopis.pl/images/artykuly-polskojezyczne/Mr_Jan-Peczkis-Review-_of_Duch_Mlodych.pdf]

Maria Ossowski
co-authors: Stanisław Ossowski

[171] Maria Ossowski, and Stanisław Ossowski 1929 *Problematyka naukoznawcza* [The problems of science studies], Paper presented at Science Studies Seminar, organized by Stanisław Michalski within the framework of the Józef Mianowski Fund. Cf. [172].

[172] — 1936 *The science of science*, Organon 1, 1936, pp. 1–12.

Julien Pacotte

[173] Julien Pacotte 1931 La Pensée technique, Librairie Félix Alcan, Paris 1931 [*Bibliothèque de philosophie contemporaine*] [Review: Bolesław Sobociński Przegląd Filozoficzny 34 (4), 1931, p. 274.] (SUDOC http://www.sudoc.fr/099077396)

Dariusz Piętka

[174] Dariusz Piętka 2008 *Stanisław Jaśkowski's logical investigations*, Organon 37 (40), 2008, pp. 39–69.

Jan Piskurewicz

[175] Jan Piskurewicz 1993 W służbie nauki i oświaty: Stanisław Michalski, 1865–1949 [In the service of science and education], Polska Akademia Nauk, Instytut Historii, Nauki, Oświaty i Techniki [IHNOiT], Warsaw 1993.

Marek Porwolik
co-authors: Kordula Świętorzecka; *see also* Marcin Łyczak

[176] Marek Porwolik 2018 *B. Sobocińskiego ujęcie powszechników jako intensjonalnych korelatów definicji przez postulaty* [B. Sobociński's account of universals as intentional correlates of definitions by postulates],
http://www.filozofia.uj.edu.pl/documents/4371778/134647442/porwolik.pdf/

[177] Marek Porwolik, and Kordula Świętorzecka 2016 *O pewnym przesądzie dotyczącym uniwersaliów. Uwagi do sformalizowanego przez Bolesława Sobocińskiego argumentu na rzecz tezy o nieistnieniu powszechników podanego przez Stanisława Leśniewskiego* [On a certain oversight concerning universals. Comments on an argument formalized by Bolesław Sobociński for the thesis of Stanisław Leśniewski that there are no universals], in: [123].

[178] — 2018 *Bolesław Sobociński. Sobociński on universals: Leśniewski's nominalism and Sobociński's metaconceptualism*, in: [139], pp. 615–632.

Richard L. Poss

[179] Richard L. Poss 1970 Weak Forms of the Axiom of Constructibility, PhD Dissertation, University of Notre Dame, Notre Dame IN, 1970 (under the direction of Bolesław Sobociński).

[180] — 1971 *Weak forms of the axiom of constructibility*, Notre Dame Journal of Formal Logic 12 (3), 1971, pp. 257–299.

Helena Radlińska
co-authors: Irena Lepalczyk

[181] Helena Radlińska, and Irena Lepalczyk 1967 Stanisława Michalskiego autobiografia i działalność oświatowa [Stanisław Michalski: The autobiography and the didactic activity] (with an introduction by Tadeusz Kotarbiński), Zakład Narodowy im. Ossolińskich, Wrocław 1967 [*Komitet Nauk Pedagogicznych Polskiej Akademii Nauk. Źródła do dziejów myśli pedagogicznej 9*].

Helena Rasiowa

[182] Helena Rasiowa 1951 *Algebraic treatment of the functional calculi of Heyting and Lewis*, Fundamenta Mathematicae 38, 1951, pp. 99–126. (Paper based on the author's PhD Dissertation at the University of Warsaw, under the supervision of Andrzej Mostowski.)

[183] — 1955 *O pewnym fragmencie implikacyjnego rachunku zdań* [On a fragment of the implicational propositional calculus], Studia Logica 3 (1), 1955, pp. 208–226. (Paper based on the author's Master's thesis, supervised first by Bolesław Sobociński, at the Warsaw Underground University, lost during a Warsaw bombing, in 1944, and re-written after the WWII, under the supervision of Andrzej Mostowski.) [Review: Johannes Bendiek, The Journal of Symbolic Logic 22 (3), [September] 2014, p. 330. 'Die Ergebnisse dieser Arbeit wurden bereits 1945 von der Verfasserin als Dissertation für den Grad des Magisters der Philosophie der Universität Warschau vorgelegt.']

Nicholas Rescher

[184] Nicholas Rescher 1968 *Many-valued logic* (*I Historical Background, II A survey of many-valued logic*), Chapter VI in: [185], I: pp. 54–62, II: pp. 63–125. (No reference to Sobociński.)

[185] — 1968a Topics in Philosophical Logic, Kluwer / D. Reidel Publishing Company, Dordrecht etc. 1968 [*Synthese Library 17*].

[186] — 1969 Many-valued Logic, McGraw-Hill Inc., New York 1969. [Review: Elliott Mendelson, The Journal of Philosophy 67 (13), 1970, pp. 457–458.]

Adrian Rezuş

[187] Adrian Rezuş 2020 Witness Theory, *Notes on λ-calculus and Logic*, College Publications, London [March] 2020 [*Studies in Logic 84*].

V. Frederick Rickey
see also Stanisław Leśniewski, Jan T. J. Srzednicki (ed.)

[188] V. Frederick Rickey 1968 An Axiomatic Theory of Syntax, PhD Dissertation, University of Notre Dame, Notre Dame IN, 1968 (under the direction of Bolesław Sobociński, cf. [189], [190].)

[189] — 1972 *Axiomatic inscriptional syntax, Part I: General syntax*, Notre Dame Journal of Formal Logic 13 (1), 1972, pp. 1–33.

[190] — 1972a *Axiomatic inscriptional syntax, Part II: The syntax of protothetic*, Notre Dame Journal of Formal Logic 14 (1), 1973, pp. 1–52. (Reprint in: [221], pp. 207–288.)

[191] — 1977 *A survey of Leśniewski's logic*, Studia Logica 36 (4), 1977, pp. 407-426. (Reprint in: [221], pp. 23–41.)

[192] — 1981 Bibliography of Bolesław Sobociński, Archives of the Warsaw University, K996 (MS sent by the author to Krysztof Tatarkiewicz, cf. [228], [229].)

[193] — 1982 *Bolesław Sobociński 1906–1980*, Proceedings and Addresses of the American Philosophical Association 55 (4), [March] 1982, pp. 498–499.

[194] — 1985 *Interpretations of Leśniewski's ontology*, Dialectica 39 (3), 1985, pp. 181–192.

[195] — 1992 *An annotated Leśniewski bibliography*, in: [160], pp. 711–785.

[196] — 2011 *Polish logic from Warsaw to Dublin, the life and work of Jan Łukasiewicz*, Proceedings of the Canadian Society for History and Philosophy of Mathematics 24 [CSHPM], 2011, pp. 93–109. ('The joint meeting of the Canadian Society for the History and Philosophy of Mathematics and the British Society for the History of Mathematics, Trinity College, Dublin, 15–17 July 2011.')

[197] — 2014 An Annotated Leśniewski Bibliography, West Point NY, February 1, 2014, 112 pp. @ https://www.academia.edu/5912188/An_Annotated_Leśniewski_Bibliography.

[198] — 2018 Bibliography of Bolesław Sobociński (1906–1980). (Herein: *Appendix III. Students of Sobociński.*) [draft as of] February 1, 2018.

[199] — 2018a *Professor Bolesław Sobociński and logic at Notre Dame*, May 30, 2018. Now published in: M. Zack, and D. Schlimm (eds.), Research in History and Philosophy of Mathematics (*Proceedings of the Canadian Society for History and Philosophy of Mathematics/ Société canadienne d'histoire et de philosophie des mathématiques*, 2019), Birkhäuser [Springer International Publishing], Basel 2020, pp. 1–24.

Jacob Ridder

[200] Jacob Ridder 1948 *Über mehrwertige Aussagenkalküle und mehrwertige engere Prädikatenkalküle*, I–III, KNAW Proceedings (Section of Sciences) 51, 1948, I: pp. 670–680, II: pp. 836–845, III: pp. 991–995 (= Indagationes Mathematicae 10, 1948, I: pp. 221–231, II: pp. 264–273, III: pp. 324–328). [Review (in French): Bolesław Sobociński, The Journal of Symbolic Logic 14 (2), 1950, p. 139 (I), and *ibid.* 14 (4), 1950, pp. 261–262 (II–III).]

[201] — 1949 *Sur quelques logiques multivalentes*, in: E. W. Beth, H. J. Pos, and J. H. A. Hollak (eds.) Proceedings of the Tenth International Congress of Philosophy (Amsterdam, August 11–18, 1948), 1, North-Holland Publishing Company, Amsterdam 1949, pp. 728–730.

Alan Rose

[202] Alan Rose 1950–1951 *Completeness of Łukasiewicz-Tarski propositional calculi*, Mathematische Annalen 122, 1950–1951, pp. 296–298. [Review: A. R. Turquette, The Journal of Symbolic Logic 16 (3), 1951, pp. 228–229.] (Online in [1], via the *Göttinger Digitalisierungszentrum*, SUB Göttingen.)

[203] — 1951 *Axiom systems for three-valued logic*, Journal of the London Mathematical Society 26 (1), 1951, pp. 50–58.

[204] — 1952 *The degree of completeness of the m-valued Łukasiewicz propositional calculus*, The Journal of the London Mathematical Society 27 (1), 1952, pp. 92–102.

[205] — 1952a *Le degré de saturation du calcul propositionnel implicatif à trois valeurs de Sobociński*, Comptes Rendus Hebdomadaires des Séances de l'Académie des Sciences (Paris) 235, 1952, pp. 1000–1002. [Review: Gene F. Rose, The Journal of Symbolic Logic 19 (1), 1954, p. 56.]

[206] — 1953 *A formalization of Sobociński's three-valued implicational propositional calculus*, The Journal of Computing Systems 1 (3), 1953, pp. 165–168. [Review: Gene F. Rose, The Journal of Symbolic Logic 19 (2), 1954, p. 144.]

[207] — 1956 *An alternative formalisation of Sobociński's three-valued implicational propositional calculus*, Zeitschrift für mathematische Logik und Grundlagen der Mathematik 2, 1956, pp. 156–162.

[208] — 1956a *Some formalisations of $\aleph_0$-valued propositional calculi*, Zeitschrift für mathematischen Logik und Grundlagen der Mathematik 2, 1956, pp. 204–209. [Review: Gene F. Rose, The Journal of Symbolic Logic 29 (4), 1964, p. 213.]

[209] — 1965 *Formalisations of certain propositional calculi with partially variable functors*, Zeitschrift für mathematische Logik und Grundlagen der Mathematik 11 (2), 1965, pp. 177–180.

J. Barkley Rosser
co-authors: Atwell Rufus Turquette

[210] J. Barkley Rosser, and Atwell Rufus Turquette 1952 Many-valued Logics, North Holland Publishing Company, Amsterdam 1952 [*Studies in Logic and the Foundations of Mathematics 7*]. [Review: Bolesław Sobociński, The Journal of Symbolic Logic 20 (1), 1955, pp. 45–50.]

George Sanford

[211] George Sanford 2003 Historical Dictionary of Poland, The Scarecrow Press Inc., Lanham MD & Oxford 2003 [second edition] [*Historical Dictionaries of Europe: European Historical Dictionaries 41*]. (Select Bibliography: *Interwar Poland* (1918–1939), pp. 261–262, *Poland in Word War II*, pp. 262–267.)

Kenneth M. Sayre

[212] Kenneth M. Sayre 2014 Adventures in Philosophy at Notre Dame, University of Notre Dame Press, Notre Dame IN [April 30] 2014.

Thomas W. Scharle

[213] Thomas W. Scharle 1962 *A diagram of the functors of the two-valued propositional calculus*, Notre Dame Journal of Formal Logic 3, 1962, pp. 243–255. (This is, essentially, the author's Master's Thesis in philosophy under the direction of Bolesław Sobociński.)

[214] — 1962a *Note to my paper: 'A diagram of the functors of the two-valued propositional calculus'*, Notre Dame Journal of Formal Logic 3, 1962, pp. 287–288.

[215] — 1965 *Axiomatization of propositional calculus with Sheffer functors*, Notre Dame Journal of Formal Logic 6 (3), 1965, pp. 209–217.

[216] — 1966 *Single axiom schemata for D and S*, Notre Dame Journal of Formal Logic 7 (4), 1966, pp. 344–348.

[217] — 1973 Axiomatization of Fragments of S5, PhD Dissertation, University of Notre Dame, Notre Dame IN, 1973 (under the direction of of Bolesław Sobociński, cf. [218]).

[218] — 1975 *Axiomatization of fragments of S5*, Notre Dame Journal of Formal Logic 16 (1), 1975, pp. 45–70.

Zbigniew Sebastian Siemaszko

[219] Zbigniew Sebastian Siemaszko 1982 Narodowe Siły Zbrojne [The National Armed Forces], Odnowa, London 1982. (First edition. Many later reprints.)

Jan T. J. Srzednicki (ed.)
co-editors: J. Czelakowski, V. Frederick Rickey, Zbigniew Stachniak

[220] Jan T. J. Srzednicki, V. Frederick Rickey, and J. Czelakowski (eds.) 1984 Leśniewski's Systems: Ontology and Mereology, Martinus Nijhoff Publishers [Kluwer Academic Publishers], The Hague, etc. & Ossolineum [The Publishing House of the Polish Academy of Sciences], Wrocław – Warsaw 1984. [*Nijhoff International Philosophy Series 13*]

[221] Jan T. J. Srzednicki, and Zbigniew Stachniak (eds.) 1998 Leśniewski's Systems: Protothetic, Kluwer, Dordrecht 1998. [Review: Arianna Betti, Studia Logica 18, 2001, pp. 401–404.]

Henryk Stonert

[222] Henryk Stonert 1955 *Sprawozdanie z I Konferencji Logików (Warszawa, grudzień 1952)* [Report on the Ist Conference of Logicians (Warsaw, December 1952], Studia Logica 2, 1955, pp. 251–266. [Review: Bolesław Sobociński, The Journal of Symbolic Logic 21 (3), 1956, p. 311.]

Wiesław Strzałkowski (ed.)
co-editors: Edward Szczepanik

[223] Wiesław Strzałkowski, and Edward Szczepanik (eds.) 1987 Filozofia polska na obczyźnie [Polish philosophy in exile], Polskie Towarzystwo Naukowe na Obczyźnie [The Polish Society of Arts and Sciences Abroad], London 1987 [*Prace Kongresu Kultury Polskiej na Obczyźnie 6 (Londyn, 14–20 września 1985)*]

Kordula Świętorzecka
see also Marcin Łyczak, Marek Porwolik

[224] Kordula Świętorzecka 2018 *Bolesław Sobociński, the ace of the second generation of the Lvov-Warsaw School*, in: [139], pp. 599–612.

Alfred Tarski

[225] Alfred Tarski 1938 *Ein Beitrag zur Axiomatik der Abelschen Gruppen*, Fundamenta Mathematicae 30 (1), 1938, pp. 253–256. (Online in [1]. Reprinted in: [226], 2.)

[226] — 1986 Collected Papers 1–4 (edited by Steven R. Givant, Ralph N. McKenzie, and Gian-Carlo Rota), Birkhäuser Verlag, Basle etc. 1986 [*Contemporary Mathematicians*]. (1: 1921–1934, 2: 1935–1944, 3: 1945–1957, 4: 1958–1979.)

Krzysztof Tatarkiewicz

[227] Krzysztof Tatarkiewicz 1996 *Trochę wspomnień, trochę refleksji* [A few memories, a few reflections], in: [147], 2, pp. 5–23.

[228] — 1998 *Profesor Sobociński i kolega Bum* [Professor Sobociński and the colleague Bum], Wiadomości Matematyczne 34, 1998 pp. 122–146.

[229] — 1998a *Logik i polityk (Bolesław Sobociński)* [The Logician and the politician (Bolesław

Sobociński)], in: [138], pp. 176–182.

[230] — 1998b *Co to są matematycy polskiego pochodzenia na obczyźnie?* [What are the mathematicians of Polish descent in exile?], in: [138], pp. 281–283.

[231] — 2005 *Korekty (dotycza wspomnien o B. Sobocińskim i K. Zorawskim)* [Corrections concerning my reminiscences about B. Sobociński and K. Zorawski], in: [233], p. 179.

Toshiharu Waragai

[232] Toshiharu Waragai 2005 *Letters and typescripts sent to Bochenski from Sobociński: Historical documents about Leśniewski's refutation of universals*, Reports of the Keio Institute of Cultural and Linguistic Studies 36 (3), 2005, pp. 203–228.
(Cf. `https://ci.nii.ac.jp/author?q=Waragai+Toshiharu`.)

Witold Więsław (ed.)

[233] Witold Więsław (ed.) 2005 Sławne dzieła matematyczne i rocznice, *Materiały XVIII Ogólnopolskiej Szkoły Historii Matematyki* [Famous mathematical works and anniversaries. Materials from the 18th National School of History of Mathematics], (Białystok – Suprasl, 31 maja – 4 czerwca 2004, Wyzsza Szkoła Matematyki i Informatyki Uzytkowej w Białymstoku), Białystok 2005.

Jan Woleński

[234] Jan Woleński 1989 Logic and Philosophy in the Lvov-Warsaw School, Kluwer Academic Publishers, Dordrecht, etc. 1989 [*Synthese Library 198*].

[235] — 1994 Philosophical Logic in Poland, Kluwer Academic Publishers, Dordrecht, etc. 1994. [*Synthese Library 228*]

[236] — *Polish attempts to modernize Thomism by logic (Bocheński and Salamucha)*, Studies in East European Thought 55, 2003, pp. 299–313. (Reprinted in: [237], **1**, pp. 51–66.)

[237] — 2013 Historico-philosophical Essays 1, Copernicus Center Press, Kraków 2013.

[238] — 2015 *Lvov-Warsaw School.* Online in [8],
`https://plato.stanford.edu/entries/lvov-warsaw/`.

[239] — 2018 *Bolesław Sobociński.* Online in [2],
`http://www.ipsb.nina.gov.pl/a/biografia/boleslaw-sobocinski`.

Joshua D. Zimmerman

[240] Joshua D. Zimmerman 2015 The Polish Underground and the Jews, 1939–1945, Cambridge University Press, New York 2015.

Leszek Żebrowski (ed.)

[241] Leszek Żebrowski (ed.) 1994–1996 Narodowe Siły Zbrojne, *Dokumenty, struktury, personalia* [The National Armed Forces: Documents, structures, *personalia*] (1–3), Wydawnictwo Burchard Edition, Warszaw 1994–1996.

Treading in Brouwer's footsteps

Wim Veldman
Institute for Mathematics, Astrophysics and Particle Physics,
Faculty of Science, Radboud University, Nijmegen
W.Veldman@science.ru.nl

Fools rush, where angels fear to tread
A. Pope

Abstract. We take a retrospective look at work that has been done in the field of intuitionistic mathematics by J. J. de Iongh and some of his students in the period 1963–1985. We go into the philosophy of intuitionistic mathematics, and sketch some results concerning the continuity of real functions, Dedekind-infinite sets, the continuum hypothesis, the perfect set theorem, the completeness of predicate logic, descriptive set theory and Ramsey's Theorem.

1 Introduction

Johan J. de Iongh (1915–1999)[1] founded the department of mathematics at Radboud University, formerly called: Katholieke Universiteit Nijmegen. He lectured on the foundations of mathematics and on the philosophy and history of mathematics. In particular, he taught intuitionistic mathematics and he had a number of master and Ph. D. students working in this field. We will pay attention to the work of Harrie de Swart, Wim Gielen, myself and Mervyn Jansen.

De Iongh had been a student of mathematics at the University of Amsterdam, a somewhat special one. Legend has it that his final examination ('*doctoraal examen*') took place, after many, many years, during a walk in the Vondelpark.

2 The philosophical basis

Brouwer famously held that mathematics, essentially, is a languageless and playful activity of the solitary mind. Language comes in if one wants to support one's own

[1] See [28].

weak memory, or if one hopes to bring some companion to mathematical activity similar to one's own. As a means of communication, either with oneself, or with someone else, language always may fail.

De Iongh shared these views. He saw a connection with Plato, who, a great writer himself, nevertheless distrusted written language and held that a moment of truth only may occur in a deeply personal conversation in which both partners put themselves at risk.

Brouwer held negative views on logic. Logic only offers a report of the linguistic accompaniment of mathematical activity. In de Iongh's formulation, the rules of logic define a social convention, a definition of *correctness*. But what matters is not correctness, but *truth*.

Brouwer further analyzed that, historically, mathematics went wrong because of its misjudgment of the rôle of logic. The rules one had observed to be valid in the domain of the finite were recklessly extended to the domain of the infinite. Recognizing this mistake, mathematicians should put themselves the task of rebuilding the mathematics of the infinite.

De Iongh hoped to be able to contribute to this program and incited his students to do so likewise.

De Iongh strongly believed in the soundness of the intuitionistic point of view. The mathematician, *'naturaliter'*, thinks constructively and only is spoiled by education to accept also indirect arguments.

He liked to tell us that the controversy about the question if mathematics is about our own constructions or about objects that exist independently of us in some other world has its roots in antiquity. As Proclus tells us, in Plato's Academy there was already discussion if every proposition essentially is a problem (that is: a construction we ourselves have to do) or if it would be better to say that every proposition essentially is a theorem (a truth we find by contemplation).[2]

He also mentioned Cusanus (1401–1464). Cusanus developed the vision that man, created by God, himself is a creator, a creator/fabricator of mathematics. Mathematics puts him on the way to understand God's creation.

De Iongh emphasized that mathematics is a *human* undertaking and not an *angelic* one. Angels may have superhuman capabilities like the one to decide, for every decidable[3] $P \subseteq \mathbb{N}$, $\exists n[P(n)] \vee \forall n[\neg P(n)]$, but we have not and should not, when doing mathematics, pretend we have.

The intuitionistic critique of mathematics shows that mathematics is not the rock of certainty one might believe it is. Every formalization is provisional only. We

[2]The 48 propositions of Euclid's First Element are divided into 14 problems and 34 theorems.

[3]$P \subseteq \mathbb{N}$ is *decidable* if and only if $\forall n[P(n) \vee \neg P(n)]$. The disjunction is taken in a constructive sense.

know only in part[4] and must begin anew again and again, trying to find together a little piece of knowledge that, tomorrow, we perhaps don't understand any more, find silly or or put into doubt.

3 The Amsterdam background

De Iongh, in Amsterdam, belonged to a group of people discussing Brouwer's philosophy and working on Brouwer's program.

De Iongh had his hesitations about Brouwer's views, although he revered him. He was influenced by G. Mannoury, a main figure in the *Signific Circle* that also counted Brouwer himself among its members. The aim of *Significs* is to study how people come to understand each other. Human beings try to influence each other's behaviour in *linguistic acts, 'taaldaden'*. A linguistic act has different meanings for the '*speaker*' than for the '*hearer*'.

Even if I am doing mathematics on my own, I have to do with such acts as I may be said to speak to myself and to hear myself. But surely, if we are doing mathematics together, the signific approach helps us to make sense of mathematics as a common undertaking and to ackowledge the rôle of applied mathematics.

It becomes difficult then to maintain Brouwer's separation of the languageless mathematical construction and its description by language. One comes to see that the linguistic formulation, in a sense, influences and may be said to be part of the mathematical activity.

In [16], de Iongh cautiously formulates his doubts. He mentions the discussion between H. Freudenthal and A. Heyting on the meaning of an implication $P \to Q$ in intuitionistic mathematics. Freudenthal had started it by bringing forward that having a method to produce a proof of Q from a proof of P only makes sense if one has a proof of P, thereby concluding that $P \to Q$ is equivalent to $P \wedge Q$. Heyting answered by suggesting that the method might start from a proof of P that is only *intended*, not actually realized.

He also mentions G. F. C. Griss, who wanted to go further than Brouwer and banish every reference to a construction that has not actually been carried out from the language of mathematics. Griss thus came to reject negation and started to work on a *negationless* intuitionistic mathematics.

De Iongh expresses his affinity with the position taken by Mannoury and D. van Dantzig. Van Dantzig judged, like Griss, that *affirmative* mathematics is the part of mathematics that makes the most sense but did not think negation should be always avoided.

[4] 1 Cor. 13,9.

From the signific point of view, the useful thing about intuitionistic mathematics, also for someone who does not share Brouwer's radical philosophy, is that it offers us a mathematical language more subtle than the usual one. It thus contributes to the study of mutual understanding in mathematics. A given classical statement has a gamut of possible translations in the language of constructive mathematics.

As an example, de Iongh treats the statement

$$(*)\ \lim_{n\to\infty} a_n = a := \forall m \exists n \forall p > n[|a_p - a| < \frac{1}{m}].$$

He claims that, using double negations, there are nine classically equivalent but intuitionistically non-equivalent ways of writing this formula.

Today, we must say that this analysis is incomplete. De Iongh does not take into account the fact that $\forall m \exists n[P(m,n)]$ is, classically as well as intuitionistically, equivalent to $\exists \alpha \forall m[P(m, \alpha(m))]$. If one does so, the number of possible translations increases substantially.

There are uncountably many sets $\mathcal{X} \subseteq \mathcal{N}$ such that $\mathbf{Fin} \subseteq \mathcal{X} \subseteq \mathbf{Fin}^{\neg\neg}$, where $\mathbf{Fin} := \{\alpha \mid \exists n \forall m > n[\alpha(m) = 0]\}$, see [27] and [31]. Using this fact, one must be able to find countably intuitionistically non-equivalent formulas that are classically equivalent to the formula $(*)$.

4 The first axiom of continuous choice and the pointwise continuity of real functions

We quote from the notes de Iongh made of a course of lectures given by Heyting, see [14, Section 8.2, p. 60]. We slightly paraphrase the text.

> One directly understands the following theorem.
>
> *Thm.* A function f from $[0, 1]$ to $\mathcal{R}$ that is defined at every point x in $[0, 1]$ is (pointwise) continuous.
>
> *Pf.* One must be able to find $f(x)$ also if x is given by a *choice sequence* of rational approximations. For such a sequence, given any n, we are able to indicate its n-th approximation $x(n) = (x'(n),\ x''(n))$ but we do not know the infinite sequence as a whole. Also if x is given by a law, one may think of a choice sequence that gives us the same rational approximations.
>
> If we want to determine an ε-approximation of $f(x)$, we must do so knowing only some $x(n) = x(n_\varepsilon)$. So for every x^* in a δ_{n_ε}-neighbourhoud of x, $f(x^*)$ will be in the given ε-neighbourhood of $f(x)$.

Surprisingly, in Heyting's book [15], carefully revised by de Iongh, this theorem and this argument do not occur. Instead, we find the stronger theorem:

> A function f from $[0,1]$ to $\mathcal{R}$ that is defined at every point x in $[0,1]$ is uniformly continuous on $[0,1]$.

This theorem is proven as a consequence of the fan theorem, see [15, Section 3.4.3]. Heyting is following Brouwer's argument in [4, §3].

De Iongh assured me that Brouwer did not accept the more simple argument for pointwise continuity just given. De Iongh once asked him about it and Brouwer sent him home with the observation: '*Do you really think, Johan, it could be that simple?*'

Why Brouwer did reject the simple argument, is not clear to me. It is connected with his distinction between *'points'* and *'point cores'*, or, in Heyting's terminology, between *'number generators'* and *'numbers'*. A point core is an equivalence class of coinciding points, where each point is an infinite sequence of rational approximations. The objection would be that, if we find δ by considering just one representative x of the point core $[x]$, it would be dangerous to claim that δ would suit the point core $[x]$.

De Iongh, probably overawed by Brouwer, agreed with his objection and even convinced C. Parsons that the simple argument could not work, see [13, page 448, footnote g].

In fact, Brouwer's own text invites the question for the simple argument. The fan theorem, as formulated in [4, §2, Theorem 2] is, in a somewhat simplified[5] version, the following statement[6]:

> If $\forall\alpha \in \mathcal{C}\exists n[\alpha R n]$, then $\exists m \forall\alpha \in \mathcal{C}\exists n \forall\beta \in \mathcal{C}[\overline{\alpha}m \sqsubset \beta \rightarrow \beta R n]$.

Brouwer starts his proof by making the following assumption:

> If $\forall\alpha\exists n[\alpha R n]$, then
> $\exists\gamma[\forall\alpha\exists n[\gamma(\overline{\alpha}n) \neq 0] \ \wedge \ \forall s[\gamma(s) \neq 0 \rightarrow \exists n \forall\alpha[s \sqsubset \alpha \rightarrow \alpha R n]]$.

He makes *this* assumption without giving an explanation. He then develops a complicated argument in order to prove:

(*): If $\forall\alpha \in \mathcal{C}\exists n[\gamma(\overline{\alpha}n) \neq 0]$, then $\exists m\forall\alpha \in \mathcal{C}\exists n \leq m[\gamma(\overline{\alpha}n) \neq 0]$.

---

[5]The simplification consists in the fact that we do not consider, like Brouwer, an arbitrary *fan* or *finitary spread*, but only Cantor space $\mathcal{C}$.

[6]In Section 13, there is a survey of the notation we use.

C. Parsons rightfully asks the question why Brouwer does not conclude the pointwise continuity of real functions from his first assumption, see [13, page 448, footnote g] and [29].

Brouwer's first assumption has been called *Brouwer's principle (for numbers)* by S. C. Kleene in [18, §7]. We call it the *First Axiom of Continuous Choice.*

Kleene tells us, in the preface to [18], that he visited Amsterdam from January to June, 1950, and that de Iongh was among those who oriented him in intuitionistic mathematics.

The name *Fan Theorem* now mostly is used for the statement $(*)$.

5 Brouwer's thoughts on $\mathcal{N}$ and spreads

Brouwer stole the term *set*, '*Menge*', from Cantor but gave the term his own meaning. Only later the term 'set' was replaced by the term *spread*, '*spreiding*'.

The difficulty Brouwer tried to solve lies with Cantor's discovery and acceptance of uncountable sets, like the set $\mathcal{N} = \mathbb{N}^{\mathbb{N}}$ of all infinite sequences $\alpha = \alpha(0), \alpha(1), \alpha(2), \ldots$ of natural numbers. Brouwer could not understand how such an uncountable set could be the result of *taking together its (previously constructed?) elements into a new whole*, as Cantor liked to formulate it. He came to see that it nevertheless could make sense to speak of *'all possible elements of $\mathcal{N}$'*, where one should emphasize the expression *possible.*

The idea is that an element α of $\mathcal{N}$ originates from a *step-by-step-construction*:

$$\textit{choose } \alpha(0),\ \textit{choose } \alpha(1),\ \textit{choose } \alpha(2),\ \ldots$$

There may be an algorithm underlying the successive choices, but not necessarily so, and, in general, one does not know if there is such an algorithm. Possibly, α is always under construction and unfinished, a *project*[7] rather than an *object.* I may change my mind, however. Even if, up to some moment n, I have ordered, step by step, by free decisions, $\alpha(0) = 0,\ \alpha(1) = 0,\ \alpha(2) = 0, \ldots, \alpha(n-1) = 0$, I may decide to 'freeze' α from now on and prescribe $\forall k[\alpha(n+k) = 1]$[8]. I also may get it in my head to make α dependent on the future course of my mathematical experience, by determining, for instance, that, for each n, $\alpha(n) \neq 0$ if and only if, at *stage* n, I find a proof of Riemann's hypothesis[9].

Grasping the set $\mathcal{N}$ as a whole, for Brouwer, means seeing the framework in which all possible infinite sequences will grow, although only a very tiny number of

[7]The use of this term is due to de Iongh.
[8]See Brouwer's argument for [4, Theorem 1].
[9]Brouwer explores this possibility in [5].

them will have been realized until now. The set is there *before* its elements, while in Cantor's picture, one first should have the elements in order to take them together into a set.

It now is important to agree on the meaning of statements of the form

$$\forall\alpha\exists n[\alpha R n].$$

For such a statement to be true, one must be able to produce the promised n in the particular case that $\alpha = \alpha(0), \alpha(1), \alpha(2), \ldots$ is given step-by-step, by free choices. If α thus comes to be known, the promised n will be produced at a moment that only finitely many values of α, say $\alpha(0), \alpha(1), \ldots, \alpha(m-1)$, have been decided upon. Now observe that *every* α, also an algorithmic one like the sequence with the constant value 0, is obtained as the result of an infinite sequence of free choices.

This line of thought makes one accept *Brouwer's Continuity Principle*:

If $\forall\alpha\exists n[\alpha R n]$, then $\forall\alpha\exists m\exists n\forall\beta[\overline{\alpha}m \sqsubset \beta \rightarrow \beta R n]$.

The Continuity Principle explains the way we think about the set $\mathcal{N}$, just as the principle of complete induction may be considered an explanation of how we think about the set $\mathbb{N}$.

The Continuity Principle may be formulated more strongly. The First Axiom of Continuous Choice that we mentioned earlier is equivalent to the statement[10]:

If $\forall\alpha\exists n[\alpha R n]$, then $\exists\varphi : \mathcal{N} \rightarrow \mathbb{N}\forall\alpha[\alpha R\varphi(\alpha)]$.

The Second Axiom of Continuous Choice is the statement:

If $\forall\alpha\exists\beta[\alpha R\beta]$, then $\exists\varphi : \mathcal{N} \rightarrow \mathcal{N}\forall\alpha[\alpha R(\varphi|\alpha)]$.

One may try to convince oneself that even these stronger versions of the Continuity Principle are reasonable starting points for our mathematical discourse, see [12, §1].

Using the Second Axiom of Continuous Choice, one may obtain a contradiction from the assumption:

$\forall\alpha\exists\beta[\alpha = \underline{0} \leftrightarrow \beta \# \underline{0}]$.

The argument is as follows. Suppose we find $\varphi : \mathcal{N} \rightarrow \mathcal{N}$ such that $\forall\alpha\exists\beta[\alpha = \underline{0} \leftrightarrow \varphi|\alpha \# \underline{0}]$. Find n such that $(\varphi|\underline{0})(n) \neq 0$.

[10]In Section 13, there is a survey of the notation we use.

Find m such that $\forall\alpha[\overline{\underline{0}}m \sqsubset \alpha \rightarrow (\varphi|\underline{0})(n) = (\varphi|\alpha)(n)]$.
Conclude: $\forall\alpha[\overline{\underline{0}}m \sqsubset \alpha \rightarrow \alpha = \underline{0}]$. Contradiction.

Brouwer suggested that, given any well-defined mathematical problem P, one may construct β such that $P \leftrightarrow \beta \# \underline{0}$.

If we follow his suggestion, the contradiction we just obtained forces us to the conclusion us that, in general, the proposition $\alpha = \underline{0}$ is not a well-defined mathematical problem. It is not difficult to see why. In the special case that α is given by an algorithm, the proposition '$\alpha = \underline{0}$' is well-defined indeed and susceptible to Brouwer's treatment. If however, α is the result of an infinite sequence of free choices, there is no mathematical question to solve, as, at every moment, only finitely many values of α are known. α then is a *project* and (perhaps) never defined completely.

The necessary restriction of what sometimes is called the *Brouwer-Kripke principle* to so-called *definite* propositions has been advocated by Johan de Iongh, see [12, §3].

Brouwer's introduction of the concept of a *spread* can be seen as an attempt to formulate more generally what he believed to be able to do with $\mathcal{N}$.

β is a *spread-law* if and only if $\forall s[\beta(s) = 0 \leftrightarrow \exists n[\beta(s * \langle n\rangle) = 0]]$. If $\beta(s) = 0$ we say that s is *admitted by the spread-law* β. β thus defines a collection of (code numbers of) finite sequences of natural numbers: the framework of the spread. If β is a spread-law, then the corresponding *spread* $\mathcal{F}_\beta$ consists of all α such that $\forall n[\beta(\overline{\alpha}n) = 0]$. The elements of $\mathcal{F}_\beta$ are infinite sequences α, possibly given by free choices. Building α in $\mathcal{F}_\beta$, we have freedom of choice within the bounds given by the spread-law β.

Brouwer enuntiates his continuity principle for spreads in general. Formally, the continuity principle for spreads in general is no stronger than the continuity principle for $\mathcal{N}$.

An example of a spread-law is τ_2, defined as follows:

$$\forall s[\tau_2(s) = 0 \leftrightarrow (s \in Bin \wedge \forall i < length(s) - 1[s(i) \leq s(i+1) \leq 1])].$$

Note: $\mathcal{T}_2 := \mathcal{F}_{\tau_2} = \{\alpha \mid \forall i[\alpha(i) \leq \alpha(i+1) \leq 1]\}$. From a classical point of view $\mathcal{T}_2$ is a countable set, as the list of its elements is the following:

$$\underline{0}, \langle 0\rangle * \underline{1}, \langle 0, 0\rangle * \underline{1}, \langle 0, 0, 0\rangle * \underline{1} \ldots$$

Intuitionistically, however, $\mathcal{T}_2$ also contains elements α for which we are unable to decide what is their place on the just-mentioned list. Brouwer's Continuity Principle enables one to derive a contradiction from the assumption that every element of α is on the just-mentioned list.

We thus see that Brouwer's way of thinking about $\mathcal{N}$ has its consequences for sets that from a classical point of view are countable sets.

6 Lawless sequences?

Reflection on the meaning and possible justification of Brouwer's continuity principle has led to the development of the notion of a *lawless sequence* by G. Kreisel and, later, also A.S. Troelstra. At first Kreisel called such sequences *absolutely free sequences*: the term *lawless* was suggested to him by K. Gödel. Lawless sequences are in sharp contrast with *lawlike* sequences. Every infinite sequence determined by an algorithm should be considered lawlike.

Lawless sequences are generated by free choices with a promise that the freedom of the choices will never be restricted, let alone that their future course, from some moment on, will be frozen by an algorithm.

Let α be such a lawless sequence and let us consider the statement: $\forall n[\alpha(n) = 0]$. I never will be able to prove this statement, as this would mean I break my promise never to restrict my freedom. Kreisel concludes: $\neg\forall n[\alpha(n) = 0]$, because, intuitionistically, the truth of a statement coincides with its provability, and, if a statement is unprovable, it must be false.

It seems to me, however, that the statement $\forall n[\alpha(n) = 0]$ should be recognized as an example of a statement that might be true, although we will never be able to prove it. The slogan *'truth is provability'* here finds its limits.

For lawless sequences, the following axiom is defended:

$$(*) \text{ If } A(\alpha), \text{ then } \exists n \forall \beta[\overline{\alpha}n \sqsubset \beta \to A(\beta)].$$

One also assumes: $\forall s \exists \alpha[s \sqsubset \alpha]$. The axiom $(*)$ then implies: for every lawless sequence α, $\neg(\alpha = \underline{0})$.

It is understood that, in the statement $A(\alpha)$ occurring in $(*)$, no lawless sequence but α is mentioned. This restriction has to do with another axiom lawless sequences are supposed to satisfy:

$$\forall \alpha \forall \beta[\alpha = \beta \lor \alpha \neq \beta],$$

One argues that, given α, β, *either* the projects α, β are the same and $\alpha = \beta$, *or* they are different. In the latter case, one never will be able to prove $\alpha = \beta$ and one concludes, going against common sense and identifying, also in this context, truth and provability: $\neg(\alpha = \beta)$.

This axiom is embarassing, also for the reason that the only known sets in constructive mathematics that have a decidable equality are (subsets of) a countable set.

It does not seem to be a fruitful idea to try to isolate the collection of the lawless sequences from the spread $\mathcal{N}$ as a whole, see also [21] and [22]. Note that the predicate *'α is lawless'* is not an *extensional* predicate on $\mathcal{N}$. As we see it, every lawlike sequence might be the result of an infinite sequence of free choices: every lawlike sequence coincides with a lawless sequence.

Troelstra considered the theory of lawless sequences as a valuable *philosophical experiment*, see [23] and [24, Vol. II, Chapter 12].

7 Infinite sets need not be Dedekind-infinite

In [14], p. 72, l. 15–17, one reads:

> *An example of an infinite set such that no 1-1 mapping on a proper subset can be indicated, is still unknown.*

A function f from $V \subseteq \mathcal{R}$ to $\mathcal{R}$ is called *(strongly) 1-1* if and only if $\forall x \in V \forall y \in V[x \#_{\mathcal{R}} y \rightarrow f(x) \#_{\mathcal{R}} f(y)]$.

A set with the above-mentioned property is called: *Dedekind-infinite.*

The question thus is:

> Is every infinite set Dedekind-infinite?

The word *'infinite'* is taken in a constructively strong sense: $V \subseteq \mathcal{R}$ is infinite if there exists a strongly 1-1 function from $\mathbb{N}$ into V.

De Iongh mentioned this problem in his Nijmegen lectures.

We now provide an example of an infinite set $V \subseteq \mathcal{R}$ such that the statement *'V is Dedekind-infinite'* is a *reckless* one.

We use *'reckless'* as a translation of the Dutch *'vermetel'*, the term de Iongh applied to statements like:

$$\exists n \forall i < 99[d(n+i) = 9] \vee \forall n \exists i < 99[d(n+i) \neq 9],$$

where $d : \mathbb{N} \rightarrow \{0, 1, \ldots, 9\}$ is the decimal expansion of π.

A statement is called *reckless* if one understands there is no constructive proof.

Let α be given. Consider

$$V_\alpha := \mathbb{N} \cup \bigcup\{(-\tfrac{1}{2} + n, n + \tfrac{1}{2}) \mid \alpha^n \# \underline{0}\}.$$

We claim:

(i) V_α is infinite, and

(ii) If $\exists f : V_\alpha \rightarrow V_\alpha[f$ *is* 1-1 *and* $\exists y \in V_\alpha \forall x \in V_\alpha[y \#_{\mathcal{R}} f(x)]]$, then $\exists m \exists n[m \neq n \wedge (\alpha^m \# \underline{0} \rightarrow \alpha^n \# \underline{0})]$.

Clearly, V_α is (positively) infinite as there exists a 1-1 mapping from $\mathbb{N}$ into V_α.

Now let $f : V_\alpha \to V_\alpha$ and y in V_α be given such that f is 1-1 and $\forall x \in V_\alpha[y \#_\mathcal{R} f(x)]$.

Define $QED := \exists m \exists n[m \neq n \wedge (\alpha^m \# \underline{0} \to \alpha^n \# \underline{0})]$.

We read QED as: *quod* est *demonstrandum, what* we (still) have to *prove.*

We now distinguish several cases.

Case 1. $y \in \bigcup\{(-\frac{1}{2}+n, n+\frac{1}{2}) \mid \alpha^n \# \underline{0}\}$.
Then $\exists n[\alpha^n \# \underline{0}]$ and: QED.

Case 2. $y \in \mathbb{N}$. Find m such that $y = n$ and distinguish two cases.

Case 2.1. $f(m) \in \bigcup\{(-\frac{1}{2}+n, n+\frac{1}{2}) \mid \alpha^n \# \underline{0}\}$.
Then $\exists n[\alpha^n \# \underline{0}]$: and QED.

Case 2.2. $f(m) \in \mathbb{N}$. Find n such that $f(m) = n$ and note: $m \neq n$.
Now assume: $\alpha^m \# \underline{0}$. Then $(-\frac{1}{2}+m, m+\frac{1}{2}) \subseteq V_\alpha$.
Note: f is continuous on $(-\frac{1}{2}+m, m+\frac{1}{2})$, see Section 4. Find $k > 1$ such that $\forall x \in (-\frac{1}{2}+n, n+\frac{1}{2})[|x-m| < \frac{1}{2^k} \to |f(x)-n| < \frac{1}{2}]$.
Conclude: $f(m+\frac{1}{2^{k+1}}) \in (-\frac{1}{2}+n, n+\frac{1}{2})$ and $(f(m+\frac{1}{2^{k+1}}) \#_\mathcal{R} n$.
Conclude: $(-\frac{1}{2}+n, n+\frac{1}{2}) \subseteq V_\alpha$ and $\alpha^n \# \underline{0}$.
Conclude: $\alpha^m \# \underline{0} \to \alpha^n \# \underline{0}$, that is: QED.

We thus see: if V_α is Dedekind-infinite, then $\exists m \exists n[m \neq n \wedge (\alpha^m \# \underline{0} \to \alpha^n \# \underline{0})]$.

One may imagine someone cooking up α such that, for all m, n satisfying $m \neq n$, the statement '$\alpha^m \# \underline{0} \to \alpha^n \# \underline{0}$' is a reckless one.

It is easier, however, to use Brouwer's Continuity Principle, as follows.

Assume $\forall\alpha[V_\alpha$ *is Dedekind-infinite*$]$.
Then $\forall\alpha\exists m \exists n[m \neq n \wedge (\alpha^m \# \underline{0} \to \alpha^n \# \underline{0})]$.
Find p, m, n such that $m \neq n$ and $\forall\alpha[\overline{\underline{0}}p \sqsubset \alpha \to (\alpha^m \# \underline{0} \to \alpha^n \# \underline{0})]$.
Note that one may define α such that $\overline{\underline{0}}p \sqsubset \alpha$ and $\alpha^m \# \underline{0} \wedge \alpha^n = \underline{0}$.
We have obtained a contradiction.

When I came at the end of my talk on this example at the Dutch Mathematical Congress in Amsterdam, 1971, Prof. Heyting announced that he just had received a manuscript from his Ph.D. student C. G. Gibson containing the same result, see [10].

8 Proving the continuum hypothesis

Cantor's continuum hypothesis may be taken to be the statement:

Every $V \subseteq \mathcal{N}$ *either* embeds into $\mathbb{N}$ *or* $\mathcal{C}$ embeds into V.

It is useful to compare this statement with the following one:

Every $V \subseteq \mathbb{N}$ *either* embeds into a natural number $n = \{0, 1, \ldots, n-1\}$ *or* $\mathbb{N}$ embeds into V.

Both statements will not be true, if taken constructively, because of the constructive interpretation of the disjunction.

Fortunately, a classical statement always has a host of possible readings in the language of constructive mathematics.

The second statement may be replaced by the following one:

For every $V \subseteq \mathbb{N}$, if V is *constructively infinite*, that is, for every n, every mapping from n to V is positively non-surjective, then $\mathbb{N}$ embeds into V.

This new statement is intuitionistically provable, as follows. Assume one has constructive evidence for the antecedent: there is a function F from $\bigcup_n V^n$ to V such that, for all n, for all s in V^n, $F(s) \in V$ and $\forall i < n[s(i) \neq F(s)]$. Then define $g : \mathbb{N} \to V$ such that, for each n, $g(n) = F(\langle g(0), g(1), \ldots, g(n-1)\rangle)$. Clearly, g embeds $\mathbb{N}$ into V.

With some hope, we make a similar version of the first statement:

If $V \subseteq \mathcal{N}$ is *constructively uncountable*, then $\mathcal{C}$ embeds into V, that is: if $\forall\alpha\exists\beta \in V\forall n[\beta \# \alpha^n]$, then $\exists\psi[\psi : \mathcal{C} \hookrightarrow V]$.

This new statement is provable, as follows. Assume $\forall\alpha\exists\beta \in V\forall n[\beta \# \alpha^n]$. Using the Second Axiom of Continuous Choice, find $\varphi : \mathcal{N} \to V$ such that $\forall\alpha\forall n[\varphi|\alpha \# \alpha^n]$.

We now prove: $(*) : \forall s\exists t\exists u[s \sqsubset u \ \wedge \ s \sqsubset u \ \wedge \ \varphi|t \perp \varphi|u]$.

Let s be given. Define $\alpha := s * \underline{0}$ and define β such that $s \sqsubset \beta$ and $\exists n[\beta^n = \varphi|\alpha]$. Note: $\forall n[\varphi|\beta \# \beta^n]$ and, therefore: $\varphi|\beta \# \varphi|\alpha$. Find p, q such that $\varphi|\overline{\beta}p \perp \varphi|\overline{\alpha}q$ and define: $t := \overline{\beta}p$ and $u := \overline{\alpha}q$.

This concludes the proof of $(*)$.

Now define $\gamma : Bin \to \mathbb{N}$ such that $\gamma(\langle\ \rangle) = \langle\ \rangle$, and, for each a in Bin, $\gamma(a * \langle 0\rangle) = t'$ and $\gamma(a * \langle 1\rangle) = t''$, where $t = \mu u[s \sqsubset u' \ \wedge \ s \sqsubset u'' \ \wedge \ \varphi|u' \perp \varphi|u'']$.

Then define $\rho : \mathcal{C} \to \mathcal{N}$ such that $\forall\alpha \in \mathcal{C}\forall n[\gamma(\overline{\alpha}n) \sqsubset \rho|\alpha]$ and define $\psi : \mathcal{C} \to \mathcal{N}$ such that $\forall\alpha \in \mathcal{C}[\psi|\alpha = \varphi|(\rho|\alpha)]$.

One easily verifies: $\psi : \mathcal{C} \hookrightarrow V$.

When I, full of joy, described this result to Gielen, he told me had proven it himself some years earlier but did not think it terribly important. He was working on other things. What he was doing will be shown in the next Section.

9 Saving the Perfect Set Theorem

9.1 Determinate subsets of ℕ are enumerable

The *Brouwer-Kripke principle* we referred to earlier is the following principle:

> *Given any* determinate *proposition P, there exists α such that $P \leftrightarrow \exists n[\alpha(n) \neq 0]$.*

The *Second Axiom of Countable Choice* is the following scheme:

> *If $\forall n \exists \alpha [nR\alpha]$, then $\exists \alpha \forall n [nR\alpha^n]$.*

This axiom seems defensible from the intuitionistic point of view, see for instance [12, §1.2]. The axiom follows from the stronger Second Axiom of Continuous Choice.

For each α, we define $E_\alpha := \{m \mid \exists p[\alpha(p) = m+1]\}$, the subset of ℕ *enumerated by* α. $X \subseteq \mathbb{N}$ is *enumerable* if and only if $\exists \alpha [X = E_\alpha]$.

Together, the Brouwer-Kripke principle and the Second Axiom of Countable Choice imply that any *determinate* subset of ℕ is enumerable.

One argues as follows. Let a determinate $X \subseteq \mathbb{N}$ be given. Every statement '$n \in X$' then is determinate, so $\forall n \exists \beta [n \in X \leftrightarrow \exists p[\beta(p) \neq 0]]$. Find β such that $\forall n[n \in X \leftrightarrow \exists p[\beta^n(p) \neq 0]]$. Define α such that, for all n, *if* $\beta^{n'}(n'') \neq 0$, then $\alpha(n) = n'$, and, *if not*, then $\alpha(n) = 0$. One easily verifies: $X = E_\alpha$.

9.2 Cantor's result reconstructed

Cantor proved that any closed subset of $\mathcal{R}$, or, for that matter, $\mathcal{N}$, may be written as the union of a perfect set and an at most countable set. This result is called *Cantor's Main Theorem* or the *Perfect Set Theorem.*

Already Brouwer himself spent a lot of thought on the possible intuitionistic content of this theorem, see [3]. He did not reach a satisfying conclusion.[11]

Gielen, not content with the result in 8, did the same. His result may be found in [12], Theorem 5.3.1. He strongly uses the enumerability of determinate subsets of ℕ, mentioned in Subsection 9.1.

We will give a free description of his result, changing even the precise formulation.

For every $X \subseteq \mathbb{N}$, we define: $\mathcal{F}(X) = \{\alpha \mid \forall n[\overline{\alpha}n \in X]\}$.

$\mathcal{X} \subseteq \mathcal{N}$ is called *closed* if and only if, for some $X \subseteq \mathbb{N}$, $\mathcal{X} = \mathcal{F}(X)$.

For every $X \subseteq \mathbb{N}$, we define: $X^* := \{s \mid \exists \alpha \in \mathcal{F}(X)[s \sqsubset \alpha]\}$, and
$D(X) := \{s \in X^* \mid \neg\neg\exists \alpha \in \mathcal{F}(X) \exists \beta \in \mathcal{F}(X)[s \sqsubset \alpha \wedge s \sqsubset \beta \wedge \alpha \# \beta]\} =$

[11] In 1952, he called the work he had done in 1919 'obsolete' and 'in need of revision'.

$\{s \in X^* \mid \neg\neg\exists t \in X^* \exists u \in X^*[s \sqsubset t \wedge s \sqsubset u \wedge t \perp u]\}$.[12]

Cantor calls the closed set $\mathcal{F}(D(X))$ the *pseudo-derivative* or the *pseudo-coherence* of $\mathcal{F}(X)$.[13] We also call the set $D(X)$, a set of natural numbers, the *pseudo-derivative* of the set X. The elements of $\mathcal{F}(D(X))$ are the *pseudo-limit points* of $\mathcal{F}(X)$. The reader should take note of the double negation occurring in the definition: we take the notion of a limit point in a constructively weak sense.

For every $X \subseteq \mathbb{N}$ we define:
$I(X) := \{s \in X^* \mid \forall \alpha \in \mathcal{F}(X) \forall \beta \in \mathcal{F}(X)[(s \sqsubset \alpha \wedge s \sqsubset \beta) \to \alpha = \beta]\}$.

One may prove intuitionistically: $I(X) = X^* \setminus D(X)$ and $D(X) = X^* \setminus I(X)$.

$\mathcal{I}(X) := \{\alpha \in \mathcal{F}(X) \mid \exists n[\overline{\alpha}n \in I(X)]\}$ is the *appendix* or the *adherence* of $\mathcal{F}(X)$, that is, the set of all *isolated* points of $\mathcal{F}(X)$.

One may prove intuitionistically:
$\mathcal{I}(X) = \mathcal{F}(X) \setminus \mathcal{F}(D(X))$ and $\mathcal{F}(D(X)) = \mathcal{F}(X) \setminus \mathcal{I}(X)$.

Assume $I(X)$ is determinate. As, according to Subsection 9.1, $I(X)$ is enumerable, find α such that $I(X) = E_\alpha$. Using the Second Axiom of Countable Choice, determine δ such that, for all n, s, if $\alpha(n) = s + 1$, then $\delta^n \in \mathcal{I}(X)$ and $s \sqsubset \delta^n$. Note: $\mathcal{I}(X) \subseteq \{\delta^n \mid n \in \omega\}$, so $\mathcal{I}(X)$ is a subset of an enumerable set.

Let $X \subseteq \mathbb{N}$ be given. We define the collection $\mathcal{D}_X$ of the iterated pseudo-derivatives of X by the following inductive[14] definition.

$\mathcal{D}_X$ is the least collection $\mathcal{Y}$ of subsets of $\mathbb{N}$ such that

(i) $X \in \mathcal{Y}$, and,

(ii) for each Z in $\mathcal{Y}$, also $D(Z) \in \mathcal{Y}$, and

(iii) for each infinite sequence $Z_0, Z_1, \ldots$ of elements of $\mathcal{Y}$, also $\bigcap_n Z_n \in \mathcal{Y}$.

Now consider $I_0(X) := \bigcup_{Z \in \mathcal{D}_X} I(Z)$.

Assume: $I_0(X)$ is determinate and find α such that $I_0(X) = E_\alpha$. Using an axiom of [15], find an infinite sequence $Z_0, Z_1, \ldots$ of elements of $\mathcal{D}_X$ such that, for all s, n,

[12]The definition of the operator Γ_A in [12, §5.2] is wrong, as wrong as it would be to replace the just-given definition by $D(X) := \{s \in X^* \mid \neg\neg\exists m \exists n[s * \langle m \rangle \in X^* \wedge s * \langle n \rangle \in X^* \wedge m \neq n]\}$. The conclusion '$\forall n[\alpha(n) = \beta(n)]$' in [12, p. 132, line 15] is wrong if one uses the definition given by the authors.

[13]We add the prefix *'pseudo'* to Cantor's terms because of the double negation.

[14]We assume such inductive definitions are intuitionistically acceptable.

[15]A more detailed version of the argumcountable choiceent would show that the Second Axiom of Countable Choice suffices.

if $\alpha(n) = s + 1$, then $s \in I(Z_n)$. Define $Z := \bigcap_n D(Z_n)$ and note: $Z \in \mathcal{D}_X$ and: $I(Z) = \emptyset$.

One may prove intuitionistically:
$\mathcal{F}(D(Z)) = \mathcal{F}(Z)$, that is: $\mathcal{F}(Z)$ is what we want to all *pseudo-perfect.*[16]

$\mathcal{F}(Z)$ is called the *pseudo-perfect kernel* $\mathcal{K}(X)$ of $\mathcal{F}(X)$.

Define $\mathcal{I}_0(X) = \bigcup_{Z \in \mathcal{D}_X} \mathcal{I}(Z)$.

Define δ such that, for all s, n, if $\alpha(n) = s + 1$, then $\delta^n \in \mathcal{I}(Z_n)$ and $s \sqsubset \delta^n$. Note: $\mathcal{I}_0(X) \subseteq \{\delta^n \mid n \in \omega\}$.

One may prove intuitionistically: $\forall \beta \in \mathcal{F}(X)[\forall n[\beta \# \delta^n] \rightarrow \beta \in \mathcal{K}(X)]$.

$\mathcal{F}(X)$ is thus seen to possess a pseudo-perfect kernel $\mathcal{K}(X)$ and there is an enumeration of points such that every point of $\mathcal{F}(X)$ positively apart from every point in that enumeration belongs to $\mathcal{K}(X)$.

9.3 Avoiding transfinite induction

Cantor's process of iterating the operation of taking the derivative transfinitely many times may be circumvented, as was observed by E. Lindelöf. Intuitionistically, this is also true. The first proof along these lines is due to J. Burgess, see [7].

For every $X \subseteq \mathbb{N}$, define
$C(X) := \{s \in X^* \mid \exists \delta \forall \beta \in \mathcal{F}(X)[\forall n[\beta \# \delta^n] \rightarrow s \perp \beta]\}$.

$C(X)$ thus consists of all s such that at least one but, in the above sense, no more than countably points of $\mathcal{F}(X)$ have s as an initial part.

We intend to prove that $X^* \setminus C(X)$ is pseudo-perfect in the sense of Subsection 9.2, but, to this end, we first establish a more positive fact about $C(X)$.

The elements of $\mathcal{F}(X^* \setminus C(X))$ are called the *condensation points* of $\mathcal{F}(X)$.

Assume that $C(X)$ is a determinate subset of $\mathbb{N}$ and find α such that $C(X) = E_\alpha$.

Using the Second Axiom of Countable Choice, find δ such that, for all m, s, if $\alpha(m) = s + 1$, then $\forall \beta \in \mathcal{F}(X)[\forall n[\beta \# \delta^{m,n}] \rightarrow s \perp \beta]$.

Note: $\forall \beta \in \mathcal{F}(X)[\forall n[\beta \# \delta^n] \rightarrow \beta \in \mathcal{F}(X^* \setminus C(X))]$.

We now prove:

all s in X^* satisfying the requirement: *for all t, u in X^* such that $s \sqsubset t$ and $s \sqsubset u$ and $t \perp u$, either $t \in C(X)$ or $u \in C(X)$*, belong to $C(X)$.

Assume $s \in X^*$ satisfies the requirement.

Then, for all t, u such that $s \sqsubset t$ and $s \sqsubset u$ and $t \perp u$, either
$\forall \beta \in \mathcal{F}(X)[\forall n[\beta \# \delta^n] \rightarrow t \perp \beta]$ or $\forall \beta \in \mathcal{F}(X)[\forall n[\beta \# \delta^n] \rightarrow u \perp \beta]$.

[16] The Brouwer-Kripke principle thus enables one to find the least fixed point of a monotone operator on subsets of $\mathbb{N}$.

Conclude: there is at most one β in $\mathcal{F}(X)$ such that $s \sqsubset \beta$ and $\forall n[\beta \# \delta^n]$. We are unable to decide if there is one.

Once more apply the Brouwer-Kripke principle and find η such that
$\exists \beta \in \mathcal{F}(X)[s \sqsubset \beta \wedge \forall n[\beta \# \delta^n]] \leftrightarrow \exists n[\eta(n) \neq 0]$.

Using the Second Axiom of Countable Choice, find ε such that,
for all n, if $\eta(n) \neq 0$, then $s \sqsubset \varepsilon^n$ and $\varepsilon^n \in \mathcal{F}(X)$ and $\forall i[\varepsilon^n \# \delta^i]$.

Define ζ such that $\forall n[\zeta^{0,n} = \delta^n \wedge \zeta^{1,n} = \varepsilon^n]$ and note:
$\forall \beta \in \mathcal{F}(X)[\forall n[\beta \# \zeta^n] \to s \perp \beta]$.

Conclude: $s \in C(X)$.

Conclude: for all s in X^*, if $s \notin C(X)$, then
$\neg \forall t \in X^* \forall t \in X^*[(s \sqsubset t \wedge s \sqsubset u \wedge t \perp u) \to (t \in C(X) \vee u \in C(X))]$.

We need the observation that, for all s in X^*, if $\neg\neg(s \in C(X))$, then $s \in C(X)$[17].

Let s in X^* be given. Note that, for each $\beta \in \mathcal{F}(X)$,
if $\neg\neg(\forall n[\beta \# \delta^n] \to s \perp \beta)$, then $\neg\neg\forall n[\beta \# \delta^n] \to \neg\neg(s \perp \beta)$,
and also: $\forall n[\beta \# \delta^n] \to s \perp \beta$.

We may conclude[18]: if $\neg\neg(\forall \beta \in \mathcal{F}(X)[\forall n[\beta \# \delta^n] \to s \perp \beta])$, then
$\forall \beta \in \mathcal{F}(X)[\forall n[\beta \# \delta^n] \to s \perp \beta]$, and therefore:
if $\neg\neg(s \in C(X))$, then $s \in C(X)$.

We thus see[19]: for all s in X^*, if $s \notin C(X)$, then
$\neg\neg\exists t \in X^* \exists u \in X^*[t \notin C(X) \wedge u \notin C(X) \wedge s \sqsubset t \wedge s \sqsubset u \wedge t \perp u]$.
Therefore: $\mathcal{K}_1(X) := \{\beta \in \mathcal{F}(X) \mid \forall n[\overline{\beta}n \notin C(X)]\}$ is pseudo-perfect.

Moreover, we constructed δ such that $\forall \beta \in \mathcal{F}(X)[\forall n[\beta \# \delta^n] \to \beta \in \mathcal{K}_1(X)]$.

9.4 Extension to analytic sets

M. Souslin noted that Cantor's result extends from closed sets to analytic sets. Burgess suggested to Gielen to extend also his result, originally formulated for closed sets only, to analytic sets. It is not difficult to do so.

Let $\varphi : \mathcal{N} \to \mathcal{N}$ be given. The set $Ran(\varphi); = \{\varphi|\alpha \mid \alpha \in \mathcal{N}\}$ then is called a *strictly*[20] *analytic set.*

Let $X \subseteq \mathbb{N}$ be given. Define
$D_\varphi(X) := \{s \in X \mid \neg\neg\exists t \in X \exists u \in X[s \sqsubset t \wedge s \sqsubset u \wedge \varphi|t \perp \varphi|u]\}$.

Arguing along the lines sketched in the previous two subsections one may find $X_0 \subseteq \mathbb{N}$ such that $D_\varphi(X_0) = X_0$.

[17] That is: $C(X)$ is a *stable* subset of X^*.

[18] The class of stable statements is closed under universal quantifications.

[19] $(\forall x[\neg\neg P(x) \to P(x)] \wedge \neg\forall x[P(x)]) \to \neg\neg\exists x[\neg P(x)]$ is a valid scheme in intuitionistic logic.

[20] We use the word 'strictly' as not every projection of a closed set is strictly analytic.

One may conclude then that the set $A_\varphi := \{\varphi|\beta \mid \forall n[\overline{\beta}n \in D_\varphi(X_0)]\}$ is pseudo-perfect and find δ such that $\forall\beta[\forall n[\varphi|\beta \# \delta^n] \to \varphi|\beta \in A_\varphi]$.

Unfortunately, if $\mathcal{K}$ is pseudo-perfect and inhabited[21], then, in general, one will not be able to construct an injective continuous $\psi : \mathcal{C} \hookrightarrow \mathcal{K}$.

9.5 Set theory?

Apart from his work on the perfect set theorem, Gielen also tried to develop *intuitionistic set theory*. Brouwer already had seen that one needs a form of *separation* or *comprehension* and introduced '*species*': definable subcollections of spreads. Gielen gave a characterization of what, in general, would be a *set* and formulated extensions of Brouwer's Continuity Principle. The manuscript [11] never led to a publication.

10 Completeness of intuitionistic predicate logic

10.1 The (weak) completeness problem

Intuitionistic logic was formalized by Heyting in 1930. The problem of its completeness was studied by E.W. Beth, who, in 1956, thought that he had solved it in a good way, also from an intuitionistic point of view, see [2].

In 1957, however, Gödel had a conversation with Kreisel, in which he explained that intuitionistic completeness of intuitionistic predicate logic would imply that, for every primitive recursive $A(n)$,

$$\text{if not-not } \exists n[A(n)], \text{ then } \exists n[A(n)].$$

The intuitionistic mathematician does not subscribe to this conclusion, that sometimes is called the *(strict) Markov Principle.*

Kreisel gave a careful exposition of the argument in [19]. We will tell the story in a slightly different way.

We first reflect on the meaning of the completeness theorem. The theorem says that, for every sentence φ,

$$\text{if } \models \varphi, \text{ then } \vdash \varphi.$$

$\models \varphi$ stands for: φ is *universally true*, that is, as Tarski defined it:
for any structure $\mathfrak{A} = (A, \ldots)$, $\mathfrak{A} \models \varphi$.

[21] $\mathcal{X} \subseteq \mathcal{N}$ is *inhabited* if and only if $\exists\alpha[\alpha \in \mathcal{X}]$, that is: one is able to effectively find an element of $\mathcal{X}$.

$\vdash \varphi$ stands for: there exists a deduction in intuitionistic first-order logic with conclusion φ.

Now assume that $\varphi = \neg\psi$ is a *negative* formula. Then, for every structure $\mathfrak{A}$, $\mathfrak{A} \models \neg\psi$ if and only if *not*: $\mathfrak{A} \models \psi$. The statement '$\mathfrak{A} \models \neg\psi$' thus is a negative statement, and, therefore, by a law of intuitionistic logic[22] already discovered by Brouwer himself:

$$\text{If not-not } \mathfrak{A} \models \neg\psi, \text{ then } \mathfrak{A} \models \neg\psi.$$

One may conclude, by another law of intuitionistic logic[23] :

$$\text{If not-not } \models \neg\psi, \text{ then } \models \neg\psi.$$

The statements '$\mathfrak{A} \models \neg\psi$' and '$\models \neg\psi$' thus are what one calls *stable* statements.[24]

If completeness holds, $\models \neg\psi$ if and only if $\vdash \neg\psi$.

We thus find:

$$\text{If not-not } \vdash \neg\psi, \text{ then } \vdash \neg\psi.$$

Note that the statement '$\vdash \neg\psi$' is of the form $\exists n[A(n)]$, with A primitive recursive.

That the conclusion extends to every statement of the form $\exists n[A(n)]$, where A is primitive recursive, may be seen as follows.

Assume $A \subseteq \mathbb{N}$ is primitive recursive and not-not $\exists n[A(n)]$. As one may learn from [17, §49, Lemma 18b], A is numeralwise expressible in Robinson's axiom system for arithmetic[25]. Let Q be the conjunction of the finitely many axioms of this system. Find a formula $\varphi = \varphi(\mathsf{x})$ in the language of arithmetic such that, for each n,

$$\text{if } A(n), \text{ then } Q \vdash \varphi(\underline{n}), \text{ and, if not } A(n), \text{ then } Q \vdash \neg\varphi(\underline{n}),$$

and, therefore,

$$A(n) \text{ if and only if } Q \vdash \varphi(\underline{n}).$$

[22] $\neg\neg\neg P \to \neg P$.

[23] $\forall x[\neg\neg P(x) \to P(x)] \to (\neg\neg\forall x[P(x)] \to \forall x[P(x)])$.

[24] The term '*stable*' has been introduced by D. van Dantzig, who hoped to be able to reconstruct 'classical', non-intuitionistic mathematics within the stable part of intuitionistic mathematics, see [8]. From the previous footnote one learns that the class of stable statements is closed under universal quantifications.

[25] We take Robinson's system of arithmetic, formulated in the language of the structure $\mathfrak{N} = (\mathbb{N}, S, +, \cdot, 0)$, as a subsystem of Heyting arithmetic HA, so with intuitionistic logic.

Here $\underline{n} := \underbrace{\mathsf{S}(\mathsf{S}(\ldots\mathsf{S}(0)\ldots))}_{n\ times}$ is the numeral for n in the language of arithmetic and $\varphi(\underline{n})$ is the result of substituting $\underline{n}$ for x in φ at every appropriate place.

Note that, if $\exists n[A(n)]$, then $Q \vdash \exists\mathsf{x}[\varphi(\mathsf{x})]$ and $\vdash \neg(Q \wedge \neg\exists\mathsf{x}[\varphi(\mathsf{x})])$.

As not-not $\exists n[A(n)]$, not-not $\vdash \neg(Q \wedge \neg\exists\mathsf{x}[\varphi(\mathsf{x})])$.

Assuming completeness, conclude: $\vdash \neg(Q \wedge \neg\exists\mathsf{x}[\varphi(\mathsf{x})])$ and $Q \vdash \neg\neg\exists\mathsf{x}[\varphi(\mathsf{x})]$.

As Q, like HA itself, is closed under *Markov's rule*, see the Friedman-Dragalin argument given for [24, Volume I, Chapter 3, Corollary 5.4], conclude: $Q \vdash \exists\mathsf{x}[\varphi(\mathsf{x})]$.

Note that, for every sentence ψ, if $Q \vdash \psi$, then $\mathfrak{N} \models \psi$.

Conclude: $\mathfrak{N} \models \exists\mathsf{x}[\varphi(\mathsf{x})]$. Find n such that $\mathfrak{N} \models \varphi[n]$ and conclude: $A(n)$.

We thus see that Completeness (of intuitionistic predicate logic) implies Markov's Principle.[26]

Beth did not work with the natural Tarskian semantics we are referring to, but introduced so-called *semi-models*, now called *Beth models*. Also in this context, validity of a negative formula is a negative and therefore a stable concept.

Gödel obtained the theorem that Markov's Principle follows from completeness of intuitionistic predicate logic as a corollary of a stronger result. He claimed that Completeness implies, for every primitive recursive $B(n)$,

$$\text{if } \forall\alpha \in \mathcal{C}\neg\neg\exists n[B(\overline{\alpha}n)], \text{ then } \forall\alpha \in \mathcal{C}\exists n[B(\overline{\alpha}n)].$$

Let us see why this claim holds true. Let a primitive recursive $B(n)$ be given such that $\forall\alpha \in \mathcal{C}\neg\neg\exists n[B(\overline{\alpha}n)]$.

Find a formula $\varphi = \varphi(\mathsf{x})$ in the language of arithmetic such that, for each n,

$$\text{if } B(n), \text{ then } Q \vdash \varphi(\underline{n}), \text{ and, if not } B(n), \text{ then } Q \vdash \neg\varphi(\underline{n}).$$

We extend the language of arithmetic with a unary function symbol $\alpha^\diamond$.

Let $\mathfrak{A}^* = (A^*, \alpha^*, S^*, +^*, \cdot^*, 0^*)$ be a structure satisfying Q.

We claim: $\mathfrak{A}^* \models \forall\mathsf{x}[\alpha^\diamond(\mathsf{x}) = \mathsf{0} \vee \alpha^\diamond(\mathsf{x}) = \mathsf{S}(\mathsf{0})] \rightarrow \neg\neg\exists\mathsf{x}[\varphi(\overline{\alpha^\diamond}\mathsf{x})]$.

In order to establish this claim, we define a sequence $0^*, 1^*, \ldots$ of elements of A such that, for each n, $(S(n))^* = S^*(n^*)$.

Assume $\mathfrak{A}^* \models \forall\mathsf{x}[\alpha^\diamond(\mathsf{x}) = \mathsf{0} \vee \alpha^\diamond(\mathsf{x}) = \mathsf{S}(\mathsf{0})]$. Define α in $\mathcal{C}$ such that, for all n, $\alpha^*(n) = (\alpha(n))^*$. Note: for each n, if $B(\overline{\alpha}n)$, then $Q \vdash \varphi[\underline{\overline{\alpha}n}]$ and $Q \vdash \varphi(\overline{\alpha^\diamond}\underline{n})$ and $\mathfrak{A} \models \exists\mathsf{x}[\varphi(\overline{\alpha^\diamond}\mathsf{x})]$. As not-not $\exists n[B(\overline{\alpha}n]$, we conclude: $\mathfrak{A} \models \neg\neg\exists\mathsf{x}[\varphi(\overline{\alpha^\diamond}\mathsf{x})]$.

We thus see that the sentence:

$$(*) : (Q \wedge \forall\mathsf{x}[\alpha^\diamond(\mathsf{x}) = \mathsf{0} \vee \alpha^\diamond(\mathsf{x}) = \mathsf{S}(\mathsf{0})]) \rightarrow \neg\neg\exists\mathsf{x}[\varphi(\overline{\alpha^\diamond}\mathsf{x})]$$

[26] One does not find this explanation of the Gödel-Kreisel argument [24, Volume II, Chapter 13, Section 3], which keeps closer to [19].

is universally true.

Assuming completeness for the sentence $(*)$ we conclude:

$$(Q \wedge \forall \mathsf{x}[\alpha^\diamond(\mathsf{x}) = \mathsf{0} \vee \alpha^\diamond(\mathsf{x}) = \mathsf{S}(\mathsf{0})]) \vdash \neg\neg\exists \mathsf{x}[\varphi(\overline{\alpha^\diamond}\mathsf{x})].$$

As the theory $Q \wedge \forall \mathsf{x}[\alpha^\diamond(\mathsf{x}) = \mathsf{0} \vee \alpha^\diamond(\mathsf{x}) = \mathsf{S}(\mathsf{0})]$ again obeys Markov's rule, we conclude:

$$(Q \wedge \forall \mathsf{x}[\alpha^\diamond(\mathsf{x}) = \mathsf{0} \vee \alpha^\diamond(\mathsf{x}) = \mathsf{S}(\mathsf{0})]) \vdash \exists \mathsf{x}[\varphi(\overline{\alpha^\diamond}\mathsf{x})].$$

For each α in $\mathcal{C}$, therefore, the structure $(\mathbb{N}, \alpha, S, +, \cdot, 0)$ satisfies the sentence $\exists \mathsf{x}[\varphi(\overline{\alpha^\diamond}\mathsf{x})]$, and: $\exists n[B(\overline{\alpha}n)]$.

Assuming completeness for the sentence $(*)$, one finds:

$$\text{if } \forall \alpha \in \mathcal{C} \neg\neg\exists n[B(\overline{\alpha}n)], \text{ then } \forall \alpha \in \mathcal{C}\exists n[B(\overline{\alpha}n)].$$

Kreisel also defines: a sentence φ satisfies *weak completeness* if and only if:

$$\text{if not} \vdash \varphi, \text{ then not} \models \varphi,$$

or, equivalently,

$$\text{if} \models \varphi \text{ then not-not} \vdash \varphi.$$

Assuming weak completeness rather than completeness for the sentence $(*)$, one is led by the above argument to the conclusion:

$$\text{if } \forall \alpha \in \mathcal{C} \neg\neg\exists n[B(\overline{\alpha}n)], \text{ then } \neg\neg\forall \alpha \in \mathcal{C}\exists n[B(\overline{\alpha}n)].$$

Kreisel judges this statement to be intuitionistically less improbable than Markov's principle itself.

10.2 The classical proof

We now ask ourselves: how does one prove, classically, the completeness of classical predicate logic?

The usual argument is *by contraposition*:

> If not $\vdash \neg\psi$, then there exists $\mathfrak{A} = (A, \ldots)$ such that A is at most countable and $\mathfrak{A} \models \psi$.

The model $\mathfrak{A}$ is obtained from a maximally consistent and exemplary set Γ of sentences, containing ψ itself, in an extension of the language of ψ by countably many new individual constants $\mathsf{c}_0, \mathsf{c}_1, \ldots$.

Let us briefly explain the notions we are using.

A set Δ of sentences is a *theory* if every sentence derivable from Δ belongs to Δ.

Δ is *consistent* if no contradiction may be derived from Δ and *maximally consistent* if, in addition, Δ has no proper consistent extension.

Δ is *disjunctive* if and only if, for every disjunctive sentence $(\varphi_0) \vee (\varphi_1)$ in Δ, either $\varphi_0 \in \Delta$ or $\varphi_1 \in \Delta$.

(Note: if Δ is maximally consistent, then Δ is disjunctive, and, for every sentence σ, either $\sigma \in \Delta$ or $\neg\sigma \in \Delta$.)

Δ is *exemplary* if and only if, for every existential sentence $\exists x[\rho(x)]$ in Δ, there exists n such that $\rho(\mathsf{c}_n) \in \Delta$.[27]

If we take the construction of $\mathfrak{A}$ from Γ for granted, the crucial fact seems to be:

> If not $\vdash \neg\psi$, then there exists a maximally consistent and exemplary theory Γ such that $\psi \in \Gamma$.

Taking the contraposition, we obtain the following statement about $\varphi = \neg\psi$:

> If φ belongs to every maximally consistent and exemplary theory Γ, then $\vdash \varphi$,

or, equivalently, as every consistent Γ has a maximally consistent and exemplary extension:

> If φ belongs to every consistent, disjunctive and exemplary theory Γ, then $\vdash \varphi$,

or, equivalently, as the condition trivially holds for an inconsistent Γ:

> (†): if φ belongs to every disjunctive and exemplary theory Γ, then $\vdash \varphi$.

10.3 The extended completeness theorem

The extended completeness theorem is the following statement:

---

[27] $\rho(\mathsf{c}_n)$ denotes the formula that results from $\rho = \rho(\mathsf{x})$ by replacing the variable x, at every place where x occurs freely in ρ, by the constant c_n.

Let $\mathcal{L}$ be a given first-order language.

For every set Γ of sentences of $\mathcal{L}$, for every sentence φ of $\mathcal{L}$, if $\Gamma \models \varphi$, then $\Gamma \vdash \varphi$.

$\Gamma \models \varphi$ stands for:
for every structure $\mathfrak{A}$, if, for all ψ in Γ, $\mathfrak{A} \models \psi$, then $\mathfrak{A} \models \varphi$.

$\Gamma \vdash \varphi$ stands for: there exists a derivation of φ from sentences in Γ.

The usual classical proof argues:

If not $\Gamma \vdash \neg\psi$, then there exists a maximally consistent and exemplary theory Δ such that $\Gamma \subseteq \Delta$ and $\psi \in \Delta$.

Arguing as in the previous Subsection, we are led to consider the following statement:

$(*)$: for every enumerable[28] theory Γ, for every sentence φ, if φ belongs to every enumerable, disjunctive and exemplary theory Δ containing Γ, then $\varphi \in \Gamma$.

Note that if we let Γ be the set of all sentences that are intuitionistically derivable, we obtain the last statement $(\dagger)$ of the previous Subsection.

In the next Subsection, we give an intuitionistic proof of $(*)$.

10.4 An intuitionistic proof

Let $\mathcal{L}$ be a given first-order relational language.

Let Γ be an enumerable theory in $\mathcal{L}$ and let $\psi_0, \psi_1, \ldots$ be an enumeration of Γ.

Let φ be a sentence of $\mathcal{L}$ that belongs to every enumerable, disjunctive and exemplary theory Δ containing Γ.

Let $\mathsf{c}_0, \mathsf{c}_1, \ldots$ be an infinite sequence of mutually different individual constants that do not occur in $\mathcal{L}$.

Let $\mathcal{L}'$ be the extension of $\mathcal{L}$ by these new individual constants.

Let $\varphi_0, \varphi_1, \ldots$ be an enumeration of the sentences of $\mathcal{L}'$ in which every sentence occurs infinitely many times.

Let $D_0, D_1, \ldots$ be an enumeration of all derivations in $\mathcal{L}'$.

We define a function $s \mapsto \Delta_s$ that associates to every s in $\bigcup_n Bin_{2n}$ a finite set of formulas.

$\Delta_{\langle\,\rangle} := \emptyset$ and, for each n, for each s in Bin_{2n}, we define:

[28] From a classical point of view, every subset of a countable set is enumerable. Intuitionistically, however, this is a non-empty extra condition.

1. if φ_n is not an existential or a disjunctive formula, then, for both $j < 2$, $\Delta_{s*\langle 1,j\rangle} = \Delta_s \cup \{\varphi_n\}$, and,

2. if $\varphi_n = (\psi_0) \vee (\psi_1)$ is a disjunctive formula, then, for both $j < 2$, $\Delta_{s*\langle 1,j\rangle} = \Delta_s \cup \{\varphi_n, \psi_j\}$, and,

3. if $\varphi_n = \exists \mathsf{x}[\rho(\mathsf{x})]$ is an existential formula, then, for both $j < 2$, $\Delta_{s*\langle 1,j\rangle} = \Delta_s \cup \{\varphi_n, \rho(\mathsf{c}_m)\}$ where m is the least k such that c_k does not occur in $\Delta_s \cup \{\varphi_n\}$, and,

4. for both $i < 2$, if there is $i \leq n$ such that *either* D_i is a deduction of φ_n from Δ_s *or* φ_n is the formula ψ_i, then $\Delta_{s*\langle 0,i\rangle} = \Delta_{s*\langle 1,i\rangle}$, and if there is no such $i \leq n$, then $\Delta_{s*\langle 0,i\rangle} = \Delta_s$.

One may prove: for each s in $\bigcup_n Bin_{2n}$,
if $\forall i < 2 \forall j < 2[\Gamma \cup \Delta_{s*\langle i,j\rangle} \vdash \varphi]$, then $\Gamma \cup \Delta_s \vdash \varphi$.
This observation will be useful in a moment.

For each α in $\mathcal{C}$, we define: $\Delta_\alpha := \bigcup_n \Delta_{\overline{\alpha}(2n)}$.
Note that, for each α in $\mathcal{C}$, Δ_α is a disjunctive and exemplary theory containing Γ.
Therefore: $\forall \alpha \in \mathcal{C}[\varphi \in \Delta_\alpha]$, that is, $\forall \alpha \in \mathcal{C} \exists n[\varphi \in \Delta_{\overline{\alpha}(2n)}]$.
Using the Fan Theorem, we find n such that $\forall \alpha \in \mathcal{C}[\varphi \in \Delta_{\overline{\alpha}(2n)}]$,
that is: $\forall s \in Bin_{2n}[\Delta_s \vdash \varphi]$.
Using the observation we just made, one proves, by backwards induction:
$\forall i \leq n \forall s \in Bin_{2i}[\Gamma \cup \Delta_s \vdash \varphi]$, and thus: $\Gamma \vdash \varphi$, and $\varphi \in \Gamma$.

We thus see:

> $(*)$: For every enumerable theory Γ, for every sentence φ,
> if φ belongs to every enumerable disjunctive and exemplary theory Δ containing Γ, then Γ contains φ.

10.5 A Kripke model

We need two consequences of the result $(*)$ of the previous Subsection. The first one is:

> $(\diamondsuit \rightarrow)$: For every enumerable disjunctive and exemplary theory Γ, for all sentences ψ_o, ψ_1, $(\psi_0) \rightarrow (\psi_1)$ belongs to Γ if and only if every enumerable disjunctive and exemplary theory Δ containing $\Gamma \cup \{\psi_0\}$ contains ψ_1.

One direction is easy: a theory Δ containing both ψ_0 and $(\psi_0) \to (\psi_1)$ will contain ψ_1.

For the other direction, we use $(*)$. Let Γ^* be the set of all sentences ψ such that $\Gamma \cup \{\psi_0\} \vdash \psi$. Note that Γ^* is enumerable. As every enumerable disjunctive and exemplary theory Δ containing Γ^* contains ψ_1, conclude: $\Gamma^* \vdash \psi$ and $\Gamma \cup \{\psi_0\} \vdash \psi_1$ and $\Gamma \vdash (\psi_0) \to (\psi_1)$ and $(\psi_0) \to (\psi_1) \in \Gamma$.

For every theory Δ, we let $IC(\Delta)$ be the set of all individual constants occurring in formulas of Δ.

The second important consequence of $(*)$ is the following statement:

> $(\Diamond\forall)$: For every enumerable disjunctive and exemplary theory Γ, for every formula $\rho = \rho(\mathsf{x})$, the sentence $\forall \mathsf{x}[\rho(\mathsf{x})]$ belongs to Γ if and only if, for every enumerable disjunctive and exemplary theory Δ containing Γ, for every individual constant c in $IC(\Delta)$, the sentence $\rho(\mathsf{c})$ belongs to Δ.

One direction is easy: a theory Δ containing $\forall \mathsf{x}[\rho(\mathsf{x})]$ will contain $\rho(\mathsf{c})$ for every c in $IC(\Delta)$.

For the other direction, we use $(*)$.

Extend the language $\mathcal{L}$ of Γ to a language $\mathcal{L}'$ containing at least one new individual constant, say c. Let Γ^* be the set all sentences ψ in $\mathcal{L}'$ such that $\Gamma \vdash \psi$. Note that Γ^* is enumerable. As every enumerable disjunctive and exemplary theory Δ containing Γ^* contains $\rho(\mathsf{c})$, conclude: $\Gamma^* \vdash \rho(\mathsf{c})$ and $\Gamma \vdash \rho(\mathsf{c})$ and, as c does not occur in Γ, $\Gamma \vdash \forall \mathsf{x}[\rho(\mathsf{x})]$ and $\forall \mathsf{x}[\rho(\mathsf{x})] \in \Gamma$.

We now construct a Kripke structure.

Let $\mathcal{L}$ be a first-order language. We assume that the language contains a propositional constant $\perp$ and no sign for negation. We then may define, for every formula φ, $\neg(\varphi) := (\varphi) \to \perp$. For every formula φ one has $\perp \vdash \varphi$ and a theory Γ is *inconsistent* if and only if $\perp \in \Gamma$.

Let C be a countable set of individual constants not occurring in $\mathcal{L}$. We let Ω be the set of all enumerable disjunctive and exemplary theories Γ in $\mathcal{L}$ such that $IC(\Gamma) \subseteq C$ and $C \setminus IC(\Gamma)$ is infinite It is very important that elements Γ of Ω may be inconsistent and that, in general, for a given Γ in Ω, we are unable to decide if Γ is inconsistent or not.[29]

Let S be the function that associates to every Γ in Ω the set $S(\Gamma)$ of all closed basic formulas occurring in Γ.

[29] Ω is a well-defined set in intuitionistic mathematics. One may define a function $s \mapsto B_s$ that associates to every s a finite set B_s of sentences such that $\Omega = \{B_\alpha \mid \alpha \in \mathcal{N}\}$ where, for each α, $B_\alpha = \bigcup_n B_{\overline{\alpha}n}$.

The structure $\mathfrak{M} = (\Omega, \subseteq, S)$ now is a *Kripke structure*.
For every formula $\varphi = \varphi(x_0, x_1, \ldots, x_{n-1})$, for every Γ in Ω, for all $a_0, \ldots, a_{n-1}$ in $IC(\Gamma)$ we define the notion

$$\mathfrak{M} \Vdash_\Gamma \varphi[a_0, a_1, \ldots, a_{n-1}]$$

as follows, by induction on the structure of the formula φ.

1. if φ is a basic formula, then $\mathfrak{M} \Vdash_\Gamma \varphi[a_0, a_1, \ldots, a_{n-1}]$ if and only if the sentence $\varphi(a_0, a_1, \ldots, a_{n-1})$ is in $S(\Gamma)$, and
2. if $\varphi = (\varphi_0) \wedge (\varphi_1)$, then: $\mathfrak{M} \Vdash_\Gamma \varphi[a_0, a_1, \ldots, a_{n-1}]$ if and only if, for both $j < 2$, $\mathfrak{M} \Vdash_\Gamma \varphi_j[a_0, a_1, \ldots, a_{n-1}]$, and
3. if $\varphi = (\varphi_0) \vee (\varphi_1)$, then: $\mathfrak{M} \Vdash_\Gamma \varphi[a_0, a_1, \ldots, a_{n-1}]$ if and only if, for at least one $j < 2$, $\mathfrak{M} \Vdash_\Gamma \varphi_j[a_0, a_1, \ldots, a_{n-1}]$ and
4. if $\varphi = (\varphi_0) \rightarrow (\varphi)_1$ then $\mathfrak{M} \Vdash_\Gamma \varphi[a_0, a_1, \ldots, a_{n-1}]$ if and only if,
for all $\Delta \supseteq \Gamma$,
if $\mathfrak{M} \Vdash_\Delta \varphi_0[a_0, a_1, \ldots, a_{n-1}]$, then $\mathfrak{M} \Vdash_\Delta \varphi_1[a_0, a_1, \ldots, a_{n-1}]$, and
5. if $\varphi = \exists \mathsf{x}[\rho(\mathsf{x})]$, then $\mathfrak{M} \Vdash_\Gamma \varphi[a_0, a_1, \ldots, a_{n-1}]$ if and only if
there exists b in $IC(\Gamma)$ such that $\mathfrak{M} \Vdash_\Gamma \rho[a_0, a_1, \ldots, a_{n-1}, b]$, and
6. if $\varphi = \forall \mathsf{x}[\rho(\mathsf{x})]$, then $\mathfrak{M} \Vdash_\Gamma \varphi[a_0, a_1, \ldots, a_{n-1}]$ if and only if, for all $\Delta \supseteq \Gamma$, for all b in $IC(\Delta)$, $\mathfrak{M} \Vdash_\Delta \rho[a_0, a_1, \ldots, a_{n-1}, b]$.

One may prove now:

> For every Γ in Ω, for every formula $\varphi = \varphi(x_0, x_1, \ldots, x_{n-1})$, for all $a_0, a_1, \ldots, a_{n-1}$ in $IC(\Gamma)$, $\mathfrak{M} \Vdash_\Gamma \varphi[a_0, a_1, \ldots, a_{n-1}]$ if and only if the formula $\varphi(a_0, a_1, \ldots, a_{n-1})$ is in Γ.

A sentence φ is called *$\mathfrak{M}$-valid* if and only if, for all Γ in Ω, $\mathfrak{M} \Vdash_\Gamma \varphi$.
Clearly, a sentence φ is $\mathfrak{M}$-valid if and only if every enumerable disjunctive and exemplary theory Γ contains φ.
Using (†), we conclude: a sentence φ is $\mathfrak{M}$-valid if and only if $\vdash \varphi$.

The crucial step making this conclusion possible is the admittance of *possibly inconsistent* disjunctive and exemplary theories Γ in Ω: $\mathfrak{M}$ sometimes is called an *exploding* Kripke model. The statement $\mathfrak{M} \models \neg\psi$ no longer is a negative statement, and the Gödel-Kreisel objection to completeness does not apply anymore.

The above completeness result may be found in [25]. A variant of the result, using Beth models rather than Kripke models, was published by de Swart, see [20]. M. Dummett, in his treatment of the semantics of intuitionistic logic in [9], introduced the expressions '*Nijmegen proof*' and '*Nijmegen school*', see also [21] and [22]. Some of us liked this predicate as a '*nom de gueux*' as they felt that others involved with intuitionistic mathematics went perhaps a wrong direction.

11 Descriptive Set Theory

In this Section, I want to give the reader some impression of the work done in [26]. The research reported in this dissertation started when de Iongh asked me if I could prove that, also intuitionistically, the class $\mathbf{\Sigma}^0_2$ is not a subclass of the class $\mathbf{\Pi}^0_2$ and the class $\mathbf{\Pi}^0_2$ is not a subclass of the class $\mathbf{\Sigma}^0_2$.

11.1 Establishing the Borel hierarchy

11.1.1 The problem

$\mathcal{X} \subseteq \mathcal{N}$ is *open* if and only if $\exists\beta\forall\alpha[\alpha \in \mathcal{X} \leftrightarrow \exists n[\beta(\overline{\alpha}n) \neq 0]]$.
$\mathcal{X} \subseteq \mathcal{N}$ is *closed* if and only if $\exists\beta\forall\alpha[\alpha \in \mathcal{X} \leftrightarrow \forall n[\beta(\overline{\alpha}n) = 0]]$.

$\mathcal{X} \subseteq \mathcal{N}$ is *(positively) Borel* if and only if $\mathcal{X}$ may be obtained from open and closed subsets of $\mathcal{N}$ by means of the operations of countable union and countable intersection.

It is an important decision to leave out the operation of taking the complement from the definition of the class of the Borel sets.

$\mathbf{\Sigma}^0_1$ is the class of the open subsets of $\mathcal{N}$, $\mathbf{\Pi}^0_1$ is the class of the closed subsets of $\mathcal{N}$, and, for each n, $\mathbf{\Sigma}^0_{n+1}$ is the class consisting of all sets $\bigcup_m \mathcal{X}_m$, where each $\mathcal{X}_m$ is in $\mathbf{\Pi}^0_n$ and $\mathbf{\Pi}^0_{n+1}$ is the class consisting of all sets $\bigcap_m \mathcal{X}_m$, where each $\mathcal{X}_m$ is in $\mathbf{\Sigma}^0_n$.

We would like to prove the *Finite Borel Hierarchy Theorem.* This Theorem says that, for each $n > 0$, the classes $\mathbf{\Sigma}^0_n \setminus \mathbf{\Pi}^0_n$ and $\mathbf{\Pi}^0_n \setminus \mathbf{\Sigma}^0_n$ are non-empty.

11.1.2 A first attempt

We shall make use of the following *reducibility relation.*

Let $\mathcal{X}, \mathcal{Y}$ be subsets of $\mathcal{N}$. $\mathcal{X} \preceq \mathcal{Y}$, ($\mathcal{X}$ *reduces to* $\mathcal{Y}$) if and only if there exists $\varphi : \mathcal{N} \to \mathcal{N}$ *reducing* $\mathcal{X}$ to $\mathcal{Y}$, that is: $\forall\alpha[\alpha \in \mathcal{X} \leftrightarrow \varphi|\alpha \in \mathcal{Y}]$.

We define an infinite sequence of Borel sets:

$\mathcal{E}_1 := \{\alpha \mid \exists n[\alpha(n) \neq 0]\}$ and $\mathcal{A}_1 := \{\alpha \mid \forall n[\alpha(n) = 0]\} = \{\underline{0}\}$, and, for each $n > 0$, $\mathcal{E}_{n+1} := \{\alpha \mid \exists m[\alpha^m \in \mathcal{A}_n]\}$ and $\mathcal{A}_{n+1} := \{\alpha \mid \forall m[\alpha^m \in \mathcal{E}_n]\}$.

Note: $\forall n > 0[\mathcal{A}_n \cap \mathcal{E}_n = \emptyset]$.

One may prove, for each $n > 0$:

$\mathcal{X} \subseteq \mathcal{N}$ is in $\mathbf{\Sigma}^0_n$ if and only if $\mathcal{X} \preceq \mathcal{E}_n$ and
$\mathcal{X} \subseteq \mathcal{N}$ is in $\mathbf{\Pi}^0_n$ if and only if $\mathcal{X} \preceq \mathcal{A}_n$.

The Finite Borel Hierarchy Theorem will be proven if we show that, for each $n > 0$, $\mathcal{E}_n \not\preceq \mathcal{A}_n$ and $\mathcal{A}_n \not\preceq \mathcal{E}_n$.

We consider an argument that *almost* gives us the conclusion we are after.

We make a first observation:
$(*) : \forall \varphi : \mathcal{N} \to \mathcal{N} \exists \alpha[\alpha \in \mathcal{E}_1 \leftrightarrow \varphi|\alpha \in \mathcal{E}_1]$.

One may see this as follows. Let $\varphi : \mathcal{N} \to \mathcal{N}$ be given. Define α such that, for each n, $\alpha(n) \neq 0$ if and only if $\varphi|\overline{\alpha}n \perp \underline{0}$.
Then $\alpha \in \mathcal{E}_1 \leftrightarrow \varphi|\alpha \in \mathcal{E}_1$, and also $\alpha \in \mathcal{A}_1 \leftrightarrow \varphi|\alpha \in \mathcal{A}_1$, as $\mathcal{A}_1 = \mathcal{N} \setminus \mathcal{E}_1$.

Now assume $\varphi : \mathcal{N} \to \mathcal{N}$ reduces $\mathcal{A}_1$ to $\mathcal{E}_1$.
Find α such that $\alpha \in \mathcal{E}_1 \leftrightarrow \varphi|\alpha \in \mathcal{E}_1$. Conclude: $\alpha \in \mathcal{E}_1 \leftrightarrow \alpha \in \mathcal{A}_1$ and: $\alpha \notin \mathcal{E}_1$ and $\alpha \notin \mathcal{A}_1$, that is: $\neg\exists n[\alpha(n) \neq 0]$ and $\neg\forall n[\alpha(n) = 0]$. Contradiction.

Conclude: $\mathcal{A}_1 \not\preceq \mathcal{E}_1$.

A similar argument leads to the conclusion: $\mathcal{E}_1 \not\preceq \mathcal{A}_1$.

We make a second observation:
$(**) : \forall \varphi : \mathcal{N} \to \mathcal{N} \exists \alpha \forall n[\alpha^n \in \mathcal{E}_1 \leftrightarrow (\varphi|\alpha)^n \in \mathcal{E}_1]$.

One may see this as follows. Let $\varphi : \mathcal{N} \to \mathcal{N}$ be given. Define α such that, for each n, for each m, $\alpha^n(m) \neq 0$ if and only if $(\varphi|\overline{\alpha}m)^n \perp \underline{0}$.

Then $\forall n[\alpha^n \in \mathcal{E}_1 \leftrightarrow (\varphi|\alpha)^n \in \mathcal{E}_1]$ and: $\forall n[\alpha^n \in \mathcal{A}_1 \leftrightarrow (\varphi|\alpha)^n \in \mathcal{A}_1]$, and therefore: $\alpha \in \mathcal{A}_2 \leftrightarrow \varphi|\alpha \in \mathcal{A}_2$ and: $\alpha \in \mathcal{E}_2 \leftrightarrow \varphi|\alpha \in \mathcal{E}_2$.

Now assume $\varphi : \mathcal{N} \to \mathcal{N}$ reduces $\mathcal{A}_2$ to $\mathcal{E}_2$. Find α such that $\alpha \in \mathcal{E}_2 \leftrightarrow \varphi|\alpha \in \mathcal{E}_2$. Conclude: $\alpha \in \mathcal{A}_2 \leftrightarrow \alpha \in \mathcal{E}_2$ and $\alpha \notin \mathcal{E}_2$ and $\alpha \notin \mathcal{A}_2$, that is $\neg\exists m \forall n[\alpha^m(n) = 0]$ and $\neg\forall m \exists n[\alpha^m(n) \neq 0]$. Unfortunately, in intuitionistic logic, these two conclusions do not lead to a contradiction.[30]

Continuing this line of argument, we may prove, for every $n > 0$:
$\mathcal{A}_n \preceq \mathcal{E}_n \to \exists \alpha[\alpha \notin \mathcal{A}_n \cup \mathcal{E}_n]$ and $\mathcal{E}_n \preceq \mathcal{A}_n \to \exists \alpha[\alpha \notin \mathcal{A}_n \cup \mathcal{E}_n]$.
Intuitionistically, if $n > 1$, this does not justify the conclusion:
$\mathcal{E}_n \not\preceq \mathcal{A}_n \ \wedge \ \mathcal{A}_n \not\preceq \mathcal{E}_n$.[31]

The classical mathematician, however, will draw this conclusion and be happy.

The argument extends to the transfinite levels of the Borel Hierarchy, see [32, Theorem 5.2].

[30] Unless one assumes the (*unrestricted*) Markov's Principle : $\forall \alpha \forall n[\neg\neg\exists n[\alpha(n) \neq (0)] \to \exists n[\alpha(n) \neq 0]$, but we do not want to do so.

[31] Markov's Principle does not enable one to obtain a contradiction from $\exists \alpha[\alpha \notin \mathcal{A}_3 \cup \mathcal{E}_3]$.

11.1.3 A solution

One may prove intuitionistically: for all $n > 0$, for all $\varphi : \mathcal{N} \to \mathcal{N}$,

if $\forall\alpha[\alpha \in \mathcal{A}_n \to \varphi|\alpha \in \mathcal{E}_n]$, then $\exists\alpha[\alpha \in \mathcal{E}_n \wedge \varphi|\alpha \in \mathcal{E}_n]$, and,

if $\forall\alpha[\alpha \in \mathcal{E}_n \to \varphi|\alpha \in \mathcal{A}_n]$, then $\exists\alpha[\alpha \in \mathcal{A}_n \wedge \varphi|\alpha \in \mathcal{A}_n]$.

These positive statements imply: $\mathcal{A}_n \not\preceq \mathcal{E}_n$ and $\mathcal{E}_n \not\preceq \mathcal{A}_n$.

We consider the cases $n = 1$ and $n = 2$.

Assume $\varphi : \mathcal{N} \to \mathcal{N}$ maps $\mathcal{A}_1$ into $\mathcal{E}_1$. Note: $\varphi|\underline{0} \in \mathcal{E}_1$. Find p such that $(\varphi|\underline{0})(p) \neq 0$. Find m such that $\overline{(\varphi|\underline{0})}(p+1) \sqsubseteq \varphi|(\overline{\underline{0}}m)$. Define $\alpha := (\overline{\underline{0}}m) * \underline{1}$ and note: $\alpha \in \mathcal{E}_1$ and $\varphi|\alpha \in \mathcal{E}_1$.

Assume $\varphi : \mathcal{N} \to \mathcal{N}$ maps $\mathcal{E}_1$ into $\mathcal{A}_1$. Suppose we find p such that $(\varphi|\underline{0})(p) \neq 0$. Find m such that $\overline{(\varphi|\underline{0})}(p+1) \sqsubseteq \varphi|(\overline{\underline{0}}m)$. Define $\alpha := (\overline{\underline{0}}m) * \underline{1}$ and note: $\alpha \in \mathcal{E}_1$ and $\varphi|\alpha \in \mathcal{E}_1$. Contradiction. Conclude: $\forall p[(\varphi|\underline{0})(p) = 0]$.
Therefore: $\varphi|\underline{0} = \underline{0}$ and: $\underline{0} \in \mathcal{A}_1$ and $\varphi|\underline{0} \in \mathcal{A}_1$.

Assume $\varphi : \mathcal{N} \to \mathcal{N}$ maps $\mathcal{A}_2$ into $\mathcal{E}_2$. Note: $\forall\gamma\forall\alpha[\forall p[\alpha^p(\gamma(p)) \neq 0] \to \alpha \in \mathcal{A}_2]$ and, therefore: $\forall\gamma\forall\alpha[\forall p[\alpha^p(\gamma(p)) \neq 0] \to \varphi|\alpha \in \mathcal{E}_2]$ and
$\forall\gamma\forall\alpha[\forall p[\alpha^p(\gamma(p)) \neq 0] \to \exists q[(\varphi|\alpha)^q \in \mathcal{A}_1]]$.

Now apply Brouwer's Continuity Principle[32] and find r, q such that
$\forall\gamma\forall\alpha[(\overline{\underline{0}}r \sqsubset \gamma \wedge \overline{\underline{1}}r \sqsubset \alpha) \to (\varphi|\alpha)^q = \underline{0}]$.

Define α such that $\forall i < r[\alpha^i = \underline{1}]$ and $\alpha^r = \underline{0}$ and $\overline{\underline{1}}r \sqsubset \alpha$.[33]

Note: $\forall m > r\exists\beta[\overline{\alpha}m \sqsubset \beta \wedge \exists\gamma[\overline{\underline{0}}r \sqsubset \gamma \wedge \forall p[\beta^p(\gamma(p)\rangle \neq 0]]]$.

Conclude: $\forall m > r\exists\beta[\overline{\alpha}m \sqsubset \beta \wedge (\varphi|\beta)^q = \underline{0}]$.

Using the continuity of the function φ, conclude: $(\varphi|\alpha)^q = \underline{0}$ and $\varphi|\alpha \in \mathcal{E}_2$. Note: also $\alpha \in \mathcal{E}_2$.

Assume: $\varphi : \mathcal{N} \to \mathcal{N}$ maps $\mathcal{E}_2$ into $\mathcal{A}_2$.

We now define an infinite sequence $s_0, s_1, \ldots$ such that $s_0 = \langle\,\rangle$ and, for each n, $s_n \sqsubset s_{n+1}$ and $(\varphi|s_{n+1})^n \perp \underline{0}$ and $(s_{n+1})^n \perp \underline{0}$.

Assume we defined s_n and have to define s_{n+1}. Define α such that $s_n \sqsubset \alpha$ and $(s_n * \alpha)^n \perp \underline{0}$. Note: $s_n * \alpha \in \mathcal{E}_2$ and therefore: $\varphi|(s_n * \alpha) \in \mathcal{A}_2$ and $\big(\varphi|(s_n * \alpha)\big)^n \in \mathcal{E}_1$. Find the least $q > 0$ such that $(s_n * \overline{\alpha}q)^n \perp \underline{0}$ and $\big(\varphi|(s_n * \overline{\alpha}q)\big)^n \perp \underline{0}$. Now define $s_{n+1} := s_n * \overline{\alpha}q$.

Finally, find α such that $\forall n[s_n \sqsubset \alpha]$ and note: $\alpha \in \mathcal{A}_2$ and $\varphi|\alpha \in \mathcal{A}_2$.

Note that the proof of the fourth and last of these four statements is elementary and constructive. In the proof of the third statement, Brouwer's Continuity Principle plays a crucial rôle.

[32]We need the observation that the set $\{(\gamma, \alpha) \mid \forall p[\alpha^p(\gamma(p)) \neq 0]\}$ is a spread.

[33]We assume: $\forall r\forall n[r \leq \langle r\rangle * n]$. There is then no conflict between the requirements '$\alpha^r = \underline{0}$' and '$\overline{\underline{1}}r \sqsubset \alpha$'.

It is possible to extend the argument to the higher levels of the Finite Borel Hierarchy.

The following two observations are then important:

for all n, for all γ, for all α,
if $\forall s \in \omega^n[\forall k[2k+1 < n \rightarrow \alpha(2k+1) = \gamma(\overline{\alpha}(2k+1))] \rightarrow \alpha(s) \neq 0]$, then $\alpha \in \mathcal{A}_n$, and:

for all n, for all γ, for all α,
if $\forall s \in \omega^n[\forall k[2k < n \rightarrow \alpha(2k) = \gamma(\overline{\alpha}(2k))] \rightarrow \alpha(s) = 0]$, then $\alpha \in \mathcal{E}_n$.

The complete proof, also for the transfinite levels of the Borel Hierarchy, may be found in [26, Chapter 9] and [32, §7].

11.2 Fine structure

Consider $\mathbb{D}^2(\mathcal{A}_1) := \{\alpha \mid \alpha^0 = \underline{0} \vee \alpha^1 = \underline{0}\}$. $\mathcal{X} \subseteq \mathcal{N}$ reduces to $\mathbb{D}^2(\mathcal{A}_1)$ if and only there exist closed sets $\mathcal{Y}_0, \mathcal{Y}_1$ such that $\mathcal{X} = \mathcal{Y}_0 \cup \mathcal{Y}_1$.

Assume $\varphi : \mathcal{N} \rightarrow \mathcal{N}$ reduces $\mathbb{D}^2(\mathcal{A}_1)$ to $\mathcal{A}_1$. Define τ such that
$\forall n \forall s \in \omega^n[\tau(s) = 0 \leftrightarrow \forall i < n \forall j < n[(s(i) \neq 0 \wedge s(j) \neq 0) \rightarrow i = j]]$. Note: τ is a spread-law. Consider the spread $\mathcal{F}_\tau := \{\alpha \mid \forall n[\tau(\overline{\alpha}n) = 0]\}$. $\mathcal{F}_\tau$ is the set of all α such that $\forall i \forall j[(\alpha(i) \neq 0 \wedge \alpha(j) \neq 0) \rightarrow i = j]$.

Note: $\forall s[\tau(s) = 0 \rightarrow \exists \alpha \in \mathbb{D}^2(\mathcal{A}_1)[s \sqsubset \alpha \wedge \varphi|\alpha = \underline{0}]]$.

Conclude, using the continuity of φ: $\forall \alpha \in \mathcal{F}_\tau[\varphi|\alpha = \underline{0}]$.

Conclude: $\forall \alpha \in \mathcal{F}_\tau[\alpha \in \mathbb{D}^2(\mathcal{A}_1)]$, that is: $\forall \alpha \in \mathcal{F}_\tau[\alpha^0 = \underline{0} \vee \alpha^1 = \underline{0}]$. Using Brouwer's Continuity Principle, find p, i such that
$\forall \alpha \in \mathcal{F}_\tau[\overline{\underline{0}}p \sqsubset \alpha \rightarrow \alpha^i = \underline{0}]$. Note: there exists α in $\mathcal{F}_\tau$ such that $\overline{\underline{0}}p \sqsubset \alpha$ and $\alpha^i \neq \underline{0}$: we reached a contradiction.

We thus see: $\mathbb{D}^2(\mathcal{A}_1) \not\preceq \mathcal{A}_1$, that is: $\mathbb{D}^2(\mathcal{A}_1)$ is not closed.
Note that $\mathbb{D}^2(\mathcal{A}_1) = \{\alpha \mid \alpha^0 = \underline{0}\} \cup \{\alpha \mid \alpha^1 = \underline{0}\}$ is a union of two closed sets.

This simple result admits of vast extensions, see [26, Chapters 4 and 12] and [34], as we briefly explain now

Let us define, for all $\mathcal{X}, \mathcal{Y} \subseteq \mathcal{N}$, $\mathcal{X} \prec \mathcal{Y} \leftrightarrow (\mathcal{X} \preceq \mathcal{Y} \wedge \mathcal{Y} \not\preceq \mathcal{X})$.

One may prove: $\forall n[\mathbb{D}^n(\mathcal{A}_1) \prec \mathbb{D}^{n+1}(\mathcal{A}_1)]$, where, for each n, for each $\mathcal{X} \subseteq \mathcal{N}$, $\mathbb{D}^n(\mathcal{X}) = \{\alpha \mid \exists i < n[\alpha^i \in \mathcal{X}]\}$.

One then may define $\mathcal{U} := \{\alpha \mid \exists n[\alpha^n \in \mathbb{D}^n(\mathcal{A}_1)]\}$ and observe: $\forall n[\mathbb{D}^n(\mathcal{A}_1) \prec \mathcal{U}]$ and $\mathcal{U} \prec \mathcal{E}_2$.

Continuing this line of thought, one comes to see that there exist uncountably many '*degrees of reducibility*', even within the class $\mathbf{\Sigma}^0_2$, see [26, Chapter 12] and [34].

11.3 The projective hierarchy and its collapse

$\mathcal{X} \subseteq \mathcal{N}$ is $\mathbf{\Sigma}_1^1$ or *analytic* if and only if there exists $\mathcal{Y}$ in $\mathbf{\Pi}_1^0$ such that $\mathcal{X} = \{\alpha \mid \exists\beta[(\alpha,\beta) \in \mathcal{Y}]\}$.
One may prove that every (positively) Borel set $\mathcal{X} \subseteq \mathcal{N}$ is analytic.
$\mathcal{E}_1^1 := \{\alpha \mid \exists\gamma\forall n[\alpha(\overline{\gamma}n) = 0]\}$ is an analytic set and a *complete* element of the class $\mathbf{\Sigma}_1^1$, i.e. $\mathcal{X} \subseteq \mathcal{N}$ is analytic if and only if $\mathcal{X} \preceq \mathcal{E}_1^1$.

$\mathcal{X} \subseteq \mathcal{N}$ is $\mathbf{\Pi}_1^1$ or *co-analytic* if and only if there exists $\mathcal{Y}$ in $\mathbf{\Sigma}_1^0$ such that $\mathcal{X} = \{\alpha \mid \forall\beta[(\alpha,\beta) \in \mathcal{Y}]\}$.
$\mathcal{A}_1^1 := \{\alpha \mid \forall\gamma\exists n[\alpha(\overline{\gamma}n) \neq 0]\}$ is a co-analytic set and a *complete* element of the class $\mathbf{\Pi}_1^1$, i.e. $\mathcal{X} \subseteq \mathcal{N}$ is co-analytic if and only if $\mathcal{X} \preceq \mathcal{A}_1^1$.

It is far from true that every (positively) Borel set is co-analytic. The set $\mathbb{D}^2(\mathcal{A}_1)$ is a counterexample, as we prove now.

Assume $\varphi : \mathcal{N} \to \mathcal{N}$ reduces $\mathbb{D}^2(\mathcal{A}_1)$ to $\mathcal{A}_1^1$.
As in Subsection 11.2, consider the spread $\mathcal{F}_\tau$ where τ satisfies
$\forall n \forall s \in \omega^n[\tau(s) = 0 \leftrightarrow \forall i < n \forall j < n[(s(i) \neq 0 \wedge s(j) \neq 0) \to i = j]]$.
Assume: $\alpha \in \mathcal{F}_\tau$. We shall prove: $\varphi|\alpha \in \mathcal{A}_1^1$.
Define β such that $\beta^0 = \underline{0}$ and $\forall n[\neg\exists p[n = \langle 0\rangle * p] \to \beta(n) = \alpha(n)]$.
Note $\beta \in \mathbb{D}^2(\mathcal{A}_1)$ and $\varphi|\beta \in \mathcal{A}_1^1$.
Let γ be given. Find p such that $(\varphi|\beta)(\overline{\gamma}p) \neq 0$.
Find q such that $\forall\delta[\overline{\beta}q \sqsubset \delta \to (\varphi|\delta)(\overline{\gamma}p) = (\varphi|\beta)(\overline{\gamma}p)]$.
If $\overline{\beta}q \sqsubset \alpha$ then $(\varphi|\alpha)(\overline{\gamma}p) \neq 0$.
If not $\overline{\beta}q \sqsubset \alpha$, then $\alpha^0 \# \underline{0}$, and, therefore, as $\alpha \in \mathcal{F}_\tau$, $\alpha^1 = \underline{0}$ and: $\alpha \in \mathbb{D}^2(\mathcal{A}_1)$ and: $\varphi|\alpha \in \mathcal{A}_1^1$ and: one may find r such that $(\varphi|\alpha)(\overline{\gamma}r) \neq 0$.
We thus see: $\forall\gamma\exists p[\varphi|\alpha)(\overline{\gamma}p) \neq 0]$, i.e. $\varphi|\alpha \in \mathcal{A}_1^1$.
Conclude: $\forall\alpha \in \mathcal{F}_\tau[\varphi|\alpha \in \mathcal{A}_1^1]$ and: $\forall\alpha \in \mathcal{F}_\tau[\alpha \in \mathbb{D}^2(\mathcal{A}_1)]$.
As in Subsection 11.2, this leads to a contradiction.

$\mathcal{X} \subseteq \mathcal{N}$ is $\mathbf{\Sigma}_2^1$ if and only if there exists $\mathcal{Y}$ in $\mathbf{\Pi}_1^1$ such that $\mathcal{X} = \{\alpha \mid \exists\beta[(\alpha,\beta) \in \mathcal{Y}]\}$.
$\mathcal{X} \subseteq \mathcal{N}$ is $\mathbf{\Pi}_2^1$ if and only if there exists $\mathcal{Y}$ in $\mathbf{\Sigma}_1^1$ such that $\mathcal{X} = \{\alpha \mid \forall\beta[(\alpha,\beta) \in \mathcal{Y}]\}$.

Using the Second Axiom of Continuous Choice, one may prove:
for all $\mathcal{X} \subseteq \mathcal{N}$, if $\mathcal{X}$ is $\mathbf{\Sigma}_2^1$, then the set $\{\alpha \mid \forall\beta[(\alpha,\beta) \in \mathcal{X}]\}$ is $\mathbf{\Sigma}_2^1$, see [26, Chapter 14] and [35, §7].

The class $\mathbf{\Sigma}_2^1$ thus is seen to be closed under the operation of universal projection. One may also see that the class $\mathbf{\Sigma}_2^1$ is closed under the operations of existential projection, countable union and countable intersection.

Every set that may obtained by these four operations from open and closed subsets of $\mathcal{N}$ thus belongs to the class $\mathbf{\Sigma}^1_2$, that is: *every projective set is* $\mathbf{\Sigma}^1_2$.

11.4 The Lusin Separation Theorem

Brouwer's *Thesis on Bars* may be formulated as a *Principle of Bar Induction*:

> For all $B, C \subseteq \omega$, if $\forall\alpha\exists n[\overline{\alpha}n \in B]$ and $B \subseteq C$ and $\forall s[s \in C \leftrightarrow \forall n[s * \langle n \rangle \in C]]$, then $\langle\,\rangle \in C$.

Brouwer introduced this principle, believing in its intuitive truth, and selling it as a theorem, in order to prove the Fan Theorem. In a way, the principle is much too strong for that, and the question arises if the principle has further going applications. Parsons mentions the fact that Lusin and Sierpinski, in 1918, used the way of arguing suggested by the principle in order to prove the Lusin Separation Theorem, see [13, p. 452]. No wonder then, there exists an intuitionistic version of this theorem.

$\mathcal{X}, \mathcal{Y} \subseteq \mathcal{N}$ are *disjoint* if and only if $\forall\alpha \in \mathcal{X}\forall\beta \in \mathcal{Y}[\alpha \# \beta]$.

$\mathcal{X}, \mathcal{Y} \subseteq \mathcal{N}$ are *Borel-separable* if and only if there exists disjoint (positively) Borel sets $\mathcal{A}, \mathcal{B}$ such that $\mathcal{X} \subseteq \mathcal{A}$ and $\mathcal{Y} \subseteq \mathcal{B}$.

Lusin's Separation Theorem is the statement:

> *Disjoint strictly analytic sets are Borel-separable.*

We sketch the proof.

Let $\varphi, \psi : \mathcal{N} \to \mathcal{N}$ be given such that $\varphi|\mathcal{N}$, $\psi|\mathcal{N}$ are disjoint.

Define $B := \{s \mid \varphi|s^0 \perp \psi|s^1\}$. Note: $\forall\alpha\exists n[\overline{\alpha}n \in B]$.

Define $C := \{s \mid \varphi|(\mathcal{N} \cap s^0), \psi|(\mathcal{N} \cap s^1)$ *are Borel-separable*$\}$.

Assume $s \in B$. Note $\varphi|(\mathcal{N} \cap s^0) \subseteq \mathcal{N} \cap \varphi|s^0$ and $\psi|(\mathcal{N} \cap s^1) \subseteq \mathcal{N} \cap \psi|s^1$ and $\mathcal{N} \cap \varphi|s^0$, $\mathcal{N} \cap \psi|s^1$ are closed-and-open and disjoint. Conclude: $s \in C$.

We thus see: $B \subseteq C$.

Let s be given such that $\forall n[s * \langle n \rangle \in C]$. We distinguish three cases.

Case (i). There exists p such that $length(s) = \langle 0 \rangle * p$. Then, for all n, $(s*\langle n \rangle)^0 = s^0 * \langle n \rangle$ and $(s * \langle n \rangle)^1 = s^1$. Using an Axiom of Countable Choice[34], find an infinite sequence $(\mathcal{A}_0, \mathcal{B}_0), (\mathcal{A}_1, \mathcal{B}_1), \ldots$ of disjoint pairs of Borel sets, such that, for each n,

[34] A more detailed version of the argument would show that the Second Axiom of Countable Choice suffices.

$\varphi|(\mathcal{N} \cap s^0 * \langle n \rangle) \subseteq \mathcal{A}_n$ and $\psi|(\mathcal{N} \cap s^1) \subseteq \mathcal{B}_n$. Define $\mathcal{A} := \bigcup_n \mathcal{A}_n$ and $\mathcal{B} := \bigcap_n \mathcal{B}_n$ and note: $\mathcal{A}, \mathcal{B}$ are disjoint and $\varphi|(\mathcal{N} \cap s^0) \subseteq \mathcal{A}$ and $\psi|(\mathcal{N} \cap s^1) \subseteq \mathcal{B}$.

We may conclude: $s \in C$.

Case (ii). There exists p such that $length(s) = \langle 1 \rangle * p$. Then, for all n, $(s * \langle n \rangle)^0 = s^0$ and $(s * \langle n \rangle)^1 = s^1 * \langle n \rangle$. Using an Axiom of Countable Choice[35], find an infinite sequence $(\mathcal{A}_0, \mathcal{B}_0), (\mathcal{A}_1, \mathcal{B}_1), \ldots$ of disjoint pairs of Borel sets, such that, for each n, $\varphi|(\mathcal{N} \cap s^0) \subseteq \mathcal{A}_n$ and $\psi|(\mathcal{N} \cap s^1 * \langle n \rangle) \subseteq \mathcal{B}_n$. Define $\mathcal{A} := \bigcap_n \mathcal{A}_n$ and $\mathcal{B} := \bigcup_n \mathcal{B}_n$ and note: $\mathcal{A}, \mathcal{B}$ are disjoint and $\varphi|(\mathcal{N} \cap s^0) \subseteq \mathcal{A}$ and $\psi|(\mathcal{N} \cap s^1) \subseteq \mathcal{B}$.

We may conclude: $s \in C$.

Case (iii) There are no i, p such that $i < 2$ and $n = \langle i \rangle * p$. Then, for both $i < 2$, $(s * \langle 0 \rangle)^i = s^i$, and, as $s * \langle 0 \rangle \in C$, also $s \in C$.

One easily sees: $\forall s \forall n[s \in C \rightarrow s * \langle n \rangle \in C]$.

Therefore: $\forall s[s \in C \leftrightarrow \forall n[s * \langle n \rangle \in C]]$.

Using the Principle of Bar Induction, one concludes:
$\langle\, \rangle \in C$, i.e., $\varphi|\mathcal{N}, \psi|\mathcal{N}$ are Borel-separable.

The theorem occurs in [26, Theorem 13.4.1], see also [35].
A slightly different theorem is proven in [1].

12 Ramsey's Theorem

12.1 A small exercise

We start with a simple case of the Finite Ramsey Theorem.

For all $l > 0$, for all $m > 0$, we define the following statement.

$R_l(m) := \forall c : [m]^2 \rightarrow l \exists i < l \exists a \in [m]^3 \forall s \in [3]^2 [c(a \circ s) = i]$.

We now prove: $\forall l > 0[R_l(3l!)]$, by induction on l.

Clearly, $R_1(3)$ holds.

Assume l is given and $R_l(3l!)$ holds. We prove: $R_{l+1}(3(l+1)!)$.

Let $c : [3(l+1)!]^2 \rightarrow l+1$ be given.

Find $i < l$ and b in $[3(l+1)!]^{3l!}$ such that
$b(0) > 0$ and $\forall k < 3l![c(\langle 0, b(k) \rangle) = i]$.

Note: *either* $\exists s \in [3l!]^2 [c(b \circ s) = i]$ *or* $\forall s \in [3l!]^2 [c(b \circ s) \neq i]$.

In the first case, find such s, define $a := \langle 0 \rangle * (b \circ s)$ and note:
$\forall s \in [3]^2 [c(a \circ s) = i]$.

[35] A more detailed version of the argument would show that the Second Axiom of Countable Choice suffices.

In the second case, use the induction hypothesis, and find $j \neq i$, g in $[3l!]^2$ such that $\forall s \in [3]^2[c(b \circ g) = j]$.

Define $a := b \circ g$ and note: $\forall s \in [3]^2[c(a \circ s) = j]$.

12.2 The Infinite Ramsey Theorem

In 1928, F. P. Ramsey proved the following statement as an exercise useful for someone who would like to prove the Finite Ramsey Theorem. We restrict ourselves to the two-dimensional case.

($\star_{class}$) For every $R \subseteq \omega$,
either $\exists \zeta \in [\omega]^\omega \forall s \in [\omega]^2[\zeta \circ s \in R]$ or $\exists \zeta \in [\omega]^\omega \forall s \in [\omega]^2[\zeta \circ s \notin R]$.

Constructively, this can not be true. The following statement is a consequence:

For every $A \subseteq \omega$,
either $\exists \zeta \in [\omega]^\omega \forall n[\zeta(n) \in A]$ or $\exists \zeta \in [\omega]^\omega \forall n[\zeta(n) \notin A]$.

This is sometimes a reckless statement, for instance, if we let A consist of the numbers n such that in the first n decimals of π, no uninterrupted sequence of 99 9's is found.

12.3 An intuitionistic version

$A \subseteq \omega$ is called 1-*almost full* if and only if $\forall \zeta \in [\omega]^\omega \exists n[\zeta(n) \in A]$.

One may prove:

If $A, B \subseteq \omega$ are 1-almost-full, then $A \cap B$ is 1-almost-full.

The argument is as follows.

Let $A, B \subseteq \omega$ be 1-almost-full. Let ζ in $[\omega]^\omega$ be given. Find, using an Axiom of Countable Choice, η in $[\omega]^\omega$ such that $\forall n[\zeta \circ \eta(n) \in A]$. Find n such that $\zeta \circ \eta(n) \in B$. Define $p := \eta(n)$ and note: $\zeta(p) \in A \cap B$.

It follows that, for each $k > 0$, for all 1-almost-full $A_0, A_1, \ldots, A_{k-1}$, also $\bigcap_{i<n} A_i$ is 1-almost-full.

$R \subseteq \omega$ is called 2-*almost-full* if and only if $\forall \zeta \in [\omega]^\omega \exists s \in [\omega]^2[\zeta \circ s \in R]$.

The two-dimensional case of the Intuitionistic Infinite Ramsey Theorem is the statement:

($\star_{int}$) If $R, T \subseteq \omega$ are 2-almost-full, then $R \cap T$ is 2-almost-full.

A classical mathematician would argue: $(\star)_{int}$ is equivalent to $(\star)_{class}$, as follows. Assuming: $(\star_{int})$ and $R \subseteq \omega$, he notes: $R \cap (\omega \setminus R) = \emptyset$, so
either R is not 2-almost-full and $\exists \zeta \in [\omega]^\omega \forall s \in [\omega]^2 [\zeta \circ s \notin R]$
or $\omega \setminus R$ is not 2-almost-full and $\exists \zeta \in [\omega]^\omega \forall s \in [\omega]^2 [\zeta \circ s \in R]$.
Conversely, assuming: $(\star_{class})$ and $R, T \subseteq \omega$ are 2-almost-full and ζ in $[\omega]^\omega$, he defines $R_\zeta = \{s \in [\omega]^2 \mid \zeta \circ s \in R\}$ and then, applying $(\star_{class})$ to R_ζ, finds η in $[\omega]^\omega$ such that $\forall s \in [\omega]^2 [\eta \circ s \in R_\zeta]$. He then notes: there must exist s in $[\omega]^2$ such that $\zeta \circ \eta \circ s \in T$ and so $\zeta \circ \eta \circ s \in R \cap T$. He thus sees: $R \cap T$ is 2-almost-full.

We now prove $(\star)_{int}$, intuitionistically, with the help of the Principle of Bar Induction. We need the following version of the latter principle:

> For all $B, C \subseteq [\omega]^{<\omega}$, if $\forall \zeta \in [\omega]^\omega \exists n [\overline{\zeta} n \in B]$ and $B \subseteq C$ and $\forall s \in [\omega]^{<\omega} [s \in C \leftrightarrow \forall n [s * \langle n \rangle \in [\omega]^{<\omega} \rightarrow s \in C]]$, then $\langle\,\rangle \in C$.

We first prove:

$$\text{If } R, T \subseteq \omega \text{ are 2-almost-full, then } \exists s \in [\omega]^2 [s \in R \cap T].$$

Let $R, T \subseteq \omega$ be 2-almost-full.

Define the proposition $QED := \exists s \in [\omega]^2 [s \in R \cap T]$.

Again, we read QED as: *quod* est *demonstrandum, what* has *to be proven.*

For each n, s in $[\omega]^n$ is called *nice* if and only if
$\exists t \in [n]^2 \exists u \in [n]^2 [t(0) = u(0) \wedge t(1) \leq u(1) \wedge s \circ t \in R \wedge s \circ u \in T]$.

Define $B := \{s \in [\omega]^{<\omega} \mid s \textit{ is nice}\}$.

We prove that B is a bar in $[\omega]^\omega$, that is: $\forall \zeta \in [\omega]^\omega \exists n [\overline{\zeta} n \in B]$.

Let ζ in $[\omega]^\omega$ be given.
Find η in $[\omega]^\omega$ such that $\forall n [\langle \zeta \circ \eta(2n), \zeta \circ \eta(2n+1) \rangle \in R]$.
Find m, n such that $m < n$ and $\langle \zeta \circ \eta(2m), \zeta \circ \eta(2n) \rangle \in T$.
Note: $\overline{\zeta}(\eta(2n) + 1) \in B]$.

Define $C := \{s \mid \forall \zeta \in [\omega]^\omega \exists n [s * \langle \zeta(n) \rangle \in B \vee QED]\}$.

Note $\forall s \in [\omega] < \omega \forall t \in [\omega]^{<\omega} [(s \in B \wedge s \sqsubseteq t) \rightarrow t \in B]$ and conclude: $B \subseteq C$.

Now let s in $[\omega]^{<\omega}$ be given such that $\forall n [s * \langle n \rangle \in [\omega]^{<\omega} \rightarrow s * \langle n \rangle \in C]$.
We prove: $s \in C$.

Let ζ in $[\omega]^\omega$ be given. Consider $l := \mathit{length}(s)$ and define $m := 3l!$.

Find $n_0, n_1, \ldots n_{m-1}$ such that $s * \langle \zeta(n_0) \rangle \in [\omega]^{<\omega}$ and
$\forall i < m \forall j < m [i < j \rightarrow s * \langle \zeta(n_i), \zeta(n_j) \rangle \in B \vee QED]$.

In order to find these numbers we repeatedly use the fact that the intersection of finitely many 1-almost-full subsets of ω is itself 1-almost-full.

Note: if $s * \langle\zeta(n_i), \zeta(n_j)\rangle \in B$ then: either $s * \langle\zeta(n_i)\rangle \in B$ or $s * \langle\zeta(n_j)\rangle \in B$ or QED or $\exists k < l[\langle s(k), \zeta(n_i)\rangle \in R \wedge \langle s(k), \zeta(n_j)\rangle \in T]$.

Conclude: for all i, j such that $i < j < 3l!$ one may determine $k < l$ such that: *either* $\exists i < m[s * \langle\zeta(n_i) \in B]$ *or* QED *or* $(\langle s(k), \zeta(n_i)\rangle \in R$ and $\langle s(k), \zeta(n_j)\rangle \in T)$.

Now, using the simple fact established in Subsection 12.1, find $k < l$ and i, j, p such that $i < j < p < m$ and *either*: $\exists i < m[s * \langle\zeta(n_i) \in B]$ *or*:

1. $(\langle s(k), \zeta(n_i)\rangle \in R$ and $\langle s(k), \zeta(n_j)\rangle \in T)$, and

2. $(\langle s(k), \zeta(n_i)\rangle \in R$ and $\langle s(k), \zeta(n_p)\rangle \in T)$, and

3. $(\langle s(k), \zeta(n_j)\rangle \in R$ and $\langle s(k), \zeta(n_p)\rangle \in T)$.

The second alternative implies: $\langle s(k), \zeta(n_j\rangle \in R \cap T$ and: QED.
Conclude: either: $\exists i < m[s * \langle\zeta(n_i) \in B]$ or QED.
We thus see: $\forall\zeta \in [\omega]^\omega \exists n[s * \langle\zeta(n) \in B \vee QED]$, that is: $s \in C$.

Using the Principle of Bar Induction, conclude: $\langle\,\rangle \in C$, and QED.

We now prove:

If $R, T \subseteq \omega$ are 2-almost-full, then $R \cap T$ is 2-almost-full.

Let $R, T \subseteq \omega$ be 2-almost-full. Let $\zeta \in [\omega]^\omega$ be given.
Define $R_\zeta := \{s \in \omega^2 \mid \zeta \circ s \in R\}$ and $T_\zeta := \{s \in \omega^2 \mid \zeta \circ s \in T\}$.
Note: R_ζ, T_ζ are 2-almost-full, find s in $R_\zeta \cap T_\zeta$, and conclude: $\zeta \circ s \in R \cap T$.
We thus see: $\forall\zeta \in [\omega]^\omega \exists s \in [\omega]^2[\zeta \circ s \in R \cap T]$, that is: $R \cap T$ is 2-almost-full.

12.4 The history

In 1974, Mervyn Jansen, as de Iongh's master student, wrote a Master's Thesis on Ramsey's Theorem and proved, using the Principle of Bar Induction, that, for each $R \subseteq \omega$ satisfying a certain condition, not both R and $\omega \setminus R$ are 2-almost-full. In 1986, M. Bezem, a Ph.D. student of D. van Dalen, independently, obtained the same result, but he could do without the condition used by Jansen. Reflection upon Bezem's proof and upon an intuitionistic proof of Dickson's Lemma:
$\forall p > 0 \forall\alpha\exists i \exists j[i < j \wedge \forall q < p[\alpha^q(i) \leq \alpha^q(j)]]$[36] then led to the more positive result treated in Subsection 12.3, see [36]. Later, in the same vein, an intuitionistic proof of Kruskal's Theorem was found, see [30].

[36] [30, §1]

13 Notation and conventions

13.1

$\mathbb{N} = \omega$ denotes the set of the natural numbers $0, 1, 2, \ldots$.

We use $a, b, \ldots m, n, \ldots p, q, r, s, \ldots$ as variables over ω.

We assume a function $J : \omega \times \omega \to \omega$ has been defined with inverse functions $K, L : \omega \to \omega$ such that $\forall n[J(K(n), L(n)) = n]$.

$\forall m \forall n[(m, n) := J(m, n)]$ and $\forall n[n' := K(n)]$ and $\forall n[n'' := L(n)]$.

Every s is the code of a finite sequence of natural numbers.

We assume a binary function $s, i \mapsto s(i)$ has been defined that gives the value of the finite sequence coded by s at i.

We assume a function $s \mapsto \mathit{length}(s)$ has been defined that gives, for each s, the length of the finite sequence coded by s.

We assume: for each s, i, if $i \geq \mathit{length}(s)$, then $s(i) = 0$.

For each k, $\omega^k := \{s \mid \mathit{length}(s) = k\}$.

$\langle\,\rangle := 0$ codes the *empty sequence* and $\omega^0 = \{0\}$.

For each n, $\langle n \rangle$ is the unique u in ω^1 such that $u(0) = n$, and $\omega^1 = \{\langle n \rangle \mid n \in \omega\}$.

For all k, for all s in ω^k, for all n, $s * \langle n \rangle$ is the unique u in ω^{k+1} satisfying $\forall i < k[u(i) = s(i)]$ and $u(k) = n$, and $\omega^{k+1} = \{s * \langle n \rangle \mid s \in \omega^k, n \in \omega\}$.

For all k, l, for all s in ω^k, for all t in ω^l, $s * t$ is the unique u in ω^{k+l} satisfying $\forall i < k[u(i) = s(i)]$ and $\forall j < l[u(k + j) = t(j)]$.

For all k, for all t in ω^k, for all s such that $\forall n < k[t(n) < \mathit{length}(s)]$, $s \circ t$ is the unique u in ω^k satisfying $\forall n < k[u(n) = s(t(n))]$.

$s \sqsubseteq t \leftrightarrow \exists u[t = s * u]$.

$s \sqsubset t \leftrightarrow (s \sqsubseteq t \ \wedge\ s \neq t)$.

$s \perp t \leftrightarrow \neg(s \sqsubseteq t \ \vee\ t \sqsubseteq s)$.

For all k, for all s in ω^k, for all $i \leq k$, $\overline{s}i$ is the unique u in ω^i such that $u \sqsubseteq s$.

For all s, n, s^n is the largest u such that $\forall m < \mathit{length}(u)[(n, m) < \mathit{length}(s) \ \wedge\ u(m) = s((n, m))]$.

$\mathit{Bin} := \{s \mid \forall n < \mathit{length}(s)[s(i) < 2]\}$.

$\mathit{Bin}_n := \{s \in \mathit{Bin} \mid \mathit{length}(s) = n\}$.

$[\omega]^k := \{s \in \omega^k \mid \forall n < k - 1[s(n) < s(n+1)]\}$.

$[\omega]^{<\omega} := \bigcup_k [\omega]^k$.

$\mu n[P(n)] = k \ \leftrightarrow\ (P(k) \ \wedge\ \forall n < k[\neg P(n)])$.

13.2

$\mathcal{N} = \omega^\omega$ is the set of all functions from ω to ω.

We use $\alpha, \beta, \ldots, \zeta, \eta, \ldots, \varphi, \psi, \ldots$ as variables over ω^ω.

$\mathcal{C} = 2^\omega = \{\alpha \in \omega^\omega \mid \forall n[\alpha(n) < 2]\}$.

For each n, $\underline{n}$ is the element β of $\mathcal{N}$ such that $\forall m[\beta(m) = n]$.

$\alpha \# \beta \leftrightarrow \alpha \perp \beta \leftrightarrow \exists n[\alpha(n) \neq \beta(n)]$.

$\overline{\alpha}k := \langle \alpha(0), \alpha(1), \ldots \alpha(k-1) \rangle$.

$s \sqsubset \alpha \leftrightarrow \overline{\alpha}(length(s)) = s$.

For all α, for all n, for all t in ω^n, $\alpha \circ t$ is the element u of ω^n such that $\forall i < n[u(i) = \alpha(t(i))]$.

For all α, for all β, $\alpha \circ \beta$ is the the element γ of $\mathcal{N}$ such that $\forall n[\gamma(n) = \alpha(\beta(n))]$.

For all α, β, (α, β) is the element γ of $\mathcal{N}$ such that $\forall n[\gamma(n) = (\alpha(n), \beta(n))]$.

$s \perp \alpha \leftrightarrow \alpha \perp s \leftrightarrow \neg(s \sqsubset \alpha)$.

$\mathcal{N} \cap s := \{\alpha \mid s \sqsubset \alpha\}$.

$\alpha^n(m) := \alpha((n, m))$.

$[\omega]^\omega = \{\alpha \in \omega^\omega \mid \forall n[\alpha(n) < \alpha(n+1)]\}$.

13.3

$\mathcal{F} \subseteq \mathcal{N}$ is a spread if and only if
$\exists \beta[\forall s[\beta(s) = 0 \leftrightarrow \exists n[\beta(s * \langle n \rangle) = 0]] \wedge \forall \alpha[\alpha \in \mathcal{F} \leftrightarrow \forall n[\beta(\overline{\alpha}n) = 0]]]$.

Let $\mathcal{F} \subseteq \mathcal{N}$ be a spread.

$\varphi : \mathcal{F} \to \mathbb{N} \quad \leftrightarrow \quad \forall \alpha \in \mathcal{F} \exists n[\varphi(\overline{\alpha}n) \neq 0]$.

If $\varphi : \mathcal{F} \to \mathbb{N}$ then, for each α in $\mathcal{F}$, $\varphi(\alpha)$ is the number q such that $\exists n[\varphi(\overline{\alpha}n) = q + 1 \ \wedge \ \forall m < n[\varphi(\overline{\alpha}m) = 0]]$.

$\varphi : \mathcal{F} \to \mathcal{N} \quad \leftrightarrow \quad \forall n[\varphi^n : \mathcal{F} \to \mathbb{N}]$.

If $\varphi : \mathcal{F} \to \mathcal{N}$, then, for each α in $\mathcal{F}$, $\varphi|\alpha$ is the element β of $\mathcal{N}$ such that $\forall n[\beta(n) = \varphi^n(\alpha)]$.

If $\varphi : \mathcal{F} \to \mathcal{N}$, then, for each s such that $\exists \delta \in \mathcal{F}[s \sqsubset \delta]$,
$\varphi|s$ is the greatest number $t \leq s$ such that
$\forall i < length(t) \exists n < length(s)[\varphi^i(\overline{s}n) = t(i) + 1 \ \wedge \ \forall j < n[\varphi^i(\overline{s}j) = 0]]$.

Note: if $\varphi : \mathcal{F} \to \mathcal{N}$, then, for all α in $\mathcal{F}$, for all β,
$\varphi|\alpha = \beta \quad \leftrightarrow \quad \forall n \exists m[\overline{\beta}n \sqsubseteq \varphi|\overline{\alpha}m]$.

For all $\varphi : \mathcal{N} \to \mathcal{N}$, for all $\mathcal{X} \subseteq \mathcal{N}$, $\varphi|\mathcal{X} := \{\varphi|\alpha \mid \alpha \in \mathcal{X}\}$.

For each $V \subseteq \mathcal{N}$, $\varphi : \mathcal{F} \hookrightarrow V$ if and only if $\varphi : \mathcal{F} \to \mathcal{N}$ and $\forall \alpha \in \mathcal{F} \forall \beta \in \mathcal{F}[\alpha \# \beta \to \varphi|\alpha \# \varphi|\beta]$ and $\forall \alpha \in \mathcal{F}[\varphi|\alpha \in V]$.

For each $\mathcal{X} \subseteq \mathcal{N}$, $\mathcal{X}^{\neg} := \{\alpha \mid \alpha \notin \mathcal{X}\}$.

$\mathcal{R}$, the set of the real numbers, may be defined as a subset of $\mathcal{N}$.

For x, y in $\mathcal{R}$, $x \#_{\mathcal{R}} y \leftrightarrow \exists n[|x - y| >_{\mathcal{R}} \frac{1}{2^n}]$.

Appendix: Johan J. de Iongh[37]

Johan J. de Iongh (1915–1999) studied mathematics, and a lot of other things, at the University of Amsterdam. He played an active rôle in Amsterdam student life and founded the students' union *Anakeion*. During the Nazi occupation of the Netherlands he spent some time in the notorious *Kamp Amersfoort*. He came under the spell of L. E. J. Brouwer and, discerning the soundness of the intuitionistic view of mathematics, he became Brouwer's associate and friend. Hearing G. Mannoury lecture on the philosophy of mathematics he developed a strong affinity for the latter's signific point of view. He discussed the foundations of physics and mathematics with D. van Dantzig. In 1952/53 he made careful notes of a course given by A. Heyting on intuitionistic mathematics. He had many conversations with S. C. Kleene when the latter visited Amsterdam in order to learn about intuitionistic mathematics.

From his early years as a student on he gave lectures on a wide variety of subjects, on all kinds of occasions. For some time, he was H. Freudenthal's assistant at the *Rijksuniversiteit Utrecht*, and, like him, he got involved with Mathematics Education, from the primary school to the University level.

In 1960, he became a Professor of Mathematics at the *Katholieke Universiteit Nijmegen*. He was the first Professor of Mathematics at this University, and founded the Department of Mathematics. While fulfilling this difficult task, he was inspired and actively supported by the Utrecht mathematicians H. Freudenthal and F. van der Blij. In the early years, he taught a meticulously thought out course on Mathematical Analysis. Later, the courses in Logic and the Foundations of Mathematics were his primary concern. He regularly treated the main subjects of the field, such as Recursive Function Theory, Axiomatic Set Theory, Introductory Model Theory and of course Intuitionistic Mathematics. He kept a watchful eye on the quality of the teaching at the Department as a whole and contributed courses on Problem Solving, School Mathematics from an Advanced Point of View, and the History and Philosophy of Mathematics. He always prepared his lectures with the utmost care.

[37] A previous version of this Appendix has been published as an obituary in [28]. The Editors of the Nieuw Archief voor Wiskunde have gracioulsy allowed us to re-use this testimony here.

When delivering them, his notes in hand, he walked enormous distances, striding up and down before the blackboard, raising and lowering his voice, slowing down and accelerating his speech, according to the needs of the subject and a good theatrical performance. He advocated a broad view and knowledge of mathematics and its applications and abhorred too narrow specialists. One of the goals he set himself and his Department was to rear excellent teachers of mathematics.

Johan de Iongh understood very well that students spend a decisive part of their lives at the university. A passionate teacher, driven by a Platonic paedagogical Eros, he wanted to know about their background and interests. He eagerly followed their attempts to organize themselves, having experienced, in his own student days, the importance of personal encounters and the possibility of making friends for life. Deeply serious about his catholic faith, he tried to create occasions where students could glimpse something of the greatness of the Christian tradition. His dreams were shattered when Paris 1968 evolved into Nijmegen 1973 and students occupied the Department for months. He felt hopeless, unable to deliver them from the false ideology in which he thought them caught. After these events, and also because of the ensuing change in the organization of the Universities directed towards "democratization", his rôle in the Department became a more modest one. He was a man of great learning and wide interests and as much a philosopher as a mathematician. Steeped in Greek culture, he loved Plato and Euclid and advertized the Elements, in particular its first book, as exciting reading, also from the perspective of modern mathematics. He pointed out that the debate in the philosophy of mathematics between Platonists and Intuitionists is foreshadowed in the difference of opinion between Menaechmos and Speusippos, as reported by Proclos. (Speusippos succeeded Plato as the head of the Academy). He also quoted Cusanus in support of the intuitionistic point of view.

His own position in the philosophy of mathematics might be described as a platonically tinged intuitionistic one. He emphasized that a proof is never to be identified with the text in which it is expressed, and that truth is not the same as correctness. Mathematics cannot be given a foundation outside itself. It comes before logic and consists in carrying out constructions in one's own mind. Language has a rôle to play when one wants to report on one's mathematics, either to oneself or to someone else, but one is never sure that it will fulfil one's expectations. It must seem that such a view forces one into solipsism. But, as Plato explains in his Seventh Letter, truth may show itself, unexpectedly and suddenly, as a spark of fire, by something like divine grace, where friends have been patiently seeking after it together, for a possibly very long time.

Johan shared Plato's hankering after such events.

He strived towards perfection and could be harsh in his judgment on other people and their work, expressing his discontent impetuously and vigorously. He was most critical of his own writings and published only a very few pages. He retired in 1984. He had problems with his health from about 1980. Somewhat lonely, and given to self-doubt, he spent his later years in his apartment across the Nijmegen Railway Station, where his loyal housekeeper, *Juffrouw* Marie, kept him company. Of the same age as Johan himself, she took care of him for 37 years, and now survives him.

He died on June 9, 1999. Let us remember him thankfully and respectfully.

References

[1] Peter Aczel *A constructive version of the Lusin Separation Theorem*, in: Sten Lindström, Erik Palmgren, Krister Segerberg, and Viggo Stoltenberg-Hansen (eds.), Logicism, Intuitionism, and Formalism, What has Become of Them?, Springer, Dordrecht 2009 [*Synthese Library, Studies in Epistemology. Logic, Methodology, and Philosophy of Science 341*], pp. 129–151.

[2] Evert Willem Beth *Semantic construction of intuitionistic logic*, Mededelingen der Koninklijke Nederlandse Akademie van Wetenschappen, Afd. Letterkunde 19 (11), 1956, pp. 357–388.

[3] Luitzen Egbertus Jan Brouwer *Begründung der Mengenlehre unabhängig vom logischen Satz vom ausgeschlossenem Dritten. Zweiter Teil: Theorie der Punktmengen*, Koninklijke Nederlandse Akademie van Wetenschappen Verhandelingen, 1e Sectie 12 (7), 1919; 33 pp.; also in: [6], pp. 191–221.

[4] — *Über Definitionsbereiche von Funktionen*, Mathematische Annalen 97, 1927, pp. 60–75; also in: [6], pp. 390–405 or in: [13], pp. 446–463.

[5] — *Essentieel-negatieve eigenschappen*, Proceedings Koninklijke Nederlandse Akademie van Wetenschappen 51, 1948, pp. 963-964; translated: *Essentially negative properties*, in: [6], pp. 478-479.

[6] — Collected Works 1, *Philosophy and Foundations of Mathematics* (edited by Arend Heyting), North-Holland Publishing Company, Amsterdam 1975.

[7] John P. Burgess *Brouwer and Souslin on transfinite cardinals*, Mathematical Logic Quarterly 26, 1980, pp. 209–214.

[8] David van Dantzig *On the principles of intuitonistic and affirmative mathematics*, Proceedings Koninklijke Nederlandse Akademie van Wetenschappen 50, 1947, pp. 918-929 and 1092-1103 = Indagationes Mathematicae 9, 1947, pp. 429-440 and 506-517.

[9] Michael Dummett Elements of Intuitionism (with the assistance of Roberto Minio), Oxford University Press, Oxford 1977 [*Oxford Logic Guides*].

[10] Christopher G. Gibson *On the definition of an infinite species*, Nieuw Archief voor Wiskunde (3) 19, 1971, pp. 196–197.

[11] Wim Gielen *The trustworthiness of the logical principles*, manuscript, 1983.

[12] Wim Gielen, Harrie de Swart, and Wim Veldman *The Continuum Hypothesis in intuitionism*, The Journal of Symbolic Logic 46, 1981, pp. 121–136.

[13] Jean van Heijenoort (ed.) From Frege to Gödel, *A Source Book in Mathematical Logic, 1879–1931*, Harvard University Press, Cambridge Mass. 1967.

[14] Arend Heyting Inleiding tot de Intuïtionistische Wiskunde, *College van Prof. Dr. A. Heyting* (Cursus 1952–1953. Uitgewerkt door J. J. de Iongh) [Introduction to Intuitionistic Mathematics, Lectures 1952–1953, Lecture Notes by J. J. de Iongh] [Dutch], University of Amsterdam 1953.

[15] — Intuitionism, an Introduction, North Holland Publishing Company, Amsterdam 1956.

[16] Johan J. de Iongh *Restricted forms of intuitionistic mathematics*, Library of the X-th International Congress of Philosophy (Amsterdam, August 11–18, 1948), 1, *Proceedings of the Congress*, North Holland Publishing Company, Amsterdam 1949, pp. 744–748.

[17] Stephen C. Kleene *Introduction to Metamathematics*, North-Holland Publishing Company, Amsterdam 1952, and D. Van Nostrand, Toronto 1952.

[18] Stephen C. Kleene, and Robert E. Vesley The Foundations of Intuitionistic Mathematics, *Especially in Relation to the Theory of Recursive Functions*, North-Holland Publishing Company, Amsterdam 1965.

[19] Georg Kreisel *On weak completeness of intuitionistic predicate logic*, The Journal of Symbolic Logic 27, 1962, pp. 139–158.

[20] Harrie de Swart *Another intuitionistic completeness proof*, The Journal of Symbolic Logic 41, 1976, pp. 644–662.

[21] — *Spreads or choice sequences?*, History and Philosophy of Logic 13, 1992, pp. 203–213.

[22] — *The Nijmegen school of intuitionism*, in: G. Alberts, L. Bergmans, and F. A. Muller (eds.), The Signific Circle and the Vienna Circle, Intersections, Springer Verlag, 2019 [to appear].

[23] Anne S. Troelstra Choice Sequences, a Chapter of Intuitionistic Mathematics, Clarendon Press, Oxford 1977.

[24] Anne S. Troelstra and Dirk van Dalen Constructivism in Mathematics, *Volume I and Volume II*, North Holland Publishing Company, Amsterdam 1988.

[25] Wim Veldman *An intuitionistic completeness theorem for intuitionistic predicate logic*, The Journal of Symbolic Logic 41, 1976, pp. 159–166.

[26] — Investigations in Intuitionistic Hierarchy Theory, PhD Dissertation, University of Nijmegen, 1981.

[27] — *Some intuitionistic variations on the notion of a finite set of natural numbers*, in: H. C. M. de Swart, and L. J. M. Bergmans (eds.), Perspectives on Negation, *Essays in Honour of Johan J. de Iongh on the Occasion of his 80th Birthday*, Tilburg University Press, Tilburg 1995, pp. 177–202.

[28] — *In Memoriam J. J. de Iongh (1915–1999)*, Nieuw Archief voor Wiskunde (5), Part 1, 1 (March 2000), pp. 16–17.

[29] — *Understanding and using Brouwer's continuity principle*, in: U. Berger, H. Osswald, and P. Schuster (eds.), Reuniting the Antipodes, *Constructive and Nonstandard Views of the Continuum* (Proceedings of a Symposium held in San Servolo / Venice, 1999), Kluwer, Dordrecht 2001, pp. 285–302.

[30] — *An intuitionistic proof of Kruskal's Theorem*, Archive for Mathematical Logic 43, 2004, pp. 215–264.

[31] — *Two simple sets that are not positively Borel*, Annals of Pure and Applied Logic 135, 2005, pp. 151–209.

[32] — *The Borel Hierarchy Theorem from Brouwer's intuitionistic perspective*, The Journal of Symbolic Logic 73, 2008, pp. 1–64.

[33] — *Some applications of Brouwer's Thesis on bars*, in: M. van Atten, P. Boldini, M. Bourdeau, and G. Heinzmann (eds.), One Hundred Years of Intuitionism (1907–2007), Birkhäuser, Basel 2008 [*Publications des Archives Henri Poincaré / Publications of the Henri Poincaré Archives*], pp. 326–340.

[34] — *The fine structure of the intuitionistic Borel Hierarchy*, The Review of Symbolic Logic 2, 2009, pp. 30–101.

[35] — *Projective sets, intuitionistically*, arXiv 1104.3077, submitted for publication to: The Review of Symbolic Logic.

[36] Wim Veldman, and Marc Bezem *Ramsey's Theorem and the Pigeonhole Principle in intuitionistic mathematics*, Journal of the London Mathematical Society 47, 1993, pp. 193–211.

II Advances in Computing

Corrado Böhm
The λ-adventure

Ariela Böhm
Rome
ariela.bohm@gmail.com

Michele Böhm
Milan
michelebohm@gmail.com

Emanuele Böhm
Florence
emanuele.bohm@gmail.com

Mariangiola Dezani-Ciancaglini
University of Turin
dezani@di.unito.it

Francesca Manfredini
Milan
manfredinibohm@gmail.com

Nora Perugia Böhm
Turin
n.perugia@gmail.com

ABSTRACT This paper is a tribute to Corrado Böhm who will live forever in the memories and hearts of those people having had the privilege of sharing in his life and work. We present a picture of Corrado Böhm not only as a scientific genius, but also as a man. Through the words of his family, the reader will be reminded of major milestones in his life from an intimate perspective. A short, factual biography will be followed by personal memories of family members and by a description of some keystones of Corrado's activity as researcher and teacher. We conclude with suggestions for further reading.

1 A Short Biography of Corrado Böhm

Corrado Böhm was born in Milan on January 17, 1923 and lived and studied there until 1942, when he relocated to Switzerland to be able to continue studying after the proclamation of the racial laws in Italy. He graduated in Electrical Engineering at the University of Lausanne and then became a research assistant at the Federal Institute of Technology in Zurich. There Corrado worked on a Zuse computing machine Z4 and obtained his PhD in Mathematics. His thesis is a milestone in the theory and development of compilers (see Section 3.1).

While Corrado was in Switzerland his parents were arrested and deported to Auschwitz where they were killed on arrival. In 2018 two commemorative blocks (part of the stumbling block project of the German artist Gunter Deming `http://www.stolpersteine.eu`), one for each of them, were placed on the sidewalk in front of their home in Milan.

In Switzerland Corrado meets Eva Romanin Jacur, an extremely sensitive and charming woman. They married in 1950 and had two sons, Michele and Emanuele, and one daughter, Ariela. Eva, a talented painter, always supported Corrado's scientific passion by making all of her husband's collaborators feel welcome and comfortable in their home, thanks to her fantastic hospitality. So many of Corrado's collaborators came to love Eva, Michele, Emanuele, and Ariela as their own family.

Corrado returned to Italy in 1951 to work at Olivetti in Ivrea. Soon after, he became a permanent researcher at the Istituto Nazionale per le Applicazioni del Calcolo (INAC) in Rome, leaving only when he was appointed as one of the first Italian full professors in computer science in 1968. Corrado was in charge of testing the computer Mark I at Manchester University. The test was successful and the machine was subsequently purchased by the Italian government for INAC.

Corrado's main research interest shifted from numerical calculus to abstract models of computation, in particular Turing Machines and λ-calculus. Corrado made keystone contributions to both of these formalisms that are discussed in Section 3.2 and 3.3, respectively.

In line with his scientific activity Corrado, along with Jaco de Bakker, Peter Landin, Dana Scott, Christopher Strachey and others, was a founding member of the IFIP (International Federation for Information Processing) Working Group 2.2 on "Formal Description of Programming Concepts". This group has been the melting pot of denotational semantics, a field that had, and continues to have, an impressive impact on the modelling and design of programming languages. The discrimination of λ-terms characterised by Corrado plays a central role in the denotational semantics of λ-calculus.

Beginning in 1960 Corrado was also teaching at the university. His lessons were considered completely unique both in their content and in Corrado's manner of working with his students. Corrado encouraged his students to engage with each other and with himself. The best students were captivated by him and selected to do their master thesis under his guidance. Many of them went on to brilliant academic careers and a short list of his students in the '60s includes Marisa Venturini-Zilli, Giorgio Ausiello, Daniel Bovet, Salvatore Caporaso, Giuseppe Jacopini and Alfonso Miola.

Corrado spent a year at the University of Modena as a full professor before being called at the University of Turin to organise an undergraduate course in computer science in 1970. The only other computer science course that existed in Italy at that time had begun just one year earlier in Pisa. In Turin, Corrado introduced a number of young people to research that later all became professors there: Mariangiola Dezani-Ciancaglini, Ines Margaria, Simona Ronchi Della Rocca, Maddalena Zacchi and Mario Coppo. In the beginning, the newly instituted department of computer science was just one small room. So it was not surprising that Corrado's spacious home, full of interesting books and papers and warmed by the kind hospitality of his entire family, became the centre of much of Corrado's research activity with his pupils.

In 1974 Corrado was appointed as a full professor at the Sapienza University of Rome, where he continued to work for the rest of his life. The following year was very successful for Corrado: he became a member of the editorial board of "Theoretical Computer Science" and also organised an international conference on "λ-calculus and Computer Science". Today "Theoretical Computer Science" is acknowledged as one of the best journals in the field. And the current international conference "Formal Structures for Computation and Deduction" has its roots in Corrado's initial conference.

At the Sapienza, Corrado continued to attract gifted students who were eager to work with him. Many of his PhD students now hold prestigious academic positions in Italy and abroad, including Alessandro Berarducci, Stefano Guerrini, Benedetto Intrigila, Ugo de'Liguoro, Silvio Micali, Adolfo Piperno, Ivano Salvo, and Enrico Tronci. It is worth noting that in 2012 Silvio Micali received the most distinguished of prizes awarded to a computer scientist, the Turing Award.

In 1994 the University of Milan awarded Corrado a "Laurea honoris causa" (honorary degree) in Computer Science.

Corrado's exceptional stature in the international field of computer science scene was critically lauded in 2001 when he received the EATCS (European Association for Theoretical Computer Science) Award just one year after its creation. The award "is given to acknowledge extensive and widely recognised contributions to theoretical computer science over a life-long scientific career."

Since October 23, 2017, Corrado is no longer with us, but the genius of his seminal work is immortal.

2 Corrado, Father and Uncle

2.1 MICHELE BÖHM: λ-Limud for Corrodi

To be the eldest son of a certified genius is not an easy feat, nor is it easy to celebrate him here, and to do so in a way that he would have liked is even more complicated. My father lived his abrasive intelligence inside his own speculative world and was sure, from the very beginning, that the computer world would be the world of the connected humanity, and that's exactly what happened.

As a child I was fascinated by his bookshelves housing volumes about games and their theory. I could not grasp the connection with Mathematics and its arcane language. Damn, right there was a connection! It was the concept of algorithm, that could be executed by a machine and had then the looks and was called a program!!! Likewise, the largest conceivable number did not carry a physical identity, it was the result of the instruction adding one to any number. And, speaking of large numbers, I wasn't 10 years old when I knew what a googol is, that one followed by a hundred zeros, that became the origin of Google's name. And also, the damned googolplex, one followed by a googol of zeroes!!! I didn't really understand what a computer was, but I already knew that a calculator, despite its incredible speed, had limits, and that some calculations would take billions of years of computation, yielding no result. I was a young kid when I learned about the experimental method, and I had already wrapped my head around the concept that, between two equally valid explanations, all scientists would embrace the shortest one. During my political years (everybody knows I was a passionate revolutionary with "Lotta Continua") those principles helped me reject any dogmatism, any denial of objective reality for the sake of ideology.

Even his Silence forged me, his never wanting to talk about the tragedy of his life, the murder of both his parents, the Shoah. His silence forcibly taught me at a very young age that the human mind is the most important battlefield, and that not speaking about something somehow means banishing that something from our brain, denying it the right to exist there. Corrodi was never bored, he was always

[0]The word Limud means "study" in Hebrew and also "a meeting of people to discuss the life of a deceased loved one". The nickname Corrodi was coined after a specific event: Eva (Corrado's wife) was waiting for Corrado to come out after discussing his dissertation in Zurich and she was asked "Sie sind die Frau von Dr. Corrodi?" ("are you Dr. Corrodi's wife?"). In the excitement of the moment she said "Ja, nein nein!" as, of course, Corrodi was someone else's surname!.

thinking and pondering, he entertained himself sketching ideas on his huge mental palette. I was so jealous!!! And in the meantime, he was also teaching, and even though I never witnessed any of his lessons, it was obvious that he was applying his peculiar teaching method, which consisted in explaining the topic of the moment to his pupils as if they were seasoned experts in the specific field, while what they were was simply eager jackasses, in dire need of some learning. I myself did some teaching in university and I can assure you that this is the best approach, the one that lets the most surprising hidden talents surface, and it's also the fastest. I used to see quaking students, whom we, the family, used to call "santini" (little saints), leave his study turned into sharp-eyed angels, ready to assist him in formulating the fundamentals of the new emerging science. Against all common sense we must look at man to understand apes, not the other way around. We must study development to understand the countless problems of underdevelopment. Therefore, I wasn't really surprised when, on my asking for a computer, he placidly answered that I would have one the day I wrote a program to play tic-tac-toe. All of which unbelievably happened.

My father taught me that inside a logical paradox a powerful idea lurks, or at least natural language's innate contradictions. He entertained me with paradoxes: Zeno's arrow which, in any one instant of time, is neither moving to where it is, nor to where it is not; or Achilles and the tortoise; or the library's catalogue which is also one of its books; or the magic of the double negative, with the opposing tribes of the Liars and the Sincere. When he explained that in functional programming a program works devouring itself down to the final result, it scared me, I thought I could see Uranus happily munching on his children!!! My father could forecast the number of pedestrians that would cross the street in one minute, and he let me understand that his secret simply lay in his counting the past in order to forecast the future, without ever mentioning Statistics, which he evidently considered as a little sister of the ultimate science, Her Majesty, Mathematics.

Without fail, he demolished me at any and all games we played, which greatly pissed me off, so much so that once I threw the chessboard and all the pieces out of the window. This didn't help me at all, since he rescued them all in a few minutes. Thank goodness such quantities of bloody rules did not apply to the world of Beauty and Art, the world to which my mother Eva introduced me, only the intensity of the vision's momentum mattered there. Dad had a fix on double-sided clothing, on self-reference, on the merciless evolutional harshness of natural selection, on radio-transmitters, and he had devised a clever scoring system to discipline us such that one could lose all points gained with a good school result just for foul language at the table. I remember that in the end I managed to carve out a prized clock/alarm clock/radio.

Corrado Böhm was one of the fathers of Computer Science, a smiling subversive radical, at war with his own followers, a truly free person, devoid of any prejudice. It is to him that we owe the modern, formidable equation: Human being = Computer + Dog. Towards the end of his long and exceptional career he became fascinated with Bioinformatics, with the speedy parallel self-organisation capability of living matter. It's basically like saying that the new computers' design is secretly modelled after Biology, and at the heart of the new Information Science is the development of a strong sensory/semantic system (first of all vision and natural language), without which speaking of Artificial Intelligence would be pompous and meaningless. Corrodi never gave up his habit of giving me small assignments, and since, as everybody knows, I deal with software myself, towards the end of the '80s he assigned me the task of developing an Optical Character Recognition software. It took me about ten years but, here too, mission accomplished! And now that he is no more, I can't be defined as "the Theorem's son" or "Professor Nulla" (Professor Nothing) rather as "Böhm the Younger" (62 years old, sic!); aware that in my life I will never attain even the anteroom of his genius, I feel a huge challenging responsibility weighing on my shoulders. Thank you, Corrodi, you've been a loving father and always helped me make the best use of my most powerful resource, my brain, to understand and grab fragments of the deep workings of reality.

Dear Dad, I miss you and always will. I will keep the Böhm's name up...

Shalom... Your son Michele

2.2 EMANUELE BÖHM: Corrado the Influencer

There is much I have absorbed from Corrado that profoundly influenced my personality and the shaping of my scientific interests. Much of what Corrado transmitted to me was very subtle yet fundamental. To him, all details of a problem were equally important. This translated into being careful never to ignore the apparently unimportant. Fascination with simplification was also an important notion that I assumed from him.

My father would regularly walk with my mother through the park in front of our house and would tell her about his research, knowing that she, being an artist, would not understand the technicalities. But I am convinced that he benefitted enormously from explaining his work to Eva. For as long as I can remember, I was utterly fascinated by the way that Corrado related to any kind of "technological gadgets", and I definitely inherited that from him. He taught me to use the slide rule but also the Kurta, a coffee grinder-shaped mechanical calculator. I also remember going on drives with my dad up to Forte Antenne (not very far from our house in Rome) to listen to my dad using his CB radio to talk to people around Europe and driving further still, to the top of the Monti Cimini at 1000m altitude, where he would talk to people all around the world.

In 1972, when I was 14, I exulted when dad wrote a sequence of instructions for the SHARP ELSI MINI pocket calculator that could be used to calculate a square root and display the result in red digits, thus overcoming the calculator limitation of being confined to just the four basic mathematical operations. In that same year, I accompanied him to the Computer Science Institute in Turin to witness a monumental event in in the evolution of microcomputers: the booting of the PDP 11 by a sequence of buttons on the front panel. Understandably, the experience is indelibly engraved in my memory.

Several years later, taking part in dad's university lectures as a student was an enriching experience that revealed to me different aspects of his personality, because until then I knew him only as a father in the context of our family interactions. This new vantage point allowed me to see how original and innovative he was as compared to all the other professors I encountered in my university training. I am pretty sure that dad never repeated a lecture twice throughout his many years of teaching "Theory and applications of computing machines". I believe he had the ability to facilitate "natural selection" of the best students by sparking their interest in the subject at hand during the course of each lecture.

I am grateful to my father for not trying to influence my professional choice (unlike his father who made him study engineering while Corrado wanted to be a mathematician) and encouraging me to both find my way and believe in my capabilities when I was less sure of myself as a young oceanographer for the first time in graduate school in the USA.

My five-year-old son Michelangelo, who bears the same name as both of my grandfathers, gives me the daily gift of expressing unique characteristics that I associate so strongly with Corrado that I can only feel, profoundly and increasingly, that my dad is still with me.

2.3 ARIELA BÖHM: **Corrado the Teacher-by-Example**

The entire scientific world is grateful to my father for his theorems and for his incredible innovative abilities. His former students are grateful to him as well for being an exemplary model of free thinking and absolute concentration in pursuing and realising abstract ideas.

His research was not only his profession, it was his existential mode. I believe that his curiosity and ability to feel amazement and wonder never changed throughout his long life, but remained as intense as they were in his childhood. These characteristics combined with his extraordinary intelligence were instrumental in making him the genius we all know.

Though life did not spare him from horror, he proceeded through it with remarkable candour and innocence. He chose to move into the dimension of abstraction and this, at times and to some, may have given the impression of a distance from reality. But his vitality and enthusiasm for even the smallest things in the real world convinced me that on some level he was fully aware of this distance and meticulously calibrated it, whether consciously or not.

I am grateful to him for having taught me by example how curiosity is the key that opens the doors of fantasy and even before that makes you look for them and find them. And for having shown me that the more open-mindedly and freely you will go through those doors, the greater the amazement that you will feel.

2.4 FRANCESCA MANFREDINI: In Awe of Corrado

I very much liked Corrado, he was fun. He had promptly defused my total awe of him through his unfazed likeability. I nurtured the idea of a non-trivial relationship with him, because of my endless admiration, and because of a supposed kinship – in name only, it goes without saying – through Math. I had loved Math a lot and had studied it with quite some success in university, including, to my utter excitement, his famous theorem! I was sure he loved me too, also because of that. He obviously tried to explain to me, as he did with just about anybody, his scientific thoughts, his theoretical results, the problems he kept tossing and turning in his mind, even while queuing at the post office, - he was never bored!!! I didn't really know about λ-calculus, but he used his signature method, he assumed that whomever he was talking to, knew and understood all about what he was saying, which would work wonders in fact, and I followed his reasonings and thought I understood, and had fun. Until he'd ask me a question and the illusion collapsed. How many crosswords have we solved together, I had inherited from my father a certain passion for word puzzles, and we had fun. We could spend hours with a "Settimana Enigmistica" magazine, I never got bored, I felt privileged in his intimacy. His erasable ball-point-pens, his little quirks.

Elegant, gentle, soft-spoken, he didn't need to raise his voice to be listened to, I heard him raise it only in his disputes with his eldest, Michele, my boyfriend, a crazy electric current fizzled between the two of them: it was their mutual jealousy over Eva, sublime wife and mother.

He was terribly stubborn in arguing, and, most of the times, maybe always, in the end he turned out to be right, he was very very unlikely to be wrong. And he knew it.

Great Corrado, we miss you!

2.5 NORA PERUGIA BÖHM: **Young Uncle Corrado**

Uncle Corrado, my father's younger brother by 20 years, was more like an older brother for me and my own brother, Sergio. But he insisted on being called "uncle". Uncle Corrado was born in 1923, I was born in 1929 and Sergio was born in 1932.

Uncle Corrado sometimes dedicated himself to us, normally during the stay at the family summer holiday house, in Maggio (a small town of less than 1000 people today) in the province of Lecco. We played one of the oldest board games in the world together: the mill game. And sometimes he invited us to hunt crickets.

We went to a little hill, laid down on the ground near the holes that he, an expert, pointed out to us. He would then skilfully insert a grass blade into the hole so that the cricket would come out and immediately be caught and placed in a special box with holes for the air. This box had a distinctive smell of crickets and later we discovered that even Uncle Corrado's clothes had this characteristic odour: neither good nor bad, just of crickets.

Uncle Corrado was very patient with us, especially when he would help us with our holiday homework, and even when, much later, we didn't understand our math, he'd never raise his voice. In contrast, his own father, our grandpa Michelangelo Böhm, was very strict with him. He didn't want Corrado to sleep late in the morning - he once woke him up at 10 AM in the fireplace room and threatened him with a stick!

Uncle Corrado studied Latin or Greek with my mother (whom Grandpa Böhm ironically called "castigamatti," an Italian expression meaning a menacing person able to obtain obedience) because she could make Uncle Corrado study ...with kindness.

Uncle Corrado loved to hunt and was oftentimes very busy preparing slingshots with scrap leather and elastic bands sewn by his mother or sister. His hunting expeditions didn't occur at fixed hours. He wouldn't want me and my brother along with him, maybe because we would have scared off the little birds. When he came home with a little bird it was a big party for him – the maid would pluck it and cook it in butter and sage for him. I remember that once I received a great prize: since there were two birds that day, he offered me "a little bird's thigh". It was delicious even though the amount of meat on the tiny leg was the size of my smallest finger nail.

I remember that my dad and Uncle Corrado practiced going up and down around the villa in Maggio with stilts!! Uncle Corrado must have been about 17 years old at the time, so my father would have been 37. I remember the two of them having fun in the meadow together, just like two small boys, flying kites. I remember them holding the strings.

The lines above were written during August 1988 in Maggio.

When we were refugees in Geneva in 1944, Uncle Corrado came to visit and would speak about high math with my dad that my brother and I couldn't follow. I remember his wedding with Eva Romanin Jacur in Padua. I also remember Eva's father, who had told her, "One day you will be the wife of a great personage." He was right.

Corrado had a great sense of humour, and I still remember having good laughs with him.

I found a letter that Uncle Corrado had written to his brother and sister on the day that he was honoured with the title of "Father of Information Technology" in Italy, telling them how much he missed them and how happy he believed that their father would have been.

When my father Arrigo passed away, his wish was to be cremated, and Corrado was the one to place the urn near my mother's and her parents' urns. He was very sad, as if it were the funeral of his own parents who had been gassed in Auschwitz.

I cared for Uncle Corrado very much and when he got sick and then passed away many years later, it was a great sorrow: a companion of my youth, a great and illustrious uncle had disappeared.

3 Corrado the Researcher

3.1 Self-Compilation

Corrado's PhD thesis [4] was singularly innovative and led to the construction of the first meta-circular compiler. Corrado described a machine, a language and a translation method for executing the programs written in the machine's language, as well as a parsing technique that was a particularly important contribution. Among parsing's features are the linearity that yields efficiency and the generation of linked binary trees in reading formulas, as today's list-processing techniques do.

Knuth and Pardo in [25] write: "Böhm's dissertation was especially remarkable because he not only described a complete compiler, he also defined that compiler in its own language! And the language was interesting in itself, because every statement (including input statements, output statements and control statements) was a special case of an assignment statement." Notably one must wait four years to get the definition of FORTRAN and seven years to get the definition of LISP.

3.2 Structured Programming

In [11] Corrado and his student Giuseppe Jacopini proved that each flow diagram can be built using only three basic diagrams: sequential composition, alternative clauses and while loop. As observed by Dijkstra [20]: "[they show] the (logical) superfluousness of the GOTO statement". This result, now known as the Böhm-Jacopini Theorem, is taught also in elementary introductions to computer science. The theorem provides a sound basis for "structured programming", the programming discipline which improves the quality and the clarity of programs by eliminating jumps between instructions [19]. All high-level modern languages allow to write only structured programs.

3.3 Discrimination of λ-terms

One of the main results in λ-calculus is the possibility of internally discriminating two different values.

The λ-calculus has only two computational rules, the β and the η rules, and each λ-term reduces to a unique irreducible λ-term (called its β-η-normal form), if any, independently from the order in which the rules are used. The β-rule evaluates the application of a λ-term $\lambda x.M$ (playing the role of a function with parameter x) to a λ-term N (playing the role of an argument) just replacing x by N inside M. The η-rule simply reduces a λ-term of the shape $\lambda x.Mx$ to M when x does not occur in M. It is then natural to consider β-η-normal forms as values of λ-calculus.

The internal discrimination means that given two different β-η-normal forms we can always apply them to suitable λ-terms to obtain two arbitrary λ-terms as results. This statement is know as Böhm Theorem and in spite of having only ever appeared as a technical report in Italian [6] it commands a large presence in the main books on λ-calculus [2, 24, 26, 23].

The Böhm Theorem has important consequences for λ-calculus semantics. It implies that no λ-calculus model can equate different β-η-normal forms. Moreover, its generalisation to approximate normal forms [29] implies the maximality of Scott's D_∞ model [28].

The proof of Böhm Theorem requires a sophisticated analysis of λ-terms structure and reduction, called Böhm-out technique by Barendregt [2]. It has been also the inspiration for a representation of β-η-normal forms as trees [8, 18]. This representation has been extended to arbitrary λ-terms by Barendregt [2], who called Böhm trees the resulting trees. Also Böhm trees play a central role in the literature on λ-calculus.

It is important to remark the clear connection between Corrado PhD thesis and

Böhm Theorem. Both these results spring from an original question which is central in Corrado's scientific activity. The question is what a system can say about itself, i.e. if and how the meta-theory of a system can be expressed in the system itself.

3.4 λ-calculus as a Programming Language

In the Sixties, while Corrado was working on the theory of Turing machines, his friend and colleague Wolf Gross pointed out to him the importance of λ-calculus. [5, 9] are the first papers which consider λ-calculus as a programming language. This language was dubbed CUCH, in honour of the logicians Haskell B. Curry [17] and Alonso Church [16], the founders of λ-calculus.

From that point, Corrado pursued a research program aimed at modelling the λ-calculus as a computer language. As a result, Corrado became the foremost authority on the connections between λ-calculus and computer science.

The set of β-η-normal forms, which, as already said in Section 3.3, can be considered the values of λ-calculus [5, 9], is frequently found at the core of Corrado's investigation.

The remainder of this section describes briefly some of the more interesting papers on λ-calculus written by Corrado in the Eighties and Nineties.

Corrado Böhm and Silvio Micali [12] looking at λ-calculus as a computational model see the reduction of λ-terms as a process losing information. They generalise the notion of β-η-normal form to that of minimal form, which are the λ-terms which either do not reduce or reduce to themselves. The minimal forms are then minimal with respect to the quasi ordering induced by the reduction relation. Reducing minimal forms there is no loss of information. Given a λ-term the property of reducing it to a minimal form is shown to be semi-decidable and a semi-algorithm is given detecting a minimal form, if any.

Corrado Böhm and Alessandro Berarducci [7] consider the β-η-normal forms typeable in Girard's system F [21] as a programming language for total functions. They call this language Λ. They give an automatic translation from the data structures defined by a free algebra (trough generators) and from the iterative functions on these data structures to programs of Λ. The key idea is that a data structure can be seen as a functional operator missing some if its arguments.

In a series of papers [14, 13, 15] Corrado Böhm, Adolfo Piperno and Enrico Tronci study systems of equations between λ-terms in which some variable play the role of unknowns. Solutions are required to be β-η-normal forms. The technical treatment uses the self-application of λ-terms in a sophisticated way. Applications include finding β-η-normal forms which are left invertible.

A fixed point equation has the shape $FX =_{\beta\eta} X$, where F is a given λ-term, X

is the unknown λ-term and $=_{\beta\eta}$ means that there is a λ-term which can be obtained by reducing both FX and X. It is well know that fixed point equations can be solved using fixed point combinators, which are λ-terms without a β-η-normal form [2]. Corrado Böhm and Benedetto Intrigila [10] find solutions for families of fixed point equations such that X is a β-η-normal form and FX is strongly normalising. i.e. every reduction of FX terminates producing X.

4 Mentor Activity

To follow Corrado's lessons has always been an unforgettable experience for students. The contents of classes were frequently inspired by the research problems Corrado was working on at the time. The relationship between Corrado and his students was extraordinary. Only a minor part of each lecture was the presentation of subject matter from the teacher's desk. During much of the lecture period students engaged each other in free discussions subtly guided by Corrado's questions.

Corrado was an exceptional research master, as demonstrated by his many pupils who became prominent researchers. His enthusiasm was compelling and he made himself available any time of the day or night to discuss technical problems with everybody. He was always correcting, but in an undeniably kind way, the (sometimes big) mistakes that the interlocutors would make. A key to Corrado's success in inspiring people to research was his total and unwavering confidence in the abilities of his pupils. This strong encouragement in turn allowed his students to trust in themselves and to express their ideas freely. Indeed, a lively discussion of λ-calculus problems resulted in a magical atmosphere where the λ-terms came to life and danced merrily about the participants, leading them along to more profound questions and discoveries. There is no doubt that Corrado was living in this magical world, a world he captured on paper for his grandchildren in the form of a fantastic novel he entitled "CurryLand".

5 Further Readings

Giorgio Ausiello has dedicated to Corrado his new book [1], in which he describes the development of computer science from the sixties to the eighties in a personal and very engaging way. Many of his memories elaborate on the Corrado's activities and his influence on Ausiello's career.

Henk Barendregt in [3] gives a personal perspective on some Corrado achievements.

A whole workshop has been dedicated to Böhm Theorem [27], and a complete list of related papers is too long to reproduce here. An informal explanation of the theorem written under Corrado's supervision can be found in [22].

A biography and a bibliography of Corrado, along with photos and words contributed by his family and friends are available at `http://www.corradobohm.it/Corrado_Bohm/Home.html`.

Acknowledgements

We wish to thank Cheryl Giordano for the impeccable English editing of the manuscript.

References

[1] Giorgio Ausiello The Making of a New Science, *A Personal Journey Through the Early Years of Theoretical Computer Science*, Springer International Publishing, 2018.

[2] Henk Barendregt The Lambda Calculus, *Its Syntax and Semantics*, North-Holland, Amsterdam 1981 (second revised edition 1984, reprinted by College Publications, London 2012).

[3] Henk Barendregt *Gems of Corrado Böhm*, arXiv 1812.02243 [December 10, 2018].

[4] Corrado Böhm *Calculatrices digitales. Du dechiffage des formules logico-mathematiques par la machine même dans la conception du programme*, Annali di Matematica Pura ed Applicata 37 (4), 1954, pp. 1–51,

[5] Corrado Böhm. *The CUCH as a formal and description language*, in: T. B. Steel (ed.), Formal Language Description Languages for Computer Programming, North Holland, Amsterdam 1966, pp. 266–294.

[6] Corrado Böhm *Alcune proprietà delle forme β-η-normali nel λ-K-calcolo*, Pubblicazioni dell'Istituto per le Applicazioni del Calcolo 696, Rome 1968.

[7] Corrado Böhm, and Alessandro Berarducci *Automatic synthesis of typed λ-programs on term algebras*, Theoretical Computer Science 39, 1985, pp. 135–154.

[8] Corrado Böhm, and Mariangiola Dezani-Ciancaglini *Combinatorial problems, combinator equations and normal forms*, in: J. Loeckx (ed.), Automata, Languages and Programming (Second Colloquium, University of Saarbrücken) Springer-Verlag, Berlin 1974. pp. 185–199 [*Lecture Notes in Computer Science 14*].

[9] Corrado Böhm, and Wolf Gross *Introduction to the CUCH*, in: Eduardo R. Caianiello (ed.), Automata Theory, Academic Press, New York 1966, pp. 35–65.

[10] Corrado Böhm, and Benedetto Intrigila *The ant-lion paradigm for strong normalization*, Information and Computation 114 (1), 1994, pp. 30–49.

[11] Corrado Böhm, and Giuseppe Jacopini *Flow diagrams, Turing machines and languages with only two formation rules*, Communications of the ACM 9 (5), 1966, pp. 366–371.

[12] Corrado Böhm, and Silvio Micali *Minimal forms in λ-calculus computations*, The Journal of Symbolic Logic 45 (1), 1980, pp. 165–171.

[13] Corrado Böhm, and Adolfo Piperno *Characterizing X-separability and one-side invertibility in λ-β-Ω-calculus*, in: Yuri Gurevich (ed.), Proceedings of the Third Annual Symposium on Logic in Computer Science (LICS '88), Edinburgh, Scotland, UK, July 5–8, 1988, IEEE Computer Society, 1988, pp. 91–101.

[14] Corrado Böhm, and Enrico Tronci *X-separability and left-invertibility in λ-calculus*, in: David Gries (ed.), Second Annual IEEE Symposium on Logic in Computer Science (LICS '87) 1987-06-22 to 1987-06-25, Ithaca, NY, IEEE Computer Society, 1987, pp. 320–328.

[15] — *About systems of equations, X-separability, and left-invertibility in the λ-calculus*, Information and Computation 90 (1), 1991, pp. 1–32.

[16] Alonzo Church *A set of postulates for the foundation of logic*, Annals of Mathematics (2) 33, 1932, pp. 346–366, and (second paper) 34, 1933, pp. 839–864.

[17] Haskell B. Curry *Grundlagen der kombinatorischen Logik*, American Journal of Mathematics 52, 1930, pp. 509–535, 789–834. (Reprinted with an English translation by Fairouz Kamareddine, and Jonathan P. Seldin, as: Foundations of Combinatory Logic / Grundlagen der kombinatorischen Logik, College Publications, London 2016 [*Logic PhDs 1*]. This edition contains also the *Errata in Curry's thesis*, pp. 167–169. Originally, *Inauguraldissertation*, Göttingen 1930, under David Hilbert.)

[18] — *On a polynomial representation of λ-β-normal forms*, in: Kuno Lorenz (ed.), Konstruktionen versus Positionen, Walter de Gruyter, Berlin 1979, pp. 94–98.

[19] Ole-Johan Dahl, Edsger Dijkstra, and Charles Hoare Structured Programming, Academic Press, New York 1972.

[20] Edsger W. Dijkstra [*Letter to the Editor*] `go to` *statement considered harmful*, Communication of the ACM 11 (3), 1968, pp. 147–148.

[21] Jean-Yves Girard Interpretation fonctionelle et elimination des coupures de l'arithmetique d'ordre superieur, PhD Dissertation, University of Paris 7.

[22] Stefano Guerrini, Adolfo Piperno, and Mariangiola Dezani-Ciancaglini *Böhm's Theorem*, in: E. Gelenbe and J.-P. Kahane (eds.), Fundamental Concepts in Computer Science, World Scientific, Singapore etc. 2009, pp. 1–15.

[23] Chris Hankin Lambda Calculi: A Guide for Computer Scientists, Oxford University Press, Oxford 1995.

[24] J. Roger Hindley, and Jonathan P. Seldin Introduction to Combinators and λ-Calculus, Cambridge University Press, Cambridge UK 1986.

[25] Donald Knuth, and Luis Pardo *The early development of programming languages*, in: J. Howlett and G. Rota (eds.) A History of Computing in the Twentieth Century, Academic Press, New York 1980, pp. 197–273.

[26] Jean-Louis Krivine λ-Calculus, Types and Models (Translated from French by René Cori), Ellis Horwood, New York 1993 [*Computers and their Applications*].

[27] Jean-Jacques Lévy (ed.) *Böhm's Theorem: Applications to Computer Science Theory*, Electronic Notes Theoretical Computer Science 50 (2), 2001. (ICALP 2001 Satellite

Workshop on 'Böhm's theorem: applications to Computer Science Theory' [BOTH 2001], held in Hersonissos, Crete, Greece, on 13 July 2001.)

[28] Dana Scott *Continuous lattices*, in: F. W. Lawvere (ed.), Toposes, Algebraic Geometry and Logic, Springer-Verlag, Berlin 1972, pp. 97–136 [*Lecture Notes in Mathematics 274*].

[29] Christopher P. Wadsworth *The relation between computational and denotational properties for Scott's D_∞-models of the λ-calculus*, SIAM Journal of Computing 5, 1976, (3) pp. 488–521.

THE REVERSE MATHEMATICS OF RAMSEY'S THEOREM FOR PAIRS

CHI TAT CHONG
Department of Mathematics, National University of Singapore
Singapore 119076
chongct@nus.edu.sg

ABSTRACT We consider the combinatorial principle RT^2_2 derived from Ramsey's Theorem for pairs, and discuss its proof-theoretic strength within the framework of reverse mathematics. Some of the techniques introduced to study this are discussed.

Let $[\mathbb{N}]^n$ denote the set of n-tuples of natural numbers. Ramsey's Theorem states that any coloring of $[\mathbb{N}^n$ in k colors (where $n, k \geq 1$) has an infinite set G all of whose n-tuples have the same color (Ramsey [37]). Such a G is said to be *homogeneous* for the coloring. This statement—otherwise called a principle—is abbreviated as RT^n_k. It is a natural generalization of the classical pigeonhole principle which is simply RT^1_k. RT^n_k is arguably the most extensively studied principle in reverse mathematics concerning combinatorics. Of particular interest is the complexity of a solution of an instance of RT^n_k, and the proof-theoretic strength of the principle over subsystems of second-order arithmetic. Our objective here is to provide an overview of this subject, disuss a number of the major mathematical problems in the area, and describe the methodologies developed to resolve them.

We will begin with fixing the notations to be used in the article, and recalling the basic notions, including those that concern subsystem of second-order atithmetic, that are central to our discussion. This will be followed by an exposition of three general problems: (i) The separation of RT^2_2 from other subsystems of second-order arithmetic, (ii) the inductive strength of RT^2_2, and (iii) conservation strength of RT^2_2. As the reader will observe, these three problems are closely related to one another, either by the method of proof or the implication of a theorem. Hence the categorization of topics into three groups is somewhat arbitrary.

Research partially supported by NUS grants C-146-000-042-001 and WBS: R389-000-040-101.

1 Preliminaries and subsystems of second-order arithmetic

1.1 Notations

The language we use is the language of second-order arithmetic, which includes both number variables $x, y, z, \ldots$. and set variables $X, Y, Z, \ldots$ It also includes a binary relation $\in$ interpreted as "element of", constant symbols 0 and 1, and function symbols $+, \times$. A structure $\mathfrak{M} = (M, S)$ in the language of second-order arithmetic consists of a first-order universe M and a second-order component S contained in the power set of M. We use ω and $\mathbb{N}$ interchangeably to denote the set of (standard) natural numbers. If $\mathfrak{M} = (\omega, S)$, then we call it an ω-model. Given $\mathfrak{M} = (M, S)$ and $X \subset M$, let $\mathfrak{M}[X] = (M, \{Z : \exists W \in S \wedge Z \leq_T X \oplus W\})$, where $\oplus$ denotes the join operation and $\leq_T$ denotes Turing reducibility.

We reserve i, j, k, m, n for numbers in $\mathbb{N}$. Given $A \subseteq M$ unbounded, let $[A]^n$ denote the set of all n-tuples $(c_0, c_1, \ldots, c_{n-1})$ selected from A where $c_0 < c_1 < \cdots < c_{n-1}$. In this paper, we are interested in colorings of $[A]^n$ in k colors, especially for $n = k = 2$.

Given $\mathfrak{M}$, we say that $K \subset M$ is $\mathfrak{M}$-finite if K is coded in $\mathfrak{M}$, i.e. there is a $c \in M$ such that for all x, $x \in K$ if and only if x divides c. Greek letters $\beta, \nu, \rho, \sigma, \tau, \ldots$ are reserved for $\mathfrak{M}$-finite binary strings, and each is identified with its characteristic function, hence an $\mathfrak{M}$-finite set (for example $\{x : \sigma(x) = 1\}$) in the obvious way. We write $\sigma \preceq \tau$ if σ is an initial segment of τ, and use $\prec$ for proper extension. The letter α is used to denote ordinals.

1.2 Z_2 and its subsystems

The system Z_2 of second-order arithmetic consists of the Peano axioms (with free number and set variables in the induction scheme), together with the full comprehension scheme, i.e.

$$\exists X \forall x (x \in X \leftrightarrow \varphi(x))$$

holds for each formula φ with possibly free number and set variables. Among the subsystems of Z_2 is RCA_0, the base system, which consists of the following axioms:

- P^-, the Peano axioms minus the induction scheme;
- The Δ^0_1-comprehension scheme:

$$\exists X \forall x (x \in X \leftrightarrow \varphi(x)),$$

  where φ is Σ^0_1 and $\varphi(x) \leftrightarrow \neg\psi(x)$ for some Σ^0_1-formula, and
- The mathematical induction scheme for Σ^0_1-formulas with free number and set variables.

The induction scheme for Σ^0_n-formulas ($n \geq 1$) is denoted $I\Sigma^0_n$. Following the proof in Paris and Kirby [34] for first-order language, one can show that $I\Sigma^0_n$ is equivalent (over P^-) to the least Σ^0_n-principle, which says that every nonempty Σ^0_n-definable set has a least member, and to the statement that every bounded Σ^0_n-definable set in a model $\mathfrak{M}$ of RCA_0 is $\mathfrak{M}$-finite. We will use these facts implicitly throughout the paper.

A scheme intermediate between $I\Sigma^0_n$ and $I\Sigma^0_{n+1}$ is $B\Sigma^0_{n+1}$, known as the Σ^0_{n+1}-bounding scheme, which states that every Σ^0_{n+1}-definable function maps a (provably) finite set to a (provably) finite set. Semantically, it means that in a model $\mathfrak{M}$ of $\mathsf{RCA}_0 + B\Sigma^0_n$, every Σ^0_n-definable function maps an $\mathfrak{M}$-finite set to an $\mathfrak{M}$-finite set. Again by Paris and Kirby [34], the following holds over P^-:

$$\cdots \to I\Sigma^0_{n+1} \to B\Sigma^0_{n+1} \to I\Sigma^0_n \to B\Sigma^0_n \to \cdots.$$

By Slaman [41], $B\Sigma^0_n$ is an induction scheme, since it is equivalent (over P^-) to induction for formulas which are provably Σ^0_n and Π^0_n.

RCA_0 is the base system for the "big five systems" studied in Simpson [40]. Two of them are particularly relevant to our main theme:

- WKL_0: RCA_0 together with the principle stating that every infinite binary tree has an infinite path.

- ACA_0: RCA_0 together with the comprehension scheme

$$\exists X \forall x (x \in X \leftrightarrow \varphi(x)),$$

  for each Σ^0_1-formula φ with number as well as set parameters.

In RCA_0 one can develop a robust theory of computation (see Chong, Li and Yang [5] for a survey). From the recursion-theoretic point of view, models of RCA_0 are structures $\mathfrak{M} = (M, S)$ such that S is closed under the join operation $\oplus$ and Turing reducibility. Furthermore, if $\mathfrak{M} \models \mathsf{ACA}_0$ then S is additionally closed under Turing jump.

It is straightforward to verify that every model of ACA_0 is a model of WKL_0, and that every model of WKL_0 is a model of RCA_0. The converse is however false. Thus $(\omega, \mathsf{REC}) \models \mathsf{RCA}_0$ but not WKL_0, where REC = the collection of recursive sets and, as shown by Hirst [22], an appljication of the Low Basis Theorem of Jockusch and Soare [24] gives a model (ω, LOW) of WKL_0 but not ACA_0, where LOW = the collection of low sets. Hence the systems RCA_0, WKL_0 and ACA_0 have increasing proof-theoretic sterngth. The discussion of Ramsey's theorem in this article is set in the background of these three systems.

1.3 $\mathsf{RCA}_0 + \mathsf{RT}^n_k$

The combinatorial principle RT^n_k is stated as follows:

> Let $k, n \in \mathbb{N}$ and $\mathfrak{M} = (M, S) \models \mathsf{RCA}_0$. Then $\mathfrak{M} \models \mathsf{RT}^n_k$ if for any $C : [M]^n \to k$ in S, there is an infinite $G \in S$ such that $C \restriction [G]^n$ is a constant, i.e. G is homogeneous for C.

A simple induction shows that for $n \geq 1$ and $k \geq 2$, $\mathsf{RT}^n_k \to \mathsf{RT}^n_{k+1}$. Thus we confine ourselves to $k = 2$. [1]

The investigation of Ramsey's Theorem from the logical point of view was inspired by two results that appeared 15 years apart. The first pointed to the case $n = 2$ as the focus for future stidy:

Theorem 1.1. (Jockusch [23])

(i) $\mathsf{RCA}_0 + \mathsf{RT}^n_2$ *implies* ACA_0 *for* $n \geq 3$.

(ii) *There is a recursive two-coloring of* $[\mathbb{N}]^2$ *with no* Σ^0_2*-definable homogeneous set.*

Sketch of proof of (i). One exhibiting a recursive coloring of $[\mathbb{N}]^3$ in 2 colors such that any set homogeneous for the coloring computes $\emptyset'$. By relativization, this implies that in any model $\mathfrak{M} = (M, S)$ of $\mathsf{RCA}_0 + \mathsf{RT}^3_2$, if $X \in S$ then its Turing jump X' is in S as well. □

It should be mentioned that the original context in which Jockush proved Theorem 1.1 was recursion-theoretic (as indicated by the title of the paper). When cast in the setting of reverse mathematics, the theorem characterizes the inductive strength of RT^3_2.

Theorem 1.1 (ii) provides evidence that solutions of instances of RT^2_2 could be beyond the computational power of $\emptyset'$. This is complemented by the next result which says that even for coloring of pairs, the inductive strength of Ramsey's Theorem is more than $I\Sigma^0_1$:

Theorem 1.2. (Hirst [22]) $\mathsf{RCA}_0 + \mathsf{RT}^2_2$ *implies* $B\Sigma^0_2$.

---

[1] Note that over RCA_0, the fact that $\mathsf{RT}^n_k \to \mathsf{RT}^n_{k+1}$ for $n \geq 1$ and $k \geq 2$ does not yield $\mathsf{RT}^n_2 \to \forall s \geq 2 \mathsf{RT}^n_s$. Indeed if we write RT^2 for $\forall s \geq 2 \mathsf{RT}^2_s$, then $\mathsf{RCA}_0 + \mathsf{RT}^2_2 \not\to \mathsf{RT}^2$. This was proved by Hirst [22]. It also follows from the combination of Theorem 3.6 below and Theorem 11.4 of Cholak, Jockusch and Slaman [3], which implies that $\mathsf{RCA}_0 + \mathsf{RT}^2 \to B\Sigma^0_3$. See also the concluding section for remarks on the principle TT^1 which is another illustration of the sharp distinction between a "fixed finite coloring" and "all finite colorings".

Sketch of proof. Suppose for the sake of contradiction that $\mathfrak{M}$ is a model of $\mathsf{RCA}_0 + \mathsf{RT}^2_2$ in which $B\Sigma^0_2$ fails. Let $f : [0, a] \to M$ be Σ^0_2-definable in $\mathfrak{M}$ witnessing this failure. Thus f is total on $[0, a]$ with unbounded range. Using f one can recursively decompose M into a-many pairwise disjoint $\mathfrak{M}$-finite sets $\{A_s : s \in [0, a]\}$. Now define a two-coloring $C : [\mathbb{M}]^2 \to 2$ by setting $C(x, y) = 0$ if x and y belong to the same A_s, and let $C(x, y) = 1$ otherwise. It is straightforward to verify that there is no unbounded set homogeneous for C that preserves $I\Sigma^0_1$. Thus $\mathsf{RCA}_0 + \mathsf{RT}^2_2$ implies $B\Sigma^0_2$. □

Theorems 1.1 amd 1.2 set the stage for the study of Ramsey's Theorem for pairs in reverse mathematics. They led to a number of fundamental questions about RT^n_k: The first group of questions concerned the strerngth of RT^2_2 *vis-à-vis* ACA_0 (which by Theorem 1.1 (i) is equivalent to $\mathsf{RCA}_0 + \mathsf{RT}^3_2$) and WKL_0, the two weakest subsystems among the "big five". The second group of questions is related to the first and second-order proof-theoretic strength of RT^2_2, e.g. how much induction or conservation does $\mathsf{RCA}_0 + \mathsf{RT}^2_2$ have? We pause for an elaboration of this question.

Let $\mathfrak{M} = (M, S)$ be a model of ACA_0. Then $\emptyset^{(n)} \in S$ for all $n \in \omega$. Now $I\Sigma^0_1$ relative to $\emptyset^{(n)}$ means $\mathfrak{M} \models I\Sigma^0_n$, and hence $\mathfrak{M} \models$ Peano arithmetic. This holds in particular for $\mathfrak{M} \models \mathsf{RCA}_0 + \mathsf{RT}^3_2$, for example. Thus whether $\mathsf{RCA}_0 + \mathsf{RT}^2_2 \to$ "$\emptyset'$ exists" is an interesting question. A negative answer would immediately imply that $\mathsf{RCA}_0 + \mathsf{RT}^2_2 \nvdash \mathsf{RT}^3_2$. The negative answer was given in Seetapun and Slaman [38].

Next, as noted by Hirst [22], $\mathsf{WKL}_0 \not\to \mathsf{RT}^2_2$. To see this, let $\mathfrak{M} = (\omega, \mathsf{LOW})$ be the model of WKL_0 introduced earlier. Theorem 1.1 (ii) says that there is a recursive two-coloring of pairs with no Σ^0_2-definable homogeneous solution in LOW. Hence $\mathfrak{M} \not\models \mathsf{RT}^2_2$, i.e. $\mathsf{WKL}_0 \not\to \mathsf{RT}^2_2$.

If conversely $\mathsf{RCA}_0 + \mathsf{RT}^2_2 \not\to \mathsf{WKL}_0$, then one proves as a corollary that there exists a combinatorial principle (indeed many, as it turns out. See Hirschfeldt and Shore [21]) outside the realm of the big five systems. This was resolved in Liu [29]).

Let Γ be the collection of sentences in the language of second-order arithmetic of a given complexity. A system T' is Γ-*conservative over* a subsystem T if every $\varphi \in \Gamma$ provable in T' is already provable in T. A result due to Harrington states that WKL_0 is Π^1_1-conservative over RCA_0 (see [3]). On the other hand, ACA_0 is not even Π^0_3-conservative over RCA_0. Here again, separating RT^2_2 from RT^3_2 can be achieved by differentiating their first-order strength, for example by showing $\mathsf{RCA}_0 + \mathsf{RT}^2_2$ to be Π^0_3-conservative over RCA_0, which was achieved by Patey and Yokoyama [36]. We will disuss these in §§2–4.

The analysis of RT^3_2 versus RT^2_2 can be approached from two different angles. The first is by comparing what each principle entails, as described above, and the second is by directly constructing an $\mathfrak{M}$ that is a model of one but not the other. The latter approach was also the first proof of $\mathsf{RCA}_0 + \mathsf{RT}^2_2 \nvdash \mathsf{RT}^3_2$ by Seetapun in Seetapun and Slaman [38]. Since every model of $\mathsf{RCA}_0 + \mathsf{RT}^3_2$ contains a set that computes the halting set, to separate RT^2_2 from RT^3_2 it is sufficient to produce a model of the former with no second-order member that computes $\emptyset'$. This was exactly what [38] accomplished. This method of "cone avoidance" has been successfully exploited to investigate other problems such as WKL_0 in Liu [29]. Before diving into the details of model construction for $\mathsf{RCA}_0 + \mathsf{RT}^2_2$, we recall the decomposition of RT^2_2 into two simpler components introduced in [3]:

Let $\mathfrak{M} = (M, S)$. SRT^2_2 is the principle of *stable Ramsey's Theorem for pairs*, which states that if $C \in S$ is a 2-coloring of pairs such that for each x, the color $C(x, y)$ is the same for all but $\mathfrak{M}$-finitely many y's. then there is a set in S that is homogeneous for C. The cohesiveness principle COH states that if $\{A_s : s \in M\}$ is an array coded as a set in S, then there is a $Z \in S$ that is *cohesive* for the array, i.e. for each s, all but $\mathfrak{M}$-finitely many members of Z are contained in A_s or its complement. These are among the most important principles known to be weaker than RT^2_2 that have been extensively studied in recent years (see Hirschfeldt and Shore [21]). The next proposition will be a basic fact we use in the discussion that follows.

Proposition 1.3. (Cholak, Jockusch and Slaman [3]) *Over* $\mathsf{RCA}_0 + B\Sigma^0_2$, RT^2_2 *is equivalent to* $\mathsf{SRT}^2_2 + \mathsf{COH}$.

2 Separating RT^2_2 from RT^3_2 and WKL_0

2.1 A model of RT^2_2 that avoids $\emptyset'$

Given a model $\mathfrak{M} = (M, S)$ of RCA_0, we call $\mathfrak{M}' = (M', S')$ an M-extension of $\mathfrak{M}$ if $M' = M$ and $S' \supseteq S$. We sketch a proof of the theorem that there is an ω-model of RT^2_2 whose second-order part does not contain the halting set. The proof proceeds in two parts. Let $\mathfrak{M}_0 = (\omega, S_0)$ be a structure such that S_0 is the collection of recursive subsets of ω. Clearly $\mathfrak{M}_0 \models \mathsf{RCA}_0$.

Lemma 2.1. *There is an ω-extension $\mathfrak{M}_1 = (\omega, S_1)$ of $\mathfrak{M}_0$ such that $\mathfrak{M}_1 \models \mathsf{RCA}_0 + \mathsf{COH}$ and $\emptyset' \notin S_1$.*

Proof. Suppose $A = \{A_s : s \in \omega\}$ is an array coded as a set in S_0. Construct Z that is cohesive for A as follows. Let $\{\Phi_e : e \in \omega\}$ be an effective list of all partial Σ^0_1-functions.

Stage 0. Suppose A_0 is finite with largest element a_0. See if there exist two finite sets σ and τ, both with least elements greater than a_0, such that for some $n \in \omega$, $\Phi^{\sigma}_0(n) \downarrow \neq \Phi^{\tau}_0(n) \downarrow$ (we call this a split of Φ_0). If the answer is "yes", denote by σ_0 the string $\eta \in \{\sigma, \tau\}$ that satisfies $\Phi^{\eta}_0(n) \neq \emptyset'(n)$ and let $X_0 = \{y : y > \max \sigma_0\}$. Otherwise, let $\sigma_0 = \emptyset$ and let $X_0 = \{y : y > a_0\}$. Note that in this case, for all infinite X with $\min X > a_0$, if Φ^X is total then the range is recursive (by the non-splitting condition) and hence $\Phi^X_0 \neq \emptyset'$.

If A_0 is infinite, let a_0 be the least member of A_0, and search for a pair of strings that split Φ_0 using numbers within A_0. If a split $\{\sigma, \tau\}$ exists, let $\sigma_0 \in \{\sigma, \tau\}$ be such that

$\Phi_0^{\sigma_0}(n) \downarrow \neq \emptyset'(n)$ for some n. Let $X_0 = \{y : y \in A_0 \wedge y > \max \sigma_0\}$. If no splitting exists, let $\sigma_0 = \emptyset$ and $X_0 = A_0$. Thus in all cases, either $\Phi_0^{\sigma_0}(n) \downarrow \neq \emptyset'(n)$ for some n, or $\sigma_0 = \emptyset$ and for all infinite $X \subset X_0$, if Φ_0^X is total, then the range is recursive. This ensures that $\sigma_0 \cup X$ does not compute $\emptyset'$ via Φ_0.

Stage $s+1$. Assume that $\sigma_0 \preceq \sigma_1 \preceq \cdots \preceq \sigma_s$ and $X_0 \supseteq X_1 \supseteq \cdots \supseteq X_s$ are defined such that

(i) X_s is infinite and $\min X_s > \max \sigma_s$;

(ii) For each $s' \leq s$, either $X_s \subset A_{s'}$ or $X_s \cap A_{s'} = \emptyset$;

(iii) For each $s' \leq s$, either there is an n such that $\Phi_{s'}^{\sigma_s}(n) \downarrow \neq \emptyset'(n)$, or for all $X \subset X_s$, whenever $\Phi_{s'}^{\sigma_s \cup X}$ is total, then the range is recursive and hence is not equal to $\emptyset'$.

Define $\sigma_{s+1} \succeq \sigma_s$ and X_{s+1} so that $\sigma_{s+1} \setminus \sigma_s$ (if nonempty) and X_{s+1} are contained in X_s, and

(iv) $\min X_{s+1} > \max \sigma_{s+1}$;

(v) Either $X_{s+1} \subset A_{s+1}$ or $X_{s+1} \cap A_{s+1} = \emptyset$;

(vi) If $\sigma_{s+1} \succ \sigma_s$, then there is an n such that $\Phi_{s+1}^{\sigma_{s+1}}(n) \downarrow \neq \emptyset'(n)$;

(vii) Suppose $\sigma_{s+1} = \sigma_s$. For all infinite $X \subset X_{s+1}$, if $\Phi_{s+1}^{\sigma_s \cup X}$ is total then the range is recursive and hence not equal to $\emptyset'$.

Note that for (vii), if there is already an n such that $\Phi_{s+1}^{\sigma_s}(n) \downarrow \neq \emptyset'(n)$, then one can simply choose $\sigma_{s+1} \succ \sigma_s$ so that $\sigma_{s+1} \setminus \sigma_s \subset X_{s+1}$.

Let $Z = \bigcup_s \sigma_s$. Then it is straightforward to verify that Z is cohesive for the array $\{A_s : s \in \omega\}$ and does not compute $\emptyset'$. Let $Z \in S_1$. Repeat the above construction for the structure $\mathfrak{M}_0[Z]$ to obtain a set $\not\geq_T \emptyset'$ that is cohesive for the next array in the expanded structure. The modification required in the construction is that, as an example, one now searches for a pair (σ, τ) that forms a split for Φ_0 in the following sense: $\Phi_0^{\sigma \oplus Z}(n) \downarrow \neq \Phi_0^{\tau \oplus Z}(n) \downarrow$ for some n. If a split exists, let σ_0 be chosen such that $\Phi_0^{\sigma_0 \oplus Z}(n) \downarrow \neq \emptyset'(n)$. Otherwise, let $\sigma_0 = \emptyset$. Define X_0 as before. Then for all infinite $X \subset X_0$, $\Phi_0^{\sigma_0 \oplus Z \oplus X} \neq \emptyset'$.

Iterating this construction ω-many times produces a model $\mathfrak{M}_1 = (\omega, S_1)$ of $\mathsf{RCA}_0 + \mathsf{COH}$ in which $\emptyset' \notin S_1$.

□

The principle D_2^2 was introduced in [3]. It states that for any Δ_2^0-function $f : M \to 2$, there is an infinite subset on which f is a constant. D_2^2 is equivalent to SRT_2^2 over the base system $\mathsf{RCA}_0 + B\Sigma_2^0$ (Chong, Lempp and Yang [4]). We will adopt this version of SRT_2^2

henceforth. A notion central to the analysis of SRT^2_2 is what is referred to as a *Seetapun disjunction* introduced in Chong, Slaman and Yang [11]. It was used in the construction of a model that separates SRT^2_2 from RT^2_2 (see Theorem 3.2 below). Although the notion was cast in the language of nonstandard arithmetic, it has a natural analog in standard arithmetic. We present an application of this notion.

2.1.1 Seetapun disjunction

Let $C:\mathbb{N}\to 2$ be a Δ^0_2-definable (possibly with parameters) coloring. Call a number x *red* (resp. *blue*) if $C(x)=0$ (resp. if $C(x)=1$). Let ρ and β denote finite binary strings which we identify with finite sets through their respective characterisic functions (ρ will usually denote a red string while β will usually denote a blue string). Given (ρ,β), an infinite set X such that $\min X > \max\{\rho,\beta\}$, and given a Σ^0_1-predicate $\varphi \equiv \exists u\psi(u,x)$ (with parameters) where x is a free variable, define the following collecction $\{o_s : s\in\omega\}$ of "blobs":

- $o_0 \subset X$ is the least finite set o such that

  $\min o > \max \rho\cup\beta$, and $\exists u(\max u \le \max o \wedge \psi(u,\rho\cup o))$

  holds;

- $o_{s+1} \subset X$ is the least finite set o enumerated such that

  $\min o > \max o_s$ and $\exists u(\max u \le \max o \wedge \psi(u,\rho\cup o))$

  holds.

Simultaneously, define the Seetapun tree T_s associated with $\{o_{s'} : s'\le s\}$ as follows: Let β be the root of T_s. Put $\beta\cup n$ at level 0 of T_s for each $n\in o_0$. Inductively, if η is a node of T_s at level s', where $s'<s$, and n is a number in $o_{s'+1}$, then $\eta\cup n$ is a node of T_s at level $s'+1$. Declare the configuration $(\vec{o}_s, T_s)$, where $\vec{o}_s=\{o_0,\dots,o_s\}$, to be a Seetapun disjunction if for every maximal parh p in T_s, there exists a subset v such that $\exists u(\max u\le \max v\wedge\psi(u,\beta\cup v))$ holds. Note that each of the enumerations of o_s, T_s and of a Seetapun disjunction is Σ^0_1 relative to X.

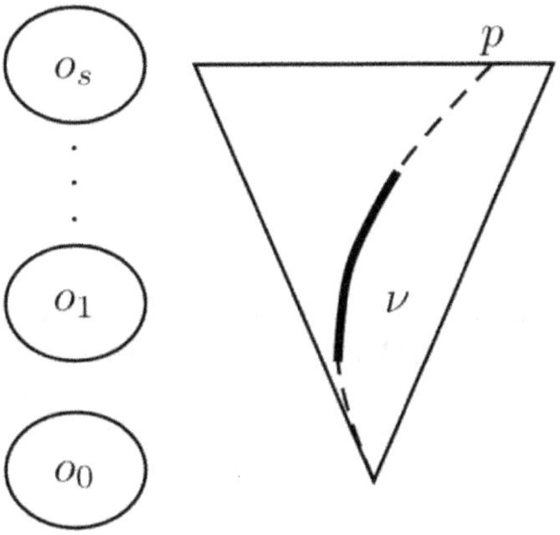

Figure 1: A Seetapun disjunction

Figure 1 is an illustration of a Seetapun disjunction consisting of finite sets $o_0, o_1, \ldots,$ a maximal path p and a finite subset v of p. The point of computing a Seetapun disjunction is encapsulated in the following claim:

Claim 2.2. *If $(\vec{o}_s, T_s)$ is a Seetapun disjunction, where ρ is red and β is blue, then either there is a red solution for φ or a blue solutioin for φ.*

A red solution is a blob $o_{s'}$, where $s' \leq s$, consisting only of red numbers, while a blue solution is a blue set v such that $\exists u(\max u \leq \max v \wedge \psi(u, \beta \cup v))$ holds. If there is no red solution on the $\vec{o}_s$ side, then it must be the case that each $o_{s'}$ contains a blue number. Hence there is a maximal path in T_s consisting only of blue members (a "blue path"). On this path there is a blue solution for φ. In other words, if a Seetapun disjunction is enumerated, then either the red side or the blue side makes progress on φ.

Now there are two ways in which one may fail to enumerate a Seetapun disjunction. One is that for some s, o_s is enumerated but not o_{s+1}. This would occur if there is no o such that $\min o > \max o_s$ and $\exists u(\max u \leq \max o \wedge \psi(u, \rho \cup o))$ holds. In particular, this applies to red o. The other way is for o_s and hence T_s to be defined for each $s \in \omega$ but there is a maximal path in T_s with no subset v for

$\exists u(\max u \leq \max v \wedge \psi(u, \beta \cup v))$ to hold. Then for any infinite path Y on $T = \bigcup_s T_s$, for any finite $v \subset Y$ such that $\min v > \max \beta$, $\neg\exists u(\max u \leq \max v \wedge \psi(u, \beta \cup v))$ holds.

In either case we ensure a "Π_1-outcome" for φ. We give the first application of Seetapun disjunction to RT^2_2:

2.1.2 $\mathsf{RCA}_0 + \mathsf{RT}^2_2 \nvdash \mathsf{RT}^3_2$

Theorem 2.3. (Seetapun and Slaman [38]) *There is a model $\mathfrak{M} = (\omega, S)$ of $\mathsf{RCA}_0 + \mathsf{RT}^2_2$ in which $\emptyset' \notin S$.*

Proof. We take as ground model $\mathfrak{M}_0 = (\omega, S_0)$ where S_0 is the collection of recursive sets. Let $C \in S$ be a two-coloring (red and blue) of pairs of numbers. We construct a G homogeneous for C such that $G \not\geq_T \emptyset'$. Let $A_s = \{y : C(s,y) = \text{ red}\}$. By the proof of Theorem 2.1, there is a Z cohesive for the array $\{A_s : s \in \omega\}$ and $Z \not\geq_T \emptyset'$. Then $C \restriction [Z]^2$ is a stable two-coloring of pairs in Z, and any $G \subset Z$ that is homogeneous for $C \restriction [Z]^2$ is homogeneoius for C. Hence it is sufficient to show that if $C \leq_T Z$ is a stable two-coloring of pairs and $Z \not\geq_T \emptyset'$, then there is a G homogeneous for C such that $G \oplus Z \not\geq_T \emptyset'$. We will sketch a poof of this.

Call a number x "red" if $\lim_{s\to\omega} C(x,s) = \text{red}$, and "blue" if $\lim_{s\to\omega} C(x,s) = \text{blue}$. Given ρ, β and reduction procedure Φ, let φ in §2.1.1 be:

$\exists n \leq \max o \exists \tilde{o}_0, \tilde{o}_1 (\tilde{o}_0, \tilde{o}_1 \subset o \wedge \Phi^{(\rho\cup\tilde{o}_0)\oplus Z}(n) \downarrow \neq \Phi^{(\rho\cup\tilde{o}_1)\oplus Z}(n) \downarrow)$.

We say in this case that $(\tilde{o}_0, \tilde{o}_1)$ "Z-splits Φ over ρ". A Seetapun disjunction is declared for $(\vec{o}_s, T_s)$ if for each maximal path p in T_s, there exists a $v \subset p$ such that $\varphi(\beta \cup v)$ holds, i.e. there exist $v_0, v_1 \subset v$ such that $v_0, v_1 \subset v$ and $\Phi^{(\beta\cup v_0)\oplus Z}(n) \downarrow \neq \Phi^{(\beta\cup v_1)\oplus Z}(n) \downarrow$ for some $n \leq \max v$. We say in this case that (v_0, v_1) "Z splits Φ over β".

Let $\{\Phi_0, \Phi_1, \ldots\}$ be an effective list of reduction procedures. The construction of the homogeneous set G proceeds in stages. Let X in the definition of Seetapun disjunction be the cohesive set Z. Hence the blobs o_s are finite sets enumerated in Z. We will repeatedlu apply the following due to Jockusch and Soare [24]:

(*) Let $Z, A \subset \omega$ such that $Z \not\geq_T A$. If T is an infinite, Z-recursively bounded, Z-recursive tree, then T has an infinite path W such that $W \oplus Z \not\geq_T A$.

At stage 0, let $\rho = \beta = \emptyset$. Enumerate $(\vec{o}_s, T_s)$ for Z-splitting of $\Phi = \Phi_0$ as described above. If a Seetapun disjunction $(\vec{o}_s, T_s)$ is enumerated, choose a red $\tilde{o}_i \subset o_{s'}$ for some $s' \leq s$ and $i < 2$, or a blue $v_i \subset v$ for some $i < 2$, as the case may be (according to the Claim), so that $\Phi_0^{\tilde{o}_i \oplus Z}(n) \downarrow \neq \emptyset'(n)$ for some $n \leq \max \tilde{o}_i$, or $\Phi_0^{v_i\oplus Z}(n) \downarrow \neq \emptyset'(n)$ for some $n \leq \max v$. If $\tilde{o}_i$ is chosen, let it be ρ_0 and let $\beta_0 = \emptyset$. If v_i is chosen, let it be β_0 and set $\rho_0 = \emptyset$.

Suppose there is no Seetapun disjunction enumerated. There are two cases to consider:

Case 1. o_s is not defined for some (least) s which we denote as s_0. Let t_0 be an upper bound of all numbers enumerated in o_{s_0-1} Let $\rho_0 = \{n_0\}$ where $n_0 > t_0$ is red, and let $\beta_0 = \emptyset$ (note that if n_0 does not exist, then all numbers in Z above t_0 are blue and we will immediately have a blue homogeneous set $\not\geq_T \emptyset'$ contained in Z).

Note 1. No infinite $V \subset Z$ such that $\min V > t_0$ computes $\emptyset'$ relative to Z via Φ_0. Suppose otherwise and W is a counterexample. Since no Z-splitting of Φ_0 exists for any $o \subset Z$ with $\min o > t_0$, it means that for any finite set $o \subset Z$ and n, if $\Phi_0^{o\oplus Z}(n) \downarrow$ then $\Phi_0^{o\oplus Z}(n) = \Phi_0^{W\oplus Z}(n) = \emptyset'(n)$. This gives an algorithm to compute $\emptyset'$ from Z, which is a contradiction.

Case 2. o_s is defined for each s. Then $T = \bigcup_s T_s$ is an infinite Z-recursively bounded Z-recursive tree. By (*) select an infinite path Y_0 in T such that $Y_0 \oplus Z \not\geq_T \emptyset'$. Let $\beta_0 = \{n_0\}$ where n_0 is a blue number in Y_0 and let $\rho_0 = \emptyset$. Again if n_0 does not exist then we will have a red homogeneous set $W \subset Y_0$ such that $W \oplus Z \not\geq_T \emptyset'$.

Note 2. No infinite subset $V$of Y_0 satisfies $\Phi_0^{V\oplus Z} = \emptyset'$. Suppose otherwise and W. is a counterexample. Then since there is no Z-splitting of Φ_0 in Y_0, for. any $v \subset Y_0$ and any n, if $\Phi_0^{v\oplus Z}(n) \downarrow$ then $\Phi_0^{v\oplus Z}(n) = \Phi_0^{W\oplus Z}(n) = \emptyset'(n)$, giving an algorithm to compute $\emptyset'$ from $Y_0 \oplus Z$, which is a contradiction.

Next, we describe the construction at stage 1. This will give an idea of the steps to take at an arbitrary stage. There are several scenarios to consider, depending on the outcome at stage 0:

- Suppose a Seetapun disjunction was enumerated at stage 0. If $\rho_0 = \emptyset$, let $\Phi = \Phi_0$. If $\rho_0 \neq \emptyset$, let $\Phi = \Phi_1$. Suppose no Seetapun disjunction was enumerated but Case 1 holds. Let $\Phi = \Phi_1$. Now do the following: Define φ as in Stage 0, upon setting $(\rho, \beta) = (\rho_0, \beta_0)$. Enumerate "blobs" $o_s \subset Z$ with $\min o_s > \max\{\rho_0, \beta_0, o_{s-1}\}$ and define the trees T_s as before, with β_0 as root. A See3tapun disjunction for $(\vec{o}_s, T_s)$ is declared if the following modified condition is satisfied: Every maximal path in T_s has a subset v that Z-splits Φ over β_0, where $\Phi = \Phi_0$ if $\beta_0 = \emptyset$ and $\Phi = \Phi_1$ otherwise. There are now three possible outcomes:

  (i) If a Seetapun disjunction is enumerated during this process and there is a red $o_{s'}$, where $s' \leq s$, such that $\varphi(\rho_0 \cup o_{s'})$ holds, let $\tilde{o}_i \subset o_{s'}$, for some $i < 2$, be chosen such that $\Phi_1^{(\rho_0\cup\tilde{o}_i)\oplus Z}(n) \downarrow\neq \emptyset'(n)$ for some n. Let $\rho_1 = \rho_0 \cup \tilde{o}_i, \beta_1 = \beta_0$. If there is a blue v such that $\varphi(\beta_0 \cup v)$ holds, let $v_i \subset v$, for some $i < 2$, be chosen such that $\Phi^{(\beta_0\cup v_i)\oplus Z}(n) \downarrow\neq \emptyset'(n)$ for some n, and let $\beta_1 = \beta_0 \cup v_i, \rho_1 = \rho_0$.

  (ii) If no disjunction is enumerated, but there is a (least) s such that o_s is not defined, denote this by s_1 and let t_1 be an upper bound of numbers in o_{s_1-1}, and let $\rho_1 = \rho_0 \cup \{n_1\}$ where $n_1 \in Z$ is a red number larger than t_1. Let $\beta_1 = \beta_0$.

Note that in either (i) or (ii), if $\rho_1 > \rho_0$ then either $\Phi^{\rho_1 \oplus Z}$ has already diagonalized against $\emptyset'$ for Φ, or for any infinite $W \subset Z$ such that $\min W > \max \rho_1$, $\Phi^{(\rho_1 \cup W) \oplus Z}$ is either partial or has range recursive in Z, hence nor equal to $\emptyset'$. The same holds if ρ_0, ρ_1 are replaced by β_0, β_1 respectively.

(iii) If no disjunction is enumerated, and o_s is defined for all s, then $T = \bigcup_s T_s$ is an infinite Z-recursively bounded, Z-recursive tree. By (*), let Y_1 be an infinite path in T such that $Y_1 \oplus Z \not\geq_T \emptyset'$. Let $\beta_1 = \beta_0 \cup \{n_1\}$, where n_1 is a blue number in Y_1 larger than $\max \beta_0$. Let $\rho_1 = \rho_0$. Then for all infinite $W \subset Y_1$ such that $\min W > \max \beta_1$, we have $\Phi^{(\beta_1 \cup W) \oplus Z}$ to be either partial or with range recursive in Z, hence not equal to $\emptyset'$.

- If Case 2 above holds at Stage 0, enumerate $o_s \subset Y_0$ after making the following modification: In φ, let $(\rho, \beta) = (\rho_0, \beta_0)$, $\Phi = \Phi_0$ and $o \subset Y_0$. Thus if o_{s-1} is defined, search for an $o \subset Y_0$ with $\min o > \max o_{s-1}, \rho_0, \beta_0$ such that there exist $\tilde{o}_0, \tilde{o}_1 \subset o$ which $Y_0 \oplus Z$-splits Φ_0 over ρ_0, i.e. $\Phi_0^{(\rho_0 \cup \tilde{o}_0) \oplus Y_0 \oplus Z}(n) \downarrow \neq \Phi_0^{(\rho_0 \cup \tilde{o}_1) \oplus Y_0 \oplus Z}(n) \downarrow$. Then o_s is the least such o enumerated. Define T_s accordingly, again with β_0 as its root. Declare $(\vec{o}_s, T_s)$ to be a Seetapun disjunction if for each maximal path p in T_s, there exists $v \subset p$ and $v_0, v_1 \subset v$ such that (v_0, v_1) $Y_0 \oplus Z$-splits Φ_1 over β_0. Define ρ_1, β_1 as follows:

(iv) A Seetapun disjunctioon is enumerated. Define ρ_1 and β_1 as in (i). Thus if there is an $s' \leq s$ such that $o_{s'}$ is red and $Y_0 \oplus Z$-splits Φ over ρ_0, let $\tilde{o}_i \subset o_{s'}$, for some $i < 2$, be the half of the split that yields $\Phi^{(\rho_0 \cup \tilde{o}_i) \oplus Y_0 \oplus Z}(n) \downarrow \neq \emptyset'(n)$ for some n. Let $\rho_1 = \rho_0 \cup \tilde{o}_i$ and $\beta_1 = \beta_0$. If there is a blue split given by a v, let $v_i \subset v$, for some $i < 2$, be that half of the split such that $\Phi_1^{(\beta_0 \cup v_i) \oplus Y_0 \oplus Z}(n) \downarrow \neq \emptyset'(n)$ for some n, and let $\rho_1 = \rho_0, \beta_1 = \beta_0 \cup v_i$.

(v) No disjunction is enumerarted and s_1, t_1 as in (ii) exist. Define ρ_1, β_1 accordingly.

(vi) No disjunction is enumerated. Then $T = \bigcup_s T_s$ is an infinite, $Y_0 \oplus Z$-recursively bounded, $Y_0 \oplus Z$-recursive tree. By (*), let Y_1 be an infinite path in T such that $Y_0 \oplus Y_1 \oplus Z \not\geq_T \emptyset'$. Let n_1 be a blue number in Y_1 larger than β_0 and set $\beta_1 = \beta_0 \cup \{n_1\}$ and $\rho_1 = \rho_0$.

Note that if (iv) holds and $\rho_1 > \rho_0$, or if (v) holds, then for all $W \subset Y_0$ with $\min W > \max \rho_1$, we have $\Phi_0^{(\rho_1 \cup W) \oplus Z} \neq \emptyset'$. Similarly, if (iv) holds and $\beta_1 > \beta_0$, then for any infinite $W \subset Y_0$ such that $\min W > \max \beta_1$, $\Phi_1^{(\beta_0 \cup \beta_1 \cup W) \oplus Z} \neq \emptyset'$. Also if (vi) holds, then for any infinite $W \subset Y_1$ such that $\min W > \max \beta_1$, $\Phi_1^{(\beta_1 \cup W) \oplus Y_0 \oplus Z} \neq \emptyset'$.

The construction at stage $k > 0$ is carried out similarly. Thus ρ_k, β_k are defined for each k and either $\rho_{k+1} \succ \rho_k$ or $\beta_{k+1} \succ \beta_k$. If $\rho_{k+1} \succ \rho_k$ for infinitely many k, let $G = \bigcup_k \rho_k$. Otherwise, let $G = \bigcup_k \beta_k$. Then G is a red or blue homogeneous set for C and $G \oplus Z \not\geq_T \emptyset'$.

Next, take $\mathfrak{M}_0[Z, G]$ and apply the argument in Theoren 2.1 to the next array A in the structure, to obtain an infinite set cohesive for A and whose join with $G \oplus Z$, denoted Z^*, does not compute $\emptyset'$. Take the structure $\mathfrak{M}_0[Z^*]$ and apply the construction above to the next instance of stable two-coloring in the structure. This produces an infinite homogeneous set (for the coloring) whose join with Z^* does not compute $\emptyset'$.

We complete the proof of the theorem by iterating this construction, arriving at a countable ω-extension $\mathfrak{M} = (\omega, S)$ of $\mathfrak{M}_0$ such that $\mathfrak{M} \models \mathsf{RCA}_0 + \mathsf{RT}^2_2$ and $\emptyset' \notin S$.

□

2.2 Weak König's Lemma WKL_0

As seen earlier, WKL_0 is stronger than RCA_0. Where or how RT^2_2 fits in the lower end of the "big five picture" is an interesting question in the reverse mathematics of Ramsey's Theorem. The question whether $\mathsf{RCA}_0 + \mathsf{RT}^2_2 \rightarrow \mathsf{WKL}_0$ was a major open problem. The solution appeared in 2012:

Theorem 2.4. (Liu [29]) $\mathsf{RCA}_0 + \mathsf{RT}^2_2 \not\rightarrow \mathsf{WKL}_0$.

To construct a model of $\mathsf{RCA}_0 + \mathsf{RT}^2_2 + \neg\mathsf{WKL}_0$, first observe that there is a recursive set which codes an infinite tree of binary strings in which every infinite path computes a completion of Peano arithmetic. This is the candidate to use for diagonalization, i.e. construct a model $\mathfrak{M} = (\omega, S)$ of $\mathsf{RCA}_0 + \mathsf{RT}^2_2$ such that no path in the tree belongs to S. A natural way to approach this is to use the cone avoidance strategy in [38]. It was clear, however, that the technique required to achieve this would have to be different and more intricate than that introduced to prove Theorem 2.3. For a start, instead of avoiding a non-recursive Turing degree, one now has to avoid uncountably many degrees in a construction to be completed in ω steps. Resolving this apparent conflict is a key challenge. Apart from [29], the reader may also refer to Hirschfeldt [20] for an excellent exposition of the proof.

In a subsequent paper [30], Liu showed that, indeed, RT^2_2 does not even imply a weak form of WKL_0 called the weak Weak König's Lemma Principle WWKL: If T is an infinite tree of binary strings such that

$$\lim_s \sum_{\sigma \in T \wedge |\sigma| = s} \frac{1}{2^s} \neq 0,$$

then there is an infinite path in T. This measure-theoretic principle was introducved in

Yu and Simpson [43] and is weaker than WKL_0 over RCA_0. It is yet another evidence of the limited proof-theoretic strength of RT^2_2. [2]

3 Separating SRT^2_2 from RT^2_2

On the face of its definition, one would naturally expect SRT^2_2 to be weaker than RT^2_2 over the base system RCA_0. However, confirming this intuition mathematically turned out to require a certain amount of effort. From the model-theoretic point of view, one can establish $\mathsf{RCA}_0 + \mathsf{SRT}^2_2 \not\rightarrow \mathsf{RT}^2_2$ by exhibiting a model $\mathfrak{M}$ for it. In particular, $\mathfrak{M}$ need not be an ω-model. This was the philosophical position and approach taken in [11].

There are two facts about two-coloring of pairs that shed light on the problem at hand:

(I) The result of Jockusch [23] stated earlier on the non-existence of a Σ^0_2-definable homogeneous solution for a particular recursive two-coloring of pairs;

(II) In contrast, every stable two-coloring of pairs has a Δ^0_2-solution.
However, Downey, Hirschfeldt, Lempp and Solomon [17] gave an example of one with no low solution.

A natural first approach, in view of (I), is to seek a model of $\mathsf{RCA}_0 + \mathsf{SRT}^2_2 + \neg\mathsf{RT}^2_2$ consisting only of Δ^0_2-sets. This approach has an obvious drawback: Suppose $\mathfrak{M} = (\omega, S)$ is such a model. Then if $G \leq_T \emptyset'$ is a member of S, a two-coloring C of pairs which is stable relative to G may not have a solution in S, i.e. one that is recursive in $\emptyset'$, since the property of being a Δ^0_2-set is not transitive under the relation "Δ^0_2 in".

A modified approach is for S to consist only of low sets. This approach would seem to have been ruled out by (II). However, a closer examination reveals that the proof in [17] applies an infinite injury priority argument that requires $I\Sigma^0_2$. It exploits the fact that under $I\Sigma^0_2$ there exist nonlow incomplete Δ^0_2-sets, a property which actually fails in some models without Σ^0_2-induction. One can trace the history of this mathematical observaion back to the development of α-recursion theory in the 1970's: Firstly, Shore [39] showed that if α is an admissible ordinal such as $\aleph^L_\omega$ or $\aleph^L_{\omega_1}$, constructible cardinals where Σ_2-replacement fails, then every incomplete α-r.e. set is low. The parallel between α-recursion theory and recursion theory in fragments of arithmetic, especially where it pertains to the jump of an incomplete r.e. set, was investigated in Mytilinaios and Slaman [32] and Chong and Yang [13]. It turns out that many of the ideas and methods in

[2] Flood [18] introduced a principle called RWKL (Ramsey like Weak König's Lemma) which was shown to be a consequence of both WKL_0 and SRT^2_2 but implies the principle DNR (Diagonally non-Recursive principle). See Bienvenue, Patey and Shafer [1] for a further study of RWKL, where among other things, it was shown that the implication $\mathsf{RCA}_0 + \mathsf{RWKL} \rightarrow \mathsf{DNR}$ is strict.

α-recursion theory are applicable in the arithmetic setting. The overall picture is that the failure of $I\Sigma_n^0$ in a model of $B\Sigma_n^0$ is the arithmetic counterpart of the failure of Σ_n-replacement in an admissible ordinal with Σ_{n-1}-replacement. In particular, there exist models of $B\Sigma_2^0 + \neg I\Sigma_2^0$ in which every incomplete r.e. set is low. To elaborate on how one might find and use such a model, we pause to recall some notions in nonstandard arithmetic (see [5]).

Given a nonstandard model $\mathfrak{M}$ of RCA_0, call $I \subset M$ a Σ_n^0-cut if it is Σ_n^0-definable, closed downwards, includes 0 and is closed under the successor operation. A set $X \subseteq I$ is *coded* by an $\mathfrak{M}$-finite set $\hat{X}$ if $\hat{X} \cap I = X$. We say in this case that X is coded (by some $\hat{X}$) for short. Now every model of $\mathsf{RCA}_0 + B\Sigma_2^0 + \neg I\Sigma_2^0$ is already endowed with a basic collection of coded sets that have a bearing on the jump of an r.e. set (see Chong and Mourad [9], Chong and Yang [13]), but a richer collection of coded sets, such as that found in a saturated structure with ω as a Σ_2^0-cut, would ensure that every incomplete r.e.-set is low ([32]. The method of proof is reminiscent of that in [39] and generalizes to showing that every incomplete Δ_2^0-set is low.

The strategy for producing a model of RCA_0 that separates SRT_2^2 from RT_2^2 is therefore (i) select as ground model an appropriate $\mathfrak{M}_0 = (M, S_0) \models \mathsf{RCA}_0 + B\Sigma_2^0 + \neg I\Sigma_2^0$ with a sufficient number of codes so that every incomplete Δ_2^0-set in $\mathfrak{M}_0$ is low, and (ii) construct an M-extension that satisfies SRT_2^2 by adding only low Δ_2^0-homogeneous sets. By (I) above, RT_2^2 fails in the M-extension.

In addition to having a sufficient number of coded sets, the model $\mathfrak{M}_0$ for (i) will also be equipped with an extra feature described below (this is a technical device introduced to make the proof work. See remarks following the proof of Theorem 3.6). Let $b \in M$ and assume that $P \subset (2^{<M})^2$ is a Σ_1^0-definable relation (with parameters) such that for each $\mathfrak{M}_0$-finite set E, $P^0(E) = \{E' : P(E, E') \text{ holds}\}$ is $\mathfrak{M}_0$-finite. Inductively for $v < b$, define $P^{v+1}(E) = \{E'' : \exists E' \in P^v(E)(P(E', E'') \text{ holds})\}$.

Definition 3.1. Let $\mathfrak{M}_0$ be a model of $\mathsf{RCA}_0 + B\Sigma_2^0$. We say that $\mathfrak{M}_0$ satisfies the bounded monotone enumeration principle (BME_1) if for all $b \in M$, Σ_1^0-definable relations P (with parameters), and any $\mathfrak{M}_0$-finite set E, there is an s^* such that $\bigcup_{v \le b} P_s^v(E) = \bigcup_{v \le b} P_{s^*}^v(E)$ for all $s > s^*$, where $P_s^v(E)$ is the set of $\mathfrak{M}_0$-finite sets enumerated into $P^v(E)$ by stage s. In other words, $\bigcup_{v \le b} P^v(E) = \bigcup_{v \le b} P_{s^*}^v(E)$ is $\mathfrak{M}_0$-finite.

In general, a model of $\mathsf{RCA}_0 + B\Sigma_2^0$ need not satisfy BME_1 (for example, the principle fails in the model of $B\Sigma_2^0 + \neg I\Sigma_2^0$ defined in [34]). By Kreutzer and Yokoyama [26], BNME_1 is equivalent over RCA_0 to the principle $P\Sigma_1^0$ (see Hájek and Pudlák [19] for a discussion of this principle), a version of the Pigeonhole principle, as wwell as totality of the Ackermann function.

Theorem 3.2. (Chong, Slaman and Yang [11]) *There is a model* $\mathfrak{M}$ *of* $\mathsf{RCA}_0 + \mathsf{SRT}_2^2 + \neg\mathsf{RT}_2^2$.

Proof. A countable model $\mathfrak{M}_0 = (M, S_0)$ of $\mathsf{RCA}_0 + B\Sigma^0_2 + \neg I\Sigma^0_2$ with the following properties was constructed in [11]:

1. ω is a Σ^0_2-cut;
2. Every arithmetically definable subset of ω is coded;
3. S_0 is the collection of all recursive sets;
4. $\mathfrak{M}_0 \models \mathsf{BME}_1$.

We construct an M-extension of $\mathfrak{M}_0$, denoted $\mathfrak{M} = (M, S)$, such that $\mathfrak{M} \models \mathsf{SRT}^2_2$ and every member of S is low. By (I) above, RT^2_2 fails in $\mathfrak{M}$. Let $C \leq_T \emptyset'$ be a two-coloring of M. Call a number $x \in M$ "red" or "blue" according to $C(x) = 0$ or $C(x) = 1$. We implement a Seetapun disjunction-type construction, but now in the nonstandard domain.

We first describe the overall procedure. Let $X \subseteq M$ be $\mathfrak{M}_0$-infinite and low, and let B be an $\mathfrak{M}_0$-finite collection of Σ^0_1-formulas (with parameters) with a free set variable $\dot{G}$. Given a pair (ρ, β) of $\mathfrak{M}_0$-finite sets (identified respectively with the binary strings which are their characteristic functions), enumerate a sequence of "blobs" $o_s \subset X$, $s \in M$, such that $\min o_s > \max_{s'<s}\{\rho, \beta, o_{s'}\}$ and $\varphi(\rho \cup o_s)$ holds for some $\varphi \in B$.[3] Simultaneously, define the Seetapun tree T_s associated with $\{o_0, \ldots, o_s\}$ as follows: The root of T_s is β. If σ is a node in $T_{s'}$ where $s' < s$, then for each $n \in o_{s'+1}$, $\sigma \cup n$ is a node in $T_{s'+1}$. Declare $(\vec{o}_s, T_s)$, where $\vec{o}_s = \{o_0, \ldots, o_s\}$, to be a Seetapun disjunction if for each maximal path p in T_s, there is a subset v such that $\varphi(\beta \cup v)$ holds for some $\varphi \in B$. In the notation of Definition 3.1, we let $E = (\rho, \beta)$, $b = |B|$ and

$$\begin{aligned} P^0_s(\rho, \beta) = &\{(\rho \cup o_{s'}, \beta) : s' \leq s \wedge \exists \varphi \in B(\varphi(\rho \cup o_{s'})\} \cup \\ &\{(\rho, \beta \cup v) : v = \text{ least subset of a maximal path in } T_s \text{ s. t. } \exists \varphi \in B(\varphi(\beta \cup v))\}, \end{aligned}$$

and call it the set of *exits* over (ρ, β). Note that if (ρ', β') is an exit, then $\rho' > \rho$ implies $\beta' = \beta$ and vice vesa. As before, one has a

Fact. Suppose ρ is red and β is blue. If $(\vec{o}_s, T_s)$ is a Seetapun disjunction, then either there is a red $o_{s'}$ for some $s' \leq s$, or there is a blue maximal path in T_s.

Upon enumeration of a Seetapun disjunction, one repeats the above steps as follows: For each exit $(\rho', \beta') \in P^0(\rho, \beta)$, replace B by $B \setminus \{\varphi\}$, where φ is satisfied by ρ' if $\rho' > \rho$ or by β' if $\beta' > \beta$ (recall if $\rho' > \rho$ then $\beta' = \beta$ and vice versa), and enumerate blobs over

[3] We adopt the convention that if $\varphi \equiv \exists u \psi(u, x)$, then to say that $\varphi(\rho \cup o_s)$ holds means that there is a u with $\max u \leq \max o_s$ such that $\psi(u, \rho \cup o_s)$ holds. The same applies to $\varphi(\beta \cup v)$.

ρ' and define simultaneously a Seetapun tree with root β'. If and when a Seetapun disjunction is enumerated, members of $P^0(\rho', \beta')$ are defined. Then $P^1(\rho, \beta) = \{P^0(\rho', \beta') : (\rho', \beta') \in P^0(\rho, \beta)\}$. In this way one obtains $P^2(\rho, \beta), P^3(\rho, \beta), \ldots$. Applying BME_1 in $\mathfrak{M}_0$, one concludes that there is an s^* such that $\bigcup_{v \le |B|} P^v(\rho, \beta) = \bigcup_{v \le |B|} P^v_{s^*}(\rho, \beta)$. In $P^v_{s^*}(\rho, \beta)$, let (ρ^*, β^*) be a maximal pair such that ρ^* is red and β^* is blue. There are two cases to consider:

Case 1. In the enumeration of blobs over ρ^*, there is a largest s, denoted s_0, such that o_{s_0} is defined.

Let $B_{\rho^*} \subset B$ be the set of formulas not satisfied by ρ^*. Then there is no set o such that $\min o > \max o_{s_0}$ and $\rho^* \cup o$ satisfies a formula in B_{ρ^*}. This means that every formula in B_{ρ^*} has a Π^0_1-outcome for any $G \subset X$ extending ρ^*. Thus for any $\varphi \in B$, $\varphi(G)$ if and only if $\varphi(\rho^*)$. We say in this case that ρ^* "decides" B.

Case 2. o_s is defined for all s.

Then the tree $T = \bigcup_s T_s$ with root β^* is $\mathfrak{M}_0[X]$-infinite. Let B_{β^*} be the set of formulas in B not satisfid by β^*. Then $U = \{\sigma \in T : \text{For all } \varphi \in B_{\beta^*} (\neg\varphi(\sigma))\}$ is an X-recursively bounded, X-recursive $\mathfrak{M}_0[X]$-infinite tree. By the Low Basis Theorem [24] which holds in models of $\mathsf{RCA}_0 + B\Sigma^0_2$, one may choose an unbounded path Z in U thar is low relative to X, and hence low as well. Then as for ρ^*, the blue set β^* decides B in the sense that for all $G \subset Z$ extending β^*, for all $\varphi \in B$, $\varphi(G)$ if and only if $\varphi(\beta^*)$.

The above outline sets the stage for the construction of our desired model. Since ω is a Σ^0_2-cut in $\mathfrak{M}_0$, there is a Σ^0_2-definable, strictly increasing and cofinal map $g : \omega \to M$.

Stage 0. Let $B = B_0 = \{\varphi : \varphi \text{ a } \Sigma^0_1\text{-formula with parameters in } g(0)\}$ and a free set variable $\dot{G}$. Let $\rho = \beta = \emptyset$ and let $X = M$. Execute the steps above to obtain $(\rho^*, \beta^*) = (\rho^-_0, \beta^-_0)$.

The next step is to preserve BME_1, for instances with parameters less than $g(0)$, relative to the generic set G we are constructing. We weave in a construction to achieve this. The construction has two versions, depending on whether (ρ^-_0, β^-_0) was obtained via Case 1 or Case 2 above.

Let $\hat{P}(x, x', \dot{G})$ be a "universal instance" of relativized BME_1 with a Σ^0_1-definition containing parameter $b < g(0)$ (the existence of a universal instance was shown in [11]). The numbers x and x' will code $\mathfrak{M}_0$-finite sets E, E; The Σ^0_1 property of $\hat{P}$ makes it amenable to a Seetapun disjunction style analysis.

First suppose Case 1 holds. Let s_0 be the largest s for which o_s is defined over ρ^-_0 in Case 1. Let $E = \emptyset$ and let $t_0 > \max o_{s_0}$. Enumerate blobs o_s (by abuse of notation, we continue to use o_s in this part of the construction), where $\min o_s > \max\{t_0, o_{s-1}\}$, such that for some E', $\hat{P}(\emptyset, E', \rho^-_0 \cup o_s)$ holds after $\max o_s$ steps of computation (write this as $E' \in \hat{P}^0_{\max o_s}(\emptyset, \rho^-_0 \cup o_s)$), and simultaneously define the Seetapun tree T_s with root β^-_0 as before. Declare $(\vec{o}_s, T_s)$ to be a Seetapun disjunction if for each maximal path in T_s there is a subset v such that an E' is enumerated in $\hat{P}^0_{\max v}(\emptyset, \beta_0^- \cup v)$, i.e. $\hat{P}(\emptyset, E', \beta^- \cup v)$ holds for some E' after max v steps of computation.

If and when this occurs, call a pair (ρ', β'), where $(\rho', \beta') = (\rho_0^- \cup o_s, \beta_0^-)$ or $(\rho', \beta') = (\rho_0^-, \beta_0^- \cup v)$, an *exit* (for uniqueness, in the latter case choose the least v for each maximal path in T_s). Note that by the definition of an exit, at least one of the exits will be a red $\rho' > \rho_0^-$ or a blue $\beta' > \beta_0^-$. Furthermore, if $\rho' > \rho_0^-$ then $\beta' = \beta_0^-$, and vice versa. If $\rho' > \rho_0^-$ then there exists an $E' \in \hat{P}^0_{\max \rho'}(\emptyset, \rho')$. If $\beta' > \beta_0^-$ then there is an $E' \in \hat{P}^0_{\max \beta'}(\emptyset, \beta')$. Given an exit (ρ', β'), let

$$\mathbb{E}_0(\rho') = \{E' : E' \in \hat{P}^0_{\max \rho'}(\emptyset, \rho')\}, \text{ where } \rho' > \rho_0^- \wedge \beta' = \beta_0^-$$
$$\mathbb{E}_0(\beta') = \{E' : E' \in \hat{P}^0_{\max \beta'}(\emptyset, \beta')\}, \text{ where } \rho' = \rho_0^- \wedge \beta' > \beta_0^-.$$

Each exit (ρ', β') and its accompanying sets $E_0(\rho')$, $E_0(\beta')$ are then used to replace

$$(\rho_0^-, \beta_0^-, \emptyset)$$

to generate the next collection of Seetapun disjunctions and their exits. More precisely, we do the following:
First suppose $\rho' > \rho_0^-$, $\beta' = \beta_0^-$. Enumerate blobs o_s, with min $o_s >$ max $\{\rho', \beta', o_{s-1}\}$, for which there is an $E' \in \mathbb{E}_0(\rho')$ and an $E'' \in \hat{P}^0_{\max o_s}(E', \rho' \cup o_s)$, i.e. $\hat{P}(E', E'', \rho' \cup o_s)$ holds. In this case we say that E'' is in $\hat{P}^1_{\max o_s}(\emptyset, \rho' \cup o_s)$. Simultaneously define an associated tree T_s with root β'. A Seetapun disjunction for $(\vec{o}_s, T_s)$ is declared if every maximal path in T_s contains a subset v for which there is an $E' \in \hat{P}^0_{\max v}(\emptyset, \beta' \cup v)$. An exit in the Seetapun disjunction is a pair (ρ'', β'') where $\rho'' = \rho' \cup o_s$ and $\beta'' = \beta'$, or $\rho'' = \rho'$ and $\beta'' = \beta' \cup v$ for some least v contained in a maximal path of T_s. Define $E_1(\rho'') = \{E'' : E'' \in \hat{P}^1_{\max \rho''}(\emptyset, \rho'')\}$ if $\rho'' > \rho'$, and $E_0(\beta'') = \{E' : E' \in \hat{P}^0_{\max \beta''}(\emptyset, \beta'')\}$ if $\beta'' > \beta'$.

Now suppose $\rho' = \rho_0^-$ and $\beta' > \beta_0^-$.

Enumerate blobs o_s such that min $o_s >$ max $\{\rho', \beta', o_{s-1}\}$ for which there is an $E' \in \hat{P}^0_{\max o_s}(\emptyset, \rho' \cup o_s)$. Simultaneousle enumerate an associated tree T_s with root β' and declare $(\vec{o}_s, T_s)$ to be a Seetapun disjunction if for every maximal path in T_s, there is a subset v for which there is an $E' \in \mathbb{E}_0(\beta')$ and an $E'' \in \hat{P}^0_{\max v}(E', \beta' \cup v)$. By definition we say that $E'' \in \hat{P}^1_{\max v}(\emptyset, \beta' \cup v)$. Define exits (ρ'', β'') as before and $E_0(\rho'')$ for $\rho'' > \rho'$ and $E_1(\beta'')$ for $\beta'' > \beta'$ similarly. In this way, uniformly in $\upsilon \le b = |B_0|$, one defines exits (ρ, β), sets $\hat{P}^\upsilon(\emptyset, \rho)$ and $\hat{P}(\emptyset, \beta)$, as well as $\mathbb{E}_\upsilon(\rho), \mathbb{E}_\upsilon(\beta)$ (where applicable). By BME_1, there is a least t, denoted t^*, such that $\hat{P}^\upsilon = \hat{P}^\upsilon_{t^*}$ for $\upsilon \le b$. Choose an exit (ρ_0, β_0) that is maximal and enumerated by t^*, such that ρ_0 is red and β_0 is blue. Note that $\rho_0 \succeq \rho_0^-$ and $\beta_0 \succeq \beta_0^-$. Assume $(\rho_0, \beta_0) \in \hat{P}^\upsilon$.

Now for (ρ_0, β_0), there are two possible outcomes:

(i) There is a largest s for which o_s is defined over ρ_0, i.e. $\hat{P}^{v+1}_{\max o_s}(\emptyset, \rho_0 \cup o_s)$ enumerates an output. Let s_0 denote the largest such s and let $t_0 > \max o_{s_0}$. Proceed to Stage 1 of the construction.

(ii) o_s is defined for each s. Let $T = \bigcup_s T_s$. Then

$$U_0 = \{\sigma : \sigma \in T \wedge \sigma \text{ contains no subset } v \text{ s. t. } \hat{P}^{v+1}(\emptyset, \beta_0 \cup v) \text{ enumerates an output}\}$$

is an $\mathfrak{M}_0$-infinite, recursively bounded recursive tree and hence contains an $\mathfrak{M}_0$-infinite low path Z_0 by the Low Basis Theorem. Proceed to Stage 1 below.

Note that in either (i) or (ii), any $\mathfrak{M}_0$-infinite G extending ρ_0 or β_0, where $\min G \setminus \rho_0 > t_0$ if (i) applies, and $G \subset Z_0$ if (ii) applies, preserves instances of BME_1 with parameters in $g(0)$.

Now suppose Case 2 holds. Let Z be an $\mathfrak{M}_0$-infinite low path in the tree U extending ρ_0^- chosen for this case. We enumerate blobs o_s, where $\min o_s > \max\{\rho_0^-, \beta_0^- o_{s-1}\}$, for which an E' is enumerated in $\hat{P}^0_{\max o_s}(\emptyset, \rho_0^- \cup o_s)$ as before, but now $o_s \subset Z$. Simultaneously define T_s over β_0^- in the same fashion. Declare $(\vec{o}_s, T_s)$ to be a Seetapun disjunction if the same conditions are met, and use each exit in the disjunction to enumerate the next Seetapun disjunction, and so on. Then BME_1 in $\mathfrak{M}_0$ ensures that there is a t^* such that $\hat{P}^v = \hat{P}^v_{t^*}$ for all $v \leq |B_0|$. Then again there are two possible outcomes similar to (i) and (ii) which we now label as (iii) and (iv) respectively. If (iv) holds, then the path Z_0 selected (as in (ii)) will be low relative to Z, hence low. This completes the construction at stage 0.

Stage 1. First suppose (i) or (iii) above holds. Let $\rho = \rho_0 \cup a$ where $a > \max o_{s_0}$ is red (and belongs to Z_0 if (iii) holds). Such a number must exist else one obtains immediately a low blue solution for the coloring.

Let $\beta = \beta_0$, $B = B_1 = \{\varphi : \varphi \text{ is } \Sigma^0_1$ with a free set variable and parameters in $g(1)\}$. Let $X = M$ if (i) holds and $X = Z_0$ if (iii) holds. Now run the machine enumerating blobs o_s in X and Seetapun trees T_s to arrive at an s^* such that $\bigcup_{v \leq |B_1|} P^v(\rho, \beta) = \bigcup_{b \leq |B_1|} P^v_{s^*}(\rho, \beta)$. As in Stage 0, let (ρ^*, β^*) be a maximal pair that is an exit, with ρ^* = red and β^* = blue. Set $(\rho_1^-, \beta_1^-) = (\rho^*, \beta^*)$ and note that there are now two scenarios (Case 1 and Case 2) to consider. In each case, proceed to the construction that enumerates outputs of the universal instance $\hat{P}$ of BME_1 with parameters less than $g(1)$. This step results in the extension of (ρ_1^-, β_1^-) to (ρ_1, β_1) which preserves BME_1 for instances with parameters less than $g(1)$. Now continue to the next stage.

If (ii) or (iv) holds, let $X = Z_0$ where $Z_0 \subset U_0$ is an $\mathfrak{M}_0$-infinite low path and extends β_0. Let $a > \max \beta_0$ be a blue number in Z_0 (which must exist else Z_0 is a red solution for the coloring). Let $B = B_1$ defioned above, $(\rho, \beta) = (\rho_0, \beta_0 \cup a)$. Enumerate o_s, T_s as before to arrive at an s^* and (ρ^*, β^*) which we accordingly denote as (ρ_1^-, β_1^-). The next step is to preserve BME_1 for instances with parameters less than $g(1)$. Again, to achieve this, there are Case 1 and Casse 2 to consider as before. Each case bifurcates into two possibilities and all result in the extension of (ρ_1^-, β_1^-) to (ρ_1, β_1) that preserves BME_1 for parameters less than $g(1)$.

In general, the construction proceeds by induction on $n \in \omega$. The countability of $\mathfrak{M}_0$ ensures that the construction is completed in ω steps. Let (ρ_n, β_n) be the pair of strings constructed at the end of stage n. Then either $\rho_n \prec \rho_{n+1}$ for infinitely many n, or $\beta_n \prec \beta_{n+1}$ for infinitely many n. Suppose it is the former. Then the set $K = \{n : \rho_n \prec \rho_{n+1}\}$ is a definable subset of ω and hence coded in $\mathfrak{M}_0$. Using the code, $\emptyset'$ can compute $g(n)$ for each $n \in K$ and thereby recover ρ_n. Then $G = \bigcup_{n \in K} \rho_n$ is a low and red homogeneous set for the coloring. The argument for the β-side is similar. Then $\mathfrak{M}_0[G]$ is a model of $\mathsf{RCA}_0 + \mathsf{BME}_1 + B\Sigma_2^0$ and solves an instance of a stable two-coloring with a low set. To obtain the model $\mathfrak{M}$, we iterate the construction above by successively adjoining a homogeneous solution preserving BME_1 every step of the way. Each solution will also be low relative to those constructed earlier.

□

Remark 3.3. It was also shown in [11] that $\mathsf{RCA}_0 + \mathsf{SRT}_2^2 + \mathsf{WKL}_0 \not\to \mathsf{RT}_2^2$.

The proof of Theorem 3.2 immediately implies

Corollary 3.4. $\mathsf{RCA}_0 + \mathsf{SRT}_2^2 \nvdash I\Sigma_2^0$.

Remark 3.5. Monin and Patey [31] have recently posted a proof of Theorem 3.2 using an ω-model. The model does not yield Corollary 3.4. It is not known if there is a model (ω, S) separating SRT_2^2. from RT_2^2 in which every member of S is low$_2$.

The limited inductive strength of SRT_2^2 extends to full scale Ramsey's Theorem for pairs. It is another manifestation of the sharp contrast between RT_2^3 and its junior sibling.

Theorem 3.6. (Chong, Slaman and Yang [12]) $\mathsf{RCA}_0 + \mathsf{RT}_2^2 \nvdash I\Sigma_2^0$.

Proof. We give sketch of the proof. By Chong, Slaman and Yang [10], every model $\mathfrak{M}$ of $\mathsf{RCA}_0 + B\Sigma_2^0$ has an M-extension that satisfies COH while preserving $B\Sigma_2^0$. This implies that $\mathsf{RCA}_0 + \mathsf{COH} + B\Sigma_2^0$ is Π_1^1-conservative over $\mathsf{RCA}_0 + B\Sigma_2^0$ (see §4). The proof can be strengthened to include preservation of BME_1 if $\mathfrak{M} \models \mathsf{BME}_1$ ([12]). By Proposition 1.3, it means that to establish the theorem, one starts with an appropriate ground model and alternately constructs sets that solve instances of COH and SRT_2^2. The resulting structure will be a model of RT_2^2.

Denote by $\mathfrak{M}_0 = (M, S_0)$ the structure in Theorem 3.2. Given a two-coloring C of pairs in $\mathfrak{M}_0$, one defines from it an array $\{A_s : s \in M\}$ such that C is stable on any set cohesive for the array. By [12] there is a solution X_0 for the array such that $\mathfrak{M}_0[X_0]$ satisfies $\mathsf{RCA}_0 + \mathsf{BME}_1 + B\Sigma^0_2$. Now C is stable on X_0 and we can implement the construction in Theorem 3.2 relative to X_0 to obtain a G homogeneous for C which preserves BME_1. We will show in fact $G \leq_T X_0'$ so that $\mathfrak{M}_0[X_0, G] \models \mathsf{BME}_1 + B\Sigma^0_2$ as well.

For the model $\mathfrak{M}_0$, the function $g : \omega \to M$ is Σ^0_2 without set parameters. A closer inspection of the construction, which is carried out relative to X_0 and follows the steps in Theorem 3.2, reveals that it is recursive in X_0''. Now the homogeneous set G is either $\bigcup_n \rho_n$ or $\bigcup_n \beta_n$, and it can be computed from X_0'' because the latter is able to compute $W_\rho = \{n : \rho_{n-1} \prec \rho_n\}$ and $W_\beta = \{n : \beta_{n-1} \prec \beta_n\}$, where we sst $\rho_{-1} = \beta_{-1} = \emptyset$. To see this, note that at each stage, the construction splits into Case 1 and Case 2, and each splits into two further cases (Case (i) and (ii) for Case 1 and Case (iii) and (iv) for Case 2). At stage n, the split into Case 1 or 2 depends on the $\Sigma^0_2(X_0)$-question: Is there a largest s where o_s is enumerated over ρ_n^-? Similarly the split into (i) or (ii), and the split into (iii) or (iv), depends on the $\Sigma^0_2(X_0)$-question: Is there a largest s where o_s is enumerated over ρ_n? Then $\rho_n \succ \rho_{n-1}$ if the answer to the first question is affirmative, and $\beta_n \succ \beta_{n-1}$ otherwise (note that $\beta_n = \beta_{n-1}$ only if Case 1 and Case (i) hold). Also, $\rho_{n+1} \succ \rho_n$ if the answer to the second question is affirmative (note that in this case one may still have $\beta_n \succ \beta_{n-1}$ if Case 2 holds at stage n). As an illustration, suppose Case 1 holds at stage n for infinitely many n, and Case (i) never holds. Consider the set

$$K_0 = \{(n, m) : \exists s(o_s \text{ is enumerated and } g(m) \leq s < g(m+1) \text{ at stage } n\}.$$

Then $K \subset \omega \times \omega$ and is $\Delta^0_2(X_0)$, hence coded by an $\mathfrak{M}_0$-finite set $\hat{K}_0$ in $\mathfrak{M}_0$ according to [9]. Now define

$$K_1 = \{n : n \in \omega \wedge \lim_{m \in \omega \wedge (n,m) \in \hat{K}_0} m \text{ exists}\}.$$

Now K_1 is definable over $\mathfrak{M}_0$ and hence coded. Then X_0' can use the code to compute the homogeneous set G. It follows that G is low relative to X_0. Thus $\mathfrak{M}[X_0, G] \models B\Sigma^0_2$.

□

Remark 3.7. Theorem 3.6 also follows from Theorem 4.5 in the next section. Namely, if $\mathsf{RCA}_0 + \mathsf{RT}^2_2$ proves $I\Sigma^0_2$, then it would imply BME_1 as well, since the latter follows from $I\Sigma^0_2$. However, since BME_1 is a Π^0_3-sentence, the Π^0_3-conservation of $\mathsf{RCA}_0 + \mathsf{RT}^2_2$ over RCA_0 would. then lead one to conclude $\mathsf{RCA}_0 \vdash \mathsf{BME}_1$, which is false.

Theorem 4.5 also implies that there is a model of $\mathsf{RCA}_0 + \mathsf{RT}^2_2$ in which BME_1 fails. It is an interesting problem to construct such a model.

4 Conservation Property of RT^2_2

The quest to undestand the relationship between infintistic mathematics and the finitistic is, in the view of many logicians, the heart of logic. It is also a major endeavour of philosophy of mathematics. In the context of the subject matter of this paper, one may pose the following question: how much light does infinitary combinatorial principles such as RT^n_k shed on finite mathematics? For $n \geq 3$, one knows that there is much more that $\mathsf{RCA}_0 + \mathsf{RT}^n_k$ proves than what RCA_0 does. The corresponding question for RT^2_2 is, however, a challenging one.

Definition 4.1. Let Γ be a class of sentences in the language of second-order arithmetic. We say that RT^n_k is Γ-conservative over RCA_0 if every $\varphi \in \Gamma$ that is provable in $\mathsf{RCA}_0 + \mathsf{RT}^2_2$ is already a theorem of RCA_0.

Since $\mathsf{RCA}_0 + \mathsf{RT}^n_2 \vdash \mathsf{ACA}_0$ for $n \geq 3$, it is not Γ-conservative over RCA_0 for $\Gamma =$ the arithmetical sentences. A long standing open question asks whether RT^2_2 is Π^1_1-conservative over RCA_0. Theorem 4.5 is a major step towards resolving this question. To discuss this result, we first recall the notion of largeness introduced by Ketonen and Solovay [25] in their analysis of the Paris-Harrington principle via the hierarchy of fast growing functions:

Definition 4.2. Let X be a finite set.

(i) X is ω-large if $|X| > \min X$;

(ii) X is ω^{n+1}-large if it $X \setminus \{\min X\}$ is the union of sets $X_1, \ldots, X_{\min X}$ such that

- $\max X_i < \min X_{i+1}$ for each $i < \min X$;
- Each X_i is ω^n-large.

(iii) X is RT^2_k-ω^n-large if every k-coloring of pairs in X has a homogeneous subset that is ω^n-large.

Clearly RT^2_k implies (iii), i.e. for any model of $\mathsf{RCA}_0 + \mathsf{RT}^2_2$ and each $k, n \in \omega$, there is an X that is RT^2_k-ω^n-large. The heart of the proof of Theorem 4.5 is in deriving a version of the converse; namely, beginning with an RT^2_2-ω^n-large set, construct a model of $\mathsf{RCA}_0 + \mathsf{RT}^2_2$.

In the absence of RT^2_k, we have the following finite version of Ramsey's theorem for pairs:

Theorem 4.3. (Ketonen and Solovay [25]) *$I\Sigma^0_1$ implies that if $k \geq 2$ and X is ω^{k+4}-large (with $\min X \geq 3$), then X is RT^2_k-ω-large.*

The next lemma is an effective generalization of this result:

Lemma 4.4. (Kołodziejczyk and Yokoyama [27]) *Let $n \in \omega$ and let X be ω^{300n}-large with min $X \geq 3$. Then X is RT^2_2-ω^n-large.*

Denote by $\forall X\Pi^0_3$ the class of sentences φ of the form $\forall X \forall x \exists y \forall z \varphi_0$, where φ_0 is bounded. Clearly $\forall X\Pi^0_3$ properly contains the class of Π^0_3-sentences.

Theorem 4.5. (Patey and Yokoyama [36], Kołodziejczyk and Yokoyama [27]) $\mathsf{RCA}_0+\mathsf{RT}^2_2+\mathsf{WKL}_0$ *is $\forall X\Pi^0_3$-conservative over* RCA_0.

Proof. We give a sketch of the simpler proof given in [27]. Suppose the theorem is false and φ is a witness to this, i.e. $\mathsf{RCA}_0 + \mathsf{RT}^2_2 + \mathsf{WKL}_0 \vdash \varphi$ while $\mathsf{RCA}_0 \nvdash \varphi$, where $\varphi \equiv \forall X \forall x \exists y \forall z \varphi_0(X \restriction z, x, y, z)$ and φ_0 is Δ^0_0. Then there is a countable model $\mathfrak{M} = (M, S)$ of RCA_0 such that

$$\mathfrak{M} \models \exists X \exists x \forall y \exists z \neg \varphi_0(X \restriction z, x, y, z).$$

Let $X_0 \in S$ and $a \in M$ be such that $\mathfrak{M} \models \forall y \exists z \neg\varphi_0(X_0 \restriction z, a, y, z)$. We will produce a cut $I \subset M$ such that $\mathfrak{I} = (I, S_I) \models \mathsf{RCA}_0 + \mathsf{RT}^2_2 + \mathsf{WKL}_0 + \neg\varphi$, where $S_I = \{E : \exists \hat{E}(E = \hat{E} \cap I \wedge \hat{E}$ is $\mathfrak{M}$-finite$\}$. This will give us the contradiction required.

First let $\{a_s : s \in M\}$ be a primitive recursive strictly increasing sequence such that for all s and $y < a_s$, there is a $z < a_{s+1}$ such that $\mathfrak{M} \models \neg\varphi_0(X_0 \mid z, a, y, z)$. Then $Y = \{a_s : s \in M\} \in S$ and is $\mathfrak{M}$-infinite. Let ω be the least infinite ordinal formalized in RCA_0. Then for each $n \in \omega$, there is an $\mathfrak{M}$-finite subset of Y that is ω^n-large. By overspill, there is a nonstandard d and an $\mathfrak{M}$-finite set $Z \subset Y$ such that Z is ω^{300^d}-large. Let $\{E_i : i \in \omega\}$ list all $\mathfrak{M}$-finite sets which are not ω-large. Let $\{C_i : i \in \omega\}$ list all $\mathfrak{M}$-finite two-colorings of pairs in Z. We construct an ω-sequence of $\mathfrak{M}$-finite sets $Z = Z_0 \supseteq Z_1 \supseteq \cdots$ such that

- Each Z_i is $\omega^{300^{d-i}}$-large;
- $C_i \restriction [Z_{2i+1}]^2$ is constant;
- $\{x : \min Z_{2i+2} \leq x \leq \max Z_{2i+2}\} \cap E_i = \emptyset$.

To define Z_1, apply Lemma 4.4 by letting $n = 300^{d-1}$ and so $Z_1 \subseteq Z_0$ is $\omega^{300^{d-1}}$-large and homogeneous for C_0. On Z_1, define a two-coloring C of pairs such that $C(u, v) =$ red if and only if $E_0 \cap \{w : u \leq w \leq v\} \neq \emptyset$. Let $Z_2 \subset Z_1$ be homogeneous for C such that Z_2 is $\omega^{300^{d-2}}$-large (by Lemma 4.4).

Claim 1. $C \restriction [Z_2]^2$ is constantly blue.
Suppose otherwise. Then for any $u < v$ in Z_2, $E_0 \cap \{w : u \leq w \leq v\} \neq \emptyset$. But this would immediately imply that E_0 is ω-large, which is not possible.

In this way one defines inductively Z_i for each $i \in \omega$. Now let $I = \{u : \exists i \in \omega(u \leq \min Z_i\}$. It is not difficult to verify that I is a cut. Let $S_I = \{E \cap I : E$ is $\mathfrak{M}$-finite$\}$.

Claim 2. $\mathfrak{I} = (I, S_I) \models \mathsf{RCA}_0 + \mathsf{RT}^2_2 + \mathsf{WKL}_0$.

We first show that $\mathfrak{I} \models I\Sigma^0_1$. For this, it is sufficient to show that every $\Sigma^0_1(\mathfrak{I})$-function defined on a bounded set in I is bounded in I. Let f be Σ^0_1 with parameter $E \in S_I$ defined on $K \subset \{x : x \le e\}$ for some $e \in I$. Suppose $f(K)$ is unbounded in I. Then $f(K) \in S_I$ and so there is an i such that $f(K) = E_i \cap I$. However, the construction ensures that E_i is disjoint from Z_{2i} and is therefore bounded in I.

Now let C be a two-coloring of $[I]^2$ that belongs to S_I. Then there is an i such that $C_i \cap I = C$. Then by constructioin Z_{2i+1} is homogeneous for C_i and hence so is $Z_{2i+1} \cap I$, which belongs to S_I. Thus $\mathfrak{I} \models \mathsf{RT}^2_2$.

To show that $\mathfrak{I} \models \mathsf{WKL}_0$, let $T \in S_I$ be an unbounded binary tree. If for each $\sigma \in T$, the collection $T_\sigma = \{\tau : \tau \succ \sigma \wedge \tau \in T\}$ is bounded in I, then by $B\Sigma^0_2$ which holds in $\mathfrak{I}$ (by Fact 1.2), T would be bounded in I. Thus there is a $\sigma_0 \in T$ such that T_{σ_0} is unbounded. By induction, one obtains an unbounded ω-sequence $\sigma_0 \prec \sigma_1 \prec \cdots \prec \sigma_i \prec \cdots$ such that $p = \bigcup_i \sigma_i$ is unbounded in I. Since $T \in S_I$, there is an $\mathfrak{M}$-finite set $\hat{T}$ such that $T = \hat{T} \cap I$ and a path $\hat{p} \in \hat{T}$ such that $p = \hat{p} \cap I$, and hence $p \in S_I$. Thus $\mathfrak{I} \models \mathsf{WKL}_0$.

Finally, since $a_s \in I$ for $s \in I$, we have $\mathfrak{I} \models \neg\varphi$, which contradicts the assumption that $\mathsf{RCA}_0 + \mathsf{RT}^2_2 + \mathsf{WKL}_0 \vdash \varphi$.

□

5 Concluding remarks

The study of the reverse mathemartics of Ramseyt's Theorem has opened up a wide area of research that connects combinatorial theory to logic. New tools and techniques have been developed to investigate problems in this fertile field. This article has considered only those directly related to Ramsey's Theorem itself. Nevertheless, studies have also been made on combinatorial principles known to be weaker than RT^2_2, such as the ascending and descending sequence principle ADS, which states that every infinite linearly ordered set has an infinite ascending or descending subsequence, and the Chain/Anti-chain principle CAC, which states that every infinte partially ordered set contains an infinite chain or anti-chain (see Hirshfeldt and Shore [21] for a discussion of these principles, Lerman, Solomon and Towsner [28] for a proof that $\mathsf{RCA}_0 + \mathsf{ADS} \not\rightarrow \mathsf{CAC}$, and Chong, Slaman and Yang [10] for Π^1_1-conservation of these principles over $\mathsf{RCA}_0 + B\Sigma^0_2$). Separately, Wang [42] showed that both the Free Set Theorem Principle FS (for every $n \ge 1$ and coloring C of $[\mathbb{N}]^n$ in infinitely many colors, there is an infinite $A \subset \mathbb{N}$ such that for any $\vec{x} = \{x_0, \ldots x_{n-1}\} \in [A]^n$, $C(\vec{x}) \in A \to C(\vec{x}) \in \vec{x}$), and the Thin Set Theorem Principle TS (for every $n \ge 1$ and infinirte coloring C of $[\mathbb{N}]^n$, there is an infinite $A \subset \mathbb{N}$ such that $C \restriction [A]^n \subsetneq \mathbb{N}$) are strictly weaker than ACA_0. A striking point about these two results is that they hold for arbitrary n-tuples—a phenomenon that is quite different from Ramsey's Theorem where $\mathsf{RT}^n_2 \to \mathsf{ACA}_0$ for $n \geq 3$.[4]

More recently, there is increasing interest in the generalization of Ramsey type combinatorial principles to other structures such as trees. Preserving the topological structure of a tree introduces a new dimension to the complexity of issues involved in constructing a homogeneous solution with the prescribed property. We give two examples below.

Let $2^{<M}$ denote the full binary tree in $\mathfrak{M} = (M, S)$.

Example 1. The TT^1 principle: Any finite coloring of the full binary tree has a homogeneous solution, i.,e. a tree isomorphic to the full binary tree all of whose nodes have the same color. TT^1 holds in any structure of $\mathsf{RCA}_0 + I\Sigma^0_2$, as can be easily verified. However, in the absence of Σ^0_2-induction, the existence of a homogeneous tree is a nontrivial problem. Corduan, Groszek and Mileti [15] showed that TT^1 is not provable in $\mathsf{RCA}_0 + B\Sigma^0_2$. Chong, Li, Wang and Yang [8] have recently shown that in the base system RCA_0, TT^1 is Π^1_1-conservative over $\mathsf{BME}_1 + B\Sigma^0_2$. As a consequence, $\mathsf{RCA}_0 + \mathsf{TT}^1$ does not prove $I\Sigma^0_2$. A rather interesting point to note is that if $I\Sigma^0_2$ fails, then there is a recursive finite coloring of the full binary tree for which there is no $\emptyset''$-computable homogeneous tree. Since little is known about the structure of Turing degrees $\not\leq \mathbf{0''}$, it may explain why solutions constructed in [8] are in general non-definable over the ground model.

Example 2. TT^n_k: Denote by $[2^{<M}]^n$ the collection of n-tuples of pairwise compatible nodes in the full binary tree. This principle states that for any coloring $C : [2^{<M}]^n \to k$, there is a tree $T \cong 2^{<M}$ in S such that $C \upharpoonright [T]^n$ is constant, where $[T]^2$ is the set of n-tuples of pairwise compatible nodes in T. Patey [35] showed that TT^2_2 is strictly stronger than RT^2_2, while Dzhafarov and Patey [16] showed that TT^2_2 is strictly weaker than TT^3_2, which is equivalent to ACA_0. Thus TT^2_2 is a Ramsey type principle whose strength lies between ACA_0 and RT^2_2. Recently Chong, Li, Liu and Yang [6] showed that despite its greater sterngth over RT^2_2, TT^2_2 still fails to imply the weak Weak Kónig's Lemma principle WWKL, and hence WKL_0 as well. In a related work [7], it is shown that TT^2_2 does not imply $I\Sigma^0_2$, generalizing Theorem 3.6. In the following figure, we give a summary of the proof-theoretic strengths of combinatorial principles discussed in the previous sections, with RT^2_2 at the center. Numbered square brackets $[n], n > 0$, refer to the papers cited. [0] refers to the fact that $\mathsf{TT}^1 \not\to \mathsf{RT}^2_2$ trivially, since the former has an ω-model whose second-order members are the recursive sets. Note that Theorem 4.5, which concerns another aspect of RT^2_2, i.e. Π^0_3-conservation over RCA_0, is not incorporated in the figure.

[4] Cholak, Gusto, Hirst and Jockusch [2] have shown that the free set theorem for pairs, i.e. when $n = 2$, is a consequence of RT^2_2.

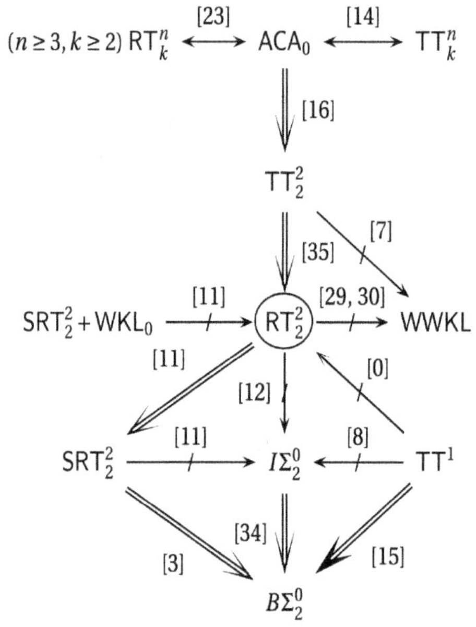

Figure 2: Combinatorial pinciples related to RT^2_2

We end this article with some questions:

Question 1. Is there an ω-model of $\mathsf{SRT}^2_2 + \neg\mathsf{RT}^2_2$ consisting only of low$_2$ sets?

The model constructed in [31] selects, roughly, paths in a $\emptyset''$-recursive tree as solutions of instances of SRT^2_2, and are therefore not necessarily low$_2$.

Question 2. Is $\mathsf{RCA}_0 + \mathsf{TT}^2_2$ or $\mathsf{RCA}_0 + \mathsf{TT}^1$ Π^0_3-conservative over RCA_0?

To follow the strategy adopted in [36], one needs an analysis of the ordinal bounds of finite versions of these principles as carried out in [25] and [27]. It is not clear how this may be achieved. A closely related problem converns Π^1_1-conservation:

Question 3. Is $\mathsf{RCA}_0 + \mathsf{RT}^2_2$ or $\mathsf{RCA}_0 + \mathsf{TT}^2_2$ Π^1_1-conservative over $\mathsf{RCA}_0 + B\Sigma^0_2$?

Question 4. What is the relative proof-theoretic strength between TT^1 and RT^2_2 or TT^2_2 over the base system RCA_0?

While TT^1 has a trivial solution over $\mathsf{RCA}_0 + I\Sigma^0_2$, so that $\mathsf{RT}^2_2 \to \mathsf{TT}^1$ (strictly) over $\mathsf{RCA}_0 + I\Sigma^0_2$, the situation is unknown without Σ^0_2-induction.

References

[1] Laurent Bienvenu, Ludovic Patey and Paul Shafer *On the logical strengths of partial solutions to mathematical problems*, Transactions of the London Mathematical Society, 2017, pp. 30–71.

[2] Peter A. Cholak, Mariagnese Giusto, Jeffry L. Hirst, and Carl G. Jockusch *Free sets and reverse mathematics*. In: Reverse Mathematics 2001, Association for Symbolic Logic, La Jolla 2001, pp. 104–119 [*Studies in Logic 21*].

[3] Peter A. Cholak, Carl G. Jockusch, and Theodore A. Slaman *On the strength of Ramsey's theorem for pairs*, The Journal of Symbolic Logic 66 (1), 2001, pp. 1–55.

[4] Chi Tat Chong, Steffen Lempp, and Yue Yang *On the role of the collection principle for Σ^0_2-formulas in second-order reverse mathematics*, Proceedings of the American Mathematical Society 138 (3), 2010, pp. 1093–1100.

[5] Chi Tat Chong, Wei Li, and Yue Yang *Nonstandard models in recursion theory and reverse mathematics*, The Bulletin of Symbolic Logic 20 (2), 2014, pp. 170–200.

[6] Chi Tat Chong, Wei Li, Lu Liu and Yue Yang. *The streength of Ramsey's Theorem for pairs over trees: I. Weak König's Lemma* [preprint].

[7] — *The streength of Ramsey's Theorem for pairs over trees: II. Σ^0_2-bounding* [in preparation].

[8] Chit Tat Chong, Wei Li, Wei Wang, and Yue Yang *On the streength of Ramsey's Theorem for trees*, Advances in Mathematics, 2020, to appear.

[9] Chi Tat Chong, and K. J. Mourad *The degree of a Σ_n-cut*, Annals of Pure and Applied Logic 48, 1990, pp. 227–235.

[10] Chi Tat Chong, Theodore A. Slaman, and Yue Yang *Π^1_1-conservation of combinatorial principles weaker than Ramsey's Theorem for pairs*, Advances in Mathematics 230 (3), 2012, pp. 1060–1077.

[11] — *The metamathematics of stable Ramsey's Theorem for pairs*, Journal of the American Mathematical Society 27 (3), 2014, pp. 863–892.

[12] — *The inductive strength of Ramsey's Theorem for pairs*, Advances in Mathematics 308, 2017, pp. 121–141.

[13] Chi Tat Chong, and Yue Yang *Σ_2-induction and infinite injury priority arguments, part II:Tame Σ_2 coding and the jump operator*, Annals of Pure and Applied Logic 87 (2), 1997. pp. 103–116.

[14] Jennifer Chubb, Jeffry L. Hirst. and Timothy H. McNicholl *Reverse mathematics, computability, and partitions of trees*, The Journal of Symbolic Logic 74 (1), 2009, pp. 201–215.

[15] Jared Corduan, Marcia J. Groszek, and Joseph R. Mileti *Reverse mathematics and Ramsey's property for trees*, The Journal of Symbolic Logic 75 (3), 2010, pp. 945–954.

[16] Damir Dzhafarov, and Ludovic Patey *Coloring trees in reverse mathematics*, Advances in Mathematics 318, 2017, pp. 497–514.

[17] Rod Downey, Denis R. Hirschfeldt, Steffen Lempp, and Reed Solomon *A Δ^0_2 set with no infinite low subset in either it or its complement*, The Journal of Symbolic Logic 66 (3), 2001, pp. 1371–1381.

[18] Stephen Flood *Reverse mathematics and a Ramsey type König's Lemma*, The Journal of Symbolic Logic 77(4), 2012, pp. 1272–1280.

[19] Petr Hájek, and Pavel Pudlák Metamathematics of First-Order Arithmetic, Springer-Verlag, Berlin 1993 [*Perspectives in Mathematical Logic*].

[20] Denis R. Hirschfeldt Slicing The Truth, *On the Computability Theoretic and Reverse Mathematical Analysis of Combinatorial Principles*, World Scientific Publishing Company, Singapore 2014 [*Lecture Notes Series, Institute for Mathematical Sciences, National University of Singapore*].

[21] Denis R. Hirschfeldt, and Richard A. Shore *Combinatorial principles weaker thanRamsey's Theorem for pairs*, The Journal of Symbolic Logic 72 (I), 2007, pp. 171–206.

[22] Jeffrey L. Hirst Combinatorics in Subsystems of second-order Arithmetic, PhD Dissertation, The Pennsylvania State University, 1987.

[23] Carl G. Jockusch Jr. *Ramsey's Theorem and recursion theory*, The Journal of Symbolic Logic 37, 1972, pp. 268–280.

[24] Carl G. Jockusch Jr., and Robert I. Soare Π^0_1 *classes and degrees of theories*, Transactions of American Mathematical Society 173, 1972. pp. 33–56.

[25] Jussi Ketonen, and Robert M. Solovay *Rapidly growing Ramsey functions*, Annals of Mathematics 113, 1981, pp. 267–314.

[26] Alexander P. Kreuzer, and Keita Yokoyama *On principles between Σ_1 - and Σ_2-induction, and monotone enumerations*, Journal of Mathematical Logic 16 (1), 2016.

[27] Leazek Aleksander Kołodziejczyk, and Keita Yokoyama *Some upper bounds on ordinal-valued Ramsey numbers for coloring of pairs*, arXiv 1807.00616.

[28] Manuel Lerman, Reed Soloman, and Henry Towner *Separating principles below Ramsey's Theorem for pairs*, Journal of Mathematical Logic 13 (2), 2013, 1350007.

[29] Jiayi Liu *RT^2_2 does not prove WKL_0*, The Journal of Symbolic Logic 77 (2), 2012, pp. 609–620.

[30] Lu Liu *Cone avoiding closed sets*, Transactions of the American Mathematical Society 367 (3), 2015, pp. 1609–1630.

[31] Benoit Monin, and Ludovic Patey SRT^2_2 *does not imply* COH *in ω-models*, arXiv 1905.08427.

[32] Michael E. Mytilinaios, and Theodore A. Slaman *Σ_2-collection and the infinite injury priority method*, The Journal of Symbolic Logic 53, 1988, pp. 212–221.

[33] Jeff B. Paris, and Leo A. Harrington *A mathematical incompleteness in Peano arithmetic*, in: John Barwise (ed.) Handbook of Mathematical Logic, North-Holland Publishing House, Amsterdam 1977, pp. 1133–1142.

[34] Jeff B. Paris, and Laurie A. S. Kirby *Σ_n-collection schemas in arithmetic*, in: Angus MacIntyre, Leszek Pacholski, and Jeff Paris (eds.) Logic Colloquium '77 (Conference Proceedings, Wrocław, 1977), North-Holland Publishing House, Amsterdam 1978, pp. 199–209 [*Studies in Logic and the Foundations of Mathematics 96*].

[35] Ludovic Patey *The strength of the tree theorem for pairs in reverse mathematics*, The Journal of Symbolic Logic 81 (4), 2016, pp. 1481–1499.

[36] Ludovic Patey, and Keita Yokoyama *The proof-theoretic strength of Ramsey's Theorem for pairs and two colors*, Advances in Mathematics 330, 2018, pp. 1034–1070.

[37] Frank Plumpton Ramsey *On a problem of formal logic*, Proceedings of the London Mathematical Society 30, 1930, pp. 264–286.

[38] David Seetapun, and Theodore A. Slaman *On the strength of Ramsey's theorem*, Notre Dame Journal of Formal Logic 36 (4), 1995, pp. 570–582.

[39] Richard A. Shore *On the jump of an α-recursively enumerable set*, Transactions of the American Mathematical Society 217, 1976, pp. 351–363.

[40] Stephen G. Simpson Subsystems of Second-order Arithmetic, Cambridge University Press, Cambridge UK 2009 (second edition) [*Perspectives in Logic*].

[41] Theodore A. Slaman *Σ_n-bounding and Δ_n-induction*, Proceedings of the American Mathematical Society 132 (8), 2004, pp. 2449–2456 (electronic).

[42] Wei Wang *Some logically weak Ramseyian theorems*, Advances in Mathematics 261 (2), 2014, pp. 1–25.

[43] Xiaokang Yu, and Stephen G. Simpson *Measure theory and weak König's lemma*, Archiv for Mathematical Logic 30, 1990, pp. 171–180.

Ramsey Theory on Infinite Structures and the Method of Strong Coding Trees

Natasha Dobrinen
University of Denver
Natasha.Dobrinen@du.edu

ABSTRACT This article discusses some recent trends in Ramsey theory on infinite structures. Trees and their Ramsey theory have been vital to these investigations. The main ideas behind the author's recent method of trees with coding nodes are presented, showing how they can be useful both for coding structures with forbidden configurations as well as those with none. Using forcing as a tool for finite searches has allowed the development of Ramsey theory on such trees, leading to solutions for finite big Ramsey degrees of Henson graphs as well as infinite dimensional Ramsey theory of copies of the Rado graph. Possible future directions for applications of these methods are discussed.

1 Introduction

Logic and Ramsey theory have a long interconnected history. In 1929, Ramsey proved his celebrated theorem in order to obtain a partial solution to Hilbert's Entscheidungsproblem. This problem, posed by Hilbert in 1928, asked for an algorithm which decides the validity of any statement of first-order logic. By Gödel's Completeness Theorem (1929), this is equivalent to asking for an algorithm which can decide whether a given formula is provable from the logical axioms. Ramsey applied his partition theorem on the natural numbers to show that the validity of formulas with only universal quantifiers in normal form is decidable [39]. In light of the Church-Turing thesis ([2] and [47]) which precludes any complete solution to Hilbert's Entscheidungsproblem, Ramsey's result is all the more striking in its success.

Over the decades, this interplay between Ramsey theory and logic has continued, each subject motivating, informing, and providing techniques to solve problems in

This work was supported by National Science Foundation Grant DMS-1600781

the other. An instance of this is seen in the 1966 result of Halpern and Läuchli [17]. While investigating the problem of whether or not the Boolean Prime Ideal Theorem (BPI) is strictly weaker than the Axiom of Choice (AC) over the Zermelo-Fraenkel Axioms (ZF), Halpern and Läuchli proved a theorem which was later interpreted to be a Ramsey theorem on products of finitely many trees. This theorem was central to the proof of Halpern and Lévy that, indeed, BPI is strictly weaker than AC over ZF [18]. In turn, the Halpern-Läuchli theorem provided means for proving Ramsey theorems for colorings of finite structures inside of infinite structures, such as the rationals as an ordered structure and the Rado graph.

Harrington later produced a proof of the Halpern-Läuchli Theorem using the set-theoretic method of forcing. His approach was novel in that the language and techniques of forcing are used to do unbounded searches for finite objects. This is in contrast to the more common use of forcing to obtain ZFC results by proving the existence of a Σ^1_2 definable object in a generic extension, and then applying Shoenfield absoluteness to deduce that the object must be in the ground model. In an interesting turn of events, Harrington's method of proof provided the backdrop for recent work of the author in [6], [5], and [4]. These results will be discussed in Section 3 and ideas of key methods developed to obtain these results will be set forth in Section 4.

This article discusses some recent trends and future directions in Ramsey theory on infinite structures. Section 2 recalls Ramsey theory on the natural numbers, and then reviews how these ideas extend to structures. Ties between Ramsey theory on structures and topological dynamics, due to Kechris, Pestov, and Todorcevic in [22] and a recent development due to Zucker in [49] provide additional motivation for these investigations. In Section 3, we present an overview of big Ramsey degrees and infinite dimensional Ramsey theory on infinite structures. As the focus here is Ramsey theory on infinite structures, and as the literature on finite structures is vast, we do not even attempt to do justice to Ramsey theory of finite structures in this article. Only a few of those results will be stated in order to provide the reader with some intuition.

The main tools used so far to obtain Ramsey results on infinite structures, aside from Ramsey's Theorem itself, have been Ramsey theorems on trees in the vein of Halpern and Läuchli, and very recently, category theory (see [29]). In Section 4, we present an overview of the recent method of trees with coding nodes, first developed by the author in [6] to prove finite big Ramsey degrees for the universal ultrahomogeneous triangle-free graph. This method was extended in [5] and [4] to determine Ramsey theory on all the Henson graphs as well as infinite dimensional Ramsey theory on the Rado graph. Forcing seemed the best first approach in the search for Ramsey theorems on these trees, and it turned out to work. An overview of some of the ideas involved in these proofs are provided. In Section 5 we point to future directions in Ramsey theory of infinite structures where these methods are likely to prove efficacious.

2 Finite and infinite dimensional Ramsey theory: from natural numbers to structures

Ramsey's theorem for the natural numbers is the following:

Theorem 2.1 (Infinite Ramsey Theorem, [39]). *Given integers $k, r \geq 1$ and a coloring $c : [\omega]^k \to r$, there is an infinite set $N \subseteq \omega$ such that $c \restriction [N]^k$ is constant.*

Here and throughout, we use the set-theoretic notation $[X]^k$ to denote the collection of all subsets of X with exactly k members. Using the Hungarian arrow notation, Theorem 2.1 is written as

$$\omega \to (\omega)^k_r. \tag{1}$$

The set N is said to be *monochromatic* or *homogeneous*. Using compactness, one obtains the finite version of Ramsey's Theorem.

Corollary 2.2 (Finite Ramsey Theorem, [39]). *Given integers $k, r, m \geq 1$ with $k \leq m$, there is an integer $n > m$ such that for any coloring $c : [n]^k \to r$, there is a subset $N \subseteq n$ of cardinality m such that $c \restriction [N]^k$ is constant.*

Theorem 2.1 and Corollary 2.2 are referred to as *finite dimensional Ramsey theory*, since the sets being colored in these theorems all have the same fixed finite size. *Infinite dimensional Ramsey theory* is concerned with coloring infinite sets of natural numbers. Although the Axiom of Choice implies there is a coloring of all infinite subsets of natural numbers into two colors for which there is no infinite monochromatic set (see Erdős-Rado [10]), for any coloring which induces a sufficiently definable partition of the Baire space, monochromatic subsets are to be found. In the context of Ramsey theory, the simplest representation of the Baire space is $[\omega]^\omega$, the set of all infinite subsets of ω, endowed with the Tychonoff topology. Nash-Williams initiated the study of infinite dimensional Ramsey theory in 1965, proving that for any clopen set $\mathcal{C}$ in the Baire space, there is some infinite set X such that $[X]^\omega$ is either contained in $\mathcal{C}$ or else is disjoint from it [32].

A few years later, Galvin and Prikry extended this to Borel sets in a strong way. The following notation is central to studies of infinite dimensional Ramsey theory: For $s \in [\omega]^\omega$ and $A \in [\omega]^\omega$, define

$$[s, A] = \{B \in [\omega]^\omega : s \sqsubset B \subseteq A\}, \tag{2}$$

where s is finite and $s \sqsubset B$ means that s is an initial segment of B, given their strictly increasing enumerations. We use the terminology of [45].

Definition 2.3. A subset $\mathcal{X} \subseteq [\omega]^\omega$ is *Ramsey* if for each non-empty set $[s, A]$, there is a $B \in [s, A]$ such that either $[s, B] \subseteq \mathcal{X}$ or else $[s, B] \cap \mathcal{X} = \emptyset$.

Theorem 2.4 (Galvin and Prikry, [13]). *Every Borel subset of the Baire space is Ramsey.*

Soon after this, Silver showed that analytic sets are Ramsey [44]. Mathias [30] and Louveau [28] attained infinite dimensional Ramsey results for the case where the infinite set comes from a Ramsey ultrafilter, a related and rich area which is not the focus of this article. The pinnacle of infinite dimensional Ramsey theory on the Baire space was achieved by Ellentuck in 1974, who provided a topological characterization of those sets which are Ramsey. The *Ellentuck topology* is the topology on $[\omega]^\omega$ generated by basic open sets of the form $[s, A]$. This topology refines the Tychonoff topology on the Baire space.

Theorem 2.5 (Ellentuck, [8]). *A subset $\mathcal{X} \subseteq [\omega]^\omega$ is Ramsey if and only if it has the property of Baire with respect to the Ellentuck topology.*

Such theorems can be notated by

$$\omega \to_* (\omega)^\omega_2, \tag{3}$$

where $\to_*$ denotes that the sets are partitioning $[\omega]^\omega$ are required to be definable in some sense with respect to some topology (see Section 11 of [22]).

Overlapping the development of infinite dimensional Ramsey theory on the Baire space, Ramsey theory on relational structures began to unfold. We review only the basics of Fraïssé theory for relational structures; more general background on Fraïssé theory can be found in Fraïssé's original paper [12], as well as [22]. We call $\mathcal{L} = \{R_i\}_{i\in I}$ a *relational signature* if is a (countable) collection of *relation symbols* R_i, where $n(i)$ denotes the *arity* of R_i, for each $i \in I$. A *structure* for $\mathcal{L}$ is a structure $\boldsymbol{A} = \langle |\boldsymbol{A}|, \{R_i^{\boldsymbol{A}}\}_{i\in I}\rangle$, where $|\boldsymbol{A}| \neq \emptyset$ is the *universe* of $\boldsymbol{A}$ and for each $i \in I$, $R_i^{\boldsymbol{A}} \subseteq A^{n(i)}$. An *embedding* between structures $\boldsymbol{A}, \boldsymbol{B}$ for $\mathcal{L}$ is an injection $\iota : |\boldsymbol{A}| \to |\boldsymbol{B}|$ such that for all $i \in I$, $R_i^{|\boldsymbol{A}|}(a_1, \dots, a_{n(i)}) \leftrightarrow R_i^{|\boldsymbol{B}|}(\iota(a_1), \dots, \iota(a_{n(i)}))$. If ι is the identity map, then we say that $\boldsymbol{A}$ is a *substructure* of $\boldsymbol{B}$. An *isomorphism* is an embedding which is onto it image. We write $\boldsymbol{A} \leq \boldsymbol{B}$ exactly when there is an embedding of $\boldsymbol{A}$ into $\boldsymbol{B}$; $\boldsymbol{A} \cong \boldsymbol{B}$ denotes that $\boldsymbol{A}$ and $\boldsymbol{B}$ are isomorphic.

A class $\mathcal{K}$ of finite structures for a relational signature $\mathcal{L}$ is called a *Fraïssé class* if it is hereditary, satisfies the joint embedding and amalgamation properties, contains (up to isomorphism) only countably many structures, and contains structures of arbitrarily large finite cardinality. These notions are recalled here for the reader's convenience. $\mathcal{K}$ is *hereditary* if whenever $\boldsymbol{B} \in \mathcal{K}$ and $\boldsymbol{A} \leq \boldsymbol{B}$, then also $\boldsymbol{A} \in \mathcal{K}$. $\mathcal{K}$ satisfies the *joint embedding property* if for any $\boldsymbol{A}, \boldsymbol{B} \in \mathcal{K}$, there is a $\boldsymbol{C} \in \mathcal{K}$ such that $\boldsymbol{A} \leq \boldsymbol{C}$ and $\boldsymbol{B} \leq \boldsymbol{C}$. $\mathcal{K}$ satisfies the *amalgamation property* if for any embeddings $f : \boldsymbol{A} \to \boldsymbol{B}$ and $g : \boldsymbol{A} \to \boldsymbol{C}$, with $\boldsymbol{A}, \boldsymbol{B}, \boldsymbol{C} \in \mathcal{K}$, there is a $\boldsymbol{D} \in \mathcal{K}$ and there are embeddings $r : \boldsymbol{B} \to \boldsymbol{D}$ and $s : \boldsymbol{C} \to \boldsymbol{D}$ such that $r \circ f = s \circ g$.

Let $\mathcal{K}$ be a Fraïssé class. For $\boldsymbol{A}, \boldsymbol{B} \in \mathcal{K}$ with $\boldsymbol{A} \leq \boldsymbol{B}$, we use $\binom{\boldsymbol{B}}{\boldsymbol{A}}$ to denote the set of all substructures of $\boldsymbol{B}$ which are isomorphic to $\boldsymbol{A}$. Given structures $\boldsymbol{A} \leq \boldsymbol{B} \leq \boldsymbol{C}$ in $\mathcal{K}$, we write

$$\boldsymbol{C} \to (\boldsymbol{B})^{\boldsymbol{A}}_k$$

to denote that for each coloring of $\binom{\boldsymbol{C}}{\boldsymbol{A}}$ into k colors, there is a $\boldsymbol{B}' \in \binom{\boldsymbol{C}}{\boldsymbol{B}}$ such that $\binom{\boldsymbol{B}'}{\boldsymbol{A}}$ is *monochromatic*, meaning every member of $\binom{\boldsymbol{B}'}{\boldsymbol{A}}$ has the same color.

Definition 2.6. A Fraïssé class $\mathcal{K}$ has the *Ramsey property* if for any two structures $\boldsymbol{A} \leq \boldsymbol{B}$ in $\mathcal{K}$ and any $k \geq 2$, there is a $\boldsymbol{C} \in \mathcal{K}$ with $\boldsymbol{B} \leq \boldsymbol{C}$ such that $\boldsymbol{C} \to (\boldsymbol{B})^{\boldsymbol{A}}_k$.

Investigations into which Fraïssé classes have the Ramsey property commenced when Graham and Rothschild proved in 1971 that the class of finite Boolean algebras has the Ramsey property [16]. Soon after this, Graham, Leeb, and Rothschild showed that the class of finite vector spaces over a finite field have the Ramsey property [14], [15]. Several years later, the class of finite ordered graphs were found to have the Ramsey property; this was proved independently by Abramson and Harrington [1] and by Nešetřil and Rödl [33], [34]. The main theorem of the papers [33] and [34] furthermore proved that all set-systems of finite ordered relational structures omitting some irreducible substructure have the Ramsey property. This includes the classes of finite ordered graphs omitting k-cliques, denoted $\mathcal{G}^<_k$, for each $k \geq 3$. Over the past several decades, more Fraïssé classes were shown to have the Ramsey property. The correspondence between the Ramsey property and extreme amenability, proved by Kechris, Pestov, and Todorcevic in 2005 in [22], has propelled a recent burst of discoveries of more Fraïssé classes with the Ramsey property.

It is interesting that while most Fraïssé classes of finite unordered structures do not have the Ramsey property, often equipping the class with an additional linear order produces the Ramsey property. In such cases, some remnant of the Ramsey property remains in the unordered reduct. Following notation in [22], given a Fraïssé class $\mathcal{K}$, for $\boldsymbol{A} \in \mathcal{K}$, $t(\boldsymbol{A}, \mathcal{K})$ denotes the smallest number t, if it exists, such that for each $\boldsymbol{B} \in \mathcal{K}$ with $\boldsymbol{A} \leq \boldsymbol{B}$ and for each $j \geq 2$, there is some $\boldsymbol{C} \in \mathcal{K}$ into which $\boldsymbol{B}$ embeds such that for any coloring $c : \binom{\boldsymbol{C}}{\boldsymbol{A}} \to j$, there is a $\boldsymbol{B}' \in \binom{\boldsymbol{C}}{\boldsymbol{B}}$ such that the restriction of c to $\binom{\boldsymbol{B}'}{\boldsymbol{A}}$ takes no more than t colors. In the arrow notation, this is written as

$$\boldsymbol{C} \to (\boldsymbol{B})^{\boldsymbol{A}}_{j,t(\boldsymbol{A},\mathcal{K})}. \tag{4}$$

A class $\mathcal{K}$ has *finite (small) Ramsey degrees* if for each $\mathrm{A} \in \mathcal{K}$ the number $t(\boldsymbol{A}, \mathcal{K})$ exists. The number $t(\boldsymbol{A}, \mathcal{K})$ is called the *Ramsey degree of* $\boldsymbol{A}$ in $\mathcal{K}$ [11]. Notice that $\mathcal{K}$ has the Ramsey property if and only if $t(\boldsymbol{A}, \mathcal{K}) = 1$ for each $\boldsymbol{A} \in \mathcal{K}$. The connection between Fraïssé classes with finite Ramsey degrees and ordered expansions was initiated in Section 10 of [22] and extended to precompact expansions in [36]. The existence of a Ramsey expansions were later shown to be equivalent to small Ramsey degrees in [48]. In particular, the Fraïssé classes of finite unordered graphs and finite unordered graphs omitting k-cliques have finite small Ramsey degrees.

At this point, it is natural to ask the following: Which infinite structures have properties similar Theorem 2.1? As it is often not possible to obtain one color for all copies of a given object, the following definition extends the notion of small Ramsey degrees (rather than Ramsey property) to infinite structures.

Definition 2.7 ([22]). Given an infinite structure $\mathcal{S}$ and a finite substructure $\boldsymbol{A} \leq \mathcal{S}$, let $T(\boldsymbol{A}, \mathcal{S})$ denote the least integer $T \geq 1$, if it exists, such that given any coloring of $\binom{\mathcal{S}}{\boldsymbol{A}}$ into finitely many colors, there is a substructure $\mathcal{S}'$ of $\mathcal{S}$, isomorphic to $\mathcal{S}$, such that $\binom{\mathcal{S}'}{\boldsymbol{A}}$ takes no more than T colors. This may be written succinctly as

$$\forall j \geq 1, \quad \mathcal{S} \to (\mathcal{S})^{\boldsymbol{A}}_{j,T(\boldsymbol{A},\mathcal{S})}. \tag{5}$$

We say that $\mathcal{S}$ has *finite big Ramsey degrees* if for each finite substructure $\boldsymbol{A} \leq \mathcal{S}$, there is an integer $T(\boldsymbol{A}, \mathcal{S})$ such that (5) holds.

By a theorem of Hjorth in [21], if $\mathcal{K}$ is a Fraïssé class with limit $\mathbb{F}$ such that $|\mathrm{Aut}(\mathbb{F})| > 1$, then there is some $\boldsymbol{A} \in \mathcal{K}$ whose big Ramsey degree is larger than one (possibly not finite). Thus any non-rigid Fraïssé structure cannot have an exact analogue of the Infinite Ramsey's Theorem, where all big Ramsey degrees are one.

Some authors prefer to color embeddings of a given $\boldsymbol{A}$ into $\mathbb{F}$. The relationship between them is simple: A structure $\boldsymbol{A} \in \mathcal{K}$ has (small or big) Ramsey degree ℓ for copies if and only if $\boldsymbol{A}$ has (small or big) Ramsey degree $\ell \cdot |\mathrm{Aut}(\boldsymbol{A})|$ for embeddings.

The following question has been investigated for several decades.

Question 2.8. Which infinite structures have finite big Ramsey degrees?

This question has been asked most often in regard to Fraïssé limits of Fraïssé classes of finite structures, which we shall call simply *Fraïssé structures.* A Fraïssé structure $\mathbb{F}$ for a Fraïssé class $\mathcal{K}$ is a countably infinite structure which is *universal* for $\mathcal{K}$ (each member of $\mathcal{K}$ embeds into $\mathbb{F}$) and *ultrahomogeneous* (any isomorphism between two finite substructures of $\mathbb{F}$ extends to an automorphism of $\mathbb{F}$). However, Question 2.8 also makes sense in the context of infinite structures which are universal for some $\mathcal{K}$ but not ultrahomogeneous, and we will see some recent progress in this direction as well in the next section.

While Question 2.8 has been of interest for many decades, it gained renewed traction in the early 2000's, both because of results on the big Ramsey degree of the Rado graph in [42] and [26], and because of Kechris, Pestov, and Todorcevic's question about finding an analogue of their correspondence for Fraïssé structures which have finite big Ramsey degrees (see Section 11 of [22]). Such an analogue was obtained by Zucker in [49]. Section 3 will give an overview of known results on big Ramsey degrees of infinite structures.

Another natural question is the following: Which infinite structures carry analogues of Theorems 2.4 or 2.5?

Question 2.9. For which infinite structures $\mathcal{S}$ does

$$\mathcal{S} \to_* (\mathcal{S})^{\mathcal{S}}_2 \tag{6}$$

hold, where $\to_*$ denotes that the partition is suitably definable, given some natural topology on the space $\binom{\mathcal{S}}{\mathcal{S}}$?

This question was brought to light in Problem 11.2 of [22], which asked for an analogue of the KPT correspondence for Fraïssé structures with some infinite dimensional Ramsey theory, while simultaneously pointing out that very little was known in this direction. In the recent paper [4], the author showed that for the collection of subcopies of the Rado graph with a certain tree-structural property, all Borel sets are Ramsey. This will be discussed in Section 4, where trees with coding nodes will be presented.

3 Big Ramsey degrees on infinite structures: An overview of previous results and methods

Ramsey theory on infinite structures seeks to find out which infinite structures carry analogues of Ramsey's Theorem. The first line of inquiry investigates which infinite structures are *indivisible*, meaning that given any partition of the universe of the structure into finitely many pieces, one of the pieces contains a copy of the structure. This is the same as saying (in the terminology of Definition 2.7) that substructures of size one have finite big Ramsey degree one.

It is straightforward to see that given a partition of the rationals into finitely many pieces, one of the pieces contains a copy of the rationals, that is, a dense linear order (without endpoints). Thus, the rationals as a linearly ordered structure is indivisible. The Rado graph, or random graph, is also indivisible. This was proved by Henson in [19], using the extension property of the Rado graph.

In contrast, proving the indivisibility of the generic k-clique-free graphs, denoted $\mathcal{H}_k$, required ideas beyond their extension properties. These *Henson graphs* were first constructed by Henson in 1971 in [19] as subgraphs of the Rado graph. In hindsight, these are seen to be Fraïssé limits of the Fraïssé classes of finite k-clique-free graphs. In [19], Henson proved that $\mathcal{H}_k$ is *weakly indivisible*: Given a partition of the vertices of $\mathcal{H}_k$ into two pieces, either the first piece of the partition contains a copy of $\mathcal{H}_k$, or else the second piece contains a copy of each finite k-clique-free graph. It is interesting to note that this was proved several years before Nešetřil and Rödl's result in [33] that the Fraïssé classes of ordered k-clique-free graphs have the Ramsey property. The first result on the indivisibility of Henson graphs was due to Komjáth and Rödl, in [23], where they proved that $\mathcal{H}_3$ is indivisible. Soon after, El-Zahar and Sauer extended this result to all Henson graphs in [7].

A sample of other notable results regarding indivisibility are the following: Hindman showed, as a consequence of his partition theorem for finite sums of natural numbers, that the vector space of countable dimension over the finite field $\mathbb{F}_2$ is indivisible [20]. However, a vector space of countable dimension over any other (nontrivial) finite field is not. A proof of this folklore theorem appears in [24], where Laflamme, Nguyen Van Thé, Pouzet, and Sauer prove that all such vector spaces retain a slightly weaker property called weak indivisibility. Nguyen Van Thé and Sauer proved that for each $m \geq 1$, the countable Urysohn metric space with distances in $\{1, \ldots, m\}$ is indivisible [37]. Later, Sauer showed that for any finite set of distances, the Urysohn space with that distance set is indivisible [43]. For more on indivisible structures, the reader is referred to the excellent Habilitation of Nguyen Van Thé [36].

While the study of indivisibility of infinite structures continues to be a rich and challenging subject, it is interesting that the nascence of the broader subject of big Ramsey degrees can actually be traced back to an example of Sierpiński. Considering the rationals as a linearly ordered structure, Sierpiński showed that there is a coloring of pairs of rationals into two colors such that any infinite subset which is again a dense linear order preserves both colors on its pairsets. His coloring plays the linear ordering of the rationals against a well-ordering of the rationals as follows, and colors pairs of rationals red if the two orders agree on the pair, and blue otherwise. Both colors persist in every subset of the rationals forming a dense linear order.

This phenomenon is perhaps best seen via trees. For $s, t \in 2^{<\omega}$, define $s \triangleleft t$ iff one of the following holds: (a) $s <_{\text{lex}} t$, (b) $s \sqsubset t$ and $t(|s|) = 1$, or (c) $t \sqsubset s$ and

$s(|t|) = 0$. Notice that $(2^{<\omega}, \lhd)$ is isomorphic to $(\mathbb{Q}, <)$. Define the coloring

$$c(\{s,t\}) = \begin{cases} 0 & \text{if } |s| \le |t| \text{ and } s \lhd t \\ 1 & \text{otherwise} \end{cases} \tag{7}$$

Then given any subset $S \subseteq 2^{<\omega}$ for which $(S, \lhd) \cong (\mathbb{Q}, <)$, both colors will persist in S; hence $T(2, \mathbb{Q}) \ge 2$.

In his PhD thesis [3], Devlin found the precise big Ramsey degrees for finite sets of rationals, building on prior unpublished work of Galvin, who proved that $T(2, \mathbb{Q}) = 2$, and of Laver, who proved existence of $T(k, \mathbb{Q})$ for all natural numbers k. Here, we will not provide much background or history, as there are already thorough expositions of Devlin's results in [46] and [45].

A key component in finding upper bounds for the numbers $T(k, \mathbb{Q})$ is a Ramsey theorem for trees, due to Milliken. This actually turns out to be at the heart of many big Ramsey degree results. Milliken's theorem utilizes the following notion of strong tree and a theorem of Halpern and Läuchli which we now briefly review. A historical record of the Halpern-Läuchli Theorem and its variants can be found in [27]. For $t \in \omega^{<\omega}$, $|t| = \mathrm{dom}(t)$ is the *length of* t. In the subject of Ramsey theory on trees, we say that a subset $T \subseteq \omega^{<\omega}$ is a *tree* if there is a subset $L \subseteq \omega$ such that $T = \{t \restriction l : t \in T,\ l \in L\}$. Thus, in this definition, a tree is closed under initial segments at levels of the tree, but it is not necessarily closed under all initial segments in $\omega^{<\omega}$. For $t \in T$, the *height of* t, $\mathrm{ht}_T(t)$, is the order-type of the set $\{u \in T : u \subset t\}$, linearly ordered by $\subseteq$. We write $T(n)$ to denote $\{t \in T : \mathrm{ht}_T(t) = n\}$. For $t \in T$, let $\mathrm{Succ}_T(t) = \{u \restriction (|t| + 1) : u \in T \text{ and } u \supset t\}$, noting that $\mathrm{Succ}_T(t)$ will not be a set of nodes in T if T does not contain any nodes of length $|t| + 1$. For $s, t \in T$ with $|s| < |t|$, the number $t(|s|)$ is called the *passing number* of t at s. Passing numbers are used to code information about the binary relations satisfied by the members of (the universe of) a given structure represented by s and t.

Definition 3.1. Let $T \subseteq \omega^{<\omega}$ be a finitely branching tree. A subset $S \subseteq T$ is a *strong subtree of* T if and only if there is an increasing sequence of natural numbers $\langle m_n : n < N \rangle$, where $N \le \omega$, such that $S = \bigcup_{n<N} S(n)$, and for each $n < N$,

1. $S(n) \subseteq T(m_n)$, and
2. $s \in S(n)$ and $u \in \mathrm{Succ}_T(s)$, there is exactly one node in $S(n+1)$ extending u.

Given $k \ge 1$, we say that S is a *k-strong subtree* of T if $N = k < \omega$.

For the next theorem, define the notation:

$$\bigotimes_{i<d} T_i := \bigcup_{n<\omega} \prod_{i<d} T_i(n). \tag{8}$$

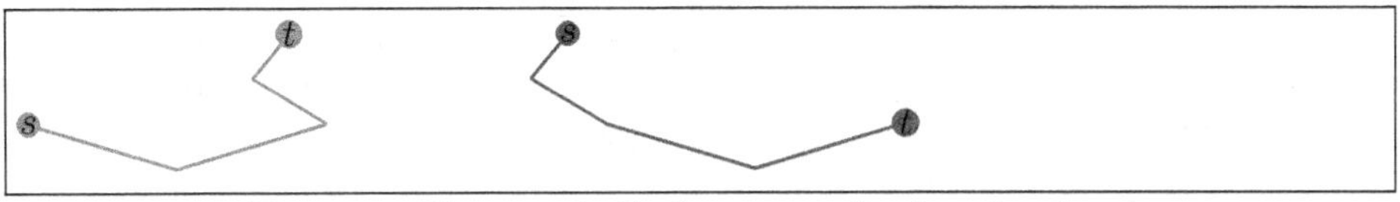

Figure 1: The two similarity types of pairs of rationals

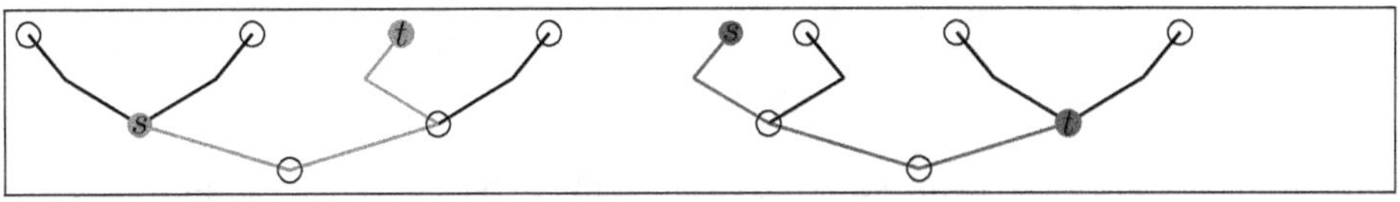

Figure 2: Strong tree envelopes for these pairs of rationals

The following is the strong tree version of the Halpern-Läuchli Theorem.

Theorem 3.2 (Halpern-Läuchli, [17]). *Let $T_i \subseteq \omega^{<\omega}$, $i < d$, be finitely branching trees with no terminal nodes and let $r \geq 2$. Given a coloring $c : \bigotimes_{i<d} T_i \to r$, there is an increasing sequence $\langle m_n : n < \omega \rangle$ and strong subtrees $S_i \leq T_i$ such that for all $i < d$ and $n < \omega$, $S_i(n) \subseteq T_i(m_n)$, and c is constant on $\bigotimes_{i<d} S_i$.*

The following theorem is proved by inductively applying Theorem 3.2 ω-many times.

Theorem 3.3 (Milliken, [31]). *Let $T \subseteq \omega^{<\omega}$ be a strong tree with no maximal nodes. Let $k \geq 1$, $r \geq 2$, and c be a coloring of all k-strong subtrees of T into r colors. Then there is a strong subtree $S \subseteq T$ such that all k-strong subtrees of S have the same color.*

Trees can be used in various ways to code structures. As mentioned above, the nodes in the tree $2^{<\omega}$ ordered by $\triangleleft$ produces a copy of the rationals. In this setting, Sierpiński's coloring can be seen structurally as giving color red to pairs of nodes $s, t \in 2^{<\omega}$ with $|s| \leq |t|$ and $s \triangleleft t$, and color blue otherwise. Figure 1. gives pairs of nodes which are colored red (left) and blue (right). These two configurations are examples of *strong similarity types*, and both of them persist in any subcopy of the rationals. Figure 2. gives examples of strong trees which *envelope* such pairs of nodes.

The proof that $T(2, \mathbb{Q}) \leq 2$ goes roughly as follows: Consider pairs of nodes s, t of different lengths such that $s(|t|) = 0$ if $|s| > |t|$, and $t(|s|) = 0$ if $|t| > |s|$. The two *strong similarity types* for such pairs are seen in Figure 1. Given a coloring of pairs of nodes in $2^{<\omega}$ into finitely many colors, fix the strong similarity type on the left in Figure 1. For each such pair $\{s, t\}$ in $2^{<\omega}$ with this strong similarity type, there are finitely many 3-strong trees enveloping s and t. Give these 3-strong trees the coloring of the pair which they envelope. Apply Milliken's theorem to this coloring of 3-strong trees to obtain an infinite strong subtree T where all pairs with this strong similarity type have the same color. Then do the same for the second strong similarity type, obtaining an infinite strong subtree T' of T. Since the rationals are coded in any infinite strong subtree of $2^{<\omega}$, one can now pull out an antichain of nodes, each of which has passing number 0 at the levels of the other nodes, so that under the ordering $\triangleleft$, the nodes in this antichain represent $\mathbb{Q}$. This produces a copy of the rationals in which all pairsets have at most two colors.

A similar but more involved strategy is behind the finite big Ramsey degrees of the Rado graph. For graphs, by interpreting the lexicographic order on ordered graphs due to Erdős, Hajnal, and Pósa in [9] in terms of trees, Sauer showed in [42] that nodes in the binary tree can code graphs. Let $\mathbf{A}$ be a graph. Enumerate the vertices of $\mathbf{A}$ as $\langle v_n : n < N\rangle$. A set of nodes $\{t_n : n < N\}$ in $2^{<\omega}$ codes $\mathbf{A}$ if and only if for each pair $m < n < N$,

$$v_n \; E \; v_m \Leftrightarrow t_n(|t_m|) = 1. \tag{9}$$

The number $t_n(|t_m|)$ is called the *passing number* of t_n at t_m. See Figure 3. where the nodes t_0, t_1, t_2 in the tree code the graph v_0, v_1, v_2, which is a path of length two.

The Rado graph, $\mathcal{R} = (R, E)$, is the Fraïssé limit of the class of all finite graphs; as such, it is ultrahomogeneous and universal for all finite graphs. Erdős, Hajnal, and Pósa launched the investigation of finite big Ramsey degrees of the Rado graph in 1975, when they showed that there is a 2-coloring of edges such that every subcopy of the Rado graph retains both colors [9], reminiscent of Sierpiński's result for pairs of rationals. Two decades later, Pouzet and Sauer proved that for each coloring of edges of $\mathcal{R}$ into finitely many colors, there is a subgraph $\mathcal{R}'$ isomorphic to $\mathcal{R}$ in which the edges have at most two colors [38]. In 2006, papers of Sauer [42] and of Laflamme, Sauer, and Vuksanovic [26] combined to the exact big Ramsey degrees for all finite graphs within the Rado graph. In fact, these papers find big Ramsey degrees for a collection of binary relational structures, including the random tournament. The strategy is similar to that outlined above for the rationals, but now the passing numbers at nodes code the edge/non-edge relation.

In [22], Kechris, Pestov, and Todorcevic asked, Which structures have finite big Ramsey degrees? In tandem with the results in [42] and [26], this sparked a new wave of interest in big Ramsey degrees of Fraïssé structures. As part of his PhD work, Nguyen Van Thé proved that the countable ultrametric Urysohn space with any finite distance set has finite big Ramsey degrees [35]. While this result used Ramsey's theorem, the next result required a new extended version of Milliken's Theorem. Laflamme, Nguyen Van Thé, and Sauer proved in [25] that the structures

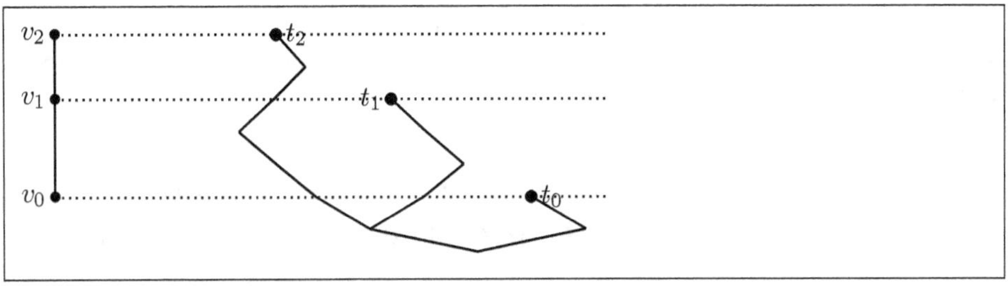

Figure 3: A path of length two coded by the nodes t_0, t_1, t_2

$\mathbb{Q}_n$ have finite big Ramsey degrees, where $\mathbb{Q}_n$ is the rationals as a linear order with an equivalence relation with n equivalence classes, each of which is dense in $\mathbb{Q}$. See [36] for an excellent exposition of these and related results.

As for the Henson graphs, Sauer proved in 1998 that the big Ramsey degree for edges in the triangle-free Henson graph is two [40]. Further results were slow in coming, mainly because of lack of analogues for k-clique free graphs of the following two fortunate facts related to the Rado graph. First, the graph $\mathcal{B}$ induced by the countable binary tree $2^{<\omega}$ is universal for countable graphs. Precisely, let the vertices of $\mathcal{B}$ be the nodes in the tree $2^{<\omega}$. Define two nodes to have an edge between them in $\mathcal{B}$ precisely when one of the nodes is longer than the other and its passing number at the level of the shorter node is one. Second, Milliken's theorem for strong subtrees of $2^{<\omega}$ can be applied to subcopies of $\mathcal{B}$, and each infinite strong subtree again codes a graph which is universal for countable graphs. After finitely many applications of Milliken's theorem to strong tree envelopes of finite antichains coding copies of a given finite graph, one can take a copy of the Rado graph where the big Ramsey degree of the finite graph under investigation is finite.

As analogues of these two facts were unknown for Henson graphs, the two main themes in the work of [6] and [5] were, first, to find means for coding the Henson graphs into trees in a way that the trees behaved like strong trees, and second, to prove Ramsey theory for such trees. In the next section, we discuss the methods developed in these two papers to handle forbidden k-cliques. These methods seem to be robust enough to handle many types of structures, including, perhaps surprisingly, infinite dimensional Ramsey theory of the Rado graph (see [4]), which will also be discussed in the next section.

We close this section by pointing out the recent work of Mašulović in [29] which uses category theory to develop transfer principals for big Ramsey degrees. In that paper, Mašulović widened the investigation of big Ramsey degrees to universal structures, regardless of ultrahomogeneity. He has transferred several of the aforementioned results to prove finite big Ramsey degrees for some new structures; most strikingly, these include Fraïssé limits of classes of finite metric spaces with finite distance sets satisfying certain properties.

4 Trees with coding nodes, their Ramsey theory, and applications to Ramsey theory of infinite graphs

In order to investigate big Ramsey degrees of the Henson graphs, the first task was to find some way of representing k-clique-free graphs via trees. Since k-cliques are forbidden, there needs to be some method for determining which nodes should be allowed to split and which ones should not, so that $\mathcal{H}_k$ is represented while the splitting is 'maximal' in some sense. In particular, the trees need to be perfect in order for Ramsey theory to have any chance of development.

To achieve this, in the construction of such trees, certain nodes are distinguished to code certain vertices of $\mathcal{H}_k$. This way, one keeps track of the finite graph that is already coded, ensuring that one knows how to branch maximally, subject to never coding a k-clique in the future. This led to the following notions of trees with coding nodes and strong coding trees.

4.1 Strong coding trees

The following definitions and theorems are taken from [6] and [5]. A *tree with coding nodes* is a structure $\langle T, N; \subseteq, <, c^T\rangle$ in the language $\mathcal{L} = \{\subseteq, <, c\}$, where $\subseteq$ and $<$ are binary relation symbols and c is a unary function symbol satisfying the following: $T \subseteq 2^{<\omega}$ and $(T, \subseteq)$ is a tree. $N \leq \omega$ and $<$ is the standard linear order on N. $c^T : N \to T$ is injective, and $m < n < N \longrightarrow |c^T(m)| < |c^T(n)|$. $c^T(n)$ is the *n-th coding node in T*, and is usually denoted c^T_n.

Notice that a collection of coding nodes $\{c^T_{n_i} : i < k\}$ in T *codes a k-clique* if and only if $c^T_{n_j}(|c^T_{n_i}|) = 1$ for all $i < j < k$. To ensure that a tree never codes a k-clique, the following branching criterion is introduced. We say that a tree T with coding nodes $\langle c^T_n : n < N\rangle$ satisfies the *K_k-Free Branching Criterion (k-FBC)* if for each non-maximal node $t \in T$, $t^\frown 0$ is always in T, and $t^\frown 1$ is in T if and only if any coding node extending $t^\frown 1$ cannot code a k-clique with coding nodes in T of length less than or equal to the length of t. It is a useful fact that given any tree T with coding nodes and no maximal nodes satisfying the k-FBC, and in which the set of coding nodes are dense, the coding nodes in T code $\mathcal{H}_k$ (Theorem 4.9, [5]).

The trees $\mathbb{S}_3$ and $\mathbb{S}_4$ in Figures 4. and 5. have coding nodes which code Henson graphs $\mathcal{H}_3$ and $\mathcal{H}_4$, respectively. These trees feature the main structural ideas behind

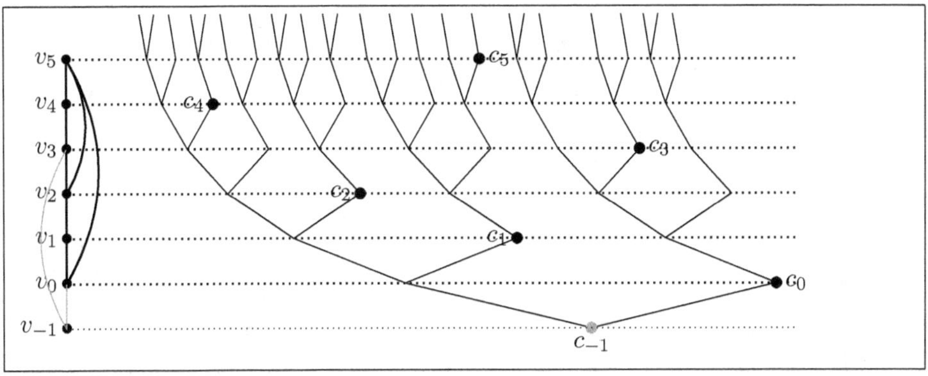

Figure 4: A strong triangle-free tree $\mathbb{S}_3$ densely coding $\mathcal{H}_3$

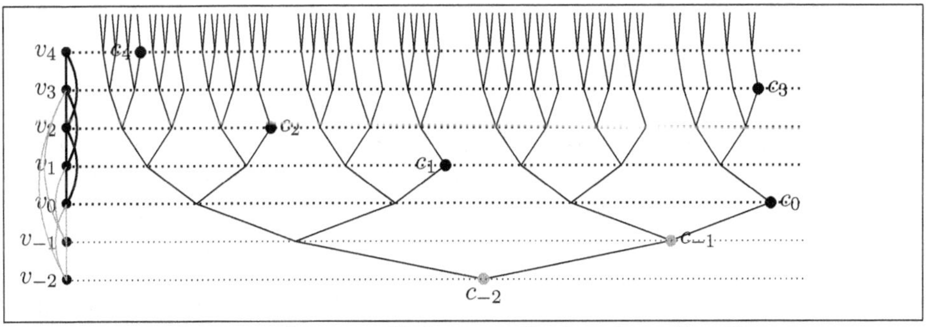

Figure 5: A strong K_4-free tree $\mathbb{S}_4$ densely coding $\mathcal{H}_4$

strong coding trees; one can extrapolate to envisage $\mathbb{S}_k$, for any $k \geq 5$ (precise constructions are given in [5]). The gray nodes $c_{-(k-2)}, \dots, c_{-1}$ are pseudo-coding nodes, which code a $(k-1)$-clique. They help to set up the tree structure so that subtrees of $\mathbb{S}_k$ coding $\mathcal{H}_k$ can be isomorphic to $\mathbb{S}_k$. The vertex v_0 is to be thought of as forming a $(k-1)$-clique with some vertices in a larger ambient copy of $\mathcal{H}_k$. This has the effect that each coding node in $\mathbb{S}_k$ does not split.

Perhaps the most illustrative way of thinking of these trees is the following: The nodes in the tree the level of c_n in the tree are coding all finite partial types over the graph on vertices $\{v_{-(k-2)}, \dots, v_{n-1}\}$. In this way, a strong coding tree is really just a means for visualizing the finite partial types over an (ordered) initial segment of the graph $\mathcal{H}_k$.

The one small but insurmountable catch to these trees is that having coding nodes and splitting nodes at the same levels prevents the development of Ramsey

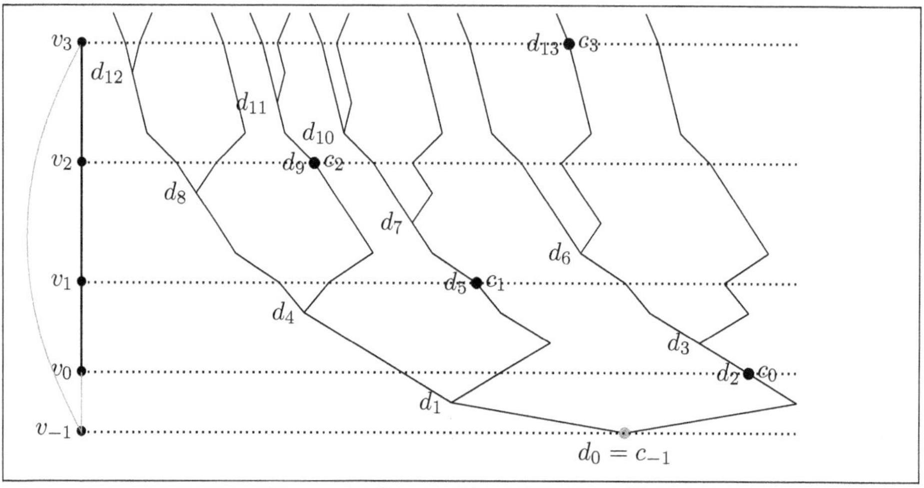

Figure 6: Strong $\mathcal{H}_3$-coding tree $\mathbb{T}_3$

theory on subtrees isomorphic to $\mathbb{S}_k$. Ironically, the failure occurs at the most basic level: There are bad colorings of coding nodes for which no subtree isomorphic to $\mathbb{S}_k$ has one color (see Example 3.18 in [6]). There turns out to be a simple solution: Skew the trees so that each level of the tree has at most one coding node or splitting node, but never both. Let $\mathbb{T}_k$ denote the skewed version of $\mathbb{S}_k$ (see Figure 6. for $\mathbb{T}_3$).

We work with a collection $\mathcal{T}_k$ of infinite subtrees of $\mathbb{T}_k$, each of which is isomorphic to $\mathbb{T}_k$ in a strong sense: Let $k \geq 3$ be given and let $S, T \subseteq \mathbb{T}_k$ be meet-closed subsets. A bijection $f : S \to T$ is a *strong similarity map* if for all nodes $s, t, u, v \in S$, the following hold:

1. f preserves lexicographic order.
2. f preserves meets, and hence splitting nodes.
3. f preserves relative lengths.
4. f preserves initial segments.
5. f preserves coding nodes.
6. f preserves passing numbers at coding nodes.

Given a subtree $T \subseteq \mathbb{T}_k$, let G_T denote the graph represented by the coding nodes in T. Notice that if $T \subseteq \mathbb{T}_k$ is strongly similar to $\mathbb{T}_k$, then G_T is isomorphic to $\mathcal{H}_k$ as ordered graphs.

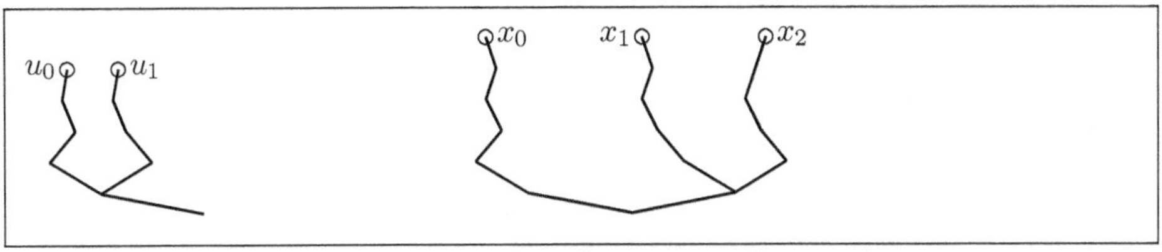

Figure 7: Two examples of unwitnessed pre-3-cliques

Essentially, a strong coding tree is a subtree of $\mathbb{T}_k$ which is strongly similar to $\mathbb{T}_k$. We say essentially, because there is one more important consideration when working with forbidden k-cliques. Any finite subtree of $\mathbb{T}_k$ which we build needs to be extendable within $\mathbb{T}_k$ to a subtree which is strongly similar to $\mathbb{T}_k$. There are many finite subtrees of $\mathbb{T}_k$ for which this is not possible, because each remembers where it came from, coding edges with the original graph represented by the coding nodes in $\mathbb{T}_k$.

Take for example $k = 3$. If A is a finite subtree of $\mathbb{T}_3$ and two nodes $s, t \in A$ have passing number 1 at some coding node c_i in $\mathbb{T}_3$, then whenever s is extended to some coding node c_m in $\mathbb{T}_3$, then any extension of t to a coding node c_n in $\mathbb{T}_3$ with length greater than $|c_m|$ cannot have passing number 1 at c_m, as that would code a triangle; no coding of a triangle occurs in $\mathbb{T}_3$. In such a situation, there is no way to extend A to a subtree of $\mathbb{T}_3$ coding all of $\mathcal{H}_3$.

To prevent this, we make a further requirement which is, roughly, as follows: Fix $a \in [3, k]$. We say that a level set $X \subseteq \mathbb{T}_k$ with nodes of length ℓ_X *has a pre-a-clique* if for some $\mathcal{I} \subseteq [\omega]^{a-2}$, letting $i_* = \max(\mathcal{I})$ and $\ell_* = |c_{i_*}|$, we have that $\ell_* \leq \ell_X$, the set $\{c_i : i \in \mathcal{I}\}$ codes an $(a-2)$-clique, and each node in X^+ has passing number 1 at c_i, for each $i \in \mathcal{I}$. The idea is that pre-a-cliques code entanglements. Essentially, we say that a subtree T of $\mathbb{T}_k$ has the *Witnessing Property* if for each pre-a-clique in T, $a \in [3, k]$, there is a set of $(a-2)$ many coding nodes $\{c_i : i \in \mathcal{I}\}$ as above, all of which are coding nodes in T. A tree $T \subseteq \mathbb{T}_k$ is a member of $\mathcal{T}_k$ iff T is strongly similar to $\mathbb{T}_k$ and has the Witnessing Property (Lemma 5.15, [5]). Some examples of unwitnessed and witnessed pre-cliques are in Figures 7 – 10.

Essentially, we define the space of *strong coding trees*, $\mathcal{T}_k$, to consist of those subtrees of $\mathbb{T}_k$ which are strongly similar to $\mathbb{T}_k$ and have the Witnessing Property. For the details, see [6] and [5]. This set of subtrees of $\mathcal{T}_k$ is the analogue of strong trees appropriate to the Henson graph $\mathcal{H}_k$

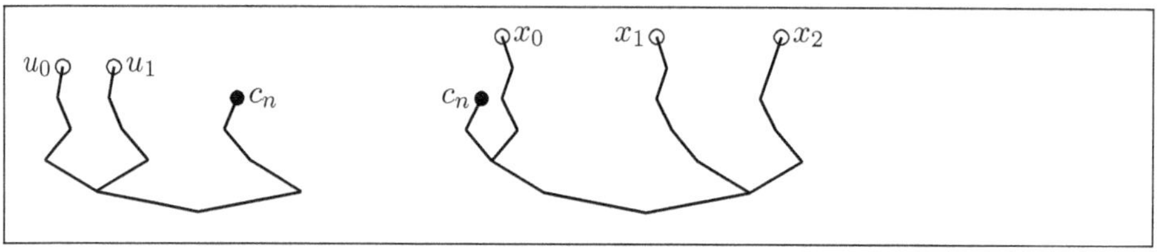

Figure 8: Two examples of witnessed pre-3-cliques

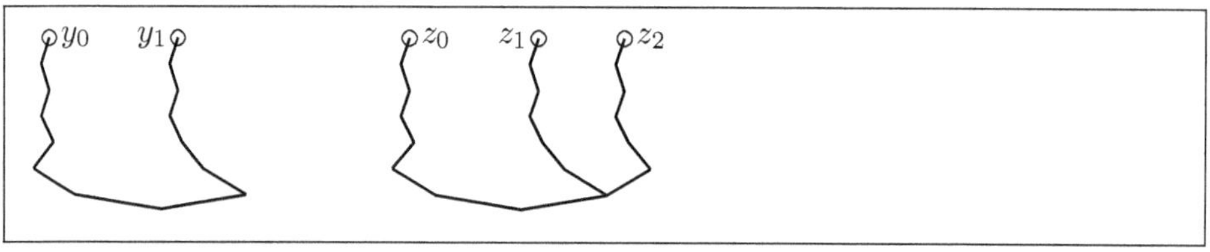

Figure 9: Two examples of unwitnessed pre-4-cliques

4.2 A Ramsey theorem for strictly similar antichains

The upper bounds for big Ramsey degrees of the Henson graphs basically come from the fact that there are only finitely many ways to code a given finite graph within a strong coding tree. Analogously to the case for the Rado graph, given any strong coding tree $T \in \mathcal{T}_k$, there is an antichain of coding nodes in T which code the Henson graph $\mathcal{H}_k$. Thus, one can restrict attention to colorings of antichains representing a given finite k-clique-free graph, say G. The relevant structural properties are those of strong similarity and new pre-cliques and their placement within the antichain. This is the idea behind *strict similarity*. For the precise definition, the reader is referred to [6] and [5].

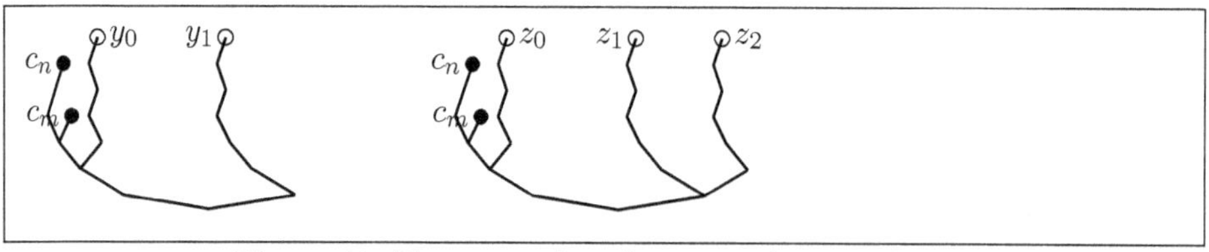

Figure 10: Two examples of witnessed pre-4-cliques

Theorem 4.1 ([6],[5]). *Let Z be a finite antichain of coding nodes in a strong $\mathcal{H}_k$-coding tree $T \in \mathcal{T}_k$, and suppose h colors of all antichains T which are strictly similar to Z into finitely many colors. Then there is an strong $\mathcal{H}_k$-coding tree $S \leq T$ such that all subsets of S strictly similar to Z have the same h color.*

The proof uses forcing to obtain a ZFC result, but not in the usual manner using absoluteness. Recall the Halpern-Läuchli Theorem 3.2. Harrington gave an insightful proof which uses Cohen forcing in the following way. Suppose we have d many infinite strong trees T_i and a coloring of the product of their level sets as in the statement of Theorem 3.2. Let κ be large enough that $\kappa \rightarrow (\aleph_1)^{2d}_{\aleph_0}$ and take $\mathbb{P}$ to be κ-Cohen forcing which adds κ new branches to each of the d many trees. Harrington gave an argument guaranteeing that there are nodes $t_i^* \in T_i$, for each $i < d$, of the same length which have some color $\varepsilon \in r$; moreover, given any level sets extending these nodes, there are further extensions to level sets so that each member in their product has the same color ε. The forcing is used to find these finite sets successively in ω many steps; the generic filter is never actually used - one never actually passes to a generic extension. Instead, the forcing language and basic facts about forcing guarantee that certain finite level sets exist, and their finiteness guarantees that they are actually in the ground model. So, this is very much not a constructive proof, but at the same time, it is a ZFC proof where one constructs each level of the subtree separately.

The author extended this idea to the context strong coding trees. However, since strong coding trees have two types of nodes, coding and non-coding, new forcings which are not equivalent to κ-Cohen forcing had to be introduced. The general approach uses Harrington's ideas as a starting point, but the implementation is much more involved. Another new element in this setting is that envelopes will be antichains of coding nodes which have a strong version of the Witnessing Property. This is quite different from envelopes for the rationals or the Rado graph being finite strong trees.

Theorem 4.1 is applied as follows to prove the finite big Ramsey degrees. Given a finite k-clique-free graph G, there are only finitely many strict similarity types of antichains of coding nodes representing G. The number of such strict similarity types is the upper bound for the big Ramsey degree of G in $\mathcal{H}_k$.

Theorem 4.2 ([6],[5]). *Suppose $k \geq 3$ and G is a finite k-clique-free graph. Let h color all copies of G in $\mathcal{H}_k$ into finitely many colors. Then there is a subgraph of $\mathcal{H}_k$ which is isomorphic to $\mathcal{H}_k$ in which the copies of G take on no more colors than the number of strict similarity types of antichains in $\mathbb{T}_k$ coding G.*

4.3 Infinite dimensional Ramsey theory of the Rado graph

We now mention a recent result of the author on infinite dimensional Ramsey theory of the Rado graph. In Problem 11.2 of [22], Kechris, Pestov, and Todorcevic ask for the topological dynamics analogue of a corresponding infinite Ramsey-theoretic result for several Fraïssé structures, in particular, the rationals, the Rado graph, and the Henson graphs. By an infinite Ramsey-theoretic result, they mean a result of the form

$$\mathbb{F} \to_* (\mathbb{F})^{\mathbb{F}}_{l,t}, \tag{10}$$

where equation (10) reads: "For each partition of $\binom{\mathbb{F}}{\mathbb{F}}$ into l many definable subsets, there is an $\mathbf{F} \in \binom{\mathbb{F}}{\mathbb{F}}$ such that $\binom{\mathbf{F}}{\mathbb{F}}$ is contained in no more than t of the pieces of the partition." Here, one assumes a natural topology on $\binom{\mathbb{F}}{\mathbb{F}}$ and *definable* refers to any reasonable class of sets definable relative to the topology, for instance, open, Borel, analytic, or property of Baire. A sub-question implicit in Problem 11.2 in [22] is the following broader version of Question 2.9:

Question 4.3. For which ultrahomogeneous structures $\mathbb{F}$ does it hold that

$$\mathbb{F} \to_* (\mathbb{F})^{\mathbb{F}}_{l,t}, \tag{11}$$

for some positive integer t?

The natural topology to give such a space is the one induced by ordering the universe F of $\mathbb{F}$ in order-type ω, and viewing $\binom{\mathbb{F}}{\mathbb{F}}$ as a subspace of the product space 2^F with the Tychonoff topology. Kechris, Pestov, and Todorcevic pointed out that very little is known about Question 4.3.

In [4], the author set out to answer this question for the Rado graph. Since the big Ramsey degrees for copies of a finite graph inside the Rado graph grow without bound as the number of vertices in the finite graph whose copies are being colored grows implies that any positive answer to Question 4.3 for the Rado graph must restrict to a collection of Rado graphs all of whose vertices are ordered in the same order. Furthermore, it is necessary that all copies of the Rado graph being colored have the same strong similarity type. Otherwise, one may use strong similarity types to make a coloring of the copies of the Rado graph to show that there is no bound t of the sort in (10), where $\mathbb{F}$ is the Rado graph.

In [4], the author answered Question 4.3 for a collection of Rado graphs, each of which has the same strong similarity type. While the Rado graph can be represented by nodes in strong trees of the kind in the Milliken Theorem 3.3, that theorem by itself does not answer this question, as it is unclear in a tree without coding nodes how the strong subtrees should be thought of as coding subcopies of a given Rado

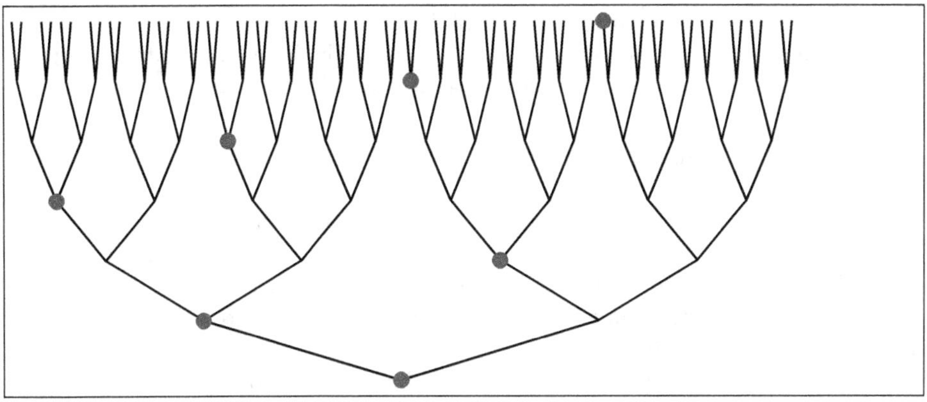

Figure 11: The Strong Rado Coding Tree $\mathbb{T}_{\mathbb{R}}$

graph. In order to make sure that the representations of the subgraphs were concrete, it turned out to be useful to work with trees with coding nodes, even though the Rado graph has no forbidden subgraphs.

Let $\mathbb{R} = (R, E)$ denote the Rado graph with vertices ordered as $\langle v_n : n < \omega \rangle$ represented by the coding nodes $\langle c_n : n < \omega \rangle$ in the tree $\mathbb{T}_{\mathbb{R}}$ in Figure 11. Let $\mathcal{T}_{\mathbb{R}}$ denote the collection of all subtrees $T \subseteq \mathbb{T}_{\mathbb{R}}$ strongly similar to $\mathbb{T}_{\mathbb{R}}$. Given $T \in \mathcal{T}_{\mathbb{R}}$, let G_T denote the ordered graph represented by the coding nodes of T, noting that each G_T is a subcopy of the Rado graph, ordered in the same way as $\mathbb{R}$. Let $\mathcal{R}$ be the collection of all G_T, where $T \in \mathcal{T}_{\mathbb{R}}$. The topology on $\mathcal{R}$ is the topology inherited from the Tychonoff topology on 2^R.

Theorem 4.4 ([4]). *If $\mathcal{X} \subseteq \mathcal{R}$ is Borel, then for each graph $\mathbb{G} \in \mathcal{R}$, either all members of $\mathcal{R}$ contained in $\mathbb{G}$ are members of $\mathcal{X}$, or else no member of $\mathcal{R}$ contained in $\mathbb{G}$ is a member of $\mathcal{X}$.*

A few remarks about this theorem are in order. First, we point out that Theorem 4.4 is the analogue of the Galvin-Prikry Theorem 2.4 for colorings of those copies of the Rado graph which have induced trees strongly similar to $\mathbb{T}_{\mathbb{R}}$. Second, the methods of proof use forcing in a similar yet simpler way than that in [6]. However, these methods were not conducive to obtaining an analogue of the Ellentuck Theorem 2.5. So it remains open whether or not, with the natural Ellentuck-like topology on $\mathcal{R}$, the subsets of $\mathcal{R}$ with the property of Baire have the Ramsey property.

Third, the strong coding tree $\mathbb{T}_{\mathbb{R}}$ is the simplest kind of tree coding $\mathbb{R}$. One could, however, fix any perfect tree (skew or not), say $\mathbb{T}'$, with coding nodes which are dense and code the Rado graph, and the same methods would produce the same result for those subcopies of the Rado graph induced by the coding nodes of subtrees of $\mathbb{T}'$ which are strongly similar to $\mathbb{T}'$. Lastly, the methods should produce a similar theorem for any of the binary relational structures in [42] and [26].

5 Future directions

As seen in the previous section, trees with coding nodes have provided means for solving problems on big Ramsey degrees for Henson graphs as well as infinite dimensional Ramsey theory for the Rado graph, which has no forbidden subgraphs. We envision several directions in which this idea can be developed to solve questions on Ramsey theory of infinite structures.

It is likely that trees with coding nodes will be useful in proving finite big Ramsey degrees for all Fraïssé limits of Fraïssé classes with finitely many binary relations and a finite constraint set satisfying the Ramsey property. Indeed, it may be enough to require finite small Ramsey degrees. Let $\mathcal{K}$ be a Fraïssé class in a finite relational language $\mathcal{L}$, where all relations are finitary. Let us call a set $\mathbf{F}$ of finite structures in $\mathcal{L}$ a *constraint set* for $\mathcal{K}$ if for any finite structure $\boldsymbol{A}$ in the language $\mathcal{L}$, $\boldsymbol{A} \in \mathcal{K}$ if and only if no induced substructure of $\boldsymbol{A}$ is in $\mathbf{F}$. In [41], Sauer gave a detailed characterization of the big Ramsey degrees of vertices in transitive free amalgamation structures in finite binary languages and provided some explicit examples of ultrahomogeneous directed graphs where the vertices do not have finite big Ramsey degrees. By the Nešetřil-Rödl Theorem, one can consider the linearly ordered versions of these classes to obtain Ramsey classes whose vertices do not have finite big Ramsey degrees.

At the conference, *Unifying Themes in Ramsey Theory* in Banff, 2018, Sauer suggested trying to move the forcing proofs from strong coding trees to a direct approach by forcing directly on the structures. Recall that the nodes in a strong coding tree represent a finite partial 1-type over a finite substructure; that is, the trees are really just a means for visualizing or making a structure out of the types. One important task is to interpret the forcings in the work related in Section 4 back into the graph setting in a way that points to the natural analogues of strong coding trees for structures with relations of any finite arity. Such an approach can hopefully lead to proving the following conjecture.

Conjecture 5.1. Let $\mathcal{K}$ be a relational Fraïssé class with finitely many relations and a finite constraint set. Suppose that $\mathcal{K}$ satisfies the Ramsey property (or has finite small Ramsey degrees). Then the Fraïssé limit of $\mathcal{K}$ has finite big Ramsey degrees.

The methods for the infinite dimensional Ramsey theory of the Rado graph are a simplified version of those developed for the big Ramsey degrees of the Henson graphs. It seems to be just a matter of double checking to see that the work in [5] will induce infinite dimensional Ramsey theory for Borel sets in any class of k-clique-free Henson graphs. It will be interesting to see if some other methods can produce analogues of the Ellentuck Theorem for these graphs, or whether there is some essential property of these graphs which prevent this. It seems likely that whatever structures have finite big Ramsey degrees will also have infinite dimensional Ramsey theorems.

Conjecture 5.2. Let $\mathcal{K}$ be a relational Fraïssé class with finitely many relations and a finite constraint set. Suppose that $\mathcal{K}$ satisfies the Ramsey property (or has finite small Ramsey degrees). Let $\mathbb{K}$ be the Fraïssé limit of $\mathcal{K}$ with some linear order of its universe in order-type ω. Then there is some notion of strong similarity type so that the collection of all subcopies of $\mathbb{K}$ having the same strong similarity type has the property that all Borel subsets are Ramsey.

References

[1] F. G. Abramson, and L. A. Harrington *Models without indiscernibles*, The Journal of Symbolic Logic 43 (3), 1978, pp. 572–600.

[2] Alonzo Church *An unsolvable problem of elementary number theory*, American Journal of Mathematics 58 (2), 1936, pp. 345–363.

[3] Dennis Devlin Some Partition Theorems for Ultrafilters on ω, PhD Dissertation, Dartmouth College, 1979.

[4] Natasha Dobrinen *Borel sets of Rado graphs and Ramsey's theorem*, 25 pp. arXiv:1904.00266v1 [submitted].

[5] — *Ramsey theory of the universal homogeneous k-clique-free graph*, 75 pp. arXiv:1901.06660 [submitted].

[6] — *The Ramsey theory of the universal homogeneous triangle-free graph*, Journal of Mathematical Logic, 75 pp. [to appear].

[7] Mohamed El-Zahar, and Norbert Sauer *The indivisibility of the homogeneous k_n-free graphs*, Journal of Combinatorial Theory (Series B) 47 (2), 1989, pp. 162–170.

[8] Erik Ellentuck *A new proof that analytic sets are Ramsey*, The Journal of Symbolic Logic 39 (1), 1974, pp. 163–165.

[9] Paul Erdős, András Hajnal, and Lajos Pósa *Strong embeddings of graphs into coloured graphs*, Colloquia Mathematica Societatis János Bolyai 10, vol. I, *Infinite and finite sets* (edited by A. Hajanal, R. Rado, and V. Sós), 10, North-Holland Publishing Company, Amsterdam 1973, pp. 585–595.

[10] Paul Erdős, and Richard Rado, *Combinatorial theorems on classifications of subsets of a given set*, Proceedings of the London Mathematical Society (3) 2, 1952, pp. 417–439.

[11] W. L. Fouché *Symmetries in Ramsey theory*, East-West Journal of Mathematics 1, 1998, pp. 43–60.

[12] Roland Fraïssé *Sur l'extension aux relations de quelques propriétés des ordres*, Annales Scientifiques de l'École Normale Supérieure 71, 1954.

[13] Fred Galvin, and Karel Prikry *Borel sets and Ramsey's Theorem*, The Journal of Symbolic Logic 38, 1973.

[14] R. L. Graham, K. Leeb, and B. L. Rothschild, *Ramsey's theorem for a class of categories*, Advances in Mathematics 8, 1972, pp. 417–433.

[15] — *Errata: "Ramsey's theorem for a class of categories"*, Advances in Mathematics 10, 1973, pp. 326–327.

[16] R. L. Graham, and B. L. Rothschild *Ramsey's theorem for n-parameter sets*, Transactions of the American Mathematical Society 159, 1971, pp. 257–292.

[17] J. D. Halpern, and H. Läuchli *A partition theorem*, Transactions of the American Mathematical Society 124, 1966, pp. 360–367.

[18] J. D. Halpern, and A. Lévy *The Boolean prime ideal theorem does not imply the axiom of choice*, in: Dana S. Scott (ed.) Axiomatic Set Theory (Proceedings of the Symposium in Pure Mathematics, Vol. XIII, Part I, University of California, Los Angeles CA 1967), American Mathematical Society, Providence RI 1971, pp. 83–134.

[19] C. Ward Henson *A family of countable homogeneous graphs*, Pacific Journal of Mathematics 38 (1), 1971, pp. 69–83.

[20] Neil Hindman *Finite sums from sequences within cells of a partition of n*, Journal of Combinatorial Theory (Series A) 17, 1974, pp. 1–11.

[21] Greg Hjorth *An oscillation theorem for groups of isometries*, Geometric and Functional Analysis 18, 2008 (2), pp. 489–521.

[22] Alexander Kechris, Vladimir Pestov, and Stevo Todorcevic *Fraïssé limits, Ramsey theory, and topological dynamics of automorphism groups*, Geometric and Functional Analysis 15 (1), 2005, pp. 106–189.

[23] Péter Komjáth, and Vojtěch Rödl *Coloring of universal graphs*, Graphs and Combinatorics 2 (1), 1986, pp. 55–60.

[24] Claude Laflamme, Lionel Nguyen Van Thé, M. Pouzet, and Norbert Sauer *Partitions and indivisibility properties of countable dimensional vector spaces*, Journal of Combinatorial Theory (Series A) 118 (1), 2011, pp. 67–77.

[25] Claude Laflamme, Lionel Nguyen Van Thé, and Norbert Sauer *Partition properties of the dense local order and a colored version of Milliken's theorem*, Combinatorica 30 (1), 2010, pp. 83–104.

[26] Claude Laflamme, Norbert Sauer, and Vojkan Vuksanovic *Canonical partitions of universal structures*, Combinatorica 26 (2), 2006, pp. 183–205.

[27] Jean Larson *Infinite combinatorics*, in: Dov M. Gabbay, Akihiro Kanamori, and John Woods (eds.) Handbook of the History of Logic 6, *Sets and Extensions in the Twentieth Century*, North-Holland Publishing Company, Amsterdam 2012, pp. 145–357.

[28] Alain Louveau *Une démonstration topologique de théorèmes de Silver et Mathias*, Bulletin des Sciences Mathématiques (2e Série) 98 (2), 1974, pp. 97–102.

[29] Dragan Mašulović *Finite big Ramsey degrees in universal structures*, Journal of Combinatorial Theory (Series A) 170, 2020, 30 pp.

[30] A. R. D. Mathias *Happy families*, Annals of Mathematical Logic 12 (1), 1977, pp. 59–111.

[31] Keith R. Milliken *A Ramsey theorem for trees*, Journal of Combinatorial Theory (Series A) 26, 1979, pp. 215–237.

[32] C. St. J. A. Nash-Williams *On well-quasi-ordering transfinite sequences*, Proceedings of the

Cambridge Philosophical Society 61, 1965, pp. 33–39.

[33] Jaroslav Nešetřil, and Vojtěch Rödl *Partitions of finite relational and set systems*, Journal of Combinatorial Theory (Series A) 22 (3), 1977, pp. 289–312.

[34] — *Ramsey classes of set systems*, Journal of Combinatorial Theory (Series A) 34 (2), 1983, pp. 183–201.

[35] Lionel Nguyen Van Thé *Big Ramsey degrees and divisibility in classes of ultrametric spaces*, Canadian Mathematical Bulletin 51 (3), 2008, pp. 413–423.

[36] — *Structural Ramsey Theory with the Kechris-Pestov-Todorcevic Correspondence in Mind*, PhD Dissertation, Université d'Aix-Marseille, 2013, 48 pp.

[37] Lionel Nguyen Van Thé, and Norbert Sauer *The Urysohn sphere is oscillation stable*, Geometric Functional Analysis 19 (2), 2009, pp. 536–557.

[38] Maurice Pouzet, and Norbert Sauer *Edge partitions of the Rado graph*, Combinatorica 16 (4), 1996, pp. 505–520.

[39] Frank P. Ramsey *On a problem of formal logic*, Proceedings of the London Mathematical Society 30, 1929, pp. 264–296.

[40] Norbert Sauer *Edge partitions of the countable triangle free homogenous graph*, Discrete Mathematics 185 (1–3), 1998, pp. 137–181.

[41] — *Canonical vertex partitions*, Combinatorics, Probability, and Computing 12 (6), 2003, pp. 671–704.

[42] — *Coloring subgraphs of the Rado graph*, Combinatorica 26 (2), 2006, pp. 231–253.

[43] — *Vertex partitions of metric spaces with finite distance sets*, Discrete Mathematics 312 (1), 2012, pp. 119–128.

[44] Jack Silver *Some applications of model theory in set theory*, Annals of Mathematical Logic 3 (1), 1971, pp. 45–110.

[45] Stevo Todorcevic Introduction to Ramsey Spaces, Princeton University Press, Princeton NJ and Oxford 2010 [*Annals of Mathematics Studies 174*].

[46] Stevo Todorcevic, and Ilijas Farah Some Applications of the Method of Forcing, Yenisei, Moskow 1995 [*Yenisei Series in Pure and Applied Mathematics*].

[47] Alan M. Turing *On computable numbers, with an application to the Entsheidungsproblem*, Proceedings of the London Mathematical Society (2) 2, 1936, pp. 230–265.

[48] Andy Zucker *Topological dynamics of automorphism groups, ultrafilter combinatorics, and the Generic Point Problem*, Transactions of the American Mathematical Society 368, 2016, pp. 6715–6740.

[49] — *Big Ramsey degrees and topological dynamics*, in: Groups, Geometry and Dynamics 13 (1), 2019, pp. 235–276.

Randomness and Computation

Rod Downey
School of Mathematics and Statistics, Victoria University, PO Box 600, Wellington, New Zealand
rod.downey@vuw.ac.nz

This article examines work seeking to understand randomness using computational tools. The focus here will be how these studies interact with classical mathematics, and progress in the recent decade. A few representative and easier proofs are given, but mainly we will refer to the literature. The article could be seen as a companion to, as well as focusing on developments since, the paper "Calibrating Randomness" from 2006 which focused more on how randomness calibrations correlated to computational ones.

1 Introduction

The great Russian mathematician Andrey Kolmogorov appears several times in this paper, First, around 1930, Kolmogorov and others founded the theory of probability, basing it on measure theory. Kolmogorov's foundation does not seek to give any meaning to the notion of an individual object, such as a single real number or binary string, being random, but rather studies the expected values of random variables. As we learn at school, all strings of length n have the same probability of 2^{-n} for a fair coin. A set consisting of a single real has probability zero. Thus there is no meaning we can ascribe to randomness of a single object.

Yet we have a persistent intuition that certain strings of coin tosses are less random than others. The goal of the theory of algorithmic randomness is to give meaning to randomness content for individual objects. Quite aside from the intrinsic mathematical interest, the *utility* of this theory is that using such objects instead distributions might be significantly simpler and perhaps giving alternative insight into what randomness might mean in mathematics, and perhaps in nature.

Downey wishes to thank the Marsden Fund of New Zealand. This article grew from earlier (and shorter) ones by the author and Denis Hirschfeldt [48, 49]. Downey thanks Hirschfeldt inclusion of some of that material as well as many excellent interactions throughout the years.

2 Historical roots

2.1 Borel and von Mises

Predating the work of Kolmogorov are early attempts to answer this kind of question by providing notions of randomness for individual objects. The modern theory of *algorithmic randomness* realizes this goal. One way to develop this theory is based on the idea that an object is random if it passes all relevant "randomness tests". That is, for a desired level of randomness, we would have computational tests for which we wold regard a real or string[1] as random it is "passed" such tests. The idea is that we could not distinguish the real passing the test from one that was "really" random.

For example, by the law of large numbers, for a random real X, we would expect the number of 1's in the binary expansion of X to have limiting frequency $\frac{1}{2}$. That is, we would expect to have

$$\lim_{n\to\infty} \frac{|\{j < n : X(j) = 1\}|}{n} = \frac{1}{2}.$$

Moreover, we would expect X to be *normal* to base 2, meaning that for any binary string σ of length k, the occurrences of σ in the binary expansion of X should have limiting frequency 2^{-k}. Since base representation should not affect randomness, we would expect X to be normal in this sense no matter what base it were written in, so that in base b the limiting frequency would be b^{-k} for a string σ of length k. Thus X should be what is known as *absolutely normal.*

The concept of (absolute) normality is due to Borel [26] in around 1909. It remains a thriving area of number theory, which has had significant advances; particularly via interactions with algorithmic randomness as we later see. In an attempt to characterize randomness, normality was extended by von Mises [131] in 1919. Von Mises suggested the following definition of randomness for individual binary sequences. A *selection function* is an increasing function $f : \mathbb{N} \to \mathbb{N}$. We think of $f(i)$ as the ith place selected in forming a subsequence of a given sequence. (For the definition of normality above, where we consider the entire sequence, $f(i) = i$.) Von Mises suggested that a sequence $a_0a_1\dots$ should be random if any selected subsequence $a_{f(0)}a_{f(1)}\dots$ is normal.

The reader will immediately notice the following problem: sequence X with infinitely many 1's, *post hoc* we could let f select the positions where 1's occur, and

[1]For this paper we consider "reals" as members of 2^ω and strings in $2^{<\omega}$ unless otherwise specified. We will denote the j-th bit of α as $\alpha(j)$, and first n bits of a real (or string) α will be denoted by $\alpha \upharpoonright n$.

X would fail the test determined by f. However, it does not seem reasonable to be able to choose the testing places *after* selecting an X. The question is then: What kinds of selection functions should be allowed, to capture the intuition that we ought not to be able to sample from a random sequence and get the wrong frequencies? The statistician Wald [132, 133] (also famous for his analysis of bullet damage on warplanes in World War 2) showed that for any *countable* collection of selection functions, we could construct a real passing the tests they generate.

It is reasonable to regard prediction as a computational process, and hence restrict ourselves to *computable* selection function[2] The reader will note that Borel's and von Mises' work predates the events the early 1930's where the notion of a computable function was clarified by Church, Kleene, Post and famously Turing [128]. We remark that it is clear that Borel had a very good intuitive understanding of what a computable process was; see Avigad and Brattka for discussion of the development of computable analysis. But there was no formal clarification until the Church-Turing Thesis.

Indeed, this suggestion to use computable selection functions was eventually made by Church [38] in 1940, and this notion is now known as *Church Stochasticity.* As we will see, von Mises' approach had a more significant flaw, but we can build on its fundamental idea: Imagine that we are judges deciding whether a sequence X should count as random. If X passes all tests we can (in principle) devise given our computational power, then we should regard X as random since, as far as we are concerned, X has all the expected properties of a random object. We will use this intuition and the apparatus of computability and complexity theory to describe notions of *algorithmic* randomness.

Aside from the intrinsic interest of such an approach, it leads to useful mathematical tools. Many processes in mathematics are computable. Indeed any process from "real life" would surely be computable. Hence the *expected behavior* of such a process should align itself with the behavior obtained by providing it with an *algorithmically random input.* Hence, instead of having to analyze the relevant distribution and its statistics, we can simply argue about the behavior of the process on a single input. For instance, the expected number of steps of a sorting algorithm should be the same as that for a single algorithmically random input. We could also be more fine-grained and seek to understand exactly "how much" randomness is needed for certain typical behaviors to arise. (See Section 5.)

As we will discuss, algorithmic randomness also goes hand in hand with other parts of algorithmic information theory, such as Kolmogorov complexity, and has

[2] Indeed for practical applications, we might restrict ourselves to *polynomial time* or even *automatic* selections.

ties with notions such as Shannon entropy and fractal dimension.

2.2 Some basic computability theory

Given that this is an expository paper in a Logic volume, we will assume that the reader is more or less cognoscent with the rudiments of classical computability theory. Thus we will give a brief reprise of the concepts we will be using Given that this is an expository paper in a Logic volume, we will assume that the reader is more or less cognoscent with the rudiments of classical computability theory. Thus we will give a brief reprise of the concepts we will be using In the 1930's, Church, Gödel, Kleene, Post, and most famously Turing [128] gave equivalent mathematical definitions capturing the intuitive notion of a computable function, leading to the *Church-Turing Thesis*, which can be taken as asserting that a function (from $\mathbb{N}$ to $\mathbb{N}$, say) is computable if and only if it can be computed by a Turing machine[3]. It has also become clear that algorithms can be treated as data, and hence that there is a *universal Turing machine*, i.e., there are a listing $\Phi_0, \Phi_1, \ldots$ of all Turing machines and a single algorithm that, on input $\langle e, n\rangle$ computes the result $\Phi_e(n)$ of running Φ_e on input n.[4]

It is important to note that a Turing machine might not halt on a given input, and hence the functions computed by Turing machines are in general *partial*. Indeed, as Turing showed, the *halting problem* "Does the eth Turing machine halt on input n?" is algorithmically unsolvable. Church and Turing famously showed that Hilbert's *Entscheidungsproblem* (the decision problem for first-order logic) is unsolvable, in Turing's case by showing that the halting problem can be coded into first-order logic. Many other problems have since been shown to be algorithmically unsolvable by similar means.

We write $\Phi_e(n)\downarrow$ to mean that the machine Φ_e eventually halts on input n. Then $\emptyset' = \{\langle e, n\rangle : \Phi_e(n)\downarrow\}$ is a set representing the halting problem. This set is an

[3]This definition can easily be transferred to other objects of countable mathematics. For instance, we think of infinite binary sequences as functions $\mathbb{N} \to \{0, 1\}$, and identify sets of natural numbers with their characteristic functions.

[4]The realization that such universal machines are possible helped lead to the development of modern computers. Previously, machines had been purpose-built for given tasks. In a 1947 lecture on his design for the Automated Computing Engine, Turing said, "The special machine may be called the universal machine; it works in the following quite simple manner. When we have decided what machine we wish to imitate we punch a description of it on the tape of the universal machine ... The universal machine has only to keep looking at this description in order to find out what it should do at each stage. Thus the complexity of the machine to be imitated is concentrated in the tape and does not appear in the universal machine proper in any way. ... [D]igital computing machines such as the ACE ... are in fact practical versions of the universal machine." From our contemporary point of view, it may be difficult to imagine how novel this idea was.

example of a noncomputable *computably enumerable* (*c.e.*) set, which means that the set can be listed (not necessarily in numerical order) by some algorithm.

Another important notion is that of *Turing reducibility* (which we define for sets of natural numbers but is similarly defined for functions), where A is Turing reducible to B, written as $A \leqslant_{\mathrm{T}} B$, if there is an algorithm for computing A when given oracle access to B. That is, the algorithm is allowed access to answers to questions of the form "Is n in B?" during its execution. This notion can be formalized using Turing machines with oracle tapes, or by adding the characteristic function of B to the Kleene partial recursive functions. If $A \leqslant_{\mathrm{T}} B$, then we regard A as no more complicated than B from a computability-theoretic perspective. We also say that A is *B-computable* or *computable relative to B*. The pre-ordering $\leq_T$ naturally leads to an equivalence relation, where A and B are *Turing equivalent* if $A \leqslant_{\mathrm{T}} B$ and $B \leqslant_{\mathrm{T}} A$. The *(Turing) degree* of A is its equivalence class under this notion. As we know, if we examine exactly how the access mechanism works we get other reducibilities refining $\leq_T$. For example, $A \leq_m B$ means that there is a computable f such that $x \in A$ iff $f(x) \in B$. The polynomial miniaturization of this is central in computational complexity theory, as per Garey and Johnson [62]. Another reducibility of relevance to algorithmic randomness is *truth table* reducibility, where $A \leq_{tt} B$ means that $A \leq_T B$ via some procedure $\Phi^B = A$ such that for all oracles X, Φ^X is total.

In general, the process of allowing access to an oracle in our algorithms is known as *relativization.* As in the unrelativized case, we can list the Turing machines $\Phi_0^B, \Phi_1^B, \ldots$ with oracle B, and let $B' = \{\langle e, n\rangle : \Phi_e^B(n)\downarrow\}$ be the relativization of the halting problem to B. This set is called the *(Turing) jump* of B. The jump operation taking B to B' is very important in computability theory, one reason being that B' is the most complicated set that is still c.e. relative to B, i.e., B' is c.e. relative to B and every set that is c.e. relative to B is B'-computable. There are several other important classes of sets that can be defined in terms of the jump. For instance, A is *low* if $A' \leqslant_{\mathrm{T}} \emptyset'$ and *high* if $\emptyset'' \leqslant_{\mathrm{T}} A'$ (where $\emptyset'' = (\emptyset')'$). Low sets are in certain ways "close to computable", while high ones partake of some of the power of $\emptyset'$ as an oracle. These properties are invariant under Turing equivalence, and hence are also properties of Turing degrees. These concepts can be iterated, for example, $A^{(2)} = (A')'$ and the hierarchy of Turing degrees $\mathbf{0}', \mathbf{0}'', \ldots$ is called the *arithmetical hierarchy*, and transfinite iterations are called the *hyperarithmetical* hierarchy. See e.g. Downey and Hirschfeldt [47], Rogers [118], or Soare [124].

2.3 Martin-Löf randomness

Von Mises approach refined by Church to consider selection functions restricted to the computable ones. However, in 1939, Ville [130] showed that von Mises' approach cannot work in its original form, no matter what *countable* collection of selection functions we choose.

Theorem 2.1 (Ville [130]). *For any countable collection of selection functions, there is a sequence X that passes all von Mises tests associated with these functions, such that for every n, there are more 0's than 1's in $X \upharpoonright n$.*

I think if you went to a casino and were told there would *always* be more heads that tails you would not think the coin to be fair. We could try to repair von Mises' definition by adding further tests, reflecting statistical laws beyond the law of large numbers. But which ones? Ville suggested ones reflecting the law of iterated logarithms, which would take care of his specific example. But how could we know that further examples along these lines—i.e., sequences satisfying both von Mises' and Ville's tests, yet failing to have some other property we expect of random sequences—would not arise? For more of this, and a modern proof of Ville's Theorem see Downey and Hirschfeldt [47], where it is also shown that the law of iterated logarithms can be defeated.

The situation was finally clarified in the 1960's by Martin-Löf [96]. In probability theory, "typicality" is quantified using measure theory, leading to the intuition that random objects should avoid null sets. Martin-Löf noticed that tests like von Mises' and Ville's can be thought of as *effectively* null sets. Instead of considering specific tests based on particular statistical laws, we should consider *all* possible tests corresponding to some precisely defined notion of effectively null set. The restriction to such a notion gets around the problem that no sequence can avoid being in *every* null set. We will see later that this idea was anticipated by Turing around 1939 in unpublished notes for work on normality.

To give Martin-Löf's definition, we work for convenience in Cantor space 2^ω, whose elements are infinite binary sequences. The choice of base is not important. For example, all of the notions of randomness we consider are enough to ensure absolute normality. The basic open sets of Cantor space are the ones of the form $[\sigma] = \{X \in 2^\omega : X \text{ extends } \sigma\}$ for $\sigma \in 2^{<\omega}$, where $2^{<\omega}$ is the set of finite binary strings. The uniform measure λ on this space is obtained by defining $\lambda([\sigma]) = 2^{-|\sigma|}$. We say that a sequence $T_0, T_1, \ldots$ of open sets in 2^ω is *uniformly c.e.* if there is a c.e. set $G \subseteq \mathbb{N} \times 2^{<\omega}$ such that $T_n = \bigcup\{[\sigma] : (n, \sigma) \in G\}$.

Definition 2.2. A *Martin-Löf test* is a sequence $T_0, T_1, \ldots$ of uniformly c.e. open sets such that $\lambda(T_n) \leqslant 2^{-n}$. A sequence X *passes* this test if $X \notin \bigcap_n T_n$. A sequence is *Martin-Löf random (ML-random)* if it passes all Martin-Löf tests.

The intersection of a Martin-Löf test is our notion of effectively null set. Since there are only countably many Martin-Löf tests, and each determines a null set in the classical sense, the collection of ML-random sequences has measure 1. It can be shown that Martin-Löf tests include all the ones proposed by von Mises and Ville, in Church's computability-theoretic versions. Indeed they include all tests that are "computably performable", which avoids the problem of having to adaptively introduce more tests as more Ville-like sequences are found.

Martin-Löf's effectivization of measure theory allowed him to consider the laws a random sequence should obey from an abstract point of view, leading to a mathematically robust definition. As Jack Lutz said in a talk at the *7th Conference on Computability, Complexity, and Randomness* (during the Alan Turing Year programme in Cambridge, 2012),

> "Placing computability constraints on a nonconstructive theory like Lebesgue measure seems *a priori* to weaken the theory, but it may strengthen the theory for some purposes. This vision is crucial for present-day investigations of individual random sequences, dimensions of individual sequences, measure and category in complexity classes, etc."

2.4 The three approaches

ML-randomness can be thought of as the *statistician's approach* to defining algorithmic randomness, based on the intuition that random sequences should avoid having statistically rare properties. There are two other major approaches:

- The *gambler's approach*: random sequences should be unpredictable.
- The *coder's approach*: random sequences should not have regularities that allow us to compress the information they contain.

2.4.1 The Gambler's Approach

The gambler's approach may be the most immediately intuitive one. It was formalized in the computability-theoretic setting by Schnorr [120], using the idea that we should not be able to make arbitrarily much money when betting on the bits of a random sequence. The following notion is a simple special case of the notion of martingale from probability theory. (See [47, Section 6.3.4] for further discussion of the relationship between these concepts.)

Definition 2.3. A *martingale* is a function $f : 2^{<\omega} \to \mathbb{R}^{\geqslant 0}$ such that

$$f(\sigma) = \frac{f(\sigma 0) + f(\sigma 1)}{2}$$

for all σ. We say that f *succeeds* on X if $\limsup_{n\to\infty} f(X \upharpoonright n) = \infty$. We call this a *supermartingale* if for all σ,

$$f(\sigma) \geq \frac{f(\sigma 0) + f(\sigma 1)}{2}$$

We think of f as the capital we have when betting on the bits of a binary sequence according to a particular betting strategy. The displayed equation ensures that the betting is fair. Success then means that we can make arbitrarily much money when betting on X, which should not happen if X is random. By considering martingales with varying levels of effectivity, we get various notions of algorithmic randomness, including ML-randomness itself, as it turns out.

For example, if we insist that f is computable, we get a notion called *computable randomness.* Defining randomness for infinite sequences, by in complexity classes we might restrict f to be polynomial time, giving *polynomial time randomness* (Lutz [91]). We call a function f *left-c.e.* if there is a computable function $g(\cdot,\cdot)$ such that

- $\lim_s g(x, s)$ exists for all x, and
- $\lim_s g(x, s) = f(x)$, and
- $g(x, s+1) \geq g(x, s)$ for all x, s.

Right-c.e. reals are defined similarly, but using approximations from above.

Theorem 2.4 (Schnorr [120]). *X is ML-random iff no left c.e. (super-)martingale succeeds on it.*

The reader might wonder why we have "left c.e." in the characterization. The way to think about this is the following. We want to bet against sequences that we think are not random. Now as time evolves we will discover more and more facts. For example, is X was $\pi - e$, and we were given the binary expansion of this thoroughly computable number, which is therefor far from random, we would probably have trouble discerning the pattern. But as time went on as we computes more and more (partial) computable functions, we might discover a predictor yielding bits of X and hence we would like to place more capital on the bits of X.

Interestingly, this leads us to some basic problems which stubbornly remain open. First we might ask what happens if we have a partial computable betting function, but instead of betting on the first bit, then the second, etc we could bet on some bit, then, depending on the outcome, we could then bit somewhere we have not yet bet upon. This notion is called *non-monotonic* randomness.

Question 2.5 (Muchnik, Semenov and Uspensky [107]). Do ML-randomness and non-monotonic randomness coincide

The essence of the question is whether partial computable, rather than left c.e. methods suffice to define ML-randomness. In some sense they do if we replace martingales by "martingales processes" (a generalization of the martingale idea) as in Merkle, Mihailovíc, and Slaman [98]. Question 2.5 has been open since 1998. Another apparently difficult problem asks if *bias* is possible, which one would suspect that this is not possible. From [47], a *Kastergale* is a supermartingale together with a partial computable function $h : 2^{<\omega} \to \{0, 1\}$. Then if $h(\sigma) \downarrow [s]$ we will promise that for all $s' > s$, $g(\sigma h(\sigma), s') > g(\sigma, 1 - h(\sigma), s')$. That is we will always henceforth bias towards, e.g., 0 is $h(\sigma) = 0$.

Question 2.6. Is Kastergale-random (i.e. random for left-c.e. kastergales) the same as ML-random?

For some recent progress on these questions, we refer the reader to Barmpalias, Fang and Lewis-Pye [9].

It is not too difficult to show that ML-randomness is strictly stronger than computable randomness, but the difference is slight as quantified by computability theory:

Theorem 2.7 (Nies, Stephan, and Terwijn [112]). *If X is a computably random real and $X' \not\geq_T \emptyset''$ (X is not high), then X is ML-random. Every high degree contains a computably random real which is not ML-random.*

We will strengthen this result soon. Clearly there are many possible variations on the notion of martingale. For example, we might ask that there is a minimum possible bet (real casinos don't allow ε as a bet), or indeed we can only bet in some set of bets like $\{n_1, \ldots, n_k\}$ or $\mathbb{N}$. This leads to the interesting notion of integer-valued betting strategies, and we refer to Bienvenu, Stephan and Teutsch [24] for more on this. Later we will see that normality can be characterized by *automatic* martingales.

There are many other interesting levels of algorithmic randomness. Schnorr also introduced another notion related to the ML-randomness definition. He defined a notion now called *Schnorr randomness*, which is like the notion of computable randomness mentioned below Definition 2.3 but with an extra effectiveness condition on the rate of success of martingales. He also showed that X is Schnorr random iff it passes all Martin-Löf tests $T_0, T_1, \ldots$ such that the measures $\lambda(T_n)$ are uniformly computable (i.e., the function $n \mapsto \lambda(T_n)$ is computable in the sense of Section 5.4

below). We remark that it is also possible to give test characterizations of computable randomness but they are somewhat counter-intuitive. (See [47].) It follows immediately from their definitions in terms of martingales that ML-randomness implies computable randomness, which in turn implies Schnorr randomness. It is more difficult to prove that none of these implications can be reversed.

Additionally to Theorem 2.7, Nies, Stephan and Terwijn [112] showed that Schnorr and computable randomness can be separated in the high degrees, but again coincide in the non-high ones.

2.4.2 The Coder's Approach

The coder's approach builds on the idea that a random string should have no short descriptions. For example, in describing 010101... (1000 times) by the brief description "print 01 1000 times", we are using regularities in this string to compress it. For a more complicated string, say the first 2000 bits of the binary expansion of e^{π}, the regularities may be harder to perceive, but are still there and can still lead to compression. A random string should have no such exploitable regularities (i.e., regularities that are not present in most strings), so the shortest way to describe it should be basically to write it out in full.

Again we see that Kolmogorov enters the picture: We formalize this using the well-known concept of Kolmogorov complexity. We can think of a Turing machine M with inputs and outputs in $2^{<\omega}$ as a description system. If $M(\tau) = \sigma$ then τ is an M-description of σ.

Definition 2.8. The *Kolmogorov complexity* $C_M(\sigma)$ of σ relative to M is the *length* of the shortest τ such that $M(\tau) = \sigma$.

Since we can enumerate all machines $\{M_e \mid e \in \mathbb{N}\}$, we can then take a universal Turing machine U, which emulates any given Turing machine with at most a constant increase in the size of programs. To wit, we would consider a machine U which on input $1^e0\sigma$ would run $M_e(\sigma)$. This is called universality by *adjugation.* We can then define the *(plain) Kolmogorov complexity* of σ as $C(\sigma) = C_U(\sigma)$. The value of $C(\sigma)$ depends on U, but only up to an additive constant independent of σ. We think of a string as random if its Kolmogorov complexity is close to its length.

For an infinite sequence X, a natural guess would be that X should be considered random if every initial segment of X is incompressible in this sense, i.e., if $C(X \upharpoonright n) \geqslant n - O(1)$[5]. However, plain Kolmogorov complexity is not quite the right notion here, because the information in a description τ consists not only of the bits of τ, but

[5] If, for example, $C(\sigma) \geq n \pm O(1)$, henceforth we will (mostly) use the economical notation $C(\sigma) \geq^{+} n$; and similarly for K below.

also its length, which can provide another $\log_2 |\tau|$ many bits of information. Indeed, Martin-Löf (see [90]) showed that it is not possible to have $C(X \upharpoonright n) \geqslant n - O(1)$: Given a long string ρ, we can write $\rho = \sigma\tau\nu$, where $|\tau|$ is the position of σ in the length-lexicographic ordering of $2^{<\omega}$. Consider the Turing machine M that, on input η, determines the $|\eta|$th string ξ in the length-lexicographic ordering of $2^{<\omega}$ and outputs $\xi\eta$. Then $N(\tau) = \sigma\tau$. For any sequence X and any k, this process allows us to compress some initial segment of X by more than k many bits.

There are several ways to get around this problem by modifying the definition of Kolmogorov complexity. The best-known one is to use prefix-free codes, which act like telephone numbers. That is, we restrict ourselves to machines M such that if $M(\tau)$ is defined (i.e., if the machine eventually halts on input τ) and μ is a proper extension of τ, then $M(\mu)$ is not defined. There are universal prefix-free machines, using the same method above, since we can enumerate the partial prefix-free machines. Then we can take such a machine U and define the *prefix-free Kolmogorov complexity* of σ as $K(\sigma) = C_U(\sigma)$. The roots of this notion be found in the work of Levin, Chaitin, and Schnorr, and in a certain sense—like the notion of Kolmogorov complexity more generally—even earlier in that of Solomonoff (see [47, 90]). As shown by Schnorr (see Chaitin [34]), it is indeed the case that the following theorem holds:

Theorem 2.9 (Schnorr). *X is Martin-Löf random if and only if $K(X \upharpoonright n) \geqslant^+ n$.*

We remark that this shows that we our definitions are reasonably robust, in that all approaches yield the same ML-random reals. We remark that it is possible to give machine characterizations of computable and Schnorr randomness. For instance, we can call a machine M a *computable domain* machine if $\lambda\text{dom}(M)$ is a computable real.

Theorem 2.10 (Downey and Griffiths [46]). *X is Schnorr random iff for all computable domain machines $K_M(X \upharpoonright n) \geq^+ n$ for all n.*

We remark that there are other methods to capture ML-randomness using incompressibility. One is to use things akin to prefix-free complexity, like *process complexity* which asks that the action be *continuous*. That is we have machines M such that if $M(\sigma)\downarrow$ and $M(\tau)\downarrow$, and $\sigma \prec \tau$, then $M(\sigma) \preceq M(\tau)$. Using this Schnorr and earlier Levin (with an analogous concept) showed that again X is random iff its segments cannot be compressed. (see [47] for a discussion about this and similar compressors.) Day [40] gave a nice machine characterization of computable randomness using a kind of process machine as described below.

There are other varieties of Kolmogorov complexity, but C and K are the main ones. For applications[6], it often does not matter which variety is used. The following surprising result establishes a fairly precise relationship between C and K. Let $C^{(1)}(\sigma) = C(\sigma)$ and $C^{(n+1)}(\sigma) = C(C^{(n)}(\sigma))$.

Theorem 2.11 (Solovay [125]). *$K(\sigma) = C(\sigma) + C^{(2)}(\sigma) \pm O(C^{(3)}(\sigma))$, and this result is tight in that we cannot extend it to $C^{(4)}(\sigma)$.*

There is a simplified version of Solovay's original proof in [47] using a suggestion of Miller and a useful result known as Symmetry of Information.

Theorem 2.12 (Symmetry of Information for K[7]-Levin and Gács [60], Chaitin [34]). $K(\sigma, \tau) =^+ K(\sigma) + K(\tau \mid \sigma, K(\sigma)) =^+ K(\tau) + K(\sigma|\tau, K(\tau))$.

Using this and other techniques, Bauwens [13] gave some simpler proofs for Theorem 2.11.

There is a vast body of research on Kolmogorov complexity and its applications. We will discuss some of these applications below; much more on the topic can be found in the books of Li and Vitányi [90] (especially for applications) and Shen, Uspenskyi and Vereshchagin [122].

One notion of compression not to be found in [47], and largely forgotten, is the following.

Definition 2.13 (Kobayashi [85]).

1. Given $f : \mathbb{N} \to \mathbb{N}$, we say that X is *f-compressible* if there exists Y which computes X via an oracle Turing machine which queries, for each n, at most the first $f(n)$ digits of Y (i.e. the Y-use) for the computation of $X \upharpoonright n$.

2. We say that a real X is *Kobayashi incompressible* if it is not f-compressible for any function f such that $n - f(n)$ is unbounded.

A recent result shows that Kobayashi incompressibility actually coincides with ML-randomness.

Theorem 2.14 (Kobayashi incompressibility and Turing reductions-Barmpalias and Downey [12], Bienvenu). *The following are equivalent:*

1. *X is Martin-Löf random;*

2. *For every Y with $X \leq_T Y$ the Y-use in any such computation of $X \upharpoonright n$ is bounded below by $n - c$ for some constant c and all n.*

---

[6]Particularly those involving effective fractal dimension we see later.

[7]There is also one for C.

Later we will look at other calibrations of randomness. For some there are characterizations along the Kobayashi-lines of the above such as for Kurtz randomness using stronger reducibilities than $\leq_T$.

3 Goals

There are several ways to explore the ideas introduced above. First, there are natural internal questions

- How do the various levels of algorithmic randomness interrelate?
- How do calibrations of randomness relate to the hierarchies of computability and complexity theory, and to relative computability?
- How should we calibrate partial randomness?
- Can a source of partial (algorithmic) randomness be amplified into a source that is fully random, or at least more random?

The books Downey and Hirschfeldt [47] and Nies [109] cover material along these lines up to about 2010.

We can also consider applications. Mathematics has many theorems that involve "almost everywhere" behavior. Natural examples come from ergodic theory, analysis, geometric measure theory, and even combinatorics. Behavior that occurs almost everywhere should occur at sufficiently random points. Using notions from algorithmic randomness, we can explore exactly *how much* randomness is needed in a given case. For example, the set of reals at which an increasing function is differentiable is null. How complicated is this null set, and hence, what level of algorithmic randomness is necessary for a real to avoid it (assuming the function is itself computable in some sense)? Is Martin-Löf randomness the right notion here? More specifically, suppose that I want to use randomness as a tool in some combinatorial algorithm. There are many such algorithms which ask for random seeds; for instance polynomial identity testing. What algorithmic level of source randomness is needed for applications to obtain results which are close to exact solutions? Also how does this theory relate to the well-developed theory of e.g. random graphs?

One recent example comes from an answer to of a question of Bollobas and of Kahane going back to 1965. In the introduction to his book [25] on random graphs, Bollobas motivates the use of probabilistic ideas in graph theory. He mentioned that earlier probabilistic application had been found in analysis via three seminal papers of Paley and Zigmund [113, 114, 115].

> "Paley and Zigmund (1930a,b,1932) had investigated random series of functions.One of their results was that if the real numbers c_n satisfy $\sum_{n=0}^{\infty} c_n^2 = \infty$ then $\sum_{n=0}^{\infty} \pm c_n \cos nx$ fails to be a Fourier-Lebesgue series for almost all choices of the signs. To exhibit a sequence of signs with this property is surprisingly difficult: indeed there is no algorithm known which constructs an appropriate sequence of signs from any sequence c_n with $\sum_{n=0}^{\infty} c_n^2 = \infty$."

We remark that an indentical question can be found even earlier in the 1968 version of Kahane's book (most recently, [80], page 47), on random trigonometric series:

> "A surprising fact is that nobody knows how to construct these signs explicitly, but a random choice works."

In recent work, Downey, Greenberg and Tangarra [45] showed that this question has a positve answer by showing that the collection of signs where the series *converges* forms a Kurtz null test (i.e the complement of a c.e. open set of measure 1)[8]. Hence by general theorems about Kurtz null tests we know that there is a computable real which succeeds on this test. There is a huge amount of largely unexplored work on random trigonometric series, some of which is explored in [45], and earlier [116].

We can also use the idea of assigning levels of randomness to individual objects to prove new theorems or give simpler proofs of known ones. We give some examples later, especially in the area of Hausdorff dimension theory.

3.0.1 The Incompressibility Method

Early examples of this method tended to use Kolmogorov complexity and what is called the "incompressibility method". For instance, in 1975, Chaitin [33] (see also [86]) famously used Kolmogorov complexity to give a proof of a version of Gödel's First Incompleteness Theorem, by showing the following:

Theorem 3.1 (Chaitin [33]; also Barzdins). *For any sufficiently strong, computably axiomatizable, consistent theory T, there is a number L such that T cannot prove that $C(\sigma) > L$ for any given string σ*[9].

Proof. (Sketch-Kritchman and Raz [86]) (For this proof, C or K are equally usable.) Let L be a large enough integer. Assume for a contradiction that, for some integer x, there is a proof for the statement "$K(x) > L$". Let w be the first proof (say,

[8] We look at Kurtz randomness, a notion of randomness weaker than ML-randomness, later in the present article.

[9] This also follows by interpreting an earlier result of Barzdins; see [90, Section 2.7]).

according to the lexicographic order) for a statement of the form "$K(x) > L$". Let z be the integer x such that w proves "$K(x) > L$". It is easy to give a computer program that outputs z : the program enumerates all possible proofs w, one by one, and for the first w that proves a statement of the form "$K(x) > L$", the program outputs x and stops. The length of this program is a $O(1) + \log L$. Thus, if L is large enough, the Kolmogorov complexity of z is less than L. Since w is a proof for "$K(x) > L$" (which is a false statement), we conclude that the theory is inconsistent. Note that the number of computer programs of length L bits is at most $2L + 1$. Hence, for any integer L, there exists an integer $0 \leq x \leq 2L+1$, such that $K(x) > L$. Thus, for some integer x, the statement "$K(x) > L$" is a true statement that has no proof. □

More recently, Kritchman and Raz [86] used these methods to give a proof of the Second Incompleteness Theorem as well.[10]

This article focuses on algorithmic randomness for infinite objects, but we should mention that there have been many applications of Kolmogorov complexity under the collective title of the *incompressibility method*, based on the observation that algorithmically random strings should exhibit typical behavior for computable processes. For example, as well as the proof of the Incompleteness Theorem above, this method can be used to give average running times for sorting, by showing that if the outcome is not what we would expect then we can compress a random input. See Li and Vitányi [90, Chapter 6] for applications of this technique to areas as diverse as combinatorics, formal languages, compact routing, and circuit complexity, among others. Many results originally proved using Shannon entropy or related methods also have proofs using Kolmogorov complexity[11]. For example, Messner and Thierauf [99] gave a constructive proof of the Lovász Local Lemma using Kolmogorov complexity.

Other applications come from the observation that in some sense Kolmogorov complexity provides an "absolute" measure of the intrinsic complexity of a string. The key is again the notion of conditional Kolmogorov complexity $C(\sigma \mid \tau)$. Then, for example, $C(\sigma \mid \sigma) = O(1)$, and σ is "independent of τ" if $C(\sigma \mid \tau) = C(\sigma) - O(1)$. Researchers comparing two sequences σ, τ representing, say, two DNA sequences, or

[10]Other recent work has explored the effect of adding axioms asserting the incompressibility of certain strings in a probabilistic way. Bienvenu, Romashchenko, Shen, Taveneaux, and Vermeeren [23] have shown that this kind of procedure does not help to prove new interesting theorems, but that the situation changes if we take into account the sizes of the proofs: randomly chosen axioms (in a sense made precise in their paper) can help to make proofs much shorter under the reasonable complexity-theoretic assumption that NP $\neq$ PSPACE.

[11]Shannon Entropy is more or less an average Kolmogorov Complexity. Hammer et al. [72] looked at how many inequalities for these concepts are interchangeable.

two phylogenetic trees, or two languages, or two pieces of music, have invented many distance metrics, such as the maximum parsimony distance on phylogenetic trees, but it is also natural to use a content-neutral measure of "information distance" like $\max\{C(\sigma \mid \tau), C(\tau \mid \sigma)\}$. There have been some attempts to make this work in practice for solving classification problems, though results have so far been mixed. Of course, C is not computable, but it can be replaced in applications by measures derived from practical compression algorithms. See [90, Sections 8.3 and 8.4]. Also see Bennett et. al. [17] for a survey of these ideas in abstract *Information Distance.* We will not give more details as this area would need a complete survey to itself.

As we will see below, a more recent line of research has used notions of effective dimension based on partial randomness to give new proofs of classical theorems in ergodic theory and obtain new results in geometric measure theory.

4 Some interactions with computability

4.1 Halting probabilities

A first question we might ask is how to generate "natural" examples of algorithmically random reals. A classic example is Chaitin's halting probability. Let U be a universal prefix-free machine and let

$$\Omega = \sum_{U(\sigma)\downarrow} 2^{-|\sigma|}.$$

This number is the measure of the set of sequences X such that U halts on some initial segment of X, which we can interpret as the halting probability of U.

Theorem 4.1 (Chaitin [35]). *Ω above is ML-random.*

Proof. (sketch) Using the Recursion Theorem we will build a prefix-free machine M which as index e within the universal machine U, and e is known in advance. Then we monitor $\Omega_s = \sum_{U(\sigma)\downarrow} 2^{-|\sigma|}[s]$, the stage s approximation. If we see some σ of length $< n - e - 2$ enter Ω_{s+1} with $U(\sigma) = \Omega_s \upharpoonright n$, then we put σ into M_{s+1} causing $1^0 0\sigma$ to enter $\Omega - \Omega_s$, and hence $\Omega_s \upharpoonright n \neq \Omega \upharpoonright n$. (This proof is from [47] and resembles the proof of the unsolvability of the Halting Problem.) □

For any prefix-free machine M in place of U we can similarly define a halting probability. In some ways, halting probabilities are the analogs of computably enumerable sets in the theory of algorithmic randomness. But, as described above, they are left-c.e. in the place of c.e.. Calude, Hertling, Khoussainov, and Wang [32] showed that every left-c.e. real is the halting probability of some prefix-free machine.

We should perhaps write Ω_U instead of Ω, to stress the dependence of its particular value on the choice of universal machine, but the fundamental properties of Ω do not depend on this choice, much as those of the halting problem do not depend on the specific choice of enumeration of Turing machines. For partial computable functions, the same could also be said. That is $\emptyset' = \{e \mid \varphi_e(e)\downarrow\}$ depends on the precise enumerations of the partial computable functions. But, of course, Myhill proved that up to m-reduction (indeed up to computable permutations) all versions of K had the same degree. So in this sense there is "only one" halting set.

In our case our halting probabilties are reals, and hence what the analogous result would entail would be a continuous version of m-reducibility. (The following is a slight generalization of the earlier version of this concept. See [47])

Definition 4.2 (Solovay [125]). We say that $X \leq_S Y$ iff there is a partial computable function f and a constant c such that for any rational $q < Y$ $f(q)\downarrow< X$ and $c(Y - q) \geq (X - f(q))$.

Using this we can effectively convert any Cauchy sequence for Y into one for X which converges just as fast. It is not hard to show that $X \leq_S Y$ means that for all n, $K(X \upharpoonright n) \leq^+ K(Y \upharpoonright n)$. That is because if I know $Y \upharpoonright n$, I can apply f to get $f(Y \upharpoonright n)$ will be within $2^{-(n-\log c)}$ of $X \upharpoonright n$, and hence if we enumerate a small diameter of strings around $f(Y \upharpoonright n)$ we will know that $X \upharpoonright n$ is one of them; and this constant is independent of n. Hence by Schnorr's Theorem, if X is ML-random and $X \leq_S Y$ then Y is ML-random also. Kučera and Slaman [88] showed that every left-c.e. real is reducible to every Ω_U up to Solovay reducibility, and hence all such Ω_U's are equivalent modulo this notion.

Theorem 4.3 (Kučera and Slaman [88]). *If X is left-c.e. then $X \leq_S \Omega$. That is Ω is Solovay complete.*

Proof. (Sketch) The proof is an illustrative "measure recycling" one. Given $X = \lim_s X_s$ if $X_{s+1} \upharpoonright n \neq X_s \upharpoonright n$, then the opponent has spent at least 2^{-n} to make this change. We would like Ω_{s+1} to make similar change of around $2^{-(n+c)}$ with c given by the recursion theorem. One way is to put a potential ML-test around $[\Omega_s \upharpoonright n]$ or alternatively issue a description $U(\tau) = \Omega_s \upharpoonright n$, where $|\tau| = n + c$ (meaning, again we build M coded in U by c). The total cost is ≤ 1 so we succeed in forcing Ω to change almost always. □

Hence all universal halting probabilities, being left-c.e. reals are essentially the same. Solovay reducibility is just one measure of relative randomness which can be used, but the reader will see that it is really quite natural. As an analogy to the study of c.e. sets under m-reducibility, we can study the structure of the left-c.e.

reals under Solovay reducibility. The following result gives some insight into this structure.

Theorem 4.4 (Downey, Hirschfeldt and Nies [51]).

1. *The Solovay degrees of left-c.e. reals forms a distributive dense upper semilattice.*
2. *The join operation is induced by +. That is $[\alpha] \vee [\beta] = [\alpha + \beta]$, where $[X]$ denotes the Solovay degree of X.*
3. *The Solovay degree of Ω is join-irreducible That is, if $[\alpha] \vee [\beta] = [\Omega]$ then one of α or β is ML-random. (Also obtained earlier by Demuth [42].)*
4. *Every incomplete Solovay degree of a left c.e. real splits over all lesser ones.*

Recently Barmpalias and Lewis-Pye [10] and Miller [102] proved some fascinating results showing that there is a calculus operating here. Is α is left-c.e. then there is a computable sequence $\alpha_s \leq \alpha_{s+1} \to \alpha$. Then Barmpalias and Lewis-Pye showed the following:

Theorem 4.5 (Barmpalias and Lewis-Pye [10]).

1. *If α and β are ML-random left-c.e. reals, then*
$$\frac{\partial \alpha}{\partial \beta} = \lim_s \frac{\alpha - \alpha_s}{\beta - \beta_s}$$
exists, and this is independent of choice of approximations for α and β.
2. *Furthermore $\alpha - \beta$ is ML-random iff $\frac{\partial \alpha}{\partial \beta} \neq 1$, and*
3. $\frac{\partial \alpha}{\partial \beta} = \sup\{c \in \mathbb{Q} \mid \alpha - c\beta$ *is left-c.e*$\} = \inf\{c \in \mathbb{Q} \mid \alpha - c\beta\}$ *is right c.e.*$\}$.

Miller [103] extended these results to what are called d.c.e. reals, being those of the form $X - Y$ with X and Y left-c.e. reals. Using this he showed that the nonrandom left-c.e. reals form a real closed field, and ∂ is derivation on this field; meaning it satisfied Leibnitz' Law:

$$\partial(\alpha\beta) = \alpha\partial\beta + \beta\partial\alpha.$$

Consequences of these results are still under active exploration.

Left-c.e. and right-c.e. reals (those of the form $1 - \alpha$ for a left-c.e. α) occur naturally in mathematics. Braverman and Yampolsky [29] showed that they arise in connection with Julia sets, and there is a striking example in symbolic dynamics: A d-dimensional *subshift of finite type* is a certain kind of collection of A-colorings of $\mathbb{Z}^d$, where A is a finite set, defined by local rules (basically saying that certain coloring patterns are illegal) invariant under the shift action

$$(S^g x)(h) = x(h+g) \text{ for } g, h \in \mathbb{Z}^d \text{ and } x \in A^{\mathbb{Z}^d}.$$

Its *(topological) entropy* is an important invariant measuring the asymptotic growth in the number of legal colorings of finite regions. It has been known for some time that entropies of subshifts of finite type for dimensions $d \geqslant 2$ are in general not computable, but the following result gives a precise characterization.

Theorem 4.6 (Hochman and Meyerovitch [75]). *The values of entropies of subshifts of finite type over $\mathbb{Z}^d$ for $d \geqslant 2$ are exactly the nonnegative right-c.e. reals.*

4.2 Algorithmic randomness and relative computability

Solovay reducibility is stronger than Turing reducibility, so Ω can compute the halting problem $\emptyset'$. Indeed Ω and $\emptyset'$ are Turing equivalent, and in fact Ω can be seen as a "highly compressed" version of $\emptyset'$. Other computability-theoretically powerful ML-random sequences can be obtained from the following remarkable result.

Theorem 4.7 (Gács [61], Kučera)[87]). *For every X there is an ML-random Y such that $X \leqslant_{\mathrm{T}} Y$.*

The proof of Theorem 4.7 uses a certain kind of weak-truth table procedure and a kind of "block coding". Recently Barmpalias and Lewis-Pye [11] have established a number of results giving a precise classification of how tightly an arbitrary X can be coded in to a random Y. This has resulted in a completely new optimal coding technique which should have other applications.

Theorem 4.7 and the Turing equivalence of Ω with $\emptyset'$ do not seem to accord with our intuition that random sets should have low "useful information". This phenomenon can be explained by results showing that, for certain purposes, the benchmark set by ML-randomness is too low. A set A has *PA degree* if it can compute a $\{0,1\}$-valued function f with $f(n) \neq \Phi_n(n)$ for all n. (The reason for the name is that this property is equivalent to being able to compute a completion of Peano Arithmetic.) If we can compute a function of this type which is not necessarily $\{0,1\}$ valued, we say that A had *DNC, diagonally noncomputable degree.* Such a function can be seen as a weak version of the halting problem, but while $\emptyset'$ has PA degree, there are sets of PA degree that are low, in the sense of Section 2.2, and hence are far less powerful than $\emptyset'$.

Theorem 4.8 (Stephan [126]). *If an ML-random sequence has PA degree then it computes* $\emptyset'$.

Thus there are two kinds of ML-random sequences. Ones that are complicated enough to somehow "simulate" randomness, and "truly random" ones that are much weaker.

On the other hand, if we remove the restriction that f must be $\{0,1\}$-valued we get the following:

Theorem 4.9 (Kučera [87]). *If A is ML-random then A computes a DNC function.*

The explanation of the apparent paradox is that if a function is $\{0,1\}$-valued, then saying it does not have value 1, means it must have value 0, and conversely. If more values are possible, saying what it is *not* does not imply what it is.

Proof. Let A be ML-random. Let $f(n)$ be the position of $A \restriction n$ in some effective listing of finite binary strings. Since A is ML-random,

$$K(f(n)) =^+ K(A \restriction n) \geq^+ n,$$

by Schnorr's Theorem, Theorem 2.9. On the other hand, if $varphi_n(n) \downarrow$, then $K(\varphi_n(n)) \leq^+ n$, so there are only finitely many n such that $f(n) = \varphi_n(n)$. By altering f at these finitely many places, we obtain an A-computable DNC function. □

Kučera's Theorem above received a lot of attention and generalizations. Crucial is the use of Schnorr's Theorem; but this also has generalizations. For example, we call a set A *complex* if there is an order h such that $K(A \restriction h(n)) \geq n$ for some computable order h, and we say that A is *autocomplex* if A is h-complex for some A-computable order.

Theorem 4.10. *(Kjos-Hanssen, Merkle, and Stephan [82]). A set is autocomplex iff it is of DNC degree.*

For more on this story, see [47], Ch. 8.

4.2.1 Stronger randomness

It is known that the class of sequences that can compute $\emptyset'$ has measure 0, so almost all ML-random sequences are in the second class. This fact is a special case of the following classical theorem

Theorem 4.11 (de Leeuw, et. al. [41], Sacks [119]). *Suppose that A is noncomputable. Then $\lambda\{X \mid A \leq_T X\} = 0$.*

Proof. This is included as it is a classic "majority vote" argument. Suppose that

$$\lambda\{X \mid A \leq_T X\} > 0.$$

Then for some fixed Φ, $\lambda\{X \mid \Phi^X = A\} > 0$, by the fact that there are only a countable number of procedures Φ. Then by Lebesgue Density, we can work in some cone of reals C where the relative density

$$\frac{\lambda\{X \mid \Phi^X = A \wedge \sigma \prec X\}}{\lambda\{Y \mid \sigma \prec Y\}} > \frac{3}{4}.$$

Then to compute $A \upharpoonright n$ for $n \geq |\sigma|$ wait till $\frac{3}{4}$ of the oracles X extending σ have $\Phi^X \upharpoonright\downarrow$ with a common value. This correctly computes $A \upharpoonright n$. □

One way to ensure that a sequence is in that class of reals with little usable information is to increase the complexity of our tests by relativizing them to non-computable oracles. It turns out that iterates of the Turing jump are particularly natural oracles to use. Let $\emptyset^{(0)} = \emptyset$ and $\emptyset^{(n+1)} = (\emptyset^{(n)})'$. We say that X is *n-random* if it passes all Martin-Löf tests relativized to to $\emptyset^{(n-1)}$. Thus the 1-random sequences are just the ML-random ones, while the 2-random ones are the ones that are ML-random relative to the halting problem. These sequences have low computational power in several ways. For instance, they cannot compute any noncomputable c.e. set, and in fact the following holds.

Theorem 4.12 (Kurtz [89]). *If X is 2-random and Y is computable relative both to $\emptyset'$ and to X, then Y is computable.*

A precise relationship between tests and the dichotomy mentioned above was established by Franklin and Ng [58] using another more technical notion of randomness.

We remark that of course, we can relativise the notion of Kolmogorov complexity, and see that as an analog of Schnorr's Theorem we have X is n-random iff $K^{\emptyset^{(n-1)}}(X \upharpoonright k) \geq^+ k$ for all k.

Quite remarkably, there is interdefinability between C and $C^{\emptyset^{(n)}}$ and similarly K and $K^{\emptyset^{(n)}}$. We will need the notion of notion of conditional Kolmogorov complexity $C(\sigma \mid \tau)$ of a string σ given another string τ (and same for K).

Theorem 4.13 (Vereshchagin [129]). $C^{\emptyset'}(\sigma) =^+ \limsup_n C(\sigma|n)$.

Proof. Let M be our fixed universal plain machine. Let $M(\tau, n) = M^{\emptyset'}(\tau)[n]$, where both the oracle and the universal machine are being approximated. If n is large enough and τ is a minimal-length $M^{\emptyset'}$-program for σ, then $M(\tau, n) = \sigma$, whence $C(\sigma \mid n) \upharpoonright |\tau| =^+ C^{\emptyset'}(\sigma) + O(1)$. Thus $\limsup_n C(\sigma|n) \leq C^{\emptyset'}(\sigma)$.

For the other direction, let $k = \limsup_n C(\sigma|n) + 1$. Let $V_n = \{\sigma \mid C(\sigma|n) < k$ We have $|V_n| < 2k$ for all n. Let $B = \{\tau \mid \exists m \forall n \geq m(|V_n \cup \{\tau\}) < 2^k\}$. Then B is $\emptyset'$-c.e. (uniformly in k), $\sigma \in B$, and $|B| < 2k$. Since we can describe σ from $\emptyset'$ by giving its position in the enumeration of B as a string of length k, we have $C^{\emptyset'}(\sigma) \leq^+ k =^+ \limsup_n C(\sigma|n)$. □

Slightly more complex methods yield the following.

Theorem 4.14 (Bienvenu, Muchnik, Shen, and Vereschagin [21]). $K^{\emptyset'}(\sigma) =^+ \limsup_n K(\sigma|n)$.

It follows that, for example, n-randomness can be defined using K alone, which is surprising in that on the face of it $K^{\emptyset'}$ would seem unrelated to K. We remark that earlier Solovay [125] has looked at relationships between K and $K^{\emptyset'}$ and some of this material can be found in [47]. Also there had been earlier results showing that 2-randomness, in particular, is naturally definable. We have seen that Martin-Löf showed that for no X is $C(X \upharpoonright n) \geq^+ n$ for all n. But it is possible that $C(X \upharpoonright n) \geq^+ n$ for *infinitely many* n, as shown by Martin-Löf [96], and he also showed that such sets were ML-random.

Theorem 4.15 (Miller [100], Nies, Stephan and Terwijn [112]). *X is 2-random iff $\exists^\infty n C(X \upharpoonright n) \geq^+ n$.*

The proof of this result in [112] uses an interesting notion called a compression function. We can also look at the K-analog. The maximal complexity n can be is $n+K(n)+O(1)$, as proved by Solovay [125]. Yu, Ding and Downey [134] proved that if a set is 3-random then it infinitely often will reach this complexity on its initial segments of length n. The final piece of the puzzle was supplied by Joe Miller:

Theorem 4.16 (Miller [101]). *X is 2-random iff $\exists^\infty n K(X \upharpoonright n) \geq^+ n + K(n)$.*

4.2.2 Computational depth

In general, among ML-random sequences, computational power (or "useful information") is inversely proportional to level of randomness. The following is one of many results attesting to this heuristic.

Theorem 4.17 (Miller and Yu [104]). *Let $X \leqslant_{\mathrm{T}} Y$. If X is ML-random and Y is n-random, then X is also n-random.*

Thus the notion that a random sequence has a "high information content" seems quite wrong. What is missing is the word "useful". There might be a lot of information but it is not accessible to a computational procedure. So when does a real contain useful in formation? One line of investigation in this area was pioneered by Bennett who defined X to be K-deep if we cannot know about A in any computable time. To wit, let K^t be a time bounded version of prefix-free complexity.

Definition 4.18. X is called *Bennett deep* or simply deep for short if for all all computable t and for all c and almost all n,

$$K^t(X \upharpoonright n) - K(X \upharpoonright n) > c.$$

If X is not deep it is called *shallow*.

The intuition is that X is deep because it contains a lot of information which is difficult to discover.

Theorem 4.19 (Bennett [18]).

1. *All computable and ML-random sets are shallow.*
2. *There are deep c.e. sets, such as the halting problem.*

The notion of depth has proven quite fruitful in giving insight into intrinsic information in languages, and several further variations on the notion, mainly involving orders[12] (in place of c) and C in place of K) have been studied. For example, Moser and Stephan [106] showed that a degree is high iff it contains a "strongly" deep set. See, for instance, [6, 7, 106], etc. As Moser [105] showed, all of these notions have a common interpretation in terms of computable time bounds and compression ratios.

All of this might lead one to suspect that Ω is a bit player in the area of algorithmic randomness. But perhaps the following two theorems say that in some sense it is a central concept.

[12] That is, a computable nondecreasing unbounded function.

Theorem 4.20 (Downey, Hirschfeldt, Miller, Nies [50]). *If X is 2-random then there is a set Y such that $X = \Omega^Y$. That is, almost all random reals are Ω-numbers relative to an oracle.*

In the same was that Ω has c.e. degree we get the following.

Theorem 4.21 (Kurtz [89]). *If X is 2-random then there is a $Z <_T X$ such that $deg_T(X)$ is c.e. relative to Z.*

The original proof of this theorem uses a technique called "*measure risking*" which allows for a procedure in a construction to be undefined on some small measure part of 2^ω, and after the fact, it is argued that the construction succeeds on all 2-randoms. Recently this idea has been portrayed using what has been dubbed by Shen to be a "fireworks" argument. We will illustrate this new method with a slightly easier proof that every 2-random bounds a 1-generic degree, where Z is called 1-generic iff for all c.e. sets of strings W, either $\exists \sigma \prec Z$ and σ has not extension in W, or there is some $\sigma \prec Z$ and $\sigma \in W$.

The fireworks metaphor is the following. We wish to purchase some fireworks from a seller who claims that all of them are good, but perhaps we are using them for an important party and it is crucial that they work when at the time. We have a lot of money so can test a large number. What we do is to ask the seller to show us, say, 100. We pick a number n between 1 and 100, randomly. We then test the first $n-1$ of the fireworks and is any fail then we reject the seller's package. If all work we accept and will use the n-th one for the party. (Of course we will need to pay for the first n, but that is another story.) For the seller to have sold us a dud, they would have to have guessed our n. The probability is at most $\frac{1}{100}$.

In computability, we use this idea for probabilistic forcing. Imagine we are meeting the requirements R_e asking that either we have some $\sigma \in W_e$ with $\sigma \prec Z$ or want to show that for some $\sigma \prec Z$, and $\tau \in W_e$ $\sigma \not\preceq \tau$.

Typically this would be done using the finite extension method, which is Cohen forcing with conditions being finite strings. Thus $\emptyset'$ can carry out such a construction. Here we need a probabilistic argument which will work for a 2-random oracle. Now in this construction we only need to do two things. If we are in a situation that we have build Z_s and there is no extension in W_e then we win by luck; the "passive guess." So, what we will do is begin a step by step construction which, for a fixed e, would pick a random $n(e)$ in some suitable interval (depending on the priority) and work with the passive guess for $n(e)$ many e-steps. That is, we assume that our guess is correct, but sometime later if we find out that it was not at that stage (i.e. Z_s actually had an extension in W_e) we would make another step. If we run out of steps say at s_1, then we will stop the construction doing the active guess seeking some $\tau \in W_e$ extending Z_{s_1}.

The random oracle is the one supplying the answers, and Γ^R would build a Martin-Löf type test, here a $\emptyset'$-computable one, such that oracles can only fail on such tests, in the sense that the construction would get stuck only inside open sets generated by the tests. For an $\emptyset'$-computable one, the construction would succeed outside it and hence a 2-random could carry the construction out. We refer the reader to Bienvenu and Patey [22] for more details and an interesting application to "bushy tree" forcing.

Of course it is impossible to combine Theorems 4.20 and 4.21. We remark that Ω showcases the difference between being c.e. relative to some set and "CEA" (c.e. relative to and *above*). In fact relativization of Ω is somewhat counter-intuitive:

Theorem 4.22 (Downey, Hirschfeldt, Miller, Nies [50]). *There are sets $A =^* B$ (i.e. the symmetric difference is finite) such that $\Omega^B|_T\Omega^A$. In fact, they are relatively random.*

For recent related work on Ω as an operator see Hölzl et. al. [76].

4.2.3 Calibrating randomness

There are many other interesting calibrations of algorithmic randomness. As we have seen, Schnorr [120] argued that his left-c.e. martingale, Theorem 2.4, characterization of ML-randomness shows that this is an intrinsically *computably enumerable* rather than *computable* notion. As well as defining computable randomness he also defined a concept now called *Schnorr randomness*, which is like the notion of computable randomness mentioned below Definition 2.3 but with an extra effectiveness condition on the rate of success of martingales. He also showed that X is Schnorr random iff it passes all Martin-Löf tests $T_0, T_1, \ldots$ such that the measures $\lambda(T_n)$ are uniformly computable (i.e., the function $n \mapsto \lambda(T_n)$ is computable in the sense of Section 5.4 below). It follows immediately from their definitions in terms of martingales that ML-randomness implies computable randomness, which in turn implies Schnorr randomness. It is more difficult to prove that none of these implications can be reversed. In fact, these levels of randomness are close enough that they agree for sets that are somewhat close to computable, as shown by the following result, where highness is as defined in Section 2.2.

Theorem 4.23 (Nies, Stephan, and Terwijn [112]). *Every high Turing degree contains a set that is computably random but not ML-random and a set that is Schnorr random but not computably random. This fact is tight, however, because every non-high Schnorr random set is ML-random.*

One last example of a variation is not called *Kurtz random* and is defined in a slightly different way.

Definition 4.24. We say that X is *Kurtz random* iff for all c.e. open sets O of measure 1, $X \in O$. The complement of O is called a Kurtz (null) test.

Kurtz randomness is a weak notion of randomness but coincides with ML-randomness on the hyperimmune-free degrees[13] We mention Kurtz randomness because it actually comes up in classifying theorems and randomness. For example, it can be shown that (with the correct definitions) a suitably computable bounded function on a closed interval has a Reimann integral iff it is undefined on Kurtz test. As we will discuss, various notions of algorithmic randomness arise naturally in applications. We already mentioned the fact that Kurtz randomness arises in the study of random (divergent) Fourier series [45]. Schnorr randomness arises a there also for convergent series.

4.3 Randomness-theoretic weakness

As mentioned above, X is ML-random iff $K(X \upharpoonright n) \geqslant n - O(1)$, i.e., X's initial segments have very high complexity. There are similar characterizations of other notions of algorithmic randomness, as well as of notions arising in other parts of computability theory, in terms of high initial segment complexity. But what if the initial segments of a sequence have *low* complexity? Such sequences have played an important role in the theory of algorithmic randomness, beginning with the following information-theoretic characterization of computability.

Theorem 4.25 (Chaitin [35]). *$C(X \upharpoonright n) \leqslant^+ C(n)$ iff X is computable.*

Proof. (Sketch) The first part of this proof is to use a combinatorial pigeonhole argument to show that

$$|\{\sigma \in 2^n \mid C(\sigma) < n + d\}| \leq O(2^d).$$

That is, only few strings have short descriptions, *and this is independent of n.* Now between lengths 2^k and 2^{k+1} there will be C-random lengths where $C(n) = \log|n|$. We can use this fact to build a computable tree of width 2^d such that if $C(A \upharpoonright n) \leq C(n) + d$ for all n, then A is a member of this tree, and hence is computable. □

It is also true that if X is computable then $K(X \upharpoonright n) \leqslant K(n) + O(1)$. Chaitin [36] considered sequences with this property, which are now called *K-trivial.* He

[13] A has hyperimmune-free degree iff for all functions $f \leq_T A$, there is a computable function g such that for all x, $f(x) < g(x)$. Some authors have called these *computably dominated* for obvious reasons. If A does not have hyperimmune-free degree it is said to have hyperimmune degree. All degrees computable from the halting problem are hyperimmune and the non-zero ones contain Kurtz random reals.

showed that every K-trivial sequence is $\emptyset'$-computable[14], and asked whether they are all in fact computable. Solovay [125] answered this question by constructing a noncomputable K-trivial sequence. Here is a simple construction of a K-trivial noncomputable real.

Theorem 4.26 (Zambella [135], after Solovay [125]). *There is a c.e. noncomputable K-trivial set.*

Proof. This proof is taken from Downey, Hirschfeldt, Nies and Stephan [52]. We define a c.e. set A as follows:

$$A_{s+1} = A_s \cup \{x \mid x \in W_{e,s} \wedge e \text{ least with } W_{e,s} \cap A_s = \emptyset \wedge \sum_{s \geq y \geq x} 2^{-K_s(y)} < 2^{-(2e+1)}\}.$$

Then A is K-trivial because we can build a machine M such that for all n, $K_M(A \restriction n) \leq K(n) + 1$. We can copy $U(n)[s]$ with σ describing $n \leq s$ by using 1σ, unless we change $A_s \restriction n$ in which case we use a string $10^e\tau$ with τ of length $\geq e+1$. The overall cost of the extra material is ≤ 1 and hence there is enough space in M to build the extra strings. □

The construction above is now in a class of "cost function" constructions which are summarized by "do what is cheap enough." In [52] it is shown that K-trivials are in fact characterized by cost functions. The class of K-trivials has several remarkable properties. It is a naturally definable *countable* class, contained in the class of low sets (as defined in Section 2.2, where we identify a set with its characteristic function, thought of as a sequence), but with stronger closure properties. (In technical terms, it is what is known as a *Turing ideal.*) Post's problem asked whether there are computably enumerable sets that are neither computable nor Turing equivalent to the halting problem. Its solution in the 1950's by Friedberg and Muchnik introduced the somewhat complex priority method, which has played a central technical role in computability theory since then. Downey, Hirschfeldt, Nies, and Stephan [52] showed that K-triviality can be used to give a simple priority-free solution to Post's problem.

Most significantly, there are many natural notions of randomness-theoretic weakness that turn out to be equivalent to K-triviality.

Theorem 4.27 (Nies [108], Nies and Hirschfeldt for (1) $\rightarrow$ (3)). *The following are equivalent.*

[14]This uses a similar argument to that of Theorem 4.25, but now note that we cannot know $K(n)$ for random n, in the same way that $C(n)$ will be $\log|n|$. Only $\emptyset'$ can know this.

1. *A is K-trivial.*

2. *A is computable relative to some c.e. K-trivial set.*

3. *A is* low for K, *meaning that A has no compression power as an oracle. i.e., that* $K^A(\sigma) \geqslant K(\sigma) - O(1)$, *where* K^A *is the relativization of prefix-free Kolmogorov complexity to A.*

4. *A is* low for ML-randomness, *meaning that A does not have any derandomization power as an oracle, i.e., any ML-random set remains ML-random when this notion is relativized to A.*

There are now over a dozen other characterizations of K-triviality. Some appear in [47, 109], and several others have emerged more recently (e.g. [64]). These have been used to solve several problems in algorithmic randomness and related areas.

Theorem 4.27 (2) above says that K-trivials are computable from c.e. K-trivials. In some sense this means that that they are intrinsically c.e. and cannot, it seems, by e.g. a forcing construction. Focussing on cost functions has allowed for a number of recent advances in the area. Nies [110] (which was available for some years before it was submitted) was the first to realize that this was a powerful abstraction in the area. Some recent examples of applications include [70, 65, 64]. This material seems tied up with derandomization power via another reducibility $A \leq_{LR} B$ meaning that $ML^A \supseteq ML^B$: everything A derandomized, B does too.

Lowness classes have also been found for other randomness notions. For Schnorr randomness, for instance, lowness can be characterized using notions of traceability related to concepts in set theory, as first explored by Terwijn and Zambella [127].

For example, we have the following.

Theorem 4.28 (Terwijn and Zambella [127]). *X is low for Schnorr tests iff X is* computably traceable. *This means that there is a computable order h such that for all* $f \leq_T X$, *we can compute an array of canonical finite sets* $\{D_{g(n)} \mid n \in \mathbb{N}\}$ *called a* trace *such that for all n,* $f(n) \in D_{g(n)}$.

Finally, using some earlier work of Bendregal and Nies, we have the following for the randomness concept:

Theorem 4.29 (Kjos-Hanssen, Nies and Stephan [83]). *X is low for Schnorr randomness iff X is low for Schnorr tests iff X is computably traceable.*

It is easy to see that if X is computably traceable it must be hyperimmune-free. And it is not hard to prove that there are continuum many computably traceable sets. The K-trivials are a countable collection of low sets below $\emptyset'$. Thus the classes are very different. In particular, it is not possible to define Schnorr randomness using K, nor ML-randomness using Schnorr tests, with any relativizable definition.

Similar characterizations were found for lowness for Kurtz randomness. Building on work of several authors, the final characterization for Kurtz randomness was the following.

Theorem 4.30 (Greenberg and Miller [66]). *A is low for Kurtz randomness iff A is hyperimmune-free and not DNC.*

Nies [108] showed that if X is low for computable randomness then X is computable.

Some of this work is related to coarse and generic computability mentioned below.

We remark that the use of traceing has become quite influential in computability theory. Can we find find a combinatorial definition of K-triviality, (i.e. not involving K) Nies and others suggested that it was related to *jump*-traceability. Let J^X denote the universal partial computable function. That is $J^X(e)$ denotes the actual value[15] (if any) of $\Phi_e(e)$.

Definition 4.31 (Figueira, Stephan and Nies [55]). If h is an order, we say that A is *jump traceable* at order h, if, for any A-partial computable function p, there is computable collection of $\{T_e \mid e \in \mathbb{N}\}$ of c.e. sets such that $p^A(e) \in T_e$ and $|T_e| \leq h(e)$ for all e.

We say that A is *strongly jump traceable* iff it is jump traceable for all orders h, iff J^A is jump traceable for all orders h.

For example, a c.e. set A is superlow (meaning $A' \equiv_{tt} \emptyset'$) iff A is jump traceable. In [55] it is shown that at order $h(n) = 2^{2n}$ there are $2^{\aleph_0}$ many h-jt sets.

Theorem 4.32 (Cholak, Downey and Greenberg [37]). *There is an order h where A being h-jt implies A is K-trivial.*

The order from [37] is around $\log \log n$, but this is certainly not optimal.

Question 4.33. Is there an order-characterization of being K-trivial. The guess would be that it would not involve a single order but a collection of a certain type, something like if $\{h_e \mid h_e \in S\}$ has property X (like some sum coverges) then A is K-trivial iff A is h_e-jump traceable for all $h_e \in S$.

Now it might seem that strong jump traceability is an artifact of stidies into randomness and now directly related but several results show that this is not the case. For example:

[15]That is not just if it halts or not.

Theorem 4.34 (Greenberg, Hirschfeldt, Nies [63]). *A is sjt iff A is computable from every superlow ML-random.*

There are similar results about ML-random reals X which are superhigh, which means that $X' \equiv_{tt} \emptyset''$ (see [63]), and characterizations involving another notion of randomness called *Demuth randomness*, which is defined via generalized ML-tests. We also remark that super jump traceable sets have been used to solve open questions in classical computability. For instance,we say that a c.e. set W is a cea-operator if for all Y, $Y <_T W^Y$. A classical theorem of Jockusch and Shore shows that for all such W there is a c.e. set Y with $Y \oplus W^Y \equiv_T \emptyset'$. This is called *pseudo-jump inversion.* Downey and Greenberg solved a longstanding question above cone avoidance by taking W to be the construction of a noncomputable strongly jump traceable set (so that W^Y was "very high" in that $\emptyset'$ would be stringly jump traceable relative to W^Y).

Theorem 4.35 (Downey and Greenberg [44]). *There is a noncomputable c.e. set B computable from all c.e. sets C with $\emptyset'$ strongly jump traceable relative to C. That is, W is a c.e. operator which cannot avoid upper cones under inversion.*

Recent unpublished work of Downey, Greenberg and Turetsky shows that B can be chosen to be superhigh. For the latest word here see Greenberg and Turetsky [71] for a survey of results and techniques.

5 Some applications

5.1 Incompressibility and information content

This article focuses on algorithmic randomness for infinite objects, but we should mention that there have been many applications of Kolmogorov complexity under the collective title of the *incompressibility method*, based on the observation that algorithmically random strings should exhibit typical behavior for computable processes. For example, this method can be used to give average running times for sorting, by showing that if the outcome is not what we would expect then we can compress a random input. See Li and Vitányi [90, Chapter 6] for applications of this technique to areas as diverse as combinatorics, formal languages, compact routing, and circuit complexity, among others. Many results originally proved using Shannon entropy or related methods also have proofs using Kolmogorov complexity. For example, Messner and Thierauf [99] gave a constructive proof of the Lovász Local Lemma using Kolmogorov complexity.

Other applications come from the observation that in some sense Kolmogorov complexity provides an "absolute" measure of the intrinsic complexity of a string. We can define a notion of conditional Kolmogorov complexity $C(\sigma \mid \tau)$ of a string σ given another string τ. Then, for example, $C(\sigma \mid \sigma) = O(1)$, and σ is "independent of τ" if $C(\sigma \mid \tau) = C(\sigma) - O(1)$. Researchers comparing two sequences σ, τ representing, say, two DNA sequences, or two phylogenetic trees, or two languages, or two pieces of music, have invented many distance metrics, such as the maximum parsimony distance on phylogenetic trees, but it is also natural to use a content-neutral measure of "information distance" like $\max\{C(\sigma \mid \tau), C(\tau \mid \sigma)\}$. There have been some attempts to make this work in practice for solving classification problems, though results have so far been mixed. Of course, C is not computable, but it can be replaced in applications by measures derived from practical compression algorithms. See [90, Sections 8.3 and 8.4].

5.2 Effective dimensions

If $X = x_0x_1\ldots$ is random, then we might expect that $x_000x_100x_200\ldots$ is "$\frac{1}{3}$-random". Making precise sense of the idea of partial algorithmic randomness has led to significant applications. Hausdorff used work of Carathéodory on s-dimensional measures to generalize the notion of dimension to possibly nonintegral values, leading to concepts such as Hausdorff dimension and packing dimension. Much like algorithmic randomness can make sense of the idea of individual reals being random, notions of partial algorithmic randomness can be used to assign dimensions to individual reals.

The measure-theoretic approach, in which we for instance replace the uniform measure λ on 2^ω by a generalized notion assigning the value $2^{-s|\sigma|}$ to $[\sigma]$ (where $0 < s \leqslant 1$), was translated by Lutz [91, 92] into a notion of *s-gale*, where the fairness condition of a martingale is replaced by $f(\sigma) = 2^{-s}(f(\sigma 0) + f(\sigma 1))$. We can view s-gales as modeling betting in a hostile environment (an idea due to Lutz), where "inflation" is acting so that not winning means that we automatically lose money. Roughly speaking, the effective fractal dimension of a sequence is then determined by the most hostile environment in which we can still make money betting on this sequence.

Mayordomo [97] and Athreya, Hitchcock, Lutz, and Mayordomo [8] found equivalent formulations in terms of Kolmogorov complexity, which we take as definitions. (Here it does not matter whether we use plain or prefix-free Kolmogorov complexity.)

Definition 5.1. [16] Let $X \in 2^\omega$. The *effective Hausdorff dimension* of X is

$$\dim(X) = \liminf_{n\to\infty} \frac{K(X \upharpoonright n)}{n}.$$

[16] Strictly speaking, this should be viewed as a Theorem, but it has become the standard definition. See [47], Chapter 13 for the full story.

The *effective packing dimension* of X is

$$\mathrm{Dim}(X) = \limsup_{n\to\infty} \frac{K(X \upharpoonright n)}{n}.$$

It is not hard to extend these definitions to elements of $\mathbb{R}^n$, yielding effective dimensions between 0 and n. They can also be relativized to any oracle A to obtain the effective Hausdorff and packing dimensions $\dim^A(X)$ and $\mathrm{Dim}^A(X)$ of X relative to A.

It is of course not immediately obvious why these notions are effectivizations of Hausdorff and packing dimension, but crucial evidence of their correctness is provided by *point to set principles*, which allow us to express the dimensions of sets of reals in terms of the effective dimensions of their elements. The most recent and powerful of these is the following, where we denote the classical Hausdorff dimension of $E \subseteq \mathbb{R}^n$ by $\dim_{\mathrm{H}}(E)$, and its classical packing dimension by $\dim_{\mathrm{p}}(E)$.

Theorem 5.2 (Lutz and Lutz [93]).

$$\dim_{\mathrm{H}}(E) = \min_{A\subseteq\mathbb{N}} \sup_{X\in E} \dim^A(X).$$

$$\dim_{\mathrm{p}}(E) = \min_{A\subseteq\mathbb{N}} \sup_{X\in E} \mathrm{Dim}^A(X).$$

For certain well-behaved sets E, relativization is actually not needed, and the classical dimension of E is the supremum of the effective dimensions of its points. In the general case, it is of course not immediately clear that the minima mentioned in Theorem 5.2 should exist, but they do. Thus, for example, to prove a lower bound of α for $\dim_{\mathrm{H}}(E)$ it suffices to prove that, for each $\varepsilon > 0$ and each A, the set E contains a point X with $\dim^A(X) > \alpha - \varepsilon$. In several applications, this argument turns out to be easier than ones directly involving classical dimension. This fact is somewhat surprising given the need to relativize to arbitrary oracles, but in practice this issue has so far turned out not to be an obstacle.

For example, Lutz and Stull [95] obtained a new lower bound on the Hausdorff dimension of generalized sets of Furstenberg type; Lutz [94] showed that a fundamental intersection formula, due in the Borel case to Kahane and Mattila, is true for arbitrary sets.

$E \subseteq \mathbb{R}^n$ is called a *Kakeya Set* for every point u on the unit sphere S^{n-1}, there is some point v such that the segment $\{su+v : s \in [0,1]\} \subseteq E$; that is E has unit lines in every direction. This is an important classical concept which has applications from harmonic analysis to extractors in complexity theory. Using the point to set principle, Lutz and Lutz [93] gave a new proof of the two-dimensional case (originally proved by Davies) of the well-known Kakeya conjecture, which states that, for all $n \geqslant 2$, if a subset of $\mathbb{R}^n$ has lines of length 1 in all directions, then it has Hausdorff dimension n. The method used by Lutz and Lutz filtered through the following result.

Theorem 5.3 (Lutz and Lutz [93]). *Let $a, b, x \in \mathbb{R}$. If a is ML-random and x is ML-random relative to the point (a, b), then the effective Hausdorff dimension of the point $(x, ax + b)$ is 2.*

The proof of Theorems 5.2 and 5.3 are familiar types of Kolmogorov complexity calculations, and are far from the classical techniques. Using Theorems 5.2 and 5.3 Lutz and Lutz gave the following proof of Davies theorem.

Proof. Let $E \subseteq \mathbb{R}^2$ be a Kakeya set, and let w be the minimizing oracle of Theorem 5.2. Let a be random ML-relative to w, and let b be such that the intersection of E with the line $y = ax + b$ contains a segment. Choose x random relative to (a, b, w) such that $(x, ax + b) \in E$. Then $\dim(E) = \sup_{z \in E} \dim^w(z) \geq \dim^w(x, ax + b)$, which is 2 by Theorem 5.3, applied relative to the oracle w. □

There had been earlier applications of effective dimension, for instance in symbolic dynamics, whose iterative processes are naturally algorithmic. For example, Simpson [123] generalized a result of Furstenberg as follows. Let A be finite and G be either $\mathbb{N}^d$ or $\mathbb{Z}^d$. A closed set $X \subseteq A^G$ is a *subshift* if it is closed under the shift action of G on A^G (see Section 4.1).

Theorem 5.4 (Simpson [123]). *Let A be finite and G be either $\mathbb{N}^d$ or $\mathbb{Z}^d$. If $X \subseteq A^G$ is a subshift then the topological entropy of X is equal both to its classical Hausdorff dimension and to the supremum of the effective Hausdorff dimensions of its elements.*

In currently unpublished work, Day has used effective methods to give a new proof of the Kolmogorov-Sinai theorem on entropies of Bernoulli shifts.

There are other applications of sequences of high effective dimension, for instance ones involving the interesting class of *shift complex* sequences. While initial segments of ML-random sequences have high Kolmogorov complexity, not all segments of such sequences do. Random sequences must contain arbitrarily long strings of consecutive 0's, for example. For example, if we knew that there was no sequence of 0's of length more than 12, but infinitely many of length 12, we could easily construct a martingales to succeed: wait for 12 0's and bet that the next bit was a 1!

However, for any $\varepsilon > 0$ there are *ε-shift complex* sequences Y such that for any string σ of consecutive bits of Y, we have $K(\sigma) \geqslant (1 - \varepsilon)|\sigma| - O(1)$. These sequences can be used to create tilings with properties such as certain kinds of pattern-avoidance, and have found uses in symbolic dynamics. See for instance Durand, Levin, and Shen [53] and Durand, Romashchenko, and Shen [54].

5.3 Randomness amplification

Many practical algorithms use random seeds. For example, the important *Polynomial Identity Testing* (*PIT*) problem takes as input a polynomial $P(x_1, \ldots, x_n)$ with coefficients from a large finite field and determines whether it is identically 0. Many practical problems can be solved using a reduction to this problem. There is a natural fast algorithm to solve it randomly: Take a random sequence of values for the variables. If the polynomial is not 0 on these values, "no" is the correct answer. Otherwise, the probability that the answer is "yes" is very high. It is conjectured that PIT has a polynomial-time deterministic algorithm,[17] but no such algorithm is known.

Thus it is important to have good sources of randomness. Some (including Turing) have believed that randomness can be obtained from physical sources, and there are now commercial devices claiming to do so. At a more theoretical level, we might ask question such as:

1. Can a weak source of randomness always be amplified into a better one?

2. Can we in fact always recover full randomness from partial randomness?

3. Are random sources truly useful as computational resources?

In our context, we can consider precise versions of such questions by taking randomness to mean algorithmic randomness, and taking all reduction processes to be computable ones. One way to interpret the first two questions then is to think of partial randomness as having nonzero effective dimension. For example, for packing dimension, we have the following negative results.

Theorem 5.5 (Downey and Greenberg [43]). *There is an X such that* $\mathrm{Dim}(X) = 1$ *and X computes no ML-random sequence. (This X can be built to be of* minimal degree, *which means that every X-computable set is either computable or has the same Turing degree as X. It is known that such an X cannot compute an ML-random sequence.)*

[17]This conjecture comes from the fact that PIT belongs to a complexity class known as BPP, which is widely believed to equal the complexity class P of polynomial-time solvable problems, since, in highly celebrated work, Impagliazzo and Wigderson [78] showed in the late 1990's that if the well-known Satisfiability problem is as hard as generally believed, then indeed BPP = P.

Theorem 5.6 (Conidis [39])). *There is an X such that* $\mathrm{Dim}(X) > 0$ *and X computes no Y with* $\mathrm{Dim}(Y) = 1$.

On the other hand, we also have the following strong positive result.

Theorem 5.7 (Fortnow, Hitchcock, Pavan, Vinochandran, and Wang [56]). *If $\varepsilon > 0$ and* $\mathrm{Dim}(X) > 0$ *then there is an X-computable Y such that* $\mathrm{Dim}(Y) > 1 - \varepsilon$. *(In fact, Y can be taken to be equivalent to X via polynomial-time reductions.)*

For effective Hausdorff dimension, the situation is quite different. Typically, the way we obtain an X with $\dim(X) = \frac{1}{2}$, say, is to start with an ML-random sequence and somehow "mess it up", for example by making every other bit a 0. This kind of process is reversible, in the sense that it easy to obtain an X-computable ML-random. However, Miller [102] showed that it is possible to obtain sequences of fractional effective Hausdorff dimension that permit no randomness amplification at all.

Theorem 5.8 (Miller [102]). *There is an X such that* $\dim(X) = \frac{1}{2}$ *and if $Y \leqslant_{\mathrm{T}} X$ then* $\dim(Y) \leqslant \frac{1}{2}$.

The proof of Theorem 5.8 is a novel forcing argument resulting in a Δ^0_2 set. The classification of such fractional dimension degrees is completely open.

Theorem 5.8 shows that effective Hausdorff dimension *cannot* in general be amplified. (In this theorem, the specific value $\frac{1}{2}$ is only an example.) Greenberg and Miller [67] also showed that there is an X such that $\dim(X) = 1$ and X does not compute any ML-random sequences.

There is one very intriguing open question here: We know that if A is random and $A_0 \oplus A_1 = A$ then A_i is relatively random to A_{1-i}. This fact is a basic result called van Lambalgen's Theorem. It implies that no random set can have minimal Turing degree.

Question 5.9. Can a set of Effective Hausdorff dimension have minimal Turing degree?

A positive answer would give a proof of a strengthening of the Greenberg-Miller Theorem.

Interestingly, Zimand [136] showed that for *two* sequences X and Y of nonzero effective Hausdorff dimension that are in a certain technical sense sufficiently independent, X and Y together can compute a sequence of effective Hausdorff dimension 1.

In some attractive recent work, it has been shown that there is a sense in which the intuition that every sequence of effective Hausdorff dimension 1 is close to an

ML-random sequence is correct. The following is a simplified version of the full statement, which quantifies how much randomness can be extracted at the cost of altering a sequence on a set of density 0. Here $A \subseteq \mathbb{N}$ has *(asymptotic) density* 0 if $\lim_{n\to\infty} \frac{|A\upharpoonright n|}{n} = 0$.

Theorem 5.10 (Greenberg, Miller, Shen, and Westrick [68]). *If* $\dim(X) = 1$ *then there is an ML-random* Y *such that* $\{n : X(n) \neq Y(n)\}$ *has density* 0.

We remark that asymptotic density has seen a lot of work recently. We say that an algorithm Φ *coarsely* computes a set X if the density of $\{n \mid \Phi(n) \neq X(n)\}$ is zero. We say that Ψ *generically* computes X if $\Psi(n)\downarrow$ implies $\Psi(n) = X(n)$ and $\{n \mid \Psi(n)\uparrow\}$ has density 1. These concepts arose in combinatorial group theory. See e.g. Kapovich, Miasnikov, Schupp and Shpilrain [81].

Here is one example applied to a word problem. (The density here is natural as measured by the words generated by the generators.)

- Let $G = \langle a, b; R\rangle$ be any 2-generator group.
- Note Any countable group is embeddable in a 2-generator group so there are uncountably many such G.
- Let $F = \langle x, y \mid \rangle$ be the free group of rank 2.
- $H = G * \langle x, y\rangle := \langle a, b, x, y; R\rangle$ be the free product of G and F.
- Then the word problem for H is generically solvable in linear time.

To see this, take a long word w on the alphabet $\{a, b, x, y\}^{\pm 1}$, e.g. $abx^{-1}bxyaxbby$. Now erase the a, b symbols, freely reduce the remaining word on $\{x, y\}^{\pm 1}$, and if any letters remain, output "no". This partial algorithm gives no incorrect answers because if the image of w under the projection homomorphism to the free group F is not 1, then $w \neq 1$ in H.

$$abx^{-1}bxyaxbby \to x^{-1}xyxy \to yxy \neq 1$$

The successive letters on $\{x, y\}^{\pm 1}$ in a long random word $w \in H$ is a long random word in F which is not equal to the identity. So the algorithm answers "No" on a generic set and gives no answer if the image in F is equal to the identity. This method is called the *quotient method* and can be used for any $G = \langle X, R\rangle$ subgroup of K of finite index for which there is an epimorphism $K \to H$ hyperbolic and not virtually cyclic, to show generically solvable word problem.

There has been a lot of work understanding coarse and generic computability, especially in group theory, but also some in computability theory such as Jockusch and Schupp [79]. One very nice theorem from that paper.

Theorem 5.11 (Jockusch and Schupp [79]). *There exists a c.e. set A of density 1 which has no computable subset of density 1.*

As well, papers such as [73] and [5] have shown that coarse computability and algorithmic randomness are very closely related. Here is one typical theorem.

Theorem 5.12 (Hirschfeldt, Kuyper, Jockusch and Schupp [73]). *If A is ML-random and Bi is computable from every coarse description D of A, then B is K-trivial. Thus, if A is in 2-random*[18] *then B is computable.*

Work is ongoing. It would also be interesting to develop this kind of analysis in the setting of computable analysis. Here generic case and coarse complexity would likely be replaced by measure. In some sense this is discussed in the material on Ergodic Theory below.

The third question above is whether sources of randomness can be useful oracles. Here we are thinking in terms of complexity rather than just computability, so results such as Theorem 4.7 are not directly relevant. Allender and others have initiated a program to investigate the speedups that are possible when random sources are queried efficiently. Let R be the set of all random finite binary strings for either plain or prefix-free Kolmogorov complexity (e.g., $R = \{x : C(x) \geqslant |x|\}$). For a complexity class $\mathcal{C}$, let $\mathcal{C}^R$ denote the relativization of this class to R. So, for instance, for the class P of polynomial-time computable functions, P^R is the class of functions that can be computed in polynomial time with R as an oracle. (For references to the articles in this and the following theorem, see [1].)

Theorem 5.13 (Buhrman, Fortnow, Koucký, and Loff [30]; Allender, Buhrman, Koucký, van Melkebeek, and Ronneburger [3]; Allender, Buhrman, and Koucký [2]).

1. $\mathrm{PSPACE} \subseteq \mathrm{P}^R$.

2. $\mathrm{NEXP} \subseteq \mathrm{NP}^R$.

3. $\mathrm{BPP} \subseteq \mathrm{P}^R_{tt}$ *(where the latter is the class of functions that are reducible to R in polynomial time via truth-table reductions, a more restrictive notion of reduction than Turing reduction).*

The choice of universal machine does have some effect on efficient computations, but we can quantify over all universal machines. In the result below, U ranges over universal prefix-free machines, and R_{K_U} is the set of random strings relative to Kolmogorov complexity defined using U.

[18] They actually proved this for A being only weakly 2-random, meaning that A passes all ML-tests for which the modulus of convergence is not necessarily computable, only that $\lambda(U_n) \to 0$.

Theorem 5.14. (Allender, Friedman, and Gasarch [4]; Cai, Downey, Epstein, Lempp, and Miller [31])

1. $\bigcap_U \mathrm{P}_{tt}^{R_{K_U}} \subseteq \mathrm{PSPACE}$.
2. $\bigcap_U \mathrm{NP}^{R_{K_U}} \subseteq \mathrm{EXPSPACE}$.

We can also say that sufficiently random oracles will always accelerate *some* computations in the following sense. Say that X is *low for speed* if for any computable set A and any function t such that A can be computed in time $t(n)$ using X as an oracle, there is a polynomial p such that A can be computed (with no oracle) in time bounded by $p(t(n))$. That is, X does not significantly accelerate any computation of a computable set. Bayer and Slaman (see [20]) constructed noncomputable sets that are low for speed, but these cannot be very random.

Theorem 5.15 (Bienvenu and Downey [20]). *If X is Schnorr random, then it is not low for speed, and this fact is witnessed by an exponential-time computable set A.*

The proof of this theorem uses a kind of speed-up technique. Interestingly, whether generic sets speed up computations depends on $P \neq NP$?. (See [20], this result is due to Bayer in his PhD Thesis.)

5.4 Analysis and Ergodic Theory

The setting for this is the area of computable analysis. For simplicity, our spaces will have dense computable bases, like the rationals in $\mathbb{R}$. Classically we know that reals are Cauchy sequences, and in computable analysis, we regard a function f as being computable iff then f is *(Type 2) computable*[19] if there is a uniform algorithm Φ taking fast converging Cauchy sequences for input x (i.e. $q_k \in B(q_n, 2^{-n})$ for all $k > n$) to fast converging Cauchy sequences for output $f(x)$. Notice that since the objects now are infinite, we won't have a finitely computable equality operator, rather we will have a computable distance function d in the sense that if x and y are reals, then uniformly in Cauchy sequences for each we can generate one for $d(x, y)$. Thus, computable analysis is an area that has developed tools for thinking about computability of objects like real-valued functions by taking advantage of separability. We can also relativize it, and it is then not difficult to see that a function is continuous iff it is computable relative to some oracle, basically because

[19] Type 1 functions take $\mathbb{N}$ to itself, and type 2 take type 1 objects to themselves, so act on infinite sequences.

to define a continuous function we need only to specify its action on a countable collection of balls. Many results in this are show that our intuition about "good" vs "bad" is realized. For example we have the following (precise definitions are not important here.)

Theorem 5.16 (Pour-E and Richards see [117]). *In this setting an operator is computable iff it is bounded.*

A consequence of this is that there is a computable ODE with computable initial conditions and having no computable solution.

Mathematics is replete with results concerning almost everywhere behavior, and algorithmic randomness allows us to to turn such results into "quantitative" ones like the following.

Theorem 5.17 (Brattka, Miller, and Nies [27], also Demuth (1975, see [27]) for (2)).

1. *The reals at which every computable increasing function $\mathbb{R} \to \mathbb{R}$ is differentiable are exactly the computably random ones.*
2. *The reals at which every computable function $\mathbb{R} \to \mathbb{R}$ of bounded variation is differentiable are exactly the ML-random ones.*

Ergodic theory is another area that has been studied from this point of view. A measure-preserving transformation T on a probability space is *ergodic* if all measurable subsets E such that $T^{-1}(E) = E$ have measure 1 or 0. Notice that this is an "almost everywhere" definition. We can make this setting computable (and many systems arising from physics will be computable). One way to proceed is to work in Cantor space without loss of generality, since Hoyrup and Rojas [77] showed that any computable metric space with a computable probability measure is isomorphic to this space in an effective measure-theoretic sense. Then we can specify a computable transformation T as a computable limit of computable partial maps $T_n : 2^{<\omega} \to 2^{<\omega}$ with certain coherence conditions. We can also transfer definitions like that of ML-randomness to computable probability spaces other than Cantor space.

The following is an illustrative result. A classic theorem of Poincaré is that if T is measure-preserving, then for all E of positive measure and almost all x, we have $T^n(x) \in E$ for infinitely many n. For a class $\mathcal{C}$ of measurable subsets, x is a *Poincaré point* for T with respect to $\mathcal{C}$ if for every $E \in \mathcal{C}$ of positive measure, $T^n(x) \in E$ for infinitely many n. An *effectively closed* set is one whose complement can be specified as a computably enumerable union of basic open sets.

Theorem 5.18 (Bienvenu, Day, Mezhirov, and Shen [19]). *Let T be a computable ergodic transformation on a computable probability space. Every ML-random element of this space is a Poincaré point for the class of effectively closed sets.*

The reader might note that the hypothesis above did not say "T is measure-preserving." This case has been analysed, and Frankin and Towsner [59] proved that the Poincaré recurrence aligns itself to yet another randomness notion called "weak 2-randomness" which is defined exactly as we did for ML-randomness, except that we only asks that $\lambda(T_n) \to 0$, without knowing the modulus of convergence. Franklin and Towsner (and others) have analysed the Birkhoff recurrence theorem, and again various interpretations align to randomness notions. In terms of the physical interpretation of these results, Bravermann, Rojas and Schneider [28] have argued that, whilst noise makes short term behaviour difficult, in fact it allows prediction easier in the long term. While many of theorems of Ergodic theory have been analysed, including the Birkhoff, Poincaré, and von Neumann ergodic theorems, but some, like Furstenberg's ergodic theorem, are yet to be understood.

In analysis, there are other theorems which concern almost everywhere behaviour. One of the problems is how to address this effectively, since a function being computable relative to an oracle imples that it must be continuous. This means even step functions with a single computable step at, for instance, 0 is not computable relative to any oracle. There seem several ways around this one especially in the case of almost everywhere behaviour. One way was suggested by Hoyrup and Rojas [77] which defines a notion of *layerwise computability*, namely if $\{U_n \mid n \in \mathbb{N}\}$ is a ML-test then demanding that f is represented by a sequence of functions f_n such that f_n acts on (names of) reals X outside of U_n, and correctly gives an answer, such that if X passes the test, then $f(X) = f_n(X)$ for all n. This is also related to an earlier suggestion of Ko and Friedman [84] who suggested that it might be reasonable to look at f defined on a Kurtz test. The latter suggestion might certainly be easier in the setting of computational complexity. Again this is largely undeveloped.

5.5 The full circle: Turing, Borel and normality again

We return to Borel's notion of normality. This is a very weak form of randomness; polynomial-time randomness is more than enough to ensure absolute normality. Schnorr and Stimm [121] showed that a sequence is normal if and only if it satisfies a notion of randomness defined using with martingales defined by certain finite state automata[20]. Building examples of absolutely normal numbers is another matter, as

[20] An finite state automaton is a constrained Turing machine which will read an input tape and according to the symbol it is reading, and its internal state, transitions to a (perhaps) new state

Borel already noted. While it is conjectured that e, π, and all irrational algebraic numbers such as $\sqrt{2}$ are absolutely normal, *none* of these have been proved to be normal to *any* base. In his unpublished manuscript "A note on normal numbers", believed to have been written in 1938, Turing built a computable absolutely normal real, which is in a sense the closest we have come so far to obtaining an explicitly-described absolutely normal real. (His construction was not published until his Collected Works in 1992, and there was uncertainty as to its correctness until Becher, Figueira, and Picchi [15] reconstructed and completed it, correcting minor errors.[21])

An interesting aspect of Turing's construction is that he more or less anticipated Martin-Löf's work by looking at a collection of computable ML-style tests sensitive enough to make a number normal in all bases, yet insensitive enough to allow computable sequences to pass all such tests. We have seen that the strong law of large implies fixed blocks of digits should occur with the appropriate frequencies in a random sequence. Translating between bases results in correlations between blocks of digits in one base and blocks of digits in the other, which is why this extension allowed Turing to construct absolutely normal numbers. Turing made enough of classical measure theory computable to generate absolute normality, yet had the tests refined enough that computable sequence could still be "random".

This approach can also be thought of in terms of effective martingales, and its point of view has brought about a great deal of progress in our understanding of normality recently. For instance, Becher, Heiber, and Slaman [16] showed that absolutely normal numbers can be constructed in low-level polynomial time, and Lutz and Mayordomo (`arXiv:1611.05911`) constructed them in "nearly linear" time. Much of the work along these lines has been number-theoretic, connected to various notions of well-approximability of irrational reals, such as that of a *Liouville number*, which is an irrational α such that for every natural number $n > 1$, there are $p, q \in \mathbb{N}$ for which $|\alpha - \frac{p}{q}| < q^{-n}$. For example, Becher, Heiber, and Slaman [16] have constructed computable absolutely normal Liouville numbers. This work has also produced results in the classical theory of normal numbers, for instance by Becher, Bugeaud, and Slaman [14].

---

and moves on to the next symbol to the right of the input tape. The automaton accepts the input if when it gets to the last symbol it is in one of the designated accept states. It is possible to have a notion of randomness using "automatic" martingales using this idea and some definitions a wee bit too technical to include in this article.

[21] See `https://www-2.dc.uba.ar/staff/becher/publications.html` for references to the papers cited here and below.

6 Summary

We have given a reasonably self-contained, if perhaps idiosyncratic, account of many of the basics of algorithmic randomness as well as a number of recent applications. Clearly the area is in a state of rapid development and we have necessarily left a lot out, but we suggest that the reader follows up the references for more details. I have definitely left a huge amount out, such as randomness lower down, in complexity classes, and also higher up, with Π^1_1-randomness, a concept going back to Martin-Löf but having a lot of interest recently, in papers such as Greenberg and Monin [69] and Hjorth-Nies [74]. Nor do I discuss the developing area of ML-randomness and quantum physics such as [111]; nor Brownian motion such as [57, 116]. These areas are all in rapid growth. The present article should at least give pointers to the aspirations and scope of such studies.

References

[1] Eric Allender *The complexity of complexity*, in: Computability and Complexity, Springer, Cham 2017 [*Lecture Notes in Computer Science 10010*], pp. 79–94.

[2] Eric Allender, Harrry Buhrman, and Michael Koucký *What can be efficiently reduced to the Kolmogorov-random strings?* Annals of Pure and Applied Logic, 138 (1–3), 2006, pp. 2–19.

[3] Eric Allender, Harry Buhrman, Michael Koucký, Dieter van Melkebeek, and Detlef Ronneburger *Power from random strings*, SIAM Journal on Computing 35 (6), 2006, pp. 1467–1493.

[4] Eric Allender, Luke Friedman, and William Gasarch *Limits on the computational power of random strings*, Information and Computation 222, 2013, pp. 80–92.

[5] Uri Andrews, Mingzhong Cai, David Diamondstone, Carl Jockusch, and Steffen Lemp *Asymptotic density, computable traceability and 1-randomness*, Fundamenta Mathematicae 234, 2016, pp. 41–53.

[6] Luis Antunes, Lance Fortnow, Dieter van Melkelbeck, and Vijay Vinochandram *Computational depth: Concept and applications*, Theoretical Computer Science 354, 2006, pp. 391–404.

[7] Luis Antunes, Armando Matos, Andre Souto, and Paul Vitanyi *Depth as randomness deficiency*, Theory of Computing Systems 45, 2009, pp. 724–739.

[8] Krishna Athreya, John Hitchcock, Jack Lutz, and Elvira Mayordomo *Effective strong dimension in algorithmic information and computational complexity*, SIAM Journal on Computing 37 (3), 2007, pp. 671–705.

[9] George Barmpalias, and Andrew Lewis-Pye *Monotonous betting strategies in warped casinos*, Information and Computation 271, 2020 [to appear].

[10] — *Differences of halting probabilities*, Journal of Computer and System Sciences 89, 2017, pp. 349–360.

[11] — *Optimal redundancy in computations from random oracles*, Journal of Computer and System Sciences 92, 2018, pp. 1–8.

[12] George Barmpalias, and Rod Downey *Kobayashi compressibility*, Theoretical Computer Science A 675, 2017, pp. 89–100.

[13] Bruno Bauwens *Relating and contrasting plain and prefix Kolmogorov complexity*, Theory of Computing Systems 58, 2015, pp. 482–501.

[14] Veronica Becher, Yves Bugeaud, and Theodore Slaman *On simply normal numbers to different bases*, Mathematische Annalen 364 (1–2), 2016 pp., 125–150.

[15] Veronica Becher, Santiago Figueira, and Rafael Picchi *Turing's unpublished algorithm for normal numbers*, Theoretical Computer Science 377 (1–3), 2007, pp. 126–138.

[16] Veronica Becher, Pablo Heiber, and Theodore Slaman *A polynomial-time algorithm for computing absolutely normal numbers*, Information and Computation 232, 2013, pp. 1–9.

[17] Charles Bennett, Peter Gács, Ming Li, Paul Vitanyi, and Wojciech Zurek *Information distance*, arxiv.org/abs/1006.3520v1.

[18] Charles Bennett *Logical depth and physical complexity*, in: Rola Herken (ed.), The Universal Turing Machine, *A Half-Century Survey*, Oxford University Press, Oxford 1992, pp. 227–257.

[19] Laurent Bienvenu, Adam Day, Ilya Mezhirov, and Alexander Shen *Ergodic-type characterizations of algorithmic randomness*, in: Programs, Proofs, Processes, Springer, Berlin 2010 [*Lecture Notes in Computer Science 6158*], pp. 49–58.

[20] Laurent Bienvenu, and Rod Downey *On Low for Speed Oracles*, in: R. Niedermeier, and B. Vallée (eds.) 35th Symposium on Theoretical Aspects of Computer Science (STACS 2018), Leibniz International Proceedings in Informatics (LIPIcs) 96, pp. 15:1–15:13, Germany, Schloss Dagstuhl–Leibniz-Zentrum für Informatik, 2018.

[21] Laurent Bienvenu, An. A. Muchnik, Alexander Shen, and Nicolai Vereshchagin *Limit complexities revisited*, in: 25th International Symposium on Theoretical Aspects of Computer Science, STACS 2008, pp. 73–84, Leibniz International Proceedings in Informatics, 2008.

[22] Laurent Bienvenu and Ludovic Patey *Diagonally non-computable functions and fireworks*, Information and Computation 253, 2017, pp. 64–77.

[23] Laurent Bienvenu, Andrei Romashchenko, Alexander Shen, Antoine Taveneaux, and Stijn Vermeeren *The axiomatic power of Kolmogorov complexity*, Annals of Pure and Applied Logic 165 (9), 2014, pp. 1380–1402.

[24] Laurent Bienvenu, Frank Stephan, and Jason Teutsch *How powerful are integer valued martingales?*, in: Proceedings CIE 2010, 2012.

[25] Béla Bollobás Random Graphs, Cambridge University Press, Cambridge 2001.

[26] Emil Borel *Les Probabilités dénombrables et leurs applications arithmétiques*, Rendiconti del Circolo Matematico di Palermo (1884–1940) *27*, 1909, pp. 247–271.

[27] Vasco Brattka, Joseph Miller, and Andre Nies *Randomness and differentiability*, Transactions of the American Mathematical Society 368, 2016, pp. 581–605.

[28] Mark Braverman, Cristobal Rojas, and Jonathan Schnieder *Space-bounded church-turing thesis and computational tractability of closed systems*, Phys. Rev. Lett. 115, Aug 28; 115 (9): 098701. Epub 2015, August 27, 2015.

[29] Mark Braverman and Michael Yampolsky *Computability of Julia sets*, Moscow Mathematical Journal 8 (2) 2008, pp. 185–231.

[30] Harry Buhrman, Michal Koucký, Lance Fortnow, and Bruno Loff *Derandomizing from random strings*, in: 25th Annual IEEE Conference on Computational Complexity CCC 2010, IEEE Computer Society, Los Alamitos CA 2010, pp. 58–63.

[31] Mingzhong Cai, Rod Downey, Rachel Epstein, Steffen Lempp, and Joseph Miller *Random strings and truth-table degrees of Turing complete c.e. sets*, Logical Methods in Computer Science, 10 (3), 2014, pp. 3-15.

[32] Cristian Calude, Peter Hertling, Bakhadyr Khoussainov, and Yongge Wang *Recursively enumerable reals and Chaitin Ω numbers*, in: M. Morvan, C. Meinel, and D. Krob (eds.), STACS 98, 15th Annual Symposium on Theoretical Aspects of Computer Science (Paris, France, February 25–27, 1998), Proceedings [*Lecture Notes in Computer Science 1373*], Springer, Berlin 1998, pp. 596–606.

[33] Gregory Chaitin *Computational complexity and Gödel's incompleteness theorem*, ACM SIGACT News 9, 1971, pp. 11–12.

[34] — *A theory of program size formally identical to information theory*, Journal of the Association for Computing Machinery 22, 1975, pp. 329–340.

[35] — *Information-theoretical characterizations of recursive infinite strings*, Theoretical Computer Science 2, 1976, pp. 45–48.

[36] — *Nonrecursive infinite strings with simple initial segments*, IBM Journal of Research and Development 21, 1977 pp. 350–359.

[37] Peter Colak, Rod Downey, and Noam Greenberg *Strong jump traceablilty I, The computably enumerable case*, Advances in Mathematics 217, 2008, pp. 2045–2074.

[38] Alonzo Church *On the concept of a random sequence*, Bulletin of the American Mathematical Society 46, 1940, pp. 130–135.

[39] Chris Conidis *A real of strictly positive effective packing dimension that does not compute a real of effective packing dimension one*, The Journal of Symbolic Logic 77 (2), 2012 pp. 447–474.

[40] Adam Day *Process and truth-table characterizations of randomness*, Theoretical Computer Science 452, 2012, pp. 47–55.

[41] Karel de Leeuw, Edward Moore, Claude Shannon, and Norman Shapiro Computability by Probabilistic Machines, *Automata Studies*, Princeton University Press, Princeton NJ 1956 [*Annals of Mathematics Studies 34*].

[42] Oswald Demuth *The differentiability of constructive functions of weakly bounded variation on pseudo numbers*, Commentationes Mathematicae Universitatis Carolinae 16 (3), 1975, pp. 583–599.

[43] Rod Downey, and Noam Greenberg *Turing degrees of reals of positive effective packing dimension*, Information Processing Letters 108, 2008, pp. 298–303.

[44] — *Pseudo-jump inversion, upper cone avoidance, and strong jump-traceability*, Advances in Mathematics 237, 2013, pp. 252–285.

[45] Rod Downey, Noam Greenberg, and Andrew Tangarra *Divergence and convergence of random series* [in preparation].

[46] Rod Downey, and Evan Griffiths *Schnorr randomness*, The Journal of Symbolic Logic 69, 2004, pp. 533–554.

[47] Rod Downey and Denis Hirschfeldt Algorithmic Randomness and Complexity, *Theory and Applications of Computability*, Springer, New York 2010.

[48] — *Algorithmic randomness*, Communications of The Association for Computing Machinery 62, 2019, pp. 70–80.

[49] — *Computability and randomness*, Notices of the American Mathematical Society 66 (7), 2019, pp. 1001–1012.

[50] Rod Downey, Denis Hirschfeldt, Joseph Miller, and Andre Nies *Relativizing Chaitin's halting probability*, Journal of Mathematical Logic 5, 2005, pp. 167–192.

[51] Rod Downey, Denis Hirschfeldt, and Andre Nies *Randomness, computability and density*, SIAM Journal of Computing 31, 2002, pp. 1169–1183.

[52] Rod Downey, Denis Hirschfeldt, Andre Nies, and Frank Stephan *Trivial reals*, in: Rod Downey, Decheng Ding, Shi Tung, Yu Qiu, and Mariko Yasugi (eds.), Proceedings of the 7th and 8th Asian Logic Conferences Held in Hsi-Tou, June 6–10, 1999, and Chongqing, August 29 – September 2, 2002), Singapore University Press and World Scientific, Singapore 2003, pp. 103–131.

[53] Bruno Durand, Leonid Levin, and Alexander Shen *Complex tilings*, The Journal of Symbolic Logic 73 (2), 2006, pp. 593–613.

[54] Bruno Durand, Andrei Romashchenko, and Alexander Shen *Fixed-point tile sets and their applications*, Journal of Computer and System Sciences 78 (3), 2012, pp. 731–764.

[55] Santiago Figueira, Frank Stephan, and Andre Nies *Lowness properties and app.roximations of the jump*, Annals of Pure and Applied Logic 152, 2008, pp. 51–66.

[56] Lance Fortnow, John Hitchcock, A. Pavan, N. Vinochandran, and Fengming Wang *Extracting Kolmogorov complexity with applications to dimension zero-one laws*, in: M. Bugliesi, B. Preneel, V. Sassone, and I. Wegener (eds.) Automata, Languages and Programming (33rd International Colloquium, ICALP 2006, Venice, Italy, July 10–14, 2006, Proceedings, Part I), Springer, Berlin 2006 [*Lecture Notes in Computer Science 4051*], pp. 335–345.

[57] Willem Fouche *Diophantine properties of Brownian motion: recursive aspects*, in: Viasco Brattka, Hannes Diener, and Dieter Spreen (eds.), Logic, Computation, Hierarchies, Springer-Verlag, 2014, pp. 139–156.

[58] Johanna Franklin, and Keng Meng Ng *Difference randomness*, Proceedings of the American Mathematical Society 139 (1), 2011, pp. 345–360.

[59] Johanna Franklin, and Henry Towsner *Randomness and non-ergodic systems*, Moscow

Mathematical Journal 14 (4), 2014, pp. 711–744.

[60] Peter Gács *On the symmetry of algorithmic information*, Soviet Mathematics Doklady 15 1974, pp. 1477–1480.

[61] — *Every set is reducible to a random one*, Information and Control 70, 1986, pp. 186–192.

[62] Michael Garey, and David Johnson. Computers and Intractability, Freeman, 1979.

[63] Noam Greenberg, Denis Hirschfeldt, and Andre Nies *Characterizing the strongly jump-traceable sets via randomness*, Advances in Mathematics 231, 2012, pp. 2252–2293.

[64] Noam Greenberg, Joseph Miller, Benoit Monin, and Daniel Turetsky *Two more characterizations of K-triviality*, Notre Dame Journal of Formal Logic 59 (2), 2018, pp. 189–195.

[65] Noam Greenberg, Joseph Miller, Andre Nies, and Daniel Turetsky *Martin-Löf reducibility and cost functions* [submitted].

[66] Noam Greenberg, and Joseph Miller *Lowness for Kurtz randomness*, The Journal of Symbolic Logic 74, 2009, pp. 665–678.

[67] — *Diagonally non-recursive functions and effective Hausdorff dimension*, Bulletin of the London Mathematical Society 43 (4), 2011, pp. 636–654.

[68] Noam Greenberg, Joseph Miller, Alexande Shen, and Linda Brown Westrick *Dimension 1 sequences are close to randoms*, Theoretical Computer Science 705, 2018, pp. 99–112.

[69] Noam Greenberg, and Benoît Monin *Higher randomness and genericity*, Forum of Mathematics: Sigma 5, 41 pp., 2017.

[70] Noam Greenberg, and Andre Nies *Benign cost functions and lowness notions*, Journal of Symbolic Logic 76, 2011, pp. 289–312.

[71] Noam Greenberg, and Daniel Turetsky *Strong jump-traceability*, The Bulletin of Symbolic Logic 24, 2018, pp. 147–164.

[72] Daniel Hammer, Andrei Romashchenko, Alexander Shen, and Nikolai Vereshchagin *Inequalities for Shannon entropies and Kolmogorov complexities*, in: IEEE Conference on Computational Complexity, 1997. (Journal version in: Journal of Computing and System Sciences 60, 2000, pp. 442–464.)

[73] Denis Hirschfeldt, Carl Jockusch, Rutger Kuyper, and Paul Schupp *Coarse reducibility and algorithmic randomness*, The Journal of Symbolic Logic 81, 2016, pp. 1028 – 1046.

[74] Gregory Hjorth, and Andre Nies *Randomness via effective descriptive set theory*, Journal of the London Mathematical Society 75, 2007, pp. 495–508.

[75] Michael Hochman, and Tom Meyerovitch *A characterization of the entropies of multidimensional shifts of finite type*, Annals of Mathematics (second series) 171 (3), 2010, pp. 2011–2038.

[76] Rupert Hölzl, Wolfgang Merkle, Frank Stephan, and Liang Yu *Chaitin's ω as a continuous function*, The Journal of Symbolic Logic 85, 2020, pp. 486–510.

[77] Mathieu Hoyrup, and Cristobal Rojas *Computability of probability measures and Martin-Löf randomness over metric spaces*, Information and Computation 207 (7), 2009,

pp. 830–847.

[78] Russel Impagliazzo, and Avi Wigderson *P = BPP if E requires exponential circuits: derandomizing the XOR lemma*, in: STOC '97 (El Paso TX), ACM, New York, 1999, pp. 220–229.

[79] Carl Jockusch, and Paul Schupp *Generic computability, Turing degrees, and asymptotic density*, Journal of the London Mathematical Society 85 (2), 2012, pp. 472–490.

[80] Jean-Pierre Kahane Some Random Series of Functions, Cambridge University Press, Cambridge UK 2003.

[81] Ilya Kapovich, Alexei Myasnikov, Paul Schupp, Vladimir Shpilrain *Generic case complexity, decision problems in group theory and random walks*, Journal of Algebra 264, 2003, pp. 665–694.

[82] Bjorn Kjos-Hanssen, Wolfgang Merkle, and Frank Stephan *Kolmogorov complexity and the recursion theorem*, in: B. Durand and W. Thomas (eds.), STACS 2006 (Proceedings of the 23rd Annual Symposium on Theoretical Aspects of Computer Science, Marseille, France, February 23–25, 2006), Springer-Verlag, Berlin 2006 [*Lecture Notes in Computer Science 3884*], pp. 149–161.

[83] Bjorn Kjos-Hanssen, Andre Nies, and Frank Stephan *Lowness for the class of Schnorr random sets*, SIAM Journal on Computing 35, 2005, pp. 647–657.

[84] Ker-I Ko, and Harvey Feideman *Computational complexity of real functions*, Theoretical Computer Science, 1982, pp. 323–352.

[85] Kojiro Kobayashi *On compressibility of infinite sequences*, Technical Report C-34, Department of Information Sciences, Tokyo Institute of Technology, Series C, 1993.

[86] Shira Kritchman, and Ran Raz *The surprise examination paradox and the second incompleteness theorem*, Notices of the American Mathematical Society 57 (11), 2010, pp. 1454–1458.

[87] Antonin Kučera *Measure, Π^0_1 classes, and complete extensions of PA*, in: H.-D. Ebbinghaus, G. H. Müller, and G. E. Sacks (eds.), Recursion Theory Week (Proceedings of the Conference Held at the Mathematisches Forschungsinstitut in Oberwolfach, April 15–21, 1984) [*Lecture Notes in Mathematics 1141*].

[88] Antonin Kučera, and Theodore Slaman *Randomness and recursive enumerability*, SIAM Journal on Computing 31, 2001, pp. 199–211.

[89] Stuart Kurtz Randomness and Genericity in the Degrees of Unsolvability, PhD Dissertation, University of Illinois at Urbana–Champaign, 1981.

[90] Ming Li, and Paul Vitányi An Introduction to Kolmogorov Complexity and its Applications. Springer, New York 2008 (third edition) [*Texts in Computer Science*].

[91] Jack Lutz *Gales and the constructive dimension of individual sequences*, in: U. Montanari, J. D. P. Rolim, and E. Welzl (eds.) Automata, Languages and Programming (27th International Colloquium, ICALP 2000, Geneva, Switzerland, July 9–15, 2000, Proceedings), Springer, Berlin 2000 [*Lecture Notes in Computer Science 1853*], pp. 902–913.

[92] — *The dimensions of individual strings and sequences*, Information and Computation

187, 2003, pp. 49–79.

[93] Jack Lutz, and Neil Lutz *Algorithmic information, plane Kakeya sets, and conditional dimension*, ACM Transactions on Computation Theory 10 (2), Art. 7, 22, 2018.

[94] Neil Lutz *Fractal intersections and products via algorithmic dimension*, in: 42nd International Symposium on Mathematical Foundations of Computer Science, LIPIcs, Leibniz Int. Proc. Inform. 83, pages Art. No. 58, 12, Schloss Dagstuhl. Leibniz-Zent. Inform., Wadern 2017.

[95] Neil Lutz, and Don Stull *Bounding the dimension of points on a line*, in: Theory and App.lications of Models of Computation, Springer, Cham 2017 [*Lecture Notes in Computer Science 10185*], pp. 425–439.

[96] Per Martin-Löf *The definition of random sequences*, Information and Control 9, 1966, pp. 602–619.

[97] Elvira Mayordomo *A Kolmogorov complexity characterization of constructive Hausdorff dimension*, Information Processing Letters 84, 2002 pp. 1–3.

[98] Wolfgang Merkle, Nenad Mihailovíc, and Theodore Slaman *Some results on effective randomness*, Theory of Computing Systems 39, 2006, pp. 707–721.

[99] Jochen Messner, and Thomas Thierauf *A Kolmogorov complexity proof of the Lovász local lemma for satisfiability*, Theoretical Computer Science 461, 2012, pp. 55–64.

[100] Joseph Miller *Kolmogorov random reals are 2-random*, The Journal of Symbolic Logic 69, 2004, pp. 907–913.

[101] — *The K-degrees, low for K-degrees, and weakly low for K sets*, Notre Dame Journal of Formal Logic 50, 2010, pp. 381–391.

[102] — *Extracting information is hard: a Turing degree of non-integral effective Hausdorff dimension*, Advances in Mathematics, 226 (1), 2011, pp. 373–384.

[103] — *On work of Barmpalias and Lewis-Pye: A derivation on the d.c.e. reals*, in: Computability and Complexity, *Essays Dedicated to Rodney G. Downey on the Occasion of His 60th Birthday*, Springer-Verlag, 2016 [*Lecture Notes in Computer Science 10010*], pp. 644–659.

[104] Joseph Miller, and Liang Yu *On initial segment complexity and degrees of randomness*, Transactions of the American Mathematical Society 360, 2008, pp. 3193–3210.

[105] Philippe Moser *On the polynomial septh of various sets of random strings*, Theoretical Computer Science 477, 2013, pp. 96–108.

[106] Philippe Moser, and Frank Stephan *Depth, highness and dnr degrees*, in: Fundamentals of Computation Theory (Twentieth International Symposium, FCT 2015, Gdansk, Poland, August 17–19, 2015, Proceedings), Springer [*Lecture Notes in Computer Science 9210*], pp. 81-94, 2015.

[107] Andrei A. Muchnik, Alexi Semenov, and Vladimir Uspensky *Mathematical metaphysics of randomness*, Theoretical Computer Science 207, 1998, pp. 263–317

[108] Andre Nies *Lowness properties and randomness*, Advances in Mathematics 197, 2005, pp. 274–305.

[109] — Computability and Randomness, Oxford University Press, Oxford, 2009 [*Oxford*

Logic Guides 51].

[110] — *Calculus of cost functions*, in: Mariya Soskova, and Barry Cooper (eds.) The Incomputable, 2017, pp. 183–216.

[111] Andre Nies, and Volkher Scholz *Martin-Löf random quantum states* [submitted arXiv: 1709.08422].

[112] Andre Nies, Frank Stephan, and Sebastiaan Terwijn *Randomness, relativization, and Turing degrees*, The Journal of Symbolic Logic 70, 2005, pp. 515–535.

[113] Raymond Paley, and Antoni Zugmund *On some series of functions (1)*, Mathematical Proceedings of the Cambridge Philosophical Society 26, 1930, pp. 337-357.

[114] — *On some series of functions (2)*, Mathematical Proceedings of the Cambridge Philosophical Society 26, 1930, pp. 459–474.

[115] — *On some series of functions* (3), Mathematical Proceedings of the Cambridge Philosophical Society 28, 1932, pp. 190–205.

[116] Paul Potgieter *Algorithmically random series and Brownian motion*, Annals of Pure and Applied Logic, 169, 2018, pp. 1210–1226.

[117] Marion Pour-El, and Ian Richards Computability in Analysis and Physics, Springer-Verlag, 1989.

[118] Hartley Rogers Jr. The Theory of Recursive Functions and Effective Computability, MIT Press, 1967.

[119] Gerald Sacks Degrees of Unsolvability, Princeton University Press, Princeton NJ 1963 [*Annals of Mathematics Studies*].

[120] Claus Schnorr Zufälligkeit und Wahrscheinlichkeit, *Eine algorithmische Begründung der Wahrscheinlichkeitstheorie*, Springer-Verlag, Berlin–New York, 1971 [*Lecture Notes in Mathematics 218*].

[121] Claus Schnorr, and H. Stimm *Endliche Automaten und Zufallsfolgen*, Acta Informatica 1 (4), 1971–1972, pp. 345–359.

[122] Alexander Shen, Vladimir Uspensky, and Nicolai Vereshchagin Kolmogorov Complexity and Algorithmic Randomness, American Mathematical Society, 2017.

[123] Stephen Simpson *Symbolic dynamics: entropy = dimension = complexity*, Theory of Computing Systems 56 (3), 2015 pp. 527–543.

[124] Robert Soare Recursively Enumerable Sets and Degrees, Springer-Verlag, 1987.

[125] Robert Solovay Draft of paper (or series of papers) on Chaitin's work, Unpublished notes, May 1975, 215 pp.

[126] Frank Stephan *Martin-Löf random sets and PA-complete sets*, in: Z. Chatzidakis, P. Koepke, and W. Pohlers (eds.), Logic Colloquium '02, pp. 342–348, Association for Symbolic Logic and A K Peters Ltd., La Jolla CA and Wellesley MA 2006 [*Lecture Notes in Logic 27*].

[127] Sebastiaan Terwijn, and Domenico Zambella *Algorithmic randomness and lowness*, The Journal of Symbolic Logic 66, 2001, pp. 1199–1205.

[128] Alan Turing *On computable numbers with an application to the Entscheidungsprob-*

lem, Proceedings of the London Mathematical Society 42, 1936, pp. 230–265. (Correction: Proceedings of the London Mathematical Society 43, 1937, pp. 544–546.

[129] Nicolai Vereshchagin *Kolmogorov complexity conditional to large integers*, Theoretical Computer Science, 271, 2002, pp. 59–67.

[130] Jean Ville Etude critique de la notion de collectif, Gauthier-Villars, Paris 1939 [*Monographies des Probabilités, Calcul des Probabilités et ses Applications*].

[131] Richard von Mises *Grundlagen der Wahrscheinlichkeitsrechnung*, Mathematische Zeitschrift, 5, 1919, pp. 52–99.

[132] Abraham Wald *Sur la notion de collectif dans la calcul des probabilités*, Comptes Rendus des Seances de l'Académie des Sciences 202, 1936, pp. 180–183.

[133] — *Die Wiederspruchsfreiheit des Kollektivbegriffes der Wahrscheinlichkeitsrechnung*, Ergebnisse eines Mathematischen Kolloquiums 8, 1937, pp. 38–72.

[134] Liang Yu, Decheng Ding, and Rod Downey *The Kolmogorov complexity of random reals*, Annals of Pure and Applied Logic, 129, 2004, pp. 163–180.

[135] Domenico Zambella *On sequences with simple initial segments*, Technical Report, The Institute for Logic, Language and Computation (ILLC), University of Amsterdam, 1990.

[136] Marius Zimand *Two sources are better than one for increasing the Kolmogorov complexity of infinite sequences*, Theory of Computing Systems 46, 2010, pp. 707–722.

Descriptive set theory and ω-powers of finitary languages

Olivier Finkel
CNRS, Université de Paris, Sorbonne Université
finkel@math.univ-paris-diderot.fr

Dominique Lecomte
Sorbonne Université, Université de Paris, CNRS,
Université de Picardie, I.U.T. de l'Oise, site de Creil
dominique.lecomte@upmc.fr

ABSTRACT The ω-power of a finitary language L over a finite alphabet Σ is the language of infinite words over Σ defined by

$$L^\infty := \{w_0 w_1 \ldots \in \Sigma^\omega \mid \forall i \in \omega \;\; w_i \in L\}.$$

The ω-powers appear very naturally in Theoretical Computer Science in the characterization of several classes of languages of infinite words accepted by various kinds of automata, like Büchi automata or Büchi pushdown automata. We survey some recent results about the links relating Descriptive Set Theory and ω-powers.

KEYWORDS Languages of finite or infinite words, context-free, one-counter automaton, ω-power, topological complexity, Borel class, complete set

1 Introduction

In the sixties, Büchi studied acceptance of infinite words by finite automata with the now called Büchi acceptance condition, in order to prove the decidability of the monadic second order theory of one successor over the integers. Since then there has been a lot of work on regular ω-languages, accepted by Büchi automata, or by some other variants of automata over infinite words, like Muller or Rabin automata, and by other finite machines, like pushdown automata, counter automata, Petri nets, Turing machines, ..., with various acceptance conditions, see [Tho90, Sta97a, PP04].

The class of regular ω-languages, those acccepted by Büchi automata, is the ω-Kleene closure of the family REG of regular finitary languages. The *ω-Kleene closure* of a class of languages of finite words over finite alphabets is the class of ω-languages of the form $\bigcup_{1\leq j\leq n} U_j \cdot V_j^{\infty}$, for some regular finitary languages U_j and V_j, $1 \leq j \leq n$, where for any finitary language $L \subseteq \Sigma^{<\omega}$ over the alphabet Σ, the *ω-power* L^{∞} of L is the set of the infinite words constructible with L by concatenation, i.e.,

$$L^{\infty} := \{ w_0 w_1 \ldots \in \Sigma^{\omega} \mid \forall i \in \omega \ \ w_i \in L \}.$$

Note that we denote here L^{∞} the ω-power associated with L, as in [Lec05, FL09], while it is often denoted L^{ω} in Theoretical Computer Science papers, as in [Sta97a, Fin01, Fin03a, FL07]. Here we reserved the notation L^{ω} to denote the Cartesian product of countably many copies of L since this will be often used in this paper.

Similarly, the operation of taking the ω-power of a finitary language appears in the characterization of the class of context-free ω-languages as the ω-Kleene closure of the family of context-free finitary languages (we refer the reader to [ABB96] for basic notions about context-free languages). And the class of ω-languages accepted by Büchi one-counter automata is also the ω-Kleene closure of the family of finitary languages accepted by one-counter automata. Therefore the operation $L \rightarrow L^{\infty}$ is a fundamental operation over finitary languages leading to ω-languages. The ω-powers of regular languages have been studied in [LT87, Sta97a].

During the last years, the ω-powers have been studied from the perspective of Descriptive Set Theory in a few papers [Fin01, Fin03a, Fin04, Lec05, DF07, FL07, FL09, FL20]. We mainly review these recent works in the present survey.

Since the set Σ^{ω} of infinite words over a finite alphabet Σ can be equipped with the usual Cantor topology, the question of the topological complexity of ω-powers of finitary languages, from the point of view of descriptive set theory, naturally arises and has been posed by Niwinski [Niw90], Simonnet [Sim92], and Staiger [Sta97a].

As the concatenation map, from L^{ω} onto L^{∞}, which associates to a given sequence $(w_i)_{i\in\omega}$ of finite words the concatenated word $w_0 w_1 \ldots$, is continuous, an ω-power is always an analytic set. It was proved in [Fin03a] that there exists a (context-free) language L such that L^{∞} is analytic but not Borel. Amazingly, the language L is very simple to describe and it is accepted by a simple one-counter automaton. Louveau has proved independently that analytic-complete ω-powers exist, but the existence was proved in a non effective way (this is non-published work).

One of our first tasks was to study the position of ω-powers with respect to the Borel hierarchy (and beyond to the projective hierarchy). A characterization of ω-powers in the Borel classes $\mathbf{\Sigma}^0_1$, $\mathbf{\Pi}^0_1$ and $\mathbf{\Pi}^0_2$ has been given by Staiger in [Sta97b].

Concerning Borel ω-powers, it was proved that, for each integer $n \geq 1$, there exist some ω-powers of (context-free) languages which are $\mathbf{\Pi}^0_n$-complete Borel sets, [Fin01]. It was proved in [Fin04] that there exists a finitary language L such that L^∞ is a Borel set of infinite rank, and in [DF07] that there is a (context-free) language W such that W^∞ is Borel above $\mathbf{\Delta}^0_\omega$. We recently proved that there are complete ω-powers of one-counter languages, for every Borel class of finite rank, [FL20]. We proved in [FL07, FL09] a result which showed that ω-powers exhibit a great topological complexity: for each countable ordinal $\xi \geq 1$, there are $\mathbf{\Pi}^0_\xi$-complete ω-powers, and $\mathbf{\Sigma}^0_\xi$-complete ω-powers. This result has an effective aspect: for each recursive ordinal $\xi < \omega_1^{\mathrm{CK}}$, where ω_1^{CK} is the first non-recursive ordinal, there are recursive finitary languages P and S such that P^∞ is $\mathbf{\Pi}^0_\xi$-complete and S^∞ is $\mathbf{\Sigma}^0_\xi$-complete.

Many questions are still open about the topological complexity of ω-powers of languages in a given class like the class of context-free languages, one-counter languages, recursive languages, or more generally languages accepted by some kind of automata over finite words. We mention some of these open questions in this paper.

This article is organized as follows. Some basic notions of topology are recalled in Section 2. Notions of automata and formal language theory are recalled in Section 3, and ω-powers of finitary languages accepted by automata are studied in this section. The study of ω-powers of finitary languages in the classical setting of descriptive set theory forms Section 4. Finally, we provide in Section 5 some complexity results about some sets of finitary languages whose associated ω-power is in some class of sets.

2 Topology

When Σ is a finite alphabet, a nonempty *finite word* over Σ is a sequence

$$w = a_0 \ldots a_{l-1},$$

where $a_i \in \Sigma$ for each $i < l$, and $l \geq 1$ is a natural number. The *length* of w is l, denoted by $|w|$. A word of length one is of the form (a). The *empty word* is denoted by λ and satisfies $|\lambda| = 0$. When w is a finite word over Σ, we write $w = w(0)w(1)\ldots w(l-1)$, and the prefix $w(0)w(1)\ldots w(i-1)$ of w of length i is denoted by $w|i$, for any $i \leq l$. We also write $u \subseteq v$ when the word u is a prefix of the finite word v. The set of finite words over Σ is denoted by $\Sigma^{<\omega}$, and Σ^+ is the set of nonempty finite words over Σ. A (finitary) *language* over Σ is a subset of $\Sigma^{<\omega}$. For $L \subseteq \Sigma^{<\omega}$, the *complement* $\Sigma^{<\omega} \setminus L$ of L (in $\Sigma^{<\omega}$) is denoted by L^-. We sometimes write a for $\{(a)\}$, for short.

The first infinite ordinal is ω. An ω-*word* over Σ is an ω-sequence $a_0a_1\ldots$, where $a_i \in \Sigma$ for each natural number i.

When σ is an ω-word over Σ, the length of σ is $|\sigma|=\omega$, and we write $\sigma=\sigma(0)\sigma(1)\ldots$, and the prefix $\sigma(0)\sigma(1)\ldots\sigma(i\text{–}1)$ of σ of length i is denoted by $\sigma|i$, for any natural number i. We also write $u\subseteq\sigma$ when the finite word u is a prefix of the ω-word σ. The set of ω-words over Σ is denoted by Σ^ω. An *ω-language* over Σ is a subset of Σ^ω. For $A\subseteq\Sigma^\omega$, the complement $\Sigma^\omega\setminus A$ of A is denoted by A^-.

The usual *concatenation* product of two finite words u and v is denoted $u^\frown v$ (and sometimes just uv). This product is extended to the product of a finite word u and an ω-word σ: the infinite word $u^\frown\sigma$ is then the ω-word such that $(u^\frown\sigma)(k)=u(k)$ if $k<|u|$, and $(u^\frown\sigma)(k)=\sigma(k-|u|)$ if $k\geq|u|$.

If E is a set, $l\in\omega$ and $(e_i)_{i<l}\in E^l$, then $^\frown{}_{i<l}\ e_i$ is the concatenation $e_0\ldots e_{l-1}$. Similarly, $^\frown{}_{i\in\omega}\ e_i$ is the concatenation $e_0e_1\ldots$ For $L\subseteq\Sigma^{<\omega}$,

$$L^\infty:=\{\sigma=w_0w_1\ldots\in\Sigma^\omega\mid\forall i\in\omega\ \ w_i\in L\}$$

is the *ω-power* of L.

We now recall some notions of topology, assuming the reader to be familiar with the basic notions, that can be found in [Mos80, Kec95, Sta97a, PP04]. The topological spaces in which we will work in this paper will be subspaces of Σ^ω, where Σ is either finite having at least two elements (like $2:=\{\mathbf{0},\mathbf{1}\}$), or countably infinite. Note that here 2 is considered as an alphabet, and we will do it also for 3,4; sometimes, we will view it as a letter, and in this case we will denote it by **2**, like we just did it for **0**,**1**. The topology on Σ^ω is the product topology of the discrete topology on Σ. For $w\in\Sigma^{<\omega}$, the set defined by $N_w:=\{\alpha\in\Sigma^\omega\mid w\subseteq\alpha\}$ is a basic clopen (i.e., closed and open) set of Σ^ω. The open subsets of Σ^ω are of the form $W^\frown\Sigma^\omega:=\{w\sigma\mid w\in W\text{ and }\sigma\in\Sigma^\omega\}$, where $W\subseteq\Sigma^{<\omega}$. When Σ is finite, this topology is called the *Cantor topology* and Σ^ω is compact. When $\Sigma=\omega$, Σ^ω is the Baire space, which is homeomorphic to $\mathbb{P}_\infty:=\{\alpha\in 2^\omega\mid\forall i\in\omega\ \exists j\geq i\ \ \alpha(j)=\mathbf{1}\}$, via the map defined on ω^ω by $h(\beta):=\mathbf{0}^{\beta(0)}\mathbf{1}\mathbf{0}^{\beta(1)}\mathbf{1}\ldots$ There is a natural metric on Σ^ω, the *prefix metric* defined as follows. For $\sigma\neq\tau\in\Sigma^\omega$, $d(\sigma,\tau):=2^{-l_{pref(\sigma,\tau)}}$, where $l_{pref(\sigma,\tau)}$ is the first natural number n such that $\sigma(n)\neq\tau(n)$. The topology induced on Σ^ω by this metric is our topology.

We now define the Borel hierarchy.

Definition 1. *Let X be a topological space, and $n\geq 1$ be a natural number. The classes $\mathbf{\Sigma}^0_n(X)$ and $\mathbf{\Pi}^0_n(X)$ of the Borel hierarchy are inductively defined as follows:*

$\mathbf{\Sigma}^0_1(X)$ is the class of open subsets of X.

$\mathbf{\Pi}^0_1(X)$ is the class of closed subsets of X.

$\mathbf{\Sigma}^0_{n+1}(X)$ is the class of countable unions of $\mathbf{\Pi}^0_n$-subsets of X.

$\mathbf{\Pi}^0_{n+1}(X)$ is the class of countable intersections of $\mathbf{\Sigma}^0_n$-subsets of X.

The Borel hierarchy is also defined for the transfinite levels. Let $\xi \geq 2$ be a countable ordinal.

$\boldsymbol{\Sigma}^0_\xi(X)$ *is the class of countable unions of subsets of X in $\bigcup_{\gamma<\xi} \boldsymbol{\Pi}^0_\gamma$.*

$\boldsymbol{\Pi}^0_\xi(X)$ *is the class of countable intersections of subsets of X in $\bigcup_{\gamma<\xi} \boldsymbol{\Sigma}^0_\gamma$.*

Suppose now that $\xi \geq 1$ is a countable ordinal and $X \subseteq Y$, where X is equipped with the induced topology. Then $\boldsymbol{\Sigma}^0_\xi(X) = \{A \cap X \mid A \in \boldsymbol{\Sigma}^0_\xi(Y)\}$, and similarly for $\boldsymbol{\Pi}^0_\xi$, see [Kec95, Section 22.A]. Note that we defined the Borel classes $\boldsymbol{\Sigma}^0_\xi(X)$ and $\boldsymbol{\Pi}^0_\xi(X)$ mentioning the space X. However, when the context is clear, we will sometimes omit X and denote $\boldsymbol{\Sigma}^0_\xi(X)$ by $\boldsymbol{\Sigma}^0_\xi$ and similarly for the dual class. The Borel classes are closed under finite intersections and unions, and continuous preimages. Moreover, $\boldsymbol{\Sigma}^0_\xi$ is closed under countable unions, and $\boldsymbol{\Pi}^0_\xi$ under countable intersections. As usual, the ambiguous class $\boldsymbol{\Delta}^0_\xi$ is the class $\boldsymbol{\Sigma}^0_\xi \cap \boldsymbol{\Pi}^0_\xi$. The class of *Borel sets* is

$$\boldsymbol{\Delta}^1_1 := \bigcup_{1 \leq \xi < \omega_1} \boldsymbol{\Sigma}^0_\xi = \bigcup_{1 \leq \xi < \omega_1} \boldsymbol{\Pi}^0_\xi,$$

where ω_1 is the first uncountable ordinal. The *Borel hierarchy* is as follows:

$$\begin{array}{ccccccccc}
 & \boldsymbol{\Sigma}^0_1 = \text{open} & & \boldsymbol{\Sigma}^0_2 & \cdots & \boldsymbol{\Sigma}^0_\omega & \cdots & & \\
\boldsymbol{\Delta}^0_1 = \text{clopen} & & \boldsymbol{\Delta}^0_2 & & & \boldsymbol{\Delta}^0_\omega & & & \boldsymbol{\Delta}^1_1 \\
 & \boldsymbol{\Pi}^0_1 = \text{closed} & & \boldsymbol{\Pi}^0_2 & \cdots & \boldsymbol{\Pi}^0_\omega & \cdots & &
\end{array}$$

This picture means that any class is contained in every class at the right of it, and the inclusion is strict in any of the spaces Σ^ω. A subset of Σ^ω is a Borel set of *rank* ξ if it is in $\boldsymbol{\Sigma}^0_\xi \cup \boldsymbol{\Pi}^0_\xi$ but not in $\bigcup_{1 \leq \gamma < \xi} (\boldsymbol{\Sigma}^0_\gamma \cup \boldsymbol{\Pi}^0_\gamma)$.

We now define completeness with respect to reducibility by continuous functions. Let Y, Σ be finite alphabets, $A \subseteq Y^\omega$ and $C \subseteq \Sigma^\omega$. We say that A is *Wadge reducible* to C if there exists a continuous function $f : Y^\omega \to \Sigma^\omega$ such that $A = f^{-1}(C)$. Now let $\boldsymbol{\Gamma}$ be a class of sets closed under continuous pre-images like $\boldsymbol{\Sigma}^0_\xi$ or $\boldsymbol{\Pi}^0_\xi$. A subset C of Σ^ω is said to be $\boldsymbol{\Gamma}$-*hard* if, for any finite alphabet Y and any $A \subseteq Y^\omega$, $A \in \boldsymbol{\Gamma}$ implies that A is Wadge reducible to C. If moreover C is in $\boldsymbol{\Gamma}(\Sigma^\omega)$, then we say that C is $\boldsymbol{\Gamma}$-*complete*. The $\boldsymbol{\Sigma}^0_n$-complete sets and the $\boldsymbol{\Pi}^0_n$-complete sets are thoroughly characterized in [Sta86]. Recall that a subset of Σ^ω is $\boldsymbol{\Sigma}^0_\xi$ (respectively $\boldsymbol{\Pi}^0_\xi$)-complete if and only if it is in $\boldsymbol{\Sigma}^0_\xi$ but not in $\boldsymbol{\Pi}^0_\xi$ (respectively in $\boldsymbol{\Pi}^0_\xi$ but not in $\boldsymbol{\Sigma}^0_\xi$), and that such sets exist (see [Kec95]). For example, the singletons of 2^ω are $\boldsymbol{\Pi}^0_1$-complete. The set $\mathbb{P}_\infty$ defined at the beginning of the present section is a well known example of a $\boldsymbol{\Pi}^0_2$-complete set. We say that $\boldsymbol{\Gamma}$ is a *Wadge class* if there is a $\boldsymbol{\Gamma}$-complete set. The *Wadge hierarchy* of Borel sets given by the inclusion of these classes is a great refinement of the Borel hierarchy of the classes $\boldsymbol{\Sigma}^0_\xi$ and $\boldsymbol{\Pi}^0_\xi$.

Among the new classes appearing in this hierarchy, we can mention the classes of transfinite differences of $\mathbf{\Sigma}^0_\xi$ sets. If η is a countable ordinal and $(A_\theta)_{\theta<\eta}$ is an increasing sequence of subsets of some set X, then we set

$$D_\eta((A_\theta)_{\theta<\eta}) := \{x \in X \mid \exists\theta<\eta \;\; x \in A_\theta \setminus \bigcup_{\theta'<\theta} A_{\theta'} \text{ and the parity of } \theta \text{ is opposite to that of } \eta\}.$$

If moreover $\xi \geq 1$ is a countable ordinal, then we set

$$D_\eta(\mathbf{\Sigma}^0_\xi) := \{D_\eta((A_\theta)_{\theta<\eta}) \mid \forall\theta < \eta \;\; A_\theta \in \mathbf{\Sigma}^0_\xi\}.$$

The class $\check{\mathbf{\Gamma}} := \{\neg A \mid A \in \mathbf{\Gamma}\}$ is the class of the complements of the sets in $\mathbf{\Gamma}$, and is called the *dual class* of $\mathbf{\Gamma}$. In particular, $\check{\mathbf{\Sigma}}^0_\xi = \mathbf{\Pi}^0_\xi$ and $\check{\mathbf{\Pi}}^0_\xi = \mathbf{\Sigma}^0_\xi$.

There are some subsets of the topological space Σ^ω which are not Borel sets. In particular, there is another hierarchy beyond the Borel hierarchy, called the projective hierarchy. The first class of the projective hierarchy is the class $\mathbf{\Sigma}^1_1$ of analytic sets. A subset A of Σ^ω is *analytic* if we can find a finite alphabet Y and a Borel subset B of $(\Sigma \times Y)^\omega$ such that $x \in A \Leftrightarrow \exists y \in Y^\omega \; (x, y) \in B$, where $(x, y) \in (\Sigma \times Y)^\omega$ means that $(x, y)(i) = (x(i), y(i))$ for each natural number i. A subset of Σ^ω is analytic if it is empty, or the image of the Baire space by a continuous map. The class $\mathbf{\Sigma}^1_1$ of analytic sets contains the class of Borel sets in any of the spaces Σ^ω. Note that $\mathbf{\Delta}^1_1 = \mathbf{\Sigma}^1_1 \cap \mathbf{\Pi}^1_1$, where $\mathbf{\Pi}^1_1 := \check{\mathbf{\Sigma}}^1_1$ is the class of *co-analytic* sets, i.e., of complements of analytic sets. Similarly, the class of projections of $\mathbf{\Pi}^1_1$ sets is denoted $\mathbf{\Sigma}^1_2$.

The ω-power of a finitary language L is always an analytic set. Indeed, if L is finite and has n elements, then L^∞ is the continuous image of the compact set $\{\mathbf{0}, \mathbf{1}, \ldots, \mathbf{n-1}\}^\omega$. If L is infinite, then there is a bijection between L and ω, and L^∞ is the continuous image of the Baire space ω^ω, [Sim92].

3 Complexity of ω-powers of languages accepted by automata

3.1 Automata

We assume the reader to be familiar with formal languages, see for example [HMU01, Tho90]. We first recall some of the definitions and results concerning automata, pushdown automata, regular and context-free languages, as presented in [ABB96, CG77, Sta97a].

Definition 2. *A pushdown automaton is a 7-tuple $\mathcal{A} = (Q, \Sigma, \Gamma, q_0, Z_0, \delta, F)$, where Q is a finite set of states, Σ is a finite input alphabet, Γ is a finite pushdown alphabet, $q_0 \in Q$ is the initial state, $Z_0 \in \Gamma$ is the start symbol which is the bottom symbol and always remains at the bottom of the pushdown stack, δ is a map from $Q \times (\Sigma \cup \{\lambda\}) \times \Gamma$ into the set of finite subsets of $Q \times \Gamma^{<\omega}$, and $F \subseteq Q$ is the set of final states.*

The automaton $\mathcal{A}$ is said to be deterministic *if δ is a map from $Q\times(\Sigma\cup\{\lambda\})\times\Gamma$ into the set of subsets of cardinal one, i.e., singletons, of $Q\times\Gamma^{<\omega}$. The automaton $\mathcal{A}$ is said to be* real-time *if there is no λ-transition, i.e., if δ is a map from $Q\times\Sigma\times\Gamma$ into the set of finite subsets of $Q\times\Gamma^{<\omega}$.*

If $\gamma\in\Gamma^+$ describes the pushdown stack content, then the leftmost symbol will be assumed to be on the "top" of the stack. A configuration *of the pushdown automaton $\mathcal{A}$ is a pair (q,γ), where $q\in Q$ and $\gamma\in\Gamma^{<\omega}$. For $a\in\Sigma\cup\{\lambda\}$, $\gamma,\beta\in\Gamma^{<\omega}$ and $Z\in\Gamma$, if (p,β) is in $\delta(q,a,Z)$, then we write $a:(q,Z\gamma)\mapsto_{\mathcal{A}}(p,\beta\gamma)$.*

Let $w=a_0\dots a_{l-1}$ be a finite word over Σ. A sequence of configurations $r=(q_i,\gamma_i)_{i<N}$ is called a run of $\mathcal{A}$ on w starting in the configuration (p,γ) *if*

(1) $(q_0,\gamma_0)=(p,\gamma)$,

(2) for each $i<N-1$, there exists $b_i\in\Sigma\cup\{\lambda\}$ satisfying $b_i:(q_i,\gamma_i)\mapsto_{\mathcal{A}}(q_{i+1},\gamma_{i+1})$ such that $a_0\dots a_{l-1}=b_0\dots b_{N-2}$.

A run r of $\mathcal{A}$ on w starting in configuration (q_0,Z_0) will be simply called a run of $\mathcal{A}$ on w*. The run is* accepting *if it ends in a final state.*

The language $L(\mathcal{A})$ accepted *by $\mathcal{A}$ is the set of words admitting an accepting run by $\mathcal{A}$. A* context-free language *is a finitary language which is accepted by a pushdown automaton. We denote by CFL the class of context-free languages.*

If we omit the pushdown stack in the definition of a pushdown automaton, we get the notion of a (finite state) automaton. Note that every finite state automaton is equivalent to a deterministic real-time finite state automaton. A regular language *is a finitary language which is accepted by a (finite state) automaton. We denote by REG the class of regular languages.*

A one-counter automaton *is a pushdown automaton with a pushdown alphabet of the form $\Gamma=\{Z_0,z\}$, where Z_0 is the bottom symbol and always remains at the bottom of the pushdown stack. A* one-counter language *is a (finitary) language which is accepted by a one-counter automaton.*

Definition 3. *Let Σ,Γ be finite alphabets.*

(a) A (Σ,Γ)-substitution *is a map $f:\Sigma\to 2^{\Gamma^{<\omega}}$.*

(b) We extend this map to $\Sigma^{<\omega}$ be setting $f(\frown_{i<l}a_i):=\{\frown_{i<l}w_i\mid\forall i<l\ \ w_i\in f(a_i)\}$, where $l\in\omega$ and $a_0,\cdots,a_{l-1}\in\Sigma$.

(c) We further extend this map to $2^{\Sigma^{<\omega}}$ by setting $f(L):=\bigcup_{w\in L}f(w)$.

(d) Let f be a (Σ,Γ)-substitution, and $\mathcal{F}$ be a family of languages. If the language $f(a)$ belongs to $\mathcal{F}$ for each $a\in\Sigma$, then the substitution f is called a $\mathcal{F}$-substitution*.*

(e) We then define the operation $\square$ on families of languages. Let $\mathcal{E}$, $\mathcal{F}$ be families of (finitary) languages. Then $\mathcal{E} \,\square\, \mathcal{F} := \{f(L) \mid L \in \mathcal{E}$ and f is a $\mathcal{F}$-substitution$\}$.

The operation of substitution gives rise to an infinite hierarchy of context-free finitary languages defined as follows.

Definition 4. *Let $OCL(0) = REG$ be the class of regular languages, $OCL(1) = OCL$ be the class of one-counter languages, and $OCL(k+1) = OCL(k) \,\square\, OCL$, for $k \geq 1$.*

It is well known that the hierarchy given by the families of languages $OCL(k)$ is strictly increasing. And there is a characterization of these languages in terms of automata.

Proposition 5 ([ABB96]). *A language L is in $OCL(k)$ if and only if L is recognized by a pushdown automaton such that, during any computation, the words in the pushdown stack remain in a language of the form $(z_{k-1})^{<\omega} \ldots (z_0)^{<\omega} Z_0$, where $\{Z_0, z_0, \ldots, z_{k-1}\}$ is the pushdown alphabet. Such an automaton is called a k-*iterated counter automaton*. The union $ICL := \bigcup_{k \geq 1} OCL(k)$ is called the family of* iterated counter languages, *which is the closure under substitution of the family OCL.*

3.2 Π^0_n-complete and Σ^0_n-complete ω-powers

Wadge first gave a description of the Wadge hierarchy of Borel sets, see [Wad83]. Duparc got in [Dup01] a new proof of Wadge's results in the case of Borel sets of finite rank, and he gave a normal form of Borel sets of finite rank, i.e., an inductive construction of a Borel set of every given degree. His proof relies on set theoretic operations which are the counterpart of arithmetical operations over ordinals needed to compute the Wadge degrees.

In fact, J. Duparc studied the Wadge hierarchy via the study of the conciliating hierarchy. He introduced in [Dup01] the conciliating sets, which are sets of finite *or* infinite words over an alphabet Σ, i.e., subsets of $\Sigma^{\leq\omega} := \Sigma^{<\omega} \cup \Sigma^{\omega}$. In particular, the set theoretic operation of exponentiation, defined over concilating sets, has been very useful in the study of context-free ω-powers.

We first recall the following.

Definition 6. *Let Σ_A be a finite alphabet, $\twoheadleftarrow$ be a letter out of Σ_A, $\Sigma := \Sigma_A \cup \{\twoheadleftarrow\}$, and x be a finite or infinite word over the alphabet Σ. Then $x^{\twoheadleftarrow}$ is inductively defined as follows.*

- $\lambda^{\twoheadleftarrow} := \lambda$.

- For a finite word $u \in \Sigma^{<\omega}$, $\begin{cases} (ua)^{\twoheadleftarrow} := u^{\twoheadleftarrow} a \text{ if } a \in \Sigma_A, \\ (u \twoheadleftarrow)^{\twoheadleftarrow} := u^{\twoheadleftarrow} \text{ with its last letter removed if } |u^{\twoheadleftarrow}| > 0, \\ (u \twoheadleftarrow)^{\twoheadleftarrow} := \lambda \text{ if } |u^{\twoheadleftarrow}| = 0. \end{cases}$

- For an infinite word σ, $\sigma^{\twoheadleftarrow} := lim_{n \in \omega} (\sigma | n)^{\twoheadleftarrow}$, where, given $(w_n) \in (\Sigma_A^{\leq\omega})^{\omega}$ and $w \in \Sigma_A^{<\omega}$, $w \subseteq lim_{n \in \omega} w_n \Leftrightarrow \exists p \in \omega\ \forall n \geq p\ w_n \| w| = w$.

Remark. For $x \in \Sigma^{\leq\omega}$, $x^{\twoheadleftarrow}$ denotes the string x, once every $\twoheadleftarrow$ occuring in x has been "evaluated" as the back space operation (the one familiar to your computer!), proceeding from left to right inside x. in: other words, $x^{\twoheadleftarrow} = x$ from which every interval of the form "$a \twoheadleftarrow$ " $(a \in \Sigma_A)$ is removed.

For example, if $x = (a \twoheadleftarrow)^n$ for some $n \geq 1$, $x = (a \twoheadleftarrow)^\omega$ or $x = (a \twoheadleftarrow\twoheadleftarrow)^\omega$ then $x^{\twoheadleftarrow} = \lambda$. If $x = (ab \twoheadleftarrow)^\omega$, then $x^{\twoheadleftarrow} = a^\omega$. If $x = bb(\twoheadleftarrow a)^\omega$, then $x^{\twoheadleftarrow} = b$.

We now can define the operation $A \mapsto A^{\sim}$ of exponentiation of conciliating sets.

Definition 7. *Let Σ_A be a finite alphabet, $\twoheadleftarrow$ be a letter out of Σ_A, $\Sigma := \Sigma_A \cup \{\twoheadleftarrow\}$, and $A \subseteq \Sigma_A^{\leq\omega}$. Then we set $A^{\sim} := \{x \in \Sigma^{\leq\omega} \mid x^{\twoheadleftarrow} \in A\}$.*

Roughly speaking, the operation $\sim$ is monotone with regard to the Wadge ordering and produces some sets of higher complexity.

The first author proved in [Fin01] that the class CFL_ω of context-free ω-languages, (i.e., those which are accepted by pushdown automata with a Büchi acceptance condition expressing that "some final state appears infinitely often during an infinite computation"), is closed under this operation $\sim$.

We now recall a slightly modified variant of the operation $\sim$, introduced in [Fin01], and which is particularly suitable to infer properties of ω-powers.

Definition 8. *Let Σ_A be a finite alphabet, $\twoheadleftarrow$ be a letter out of Σ_A, $\Sigma := \Sigma_A \cup \{\twoheadleftarrow\}$, and $A \subseteq \Sigma_A^{\leq\omega}$. Then we set $A^{\approx} := \{x \in \Sigma^{\leq\omega} \mid x^{\twoheadleftarrow} \in A\}$, where $x^{\twoheadleftarrow}$ is inductively defined as follows.*

- $\lambda^{\twoheadleftarrow} := \lambda$.
- *For a finite word $u \in \Sigma^{<\omega}$,* $\begin{cases} (ua)^{\twoheadleftarrow} := u^{\twoheadleftarrow} a \text{ if } a \in \Sigma_A, \\ (u \twoheadleftarrow)^{\twoheadleftarrow} := u^{\twoheadleftarrow} \text{ with its last letter removed if } |u^{\twoheadleftarrow}| > 0, \\ (u \twoheadleftarrow)^{\twoheadleftarrow} \text{ is undefined if } |u^{\twoheadleftarrow}| = 0. \end{cases}$
- *For an infinite word σ, $\sigma^{\twoheadleftarrow} := lim_{n \in \omega} (\sigma | n)^{\twoheadleftarrow}$.*

The only difference is that here $(u \twoheadleftarrow)^{\twoheadleftarrow}$ is undefined if $|u^{\twoheadleftarrow}| = 0$. It is easy to see that if $A \subseteq \Sigma_A^\omega$ is a Borel set such that $A \neq \Sigma_A^\omega$, i.e., $A^- \neq \emptyset$, then $A^{\approx}$ is Wadge equivalent to $A^{\sim}$ (see [Fin01]) and this implies the following result:

Theorem 9. *Let Σ_A be a finite alphabet, and $n \geq 2$ be a natural number. If $A \subseteq \Sigma_A^\omega$ is $\mathbf{\Pi}^0_n$-complete, then $A^{\approx}$ is $\mathbf{\Pi}^0_{n+1}$-complete.*

Notation. Let Σ_A be a finite alphabet, $\twoheadleftarrow$ be a letter out of Σ_A, and $\Sigma := \Sigma_A \cup \{\twoheadleftarrow\}$.

The language L_3 over Σ is the context-free language generated by the context-free grammar with the following production rules:

$$S \to aS \twoheadleftarrow S \text{ with } a \in \Sigma_A,$$
$$S \to a \twoheadleftarrow S \text{ with } a \in \Sigma_A,$$
$$S \to \lambda$$

(see [HMU01] for the basic notions about grammars). This language L_3 corresponds to the words where every letter of Σ_A has been removed after using the backspace operation. It is easy to see that L_3 is a deterministic one-counter language, i.e., L_3 is accepted by a deterministic one-counter automaton. Moreover, for $a \in \Sigma_A$, the language L_3a is also accepted by a deterministic one-counter automaton.

We can now state the following result, which implies that the class of ω-powers is closed under the operation $A \to A^{\approx}$.

Lemma 10 (see [Fin01]). *Whenever $A \subseteq \Sigma_A^\omega$, the ω-language $A^{\approx} \subseteq \Sigma^\omega$ is obtained by substituting in A the language L_3a for each letter $a \in \Sigma_A$.*

An ω-word $\sigma \in A^{\approx}$ may be considered as an ω-word $\sigma^{\twoheadleftarrow} \in A$ to which we possibly add, before the first letter $\sigma^{\twoheadleftarrow}(0)$ of $\sigma^{\twoheadleftarrow}$ (respectively, between two consecutive letters $\sigma^{\twoheadleftarrow}(n)$ and $\sigma^{\twoheadleftarrow}(n+1)$ of $\sigma^{\twoheadleftarrow}$), a finite word belonging to the context-free (finitary) language L_3.

Corollary 11. *Whenever $A \subseteq \Sigma_A^\omega$ is an ω-power of a language L_A, i.e., $A = L_A^\infty$, then $A^{\approx}$ is also an ω-power, i.e., there exists a (finitary) language E_A such that $A^{\approx} = E_A^\infty$. Moreover, if the language L_A is in the class $OCL(k)$ for some natural number k, then the language E_A can be found in the class $OCL(k+1)$.*

Proof. Let $h : \Sigma_A \to 2^{\Sigma^{<\omega}}$ be the substitution defined by $a \mapsto L_3a$, where L_3 is the context-free language defined above. Then it is easy to see that now $A^{\approx}$ is obtained by substituting in A the language L_3a for each letter $a \in \Sigma_A$. Thus $E_A = h(L_A)$ satisfies the statement of the theorem. □

The following result was proved in [Fin01].

Theorem 12. *For each natural number $n \geq 1$, there is a context-free language P_n in the subclass of iterated counter languages such that P_n^∞ is $\mathbf{\Pi}_n^0$-complete.*

Proof. Let $B_1 = \{\sigma \in \{0,1\}^\omega \mid \forall i \in \omega \;\; \sigma(i) = 0\} = 0^\infty$. B_1 is a $\mathbf{\Pi}_1^0$-complete set of the form P_1^∞ where P_1 is the singleton containing only the word (0). Note that $P_1 = \mathbf{0}$ is a regular language, hence in the class $OCL(0)$.

Let then $B_2 = \mathbb{P}_\infty$ be the well known $\mathbf{\Pi}_2^0$-complete regular ω-language. Note that $B_2 = (\mathbf{0}^{<\omega}\mathbf{1})^\infty$. Let $P_2 := \mathbf{0}^{<\omega}\mathbf{1}$. Then P_2 is a regular language, hence in the class $OCL(0)$.

We can now use iteratively Corollary 11 to end the proof □

Note that P_1 and P_2 are regular, hence accepted by some (real-time deterministic) finite automata (without any counter). On the other hand, the language P_3 is accepted by a one-counter automaton. Notice that the ω-powers of regular languages are regular ω-languages, and thus are boolean combination of $\mathbf{\Pi}^0_2$-sets, hence $\mathbf{\Delta}^0_3$-sets. Therefore there are no $\mathbf{\Pi}^0_3$-complete or $\mathbf{\Sigma}^0_3$-complete (or even higher in the Borel hierarchy) ω-powers of regular languages.

For the classes $\mathbf{\Sigma}^0_n$, we first give an example of a $\mathbf{\Sigma}^0_n$-complete ω-power for $n=1,2$. Consider the finitary language $S_1:=\{s\in 2^{<\omega} \mid 0\subseteq s \text{ or } \exists k\in\omega\ 10^k1\subseteq s\}$ which is regular. Then the ω-power $S_1^\infty = 2^\omega\setminus\{10^\infty\}$ is open and not closed, and thus $\mathbf{\Sigma}^0_1$-complete.

Using another modification of the operation of exponentiation, we proved in [FL09] that there exists a one counter language $L\subseteq 2^{<\omega}$ such that L^∞ is $\mathbf{\Sigma}^0_2$-complete. It is enough to find a finitary language $S_2\subseteq 3^{<\omega}$, where $3=\{\mathbf{0},\mathbf{1},\mathbf{2}\}$. We set, for $j\in 3$ and $s\in 3^{<\omega}$,

$$n_j(s) := \text{Cardinality}(\{i<|s| \mid s(i)=j\}),$$

$T:=\{\alpha\in 3^{\leq\omega} \mid \forall l<1+|\alpha|\ n_{\mathbf{2}}(\alpha|l)\leq n_{\mathbf{1}}(\alpha|l)\}$. We inductively define, for $s\in T\cap 3^{<\omega}$, a "back space" sequence $s^{\hookleftarrow}\in 2^{<\omega}$ as follows:

$$s^{\hookleftarrow} := \begin{cases} \emptyset \text{ if } s=\emptyset, \\ t^{\hookleftarrow}\varepsilon \text{ if } s=t\varepsilon \text{ and } \varepsilon\in 2, \\ t^{\hookleftarrow}, \text{ except that its last } \mathbf{1} \text{ is replaced with } \mathbf{0}, \text{ if } s=t\mathbf{2}. \end{cases}$$

We then set $E:=\mathbf{0}\cup\{s\in T\cap 3^{<\omega}\setminus\{\emptyset\} \mid n_{\mathbf{2}}(s)=n_{\mathbf{1}}(s) \text{ and } \mathbf{1}\subseteq (s|(|s|-1))^{\hookleftarrow}\}$, and

$$E^* := \{\frown_{i<l}\, s_i\in 3^{<\omega} \mid l\in\omega \text{ and } \forall i<l\ \ s_i\in E\}.$$

We put $S_2 := E\cup\{\frown_{j\leq k}\,(c_j\mathbf{1})\in 3^{<\omega} \mid k\in\omega \text{ and } (k=0\Rightarrow c_0\neq\emptyset) \text{ and } \forall j\leq k\ \ c_j\in E^*\}$, and S_2^∞ is $\mathbf{\Sigma}^0_2$-complete. Note that S_2 is accepted by a one-counter automaton.

Finally, we recently proved in [FL20] the following result giving some complete ω-powers of a one-counter language, for any Borel class of finite rank.

Theorem 13. *Let $n\geq 1$ be a natural number.*

(a) There is a finitary language P_n which is accepted by a one-counter automaton and such that the ω-power P_n^∞ is $\mathbf{\Pi}^0_n$-complete.

(b) There is a finitary language S_n which is accepted by a one-counter automaton and such that the ω-power S_n^∞ is $\mathbf{\Sigma}^0_n$-complete.

Moreover, for any given integer $n\geq 1$, one can effectively construct some one-counter automata accepting such finitary languages P_n and S_n (here a construction is effective if there is an algorithm allowing it).

3.3 Borel ω-powers of infinite rank

A first example of an ω-power which is a Borel set of infinite rank was obtained in [Fin04]. The idea was to iterate the operation $L \to L^{\approx}$, using an infinite number of erasers.

We can first iterate k times this operation $A \to A^{\approx}$. More precisely, we define, for a set $A \subseteq \Sigma^\omega$, where Σ is a finite alphabet.

- $A_k^{\approx.0} := A$,
- $A_k^{\approx.1} := A^{\approx}$,
- $A_k^{\approx.2} := (A_k^{\approx.1})^{\approx}$,

....

- $A_k^{\approx.(k)} := (A_k^{\approx.(k-1)})^{\approx}$,

where we apply k times the operation $A \to A^{\approx}$ with different new letters $\twoheadleftarrow_k$, $\twoheadleftarrow_{k-1}$, $\ldots, \twoheadleftarrow_3$, $\twoheadleftarrow_2$, $\twoheadleftarrow_1$, in such a way that we successively have

$A_k^{\approx.0} = A \subseteq \Sigma^\omega$,

$A_k^{\approx.1} \subseteq (\Sigma \cup \{\twoheadleftarrow_k\})^\omega$,

$A_k^{\approx.2} \subseteq (\Sigma \cup \{\twoheadleftarrow_k, \twoheadleftarrow_{k-1}\})^\omega$,

....

$A_k^{\approx.(k)} \subseteq (\Sigma \cup \{\twoheadleftarrow_k, \twoheadleftarrow_{k-1}, \ldots, \twoheadleftarrow_1\})^\omega$.

and we set $A^{\approx.(k)} = A_k^{\approx.(k)}$.

Note that the choice of the erasers $\twoheadleftarrow_k$, $\twoheadleftarrow_{k-1}$, $\ldots$, $\twoheadleftarrow_2$, $\twoheadleftarrow_1$ in this precise order is important in the proof in [Fin04]. We can now describe the operation $A \to A^{\approx.(k)}$ in a manner similar to the case of the operation $A \to A^{\approx}$, using the notion of a substitution. Let $T_k \subseteq (\Sigma \cup \{\twoheadleftarrow_k, \twoheadleftarrow_{k-1}, \ldots, \twoheadleftarrow_1\})^{<\omega}$ be the language containing the finite words u over the alphabet $\Sigma \cup \{\twoheadleftarrow_k, \twoheadleftarrow_{k-1}, \ldots, \twoheadleftarrow_1\}$ such that one gets the empty word after applying to u the successive erasing operations with the erasers $\twoheadleftarrow_1, \twoheadleftarrow_2, \ldots, \twoheadleftarrow_{k-1}, \twoheadleftarrow_k$. More precisely, $u \in T_k$ if when we start with u, we evaluate $\twoheadleftarrow_1$ as an eraser, and obtain $u_1 = u^{\twoheadleftarrow_1}$ (following Definition 8, i.e., every occurrence of a symbol $\twoheadleftarrow_1$ does erase a letter of Σ or an eraser $\twoheadleftarrow_i$ for $i > 1$). Then we start again with u_1, this time we evaluate $\twoheadleftarrow_2$ as an eraser, which yields $u_2 = u_1^{\twoheadleftarrow_2}$, and so on. When there is no more symbol $\twoheadleftarrow_i$ to be evaluated, then there remains $u_k \in \Sigma^{<\omega}$. By definition, $u \in T_k$ if and only if $u_k = \lambda$. It is easy to see that T_k is a context free language belonging to the subclass of iterated counter languages. Now let h_k be the substitution $\Sigma \to 2^{((\Sigma \cup \{\twoheadleftarrow_k, \twoheadleftarrow_{k-1}, \ldots, \twoheadleftarrow_1\})^{<\omega})}$ defined by $h_k(a) := T_k {}^\frown a$ for every letter $a \in \Sigma$. It holds that $A^{\approx.(k)} = h_k(A)$, for every $A \subseteq \Sigma^\omega$.

We now set $\Sigma = \{\mathbf{0}, \mathbf{1}\}$. Consider now the ω-language $B_2 := (\mathbf{0}^{<\omega}\mathbf{1})^\infty = P_2^\infty$, where P_2 is the language $\mathbf{0}^{<\omega}\mathbf{1}$. B_2 is $\mathbf{\Pi}_2^0$-complete. Then, as in the proof of Theorem 12, $h_p(P_2^\infty) = (h_p(P_2))^\infty$ is a $\mathbf{\Pi}_{p+2}^0$-complete set, for each integer $p \geq 1$.

On the other hand, the languages T_k, for $k \geq 1$, form a sequence which is strictly increasing for the inclusion relation:

$$T_1 \subsetneq T_2 \subsetneq T_3 \subsetneq \ldots \subsetneq T_i \subsetneq T_{i+1} \ldots$$

In order to construct an ω-power which is Borel of infinite rank, the first idea is to substitute the language $\bigcup_{k\geq 1} T_k{}^\frown a$ to each letter $a \in \Sigma = \{\mathbf{0}, \mathbf{1}\}$ in the language P_2^∞. But this way we would get a language over the *infinite* alphabet $\Sigma \cup \{\twoheadleftarrow_1, \twoheadleftarrow_2, \twoheadleftarrow_3, \ldots\}$. in: order to obtain a finitary language over a *finite* alphabet, every eraser $\twoheadleftarrow_j$ can be coded by a finite word $\alpha.\beta^j.\alpha$ over the alphabet $\{\alpha, \beta\}$, where α and β are two new letters.

We defines the substitution $\varphi_k : (\Sigma \cup \{\twoheadleftarrow_1, \ldots, \twoheadleftarrow_k\})^{<\omega} \to 2^{(\Sigma \cup \{\alpha, \beta\})^{<\omega}}$ by $\varphi_p(c) := \{c\}$ for each $c \in \Sigma$ and $\varphi_k(\twoheadleftarrow_j) = \{\alpha.\beta^j.\alpha\}$ for each integer $j \in [1, k]$. Now let

$$\mathcal{L} := \bigcup_{k\geq 1} \varphi_k(T_k),$$

and $h : \Sigma \to 2^{((\Sigma \cup \{\alpha, \beta\})^{<\omega})}$ be the substitution defined by $h(a) := \mathcal{L}^\frown a$, for each $a \in \Sigma$.

Theorem 14. *Let $P_2 := \mathbf{0}^{<\omega}\mathbf{1}$. Then the ω-power $(h(P_2))^\infty \subseteq \{\mathbf{0}, \mathbf{1}, \alpha, \beta\}^\omega$ is a Borel set of infinite rank.*

The language $(h(P_2))$ is a simple recursive language but it is not context-free. Later, with a modification of the construction, and using a coding of an infinity of erasers previously defined in [Fin03b], Finkel and Duparc got a context-free language W such that W^∞ is a Borel set of infinite rank [DF07].

Theorem 15. *There exists a context-free finitary language $W \subseteq \Gamma^{<\omega}$, where Γ is a finite alphabet, such that W^∞ is a Borel set of infinite rank. Moreover W^∞ is above the class $\mathbf{\Delta}_\omega^0$.*

The coding of the infinity of erasers $\twoheadleftarrow_n$ is given by $\Phi(\twoheadleftarrow_n) = \alpha B^n C^n D^n E^n \beta$ with new letters $\alpha, B, C, D, E, \beta$. Actually the pushdown automaton constructed in order to accept the language W must be able to read the number n identifying the eraser four times.

The ω-power W^∞ is above the class $\mathbf{\Delta}_\omega^0$, i.e., it is not in the Borel class $\mathbf{\Delta}_\omega^0$. Note that the ω-power $(h(P_2))^\infty$ was actually also above the class $\mathbf{\Delta}_\omega^0$ but this was not shown in [Fin04]. We give the argument in this latter case, where the language $h(P_2)$ is simpler than W.

This follows from the fact that $((h(P_2))^\infty)^\approx$ is Wadge equivalent to $(h(P_2))^\infty$, which is due to the precise way we ordered the erasers, as described above. On the other side the operation $A \to A^\approx$ is strictly increasing for the Wadge ordering inside $\mathbf{\Delta}^0_\omega$ (see [Dup01]). This implies that $(h(P_2))^\infty$, and also W^∞, are not in the class $\mathbf{\Delta}^0_\omega$.

Note that the language W is context-free but it cannot be accepted by a one-counter automaton.

3.4 Non-Borel ω-powers which are even $\mathbf{\Sigma}^1_1$-complete

A first example of language L such that L^∞ is not Borel, and even $\mathbf{\Sigma}^1_1$-complete, was obtained in [Fin03a]. It turned out that the language L may be described in a very simple way. Surprisingly it is actually accepted by a one-counter automaton. It was obtained via a coding of infinite labelled binary trees. We now recall the construction of this language L using the notion of a substitution.

Let d be a letter not in 2 and $D := \{ u \cdot d \cdot v \mid u, v \in 2^{<\omega}$ and $|v| = 2|u|$ or $|v| = 2|u|+1 \}$. It is easy to see that the language $D \subseteq (2 \cup \{d\})^{<\omega}$ is a context-free language accepted by a one-counter automaton.

Let $g : \Sigma \to 2^{(2 \cup \{d\})^{<\omega}}$ be the substitution defined by $g(a) = a \cdot D$. Since $W := \{\mathbf{0}\}^{<\omega} \cdot \mathbf{1}$ is a regular language, $L := g(W)$ is a context-free language and it is accepted by a one-counter automaton. Moreover, it is proved in [Fin03a] that $(g(W))^\infty$ is $\mathbf{\Sigma}^1_1$-complete, and thus non-Borel. This is done by reducing to this ω-language a well-known example of a $\mathbf{\Sigma}^1_1$-complete set: the set of infinite binary trees labelled in the alphabet 2 which have an infinite branch in the $\mathbf{\Pi}^0_2$-complete set W^∞.

4 Classical and effective complexity of the ω-powers

In [FL07], we prove that there are some ω-powers of any Borel rank. More precisely, Theorem 2 in [FL07] is as follows.

Theorem 16. *Let $\xi \geq 1$ be a countable ordinal.*

(a) There is a finitary language $P_\xi \subseteq 2^{<\omega}$ such that the ω-power P_ξ^∞ is $\mathbf{\Pi}^0_\xi$-complete.

(b) There is a finitary language $S_\xi \subseteq 2^{<\omega}$ such that the ω-power S_ξ^∞ is $\mathbf{\Sigma}^0_\xi$-complete.

In fact, we provide a general method proving this when $\xi \geq 3$. Examples of such finitary languages were given in Section 3.2 when $\xi \leq 2$.

We now turn to the general case. Let $\mathbf{\Gamma}$ be a class of sets of the form $\mathbf{\Sigma}^0_\xi$ or $\mathbf{\Pi}^0_\xi$, with $\xi \geq 3$. Fix a $\mathbf{\Gamma}$-complete set $B \subseteq 2^\omega$, so that $B \in \mathbf{\Pi}^0_{\xi+1}$.

A result due to Kuratowski provides a closed subset C of ω^ω and a continuous bijection $f:C\to B$ with the property that f^{-1} is $\mathbf{\Sigma}^0_\xi$-measurable (i.e., $f[O]$ is a $\mathbf{\Sigma}^0_\xi$ subset of B if O is an open subset of C, see [Kur66]). This result is a level by level version of a result, due to Lusin and Souslin, asserting that every Borel subset of 2^ω is the image of a closed subset of ω^ω by a continuous bijection. By Proposition 11 in [Lec05], it is enough to find a finitary language $A\subseteq 4^{<\omega}$, where $4:=\{\mathbf{0},\mathbf{1},\mathbf{2},\mathbf{3}\}$, such that A^∞ is $\mathbf{\Gamma}$-complete.

The language A will be made of two pieces: $A=\mu\cup\pi$. The set π will code f, and π^∞ will look like B on some compact sets $K_{N,j}$. Outside this countable family of compact sets we will hide f, so that A^∞ will be the simple set μ^∞.

The Lusin-Souslin theorem has been used by Arnold in [Arn83] to prove that every Borel subset of Σ^ω, where Σ is a finite alphabet, is accepted by a non-ambiguous finitely branching transition system with a Büchi acceptance condition, and our first idea was to code the behaviour of such a transition system.

Definition 17. *A Büchi transition system is a 5-tuple $\mathcal{T}=(Q,\Sigma,q_0,\Delta,F)$, where Q is a (possibly infinite) countable set of states, Σ is a finite input alphabet, $q_0\in Q$ is the initial state, $\Delta\subseteq Q\times\Sigma\times Q$ is the transition relation, and $F\subseteq Q$ is the set of final states.*

Let $\sigma=a_0a_1\ldots$ be an ω-word over Σ. An ω-sequence of states $r=(t_i)_{i\in\omega}$ is called a run of $\mathcal{T}$ on σ if

(1) $t_0=q_0$,

(2) for each $i\in\omega$, $(t_i,\sigma(i),t_{i+1})\in\Delta$.

The run r is said to be accepting when $t_i\in F$ for infinitely many i's. The transition system $\mathcal{T}$ is said to be

- non-ambiguous if each infinite word $\sigma\in\Sigma^\omega$ has at most one accepting run by $\mathcal{T}$,

- finitely branching if for each state $q\in Q$ and each $a\in\Sigma$, there are only finitely many states q' such that $(q,a,q')\in\Delta$.

The ω-language accepted by $\mathcal{T}$ is

$$A(\mathcal{T}):=\{\sigma\in\Sigma^\omega\mid \textit{there exists an accepting run } r \textit{ of } \mathcal{T} \textit{ on } \sigma\}.$$

We will code the behaviour of a transition system coming from f.

- The set of states is $Q:=\{(s,t)\in 2^{<\omega}\times 2^{<\omega}\mid |s|=|t|\}$, which is countably infinite. We enumerate Q as follows. We start with $q_0:=(\emptyset,\emptyset)$. Then we put the sequences of length 1 of elements of 2×2, in the lexicographical ordering: $q_1:=(0,0)$, $q_2:=(0,1)$, $q_3:=(1,0)$, $q_4:=(1,1)$. Then we put the 16 sequences of length 2: $q_5:=(0^2,0^2)$, $q_6:=(0^2,01)$, ... And so on.

We will sometimes use the coordinates of $q_n := (q_n^0, q_n^1)$. We put $M_j := \Sigma_{i<j}\, 4^{i+1}$. Note that the sequence $(M_j)_{j\in\omega}$ is strictly increasing, and that q_{M_j} is the last sequence of length j of elements of 2×2. We define, for $N, j\in\omega$ with $N\leq M_j$, the compact set

$$K_{N,j} := \{\ \mathbf{2}^N(\frown_{i\in\omega}\ m_i\ \mathbf{2}^{M_{j+i+1}}\ \mathbf{3}\ \mathbf{2}^{M_{j+i+1}}) \in 4^\omega \mid \forall i\in\omega\ \ m_i\in 2\ \}.$$

- The input alphabet is 2.

- The initial state is $q_0 := (\emptyset, \emptyset)$.

- If $m\in 2$ and $n, p\in\omega$, then we write $n \xrightarrow{m} p$ if $q_n^0 \subseteq q_p^0$ and $q_p^1 = q_n^1 m$. As f is continuous on C, the graph Graph(f) of f is a closed subset of $C\times 2^\omega$. As C is a closed subset of $\mathbb{P}_\infty$, Graph(f) is also a closed subset of $\mathbb{P}_\infty\times 2^\omega$. So there is a closed subset P of $2^\omega\times 2^\omega$ with the property that Graph$(f) = P\cap(\mathbb{P}_\infty\times 2^\omega)$. We identify $2^\omega\times 2^\omega$ with $(2\times 2)^\omega$, i.e., we view (β,α) as $(\beta(0),\alpha(0)), (\beta(1),\alpha(1))$,... By Proposition 2.4 in [Kec95], there is $R\subseteq (2\times 2)^{<\omega}$, closed under initial segments, such that

$$P = \{(\beta,\alpha)\in 2^\omega\times 2^\omega \mid \forall k\in\omega\ \ (\beta,\alpha)|k\in R\};$$

note that R is a tree whose infinite branches form the set P. In particular, we get

$$(\beta,\alpha)\in \text{Graph}(f) \Leftrightarrow \beta\in\mathbb{P}_\infty \text{ and } \forall k\in\omega\ \ (\beta,\alpha)|k\in R.$$

The transition relation $\Delta\subseteq Q\times 2\times Q$ is given by $(q_n, m, q_p)\in\Delta \Leftrightarrow n\xrightarrow{m} p$, for $m\in 2$ and $n, p\in\omega$.

- The set of final states is $F := \{(t,s)\in R \mid t\neq\emptyset \text{ and } t(|t|-1) = 1\}$. Note that F is simply the set of pairs $(t,s)\in R$ such that the last letter of t is a 1.

Recall that a run of $\mathcal{T}$ is said to be Büchi accepting if final states occur infinitely often during this run. Then the set of ω-words over the alphabet 2 which are accepted by the transition system $\mathcal{T}$ from the initial state q_0 with Büchi acceptance condition is exactly the Borel set B.

We are now ready to define the finitary language π. We set

$$\pi := \left\{ \begin{array}{l} s\in 4^{<\omega} \mid \exists j, l\in\omega\ \exists (m_i)_{i\leq l}\in 2^{l+1}\ \ \exists (n_i)_{i\leq l}, (p_i)_{i\leq l}, (r_i)_{i\leq l}\in\omega^{l+1} \\ \\ \qquad\qquad n_0\leq M_j \\ \qquad\qquad\quad \text{and} \\ \qquad\qquad \forall i\leq l\ \ n_i \xrightarrow{m_i} pi \text{ and } p_i + r_i = M_{j+i+1} \\ \qquad\qquad\quad \text{and} \\ \qquad\qquad \forall i< l\ \ p_i = n_{i+1} \\ \qquad\qquad\quad \text{and} \\ \qquad\qquad q_{p_l}\in F \\ \qquad\qquad\quad \text{and} \\ \qquad\qquad s = \frown_{i\leq l}\ \mathbf{2}^{n_i}\ m_i\ \mathbf{2}^{p_i}\ \mathbf{2}^{r_i}\ \mathbf{3}\ \mathbf{2}^{r_i} \end{array} \right\}.$$

We are also ready to define μ. The idea is that an infinite sequence containing a word in μ cannot be in the union of the $K_{N,j}$'s. We set

$$\mu_0 := \left\{ \begin{array}{l} s \in 4^{<\omega} | \exists l \in \omega \ \exists (m_i)_{i \leq l+1} \in 2^{l+2} \ \exists N \in \omega \ \exists (P_i)_{i \leq l+1}, (R_i)_{i \leq l+1} \in \omega^{l+2} \\ \qquad \forall i \leq l+1 \ \exists j \in \omega \ P_i = M_j \\ \qquad \text{and} \\ \qquad P_l \neq R_l \\ \qquad \text{and} \\ \qquad s = \mathbf{2}^N (\frown_{i \leq l+1} m_i \ \mathbf{2}^{P_i} \ \mathbf{3} \ \mathbf{2}^{R_i}) \end{array} \right\},$$

$$\mu_1 := \left\{ \begin{array}{l} s \in 4^{<\omega} | \exists l \in \omega \ \exists (m_i)_{i \leq l+1} \in 2^{l+2} \ \exists N \in \omega \ \exists (P_i)_{i \leq l+1}, (R_i)_{i \leq l+1} \in \omega^{l+2} \\ \qquad \forall i \leq l+1 \ \exists j \in \omega \ P_i = M_j \\ \qquad \text{and} \\ \qquad \exists j \in \omega \ (P_l = M_j \text{ and } P_{l+1} \neq M_{j+1}) \\ \qquad \text{and} \\ \qquad s = \mathbf{2}^N (\frown_{i \leq l+1} m_i \ \mathbf{2}^{P_i} \ \mathbf{3} \ \mathbf{2}^{R_i}) \end{array} \right\},$$

and $\mu := \mu_0 \cup \mu_1$. Recall that $A = \mu \cup \pi$.

We just described how to get the finitary languages in the statement of Theorem 16. For the other Borel classes $\mathbf{\Delta}^0_\xi$, only $\mathbf{\Delta}^0_1$ is a Wadge class, and $A := \{s \in 2^{<\omega} \mid 0 \subseteq s \text{ or } 1^2 \subseteq s\}$ has the property that $A^\infty = 2^\omega \setminus N_{10}$ is $\mathbf{\Delta}^0_1$-complete (see [FL09]). In [FL09], we provide some complete sets for some other Wadge classes of Borel sets, in fact some dual classes of classes of differences of $\mathbf{\Sigma}^0_\xi$ sets (see also [Lec05]). It is worth noting that Theorem 16 may seem to indicate that ω-powers can be arbitrarily complex, but its proof uses closure properties of the classes of the Borel hierarchy that are not shared by all the Wadge classes of Borel sets, such as the closure by finite unions. The extension of Theorem 16 to all Wadge classes of Borel sets is an open problem.

An important result in [FL09] shows that Theorem 16 is as effective as it can be, in the context of effective descriptive set theory. in: order to state it, we must recall some notions about this theory. Effective descriptive set theory is based on the notion of a recursive function. A function from ω^k to ω is said to be *recursive* if it is total and computable. By extension, a relation is called *recursive* if its characteristic function is recursive.

Definition 18. *A recursive presentation of a topological space X is a pair $((x_n)_{n \in \omega}, d)$ such that*

1. $(x_n)_{n \in \omega}$ *is dense in X,*

2. *d is a compatible complete distance on X such that the following relations P and Q are recursive:*

$$P(i,j,m,k) \iff d(x_i,x_j) \leq \frac{m}{k+1},$$
$$Q(i,j,m,k) \iff d(x_i,x_j) < \frac{m}{k+1}.$$

A topological space X is recursively presented *if it is given with a recursive presentation of it.*

Note that every recursively presented space is *Polish* (i.e., separable and completely metrizable). For example, one can check that the spaces ω and Σ^ω have a recursive presentation. Moreover, a product of two recursively presented spaces has a recursive presentation.

Note that the formula $(p,q)\mapsto 2^p(2q+1)-1$ defines a recursive bijection $\omega^2\to\omega$. One can check that the coordinates of the inverse map are also recursive. They will be denoted $n\mapsto(n)_0$ and $n\mapsto(n)_1$ in the sequel. These maps will help us to define some of the basic effective classes.

Definition 19. *Let $((x_n)_{n\in\omega},d)$ be a recursive presentation of a topological space X.*

1. *We fix a countable basis of X: $B(X,n)$ is the open ball $B_d(x_{(n)_0},\frac{((n)_1)_0}{((n)_1)_1+1})$.*

2. *A subset S of X is* semirecursive, *or* effectively open *(denoted $S\in\Sigma^0_1$) if*

$$S=\bigcup_{n\in\omega} B(X,f(n)),$$

for some recursive function f.

3. *If $n\geq 1$ is a natural number, then Π^0_n is the class of complements of Σ^0_n sets. We say that $B\in\Sigma^0_{n+1}$ if there is $C\in\Pi^0_n(\omega\times X)$ such that*

$$B=\exists^\omega C:=\{x\in X\mid \exists i\in\omega\ (i,x)\in C\}.$$

We also set $\Delta^0_n:=\Sigma^0_n\cap\Pi^0_n$.

4. *A subset S of X is* effectively analytic *(denoted $S\in\Sigma^1_1$) if there is a Π^0_1 subset C of $X\times\omega^\omega$ such that $S=proj_X[C]:=\{x\in X\mid \exists\alpha\in\omega^\omega\ \ (x,\alpha)\in C\}$. A subset S of X is* effectively co-analytic *(denoted $S\in\Pi^1_1$) if its complement $\neg S$ is effectively analytic, and* effectively Borel *if it is in Σ^1_1 and Π^1_1 (denoted $S\in\Delta^1_1$). We also set $\Sigma^1_2:=\{\exists^{\omega^\omega}C\mid C\in\Pi^1_1\}$, $\Pi^1_2:=\check{\Sigma}^1_2$ and $\Delta^1_2:=\Sigma^1_2\cap\Pi^1_2$.*

5. *We will consider the relativized classes: if Y is a recursively presented space and $y \in Y$, then we say that $A \subseteq X$ is in $\Sigma_1^1(y)$ if there is $S \in \Sigma_1^1(Y \times X)$ such that*
$$A = S_y := \{x \in X \mid (y, x) \in S\}.$$
The class $\Pi_1^1(y)$ is defined similarly. We also set $\Delta_1^1(y) := \Sigma_1^1(y) \cap \Pi_1^1(y)$.

6. *Let $\gamma \in \omega^\omega$. We say that $\gamma \in \Sigma_1^0$ if $\{k \in \omega \mid \gamma \in B(\omega^\omega, k)\} \in \Sigma_1^0(\omega)$. A countable ordinal ξ is a recursive ordinal if there is $\gamma \in \Sigma_1^0$ coding a well-ordering on ω of order type ξ.*

7. *There is a good parametrization in Σ_1^0 for $\mathbf{\Sigma}_1^0$ (see 3E.2, 3F.6 and 3H.1 in [Mos80]). This means that there is a system of sets $G^{\Sigma_1^0, Y} \in \Sigma_1^0(\omega^\omega \times Y)$ such that, for each recursively presented space Y and for each $P \subseteq Y$,*
$$P \in \mathbf{\Sigma}_1^0 \Leftrightarrow \exists \gamma \in \omega^\omega \ P = G_\gamma^{\Sigma_1^0, Y},$$
$$P \in \Sigma_1^0 \Leftrightarrow \exists \gamma \in \Sigma_1^0 \ P = G_\gamma^{\Sigma_1^0, Y}.$$
Moreover, if Z is a recursively presented space of type at most 1 (i.e., a finite product of spaces equal to ω, ω^ω or 2^ω), and Y is a recursively presented space, then there is $S_{\Sigma_1^0}^{Z,Y} : \omega^\omega \times Z \to \omega^\omega$ recursive such that
$$(\gamma, z, y) \in G^{\Sigma_1^0, Z \times Y} \Leftrightarrow (S_{\Sigma_1^0}^{Z,Y}(\gamma, z), y) \in G^{\Sigma_1^0, Y}$$
(here, by $S_{\Sigma_1^0}^{Z,Y}$ recursive we mean that the relation defined by
$$R(\gamma, z, k) \Leftrightarrow S_{\Sigma_1^0}^{Z,Y}(\gamma, z) \in B(\omega^\omega, k)$$
defines a Σ_1^0 subset of $\omega^\omega \times Z \times \omega$).

8. *We can code the partial recursive functions. Let Y be a recursively presented space, $f : X \to Y$ be a partial function, $D \subseteq Domain(f)$ and $P \subseteq X \times \omega$. Then P computes f on D if*
$$x \in D \Rightarrow \forall k \in \omega \ \left(f(x) \in B(Y, k) \Leftrightarrow (x, k) \in P\right).$$
If P is in Σ_1^0 and computes f on D, then we say that f is recursive on D. This means that $f^{-1}(B(Y, k)) \in \Sigma_1^0$, uniformly in k.

We now define a partial function $U : \omega^\omega \times X \to Y$ by
$$U(\gamma, x)\downarrow \ \Leftrightarrow U(\gamma, x) \text{is defined}$$
$$\Leftrightarrow \exists y \in Y \ \forall k \in \omega \ \left(y \in B(Y, k) \Leftrightarrow (\gamma, x, k) \in G^{\Sigma_1^0, X \times \omega}\right),$$
$$U(\gamma, x) := \text{the unique } y \in Y \text{ such that}$$
$$\forall k \in \omega \ \left(y \in B(Y, k) \Leftrightarrow (\gamma, x, k) \in G^{\Sigma_1^0, X \times \omega}\right).$$

Now let $\gamma \in \omega^\omega$*. The function* $\{\gamma\}^{X,Y} : X \to Y$ *is defined by* $\{\gamma\}^{X,Y}(x) := U(\gamma, x)$*. Then a partial function* $f : X \to Y$ *is recursive on its domain if and only if there is* $\gamma \in \Sigma^0_1$ *such that* $f(x) = \{\gamma\}^{X,Y}(x)$ *when* $f(x)$ *is defined. More generally, the functions of the form* $\{\gamma\}^{X,Y}$ *are the partial continuous functions from a subset of* X *into* Y*. In order to simplify the notation, we will write* $\{\gamma\}$ *instead of* $\{\gamma\}^{X,Y}$ *when* $Y = \omega^\omega$*.*

9. *We now define, by induction on the countable ordinal* $\xi \geq 1$*, the set* BC_ξ *of Borel codes for* $\mathbf{\Sigma}^0_\xi$ *as follows. If* $\gamma \in \omega^\omega$*, then we define* $\gamma^* \in \omega^\omega$ *by* $\gamma^*(i) := \gamma(i+1)$*. We set*

$$BC_1 := \{\, \gamma \in \omega^\omega \mid \gamma(0) = 0 \,\},$$
$$BC_\xi := \{\, \gamma \in \omega^\omega \mid \gamma(0) = 1 \text{ and } \forall i \in \omega \;\; \{\gamma^*\}(i)\!\downarrow \text{ and } \{\gamma^*\}(i) \in \textstyle\bigcup_{1 \leq \eta < \xi} BC_\eta \,\} \quad \text{if } \xi \geq 2.$$

The set of Borel codes is $BC := \bigcup_{1 \leq \xi < \omega_1} BC_\xi$*. We also set*

$$BC^* := \bigcup_{2 \leq \xi < \omega_1} \uparrow BC_\xi.$$

We define $\rho^X : BC \to \mathbf{\Delta}^1_1(X)$ *by induction:*

$$\rho^X(\gamma) := \begin{cases} \bigcup_{i \in \omega} B(X, \gamma^*(i)) \text{ if } \gamma \in BC_1, \\ \bigcup_{i \in \omega} X \setminus \rho^X(\{\gamma^*\}(i)) \text{ if } \gamma \in BC^*. \end{cases}$$

Clearly, $\rho^X[BC_\xi] = \mathbf{\Sigma}^0_\xi(X)$*, by induction on* ξ*.*

10. *We can now define the* hyperarithmetical hierarchy*. Let* $\xi \geq 1$ *be a countable ordinal. Then*

$$\Sigma^0_\xi(X) = \{\rho^X(\gamma) \mid \gamma \in \Sigma^0_1 \cap BC_\xi\},$$

$$\Pi^0_\xi(X) = \check{\Sigma}^0_\xi(X),$$

$$\Delta^0_\xi(X) = \Sigma^0_\xi(X) \cap \Pi^0_\xi(X).$$

This definition is compatible with the item 3.

The crucial link between the effective classes and the classical corresponding classes is as follows: the class of analytic (resp., co-analytic, Borel) sets is equal to $\bigcup_{\alpha \in \omega^\omega} \Sigma^1_1(\alpha)$ (resp., $\bigcup_{\alpha \in \omega^\omega} \Pi^1_1(\alpha)$, $\bigcup_{\alpha \in \omega^\omega} \Delta^1_1(\alpha)$). This allows to use effective descriptive set theory to prove results of classical type.

Theorem 20. *Let $\xi \geq 1$ be a recursive ordinal.*

(a) There is a finitary language $P_\xi \subseteq 2^{<\omega}$, that can be coded by a Δ_1^0 subset of ω, such that the ω-power P_ξ^∞ is in the effective class Π_ξ^0 but not in $\mathbf{\Sigma}_\xi^0$.

(b) There is a finitary language $S_\xi \subseteq 2^{<\omega}$, that can be coded by a Δ_1^0 subset of ω, such that the ω-power S_ξ^∞ is in the effective class Σ_ξ^0 but not in $\mathbf{\Pi}_\xi^0$.

5 Complexity of some sets of finitary languages related to the ω-powers

In [Lec05], the following question is raised. What is the topological complexity of the set of finitary languages whose associated ω-power is of a given level of complexity?

This question arises naturally when we look at the characterizations of closed, $\mathbf{\Pi}_2^0$ and open ω-powers obtained in [Sta97b] (see Corollary 14 and Lemmas 25, 26). This leads to set, for a class of sets $\mathbf{\Gamma}$, $\mathcal{L}_{\mathbf{\Gamma}} := \{L \subseteq 2^{<\omega} \mid L^\infty \in \mathbf{\Gamma}\}$. It is proved in [Lec05] (see Theorem 4) that $\mathcal{L}_{\{\emptyset\}}$ is $\mathbf{\Pi}_1^0$-complete, $\mathcal{L}_{\check{\{\emptyset\}}}$ is $\mathbf{\Sigma}_1^0$-complete, and

Theorem 21. *The set $\mathcal{L}_{\mathbf{\Delta}_1^0}$ is $\mathbf{\Sigma}_2^0$-complete.*

For the next classes of the Borel hierarchy, it is proved in [Lec05] that $\mathcal{L}_{\mathbf{\Sigma}_\xi^0}$ are $\mathcal{L}_{\mathbf{\Pi}_\xi^0}$ are $\mathbf{\Sigma}_2^1$ (see Proposition 16). A consequence of Theorem 20 is that these sets are $\mathbf{\Pi}_1^1$-hard if $\xi \geq 3$ (see Corollary 6.4 in [FL09]). It is proved in [Fin10] that for every integer $k \geq 2$ (respectively, $k \geq 3$) the set $\mathcal{L}_{\mathbf{\Pi}_{k+1}^0}$ (respectively, $\mathcal{L}_{\mathbf{\Sigma}_{k+1}^0}$) is "more complex" than the set $\mathcal{L}_{\mathbf{\Pi}_k^0}$ (respectively, $\mathcal{L}_{\mathbf{\Sigma}_k^0}$), with respect to the Wadge reducibility. The following result is proved in [Lec05, Fin10].

Theorem 22. *The set $\mathcal{L}_{\mathbf{\Delta}_1^1}$ is in $\mathbf{\Sigma}_2^1 \setminus \mathbf{\Pi}_2^0$.*

Along similar lines, some other results of effective nature are available in [Lec05, FL09]. For instance, we set $\mathcal{L}_\Delta := \{L \subseteq 2^{<\omega} \mid L^\infty \in \Delta_1^1(L)\}$. The following is proved in [Lec05] and [FL09].

Theorem 23. *The following sets are co-analytic and not Borel.*

(a) $\mathcal{L}_\Delta$,

(b) $\mathcal{L}_{\mathbf{\Sigma}_\xi^0} \cap \mathcal{L}_\Delta$ ($\mathbf{\Pi}_1^1$-complete if $\xi \geq 3$),

(c) $\mathcal{L}_{\mathbf{\Pi}_\xi^0} \cap \mathcal{L}_\Delta$ if $\xi \geq 2$ ($\mathbf{\Pi}_1^1$-complete if $\xi \geq 3$).

There is a very natural subset of $\mathcal{L}_{\mathbf{\Pi}^0_1}$, namely the set of finitely generated ω-powers. If we set $\mathbf{\Gamma}_f := \{L^\infty \mid L \text{ is finite}\}$, then this is $\mathcal{L}_{\mathbf{\Gamma}_f}$. We can decompose $\mathbf{\Gamma}_f$ with respect to the cardinality, setting, for $p \in \omega$, $\mathbf{\Gamma}_p := \{L^\infty \mid \text{Cardinality}(L) = p\}$, so that

$$\mathbf{\Gamma}_f = \bigcup_{p \in \omega} \mathbf{\Gamma}_p.$$

Note that $\mathbf{\Gamma}_0 = \mathcal{L}_{\{\emptyset\}}$, and we can prove that $\mathbf{\Gamma}_1$ is $\mathbf{\Pi}^0_1$-complete (see Proposition 6 in [Lec05]). The complexity of $\mathbf{\Gamma}_2$ is very surprising since it is not clear at all on its definition (see Corollary 10 in [Lec05]).

Theorem 24. *The set* $\mathbf{\Gamma}_2$ *is* $\check{D}_\omega(\mathbf{\Sigma}^0_1)$*-complete.*

6 Open questions

It is still open to determine all the infinite Borel ranks of the ω powers of context-free languages. However the results of [Fin06] suggest that the ω-powers of context-free languages or even of languages accepted by one-counter automata exhibit also a great topological complexity. Indeed, there are ω-languages accepted by Büchi one-counter automata of every Borel rank (and even of every Wadge degree) of an effective analytic set.

In particular, for each recursive ordinal $\xi < \omega_1^{\mathrm{CK}}$, there are some ω-languages P_ξ and S_ξ in the class Δ^1_1 such that P_ξ is $\mathbf{\Pi}^0_\xi$-complete and S_ξ is $\mathbf{\Sigma}^0_\xi$-complete. But effective analytic sets are much more complicated than Δ^1_1 sets: Kechris, Marker and Sami proved in [KMS89] that the supremum of the set of Borel ranks of (effective) Σ^1_1 sets is the ordinal γ^1_2. This ordinal is proved to be strictly greater than the ordinal δ^1_2 which is the first non Δ^1_2 ordinal. In particular, the ordinal γ^1_2 is strictly greater than the ordinal ω_1^{CK} (note that the exact value of the ordinal γ^1_2 may depend on axioms of set theory).

Moreover each ω-language $L \subseteq \Sigma^\omega$ accepted by a Büchi one-counter automaton is of the form $L = \bigcup_{1 \le j \le n} U_j \cdot V_j^\infty$, for some one-counter finitary languages U_j and V_j, $1 \le j \le n$.

Therefore it seems plausible that there exist complete ω-powers of a one-counter language, for each Borel class of recursive rank, and we can even conjecture that there exist some ω-powers of languages accepted by one-counter automata which have Borel ranks up to the ordinal γ^1_2, although these languages are located at the very low level in the complexity hierarchy of finitary languages.

References

[Arn83] André Arnold *Topological characterizations of infinite behaviours of transition systems*, in: Josep Díaz (ed.), Automata, Languages and Programming, Springer, 1983 [*Lecture Notes in Computer Science 154*] pp. 28–38.

[ABB96] Jean-Michel Autebert, Jean Berstel, and Luc Boasson *Context free languages and pushdown automata*, in: Grzegorz Rozenberg and Arto Salomaa (eds.), Handbook of Formal Languages 1, 111-174, Springer-Verlag, 1997.

[CG77] Rina S. Cohen, and Arie Y. Gold *Theory of ω-languages, parts one and two*, Journal of Computer and System Science 15, 1977, pp.169–208.

[DF07] Jacques Duparc, and Olivier Finkel *An ω-power of a context free language which is Borel above* $\boldsymbol{\Delta}^0_\omega$, in: Stefan Bold, Benedikt Löwe, Thoralf Räsch and Johan van Benthem (eds.), Infinite Games (Proceedings of the International Conference Foundations of the Formal Sciences V, November 26th to 29th, 2004, Bonn, Germany), London, 2007 [*College Publications at King's College 11 (Studies in Logic)*], pp. 109–122.

[Dup01] Jacques Duparc *Wadge hierarchy and Veblen hierarchy: Part 1: Borel sets of finite rank*, The Journal of Symbolic Logic 66 (1), 2001, pp. 56–86.

[Fin01] Olivier Finkel *Topological properties of omega context free languages*, Theoretical Computer Science 262 (1–2), 2001, pp. 669–697.

[Fin03a] — *Borel hierarchy and omega context free languages*, Theoretical Computer Science 290 (3), 2003, pp. 1385–1405.

[Fin03b] — *On omega context free languages which are Borel sets of infinite rank*, Theoretical Computer Science 299 (1-3), 2003, pp. 327–346.

[Fin04] — *An omega-power of a finitary language which is a Borel set of infinite rank*, Fundamenta Informaticae 62 (3–4), 2004, pp. 333–342.

[Fin06] — *Borel ranks and Wadge degrees of omega context free languages*, Mathematical Structures in Computer Science 16 (5), 2006, pp. 813–840.

[Fin10] — *On some sets of dictionaries whose* omega-*powers have a given complexity*, Mathematical Logic Quarterly 56 (5), 2010, pp. 452–460.

[FL07] Olivier Finkel, and Dominique Lecomte *There exist some ω-powers of any Borel rank*, in: Jacques Duparc and Thomas A. Henzinger (eds.), Proceedings of the 16th EACSL Annual International Conference on Computer Science and Logic, CSL 2007 (Lausanne, Switzerland, September 11-15, 2007), Springer, 2007 [*Lecture Notes in Computer Science 4646*], pp. 115–129.

[FL09] — *Classical and effective descriptive complexities of omega-powers*, Annals of Pure and Applied Logic 160 (2), 2009, pp. 163–191.

[FL20] — *Some complete ω-powers of a one-counter language, for any Borel class of finite rank*, Archive for Mathematical Logic, 2020 [to appear].

[HMU01] John E. Hopcroft, Rajeev Motwani, and Jeffrey D. Ullman Introduction to Automata Theory, Languages, and Computation, Addison-Wesley Publishing Company, Reading MA 2001 [*Addison-Wesley Series in Computer Science*].

[Kec95] Alexander S. Kechris Classical Descriptive Set Theory, Springer-Verlag, New York 1995.

[KMS89] Alexander S. Kechris, David Marker, and Ramez Labib Sami Π^1_1 *Borel sets*, The Journal of Symbolic Logic 54 (3), 1989, pp. 915–920.

[Kur66] Kazimierz Kuratowski Topology 1, Academic Press, New York 1966.

[Lec05] Dominique Lecomte *Omega-powers and descriptive set theory*, The Journal of Symbolic Logic 70 (4), 2005, pp. 1210–1232.

[LT87] Igor Litovsky and Erick Timmerman *On generators of rational omega-power languages*, Theoretical Computer Science 53, 1987, pp. 187–200.

[Mos80] Yiannis N. Moschovakis Descriptive Set Theory, North-Holland Publishing Company, Amsterdam 1980.

[Niw90] Damian Niwinski A problem on ω-powers, in: 1990 Workshop on Logics and Recognizable Sets, University of Kiel, 1990.

[PP04] Dominique Perrin, and Jean-Eric Pin Infinite Words, Automata, Semigroups, Logic and Games, Elsevier, 2004 [*Pure and Applied Mathematics 141*].

[Sim92] Pierre Simonnet Automates et théorie descriptive, PhD Dissertation, University of Paris 7, 1992.

[Sta86] Ludwig Staiger *Hierarchies of recursive ω-languages*, Elektronische Informationsverarbeitung und Kybernetik 22 (5–6), 1986, pp. 219–241.

[Sta97a] — *ω-languages*, in: Grzegorz Rozenberg and Arto Salomaa (eds.), Handbook of Formal Languages 3, Springer, Berlin 1997, pp. 339–387.

[Sta97b] — *On ω-power languages*, in: Gheorghe Păun and Arto Salomaa (eds.), New Trends in Formal Languages, Control, Cooperation and Combinatorics, Springer-Verlag, 1997 [*Lecture Notes in Computer Science 1218*], pp. 377–393.

[Tho90] Wolfgang Thomas *Automata on infinite objects*, in: Jan van Leeuwen (ed.), Handbook of Theoretical Computer Science, *B Formal Models and Semantics*, Elsevier, 1990, pp. 133–191.

[Wad83] William W. Wadge Reducibility and Determinateness in the Baire space, PhD Dissertation, University of California, Berkeley CA 1983.

Low, Superlow, and Superduperlow Sets:

An Exposition of a Known But Not Well-Known Result

William Gasarch
University of Maryland
gasarch@cs.umd.edu

1 Introduction

We use the following standard notation.

Notation 1.1.

1. $M_0, M_1, \ldots$ is be a standard list of Turing Machines.
2. W_e is the domain of M_e. Hence $W_0, W_1, \ldots$ is a list of all c.e. sets.
3. $M_0^{()}, M_1^{()}, \ldots$ is a standard list of oracle Turing Machines.
4. K is the set $\{e : M_e(e) \downarrow\}$.
5. If A is a set then $A' = \{e : M_e^A(e) \downarrow\}$.

We use the following definitions. *Low* and *Superlow* are standard; however, *Superduperlow* seems to be a new term.

Def 1.2.

1. A set A is *Low* if $A' \leq_{\mathrm{T}} K$.
2. A set A is *Superlow* if $A' \leq_{\mathrm{tt}} K$.
3. A set A is *Superduperlow* if $A' \leq_{\mathrm{btt}} K$.

By a finite injury priority argument one can construct a noncomputable superlow c.e. set A (we include this proof in the appendix). This raises the question: Is there a noncomputable superduperlow set A?

We asked about this at a logic conference and found out:

1. Four prominent computability theorists thought that there was no such set; however, none knew of a proof or reference.

2. Carl Jockusch, also a prominent computability theorist, knew of unpublished proofs by: (1) Bickford and Mills, (2) Phillips, (3) himself, and (4) Stephan. He also knew of a more complicated published proof by Mohrerr [5]. She actually proved the stronger result that if $A^{\mathrm{tt}} \leq_{\mathrm{btt}} B'$ then $A \leq_{\mathrm{T}} B$, as did Bickford and Mills.

It is our opinion that the proofs of Jockusch and Stephan are beautiful and should be better known. Hence we present them.

We will need the following two standard Lemmas. The first one is *the Shoenfield Limit Lemma.*

Lemma 1.3. *$A \leq_{\mathrm{T}} K$ iff there exists a computable function $h : \mathsf{N} \times \mathsf{N} \to \mathsf{N}$ such that*

$$A(x) = \lim_{s \to \infty} h(x, s).$$

Lemma 1.4. *If $A \leq_{\mathrm{T}} B$ then $A \leq_{\mathrm{m}} B'$.*

2 Bi-immune and Hyperimmune Sets

Def 2.1.

1. A set C is *immune* if C is infinite and has no infinite c.e. subsets

2. A set C is *bi-immune* if both C and $\overline{C}$ are immune.

3. If B is an infinite set then *the principal function of B*, denoted p_B, is defined as

$$p_B(s) = \text{ the } s^{\text{th}} \text{ element of } B.$$

4. A function f is *majorized* by a function g if, for all x, $f(x) < g(x)$.

5. A set B is *hyperimmune* if p_B is not majorized by any computable function. (There is an alternative definition of hyperimmune which is a variant of immune, hence the name. We do not need that alternative definition.)

The following Lemma is due to Miller and Martin [4]. We include the proof since the original article is behind a paywall.

Lemma 2.2. *If $\emptyset <_T A \le_T K$ then there exists a hyperimmune set B such that $B \equiv_T A$. (We will only use the $B \le_T A$ part.)*

Proof. Since $A \le_T K$ there exists, by Lemma 1.3, a computable h such that

$$A(x) = \lim_{s\to\infty} h(x, s).$$

Let

$$f(x) = \text{ the least } s \ge \max\{x, f(x-1)\} \text{ such that } (\forall y \le x)[A(y) = h(y, s)].$$

Let $s = f(x)$. Then, for all $y \le x$, $h(y, s) = A(y)$. One might think that the limit has settled down on all $y \le x$; however, it may be that there is some $y \le x$ and $t > s$ such that $h(y, t) \ne A(y)$. If so then there will be some $t' > t$ with $h(y, t') = A(y)$. And $h(y, -)$ may even change its mind again! But eventually it will always be $A(y)$.

Let B be the image of f. Clearly $B \le_T A$. We show B is hyperimmune. Since $f = p_B$ is the principal function of B, it suffices to show that f cannot be majorized by any computable function.

Assume, by way of contradiction, that there is a computable g such that, for all x, $f(x) < g(x)$. We use this to obtain an algorithm for A. Given x, we want to determine $A(x)$.

Note that, for all y,

$$y \le f(y) < g(y)$$

Let $y \ge x$. Imagine what would happen if

$$h(x, y) = h(x, y+1) = \cdots = h(x, f(y)) = \cdots = h(x, g(y)) = b.$$

We would have:

$$A(x) = h(x, f(y)) = b.$$

Hence we would know $A(x)$. Therefore we need to find such a y. If we knew that one existed we could just look for it.

One does exist! Let y be such that

$$h(x, y) = h(x, y+1) = \cdots =$$

Such a y exists since h reaches a limit. This y clearly suffices. We cannot find this particular y but we do not need to. We need only find *some* y such that

$$h(x, y) = h(x, y+1) = \cdots = h(x, g(y)) = b.$$

Here is the formal algorithm for A.

1. Input(x)

2. Find a $y \geq x$ such that

$$h(x,y) = h(x,y+1) = \cdots = h(x,g(y)) = b$$

3. Output the value b.

Thus A is computable— a contradiction. Hence f cannot be majorized by any computable function.

Therefore we have a set $B \leq_T A$ such that B is hyperimmune.

We leave the proof that $A \leq_T A$ to the reader.

□

The following is a result of Carl Jockusch [2].

Lemma 2.3. *For every hyperimmune B there exists a bi-immune $C \leq_T B$.*

Proof. Let B be hyperimmune. Since B is hyperimmune, p_B is not majorized by any computable function. We use this to construct a bi-immune $C \leq_T B$. To ensure that C is bi-immune we make sure that C satisfies the following requirements:

$$R_e : W_e \text{ infinite } \rightarrow (W_e \cap C \neq \emptyset \wedge W_e \cap \overline{C} \neq \emptyset).$$

CONSTRUCTION

Stage 0: For all e, R_e is not satisfied.

Stage s: Find the least $e \leq s$, if it exists, such that R_e is not satisfied and $W_{e,p_B(s)}$ has at least two elements $x_1, x_2 \geq s$ which have not yet been put into C or $\overline{C}$. Put x_1 into C, x_2 into $\overline{C}$, and declare R_e *satisfied.* (it will never become unsatisfied). We also say that R_e has *acted.* If there is no such e, do nothing.

END OF CONSTRUCTION

We have $C \leq_T B$ since (1) the only noncomputable part of the construction is computing $p_B \leq_T B$, and (2) $C(n)$ is decided by stage n . (For definiteness, a number is in C iff the construction puts it into C.)

We show that C is bi-immune by showing that it satisfies each requirement. We assume that $R_1, \ldots, R_{e-1}$ are satisfied and show that R_e is satisfied. There are two cases.

1. W_e is finite. Then clearly R_e is satisfied.

2. W_e is infinite. Assume, by way of contradiction, that W_e is not satisfied. From this we will construct a computable function g that majorizes f which will be the contradiction. Let s_0 be such that by state s_0 all of $R_1, \ldots, R_{e-1}$ that are going to act have acted. So for all $s \geq s_0$ R_e is not satisfied yet fails to act! Why!?

   Let $g(s)$ be the least $t \geq s_0$ such that $W_{e,t}$ has at least $2s+2$ elements $\geq s$. Note that g is computable. Lets look at the construction at a stage $s \geq s_0$. R_e did not act. Why? It must be that *there is no* x_1, x_2 such that (1) $x_1, x_2 \in W_{e,p_B(s)}$, (2) $x_1, x_2 \geq s$, and (3) x_1, x_2 were not used by any of $P_0, \ldots, P_{e-1}$. By the definition of g *there is* x_1, x_2 such that (1) $x_1, x_2 \in W_{e,g(s)}$, (2) $x_1, x_2 \geq s$, and (3) x_1, x_2 were not used by any of $P_0, \ldots, P_{e-1}$. Hence we have that,

   $$(\forall s \geq s_0)[p_B(s) < g(s)].$$

   Recall that g is computable. We want to say g *majorizes* p_B but this is not quite true; however, a finite variant of g majorizes p_B and is clearly computable. Hence there is a computable function that majorizes p_B. This contradicts B being hyperimmune.

□

Lemma 2.4. *Let $\emptyset \leq_T A \leq_T K$.*

1. *If A is not computable then there exists C bi-immune such that $C \leq_T A$.*
2. *If there is no bi-immune set $C \leq_T A$ then A is computable (this is the contrapositive of Part 1).*

Proof. 1) By Lemma 2.2 there is a hyperimmune set $B \leq_T A$. By Lemma 2.3 there is a bi-immune set $C \leq_T B$. Hence there is a bi-immune set $C \leq_T B \leq_T A$. □

3 Superduperlow implies Decidable: Proof One

Def 3.1. Let $n \geq 0$. A set D is *weakly n-c.e.* if there exists a function h such that

- $D(x) = \lim_{s \to \infty} h(x, s)$
- $|\{s : h(x, s) \neq h(x, s+1)\}| \leq n$.

Note 3.2. Let D be weakly n-c.e. and $x \in \mathsf{N}$. Our view: initially x thinks $D(x) = h(x, 0)$; however, it can change its mind $\leq n$ times.

The following easy lemma we leave to the reader.

Lemma 3.3. *If* $\emptyset <_{\mathrm{T}} D \leq_{\mathrm{btt}} K$ *then there exists an* $n \geq 1$ *such that* D *is* n*-c.e. but not* $(n-1)$*-c.e.*

The following is an unpublished Theorem of Jockusch.

Theorem 3.4. *If* A *is superduperlow then* A *is decidable.*

Proof. Since A is superduperlow $A' \leq_{\mathrm{btt}} K$.

Let $D \leq_{\mathrm{T}} A$. We will show that D is not bi-immune then apply Lemma 2.4 to deduce that A is computable. If D is decidable then D is not bi-immune; hence we assume D is undecidable.

Since $D \leq_{\mathrm{T}} A$, by Lemma 1.4, $D \leq_{\mathrm{m}} A'$. Since $A' \leq_{\mathrm{btt}} K$ we have

$$D \leq_{\mathrm{m}} A' \leq_{\mathrm{btt}} K.$$

Hence $D \leq_{\mathrm{btt}} K$. Since D is not decidable, by Lemma 3.3, there exists $n \geq 1$ such that D is weakly n-c.e. but not weakly $(n-1)$-c.e. Let h be such that

- $D(x) = \lim_{s\to\infty} h(x,s)$
- $|\{s : h(x,s) \neq h(x,s+1)\}| \leq n$.

Let

$$E = \{x : |\{s : h(x,s) \neq h(x,s+1)| = n\}\}.$$

E is infinite, else D is weakly $(n-1)$-c.e. Let

$$E_0 = \{x : h(x,0) = 0 \wedge |\{s : h(x,s) \neq h(x,s+1)| = n\}\}.$$

$$E_1 = \{x : h(x,0) = 1 \wedge |\{s : h(x,s) \neq h(x,s+1)| = n\}\}.$$

Clearly both E_0 and E_1 are c.e: Clearly at least one of E_0 or E_1 is infinite. There are four cases depending on (1) which of E_0, E_1 is infinite, and (2) the parity of n. For all four cases keep in mind that E_0 and E_1 are c.e.

Case E_0 infinite, n even: Every $x \in E_0$ starts out thinking it's not in D and changes its mind an even number of times. Hence E_0 is an infinite c.e. subset of $\overline{D}$, so D is not bi-immune.

Case E_0 infinite, n odd: Every $x \in E_0$ starts out thinking it's not in D and changes its mind an odd number of times. Hence E_0 is an infinite c.e. subset of D, so D is not bi-immune.

Case E_1 infinite, n even: Every $x \in E_1$ starts out thinking it's in D and changes its mind an even number of times. Hence E_1 is an infinite c.e. subset of D, so D is not bi-immune.

Case E_1 infinite, n odd: Every $x \in E_1$ starts out thinking it's in D and changes its mind an odd number of times. Hence E_1 is an infinite c.e. subset of $\overline{D}$, so D is not bi-immune.

The upshot is that, for *every* set $D \leq_T A$, D is not bi-immune. By Lemma 2.4, A is computable. □

4 Superduperlow implies Decidable: Proof Two

In this section we present a proof by Frank Stephan that uses concepts from Bounded Queries in Computability Theory. We will provide all that you need; however, for more information, see the survey by Gasarch [1].

Def 4.1. Let $A \subseteq \mathsf{N}$ and $n \in \mathsf{N}$.

1. $\chi_n^A : \mathsf{N}^n \to \{0,1\}^n$ is the following function:

$$\chi_n^A(x_1, \dots, x_n) = A(x_1) \cdots A(x_n).$$

2. $\#_n^A : \mathsf{N}^n \to \mathsf{N}$ is the following function:

$$\#_n^A(x_1, \dots, x_n) = |\{x_1, \dots, x_n\} \cap A|.$$

We give an intuition for the next definition. Given 3 numbers a, b, c we want to know $\chi_3^K(a, b, c)$. We could just output (0,0,0), (0,0,1), ..., (1,1,1) and be happy that *one of them is* $\chi_3^K(a, b, c)$. Can we output ≤ 7 tuples, one of which is $\chi_3^K(a, b, c)$? Yes if we are willing to not know when the process has output its last candidate[1]. We can do the following: Output (0,0,0) since it is certainly possible that $\chi_3^K(a, b, c) = (0, 0, 0)$. Then run $M_a(a)$, $M_b(b)$, and $M_c(c)$ at the same time until (if it happens) one of them halts: (1) if it's a then output $(1, 0, 0)$, (2) if it's b then output $(0, 1, 0)$, (3) if it's c then output $(0, 0, 1)$. Then run all those that haven't halted until (if it happens) one of them halts. We leave it to the reader to finish this up. Throughout this process you will output at most 4 tuples *one of which is* $\chi_3^K(a, b, c)$. We do not know when the process has output its last candidate. This motivates the next definition:

[1]It is known that there is no algorithm that will, on input (a, b, c), always output 7 3-tuples one of which is $\chi_3^K(a, b, c)$. The proof uses the recursion theorem.

Def 4.2. A function f is in $\mathrm{EN}(m)$ if there exists a Turing machine M that will, on input x, over time, output at most m numbers, one of which is $f(x)$.

Clearly, for all A, $\#_n^A \in \mathrm{EN}(n+1)$. Kummer [3] (see Gasarch [1] for an alternative proof) showed that, for undecidable sets, this is the best one can do.

Theorem 4.3. *For all A, if there exists n such that $\#_n^A \in \mathrm{EN}(n)$, then A is computable.*

We use Theorem 4.3 to show that all superduperlow sets are decidable.

We need a known lemma. We give the proof for completeness.

Lemma 4.4. $\chi_n^K \in \mathrm{EN}(n+1)$.

Proof. On input $(x_1, \ldots, x_n)$:

1. Output $(0, \ldots, 0)$ as a possible answer. We call $(0, \ldots, 0)$ *the current tuple.*
2. Run $M_{x_1}(x_1), \ldots, M_{x_n}(x_n)$ at the same time. If the ith one halts then change the ith bit of the current tuple to 1, output the new current tuple, and keep running the machines until another one halts (which may never happen).

If j of them halt then this process will output $j+1$ tuples, of which the last one is $\chi_n^K(x_1, \ldots, x_n)$. Hence this process outputs at most $n+1$ tuples, one of which is $\chi_n^K(x_1, \ldots, x_n)$. □

Theorem 4.5. *If A is superduperlow then A is decidable.*

Proof. Assume that A is superduperlow. Let k be such that $A' \leq_{\text{k-tt}} K$. We show that, for some (large enough) n, $\#_{2^n-1}^A \in \mathrm{EN}(2^n - 1)$, hence A is decidable.

Let $A_1, \ldots, A_n$ be the following sets.

$$A_i = \{(x_1, \ldots, x_{2^n-1}) : \text{ the } i\text{th bit of } \#_{2^n-1}^A(x_1, \ldots, x_n) \text{ is } 1\ \}.$$

For each i, $A_i \leq_{\mathrm{T}} A$ (actually $A_i \leq_{\mathrm{tt}} A$ but we do not need this). Hence, by Lemma 1.4, $A_i \leq_{\mathrm{m}} A'$. Since $A' \leq_{\text{k-tt}} K$ we get $A_i \leq_{\text{k-tt}} K$. We use this to obtain a procedure for

$$\#_{2^n-1}^A \in \mathrm{EN}(kn+1).$$

1. Input $(x_1, \ldots, x_{2^n-1})$.
2. For $1 \leq i \leq n$ do the following: Using $A_i \leq_{\text{k-tt}} K$ find k numbers $y_{i1}, \ldots, y_{ik}$ such that if we knew $\chi_{ik}^K(y_{i1}, \ldots, y_{ik})$ then we would know if $(x_1, \ldots, x_{2^n-1})$ is in A_i.

3. We now have kn numbers $(y_{11}, \ldots, y_{1k}, y_{21}, \ldots, y_{2k}, \ldots, y_{n1}, \ldots, y_{nk})$ such that if we knew $\chi_{kn}^{K}(y_{11}, \ldots, y_{1k}, y_{21}, \ldots, y_{2k}, \ldots, y_{n1}, \ldots, y_{nk})$ we would, for $1 \le i \le n$, know if $(x_1, \ldots, x_{2^n-1}) \in A_i$.

4. By Lemma 4.4 $\chi_{kn}^{K} \in \mathrm{EN}(kn+1)$. Run this enumeration algorithm on

   $(y_{11}, \ldots, y_{1k}, y_{21}, \ldots, y_{2k}, \ldots, y_{n1}, \ldots, y_{nk})$.

   Every time a candidate is enumerated, use it to obtain a candidate for $\#_{2^n-1}^{A}$. Output that candidate and continue.

By the above enumeration algorithm $\#_{2^n-1}^{A} \in \mathrm{EN}(kn+1)$. Take n large enough so that $kn+1 \le 2^n - 1$ to obtain that $\#_{2^n-1}^{A} \in \mathrm{EN}(2^n-1)$. By Theorem 4.3 A is computable. □

5 Acknowledgment

We thank Carl Jockusch and Frank Stephan for supplying most of the material for this paper and for helpful discussions.

A There exists an undecidable c.e. Superlow Set

We give the standard construction of a noncomputable low c.e. set A; however, we analyze the construction carefully to show that A is actually superlow.

Theorem A.1. *There exists an undecidable c.e. superlow set.*

Proof. We construct a c.e. set A that satisfies the following requirements:

$P_e : W_e$ infinite $\implies W_e \cap A \neq \emptyset$.

These are called *positive requirements* since they act by *putting numbers into A*. It is easy to show that, if A is co-infinite and all of the P_e's are satisfied, then A is undecidable. (We will also make A co-infinite, though we do not state it as a formal requirement.)

$N_e : (\exists^\infty s)[M_{e,s}^{A_s}(e)\downarrow] \implies M_e^A(e)\downarrow$.

These are called *negative requirements* since they act by *keeping numbers out of A*. We will show that, if all N_e are satisfied, then A is superlow. These requirements protect a computation from being injured. By this we mean that, if $M_{e,s}^{A_s}(e)\downarrow$ then this requirement will try to make sure that no numbers enter A that might make this computation diverge (N_e will try to keep *all* numbers $\le$ the max number queried from going into A). Associated to every N_e will be a restraint function $r(e,s)$. This is N_e saying *you cannot put an element into* A *that is* $\le r(e,s)$. This restraint will be respected by the lower priority positive requirements (P_e, P_{e+1}, etc.) but not by the higher priority positive requirements (P_0, P_1, ..., P_{e-1}).

The requirements are in the following priority ordering

$$N_0, P_0, N_1, P_1, \ldots$$

CONSTRUCTION
Stage 0: $A_0 = \emptyset$. $(\forall e)[r(e,0) = 0]$. For all e, P_e is not satisfied.

Stage s**:** Visit each requirement in turn, via the priority ordering, up to P_s.
Case 1: A positive requirement P_e. If (a) P_e is not satisfied, and (b) there exists $x \in W_{e,s}$ such that $x \ge 2e$ and $x > \max_{i \le e} r(e,s)$, then P_e acts by putting x into A. P_e is declared satisfied. We say that N_i is *injured.* Note that P_e will never become unsatisfied.
Case 2: A negative requirement N_e. If $M_{e,s}^{A_s}(e) \downarrow$ then set $r(e,s)$ to be the largest number that is queried in this computation. Note that if no number ever enters A that is $\le r(e,s)$ then $M_e^A(e) \downarrow$.
END OF CONSTRUCTION

Claim 1: Every P_e acts finitely often.
Proof of Claim 1: If P_e never acts then clearly P_e acts finitely often.

If P_e ever acts then it is satisfied, will never be injured, and will never acts again. Hence P_e acts finitely often.
End of Proof of Claim 1:

Claim 2: For all e, N_e is satisfied and $\lim_{s\to\infty} r(e,s) < \infty$.
Proof of Claim 2: Let s_0 be such that, for all $i < e$, P_e will never act past stage s_0. Note that s_0 exists by Claim 1.

If $(\forall^\infty s)[M_{e,s}^{A_s}(e) \uparrow]$ then N_e is satisfied since its premise is false. Past some point N_e will never act, hence its restraint changes only finitely often, so
$\lim_{s\to\infty} r(e,s) < \infty$.

If $(\exists^\infty s)[M_{e,s}^{A_s}(e) \downarrow]$ then there is some $s_1 \ge s_0$ such that $M_{e,s_1}^{A_s}(e) \downarrow$. When that happens N_e will act and a restraint $r(e,s_1)$ will be set. Since no higher priority positive requirement ever acts, N_e is never injured, and hence is satisfied. Since N_e never acts again, $\lim_{s\to\infty} r(e,s) = r(e,s_1) < \infty$.
End of Proof of Claim 2:

Claim 3: Every P_e is satisfied.

Proof of Claim 3: If W_e is finite then P_e is satisfied. Hence we assume that W_e is infinite. Let s_0 be such that for all $i < e$ $\lim_{s\to\infty} r(i,s) = r(i,s_0)$. Let $R(e) = \max_{i\le e} r(i,s_0)$. Since W_e is infinite there will be an $x > \max\{2e, R(e)\}$ that is enumerated into W_e at some stage $s > \max\{s_0, e\}$. If P_e is not yet satisfied then P_e will act at stage s and be satisfied.
End of Proof of Claim 3:

Claim 4: A is co-infinite.
Proof of Claim 4: Look at the numbers $S_e = \{1, 2, \ldots, 2e, 2e+1\}$. Since P_{e+1} only uses numbers $\ge 2e+2$, the only positive requirements that will use elements of S_e are $P_0, \ldots, P_e$. Hence at most $e+1$ of the elements of S_e will enter A. Hence at least e of the elements of S_e will not enter A. Since this is true for all e, A is co-infinite.
End of Proof of Claim 4:

Claim 5: A is superlow.
Proof of Claim 5:

Note that the only requirements that can injure N_e are $P_0, P_1, \ldots, P_{e-1}$. These requirements act at most once. Hence N_e is injured at most e times. We can determine $e \in A'$ by asking the following questions at the same time (which is why we get $A' \le_{\text{tt}} K$). For each $i \le e$ we have the following two questions to K.

- Is N_e injured at least i times?
- Is there a stage s that occurs after N_e is injured i times where $M_{e,s}^{A_s}(e)\downarrow$?

From the answers (1) determine the largest i_0 such that N_e is injured exactly i_0 times, and (2) determine if there is some stage s after the i_0 injuries such that $M_{e,s}^{A_s}(e)\downarrow$. If the answer to (2) is YES then $e \in A'$ since that computation will never be injured. If the answer is NO then $e \notin A'$ since for almost all s, $M_{e,s}^{A_s}(e)\uparrow$. □

References

[1] William Gasarch *Gems in the field of bounded queries*, in: Barry S. Cooper, and Sergey Goncharov (eds.) Computability and Models, *Perspectives East and West*, Kluwer Academic / Plenum Publisheres, New York 2003 [*University Series in Mathematics*]. (Cf. `http://www.cs.umd.edu/~gasarch/papers/papers.html`.)

[2] Carl G. Jockusch Jr. *The degrees of bi-immune sets*, Zeitschrift für Logik and Grundlagen der Mathathematik [now: Mathematical Logic Quarterly] 15, 1986, pp.135–140.

[3] Martin Kummer *A proof of Beigel's cardinality conjecture*, The Journal of Symbolic Logic 57 (2), [June] 1992, pp. 677–681.

[4] Webb Miller, and D. A. Martin *The degree of hyperimmune sets*, Zeitschrift für Logik and Grundlagen der Mathematik 14, 1968, pp. 159–166.

[5] Jeanleah Mohrherr *A refinement of* low_n *and* high_n *for the r.e. degrees*, Zeitschrift für Logik and Grundlagen der Mathematik 32, 1986, pp. 5–12.

Small NFA's for Cofinite Unary Languages

William Gasarch
University of Maryland
gasarch@cs.umd.edu

Erik Metz
University of Maryland
emetz1618@gmail.com

Yuang Shen
University of Maryland
eric.shen2000@gmail.com

Zan Xu
University of Maryland
zanturtle@gmail.com

Sam Zbarsky
Princeton University
zbarskysam@gmail.com

ABSTRACT For all n there is a DFA for $\{a^i : i \neq n\}$ of size $n+2$; however there is no smaller DFA. What about NFA's? We show that there is an NFA for $\{a^i : i \neq n\}$ of size $\sqrt{n} + \widetilde{O}(1)$. We also find small NFA's for many other cofinite unary sets. How small can we go? We show that any NFA for $\{a^i : i \neq n\}$ must have at least $\sqrt{n}$ states.

1 Introduction

Consider the language

$$\mathrm{MN}(n) = \{a^i : i \neq n\}.$$

(MN stands for *Missing Number.*)

It is easy to show that (1) there is a DFA for MN(n) with $n+2$ states, and (2) any DFA for MN(n) has at least $n+2$ states. What about an NFA for MN(n)? We show that there is an NFA for MN(n) that has substantially fewer than n states. We also obtain small NFA's for many other cofinite unary languages.

Notation 1.1. N is $\{0, 1, 2, \ldots\}$ (that is, we include 0).

Def 1.2. If $A \subseteq \mathsf{N}$ then

$$\mathrm{MN}(A) = \{a^i : i \notin A\}.$$

We will only use this definition when A is finite. We will write MN(a, b, c) instead of the formally correct MN($\{a, b, c\}$).

Notation 1.3. If f and g are functions then, informally, $f \leq \widetilde{O}(g)$ means that f is less than g if we ignore polylog factors. Formally it means that

$$(\exists n_0)(\exists c)(\forall n \geq n_0)[f(n) \leq c(\log n)^c g(n)].$$

1. In Section 3 we show that (1) there is an NFA for MN(100) on 29 states, and (2) for all n there is an NFA for MN(n) with $\leq n^{1/2} + \widetilde{O}(1)$ states.

2. In Section 4 we show that (1) there is an NFA for MN(998, 999, 1000) on 104 states, (2) for any $A \subseteq \{998, 999, 1000\}$ there is an NFA for MN(A) on 104 states, (3) for all n, for all $0 < \delta < 1$ there is an NFA for MN($n - n^\delta, \ldots, n$) on $5n^{\max\{1/2,\delta\}} + \widetilde{O}(1)$ states, and (4) for any $A \subseteq \{n - n^\delta, \ldots, n\}$ there is an NFA for MN(A) on $5n^{\max\{1/2,\delta\}} + \widetilde{O}(1)$ states.

3. In Section 5 we show that, for all n, for all $0 < \alpha < 1$ such that $\alpha n \in \mathsf{N}$, there is an NFA for MN($\alpha n, n$) on $2n^{1/2}\ln(n) + \widetilde{O}(1)$ states.

4. In Section 6 we prove a general theorem about unary sets with big gaps. We obtain the following corollary: for all $0 < \delta < 1$ there is an NFA for MN(n^δ, n) on $n^{1/2} + n^\delta + \widetilde{O}(1)$ states.

5. In Section 7 we show that any NFA for MN(n) requires at least $n^{1/2}$ states.

6. In Section 8 we discuss our empirical results.

7. In Section 9 we state open problems.

Def 1.4. A set X has a *small NFA* if there is an NFA that accepts it that is much smaller than any DFA for it. We do not define the term *much smaller* rigorously. However, all of our results are about small NFA's.

All of our general results are asymptotic; however, we will present empirical evidence that indicates the results hold for small n as well.

2 Needed Lemma

The following problem is attributed to Frobenius:

Given a set of relatively prime positive integers $\{a_1, \ldots, a_m\}$ *find the set* $\{\sum_{i=1}^{n} a_i x_i : x_1, \ldots, x_m \in \mathsf{N}\}$.

It is known that this set is always cofinite. The $m = 2$ case was solved by James Joseph Sylvester in 1884:

Lemma 2.1. *Let* $c, d \in \mathsf{N}$ *be relatively prime.*

1. *For all* $i \geq cd - c - d + 1$ *there exists* $x, y \in \mathsf{N}$ *such that* $i = cx + dy$.

2. *There is no* $x, y \in \mathsf{N}$ *such that* $cd - c - d = cx + dy$.

3. *There is no* $x, y, C, D \in \mathsf{N}$ *such that* $cd - c - d - Cc - Dd = cx + dy$. *(If there was then* $cd - c - d = (C + x)c + (D + y)d$.*) We use this part in Section 4.*

3 Small NFA's for MN(100) and MN(n)

3.1 Small NFA for MN(100)

Theorem 3.1.

1. *For all* $i \geq 96$ *there exists* $x, y \in \mathsf{N}$ *such that* $i = 13x + 9y$.

2. *There does not exist* $x, y \in \mathsf{N}$ *such that* $95 = 13x + 9y$.

3. *For all* $i \geq 101$ *there exists* $x, y \in \mathsf{N}$ *such that* $i = 13x + 9y + 5$.

4. There does not exist $x, y \in \mathsf{N}$ *such that* $100 = 13x + 9y + 5$.

5. There exists an NFA M *such that the following are true:*

   (a) For all $i \geq 101$, M *accepts* a^i.

   (b) M *rejects* a^{100}.

   (c) We have no comment on the behavior of M *on other* a^i.

   (d) M *has 13 states.*

6. There exists an NFA on 29 states that accepts MN(100).

Proof. 1,2) These follow from Lemma 2.1, though they can be proven directly by an easy induction.
3,4) These follow from Parts 1 and 2
5) The NFA is constructed as follows: (also see Figure 1, the caption will be explained later).

- M has states $0, \ldots, 12$, 0 is the start state, and 5 is the only final state. For $0 \leq j \leq 12$, $\delta(j, a) = j + 1 \pmod{13}$. ($\delta$ is not fully defined yet.)
- If we go no further then M accepts $\{a^{13x+5} : x \in \mathsf{N}\}$.
- We put in an e-transition from state 5 to state 9. Now M accepts

$$\{a^{13x+9y+5} : x, y \in \mathsf{N}\}.$$

  (The $9y$ is not because the e-transition went to state 9. It is because the distance from state 9 back to state 5 is 9.)

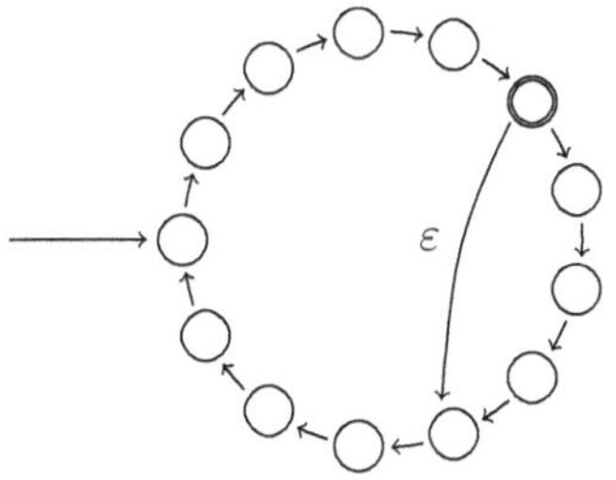

Figure 1: LOOP(9,13,5) Case 1

By Parts 3,4 M satisfies 5a and 5b. M clearly has 13 states, so it satisfies 5d.

6) Let $Q = \{3, 5, 7\}$. Note that $3 \times 5 \times 7 = 105 > 100$. For each $p \in Q$ let M_p be the DFA that accepts $\{a^i : i \not\equiv 100 \pmod{p}\}$.

The NFA is constructed as follows: (see also Figure 2)

1. The NFA M is part of our new NFA. We create a new start state, and then put an e-transition from this new state to M's original start state. Note that M (a) accepts all a^i with $i \geq 101$ (it also accepts other strings), (b) rejects a^{100}, and (c) has 13 states.

2. For each $p \in Q$ put an e-transition from our new start state to the start state of M_p. Note that M_p (a) accepts all a^i with $i \not\equiv 100 \pmod{p}$, (b) rejects a^{100}, and (c) has p states.

Clearly the NFA has $13 + 3 + 5 + 7 + 1 = 29$ states and rejects a^{100}. We show that it accepts everything else.

Let a^i be rejected by this NFA.

- Since the M part rejects a^i, $i \leq 100$ (note, hence $i \leq 3 \times 5 \times 7 = 105$).
- Since the M_3 part rejects a^i, $i \equiv 100 \pmod{3}$
- Since the M_5 part rejects a^i, $i \equiv 100 \pmod{5}$
- Since the M_7 part rejects a^i, $i \equiv 100 \pmod{7}$

By the Chinese Remainder Theorem there is a unique number $0 \leq z \leq 3 \times 5 \times 7 = 105$ such that, for every $p \in \{3, 5, 7\}$, $z \equiv 100 \pmod{p}$. Since both i and 100 satisfies these criteria, $i = n$. □

3.2 Small NFA for MN(n)

We generalize the construction of a small NFA for MN(100) to get a small NFA for MN(n).

Def 3.2. Let $c, d, e \in \mathsf{N}$ be such that $c < d$ and c, d are relatively prime. LOOP(c, d, e) is the NFA defined as follows. There are two cases.
Case 1: $e \leq d - 1$.

1. The NFA has states $0, \ldots, d - 1$, with 0 as the start state and e as the only final state. For $0 \leq j \leq d - 1$, $\delta(j, a) = j + 1 \pmod{d}$.

2. So far this NFA accepts $\{a^{dx+e} : x \in \mathsf{N}\}$.

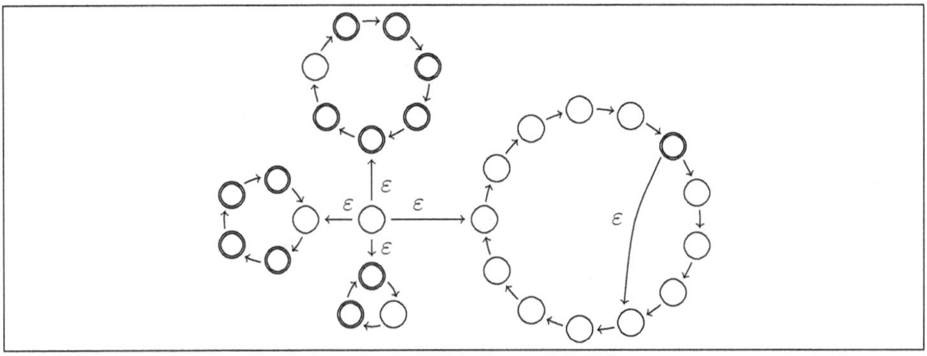

Figure 2: NFA for MN(100)

3. We put in an e-transition from state e to state $e - c \pmod{d}$. Note that the distance from state $e - c \pmod{d}$ to state e is c. Now the NFA accepts

$$\{a^{cx+dy+e} : x, y \in \mathsf{N}\}.$$

4. This NFA has d states.

Note that Figure 1 is LOOP(9, 13, 5) which is an example of a Case 1 LOOP.

Case 2: $e \geq d$

1. The NFA has states $s_0, s_1, \ldots, s_{e-d+2}$ such that s_0 is the start state. For $0 \leq j \leq e - d + 1$, $\delta(s_j, a) = s_{j+1}$.

2. The NFA has states $0, \ldots, d - 1$, with $d - 1$ as the only final state. For $0 \leq j \leq d - 1$, $\delta(j, a) = j + 1 \pmod{d}$. The state 0 is identical to the state s_{e-d+2}.

3. In total there are $(e - d + 1) + (d - 1) = e$ transitions to get to the final state the first time, after which each loop of length d brings you back to the same state, so the NFA accepts $\{a^{dx+e} : x \in \mathsf{N}\}$.

4. We put in an e-transition from state $d - 1$ to state $d - c - 1$. Note that the distance from state $d - c - 1$ to state $d - 1$ is c. Now the NFA accepts

$$\{a^{cx+dy+e} : x, y \in \mathsf{N}\}.$$

5. This NFA has $e + 1$ states.

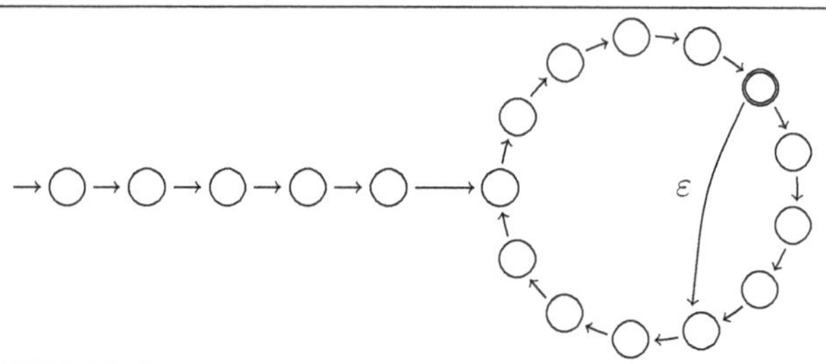

Figure 3: NFA LOOP(9,13,17) Case 2

Figure 3 is LOOP(9, 13, 17) which is an example of a Case 2 LOOP.

The following is clear:

Lemma 3.3. *Let $c, d, e \in \mathsf{N}$ be such that $e, c < d$ and c, d are relatively prime.*

1. LOOP(c, d, e) *accepts* $\{a^i : i \geq cd - c - d + e + 1\}$
2. LOOP(c, d, e) *rejects* $\{a^{cd-c-d+e}\}$.
3. LOOP(c, d, e) *rejects* $\{a^{cd-Cc-Dd+e} : C, D \in \mathsf{N}\}$ *(since if $a^{cd-Cc-Dd+e}$ can reach the accept state then by adding C c's and D d's the NFA gets back to the accept state). We will use this part in Section 4.*
4. *If we used Case 1 then* LOOP(c, d, e) *has d states.*
5. *If we used Case 2 then* LOOP(c, d, e) *has $e + 1$ states.*

Note 3.4. Below, we use o(1) to denote a number that may be positive or negative, but that goes to 0 as our variable of interest (N in Lemma 3.5), goes to infinity. This may be non-standard.

Lemma 3.5. *Let $N \in \mathsf{N}$. Let Q_N be the set of the first N primes.*

1. $\prod_{p \in Q_N} p \sim e^{((1+o(1))N \log N}$. *(This is well known.)*
2. $\sum_{p \in Q_N} p \sim O(N^2 \log N) = \tilde{O}(N^2)$. *(For references and more precise estimates see Axler [1].)*

3. *Let $n \in \mathsf{N}$. The product of the first $\Omega(\log n)$ primes is $\geq n$. The sum of the first $O(\log n)$ primes is $\leq O((\log n)^2 \log\log n) \leq \tilde{O}(1)$. (This follows from parts 1 and 2.)*

4. *$\prod_{p \leq N,\ p \text{ prime}} p \sim e^{(1+o(1))N}$. (This is well known.)*

Theorem 3.6. *Let $n \in \mathsf{N}$.*

1. *There exists an NFA M such that the following are true:*
   (a) For all $i \geq n + 2\left\lceil n^{1/2} \right\rceil$, M accepts a^i.
   (b) M rejects a^n.
   (c) We have no comment on the behavior of M on other a^i.
   (d) M has $\leq n^{1/2} + O(1)$ states.

2. *There exists an NFA on $\leq n^{1/2} + \tilde{O}(1)$ states that accepts* MN(n).

Proof. 1) Let $c = \left\lceil n^{1/2} \right\rceil + 1$ and $e = n + 1 \pmod{c}$. Note that $e \leq c$. Let M be LOOP$(c, c+1, e)$. Note that

$$c(c+1) - c - c - 1 + e = c^2 - c - 1 + e \leq c^2 - c - 1 + c = c^2 - 1$$

By Lemma 3.3 M accepts a^i where $i \geq c^2 - 1 + 1 = c^2 \geq n + 2\left\lceil n^{1/2} \right\rceil$.

We show that M rejects a^n. Assume, by way of contradiction, that M accepts a^n. Then there exists $x, y \geq 0$ such that

$$cx + (c+1)y + e = n$$

Take this equation mod c. Then

$$0x + 1 \times y + (n+1) \equiv n \pmod{c}$$

$$y + 1 \equiv 0 \pmod{c}$$

$$y \equiv -1 \pmod{c}.$$

Since $y \geq 0$, $y \geq c - 1$. Hence

$$n = cx + (c+1)y + e \geq (c+1)(c-1) = c^2 - 1 = (\left\lceil n^{1/2} \right\rceil + 1)^2 - 1 = n + 2n^{1/2} - 1.$$

This is a contradiction.

Since $e \le c$, M has $c + 1 = n^{1/2} + O(1)$ states.

2) By Lemma 3.5.3 there is a set of primes Q such that

- $\prod_{p \in Q} p \ge n + 2\lceil n^{1/2} \rceil$.
- $\sum_{p \in Q} p \le \tilde{O}(1)$.

For each $p \in Q$ let M_p be the DFA that accepts $\{a^i : i \not\equiv n \pmod{p}\}$.

The NFA is constructed as follows:

1. The NFA M is part of our NFA. We create a new start state, and then put an e-transition from this new state to M's original start state. Note that (1) M accepts a^i for $i \ge n + 2\lceil n^{1/2} \rceil$ (it also accepts other strings), (2) M rejects a^n, and (3) M has $\le n^{1/2} + O(1)$ states.
2. For each $p \in Q$ there is an e-transition from our new start state to the start state of M_p. Note that (1) M_p accepts a^i if $i \not\equiv n \pmod{p}$, (2) M_p rejects a^n, and (3) M_p has p states.

Clearly the NFA has $\le n^{1/2} + \sum_{p \in Q} p \le n^{1/2} + \tilde{O}(1)$ states and rejects a^n. We show that it accepts everything else.

Let a^i be rejected by this NFA.

- Since the M part rejects a^i, $i \le n + 2\lceil n^{1/2} \rceil$ (note, hence $i \le \prod_{p \in Q} p$).
- For each $p \in Q$, since the M_p part rejects a^i, $i \equiv n \pmod{p}$.

By the Chinese Remainder Theorem there is a unique number $0 \le z < \prod_{p \in Q} p \ge n + 2\lceil n^{1/2} \rceil$ such that, for every $p \in Q$, $z \equiv n \pmod{p}$. Since both i and n satisfy the criteria, $i = n$. □

4 Small NFA's for MN(998, 999, 1000) and MN(A)

4.1 Small NFA for MN(998, 999, 1000) and MN(998, 1000)

Theorem 4.1.

1. *There exists an NFA M such that the following are true:*

   (a) *For all $i \ge 1067$, M accepts a^i.*

(b) For all $i \in \{998, 999, 1000\}$ M rejects a^i.

(c) We have no comment on the behavior of M for other a^i's.

(d) M has 34 states.

2. *There exists an NFA with 104 states that accepts* $\mathrm{MN}(998, 999, 1000)$.

3. *There exists an NFA with 104 states that accepts* $\mathrm{MN}(A)$ *where* $A \subseteq \{998, 999, 1000\}$. *(For $A = \emptyset$ this is trivial.)*

Proof. 1) Let M be $\mathrm{LOOP}(33, 34, 11)$. From Lemma 3.3 we know the following:

a) For all $i \geq 1067$, M accepts a^i.

b) For all $C, D \in \mathsf{N}$, M rejects $a^{1066-33C-34D}$ which we write as $a^{1066-33(C+D)-D}$.

We set (C, D) carefully to obtain, using item b, strings that M rejects.

- If $(C, D) = (2, 0)$ then we get $1066 - 33 \times 2 - 0 = 1000$
- If $(C, D) = (1, 1)$ then we get $1066 - 33 \times 2 - 1 = 999$
- If $(C, D) = (0, 2)$ then we get $1066 - 33 \times 2 - 2 = 998$

Clearly M has 34 states.

2) We will once again use primes and mods. We can't use mod 2 or mod 3 since then one of a^{998}, a^{999}, a^{1000} will be accepted.

We can use any mod from 5 up. We need another trick, as you will see.

Let $Q = \{5, 7, 11\}$. Note that $3 \times 5 \times 7 \times 11 = 1155 > 1066$ (That is not a typo. We really do mean to multiply by 3. We chose 3 because $|\{998, 999, 1000\}| = 3$. We chose $\{5, 7, 11\}$ since none of them divide 3 and the product $3 \times 5 \times 7 \times 11 > 1066$. The fact that 3 is a prime is not important.)

For each $p \in Q$ let M_{3p} be the DFA that accepts

$$\{a^i : i \not\equiv 998, 999, 1000 \pmod{3p}\}$$

Note that $\mathrm{LCM}(3 \times 5, 3 \times 7, 3 \times 11) = 3 \times 5 \times 7 \times 11 = 105 > 100$.

The NFA is constructed as follows:

1. The NFA M is part of our new NFA. We create a new start state, and then put an e-transition from this new state to M's original start state. Note that M (1) accepts all a^i with $i \geq 1067$ (it also accepts other strings), (2) rejects any of a^i with $i \in \{998, 999, 1000\}$, and (3) has 34 states.

2. For each $p \in Q$, there is an e-transition from our new start state to the start state of M_{3p}. Note that M_{3p} (1) accepts a^i when $i \not\equiv 998, 999, 1000 \pmod{3p}$, (2) rejects any a^i with $i \in \{998, 999, 1000\}$, and (3) has $3p$ states.

This NFA has $34 + 3(5 + 7 + 11) + 1 = 104$ states and rejects any a^i with $i \notin \{998, 999, 1000\}$. We show that it accepts everything else.

Let a^i be rejected by this NFA.

- Since the M part rejects a^i, $i \leq 1066$ (note, hence $i \leq 3 \times 5 \times 7 \times 11 = 1155$).
- For all $p \in Q$, since the M_{3p} part rejects a^i, there exists $x \in \{998, 999, 1000\}$ such that $i \equiv x \pmod{3p}$.

We cannot use the Chinese Remainder Theorem (yet) since it is possible that, say $i \equiv 998 \pmod{3 \times 7}$ but $i \equiv 1000 \pmod{3 \times 11}$. We need that i is equivalent to the same x with all of those mods.

Let $i \equiv x \pmod 3$ where $x \in \{998, 999, 1000\}$. Note that x is unique. Let $p \in Q$. Let $y \in \{998, 999, 1000\}$ be such that $i \equiv y \pmod{3p}$. Note that y is unique since $3 < 3p$.

We show that $x = y$.

Since $i \equiv x \pmod 3$ there exists $a \in \mathsf{Z}$ such that

$$\text{Eq 1 } i = x + 3a.$$

Since $i \equiv y \pmod{3p}$ there a $b \in \mathsf{Z}$ such that

$$\text{Eq 2 } i = y + 3pb.$$

By subtracting Eq 2 from Eq 1 we get

$$x - y = 3pb - 3a \equiv 0 \pmod 3$$

Since $x, y \in \{998, 999, 1000\}$ and $x \equiv y \pmod 3$, $x = y$. To recap we now have that there exists $x \in \{998, 999, 1000\}$ such that, for all $p \in Q$, $i \equiv x \pmod{3p}$.

By the Chinese Remainder Theorem there is a unique number $0 \leq z \leq \text{LCM}(3 \times 5, 3 \times 7, 3 \times 11) = 1155$ such that, for all $p \in Q$, $z \equiv x \pmod{3p}$. Since both i and x satisfy those criteria, $i = x$.

3) We look at MN(998, 1000) as an example. The construction is similar to the one for MN(998, 1000) except that, at the end, use the DFA for $\{a^i : i \not\equiv 998, 1000 \pmod{3p}\}$ instead of $\{a^i : i \not\equiv 998, 999, 1000 \pmod{3p}\}$. The other cases are similar. □

Theorem 4.2. *Let $0 < \delta < 1$. Let $n \in \mathsf{N}$. (We will assume $n^\delta \in \mathsf{N}$ and leave it to the reader to adjust the statement and the proof for when $n^\delta \notin \mathsf{N}$.) Assume $n = c^2 + f$ where $0 \le f \le 2c$.*

1. *There exists an NFA M such that the following are true:*
   (a) *For all $i \ge n + n^{1/2+\delta} + n^{2\delta} + 1$, M accepts a^i.*
   (b) *For all $i \in \{n - n^\delta, n - n^\delta + 1, \ldots, n\}$, M rejects a^i.*
   (c) *We have no comment on the behavior of M for other a^i's.*
   (d) *M has $\le 5n^{\max\{1/2,\delta\}} + O(1)$ states.*
2. *There exists an NFA on*
   $$\le 5n^{\max\{1/2,\delta\}} + \tilde{O}(1) \text{ states}$$
   that accepts $\mathrm{MN}(n - n^\delta, n - \delta + 1, \ldots, n)$.
3. *Let $A \subseteq \{n - n^\delta, \ldots, n\}$. There exists an NFA on*
   $$5n^{\max\{1/2,\delta\}} + \tilde{O}(1) \text{ states}$$
   that accepts $\mathrm{MN}(A)$. (For $A = \emptyset$ this is trivial.)

Proof. 1) Let M be the NFA $\mathrm{LOOP}(c+k, c+k+1, f+1+x(k))$ where we determine k and $x(k)$ later.
Claim:

1. If M rejects $a^{n+n^\delta(c+k)}$ then, for $i = n - n^\delta, \ldots, n$, M rejects a^i.
2. If $x(k) = n^\delta(c+k) - k^2 - 2ck + c + k$ then M rejects $a^{n+n^\delta(c+k)}$.
3. If $x(k) = n^\delta(c+k) - k^2 - 2ck + c + k$ then, for $i = n - n^\delta, \ldots, n$, M rejects a^i (this follows from parts 1 and 2).

Proof of Claim:
1) Assume M rejects $n + n^\delta(c+k)$. Then it also rejects everything of the form

$$n + n^\delta(c+k) - (c+k)C - (c+k+1)D = n + n^\delta(c+k) - (C+D)(c+k) - D$$

(since otherwise M would accept $n + n^\delta(c+k) - (c+k)C - (c+k+1)D + (c+k)C + (c+k+1)D = n + n^\delta(c+k)$).

We set (C, D) as follows:

- If $(C, D) = (n^\delta, 0)$ then we get $n + n^\delta(c + k) - n^\delta(c + k) - 0 = n$.
- If $(C, D) = (n^\delta - 1, 1)$ then we get $n + n^\delta(c + k) - n^\delta(c + k) - 1 = n - 1$.
- $\vdots$
- If $(C, D) = (0, n^\delta)$ then we get $n + n^\delta(c + k) - n^\delta(c + k) - n^\delta = n - n^\delta$.

2) For $k \in \mathsf{N}$ we need $x(k) \in \mathsf{N}$ such that $\mathrm{LOOP}(c+k, c+k+1, f+1+x(k))$ rejects $n + n^\delta(c + k)$. Note that this NFA rejects

$$(c + k)(c + k + 1) - (c + k) - (c + k + 1) + f + 1 + x(k)$$
$$= c^2 + k^2 + 2ck + c + k - 2c - 2k - 1 + f + 1 + x(k)$$
$$= n + k^2 + 2ck - c - k + x(k)$$

Hence we find $x(k)$ via:

$$n + k^2 + 2ck - c - k + x(k) = n + n^\delta(c + k)$$

or equivalently

$$x(k) = n^\delta(c + k) - k^2 - 2ck + c + k$$

End of Proof of Claim

We choose k such that the max of $\{c + k + 1, f + 1 + x(k)\}$ is small. We look at what happens to $x(k)$ for $k \in \{0, \ldots, n^\delta\}$. We consider only when $n \geq 9$, with smaller n being expressed within the $O(1)$ term. Note that

- $x(0) = n^\delta c + c > 0$.
- $x(n^\delta) = n^{2\delta} + n^\delta c - n^{2\delta} - 2cn^\delta + c + n^\delta = c + n^\delta - cn^\delta < 0$ (since $n \geq 9$).
- there exists k_o such that $x(k_o) \geq 0$ and $x(k_o + 1) \leq 0$ (this follows from the first two points).

Note that

$$x(k_o) \leq x(k_o) - x(k_o + 1) \leq |-2c + n^\delta - 2k_o| \leq 2c + n^\delta.$$

Let $M = \mathrm{LOOP}(c+k_o, c+k_o, f+1+x(k_o))$. Since $c \leq n^{1/2}$, $f \leq 2n^{1/2}$, $k_o \leq n^\delta$, and $x(k_o) \leq 2c+n^\delta \leq 3n^{\max\{1/2,\delta\}}$. $e = x(k_o)+f+1$ has $\leq 5n^{\max\{1/2,\delta\}}+O(1)$ states, while $c+k_o+1$ has $\leq 3n^{\max\{1/2,\delta\}}+O(1)$ states, so M must have $\leq 5n^{\max\{1/2,\delta\}}+O(1)$ states overall.

M satisfies conditions of what to reject and how many states it has. We now consider what it accepts. Note that $x(k)$ was chosen so that the largest number M (with $k = k_o$) rejects is $a^{n+n^\delta(c+k_o)}$. We need to estimate this.

$$n + n^\delta(c + k_o) \leq n + n^\delta(n^{1/2} + n^\delta) \leq n + n^{1/2+\delta} + n^{2\delta}.$$

By Lemma 3.3 M accepts what it should.

2) To simplify the algebra we just use that the NFA in Part 1 accepts $\{a^i : i \geq n^2\}$.

By Lemma 3.5 there is a set of primes Q' such that (1) $\prod_{p\in Q'} p \geq n^2$, (2) $\sum_{p\in Q'} p \leq \widetilde{O}(1)$. We form Q as follows: (1) remove from Q' all of the primes that divide n^δ, (2) add in the smallest primes possible that do not divide n^δ so that $n^\delta \prod_{p\in Q} p \geq n^2$.

One can show that $\sum_{p\in Q} p \leq O(\sum_{p\in Q'} p) \leq \widetilde{O}(1)$. Hence we have a set Q such that (1) $n^\delta \prod_{p\in Q} p \geq n^2$ (2) $\sum_{p\in Q} p \leq \widetilde{O}(1)$, and (3) for all $p \in Q$, p does not divide n^δ. For each $p \in Q$ let $M_{n^\delta p}$ be the DFA that accepts $\{a^i : i \not\equiv n \pmod{n^\delta p}\}$.

Note that the LCM$\{n^\delta p : p \in Q\} = n^\delta \prod_{p\in Q} p \geq n$.

The NFA is constructed as follows:

1. The NFA M is part of our new NFA . We create a new start state, and then put an e-transition from this new state to M's original start state. Note that M (1) accepts all a^i with $i \geq n^2$ (it also accepts other strings), (2) rejects any of a^i with $i \in \{n - n^\delta, \ldots, n\}$, and (3) has $\leq 5n^{\max\{1/2,\delta\}} + O(1)$ states.

2. For each $p \in Q$ put an e-transition from our new start state to the start state of $M_{n^\delta p}$. Note that $M_{n^\delta p}$ (1) accepts a^i with $i \not\equiv n - n^\delta, \ldots, n \pmod{n^\delta p}$, (2) rejects any of a^i with $i \in \{n - n^\delta, \ldots, n\}$, and (3) has $n^\delta p$ states.

This NFA has $5n^{\max\{1/2,\delta\}} + \widetilde{O}(1)$ states and rejects any a^i with $i \in \{n - n^\delta, \ldots, n\}$. We show that it accepts everything else.

Let a^i be rejected by this NFA.

- Since the M part rejects a^i, $i \leq n^2$ (note, hence $i \leq n^\delta \prod_{p\in Q} p$).

- For each $p \in Q$, since the $M_{n^\delta p}$ part rejects a^i, there exists $x \in \{n - n^\delta, \ldots, n\}$ such that $i \equiv x \pmod{n^\delta p}$.

We cannot use the Chinese Remainder Theorem (yet) since it is possible that, say $i \equiv 95 \pmod{n^\delta \times 7}$ but $i \equiv 92 \pmod{n^\delta \times 11}$. We need that a^i is equivalent to the same $n^\delta p$ with all those mods.

Let $i \equiv x \pmod{n^\delta}$ where $x \in \{n - n^\delta, \ldots, n\}$. Note that x is unique. Let $n^\delta p \in n^\delta Q$. Let $y \in \{n - n^\delta, \ldots, n\}$ be such that $i \equiv y \pmod{n)^\delta p}$. Note that y is unique since $n^\delta < n^\delta p$. We show that $x = y$.

Since $i \equiv x \pmod{n^\delta}$ there exists $a \in \mathsf{Z}$ such that:

$$\text{Eq } 1\ i = x + n^\delta a.$$

Since $i \equiv y \pmod{n^\delta}$ there exists a $b \in \mathsf{Z}$ such that

$$\text{Eq } 2\ i = y + n^\delta pb.$$

By subtracting Eq 2 from Eq 1 we get

$$x - y = n^\delta pb - n^\delta a \equiv 0 \pmod{n^\delta}$$

Since $x, y \in \{n - \delta, \ldots, n\}$ and $x \equiv y \pmod{n^\delta}$, $x = y$. To recap we now have that there exists $x \in \{n - \delta, \ldots, n\}$ such that, for all $n^\delta p \in Q$, $i \equiv x \pmod{n^\delta p}$. By the Chinese Remainder Theorem there is a unique number $0 \le z \le \mathrm{LCM}\{n^\delta p : p \in Q\} \ge n$ such that, for all $p \in Q$, $z \equiv x \pmod{n^\delta p}$. Since both i and x satisfy those criteria, $i = x$.

3) This is an easy modification of Part 2 which we leave to the reader. □

5 Small NFA's for $\mathrm{MN}(\alpha n, n)$

Lemma 5.1. *Let $x, x', y, y', c \in \mathsf{N}$ with $c \ge 1$ be such that the following hold.*

1. $c(x - x') + (c + 1)(y - y') = 0$.
2. $|x - x'| \le c$.

Then $x = x'$ and $y = y'$.

Proof. Since $c + 1$ divides $c(x - x')$ and $c + 1$ is rel prime to c we have that $c + 1$ divides $x - x'$. Since $|x - x'| \le c$, $x = x'$. Hence $y = y'$. □

Theorem 5.2. *Let $n \in \mathsf{N}$ and $0 < \alpha < 1$ be such that $\alpha n \in \mathsf{N}$.*

1. *There exists an NFA M such that the following are true:*
   (a) *For all $i \ge 2n \ln n$, M accepts a^i.*
   (b) *For all $i \in \{\alpha n, n\}$, M rejects a^i.*

(c) We have no comment on the behavior of M for any other a^i's.

(d) M has $\le 2n^{1/2}\ln n + \tilde{O}(1)$ states.

2. *There exists an NFA on $\le 2n^{1/2}\ln n + \tilde{O}(1)$ states that accepts* $\mathrm{MN}(\alpha n, n)$.

Proof. Let $c = \lceil n^{1/2} \rceil + 1$ and $e = n+1 \pmod{c}$. Note that $e \le c$.
1) Let M' be $\mathrm{LOOP}(c, c+1, e)$. By the proof of Theorem 3.6 we have:

1. For all $i \ge n+2\lceil n^{1/2} \rceil$, M' accepts a^i. Note that for all $i \ge 2n\ln n$, M' accepts a^i.
2. M' rejects a^n.
3. We have no comment on the behavior of M' on other a^i.
4. M' has $\le n^{1/2} + O(1)$ states.

Case 1: M' rejects $a^{\alpha n}$. Then take M to be M'.
Case 2: M' accepts $a^{\alpha n}$. We use the very acceptance of $a^{\alpha n}$ to find an NFA M that satisfies the theorem.
Claim 1: There exists a unique x, y, such that $cx + (c+1)y + e = \alpha n$. Both x, y are $\le c - 1 \le n^{1/2}$.
Proof of Claim: Since M' accepts $a^{\alpha n}$ there is at least one such x, y such that:

$$\text{Eq 1 } cx + (c+1)y + e = \alpha n$$

$x \le c-1$ since otherwise Eq 1 implies:

$$\alpha n = cx + (c+1)y + e \ge c^2 + (c+1)y + e \ge c^2 = n.$$

$y \le c-1$ by a similar argument.
Assume that x', y' also works.

$$\text{Eq 2 } cx' + (c+1)y' + e = \alpha n$$

By the same reasoning that $x \le c-1$, we have $x' \le c-1$, so $|x - x'| \le c-1$. Subtract the second equation from the first to obtain:

$$c(x - x') + (c+1)(y - y') = 0$$

By Lemma 5.1, $x = x'$ and $y = y'$.
End of Proof of Claim 1

Let p be the least prime that does not divide yc (hence does not divide y or c). Since $y \le c^{1/2}$ and $c = \lceil n^{1/2} \rceil + 1$, $yc \le n + n^{1/2} + O(1)$. By Lemma 3.5.3, $p \le (1+o(1))\ln(n+n^{1/2}+O(1)) \le 2\ln(n)+O(1)$. Let M be LOOP$(c, p(c+1), e)$. Note that M has $\le 2n^{1/2}\ln n + \widetilde{O}(1)$ states. We need to show that (1) for all $i \ge n \ln n$, M accepts a^i, and (2) M rejects $a^{\alpha n}$ and a^n. Note that

$$cp(c+1) - c - cp - p + e = c^2p + cp - c - cp - p + e = c^2p - c - p + e \le 2n\ln n - 1$$

By Lemma 3.3

- For all $i \ge 2n\ln n$, M accepts a^i.
- For all C, D, M rejects a^i where $i = c^2p - c - p + e - Cc - D(p(c+1))$. (We will not be using this.)

Claim 2: M rejects $a^{\alpha n}$.
Proof of Claim 2:
Assume, by way of contradiction, that M accepts αn. Then there exists x', y' such that

$$\text{Eq 1 } \alpha n = cx' + (c+1)y' + e$$

Recall that from Claim 1 there exists unique x, y such that

$$\text{Eq 2 } \alpha n = cx + (c+1)y + e.$$

Hence $x = x'$ and $y = y'p$. This contradicts that p does not divide y.
End of Proof of Claim 2
Claim 3: M rejects a^n.
Proof of Claim 3:
Assume, by way of contradiction, that M accepts n. Then there exists x', y' such that

$$n = cx' + p(c+1)y' + e$$

Then a^n is accepted by LOOP$(c, c+1, e)$, which is a contradiction.
End of Proof of Claim 3

2) This proof is similar to that of Theorem 4.2.2.

□

6 Small NFA's for MN(A) where A has large gaps

Theorem 6.1. *Let $A \subseteq \mathsf{N}$ with maximum element n'. Let $n' < n$. Then there is an NFA for* $\mathrm{MN}(A \cup \{n\})$ *of size* $n^{1/2} + n' + \widetilde{O}(1)$.

Proof. By Theorem 3.6 there exists an NFA M' for $\mathrm{MN}(n - n')$ of size $(n - n')^{1/2} + \widetilde{O}(1) \leq n^{1/2} + \widetilde{O}(1)$. We form M as follows:

1. Add states $0, 1, \ldots, n'$ where state n' is the start state of M'.
2. 0 is the start state of M.
3. For $0 \leq i \leq n' - 1$ we have transitions $\delta(i, a) = i + 1$.
4. For all $0 \leq i \leq n'$, make i an accept state iff $i \in A$.

Clearly M'' accepts $A \cup \{n\}$ and has $n^{1/2} + n' + \widetilde{O}(1)$ states. □

Corollary 6.2.

1. *For all n, for all $0 < \delta < 1$, there is an NFA for $MN(n^{\delta}, n)$ of size $n^{1/2} + n^{\delta} + \widetilde{O}(1)$.*
2. *For all n, for all $\beta \in \mathsf{R}^{+}$, there is an NFA for $MN((\log n)^{\beta}, n)$ of size $n^{1/2} + \widetilde{O}(1)$.*

7 Every NFA for $MN(n)$ has $\geq n^{1/2}$ States

This section is due to Jeff Shallit who shared it with us.

Chroback [2] proved the following.

Theorem 7.1. *Let L be a cofinite unary regular language. If there is an NFA for L with n states then there is an NFA for L of the following form:*

- *There is a sequence of $\leq n^2$ states from the start state to a state we will call X. Note that there is no nondeterminism involved yet.*
- *From X there are e-transitions to $X_1, \ldots, X_m$. (This is nondeterministic.)*
- *Each X_i is part of a cycle C_i. All of the C_i are disjoint.*

Theorem 7.2.

1. *Let L be a cofinite unary language where the shortest string that is not in L is of length n. Then any NFA for L requires $n^{1/2}$ states*

2. *Any NFA for $MN(n)$ has $n^{1/2}$ states (this follows from part 1).*

Proof. Assume there was an NFA with $< n^{1/2}$ states for L. Then by Theorem 7.1 there would be an NFA for L with a path from the start state to a state X of length $< n$ and then from X a branch to many cycles. Let X_i and cycle's C_i as described in Theorem 7.1.

Run a^n through the NFA and try out all paths. For each i there will be a point in C_i that you end up at. Let n_i be the length of C_i. For every i there is a state on C_i that rejects. Hence the strings $a^{n+Kn_1n_2\cdots n_m}$ are all rejected. This is an infinite number of strings. This is a contradiction. □

8 Empirical Results

We have written a program that, given n, tries to find the smallest NFA for $\mathrm{MN}(n)$. We first set $c = \lceil\sqrt{n}\rceil$, $d = c+1$, and e such that $\mathrm{LOOP}(c,d,e)$ (1) rejects a^n and (2) for all $i \geq n+1$, accepts a^i. We then looked at sets of prime powers (these work as well as primes) so that the usual M_{p^b} machines will accept all a^i such that $i \leq n-1$. We took the smallest NFA among all of these choices. We ran this program for $1 \leq n \leq 10^{27}$. Here is what we discovered:

1. The smallest NFA for $\mathrm{MN}(n)$ was around $n^{1/2} + (\ln n)^g$ where g had the following values:

   (a) For $1 \leq n \leq 10^6$ g decreases from 2 to around 1.55.

   (b) For $10^6 \leq n \leq 10^{27}$ g fluctuates around 1.55 but slowly increases.

The log-term is actually the inverse of Landau's function. As such, it is known that g is bounded by 2.

We also wrote a second program that tries to find smaller NFAs than the first program. How? Note that, in the first program, we found a set of M_{p^b} machines to accept all a^i such that $i \leq n-1$. However, $\mathrm{LOOP}(c,d,e)$ already accepts some of those strings. Our second program finds a set of M_{p^b} machines that accepts all a^i that $\mathrm{LOOP}(c,d,e)$ did not accept. We ran this program for $1 \leq n \leq 1700$ (this program took much longer to run then the first one). Here is what we discovered:

1. For slightly more than half of the n, we found a smaller NFA this way.

2. The most common improvement was 1. Then 2. ... Then 6. There were no improvements bigger than 6. There was a slight tendency of getting bigger improvements for bigger n.

We conjecture that, for all L, there exists n, so that the second program will produce an NFA that has at least L states fewer than the first program.

9 Open Questions

We conjecture that every cofinite unary language has a small NFA; however, this is hard to state rigorously.

The NFA for MN(n) is optimal up to $\widetilde{O}(1)$ terms. We would like to know if the other NFA's we have presented are optimal up to $\widetilde{O}(1)$ terms.

10 Acknowledgments

We would like to thank Jeff Shallit for his slides [3] that got us started on this subject, his sharing the proof of Theorem 7.2 with us, and for his encouragement to pursue this work.

References

[1] Christian Axler *On the sum of the first n prime numbers*, 2014. `https://arxiv.org/pdf/1409.1777.pdf`.

[2] Marek Chrobak *Finite automata and unary languages*, Theoretical Computer Science 47, 1986, pp. 149–158. Erratum for paper is at `https://dl.acm.org/citation.cfm?id=860232`.

[3] Jeff Shallit *The Frobenius problem and its generalization*, Slides: `https://cs.uwaterloo.ca/~shallit/Talks/frob14.pdf`.

AN INTRODUCTION TO A MODEL OF ABSTRACT COMPUTATION: THE BSS-RAM MODEL

CHRISTINE GASSNER
University of Greifswald, Germany
gassnerc@uni-greifswald.de

ABSTRACT We give a detailed definition of BSS RAM's for the sequential computation over first-order structures. These random access machines are abstract machines with an input procedure and an output procedure. Each program of a machine is an element of a formal language whose syntax is defined by syntactic rules. This means that the programs are only strings that can be formed by syntactic rules. For giving these strings a semantic meaning, we use transition systems. Any random access machine determines such a system for transforming one configuration of the machine into the next configuration. The semantics of any program is dependent on such a transition system that results from the program and the interpretation of the symbols in the program by means of the underlying structure. For this purpose, all operations and relations of the underlying structure are available. Since any machine has its own program, it is not necessary to have a universal machine that is able to execute all programs. Here, we give the basic definitions and discuss the possibility to define, for certain mathematical structures, universal machines or programs that are useful for further investigations on this field.

1 Introduction

There are a lot of models used for describing the sequential processing of objects and data. In general, there is a finite pool of elementary operations and tests that can be used. In most cases, the processing procedures are described by texts that are similar to programs over abstract structures. Examples for such processes can be found in the areas of programming as well as in mathematics and natural sciences. In many cases, the procedures are algorithms in the widest sense of the word.

They can be executed by using existing basic routines or certain substances without knowing the details of the implementation of the basic procedures or the details of producing such substances. One of the attempts to provide a suitable framework for the formalisation of all uniform processes of this kind — where "uniform" means here to consider inputs of any length — was realized by defining the BSS-RAM model of computation over several structures. It is a generalisation of models that are the result of combining properties of the BSS model and the real RAM's. It was used exactly in this form since 1994 (in publications since 1996, see e.g. [13]) without explaining all details. In principle, we can say that every mathematician has an idea of what an algorithm is (see e.g. [19]). But, for developing theories that help to answer questions about the computability and complexity of various problems, we need a formal concept. The BSS-RAM model is a logic-based concept that provides a mathematical framework for collecting results referring to several computational models, generalising results, transferring results, and finding new relationships at a higher level of abstraction. The definition of the BSS-RAM model presented here is the result of a long formalisation process. Many different concepts go into defining the BSS-RAM model. A consequence is that, on the one hand, Turing machines and the known BSS machines can be simulated by BSS RAM's and, moreover, for several classes of structures, the classical concepts such as the concept of the universal machine (cf. [28, pp. 241–246]) can be transferred in order to simulate abstract computing processes. On the other hand, the programs of BSS-RAM's have features suitable for modelling the object-oriented programming. This makes it easy for beginners to start their studies with the BSS-RAM model in order to gain insights into the classical recursion theory, the classical complexity theory, and the possibility to deal with Turing machines or to investigate BSS machines and the corresponding complexity classes. The BSS-RAM model brings together the basic concepts given in [5], [3], and [24]. Computational models as given by Dana Scott [25] together with the mathematical description of this type of computations by the transition systems discussed in detail by Egon Börger [5] offer a framework under which a standardisation and a classification of a lot of computational models is possible. In the first BSS model developed by Lenore Blum, Micheal Shub, and Steve Smale in [3] the change of values in sequences of real numbers was possible by means of programs given in form of flow charts. It was the starting point for discussing the similarities between the known real-RAM model and uniform computational models such as the Turing machine. Whereas the real RAM's equipped with registers for storing real numbers were introduced for investigating the complexity of single algorithms on the basis of different elementary operations und tests by Franco P. Preparata and Michael I. Shamos [24], the BSS model is a uniform model that allows to create a complexity theory over the reals on the basis of elementary operations and certain test procedures as basic relations. The BSS-RAM model is a uniform register-machine-model which combines properties of the RAM's and the BSS model. In general, the form of instructions for building programs are only shortly defined before new observations and results are presented, as in [17, 18]. However, small differences in the definitions can influence the capability of the machines and the complexity of the executed algorithms. Therefore, we want to start with the definition of all details of our $\mathcal{A}$-machines and the BSS-RAM model for several structures $\mathcal{A}$ in order to make visible the relationships between several models, on the one hand, and the fact that programs of machines over mathematical structures need an interpretation in a similar way as all logical formulas, on the other hand. Moreover, by defining the transition systems in detail, we provide a good basis for comparing the $\mathcal{A}$-machines and the BSS RAM's with other computational models over several structures that are closer related to the classical recursion theory (see e.g. [22]). The details here presented were elaborated in [16]. Here we give first examples that show how known models can be discussed, studied, and compared in this framework. We hope that this article can be the starting point for many comparisons of known results, generalising results, and transferring results from one theory to another theory. The discussion started here can be continued with an analysis of the structural complexity of computational problems.

We want to mention that there are a lot of machine-oriented models such as the uniform models presented in [23, 20, 15] and the machine-independent models of abstract computation studied in generalised recursion theory (see e.g. [22]) for modelling the computation over algebraic structures supplemented with some relations. For structures over the real numbers, we know machine-oriented models such as the real RAM's introduced on the basis of the concept presented in [26, 1], the modified real-RAM model considered in [7], and the uniform Blum-Shub-Smale (BSS) machines introduced in [3]. A common feature of these machines is that each real number can be stored as a single unit during the computation. They are suitable for describing algorithms and analysing their complexity under the assumption that some operations and relations can be used as primitive operations and relations at unit cost. However, the comparability of the results is difficult since, for many models, the possibilities of computation are not sufficiently investigated. [5, 4, 20, 7, 6] can help to get an overview of the great variety of models. For example, there are also a lot of different real-RAM models. The first real RAM's were considered by Preparata and Shamos in [24]. The authors adopted the definition of the random access machine that was introduced for processing integers in [1]. In analogy with this model, many real RAM's work with an instruction for reading one input value and copying the real number to a register and an instruction for writing the content of one register on an output tape. Consequently, the evaluation of the complexity of algorithms on the basis of real RAM's refers often to an arbitrary fixed number of input values since the uniform treatment of all sequences of input values by one machine or one program is neither required nor expected. In contrast to that, the BSS machines use an input procedure that allows an exact mathematical description of the single steps of a uniform computational process for all possible finite sequences of real numbers without any restriction of the input space. It has been confirmed that the BSS model is suitable for developing a theory of computation and complexity over the real numbers. That is one reason why we continue this approach.

We present a random-access-machine model of abstract computation over a first-order structure including the uniform machines. These machines, the so-called BSS RAM's, can use every operation of the underlying structure as a primitive operation and the relations of the structure for changing their state. As above mentioned, our model of computation has been developed on the basis of the concept presented by Scott in [25] and the transition systems defined in [5]. It is a generalisation of the BSS model of computation over a ring and it was used, for instance, also in [14, 15]. More precisely, we extended the model considered by Börger [5] in order to get a uniform model in the sense of the BSS model. This means that several types of instructions are now available for accessing memory and for using the indirect addressing techniques. However, the capabilities with respect to indirect addressing are so limited that the BSS RAM's over a finite structure (containing only a finite number of elements) have a power that is comparable to the power of the Turing machine. Over the ordered ring $(\mathbb{R};\mathbb{R};+,-,\cdot;\leq)$, where all real numbers can be used as constants, the BSS RAM's have the same computational power as the BSS machines over this ring.

Here, we continue the investigation in [16] and discuss a concept of universal programs and machines for several structures. We present a model of abstract computation without taking questions about a possible digitisation into consideration and we formalise the concept of *algorithm* as precisely and extensively as possible in our framework. This means that we provide a general framework for describing sequential computations by a uniform transition model and investigating the power of machines that can perform each operation and use each relation of the underlying structures, by assumption, in the same time unit. We hope this model helps to better understand the different approaches to solve or analyse algorithmic problems, the meaning of different models of sequential computation, and the common properties and the differences of these models.

In Section 2 we give a detailed definition and description of $\mathcal{A}$-machines and BSS RAM's over any first-order structure $\mathcal{A}$ of signature σ. The discussion includes σ-programs, the use of pseudo instructions, and multi-tape machines. Section 3 presents a concept of universal σ-programs and universal BSS RAM's and deals with questions concerning the importance of universal BSS RAM's for the completeness of decision problems.

We use the usual mathematical notations. $\mathbb{N}$, $\mathbb{N}_+$, and $\mathbb{R}$ are the sets of non-negative integers, positive integers, and real numbers, respectively. $=_{\mathrm{df}}$ means "is equal by definition". In defining sets we use the usual symbols such as $\forall, \exists, \&, \in$, $\max, \leq, \ldots$ where & stands for "and". $\{x \mid \phi(x)\}$ is the set of all objects x satisfying the property described by $\phi(x)$. The strings $^{(0)}$, $^{(1)}$, $^{(i)}$, $_i, \ldots$ are indices and placeholders for indices, respectively. For arithmetic expressions t such as $i+1$, $_t$ is usually the integer index resulting from the evaluation of t. $i \in \{1, \ldots, n\}$ and $i \leq n$ means that i is an integer with $1 \leq i \leq n$ and $i \geq 1$ means $i \in \mathbb{N}_+$, and so on. $f :\subseteq M \to N$ means that f is a partially defined function from M into N. We write $f : M \to N$ only if f is a totally defined function. For $f : M \to N$, let $f_i = f(i)$ for all $i \in M$ and, for $M = \mathbb{N}$ and $n \geq 1$, let $f_1, \ldots, f_n$ be the list of values assigned to the integers 1 to n by f in this order. Let $M^\infty = \bigcup_{n \geq 1} M^n$ be the set of all tuples $\vec{x} =_{\mathrm{df}} (x_1, \ldots, x_n)$ with $x_i \in M$ $(i \leq n)$ and $n \geq 1$ and let M^ω be the set of all infinite sequences $\bar{u} =_{\mathrm{df}} (u_1, u_2, \ldots)$ with $u_i \in M$ for all $i \geq 1$. For $n = 0$, let $x_1, \ldots, x_n$ and $\vec{x}$ be the empty list and, thus, $\sum_{i=1}^n x_i = 0$, and so on. For all $\vec{x} \in M^n$, $\vec{y} \in N^m$, and $\bar{u} \in N^\omega$, let $(\vec{x} \,.\, \vec{y}) = (x_1, \ldots, x_n, y_1, \ldots, y_m) \in (M \cup N)^{n+m}$ and $(\vec{x} \,.\, \bar{u}) = (x_1, \ldots, x_n, u_1, u_2, \ldots) \in (M \cup N)^\omega$. We use, for $d \geq 1$, $(\vec{x}^{(j)})_{j=1..d} =_{\mathrm{df}} (\vec{x}^{(1)}, \ldots, \vec{x}^{(d)})$ and $(\bar{u}^{(j)})_{j=1..d} =_{\mathrm{df}} (\bar{u}^{(1)}, \ldots, \bar{u}^{(d)})$ and, for all $j \geq 1$, $\vec{x}^{(j)} \in M^n$, and $\bar{u}^{(j)} \in N^\omega$, let $\vec{x}^{(j)} = (x_{j,1}, \ldots, x_{j,n})$ and $\bar{u}^{(j)} = (u_{j,1}, u_{j,2}, \ldots)$. For any $f :\subseteq M^\infty \to N$ and any $\vec{x} \in M^\infty$, let $f(x_1, \ldots, x_n) = f(\vec{x})$. For any $f :\subseteq M^\omega \to N$ and any $\bar{u} \in M^\omega$, let $f(u_1, u_2, \ldots) = f(\bar{u})$. For $g :\subseteq M \to N$ and $f :\subseteq N \to O$, $f \circ g :\subseteq M \to O$ is given by $(f \circ g)(x) = f(g(x))$ for all $x \in M$ and, thus, defined for those $x \in M$ for which both $g(x)$ and $f(g(x))$ are defined. Let $f_{g_x} = f(g(x))$. $\{\alpha_1, \ldots, \alpha_m\}^*$ $(m \geq 1)$ consists of the *empty string* Λ and all *strings* $\alpha_{i_1} \cdots \alpha_{i_l}$ with $l \geq 1$ and $i_1, \ldots, i_l \leq m$.

2 The BSS-RAM model

2.1 The idea and the background

For any algebraic structure $\mathcal{A}$, every $\mathcal{A}$-machine has its own program. It can perform each operation of the underlying structure $\mathcal{A}$ in a single step of computation. The BSS RAM's over a structure $\mathcal{A}$ are $\mathcal{A}$-machines that get as inputs arbitrary finite sequences of elements of $\mathcal{A}$. Figure 1 shows the essential processing steps for the computation of a function. The inputted values are assigned to registers by an input procedure. Afterwards, the machine executes its program defined by a finite sequence of labelled instructions until a stop criterion is satisfied. If the stop criterion is reached, then the machine halts and the computed values can be outputted by means of an output procedure. If $\mathcal{A}$ is a structure with the universe $U_{\mathcal{A}}$, we will say that a partially defined function f from $U_{\mathcal{A}}^{\infty}$ into $U_{\mathcal{A}}^{\infty}$ is computable by a BSS RAM over $\mathcal{A}$ if there is a BSS RAM over $\mathcal{A}$ that halts on input $(x_1, \ldots, x_n) \in U_{\mathcal{A}}^{\infty}$ if and only if the value of f is defined for $(x_1, \ldots, x_n)$, and then it outputs the value $f(x_1, \ldots, x_n)$.

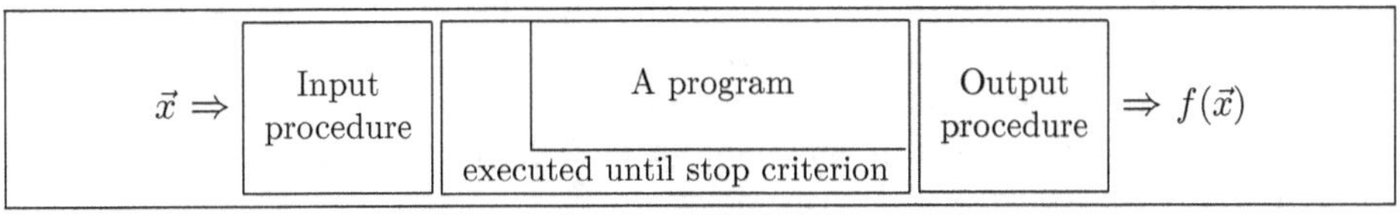

Figure 1: The BSS-RAM model

In general, we use *infinite dimensional* machines $\mathcal{M}$ over $\mathcal{A}$ that are equipped with an infinite number of registers $Z_1, Z_2, \ldots$ for storing elements of $U_{\mathcal{A}}$, a finite number of registers $I_1, I_2, \ldots, I_{k_{\mathcal{M}}}$ for storing indices in $\mathbb{N}_+$, and an auxiliary register B for using it as instruction counter, the label of the current instruction (see Figure 2). $Z_1, Z_2, \ldots$ are called *Z-registers* and $I_1, I_2, \ldots, I_{k_{\mathcal{M}}}$ are called *index registers*. But sometime we use *finite dimensional* machines $\mathcal{M}$ over $\mathcal{A}$ that are only equipped with a finite number of Z-registers $Z_1, Z_2, \ldots, Z_{n_{\mathcal{M}}}$ for storing the elements of $U_{\mathcal{A}}$ and a register B for the instruction counter. Such machines $\mathcal{M}$ can compute functions from $U_{\mathcal{A}}^n$ into $U_{\mathcal{A}}^m$ for some fixed $n, m \leq n_{\mathcal{M}}$.

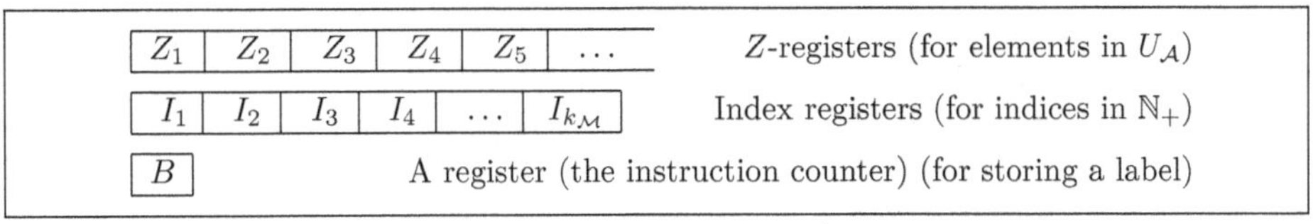

Figure 2: The registers of an $\mathcal{A}$-machine

The index registers are important if a machine uses an infinite number of Z-registers. They are used for storing the length of the input, for determining the length of the output, and in copy instructions. The *infinite dimensional machines* can copy the content of one Z-register to another Z-register by means of indirect addressing where each address that may be used in the copy process must be stored in an index register. The content of an index register can only be changed with the help of several types of index instructions.

The underlying structure $\mathcal{A}$ can be an arbitrary first-order structure as defined e.g. in [2, p. 22]. Let $\mathcal{A} = (U; (c_i)_{i \in N_1}; (f_i)_{i \in N_2}; (r_i)_{i \in N_3})$, where U is a non-empty set of individuals and, for suitable sets N_1, N_2, and N_3, $(c_i)_{i \in N_1}$ is a family of constants $c_i \in U$, $(f_i)_{i\in N_2}$ is a family of operations f_i of arity m_i, and $(r_i)_{i\in N_3}$ is a family of relations $r_i \subseteq U^{k_i}$ of arity k_i. Let $U_\mathcal{A} = U$. The *signature of* $\mathcal{A}$ is $(|N_1|;(m_i)_{i\in N_2};(k_i)_{i\in N_3})$. In general, any operation f_i is an everywhere defined function from $U_\mathcal{A}^{m_i}$ into $U_\mathcal{A}$. But it is also possible to permit partial functions $f_i :\subseteq U_\mathcal{A}^{m_i} \to U_\mathcal{A}$ and the characteristic functions $\chi_{r_i} : D_1 \times \cdots \times D_{k_i} \to \{0,1\}$ of relations $r_i \subseteq D_1 \times \cdots \times D_{k_i}$ with $D_1, \times \cdots \times D_{k_i} \subseteq U_\mathcal{A}^{k_i}$. This could be interesting if we wanted to use certain computable functions (computable e.g. by type-2 Turing machines defined in [29] or BSS machines) as primitive operations or a relation $<_k$ that makes it possible to execute a finite precision test up to an accuracy of $1/(k+1)$ for real numbers and that is important e.g. for the modified real-RAM model defined in [7]. Any structure $\mathcal{B} = (U_\mathcal{A};(c_i)_{i\in L_1};(f_i)_{i\in L_2};(r_i)_{i\in L_3})$ with $L_1 = \{i_1,\ldots,i_{n_1}\} \subseteq N_1$, $L_2 = \{j_1,\ldots,j_{n_2}\} \subseteq N_2$, and $L_3 = \{l_1,\ldots,l_{n_3}\} \subseteq N_3$ (for some $n_1,n_2,n_3 \geq 0$) is a *reduct* of $\mathcal{A}$ and a structure of signature $\sigma =_{\mathrm{df}} (n_1;m_{j_1},\ldots,m_{j_{n_2}};k_{l_1},\ldots,k_{l_{n_3}})$. For such a structure, we use also the notation $(U_\mathcal{A};c_{i_1},\ldots,c_{i_{n_1}};f_{j_1},\ldots,f_{j_{n_2}};r_{l_1},\ldots,r_{l_{n_3}})$. σ is a *finite signature* and a *subsignature* of the signature of $\mathcal{A}$.

Since each program is a finite string that contains only a finite number of symbols, we define a set of programs for every finite signature. Now, let $\sigma = (n_1;m_1,\ldots,m_{n_2};k_1,\ldots,k_{n_3})$ and $n_1,n_2,n_3 \geq 0$. Let P_σ be the set of all σ-programs defined as follows. Every *σ-program* $\mathcal{P}$ has the following form.

$$1 : \mathsf{instruction}_1;\ 2 : \mathsf{instruction}_2; \ldots;\ \ell_\mathcal{P} - 1 : \mathsf{instruction}_{\ell_\mathcal{P}-1};\ \ell_\mathcal{P} : \mathsf{stop}.$$

It is a string that consists of a finite number of substrings, each of which starts with a *label* $i \in \{1,\ldots,\ell_\mathcal{P}\}$ followed by the symbol : and a string denoted by $\mathsf{instruction}_i$ and called *σ-instruction*. Here and in the following, each arithmetic expression before the symbol : such as $\ell_\mathcal{P} - 1$ stands for the integer resulting from the evaluation of this expression. For each $\mathcal{P}$, there is some $l \geq 1$ such that $\mathcal{P}$ results from the application of a syntactic rule to each of the placeholders $\langle instruction\rangle$ in a string of the form

$$1 : \langle instruction\rangle;\ 2 : \langle instruction\rangle;\ \ldots;\ l-1 : \langle instruction\rangle;\ l : \mathsf{stop}. \qquad (*)$$

Let $\ell_\mathcal{P} = l$. We distinguish 8 types of σ-instructions. $\langle instruction\rangle$ can be replaced by an instruction of one of the types (1), ..., (7) given after the labels in Overview 1 where all $j, k, j_1, j_2, \ldots$ are placeholders for positive integers, each ℓ is a placeholder for a label in $\{1,\ldots,l-1\}$, ℓ_1 and ℓ_2 are placeholders for labels in $\{1,\ldots,l\}$, and every i stands for a positive integer that is less than or equal to n_1, n_2, and n_3, respectively. The names of the types are given in parentheses at the end of the lines.

Purely formally, the instructions are only strings and we must distinguish between the symbols in a program as presented in Overview 1 and their interpretation determined by a structure of signature σ. This means that one symbol for a binary operation, f_i^2, can stand for the addition or for the multiplication, and so on. Strictly speaking, we always distinguish between each symbol $f_i^{m_i}$ for an m_i-ary operation in a program and its interpretation, between the symbols $r_i^{k_i}$ and the used k_i-ary relations, and between c_i^0 and the constants themselves.

Overview 1 (σ-instructions).

Computation instructions		
(1)	$\ell : Z_j := f_i^{m_i}(Z_{j_1}, \ldots, Z_{j_{m_i}})$	(F-instructions)
(2)	$\ell : Z_j := c_i^0$	(F_0-instructions)
Copy instructions		
(3)	$\ell : Z_{I_j} := Z_{I_k}$	(C-instructions)
Branching instructions		
(4)	$\ell :$ if $r_i^{k_i}(Z_{j_1}, \ldots, Z_{j_{k_i}})$ then goto ℓ_1 else goto ℓ_2	(T-instructions)
Index instructions		
(5)	$\ell :$ if $I_j = I_k$ then goto ℓ_1 else goto ℓ_2	(H_T-instructions)
(6)	$\ell : I_j := 1$	(H_1-instructions)
(7)	$\ell : I_j := I_j + 1$	(H_{+1}-instructions)
Stop instruction		
(8)	$l :$ stop	(S-instruction)

If a function can be computed by means of a finite number of Z-registers, then *copy instructions* of the form $Z_j := Z_k$ with fixed addresses j and k could be sufficient. The *σ-programs for finite dimensional machines* are restricted to instructions of the form $Z_j := f_i^{m_i}(Z_{j_1}, \ldots, Z_{j_{m_i}})$, $Z_j := c_i^0$, $Z_j := Z_k$, and if $r_i^{k_i}(Z_{j_1}, \ldots, Z_{j_{k_i}})$ then goto ℓ_1 else goto ℓ_2.

2.2 The formal definition of $\mathcal{A}$-machines

For any structure $\mathcal{A}$, a machine $\mathcal{M}$ over $\mathcal{A}$ equipped with $k_{\mathcal{M}}$ index registers should be able to execute a σ-program $\mathcal{P}$ if $\mathcal{A}$ includes a reduct for interpreting all symbols of constants, operations, and relations occurring in $\mathcal{P}$ and if there is a $k \leq k_{\mathcal{M}}$ such that any index after a substring "I" in $\mathcal{P}$ is a number less than or equal to k. Let $k_{\mathcal{P}}$ be the smallest such k. Consequently, we can formally define an $\mathcal{A}$-machine as follows.

Definition 1 (Infinite dimensional $\mathcal{A}$-machine [16]). Let a structure $\mathcal{A}$, any sets I and O, and a signature $\sigma = (n_1; m_1, \ldots, m_{n_2}; k_1, \ldots, k_{n_3})$ be given. A tuple $(U^\omega, (\mathbb{N}_+)^k, \mathcal{L}, \mathcal{P}, \mathcal{B}, \mathrm{In}, \mathrm{Out})$ consisting of

- a space of memory states, U^ω,
- a space of addresses (or indices), $(\mathbb{N}_+)^k$, with $k \geq 1$,
- a set of labels, $\mathcal{L} = \{1, \ldots, l\}$, with $l \geq 1$,
- a program $\mathcal{P} \in \mathsf{P}_\sigma$ with $\ell_\mathcal{P} = l$ and $k_\mathcal{P} \leq k$,
- a structure $\mathcal{B} = (U; c_1, \ldots, c_{n_1}; f_1, \ldots, f_{n_2}; r_1, \ldots, r_{n_3})$ of signature σ,
- an input function $\mathrm{In} : \mathsf{I} \to \{(\vec{\nu} \,.\, \bar{u}) \mid (\vec{\nu}, \bar{u}) \in (\mathbb{N}_+)^k \times U^\omega\}$,
- an output function $\mathrm{Out} : \{(\vec{\nu} \,.\, \bar{u}) \mid (\vec{\nu}, \bar{u}) \in (\mathbb{N}_+)^k \times U^\omega\} \to \mathsf{O}$

is an *infinite dimensional machine over* $\mathcal{A}$ (or $\mathcal{A}$*-machine*) with the machine constants $c_1, \ldots, c_{n_1}$, the input space I, and the output space O if $\mathcal{B}$ is a reduct of $\mathcal{A}$.

Let $\mathsf{IM}_\mathcal{A}$ be the class of all infinite dimensional $\mathcal{A}$-machines. For any machine $\mathcal{M} = (U^\omega, (\mathbb{N}_+)^k, \mathcal{L}, \mathcal{P}, \mathcal{B}, \mathrm{In}, \mathrm{Out})$ with $\mathcal{L} = \{1, \ldots, \ell_\mathcal{P}\}$, let $k_\mathcal{M} = k$, $\mathcal{L}_\mathcal{M} = \mathcal{L}$, $\mathcal{P}_\mathcal{M} = \mathcal{P}$, $\mathcal{B}_\mathcal{M} = \mathcal{B}$, $\mathrm{In}_\mathcal{M} = \mathrm{In}$, and $\mathrm{Out}_\mathcal{M} = \mathrm{Out}$. Moreover, let $\mathsf{I}_\mathcal{M} = \mathsf{I}$, $\mathsf{O}_\mathcal{M} = \mathsf{O}$, and $\vec{c}^{(\mathcal{M})}$ be the tuple $(c_1, \ldots, c_{n_1})$ of the constants of $\mathcal{B}_\mathcal{M}$. Note that — similarly to Turing machines if they are described by tuples in the usual way — $\mathcal{A}$-machines are over defined by $(U^\omega, (\mathbb{N}_+)^k, \mathcal{L}, \mathcal{P}, \mathcal{B}, \mathrm{In}, \mathrm{Out})$ and the inclusion of the first three components and the components $\mathcal{P}$ and $\mathcal{B}$ implies redundancy. However, it could be useful to have all components if further types of machines are taken into consideration.

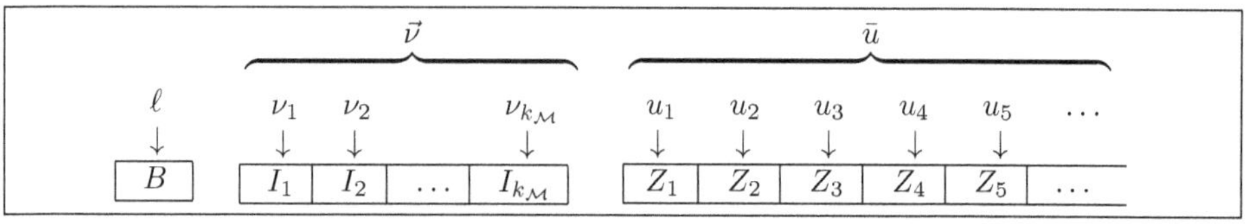

Figure 3: The assignment of values to the registers of an $\mathcal{A}$-machine $\mathcal{M}$

The overall state of $\mathcal{M} \in \mathsf{IM}_\mathcal{A}$ results from the values assigned to the registers as presented in Figure 3. It is given by a configuration and can be changed by executing the program $\mathcal{P}_\mathcal{M}$.

Definition 2 (Configuration). Any possible *configuration* of an $\mathcal{A}$-machine $\mathcal{M}$ is given by a sequence $(\ell \,.\, \vec{\nu} \,.\, \bar{u})$ where ℓ is a label in $\mathcal{L}_\mathcal{M}$ and $\vec{\nu} = (\nu_1, \ldots, \nu_{k_\mathcal{M}}) \in (\mathbb{N}_+)^{k_\mathcal{M}}$ and $\bar{u} \in U_\mathcal{A}^\omega$ hold. If $\ell = 1$, then we call $(\ell \,.\, \vec{\nu} \,.\, \bar{u})$ an *initial configuration*. If $\ell = \ell_\mathcal{P}$, then we call $(\ell \,.\, \vec{\nu} \,.\, \bar{u})$ a *stop configuration*.

For any $\mathcal{M} \in \mathsf{IM}_{\mathcal{A}}$, let $\mathsf{S}_{\mathcal{M}} = \{(\ell\,.\,\vec{\nu}\,.\,\bar{u}) \mid \ell \in \mathcal{L}_{\mathcal{M}}\ \&\ \vec{\nu} \in (\mathbb{N}_+)^{k_{\mathcal{M}}}\ \&\ \bar{u} \in U_{\mathcal{A}}^{\omega}\}$ be the space of all possible configurations of $\mathcal{M}$. For any register R, let the *content of* R be the value stored in R and let it be denoted by $c(R)$. Let $\mathcal{M} \in \mathsf{IM}_{\mathcal{A}}$. In case that, for any Z-register Z_i of $\mathcal{M}$ (with $i \geq 1$), the current content $c(Z_i)$ of Z_i is u_i and, for any register I_j of $\mathcal{M}$ (with $j \leq k_{\mathcal{M}}$), the current content $c(I_j)$ of I_j is ν_j, and the current content $c(B)$ of the register B of $\mathcal{M}$ is ℓ, the configuration $(\ell\,.\,\vec{\nu}\,.\,\bar{u}) \in \mathsf{S}_{\mathcal{M}}$ describes the current state of $\mathcal{M}$. $\mathcal{M}$ should work similarly as a usual finite machine equipped with a finite number of Z-registers, however there are a few things one needs to make clear if one wants to know the complete current state at any given time $t \geq 1$. That is the reason why we now define a transition system $\mathcal{S}_{\mathcal{M}}$ for transforming one configuration in $\mathsf{S}_{\mathcal{M}}$ into the next configuration. For any $\mathcal{M} \in \mathsf{IM}_{\mathcal{A}}$, let $\mathcal{L}_{\mathcal{M}} = \mathcal{L}_{\mathcal{M},\mathrm{F}} \cup \mathcal{L}_{\mathcal{M},\mathrm{F}_0} \cup \mathcal{L}_{\mathcal{M},\mathrm{C}} \cup \mathcal{L}_{\mathcal{M},\mathrm{T}} \cup \mathcal{L}_{\mathcal{M},\mathrm{H}_\mathrm{T}} \cup \mathcal{L}_{\mathcal{M},\mathrm{H}_1} \cup \mathcal{L}_{\mathcal{M},\mathrm{H}_{+1}} \cup \mathcal{L}_{\mathcal{M},\mathrm{S}}$ be the union of the pairwise disjoint sets $\mathcal{L}_{\mathcal{M},\mathrm{F}}, \mathcal{L}_{\mathcal{M},\mathrm{F}_0}, \ldots$ consisting of the labels of all F-instructions in $\mathcal{P}_{\mathcal{M}}$, the labels of all F_0-instructions in $\mathcal{P}_{\mathcal{M}}$, and so on, respectively, and let $\mathcal{L}_{\mathcal{M},\mathrm{S}} = \{\ell_{\mathcal{P}_{\mathcal{M}}}\}$. For determining the semantics of $\mathcal{P}_{\mathcal{M}}$, the following family $\mathcal{F}_{\mathcal{M}}$ of totally defined functions for changing the labels and the values in the registers of an $\mathcal{A}$-machine $\mathcal{M} \in \mathsf{IM}_{\mathcal{A}}$ are introduced. The functions are determined by $k_{\mathcal{M}}$, $\mathcal{P}_{\mathcal{M}}$, and $\mathcal{B}_{\mathcal{M}} = (U_{\mathcal{A}}; c_1, \ldots, c_{n_1}; f_1, \ldots, f_{n_2}; r_1, \ldots, r_{n_3})$. For $\ell \in \mathcal{L}_{\mathcal{M}}$, let $\phi_{\mathcal{M},\ell}$ be a description such that the ℓth instruction of $\mathcal{P}_{\mathcal{M}}$ has exactly the form (including the indices) presented for this type of instructions in Overview 1 if $\phi_{\mathcal{M},\ell}$ holds. For instance, for the instruction $Z_1 := f_2^3(Z_3, Z_1, Z_2)$ labelled by 2, let $\phi_{\mathcal{M},2}$ be the expression $i = 2\ \&\ j = 1\ \&\ j_1 = 3\ \&\ j_2 = 1\ \&\ j_3 = 2$.

Definition 3 (Operations and functions in $\mathcal{F}_{\mathcal{M}}$). For any $\mathcal{M} \in \mathsf{IM}_{\mathcal{A}}$, let the functions in $\mathcal{F}_{\mathcal{M}}$ be defined as follows if $\phi_{\mathcal{M},\ell}$ holds for the considered label ℓ.

- *Elementary operations* $F_\ell : U_{\mathcal{A}}^{\omega} \to U_{\mathcal{A}}^{\omega}$

  $F_\ell(\bar{u}) = (u_1, \ldots, u_{j-1}, f_i(u_{j_1}, \ldots, u_{j_{m_i}}), u_{j+1}, \ldots)$ for $\ell \in \mathcal{L}_{\mathcal{M},\mathrm{F}}$
  $F_\ell(\bar{u}) = (u_1, \ldots, u_{j-1}, c_i, u_{j+1}, \ldots)$ for $\ell \in \mathcal{L}_{\mathcal{M},\mathrm{F}_0}$

- *Copy functions* $C_\ell : (\mathbb{N}_+)^{k_{\mathcal{M}}} \times U_{\mathcal{A}}^{\omega} \to U_{\mathcal{A}}^{\omega}$

  $C_\ell(\vec{\nu}, \bar{u}) = (u_1, \ldots, u_{\nu_j - 1}, u_{\nu_k}, u_{\nu_j + 1}, \ldots)$ for $\ell \in \mathcal{L}_{\mathcal{M},\mathrm{C}}$

- *Test functions* $T_\ell : U_{\mathcal{A}}^{\omega} \to \mathcal{L}_{\mathcal{M}}$ and $T_\ell : (\mathbb{N}_+)^{k_{\mathcal{M}}} \to \mathcal{L}_{\mathcal{M}}$

  $$T_\ell(\bar{u}) = \begin{cases} \ell_1 \text{ if } (u_{j_1}, \ldots, u_{j_{k_i}}) \in r_i \\ \ell_2 \text{ if } (u_{j_1}, \ldots, u_{j_{k_i}}) \notin r_i \end{cases} \quad \text{for } \ell \in \mathcal{L}_{\mathcal{M},\mathrm{T}}$$

  $$T_\ell(\vec{\nu}) = \begin{cases} \ell_1 \text{ if } \nu_j = \nu_k \\ \ell_2 \text{ if } \nu_j \neq \nu_k \end{cases} \quad \text{for } \ell \in \mathcal{L}_{\mathcal{M},\mathrm{H}_\mathrm{T}}$$

- *Auxiliary functions* $H_\ell : (\mathbb{N}_+)^{k_{\mathcal{M}}} \to (\mathbb{N}_+)^{k_{\mathcal{M}}}$

$$(\ell_0 + 1 \,.\, \vec{\nu} \,.\, \underbrace{(u_1, \ldots, u_{j-1}, f_i(u_{j_1}, \ldots, u_{j_{m_i}}), u_{j+1}, \ldots)}_{F_{\ell_0}(\bar{u})}) \quad \text{if } \ell_0 \in \mathcal{L}_{\mathcal{M},\mathrm{F}}$$

$$(\ell_0 + 1 \,.\, \underbrace{(\nu_1, \ldots, \nu_{j-1}, 1, \nu_{j+1}, \ldots, \nu_{k_{\mathcal{M}}})}_{H_{\ell_0}(\vec{\nu})} \,.\, \bar{u}) \quad \text{if } \ell_0 \in \mathcal{L}_{\mathcal{M},\mathrm{H}_1}$$

Figure 4: Examples for a possible successor of the configuration $(\ell_0 \,.\, \vec{\nu} \,.\, \bar{u})$

$$H_\ell(\vec{\nu}) = (\nu_1, \ldots, \nu_{j-1}, 1, \nu_{j+1}, \ldots, \nu_{k_{\mathcal{M}}}) \qquad \text{for } \ell \in \mathcal{L}_{\mathcal{M},\mathrm{H}_1}$$
$$H_\ell(\vec{\nu}) = (\nu_1, \ldots, \nu_{j-1}, \nu_j + 1, \nu_{j+1}, \ldots, \nu_{k_{\mathcal{M}}}) \qquad \text{for } \ell \in \mathcal{L}_{\mathcal{M},\mathrm{H}_{+1}}$$

In order to complete the concept of computation with respect to our model, we define a computational system by means of $\mathcal{F}_{\mathcal{M}}$ for any machine $\mathcal{M} \in \mathsf{IM}_{\mathcal{A}}$.

Definition 4 (Computational system). For any $\mathcal{M} \in \mathsf{IM}_{\mathcal{A}}$, the *computational system* $\mathcal{R}_{\mathcal{M}} = (\mathcal{S}_{\mathcal{M}}, \mathrm{Input}_{\mathcal{M}}, \mathrm{Output}_{\mathcal{M}}, \mathrm{Stop}_{\mathcal{M}})$ is given by

- the *transition system* $\mathcal{S}_{\mathcal{M}} = (\mathsf{S}_{\mathcal{M}}, \rightarrow_{\mathcal{M}})$ defined by

$$\begin{aligned}\rightarrow_{\mathcal{M}} = \; & \{((\ell.\vec{\nu}.\bar{u}), (\ell+1.\vec{\nu}.F_\ell(\bar{u}))) \in \mathsf{S}^2_{\mathcal{M}} \mid \ell \in \mathcal{L}_{\mathcal{M},\mathrm{F}} \cup \mathcal{L}_{\mathcal{M},\mathrm{F}_0}\} \\ \cup & \{((\ell.\vec{\nu}.\bar{u}), (\ell+1.\vec{\nu}.C_\ell(\vec{\nu},\bar{u}))) \in \mathsf{S}^2_{\mathcal{M}} \mid \ell \in \mathcal{L}_{\mathcal{M},\mathrm{C}}\} \\ \cup & \{((\ell.\vec{\nu}.\bar{u}), (T_\ell(\bar{u}).\vec{\nu}.\bar{u})) \in \mathsf{S}^2_{\mathcal{M}} \mid \ell \in \mathcal{L}_{\mathcal{M},\mathrm{T}}\} \\ \cup & \{((\ell.\vec{\nu}.\bar{u}), (T_\ell(\vec{\nu}).\vec{\nu}.\bar{u})) \in \mathsf{S}^2_{\mathcal{M}} \mid \ell \in \mathcal{L}_{\mathcal{M},\mathrm{H_T}}\} \\ \cup & \{((\ell.\vec{\nu}.\bar{u}), (\ell+1.H_\ell(\vec{\nu}).\bar{u})) \in \mathsf{S}^2_{\mathcal{M}} \mid \ell \in \mathcal{L}_{\mathcal{M},\mathrm{H}_1} \cup \mathcal{L}_{\mathcal{M},\mathrm{H}_{+1}}\} \\ \cup & \{((\ell.\vec{\nu}.\bar{u}), (\ell.\vec{\nu}.\bar{u})) \in \mathsf{S}^2_{\mathcal{M}} \mid \ell = \ell_{\mathcal{P}_{\mathcal{M}}}\},\end{aligned}$$

- the *input procedure* $\mathrm{Input}_{\mathcal{M}} : \mathsf{I}_{\mathcal{M}} \to \mathsf{S}_{\mathcal{M}}$ defined by $\mathrm{Input}_{\mathcal{M}}(\mathsf{i}) = (1 \,.\, \mathrm{In}_{\mathcal{M}}(\mathsf{i}))$ for all $\mathsf{i} \in \mathsf{I}_{\mathcal{M}}$,
- the *output procedure* $\mathrm{Output}_{\mathcal{M}} : \mathsf{S}_{\mathcal{M}} \to \mathsf{O}_{\mathcal{M}}$ defined by $\mathrm{Output}_{\mathcal{M}}(\ell \,.\, \vec{\nu} \,.\, \bar{u}) = \mathrm{Out}_{\mathcal{M}}(\vec{\nu} \,.\, \bar{u})$ for all $(\ell \,.\, \vec{\nu} \,.\, \bar{u}) \in \mathsf{S}_{\mathcal{M}}$,
- the *stop criterion* $\mathrm{Stop}_{\mathcal{M}}(\ell \,.\, \vec{\nu} \,.\, \bar{u}) = 0$ satisfied if $\ell = \ell_{\mathcal{P}_{\mathcal{M}}}$.

The relation $\rightarrow_{\mathcal{M}}$ determines the transition rules $\mathsf{conf}_1 \rightarrow_{\mathcal{M}} \mathsf{conf}_2$ that are permitted for all $(\mathsf{conf}_1, \mathsf{conf}_2) \in \mathsf{S}^2_{\mathcal{M}}$ if $(\mathsf{conf}_1, \mathsf{conf}_2) \in \rightarrow_{\mathcal{M}}$. Each instruction causes the change of a configuration by applying a transition rule. For instance, the instruction $Z_{I_j} := Z_{I_k}$ labelled by ℓ_0 implies the transformation of $(\ell_0 \,.\, \vec{\nu} \,.\, \bar{u})$ into $(\ell_0 + 1 \,.\, \vec{\nu} \,.\, (u_1, \ldots, u_{\nu_j - 1}, u_{\nu_k}, u_{\nu_j + 1}, \ldots))$ for all $\bar{u} = (u_1, u_2, \ldots) \in U^\infty_{\mathcal{A}}$ (for further examples of transformations see Figure 4).

For any $\mathcal{M} \in \mathsf{IM}_{\mathcal{A}}$ and each input $\mathsf{i} \in \mathsf{I}_{\mathcal{M}}$, the system $(\mathsf{S}_{\mathcal{M}}, \rightarrow_{\mathcal{M}})$ can be used to generate finite sequences $(\ell_t \,.\, \vec{\nu}^{(t)} \,.\, \bar{u}^{(t)})_{t=1..s}$ of configurations or an infinite sequence of configurations $(\ell_t \,.\, \vec{\nu}^{(t)} \,.\, \bar{u}^{(t)})$ where we have $(\ell_1 \,.\, \vec{\nu}^{(1)} \,.\, \bar{u}^{(1)}) = \mathrm{Input}_{\mathcal{M}}(\mathsf{i})$

and $(\ell_t \,.\, \vec{\nu}^{(t)} \,.\, \bar{u}^{(t)}) \to_{\mathcal{M}} (\ell_{t+1} \,.\, \vec{\nu}^{(t+1)} \,.\, \bar{u}^{(t+1)})$ and $\ell_t \neq \ell_{\mathcal{P}_{\mathcal{M}}}$ hold for all $t < s$ and $t \geq 1$, respectively. We call the longest such sequences *computation paths.* Informally, we say that $\mathcal{M}$ *goes through a computation path* that is given by the sequence of instructions executed in this order by $\mathcal{M}$ and that $\mathcal{M}$ does it *in exactly* t_0 *steps* if this path is a sequence of $t_0 + 1$ instructions. We say that $\mathcal{M}$ *halts* on $\mathsf{i} \in \mathsf{I}_{\mathcal{M}}$ if the corresponding maximal sequence contains the S-instruction labelled by $\ell_{\mathcal{P}_{\mathcal{M}}}$. Only in this case, the machine outputs an element of $\mathsf{O}_{\mathcal{M}}$.

Since $\to_{\mathcal{M}}$ is a single-valued mapping, it is possible to write $(\to_{\mathcal{M}})(\mathsf{conf}_1) = \mathsf{conf}_2$ and $(\to_{\mathcal{M}})^1(\mathsf{conf}_1) = \mathsf{conf}_2$ instead of $\mathsf{conf}_1 \to_{\mathcal{M}} \mathsf{conf}_2$ and to use the composition $(\to_{\mathcal{M}})^{t+1}$ defined by $(\to_{\mathcal{M}})^{t+1}(\mathsf{conf}) = (\to_{\mathcal{M}})((\to_{\mathcal{M}})^t(\mathsf{conf}))$ for all $\mathsf{conf} \in \mathsf{S}_{\mathcal{M}}$ and $t \geq 1$. Moreover, let the function $(\to_{\mathcal{M}})_{\mathsf{Stop}_{\mathcal{M}}} :\subseteq \mathsf{S}_{\mathcal{M}} \to \mathsf{S}_{\mathcal{M}}$ be given by $(\to_{\mathcal{M}})_{\mathsf{Stop}_{\mathcal{M}}}(\mathsf{conf}) = (\to_{\mathcal{M}})^{t_0}(\mathsf{conf})$ if $h_{\mathsf{conf}} = \{t \mid \mathrm{Stop}_{\mathcal{M}}((\to_{\mathcal{M}})^t(\mathsf{conf})) = 0\}$ is not empty and $t_0 = \min h_{\mathsf{conf}}$ holds and otherwise let $(\to_{\mathcal{M}})_{\mathsf{Stop}_{\mathcal{M}}}(\mathsf{conf})$ be undefined. Thus, for every finite computation path $(\ell_t \,.\, \vec{\nu}^{(t)} \,.\, \bar{u}^{(t)})_{t=1..s}$, the stop configuration $(\ell_s \,.\, \vec{\nu}^{(s)} \,.\, \bar{u}^{(s)})$ is given by $(\to_{\mathcal{M}})_{\mathsf{Stop}_{\mathcal{M}}}(\ell_1 \,.\, \vec{\nu}^{(1)} \,.\, \bar{u}^{(1)}) = (\ell_s \,.\, \vec{\nu}^{(s)} \,.\, \bar{u}^{(s)})$.

Definition 5 (Result function). For every $\mathcal{A}$-machine $\mathcal{M}$, the partial function $\mathrm{Res}_{\mathcal{M}} :\subseteq \mathsf{I}_{\mathcal{M}} \to \mathsf{O}_{\mathcal{M}}$, defined by

$$\mathrm{Res}_{\mathcal{M}}(\mathsf{i}) = \mathrm{Output}_{\mathcal{M}} \circ (\to_{\mathcal{M}})_{\mathsf{Stop}_{\mathcal{M}}} \circ \mathrm{Input}_{\mathcal{M}}(\mathsf{i})$$

for all $\mathsf{i} \in \mathsf{I}_{\mathcal{M}}$, is called *the result function of* $\mathcal{R}_{\mathcal{M}}$ and *computable over* $\mathcal{A}$.

We say that $\mathcal{M}$ *computes* $\mathrm{Res}_{\mathcal{M}}$ and that $\mathcal{M}$ *executes the program* $\mathcal{P}_{\mathcal{M}}$ *on input* i in order to compute $\mathrm{Res}_{\mathcal{M}}(\mathsf{i})$. For any $\mathcal{A}$-machine $\mathcal{M}$, the input function $\mathrm{In}_{\mathcal{M}}$ and the output function $\mathrm{Out}_{\mathcal{M}}$ can be defined in several ways where the input space and the output space can, for instance, be $U_{\mathcal{A}}^n$ for some $n \geq 1$, $U_{\mathcal{A}}^{\infty}$, $(U_{\mathcal{A}}^{\omega,u})_{\mathrm{fin}} =_{\mathrm{df}} \{(u_1, u_2, \ldots) \in U_{\mathcal{A}}^{\omega} \mid (\exists j \in \mathbb{N}_+)(u = u_j = u_{j+1} = \cdots)\}$ for some $u \in U_{\mathcal{A}}$, and $U_{\mathcal{A}}^{\omega}$, respectively, or the result of a combination of these or other sets. The following machine with the input space $\mathbb{R}^{\infty}$ is a variant of the BSS machine over $\mathbb{R}^{\geq 0} =_{\mathrm{df}} (\mathbb{R}; \mathbb{R}; +, -, \cdot, /; \{r \mid r \geq 0\})$ that essentially corresponds to the real Turing machines described in [12, pp. 455–456]. As described in [11, Remark 1, p. 367], we say that $r/0$ is here defined for all real numbers r and a machine returns zero if the computation of $r/0$ is necessary.

Example 6 (BSS $\mathbb{R}^{\geq 0}$-machine). Let $\mathcal{P}$ be an $(n_1; 2, 2, 2, 2; 1)$-program each of whose instructions can be an F-instruction, an F_0-instruction, an index instruction of the type (6) or (7) for the two index register I_1 and I_2, the C-instruction $Z_{I_2} := Z_{I_1}$, or a T-instruction with the condition $r_1^1(Z_1)$ for evaluating $c(Z_1) \geq 0$. Let $\mathcal{B} = (\mathbb{R}; c_1, \ldots, c_{n_1}; +, -, \cdot, /; \{r \mid r \geq 0\})$ for some $c_1, \ldots, c_{n_1} \in \mathbb{R}$. Then, the following tuple is a BSS $\mathbb{R}^{\geq 0}$-machine.

$\mathcal{M}^{\mathrm{BSS}}_{\mathbb{R}^{\geq 0}} = (\mathbb{R}^{\omega}, (\mathrm{N}_{+})^2, \mathcal{L}, \mathcal{P}, \mathcal{B}, \mathrm{In}, \mathrm{Out}),$	$(x_1, \ldots, x_n) \in \mathbb{R}^{\infty}$
$\mathrm{In}(x_1, \ldots, x_n) = (1, 1, x_1, n, x_2, 0, x_3, \ldots, 0, x_n, 0, 0, \ldots)$	
$\mathrm{Out}(\nu_1, \nu_2, u_1, u_2, u_3, \ldots) = (u_1, u_2, u_3, \ldots)$	

Note, that the usual division can be computed by a BSS $\mathbb{R}^{\geq 0}$-machine $\mathcal{M}$ with the following program and constants given by $\vec{c}_{\mathcal{M}} = (-1, \ldots)$. 1 : if $r_1^1(Z_2)$ then goto 2 else goto 5; $2 : Z_3 := c_1^0$; $3 : Z_3 := f_3^2(Z_3, Z_2)$; 4 : if $r_1^1(Z_3)$ then goto 4 else goto 5; $5 : Z_3 := f_4^2(Z_1, Z_2)$; $6 : Z_1 := Z_3$; $7 : I_1 := 1$; 8 : stop.

By [3], every BSS machine is a graph such that, for any input, the computational process of this machine can be represented by a computation path as usual in flow charts where the nodes correspond to the labelled instructions of some BSS $\mathbb{R}^{\geq 0}$-machine.

2.3 Subprograms and pseudo instructions

Here, any *σ-subprogram* is a substring of a σ-program (a macro) that has the form instruction$_{\ell_1}$; ℓ_1+1 : instruction$_{\ell_1+1}$; $\ldots$; ℓ_2 : instruction$_{\ell_2}$ where every instruction$_\ell$ ($\ell_1 \leq \ell \leq \ell_2$) denotes a σ-instruction. Thus, any $\mathcal{P} \in \mathsf{P}_\sigma$ can be represented by certain strings of the form

$$1 : \mathsf{subprogram}_1;\ \ell_2 : \mathsf{subprogram}_2; \ldots;\ \ell_s : \mathsf{subprogram}_s;\ \ell_{\mathcal{P}} : \mathsf{stop}.$$

with strings subprogram$_m$ ($1 \leq m \leq s$) denoting σ-subprograms and resulting from replacing the placeholder ⟨*subprogram*⟩ with the help of syntactic rules given in Table 1 (for suitable $\ell \geq 1$). Any pair (A_1, A_2) given in one row of Table 1 means that the placeholder A_1 given in the first column can be replaced by the string A_2 given in the second column by the application of the rule $A_1 \to A_2$. In the replacing process, all placeholders marked by angle brackets ⟨ and ⟩ must be replaced step-by-step. Accordingly, all placeholders ⟨*pseudo_instr*⟩ and ⟨*condition*⟩ may also be replaced by new pseudo instructions and new conditions condition$_j$, respectively, introduced later. In this way we get new strings for denoting subprograms. For example, any resulting string "ℓ_1 : if condition$_j$ then subprogram$_m$; ℓ_2 :" corresponds to "ℓ_1 : if condition$_j$ then goto ℓ_1+1 else goto ℓ_2; ℓ_1+1 : subprogram$_m$; ℓ_2 :" where subprogram$_m$ can stand, e.g., for "instruction$_{\ell_1+1}$; ℓ_1+2 : instruction$_{\ell_1+2}$; $\ldots$; ℓ_1+s : instruction$_{\ell_1+s}$" if, for the arithmetic expressions, $\ell_2 = \ell_1+s+1$ holds. In the following, some of the labels before an instruction that are not used in branching instructions (and the symbol : after these labels) are omitted. Summarizing, we can say that an application of a syntactic rule is only permitted if the resulting string does not contain two (pseudo) instructions with the same label and the result

A_1	A_2
$\langle condition \rangle$	$I_j = I_k$
$\langle condition \rangle$	$r_i^{k_i}(Z_{j_1}, \ldots, Z_{j_{k_i}})$
$\langle condition \rangle$	$\langle condition \rangle$ & $\langle condition \rangle$
$\langle condition \rangle$	$\langle condition \rangle$ or $\langle condition \rangle$
$\langle instructions \rangle$	$\langle instruction \rangle$
$\langle instructions \rangle$	$\langle pseudo\_instr \rangle$
$\langle instructions \rangle$	$\langle instructions \rangle$; ℓ: $\langle instructions \rangle$
$\langle pseudo\_instr \rangle$	goto ℓ
$\langle pseudo\_instr \rangle$	if $\langle condition \rangle$ then $\langle subprogram \rangle$
$\langle pseudo\_instr \rangle$	if $\langle condition \rangle$ then $\langle subprogram \rangle$ else $\langle subprogram \rangle$
$\langle subprogram \rangle$	$\langle instruction \rangle$
$\langle subprogram \rangle$	$\{\langle instructions \rangle\}$

Table 1: Syntactic rules $A_1 \to A_2$

of all applications is — after a possible renumbering of all instructions and the corresponding renaming of the labels — a σ-program. The braces can be omitted if no ambiguity is possible. Therefore, the braces in $\{\langle instructions \rangle\}$ are omitted if $\langle instructions \rangle$ is replaced by a simple pseudo instruction without an if part (such as some of the pseudo instructions given e.g. in Overview 2).

2.4 Meaning of the space of indices and properties

The BSS machines introduced in [3] work with only two index registers and they have been defined for a large class of commutative rings that includes the ring over the real numbers and integers, respectively. Our generalisation of this kind of machines leads to the concept of $\mathcal{A}$-machines. Accordingly, all Z-registers have an address such that their contents can definitely be read by means of index registers. But, we clearly distinguish between the values used as addresses of storage locations and the elements of the underlying structure $\mathcal{A}$. We want to have the possibility to store some additional information such as the length of the input regardless of whether the encoding of the information by the inputted individuals or other elements of the structure is possible. Therefore, the space of addresses contains, in general, tuples of more than two components. The addresses belong to a second structure, to the Peano structure $\mathcal{A}_{\mathbb{N}} = (\mathbb{N}_+; 1; succ; =)$ where $succ(n) = n + 1$ for all $n \geq 1$. Consequently, any $\mathcal{A}$-machine is able to use the functions and the relations of both structures $\mathcal{A}$ and $\mathcal{A}_{\mathbb{N}}$ and to compute several arithmetic functions $f : \mathbb{N}_+^s \to \mathbb{N}_+$ by manipulating indices. With respect to processing indices, we allow only the interpretation of the symbols as usual in arithmetic. Therefore, we do not distinguish between the arithmetic functions and relations and their symbols in arithmetic expressions occurring in pseudo instructions and comparisons if they refer to indices or index registers. Hence, the semantics is easy to understand without explaining all details and the meaning of each single pseudo instruction. Accordingly, we use further pseudo instructions and the like and further conditions A_2 in syntactic rules of the form $\langle condition \rangle \to A_2$. In Overview 2, let $\nu, \mu \geq 1$ or, in the third line, let $\mu - \nu > 0$ and $\mu \geq \nu \geq 0$, respectively, and, in the fourth line, let $\nu + \mu \geq 1$. Any expression $t + 0$ may be replaced by t.

For $\nu = 1$, the third pseudo instruction can be realised as follows where K_1 and K_2 are new index registers. $\mathsf{if}\ I_j > 1\ \mathsf{then\ goto}\ \ell_1\ \mathsf{else\ goto}\ \ell_4;\ \ell_1 : K_1 := 1;$ $K_2 := 1;\ \ell_2 : K_1 := K_1 + 1;\ \mathsf{if}\ K_1 = I_j\ \mathsf{then\ goto}\ \ell_3\ \mathsf{else}\ K_2 := K_2 + 1;\ \mathsf{goto}\ \ell_2;$ $\ell_3 : I_j := K_2;\ \ell_4 : \ldots$ The fourth last pseudo instruction corresponds to $I_m := I_j + \nu;$ $\mathsf{if}\ I_k > I_m\ \mathsf{then}\ \{Z_{I_m} := Z_{I_k};\ K_1 := 1;\ \mathsf{goto}\ \ell_1\}\ \mathsf{else}\ \{K_1 := \mu - \nu;\ \mathsf{goto}\ \ell_2\};\ \ell_1 : \mathsf{if}$ $K_1 \leq \mu - \nu\ \mathsf{then}\ \{Z_{I_m+K_1} := Z_{I_k+K_1};\ K_1 := K_1 + 1;\ \mathsf{goto}\ \ell_1\}\ \mathsf{else\ goto}\ \ell_3;\ \ell_2 : \mathsf{if}$ $K_1 \geq 1\ \mathsf{then}\ \{Z_{I_m+K_1} := Z_{I_k+K_1};\ K_1 := K_1 - 1;\ \mathsf{goto}\ \ell_2\};\ Z_{I_m} := Z_{I_k};\ \ell_3 : \ldots.$

Overview 2 (Pseudo instructions).

$I_j := \nu, \quad I_j := I_j + \nu, \quad \mathsf{if}\ I_j > \nu\ \mathsf{then}\ I_j := I_j - \nu$

$I_j := I_k + \nu, \quad I_j := I_k + I_m, \quad Z_{I_j} := Z_{I_k+I_m}, \quad Z_{I_j+I_k} := Z_{I_l+I_m}$

$I_j := \mu - \nu, \quad (Z_{I_j+\nu}, \ldots, Z_{I_j+\mu}) := (Z_{I_k}, \ldots, Z_{I_k+\mu-\nu})$

$(Z_{I_j+\nu}, \ldots, Z_{I_k+\nu}) := (Z_{I_j+\mu}, \ldots, Z_{I_k+\mu})$

(permitted if $c(I_j) \leq c(I_k)$ is ensured)

$(Z_1, Z_{\mu+1}, \ldots, Z_{\nu\cdot\mu+1}) := (Z_1, Z_2, \ldots, Z_{\nu+1}), \quad (Z_{I_j}, Z_{I_j+1}) := (c_{i_1}, c_{i_2})$

Further pseudo conditions: $(Z_{I_j}, Z_{I_j+1}) = (c_{i_1}, c_{i_2})$

$I_j = \nu,\ I_j > \nu,\ I_j \geq \nu,\ I_j \leq \mu - \nu,\ I_j \geq I_k,\ I_j = I_k + \nu,\ I_j > I_k + \nu$

In any program, every Z_i is a variable to which values can be assigned. However, informally we can associate $Z_1, Z_2, \ldots$ with places for storing objects. These places are boxes, cells or so-called registers. Since each of the Z-registers has a number (an address), this corresponds to the idea that the Z-registers form a tape infinite to the right and that any value on the tape can be read by a head whose position is stored in one index register.

Let $\mathcal{A}_0 = (\{0,1\}; 0, 1; ; =)$. Then, any simple 1-tape Turing machine M (for the definition see, e.g., [8, p. 159 in the German version]) computing a function $f :\subseteq \{0,1\}^* \to \{0,1\}^*$ corresponds to the following Turing $\mathcal{A}_0$-machine $\mathcal{M}^{\mathrm{T}}(M)$ with the input space $\mathsf{I}_{\mathcal{M}^{\mathrm{T}}(M)} = \{0,1\}^*$. Since every positive integer has a unique binary representation, it is also possible that M computes a function $f :\subseteq \mathbb{N}_+ \to \mathbb{N}_+$. For this purpose, we define a Turing $\mathcal{A}_0$-machine $\mathcal{M}^{\mathrm{T}}_{\mathbb{N}}(M)$ with $\mathsf{I}_{\mathcal{M}^{\mathrm{T}}_{\mathbb{N}}(M)} = \mathbb{N}_+$ and new input and output functions, $\mathrm{In}_{\mathcal{M}^{\mathrm{T}}_{\mathbb{N}}(M)}$ and $\mathrm{Out}_{\mathcal{M}^{\mathrm{T}}_{\mathbb{N}}(M)}$, derived from $\mathrm{In}_{\mathcal{M}^{\mathrm{T}}(M)}$ and $\mathrm{Out}_{\mathcal{M}^{\mathrm{T}}(M)}$.

Example 7 (Turing $\mathcal{A}_0$-machines). Let M work with the symbols $0, 1, \square$ and a finite number of states in $Q = \{q_1, \ldots, q_{i_0}\}$ for some $i_0 \geq 1$. Here, the symbols 0 and 1 are encoded by $(0,0)$ and $(0,1)$, respectively, and the symbol $\square$ by $(1,1)$. The content of the relevant cells $\tau_{-s_1}, \ldots, \tau_{s_2}$ of M (that do not contain the symbol $\square$) is stored in the Z-registers $Z_1, \ldots, Z_{2(s_1+s_2+1)}$, the head position s is given by the index register I_2 such that $c(I_2) = 2(s_1 + s + 1) - 1$. $c(I_1) = 2(s_1 + s_2 + 1)$ is the address of the last relevant Z-register whereas the first relevant cell τ_{-s_1} corresponds to Z_1 and Z_2. Any state $q_i \in Q$ is represented by one label $\ell_i \in \mathcal{L}_{\mathcal{M}^{\mathrm{T}}(M)}$. The program $\mathcal{P} = \mathcal{P}_{\mathcal{M}^{\mathrm{T}}(M)}$ contains, for any label ℓ_i ($i \leq i_0$), a pseudo instruction of the following form with $(c_{i_1}, c_{i_2}) \in \{(0,1), (0,1), (1,1)\}$ and $j \leq i_0$.

$$\ell_i : \textsf{if } (Z_{I_2}, Z_{I_2+1}) = (c_{i_1}, c_{i_2}) \textsf{ then } \{\textsf{subprogram}_{i,i_1,i_2};\ \textsf{goto } \ell_j\}$$

The Turing machine M can only change the content of a cell τ_s or make a move to a neighbour cell τ_{s-1} or τ_{s+1} in dependence on the current state and the content of τ_s. Hence, any subprogram $\textsf{subprogram}_{i,i_1,i_2}$ is one of the following pseudo instructions. (a) means to write the symbol encoded by (c_{i_1}, c_{i_2}). (b) means to go to the right. (c) means to go to the left.

(a)	$(Z_{I_2}, Z_{I_2+1}) := (c_{i_1}, c_{i_2})$
(b)	$\textsf{if } I_1 = I_2 + 1 \textsf{ then } I_1 := I_1 + 2;\ I_2 := I_2 + 2$
(c)	$\textsf{if } I_2 > 2 \textsf{ then } I_2 := I_2 - 2 \textsf{ else } \{I_1 := I_1 + 2;$ $(Z_{I_2+2}, \ldots, Z_{I_1+2}) := (Z_{I_2}, \ldots, Z_{I_1});\ (Z_{I_2}, Z_{I_2+1}) := (1,1)\}$

Without loss of generality, let Q contain only one final state. Therefore, let the final state be represented by $\ell_{\mathcal{P}} \in \mathcal{L}_{\mathcal{M}^{\mathrm{T}}(M)}$. Formally, we describe $\mathcal{M}^{\mathrm{T}}(M)$ as follows. $\mathcal{L}$ contains all ℓ_i for $i \leq i_0$ and the labels used in the subprograms. $k > 2$ is also dependent on the subprograms. For each stop configuration, we define two integers. Let λ_0 be defined by $\lambda_0 = \min\{\nu \mid (2\nu - 1 \geq \nu_2 \ \&\ (u_{2\nu-1}, u_{2\nu}) \neq (1,1)) \text{ or } \nu = \frac{\nu_1}{2}\}$ and let λ_1 be defined by $\lambda_1 = \min\{\nu \mid \nu \geq \lambda_0 \ \&\ (u_{2\nu+1}, u_{2\nu+2}) = (1,1)\}$. Then, λ_0 and λ_1 give, if possible, the positions for an output without the symbol $\square$.

$\mathcal{M}^{\mathrm{T}}(M) = (\{0,1\}^\omega, (\mathbb{N}_+)^k, \mathcal{L}, \mathcal{P}, \mathcal{A}_0, \mathrm{In}_*, \mathrm{Out}_*),$	$x_1 \cdots x_n \in \{0,1\}^*$
$\mathrm{In}_*(x_1 \cdots x_n) = ((2n, 1, \ldots, 1) \,.\, (0, x_1, 0, x_2, \ldots, 0, x_n, 1, 1, 1, 1, \ldots))$	
$\mathrm{Out}_*(\nu_1, \ldots, \nu_k, u_1, u_2, u_3, \ldots) = \Lambda$	if $u_{2\lambda_0 - 1} = 1$
$\mathrm{Out}_*(\nu_1, \ldots, \nu_k, u_1, u_2, u_3, \ldots) = u_{2\lambda_0} u_{2\lambda_0+2} \cdots u_{2\lambda_1}$	otherwise

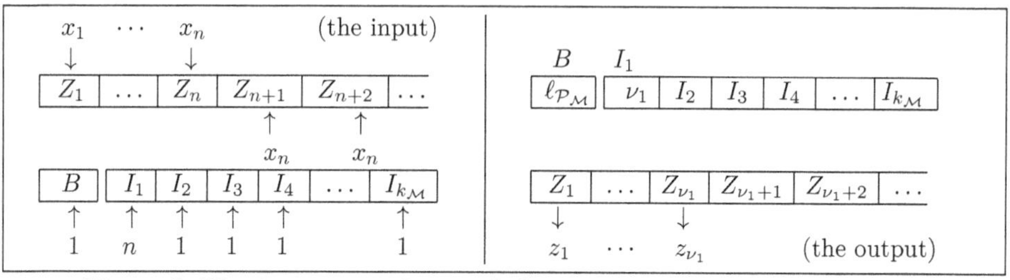

Figure 5: The input procedure and the output procedure for $\mathcal{M} \in \mathsf{M}_{\mathcal{A}}$

Let bin : $\mathbb{N}_+ \to \{0,1\}^*$ be defined by $\mathrm{bin}(\sum_{i=1}^{n} x_i \cdot 2^{i-1}) = x_n \cdots x_1$ for all $\vec{x} \in \{0,1\}^\infty$ with $x_n = 1$ and let bin^{-1} be the partial inverse of bin.

$\mathcal{M}^{\mathrm{T}}_{\mathbb{N}}(M) = (\{0,1\}^\omega, (\mathbb{N}_+)^k, \mathcal{L}, \mathcal{P}, \mathcal{A}_0, \mathrm{In}_{\mathbb{N}}, \mathrm{Out}_{\mathbb{N}})$,	$m \in \mathbb{N}_+$
$\mathrm{In}_{\mathbb{N}}(m) = \mathrm{In}_* \circ \mathrm{bin}(m)$	
$\mathrm{Out}_{\mathbb{N}}(\nu_1, \ldots, \nu_k, u_1, u_2, u_3, \ldots) = \mathrm{bin}^{-1} \circ \mathrm{Out}_*(\nu_1, \ldots, \nu_k, u_1, u_2, u_3, \ldots)$	

2.5 BSS RAM's over $\mathcal{A}$

One of the main characteristics of the infinite dimensional $\mathcal{A}$-machines is that any $\mathcal{M} \in \mathsf{IM}_{\mathcal{A}}$ has its own program. This means that the code of the program $\mathcal{P}_{\mathcal{M}}$ does not need to be part of the input of $\mathcal{M}$ and it is not necessary to have a universal algorithm for executing a program whose code is only given as input. Moreover, $\mathcal{A}$-machines have features suitable for modelling the object-oriented programming where provided data structures, the so-called classes, including the operations for processing the data, the so-called methods, can be used without knowing the implementation of the methods. Besides these properties, a further key feature of BSS RAM's over $\mathcal{A}$ should be the ability to process all tuples of elements of $U_{\mathcal{A}}$ uniformly.

Definition 8 (BSS RAM). For any structure $\mathcal{A}$, any $\mathcal{M} \in \mathsf{IM}_{\mathcal{A}}$ with the spaces $\mathsf{I}_{\mathcal{M}} = \mathsf{O}_{\mathcal{M}} = U_{\mathcal{A}}^\infty$ is called *BSS RAM over* $\mathcal{A}$ if $\mathrm{In}_{\mathcal{M}}$ and $\mathrm{Out}_{\mathcal{M}}$ are defined for all $(x_1, \ldots, x_n) \in U_{\mathcal{A}}^\infty$ and $((\nu_1, \ldots, \nu_{k_{\mathcal{M}}}), (u_1, u_2, \ldots)) \in \mathbb{N}_+^{k_{\mathcal{M}}} \times U_{\mathcal{A}}^\omega$ as follows (see also Figure 5).

$$\mathrm{In}_{\mathcal{M}}(x_1, \ldots, x_n) = (\underbrace{n, 1, \ldots, 1}_{k_{\mathcal{M}} \text{ indices}}, x_1, \ldots, x_n, x_n, x_n, \ldots)$$

$$\mathrm{Out}_{\mathcal{M}}(\nu_1, \ldots, \nu_{k_{\mathcal{M}}}, u_1, u_2, u_3, \ldots) = (u_1, \ldots, u_{\nu_1})$$

Let $\mathsf{M}_{\mathcal{A}}$ be the class of all BSS RAM's over $\mathcal{A}$. Consequently, for any $\mathcal{M} \in \mathsf{M}_{\mathcal{A}}$ and each input $(x_1, \ldots, x_n) \in U_{\mathcal{A}}^\infty$, the input procedure $\mathrm{Input}_{\mathcal{M}}$ of $\mathcal{R}_{\mathcal{M}}$ provides the initial configuration $(1 . (n, 1, \ldots, 1) . (x_1, \ldots, x_n, x_n, x_n, \ldots))$. Here, to simplify matters, the registers $Z_{n+1}, Z_{n+2}, \ldots$ get the value x_n. Therefore, for the input $(x_1, \ldots, x_n)$, $\mathcal{M}$ uses only the subspace $(U_{\mathcal{A}}^{\omega, x_n})_{\text{fin}}$ of the space $U_{\mathcal{A}}^{\omega}$. Note that the concept could also be changed such that the initialisation of Z_j for $j > n$ would be done during the computation before Z_j should be used. Moreover, a constant c_i could be assigned to $Z_{n+1}, Z_{n+2}, \ldots$ if $\mathcal{A}$ is a structure with a constant and c_i is one of the constants of $\mathcal{A}$. If the machine halts, then the values $c(Z_1), \ldots, c(Z_{c(I_1)})$ are outputted. The following example helps to understand that there are differences between the BSS RAM's and the real RAM's working with a read instruction and a write instruction. It remains unclear how the following functions could be computed by means of real RAM's of the latter form.

Example 9 (Functions computable over $\mathbb{R}^{\geq 0}$). Since the register I_1 of any BSS RAM contains the length of any input at the beginning, for $i \in \{1, 2, 3\}$, there are BSS RAM's $\mathcal{M}_i$ over $\mathbb{R}^{\geq 0}$ computing the functions $f_i : \mathbb{R}^{\infty} \to \mathbb{R}^{\infty}$ everywhere defined by $f_1(x_1, \ldots, x_n) = \sum_{i=1}^{n} x_i$, $f_2(x_1, \ldots, x_n) = x_n$, and $f_3(x_1, \ldots, x_n) = n$, respectively, for any $n \geq 1$ and all $(x_1, \ldots, x_n) \in \mathbb{R}^{\infty}$. More precisely, we have $\mathrm{Res}_{\mathcal{M}_i} = f_i$ for each $i \in \{1, 2, 3\}$.

2.6 Multi-tape machines and the multi-tape mode

Now, we consider machines that work with a finite number of tapes. For any $d \geq 1$, a *d-tape $\mathcal{A}$-machine* is a tuple $((U_{\mathcal{A}}^{\omega})^d, \mathbb{N}_+^{\kappa_1} \times \cdots \times \mathbb{N}_+^{\kappa_d}, \mathcal{L}, \mathcal{P}, \mathcal{B}, \mathrm{In}, \mathrm{Out})$ that can be defined analogously to the usual $\mathcal{A}$-machines in $\mathsf{IM}_{\mathcal{A}}$. It is equipped, for all $j \leq d$, with the Z-registers $Z_{j,1}, Z_{j,2}, \ldots$ forming the jth tape and a finite number of index registers $I_{j,1}, \ldots, I_{j,\kappa_j}$. For $\ell \in \mathcal{L}$, $(\vec{\nu}^{(j)})_{j=1..d} \in \mathbb{N}_+^{\kappa_1} \times \cdots \times \mathbb{N}_+^{\kappa_d}$, and $(\bar{u}^{(j)})_{j=1..d} \in (U_{\mathcal{A}}^{\omega})^d$, the tuple $(\ell, (\vec{\nu}^{(j)} . \bar{u}^{(j)})_{j=1..d})$ is a *configuration of this machine*. The overall state of such a machine $\mathcal{M}$ is described by $(\ell, (\vec{\nu}^{(j)} . \bar{u}^{(j)})_{j=1..d})$ if ℓ is the label $c(B)$ stored in the register B of $\mathcal{M}$, every component $\nu_{j,k}$ of $\vec{\nu}^{(j)}$ is the content $c(I_{j,k})$ of the index register $I_{j,k}$ of $\mathcal{M}$, and the component $u_{j,k}$ of $\bar{u}^{(j)}$ is the content $c(Z_{j,k})$ of the Z-register $Z_{j,k}$ of $\mathcal{M}$. Let $\mathsf{IM}_{\mathcal{A}}^{(d)}$ be the class of all d-tape $\mathcal{A}$-machines. For signature $\sigma = (n_1; m_1, \ldots, m_{n_2}; k_1, \ldots, k_{n_3})$, any *d-tape σ-program* results from replacing each placeholder $\langle instruction \rangle$ in a string of the form (*) by one of the *d-tape σ-instructions* whose form is given in Overview 3 where i is a positive integer less than or equal to n_1, n_2, and n_3, respectively, $d_0, d_1, d_2, \ldots$ stand for positive integers $\leq d$, and all $j, k, j_1, j_2, \ldots$ are placeholders for positive integers. Let $\mathsf{P}_{\sigma}^{(d)}$ be the set of all d-tape σ-programs.

Overview 3 (d-tape σ-instructions).

$Z_{d_0,j} := f_i^{m_i}(Z_{d_1,j_1},\ldots,Z_{d_{m_i},j_{m_i}})$,	$Z_{d_0,j} := c_i^0$,	$Z_{d_1,I_{d_1,j}} := Z_{d_2,I_{d_2,k}}$
if $r_i^{k_i}(Z_{d_1,j_1},\ldots,Z_{d_{k_i},j_{k_i}})$ then goto ℓ_1 else goto ℓ_2		
if $I_{d_1,j} = I_{d_2,k}$ then goto ℓ_1 else goto ℓ_2,	$I_{d_1,j} := 1$,	$I_{d_1,j} := I_{d_1,j} + 1$

Each execution of an instruction in $\mathcal{P}_\mathcal{M} \in \mathsf{P}_\sigma^{(d)}$ by a machine $\mathcal{M} \in \mathsf{IM}_\mathcal{A}^{(d)}$ can cause the change of the current configuration. For instance, $\ell_0 : Z_{1,I_{1,j}} := Z_{2,I_{2,k}}$ means $(\ell_0, (\vec{\nu}^{(j)} . \bar{u}^{(j)})_{j=1..d}) \to_\mathcal{M} (\ell_0 + 1, (\vec{\nu}^{(j)} . \bar{w}^{(j)})_{j=1..d})$ where $\bar{w}^{(j)} = \bar{u}^{(j)}$ for $j \in \{2,\ldots,d\}$ and $\bar{w}^{(1)} = (u_{1,1},\ldots,u_{1,\nu_{1,j}-1}, u_{2,\nu_{2,k}}, u_{1,\nu_{1,j}+1}, u_{1,\nu_{1,j}+2},\ldots)$. Accordingly, let the result function be defined.

A machine $\mathcal{M} \in \mathsf{IM}_\mathcal{A}^{(d)}$ is a *d-tape BSS RAM* if, for all inputs $\vec{x} = (x_1,\ldots,x_n)$, the values $\mathrm{In}_\mathcal{M}(\vec{x}) = (\vec{\nu}^{(j)} . \bar{u}^{(j)})_{j=1..d}$ are determined by $\vec{\nu}^{(1)} = (n,1,\ldots,1)$ and $\bar{u}^{(1)} = (x_1,\ldots,x_n,x_n,\ldots)$ and by $\vec{\nu}^{(j)} = (1,\ldots,1)$ and $\bar{u}^{(j)} = (x_n,x_n,\ldots)$ for all $j \in \{2,\ldots,d\}$ and $\mathrm{Out}_\mathcal{M}$ is defined by $\mathrm{Out}_\mathcal{M}((\vec{\nu}^{(j)} . \bar{u}^{(j)})_{j=1..d}) = (u_{1,1},\ldots,u_{1,\nu_{1,1}})$. Let $\mathsf{M}_\mathcal{A}^{(d)}$ be the class of all d-tape BSS RAM's.

Since the Z-registers of the first tape of a machine in $\mathsf{M}_\mathcal{A}^{(d)}$ get the input values and provide the output in the same way as the registers of a machine in $\mathsf{M}_\mathcal{A}$, we have $\mathsf{M}_\mathcal{A}^{(1)} = \mathsf{M}_\mathcal{A}$. On the other hand, any program of a machine $\mathcal{M} \in \mathsf{IM}_\mathcal{A}^{(d)}$ can be simulated by a machine $\mathcal{N}_\mathcal{M} \in \mathsf{IM}_\mathcal{A}^{(1)}$. Let $\mathcal{M}$ be any machine in $\mathsf{IM}_\mathcal{A}^{(d)}$ and, for all $j \leq d$, let $I_{j,1},\ldots,I_{j,\kappa_j}$ be the index registers of $\mathcal{M}$ allowing the access to the jth tape $Z_{j,1}, Z_{j,2},\ldots$ of $\mathcal{M}$. Then, let $\mathcal{N}_\mathcal{M}$ be a machine in $\mathsf{IM}_\mathcal{A}^{(1)}$ with $k_{\mathcal{N}_\mathcal{M}} = \kappa_1 + \cdots + \kappa_d$ working in *d-tape mode* with d tracks by using the jth *track* $Z_j, Z_{d+j}, Z_{2d+j},\ldots$ instead of the jth tape $Z_{j,1}, Z_{j,2},\ldots$ of $\mathcal{M}$. If $\mathrm{Input}_\mathcal{M}(\mathrm{i}) = (1, (\vec{\nu}^{(j)} . \bar{u}^{(j)})_{j=1..d})$ holds, then let $\mathrm{In}_{\mathcal{N}_\mathcal{M}}(\mathrm{i}) = ((\vec{\nu}^{(1)} . \ldots . \vec{\nu}^{(d)}) . \bar{u})$ with $(u_j, u_{d+j}, u_{2d+j},\ldots) = \bar{u}^{(j)}$ for all $j \leq d$. However, if $\mathcal{M}$ and $\mathcal{N}_\mathcal{M}$ should be BSS RAM's, then, after the input, let $\mathcal{N}_\mathcal{M}$ arrange its input $\vec{x} \in U_\mathcal{A}^\infty$ on the first track $Z_1, Z_{d+1}, Z_{2d+1},\ldots$ and execute $I_1 := (I_1 - 1) \cdot d + 1$ before the simulation starts with a suitable configuration.

Proposition 10 (Simulation of d-tape $\mathcal{A}$-machines). *For any structure $\mathcal{A}$ and all $t \geq 1$, the work of any $\mathcal{M} \in \mathsf{IM}_\mathcal{A}^{(d)}$ can be done by a machine $\mathcal{N}_\mathcal{M} \in \mathsf{IM}_\mathcal{A}^{(1)}$ where t steps of $\mathcal{M}$ can be realised by $\mathcal{N}_\mathcal{M}$ within dt steps.*

Proof. Let $\mathcal{P}_\mathcal{M}$ contain only instructions given in Overview 3. To get the corresponding program $\mathcal{P}_{\mathcal{N}_\mathcal{M}}$, we replace every variable $Z_{j,i}$ ($j \leq d, i \geq 1$) in $\mathcal{P}_\mathcal{M}$ by the variable $Z_{(i-1)d+j}$ so that we have to consider the register blocks $Z_{(i-1)d+1},\ldots,Z_{id}$ of $\mathcal{N}_\mathcal{M}$ instead of the registers $Z_{1,i},\ldots,Z_{d,i}$. The variables for index registers $I_1,\ldots,I_{\kappa_1+\cdots+\kappa_d}$ are substituted for $I_{1,1},\ldots,I_{1,\kappa_1}, \ldots I_{d,1},\ldots,I_{d,\kappa_d}$ in this order. Consequently, in all copy instructions in $\mathcal{P}_\mathcal{M}$, the indices of Z-registers including the index register variables must be replaced as follows. $Z_{I_{\kappa_1+\cdots+\kappa_{j-1}+i}}$ must be substituted for $Z_{j,I_{j,i}}$. Moreover, we replace any index instruction $I_{j,i} := 1$ in $\mathcal{P}_{\mathcal{M}}$ by $I_{\kappa_1+\cdots+\kappa_{j-1}+i} := j$ and $I_{j,i} := I_{j,i}+1$ by $I_{\kappa_1+\cdots+\kappa_{j-1}+i} := I_{\kappa_1+\cdots+\kappa_{j-1}+i}+d$. Then, the execution of each of these pseudo instructions in $\mathcal{P}_{\mathcal{N}_{\mathcal{M}}}$ can be realised within d transformation steps defined by $\rightarrow_{\mathcal{N}_{\mathcal{M}}}$. Any other instruction of $\mathcal{P}_{\mathcal{M}}$ can be simulated in one step. □

3 Universal machines and consequences

3.1 Universal programs and machines: the definitions

We want to transfer the classical theorems about the existence of a universal partial recursive function (see e.g. the Enumeration Theorem in [27, p. 18]) and the existence of a universal Turing machine (see e.g. [9, Theorem IV, p. 22]) and generalise a result about the existence of a universal BSS machine presented by a flow chart in Figure 15 in [3, p. 35]. The following definition also includes the case considered for a similar model in [20, p. 39]. Let a and b be any two distinct objects and let λ be an integer such that it is possible to encode the characters in any string $\mathcal{P} \in \mathsf{P}_\sigma$ (including all symbols such as r_{21}^3 after their replacement, e.g., by r_21_3) by tuples in $\{a,b\}^\lambda$. Then, for any two distinct objects a and b and for any σ-program $\mathcal{P}$, let $\mathrm{code}_{(a,b)}(\mathcal{P}) \in \{a,b\}^\infty$ result from the concatenation of an a and the sequence of the codes of all characters in $\mathcal{P}$ in this order. In the following definition, let $a = 1$ and $b = 0$. Let $\mathrm{code}^*(\mathcal{P})$ be the string $s_1 \cdots s_m \in \{0,1\}^*$ if $\mathrm{code}_{(1,0)}(\mathcal{P}) = (s_1,\ldots,s_m)$ holds and let $\mathrm{code}_{\mathbb{N}}(\mathcal{P}) = \mathrm{bin}^{-1}(\mathrm{code}^*(\mathcal{P}))$. Let $\mathcal{A}$ be a structure. For any $\mathcal{M} \in \mathsf{IM}_{\mathcal{A}}$ with the constants in $\vec{c}^{\,(\mathcal{M})} = (c_{j_1},\ldots,c_{j_{n_1}})$, let $\mathrm{code}_{(a,b)}(\mathcal{M}) = (\mathrm{code}_{(a,b)}(\mathcal{P}_{\mathcal{M}}) \,.\, \vec{a}^{\,(\mathcal{M},a)})$ where $\vec{a}^{\,(\mathcal{M},a)} = (a_1,\ldots,a_{\ell_{\mathcal{P}_{\mathcal{M}}}})$ is defined as follows. For any $\ell \le \ell_{\mathcal{P}_{\mathcal{M}}}$, let a_ℓ be the ith component c_{j_i} in $\vec{c}^{\,(\mathcal{M})}$ if the ℓth instruction of $\mathcal{P}_{\mathcal{M}}$ is the instruction $Z_j := c_i^0$ for some j and otherwise let $a_\ell = a$.

Definition 11 (Universal machines of type 1 for BSS RAM's). Let a and b be two constants of $\mathcal{A}$. We say that $\mathcal{M}_0 \in \mathsf{M}_{\mathcal{A}}$ is a *universal BSS RAM* over $\mathcal{A}$ if $\mathrm{Res}_{\mathcal{M}_0}((\mathrm{code}_{(a,b)}(\mathcal{M}) \,.\, \vec{x})) = \mathrm{Res}_{\mathcal{M}}(\vec{x})$ holds for all $\vec{x} \in U_{\mathcal{A}}^\infty$ and any $\mathcal{M} \in \mathsf{M}_{\mathcal{A}}$.

The following definition includes also the definition of a further variant of a universal BSS RAM for several structures. It is in particular suitable for structures of finite signature without constants. For this purpose, let $p_1, p_2, \ldots$ be the sequence of all prime numbers with $p_i < p_{i+j}$ for all $i,j \ge 1$. For any $\mathcal{A}$-machine $\mathcal{M}_0$, let $\mathsf{ireg} : \mathbb{N}_+^\infty \to (\mathbb{N}_+^{k_{\mathcal{M}_0}} \cap (\mathbb{N}_+ \times \{1\} \times \cdots \times \{1\}))$ be any bijective function such that this function as well as its partial inverse can be computed by BSS RAM's over $\mathcal{A}_{\mathbb{N}}$. Moreover, for any $k \ge 1$, let the function $\mathsf{end}_{\mathcal{M}_0,k} : \mathsf{S}_{\mathcal{M}_0} \to \{(\vec{\nu} \,.\, \bar{u}) \mid (\vec{\nu}, \bar{u}) \in$

$\mathbb{N}_+^k \times U_{\mathcal{A}}^\omega\}$ be defined, for all $(\ell\,.\,\vec{\mu}\,.\,\bar{u}) \in \mathsf{S}_{\mathcal{M}_0}$, by $\mathsf{end}_{\mathcal{M}_0,k}(\ell\,.\,\vec{\mu}\,.\,\bar{u}) = (\vec{\nu}\,.\,\bar{u})$ where $\vec{\nu} = (\mu_1, \nu_2, \ldots, \nu_k)$ holds if $\mu_2 = 2^{\nu_2} \cdots p_{k-1}^{\nu_k}$ and otherwise $\vec{\nu} = (\mu_1, 1, \ldots, 1) \in \mathbb{N}_+^k$ holds.

Definition 12 (Universal programs and machines of type 2). Let $\mathcal{M}_0$ be an $\mathcal{A}$-machine and N be a class of $\mathcal{A}$-machines. We say that $\mathcal{P}_{\mathcal{M}_0}$ is N-*universal* (or *universal*) if, for all $\mathcal{M} \in \mathsf{N}$ and $\mathsf{i} \in \mathsf{I}_{\mathcal{M}}$,

$$\mathrm{Res}_{\mathcal{M}}(\mathsf{i}) = \mathrm{Out}_{\mathcal{M}} \circ \mathsf{end}_{\mathcal{M}_0,k_{\mathcal{M}}} \circ (\rightarrow_{\mathcal{M}_0})_{\mathrm{Stop}_{\mathcal{M}_0}}(\mathsf{init}_{\mathcal{M}_0}(\mathcal{M}, \mathsf{i}))$$

holds for the initial configuration $\mathsf{init}_{\mathcal{M}_0}(\mathcal{M}, \mathsf{i}) \in \mathsf{S}_{\mathcal{M}_0}$ that is given by

$$\mathsf{init}_{\mathcal{M}_0}(\mathcal{M}, \mathsf{i}) = (1\,.\,\mathsf{ireg}(\vec{\nu}\,.\,\mathrm{code}_{\mathbb{N}}(\mathcal{P}_{\mathcal{M}}))\,.\,\vec{a}^{\,(\mathcal{M},u_1)}\,.\,\bar{u})$$

and $(\vec{\nu}\,.\,\bar{u}) = \mathrm{In}_{\mathcal{M}}(\mathsf{i})$ with $\vec{\nu} \in \mathbb{N}_+^{k_{\mathcal{M}}}$. $\mathcal{M}_0$ is called ($\mathsf{M}_{\mathcal{A}}$-)*universal BSS RAM* if $\mathcal{M}_0$ is a BSS RAM and $\mathcal{P}_{\mathcal{M}_0}$ is $\mathsf{M}_{\mathcal{A}}$-universal.

If $\mathsf{I}_{\mathcal{M}_0} = \{(\mathcal{M}, \mathsf{i}) \mid \mathcal{M} \in \mathsf{N}\ \&\ \mathsf{i} \in \mathsf{I}_{\mathcal{M}}\}$ is the input space of a machine $\mathcal{M}_0$ with an N-universal program, then we will consider only the case $\mathrm{Input}_{\mathcal{M}_0} = \mathsf{init}_{\mathcal{M}_0}$. If $\mathcal{M}$ is a BSS RAM and $\mathcal{M}_0$ is an $\mathsf{M}_{\mathcal{A}}$-universal BSS RAM, then the function ireg is determined by a binary function. This means that there is some $\mathsf{idx} : \mathbb{N}_+^2 \to \mathbb{N}_+$ such that $\mathsf{ireg}(\vec{\nu}\,.\,\mu) = (\mathsf{idx}(\nu_1, \mu), 1, \ldots, 1)$ holds for all $\vec{\nu} = (\nu_1, \ldots, \nu_{k_{\mathcal{M}}}) \in \mathbb{N}_+^\infty$ and all $\mu \in \mathbb{N}_+$.

In the following, we want to consider universal $\mathcal{A}$-machines that are able to simulate the execution of every $\mathcal{M} \in \mathsf{M}_{\mathcal{A}}$ instruction-by-instruction. Let $\sigma = (n_1; m_1, \ldots, m_{n_2}; k_1, \ldots, k_{n_3})$ and $\mathcal{P}_{\mathcal{M}} \in \mathsf{P}_\sigma$. Since we want to make the important information about the instructions of $\mathcal{P}_{\mathcal{M}}$ available for easy access, we use strings $\alpha_{e_1} \cdots \alpha_{e_{\max\{m_1,\ldots,m_{n_2},k_1,\ldots,k_{n_3}\}}} \in \{\alpha_0, \ldots, \alpha_{h_{\mathcal{P}_{\mathcal{M}}}}\}^*$ for some $h_{\mathcal{P}_{\mathcal{M}}} \geq 1$ and Gödel numberings (for the definition see e.g. [21, p. 183]) for storing the information about a single instruction labelled by ℓ in a Gödel number $\mu_{\mathcal{P}_{\mathcal{M}},\ell}$. These numbers are dependent on the type of instructions and $\phi_{\mathcal{P}_{\mathcal{M}},\ell} =_{\mathrm{df}} \phi_{\mathcal{M},\ell}$ (used also in Definition 3) for $\ell \leq \ell_{\mathcal{P}_{\mathcal{M}}}$. Table 2 shows the details for each type. Accordingly, for all $s \geq 1$, let $e_s = 0$ if there is no other information for e_s in the table. Thus, for any $\ell \leq \ell_{\mathcal{P}_{\mathcal{M}}}$, there are numbers $e_1, e_2, \ldots$ such that $\mu_{\mathcal{P}_{\mathcal{M}},\ell} = p_1^{e_1} \cdot p_2^{e_2} \cdot \cdots$. Moreover, let a be any element of $\mathcal{A}$. For any $\mathcal{P} \in \mathsf{P}_\sigma$, let the program information be given by $\mu_\ell = \mu_{\mathcal{P},\ell}$ for $\ell \leq \ell_{\mathcal{P}}$ and

$$\mathrm{code}_a(\mathcal{P}) = (a, \ldots, a) \in \{a\}^{\nu_{\mathcal{P}}} \subseteq \{a\}^\infty \text{ with } \nu_{\mathcal{P}} = 2^{\mu_1} 3^{\mu_2} \cdots p_{\ell_{\mathcal{P}}-1}^{\mu_{\ell_{\mathcal{P}}-1}} p_{\ell_{\mathcal{P}}} \in \mathbb{N}_+.$$

For any $\mathcal{M} \in \mathsf{IM}_{\mathcal{A}}$, let $\ell_{\mathcal{M}} = \ell_{\mathcal{P}_{\mathcal{M}}}$ and $\nu_{\mathcal{M}} = \nu_{\mathcal{P}_{\mathcal{M}}}$. Consequently, every machine $\mathcal{M} \in \mathsf{IM}_{\mathcal{A}}$ could be encoded by $\vec{a}^{\,(\mathcal{M},a)}$ and a tuple $(a, \ldots, a)$ whose length is dependent

Type	e_1	e_2	e_3	e_4	e_5	e_6	e_7	$\cdots$	$e_{\ldots}$	$\mu_{\mathcal{P},\ell}$
(1)	1	i			j	j_1	j_2		j_{m_i}	$2^1 \cdot 3^i \cdot 11^j \cdot 13^{j_1} \cdot 17^{j_2} \cdots p_{m_i+5}^{j_{m_i}}$
(2)	2				j					$2^2 \cdot 11^j$
(3)	3				j	k				$2^3 \cdot 11^j \cdot 13^k$
(4)	4	i	ℓ_1	ℓ_2		j_1	j_2		j_{k_i}	$2^4 \cdot 3^i \cdot 5^{\ell_1} \cdot 7^{\ell_2} \cdot 13^{j_1} \cdot 17^{j_2} \cdots p_{k_i+5}^{j_{k_i}}$
(5)	5		ℓ_1	ℓ_2	j	k				$2^5 \cdot 5^{\ell_1} \cdot 7^{\ell_2} \cdot 11^j \cdot 13^k$
(6)	6				j					$2^6 \cdot 11^j$
(7)	7				j					$2^7 \cdot 11^j$
(8)	8									1

Table 2: The codes $\mu_{\mathcal{P},\ell}$ for the instruction with label ℓ in $\mathcal{P}$

on the Gödel number $\nu_{\mathcal{M}} = |\text{code}_a(\mathcal{P}_{\mathcal{M}})|$ and the length of the input. However, in order to complement any possible input $(x_1, \ldots, x_n) \in U_{\mathcal{A}}^{\infty}$ of an $\mathcal{M} \in \mathsf{M}_{\mathcal{A}}$ with a code of $\mathcal{M}$ that is computable from the input by some machine in $\mathsf{M}_{\mathcal{A}}$, we take $\vec{a}^{(\mathcal{M},x_1)}$ and a suitable tuple in $\{x_n\}^{\infty}$. For computing the total length of this tuple we use the Cantor pairing function $cantor : \mathbb{N}^2 \to \mathbb{N}$ defined by $cantor(\mu_1, \mu_2) = \frac{1}{2}((\mu_1 + \mu_2)^2 + 3\mu_1 + \mu_2)$. Now, let $\text{code}_{n,a}(\mathcal{P}_{\mathcal{M}}) = (a, \ldots, a) \in \{a\}^{cantor(n,\nu_{\mathcal{M}})-\ell_{\mathcal{M}}-n}$ for $n \geq 1$ and any possible $a \in U_{\mathcal{A}}$. In this way, for all $\vec{x} \in U_{\mathcal{A}}^n$ and $\mathcal{M} \in \mathsf{M}_{\mathcal{A}}$, we get an input

$$(\vec{a}^{(\mathcal{M},x_1)} \,.\, \vec{x} \,.\, \text{code}_{n,x_n}(\mathcal{P}_{\mathcal{M}})) \in U_{\mathcal{A}}^{cantor(n,\nu_{\mathcal{M}})} \qquad (**)$$

for a universal BSS RAM $\mathcal{M}_0$ of type 2 such that we have

$$\mathsf{init}_{\mathcal{M}_0}(\mathcal{M}, \vec{x}) = (1 \,.\, (cantor(n, \nu_{\mathcal{M}}), 1, \ldots, 1) \,.\, (\vec{a}^{(\mathcal{M},x_1)} \,.\, \vec{x} \,.\, (x_n, x_n, \ldots))). \qquad (***)$$

3.2 Simulation of BSS RAM's by universal machines

Here, we will describe the work of a universal BSS RAM $\mathcal{M}_0 \in \mathsf{M}_{\mathcal{A}}$ that can simulate any machine $\mathcal{M} \in \mathsf{M}_{\mathcal{A}}$ by means of three tracks after assigning its own input to the first track in case that the input has the form (**). For explaining the algorithm we only construct a BSS RAM $\mathcal{M}_0^{(3)} \in \mathsf{M}_{\mathcal{A}}^{(3)}$ that is able to simulate any $\mathcal{M} \in \mathsf{M}_{\mathcal{A}}$ by a program that is also useful for other investigations. To simplify matters, we use further pseudo instructions. The strings in square brackets are optional.

Overview 4 (Pseudo instructions for d tapes).

$$Z_{1,I_{1,k}} := f_i^{m_i}(Z_{1,I_{1,k+1}}, \ldots, Z_{1,I_{1,k+m_i}})$$
$$(Z_{d_1,[I_{d_1,j}+]1}, \ldots, Z_{d_1,[I_{d_1,j}+]I_{1,m}}) := (Z_{1,[I_{1,k}+]1}, \ldots, Z_{1,[I_{1,k}+]I_{1,m}})$$

Further pseudo conditions: $r_i^{k_i}(Z_{1,I_{1,k+1}}, \ldots, Z_{1,I_{1,k+k_i}})$

In Overview 5, J_i, J_k, J_l, and J_m stand for $I_{1,i}, I_{1,k}, I_{1,l}$, and $I_{1,m}$. The operator $\div$ denotes the integer division defined by $\mu \div \nu = \max\{s \in \mathbb{N}_+ \mid s \cdot \nu \leq \mu\}$ for all $\nu, \mu \in \mathbb{N}_+$ with $\nu \leq \mu$, and $\nu|\mu$ means that ν is a divisor of μ. Moreover, let $cantor_1(\mu) = \mu_1$ and $cantor_2(\mu) = \mu_2$ if $\mu = cantor(\mu_1, \mu_2)$.

Overview 5 (Pseudo instructions for decoding numbers).

$J_i := cantor_1(J_k)$, $\quad J_i := cantor_2(J_k)$, $\quad J_i := p_m^{J_k}$, $\quad J_i := J_k \cdot p_{J_m}$	
$J_i := \max\{s \mid p_m^s \| J_k\}$	*(permitted if $p_m \| c(J_k)$ is ensured)*
$J_i := [J_l+]\max\{s \mid p_{J_m}^s \| J_k\}$	*(permitted if $p_{c(J_m)} \| c(J_k)$ is ensured)*
$J_i := \max\{m \mid p_m \| J_k\}$	*(permitted if $c(J_k) > 1$ is ensured)*
$J_i := J_i \div p_{J_m}^{J_l - J_k - 1}$	*(permitted if $c(J_i) \geq p_{c(J_m)}^{c(J_l)-c(J_k)-1} > 1$ is ensured)*
Further pseudo conditions: $p_m \| J_k$, $p_{J_m}^s \| J_k$	

If $d_1 \leq 3$, then the instructions in Overview 4 can be executed by a 3-tape machine. All values that are necessary for executing the first pseudo instruction and checking the new condition are firstly copied on a second tape where the new values are computed and the tests are performed. For this purpose, a finite number of instructions of the form $I_{2,1} := 1, I_{2,2} := 2, \ldots, Z_{2,I_{2,1}} := Z_{1,I_{1,k+1}}, Z_{2,I_{2,2}} := Z_{1,I_{1,k+2}}, \ldots, Z_{2,1} := f_i^{m_i}(Z_{2,1}, \ldots, Z_{2,m_i}), Z_{1,I_{1,k}} := Z_{2,I_{2,1}}$, and if $r_i^{k_i}(Z_{2,1}, \ldots, Z_{2,k_i})$ then goto ℓ_1 else goto ℓ_2, respectively, can be used. As known from the classical recursion theory, the pseudo instructions listed in Overview 5 can be used for evaluating Gödel numbers, and they can be replaced by subprograms consisting only of index instructions of the types (5) to (7). For example, since $J_k|J_i$ holds if and only if $J_i \geq J_k$ & $(J_i \div J_k) \cdot J_k = J_i$ holds, the pseudo instruction if $J_k|J_i$ then goto ℓ_1 else goto ℓ_2 can be performed as follows. One goes to ℓ_2 if $c(J_i) < c(J_k)$. Otherwise one takes a new index register J_l, executes $J_l := J_k$, and repeats the execution of $J_l := J_l + J_k$ until $c(J_l) \geq c(J_i)$. If $c(J_l) = c(J_i)$, then the execution can be continued with the instruction labelled by ℓ_1 and, otherwise, with the instruction labelled by ℓ_2. If $c(J_i) \geq c(J_k)$, then it is also possible to realise the pseudo instruction $J_j := J_i \div J_k$ that allows to compute $c(J_j) = \max\{s \in \mathbb{N}_+ \mid s \cdot c(J_k) \leq c(J_i)\}$. In this case, one takes two new index registers J_{l_1} and J_{l_2}, executes $J_{l_1} := J_k$ and $J_{l_2} := 1$, and repeats the execution of $J_{l_1} := J_{l_1} + J_k$ and $J_{l_2} := J_{l_2} + 1$ until $c(J_{l_1}) \geq c(J_i)$.

If $c(J_{l_1}) = c(J_i)$, then $c(J_i) \div c(J_k)$ is $c(J_{l_2})$. If $c(J_{l_1}) > c(J_i)$, then $c(J_i) \div c(J_k)$ is the number $c(J_{l_2}) - 1$. $c(J_k) = p_m$ can be computed by a program as follows where the index register K_1 is used for storing the index of the next searched prime number after $p_{c(K_1)-1}$ (if $c(K_1) - 1 \geq 1$), the index register K_2 is used for storing the integers greater than $p_{c(K_1)-1}$ and checking whether the stored number $c(K_2)$ is the next prime number, and the index register K_3 is used for storing a possible non-trivial factor of $c(K_2)$. $\ell_1 : K_1 := 1;\ K_2 := 2;\ \ell_2 :$ if $K_1 = m$ then goto ℓ_7; $\ell_3 \colon K_1 := K_1 + 1;\ \ell_4 \colon K_2 := K_2 + 1;\ K_3 := 1;\ \ell_5 \colon K_3 := K_3 + 1;$ if $K_2 = K_3$ then goto ℓ_2; $\ell_6\colon$ if $K_3 | K_2$ then goto ℓ_4 else goto ℓ_5; $\ell_7 \colon J_k := K_2$.

Now, we assume that corresponding to (**) the registers $Z_{1,1}, Z_{1,2}, \ldots$ of the first tape $\mathcal{M}_0^{(3)}$ contain the values $a_1, \ldots, a_{\ell_\mathcal{M}}, x_1, \ldots, x_n, x_n, \ldots$ where $(a_1, \ldots, a_{\ell_\mathcal{M}}) = \vec{a}^{(\mathcal{M}, x_1)}$ before $\mathcal{M}_0^{(3)}$ executes the $(0; m_1, \ldots, m_{n_2}; k_1, \ldots, k_{n_3})$-program $\mathcal{P}_0^{(3)}$ given in Figure 6. Let $\mathcal{P}_{\mathcal{M}_0^{(3)}} = \mathcal{P}_0^{(3)}$. In $\mathcal{P}_0^{(3)}$, the registers $I_{1,1}, \ldots, I_{1,2m_0+15}$ with $m_0 = \max\{m_1, \ldots, m_{n_2}, k_1, \ldots, k_{n_3}\}$ are denoted by $N_0, N, C, L, L_P, V, S,\ E_1, \ldots, E_{m_0+5}$, $J_0, \ldots, J_{m_0},\ H_1$, and H_2. $c(N_0)$ is firstly the length n_0 of the input of $\mathcal{M}_0^{(3)}$ and later the length of the output of $\mathcal{M}_0^{(3)}$. $c(N)$ is firstly the length n of the input of $\mathcal{M}$. C is used for storing the Gödel number $\nu_\mathcal{M}$, and we use $c(L_P) = \ell_\mathcal{M}$. Since any machine can contain only a finite number of index registers, $\mathcal{M}_0^{(3)}$ cannot store all values $c(I_1), \ldots, c(I_{k_\mathcal{M}})$ for every possible $k_\mathcal{M} \geq 1$ in different index registers. Thus, we use the possibility to store all values of the registers $I_1, \ldots, I_{k_\mathcal{M}}$ of $\mathcal{M}$ in one index register of $\mathcal{M}_0^{(3)}$. V contains the product $p_{s_1}^{c(I_{s_1})} \cdots p_{s_\mu}^{c(I_{s_\mu})}$ for the registers $I_{s_1}, \ldots, I_{s_\mu}$ $(s_1 < \cdots < s_\mu \leq k_\mathcal{M})$ of $\mathcal{M}$ having already been considered during the simulation of $\mathcal{M}$ by $\mathcal{M}_0^{(3)}$ until that time. L and S contain the label and the code, respectively, of the current instruction of $\mathcal{P}_\mathcal{M}$, $E_1, E_2, \ldots$ contain the current values $e_1, e_2, \ldots$. $J_0, J_1, \ldots$ contain the values $n_0 + e_5, n_0 + e_6, \ldots$. Since the Z-registers $Z_{1,n_0+1}, Z_{1,n_0+2}, \ldots$ are used for storing all values $c(Z_1), c(Z_2), \ldots$ of the Z-registers of $\mathcal{M}$ during the simulation, the universal machine $\mathcal{M}_0^{(3)}$ assigns the start values $x_1, \ldots, x_n$ to its registers $Z_{1,n_0+1}, \ldots, Z_{1,n_0+n+1}$ and the start values $a_1, \ldots, a_{\ell_\mathcal{M}}$ to the registers $Z_{3,1}, \ldots, Z_{3,\ell_\mathcal{M}}$ by the subprogram of $\mathcal{P}_0^{(3)}$ that is labelled by 1 at the beginning. $Z_{1,n_0+n+1}, Z_{1,n_0+n+2}, \ldots$ have got the value x_n by the input procedure of $\mathcal{M}_0^{(3)}$. For any $e_1 \leq 7$, $\mathcal{P}_0^{(3)}$ contains a subprogram labelled by $\tilde{\ell}_{e_1}$ for simulating the instructions of type (e_1). In addition to the pseudo instructions listed in Overview 5, the subprograms contain 3-tape pseudo instructions that can be introduced in analogy with the pseudo instructions given in Overview 2 and the like. Since $\mathcal{P}_0^{(3)}$ can also contain only a finite number of strings denoting indices of Z-registers, the F-instructions $Z_j := f_i^{m_i}(Z_{j_1}, \ldots, Z_{j_{m_i}})$ cannot in general be simulated

The program $\mathcal{P}_0^{(3)}$

1 : $N := cantor_1(N_0)$; $C := cantor_2(N_0)$;
$L_P := \max\{m \mid p_m|C\}$; $(Z_{3,1},\ldots,Z_{3,L_P}) := (Z_{1,1},\ldots,Z_{1,L_P})$;
$(Z_{1,N_0+1},\ldots,Z_{1,N_0+N}) := (Z_{1,L_P+1},\ldots,Z_{1,L_P+N})$; $V := p_1^N$; $L := 1$;

ℓ_1 : if $L = L_P$ then goto ℓ_7 else goto ℓ_2;

ℓ_2 : $S := \max\{s \mid p_L^s|C\}$;
if $p_1|S$ then $E_1 := \max\{e \mid p_1^e|S\}$; $\ldots$;
if $p_{m_0+5}|S$ then $E_{m_0+5} := \max\{e \mid p_{m_0+5}^e|S\}$;
if $E_1 = 1$ or $E_1 = 2$ then $J_0 := N_0 + E_5$;
if $E_1 = 1$ or $E_1 = 4$ then $\{J_1 := N_0 + E_6; \ldots; J_{m_0} := N_0 + E_{m_0+5}\}$;
if $E_1 = 3$ or $E_1 = 5$ then $\{H_1 := \mathrm{comp}_{E_5}; H_2 := \mathrm{comp}_{E_6}\}$;
if $E_1 = 6$ or $E_1 = 7$ then $H_1 := \mathrm{comp}_{E_5}$;

ℓ_3 : if $E_1 = 1$ then goto $\tilde{\ell}_1$; $\ldots$; if $E_1 = 7$ then goto $\tilde{\ell}_7$;

$\tilde{\ell}_1$: if $E_2 = 1$ then $Z_{1,J_0} := f_1^{m_1}(Z_{1,J_1},\ldots,Z_{1,J_{m_1}})$; $\ldots$;
if $E_2 = n_2$ then $Z_{1,J_0} := f_{n_2}^{m_{n_2}}(Z_{1,J_1},\ldots,Z_{1,J_{m_{n_2}}})$; goto ℓ_4;

$\tilde{\ell}_2$: $Z_{1,J_0} := Z_{3,L}$; goto ℓ_4;

$\tilde{\ell}_3$: $Z_{1,H_1} := Z_{1,H_2}$; goto ℓ_4;

$\tilde{\ell}_4$: if $E_2 = 1$ & $r_1^{k_1}(Z_{1,J_1},\ldots,Z_{1,J_{k_1}})$ then goto ℓ_5 else goto ℓ_6; $\ldots$;
if $E_2 = n_3$ & $r_{n_3}^{k_{n_3}}(Z_{1,J_1},\ldots,Z_{1,J_{k_{n_3}}})$ then goto ℓ_5 else goto ℓ_6;

$\tilde{\ell}_5$: if $H_1 = H_2$ then goto ℓ_5 else goto ℓ_6;

$\tilde{\ell}_6$: if $H_1 > N_0 + 1$ then $V := V \div p_{E_5}^{H_1 - N_0 - 1}$; goto ℓ_4;

$\tilde{\ell}_7$: $V := V \cdot p_{E_5}$;

ℓ_4 : $L := L + 1$; goto ℓ_1;

ℓ_5 : $L := E_3$; goto ℓ_1;

ℓ_6 : $L := E_4$; goto ℓ_1;

ℓ_7 : $N := \max\{j \mid p_1^j|V\}$; $H_1 := 1$;

ℓ_8 : if $N \geq H_1$ then $\{Z_{1,H_1} := Z_{1,N_0+H_1}; H_1 := H_1 + 1;$ goto $\ell_8\}$; $N_0 := N$;

ℓ_9 : stop.

The subprogram $H_\nu := \mathrm{comp}_{E_\mu}$ (given in form of a pseudo instruction)
if $p_{E_\mu}|V$ then $H_\nu := N_0 + \max\{s \mid p_{E_\mu}^s|V\}$ else $\{V := V \cdot p_{E_\mu}; H_\nu := N_0 + 1\}$

Figure 6: The program $\mathcal{P}_0^{(3)}$ of the universal machine $\mathcal{M}_0^{(3)}$

by executing F-instructions of the form $Z_{1,n_0+j} := f_i^{m_i}(Z_{1,n_0+j_1}, \ldots, Z_{1,n_0+j_{m_i}})$. The parameters $n_0 + j, n_0 + j_1, \ldots, n_0 + j_{m_i}$ may vary depending on the machines that should be simulated. Therefore, the index registers $J_0, J_1, \ldots$ get these values before, for $k = m_0 + 13$, the first pseudo instruction given in Overview 4 or a branching instruction with a pseudo condition will be executed. Moreover, further pseudo instructions given in Overview 5 are used before some registers of the second tape $Z_{2,1}, Z_{2,2}, \ldots$ can be used by $\mathcal{M}_0^{(3)}$ for the simulation of an F-instruction or a T-instruction by $\mathcal{M}$. To search an information in $c(C) = |\mathrm{code}_a(\mathcal{P}_\mathcal{M})|$ necessary for the simulation of the considered instruction of $\mathcal{M}$, the machine $\mathcal{M}_0^{(3)}$ uses L for storing the current label. The instruction labelled by ℓ_2 in Figure 6 means to search the code of the instruction with label $c(L)$. During computing the code $c(S)$ of the instruction with the current label $c(L)$ from $c(C)$, $p_{c(L)}$ is the corresponding prime number. The subprogram labelled by ℓ_2 in Figure 6 also allows to compute the values $c(E_1) = e_1, c(E_2) = e_2, \ldots$ (as given in Table 2) for the current instruction with label $c(L)$. The index register E_1 stands for the type of the current instruction and E_2, E_3, E_4 for the used indices i and the labels ℓ_1 and ℓ_2, respectively, in the current instruction of $\mathcal{P}_\mathcal{M}$. Then, depending on the type of the instruction, further values $e_5, e_6, \ldots$ can be made available before simulating the current instruction. The subprograms labelled by ℓ_4, ℓ_5, and ℓ_6 in Figure 6 are intended for determining the label of the next relevant instruction in $\mathcal{P}_\mathcal{M}$. The subprograms labelled by ℓ_7 and ℓ_8 are used for preparing the output of $\mathcal{M}_0^{(3)}$. The subprogram $H_\nu := \mathrm{comp}_{E_\mu}$ allows to compute the values $c(I_{c(E_5)})$ and $c(I_{c(E_6)})$ of the relevant index registers of $\mathcal{M}$ from $c(V)$.

Consequently, by Proposition 10, we have shown the following theorem.

Theorem 13. *For any structure $\mathcal{A}$ with a finite number of operations and relations, there exists an $\mathsf{M}_\mathcal{A}$-universal machine $\mathcal{M}_0$ of type 2 such that*

$$\mathrm{Res}_\mathcal{M}(x_1, \ldots, x_n) = \mathrm{Res}_{\mathcal{M}_0}(\vec{a}^{\,(\mathcal{M},x_1)} \,.\, (x_1, \ldots, x_n) \,.\, \mathrm{code}_{n,x_n}(\mathcal{P}_\mathcal{M}))$$

for any $\mathcal{M} \in \mathsf{M}_\mathcal{A}$ and any $(x_1, \ldots, x_n) \in U_\mathcal{A}^\infty$.

3.3 Complete problems

The here considered *decision problems* are algorithmic problems completely defined by a decision question for which we want to get the answer yes or the answer no only. For instance, for any structure $\mathcal{A}$ and $\mathsf{I} = \mathsf{M}_\mathcal{A} \times U_\mathcal{A}^\infty$, the halting problem $\mathrm{HP}_\mathcal{A}^t = \{(\mathcal{M}, \vec{x}) \in \mathsf{I} \mid \mathcal{M}(\vec{x}) \downarrow^t\}$ where $\mathcal{M}(\vec{x}) \downarrow^t$ stands for the fact that $\mathcal{M}$ halts (stops) on input $\vec{x}$ after the execution of t transformation steps is a decision problem.

Let $\mathcal{A}$ be a structure of signature $(|N_1|; m_1, \ldots, m_{n_2}; k_1, \ldots, k_{n_3})$. Then, $\mathrm{HP}^t_{\mathcal{A}}$ is decidable by a machine $\mathcal{M}_0 \in \mathsf{IM}_{\mathcal{A}}$ with $\mathsf{I}_{\mathcal{M}_0} = \mathsf{I}$, $\mathsf{O}_{\mathcal{M}_0} = \{\mathsf{yes}, \mathsf{no}\}$, $\mathrm{Input}_{\mathcal{M}_0} = \mathsf{init}_{\mathcal{M}_0}$ (as defined by (***)), and the function $\mathrm{Out}_{\mathcal{M}_0}$ defined, for all $(\vec{\nu}, \bar{u}) \in \mathbb{N}^{k_{\mathcal{M}_0}} \times U^{\omega}_{\mathcal{A}}$, by $\mathrm{Out}_{\mathcal{M}_0}(\vec{\nu}\,.\,\bar{u}) = \mathsf{yes}$ if $\nu_1 = 1$ and otherwise by $\mathrm{Out}_{\mathcal{M}_0}(\vec{\nu}\,.\,\bar{u}) = \mathsf{no}$. Starting from a configuration given by (***), the execution of $\mathcal{M}$ on $\vec{x}$ can be simulated by $\mathcal{M}_0$ with the help of an $\mathsf{M}_{\mathcal{A}}$-universal program derived from $\mathcal{P}_0^{(3)}$ where the simulated steps are simultaneously counted by means of an additional index register. After simulating t steps, $\mathcal{M}_0$ can output yes if the S-instruction of $\mathcal{M}$ is reached and otherwise no. Thus, the corresponding result function $\mathrm{Res}_{\mathcal{M}_0}$ is totally defined on $\mathsf{I}_{\mathcal{M}_0}$.

With respect to BSS RAM's over $\mathcal{A}$, any *decision problem* is a subset $P \subseteq U^{\infty}_{\mathcal{A}}$ connected with the question whether $\vec{x} \in U^{\infty}_{\mathcal{A}}$ is in P. If $\mathcal{A}$ has at least two constants, denoted here by a and b, we want to assume that a stands for the answer yes and the second constant b stands for the answer no. Under this condition, an $\mathcal{M} \in \mathsf{M}_{\mathcal{A}}$ *decides* the problem P if the computed function is the characteristic function $\chi_P : U^{\infty}_{\mathcal{A}} \to \{a, b\}$ such that $\mathrm{Res}_{\mathcal{M}}(\vec{x}) = \chi_P(\vec{x}) = a$ holds for all $\vec{x} \in P$ and $\mathrm{Res}_{\mathcal{M}}(\vec{x}) = \chi_P(\vec{x}) = b$ holds for all $\vec{x} \in U^{\infty}_{\mathcal{A}} \setminus P$. In more general terms, we can define the notion decidable *by a BSS RAM* as follows. A set $P \subseteq U^{\infty}_{\mathcal{A}}$ is a *halting set* (*over* $\mathcal{A}$) if it is the domain of definition of the result function $\mathrm{Res}_{\mathcal{M}}$ for some $\mathcal{M} \in \mathsf{M}_{\mathcal{A}}$. A problem $P \subseteq U^{\infty}_{\mathcal{A}}$ is *semi-decidable* (*over* $\mathcal{A}$) if it is the halting set of a BSS RAM. Consequently, a semi-decidable set $P \subseteq U^{\infty}_{\mathcal{A}}$ does not have to be recursively enumerable by a function $f : U^{\infty}_{\mathcal{A}} \to U^{\infty}_{\mathcal{A}}$ with the properties that f is computable over $\mathcal{A}$, P is the image of f, and $f(\vec{x}) = f(\vec{y})$ holds for all $\vec{x} \in U^n_{\mathcal{A}}$ and $\vec{y} \in U^m_{\mathcal{A}}$ if $n = m$. $P \subseteq U^{\infty}_{\mathcal{A}}$ is *decidable* (*over* $\mathcal{A}$) if P and its complement $U^{\infty}_{\mathcal{A}} \setminus P$ are semi-decidable over $\mathcal{A}$ and, thus, halting sets of BSS RAM's.

In [18], two hierarchies of decision problems over algebraic structures are defined such that both definitions coincide with the usual descriptions of the arithmetical hierarchy over $(\mathbb{N}; \mathbb{N}; +; =)$ and the definitions given in [10]. For any $Q \subseteq U^{\infty}_{\mathcal{A}}$, let $\mathsf{M}^Q_{\mathcal{A}}$ be the class of all oracle BSS RAM's over $\mathcal{A}$ that are able to execute, additionally to the instructions of the types $(1), \ldots, (8)$, all oracle instructions of the form $\mathsf{if}\ (Z_1, \ldots, Z_{I_1}) \in \mathcal{O}\ \mathsf{then\ goto}\ \ell_1\ \mathsf{else\ goto}\ \ell_2$ by evaluating the query $(c(Z_1), \ldots, c(Z_{c(I_1)})) \in Q$?. Then, the first hierarchy consists of the class $\mathcal{A}\text{-}\Sigma^0_0 = \mathrm{DEC}_{\mathcal{A}}$ of all problems decidable by a machine in $\mathsf{M}_{\mathcal{A}}$ and the classes $\mathcal{A}\text{-}\Sigma^0_n = \bigcup_{Q \in \mathcal{A}\text{-}\Sigma^0_{n-1}} \mathrm{SDEC}^Q_{\mathcal{A}}$ of all problems semi-decidable by a machine in $\mathsf{M}^Q_{\mathcal{A}}$ for some $Q \in \mathcal{A}\text{-}\Sigma^0_{n-1}$. By generalising Theorem 13, we get the following where the oracle instructions are encoded by $2^9 \cdot 5^{\ell_1} \cdot 7^{\ell_2}$

Theorem 14. *For any structure $\mathcal{A}$ with a finite number of operations and relations and any oracle $Q \subseteq U^{\infty}_{\mathcal{A}}$, there exists an $\mathsf{M}^Q_{\mathcal{A}}$-universal machine $\mathcal{M}_0$ of type 2* *satisfying*

$$\mathrm{Res}_{\mathcal{M}}(x_1,\ldots,x_n) = \mathrm{Res}_{\mathcal{M}_0}(\vec{a}^{\,(\mathcal{M},x_1)}\,.\,(x_1,\ldots,x_n)\,.\,\mathrm{code}_{n,x_n}(\mathcal{P}_{\mathcal{M}}))$$

for any $\mathcal{M} \in \mathsf{M}_{\mathcal{A}}^{Q}$ *and any* $(x_1,\ldots,x_n) \in U_{\mathcal{A}}^{\infty}$.

Whereas in [18], the halting problems are considered only for structures with two constants a and b that are effectively distinguishable, we can now define further *halting problems* as follows, where $\mathcal{M}(\vec{x})\downarrow$ means $\mathcal{M}(\vec{x})\downarrow^{t}$ for some $t \geq 1$.

$$\mathbb{H}_{\mathcal{A}}^{Q} =_{\mathrm{df}} \{(\vec{a}^{\,(\mathcal{M},x_1)}\,.\,\vec{x}\,.\,\mathrm{code}_{n,x_n}(\mathcal{P}_{\mathcal{M}})) \mid \mathcal{M} \in \mathsf{M}_{\mathcal{A}}^{Q} \ \&\ \vec{x} \in U_{\mathcal{A}}^{\infty} \ \&\ \mathcal{M}(\vec{x})\downarrow\}$$

If $(|N_1|; m_1,\ldots,m_{n_2}; k_1,\ldots,k_{n_3})$ is the signature of $\mathcal{A}$, then let $\mathcal{M}_0^{Q} \in \mathsf{M}_{\mathcal{A}}^{Q}$ be the universal oracle machine defined analogously to the universal BSS RAM considered in Section 3.2. If the set of constants of $\mathcal{A}$ is decidable, then the set $\mathsf{I} =_{\mathrm{df}} \{(\vec{a}^{\,(\mathcal{M},x_1)}\,.\,\vec{x}\,.\,\mathrm{code}_{n,x_n}(\mathcal{P}_{\mathcal{M}})) \mid (\mathcal{M},\vec{x}) \in \mathsf{M}_{\mathcal{A}}^{Q} \times U_{\mathcal{A}}^{\infty}\}$ is decidable and $\mathbb{H}_{\mathcal{A}}^{Q} \subseteq U_{\mathcal{A}}^{\infty}$ is semi-decidable by a machine in $\mathsf{M}_{\mathcal{A}}^{Q}$ that uses $\mathcal{P}_{\mathcal{M}_0^{Q}}$ for all inputs in I and does not halt for all inputs in $U_{\mathcal{A}}^{\infty} \setminus \mathsf{I}$. By Theorem 14 we get the following result.

Theorem 15. *For any structure* $\mathcal{A}$ *with a decidable set of constants and a finite number of operations and relations, the Halting problem* $\mathbb{H}_{\mathcal{A}}^{Q}$ *is semi-decidable by a universal oracle machine in* $\mathsf{M}_{\mathcal{A}}^{Q}$.

The decision problems $\mathbb{H}_{\mathcal{A}}^{(n)} =_{\mathrm{df}} \mathbb{H}_{\mathcal{A}}^{\mathbb{H}_{\mathcal{A}}^{(n-1)}}$ (for $n > 0$) form a sequence of halting problems resulting from the so-called jumps where $\mathbb{H}_{\mathcal{A}}^{(0)} =_{\mathrm{df}} \emptyset$ and each $\mathbb{H}_{\mathcal{A}}^{(n)}$ is semi-decidable by a machine in $\mathsf{M}_{\mathcal{A}}^{\mathbb{H}_{\mathcal{A}}^{(n-1)}}$. Let $P \subseteq U_{\mathcal{A}}^{\infty}$ and $Q \subseteq U_{\mathcal{A}}^{\infty}$ be given. Then, by analogy with the classical case (cf. [27], p. 50 and p. 19), we say that P is *Turing reducible to* Q (denoted by $P \preceq_{\mathcal{A},T} Q$) if P is decidable by a machine in $\mathsf{M}_{\mathcal{A}}^{Q}$ and P *is one-one reducible to* Q (denoted by $P \preceq_{\mathcal{A},1} Q$) if there is an $\mathcal{M} \in \mathsf{M}_{\mathcal{A}}$ computing a total and injective function $\mathrm{Res}_{\mathcal{M}}$ such that, for all $\vec{x} \in U_{\mathcal{A}}^{\infty}$, $\mathrm{Res}_{\mathcal{M}}(\vec{x}) \in Q$ holds if and only if $\vec{x} \in P$ holds. $\mathbb{H}_{\mathcal{A}}^{(n)}$ is called *complete* in $\mathcal{A}$-Σ_n^0 since any problem of this class is one-one-reducible to it. This means that

$$\mathcal{A}\text{-}\Sigma_n^0 = \{P \subseteq U_{\mathcal{A}}^{\infty} \mid P \preceq_{\mathcal{A},1} \mathbb{H}_{\mathcal{A}}^{(n)}\}$$

and

$$(\mathcal{A}\text{-}\Sigma_n^0) \cap \{U_{\mathcal{A}}^{\infty} \setminus P \mid P \in \mathcal{A}\text{-}\Sigma_n^0\} = \{P \subseteq U_{\mathcal{A}}^{\infty} \mid P \preceq_{\mathcal{A},T} \mathbb{H}_{\mathcal{A}}^{(n-1)}\}$$

hold.

The described universal oracle $\mathcal{A}$-machines allow to extend the results from [18]. Note that the discussion about the existence of universal BSS RAM's is also very helpful for characterising complexity classes by complete problems, in particular, for structures without constants and for groups (see e.g. [14]).

4 Acknowledgement

I would like to thank the anonymous referees for the valuable remarks and suggestions. Moreover, my thanks go to Adrian Rezuş for preparing the final format.

References

[1] Alfred V. Aho, John E. Hopcroft, and Jeffrey D. Ullman The Design and Analysis of Computer Algorithms, Addison-Wesley 1974.

[2] Günter Asser Einführung in die mathematische Logik, Teil 2, Prädikatenkalkül der ersten Stufe, BSB B. G. Teubner Verlagsgesellschaft, Leipzig 1975.

[3] Lenore Blum, Michael Shub, and Steve Smale, *On a theory of computation and complexity over the real numbers: NP-completeness, recursive functions and universal machines*, Bulletin of the American Mathematical Society 21 1989, pp. 1–46.

[4] Lenore Blum, Felipe Cucker, Michael Shub, and Steve Smale Complexity and Real Computation, Springer 1998.

[5] Egon Börger Berechenbarkeit, Komplexität, Logik, Vieweg 1992. (In English: Computability, Complexity, Logic, Elsevier 1989.)

[6] Olivier Bournez, and Amaury Pouly *A survey on analog models of computation*, arXiv:1805.05729, 2018.

[7] Vasco Brattka, and Peter Hertling *Feasible real random access machines*, Journal of Complexity 14, 1998, pp. 490–526.

[8] John E. Hopcroft, and Jeffrey D. Ullman *Einführung in die Automatentheorie, formale Sprachen und Komplexitätstheorie*, Addison-Wesley 1993. (In English: Introduction to Automata Theory, Languages, and Computation, Addison-Wesley 1979.)

[9] Hartley Rogers Theory of Recursive Functions and Effective Computability, McGraw-Hill 1967.

[10] Felipe Cucker *The arithmetical hierarchy over the reals*, Journal of Logic and Computation 2 (3), 1992, pp. 375–395.

[11] Felipe Cucker, and Klaus Meer *Logics which capture complexity classes over the reals*, The Journal of Symbolic Logic 64 (1), 1999, pp. 363–390.

[12] Felipe Cucker, and A. Torrecillas *Two P-complete problems in the theory of the reals*, Journal of Complexity 8, 1992, pp. 454–466.

[13] Christine Gaßner *An NP-complete problem for linear real machines*, Workshop on Computability, Complexity and Logic (WCCL'96) Zinnowitz, 1996, 37–38.

[14] — *Computation over groups*, in: Arnold Beckmann, Costas Dimitracopoulos, and Benedikt Löwe (eds.) Logic and Theory of Algorithms, 2008, pp. 147–156.

[15] — *Oracles and relativizations of the P =? NP question for several structures*, Journal of Universal Computer Science 15 (6), 2009, pp. 1186–1205.

[16] — Gelöste und offene P-NP-Probleme über verschiedenen Strukturen, Habilitationsschrift, Greifswald 2011.

[17] — *Strong Turing degrees for additive BSS RAM's*, Logical Methods in Computer Science 9 (4:25), 2013, pp. 1–18.

[18] — *Computation over algebraic structures and a classification of undecidable problems*, Mathematical Structures in Computer Science 27 (8), 2017, pp. 1386–1413.

[19] W. Gellert, H. Küstner, M. Hellwich, and H. Kästner (eds.): Kleinen Enzyklopädie Mathematik, VEB Interdruck Leipzig (1968).

[20] Armin Hemmerling *Computability of string functions over algebraic structures*, Mathematical Logic Quarterly 44, 1998, pp. 1–44.

[21] Nikolai I. Kondakow Wörterbuch der Logik (The editors of the German edition: Erhard Albrecht and Günter Asser), Bibliographisches Institut, Leipzig 1983.

[22] Yiannis N. Moschovakis *Abstract first order computability I*, Transactions of the American Mathematical Society 138, 1969, pp. 427–464.

[23] Bruno Poizat Les Petits cailloux, *Une approche modèle-théorique de l'Algorithmie*, Aléas, Lyon 1995 [*Nur al-Mantiq wal-Ma'rifah 3*]. (Paperback 2002.)

[24] Franco P. Preparata, and Michael I. Shamos Computational Geometry, *An Introduction*, Springer 1985.

[25] Dana Scott *Some definitional suggestions for automata theory*, Journal of Computer and System Sciences 1, 1967, pp. 187–212.

[26] John C. Shepherdson, and Howard E. Sturgis *Computability of recursive functions*, Journal of the Association for Computing Machinery 10, 1963, pp. 217–255.

[27] Robert I. Soare Recursively Enumerable Sets and Degrees, *A Study of Computable Functions and Computably Generated Sets*, Springer, 1987.

[28] Alan M. Turing *On computable numbers, with an application to the Entscheidungsproblem*, Proceedings of the London Mathematical Society 42 (1), 1937, pp. 230–265.

[29] Klaus Weihrauch Computable Analysis, Springer 2000.

Two applications of admissible computability

Noam Greenberg
School of Mathematics and Statistics, Victoria University of Wellington, Wellington, New Zealand
greenberg@msor.vuw.ac.nz

ABSTRACT We discuss two applications of admissible computability: to higher randomness, and computability of uncountable structures.

Admissible computability, as formalised by Kripke [62] and Platek [73], is a common generalisation of metarecursion theory (Kreisel and Sacks [61, 60]) and Takeuti's approach to constructibility via recursion on the ordinals [88, 89]. Takeuti showed how to recover constructibility by considering the ordinals first and using a class of partial functions on the ordinals, resembling Kleene's partial recursive functions. Kreisel and Sacks were motivated by Church and Kleene's [20, 19, 56] development of the computable ordinals and the hyperarithmetic sets. The end result was a generalisation of computability to domains beyond the natural numbers, namely some ordinals greater than ω.

The first line of enquiry in the field known at the time as *α-recursion theory* was the attempt to lift to the admissible setting the constructions of classical computability, in particular in the areas of the lattice of c.e. sets and the Turing degrees of c.e. sets. For example, Sacks and Simpson [77] showed that the Friedberg-Muchnik resolution of Post's problem holds for every admissible ordinal, and later Shore extended Sacks's density theorem to all admissible ordinals [82]. On the other hand, some ordinals were shown to have unusual computable structure, for example, for some α, all incomplete c.e. degrees are low [81]. The techniques of α-recursion theory were later used in the study of nonstandard models of arithmetic and their computability, via the resemblance of failure of definable regularity of some singular ordinals and failure of the bounding principle in models of arithmetic. A more recent application of these investigations is Chong, Slaman and Yang's [15] construction of a non-standard model separating the stable and general forms of Ramsey's theorem for pairs, where again a crucial property is that incomplete c.e. degrees are low.

The author is supported by a Marsden Fund grant #17-VUW-090.

In this chapter we survey a couple of more recent applications of admissible computability, namely to the study of higher randomness and the study of uncountable computable structure theory. The methods of α-recursion theory showed that the generalistion of computability allows us to elucidate the underlying nature of basic notions and constructions of classical computability. This is the main theme of the work that we present. By contrasting classical computability with its generalisations, we can separate between the fundamental and the accidental. What is common to all generalisations, and thus can be considered necessary to computability, and what is special to the natural numbers?

An example is given by the work in [39], which exhibits some of the role that finiteness plays in computability. Kreisel [61] studied the analogy between Π^1_1 sets of numbers and computably enumerable ones, and noted that the correct analogue of hyperarithmetic is not computable but *finite* (for example, the image of a hyperarithmetic function is hyperarithmetic). Thus, in admissible computability, it is not only the notion of computability that is generalised, but the notion of finiteness. If α is an admissible ordinal, then α-computable processes are those which take up to α many steps to perform, and each ordinal $\beta < \alpha$ is in this context considered as finite. However in [39] it is shown that some constructions of computability theory rely on the fact that finite ordinals have predecessors. In particular, it is shown that Lachlan's [64] continuous tracing technique is necessary for his embedding of the 1-3-1 lattice into the c.e. degrees, and its success relies on the "true finiteness" of the natural numbers.

Similarly, studying higher randomness allows us to observe how "time tricks" are heavily utilised in classical algorithmic randomness; and studying uncountable linear orderings and free groups shows how to properly generalise classical results such as the Dzgoev-Remmel characterisation of computably categorical linear orderings.

Below, we first give a brief development of admissible computability, and then discuss the two applications mentioned. For more details on admissible computability we refer the reader to the classic [79] and to [8, 18]. For more on α-recursion theory see [14] and [83].

1 Admissible computability

There are several equivalent ways for defining admissible computability. Kripke, following Takeuti and Kleene, used an equation calculus. Platek [73] and later Köpke and Seyfferth [59] gave a more intuitive definition in terms of idealised computers or Turing machines with ordinal-length tape. The approach used most frequently appeals to set theory. The motivating example here is computability as definability in the structure HF, the collection of hereditarily finite sets. The structure (HF; $\in$) is effectively bi-interpretable with the standard model ($\mathbb{N}; +, \times$) of arithmetic: in one direction, the set $\mathbb{N}$ and the (graphs of the) functions $+$ and $\times$ are Δ_1-definable in HF; and via the *Ackermann interpretation* [1], the structure (HF; $\in$) is interpretable in $\mathbb{N}$ by a computable relation. Further, the map sending $n \in \mathbb{N}$ to the number coding n in this interpretation is computable, with computable range. It follows that a function $f\colon \mathbb{N} \to \mathbb{N}$ is computable if and only if it is Δ_1-definable in (HF; $\in$), and a set $A \subseteq \mathbb{N}$ is computably enumerable if and only if it is Σ_1-definable in (HF; $\in$). Here by Σ_1 we refer to the Levy hierarchy of formulas in the language of set theory, which is built up from formulas which only use bounded quantifiers.

We can therefore redevelop the theory of computability by *defining* a set $A \subseteq$ HF to be c.e. exactly if it is Σ_1-definable in (HF; $\in$), and a partial function $f\colon$ HF $\to$ HF to be partial computable if its graph is c.e. As a kind of motivation, we consider that once we agree that the relations $\in$ and equality should be considered computable, then bounded quantifiers correspond to a bounded search, which should also be admitted as computable; and so all Δ_0-definable sets should be considered computable. The external existential quantifier in a Σ_1 formula then corresponds to an unbounded search.

Proceeding with our defintions, a we call a set $A \subseteq$ HF computable if it is Δ_1-definable, that is, if it is c.e. and co-c.e. A partial function $f\colon$ HF $\to$ HF is computale if it is partial computable and its domain is computable. We can then proceed to prove the basic facts about computability from these definitions. For example:

Proposition 1.1. *A set $A \subseteq$ HF is computable if and only if its characteristic function 1_A is computable.*

To see this, assuming that A is computable, we observe that the relation $1_A(x) = y$ is

$$((y = 0)\ \&\ (x \notin A)) \vee ((y = 1)\ \&\ (x \in A)),$$

which is Δ_1-definable as well. Similarly:

Proposition 1.2. *The composition of partial computable functions is partial computable.*

For if the relations $f(x) = y$ and $g(y) = z$ are both Σ_1-definable, then so is the binary relation

$$(\exists y)\ f(x) = y\ \&\ g(y) = z.$$

Proposition 1.3. *A set $A \subseteq$ HF is c.e. if and only if it is the range of an injective partial computable function.*

To see this, in the harder direction, let $A \subseteq \mathrm{HF}$ be c.e.; let $R \subseteq \mathrm{HF}^2$ be a Δ_0-definable relation such that $x \in A \iff (\exists y)\, R(x,y)$. There is a computable ordering $<_{\mathrm{HF}}$ of HF of order-type ω. We let C be the collection of pairs (x,y) such that y is $<_{\mathrm{HF}}$-least such that $R(x,y)$ holds; then C is computable, and the map $(x,y) \mapsto x$ restricted to C is injective, computable and its range is A.

We should say a little more on the ordering $<_{\mathrm{HF}}$. It is the image of the natural ordering on $\mathbb{N}$ under the isomorphism between HF and $\mathbb{N}$ given by the Ackermann interpretation. A direct construction of $<_{\mathrm{HF}}$, not appealing to arithmetic, is done by *recursion*. In the context of HF, defining computable functions by recursion is stated as follows:

Proposition 1.4. *Suppose that* $I\colon \mathrm{HF} \to \mathrm{HF}$ *is computable. Then there is a unique function* $g\colon \mathbb{N} \to \mathrm{HF}$ *satisfying* $g(n) = I(g \upharpoonright n)$ *for all* $n \in \mathbb{N}$*; and this function is computable.*

Since $\mathrm{HF} = V_\omega = \bigcup_n V_n$, we can construct $<_{\mathrm{HF}}$ as the union of linear orderings $<_n$ of V_n such that each $<_{n+1}$ is an end-extension of $<_n$; to do so, all we need to do is to define a computable operation taking a linear ordering $<_X$ of a finite transitive set X and producing an ordering $<_{\mathscr{P}(X)}$ of $\mathscr{P}(X)$ which is an end-extension of $<_X$. One way to do this is to let $<_{\mathscr{P}(X)}$ be the right-lexicogrpahic ordering of $\mathscr{P}(X)$ based on $<_X$.[1] We just need to check that the map taking $<_X$ to $<_{\mathscr{P}(X)}$ is Σ_1-definable in HF, and then use recursion.

We remark that the recursion scheme given by Proposition 1.4 can be extended to define computable functions on other well-founded relations, for example $\in$; that type of recursion shows, for example, that the function taking $x \in \mathrm{HF}$ to its transitive closure is computable. Another use of recursion allows us to formalise first-order logic in HF. Under any reasonable formalisation of formulas, the collection of formulas is computable, and the satisfaction relation between finite structures and formulas is also computable, as all involve "bounded search". Now a Σ_1 sentence ψ (with parameters in HF) holds in HF if and only if there is some transitive $M \in \mathrm{HF}$ such that $M \models \psi$. This is because of absoluteness for Δ_0 predicates. This shows that the global Σ_1 satisfaction relation is c.e. We can then fix a computable numbers $\langle \psi_e \rangle$ of all Σ_1 formulas, and let

$$W_e = \{x \in \mathrm{HF} \,:\, \mathrm{HF} \models \psi_e(x)\}\,;$$

then the list $\langle W_e \rangle$ is a list of all c.e. sets; it is *uniformly* c.e., in that the set $\bigoplus_e W_e$ is c.e. (because the global Σ_1 satisfaction relation is Σ_1-definable). It is also *acceptable*,

[1] $a <_{\mathscr{P}(X)} b$ if $x \in b$ for the $<_X$-greatest element of $a \triangle b$

which means that whenever $\langle A_e \rangle$ are uniformly c.e. sets, then there is a computable function f such that for all e, $A_e = W_{f(e)}$. For if $\theta(x, y)$ is a Σ_1 formula defining $\bigoplus_e A_e$, then the function f takes e to the code of the formula $\theta(e, -)$.[2]

1.1 Admissible sets

The definition of an admissible set aims to answer the question: what is the minimal amount of set-theoretic closure required of a set M so that we can mimic the definition of computability above with satisfactory result? That is, if we: consider the elements of M to be "finite"; the ordinals of M to be "numbers"; and interpret "c.e." as Σ_1-definable in M — would this give us a reasonable theory of computability?

Some minimal amount of closure is certainly required. Take, for example, the structure $M = V_{\omega+\omega}$. The ordinals of this structure are $\omega+\omega$. Under any reasonable definition, ordinal addition should be a computable operation. However, $\omega \in M$ but $\omega + \omega \notin M$. That is, the sum of two "finite numbers" is "infinite", which should not be the case. So for a reasonable theory of computability, it should be the case that the ordinals of M are closed under addition. We could make a longer and longer list of similar operations (multiplication, exponentiation,...) but it is not clear where to stop. Rather, we (i) require some very basic amount of closure, so that definability of M makes any kind of sense; and (ii) then, anticipating the definition of M-computable functions, we require the image of a finite object under an M-computable function to be finite, or at least bounded. Here are the formal details.

Definition 1.5. A nonempty transitive set M is *amenable* if:

- For all $x, y \in M$, $\{x, y\} \in M$, $\bigcup x \in M$, and $x \times y \in M$;
- For every $\Delta_0(M)$ predicate R and every set $a \in M$, $a \cap R \in M$.

The second condition is referred to as Δ_0-*comprehension.*

Definition 1.6. Let M be an amenable set. A set $A \subseteq M$ is M*-computably enumerable* if it is $\Sigma_1(M)$. It is M*-computable* if it is $\Delta_1(M)$, i.e., M-c.e. and M-co-c.e.

A partial function from M to M is M*-partial computable* if its graph is M-c.e. An M-partial computable function is M*-computable* if its domain is M-computable.

Some basic facts about computability hold for all amenable sets, with exactly the same proofs. For example:

- The graph of an M-computable function is M-computable.

[2] The acceptability property is also known as the "s-m-n theorem".

- A set is M-computable if and only if its characteristic function is M-computable.
- The composition of M-partial computable functions is M-partial computable.

However, as was observed above, some amenable sets, such as $V_{\omega+\omega}$, are poor choices for computability purposes. The seond step consists of the following definition:

Definition 1.7. A nonempty transitive set M is *admissible* if it is amenable, and it satisfies Δ_0 *collection*: If $R \subseteq M^2$ is a $\Delta_0(M)$ relation, then for all $a \in M$ such that $a \subseteq \operatorname{dom} R$ there is some $b \in M$ such that for all $x \in a$ there is some $y \in b$ such that $R(x, y)$.

The definition is made to be minimal, so that it is easier to verify that certain sets are admissible; however it implies more:

Proposition 1.8. *Every admissible set satisfies Δ_1 comprehension and Σ_1 collection.*

Recalling our intentions, it is common to refer to the elements of an admissible set M as "M-finite". Thus, Δ_1-comprehension says: the intersection of an M-finite set with an M-computable set is M-finite. And Σ_1-collection means: if R is an M-c.e. relation and $a \subseteq \operatorname{dom} R$ is M-finite, then there is an M-finite $b \subseteq \operatorname{range} M$ which contains R-images for all $x \in a$. In particular, the image of an M-finite set under an M-computable function is contained in an M-finite set, and by comprehension, is in fact M-finite. A key fact used is that M-c.e. relations are closed under bounded quantification: if R is M-c.e., then so is $(\forall x \in y)\, R$.

Example 1.9. If κ is a cardinal, then

$$H_\kappa = \{x : |\operatorname{tc}(x)| < \kappa\}$$

(where $\operatorname{tc}(x)$ is the transitive closure of x) is an admissible set. This is clear if κ is regular, and uses a reflection argument when κ is singular. In particular, $\mathrm{HF} = H_\omega$ is admissible, and HF-computability is classical computability. On the other hand, for every admissible set M, we have $\mathrm{HF} \subseteq M$, and if $M \neq \mathrm{HF}$ then $\mathrm{HF} \in M$.

The key to admissibility is that it is precisely what is required to be able to define M-computable functions by recursion. The analogue of Proposition 1.4 is:

Proposition 1.10. *Let M be an admissible set. Let $\alpha = o(M) = M \cap \mathrm{On}$ (the ordinals of M). Suppose that $I\colon M \to M$ is M-computable. Then there is a unique function $g\colon \alpha \to M$ satisfying:*

- *For all $\beta < \alpha$, $g \upharpoonright \beta$ is M-finite and $g(\beta) = I(g \upharpoonright \beta)$.*

The unique such function g is M-computable.

Given Proposition 1.4 (and its generalisations to other M-computable well-founded relations other than $(\alpha; <)$), we can proceed with the development of computability theory as above, with proofs copied over nearly verbatim. For example, the formalisation of first-order logic proceeds in the same way, giving us a universal M-c.e. set: an M-c.e. set W such that letting, for $x \in M$, $W^{[x]} = \{y : (x, y) \in W\}$ be the x-section of W, the collection $\left\{W^{[x]} : x \in M\right\}$ is the collection of all M-c.e. sets. All standard proofs of the Kleene fixed point ("recursion") theorem hold in all admissible sets, and so on. An admissible set M is closed under basic ordinal arithmetic (addition, multiplication, exponentiation), and these operations are M-computable.

Further, some set-theoretic concepts have admissible effectivisations. A set-theoretic way of viewing admissibility is by saying that the ordinal $\alpha = o(M)$ for an admissible set M is "M-effectively regular"; it may fail to be a regular cardinal (indeed may not be a cardinal at all), but M-computable functions cannot witness this fact: there is no M-computable sequence of order-type $< \alpha$, unbounded in α. Some properties of regular cardinals then carry over to admissible ordinals, once we restrict to M-computable objects. For example:

Lemma 1.11. Let $M \neq \mathrm{HF}$ be admissible, let $\gamma < o(M)$ and let $\langle C_\alpha \rangle_{\alpha<\gamma}$ be a uniformly M-computable sequence of closed and unbounded subsets of $o(M)$. Then $\bigcap_{\alpha<\gamma} C_\alpha$ is M-computable, closed and unbounded in $o(M)$.

Similarly, if $M \neq \mathrm{HF}$ is admissible, and $f\colon o(M) \to o(M)$ is M-computable, then

$$\{\beta < o(M) : f[\beta] \subseteq \beta\}$$

is closed and unbounded in $o(M)$.

1.2 Constructibility

One part of classical computability which we developed above but have not generalised yet is $<_{\mathrm{HF}}$, the computable well-ordering of the "universe". This is because in general, there is no reason to assume such a well-ordering exists. For example, $M = H_{\omega_1}$ may not have any definable well-ordering, as the reals may fail to have such an ordering. To make computability "linear", we restrict ourselves to the constructible universe.

Just as for HF, if M is admissible, then satisfaction for structures inside M is M-computable (this was used to get a universal M-c.e. set). Further applications of the recursion principle (analgoues of Proposition 1.10) shows that if $A \in M$ then $\mathscr{P}_{\mathrm{DEF}}(A)$, the collection of A-definable subsets of A, is also an element of M, and the map $A \mapsto \mathscr{P}_{\mathrm{DEF}}(A)$ is M-computable. Applying recursion once more, we get:

Proposition 1.12. *Let M be admissible.*

1. *For all $\alpha < o(M)$, $L_\alpha \in M$, and the map $\alpha \to L_\alpha$ is M-computable.*

2. *$L_{o(M)} = L^M = \bigcup_{\alpha<o(M)} L_\alpha$ is M-c.e.*

Further, recall that $<_L$, the well-ordering of L, is defined recursively, with the ordering of $L_{\alpha+1}$ being an end-extension of the ordering of L_α; again, an examination shows that this operation can be defined in a Σ_1 way, and so the restriction of $<_L$ to an admissible set M is its restriction to $L_{o(M)}$, and is M-c.e.; the map taking α to $<_L\restriction L_\alpha$ is M-computable, and the map taking $x \in L_{o(M)}$ to $\{y \in L : y <_L x\}$ is M-partial computable.

A key fact is that L inherits admissibility:

Proposition 1.13. *Let M be an admissible set. Then $L_{o(M)}$ is admissible as well.*

There are some details to the argument, which concern the development of L inside L, but the crux of the proof is in showing that if $\beta < o(M)$ and $f\colon \beta \to o(M)$ is $L_{o(M)}$-computable, then it is bounded below $o(M)$; and the point is that since L^M is itself M-c.e., any Σ_1-definition within L^M can be translated to a Σ_1-definition in M, replacing the unrestricted quantifiers by quantifiers ranging over L^M. Hence f is also M-computable, and hence bounded. We thus define:

Definition 1.14. An ordinal α is admissible if L_α is an admissible set.

By Proposition 1.13, α is admissible if and only if there is an admissible set M such that $\alpha = o(M)$. We say that a set is α-c.e. if it is L_α-c.e., and α-computable if it is L_α-computable.

Working in initial segments of L, we utilise the following:

Proposition 1.15. *If α is admissible, then there is an α-computable bijection between α and L_α.*

Indeed, if $j(\beta)$ is the βth element of L according to $<_L$, then j restricts to a bijection between α and L_α, as j can be defined by recursion, and so $j[\alpha] \subseteq L_\alpha$ and the map $j \upharpoonright \alpha$ is L_α-computable. Similarly, a recursion on $<_L \upharpoonright L_\alpha$ defines $j^{-1} \upharpoonright L_\alpha$ by recursion inside L_α, and so $j \upharpoonright \alpha$ is the required bijection.[3]

Proposition 1.15 allows us to "linearize" α-computability. For example, we now get a numbering $\langle W_\beta \rangle_{\beta<\alpha}$ of all α-c.e. sets, rather than a numbering indexed only by the elements of L_α. Similarly, when performing priority arguments in α-computability, we can order all requirements in order-type α (rather than just indexed by elements of the admissible set), and so can set a priority ordering between them. We can regard every α-computable process as being recursively defined along α. More informally, we think of such processes as taking α many steps. In general, working in α-computability, with experience, we apply some kind of Church-Turing thesis to α-computable functions. Just as in classical computability, we eventually describe computable processes informally, rather than writing computer programs in detail, in admissible computability, we eventually cease to write down precise Σ_1 formulas defining the functions we are interested in. Instead, we develop an intuition as to what constitutes "legal" α-computable manipulations of α-finite objects (elements of L_α), and get a sense of the "time" that a process takes; if it takes fewer than α steps, then it "halts".

1.3 The least admissible ordinal (beyond ω)

So far we have given only one kind of example of admissible ordinals, namely the cardinals (Example 1.9). By a reflection argument (collapsing elementary substrctures), we see that there are many admissible ordinals which are not cardinals, indeed many countable ones. We can give a concrete description of the least admissible ordinal beyond ω. Interestingly, this ordinal arises from Church and Kleene's theory of the computable ordinals. An ordinal β is called computable if there is a computable well-ordering of $\mathbb{N}$ of order-type β. The computable ordinals form a countable initial segment of the ordinals, and the least non-computable ordinal is denoted by ω_1^{ck} (Church-Kleene ω_1). Now a computable ordinal $\beta > \omega$ cannot be admissible: if $<_\beta$ is some computable well-ordering of $\mathbb{N}$ of order-type β, then the ordering $<_\beta$ is an element of $L_{\omega+1}$, and so of L_β; if L_β were admissible, then by β-recursion, we would see that the isomorphism from $(\mathbb{N}; <_\beta)$ to $(\beta; <)$ would be β-computable, contradicting admissibility. However:

Proposition 1.16. *ω_1^{ck} is admissible.*

The reason for this is Σ_1^1 bounding, a key aspect of the theory of computable ordinals and hyperarithmetic sets. We give a quick review. Σ_1^1 bounding says that

[3]We remark that in fact, for every limit ordinal α, L_α is amenable and there is an L_α-computable bijection between α and L_α, but the definition of this bijection is not uniform in α.

if A is a Σ^1_1 collection of well-orderings of $\mathbb{N}$, then there is a computable bound on the order-types of all the orderings in A. One way to see this is to note that the collection of computable well-orderings is Π^1_1-complete, and so, by Cantor's diagonal argument, cannot be Σ^1_1; however, if the order-types of the orderings in A are not computably bounded, then we can give a Σ^1_1 definition of the computable well-orderings by asking for an embedding into some element of A. A more constructive approach (which also gives uniformity) is as follows: by a normal form argument, A is the projection of an effectively closed set P in Baire space. We let Q be the closed set of triples (L, f, g), where $(L, f) \in P$ (so $L \in A$) and g is an infinite descending sequence in L. Since every $L \in A$ is a well-ordering, Q is actually empty. This means that the tree S associated with the definition we gave Q is well-founded. For every $L \in A$, the tree of finite descending sequences in L is embeddable into S, and so the rank of S, which is computable, bounds the order-types of all elements of A.

Spector [85] showed, essentially, that if you take an iteration of the Turing jump along any computable well-ordering, then the result depends only on the order-type of the ordering. Thus, for every computable ordinal α, there is a well-defined Turing degree $\mathbf{0}^{(\alpha)}$, which contains all computable iterations of the Turing jump of length α. This is an increasing hierarchy of Turing degrees of length ω_1^{ck}; a set of numbers is defined to be *hyperarithmetic* if its Turing degree lies below some $\mathbf{0}^{(\alpha)}$. Kleene [57] used the Σ^1_1 bounding principle to show that the hyperarithmetic sets coincide with the Δ^1_1 sets, an effective analogue of the coincidence of the Borel sets with the $\boldsymbol{\Delta}^1_1$ ones (due to Suslin).

The next step is the Spector-Gandy theorem [86, 35], which analyses the quantifiers ranging over the hyperarithmetic reals.

Theorem 1.17. *A set $A \subseteq \mathbb{N}$ is Π^1_1 if and only if it is of the form "there exists a hyperarithmetic x such that $Q(-, x)$", where Q is Π^0_2.*

One direction is easier. Suppose that Q is an arithmetical predicate. The map taking a (computable index of a) computable well-ordering K to the iteration $\emptyset^{(K)}$ of the Turing jump along K is Π^0_2-definable, and so $\emptyset^{(K)}$ is Δ^1_1 uniformly in K. There is a hyperarithmetic x such that $Q(-, x)$ if and only if there is a computable well-ordering K and a Turing reduction Φ such that $Q(-, \Phi(\emptyset^{(K)}))$ holds. The search for indices for K and Φ is arithmetic; the main complexity is asking whether K is well-founded or not.

In the other direction, we start with an analysis of *pseudo-ordinals.* A (computable) pseudo-ordinal is a computable ordering of $\mathbb{N}$ which is not well-founded, however it has no hyperarithmetic infinite descending sequences. The existence of such objects can be concluded using Σ^1_1 bounding, in its guise as a "overspill" argument. By the easy direction of the Spector-Gandy theorem, the collection of computable well-orderings together with the computable pseudo-ordinals is Σ^1_1; since it contains all computable well-orderings, and it cannot coincide with the collection of computable well-orderings, a pseudo-ordinal must exist.[4] The length of the well-founded part of any pseudo-ordinal must be precisely ω_1^{ck}; it cannot be longer, as then a principal initial segment would give a computable copy of ω_1^{ck}. And it cannot be shorter, because otherwise, an argument using effective transfinite recursion shows that the well-founded part would be hyperarithmetic, and so we would be able to hyperarithmetically define an infinite descending sequence by avoiding the well-founded part. It follows that any iteration of the Turing jump along a pseudo-ordinal must compute all hyperarithmetic sets (and in particular will not be hyperarithmetic). Note that the existence of such an iteration is not automatic, as the pseudo-ordinal is, in fact, ill-founded; however, an overspill argument shows that there is a pseudo-ordinal L^* with a jump hiearchy along L^*.

We can now prove the harder direction of the Spector-Gandy theorem. Let A be Π^1_1. Membership of some n in A can be translated to the question of whether some computable linear ordering K is well-founded or not. K is well-founded if and only for some e in the well-founded part of L^*, the unique interation of the Turing jump along L^* up to e computes an embedding of K into L^* up to e; and this happens if and only if for some $e \in L^*$, a hyperaeithmetic iteration of the jump along L^* up to e computes such an embedding.

In essence, a similar argument can be now used to show that ω_1^{ck} is admissible. After learning some general facts about L_α for limit α, and about $L_{\omega_1^{\mathrm{ck}}}$ in particular, we can show that for admissibility, it suffices to show that every function $f\colon \omega \to \omega_1^{\mathrm{ck}}$ which is Σ_1-definable in $L_{\omega_1^{\mathrm{ck}}}$ (without parameters) is bounded below ω_1^{ck}. By effective transfinite recursion, we show that for all $\alpha < \omega_1^{\mathrm{ck}}$, there is a hyperarithmetic copy of L_α, in a uniform way: essentially, $\mathbf{0}^{(\omega\alpha)}$ computes such a copy. Now consider the copies of various L_α's computed by an iteration of the jump along L^*. At the well-founded levels we get L_α for all $\alpha < \omega_1^{\mathrm{ck}}$. At the ill-founded levels, we get ill-founded models behaving like L_α's, whose well-founded part includes $L_{\omega_1^{\mathrm{ck}}}$. Observing the interpretation of the function $f\colon \omega \to \omega_1^{\mathrm{ck}}$ in these models, we see that in each copy we get a restriction of f to (possibly) a subset of ω, but that in the ill-founded models, by upward absoluteness, we get f itself. That is, the ill-founded models believe that f is total. By "underspill", there must be some well-founded model which believes that f is total, that is, some L_α for $\alpha < \omega_1^{\mathrm{ck}}$ believes that f is total, whence f is bounded by α.

[4]A completely different way to obtain pseudo-ordinals is by the Gandy basis theorem, from which we get a countable, ill-founded model of ZFC whose well-founded part has height ω_1^{ck}; that model contains ill-founded "ordinals" which the model believes are computable ordinals.

Yet another similar argument gives a set-theoretic interpretation of the Spector-Gandy theorem: a set $A \subseteq \omega^\omega$ is Π^1_1 if and only if there is some Σ_1 formula in the language of set theory such that for all $y \in \omega^\omega$, $y \in A$ if and only if $L_{\omega_1^y}[y] \models \varphi(y)$. Here ω_1^y is ω_1^{ck} relativised to y, that is, the least ordinal which does not have a y-computable copy. The set $L_{\omega_1^y}[y]$ is the smallest admissible set containing y as an element. Restricting to subsets of $\mathbb{N}$, and recalling the definition above of α-c.e. sets for admissible ordinals α, we obtain:

Proposition 1.18. *A set $A \subseteq \mathbb{N}$ is Π^1_1 if and only if it is ω_1^{ck}-c.e.*

Also, a set $A \subseteq \mathbb{N}$ is Δ^1_1 (hyperarithmetic) if and only if it is ω_1^{ck}-finite (an element of $L_{\omega_1^{\mathrm{ck}}}$). As mentioned above, because of the strong analogy between Π^1_1 sets and c.e. sets, which is exemplified by Proposition 1.18, one would be led to believe that Δ^1_1 should be analogous to "computable". Kreisel and Sacks realised that the correct analogue is "finite", and so turned to investigate the complexity of subsets of ω_1^{ck}, which may be ω_1^{ck}-computable and not ω_1^{ck}-finite. Nonetheless, ω_1^{ck}-computability is very useful in the investigation of Π^1_1 sets.

1.4 Higher computability and effective descriptive set theory

As an aside, we give an example for how computability can be used to prove theorems of descriptive set theory. This relies on the connection between the "boldface" set-theoretic notions (Borel, $\boldsymbol{\Pi}^1_1$) and their "lightface" computable analogue (hyperarithmetic, Π^1_1), which was mentioned above (Addison [2]). We note that the coincidence of hyperarithmetic and Δ^1_1 holds for sets of reals as well as of numbers; a set $A \subseteq \omega^\omega$ is hyperarithmetic if for some computable α, $y \in A \iff y^{(\alpha)} \in B$ for some computable set B. We show:

Proposition 1.19. *If $B \subseteq \omega^\omega$ is Δ^1_1 and $f \colon B \to \omega^\omega$ is computable and injective, then $f[B]$ is Δ^1_1.*

Proof. Since the map $y \mapsto y^{(\alpha)}$ is Π^0_2-definable, for every $x \in B$, x is a $\Pi^0_2(f(x))$-singleton, which in turn implies that x is $\Delta^1_1(f(x))$, and so by Kleene's theorem, hyperarithmetic relative to $f(x)$. Now $f[B]$ is naturally Σ^1_1 (as B is Σ^1_1); it is also Π^1_1, as $y \in f[B]$ if and only if there is some $x \in B$, hyperarithmetic relative to y, such that $f(x) = y$; we apply the easy direction of the Spector-Gandy theorem. □

Since a function is contiuous if and only if it is computable relative to an oracle, we obtain:

Corollary 1.20. *If $B \subseteq \omega^\omega$ is Borel and $f \colon B \to \omega^\omega$ is continuous and injective, then $f[B]$ is Borel.*

There are several sophisticated uses of computability in descriptive set theory. For instance, the Glimm-Effros dichotomy (Harrington, Kechris and Louveau [48]) can be deduced from an effective version. Most recently, Day and Marks used a variety of effective considerations in their recent resolution of the decomposability conjecture (in preparation). Work leading up to their resolution also used surprising tools, such as the Shore-Slaman join theorem [84] used by Kihara [55].

2 Higher computability and randomness

In this section we give various examples of how the contrast and comparison between ω_1^{ck}-computability and classical computability give us new insights into the nature of computability itself. Some of these examples arise when we consider computability relative to an oracle; others, when we consider the interaction of computability and randomness. A main theme is the failure of "time tricks". Guided by questions of higher algorithmic randomness, we use ω_1^{ck}-computability to study reals, i.e., subsets of ω, rather than subsets of ω_1^{ck}. That is, the objects that we study have "height" ω. On the other hand, the computable processes that we use take up to ω_1^{ck} many "steps". This discrepancy between height and time exposes many instances of classical computability in which the coincidence of height and time in the lower setting is used in arguments. In the higher setting (of Π_1^1 sets and ω_1^{ck}-computability), some of the classical results still hold, but we need to devise new proofs; and some of the classical results fail. This tells us when time tricks are essential.

Algorithmic randomness attempts to answer the question "what does it means for a (finite or infinite) binary sequence to be random?" Using the tools of computability, it gives a hierarchy of randomness notions, based on ever more complicated null sets. For a detailed account, see Downey's chapter in this volume, or [27, 26, 72].

While most null sets considered in algorithmic randomness are around the level of effectively open and effectively G_δ sets, very early in the development of the theory, Martin-Löf [67] and Sacks [78, 79], and later Stern [87], introduced notions of randomness at a "higher setting", i.e., the level of arithmetic, hyperarithmetic and Π_1^1 sets. They defined the notion of Δ_1^1-randomness, which means avoiding all Δ_1^1 null sets; Sacks and Stern also introduced the notion of Π_1^1-randomness, with a similar definition.

A different approach was taken by Hjorth and Nies [50]. They relied on the analogy between Π_1^1 and c.e., and used the main definition of "classical" algorithmic randomness, namely ML-randomness, and simply replaced every instance of "c.e."

by Π^1_1. They showed that the resulting notion of Π^1_1-ML randomness[5] shared many of the properties held by the classical notion. For example, it can be charaterised using a Π^1_1 version of Kolmogorov complexity, and a higher analogue of the Kučera-Gács theorem [63, 34] holds.[6] Along with this notion of randomness, they introduced a continuous reducibility which interacts well with the study of randomness.

This last point was taken up and studied in detail in [9]. The Kučera-Gács theorem is just one of many ways that randomness and Turing reducibility interact.[7] When studying the "higher" (Π^1_1) analogue of randomness, we therefore also need to understand what is the correct higher analogue of Turing reducibility. The first guess would be relative hyperarithmetic reducibility. However the main drawback of this reducibility is that it is not given by continuous maps. This is why Hjorth and Nies introduced $\leqslant_{\text{fin-h}}$, which is a continuous version of hyperarithmetic reducibility. For example, this is the reducibility that they use in their version of the Kučera-Gács theorem.

2.1 Choosing the correct higher continuous reducibility

In [9], the authors showed that there are inequivalent ways of generalising Turing reducibility to the higher setting, and argued that one more general than $\leqslant_{\text{fin-h}}$ is the correct one for studying higher randomness. The issue revolves around *consistency* of functionals. Let us give some details. A *functional* is a set $\Phi \subseteq 2^{<\omega} \times 2^{<\omega}$. If Φ is a functional and $x \in 2^\omega$, then we let $\Phi(x)$ be the union of all σ such that for some $\tau \prec x$ we have $(\tau, \sigma) \in \Phi$. The motivation is that for any $x, y \in 2^\omega$ we have $y \leqslant_T x$ if and only if there is some c.e. functional Φ such that $\Phi(x) = y$. The pair (τ, σ) being in a functional Φ says that in the oracle machine coded by Φ, for any oracle x extending τ and any $k < |\sigma|$, on input k with oracle x the machine halts and outputs $\sigma(k)$.

For a functional Φ and $x \in 2^\omega$, there may be two reasons that $\Phi(x)$ would not be properly defined. One is partialness; in our formulation, $\Phi(x)$ may be a finite binary string rather than an infinite one. Another is that $\Phi(x)$ may be inconsistent, i.e.,

[5]A Π^1_1-ML null set is a G_δ set of the form $\bigcap_n U_n$, where the sets U_n are uniformly Π^1_1 open, and $\lambda(U_n) \leqslant 2^{-n}$; here λ denotes the fair coin measure on 2^ω. Using Π^1_1 open sets is made easy by the fact that a Π^1_1 set is open if and only if it is generated by a Π^1_1 set of string, that is, if it is of the form $\{x \in 2^\omega : (\exists \sigma \in W)\, \sigma \prec x\}$ for some Π^1_1 set $W \subseteq 2^{<\omega}$.

[6]The Kučera-Gács theorem says that every real is computable from some random sequence.

[7]Another very well-known example is van-Lambalgen's effective "2-step iteration" theorem [65], which says that the join $X \oplus Y$ is ML-random if and only if X is ML-random and Y is ML-random relative to X. Yet another example is the Miller-Yu theorem [68], which states that if $X \leqslant_T Y$ are both ML-random, and Y has one of many stronger randomness properties (for example, being 2-random), then X must share this property as well.

not a function: we could have (τ_0, σ_0) and (τ_1, σ_1) both in Φ where τ_0, τ_1 are both prefixes of x, but σ_0 and σ_1 are incomparable. This last point is often ignored in classical computability, because inconsistencies can be fixed: if $y \leqslant_{\mathrm{T}} x$ then in fact there is a consistent c.e. functional Φ such that $\Phi(x) = y$; Φ being consistent simply means that the situation above cannot occur, in other words, that it is consistent on all oracles.

Let us consider how we can remove inconsistencies. Suppose that Ψ is a c.e. functional; we can produce a consistent c.e. functional Φ such that for all x, if $\Psi(x)$ is consistent then $\Phi(x) = \Psi(x)$. How do we do this? we enumerate the "axioms" of Ψ (the pairs of strings in Ψ). Suppose that at stage s we have already enumerated Φ_s (a finite set of axioms), and see that a new axiom (τ, σ) is now enumerated into Ψ. It is possible that (τ, σ) is inconsistent with some axioms already in Φ_s, but it is also possible that the axiom applies to some oracles on which Ψ is consistent. What we do is look at every extension $\bar{\tau}$ of τ of length s, and enumerate the axiom $(\bar{\tau}, \sigma)$ into Φ_{s+1} only if it does not contradict another axiom already in Φ_s.

This was a time trick: at stage s, we used strings of length s, which were "fresh", in that they are longer than all strings that we dealt with so far. Now suppose that we work in the higher setting, with Π^1_1 functionals, which are enumerated in ω_1^{ck} many stages. We would like to mimic the argument, but now at stage $s \geqslant \omega$ we may be in bad shape. Suppose, for example, that at stage $n < \omega$ we see the axiom $0^n1 \mapsto 0$ in Ψ and copy it over to Φ; all these axioms are pairwise consistent. However at stage ω we see that Ψ maps some 0^k to 1, and in fact it is possible that $\Psi(0^\omega) = 1\cdots$ is total and consistent. However enumerating $0^m \to 1$ into Φ for any m after stage ω will make Φ inconsistent.

In fact, this argument is turned around in [10] to show that there are $x, y \in 2^\omega$ such that $\Phi(x) = y$ for some Π^1_1 functional Φ, but there is no consistent Π^1_1 functional Ψ such that $\Psi(x) = y$. That is, the time trick is essential in the previous argument. In [9], the authors argued that the relation $\leqslant_{\omega_1^{\mathrm{ck}}\mathrm{T}}$, defined by:

Definition 2.1. Let $x, y \in 2^\omega$. We say that x is *higher computable from* x, and write $y \leqslant_{\omega_1^{\mathrm{ck}}\mathrm{T}} x$, if there is some Π^1_1 functional Φ such that $\Phi(x) = y$.

is the correct definition to use. One piece of evidence is the relationship between computability and enumerability. There is only one reasonable definition for the relation "continuously relatively higher-x-c.e.": an *enumeration functional* is a set $W \subseteq 2^{<\omega} \times \omega$. For an enumeration functional W and an oracle $x \in 2^\omega$, we let W^x be the collection of all $n \in \mathbb{N}$ such that for some $\tau \prec x$ we have $(\tau, n) \in W$. Again the point is that for all x and $A \subseteq \mathbb{N}$, the set A is c.e. relative to x if and only if there is a c.e. functional W such that $W^x = A$. With enumeration functionals there are no issues of partialness or consistency, and so we define:

Definition 2.2. Let $x \in 2^\omega$. A set $A \subseteq \mathbb{N}$ is *higher x-c.e.* if $A = W^x$ for some Π^1_1 enumeration functional W.

The standard classical argument shows that that for any x and A, the set A is higher x-computbale (that is, $A \leqslant_{\omega_1^{\text{ck}}\text{T}} x$) if and only if A is both higher x-c.e. and higher co-x-c.e. Another piece of evidence for using $\leqslant_{\omega_1^{\text{ck}}\text{T}}$ is that some basic theorems about ML-random sequences, such as van-Lambalgen's theorem, hold for Π^1_1-ML sequences, with Turing replaced with $\leqslant_{\omega_1^{\text{ck}}\text{T}}$, rather than the consistent version of higher Turing reducibility.

Hjorth and Nies's definition is stricter than "consistent higher Turing". They defined $y \leqslant_{\text{fin-h}} x$ if there is a Π^1_1 functional Φ such that $\Phi(x) = y$, and such that Φ (as a set of pairs) is a monotone function defined on a subtree of $2^{<\omega}$. That is, not only is Φ consistent, but when we state that Φ maps τ to σ, we have already stated what Φ does on all of its initial segments. This appears to be a significant restriction. In the lower setting, an argument such as above uses a time trick to take a consistent functional and turn it into a c.e. functional of this type. It was therefore a little suprising to learn the following:

Proposition 2.3. *$y \leqslant_{\text{fin-h}} x$ if and only if there is a consistent Π^1_1 functional Φ such that $\Phi(x) = y$.*

That is, the time trick is not as essential to the result. Of course we need to make a new argument, and this argument is non-uniform (and must be so). Roughly, it goes as follows. Suppose that Φ is a consistent functional and that $y = \Phi(x)$. Define a Π^1_1 tree $T \subseteq 2^{<\omega}$: at stage $s < \omega_1^{\text{ck}}$, enumerate into T_s all strings ρ such that for all n, some extension of ρ is mapped by Φ_s to a string of length $\geqslant n$. Since $\Phi(x)$ is total, an admissibility argument show that $x \in [T]$.[8] Now there are two cases. If there is no s such that $x \in [T_s]$, then x collapses ω_1^{ck} in a continuous way: the map taking n to the least s such that $x \upharpoonright n \in T_s$ is cofinal in ω_1^{ck} and higher computable from x. Now we can transform Φ to a fin-h functional by copying $\Phi_s(\tau)$ if $\tau \in T_{s+1} \setminus T_s$. Otherwise, $x \in [T_s]$ for some $s < \omega_1^{\text{ck}}$. We can then use Φ_s to give a fin-h functional Ψ with $\Psi(x) = y$, as we can examine Φ_s in a hyperarithmetic way and let Ψ map ρ to the longest string compatible with $\Phi_s(\tau)$ for all extensions τ of ρ.

Another comparison between higher and classical computability is done by examining relative effectively closed sets. For every x, there is a $\Pi^0_1(x)$ class which

[8]For every $\tau \prec x$, for any n we know that Φ maps some extension of τ to some string of length $\geqslant n$ (namely some initial segment of x); the map taking n to the least s at which such an extension appears is ω_1^{ck}-computable, and so bounded.

contains no x-computable points. The usual argument is a time trick, but nonetheless, by a nonuniform argument, we can show that for every x, there is a higher x-effectively closed set containing no $y \leqslant_{\omega_1^{\mathrm{ck}}\mathrm{T}} x$ [10]. We can again show that that the nonuniformity is necessary.[9]

2.2 A deeper look into ML randomness

The idea of replacing "c.e." by Π^1_1 can be now relativised using Definition 2.2. We thus define:

- For every x, a higher x-ML null set is a set $\bigcap_n U_n$ such that $\lambda(U_n) \leqslant 2^{-n}$, and the sets U_n are uniformly higher x-c.e. open (generated by higher x-c.e. sets of strings).

Classically, the centrality of ML-randomness is witnessed by its robustness: many equivalent definitions coincide. Which of the implications are necessary, and which coincidental? Consider, for example, discrete measures. A *discrete measure* is a function $\mu\colon \omega \to \mathbb{R}^{\geqslant 0}$ such that $\mu(\omega) = \sum_n \mu(n)$ is finite. After identifying between numbers and finite binary strings, each discrete measure determines a co-null set: the set R_μ of x such that $\mu(x \upharpoonright n) \geqslant^\times 2^{-n}$.[10] Classically, a real x is ML-random if and only if for every left-c.e. (lower semicomputable) discrete measure μ, x is in the associated co-null set R_μ. Let us recall how to show this.

In one direction, let μ be a left-c.e. discrete measure. We show that there is a ML null set $\bigcap_n U_n$ continaing the complement of R_μ, so that every real "captured" by μ (in the sense that $x \notin R_\mu$) is captured by the ML-null set; this will show that if x is ML-random then it is random for discrete measures (meaning $x \in R_\mu$ for all left-c.e. μ). For each n, we let S_n be the set of strings σ such that $\mu(\sigma) > 2^n 2^{-|\sigma|}$, and let U_n be the open set generated by S_n. The sets U_n are uniformly c.e., and $\lambda(U_n) \leqslant^\times 2^{-n}$, the point being that $\lambda(U_n)$ is bounded by the *weight* $\sum_{\sigma \in S_n} 2^{-|\sigma|}$ of S_n, which in turn is bounded by $2^{-n}\mu(\omega)$.

In the other direction, let $\bigcap_n U_n$ be an ML-null set; we find a left-c.e. discrete measure μ which captures every $x \in \bigcap_n U_n$. To do this, for every n (uniformly) we find a c.e. antichain A_n of strings which generates U_n. We then define $\mu_n(\sigma) = 2^n 2^{-|\sigma|}$ for every $\sigma \in A_n$, and $\mu_n(\sigma) = 0$ otherwise; then $\mu_n(\omega)$ is bounded by the weight of A_n, which is the measure of U_n, and so $\mu = \sum_n \mu_n$ is as required.

[9] An interesting contrast is given by work of J. Miller and M. Soskova (in preparartion), who examine relativised randomness and related notions working with enumeration oracles, that is, using enumeration reducibility. One of their results is the construction of a "self-PA" oracle in their context, an oracle for which the property we discussed above fails.

[10] This means: for some constant $\delta > 0$, for all n, $\mu(x \upharpoonright n) \geqslant \delta 2^{-n}$.

Now the first implication is completely natural, and the same argument shows:

Proposition 2.4. *For every $x \in 2^\omega$, every sequence r which is higher x-ML random is random for higher x-c.e. discrete measures, that is, for every higher x-left-c.e. discrete measure μ, $r \in R_\mu$.*

In the other direction, one step is problematic: finding an antichain A_n generating U_n. This relies on a time trick. Let W be a c.e. set of strings generating an effectively open set W. We enumerate an antichain A in stages; at stage s, when we see a string σ enter S, we enumerate into A all extensions of σ of length s which do not extend any string previously enumerated into A. As above, we use the fact that at stage s, all strings in A have length $< s$. We can, in fact, show that this time trick is necessary:

Proposition 2.5 ([10]). *There is a higher effectively open set (a set generated by a Π^1_1 set of strings) which is not generated by any Π^1_1 antichain.*

Sketch of proof. We enumerate a higher c.e. set of strings V, ensuring that for all e, the eth higher c.e. set of strings W_e is not an antichain, or it does not generate $[V]^\prec$ (the open set generated by V). Let $\sigma_e = 0^e1$. For each e, let $A_e = \{\sigma_e\hat{}\sigma_n : n \in \mathbb{N}\}$. We let $V_0 = \bigcup_e A_e$. At stage $s < \omega_1^{\text{ck}}$, for every e, we check if both $[A_e]^\prec \subseteq [W_{e,s}]^\prec$ and $\sigma_e\hat{}0^\infty \notin [W_{e,s}]^\prec$. If so, then we enumerate σ_e into V_{s+1}.

Fix e, and suppose that $[W_e]^\prec = [V]^\prec$. By compactness of 2^ω, every $\tau \in A_e$ is covered by finitely many strings in W_e, and so there is some s such that $[\tau] \subseteq [W_{e,s}]^\prec$; this is recognised computably. By admissibility of ω_1^{ck}, as A_e is ω_1^{ck}-finite, there is some s such that $[A_e]^\prec \subseteq [W_{e,s}]^\prec$. Let s be least such. If $\sigma_e\hat{}\infty \in [W_{e,s}]^\prec$ then by our instructions, $\sigma_e\hat{}0^\infty \notin [V]^\prec$, contradicting $[W_e]^\prec = [V]^\prec$. Hence $\sigma_e\hat{}0^\infty \notin [W_{e,s}]^\prec$, whence by our actions, $\sigma_e\hat{}0^\infty \in [V]^\prec$, implying that $\sigma_e\hat{}0^\infty \in [W_e]^\prec$; an initial segment of $\sigma_e\hat{}0^\infty$ is enumerated into W_e at some stage $t > s$, and so must be incomparable with some string already in $W_{e,s}$, as $[A_e]^\prec$ is dense along $\sigma_e\hat{}0^\infty$. □

Thus, it is not clear that the result holds in all settings, and indeed, it does not; in upcoming work, the author, together with J. Miller and B. Monin, show that there is some oracle x relative to which there is a sequence r, which is random for higher x-c.e. discrete measures, but is not higher x-ML random.

Is the oracle necessary? the answer is positive; indeed, Hjorth and Nies showed the equivalence of higher ML-randomness with the even weaker property of being random for higher prefix-free complexity. To overcome the reliance on time tricks, they used the next lemma below. In the argument above, it was not actually important that the sets A_n were antichains; what we really used are the properties $U_n \subseteq [A_n]^\prec$ and $\text{wt}(A_n) \leqslant 2^{-n}$, where again $\text{wt}(A) = \sum_{\tau \in A} 2^{-|\tau|}$; we used an antichain because if A is an antichain then $\mathrm{wt}(A) = \lambda(A)$.[11] And the condition $\mathrm{wt}(A_n) \leqslant 2^{-n}$ is not fundamental either; all we need is $\sum_n \mathrm{wt}(A_n)$ to be finite. The following lemma then suffices.

Lemma 2.6. For every Π^1_1 open set U and every $\varepsilon > 0$ there is a Π^1_1 set of strings V such that $U \subseteq [V]^{\prec}$ and $\mathrm{wt}(V) \leqslant \lambda(U) + \varepsilon$. This is uniform in U and ε.

Sketch of proof. We give a proof slightly different to the one in [50], by introducing a new tool: the *projectum function.* There is an ω_1^{ck}-computable injective function $p\colon \omega_1^{\mathrm{ck}} \to \omega$ (essentially, take α to some index for a computable copy of α). We use p to "distribute mass" along the stages $s < \omega_1^{\mathrm{ck}}$. By stage s, we will have enumerated V_s with $U_s \subseteq [V_s]^{\prec}$. Suppose that $U_{s+1} = U_s \cup [\tau_s]$. Since $[\tau_s] \setminus [V_s]^{\prec}$ is hyperarithmetic, we can effectively (in the sense of $L_{\omega_1^{\mathrm{ck}}}$) find an antichain C_s (indeed a finite one) satisfying $[\tau_s] \subseteq [V_s]^{\prec} \cup [C_s]^{\prec}$ and $\lambda(C_s) \leqslant \lambda([\tau_s] \setminus [V_s]^{\prec}) + \varepsilon 2^{-p(s)}$; we add C_s to V_{s+1}. The sum of all the "extra errors" $\sum \varepsilon 2^{-p(s)}$ is bounded by 2ε, as p is injective. □

Along these lines, Hjorth and Nies showed that there is a universal Π^1_1-ML test (a largest Π^1_1-ML null set); the standard argument applies. However, when relativising, in the lower setting, a time trick is used to construct a *uniform* oracle universal ML-test; oracle effectively open operators U_n such that for every oracle x, $\langle U_n^x \rangle$ is a universal test for x-ML-randomness. In the higher setting, in [9] it is shown that this cannot be done in the higher setting, and in fact, with more work, an oracle x is constructed for which there is no universal higher x-ML test at all.

Further work, however, is required to completely elucidate the relationships between all variants of ML-randomness, for example, those that rely on c.e. martingales (equivalently, left-c.e. continuous measures), prefix-free Kolmogorov complexity, and Schnorr tests. The necessary implications are clear, but constructions of oracles witnessing the failure of other implications appear to be hard.

2.3 The higher limit lemma

Recall that Shoenfield's limit lemma states that a function f is $\emptyset'$-computable functions iff it has a *computable approximation*: a uniformly computable sequence $\langle f_s \rangle$ such that $f = \lim_s f_s$ (in the discrete topology). In the higher setting, any complete Π^1_1 (such as Kleene's $\mathcal{O}$, or the set of indices of computable well-orderings) plays the role of $\emptyset'$. For computable approximations, though, we need approximations

[11] We write $\lambda(A)$ for $\lambda([A]^{\prec})$.

of length ω_1^{ck} (the limit of a hyperarithmetic ω-sequence of functions is hyperarithmetic, as Δ_1^1 is closed under taking the Turing jump). Indeed, fixing a complete Π_1^1 set O, the following are equivalent for a function f:

- $f \leqslant_{\omega_1^{\mathrm{ck}}\mathrm{T}} O$;[12]
- there is a ω_1^{ck}-computable sequence $\langle f_s \rangle_{s<\omega_1^{\mathrm{ck}}}$ (of functions $f_s \colon \omega \to \omega$, each necessarily being hyperarithmetic) such that $f = \lim_s f_s$, in the sense that for all n, for some $s < \omega_1^{\mathrm{ck}}$, $f_t(n) = f(n)$ for all $t \in [s, \omega_1^{\mathrm{ck}})$.

The fact that the approximation has limit stages allows us to define and investigate subclasses of the O-computable functions which have no classical analogue. For example, a *finite-change approximation* is an approximation $\langle f_s \rangle$ such that for all n, there is no infinite increasing sequence $\langle s_k \rangle$ of stages such that for all k, $f_{s_{k+1}}(n) \neq f_{s_k}(n)$.[13] Not all functions $f \leqslant_{\omega_1^{\mathrm{ck}}\mathrm{T}} O$ have finite change approximations.

Such subclasses can be used to answer questions about notions of randomness that lie between higher ML randomness and Π_1^1-randomness, prime among them the higher version of weak 2-randomness (or strong 1-randomness): a higher test for weak 2-randomness is a set $\bigcap_n U_n$ which is null, where the sets U_n are uniformly Π_1^1 open. Each such null set is Π_1^1, and so we get the implications:

$$\Pi_1^1\text{-random} \quad \Longrightarrow \quad \text{higher weakly 2-random} \quad \Longrightarrow \quad \text{higher ML random.}$$

Some basic facts about weak 2-randomness rely on time tricks. Consider, for example, the fact [23] that no weakly 2-random sequence can be Δ_2^0: let $\langle x_s \rangle$ be a computable approximation of a Δ_2^0 sequence x. We let U_n be the open set generated by $\{x_s \upharpoonright n : s > n\}$; note how stages and lengths are compared. Then $\bigcap_n U_n = \{x\}$, which is null, and the sets U_n are uniformly c.e. We know that to some extent the time trick is necessary: Gandy's basis theorem implies that there is a Π_1^1-random sequence $x \leqslant_{\mathrm{T}} O$.

To what extent can O-computable sequences be higher weak 2-random? Chong and Yu [16] showed, using the Lebesgue density theorem, that no higher weakly 2-random sequence can be higher left-c.e. (left-Π_1^1). In [44], the following is shown:

Proposition 2.7. *No sequence which has a finite-change approximation can be higher weakly-2 random.*

[12] We remark that $f \leqslant_{\omega_1^{\mathrm{ck}}\mathrm{T}} O$ if and only if $f \leqslant_{\mathrm{T}} O$.

[13] Note that it is not enough to require that $f_{s+1}(n) \neq f_s(n)$ for only finitely many $s < \omega_1^{\mathrm{ck}}$; to that, we need to add that for all limit $s < \omega_1^{\mathrm{ck}}$, the limit $f_{<s} = \lim_{t<s} f_t$ exists and for all n, $f_s(n) \neq f_{<s}(n)$ for only finitely many limit s. In general, in approximations for which the limits $f_{<s}$ exist for all limit s, by reindexing, we assume that $f_s = f_{<s}$.

Sketch of proof. Let $\langle x_s \rangle$ be a finite-change approximation of $x = x_{\omega_1^{ck}}$. The property that we use about such approximations is that (after perhaps modifying the limit steps) the set $X = \left\{x_s : s \leqslant \omega_1^{ck}\right\}$ is *closed*. Let U_n be the open set generated by $\left\{x_s \upharpoonright n : s < \omega_1^{ck}\right\}$. Then every $y \in \bigcap_n U_n$ lies in the closure of X, and hence in X. Since X is countable, it is null. □

Proposition 2.7 is used, for example, to show that the two halves of the higher version of Chaitin's Ω are not Π^1_1-random, indeed, they are not higher weakly 2-random. Another property, weaker than having a finite-change approximation, can be used to separate Π^1_1-randomness from higher weak 2-randomness. This relies on Stern's result [87], rediscovered by Chong, Nies and Yu [17], that x is Π^1_1-random if and only if it is Δ^1_1-random and $\omega_1^x = \omega_1^{ck}$. Thus, it suffices to constuct a higher weakly 2-random sequence x such that $\omega_1^x > \omega_1^{ck}$. For example, every non-hyperarithmetic x with a finite-change approximation satisfies $\omega_1^x > \omega_1^{ck}$: the function mapping n to the least s such that $x \upharpoonright n = x_s \upharpoonright n$ is unbounded in ω_1^{ck}. Unfortunately, Proposition 2.7 says that this cannot be used for the desired separation. However a weaker notion is compatible with being higher weakly 2-random: having an approximation $\langle x_s \rangle$ such that for all n, the set $\left\{s < \omega_1^{ck} : x_s \upharpoonright n = x \upharpoonright n\right\}$ is closed (and neccessarily unbounded). Again, if x has such an approximation and is not hyperarithmetic, then the map $n \mapsto$the least s such that $x_s \upharpoonright n = x \upharpoonright n$ is unbounded in ω_1^{ck} (if $t \leqslant \omega_1^{ck}$ is a bound, then by closure, $x_t = x$; for all $t < \omega_1^{ck}$, x_t is hyperarithmetic).

Finally, we remark that subclasses of the O-computable functions can be used to give Demuth-style characterisations of higher weak 2-randomness. This is related to the class $\mathtt{MLR}[\emptyset']$, determined by null sets $\bigcap_n U_n$ for which $\lambda(U_n) \leqslant 2^{-n}$, each U_n is effectively open, but the sequence itself is not necessarily uniformly so; rather, $\emptyset'$ can be used to find a c.e. open index for U_n. In the lower setting, $\mathtt{MLR}[\emptyset']$ is equivalent to weak 2-randomness, but in the higher setting, the class $\mathtt{MLR}[O]$ (modified so the indices are of Π^1_1 open sets) is very strong, strictly stronger than Π^1_1-randomness, as it is incompatible with being O-computable. A time trick similar to the one above is used in the lower setting. In the higher setting, we restrict the kind of functions that give indices.

For example, letting $\langle W_e \rangle_{e<\omega}$ be an effective enumeration of all Π^1_1 open sets, we say that a *finite-change null set* is a null set of the form $\bigcap \langle W_{f(n)} \rangle_{n<\omega}$ which is nested (meaning $W_{f(n+1)} \subseteq W_{f(n)}$), satisfies $\lambda(W_{f(n)}) \leqslant 2^{-n}$, and such that f has a finite-change approximation. Avoiding all finite-change null sets is equivalent to higher weak 2-randomness. In the direction which requires a new argument, curiously, we are informed by the proof of Proposition 2.7. Suppose that $\langle W_{f(n)} \rangle$ determines a finite-change null set, which we want to cover by a higher weak 2-test; let $\langle f_s \rangle$ be a finite-change approximation of f. By fiddling, we may assume that for all s, $\langle W_{f_s(n)}\rangle$ is nested and $\lambda(W_{f_s(n)}) \leqslant 2^{-n}$. Let $U_n = \bigcup_{s<\omega_1^{\mathrm{ck}}} W_{f_s(n)}$; we show that $\bigcap U_n$ is null. For $s \leqslant \omega_1^{\mathrm{ck}}$, let $A_s = \bigcap_n W_{f_s(n)}$, and let $A = \bigcup_{s\leqslant\omega_1^{\mathrm{ck}}} A_s$. Each A_s is null, and ω_1^{ck} is countable, so A is null. And $\bigcap_n U_n \subseteq A$; this uses the fact that $\left\{f_s : s \leqslant \omega_1^{\mathrm{ck}}\right\}$ is compact.

2.4 Other work

There is much more to say about higher randomness. There is extensive work on lowness notions for higher randomness (the higher analogues of K-trivial sets) in [50, 9, 3], work on the Borel complexity of the set of Π^1_1-randoms [69], and a higher analogue of the Miller-Hirschfeldt theorem saying that a ML-random is weak 2 random if and only if it forms a minimal pair with $\emptyset'$ [44]. In a different direction, recently researchers have been studying randomness defined using infinite-time Turing machines [11, 70], in which admissibility plays an important role as well.

3 Uncountable computable structure theory

Effective considerations of rings and fields were made even before the formalisation of computability itself [21, 49, 90]. The modern incarnation of these is the field of computable structure theory (or computable algebra). The aim is to understand the relationship between the algebraic structure and the information stored in that structure; see the surveys [58, 47, 22], the books [6, 31] and the upcoming book by Montalbán.

By the nature of the tools involved, such considerations are restricted to countable structures. It is natural to attempt to study effective properties of uncountable structures as well, and several approaches have been suggested (see [40]). In [42], J. Knight and the author suggested using admissible computability on cardinals for this purpose (recalling Example 1.9 saying that every cardinal is an admissible ordinal). For example, for any cardinal κ, we say that a group G with universe κ is *κ-computable* if its group operation is a κ-computable function, and similarly a linear ordering with universe κ is κ-computable if the ordering relation is κ-computable. We can then attempt to answer the same questions asked in the countable setting: which linear orderings of size κ have κ-computable copies? What does it take to compute isomorphisms between two κ-computable copies of some structure?

This last question gives rise to the concept of a *κ-computably categorical* structure: a structure M such that for any two κ-computable copies of M, there is a κ-computable isomorphism between them. The classical notion of ω-computably categorical structres, which was introduced by Mal'cev [66], has been studied extensively. There are two lines of inquiry. One is general; important results, for example, are the characterisation of relative computable categoricity by syntactic means of an effective family of formulas defining the orbits of the structure [13, 7]; Goncharov's result saying that computable categoricity is equivalent to its relative version in 2-decidable structures [36]; work on uniform computable categoricity [24]; Goncahrov's construction of structures with finite computable dimension (the number of computable copies up to computable isomorphism) [37]; and more recent work on the complexity of being computably categorical [28]. In [43], some work along these lines is carried out in the higher setting.

Another line of inquiry is trying to understand computable categoricity in particular classes of structures. An important example is linear orderings, where the fundamental result is due to Dzgoev (see [38]) and Remmel [75]:

Theorem 3.1. *A computable linear ordering is computably categorical if and only if it has finitely many elements with successors.*

Theorem 3.1 formalises the idea that the only way that a linear ordering can be computably categorical is if Cantor's back-and-forth technqiue can be used to construct isomorphisms between any two of its copies. The point about having finitely many successors is that after matching up finitely many elements (the successor pairs), which can be done non-uniformly, the rest is broken up to finitely many dense pieces, which we can match with a back-and-forth process.

In [41], the authors characterise ω_1-computably categorical linear orderings. As we shall see, this sheds light on the classical result as well. The naive attempt at generalising Theorem 3.1 to the higher setting would be to guess that an ω_1-computable linear ordering is ω_1-computably categorical if and only if it has only countably many successor pairs; recall that in L, $L_\kappa = H_\kappa$, so the ω_1-finite sets are precisely the hereditarily countable ones.[14] This, however, fails in both directions.

In one direction, we observe that $2\cdot\mathbb{R}$, the linear ordering obtained from the reals by replacing each real by a successor pair, is ω_1-computably categorical. If we are given two copies of $2\cdot\mathbb{R}$, then non-uniformly, we fix two copies of $2\cdot\mathbb{Q}$ inside them. Then we can match the rest. The important point is that if A is a countable subset of an ω_1-computable linear ordering K, then for every $x \in K\setminus A$, we can ω_1-effectively find the left and the right cuts that x defines in A, that is, the sets $\{a \in A : a <_K x\}$ and $\{a \in A : x <_K a\}$. So suppose that K and K' are two ω_1-computable copies of $2\cdot\mathbb{R}$; let A and A' be the images of $2\cdot\mathbb{Q}$ in K and K', respectively. As stated, an order-preserving bijection $\pi\colon A \to A'$ can be fixed, as it is a countable object.

[14]Throughout, we assume, for simplicity, that $V = L$.

Now given any $p \in K \setminus A$, we compute the A-cut (B_1, B_2) determined by p. We first wait for another point q which determines the same A-cut. Then, we find two points p' and q' which in K' determine the A'-cut $(\pi[B_1], \pi[B_2])$, and match $\{p, q\}$ with $\{p', q'\}$ in an order-preserving way.

What this second example tells us is that the Dzgoev-Remmel result relies, to put it flippantly, on the fact that $\aleph_0$ is a strongly inaccessible cardinal: in more serious language, on the fact that a finite subset of a linear ordering can be split into only finitely many cuts, and so determines only finitely many intervals in the linear ordering. In any case, the analysis so far shows that the phrasing of the result in terms of successor pairs is a little misleading. Rather, what it should say, is that a countable linear orderin K is computably categorical if and only if there is a finite set $A \subset K$ such that for every cut (A_1, A_2) of A, the order-type of the K-interval determined by that cut allows us to effectively match it to its copy.[15] The example $2 \cdot \mathbb{R}$ shows that in the collection of these "effectively matchable" order-types, we must include the finite ones, as well as, in the countable setting, the rationals.

Under this reconsideration, we could guess that an ω_1-computable linear ordering K is ω_1-computably categorical if and only if there is a countable set $A \subset K$ such that every K-interval determined by a cut of A is either finite or dense. However, in the uncountable setting, density is insufficient. For we note that the linear ordering $\mathbb{Q} \cdot \omega_1$ (the result of replacing each point in the linear ordering ω_1 by a copy of the rationals) is dense, but not ω_1-computably categorical. We can construct two ω_1-computable copies K and K' of this linear ordering and at the same time diagonalise against all ω_1-computable attempts at an isomorphism between them. This is done with a priority argument. For each $e < \omega_1$, we ensure that the eth partial ω_1-computable function φ_e is not an isomorphism from K to K'. To do that, we fix an interval C of K of order-type $\mathbb{Q}$; when we see that φ_e has halted on every point of C, we add a new point to K', between points of $\varphi_e[C]$. Note that if φ_e is total, we will witness $\varphi_e \restriction C$ at a countable stage, as C is countable. The requirement then imposes restraint on weaker ones. This only restrains countable pieces of K and K', and so the regularity of ω_1 implies that each requirement can find some unrestrained space.

What underlies this argument is that in the original plan of carrying out the back-and-forth process, density was not really the important property of $\mathbb{Q}$; what is important is that $\mathbb{Q}$ is *saturated*. There is a unique saturated ω_1-linear ordering, denoted by η_1, and we can construct ω_1-computable copies of it. In any case, this

[15]Here a cut (A_1, A_2) of A is a partition of A into an initial and final segment (one of which may be empty), and the K-interval determined by this cut is $\{x \in K : A_1 < x < A_2\}$, where as expected $A_1 < x$ means $(\forall a \in A_1)\, a <_K x$, and similarly for $x < A_2$. We sometimes write $(A_1, A_2)_K$ for this K-interval.

shows that perhaps a reformulation of the Dzgoev-Remmel theorem which grasps at its "real essence" is as follows:

Theorem 3.2. *A computable linear ordering K is computably categorical if and only if there is a finite set $A \subset K$ such that every K-interval determined by a cut A is either finite or saturated.*

We almost have a good guess for how to characterise ω_1-computably categorical linear orderings; we just require the parameter set A to be countable. But there is one last issue. Given two linear orderings of finite size n, we can effectively match between them by waiting for all n points to appear. However, this assumes that *we know* n. If there are infinitely many intervals, we could have some, many, or all sizes appear as intervals. To complete a construction for all intervals together, we need to know which is which. In the lower setting, this effective aspect of the characterisation is missing, because adding a finite amount of information costs nothing. This hidden aspect is revealed in the eventual characterisation:

Theorem 3.3 ([41]). *An ω_1-computable linear ordering K is ω_1-computably categorical if and only if there is a countable set $A \subset K$ such that every K-interval determined by a cut of A is either finite or saturated, and further, there is a partial ω_1-computable function which for every cut (A_1, A_2) of A for which the interval $(A_1, A_2)_K$ is finite, maps the cut to the size of that interval.*[16]

Now one may ask why we restricted ouselves to $\kappa = \omega_1$. What about ω_2, and other cardinals? The answer lies in the proof of Theorem 3.3. It is beyond the scope of this survey to give a detailed account. We only mention that the argument relies on the fact that every countable linear ordering has a proper self-embedding (this is how we force an opponent to add points where we want them). Further, it makes use of Hausdorff's separation between scattered and nonscattered countable linear orderings, and so in contrast with the classical case, the construction on an interval will have two pahses: first, we must force the opponent's interval to be nonscattered; then we make it saturated. This special analysis of countable linear orderings does not generalise to uncountable ones; indeed, there are uncountable linear orderings with no proper self-embeddings. To date, there is no charaterisation of ω_2-computably categorical linear orderings.

[16]On cuts defining infinite outputs, the output of this function may be anything. The point being that during the construction of an isomorphism, we may believe that $(A_1, A_2)_K$ has some size n, and match accordingly. If this guess is false, then $(A_1, A_2)_K$ is saturated, so we can extend the matching that we already made to an isomorphism.

3.1 Free abelian groups

When a structure is not computably categorical, it still makes sense to ask how much information is required to compute isomorphisms between any two computable copies. In the special case of the countable free abelian group $\mathbb{Z}^{(\omega)}$, computing an isomorphism with the standard copy is the same as computing a basis for the group. In the countable case, Downey and Melnikov [25] showed that the free abelian group is Δ^0_2-categorical; equivalently, if we are given a copy of the countable free abelian group (via its groups table), then with one Turing jump we can construct a basis. The construction is recursive: at a finite stage we have a finite piece of the basis, and then we can extend it to a larger finite piece, making sure that the next element of the group is generated.

Of course, this resembles the construction of a basis of a vector space: add a vector not spanned by the ones chosen so far. In the case of groups, the fact that we cannot always divide means that some finite linearly independent sets cannot be part of a basis (think the subset $\{2\}$ of the group $\mathbb{Z}$). At the minimum, they need to span a *pure subgroup*: roughly, one in which every division which occurs in the larger group, already occurs within the subgroup. Pontryagin [74] essentially showed that a finite set generating a pure subgroup can be extended to a basis; in other words, a finitely generated pure subgroup H of a free group G *detaches* in G, in the sense that $G = H \oplus K$ for some subgroup K of G. This allowed Downey and Melnikov to give to construct a basis by recursion.

What happens in the uncountable setting? The situation for vector spaces is the same as in the countable world: we may need one jump to determine linear independence, but once we have that, we can construct bases by transfinite recursion, adding one element at a time. With free abelian groups, the situation is very different. Fuchs (see [33]) showed that there is a pure subgroup H of $\mathbb{Z}^{(\omega)}$ which does not detach in $\mathbb{Z}^{(\omega)}$. Suppose that this $\mathbb{Z}^{(\omega)}$ sits inside an uncountable free group G (all groups henceforth are abelian). The following situation may occur: we try to build a basis for G by recursion. We keep adding elements, all within H, each generating a pure subgroup. At each finite step, we have a finite set; by Pontryagin, it can be extended to a basis of G. But if we happened to work within H, at stage ω we may have an infinite set which cannot be extended to a basis of G. New complications are introduced at limit steps that do not occur in the countable world.

In fact, the situation is dire. Not only does not a jump or two suffice, no transfinite iteration suffices, and beyond:

Theorem 3.4 ([45]). *For every uncountable regular cardinal κ, and every set $X \subseteq \kappa$ which is $\Delta^1_1(L_\kappa)$-definable, there is a κ-computable copy G of the free abelian group $\mathbb{Z}^{(\kappa)}$ which has no X-computable basis.*

The class $\Delta^1_1(L_\kappa)$ is huge, much bigger than any definition of "hyperarithmetic" in the context of κ-computability. So in particular, Theorem 3.4 implies that there is a κ-computable copy of $\mathbb{Z}^{(\kappa)}$ which has no basis which is first-order definable over L_κ. Essentially, no process of recursion can be designed that builds bases of uncountable free groups "from below". A basis needs to be given in its entirety.

The work in [45] continues work in [46], in which Theorem 3.4 was first proved for all successor cardinals κ, in fact for all regular cardinals which are not weakly compact. For such cardinals, it is shown in [46] that the problem of *identifying* free groups is as complicated as possible: it is $\mathbf{\Sigma}^1_1(L_\kappa)$-complete.

In general, the structure of uncountable torsion-free abelian groups is "sufficiently thin" so that set-theoretic considerations play a major role in their investigation. This is most famously seen in Shelah's independence result for the Whitehead problem (asking whether every abelian group satisfying $\mathrm{Ext}(G, \mathbb{Z}) = 0$ is free) [80]; for more on this extensive body of work see the book [30]. The results mentioned above heavily rely on set-theoretic methods as well.

For example, the identification of free groups is tied to the problem of telling whether a given subset of a regular cardinal κ is nonstationary. Suppose that G is an abelian group with universe κ. A *filtration* of G is a sequence $\langle G_\alpha \rangle_{\alpha \leqslant \kappa}$ which is increasing ($\alpha < \beta$ implies $G_\alpha \subseteq G_\beta$), continuous ($G_\beta = \bigcup_{\alpha<\beta} G_\alpha$ for all limit ordinals $\beta \leqslant \kappa$), satisfying $G = G_\kappa$ and $|G_\alpha| < \kappa$ for all $\alpha < \kappa$. If $\bar{G} = \langle G_\alpha \rangle$ is a filtration of G, then we let the *detachment set* of $\bar{G}$ to be

$$\mathrm{Div}(\bar{G}) = \{\alpha < \kappa \,:\, (\forall \beta \in (\alpha, \kappa))\ G_\alpha \mid G_\beta\},$$

where $H \mid K$ denotes that H detaches in K as a direct summand.[17] The detachment set depends on the choice of filtration, however any two choices result in detachment sets that are equivalent modulo the nonstationary ideal on κ; any two filtrations agree on a club (closed and unbounded subset of κ). If every G_α for $\alpha < \kappa$ is free, then G is free if and only if $\mathrm{Div}(\bar{G})$ contains a club [29].

Fokina et al. [32] showed that the nonstationary ideal on ω_1 is $\Sigma^1_1(L_{\omega_1})$-complete. For most regular cardinals, we can show that telling which sets contain clubs is computationally equivalent, in the sense of κ-computability, to the problem of identifying which groups are free. In the harder direction, given a set $X \subseteq \kappa$, we want to κ-effectively construct a filtration $\bar{G}$ of a group $G = G(X)$ which is free if and only if X contains a club. In set theory, this is usually done "statically", but when effective considerations are applied, it is useful to think of the construction as being done recursively. At step α of the construction we will have defined G_γ for all $\gamma \leqslant \alpha$, and need to define $G_{\alpha+1}$, depending on whether $\alpha \in X$ or not. If $\alpha \in X$ then we want

[17] When G_β is free this is equivalent to G_β / G_α being free.

to arrange that $\alpha \in \mathrm{Div}(\bar{G})$, by adding copies of $\mathbb{Z}$ as direct summands. If $\alpha \notin X$ then we want to "twist" G_α inside $G_{\alpha+1}$ so that $\alpha + 1$ witnesses that $\alpha \notin \mathrm{Div}(\bar{G})$. In doing so, we need to ensure that we do not destroy past detachments of G_β's for $\beta < \alpha$.

It would seem that we want to ensure that $\mathrm{Div}(\bar{G}) = X$. However, this ignores an important point: we need to ensure that for all limit $\alpha < \kappa$, G_α is free. In other words, we want to ensure that we haven't twisted too much, even when X is sparse. To do that, we restrict our twisting further, to a set $E \subseteq \kappa$ which is stationary in κ, but does not reflect (for all $\alpha < \kappa$, $E \cap \alpha$ is nonstationary in α). In the case that κ is a successor cardinal, such a set E is given by Jensen's elaborate machinery [51] which he used to define a global square sequence in L. When κ is inaccessible, we need to thin E further, to a set which witnesses the describability of κ, when such a set exists.[18] When such a set does not exist, the cardinal κ is weakly compact, and then this very compactness tells us that the problem of determining which group is free is actually relatively simple: a group G of size κ is free if and only if every subgroup of size $< \kappa$ is free, and so in this case the collection of free groups is $\Pi^0_2(L_\kappa)$.

3.2 Further work

Ash et al. [4, 5] generalised Watnick's result [91, 76] (independently discovered by Downey (see [22]) and by Ash et al.) by showing that for any computable ordinal α and any $\mathbf{0}^{(2\alpha)}$-computable linear ordering K, there is a computable copy of $\mathbb{Z}^\alpha \cdot K$.[19] This shows that 2α jumps are not only sufficient but required to compute the iteration of the Hausdorff drivative of a linear ordering (the derivative is taken by identifying points which are finitely far apart). Ash's technique is related to Montalbán's "true stage" approximation of the iterations of the jump [71]. Work in progress by Turetsky and the author examines the situation in the uncountable setting, again revealing what is special about the countable case and what is general. For example, the work seems to imply that the fundamental operation is not actually Hausdorff's derivative, but rather, one closer to that of Cantor-Bendixson.

Further extensive work on thin trees of uncountable height was undertaken by Johnston [54], who again utilised set-theoretic tools in novel ways. For example, he uses Suslin trees when building Π^0_1 classes (in the uncountable sense); they allow us to work toward an empty class but later recover the construction if we need to change our mind. Further work by Johnson [52, 53] considers fields and uses abstract

[18] An interesting aspect of the construction in these cases is that given X, we construct a group G which is X-computable, but the filtration $\bar{G}$ will not be X-computable but merely left-κ-c.e.; this kind of distinction does not of course come up in set theory.

[19] When $\alpha \geqslant \omega$ we need to replace 2α by $2\alpha + 1$.

elementary classes. On the arithmetical hierarchy in the uncountable setting see [12].

Many questions abound; a particular one is what happens when we drop the assumption $V = L$.

References

[1] Wilhelm Ackermann *Die Widerspruchsfreiheit der allgemeinen Mengenlehre*, Mathematische Annalen 114 (1), 1937, pp. 305–315.

[2] J. W. Addison *Separation principles in the hierarchies of classical and effective descriptive set theory*, Fundamenta Mathematicae 46, 1959, pp. 123–135.

[3] Paul-Elliot Angles d'Auriac, and Benoît Monin *Another characterization of the higher K-trivials*, in: 42nd International Symposium on Mathematical Foundations of Computer Science, *LIPIcs. Leibniz Int. Proc. Inform. 83*, pages Art. No. 34, 13. Schloss Dagstuhl, Leibniz-Zentrum Informatik, Wadern 2017.

[4] Chris J. Ash *A construction for recursive linear orderings*, The Journal of Symbolic Logic 56 (2), 1991, pp. 673–683.

[5] Chris J. Ash, C. G. Jockusch, Jr., and J. F. Knight *Jumps of orderings*, Transactions of the American Mathematical Society 319 (2), 1990, pp. 573–599.

[6] Chris J. Ash and Julia F. Knight Computable Structures and the Hyperarithmetical Hierarchy, North-Holland Publishing Company, Amsterdam 2000 [*Studies in Logic and the Foundations of Mathematics 144*].

[7] Chris J. Ash, Julia F. Knight, Mark Manasse, and Theodore A. Slaman *Generic copies of countable structures*, Annals of Pure and Applied Logic 42 (3), 1989, pp. 195–205.

[8] Jon Barwise Admissible Sets and Structures, *An Approach to Definability Theory*, Springer-Verlag, Berlin & New York 1975 [*Perspectives in Mathematical Logic*].

[9] Laurent Bienvenu, Noam Greenberg, and Benoît Monin *Continuous higher randomness*, Journal of Mathematical Logic 17 (1), 2017, 1750004, 53.

[10] — *Bad oracles in higher computability and randomness*, Israel Journal of Mathematics [to appear].

[11] Merlin Carl, and Philipp Schlicht *Randomness via infinite computation and effective descriptive set theory*, The Journal of Symbolic Logic 83 (2), 2028, pp. 766–789.

[12] Jacob Carson, Jesse Johnson, Julia Knight, Karen Lange, Charles McCoy, and John Wallbaum *The arithmetical hierarchy in the setting of ω_1*, Computability 2 (2), 2013, pp. 93–105.

[13] John Chisholm *Effective model theory vs. recursive model theory*, The Journal of Symbolic Logic 55 (3), 1990, pp. 1168–1191.

[14] Chi Tat Chong Techniques of Admissible Recursion Theory, Springer-Verlag, Berlin 1984. [*Lecture Notes in Mathematics 1106*].

[15] Chi Tat Chong, Theodore A. Slaman, and Yue Yang *The inductive strength of Ramsey's Theorem for pairs*, Advances in Mathematics 308, 2017, pp. 121–141.

[16] Chi Tat Chong, and Liang Yu *Randomness in the higher setting*, The Journal of Symbolic Logic 80 (4), 2015, pp. 1131–1148.

[17] Chi Tat Chong, André Nies, and Liang Yu *Lowness of higher randomness notions*, Israel Journhal of Mathematics 166, 2008, pp. 39–60.

[18] Chi Tat Chong, and Liang Yu Recursion Theory, De Gruyter, Berlin, 2015 [De Gruyter Series in Logic and its Applications 8]. (Computational aspects of definability, With an interview with Gerald E. Sacks.)

[19] Alonzo Church *The constructive second number class*, Bulletin of the American Mathematical Society 44, 1938, pp. 224–232.

[20] Alonzo Church, and Stephen Cole Kleene *Formal definitions in the theory of ordinal numbers*, Fundamenta Mathematicae 28, 1937, pp. 11–21.

[21] Max Dehn *Über unendliche diskontinuierliche Gruppen*, Mathematische Annalen 71 (1), 1911, pp. 116–144.

[22] Rod G. Downey *Computability theory and linear orderings*, in: Handbook of Recursive Mathematics 2, North-Holland Publishing Company, Amsterdam 1998 [*Studies in Logic and the Foundations of Mathematics 139*], pp. 823–976.

[23] Rod G. Downey, Andre Nies, Rebecca Weber, and Liang Yu *Lowness and* Π^0_2 *nullsets*, The Journal of Symbolic Logic 71 (3), 2006, pp. 1044–1052.

[24] Rod G. Downey, Denis R. Hirschfeldt, and Bakhadyr M. Khoussainov *Uniformity in the theory of computable structures*, Algebra i Logika 42 (5), 2003, pp. 566–593, 637.

[25] Rodney Downey, and Alexander G. Melnikov *Effectively categorical abelian groups*, Journal of Algebra 373, 2023, pp. 223–248.

[26] Rodney G. Downey, and Denis R. Hirschfeldt Algorithmic Randomness and Complexity, *Theory and Applications of Computability*, Springer, New York 2010.

[27] Rodney G. Downey, Denis R. Hirschfeldt, André Nies, and Sebastiaan A. Terwijn *Calibrating randomness*, The Bulletin of Symbolic Logic 12 (3), 2006, pp. 411–491.

[28] Rodney G. Downey, Asher M. Kach, Steffen Lempp, Andrew E. M. Lewis-Pye, Antonio Montalbán, and Daniel D. Turetsky *The complexity of computable categoricity*, Advances in Mathematics 268, 2015, pp. 423–466.

[29] Paul C. Eklof *Methods of logic in abelian group theory*, in: Abelian Group Theory (Proceedings of the Second New Mexico State University Conference, Las Cruces NM 1976) [*Lecture Notes in Mathematics 616*], 1977, pp. 251–269.

[30] Paul C. Eklof, and Alan H. Mekler Almost Free Modules, North-Holland Publishing Company, Amsterdam 2002 [*North-Holland Mathematical Library 65*], revised edition. (Set-theoretic methods.)

[31] Yuri L. Ershov, and Sergei S. Goncharov Constructive Models (Siberian School of Algebra and Logic), Consultants Bureau, New York 2000.

[32] Ekaterina Fokina, Sy-David Friedman, Julia Knight, and Russell Miller *Classes of structures with universe a subset of* ω_1, Journal of Logic and Computation 23 (6), 2013, pp.1249–1265.

[33] László Fuchs Infinite Abelian Groups 1, Academic Press, New York and London 1970

[*Pure and Applied Mathematics 36*].

[34] Péter Gács *Every sequence is reducible to a random one*, Information and Control 70 (2–3), 1986, pp. 186–192.

[35] Robin O. Gandy *Proof of Mostowski's conjecture*, Bulletin de l'Académie Polonaise des Sciences (Sér. Sci. Math. Astronom. Phys.) 8, 1960, pp. 571–575.

[36] Sergei S. Gončarov *Selfstability, and computable families of constructivizations*, Algebra i Logika 14 (6), 1975, pp. 647–680, 727.

[37] — *The problem of the number of nonautoequivalent constructivizations*, Algebra i Logika 19 (6), 1980, pp. 621–639, 745.

[38] Sergei S. Gončarov, and V. D. Dzgoe *Autostability of models*, Algebra i Logika 19 (1), 1980, pp. 45–58, 132.

[39] Noam Greenberg *The role of true finiteness in the admissible recursively enumerable degrees*, Memoirs of the American Mathematical Society 18 (854), 2006 [vi + 99 pp.]

[40] Noam Greenberg, Joel David Hamkins, Denis Hirschfeldt, and Russell Miller (eds.). Effective Mathematics of the Uncountable, Association for Symbolic Logic, La Jolla CA and Cambridge University Press, Cambridge 2013 [*Lecture Notes in Logic 41*].

[41] Noam Greenberg, Asher M. Kach, Steffen Lempp, and Daniel D. Turetsky *Computability and uncountable linear orders I: Computable categoricity*, The Journal of Symbolic Logic 80 (1), 2015, pp. 116–144.

[42] Noam Greenberg, and Julia F. Knight *Computable structure theory on ω_1 using admissibility* in: Effective Mathematics of the Uncountable, Association for Symbolic Logic, La Jolla CA 2013 [*Lecture Notes in Logic 41*], pp. 50–80.

[43] Noam Greenberg, Alexander G. Melnikov, Julia F. Knight, and Daniel Turetsky *Uniform procedures in uncountable structures*, The Journal of Symbolic Logic 83 (2), 2018, pp. 529–550.

[44] Noam Greenberg, and Benoît Monin *Higher randomness and genericity*, Forum of Mathematics, Sigma 5, e31, 41, 2017.

[45] Noam Greenberg, Linus Richter, Saharon Shelah, and Dan Turetsky *More on bases of uncountable free abelian groups* [submitted].

[46] Noam Greenberg, Dan Turetsky, and Linda Brown Westrick *Finding bases of uncountable free abelian groups is usually difficult*, Transactions of the American Mathematical Society 370 (6), 2018, pp. 4483–4508.

[47] Valentina S. Harizanov *Pure computable model theory*, in: Handbook of Recursive Mathematics, 1, North-Holland, Amsterdam 1998 [*Studies in Logic and the Foundations of Mathematics 138*], pp. 3–114.

[48] L. A. Harrington, A. S. Kechris, and A. Louveau *A Glimm-Effros dichotomy for Borel equivalence relations*, Journal of the American Mathematical Society 3 (4), 1990, pp. 903–928.

[49] Grete Hermann *Die Frage der endlich vielen Schritte in der Theorie der Polynomideale*, Mathematische Annalen 95 (1), 1926, 736–788.

[50] Greg Hjorth, and André Nies *Randomness via effective descriptive set theory*, Journal

of the London Mathematical Society (2), 5 (2), 2007, pp. 495–508.

[51] Ronald Björn Jensen *The fine structure of the constructible hierarchy*, Annals of Mathematical Logic 4, 1972, pp. 229–308; *erratum*, ibid. 4, 1972, p. 443. (With a section by Jack Silver.)

[52] Jesse Johnson *Computable categoricity for pseudo-exponential fields of size* $\aleph_1$, Annals of Pure and Applied Logic 165 (7–8), 2014, pp. 1301–1317.

[53] Jesse W. Johnson Computable Model Theory for Uncountable Structures, PhD Dissertation, University of Notre Dame, 2013, ProQuest LLC, Ann Arbor MI 2013.

[54] Reese Johnston *Computability in uncountable binary trees*, The Journal of Symbolic Logic 84 (3), 2029, pp. 1049–1098.

[55] Takayuki Kihara *Decomposing Borel functions using the Shore-Slaman join theorem*, Fundamenta Mathematicae 230 (1), 2015, pp. 1–13.

[56] Stephen Cole Kleene *On notations for ordinal numbers*, The Journal of Symbolic Logic 3, 1938, pp. 150–155.

[57] — *Hierarchies of number-theoretic predicates*, Bulletin of the American Mathematical Society 61, 1955, pp. 193–213.

[58] Julia F. Knight *Degrees of models*, in: Handbook of Recursive Mathematics, 1, North-Holland Publishing Company, Amsterdam 1998 [*Studies in Logic and the Foundations of Mathematics 138*], pp. 289–309.

[59] Peter Koepke, and Benjamin Seyfferth *Ordinal machines and admissible recursion theory*, Annals of Pure and Applied Logic 160 (3), 2009, pp. 310–318.

[60] Georg Kreisel, and Gerald E. Sacks *Metarecursive sets*, The Journal of Symbolic Logic 30, 1965, pp. 318–338.

[61] Georg Kreisel *Set theoretic problems suggested by the notion of potential totality*, in: Infinitistic Methods (Proc. Sympos. Foundations of Mathematics, Warsaw, 1959), Pergamon, Oxford, and Państwowe Wydawnictwo Naukowe, Warsaw 1961, pp. 103–140.

[62] Saul Kripke *Transfinite recursion on admissible ordinals i, ii* (abstracts), The Journal of Symbolic Logic 29, 1964, pp. 161–162.

[63] Antonín Kučera *Measure,* Π^0_1*-classes and complete extensions of* PA, in: Recursion Theory Week (Oberwolfach, 1984), Springer, Berlin 1985 [*Lecture Notes in Mathematics 1141*], pp. 245–259.

[64] Alistair H. Lachlan *Embedding nondistributive lattices in the recursively enumerable degrees*, in: Wilfrid Hodges (ed.) Conference in Mathematical Logic, London '70 (Proc. Conf., Bedford College, London, 1970), Springer, Berlin 1972 [*Lecture Notes in Mathematics 255*] pp. 149–177.

[65] Michiel van Lambalgen *The axiomatization of randomness*, The Journal of Symbolic Logic 55 (3), 1990, pp. 1143–1167.

[66] Anatoly I. Mal'cev *On recursive Abelian groups*, Doklady Akademii Nauk SSSR 146, 1962, pp. 1009–1012.

[67] Per Martin-Löf *On the notion of randomness*, in: A. Kino, J. Myhill, and R. E. Vesley (eds.) Intuitionism and Proof Theory (Proceedings of the Summer Conference, Buffalo

NY, 1968), North-Holland Publishing Company, Amsterdam 1970, pp. 73–78 [*Studies in Logic and the Foundations of Mathematics 60*].

[68] Joseph S. Miller, and Liang Yu *On initial segment complexity and degrees of randomness*, Transactions of the American Mathematical Society 360 (6), 2008, pp. 3193–3210.

[69] Benoît Monin *Higher randomness and forcing with closed sets*, Theory of Computing Systems 60 (3), 2027, pp. 421–437.

[70] Benoît Monin, and Paul-Elliot Anglès d'Auriac *Genericity and randomness with ITTMs*, The Journal of Symbolic Logic 84 (4), 2029, pp. 1670–1710.

[71] Antonio Montalbán *Priority arguments via true stages*, The Journal of Symbolic Logic 79 (4), 2014, pp. 1315–1335.

[72] André Nies Computability and Randomness, Oxford University Press, Oxford 2009 [*Oxford Logic Guides 51*].

[73] Richard Platek Foundations of Recursion Theory, PhD Dissertation, Stanford University, Stanford CA, 1965.

[74] L. Pontrjagin *The theory of topological commutative groups*, Annals of Mathematics (2), 35 (2), 1934, pp. 361–388.

[75] Jeffery B. Remmel *Recursively categorical linear orderings*, Proceedings of the American Mathematical Society 83 (2), 1981, pp. 387–391.

[76] Dev K. Roy, and Richard Watnick *Finite condensations of recursive linear orders*, Studia Logica 47 (4), 1989, pp. 311–317.

[77] Gerald E. Sacks, and S. G. Simpson *The α-finite injury method*, Annals of Mathematical Logic 4, 1972, pp. 343–367.

[78] Gerald E. Sacks *Measure-theoretic uniformity in recursion theory and set theory*, Transactions of the American Mathematical Society 142, 1969, pp. 381–420.

[79] — Higher Recursion Theory, Springer-Verlag, Berlin 1990 [*Perspectives in Mathematical Logic*]

[80] Saharon Shelah *Infinite abelian groups, Whitehead problem and some constructions*, Israel Journal of Mathematics 18, 1974, pp. 243–256.

[81] Richard A. Shore *On the jump of an α-recursively enumerable set*, Transactions of the American Mathematical Society 217, 1976, pp. 351–363.

[82] — *The recursively enumerable α-degrees are dense*, Annals of Mathematical Logic 9 (1–2), 1976, pp. 123–155.

[83] — *α-recursion theory*, in: Handbook of Mathematical Logic, North-Holland Publishing Company, Amsterdam, 1977 [*Studies in Logic and Foundations of Mathematics 90*], pp. 653–680.

[84] Richard A. Shore, and Theodore A. Slaman *Defining the Turing jump*, Mathematical Research Letters 6 (5–6), 1999, pp. 711–722.

[85] Clifford Spector *Recursive well-orderings*, The Journal of Symbolic Logic 20, 1955, pp. 151–163.

[86] — *Hyperarithmetical quantifiers*, Fundamenta Mathematicae 48, 1959–1960, pp. 313–

320.

[87] Jacques Stern *Some measure theoretic results in effective descriptive set theory*, Israel Journal of Mathematics 20 (2), 1975, pp. 97–110.

[88] Gaisi Takeuti *On the recursive functions of ordinal numbers*, Journal of the Mathematical Society of Japan 12, 1960, pp. 119–128.

[89] — *Recursive functions and arithmetical functions of ordinal numbers*, in: Logic, Methodology and Philosophy of Sciences (Proc. 1964 Internat. Congr.), North-Holland Publishing Company, Amsterdam 1965. pp. 179–196.

[90] Bartel L. van der Waerden *Eine Bemerkung über die Unzerlegbarkeit von Polynomen*, Mathematische Annalen 102 (1), 1930, pp. 738–739.

[91] Richard Watnick *A generalization of Tennenbaum's theorem on effectively finite recursive linear orderings*, The Journal of Symbolic Logic 49 (2), 1984, pp. 563–569.

THE CONSTRUCTIVE HAHN-BANACH THEOREM, REVISITED

HAJIME ISHIHARA
Japan Advanced Institute of Science and Technology
ishihara@jaist.ac.jp

ABSTRACT After reviewing results in constructive analysis and functional analysis, we consider the approximate versions of the separation theorem and the continuous extension theorem for a separable space due to Bishop [3], and exact versions of them for an inseparable space with some geometric properties due to the author [14]. We give a new proof of an approximate version of the separation theorem by using the Baire category theorem, and a proof of an approximate version of the (1-dimensional) dominated extension theorem.

KEYWORDS Hahn-Banach theorem, separation theorem, continuous extension theorem, dominated extension theorem, constructive mathematics

1 Introduction

The Hahn-Banach theorem is a central tool in functional analysis, and is named after the mathematicians *Hans Hahn* and *Stefan Banach* who proved it independently in the late 1920s. Nowadays the theorem has several forms.

- The *continuous extension theorem* which allows an extension of a bounded linear functional defined on a subspace of a normed space to the whole space;
- The *separation theorem* which allows a separation of two convex subsets of a normed space by a hyperplane, and has numerous uses in convex analysis (geometry);
- The *dominated extension theorem* which is the most general form of the theorem, and allows an extension of a linear functional defined on a subspace of a normed space, and dominated by a sublinear function, to the whole space.

The standard proof of, say, the dominated extension theorem consists of two parts; the first part contains computing the least upper bound of a set of real numbers with upper bound; the second part makes use of transfinite induction such as well-ordering, Zorn's lemma and Hausdorff's maximality theorem; see, for example, [21, 3.2].

Bishop's constructive mathematics [3] has been called mathematics with intuitionistic logic and neutral constructive mathematics in contrast with Brouwer's intuitionistic mathematics [9, 13] and constructive recursive mathematics of the Russian school of Markov [18], as it is consistent with classical mathematics. Intuitionistic logic is a logic weaker than classical logic in which classical principles, such as the principle of excluded middle PEM: $\varphi \vee \neg\varphi$, De Morgan's law DML: $\neg(\varphi \wedge \psi) \rightarrow \neg\varphi \vee \neg\psi$, and the double negation elimination DNE: $\neg\neg\varphi \rightarrow \varphi$, are not deducible; see [11, 23]. Note that PEM and DNE, as general schemata, are equivalent (not necessarily, of course, if they are restricted to some class of formulae).

Since the least upper bound of a set of real numbers does not always exists constructively (see Proposition 11 below) and transfinite induction is highly non-constructive, the standard proof of the dominated extension theorem does not work in constructive mathematics. In 1967, Bishop first proved the separation theorem [3, Chapter 9, Theorem 3], and then the continuous extension theorem [3, Chapter 9, Theorem 4] as a corollary. He overcame the difficulties in the standard proof by proving an approximate version, instead of the exact version, of the separation theorem and assuming separability of the space.

In this paper, we review results in constructive analysis and functional analysis which are assumed in the rest of the paper in Section 2. Then, in Section 3, we consider the approximate versions of the separation theorem (for two *bounded* convex subsets) and the continuous extension theorem for a separable space by Bishop [3], and then exact versions of them for an inseparable space with some geometric properties by the author [14]. Thus far, no version of the dominated extension theorem has been proven constructively. Finally, in Section 4, we give a new proof of an approximate version of the separation theorem (for two *unbounded* convex subsets) by using the Baire category theorem, and a proof of an approximate version of the (1-dimensional) dominated extension theorem.

2 Preliminaries

In this section, we review results in constructive analysis and functional analysis which are assumed in the rest of the paper; see also [3, 5, 4, 7, 8, 16].

The material in the following could be formalised in *constructive Zermelo-Fraen-*

kel set theory **CZF**, founded by Aczel as a foundation of constructive mathematics, together with the axiom of countable choice and the axiom of dependent choice; see [1, 2].

Principles of omniscience are principles which are provable in classical logic, but unprovable in intuitionistic logic and hence constructively unacceptable. A theorem is underivable in constructive mathematics if it implies an omniscience principle. Most of them are instances of classical principles, PEM, DML and DNE, and we list some of them below. Here and in the following, we follow the notational conventions in [23]: α, β are supposed to range over $\mathbf{N}^{\mathbf{N}}$; $\mathbf{0} = \lambda n.0$ and $\alpha \# \beta \Leftrightarrow \exists n(\alpha(n) \neq \beta(n))$.

1. The limited principle of omniscience (LPO, Σ^0_1-PEM):
$$\forall\alpha[\alpha \# \mathbf{0} \vee \neg\alpha \# \mathbf{0}]$$
2. The lesser limited principle of omniscience (LLPO, Σ^0_1-DML):
$$\forall\alpha\beta[\neg(\alpha \# \mathbf{0} \wedge \beta \# \mathbf{0}) \to \neg\alpha \# \mathbf{0} \vee \neg\beta \# \mathbf{0}]$$
3. Markov's principle (MP, Σ^0_1-DNE):
$$\forall\alpha[\neg\neg\alpha \# \mathbf{0} \to \alpha \# \mathbf{0}]$$

Definition 1. A *real number* is a sequence $(p_n)_n$ of rationals such that
$$\forall mn\left(|p_m - p_n| < 2^{-m} + 2^{-n}\right).$$
The *ordering relation* $<$ between real numbers $x = (p_n)_n$ and $y = (q_n)_n$ is defined by
$$x < y \Leftrightarrow \exists n\left(2^{-n+2} < q_n - p_n\right).$$
We shall write $\mathbf{R}$ for the set of real numbers as usual. Note that the set $\mathbf{Q}$ of rationals is embedded into $\mathbf{R}$ by the mapping $p \mapsto p^* = \lambda n.p$.

Proposition 2. *Let $x, y, z \in \mathbf{R}$. Then*

1. $\neg(x < y \wedge y < x)$,

2. $x < y \to x < z \vee z < y$.

Proof. See [3, Chapter 2, Corollary], [4, Chapter 2, (2.17)] and [16, Proposition 1]. □

Definition 3. We define the *apartness* $\#$, the *equality* $=$, and the ordering relation $\leq$ between real numbers x and y by

1. $x \# y \Leftrightarrow (x < y \vee y < x)$,
2. $x = y \Leftrightarrow \neg(x \# y)$,
3. $x \leq y \Leftrightarrow \neg(y < x)$.

Lemma 4. *For each binary sequence α, there exists $x \in \mathbf{R}$ such that*

$$\alpha \# \mathbf{0} \leftrightarrow 0 < x.$$

Conversely, for each $x \in \mathbf{R}$, there exists a binary sequence α such that

$$0 < x \leftrightarrow \alpha \# \mathbf{0}.$$

Proof. Suppose that α is a binary sequence, and define a sequence $(p_n)_n$ of rationals by

$$p_n = \sum_{k=0}^{n} \alpha(k) \cdot 2^{-(k+1)}.$$

Then $x = (p_n)_n \in \mathbf{R}$, and $\alpha \# \mathbf{0} \leftrightarrow 0 < x$.

Conversely, suppose that $x = (p_n)_n \in \mathbf{R}$, and define a binary sequence α by

$$\alpha(n) = 1 \Leftrightarrow 2^{-n+2} < p_n.$$

Then $0 < x \leftrightarrow \alpha \# \mathbf{0}$. □

We have the following correspondences between relations on real numbers and omniscience principles.

Proposition 5. *1.* $\forall x \in \mathbf{R}(0 < x \vee x \leq 0) \Leftrightarrow$ LPO,

2. $\forall x \in \mathbf{R}(x \leq 0 \vee 0 \leq x) \Leftrightarrow$ LLPO,

3. $\forall x \in \mathbf{R}(\neg x \leq 0 \rightarrow 0 < x) \Leftrightarrow$ MP.

Proof. Straightforward by Lemma 4. □

It is straightforward to define the arithmetical operations on $\mathbf{R}$ in terms of operations on sequences of rational numbers; see [3, Chapter 2, Proposition 2], [4, Chapter 2, (2.4) and (2.5)], [7, Chapter 1, (5.1)] and [23, Chapter 5] for details.

Proposition 6. *Let $x_0, \ldots, x_n$ be real numbers. If $0 < \sum_{k \leq n} x_k$, then $0 < x_k$ for some $k \leq n$.*

Proof. We proceed by induction on n. It is trivial for $n = 0$. If $0 < \sum_{k \leq n+1} x_k$, then either $0 < \sum_{k \leq n} x_k$ or $\sum_{k \leq n} x_k < \sum_{k \leq n+1} x_k$, by Proposition 2 (2); in the former case, $0 < x_k$ for some $k \leq n$, by the induction hypothesis; in the latter case, we have $0 < x_{n+1}$. See also [3, Chapter 2, Proposition 7] and [4, Chapter 2, (2.16)]. □

Definition 7. A sequence (x_n) of real numbers is a *Cauchy sequence* if

$$\forall \epsilon > 0 \exists N \forall mn \geq N[|x_m - x_n| \leq \epsilon];$$

converges to a limit $x \in \mathbf{R}$ if

$$\forall \epsilon > 0 \exists N \forall n \geq N[|x_n - x| \leq \epsilon],$$

and we then write $x_n \to x$ as $n \to \infty$ or $x = \lim_{n \to \infty} x_n$.

Theorem 8. *A sequence of real numbers converges if and only if it is a Cauchy sequence.*

Proof. See [3, Chapter 2, Theorem 2], [4, Chapter 2, (3.3)] and [7, Chapter 2, (5.1)]. □

Remark 9. A proof of the Cauchy completeness without countable axiom of choice is given in [23, 5.4.2]. A treatment of Dedekind reals and the difference between the Cauchy completion and the order completion can be found in [23, 5.5]. A construction of the real numbers using a form of interval arithmetic is given in [8].

Definition 10. Let S be an inhabited [1] subset of $\mathbf{R}$. A real number z is an *upper bound* of S if $s \leq z$ for each $s \in S$; *supremum* of S if it is an upper bound of S, and for each $\epsilon > 0$ there exists $s \in S$ such that $z < s + \epsilon$, and we then write $z = \sup S$. A *lower bound* and *infimum* of S are defined similarly.

Classically, every subset with an upper bound has a least upper bound. However, constructively, a subset with an upper bound does not always have a supremum.

Proposition 11. *If every inhabited subset S of $\mathbf{R}$ with an upper bound has a supremum, then* LPO *holds.*

[1] A set S is *inhabited* if there exists an element of S. Note that if a set S is inhabited, then S is nonempty, but the converse does not hold constructively.

Proof. Let α be a binary sequence. Then $S = \{\alpha(n) \mid n \in \mathbf{N}\}$ is an inhabited subset of $\mathbf{R}$ with an upper bound 1. If $\sup S$ exists, then either $0 < \sup S$ or $\sup S < 1$; in the former case, we have $\alpha \# \mathbf{0}$; in the latter case, we have $\neg\alpha \# \mathbf{0}$. □

Remark 12. Note that, in Proposition 11 (and in Proposition 20 below), taking a formula φ such that the class

$$S = \{0\} \cup \{1 \mid \varphi\}$$

is a set, we can show a stronger consequence: PEM for the formula φ.

As an application of the completeness of the real numbers, we have the following *constructive supremum principle.*

Theorem 13. *Let S be an inhabited subset of $\mathbf{R}$ with an upper bound. If for each $a, b \in \mathbf{R}$ with $a < b$, either $\exists s \in S(a < s)$ or $\forall s \in S(s < b)$, then a supremum of S exists.*

Proof. See [4, Chapter 2, (4.3)], [7, Chapter 2, (1.2)] and [16, Theorem 2]. □

Definition 14. A *metric space* is a set X equipped with a *metric* $d : X \times X \to \mathbf{R}$ such that

1. $d(x, y) = 0 \leftrightarrow x = y$,
2. $d(x, y) = d(y, x)$,
3. $d(x, y) \leq d(x, z) + d(z, y)$,

for each $x, y, z \in X$. For $x, y \in X$ and $r > 0$, we write $x \# y$ for $0 < d(x, y)$, and $B(x, r)$ to denote the *open ball* with *centre* x and *radius* r, that is,

$$B(x, r) = \{y \in E \mid d(x, y) < r\}.$$

Definition 15. A subset S of a metric space X is *open* if

$$\forall x \in S \exists \epsilon > 0 \forall y \in B(x, \epsilon)[y \in S];$$

closed if

$$\forall x \in X[\forall \epsilon > 0 \exists y \in B(x, \epsilon)(y \in S) \to x \in S].$$

Let S be a subset of a metric space X. Then the *closure* $\overline{S}$ of S is defined by

$$\overline{S} = \{x \in X \mid \forall \epsilon > 0 \exists y \in B(x, \epsilon)[y \in S]\}.$$

A subset S of a metric space X is *dense* in X if $\overline{S} = X$; a metric space is *separable* if there exists a countable dense subset.

Definition 16. A sequence $(x_n)_n$ in a metric space is a *Cauchy sequence* if

$$\forall \epsilon > 0 \exists N \forall mn \geq N[d(x_m, x_n) \leq \epsilon];$$

converges to a limit $x \in X$ if

$$\forall \epsilon > 0 \exists N \forall n \geq N[d(x_n, x) \leq \epsilon],$$

and we then write $x_n \to x$ as $n \to \infty$ or $x = \lim_{n \to \infty} x_n$.

A metric space is *complete* if every Cauchy sequence converges.

We have the following constructive version of the Baire category theorem.

Theorem 17. *If* $(U_n)_n$ *is a sequence of open dense subsets of a complete metric space* X, *then* $\bigcap_{n=0}^{\infty} U_n$ *is dense in* X.

Proof. See [3, Chapter 4, Theorem 4], [4, Chapter 4, (3.9)] and [7, Chapter 2, (1.3)]. □

Remark 18. A detailed analysis of the constructive Baire theorem is given in [7, Chapter 2, 2], and its applications can be found in [8, 6.6].

Definition 19. An inhabited subset S of a metric space X is *located* if

$$d(x, S) = \inf\{d(x, y) \mid y \in S\}$$

exists for each $x \in X$.

Proposition 20. *If every inhabited subset* S *of* $\mathbf{R}$ *is located, then* LPO *holds.*

Proof. Let α be a binary sequence. Then $S = \{\alpha(n) \mid n \in \mathbf{N}\}$ is an inhabited subset of $\mathbf{R}$. If $d(1, S)$ exists, then either $0 < d(1, S)$ or $d(1, S) < 1$; in the former case, we have $\neg \alpha \# \mathbf{0}$; in the latter case, we have $\alpha \# \mathbf{0}$. □

Remark 21. The definition of a located set in Definition 19 is the notion of a *metrically* located set, and the notion of a *topologically* located set can be found in [24, Chapter 7].

Definition 22. A mapping f between metric spaces X and Y is *continuous at* $x \in X$ if

$$\forall \epsilon > 0 \exists \delta > 0 \forall y \in X[d(x, y) < \delta \to d(f(x), f(y)) \leq \epsilon];$$

continuous if it is continuous at each point of X; *uniformly continuous* if

$$\forall \epsilon > 0 \exists \delta > 0 \forall xy \in X[d(x, y) < \delta \to d(f(x), f(y)) \leq \epsilon].$$

Definition 23. A sequence $x = (x_n)_n$ of a metric space X is *regular* if

$$\forall mn \left(d(x_m, x_n) \leq 2^{-m} + 2^{-n} \right).$$

Proposition 24. *Let $\tilde{X}$ be the set of regular sequences of a metric space X, and define*

$$\tilde{d}(x, y) = \lim_{n \to \infty} d(x_n, y_n)$$

for each $x = (x_n)_n$ and $y = (y_n)_n$ in $\tilde{X}$. Then $\tilde{X}$ with the equality relation $x =_{\tilde{X}} y \Leftrightarrow \tilde{d}(x, y) = 0$ is a complete metric space with the metric $\tilde{d}$, called the completion of X. Moreover, the inclusion map $\iota : X \to \tilde{X}$ defined by $\lambda n.x$ for each $x \in X$ satisfies $\tilde{d}(\iota(x), \iota(y)) = d(x, y)$ for each $x, y \in X$, and realises X as a dense subset of $\tilde{X}$.

Proof. See [3, Chapter 4, Theorem 1 and Theorem 2] and [4, Chapter 4, (3.2) and (3.4)]. □

Lemma 25. *If S is a located subset of a metric space X, then it is a located subset of the completion $\tilde{X}$.*

Proof. Note that $|d(x, S) - d(y, S)| \leq d(x, y)$ for each $x, y \in X$. Then for each $x = (x_n)_n \in \tilde{X}$, it is straightforward to see that $(d(x_n, S))_n$ is a Cauchy sequence of $\mathbf{R}$, and that $\tilde{d}(x, S) = \lim_{n \to \infty} d(x_n, S)$. □

Lemma 26. *If f is a uniformly continuous function of a metric space into a complete metric space Y, then there exists a uniformly continuous extension $\tilde{f} : \tilde{X} \to Y$ of f.*

Proof. For each $x = (x_n)_n \in \tilde{X}$, it is straightforward to see that $(f(x_n))_n$ is a Cauchy sequence of Y, and that $\tilde{f}(x) = \lim_{n \to \infty} f(x_n)$ defines a uniformly continuous extension of f. □

Definition 27. Let E be a linear space. Then a subset C of E is *convex* if

$$\lambda x + (1 - \lambda) y \in C$$

for each $x, y \in C$ and $\lambda \in [0, 1]$; a *subspace* M of E is a linear subset of E.

Definition 28. A *normed space* is a linear space E equipped with a *norm* $\| \cdot \| : E \to \mathbf{R}$ such that

1. $\|x\| = 0 \leftrightarrow x = 0$,

2. $\|ax\| = |a| \|x\|$,

3. $\|x+y\| \leq \|x\| + \|y\|$,

for each $x, y \in E$ and $a \in \mathbf{R}$. Note that a normed space E is a metric space with the metric

$$d(x,y) = \|x-y\|.$$

A *Banach space* is a normed space which is complete with respect to the metric.

Example 29. Let ξ be a sequence of real numbers. Then we write ξ_n for $\xi(n)$. For $1 \leq p < \infty$, let

$$l_p = \{\xi \in \mathbf{R}^{\mathbf{N}} \mid \textstyle\sum_{n=0}^{\infty} |\xi_n|^p < \infty\}$$

and define a norm by $\|\xi\| = (\sum_{n=0}^{\infty} |\xi_n|^p)^{1/p}$. Then l_p is a (separable) Banach space. Classically the normed space

$$l_\infty = \{\xi \in \mathbf{R}^{\mathbf{N}} \mid (\xi_n) \text{ is bounded}\}$$

with the norm $\|\xi\| = \sup_n |\xi_n|$ is an *inseparable* Banach space. However, constructively, it is *not* a normed space.

Remark 30. The completion $\tilde{E}$ of a normed space E as a metric space is a Banach space with the norm

$$\|x\| = \lim_{n\to\infty} \|x_n\|,$$

where $x = (x_n)_n \in \tilde{E}$. The inclusion map $\iota : E \to \tilde{E}$ preserves norms and realises E as a dense subspace of $\tilde{E}$; see [3, Chapter 9, Proposition 4] and [4, Chapter 7, (1.12)].

Proposition 31. *Let Y be a located subspace of a normed space E, and define*

$$\|x\|_{E/Y} = d(x, Y).$$

Then E with the equality relation $x =_{E/Y} y \Leftrightarrow \|x-y\|_{E/Y} = 0$ is a normed space with the norm $\|\cdot\|_{E/Y}$, called the quotient space of E by Y, and written E/Y. Moreover, if E is a Banach space, then E/Y is a Banach space.

Proof. See [4, Chapter 7, (2.9)], [8, Proposition 2.3.8] and [16, Proposition 14]. □

Definition 32. A normed space E is *uniformly convex* if for each $\epsilon > 0$ there exists $\delta > 0$ such that

$$\epsilon < \|x-y\| \to \|(x+y)/2\| \leq 1 - \delta$$

for each $x, y \in B(0,1)$.

Remark 33. l_p is uniformly convex for $1 < p < \infty$; see [3, Chapter 9, Theorem 1 and Corollary] and [4, Chapter 7, (3.22)].

Proposition 34. *Let C be a closed convex subset of a uniformly convex Banach space E, and let x be an element of E such that*

$$d(x, C) = \inf\{\|x - z\| \mid z \in C\}$$

exists. Then there exists a unique element y of C such that $d(x, C) = \|x - y\|$.

Proof. See [3, Chapter 9, Problem 5], [4, Chapter 7, Problem 11] and [16, Proposition 22]. □

Proposition 35. *Let f be a nonzero normable linear functional on a uniformly convex Banach space E. Then there exists $x \in E$ such that $f(x) = \|f\|$ and $\|x\| = 1$.*

Proof. See [4, Chapter 7, (3.23)]. □

Definition 36. A function p of a linear space E into $\mathbf{R}$ is *convex* if

$$p(\lambda x + (1 - \lambda)y) \leq \lambda p(x) + (1 - \lambda)p(y)$$

for each $x, y \in E$ and $\lambda \in [0, 1]$; *sublinear* if

1. $p(ax) = ap(x)$,
2. $p(x + y) \leq p(x) + p(y)$

for each $x, y \in E$ and $a \in \mathbf{R}$ with $a \geq 0$; *linear* if

1. $p(ax) = ap(x)$,
2. $p(x + y) = p(x) + p(y)$

for each $x, y \in E$ and $a \in \mathbf{R}$.

A *linear functional* f on a linear space E is a linear function of E into $\mathbf{R}$. The *kernel* $\ker(f)$ of a linear functional f on a linear space E is defined by

$$\ker(f) = \{x \in E \mid f(x) = 0\}.$$

Definition 37. A sublinear function p on a normed space E is *bounded* if there exists $c > 0$ such that

$$|p(x)| \leq c\|x\|$$

for each $x \in E$.

Proposition 38. *Let p be a sublinear function on a normed space E. Then the following are equivalent.*

1. *p is continuous,*
2. *p is uniformly continuous,*
3. *p is bounded.*

Proof. Note that

$$p(x) - p(y) \leq p(x-y) \leq |p(x-y)|, \qquad p(y) - p(x) \leq p(y-x) \leq |p(y-x)|.$$

(1) $\Rightarrow$ (2): Given an $\epsilon > 0$, since p is continuous at 0, there exists $\delta > 0$ such that for each $z \in E$

$$\|z\| < \delta \rightarrow |p(z) - p(0)| = |p(z)| \leq \epsilon.$$

Therefore, for each $x, y \in E$, if $\|x - y\| < \delta$, then $|p(x-y)| \leq \epsilon$ and $|p(y-x)| \leq \epsilon$, and hence $|p(x) - p(y)| \leq \epsilon$.

(2) $\Rightarrow$ (3): Since p is uniformly continuous, there exists $\delta > 0$ such that for each $z \in E$

$$\|z\| < \delta \rightarrow |p(z)| \leq 1.$$

For each $x \in E$ and $\epsilon > 0$, since $\|\delta x/(\|x\| + \epsilon)\| < \delta$, we have

$$\frac{\delta}{\|x\| + \epsilon}|p(x)| = \left|p\left(\frac{\delta x}{\|x\| + \epsilon}\right)\right| \leq 1,$$

and hence $|p(x)| \leq (\|x\| + \epsilon)/\delta$. Therefore, since ϵ is arbitrary, we have

$$|p(x)| \leq c\|x\|$$

for each $x \in E$, where $c = 1/\delta$.

(3) $\Rightarrow$ (1): Let $c > 0$ be such that $|p(z)| \leq c\|z\|$ for each $z \in E$, and given an $x \in E$ and an $\epsilon > 0$, set $\delta = \epsilon/c$. Then for each $y \in E$ with $\|x - y\| < \delta$, we have

$$|p(x-y)| \leq c\|x - y\| < \epsilon, \qquad |p(y-x)| \leq c\|y - x\| < \epsilon,$$

and hence $|p(x) - p(y)| < \epsilon$.

See also [3, Chapter 9, Proposition 2], [4, Chapter 7, (1.5)] and [8, Proposition 2.3.3]. ☐

Definition 39. A linear functional f on a normed space E is *normable* if

$$\|f\| = \sup\{|f(x)| \mid x \in B(0,1)\}$$

exists.

Classically, every bounded linear functional is normable. However, constructively, we can construct a bounded linear functional which is *not* normable.

Proposition 40. *If every bounded linear functional on l_2 is normable, then* LPO *holds.*

Proof. Let α be a binary sequence with at most one nonzero term, and define a linear functional f on l_2 by

$$f(\xi) = \sum_{k=0}^{\infty} \alpha(k)\xi_k.$$

Then f is bounded. If f is normable, then either $0 < \|f\|$ or $\|f\| < 1$; in the former case, we have $\alpha \# \mathbf{0}$; in the latter case, we have $\neg\alpha \# \mathbf{0}$. □

Proposition 41. *A nonzero bounded linear functional f on a normed space E is normable if and only if its kernel* $\ker(f)$ *is located.*

Proof. See [3, Chapter 9, Proposition 8], [4, Chapter 7, (1.10)] and [8, Proposition 2.3.6]. □

3 The constructive Hahn-Banach theorem

In this section, we present the approximate versions of the separation theorem (Theorem 45) and the continuous extension theorem (Theorem 47) for a separable space by Bishop [3], and then exact versions of the continuous extension theorem (Theorem 52) and the separation theorem (Theorem 53) for an inseparable space with some geometric properties by the author [14].

Classically, the following Hahn-Banach theorem (the continuous extension theorem) and its corollary hold.

Theorem 42 (classical)**.** *Let M be a subspace of a normed space E, and let f be a bounded linear functional on M. Then there exists a bounded linear functional g on E such that $g(x) = f(x)$ for each $x \in M$ and $\|g\| = \|f\|$.*

Corollary 43 (classical)**.** *Let x be a nonzero element of a normed space E. Then there exists a bounded linear functional f on E such that $f(x) = \|x\|$ and $\|f\| = 1$.*

However, constructively, the corollary implies LLPO.

Proposition 44. *The classical Hahn-Banach theorem implies* LLPO.

Proof. Let $(1, a)$ be a nonzero element of the normed space $\mathbf{R}^2$ with a norm $\|(x, y)\| = |x|+|y|$. Then there exists a bounded linear functional f such that $f(1, a) = 1+|a|$ and $\|f\| = 1$. Since $|f(1,0)| \leq 1$ and $|f(0,1)| \leq 1$, we have

$$1 + |a| = f(1, a) = f(1,0) + af(0,1) \leq f(1,0) + |a|,$$

and therefore $f(1,0) = 1$ and $af(0,1) = |a|$. Either $-1 < f(0,1)$ or $f(0,1) < 1$. In the former case, we have $0 \leq a$; in the latter case, we have $a \leq 0$. See also [15] and [20] for a recursive counterexample to the classical Hahn-Banach theorem. □

Bishop [3] first gave a constructive proof of the following approximate version of the separation theorem. Note that, in Bishop's proof, the boundedness of convex subsets is crucial.

Theorem 45. *Let A and B be bounded convex subsets of a separable normed space E such that the algebraic difference*

$$B - A = \{y - x \mid x \in A, y \in B\}$$

is located and $d = d(0, B - A) > 0$. Then for each $\epsilon > 0$ there exists a normable linear functional f on E such that $\|f\| = 1$ and

$$f(x) + d \leq f(y) + \epsilon$$

for each $x \in A$ and $y \in B$.

Proof. See [3, Chapter 9, Theorem 3], [5, Chapter 3, 3.3], [4, Chapter 7, (4.3)] or [8, Theorem 5.2.9]. □

Corollary 46. *Let x_0 be a nonzero element of a separable normed space E. Then for each $\epsilon > 0$ there exists a normable linear functional f on E such that $\|f\| = 1$ and*

$$\|x_0\| \leq f(x_0) + \epsilon.$$

Proof. Apply Theorem 45 for $A = \{0\}$ and $B = \{x_0\}$; see [3, Chapter 9, Corollary of Theorem 3], [5, Chapter 3, 3.4] and [4, Chapter 7, (4.5)] for details. □

Theorem 47. *Let M be a subspace of a separable normed space E, and let f be a nonzero linear functional on M such that the kernel* $\ker(f)$ *is located in E. Then for each $\epsilon > 0$ there exists a normable linear functional g on E such that $g(x) = f(x)$ for each $x \in M$ and*

$$\|g\| \leq \|f\| + \epsilon.$$

Proof. Apply Corollary 46 for x_0 as a point of the quotient space $E/\ker(f)$ (see Proposition 31) with $f(x_0) = 1$; see [3, Chapter 9, Theorem 4], [5, Chapter 3, 3.5] and [4, Chapter 7, (4.6)] for details. □

Remark 48. A constructive proof of Theorem 47 (the continuous extension theorem) following the lines of a standard classical proof, for example, [21, 3.2], was given in [7, Chapter 2, (5.9) and (5.10)].

The author [14] showed, with a help of geometric properties of a Banach space such as uniform convexity and Gâteaux differentiability of the norm, exact versions of the separation theorem and the continuous extension theorem for an inseparable space without invoking transfinite induction.

Definition 49. Let E be a linear space. Then a convex function $p : E \to \mathbf{R}$ is *Gâteaux differentiable at $x \in E$* with the *derivative* $f : E \to \mathbf{R}$ if for each $y \in E$ and $\epsilon > 0$ there exists $\delta > 0$ such that

$$\forall t \in \mathbf{R}(0 < |t| < \delta \to |p(x + ty) - p(x) - tf(y)| < \epsilon|t|).$$

Note that the derivative f is given by

$$f(y) = \lim_{t \to 0} \frac{p(x + ty) - p(x)}{t},$$

and is linear; see Lemma 59 below.

The norm of a normed space E is *Gâteaux differentiable* if it is Gâteaux differentiable, as a convex function, at each $x \in E$ with $\|x\| = 1$.

Remark 50. The norm of l_p for $1 < p < \infty$ and the norm of a Hilbert space are Gâteaux differentiable at each $x \in E$ with $x \# 0$; see [17] and [6, Proposition 3.3].

Proposition 51. *Let x be a nonzero element of a normed linear space E whose norm is Gâteaux differentiable at x. Then there exists a unique normable linear functional f on E such that $f(x) = \|x\|$ and $\|f\| = 1$.*

Proof. Take the derivative f of the norm at x; see [14, Lemma 1] for details. □

Theorem 52. *Let M be a subspace of a uniformly convex Banach space E with a Gâteaux differentiable norm, and let f be a normable linear functional on M. Then there exists a unique normable linear functional g on E such that $g(x) = f(x)$ for each $x \in M$ and $\|g\| = \|f\|$.*

Proof. We may assume without loss of generality that $\|f\| = 1$. Let $\overline{M}$ be the closure of M. Then there exists a normable extension $\overline{f}$ of f on $\overline{M}$. Since $\overline{M}$ is a uniformly convex Banach space, there exists $x \in \overline{M}$ such that $\overline{f}(x) = \|x\| = 1$, by Proposition 35. Take the derivative g of the norm at x; see [14, Theorem 1] for details. □

Theorem 53. *Let A and B be convex subsets of a uniformly convex Banach space E with a Gâteaux differentiable norm such that*

$$d = d(0, B - A) = \inf\{y - x \mid x \in A, y \in B\}$$

exists. Then there exists a normable linear functional f on E such that $\|f\| = 1$ and

$$f(x) + d \leq f(y)$$

for each $x \in A$ and $y \in B$.

Proof. Let $z \in \overline{B - A}$ be such that $\|z\| = d(0, B - A)$ (such a z exist by Proposition 34), and let f be the derivative of the norm at z. Then $f(x) + d \leq f(y)$ for each $x \in A$ and $y \in B$. See [14, Theorem 2] for details. □

Remark 54. A treatment of the Hahn-Banach theorem in the setting of formal topology [22] can be found in [10].

4 A new proof

In this section, we give yet another proof of an approximate version of the separation theorem (Theorem 62) by using the Baire category theorem, and a proof of an approximate version of the (1-dimensional) dominated extension theorem (Theorem 66).

Lemma 55. *Let E be a linear space, and let $p : E \to \mathbf{R}$ be a convex function. Then for each $u, v \in E$ and $t, s \in \mathbf{R}$ with $0 < t \leq s$,*

$$\frac{p(u - sv) - p(u)}{-s} \leq \frac{p(u - tv) - p(u)}{-t} \leq \frac{p(u + tv) - p(u)}{t} \leq \frac{p(u + sv) - p(u)}{s}.$$

Proof. Since

$$\begin{aligned} p(u+tv)-p(u) &= p\left(\frac{t}{s}(u+sv)+\left(1-\frac{t}{s}\right)u\right)-p(u) \\ &\leq \frac{t}{s}p(u+sv)+\left(1-\frac{t}{s}\right)p(u)-p(u) \\ &= \frac{t}{s}(p(u+sv)-p(u)), \end{aligned}$$

we have

$$\frac{p(u+tv)-p(u)}{t} \leq \frac{p(u+sv)-p(u)}{s}.$$

Similarly, we have

$$\frac{p(u-sv)-p(u)}{-s} \leq \frac{p(u-tv)-p(u)}{-t}.$$

Since

$$p(u) = p\left(\frac{(u+tv)+(u-tv)}{2}\right) \leq \frac{1}{2}p(u+tv)+\frac{1}{2}p(u-tv),$$

we have $2p(u) \leq p(u+tv)+p(u-tv)$, and hence

$$\frac{p(u-tv)-p(u)}{-t} \leq \frac{p(u+tv)-p(u)}{t}.$$

□

Lemma 56. *Let E be a normed space, and let $p : E \to \mathbf{R}$ be a convex function. If p is continuous at $x \in E$, then there exists $r > 0$ and $L > 0$ such that*

$$|p(u)-p(v)| \leq L\|u-v\|$$

for each $u, v \in B(x, r)$.

Proof. Suppose that p is continuous at $x \in E$. Then there exists $\delta > 0$ such that $|p(x)-p(y)| \leq 1$ for each $y \in B(x,\delta)$, and hence

$$p(x)-1 \leq p(y) \leq p(x)+1$$

for each $y \in B(x,\delta)$. Let $r = \delta/2$. Then for each $u, v \in B(x,r)$ and $\epsilon > 0$, by setting

$$a = \frac{r}{\|v-u\|+\epsilon}, \qquad w = v+a(v-u), \qquad \lambda = \frac{1}{1+a},$$

we have

$$\|w-x\| \leq \|v-x\|+a\|v-u\| < r+r = \delta$$

and

$$\lambda w + (1-\lambda)u = \frac{1}{1+a}((1+a)v - au) + \frac{a}{1+a}u = v,$$

and, since p is convex, we have

$$p(v) \leq \lambda p(w) + (1-\lambda)p(u) = \lambda(p(w) - p(u)) + p(u).$$

Therefore

$$\begin{aligned} p(v) - p(u) &\leq \frac{1}{1+a}(p(w) - p(u)) < \frac{(p(x)+1) - (p(x)-1)}{a} \\ &= \frac{2}{r}(\|v-u\| + \epsilon), \end{aligned}$$

and similarly, we have $p(u) - p(v) \leq (2/r)(\|v-u\| + \epsilon)$. Thus, since ϵ is arbitrary, we have

$$|p(u) - p(v)| \leq L\|v-u\|,$$

where $L = 2/r$. □

Definition 57. A linear functional f on a linear space E is a *subderivative* of a convex function $p : E \to \mathbf{R}$ at $x \in E$ if

$$f(y-x) \leq p(y) - p(x)$$

for each $y \in E$; a convex function $p : E \to \mathbf{R}$ is *subdifferentiable* at $x \in E$ if it has a subderivative at x.

Lemma 58. *Let E be a normed space, and let $p : E \to \mathbf{R}$ be a convex function. If p is continuous at $x \in E$, then there exist $r > 0$ and $L > 0$ such that*

$$|f(y)| \leq L\|y\|$$

for each subderivative f of p at $z \in B(x,r)$ and $y \in E$.

Proof. Suppose that p is continuous at $x \in E$. Then there exist $r > 0$ and $L > 0$ such that $|p(u) - p(v)| \leq L\|u-v\|$ for each $u, v \in B(x, 2r)$, by Lemma 56. Let f be a subderivative of p at $z \in B(x,r)$, and consider $y \in E$ and $t > 0$ with $t\|y\| < r$. Then, since $z, z+ty, z-ty \in B(x,2r)$, we have

$$tf(y) = f((z+ty) - z) \leq p(z+ty) - p(z) \leq tL\|y\|$$

and

$$-tf(y) = f((z-ty) - z) \leq p(z-ty) - p(z) \leq tL\|y\|.$$

Therefore

$$-L\|y\| \leq f(y) \leq L\|y\|.$$

□

Lemma 59. *Let E be a linear space, and let $p : E \to \mathbf{R}$ be a convex function. If p is Gâteaux differentiable at $x \in E$, then the derivative f is linear, and is a subderivative of p at x.*

Proof. Suppose that a convex function $p : E \to \mathbf{R}$ is Gâteaux differentiable at $x \in E$ with the derivative $f : E \to \mathbf{R}$. Then, for each $t > 0$, we have

$$\begin{aligned}\frac{p(x+t(y+z))-p(x)}{t} &= \frac{p\left(\frac{1}{2}(x+2ty)+\frac{1}{2}(x+2tz)\right)-p(x)}{t}\\ &\le \frac{\frac{1}{2}p(x+2ty)-\frac{1}{2}p(x)+\frac{1}{2}p(x+2tz)-\frac{1}{2}p(x)}{t}\\ &= \frac{p(x+2ty)-p(x)}{2t}+\frac{p(x+2tz)-p(x)}{2t}\end{aligned}$$

and similarly, we have

$$\frac{p(x-2ty)-p(x)}{-2t}+\frac{p(x-2tz)-p(x)}{-2t} \le \frac{p(x-t(y+z))-p(x)}{-t}.$$

Hence

$$\begin{aligned}f(y+z) &= \lim_{t\to 0}\frac{p(x+t(y+z))-p(x)}{t}\\ &= \lim_{t\to 0}\frac{p(x+ty)-p(x)}{t}+\lim_{t\to 0}\frac{p(x+tz)-p(x)}{t} = f(y)+f(z).\end{aligned}$$

Consider $a \in \mathbf{R}$ with $a \# 0$. Then, since

$$\frac{p(x+tay)-p(x)}{t} = a\frac{p(x+tay)-p(x)}{at}$$

for each $t > 0$, we have

$$\begin{aligned}f(ay) &= \lim_{t\to 0}\frac{p(x+tay)-p(x)}{t} = a\lim_{t\to 0}\frac{p(x+tay)-p(x)}{at}\\ &= a\lim_{t\to 0}\frac{p(x+ty)-p(x)}{t} = af(y).\end{aligned}$$

Given an $a \in \mathbf{R}$, suppose that $f(ay) \# af(y)$. If $a \# 0$, then $f(ay) = af(y)$, a contradiction. Therefore $a = 0$, and so $f(ay) = 0 = af(y)$, again a contradiction. Thus $f(ay) = af(y)$.

For each $y \in E$ and t with $0 < t < 1$, we have

$$\begin{aligned}\frac{p(x+t(y-x))-p(x)}{t} &= \frac{p((1-t)x+ty)-p(x)}{t}\\ &\le \frac{(1-t)p(x)+tp(y)-p(x)}{t} = p(y)-p(x),\end{aligned}$$

and hence

$$f(y-x)=\lim_{t\to 0}\frac{p(x+t(y-x))-p(x)}{t}\leq p(y)-p(x).$$

Therefore f is a subderivative of p at x. □

Proposition 60. *Let C be a located convex subset of a normed space E, and let $x \in E$ be such that $d = d(x, C) > 0$. Then the following are equivalent.*

1. *The convex function $d(\cdot, C)$ is subdifferentiable at x.*

2. *There exists a normable linear functional f on E such that $\|f\| = 1$ and $f(x) + d \leq f(y)$ for each $y \in C$.*

Proof. We may assume without loss of generality that $x = 0$.

(1 $\Rightarrow$ 2): Suppose that $d(\cdot, C)$ is subdifferentiable at 0. Then there exists a linear functional g on E such that $g(z) \leq d(z, C) - d$ for each $z \in E$. Let $f = -g$. Then since $g(z) \leq d(z, C) - d \leq \|z\|$ for each $z \in E$, we have $|f(z)| \leq \|z\|$ for each $z \in E$. Let (z_n) be a sequence in C such that $\|z_n\| \to d$ as $n \to \infty$, and note that $0 < d \leq \|z_n\|$ for each n. Then

$$f(z_n)/\|z_n\| \geq (d - d(z_n, C))/\|z_n\| = d/\|z_n\| \to 1$$

as $n \to \infty$. Therefore f is normable with $\|f\| = 1$, and, since $d(y, C) = 0$ for each $y \in C$, we have

$$f(0) + d = d = d - d(y, C) \leq f(y)$$

for each $y \in C$.

(2 $\Rightarrow$ 1): Suppose that there exists a normable linear functional f such that $\|f\| = 1$ and $d \leq f(y)$ for each $y \in C$. Then given a $z \in E$, we have $d - f(z) \leq f(y) - f(z) \leq \|y - z\|$ for each $y \in C$, and hence $d - f(z) \leq d(z, C)$. Therefore $-f(z) \leq d(z, C) - d$ for each $z \in E$, and so $-f$ is a subderivative of $d(\cdot, C)$ at 0. □

The following theorem, a variant of Mazur's theorem [19], states that a continuous convex function on a separable Banach space has sufficiently many points at which it is Gâteaux differentiable; see also [12, V.9.8]

Proposition 61. *Let E be a separable Banach space, and let $p : E \to \mathbf{R}$ be a continuous convex function. Then p is Gâteaux differentiable at each point of a dense subset of E.*

Proof. Given $y \in E$ and $\epsilon > 0$, let

$$U_{y,\epsilon} = \{x \in E \mid \exists t \in \mathbf{Q}^+(p(x+ty) + p(x-ty) - 2p(x) < \epsilon t)\},$$

where $\mathbf{Q}^+ = \{t \in \mathbf{Q} \mid 0 < t\}$. Then, since p is continuous, $U_{y,\epsilon}$ is open. We show that $U_{y,\epsilon}$ is dense in E. Consider $x \in E$ and $\delta > 0$. Then there exist $r > 0$ and $L > 0$ such that $|p(u) - p(v)| \leq L\|u - v\|$ for each $u, v \in B(x, r)$, by Lemma 56. Choose n so that $2L\|y\| < n\epsilon$, and then choose $s \in \mathbf{Q}^+$ so that $(n+1)s\|y\| < \min\{r, \delta\}$. For each $k < n$, let

$$a_k = p(x + (k+1)sy) - p(x + ksy).$$

Then, since $x + ksy \in B(x, r)$ for each $k \leq n+1$, we have

$$\begin{aligned}\sum_{k<n}(a_{k+1} - a_k) &= a_n - a_0 \\ &= p(x + (n+1)sy) - p(x + nsy)) - (p(x + sy) - p(x)) \\ &< 2Ls\|y\| < n\epsilon s,\end{aligned}$$

and hence $a_{k+1} - a_k < \epsilon s$ for some $k < n$, by Proposition 6. Let $z = x + (k+1)sy$. Then

$$p(z + sy) + p(z - sy) - 2p(z) = a_{k+1} - a_k < \epsilon s$$

and $\|x - z\| = (k+1)s\|y\| < \delta$. Therefore $U_{y,\epsilon}$ is dense in E.

Let $(y_n)_n$ be a dense sequence of E, and let $\epsilon_m = 2^{-m}$ for each m. Then, by Theorem 17, the subset

$$U = \bigcap_{n,m} U_{y_n,\epsilon_m}$$

is also dense in E. Let $x \in U$. Then there exist $r > 0$ and $L > 0$ such that $|p(u) - p(v)| \leq L\|u - v\|$ for each $u, v \in B(x, r)$, by Lemma 56. Given $y \in E$ and $\epsilon > 0$, choose n and m so that $\epsilon_m + 2L\|y - y_n\| < \epsilon$. Then there exists $s \in \mathbf{Q}^+$ such that $p(x + sy_n) + p(x - sy_n) - 2p(x) < \epsilon_m s$, or

$$\frac{p(x + sy_n) - p(x)}{s} - \frac{p(x - sy_n) - p(x)}{-s} < \epsilon_m.$$

Choose $\delta > 0$ so that $\delta < s$ and $\delta \max\{\|y\|, \|y_n\|\} < r$. Then for each t with $0 < t < \delta$, since

$$x + ty, x + ty_n, x - ty, x - ty_n \in B(x, r),$$

we have

$$\begin{aligned}
&\frac{p(x+ty)-p(x)}{t} - \frac{p(x-ty)-p(x)}{-t} \\
&\leq \frac{p(x+ty_n)-p(x)+tL\|y-y_n\|}{t} - \frac{p(x-ty_n)-p(x)+tL\|y-y_n\|}{-t} \\
&= \frac{p(x+ty_n)-p(x)}{t} - \frac{p(x-ty_n)-p(x)}{-t} + 2L\|y-y_n\| \\
&\leq \frac{p(x+sy_n)-p(x)}{s} - \frac{p(x-sy_n)-p(x)}{-s} + 2L\|y-y_n\| \\
&< \epsilon_m + 2L\|y-y_n\| < \epsilon,
\end{aligned}$$

by Lemma 55. Therefore the limit

$$f(y) = \lim_{t\to 0} \frac{p(x+ty)-p(x)}{t},$$

exists. □

The following theorem is an approximate version of the separation theorem for two *unbounded* convex subsets.

Theorem 62. *Let A and B be convex subsets of a separable Banach space E such that the algebraic difference*

$$B - A = \{y - x \mid x \in A, y \in B\}$$

is located and $d = d(0, B-A) > 0$. Then for each $\epsilon > 0$ there exists a normable linear functional f on E such that $\|f\| = 1$ and

$$f(x) + d \leq f(y) + \epsilon$$

for each $x \in A$ and $y \in B$.

Proof. Let $z \in E$ be such that $\|z\| < \epsilon$ and $d(\cdot, B-A)$ is Gâteaux differentiable, and hence subdifferentiable at z (such a z exists by Proposition 61). Then, by Proposition 60, there exists a normable linear functional f on E such that $\|f\| = 1$ and

$$f(z) + d \leq f(y - x)$$

for each $x \in A$ and $y \in B$. Therefore

$$-\epsilon + d \leq f(z) + d \leq f(y) - f(x)$$

for each $x \in A$ and $y \in B$. □

Corollary 63. *Let A and B be convex subsets of a separable normed space E such that the algebraic difference*

$$B - A = \{y - x \mid x \in A, y \in B\}$$

is located and $d = d(0, B - A) > 0$. Then for each $\epsilon > 0$ there exists a normable linear functional f on E such that $\|f\| = 1$ and

$$f(x) + d \leq f(y) + \epsilon$$

for each $x \in A$ and $y \in B$.

Proof. Note that A and B are convex subsets of the Banach space $\tilde{E}$ such that $B - A$ is located in $\tilde{E}$ and $d = d(0, B - A) > 0$, by Remark 30 and Lemma 25, and apply Theorem 62. □

Since, by Theorem 62, we can separate two unbounded convex subsets, we have the following proof of the continuous extension theorem.

Theorem 64. *Let M be a subspace of a separable normed space E, and let f be a nonzero linear functional on M such that the kernel* $\ker(f)$ *is located in E. Then for each $\epsilon > 0$ there exists a normable linear functional g on E such that $g(x) = f(x)$ for each $x \in M$ and*

$$\|g\| \leq \|f\| + \epsilon.$$

Proof. Note that, since $\ker(f)$ is located in M, f is a normable linear functional with $0 < \|f\|$. Let $x_0 \in M$ be such that $f(x_0) = 1$, and let $d = d(0, \ker(f) - \{x_0\})$. Then, since $1 \leq |f(y - x_0)| \leq \|f\| \|y - x_0\|$ for each $y \in \ker(f)$, we have $1 \leq \|f\| d$, and hence $0 < d$. Given an $\epsilon > 0$, choose $\delta > 0$ so that $\delta < d$ and $\|f\| \delta / (d - \delta) \leq \epsilon$. Then, by Theorem 62, there exists a normable linear functional g_0 on E such that $\|g_0\| = 1$ and

$$g_0(x_0) + d \leq g_0(y) + \delta$$

for each $y \in \ker(f)$. Note that, since $0 \in \ker(f)$, we have $g_0(x_0) \leq \delta - d < 0$. Define a normable linear functional g on E by $g(x) = g_0(x)/g_0(x_0)$. Then

$$\|g\| = 1/|g_0(x_0)| \leq 1/(d - \delta) \leq \|f\| d/(d - \delta) \leq \|f\| + \|f\| \delta/(d - \delta) \leq \|f\| + \epsilon$$

and

$$1 + (d - \delta)/g_0(x_0) \geq g(y)$$

for each $y \in \ker(f)$. For each $x \in M$ and $t > 0$, since

$$t(x - f(x)x_0), -t(x - f(x)x_0) \in \ker(f),$$

we have
$$\frac{1+(d-\delta)/g_0(x_0)}{t} \geq g(x) - f(x) \geq -\frac{1+(d-\delta)/g_0(x_0)}{t},$$
and hence, letting $t \to \infty$, we have $g(x) = f(x)$. □

Proposition 65. *Let E be a linear space, let $p : E \to \mathbf{R}$ be a sublinear function, and let $x \in E$. Then the following are equivalent.*

1. *p is subdifferentiable at x.*
2. *There exists a linear functional f on E such that $f(x) = p(x)$ and $f(y) \leq p(y)$ for each $y \in E$.*

Proof. (1) ⇒ (2): Suppose that p is subdifferentiable at x with a subderivative f. Then
$$f(y-x) \leq p(y) - p(x)$$
for each $y \in E$, and hence
$$f(y) = f((x+y)-x) \leq p(x+y) - p(x) \leq p(x) + p(y) - p(x) = p(y)$$
for each $y \in E$. Since
$$-f(x) = f(0-x) \leq p(0) - p(x) = -p(x),$$
we have $p(x) \leq f(x)$, and so $f(x) = p(x)$.

(2) ⇒ (1): Suppose that f is a linear functional on E such that $f(x) = p(x)$ and $f(y) \leq p(y)$ for each $y \in E$. Then
$$f(y-x) = f(y) - f(x) \leq p(y) - p(x).$$
□

The following theorem is a constructive (approximate) version of the (1-dimensional) dominated extension theorem.

Theorem 66. *Let E be a separable Banach space E, let $p : E \to \mathbf{R}$ be a continuous sublinear function, and let $x \in E$. Then for each $\epsilon > 0$ there exists a bounded linear functional f on E such that $p(x) \leq f(x) + \epsilon$ and*
$$f(y) \leq p(y)$$
for each $y \in E$.

Proof. Since p is continuous at x, there exists $r > 0$ and $L > 0$ such that

$$|p(u) - p(v)| \leq L\|u - v\|$$

for each $u, v \in B(x, r)$, and

$$|f(y)| \leq L\|y\|$$

for each $y \in E$ and subderivative f of p at $z \in B(x, r)$, by Lemma 56 and Lemma 58. Choose $\delta > 0$ so that $\delta < r$ and $2L\delta < \epsilon$, and then choose $z \in B(x, \delta)$ so that p is Gâteaux differentiable, and hence subdifferentiable at z (such a z exists by Proposition 61). Let f be a subderivative of p at z. Then, as in the proof of Proposition 65, we have $f(z) = p(z)$ and $f(y) \leq p(y)$ for each $y \in E$. Since $z \in B(x, r)$, we have

$$\begin{aligned} p(x) &\leq p(z) + L\|x - z\| = f(z) + L\|x - z\| \\ &\leq f(x) + 2L\|x - z\| < f(x) + 2L\delta < f(x) + \epsilon. \end{aligned}$$

□

Corollary 67. *Let E be a separable normed space E, let $p : E \to \mathbf{R}$ be a continuous sublinear function, and let $x \in E$. Then for each $\epsilon > 0$ there exists a bounded linear functional f on E such that $p(x) \leq f(x) + \epsilon$ and*

$$f(y) \leq p(y)$$

for each $y \in E$.

Proof. Note that, since p is uniformly continuous, by Proposition 38, $\tilde{p}$ is a uniformly continuous sublinear function on the Banach space $\tilde{E}$, by Lemma 26, and apply Theorem 66. □

Acknowledgment

The author thanks the Japan Society for the Promotion of Science (JSPS), Core-to-Core Program (A. Advanced Research Networks) and JSPS KAKENHI Grant Number JP16K05251 for supporting the research.

References

[1] Peter Aczel, and Michael Rathjen Notes on Constructive Set Theory, Report No. 40, Institut Mittag-Leffler, The Royal Swedish Academy of Sciences, 2001.

[2] — CST Book [draft], August 19, 2010
`http://www1.maths.leeds.ac.uk/~rathjen/book.pdf`.

[3] Errett Bishop Foundations of Constructive Analysis, McGraw-Hill Book Company, New York, Toronto & London 1967.

[4] Errett Bishop, and Douglas Bridges Constructive Analysis, Springer-Verlag, Berlin 1985 [*Grundlehren der Mathematischen Wissenschaften 279*].

[5] Douglas S. Bridges Constructive Functional Analysis, Pitman, Boston MA & London 1979 [*Advanced Publishing Program – Research Notes in Mathematics 28*].

[6] Douglas Bridges, Hajime Ishihara, and Luminiţa Vîţă *Computing infima on convex sets, with applications in Hilbert spaces*, Proceedings of the American Mathematical Society 132, 2004, pp. 2723–2732.

[7] Douglas Bridges, and Fred Richman Varieties of Constructive Mathematics, Cambridge University Press, Cambridge 1987 [*London Mathematical Society Lecture Note Series 97*].

[8] Douglas S. Bridges, and Luminiţa Simona Vîţă Techniques of Constructive Analysis, Springer, New York 2006 [*Universitext*].

[9] Luitzen E. J. Brouwer Brouwer's Cambridge Lectures on Intuitionism, Edited and with a preface by Dirk van Dalen, Cambridge University Press, Cambridge, New York 1981.

[10] Jan Cederquist, Thierry Coquand, and Sara Negri *The Hahn-Banach theorem in type theory*, in: Giovanni Sambin, and Jan Smith (eds.) Twenty-five Years of Constructive Type Theory (Proceedings of a Congress held in Venice, October 1995), Oxford University Press, New York 1998 [*Oxford Logic Guides 36*], pp. 57–72.

[11] Dirk van Dalen Logic and Structure, Springer, London 2013 (fifth edition) [*Universitext*].

[12] Nelson Dunford, and Jacob T. Schwartz Linear Operators, *Part I, General Theory* [With the assistance of William G. Bade and Robert G. Bartle], Interscience Publishers, John Wiley & Sons, Inc., New York 1988 [*Wiley Classics Library*]. (A reprint of the 1958 original.)

[13] Arend Heyting Intuitionism, *An Introduction*, North-Holland Publishing Company, Amsterdam 1966 [second revised edition].

[14] Hajime Ishihara *On the constructive Hahn-Banach theorem*, Bulletin of the London Mathematical Society 21, 1989, pp. 79–81.

[15] — *An omniscience principle, the König lemma and the Hahn-Banach theorem*, Zeitschrift für mathematische Logik und Grundlagen der Mathematik 36, 1990, pp. 237–240.

[16] — *Constructive functional analysis: an introduction*, in: K. Mainzer, P. Schuster, and H. Schwichtenberg (eds.), Proof and Computation, *Digitization in Mathematics, Computer Science, and Philosophy*, World Scientific, Singapore 2018, pp.109–165.

[17] David L. Johns, and Christopher G. Gibson *A constructive approach to the duality theorem for certain Orlicz spaces*, Mathematical Proceedings of the Cambridge Philosophical Society 89, 1981, pp. 49–69.

[18] Boris A. Kushner Lectures on Constructive Mathematical Analysis [Translated from Rus-

sian by Eliott Mendelson, and edited by Lev J. Leifman], American Mathematical Society, Providence RI 1984 [*Translations of Mathematical Monographs 60*]. (Russian original: Lekcii po konstructivnomu matematičeskomu analizu, Nauka, Moskow 1973.)

[19] Stanisław Mazur *Über konvexe Mengen inlinearen normierte Räumen*, Studia Mathematica 4 (1), 1933, pp. 70–84.

[20] George Metakides, Anil Nerode, and Richard A. Shore *Recursive limits on the Hahn-Banach theorem*, in: Murray Rosenblatt (ed.) Errett Bishop: Reflections on Him and His Research (Proceedings of the Memorial Meeting for Errett Bishop: September 24, 1983, University of California, San Diego), American Mathematical Society, Providence RI 1985 [*Contemporary Mathematics. 39*], pp. 85–91.

[21] Walter Rudin Functional Analysis, McGraw-Hill, Inc., New York 1991 (second edition) [*International Series in Pure and Applied Mathematics*].

[22] Giovanni Sambin *Some points in formal topology*, Theoretical Computer Science 305 (1–3), 2003, pp. 347–408.

[23] Anne S. Troelstra, and Dirk van Dalen Constructivism in Mathematics, *An introduction*, vol. 1, North-Holland Publishing Company, Amsterdam 1988 [*Studies in Logic and the Foundations of Mathematics 121*].

[24] — Constructivism in Mathematics, *An Introduction*, vol. 2, North-Holland Publishing Company, Amsterdam 1988 [*Studies in Logic and the Foundations of Mathematics 123*]

METAFINITE MODEL THEORY AND REAL NUMBER COMPUTATIONS

KLAUS MEER
Brandenburg University of Technology, Cottbus
meer@b-tu.de

ABSTRACT In this article we survey research done in the area of descriptive complexity theory in relation with the Blum-Shub-Smale model of real number computations. Parallel to the classical development of discrete complexity theory based on the Turing machine model since the 1970s, descriptive complexity has been an inspiring research field in relation to finite model theory. For dealing with real number problems, this framework has to be enlarged to *metafinite* model theory. We describe this approach and show how it leads to capturing most of the important classes in real number complexity theory.

KEYWORDS Descriptive complexity theory, real number computations, meta-finite model theory

1 Introduction

There has been and still is a tremendous interplay between logic and computation. One prominent early example is due to Büchi [12] and, independently, Elgot [31] and Trakhtenbrot [60]. It relates the most fundamental computational concept of a finite state automaton to certain logics defined over the natural numbers. More precisely, viewing each word $w \in \Sigma^*$ built by letters from a finite set (called alphabet) Σ - the central objects finite automata compute with, as a relational structure $\mathcal{S}_w$ whose signature includes a number of built-in relations - the central objects on which logics are defined, Büchi showed the following: A set $L \subseteq \Sigma^*$ of words (also called a language) can be accepted by a finite state automaton, i.e., is a regular language, if and only if there exists a sentence ϕ in weak monadic second-order MSO logic over the signature $\langle \mathbb{N}, =, succ, 0 \rangle$ which is precisely satisfied by those structures resulting from words in $L : w \in L \Leftrightarrow \mathcal{S}_w \models \phi$. Here, *succ* is the binary successor relation and 'weak' reflects that monadic second-order quantifiers range over finite subsets of $\mathbb{N}$ only. Büchi also proved a generalization for full MSO logic. These results implied the decidability of certain fragments of arithmetic by using automata theoretic methods. It is one of the first examples connecting finite model theory and the theory of computation.

About a decade later, the beginning 1970s documented by the work of Cook [19] and many others the ever increasing interest in the general study of complexity of problems and solution algorithms over finite alphabets. The latter are formalized in classical computer science by the most fundamental concept of a Turing machine. It lead to the very successful theory of NP-completeness and the still unsolved millennium problem asking whether the two complexity classes P and NP coincide or not. Studying relations between finite model theory and computational complexity in the Turing machine framework started soon after and is the core topic of interest in descriptive complexity theory. Whereas complexity theory analyses the necessary resources to solve a problem algorithmically, descriptive complexity theory studies the logical resources required for expressing a problem in a logical manner like done in Büchi's theorem. One of its first important results being very influential for the further development of this research direction was Fagin's theorem [32].[1] It states that the languages in NP, after identifying words with finite structures over suitable signatures similarly as done above in Büchi's theorem, are precisely the models of sentences expressed in existential second-order logic over the considered signature.

Both above results witness the strong relation between logic and computability and complexity theory, respectively; this relation has emerged into an own research direction. Important applications on the computer science side are, for example, the design of efficient algorithms for subclasses of hard problems in the field of parameterized algorithms [35], the relations to database theory [1] or to the area of model checking and software verification [18]. All this is located in the framework of discrete structures as they foremost are present when designing and analysing algorithms in computer science in the Turing machine model.

Computational models other than the Turing machine have become more and more objects of study in the last decades; the latter includes such diverse models as Neural Networks [44], Quantum Computers [54], Analog Computers [11] and Biological Computing [36]. An approach reflecting the algorithm design in classical areas of mathematics is Algebraic Complexity Theory [14]. Here, problems are formalized over uncountable structures like $\mathbb{R}$ and $\mathbb{C}$ and algorithms can perform the basic arithmetic operations $+, -, \times, :$ of these structures, including binary test

[1] Close variants of Fagin's theorem have been proved independently by Christen [17] and Jones and Selman [46], see [8] for a historical account.

operations using the order relation $\geq$ over $\mathbb{R}$ or equality $=$ over $\mathbb{C}$. Models of computations used in this approach are algebraic circuits and straight-line programs. In 1989, Blum, Shub, and Smale [7] introduced a model of computation over arbitrary structures, now called the BSS model. Especially when considered over the real or complex numbers it gives a uniform version of algebraic computations. One main focus of the approach was to introduce a complexity theory over $\mathbb{R}$ similar to the classical one for discrete structures. In the three decades passed since then a large pool of complexity results in this framework has been obtained, see for example the collections [55, 52]. Among the different directions related to the BSS approach one is descriptive complexity. In this article we shall outline the approach and survey some of its results, focussing on the real numbers as underlying structure.

Section 2 will briefly recall the BSS model over $\mathbb{R}$ and the definition of major complexity classes. Then, we describe the background of meta-finite model theory introduced by Grädel and Gurevich [39], putting the focus on its use for real number computations as developed in [41]. Section 3.2 overviews some of the main results obtained characterizing most real number complexity classes by logical means. The final section will discuss some further results, including an outline of interesting research directions.

The paper will not at all try to survey the huge field of 'classical' descriptive complexity. There are many excellent textbooks which could serve as starting point and orientation about work in finite model theory and its relations to computation and complexity. The interested reader is referred to [45, 40]. In [35], as well the relation between logic and the design of efficient algorithms is addressed; it also contains a discussion of Büchi's result mentioned at the beginning of this introduction.

2 The real number model of computation

We first give a brief account of the Blum-Shub-Smale (for short: BSS) computational model and define the complexity classes used in subsequent sections. We solely deal with $\mathbb{R}$ as underlying structure. For a much deeper treatment see [6].

Definition 1. A (real number) Blum-Shub-Smale (shortly: BSS) machine is a Random Access Machine over $\mathbb{R}$. Such a machine has a countable number of registers each storing a real number. It is able to perform the basic arithmetic operations $\{+,-,\times,:\}$ together with branch instructions of the form: is a real number $x \geq 0$? These operations are performed on the input components and the intermediate results; moreover, there is a finite number of machine constants from the underlying domain used by the machine. In addition, there are instructions for direct and indirect addressing of registers. A BSS machine M now can be defined as a directed graph. Each node of the graph corresponds to an instruction. An outgoing edge points to the next instruction to be performed; a branch node has two outgoing edges related to the two possible outcomes of the test. The algorithm performed by M starts with one designated start instruction. It stops if a designated stop instruction is reached. Such a machine handles finite sequences of real numbers as inputs, i.e., elements from the set $\mathbb{R}^\infty := \bigsqcup_{k\in\mathbb{N}} \mathbb{R}^k$. In case of termination of a computation the machine outputs an element from $\mathbb{R}^\infty$ as result.

Definition 2. A pair (A, A^+) with sets $A^+ \subseteq A \subseteq \mathbb{R}^\infty$ is called *(real number) decision problem.* It is *decidable* if there is a BSS algorithm over $\mathbb{R}$ that, given an input $x \in A$, terminates with result 1 in case $x \in A^+$ and with result 0 otherwise. We then say that the algorithm *decides* (A, A^+).

In the sequel we shall only consider decidable problems. For sake of simplicity we also disregard division as an operation since it neither has an influence on decidability nor on polynomial time complexity as defined below.

Next, algorithms should be equipped with a time measure for their execution. As usual, in order to then classify problems with respect to the running time needed to solve them one also has to define the size of an instance. The time consumption is considered as function in the input size. The intuitive approach for measuring the algebraic complexity of a problem described at the beginning of this section is now made more precise as follows.

Definition 3. Let M be a BSS machine. The *size* of an element $x \in \mathbb{R}^k$ is $size_{\mathbb{R}}(x) := k$. The *cost* of each basic operation and of a branching instruction is 1. The cost of an entire computation is the number of operations (including branching instructions) performed until the machine halts, or ∞ in case it computes forever. The (partial) function from $\mathbb{R}^\infty$ to $\mathbb{R}^\infty$ computed by M is denoted by Φ_M. The cost of M's computation on input $x \in \mathbb{R}^\infty$ is also called its *running time* and denoted by $T_M(x)$.

Most of the well known Boolean time-complexity classes can now be defined analogously over the reals. We first give a precise definition of the two main such classes. Note that below we once in a while also consider more general input spaces, for example pairs of elements from $\mathbb{R}^\infty$. This easily can be handled in the above framework as well by using an embedding of spaces like $\mathbb{R}^\infty \times \mathbb{R}^\infty$ into $\mathbb{R}^\infty$; for example, one could use an encoding which explicitly contains information about the length of the two parts of the input from $\mathbb{R}^\infty \times \mathbb{R}^\infty$. Clearly, the size of an element in $\mathbb{R}^\infty \times \mathbb{R}^\infty$ is the sum of the sizes of both components from $\mathbb{R}^\infty$ plus a constant.

Definition 4. a) A decision problem (A, A^+), $A^+ \subseteq A \subseteq \mathbb{R}^\infty$ is in class $\mathrm{P}_\mathbb{R}$ (*decidable in polynomial time over* $\mathbb{R}$) iff there exist a polynomial p and a BSS machine M over $\mathbb{R}$ deciding the problem such that $T_M(x) \leq p(size_\mathbb{R}(x))\ \forall x \in \mathbb{R}^\infty$. b) (A^+, A) is in class $\mathrm{NP}_\mathbb{R}$ (*verifiable in non-deterministic polynomial time over* $\mathbb{R}$) iff there exist a polynomial p and a BSS machine M over $\mathbb{R}$ working on input space $A \times \mathbb{R}^\infty$ such that

(i) $\Phi_M(x, y) \in \{0, 1\}\ \forall x \in A, y \in \mathbb{R}^\infty$

(ii) $\Phi_M(x, y) = 1 \Rightarrow x \in A^+$

(iii) $\forall x \in A^+\ \exists y \in \mathbb{R}^\infty\ \Phi_M(x, y) = 1$ and $T_M(x, y) \leq p(size_\mathbb{R}(x))$.

c) A problem $(A, A^+) \in \mathrm{NP}_\mathbb{R}$ is $\mathrm{NP}_\mathbb{R}$-*complete* iff every other problem in $\mathrm{NP}_\mathbb{R}$ can be reduced to it in polynomial time. *Polynomial time reducibility* of problem (B, B^+) to problem (A, A^+) means: There is a polynomial time computable function $f : B \to A$ which satisfies $\forall x \in B : x \in B^+ \Leftrightarrow f(x) \in A^+$. This type of reduction is also called *polynomial time many one reduction.*

When talking about a problem $(A, A^+) \in \mathrm{NP}_\mathbb{R}$, for an input $x \in A^+$ the y whose existence is required in part b,ii) above can be seen as a proof of x's membership in A^+. The definition then requires that correctness of this proof can be checked efficiently, i.e., in polynomial time in the size of x. The definition directly implies that $\mathrm{P}_\mathbb{R}$ is included in $\mathrm{NP}_\mathbb{R}$. Just as in discrete complexity theory, the currently most important open question in real (and also complex) number complexity theory is whether this inclusion is strict. It is easily seen that $\mathrm{P}_\mathbb{R} = \mathrm{NP}_\mathbb{R}$ is equivalent to the existence of already one single $\mathrm{NP}_\mathbb{R}$-complete problem which belongs to class $\mathrm{P}_\mathbb{R}$.

Example 5. Let 4-FEAS denote the set of real polynomials with arbitrary number of variables and of degree four, and let 4-FEAS$^+$ be those which have a real root. Here, given n and a polynomial $f \in \mathbb{R}[z_1, \ldots, z_n]$ we represent f as an element in $\mathbb{R}^\infty$ coding it by the sequence of its coefficients. Note that the size of such an encoding is $\mathcal{O}(n^4)$ because this bounds the number of coefficients. As decision problem (4-FEAS, 4-FEAS$^+$) belongs to $\mathrm{NP}_\mathbb{R}$: Given an f (playing the role of x in Definition 4, b) a machine verifying that $f \in$ 4-FEAS$^+$ expects the second part y of its input to represent a real zero of f in $\mathbb{R}$ and checks $f \in 4\text{-}FEAS^+$ by evaluating $f(y)$ in polynomial time in the input size $\mathcal{O}(n^4)$.

One of the main results in [7] shows that (4-FEAS, 4-FEAS$^+$) is an $\mathrm{NP}_\mathbb{R}$-complete problem.

For formalizing some of the results to be presented later on we additionally need to recall the definition of the well known computational model of algebraic circuits. When equipped with a uniformity condition it is known to be equivalent to real number BSS machines.

Definition 6. An *algebraic circuit* is a connected and directed acyclic graph having nodes either of indegree 0 (input nodes or constant nodes) , 1 (test nodes) or 2 (computation nodes). Nodes of indegree 0 are labelled either as input nodes or with a real constant; nodes of indegree 2 are labelled with one of the operations $+, -, \times$, nodes of indegree 1 are labelled with $' \geq 0?'$ A circuit has one output node of outdegree 0. The *size* of a circuit is the number of its nodes, its *depth* is the length of the longest path from an input node to the output node. A circuit C_n with n input nodes computes in the straightforward manner a function from $\mathbb{R}^n \to \mathbb{R}$; on input $x \in \mathbb{R}^n$ it propagates values along the labels of nodes in the obvious way. The value of a test node is either 1 or 0, depending on whether its incoming value is ≥ 0 or not. We only consider circuits with one output node which is a test node. Thus, our circuits compute characteristic functions.

A family $\{C_n\}_{n\in\mathbb{N}}$ of algebraic circuits then computes the characteristic function of a real language $L \subseteq \mathbb{R}^\infty$. However, in order to relate such families to real complexity classes in the BSS model or to use them for defining new ones it makes sense to impose some uniformity condition on the circuits. One such condition refers to the constants used by the different circuits. They should be the same for all input dimensions. Another addresses the structure of circuits for different input dimensions. Depending on the uniformity condition different complexity classes are obtained. Note that each node of a circuit is uniquely coded by its type and the incoming node(s) in the underlying acyclic graph.

Definition 7. Let $s \in \mathbb{N}, c := (c_1, \ldots, c_s) \in \mathbb{R}^s$ be fixed and let $\{C_n\}_{n\in\mathbb{N}}$ be a family of algebraic circuits which for every $n \in \mathbb{N}$ have n input nodes; all circuits are assumed to have s constant nodes labelled with $c_1, \ldots, c_s$, respectively.

a) The family is *L-uniform* if there exists a Turing machine computing for each pair (n, i) the (discrete) description of its i-th node in time $\mathcal{O}(\log n)$. If there exists a Turing machine running in polynomial time in n the family is called *P-uniform.*

b) Define $\mathrm{NC}^k_{\mathbb{R}}$ for $k \geq 1$ to be the class of decision problems (S, S^+) such that there is an $s \in \mathbb{N}, c \in \mathbb{R}^s$ and a L-uniform family of algebraic circuits $\{C_n\}_{n\in\mathbb{N}}$ having size polynomial in n and depth $\mathcal{O}(\log^k n)$ that computes the characteristic function of S^+ in S when using c as labels of its constant nodes for all input dimensions n. The union of all $\mathrm{NC}^k_{\mathbb{R}}$ is denoted by $\mathrm{NC}_{\mathbb{R}}$.

c) Similarly, we define $\mathrm{PAR}_{\mathbb{R}}$ to be the class of all decision problems (L, L^+) such that there is an $s \in \mathbb{N}, c \in \mathbb{R}^s$ and a P-uniform family of algebraic circuits $\{C_n\}_{n\in\mathbb{N}}$

having depth polynomial (and therefore size exponential) in n that computes the characteristic function of L^+ in L when using c as labels of its constant nodes for all n.

It is not hard to see that such uniform circuit families can simulate BSS algorithms and vice versa in such a way, that the necessary resources are closely related. For example, the class $P_{\mathbb{R}}$ is characterized as those problems being decidable by a P-uniform family of circuits of polynomial size (and where the labelling of the constant nodes by reals is the same for all n).

Example 8. Consider an algebraic circuit C being a decision circuit, i.e., having only a single output node which is a sign node. The Circuit Satisfiability Problem asks for the existence of a real input $x \in \mathbb{R}^n$ of suitable dimension n such that $C(x) = 1$. Due to the above remarks concerning the simulation of BSS machines by algebraic circuits this problem is $NP_{\mathbb{R}}$-complete as well. The Circuit Evaluation Problem gets as input such a decision circuit C together with a point $x \in \mathbb{R}^n$ and asks whether $C(x) = 1$. In this restricted form the problem is in $P_{\mathbb{R}}$. Here, it is complete under a restricted version of polynomial reductions using $NC_{\mathbb{R}}$-computations [27]. This implies that it cannot be parallelized because it is known that $NC_{\mathbb{R}} \neq P_{\mathbb{R}}$, see [24].

This only gives a sketchy overview of complexity classes defined and results obtained in the BSS framework. The treated classes and examples are used below in relation to the main topic of this article.

3 Metafinite model theory and descriptive complexity over $\mathbb{R}$

In finite model theory it is not really possible to handle situations in which the description of problems at least partially involves objects from infinite numerical domains which are used in arithmetic operations. Such situations, however, frequently occur in computer science, for example in areas like database theory, problems on weighted graphs, fault-tolerant networks etc. So one might look for an extension going beyond finite models in order to be able to reason about such problems adequately, both preserving and extending the methods of finite model theory. Obviously, given a computational model for such infinite domains together with a complexity theory, as it is the case with the BSS model over $\mathbb{R}$, we can ask for the relations between the latter and logical properties of problems in such an extended 'metafinite' model theoretic framework. If we would consider the reals as underlying structure with function and relation symbols interpreting the arithmetic operations and the order, then already first-order logic - typically a very weak logic in the finite setting - will result in a much too strong logic for describing real complexity classes. It is not only able to express all problems in $\mathrm{NP}_{\mathbb{R}}$, since the $\mathrm{NP}_{\mathbb{R}}$-complete (4-FEAS,4-FEAS$^+$) problem can be expressed as a first-order sentence with existential quantifiers only, but even the potentially much larger so called polynomial hierarchy PH, see [6]. If other infinite domains like the integers are considered, a careless approach even would lead to typical problems of such structures like the undecidability of arithmetic. This implies that an approach extending finite model theory to infinite domains has to proceed more carefully with respect to the issue which parts of the structure can be accessed in which manner by a logic. This program was initiated by Grädel and Gurevich in [39], where metafinite model theory was introduced. Its basic idea is a clean split of the discrete and the numerical parts of a structure; then suitable logics integrate typical constructs already used in finite model theory together with a potentially limited access to the infinite numerical domain of a structure. In [39] this approach is developed and applied to a number of metafinite structures important for different areas of computer science. A more recent survey is given in [38]. Here, we want to describe this approach when applied to real number computations and BSS complexity theory. This was started in [41].

3.1 $\mathbb{R}$-structures; first-order and second-order logic on $\mathbb{R}$-structures

Consider a typical real number decision problem like (4-FEAS,4-FEAS$^+$). Though the question can be formalized using an existential first-order sentence over the structure $\langle \mathbb{R}, +, -, \times, \geq, 0, 1, \{c_r\}_{r\in\mathbb{R}}\rangle$, where c_r stands for a constant symbol representing the real r, a closer look identifies both discrete and real number data underlying a description of the task. Typical for the former are the set of variables and the monomials built over them up to degree 4. The 'only' real number part of the data is the vector of coefficients which can be seen as a function mapping the (discrete) set of monomials to $\mathbb{R}$. Of course, also the decision question itself asks something about the existence of certain real numbers.The logics we are going to define can separate much more sensitively properties of the discrete and the real number part. This split is the underlying idea for defining metafinite model theory in general. Over the reals it leads to the concept of $\mathbb{R}$-structures and logics for them. For basic terminology in logic such as vocabulary, structures, first-order formulas and sentences, interpretation etc. see, for example, [30].

Definition 9. Let L_s, L_f be finite vocabularies, where L_s may contain relation and function symbols, and L_f contains function symbols only.

a) An $\mathbb{R}$-*structure of signature* $\sigma = (L_s, L_f)$ is a pair $\mathfrak{D} = (\mathcal{A}, \mathcal{F})$ consisting of

(i) a finite structure $\mathcal{A}$ of vocabulary L_s, called the *skeleton* of $\mathfrak{D}$, whose universe A will also be said to be the *universe* of $\mathfrak{D}$, and

(ii) a finite set $\mathcal{F}$ of functions $X : A^k \to \mathbb{R}$ interpreting the function symbols in L_f (where k depends on the respective symbol only).

We shall denote the set of all $\mathbb{R}$-structures of signature σ by $\text{Struct}_{\mathbb{R}}(\sigma)$.

b) For an $\mathbb{R}$-structure $\mathfrak{D}$ with skeleton $\mathcal{A}$ we denote by $|\mathfrak{D}|$ the cardinality of its universe A. This number is called the *size* of the structure $\mathfrak{D}$.

c) An $\mathbb{R}$-structure $\mathfrak{D} = (\mathcal{A}, \mathcal{F})$ is *ranked* if there is a unary function symbol $r \in L_f$ whose interpretation $r_{\mathfrak{D}}$ in $\mathfrak{D}$ bijects A with $\{0, 1, \ldots, |A| - 1\}$. The function $r_{\mathfrak{D}}$ is called *ranking*. We will write $i < j$ for $i, j \in A$ iff $r_{\mathfrak{D}}(i) < r_{\mathfrak{D}}(j)$. A k-ranking on A is a bijection between A^k and $\{0, 1, \ldots, |A|^k - 1\}$. It can easily be defined if a ranking is available. We denote by $r^k_{\mathfrak{D}}$ the interpretation of the k-ranking induced by $r_{\mathfrak{D}}$.

Unless otherwise stated we only consider ranked $\mathbb{R}$-structures and will thus most of the time suppress notationally the order relation in the vocabularies considered.

A first observation is that $\mathbb{R}^\infty$ easily can be coded as a set of $\mathbb{R}$-structures over $\sigma = (L_s, L_f)$, where $L_s = \emptyset$ and $L_f = \{r, C\}$ consists of two unary function symbols. Now an $x = (x_0, x_1, \ldots, x_{n-1})$ is coded as an $\mathbb{R}$-structure $\mathfrak{D}_x$ over σ with universe $A = \{0, 1, \ldots, n-1\}$, where the interpretation of r is a ranking on A and the interpretation of C gives the components of x in the order induced by the ranking. Conversely, given an $\mathbb{R}$-structure $\mathfrak{D} = (\mathcal{A}, \mathcal{F})$ we can encode it as an element $e(\mathfrak{D}) \in \mathbb{R}^\infty$ as follows: Fix a ranking on the underlying universe and write all the functions and relations from L_s as characteristic function of form $\chi : A^k \to \{0, 1\} \subset \mathbb{R}$ for a suitable k. Next, identify all functions $X : A^k \to \mathbb{R}$ as a point in $\mathbb{R}^\infty$. The ordering of the components is determined by the ranking, the concatenation of all resulting vectors gives the encoding $e(\mathfrak{D})$ of $\mathfrak{D}$. Note that the size of $e(\mathfrak{D})$ is polynomially bounded in $|\mathfrak{D}|$. Below, when we speak about the complexity of a decision problem (F, F^+) defined on $\mathbb{R}$-structures we mean the complexity of the respective encoded problem $(e(F), e(F^+))$ over $\mathbb{R}^\infty$, where $e(F) = \{e(\mathfrak{D}) | \mathfrak{D} \in F\}$.

The decisive aspect when combining $\mathbb{R}$-structures and logics for them with real number complexity theory is how the logics can access the different parts of such a structure. The first logic we introduce on $\mathbb{R}$-structures is first-order logic. Just as in the finite model theoretic framework it turns out to be a relatively weak logic, the reason being that first-order variables only range over the skeleton. Let $V = \{v_0, v_1, \ldots\}$ denote a fixed countable set of variables.

Definition 10. The language $\mathrm{FO}_{\mathbb{R}}$ contains, for each signature $\sigma = (L_s, L_f)$, a set of formulas and terms. Each term t takes, when interpreted in some $\mathbb{R}$-structure, values in either the skeleton, in which case we call it an *index term*, or in $\mathbb{R}$, in which case we call it a *number term*. Terms are defined inductively as follows

(i) The set of index terms is the closure of the set V of variables under applications of function symbols of L_s.

(ii) Any real number is a number term.

(iii) If $h_1, \ldots, h_k$ are index terms and X is a k-ary function symbol of L_f, then $X(h_1, \ldots, h_k)$ is a number term.

(iv) If t, t' are number terms, then so are $t + t'$, $t - t'$, $t \times t'$, and $\mathrm{sign}\,(t)$. Here, the sign function is defined as $\mathrm{sign}\,(t) := \begin{cases} 1 & \text{if } t \geq 0 \\ 0 & \text{if } t < 0 \end{cases}$.

Atomic formulas are equalities $h_1 = h_2$ of index terms, equalities $t_1 = t_2$ and inequalities $t_1 < t_2, t_1 \leq t_2$ of number terms, and expressions $P(h_1, \ldots, h_k)$, where P is a k-ary predicate symbol in L_s and $h_1, \ldots, h_k$ are index terms.

The set of formulas of $\mathrm{FO}_{\mathbb{R}}$ is the smallest set containing all atomic formulas and which is closed under Boolean connectives and quantification $(\exists v)\psi$ and $(\forall v)\psi$.

Example 11. If $\mathfrak{D}$ is an $\mathbb{R}$-structure of signature (L_s, L_f) and $r \in L_f$ is a unary function symbol we can express in first-order logic the requirement that r is interpreted as a ranking in $\mathfrak{D}$ using the sentence

$$inj(r) \wedge \exists \mathtt{o} \in A\; r(\mathtt{o}) = 0 \;\wedge \forall u \in A\; [u \neq \mathtt{o} \Rightarrow (0 < r(u) \;\wedge\; \exists v \in A\; r(u) = r(v) + 1)].$$

Here, $inj(r) \equiv \forall x, y : \; x \neq y \Rightarrow r(x) \neq r(y)$ expresses injectivity of r. Moreover, such an r can be used to define two elements $\mathtt{o}, \mathtt{1} \in A$ such that $r(\mathtt{o}) = 0$ and $r(\mathtt{1}) = |A| - 1$. These symbols thus define a first and a last element of A. This of course only makes sense when $|A| \geq 2$. Similarly, any ranking r induces, for all $k \geq 1$, a k-ranking r^k on A by lexicographical ordering. Note that r^k is definable in the sense that for all $(v_1, \ldots, v_k) \in A^k$

$$r^k(v_1, \ldots, v_k) = r(v_1)|A|^{k-1} + r(v_2)|A|^{k-2} + \ldots + r(v_k).$$

This can be used to express logically the process of cycling through A^k.

Example 12. Let us see how to describe the structure of algebraic circuits using first-order logic. To do so, we consider the vocabularies $L_s = \{f_l, f_r\}$, where f_l and f_r

are function symbols of arity one, and $L_f = \{r, C, F_t\}$, whose three symbols are also of arity one. If $\mathfrak{D}$ is an $\mathbb{R}$-structure with signature (L_s, L_f) we intend to interpret its universe A as the set of nodes of an algebraic circuit. Left and right predecessors are to be given by the functions f_l and f_r, respectively. There are finitely many types of nodes (such as input nodes, summation nodes etc.), each coded by a natural number; for example, below we code input nodes by '1' and constant nodes by '2'. The type of each node will be given by F_t and the real numbers associated to the constant nodes by the function C. Finally, we will require r to be interpreted as a ranking.

We now express all these requirements with first-order sentences. To interpret r as a ranking we use the sentence in Example 11. The sentence $\forall v\,[\bigvee_k F_t(v) = k]$ ensures that the elements of A are of one of the finitely many types of a node.

The sentence

$$\forall v \forall w\ [(F_t(v) = 1\ \wedge\ F_t(w) \neq 1) \Rightarrow r(v) < r(w)]$$

requires that input nodes are the first nodes of the circuit. Finally, the sentence

$$\forall v\ [(F_t(v) \neq 1\ \wedge\ F_t(v) \neq 2) \Rightarrow (r(f_l(v)) < r(v)\ \&\ r(f_r(v)) < r(v))]$$

requires that, when a gate v is not of input or constant type, its predecessors are really so, that is, they are ranked before v. It also ensures that A, considered as a directed graph, is acyclic.

The expressive power of first-order logic is relatively weak. Given a first-order sentence in prenex form we can evaluate the quantifier free part in constant time (only depending on the fixed formula size) for any assignment of the variables. Then validity of the quantifier prefix can be checked in parallel. This implies

Proposition 13 ([26]). *Let σ be a signature, $S = \mathrm{Struct}_{\mathbb{R}}(\sigma)$ and $S^+ = \{\mathfrak{D} \in S \mid \mathfrak{D} \models \varphi\}$, where φ is a first-order sentence. Then $(S, S^+) \in \mathrm{NC}^1_{\mathbb{R}}$.*

The above result recently has been strengthened to a characterization of $\mathrm{FO}_{\mathbb{R}}$ via the class $\mathrm{AC}^0_{\mathbb{R}}$ of real number decision problems decidable by a uniform family of algebraic circuits of polynomial size, constant depth and with *unbounded* fan-in of its gates.

Proposition 14 ([3]). *It holds* $\mathrm{AC}^0_{\mathbb{R}} = \mathrm{FO}_{\mathbb{R}}$

Given these propositions together with Example 8 one sees that the Circuit Evaluation Problem cannot be expressed in $\mathrm{FO}_{\mathbb{R}}$-logic.

Consequently, in order to capture classes like $\mathrm{P}_{\mathbb{R}}$ and $\mathrm{NP}_{\mathbb{R}}$ by logics we have to extend first-order logic. For describing the existence of real number objects like zeros of polynomials it is necessary to allow quantification over function symbols from the finite universe to the real number part of an $\mathbb{R}$-structure. This leads to second-order logic and its existential fragment.

Definition 15. Second-order logic $SO_{\mathbb{R}}$ on $\mathbb{R}$-structures is obtained starting from $FO_{\mathbb{R}}$ logic by adding function variables and quantifiers for these variable. More precisely, given a vocabulary $\sigma = (L_s, L_f)$, where L_f contains a function symbol Y interpreted as a function from some $A^k \to \mathbb{R}$, together with a first-order formula ϕ over σ, both $\exists Y \phi(Y)$ and $\forall Y \phi(Y)$ are second-order formulas. If all quantified function symbols are existentially quantified we get *existential second-order logic* $\exists SO_{\mathbb{R}}$.

Example 16. Let us explain how to encode instances from 4-FEAS as $\mathbb{R}$-structures and how to express membership in 4-FEAS$^+$ via an existential second-order sentence. Consider the signature $(\emptyset, \{r, C\})$, where the arities of r and C are 1 and 4, respectively, and require that r is interpreted as a ranking. Let $\mathfrak{D} = (\mathcal{A}, \mathcal{F})$ be any $\mathbb{R}$-structure where $\mathcal{F}$ consists of interpretations $C_{\mathfrak{D}} : A^4 \to \mathbb{R}$ and $r_{\mathfrak{D}} : A \to \mathbb{R}$ of C and r. Let $n = |A| - 1$ so that $r_{\mathfrak{D}}$ bijects A with $\{0, 1, \dots, n\}$. Then $\mathfrak{D}$ defines a homogeneous polynomial $\widehat{g} \in \mathbb{R}[X_0, \dots, X_n]$ of degree four, namely

$$\widehat{g} = \sum_{(i,j,k,\ell) \in A^4} C(i,j,k,\ell) X_i X_j X_k X_\ell.$$

An arbitrary, that is, not necessarily homogeneous polynomial $g \in \mathbb{R}[X_1, \dots, X_n]$ of degree four can now be obtained by setting $X_0 = 1$ in $\widehat{g}$. As explained in Example 11 denote by $\mathtt{o}$, $\mathtt{1}$, $\bar{\mathtt{o}}$ and $\bar{\mathtt{1}}$ the first and last elements of A and A^4 with respect to $r_{\mathfrak{D}}$ and $r^4_{\mathfrak{D}}$, respectively. The following sentence quantifies two functions $X : A \to \mathbb{R}$ and $Y : A^4 \to \mathbb{R}$:

$$\begin{aligned} \psi \equiv\ & (\exists X)(\exists Y) \Big(Y(\bar{\mathtt{o}}) = C(\bar{\mathtt{o}}) \ \wedge\ Y(\bar{\mathtt{1}}) = 0 \ \wedge\ X(\mathtt{o}) = 1 \ \wedge \\ & \wedge\ \forall u_1 \dots \forall u_4\ [u \neq \bar{\mathtt{o}} \Rightarrow\ \exists v_1 \dots \exists v_4\ (r^4_{\mathfrak{D}}(u) = r^4_{\mathfrak{D}}(v) + 1) \\ & \wedge\ Y(u) = Y(v) + C(u) X(u_1) X(u_2) X(u_3) X(u_4)] \Big). \end{aligned}$$

Here, if $a_i = r_{\mathfrak{D}}^{-1}(i)$ for $i = 1, \dots, n$ then, $(X(a_1), \dots, X(a_n)) \in \mathbb{R}^n$ describes the zero of g and $Y(u)$ is the partial sum of all its monomials up to $u = (u_1, \dots, u_4) \in A^4$ evaluated at the point $(X(a_1), \dots, X(a_n))$.

The sentence ψ describes 4-FEAS$^+$ in the sense that for any $\mathbb{R}$-structure $\mathfrak{D}$ it holds $\mathfrak{D} \models \psi$ if and only if the polynomial g of degree four defined by $\mathfrak{D}$ has a real zero.

3.2 Capturing $\mathrm{P}_{\mathbb{R}}$ and $\mathrm{NP}_{\mathbb{R}}$

The first two major results in descriptive complexity theory over the reals in [41] give logical characterizations of both $\mathrm{NP}_{\mathbb{R}}$ and $\mathrm{P}_{\mathbb{R}}$, thus providing a real number analogue of Fagin's result. The presentation below closely follows the original source.

Theorem 17 ([41]). *Let (F, F^+) be a decision problem of $\mathbb{R}$-structures. Then $(F, F^+) \in \mathrm{NP}_{\mathbb{R}}$ if and only if there exists an $\exists\mathrm{SO}_{\mathbb{R}}$-sentence ψ such that $F^+ = \{\mathfrak{D} \in F | \mathfrak{D} \models \psi\}$.*

Proof. The if-part is easy using Proposition 13. Given an $\exists\mathrm{SO}_{\mathbb{R}}$-formula $\psi = \exists Y_1 \ldots \exists Y_s \phi$ and an input $\mathbb{R}$-structure $\mathfrak{D} \in F$, an $\mathrm{NP}_{\mathbb{R}}$ machine non-deterministically guesses an interpretation for the functions $Y_1, \ldots, Y_s$. Then, it evaluates $\phi(Y_1, \ldots, Y_s)$ on the given structure and checks its truth in $\mathrm{NC}^1_{\mathbb{R}}$.

For the only-if part let $(G, G^+) \in \mathrm{P}_{\mathbb{R}}$ be such that the given $\mathrm{NP}_{\mathbb{R}}$-problem (F, F^+) satisfies $G = \{(\mathfrak{D}, Y) | \mathfrak{D} \in F \text{ and } Y \text{ is a } k\text{-ary function}\}$ and $F^+ = \{\mathfrak{D} \in F | \exists Y (\mathfrak{D}, Y) \in G^+\}.(G, G^+)$ is easily obtained from the definition of $\mathrm{NP}_{\mathbb{R}}$ by expanding structures in F with the verification proof Y; the latter by definition has polynomial size in $|\mathfrak{D}|$ and thus can be modelled as k-ary function from the universe of a given input structure to $\mathbb{R}$. Here, k only depends on the $\mathrm{NP}_{\mathbb{R}}$-machine for (F, F^+).

Since $(G, G^+) \in \mathrm{P}_{\mathbb{R}}$ there is a BSS machine M and an $m \in \mathbb{N}$ such that M on inputs $(\mathfrak{D}, Y)$ with $|\mathfrak{D}| = n$ stops in at most n^m steps and uses at most $n^m - 1$ registers during its computation.[2] The idea now is to describe the computation of M on $(\mathfrak{D}, Y)$ by a first-order formula. Then, the $\mathrm{NP}_{\mathbb{R}}$-question whether $\mathfrak{D} \in F^+$ is expressed adding to this formula an existential quantification of Y. By Example 11 the existence of a ranking $r^m_{\mathfrak{D}} : A^m \to \mathbb{R}$ can easily be expressed in $\exists\mathrm{SO}_{\mathbb{R}}$, so without loss of generality we suppose $r^m_{\mathfrak{D}}$ to be available and use it to identify A^m with $\{0, 1, \ldots |A|^m - 1\}$. This way, tuples $t \in A^m$ can be thought of ranging over the natural numbers $< n^m$ and also occur in comparisons and formulas involving arithmetic operations. Below, we therefore denote by t also the corresponding integer $r^m_{\mathfrak{D}}(t)$. The computation of M on $(\mathfrak{D}, Y)$ now is described using a function $Z : A^{2m} \to \mathbb{R}$. Here, the first m components of an argument encode the registers used by M, and the second m components encode a time step. More precisely, at time $t \in A^m$ of the computation $Z(0, t)$ encodes the instruction performed by M. Similarly, $Z(j, t)$ encodes the content of register $j \in A^m$ at time t. The goal now is to construct a first-order formula ψ over the vocabulary of $\mathfrak{D}, Y$, and Z which expresses accepting computations of $M : (\mathfrak{D}, Y, Z) \models \psi$ iff M accepts $\mathfrak{D}$ using guess Y. The construction

[2] Note here again that the size of $Y : A^k \to \mathbb{R}$ is n^k, so it can be subsumed in some n^m for a suitable choice of m.

has to guarantee that $Z(i,0)$ encodes the input $(\mathfrak{D},Y)$ for M's computation, that $Z(i,n^m-1)$ codes an accepting configuration and that for all time steps t the values of $Z(i,t)$ and $Z(i,t+1)$ for all $i \in A^m$ reflect a correct computation step. To grasp the general idea it suffices to describe a formula for the latter situation. Machine M has finitely many instructions, each coded by a number in $\{1,\ldots,N\}$ for some fixed $N \in \mathbb{N}$ only depending on M. Suppose then, for example, that instruction 2 is of form $x_4 \leftarrow x_5 \times x_7$.[3] Executing instruction 2 means that the value of register 4 is changed to that of the product of the values in registers 5 and 7; all other register values remain unchanged. The next instruction (by convention) has number 3. This leads to the following first-order formula:

$$\begin{aligned}\forall t,i < n^m: \quad & \{Z(0,t)=2 \Rightarrow (Z(0,t+1)=3 \wedge Z(4,t+1)=Z(5,t)\times Z(7,t))\}\\ & \wedge \{i \neq 4 \Rightarrow Z(i,t+1)=Z(i,t)\}\,.\end{aligned}$$

A corresponding formula has to be constructed for all of the finitely many instructions of M. First-order formulas expressing that $Z(\bullet,0)$ codes a correct starting configuration as well as $Z(\bullet,n^m-1)$ codes an accepting final configuration are obtained similarly. The conjunction of all these formulas gives a formula $\tilde{\psi}$. It then follows that $(\mathfrak{D},Y,Z) \models \psi$ iff M has an accepting computation on $\mathfrak{D}$ using Y as guess; Z represents the evolvement of register assignments during this computation. The conjunction $\tilde{\psi} \wedge rank$, where $rank$ is the $\exists\mathrm{SO}_{\mathbb{R}}$-formula defining a ranking, then gives the desired ψ. □

The proof immediately implies a characterization of further real number complexity classes. Full $\mathrm{SO}_{\mathbb{R}}$-logic - which corresponds to first-order logic on the structure $\langle\mathbb{R},+,-,\times,0,1,$
$\{c_r\}_r\rangle$ - captures the so-called polynomial hierarchy $\mathrm{PH}_{\mathbb{R}}$. This is an analogue of the polynomial hierarchy in the Turing model [59]. It can be defined by means of oracle machines; starting with classes $\Sigma_{0,\mathbb{R}} := \mathrm{P}_{\mathbb{R}}, \Sigma_{1,\mathbb{R}} := \mathrm{NP}_{\mathbb{R}}$ one recursively defines $\Sigma_{k+1,\mathbb{R}} := \mathrm{NP}_{\mathbb{R}}^{\Sigma_{k,\mathbb{R}}}$, where $\mathrm{NP}_{\mathbb{R}}^{\Sigma_{k,\mathbb{R}}}$ stands for problems decidable by an $\mathrm{NP}_{\mathbb{R}}$-algorithm which can access at unit cost an oracle answering membership questions to an oracle set $A \in \Sigma_{k,\mathbb{R}}$. An equivalent definition of problems in $\Sigma_{k,\mathbb{R}}$ is obtained starting from a problem $(L,L^+) \in \mathrm{P}_{\mathbb{R}}$. Now, $A \in \Sigma_{k,\mathbb{R}}$ iff there exists a polynomial p such that $A = \{x \in \mathbb{R}^\infty | \exists y_1 \in \mathbb{R}^\infty \forall y_2 \in \mathbb{R}^\infty \ldots Q_k y_k \in \mathbb{R}^\infty\ (x,y_1,\ldots,y_k) \in L^+\}$,

[3]Note that for a BSS program the indexes 4, 5, 7 are fixed for an instruction. The machine can apply such an instruction to arbitrary registers by first copying their content into the respective registers used by the instruction. Thus, in the formulas the indexes 4, 5, 7 are fixed. In a complete description one has to express as well the action of a copy instruction in first-order logic. This, however, is straightforward once the general idea has been grasped, so we exclude it from our description for sake of simplicity.

where the algebraic size of each block y_i is at most $p(|x|)$ and $Q_k \in \{\exists, \forall\}$ depending on k. Each level $\Sigma_{k,\mathbb{R}}$ of the hierarchy is thus characterized in at least three different ways: through oracle machines of the above structure, through quantification of $\mathrm{P}_{\mathbb{R}}$-properties with a quantifier format $\exists\forall \ldots Q_k$, and through formulas in $\mathrm{SO}_{\mathbb{R}}$-logic on $\mathbb{R}$-structures with the same limited quantifier format for functions from the finite universe to the real part of the structure. The entire polynomial hierarchy $\mathrm{PH}_{\mathbb{R}} = \bigcup_{k\geq 0} \Sigma_{k,\mathbb{R}}$ thus is captured by $\mathrm{SO}_{\mathbb{R}}$-logic:

Theorem 18 ([41]). *Let (F, F^+) be a decision problem of $\mathbb{R}$-structures. Then $(F, F^+) \in \mathrm{PH}_{\mathbb{R}}$ if and only if there exists a $\mathrm{SO}_{\mathbb{R}}$-sentence ψ such that $F^+ = \{\mathfrak{D} \in F | \mathfrak{D} \models \psi\}$.*

Proposition 13 showed that $\mathrm{NC}^1_{\mathbb{R}}$, being a proper subclass of $\mathrm{P}_{\mathbb{R}}$, contains all decision problems expressible in $\mathrm{FO}_{\mathbb{R}}$-logic. A logic for $\mathrm{P}_{\mathbb{R}}$ thus must extend $\mathrm{FO}_{\mathbb{R}}$-logic; on the other hand, assuming that $\mathrm{P}_{\mathbb{R}} \neq \mathrm{NP}_{\mathbb{R}}$ the logic $\exists\mathrm{SO}_{\mathbb{R}}$ is too strong. As in finite model theory, adding to $\mathrm{FO}_{\mathbb{R}}$-logic a so-called fixed-point operator for functions from the finite universe to $\mathbb{R}$ will lead to the desired logic on ranked structures. The necessary extension can be done by adding two grammatical rules for defining further number terms. Without loss of generality suppose all signatures to contain functional symbols only, i.e., $\sigma = (\emptyset, L_f)$. Each discrete relation $R \subseteq A^k$ is replaced by its characteristic function $\chi_R : A^k \to \{0,1\} \subset \mathbb{R}$. The first rule for building new number terms, called maximization and denoted by $\mathrm{MAX}_{\mathbb{R}}$, can be applied to number terms $F(s, \bar{t})$ with free variables $s, \bar{t} = (t_1, \ldots, t_r)$. Then, $\max_s F(s, \bar{t})$ is a number term with the obvious interpretation on $\mathfrak{D}$-structures. It is not hard to see that for each $\mathrm{FO}_{\mathbb{R}}$-formula $\phi(x_1, \ldots, x_\ell)$ with free variables $x_1, \ldots x_\ell$, its characteristic function can be described by a number term built from those in $\mathrm{FO}_{\mathbb{R}}$-logic together with the maximization rule. The second rule for extending $\mathrm{FO}_{\mathbb{R}}$-logic is one for defining fixed-points of functions represented by number terms. The intuitive idea is to start from a partially defined function related to a number term over an extended signature; then, this function is updated iteratively at those arguments for which it was not yet defined. This is done until a stable function, the fixed-point, is obtained. Let us make the idea more precise. Consider a signature $\sigma = (\emptyset, L_f)$ and an additional function symbol Z of arity r. Let $F(Z, \bar{x})$ be a number term in $\mathrm{FO}_{\mathbb{R}}+\mathrm{MAX}_{\mathbb{R}}$ logic of signature $L_s \cup \{Z\}$ with free variables $\bar{x} = (x_1, \ldots, x_\ell)$. Note that despite Z having arity r it might have other arguments than just the x_i's. Given an $\mathbb{R}$-structure $\mathfrak{D}$ over σ with universe A, the number term F can be used to iteratively update a function interpreting Z as follows: start by setting $Z^0(\bar{x}) := undef \ \forall \bar{x} \in A^r$. Here, *undef* is a new element added to $\mathbb{R}$ such that $a + undef := a - undef := undef \ \forall a \in \mathbb{R}, 0 \times undef := 0$ and $a \times undef :=$

undef $\forall a \in \mathbb{R} \setminus \{0\}$. Now define iteratively

$$Z^{i+1}(\bar{x}) := \begin{cases} F(Z^i, \bar{x}) & \text{if } Z^i(\bar{x}) = undef \\ Z^i(\bar{x}) & \text{else} \end{cases}$$

This process becomes stable because Z^{i+1} potentially only updates values for arguments $\bar{x}$ where Z^i equals *undef*. If at all such an update is performed, this has to happen after at most $|A|^r$ many iterations. The resulting function is the fixed-point of $F(Z, \bar{x})$ on $\mathfrak{D}$ and denoted by Z^∞. It might still attain the value *undef* at some arguments. Whether this happens for $\bar{x}$ can be figured out by computing Z^∞ along the above update formula. The fixed-point rule now defines for all $\bar{u} \in A^r$ the term $Z^\infty(\bar{u})$ as further number term.

Definition 19. Functional fixed-point logic $\mathrm{FFP}_{\mathbb{R}}$ is obtained as closure of the set of first-order number terms under the maximization rule and the fixed-point rule. $\mathrm{FP}_{\mathbb{R}}$ denotes the class of characteristic functions definable in functional fixed-point logic.

Example 20. The interested reader might work out a description of the Circuit Evaluation Problem from Example 8 in $\mathrm{FP}_{\mathbb{R}}$-logic. Towards this goal use a suitable vocabulary to model along the way it is done in Example 12 the type of nodes of a circuit, the assignment of reals to constant nodes, the structure of the directed graph underlying the circuit, and the input $x \in \mathbb{R}^n$ for which the circuit should be evaluated. Now define a function E which assigns to each node of the graph its output value once the input has been propagated through the circuit in the obvious manner. Then E can be obtained as a fixed-point resulting from the updates of nodes which are input nodes, nodes of maximal distance one to input nodes, nodes of maximal distance 2 to input nodes etc.

Maximization obviously can be computed in polynomial time because there are only polynomially many candidates. Since each update in the fixed-point rule above needs only to cycle once through the $|A|^r$ many arguments and since $|A|^r - 1$ rounds are sufficient to compute Z^∞, also the fixed-point can be computed in polynomial time. This implies that any function in $\mathrm{FFP}_{\mathbb{R}}$ is computable in polynomial time in the BSS model. Actually, the converse holds as well.

Theorem 21 ([41]). *On ranked* $\mathbb{R}$*-structures the functions in* $\mathrm{FFP}_{\mathbb{R}}$ *are exactly those computable in polynomial time.*[4] *In particular, the characteristic functions in* $\mathrm{FFP}_{\mathbb{R}}$ *are those of polynomial time solvable decision problems , i.e.* $\mathrm{FP}_{\mathbb{R}} = \mathrm{P}_{\mathbb{R}}$.

[4] If $f(x) = undef$ for a function we mean by polynomial time computability that an algorithm detects it in polynomial time. This should not be confused with non-terminating algorithms as considered in general computability theory. A slightly varied fixed-point rule using only functions with real values everywhere was used in [26].

Proof. Similarly to the proof of Theorem 17 a computation of a BSS machine M on a structure $\mathfrak{D}$ deciding a problem $(L, L^+) \in \mathrm{P}_{\mathbb{R}}$ is represented as function $Z : A^{2m} \to \mathbb{R}$ for suitable m. We just outline how the fixed-point rule is used for defining the correct Z. The crucial step is to define a term $F(Z, \bar{j}, \bar{t})$ describing the updates $Z^i(\bar{j}, \bar{t})$, starting from a function Z^{-1} being undefined everywhere. The intended interpretation is that for all $i \leq |A|^m, \bar{j}, \bar{t} \in A^m$ the value $Z^i(\bar{j}, \bar{t})$ gives the register content of register $\bar{j}$ at time $\bar{t} \leq i$, where $\bar{j}$ and $\bar{t}+1$ as above code numbers in $\{0, 1, \ldots, |A|^m\}$. For values $\bar{t} > i$ it is $Z^i(\bar{j}, \bar{t}) = \mathit{undef}$. In order to achieve this, $F(Z, \bar{j}, \bar{t})$ is defined as a $\mathrm{FFP}_{\mathbb{R}}$-term describing the updates appropriately. More precisely,

$$F(Z, \bar{j}, \bar{t}) = \chi(\bar{t} = 0) \cdot F_{input}(\bar{j}) + \max_{\bar{s}} \chi(\bar{t} = \bar{s} + 1) \cdot \sum_{k=1}^{N} \chi(Z(0, \bar{s}) = k) \cdot F_k(Z, \bar{j}, \bar{s}).$$

Here, χ denotes the characteristic function of its respective argument predicate; the instructions of M are labeled $\{1, \ldots, N\}, F_{input}$ is a first-order number term describing the input encoding, and the F_k are $\mathrm{FFP}_{\mathbb{R}}$-terms (actually, terms definable in $\mathrm{FO}_{\mathbb{R}} + \mathrm{MAX}_{\mathbb{R}}$) describing the action of instruction k of M. The precise format of those terms can easily be derived similarly to the proof of Theorem 17. The fixed-point rule, when applied to F, updates $Z(\bar{j}, \bar{t})$ for the current time step $\bar{t}$. Machine M terminates after at most $|A|^m$ steps , so Z^∞ is obtained after at most that many updates. It is defined for all arguments yielding the desired interpretation. In particular, the result of M on an input is represented by the number term $Z^\infty(\bar{j}_{res}, |A|^m - 1)$, where register $\bar{j}_{res}$ is supposed to contain M's result of the question to decide. □

4 Further results and research directions

In this final section we outline further work that has been done in descriptive complexity theory over the reals and identify some interesting research directions.

Following [41], in [26] many other important real number complexity classes are captured by suitable logics. Given the results from the previous section logics for the parallel classes $\mathrm{NC}^k_{\mathbb{R}}$ from Definition 7 have to be weaker extensions of $\mathrm{FO}_{\mathbb{R}}$ than the one given by $\mathrm{FFP}_{\mathbb{R}}$. The way to obtain such logics is to restrict the number of updates allowed to perform when applying the fixed-point rule. For a function class $\mathcal{U}$ denote by $\mathrm{FP}_{\mathbb{R}}[\mathcal{U}]$ the characteristic functions that can be obtained from $\mathrm{FO}_{\mathbb{R}} + \mathrm{MAX}_{\mathbb{R}}$ when the fixed-point operator is applied only $f(n)$ many times, where $f \in \mathcal{U}$ and n denotes the size of an input $\mathbb{R}$-structure. Then the following holds:

Theorem 22 ([26]). $\mathrm{FP}_{\mathbb{R}}[O((\log^k(n))] = \mathrm{NC}^k_{\mathbb{R}}$.

There are as well important complexity classes above $NP_{\mathbb{R}}$. A characterization of the problems in the polynomial hierarchy by $SO_{\mathbb{R}}$ has been discussed in the previous section. A class provably larger than $NP_{\mathbb{R}}$ is $EXP_{\mathbb{R}}$; it contains all real number decision problems solvable in simple exponential time. In order to capture $EXP_{\mathbb{R}}$ by a metafinite logic one can change the fixed-point rule, this time increasing its power. In a first step one structurally extends number terms so that they can take as well *subsets* of some A^r as arguments, where A as usual denotes the universe of an $\mathbb{R}$-structure. Now, maximization is extended to search for maxima among such subsets and the fixed-point rule is extended to deal with new function symbols that can receive subsets as arguments. Then the corresponding update scheme when (re-)defining a fixed-point rule causes that the number of iterations before a fixed-point is obtained can be simple exponential. If we denote the corresponding logic by $FP^2_{\mathbb{R}}$ (where the superscript '2' indicates the presence of subsets as second-order objects opposed to first-order variables) a careful elaboration of the above intuitive arguments leads to

Theorem 23 ([26]). $FP^2_{\mathbb{R}} = EXP_{\mathbb{R}}$.

Similar ideas are used in [26] to capture the class $PAR_{\mathbb{R}}$ of problems solvable in parallel polynomial time, see Definition 7. Here, the basic idea is to use once more the second-order fixed-point rule, but restricting again the number of updates to be polynomial, similar to the construction of a logic for $NC^k_{\mathbb{R}}$ from $FP_{\mathbb{R}}$ indicated above in Theorem 22.

Another area of problems in relation to which descriptive complexity has been studied are counting problems. This has been considered both in finite and metafinite model theory. Typical examples of counting problems in the two frameworks ask for the number of satisfying assignments of a propositional formula and computing the number of zeros of a real polynomial. The relation of such problems to descriptive complexity is made by considering formulas in certain logics with free variables of different types. Then count for a given structure the number of assignments for those free variables such that the given structure together with the assignment is a model for the formula. Consider once more the 4-FEAS Example 16 and the formula ψ defined there to express existence of a real zero. Studying the counting version denoted by #4-FEAS we put focus on the subformula $\rho(X,Y)$ such that $\psi \equiv \exists X \exists Y \rho(X,Y)$. Then for a given $\mathbb{R}$-structure $\mathfrak{D}$ representing a polynomial of degree 4 one is interested in the cardinality of the set $\{(X,Y) | (\mathfrak{D}, X, Y) \models \rho(X,Y)\}$. Note that ρ is a first-order formula over the extended vocabulary $(\emptyset, \{r, C\} \cup \{X, Y\})$ having as free variables the second-order variables X, Y and a first-order quantifier structure $\forall\exists$. Now [50] studies a hierarchy of counting classes being logically defined by varying the first-order quantifier structure of formulas. It turns out that this hierarchy is finite consisting of five different levels, the largest class being the one defined by the above first-order quantifier structure $\forall\exists$. The results are analogue to the ones in finite model theory given in [57]. An extension of defining counting problems in the metafinite framework being based on generating functions of graph properties in the spirit of [13] is studied in [48]. And in [34] a logical hierarchy similar to the above mentioned one is presented for a certain class of optimization problems over $\mathbb{R}$. Let us mention that a deep study of real number counting problems under a complexity theoretic perspective (which the above mentioned logical hierarchy does not provide) can be found in a series of papers by Bürgisser, Cucker and coauthors, see [15, 16]. Real and complex number variants of a famous result by Toda on the complexity of counting problems have been obtained in [4, 5].

Very recent research in [43] studies the relation between descriptive complexity for $\mathbb{R}$-structures and so-called *probabilistic independence logic* introduced in [29]. The latter is a logic studied in the area of team semantics, see [61], in which formulas are evaluated with respect to sets of assignments called teams. In probabilistic independence logic probabilistic teams are modeled by discrete probability distributions with real number probability values. This gives a natural embedding into $\mathbb{R}$-structures by modeling such distributions as functions from the skeleton to the numerical part. Towards the above aim, in [43] the authors introduce both a restricted version of deterministic and non-deterministic BSS machines called *separate BSS machines* and a restricted version of the existential second order logic $\exists\mathrm{SO}_{\mathbb{R}}$ introduced in Definition 15. On the algorithmic side the restrictions are twofold. First, test instructions are associated with two fixed reals $\epsilon_i < \epsilon_+$; a test checks, whether for a real x it either holds $x \leq \epsilon_-$ or $x \geq \epsilon_+$. The computation then continuous depending on the result. If, however, $x \in (\epsilon_-, \epsilon_+)$ the machine stops its computation and rejects the input. Secondly, for non-deterministic computations only real numbers in the compact set $[0,1]$ can be guessed, i.e., all components of the point y in Definition 4, b) must belong to $[0,1]$. On the side of defining suitable logics $\exists\mathrm{SO}_{\mathbb{R}}$ is restricted accordingly by only allowing inequalities of type $\leq$ to occur in formulas and by restricting existential second order quantifiers to function symbols $Y : A^k \to [0,1]$, see Definition 15. The resulting logic is called *existential loose* $[0,1]$*-guarded real arithmetic*. One of the main results in [43] now shows that the complexity class of real number problems defined via non-deterministic polynomial time separate BSS machines is precisely captured by the above logic for existential loose $[0,1]$-guarded real arithmetic, thus complementing Theorem 17. As an interesting by-product, the authors as well derive a logic capturing a subclass of the Boolean complexity class $\exists\mathbb{R}$, the class of problems being polynomial time many-one reducible (in the Turing sense) to the existential theory of the reals. The class $\exists\mathbb{R}$ has been first isolated as a complexity class in its own right in [58] and has raised increasing interest in recent years.

Research combining descriptive complexity with algorithmic complexity that has turned out to be important in the last two decades is related to the field of parameterized complexity theory. Here, the basic idea is as follows: Consider the logical description of a decision problem (F, F^+) in the way done above, and let φ be a formula defining the instances of F^+. If the initial problem is (supposed to be) hard, for example NP-complete, the idea of parameterized complexity theory is to identify significant problem parameters such that restricting problem instances to a subclass where such a parameter is constant yields the existence of efficient algorithms solving the problem. With respect to such a parameter the problem then is called fixed-parameter tractable. For introductions into the field of parameterized complexity theory see [28, 35, 53]. Now descriptive complexity theory combines this idea of identifying suitable parameters with the logical expressibility of a problem. In the ideal situation one obtains meta-results of the following form: if a (hard) problem is expressible in a certain logic and inputs are restricted to the subclass of instances on which the respective parameter is fixed, then efficient algorithms exist for this subclass. The general proof idea is to decompose both the formula expressing the problem and the instance into smaller parts. This decomposition heavily relies on both the problem parameter and the logic. Bounding the former implies that a typical backtracking algorithm does not lead to an exploding running time. Of course, this approach also relies on the logic in which the problem is expressible. It has been applied successfully to the so-called treewidth of a graph, introduced independently in [42] and [56]. For many problems there is attached a graph in a natural way. The underlying intuition then is that many otherwise hard problems turn out to be efficiently solvable on trees. Treewidth now in a certain sense measures how close a graph is to a tree. In the ideal case it comes with a decomposition of the graph into pieces each of which corresponds to a subgraph whose size is determined by the treewidth. Solving the original problem then reduces to performing work on the (small) subgraphs and glueing the results together along the tree decomposition. The main gain with respect to complexity is that backtracking is significantly reduced when dealing with the subgraphs. Whereas the above foregoing seems to be designed particularly for the respective problem, descriptive complexity theory allows to apply it in a much more general context. If the original problem can be expressed in a suitable logic such that the decomposition of the graph into subgraphs goes hand in hand with a similar decomposition of the sentence on substructures expressing the resulting problems on subgraphs, then efficient algorithms are obtained for a much broader class of problems. The most important logic in this respect for dealing with problems in discrete complexity theory is existential monadic second-order logic $\exists$MSO. It is the fragment of $\exists$SO-logic where SO quantifiers only can range over unary relations, i.e., subsets of the universe. In the papers [20, 23, 2] it was shown that graph properties expressible in ∃MSO logic (considering graphs as two-sorted structures) are fixed-parameter tractable on graphs of bounded treewidth. This kind of result can be obtained as well for other important graph parameters; in [21, 22] it is shown for one-sorted graph-structures of bounded clique-width. The deep logical background for such results is the Feferman-Vaught theorem [33]. It allows the decomposition of a given problem along a decomposition of the instance, in case of treewidth as parameter the latter being the above mentioned tree decomposition. Its algorithmic features are thoroughly described in the paper [47]. The above ideas as well can be applied to algebraic problems. A typical example is the computation of the permanent of a real square matrix, a well known and potentially hard problem [62]. A graph naturally attached to such a matrix is defined by including edges for non-zero entries (where row and column numbers correspond to vertices). In [22] it is shown that this and many other counting problems can be solved efficiently on matrices of bounded treewidth of the attached graph; similar ideas are already present in [2].

Given the success of this approach relating discrete complexity theory and finite model theory it is a natural question whether similar meta-theorems can be derived using metafinite model theory and suitable parameters for real number problems. In [49] a first step into this direction was made. Since polynomial equations play a central role in BSS complexity theory, a system of such equations is related to a hypergraph; the variables of the system build its vertices, the monomials with non-zero coefficients constitute the hyperedges. Typical graph parameters now can be studied for this graph. For example, that way the treewidth parameter defines the *algebraic treewidth* of a polynomial system. The main task is to define a reasonable logic expressing interesting computational problems over $\mathbb{R}$ such that decomposing the above hypergraph with respect to typical graph parameters corresponds to a similar decomposition of formulas in the logic. In [49] one such logic, denoted by $\exists\mathrm{MSO}_{\mathbb{R}}$ is constructed. It covers usual MSO logic on the discrete part (the hypergraph as two-sorted structure) of an $\mathbb{R}$-structure. The more delicate question is which algebraic features to include in the logic by involving (real) number terms and functions from the universe to $\mathbb{R}$. This refers both to the 'monadic' quantifiers allowed and to the structure of terms that can be decomposed in a suitable manner along decompositions of the underlying hypergraph. Due to the limited space we do not review here the precise definition of $\exists\mathrm{MSO}_{\mathbb{R}}$ logic. It can, for example, express properties like the solvability of a real polynomial system over a given finite subset $A \subset \mathbb{R}^n$. This implies the existence of efficient algorithms for this problem if instances of such systems are restricted to be of constant algebraic treewidth in the above sense. Though this shows that the design of efficient algorithms using metafinite model theory is possible, it is fair to say that in the algebraic framework this has not yet lead to the multitude of important results and consideration of several different parameters as in finite model theory. One reason for this seems to be the variety of potential extensions leading to a reasonable definition of a monadic second-order logic over $\mathbb{R}$, especially with respect to the structure of number terms allowed in formulas. In our point of view this is a very interesting research area with many open problems.

Closing the circle by returning to Büchi's result mentioned at the beginning another version of a monadic second-order logic on $\mathbb{R}$-structures recently has been studied in [51]. There, an extension of the classical finite automata model and of regular languages to the real number world is studied, based on a model of generalized finite automata introduced in [37]. For a slight variation of this model real regular languages can be defined sharing a lot of structural properties of classical regular languages. Among those properties one is a logical characterization of the languages accepted by such generalized automata over $\mathbb{R}$ in the spirit of Büchi's result using a metafinite logic on $\mathbb{R}$-structures which consists of monadic second-order constructs, however being different from the logic $\exists\mathrm{MSO}_{\mathbb{R}}$ mentioned above.

Let us finally briefly mention another research area loosely connected to the topics of this paper. As shown above, descriptive complexity provides a machine-independent characterization of many real number complexity classes using logic. Another approach yielding such characterizations in the BSS model is *implicit complexity.* It is based on the idea of recursion used to define hierarchies of sets of functions characterizing real number complexity classes. For more on this topic we just refer to [9, 10, 25].

References

[1] Serge Abiteboul, Richard Hull, and Victor Vianu Foundations of Databases, Addison Wesley, Reading MA 1995.

[2] Stefan Arnborg, Jens Lagergren, and Detlef Seese *Easy problems for tree-decomposable graphs*, Journal of Algorithms 12 (2), 1991, pp. 308–340.

[3] Timon Barlag, and Heribert Vollmer *A Logical Characterization of Constant-Depth Circuits over the Reals*, Preprint 2020 [arXiv:2005.04916].

[4] Saugata Basu *A complex analogue of Toda's theorem*, Foundations of Computational Mathematics 12, 2012, pp. 327–362.

[5] Saugata Basu, and Thierry Zell *Polynomial hierarchy, Betti numbers, and a real analogue of Toda's theorem*, Foundations of Computational Mathematics 10 (4), 2010, pp. 429–454.

[6] Lenore Blum, Felipe Cucker, Mike Shub, and Steve Smale Complexity and Real Computation, Springer 1998.

[7] Lenore Blum, Mike Shub, and Steve Smale *On a theory of computation and complexity over the real numbers: NP-completeness, recursive functions and universal machines*, Bulletin of the American Mathematical Society 21, 1989, pp. 1–46.

[8] Egon Börger *Decision problems in predicate logic*, in: Gabriele Lolli, Giuseppe Longo, and Annalisa Marcja (eds.) Logic Colloquium '82, Elsevier / North-Holland Publishing Company 1984, pp. 263–301.

[9] Olivier Bournez, Felipe Cucker, Paulin Jacobé de Naurois, and Jean-Yves Marion *Computability over an arbitrary structure: sequential and parallel polynomial time*, in: Andrew D. Gordon (ed.) Proc. 6th FOSSACS, Springer 2003 [*Lecture Notes in Computer Science 2620*], pp. 185–199.

[10] — *Implicit complexity over an arbitrary structure: quantifier alternations*, Information and Computation 204, 2006, pp. 210–230.

[11] Olivier Bournez, and Amaury Pouly *A Survey on Analog Models of Computation*, in: Vasco Brattka, and Peter Hertling (eds.) Handbook of Computability and Complexity in Analysis, Springer [to appear].

[12] Julius Richard Büchi *Weak second-order arithmetic and finite automata*, Zeitschrift für mathematische Logik und Grundlagen der Mathematik 6, 1960, pp. 66–92.

[13] Peter Bürgisser Completeness and Reduction in Algebraic Complexity Theory, Springer 2000 [*Algorithms and Computation in Mathematics 7*].

[14] Peter Bürgisser, Michael Clausen, and Mohammad Amin Shokrollahi Algebraic Complexity Theory, Springer 1997 [*Grundlehren der mathematischen Wissenschaften 315*].

[15] Peter Bürgisser, and Felipe Cucker *Variations by complexity theorists on three themes of Euler, Bézout, Betti, and Poincaré*, in: Jan Krajicek (ed.) Complexity of computations and proofs 13, *Quaderni di Matematica*, 2005, pp. 73–152.

[16] — *Counting complexity classes for numeric computations, II: Algebraic and semialgebraic sets*, Journal of Complexity 22 (2), 2006, pp. 147–191.

[17] Claude-André Christen Spektren and Klassen elementarer Funktionen, PhD Dissertation, ETH Zürich 1974.

[18] Edmund M. Clarke, Thomas A. Henzinger, Helmut Veith, and Roderick Bloem (eds.) Handbook of Model Checking, Springer 2018.

[19] Stephen A. Cook *The complexity of theorem-proving procedures*, in: Michael A. Harrison, Ranan B. Banerji, and Jeffrey D. Ullman (eds.) Proc. 3rd ACM Symposium on Theory of Computing STOC, ACM 1971, pp. 151–158.

[20] Bruno Courcelle *The monadic second-order logic of graphs, I: Recognizable sets of finite graphs*, Information and Computation 85 (1), 1990, pp. 12–75.

[21] Bruno Courcelle, Johann A. Makowsky, and Udi Rotics *Linear time solvable optimization problems on graphs of bounded clique-width*, Theory of Computing Systems 33 (2), 2000, pp. 125–150.

[22] — *On the fixed parameter complexity of graph enumeration problems definable in*

monadic second-order logic, Discrete Applied Mathematics 108 (1–2), 2001, pp. 23–52.

[23] Bruno Courcelle, and Mohamed Mosbah *Monadic second-order evaluations ontree decomposable graphs*, Theoretical Computer Science 109, 1993, pp. 49–82.

[24] Felipe Cucker $NC_{\mathbb{R}} \neq P_{\mathbb{R}}$, Journal of Complexity 8, 1992, pp. 230–238.

[25] — *Complexity classes over the reals: A logician's viewpoint*, in: Rodney G. Downey, Decheng Ding, Tung Shih Ping, Qiu Yu Hui, and Mariko Yasugi (eds.) Proceedings of the 7th and 8th Asian Logic Conferences, World Scientific 2003, pp. 39–62.

[26] Felipe Cucker, and Klaus Meer *Logics which capture complexity classes over the reals*, The Journal of Symbolic Logic 64 (1), 1999, pp. 363–390.

[27] Felipe Cucker, and A. Torecillas *Two P-Complete Problems in the Theory of the Reals*, in: Javier Leach Albert, Burkhard Monien, and Mario Rodríguez-Artalejo (eds.) Automata, Languages and Programming (18th International Colloquium ICALP91), Springer 1991 [*Lecture Notes in Computer Science 510*], pp. 556–565.

[28] Rodney G. Downey, and Michael R. Fellows Parameterized Complexity, Springer, New York 1999 [*Monographs in Computer Science*].

[29] Arnaud Durand, Miika Hannula, Juha Kontinen, Arne Meier, and Jonni Virtema *Probabilistic Team Semantics*, in: Flavio Ferrarotti, and Stefan Woltran (eds.) Foundations of Information and Knowledge Systems – 10th International Symposium FoIKS, Springer 2018 [*Lecture Notes in Computer Science 10833*], pp. 186–206.

[30] Heinz-Dieter Ebbinghaus, and Jörg Flum Finite Model Theory, Springer 1995.

[31] Calvin C. Elgot *Decision problems of finite automata design and related arithmetics*, Transactions of the American Mathematical Society 98, 1961, pp. 21–52.

[32] Ronald Fagin *Generalized first-order spectra and polynomial-time recognizable sets*, in: Richard M. Karp (ed.) Complexity of Computation, American Mathematical Society, Providence RI 1974 [*Proceedings SIAM-AMS 7*], pp. 43–73.

[33] Solomon Feferman, and Robert L. Vaught *The first-order properties of products of algebraic systems*, Fundamenta Mathematicae 47, 1959, pp. 57–103.

[34] Uffe Flarup, and Klaus Meer *Two logical hierarchies of optimization problems over the real numbers*, Mathematical Logic Quarterly 52 (1), 2006, pp. 37-50.

[35] Jörg Flum, and Martin Grohe Parameterized Complexity Theory, Springer 2006 [*Springer Texts in Theoretical Computer Science*].

[36] Pierluigi Frisco Computing with Cells – Advances in Membrane Computing, Oxford University Press, Oxford 2009.

[37] Aniruddh Gandhi, Bakhadyr Khoussainov, and Jiamou Liu *Finite Automata over Structures*, in: Manindra Agrawal, S. Barry Cooper, and Angsheng Li (eds.) Proc. Theory and Applications of Models of Computation, Springer 2012 [*Lecture Notes in Computer Science 7287*], pp. 373–384.

[38] Erich Grädel *Finite Model Theory and Descriptive Complexity*, in: Erich Grädel, Phokion G. Kolaitis, Leonid Libkin, Maarten Marx, Joel Spencer, Moshe Y. Vardi, Yde Venema, and Scott Weinstein (eds.) Finite Model Theory and its Applications, Springer 2007 [*Texts in Theoretical Computer Science*], pp. 125–230.

[39] Erich Grädel, and Yuri Gurevich *Metafinite model theory*, Information and Computation 140, 1998, pp. 26–81.

[40] Erich Grädel, Phokion G. Kolaitis, Leonid Libkin, Maarten Marx, Joel Spencer, Moshe Y. Vardi, Yde Venema, and Scott Weinstein (eds.) Finite Model Theory and its Applications, Springer 2007 [*Texts in Theoretical Computer Science*].

[41] Erich Grädel, and Klaus Meer *Descriptive Complexity Theory over the Real Numbers*, in: James Renegar, Mike Shub, and Steve Smale (eds.) Proc. of the AMS Summer Seminar on Mathematics of Numerical Analysis: Real Number Algorithms, American Mathematical Society, Providence RI 1996 [*Lectures in Applied Mathematics*], pp. 381–404.

[42] Rudolf Halin *S-functions for graphs*, Journal of Geometry 8 (1-2), 1976, pp. 171–186.

[43] Miika Hannula, Juha Kontinen, Jan Van den Bussche, and Jonni Virtema *Descriptive complexity of real computation and probabilistic independence logic*, in: Proc. 35th Annual ACM/IEEE Symposium on Logic in Computer Science, 2020 [to appear].

[44] Simon Haykin Neural Networks, *A Comprehensive Foundation*, Prentice Hall, Upper Saddle River NJ 1999 [2nd edition].

[45] Neil Immerman, and Phokion G. Kolaitis (eds.) Descriptive Complexity and Finite Models, American Mathematical Society, Providence RI 1997 [*DIMACS Series in Discrete Mathematics and Theoretical Computer Science 31*].

[46] Neil D. Jones, and Alan L. Selman *Turing machines and the spectra of first-order formulas*, The Journal of Symbolic Logic 39, 1974, pp. 139–150.

[47] Johann A. Makowsky *Algorithmic uses of the Feferman-Vaught theorem*, Annals of Pure and Applied Logic 126 (1–3), 2004, pp. 159–213.

[48] Johann A. Makowsky, and Klaus Meer *On the complexity of combinatorial and metafinite generating functions*, in: Peter Clote and Helmut Schwichtenberg (eds.) Proc. of 14th Computer Science Logic conference CSL, Springer 2000 [*Lecture Notes in Computer Science 1862*], pp. 399-410.

[49] — *Polynomials of bounded tree-width*, in: Felipe Cucker and J. Maurice Rojas (eds.) Foundations of Computational Mathematics (Proceedings of the Smalefest 2000), World Scientific, New Jersey 2002, pp. 211–250.

[50] Klaus Meer *Counting Problems over* $\mathbb{R}$, Theoretical Computer Science 242 (1–2), 2000, pp. 41–58.

[51] Klaus Meer, and Ameen Naif *Automata over infinite sequences of reals*, in: Carlos Martín-Vide, Alexander Okhotin, and Dana Shapira (eds.) Proc. 13th International Conference on Language and Automata Theory and Applications LATA, Springer 2019 [*Lecture Notes in Computer Science 1862*], pp. 121–133.

[52] José Luis Montaña, and Luis Miguel Pardo (eds.) Recent Advances in Real Complexity and Computation, American Mathematical Society, 2013 [*Contemporary Mathematics 604*].

[53] Rolf Niedermeier Invitation to Fixed-parameter Algorithms, Oxford University Press, Oxford 2006 [*Oxford Lecture Series in Mathematics and its Applications 31*].

[54] Michael A. Nielsen, and Isaac L. Chuang Quantum Computation and Quantum Information, Cambridge University Press 2000.

[55] James Renegar, Mike Shub, and Steve Smale (eds.) Proceedings of the AMS Summer Seminar on Mathematics of Numerical Analysis, American Mathematical Society, Providence RI 1996 [*Lectures in Applied Mathematics*].

[56] Neil Robertson, and Paul D. Seymour *Graph minors, II: Algorithmic aspects of treewidth*, Journal of Algorithms 7 (3), 1986, pp. 309–322.

[57] Sanjeev Saluja, K. V. Subrahmanyam, and Madhukar N. Thakur *Descriptive complexity of #P functions*, Journal of Computer and System Sciences 50 (3), 1995, pp. 493–505.

[58] Marcus Schaefer *Complexity of some geometric and topological problems*, in: David Eppstein, and Emden R. Gansner (eds.) Proc. 17th International Symposium Graph Drawing [*Lecture Notes in Computer Science 5849*], Springer 2009, pp. 334–344.

[59] Larry Stockmeyer *The polynomial hierarchy*, Theoretical Computer Science 3, 1976, pp. 1–22.

[60] Boris A. Traxtenbrot [Trakhtenbrot] *Konečnye avtomaty i logika ondomestnyx predikatov* [Finite automata and the logic of monadic predicates] [Russian], Doklady Akademii Nauk SSSR 140 (2), 1961, pp. 326–329.

[61] Jouko A. Väänänen Dependence Logic, *A New Approach to Independence Friendly Logic*, Cambridge University Press, Cambridge UK 2007 [*London Mathematical Society Student Texts 70*].

[62] Leslie G Valiant *The complexity of computing the permanent*, Theoretical Computer Science 8 (2), 1979, pp. 189–201.

Moschovakis extension of multi-represented spaces

Dimiter Skordev
Sofia University, Faculty of Mathematics and Informatics
Sofia, Bulgaria
skordev@fmi.uni-sofia.bg

ABSTRACT Given a multi-represented space (in the sense of TTE theory), an appropriate multi-representation is constructed for the Moschovakis extension of its carrier. Some results are presented about TTE computability in the multi-represented space obtained in this way. For partial multiple-valued functions in the Moschovakis extension, a relative computability notion called neat computability is considered which is equivalent to absolute prime computability in the case of partial single-valued functions. We prove the TTE computability of any function which is neatly computable in some TTE computable functions (computability via single-valued realizations and computability via Brattka style ones are considered).

KEYWORDS Moschovakis extension, multi-represented space, TTE, single-valued function, multiple-valued function, realization, abstract first order computability, computable, absolutely prime computable, composition, juxtaposition, iteration, branching, first recursion theorem.

Introduction

The present paper generalizes and extends some of the results from [7].

The Moschovakis extension of a set B is the set B^* defined in [4]. Assuming without loss of generality that no element of the set B is an ordered pair, one builds B^* as the closure of $B \cup \{\boldsymbol{o}\}$ under formation of ordered pairs, where $\boldsymbol{o}$ (denoted by 0 in [4]) is some object which does not belong to B and also is not an ordered pair.[1] Functions π and δ from B^* to B^* are considered which are defined as follows:

[1]The bold symbol $\boldsymbol{o}$ used here should not be confused with the normal text symbol o used elsewhere by the present author for the least elements of a certain kind of partially ordered structures studied by him.

$\pi(u) = s$ and $\delta(u) = t$ for $u = (s,t)$, $\pi(\boldsymbol{o}) = \delta(\boldsymbol{o}) = \boldsymbol{o}$, $\pi(u) = \delta(u) = (\boldsymbol{o},\boldsymbol{o})$ for all u in B.

Certain relative computability notions for functions in B^* are introduced and studied in [4]. The functions considered there are, in general, multiple-valued. In the case of single-valued functions, one of these notions, namely absolute prime computability, seems to be able to cover any reasonable kind of computability by means of deterministic programs using some given functions. In the general case, however, a modification of absolute prime computability coinciding with it in the case of single-valued functions also deserves attention. This modification (introduced in [6] and studied in [7]) is called neat computability in the present paper.

As *e.g.* in [2, Definition 2.1], a *representation* of a set X is a partial mapping of $\mathbb{N}^{\mathbb{N}}$ onto X (where, of course, $\mathbb{N}^{\mathbb{N}}$ is the set of all total functions from $\mathbb{N}$ to $\mathbb{N}$), and a *represented space* is an ordered pair whose terms are a set and a representation of this set. By modifying this definition we get the following one in the spirit of [9, Definition 16]: a *multi-representation* of a set X is a partial mapping ϱ of $\mathbb{N}^{\mathbb{N}}$ into the set of the nonempty subsets of X such that $\bigcup\{\varrho(p)\,|\,p \in \mathrm{dom}(\varrho)\} = X$; a *multi-represented space* is an ordered pair whose terms are a set and a multi-representation of this set.[2]

Given a multi-represented space, we construct an appropriate multi-representation of the Moschovakis extension of its carrier, and we consider TTE computability via single-valued realizations for functions in the multi-represented space obtained in this way. We establish some statements which are in the same vein as [1, Theorem 31], [8, Theorems 3.1.6 and 3.1.7], [2, Theorem 8.3] and the results in [9] by proving the TTE computability of any function which is neatly computable in some TTE computable functions (we consider both TTE computability of the kind studied in [8, 9], as well as Brattka style TTE computability, i.e. computability in the sense of [2], appropriately generalized for the case of multi-representations).

In contrast to the presentation in [7], the author does almost not touch here a background for his considerations, namely the generalization of computability theory from [5] and other publications of him.

1 Neat computability

When considering functions in B^*, it is not an essential restriction to confine oneself to unary ones. Let $\mathcal{F}$ be the set of all unary partial multiple-valued functions in B^* (the set of the single-valued ones can be regarded as a subset of $\mathcal{F}$). We will consider the elements of $\mathcal{F}$ as the partial functions from B^* to the set of the nonempty

[2] Argumentation for using multi-representations can be found *e.g.* in [9, Section 6].

subsets of B^*, whereas, according to [4, pp. 429–430], the unary partial multiple-valued functions in B^* are the functions from the whole B^* to the set of all subsets of B^*. However, there is an obvious one-to-one correspondence between these two kinds of mathematical objects, and we will identify the objects that correspond one to the other. Another equipollent option is (as, for example, in [5, Ch. 1, Sect. 5]) to consider the partial multiple-valued functions as relations.

It is shown in [5, Ch. 1, Sect. 7] that a function $\varphi \in \mathcal{F}$ is absolutely prime computable in some given functions $\psi_1, \ldots, \psi_l \in \mathcal{F}$ iff φ can be obtained from $\psi_1, \ldots, \psi_l$ and the functions π and δ by finitely many applications of three natural operations, namely:

(a) The usual "relational" composition in $\mathcal{F}$ transforming any φ and ψ from $\mathcal{F}$ into the corresponding element θ of $\mathcal{F}$ (denoted by $\varphi\psi$) such that

$$\mathrm{dom}(\theta) = \{z \in \mathrm{dom}(\psi) \mid \psi(z) \cap \mathrm{dom}(\varphi) \neq \varnothing\},$$
$$\theta(z) = \bigcup\{\varphi(u) \mid u \in \psi(z) \cap \mathrm{dom}(\varphi)\} \text{ for all } z \in \mathrm{dom}(\theta).$$

(b) The operation $\varphi, \psi \mapsto \lambda x.\varphi(x) \times \psi(x)$. This operation is called *combination* in [5] and *juxtaposition* in [1, 2, 9]; we will use the second of these names here.

(c) The operation $\varphi, \chi \mapsto \iota$, where $x \in \mathrm{dom}(\iota)$ and $y \in \iota(x)$ iff a finite sequence $z_0, z_1, \ldots, z_n$ of elements of B^* exists such that $z_0 = x$, $z_n = y$,

$$z_i \in \mathrm{dom}(\varphi) \cap \mathrm{dom}(\chi) \;\&\; \chi(z_i) \setminus (B \cup \{\boldsymbol{o}\}) \neq \varnothing \;\&\; z_{i+1} \in \varphi(z_i) \qquad (1)$$

for all $i < n$, and $z_n \in \mathrm{dom}(\chi)$, $\chi(z_n) \cap (B \cup \{\boldsymbol{o}\}) \neq \varnothing$. As in [5], the function ι will be called *the iteration of φ controlled by χ*. Any sequence (finite or infinite) $z_0, z_1, \ldots$ of elements of B^* such that (1) holds whenever there are terms z_i and z_{i+1} in the sequence will be called a *φ, χ-path*.

Remark 1. *If φ and χ are single-valued then the corresponding function ι can be represented in Pascal-like notations by means of the declaration*

```
function ι(x : B*) : B*;
  var z : B*;
begin
  z := x;
  while χ(z) ∈ B* \ (B ∪ {o}) do z := φ(z);
  ι := z
end
```

Remark 2. *Among much others, the equivalence proof for the above-mentioned characterization of absolute prime computability makes use of the fact that the constant function with domain B^* and range $\{\boldsymbol{o}\}$ as well as the identity function of B^* can be obtained from the function π by means of iteration and composition. This is seen by observing that the iteration of π controlled by π maps any element of B^* into some element x with $\pi\pi\pi(x) = \boldsymbol{o}$ and the iteration of any function in $\mathcal{F}$ controlled by the constant function in question equals the identity function of B^*.*

By appropriately modifying the first and the third of the operations in the above-mentioned characterization of absolute prime computability we get the definition of another relative computability notion in $\mathcal{F}$ which is equivalent to absolute prime computability in the case of single-valued functions, but turns out to be more useful for the study of TTE computability in the general case (the intuitive idea of the modification can be described as follows: the modified operations must produce functions with evaluation procedures that always terminate for argument values belonging to the function domain, provided the operations are applied to functions with such evaluation procedures).

The *modified composition* $\varphi \circ \psi$ of two elements φ and ψ of $\mathcal{F}$ (sometimes called their *functional composition*) is the restriction of $\varphi\psi$ to the set

$$\{z \in \mathrm{dom}(\psi) \mid \psi(z) \subseteq \mathrm{dom}(\varphi)\}.$$

Such a kind of compositions are used for instance in [8, 2, 9]), whereas in [1] such a composition is additionally modified by the formation of metric space closure[3]. Clearly $\varphi \circ \psi = \varphi\psi$ in the case when ψ is single-valued, as well in the case when $\mathrm{dom}(\varphi) = B^*$.

The *modified iteration* operation is $\varphi, \chi \mapsto \iota \restriction E$, where ι is the same as above, and E is the set of the elements x of B^* such that

- no infinite φ, χ-path $z_0, z_1, z_2, \ldots$ exists with $z_0 = x$;
- whenever $z_0, z_1, \ldots, z_n$ is a finite φ, χ-path with $z_0 = x$,

$$z_n \in \mathrm{dom}(\chi) \;\&\; (\chi(z_n) \setminus (B \cup \{\boldsymbol{o}\}) \neq \varnothing \Rightarrow z_n \in \mathrm{dom}(\varphi)). \tag{2}$$

It is easy to see that the modified iteration operation coincides with the initially considered one when applied to single-valued elements of $\mathcal{F}$.

Definition 1. *A function $\varphi \in \mathcal{F}$ will be called* neatly computable *in some given functions $\psi_1, \ldots, \psi_l$ from $\mathcal{F}$ if φ can be obtained from $\psi_1, \ldots, \psi_l$ and the functions π and δ by means of finitely many applications of modified composition, juxtaposition and modified iteration.*

[3]Thus the mentioning of the paper [1] in the statement "The modified composition is one used in [Br96, We00, Br03, We08]" on the second page of the abstract [6] needs to be made more precise).

Remark 3. *The constant function with domain B^* and range $\{\boldsymbol{o}\}$ as well as the identity function of B^* are neatly computable in any given functions from $\mathcal{F}$ because the iterations and the composition $\pi\pi\pi$ used in Remark 2 coincide with the corresponding modified ones.*

Of course, if $\psi_1, \ldots, \psi_l$ are single-valued then the neat computability in them is equivalent to absolute prime computability in $\psi_1, \ldots, \psi_l$. As seen from [7, Example 1.5], it is not so in the general case (although each function neatly computable in $\psi_1, \ldots, \psi_l$ is a restriction of some function prime computable in $\psi_1, \ldots, \psi_l$).

Besides the operations composition, juxtaposition and iteration used in the characterizations of the absolute prime computability and in the definition of neat computability, there is another important operation on functions from $\mathcal{F}$, namely branching (the conditional operator). Its version we will consider is a ternary operation in $\mathcal{F}$ whose action on a triple of single-valued functions χ, φ, ψ from $\mathcal{F}$ is representable in Pascal-like notations as follows:

function $\theta(x : B^*) : B^*$;
begin if $\chi(x) \in B^* \setminus (B \cup \{\boldsymbol{o}\})$ then $\theta := \varphi(x)$ else $\theta := \psi(x)$ end

As in the case of composition and iteration, two variants of this operation will be considered. The first one is $\chi, \varphi, \psi \mapsto \theta$, where $\mathrm{dom}(\theta)$ is the set of the elements z of $\mathrm{dom}(\chi)$ satisfying some of the conditions

$$\chi(z) \setminus (B \cup \{\boldsymbol{o}\}) \neq \varnothing \ \&\ z \in \mathrm{dom}(\varphi), \tag{3}$$
$$\chi(z) \cap (B \cup \{\boldsymbol{o}\}) \neq \varnothing \ \&\ z \in \mathrm{dom}(\psi), \tag{4}$$

and, for any z in $\mathrm{dom}(\theta)$, the set $\theta(z)$ consists of all u such that $u \in \varphi(z)$ with (3) holding or $u \in \psi(z)$ with (4) holding. The second variant of the operation is $\chi, \varphi, \psi \mapsto \theta \restriction E$, where θ is the same as above, and E is the set of the elements z of $\mathrm{dom}(\chi)$ satisfying the following two conditions:

$$\chi(z) \setminus (B^* \cup \{\boldsymbol{o}\}) \neq \varnothing \Rightarrow z \in \mathrm{dom}(\varphi),$$
$$\chi(z) \cap (B^* \cup \{\boldsymbol{o}\}) \neq \varnothing \Rightarrow z \in \mathrm{dom}(\psi).$$

It turns out that the first of the above variants of branching preserves absolute prime computability, and the second one preserves neat computability. This follows from the fact that the result of applying them to a triple χ, φ, ψ of elements of $\mathcal{F}$ equals, respectively, $\delta\upsilon\delta\iota\alpha$ and $\delta \circ \mathring{\upsilon} \circ \delta \circ \mathring{\iota} \circ \alpha$, where:

- $\alpha = \lambda z.\chi(z) \times \{(\boldsymbol{o}, \boldsymbol{o}), z)\}$,

- ι and $\mathring{\iota}$ are, respectively, the iteration and the modified iteration of the function $\lambda z.\{\boldsymbol{o}\} \times (\{\boldsymbol{o}\} \times \varphi\delta\delta(z))$ controlled by π,
- υ and $\mathring{\upsilon}$ are, respectively, the iteration and the modified iteration of the function $\lambda z.\{\boldsymbol{o}\} \times \psi\delta(z)$ controlled by π

(taking also into account that $\varphi\delta\delta = \varphi \circ \delta \circ \delta$ and $\psi\delta = \psi \circ \delta$ because δ is single-valued).[4]

An essential thing concerning neat computability is that the first recursion theorem remains valid for it. The precise formulation and the proof of this can be found in [5] (*cf.* [5, Ch. III, Sect. 4, and Appendix, Sect. 8, paying attention to Remark 6 there]).

2 Moschovakis extension of a multi-represented space

Let (B, ϱ) be a multi-represented space. To turn the Moschovakis extension B^* of B into a multi-represented space, we choose some computable bijection J from $\mathbb{N}^2$ to $\mathbb{N}$ such that the inequality $2J(m,n) \geq \max(m,n)$ holds for all $m, n \in \mathbb{N}$, and we define a partial mapping ϱ^* of $\mathbb{N}^\mathbb{N}$ into the set of the nonempty subsets of B^* in the following inductive way, where k ranges over $\mathbb{N}$:

(i) $\lambda k.2q(k) + 2 \in \mathrm{dom}(\varrho^*)$ and $\varrho^*(\lambda k.2q(k) + 2) = \varrho(q)$ for any $q \in \mathrm{dom}(\varrho)$.

(ii) $\lambda k.0 \in \mathrm{dom}(\varrho^*)$ and $\varrho^*(\lambda k.0) = \{\boldsymbol{o}\}$.

(iii) If $q, r \in \mathrm{dom}(\varrho^*)$ then $\lambda k.2J(q(k), r(k)) + 1 \in \mathrm{dom}(\varrho^*)$ and

$$\varrho^*(\lambda k.2J(q(k), r(k)) + 1) = \varrho^*(q) \times \varrho^*(r).$$

Clearly $\bigcup\{\varrho^*(p) \mid p \in \mathrm{dom}(\varrho^*)\} = B^*$, hence (B^*, ϱ^*) is a multi-represented space. We will consider it as a Moschovakis extension of (B, ϱ).

Of course, the multi-representation ϱ^* depends on the choice of the computable bijection J. However, if J' is an arbitrary computable bijection from $\mathbb{N}^2$ to $\mathbb{N}$ such that we have $2J'(m,n) \geq \max(m,n)$ for all $m, n \in \mathbb{N}$ then the corresponding multi-representation $\varrho^{*\prime}$ is connected with ϱ^* in a simple way, namely a computable bijection h from $\mathbb{N}$ to $\mathbb{N}$ exists such that $\mathrm{dom}(\varrho^{*\prime}) = h^{-1}(\mathrm{dom}(\varrho^*))$ and the equality

[4]The easiest way to prove the above statements about representability of the two variants of branching in the forms $\delta\upsilon\delta\iota\alpha$ and $\delta \circ \mathring{\upsilon} \circ \delta \circ \mathring{\iota} \circ \alpha$, respectively, is by direct verification. However, it is worth noting that, in view of Proposition 1 in [5, Ch. II, Sect. 5], the representability in question is a particular instance of Proposition 4 there (the mentioned Proposition 1 is based on a result from [3]).

$\varrho^{*\prime}(p) = \varrho^*(hp)$ holds for any $p \in \mathrm{dom}(\varrho^{*\prime})$ (where, of course, $hp = \lambda k.h(p(k))$). In fact, we may define h in the following inductive way:

$$h(2l) = 2l, \quad h(2J'(m,n)+1) = 2J(h(m),h(n))+1.$$

3 On the TTE computability in the Moschovakis extension of a multi-represented space

In this section, an arbitrary multi-represented space (B, ϱ) will be supposed to be given. TTE computability in the multi-represented space (B^*, ϱ^*) constructed as in the previous section will be considered.

3.1 Computable elements of the Moschovakis extension

As well known, the values of the second component of a represented space at computable sequences of natural numbers are called *computable elements* of the space. The natural corresponding definition for multi-represented spaces is: if (X, ϱ) is a multi-represented space then an element of X is called *computable* if it belongs to $\varrho(p)$ for some computable element p of $\mathrm{dom}(\varrho)$ (cf. [9, Definition 24]). The following three statements are easily verifiable, and they entail the corollary after them:

Proposition 4. *The element* $\boldsymbol{o}$ *is a computable element of* (B^*, ϱ^*).

Proposition 5. *If* $x \in B$ *then* x *is a computable element of* (B^*, ϱ^*) *iff* x *is a computable element of* (B, ϱ).

Proposition 6. *If* u *and* v *belong to* B^* *then* (u, v) *is a computable element of* (B^*, ϱ^*) *iff* u *and* v *are computable elements of* (B^*, ϱ^*).

Corollary 7. *The set of the computable elements of the multi-represented space* (B^*, ϱ^*) *is the closure under the pairing operation of the set consisting of the element* $\boldsymbol{o}$ *and the computable elements of the multi-represented space* (B, ϱ).

3.2 On computability of functions in B

Having in mind [9, Definitions 17 and 24], we accept the following definition: if (X_1, ϱ_1) and (X_2, ϱ_2) are multi-represented spaces, then a *single-valued* (ϱ_1, ϱ_2)-*realization* of a partial function φ from X_1 to the set of the nonempty subsets of X_2 is any partial mapping γ of $\mathbb{N}^{\mathbb{N}}$ into $\mathbb{N}^{\mathbb{N}}$ such that $\gamma(p) \in \mathrm{dom}(\varrho_2)$ and $\varrho_2(\gamma(p)) \cap \varphi(x) \neq \varnothing$ whenever $p \in \mathrm{dom}(\varrho_1)$ and $x \in \varrho_1(p) \cap \mathrm{dom}(\varphi)$; the function φ is said to be (ϱ_1, ϱ_2)-*computable* if a computable single-valued (ϱ_1, ϱ_2)-realization of φ exists.[5]

If φ is a partial function from B to the set of the nonempty subsets of B then, of course, φ is also a partial function from B to the set of the nonempty subsets of B^*, as well as a partial function from B^* to the set of the nonempty subsets of B and a partial function from B^* to the set of the nonempty subsets of B^*.

Lemma 8. *Let θ be the function with domain B such that $\theta(x) = \{x\}$ for any x in B, and let $\varrho_1, \varrho_2 \in \{\varrho, \varrho^*\}$. Then θ is (ϱ_1, ϱ_2)-computable.*

Proof. If $\varrho_1 = \varrho_2$ then the identity mapping of $\mathrm{dom}(\varrho_1)$ is obviously a computable single-valued (ϱ_1, ϱ_2)-realization of θ. In the case of $\varrho_1 = \varrho$, $\varrho_2 = \varrho^*$ and in the case of $\varrho_1 = \varrho^*$, $\varrho_2 = \varrho$, so are, respectively, the mapping $q \mapsto \lambda k.2q(k) + 2$ and the mapping $p \mapsto \lfloor p(k)/2 \rfloor \dot{-} 1$ of $\mathrm{dom}(\varrho_1)$ into $\mathrm{dom}(\varrho_2)$. □

Proposition 9. *Let φ be a partial function from B to the set of the nonempty subsets of B. Then the following conditions are equivalent:*

(i) φ is (ϱ, ϱ)-computable;
(ii) φ is (ϱ, ϱ^*)-computable;
(iii) φ is (ϱ^*, ϱ)-computable;
(iv) φ is (ϱ^*, ϱ^*)-computable.

Proof. Let θ be as in the above lemma. Then the equalities

$$\varphi = \theta \circ \varphi = \varphi \circ \theta = \theta \circ \varphi \circ \theta$$

hold. The implication from (i) to (ii) follows from the equality $\varphi = \theta \circ \varphi$ because if γ is a computable single-valued (ϱ, ϱ)-realization of φ, and τ is a computable single-valued (ϱ, ϱ^*)-realization of θ then the function $q \mapsto \tau(\gamma(q))$ will be a computable single-valued (ϱ, ϱ^*)-realization of $\theta \circ \varphi$, *i.e.* of φ. The other implications from some of the conditions (i)–(iv) to another one can be established in a similar way. □

3.3 Neat computability preserves the usual TTE computability

Theorem 10. *Let $\mathcal{F}_\varrho$ be the set of all functions from $\mathcal{F}$ which are (ϱ^*, ϱ^*)-computable. If $\psi_1, \dots, \psi_l$ belong to $\mathcal{F}_\varrho$ then all functions neatly computable in $\psi_1, \dots, \psi_l$ also belong to $\mathcal{F}_\varrho$.*

[5] In [9] also multi-valued realizations are considered, and there is no requirement in Definition 24 about single-valuedness of the realization. We impose this requirement in order to avoid specifying a computability notion for multi-valued mappings of $\mathbb{N}^{\mathbb{N}}$ into $\mathbb{N}^{\mathbb{N}}$.

Proof. First of all, we will show that the functions π and δ belong to $\mathcal{F}_\varrho$. A computable single-valued (ϱ^*, ϱ^*)-realization γ of π can be defined by setting

$$\gamma(p)(k) = g(p(k)),$$

where the function $g : \mathbb{N} \to \mathbb{N}$ is defined by means of the equalities

$$g(0) = 0, \;\; g(2i+2) = 1,^{6} \; g(2J(m,n)+1) = m.$$

For δ, the construction of a computable single-valued (ϱ^*, ϱ^*)-realization is similar – with n instead of m in the right-hand side of the third equality for g.

The rest of the proof consists of showing that $\mathcal{F}_\varrho$ is closed under modified composition, juxtaposition and modified iteration. Let $\varphi_1, \varphi_2 \in \mathcal{F}_\varrho$. Then there are computable single-valued (ϱ^*, ϱ^*)-realizations γ_1 and γ_2 of φ_1 and φ_2, respectively. The composition $\varphi_1 \circ \varphi_2$ and the juxtaposition of φ_1 and φ_2 belong to $\mathcal{F}_\varrho$ because the functions $p \mapsto \gamma_1(\gamma_2(p))$ and $p \mapsto 2J(\gamma_1(p), \gamma_2(p)) + 1$ are computable single-valued (ϱ^*, ϱ^*)-realizations of them, respectively. Let now ι be the modified iteration of φ_1 controlled by φ_2. We set

$$D = \{(q,z) \mid q \in \mathrm{dom}(\varrho^*), \; z \in \varrho^*(q) \cap \mathrm{dom}(\varphi_1)\}.$$

Whenever $(q,z) \in D$,

$$q \in \mathrm{dom}(\gamma_1), \;\; \gamma_1(q) \in \mathrm{dom}(\varrho^*), \;\; \varrho^*(\gamma_1(q)) \cap \varphi_1(z) \neq \varnothing.$$

Therefore a mapping σ of D into B^* exists such that

$$\sigma(q,z) \in \varrho^*(\gamma_1(q)) \cap \varphi_1(z)$$

whenever $(q,z) \in D$. For any n in $\mathbb{N}$, we define the partial mapping γ_1^n of $\mathbb{N}^\mathbb{N}$ into $\mathbb{N}^\mathbb{N}$ by setting $\gamma_1^0(p) = p$ for all p in $\mathbb{N}^\mathbb{N}$, $\gamma_1^{n+1}(p) = \gamma_1(\gamma_1^n(p)$. Let $p \in \mathrm{dom}(\varrho^*)$, and H_p be the set of all natural numbers n such that

(i) $p \in \mathrm{dom}(\gamma_1^n)$;

(ii) $\gamma_1^i(p) \in \mathrm{dom}(\gamma_2)$ and $\gamma_2(\gamma_1^i(p))(0)$ is odd for all $i < n$.

Clearly $0 \in H_p$, and all natural numbers which are less than a number from H_p also belong to H_p. Let $z \in \varrho^*(p) \cap \mathrm{dom}(\iota)$, and $H_{p,z}$ be the set of the numbers $n \in H_p$ such that a φ_1, φ_2-path $z_0, z_1, \ldots, z_n$ with $z_0 = z$ exists satisfying the conditions

$$(\gamma_1^i(p), z_i) \in D, \;\; z_{i+1} = \sigma(\gamma_1^i(p), z_i) \tag{5}$$

[6]Note that $1 = 2J(0,0) + 1$ because the inequalities $2J(m,n) \geq m$, $2J(m,n) \geq n$ and the surjectivity of J imply $J(0,0) = 0$.

for all $i < n$. We will show that actually $H_{p,z} = H_p$, *i.e.* that $n \in H_p$ implies $n \in H_{p,z}$. This will be done by induction. The implication obviously holds for $n = 0$ since the statement that $0 \in H_{p,z}$ is trivially true.

Before taking the inductive step, we will show that, whenever $n \in H_{p,z}$ and $z_0, z_1, \ldots, z_n$ is a φ_1, φ_2-path with $z_0 = z$ satisfying the conditions (5) for all $i < n$, the following statements hold:

$$\gamma_1^n(p) \in \mathrm{dom}(\varrho^*),\ z_n \in \varrho^*(\gamma_1^n(p)), \tag{6}$$

$$z_n \in \mathrm{dom}(\varphi_2) \,\&\, (\varphi_2(z_n) \setminus (X \cup \{o\}) \neq \varnothing \Rightarrow z_n \in \mathrm{dom}(\varphi_1)). \tag{7}$$

Suppose $n \in H_{p,z}$ and $z_0, z_1, \ldots, z_n$ is a φ_1, φ_2-path with $z_0 = z$ satisfying the conditions (5) for all $i < n$. The statements (6) hold because $\gamma_1^n(p) = p$, $z_n = z$ if $n = 0$, and

$$(\gamma_1^{n-1}(p), z_{n-1})) \in D,\ \gamma_1^n(p) = \gamma_1(\gamma_1^{n-1}(p)),\ z_n = \sigma(\gamma_1^{n-1}(p), z_{n-1})$$

otherwise. As to the conjunction (7), it is the conjunction (2) with φ_1, φ_2 in the roles of φ, χ, respectively.

We note also that, whenever $n \in H_{p,z}$,

$$\gamma_1^n(p) \in \mathrm{dom}(\gamma_2) \,\&\, \gamma_2(\gamma_1^n(p)) \in \mathrm{dom}(\varrho^*) \,\&\, \varrho^*(\gamma_2(\gamma_1^n(p))) \cap \varphi_2(z_n) \neq \varnothing, \tag{8}$$

and if, in addition, the number $\gamma_2(\gamma_1^n(p))(0)$ is odd, then $n+1 \in H_{p,z}$ too. Indeed, let $n \in H_{p,z}$. Then a φ_1, φ_2-path $z_0, z_1, \ldots, z_n$ with $z_0 = z$ exists which satisfies the conditions (5) for all $i < n$, and the statements (6), (7) hold for it. Since $z_n \in \varrho^*(\gamma_1^n(p))$, the first term of the conjunction (7) implies (8). Suppose additionally that $\gamma_2(\gamma_1^n(p))(0)$ is odd. Then

$$\varrho^*(\gamma_2(\gamma_1^n(p))) \cap (X \cup \{o\}) = \varnothing,$$

therefore $\varphi_2(z_n) \setminus (X \cup \{o\}) \neq \varnothing$, hence $z_n \in \mathrm{dom}(\varphi_1)$. Thus $(\gamma_1^n(p), z_n) \in D$. But then the sequence $z_0, z_1, \ldots, z_n, z_{n+1}$, where $z_{n+1} = \sigma(\gamma_1^n(p), z_n)$ witnesses that $n+1 \in H_{p,z}$.

For the inductive step, let now n be a natural number such that $n \in H_p$ implies $n \in H_{p,z}$. Suppose that $n+1 \in H_p$. Then $n \in H_p$, hence $n \in H_{p,z}$. The number $\gamma_2(\gamma_1^n(p))(0)$ is odd, since $n < n+1 \in H_p$, hence $n+1 \in H_{p,z}$.

It is easy to see now that H_p is a finite set – otherwise p would belong to $\mathrm{dom}(\gamma_1^i)$ for all i and an infinite φ_1, φ_2-path $z_0, z_1, z_2, \ldots$ with $z_0 = z$ satisfying the conditions (5) for all i would exist, but this is impossible because z belongs to $\mathrm{dom}(\iota)$. Let n be the maximal number in H_p, and let $z_0, z_1, \ldots, z_n$ be a φ_1, φ_2-path with $z_0 = z$ satisfying the conditions (5) for all $i < n$. The maximality of n implies that

$n+1 \notin H_p$, hence the number $\gamma_2(\gamma_1^n(p))(0)$ is even. Therefore $\varphi_2(z_n) \cap (X \cup \{o\}) \neq \varnothing$ and consequently $z_n \in \iota(z)$.

Thus we proved the following statement: if

$$p \in \mathrm{dom}(\varrho^*) \text{ and } z \in \varrho^*(p) \cap \mathrm{dom}(\iota) \tag{9}$$

then a natural number n exists with the following properties: $p \in \mathrm{dom}(\gamma_1^n)$, $\gamma_1^i(p) \in \mathrm{dom}(\gamma_2)$ for all $i \leq n$, $\gamma_2(\gamma_1^i(p))(0)$ is odd for all $i < n$, $\gamma_2(\gamma_1^n(p))(0)$ is even, $\gamma_1^n(p) \in \mathrm{dom}(\varrho^*)$ and $\varrho^*(\gamma_1^n(p)) \cap \iota(z) \neq \varnothing$.

The mappings γ_1 and γ_2 can be extended to computable operators Γ_1 and Γ_2 in the set of all unary partial functions in $\mathbb{N}$. If condition (9) holds then

$$\iota(z) = \varrho^*(\Gamma_1^n(p)) \text{ with } n = \mu i[\Gamma_2(\Gamma_1^i(p))(0) \bmod 2 = 0].$$

Let $\Gamma_1^\dagger$ be the operator transforming unary partial functions in $\mathbb{N}$ into binary ones which is defined by the equality

$$\Gamma_1^\dagger(f)(i,k) = \Gamma_1^i(f)(k).$$

Let the operators $\Gamma_2^\dagger$ and Γ transforming unary partial functions in $\mathbb{N}$ into unary ones be defined as follows:

$$\Gamma_2^\dagger(f)(i) = \Gamma_2(\Gamma_1^i(f))(0),$$
$$\Gamma(f)(k) = \Gamma_1^\dagger(f)(\mu i[\Gamma_2^\dagger(f)(i) \bmod 2 = 0], k).$$

One consecutively shows that $\Gamma_1^\dagger, \Gamma_2^\dagger, \Gamma$ are computable (this can be done, for instance, by using their continuity and the effectiveness of their restrictions to partial recursive functions). Whenever (9) holds,

$$\Gamma(p) \in \mathrm{dom}(\varrho^*) \text{ and } \iota(z) \in \varrho^*(\Gamma(p)).$$

Therefore some restriction of Γ will be a computable single-valued (ϱ^*, ϱ^*)-realization of ι. □

3.4 Neat computability preserves TTE computability in Brattka's sense (generalized for the case of multi-representations)

We accept the following convention:

Definition 2. *Let (X_1, ϱ_1) and (X_2, ϱ_2) be multi-represented spaces, and φ be a partial function from X_1 to the set of the nonempty subsets of X_2. A* Brattka style (ϱ_1, ϱ_2)-realization of φ *is any partial mapping γ of $\mathbb{N}^{\mathbb{N}} \times \mathbb{N}^{\mathbb{N}}$ into $\mathbb{N}^{\mathbb{N}}$ such that the next conditions are satisfied whenever $p \in \mathrm{dom}(\varrho_1)$ and $z \in \varrho_1(p) \cap \mathrm{dom}(\varphi)$:*

(i) $(p, q) \in \mathrm{dom}(\gamma)$, $\gamma(p, q) \in \mathrm{dom}(\varrho_2)$, $\varrho_2(\gamma(p, q)) \cap \varphi(z) \neq \varnothing$ for all q in $\mathbb{N}^{\mathbb{N}}$;

(ii) $\varphi(z) \subseteq \bigcup \left\{ \varrho_2(\gamma(p, q)) \,\middle|\, q \in \mathbb{N}^{\mathbb{N}} \right\}$.

The function φ will be called (ϱ_1, ϱ_2)-computable in Brattka's sense *if there is a computable Brattka style (ϱ_1, ϱ_2)-realization of φ.*

Thanks to condition (i), the (ϱ_1, ϱ_2)-computability of the function φ in Brattka's sense implies its computability in the usual sense. If φ is single-valued then the converse implication also holds. In the case of single-valued ϱ_2, a partial mapping γ of $\mathbb{N}^{\mathbb{N}} \times \mathbb{N}^{\mathbb{N}}$ into $\mathbb{N}^{\mathbb{N}}$ is a Brattka style (ϱ_1, ϱ_2)-realization of φ iff $(p, q) \in \mathrm{dom}(\gamma)$, $\gamma(p, q) \in \mathrm{dom}(\varrho_2)$ for all q in $\mathbb{N}^{\mathbb{N}}$, and

$$\varphi(z) = \left\{ \varrho_2(\gamma(p, q)) \,\middle|\, q \in \mathbb{N}^{\mathbb{N}} \right\}$$

whenever $p \in \mathrm{dom}(\varrho_1)$ and $z \in \varrho_1(p) \cap \mathrm{dom}(\varphi)$. Therefore the (ϱ_1, ϱ_2) computability in Brattka's sense is equivalent to the one introduced by Definition 7.1 in [2] if (X_1, ϱ_1) and (X_2, ϱ_2) are represented spaces.[7]

Remark 11. *Proposition 9 remains valid after replacement of "computable" with "computable in Brattka's sense".*

Theorem 12. *Let $\mathcal{F}_\varrho^{\mathrm{B}}$ be the set of all functions from $\mathcal{F}$ which are (ϱ^*, ϱ^*)-computable in Brattka's sense. If $\psi_1, \ldots, \psi_l$ belong to $\mathcal{F}_\varrho^{\mathrm{B}}$ then all functions neatly computable in $\psi_1, \ldots, \psi_l$ belong to $\mathcal{F}_\varrho^{\mathrm{B}}$.*

Proof. A single-valued function from $\mathcal{F}$ belongs to $\mathcal{F}_\varrho^{\mathrm{B}}$ iff it belongs to the set $\mathcal{F}_\varrho$ from Theorem 10. Hence π and δ belong to $\mathcal{F}_\varrho^{\mathrm{B}}$. It remains to show that $\mathcal{F}_\varrho^{\mathrm{B}}$ is closed under modified composition, juxtaposition and modified iteration. To this end, suppose $\varphi_1, \varphi_2 \in \mathcal{F}_\varrho^{\mathrm{B}}$. Let γ_1 and γ_2 be computable Brattka style (ϱ^*, ϱ^*)-realizations of φ_1 and φ_2, respectively. It is easy to verify that the following two mappings of $\mathbb{N}^{\mathbb{N}} \times \mathbb{N}^{\mathbb{N}}$ into $\mathbb{N}^{\mathbb{N}}$ are computable Brattka style (ϱ^*, ϱ^*)-realizations of

[7]It may seem that a more natural way to generalize that definition to the case of multi-represented spaces would be by leaving out the condition (i) of Definition 2 and replacing the inclusion sign by an equality sign in condition (ii). Unfortunately this would not guarantee the equivalence of computability in Brattka's sense of single-valued functions to their computability in the usual sense.

$\varphi_1 \circ \varphi_2$ and of the juxtaposition of φ_1 and φ_2), respectively:

$$(p,q) \mapsto \gamma_1(\gamma_2(p, \lambda k.q(2k)), \lambda k.q(2k+1)),$$
$$(p,q) \mapsto 2J(\gamma_1(p, \lambda k.q(2k+1)), \gamma_2(p, \lambda k.q(2k)) + 1.$$

Let now ι be the modified iteration of φ_1 controlled by φ_2. We consider the partial mapping δ of $\mathbb{N}^{\mathbb{N}} \times \mathbb{N}^{\mathbb{N}} \times \mathbb{N}$ into $\mathbb{N}^{\mathbb{N}} \times \mathbb{N}^{\mathbb{N}}$ and the partial mapping γ of $\mathbb{N}^{\mathbb{N}} \times \mathbb{N}^{\mathbb{N}}$ into $\mathbb{N}^{\mathbb{N}}$ which are defined as follows:

$$\delta(p,q,0) = p, \tag{10}$$
$$\delta(p,q,i+1) = \gamma_1(\delta(p,q,i), \lambda k.q(J(2i,k))), \tag{11}$$
$$\gamma(p,q) = \delta(p,q, \mu i[\gamma_2(\delta(p,q,i), \lambda k.q(J(2i+1,k)))(0) \bmod 2 = 0]). \tag{12}$$

We will show that γ is a computable Brattka style (ϱ^*, ϱ^*)-realization of ι.

The computability of γ can be proved by showing that an appropriate continuous extension of γ transforming arbitrary pairs of unary partial functions in $\mathbb{N}$ into such functions is effective when considered on partial recursive functions. For the proof that γ is a Brattka style (ϱ^*, ϱ^*)-realization of ι, suppose $p \in \mathrm{dom}(\varrho^*)$ and $z \in \varrho^*(p) \cap \mathrm{dom}(\iota)$. We have to show that

$$(p,q) \in \mathrm{dom}(\gamma), \;\; \gamma(p,q) \in \mathrm{dom}(\varrho^*), \;\; \varrho^*(\gamma(p,q)) \cap \iota(z) \neq \varnothing \tag{13}$$

for all $q \in \mathbb{N}^{\mathbb{N}}$, and that

$$\iota(z) \subseteq \bigcup \left\{ \varrho^*(\gamma(p,q)) \;\middle|\; q \in \mathbb{N}^{\mathbb{N}} \right\}. \tag{14}$$

To prove that we have (13) for all $q \in \mathbb{N}^{\mathbb{N}}$, suppose q is an arbitrary function from $\mathbb{N}^{\mathbb{N}}$. Let H be the set of all natural numbers n such that

(i) $(p,q,n) \in \mathrm{dom}(\delta)$;

(ii) for all $i < n$, $(\delta(p,q,i), \lambda k.q(J(2i+1,k))) \in \mathrm{dom}(\gamma_2)$ and the number $\gamma_2(\delta(p,q,i), \lambda k.q(J(2i+1,k)))(0)$ is odd.

Clearly $0 \in H$ and all natural numbers which are less than a number from H also belong to H. Let

$$D = \{(r,u) \mid r \in \mathrm{dom}(\varrho^*), \; u \in \varrho^*(r) \cap \mathrm{dom}(\varphi_1)\}.$$

If $(r,u) \in D$ and $i \in \mathbb{N}$ then, by the fact that γ_1 is a Brattka style (ϱ^*, ϱ^*)-realization of φ_1, we have

$$(r, \lambda k.q(J(2i,k)) \in \mathrm{dom}(\gamma_1), \;\; \gamma_1(r, \lambda k.q(J(2i,k)) \in \mathrm{dom}(\varrho^*),$$
$$\varrho^*(\gamma_1(r, \lambda k.q(J(2i,k))) \cap \varphi_1(u) \neq \varnothing.$$

Let σ be a mapping of $D \times \mathbb{N}$ into X^* such that

$$\sigma(r,u,i) \in \varrho^*(\gamma_1(r, \lambda k.q(J(2i,k)))) \cap \varphi_1(u)$$

whenever $(r,u,i) \in D \times \mathbb{N}$. Let $\hat{H}$ be the set of the numbers $n \in H$ such that a φ_1, φ_2-path $z_0, z_1, \ldots, z_n$ with $z_0 = z$ exists satisfying the conditions

$$(\delta(p,q,i), z_i) \in D, \quad z_{i+1} = \sigma(\delta(p,q,i), z_i, i) \tag{15}$$

for all $i < n$. We will show that actually $\hat{H} = H$, *i.e.* that $n \in H$ implies $n \in \hat{H}$. This will be done by induction. The implication obviously holds for $n = 0$ since the statement that $0 \in \hat{H}$ is trivially true. Before taking the inductive step, we will show that, whenever $n \in \hat{H}$ and $z_0, z_1, \ldots, z_n$ is a φ_1, φ_2-path with $z_0 = z$ satisfying the conditions (15) for all $i < n$, the following statements hold:

$$\delta(p,q,n) \in \mathrm{dom}(\varrho^*), \quad z_n \in \varrho^*(\delta(p,q,n)), \tag{16}$$

$$z_n \in \mathrm{dom}(\varphi_2) \,\&\, (\varphi_2(z_n) \setminus (X \cup \{o\}) \neq \varnothing \Rightarrow z_n \in \mathrm{dom}(\varphi_1)). \tag{17}$$

Suppose $n \in \hat{H}$ and $z_0, z_1, \ldots, z_n$ is a φ_1, φ_2-path with $z_0 = z$ satisfying the conditions (15) for all $i < n$. The statements (16) hold because $\delta(p,q,n) = p$, $z_n = z$ if $n = 0$, and

$$(\delta(p,q,n-1), z_{n-1}) \in D, \ \delta(p,q,n) = \gamma_1(\delta(p,q,n-1), \lambda k.q(J(2n-2,k))),$$
$$z_n = \sigma(\delta(p,q,n-1), z_{n-1}, n-1) \in \varrho^*(\gamma_1(\delta(p,q,n-1), \lambda k.q(J(2n-2,k))))$$

otherwise. As to the conjunction (17), it is the conjunction (2) with φ_1, φ_2 in the roles of σ, χ, respectively.

We note also that, whenever $n \in \hat{H}$,

$$(\delta(p,q,n), \lambda k.q(J(2n+1,k))) \in \mathrm{dom}(\gamma_2), \tag{18}$$

$$\gamma_2(\delta(p,q,n), \lambda k.q(J(2n+1,k))) \in \mathrm{dom}(\varrho^*), \tag{19}$$

$$\varrho^*(\gamma_2(\delta(p,q,n), \lambda k.q(J(2n+1,k)))) \cap \varphi_2(z_n) \neq \varnothing, \tag{20}$$

$$\gamma_2(\delta(p,q,n), \lambda k.q(J(2n+1,k)))(0) \text{ is odd } \Rightarrow \ n+1 \in \hat{H}. \tag{21}$$

Indeed, let $n \in \hat{H}$. Then a φ_1, φ_2-path $z_0, z_1, \ldots, z_n$ with $z_0 = z$ exists which satisfies the conditions (15) for all $i < n$, and the statements (16), (17) hold for it. Since $z_n \in \varrho^*(\delta(p,q,n))$ and γ_2 is a Brattka style (ϱ^*, ϱ^*)-realization of φ_2, the first term of the conjunction (17) implies the statements (18)–(20). For the proof of (21), suppose that $\gamma_2(\delta(p,q,n), \lambda k.q(J(2n+1,k)))(0)$ is odd. Then

$$\varrho^*(\gamma_2(\delta(p,q,n), \lambda k.q(J(2n+1,k)))) \cap (X \cup \{o\}) = \varnothing,$$

hence $\varphi_2(z_n) \setminus (X \cup \{o\}) \neq \varnothing$, and therefore $z_n \in \mathrm{dom}(\varphi_1)$. Thus

$$(\delta(p, q, n), z_n) \in D.$$

But then the sequence $z_0, z_1, \dots, z_n, z_{n+1}$, where $z_{n+1} = \sigma(\delta(p, q, n), z_n, n)$ turns out to be a φ_1, φ_2-path starting at z and satisfying the conditions (15) for all $i < n+1$, hence $n + 1 \in \hat{H}$.

Let now n be a natural number such that $n \in H$ implies $n \in \hat{H}$, and suppose that $n+1 \in H$. Then $n \in H$, hence $n \in \hat{H}$ and therefore the statements (18)–(21) hold. The number

$$\gamma_2(\delta(p, q, n), \lambda k.q(J(2n+1, k)))(0) \tag{22}$$

is odd, since $n < n+1 \in H$, hence $n+1 \in \hat{H}$.

It is easy to see now that H is a finite set – otherwise the triple (p, q, i) would belong to $\mathrm{dom}(\delta)$ for all i and an infinite φ_1, φ_2-path $z_0, z_1, z_2, \dots$ with $z_0 = z$ satisfying the conditions (15) for all i would exist, but this is impossible because $z \in \mathrm{dom}(\iota)$. Let n be the maximal number in H, and let $z_0, z_1, \dots, z_n$ be a φ_1, φ_2-path with $z_0 = z$ satisfying the conditions (15) for all $i < n$. The maximality of n implies that $n+1 \notin H$, hence the number (22) is even. Therefore

$$\varphi_2(z_n) \cap (X \cup \{o\}) \neq \varnothing$$

and consequently $z_n \in \iota(z)$, hence $\varrho^*(\delta(p, q, n) \cap \iota(z) \neq \varnothing$. This implies the statements (13). Indeed, we have the equality

$$n = \mu i[\gamma_2(\delta(p, q, i), \lambda k.q(J(2i+1, k)))(0) \bmod 2 = 0]$$

because $n \in H$ and the number (22) is even, hence $(p, q) \in \mathrm{dom}(\gamma)$ and the equality $\gamma(p, q) = \delta(p, q, n)$ holds.

It remains to prove the inclusion (14). Suppose $v \in \iota(z)$. Then a finite φ_1, φ_2-path $u_0, u_1, \dots, u_{n-1}, u_n$ exists such that $u_0 = z$, $u_n = v$ and

$$\varphi_2(v) \cap (X \cup \{o\}) \neq \varnothing.$$

Making use again of the fact that γ_1 and γ_2 are Brattka style (ϱ^*, ϱ^*)-realizations of φ_1 and φ_2, respectively, we see that functions $p_0, p_1, \dots, p_n \in \mathrm{dom}(\varrho^*)$ and functions $q_0, q_1, \dots, q_{n-1}, q_n, q'_0, q'_1, \dots, q'_{n-1}$ in $\mathbb{N}^{\mathbb{N}}$ with the following properties can be found:

$$p_0 = p, \;\; u_i \in \varrho^*(p_i) \text{ for } i \leq n,$$
$$\varrho^*(\gamma_2(p_i, q_i)) \setminus (X \cup \{o\}) \neq \varnothing \text{ and } p_{i+1} = \gamma_1(p_i, q'_i) \text{ for } i < n,$$
$$\varrho^*(\gamma_2(p_n, q_n)) \cap (X \cup \{o\}) \neq \varnothing.$$

Then $v \in \varrho^*(p_n)$ and

$$\gamma_2(p_i, q_i)(0) \bmod 2 = \begin{cases} 1 & \text{if } i < n, \\ 0 & \text{if } i = n. \end{cases}$$

Let $q \in \mathbb{N}^{\mathbb{N}}$ be such that

$$\lambda k.q(J(2i+1,k)) = q_i \text{ for } i \le k,$$
$$\lambda k.q(J(2i,k)) = q_i' \text{ for } i < k.$$

Then $\delta(p, q, i) = p_i$ for $i \le n$, hence

$$\gamma(p, q) = \delta(p, q, \mu i[\gamma_2(p_i, q_i)(0) \bmod 2 = 0]) = \delta(p, q, n) = p_n,$$

and therefore $(p, q) \in \mathrm{dom}(\gamma)$, $\gamma(p, q) \in \mathrm{dom}(\varrho^*)$, $v \in \varrho^*(\gamma(p, q))$ for the function q chosen in the above way. □

References

[1] Vasko Brattka *Recursive characterization of computable real-valued functions and relations*, Theoretical Computer Science 162, 1996, pp. 45–77.

[2] — *Computability over topological structures*, in: Barry S. Cooper and Sergey Goncharov (eds.), Computability and Models, *Perspectives East and West*, Kluwer Academic/Plenum Publishers, Dordrecht/New York 2003 [*University Series in Mathematics*], pp. 93–136.

[3] Lyubomir Ivanov Algebraic Recursion Theory, Ellis Horwood, Chichester [UK] 1986.

[4] Yiannis Moschovakis *Abstract first order computability*, I, Transactions of the American Mathematical Society 138, 1969, pp. 427–464.

[5] Dimiter Skordev Computability in Combinatory Spaces, *An Algebraic Generalization of Abstract First Order Computability*, Kluwer Academic Publishers, Dordrecht 1992.[8]

[6] — *Moschovakis extension of multi-represented spaces*, in: Fifteenth International Conference on Computability & Complexity in Analysis, 5–8 August 2018, Lake Kochel, Germany (CCA 2018 Abstract Booklet).

[7] — *Moschovakis extension of represented spaces*, Logical Methods in Computer Science 15 (1), 2019, pp. 35:1–35:21.

[8] Klaus Weihrauch Computable Analysis, *An Introduction*, Springer-Verlag, Berlin, Heidelberg etc. 2000.

[9] — *The computable multi-functions on multi-represented sets are closed under programming*, Journal of Universal Computer Science 14, 2008, pp. 801–844.

[8]Errata at `https://www.fmi.uni-sofia.bg/fmi/logic/skordev/errata_combinatory.htm`

THE THEORY OF THE ENUMERATION DEGREES, DEFINABILITY, AND AUTOMORPHISMS

MARIYA I. SOSKOVA
Department of Mathematics
University of Wisconsin
Madison, WI 53706, USA
msoskova@math.wisc.edu

1 Introduction

Enumeration reducibility captures a natural relationship between sets of natural numbers in which positive information about the first set is used to produce positive information about the second set. It was introduced independently several times in works by Friedberg and Rogers [10], Myhill [39], and Selman [45], who were searching for a natural way to extend the notion of relative Turing computability from total functions to partial functions. Informally, $A \subseteq \omega$ is enumeration reducible to $B \subseteq \omega$ if there is a uniform way to compute an enumeration of A from an enumeration of B. The formal definition that we give below is the one by Friedberg and Rogers [10].

Definition 1.1. $A \subseteq \omega$ is *enumeration reducible* to a set $B \subseteq \omega$ ($A \leq_e B$) if there is a c.e. set W such that

$$A = \{n \colon (\exists e)[\langle n, e\rangle \in W \text{ and } D_e \subseteq B]\},$$

where D_e is the eth finite set in a canonical enumeration.

By identifying sets that are reducible to each other we obtain an algebraic representation of this reducibility as a partial order: the structure of the enumeration degrees $\mathcal{D}_e$. The degree structure $\mathcal{D}_e$ is an upper semi-lattice with least upper bound induced by the effective join operation $A \oplus B = \{2n \mid n \in A\} \cup \{2n+1 \mid n \in B\}$ and a least element $\mathbf{0}_e$, the degree of all c.e. sets.

Motivation for the interest in the enumeration degrees comes from its nontrivial connections to the study of the Turing degrees. In Turing reducibility, $\leq_T$, we use

The author was partially supported by National Science Foundation grant DMS1762648.

membership information, both positive and negative, from a given oracle set B to obtain the same type of membership information about a reduced set A. Enumeration reducibility, restricts us to both using and producing only positive information. There is a further relation that sits between Turing and enumeration reducibility. The relation *relative computable enumerability* (c.e. in) uses positive and negative information about an oracle set B to produce only positive information about a set A. We can express the positive and negative information about a set A in a positive way by replacing it with $A \oplus \overline{A}$. This gives the following relationship between the three reducibilities:

Proposition 1.2. $A \leq_T B \Leftrightarrow A \oplus \overline{A}$ *is* B*-c.e.* $\Leftrightarrow A \oplus \overline{A} \leq_e B \oplus \overline{B}$.

Myhill [39] used this relationship to define a natural embedding of the Turing degrees into the enumeration degrees. He proved that the embedding $\iota\colon \mathcal{D}_T \to \mathcal{D}_e$, defined by

$$\iota(d_T(A)) = d_e(A \oplus \overline{A}),$$

preserves the order and the least upper bound. Thus we have a copy of the Turing degrees sitting inside the enumeration degrees. Medvedev [35] observed that sufficiently generic sets have enumeration degrees outside of the range of this embedding and so the enumeration degrees properly extend the Turing degrees.

In this article we outline some of the more recent results in the study of the enumeration degrees. We will focus on three aspects of the structure of the enumeration degrees:

I. The first order theory of $\mathcal{D}_e$ and its fragments;

II. First order definability;

III. Automorphisms and automorphism bases.

We will outline the current state of the art and discuss open problems on each topic that we believe mark important goals for the advancement of knowledge in the field. We will see that enumeration reducibility and the enumeration degrees have nontrivial interactions with other parts of mathematics, most prominently with topology.

The structure of this article is motivated by a theorem that was proved by Slaman and Woodin [53] for the Turing degrees and extended by Soskova [54] to the enumeration degrees. We state it here and elaborate on it throughout the paper:

Theorem 1.3 (Slaman and Woodin, Soskova). *The following are equivalent:*

1. $\mathcal{D}_e$ *is* biinterpretable *with second order arithmetic.*

2. *The definable relations in $\mathcal{D}_e$ are exactly the ones induced by degree invariant definable relations in second order arithmetic.*

3. *$\mathcal{D}_e$ is a rigid structure.*

Thus, the three aspects we consider are related and provide different outlooks on one main problem.

Question 1.4. Are there nontrivial automorphisms of the enumeration degrees?

2 The first order theory of $\mathcal{D}_e$ and its fragments

We start our explorations with the theory of the enumeration degrees $Th(\mathcal{D}_e)$, the set of sentences in the language of partial orders that are true in the structure. Our first observation is that enumeration reducibility is a relation that can be defined in second order arithmetic Z_2. Assuming some basic facts from classical recursion theory, it is straightforward to check that enumeration reducibility is a Σ^0_3 relation. Therefore, the theory $Th(\mathcal{D}_e)$ can be effectively interpreted in second order arithmetic. This does not really help us with understanding the complexity of the set, except for providing the evident upper bound $Th(\mathcal{D}_e) \leq_T Th(Z_2)$. It does, however, set the stage for our investigations and hints to the connections between $\mathcal{D}_e$ and Z_2 outlined in Theorem 1.3. We will first consider fragments of the theory that we obtain by restricting the quantifier complexity of the sentences.

2.1 The existential theory of $\mathcal{D}_e$

The simplest fragment of the theory of $\mathcal{D}_e$ is the existential theory, the set of existential sentences true in the structure. We denote this set by $\exists$-$Th(\mathcal{D}_e)$. An existential statement has the form

$$(\exists \mathbf{x}_1)\dots(\exists \mathbf{x}_n)[\varphi(\mathbf{x}_1,\dots \mathbf{x}_n)],$$

where φ is either obviously false because it contradicts the axioms of partial orderes or else it is a disjunction of quantifier free formulas that partially describe a finite partial order. In order for an existential sentence to be true in $\mathcal{D}_e$, it first must comply with the axioms of partial order. This is something that can be effectively checked. It follows that in order to decide whether an existential sentence is true in $\mathcal{D}_e$ we must understand which finite partial orders can be embedded in $\mathcal{D}_e$. The answer is simple: all partial orders can be embedded. This result can be traced back to Sacks [43], who showed that every countable partial order can be embedded in

$\mathcal{D}_T$ (in fact, he showed it for $\mathcal{D}_T(\leq_T \mathbf{0}')$, the initial interval of the Turing degrees bounded by $\mathbf{0}'$), in combination with the embedding of $\mathcal{D}_T$ in $\mathcal{D}_e$. In the enumeration degrees this result was extended first by Lagemann [26], who showed that every countable partial order can be embedded below any nonzero Δ^0_2 enumeration degree, then by Bianchini [4] who found such embeddings in any nonempty interval of Σ^0_2 enumeration degrees, then by Soskov and Soskova [55], who replaced Σ^0_2 with *good*. A good enumeration degree is a degree that contains a set with a *good approximation*. The good approximations were introduced by Lachlan and Shore [24]. They use them to show density for the n-c.e.a. degrees: a hierarchy of enumeration degrees based on the relation c.e. in. Most recently this series of results have been extended by Slaman and Sorbi [48].

Theorem 2.1 (Slaman, Sorbi). *Every countable partial order can be embedded below any nonzero enumeration degree.*

Note that the statement of the theorem above, reveals an important structural property of the enumeration degrees, initially proved by Gutteridge [16]: the enumeration degrees are downwards dense. We will see that this statement will play a trick on us when we consider more complex fragments of the theory. It also provides an example of a structural difference between $\mathcal{D}_e$ and $\mathcal{D}_T$, where minimal degrees exist.

In any case, we now know that the existential theory of the enumeration degrees is decidable. We move on to the next quantifier complexity level, where the situation is less clear.

2.2 The two quantifier theory of $\mathcal{D}_e$

We only have partial understanding of the two quantifier theory of the enumeration degrees $\forall\exists\text{-}Th(\mathcal{D}_e)$. In order to describe this, let us again consider the problem in more detail. A two quantifier statement has the form:

$$(\forall \mathbf{x}_1)\dots(\forall \mathbf{x}_k)(\exists \mathbf{y}_1)\dots(\exists \mathbf{y}_n)[\varphi(\mathbf{x}_1,\dots \mathbf{x}_k,\mathbf{y}_1,\dots \mathbf{y}_n)],$$

where φ is once again a disjunction of conjunctions. Let $\psi_1,\dots\psi_r$ be formulas that describe the complete quantifier free type of the variables $\mathbf{x}_1,\dots \mathbf{x}_k$. The statement above is equivalent to the following:

$$\bigwedge_{i<r}(\forall \mathbf{x}_1)\dots(\forall \mathbf{x}_k)[\psi_i(\mathbf{x}_1,\dots \mathbf{x}_k) \to (\exists \mathbf{y}_1)\dots(\exists \mathbf{y}_n)\varphi(\mathbf{x}_1,\dots \mathbf{x}_k,\mathbf{y}_1,\dots \mathbf{y}_n)].$$

And so it is sufficient to decide statements of the form:

$$(\forall \mathbf{x}_1)\dots(\forall \mathbf{x}_k)[\psi(\mathbf{x}_1,\dots \mathbf{x}_k) \to \bigvee_{i<t}(\exists \mathbf{y}_1)\dots(\exists \mathbf{y}_n)\varphi_i(\mathbf{x}_1,\dots \mathbf{x}_k,\mathbf{y}_1,\dots \mathbf{y}_n),$$

where ψ describes the quantifier free type of the variables $\mathbf{x}_1, \dots \mathbf{x}_k$ and each φ_i describes one possible quantifier free type of the variables $\mathbf{x}_1, \dots, \mathbf{x}_k, \mathbf{y}_1, \dots, \mathbf{y}_n$ that is consistent with the type described by ψ. We can restate this question in a structural way as follows:

Problem 2.2. *If P is a finite partial order and $Q_1, \dots Q_t$ are finite extensions of P, decide whether every embedding of P into the enumeration degrees can be extended to an embedding of one of the Q_i.*

The simpler problem when $t = 1$ is the *the extension of embeddings problem.*

For $\mathcal{D}_T$ the problem above is decidable. Lerman [30] showed that every finite lattice P can be embedded as an initial segment of $\mathcal{D}_T$. Thus, if P is a lattice then this embedding of P can be extended to an embedding of Q only if no new element in $Q \setminus P$ is below any element of P. In addition, Q must respect least upper bounds: i.e. if $x \in Q \setminus P$ is above two old elements $u, v \in P$ then x must be above $u \vee v$. If P is not a lattice then points in $Q \setminus P$ can also take the place of least upper bounds that need to be added just because we are embedding in an upper semi-lattice. Shore [46] and Lerman [31] then proved that these are the only obstacles and so the decision problem is computable. The algorithm does not even use the possibility of selecting different possible extensions in different situations: the decision problem is reduced to its simplest case, it is equivalent to the extension of embeddings problem.

In $\mathcal{D}_e$ the situation is very interesting for the following reasons. As we mentioned earlier, Gutteridge [16] showed that the enumeration degrees are downwards dense and so no finite lattice can be embedded as an initial segment. Cooper [8] proved, however, that the enumeration degrees are not dense and Slaman and Calhoun [6] extended Cooper's results by showing that there are empty intervals in the Π^0_2-enumeration degrees. Kent, Lewis-Pye, and Sorbi [21] showed that there are strong minimal covers in the enumeration degrees:

Definition 2.3. The degree $\mathbf{b}$ is a *strong minimal* cover of $\mathbf{a}$ if $\mathbf{a} < \mathbf{b}$ and every degree $\mathbf{x} < \mathbf{b}$ is also bounded by $\mathbf{a}$.

Consider the two-element lattice P consisting of two elements $u < v$. The embedding of P to degrees $\mathbf{a} < \mathbf{b}$ such that $\mathbf{b}$ is a strong minimal cover of $\mathbf{a}$ extends to an embedding of Q only if new elements $x \in Q \setminus P$ that are strictly below v are also below u. The embedding of P to degrees $\mathbf{0}_e < \mathbf{b}$, on the other hand, extends to an embedding of Q only if new elements $x \in Q \setminus P$ are above u. Using Theorem 2.1 and a fairly standard forcing construction we can conclude that these are the only obstacles. Thus for this lattice P we can decide the problem of extending to one of many Q_i's: every embedding of P extends to an embedding $Q_1, \dots Q_n$, if and only if there is one Q_i that places elements strictly below v also below u and there is another Q_j that places new elements above u. The decision procedure is already slightly more complicated than that for the same lattice in $\mathcal{D}_T$ and is not equivalent to the extension of embeddings problem.

Towards a possible decision procedure for the more general problem Lempp, Slaman, and Soskova [28] prove the following

Theorem 2.4 (Lemmp, Slaman, Soskova). *Every finite distributive lattice can be embedded as an interval* $[\mathbf{a}, \mathbf{b}]$, *so that if* $\mathbf{x} \leq \mathbf{b}$ *then* $\mathbf{x} \in [\mathbf{a}, \mathbf{b}]$ *or* $\mathbf{x} < \mathbf{a}$.

Note that in the theorem above the range of our embedding is the whole interval $[\mathbf{a}, \mathbf{b}]$, and so this is an extension of the existence of strong minimal covers in the enumeration degrees. We will say that this is a *strong interval lattice embedding.* This turns out to be sufficient to decide the extension of embeddings problem:

Theorem 2.5 (Lemmp, Slaman, Soskova). *The extension of embeddings problem for* $\mathcal{D}_e$ *is decidable.*

Some very important questions remain open. First of all, we do not know, whether we can remove the distributivity restriction in the theorem above.

Question 2.6. Does every finite lattice have a strong interval embedding in $\mathcal{D}_e$?

Even if we had a positive answer to the question above, we still do not know, whether we can decide the two quantifier theory of $\mathcal{D}_e$. Before we can answer that question, we need to understand more about the structure of $\mathcal{D}_e$. One particularly difficult structural questions concerns the existence of a strong minimal pair.

Definition 2.7. A pair of degrees $\mathbf{a}$ and $\mathbf{b}$ form a *strong minimal pair* if

- $\mathbf{a}$ and $\mathbf{b}$ are incomparable degrees with only $\mathbf{0}_e$ as their common lower bound.
- if $\mathbf{x} \leq \mathbf{b}$ then $\mathbf{a} \vee \mathbf{x} \geq \mathbf{b}$, and, similarly, if $\mathbf{x} \leq \mathbf{a}$ then $\mathbf{b} \vee \mathbf{x} \geq \mathbf{b}$.

Question 2.8. Are there strong minimal pairs in $\mathcal{D}_e$?

2.3 The three quantifier theory of $\mathcal{D}_e$

Finite distributive lattices are already fairly complicated structures. Nies [40] showed that their $\forall\exists\forall$-theory in the language of partial orders is hereditarily undecidable. He also gave a way to transfer this undecidability to structures in which we can define finite distributive lattices.

Definition 2.9. Let $\mathcal{C}$ be a class of structures in a finite relational language $L = \{R_1, \dots R_n\}$. We say that $\mathcal{C}$ is Σ_k*-elementary definable* in $\mathcal{D}_e$ if there are Σ_k formulas φ_U, φ_{R_i}, and $\varphi_{\neg R_i}$ for $i \leq n$ such that for every $C \in \mathcal{C}$ there are parameters $\vec{\mathbf{p}} \in \mathcal{D}_e$ that make the structure with universe $U = \{\mathbf{x} \mid \mathcal{D}_e \models \varphi_U(\mathbf{x}, \vec{\mathbf{p}})\}$ and relations R_i defined as $\{\vec{\mathbf{x}} \mid \mathcal{D}_e \models \varphi_{R_i}(\vec{\mathbf{x}}, \vec{\mathbf{p}})\} = \{\vec{\mathbf{x}} \mid \mathcal{D}_e \models \neg\varphi_{\neg R_i}(\vec{\mathbf{x}}, \vec{\mathbf{p}})\}$ isomorphic to C.

Theorem 2.4 implies that the class of finite distributive lattices is Σ_1-elementary definable in the partial order $\mathcal{D}_e$ with two parameters: $\varphi_U(\mathbf{x}, \mathbf{a}, \mathbf{b})$ is the formula $\mathbf{a} \leq \mathbf{x}$ & $\mathbf{x} \leq \mathbf{b}$ and $=$, $\neq$, $\leq$ and $\nleq$ are interpreted via $=$, $\neq$, $\leq$ and $\nleq$. We next apply the Nies Transfer Lemma [40]:

Lemma 2.10 (Nies Transfer Lemma). *Let $r \geq 2$ and $k \geq 1$. If a class of models $\mathcal{C}$ is Σ_k-elementarily definable in $\mathcal{D}_e$ and the $r+1$-quantifier fragment of $\mathcal{C}$ is hereditarily undecidable then the $k+r$-quantifier fragment of $\mathcal{D}_e$ is hereditarily undecidable.*

We can now conclude:

Corollary 2.11. *The three quantifier theory of $\mathcal{D}_e$ is (hereditarily) undecidable.*

This makes the question of the decidability of the $\forall\exists$-theory of $\mathcal{D}_e$ all the more interesting, as it would exactly give us the quantifier complexity where decidability breaks down.

2.4 The full theory of $\mathcal{D}_e$

Let us now turn to the full theory of the enumeration degrees. It follows from what we have said so far that this theory is not decidable. But how complicated is it? Well, first of all, as we already discussed, enumeration reducibility is arithmetically definable in Z_2 and so Z_2 can interpret $\mathcal{D}_e$. This sets an upper bound to the complexity of $Th(D_e)$, namely we see that it is 1-reducible to $Th(Z_2)$. Slaman and Woodin [52] prove that the reverse is true as well.

Theorem 2.12 (Slaman, Woodin). *The first order theory of the enumeration degrees is computably isomorphic to the theory of second order arithmetic.*

In other words, they show that there is an algorithm that allows us to translate a sentence φ in the language of second order arithmetic to a sentence ψ in the language of partial orders so that $Z_2 \models \varphi$ if and only if $\mathcal{D}_e \models \psi$. The main tool that they use is their *Coding Theorem*:

Theorem 2.13 (Coding Theorem). *There is a uniform way to define every countable relation on $\mathcal{D}_e$ using parameters. In other words, for every n there is a formula φ_n such that for every countable relation $R \subseteq \mathcal{D}_e^n$ there are parameters $\vec{\mathbf{p}}$ such that $R(\mathbf{a}_1, \dots \mathbf{a}_n)$ is true if and only if $\mathcal{D}_e \models \varphi_n(\mathbf{a}_1, \dots \mathbf{a}_n, \vec{\mathbf{p}})$.*

The Coding Theorem lets us pick out in a definable way a tuple of parameters $\vec{\mathbf{p}}$ that codes unary relations $N_{\vec{\mathbf{p}}}$ and $C_{\vec{\mathbf{p}}}$, 3-ary relations $R_{+,\vec{\mathbf{p}}}$ and $R_{*,\vec{\mathbf{p}}}$, such that the structure $M_{\vec{\mathbf{p}}} = (N_{\vec{\mathbf{p}}}; +_{\vec{\mathbf{p}}}, *_{\vec{\mathbf{p}}}, C_{\vec{\mathbf{p}}})$, where $\mathbf{a}_1 +_{\vec{\mathbf{p}}} \mathbf{a}_2 = \mathbf{a}_3$ if and only if $R_{+,\vec{\mathbf{p}}}(\mathbf{a}_1, \mathbf{a}_2, \mathbf{a}_3)$ and, similarly, $\mathbf{a}_1 *_{\vec{\mathbf{p}}} \mathbf{a}_2 = \mathbf{a}_3$ if and only if $R_{*,\vec{\mathbf{p}}}(\mathbf{a}_1, \mathbf{a}_2, \mathbf{a}_3)$, is isomorphic to true arithmetic with a predicate for a set C. We say that $\vec{\mathbf{p}}$ codes the model $(\mathbb{N}, +, *, C)$.

The *biinterpretability conjecture* suggests that the relationship between Z_2 and $\mathcal{D}_e$ is even stronger:

Conjecture 2.14 (Biinterpretability conjecture). *The relation $Bi(\vec{\mathbf{p}}, \mathbf{c})$, true when $\vec{\mathbf{p}}$ codes a model $(\mathbb{N}, +, *, C)$ and $\deg_e(C) = \mathbf{c}$, is first order definable in $\mathcal{D}_e$.*

Slaman and Woodin formulate this conjecture for a number of degree structures, including $\mathcal{D}_T$, $\mathcal{D}_e$, the arithmetical degrees $\mathcal{D}_a$, the hyperarithmetical degrees $\mathcal{D}_h$. In their fundamental work [53] on the analysis on the automorphism group of $\mathcal{D}_T$ they prove that the conjecture is true for the hyperarithmetical degrees. For $\mathcal{D}_T$ they are only able to show that it is true modulo the use of a single parameter. Soskova [54], extends their work and shows that, as anticipated, the same is true for $\mathcal{D}_e$.

Theorem 2.15 (Slaman and Woodin, Soskova). *There is a single parameter $\mathbf{g}$ and a formula φ such that $Bi(\vec{\mathbf{p}}, \mathbf{c})$ is true if and only if $\mathcal{D}_e \models \varphi(\mathbf{g}, \vec{\mathbf{p}}, \mathbf{c})$.*

Using the equivalence proved in Theorem 1.3 we can infer something about our next theme: first order definability. However, in this case as well, we need to use a parameter.

Corollary 2.16. *Every relation on $\mathcal{D}_e$ that is induced by a degree invariant definable relation in second order arithmetic can be defined in $\mathcal{D}_e$ using a single parameter $\mathbf{g}$.*

3 First order definability in $\mathcal{D}_e$

One of the most celebrated first order definability results in the structure of the Turing degrees is the first order definability of the Turing jump. Recall, that the Turing jump is an operator on $\mathcal{D}_T$ that maps a degree $\mathbf{a}$ to $\mathbf{a}' > \mathbf{a}$ and is defined by relativizing the halting problem to an arbitrary set. Slaman and Woodin's [53] analysis of the automorphism group of $\mathcal{D}_T$ allows them to prove that for every automorphism on the Turing degrees π and every degree $\mathbf{x}$ we have that $\pi(\mathbf{x})'' = \pi(\mathbf{x}'')$. In other words, the double jump operator is preserved by automorphisms. They also show that if a relation is invariant under all automorphisms then it must be definable without parameters. As a result they get the first order definability of the double jump operator. Shore and Slaman [47] then build on top of that result to prove that the jump operator is also definable. The proof uses the Kumabe-Slaman forcing method. As you can probably guess, the first order definition of the jump operator that comes out of this elaborate proof is not intuitive and has fairly high quantifier complexity. We will see in this section that definability in the enumeration degrees is quite different.

3.1 The enumeration jump and the total enumeration degrees

Before we can illustrate how definability differs in the enumeration degrees, we need to isolate interesting relations on $\mathcal{D}_e$ whose first order definability would be informative. And what better way to start than with an enumeration degree analog of the jump operator. Recall that the halting set K^A relative to a set A is the uniform join of all c.e. in A sets. When we try to adapt the definition to the world of enumeration degrees we naturally consider the set $K_A = \bigoplus_{e<\omega} \Gamma_e(A)$—the uniform join of the sets that are enumeration reducible to A. Unfortunately, this does not give rise to a very interesting operator, because $K_A \equiv_e A$. In the proof that $A \not\leq_T K^A$ we actually use the fact that $\overline{K^A}$ is not c.e. in A. This idea gives rise to the following definition of the enumeration jump operator introduced by Cooper [7].

Definition 3.1. The *enumeration jump* of a set A is the set $A' = K_A \oplus \overline{K}_A$. The jump of a degree is $\deg_e(A)' = \deg_e(A')$.

The enumeration jump operator has many of the properties that we expect from a jump operator: for instance for all $\mathbf{a}$ we have that $\mathbf{a} < \mathbf{a}'$ and $\mathbf{a} \leq \mathbf{b}$ implies $\mathbf{a}' \leq \mathbf{b}'$. It also agrees with the Turing jump under the standard embedding ι. So naturally we may wonder: Is the enumeration jump operator first order definable?

Another, possibly more important, class of enumeration degrees is the class of all total enumeration degrees.

Definition 3.2. A set A is *total* if $\overline{A} \leq_e A$. An enumeration degree is *total* if it contains a total set.

To understand where the name *total* note that the graph of a total function is total and that every total degree contains the graph of a total function. An equivalent way of defining total degrees is as the enumeration degrees of sets of the form $A \oplus \overline{A}$. In other words, the total enumeration degrees are exactly the degrees that are images of Turing degrees under the standard embedding ι. It was Rogers [42] who asked first whether the total degrees are first order definable in $\mathcal{D}_e$. In fact, Rogers [42] had a list of questions among which were whether $\mathcal{D}_e$ and $\mathcal{D}_T$ are rigid, whether the Turing jump is first order definable and whether definability is equivalent to invariance under automorphisms.

We will see that in $\mathcal{D}_e$ both the enumeration jump and the total enumeration degrees have natural, simple first order definitions. At the heart of these definition is a notion introduced by Jockusch [18] in his thesis.

Definition 3.3. A set A is semi-computable if and only if there is a total computable selector function $s_A\colon \omega^2 \to \omega$—a function such that $\forall x, y \in \omega$ we have that $s_A(x, y) \in \{x, y\}$ and whenever $\{x, y\} \cap A \neq \emptyset$ we have that $s_A(x, y) \in A$.

Jockusch [18] characterized semi-computable sets as left cuts in computable linear orderings on ω. One direction of this characterization is straightforward: if $\leq_L$ is a computable linear ordering then the function $s(x, y)$ that compares its inputs and outputs the one that is smaller with respect to $\leq_L$ witnesses that all left cuts in that linear ordering are semi-computable. Jockusch also proved that semi-computable sets are far from computable. In fact, every Turing degree contains a semi-computable set that is neither c.e. nor co-c.e. Translated through the embedding ι into enumeration degree theoretic terms this shows that every total enumeration degrees is the nontrivial join of the enumeration degrees of a semi-computable set and its complement. Here by nontrivial, we mean that neither of these two degrees is $\mathbf{0}_e$. Arslanov, Cooper and Kalimullin [3] realized that the enumeration degrees of a semi-computable set and its complement satisfy an unusual structural property:

Definition 3.4. A pair of enumeration degrees $\{\mathbf{a}, \mathbf{b}\}$ is a *robust minimal pair* if and only if:

$$(\forall \mathbf{x})[(\mathbf{a} \vee \mathbf{x}) \wedge (\mathbf{b} \vee \mathbf{x}) = \mathbf{x}.]$$

Note that $\mathbf{0}$ forms a robust minimal pair with any other degree. We will call this a *trivial* robust minimal pair. The reason that we called the property above unusual is once again rooted in intuition coming from the Turing degrees. Posner and Robinson [41] prove that if $D \geq_T \emptyset'$ and $\{A_i\}_{i<\omega}$ is a sequence of uniformly D-computable incomputable sets then there is a set G such that

$$(\forall i)(A_i \oplus G \equiv_T G' \equiv_T D)$$

As a consequence we get that for any pair of nonzero Turing degrees $\{\mathbf{a}, \mathbf{b}\}$ there is a Turing degree $\mathbf{g}$ such that $\mathbf{a} \vee \mathbf{g} = \mathbf{b} \vee \mathbf{g} = \mathbf{g}'$. As $\mathbf{g} < \mathbf{g}'$, the degree $\mathbf{g}$ witnesses that the pair $\{\mathbf{a}, \mathbf{b}\}$ is not a robust minimal pair. So there are no nontrivial robust minimal pairs in the Turing degrees.

It is alluring to hope that the robust minimal pairs define semi-computable pairs, as that would give a fairly simple definition of the nonzero total enumeration degrees: joins of nontrivial robust minimal pairs. Kalimullin [19] showed that this is, unfortunately, not the case. He gave a combinatorial characterization of the pairs of sets whose degrees form robust minimal pairs:

Definition 3.5. Sets A and B form a *Kalimullin pair* ($\mathcal{K}$-pair) relative to a set U if and only if there is a set $W \leq_e U$ such that $A \times B \subseteq W$ and $\overline{A} \times \overline{B} \subseteq \overline{W}$.

Theorem 3.6 (Kalimullin). *A pair of sets A and B are a $\mathcal{K}$-pair relative to a set U if and only if their enumeration degrees* $\mathbf{a}$, $\mathbf{b}$ *and* $\mathbf{u}$ *satisfy*

$$(\forall \mathbf{x} \geq \mathbf{u})[(\mathbf{a} \vee \mathbf{u} \vee \mathbf{x}) \wedge (\mathbf{b} \vee \mathbf{u} \vee \mathbf{x}) = \mathbf{x}$$

We will say that $\mathbf{a}$ and $\mathbf{b}$ are a *robust minimal pair relative to* $\mathbf{u}$ if they satisfy the formula from the definition above. And so, the robust minimal pairs are exactly the degrees of $\mathcal{K}$-pairs relative to any c.e. set. We call such pairs simply $\mathcal{K}$-pairs. It is not difficult to show that this class is much larger than the class of semi-computable pairs. Nevertheless, Kalimullin [19] was able to show that they are extremely useful for definability results in $\mathcal{D}_e$. He proved that the enumeration jump can be characterized via robust minimal pairs.

Theorem 3.7 (Kalimullin). *The jump of an enumeration degree* $\mathbf{u}$ *is the greatest degree that can be represented as* $\mathbf{a} \vee \mathbf{b} \vee \mathbf{c}$ *where* $\{\mathbf{a}, \mathbf{b}\}$, $\{\mathbf{b}, \mathbf{c}\}$, *and* $\{\mathbf{a}, \mathbf{c}\}$ *all form robust minimal pairs relative to* $\mathbf{u}$.

Ganchev and Soskova [14] gave an alternative definition of the enumeration jump which only relies on unrelativized robust minimal pairs:

Theorem 3.8 (Ganchev, Soskova). *The jump of an enumeration degree* $\mathbf{u}$ *is the greatest degree that can be represented as* $\mathbf{a} \vee \mathbf{b}$ *for a nontrivial robust minimal pair* $\{\mathbf{a}, \mathbf{b}\}$ *such that* $\mathbf{a} \leq \mathbf{u}$.

It took several more years to arrive at the correct approach to the first order definability of the total enumeration degrees. Ganchev and Soskova [14] realized that semi-computable pairs satisfy a stronger structural feature: they are *maximal* robust minimal pairs.

Definition 3.9. A robust minimal pair $\{\mathbf{a}, \mathbf{b}\}$ is *maximal* if and only if whenever $\{\mathbf{c}, \mathbf{d}\}$ is a robust minimal pair such that $\mathbf{a} \leq \mathbf{c}$ and $\mathbf{b} \leq \mathbf{d}$, we have that $\mathbf{a} = \mathbf{c}$ and $\mathbf{b} = \mathbf{d}$.

In other words, these are robust minimal pairs such that neither side can be further lifted to form a higher robust minimal pair. The final piece of the puzzle was to show that for every $\mathcal{K}$-pair $\{A, B\}$ there is a semi-computable set C such that $A \leq_e C$ and $B \leq_e \overline{C}$. Ganchev and Soskova [14] showed that this holds for $\mathcal{K}$-pairs bounded by $\mathbf{0}_e'$. The full result was then obtained by Cai, Ganchev, Lempp, Miller and Soskova [5]:

Theorem 3.10 (Cai, Ganchev, Lempp, Miller and Soskova). *A nonzero enumeration degree is total if and only if it can be represented as* $\mathbf{a} \vee \mathbf{b}$, *for a* maximal *robust minimal pair.*

The first order definability of the total enumeration degrees clarified a lot of the parallels that we were observing between $\mathcal{D}_T$ and $\mathcal{D}_e$. For example, Theorem 2.12, Theorem 2.13, and Theorem 2.15 are now a direct consequences of the corresponding facts true of the Turing degrees.

This definition of the total enumeration degrees lead to an additional surprising consequence: the first order definability of the image of the relation "c.e. in". Recall that for Turing degrees $\mathbf{a}$ and $\mathbf{b}$, we say that $\mathbf{a}$ is c.e. in $\mathbf{b}$ if and only if there is some set $A \in \mathbf{a}$ which is c.e. in some set (or equivalently all sets) $B \in \mathbf{b}$.

Theorem 3.11 (Cai, Ganchev, Lempp, Miller and Soskova [5]). *The relation*

$$\{(\mathbf{a}, \mathbf{b}) \in \mathcal{D}_e^2 \mid \mathbf{a}, \mathbf{b} \textit{ are total } \& \ \iota^{-1}(\mathbf{a}) \textit{ is c.e. in } \iota^{-1}(\mathbf{b})\}$$

is first order definable in $\mathcal{D}_e$.

So far we have only looked at relations and classes that are closely related to our study and understanding of the Turing degrees. We next take a look at a class that arises differently: from the study of effective mathematics.

3.2 The continuous degrees

Computable analysis allows us to lift computability theoretic notions from sets of natural numbers to more complex mathematical objects, such as real numbers, continuous functions, elements of the Hilbert cube. All of these are examples of points in computable metric spaces, a notion introduced by Lacombe [25].

Definition 3.12. A *computable metric space* is a metric space $\mathcal{M}$ together with a countable dense sequence $Q^{\mathcal{M}} = \{q_n^{\mathcal{M}}\}_{n\in\omega}$ on which the metric is computable, i.e. there is a computable function that maps a pair of indices i, j and a precision $\varepsilon \in \mathbb{Q}^+$ to a rational that is within ε of $d_{\mathcal{M}}(q_i, q_j)$.

The canonical example of a computable metric space is $\mathbb{R}$ with $Q^{\mathbb{R}}$ some computable listing of the rational numbers $\mathbb{Q}$. But many other second countable metric spaces (metric spaces with a countable base) can be supplied with a listing of a dense sequence to make them computable: for example, Cantor space 2^ω and Baire space ω^ω with the usual metric, the continuous functions on the unit interval $\mathbf{C}[0, 1]$ with the uniform metric, the Hilbert cube $[0, 1]^\omega$ with metric $d(\alpha, \beta) = \sum_{n\in\omega} \frac{|\alpha(n)-\beta(n)|}{2^n}$.

To every member x of a computable metric space we associate a set of *names*—discrete objects that give us a way to approximate x with arbitrary precision using the distinguished dense sequence:

Definition 3.13. $\lambda\colon \mathbb{Q}^+ \to \omega$ is a *name* of a point x in a computable metric space $\mathcal{M}$ if for all rationals $\varepsilon > 0$ we have $d_{\mathcal{M}}(x, q^{\mathcal{M}}_{\lambda(\varepsilon)}) < \varepsilon$.

We think of the set of names for x as carrying the algorithmic content of x. In particular, we can define a computable function on computable metric spaces $f\colon \mathcal{M} \to \mathcal{N}$ as (represented by) a computable functional Ψ that takes names of a point $x \in \mathcal{M}$ to names of $f(x) \in \mathcal{N}$. We can also talk about the *Turing degree* of a point: the least Turing degree of a name for that point. It is fairly easy to see that every real number r has a Turing degree, the Turing degree of its Dedekind cut $\{q \in \mathbb{Q} \mid q < r\} \oplus \{q \in \mathbb{Q} \mid q > r\}$. Pour El and Lempp asked whether this is also true for continuous functions on the real numbers. To answer their question, Miller [36] introduced a way to compare the computable strength of points in arbitrary computable metric spaces.

Definition 3.14. If x and y are members of (possibly different) computable metric spaces, then $x \leq_r y$ if there is a uniform way to compute a name for x from a name for y.

This reducibility induces a degree structure, which Miller [36] called the *continuous degrees*. His reason for the choice of name comes from the following characterization:

Theorem 3.15 (Miller). *Every continuous degree contains a point from* $[0,1]^\omega$ *and a point from* $\mathbf{C}[0,1]$.

In other words, we can think of the continuous functions on the unit interval and of the Hilbert cube as universal spaces. Using the universality of $[0,1]^\omega$ Miller [36] was able to show that the continuous degrees embed into the enumeration degrees. For $\alpha \in [0,1]^\omega$, let

$$C_\alpha = \bigoplus_{i\in\omega} \{q \in \mathbb{Q} \mid q < \alpha(i)\} \oplus \{q \in \mathbb{Q} \mid q > \alpha(i)\}.$$

Enumerating C_α is exactly as hard as computing a name for α. So $\alpha \mapsto C_\alpha$ induces the aforementioned embedding. Each element of 2^ω, ω^ω, and $\mathbb{R}$ is mapped onto the total degree of its least Turing degree name (i.e., the image of its Turing degree). Lempp and Pour El's question can be restated in terms of this embedding as: is there a continuous degree that is non-total. Miller [36] answered this question:

Theorem 3.16 (Miller). *There is a nontotal continuous degree.*

It is worth pointing out that every known proof of this result uses nontrivial topological facts: Miller [36] used a variant of Brouwer's fixed point theorem for multivalued functions on an infinite dimensional space. Day and Miller [9] gave an alternative proof that relies on neutral measures. Levin [33] used Sperner's lemma to construct such measures. More recently, Kihara and Pauly [23], and independently Hoyrup (unpublished) used results from topological dimension theory—that $[0, 1]^\omega$ is strongly infinite dimensional and therefore not the countable union of finite dimensional spaces.

The continuous degrees therefore constitute an interesting class of enumeration degrees. They properly extend the Turing degrees. Miller [36] proved that no continuous degree can be quasiminimal, so they are a proper subclass of the enumeration degrees. Are they first order definable? Surprisingly, the answer turns out to be: yes and they have a very natural first order definition. Andrews, Igusa, Miller and Soskova [2] use an effective version of Urysohn's metrization theorem due to Schröder [44] to show the following:

Theorem 3.17 (Andrews, Igusa, Miller, and Soskova). *An enumeration degree* $\mathbf{a}$ *is continuous if and only if it is* almost total*: if* $\mathbf{x} \not\leq \mathbf{a}$ *and* $\mathbf{x}$ *is total then* $\mathbf{a} \vee \mathbf{x}$ *is total.*

It follows from the definability of the total enumeration degrees that the continuous degrees are first order definable. In this case as well, the definability of the continuous degrees has a pleasing further consequence. Recall, that a Turing degree $\mathbf{a}$ is *PA above* a Turing degree $\mathbf{b}$ if $\mathbf{a}$ computes a path in every infinite $\mathbf{b}$-computable tree. Using the embedding ι we can transfer this relation to total degrees. Miller [36] proved that nontotal continuous degrees can be used to characterize this relation.

Theorem 3.18. *For total degrees* $\mathbf{a}$ *is* PA above $\mathbf{b}$ *if and only if there is a nontotal continuous degree* $\mathbf{c}$ *such that* $\mathbf{b} < \mathbf{c} < \mathbf{a}$.

The definability of the non-total continuous degrees now yields:

Corollary 3.19. *The image of the relation "PA above" is first order definable in* $\mathcal{D}_e$.

Ganchev, Kalimullin, Miller, and Soskova [11] give an alternative first order definition of the continuous degrees that relies only on $\mathcal{K}$-pairs and avoids invoking the definability of the total degrees. They show that an enumeration degree is continuous if and only if it is not half of any nontrivial relativized $\mathcal{K}$-pair. This gives a structural dichotomy in the enumeration degrees:

Theorem 3.20 (Ganchev, Kalimullin, Miller, and Soskova). *For every enumeration degree* $\mathbf{a}$, *exactly one of the following two properties holds:*

1. *The degree* $\mathbf{x}$ *is continuous, so for every total enumeration degree* $\mathbf{x} \not\leq \mathbf{a}$, $\mathbf{a} \vee \mathbf{x}$ *is total.*

2. *There is a total enumeration degree* $\mathbf{x} \not\leq \mathbf{a}$ *such that* $\mathbf{a} \vee \mathbf{x}$ *is a strong quasi-minimal cover of* $\mathbf{x}$.

3.3 The skip operator and the cototal enumeration degrees

Before we can explore first order definability in the enumeration degrees further, we must accumulate a collection of classes and relations on the enumeration degrees and understand their interactions with classes that we have already explored. Andrews, Ganchev, Kuyper, Lempp, Miller, A. Soskova and M. Soskova [1] initiate the study of a natural operator on the enumeration degrees: *the skip operator*, and the related class of the cototal degrees. Recall that by K_A we denote the uniform join of all set that are enumeration reducible to A.

Definition 3.21. The *skip* of a set A is the set $A^\diamond = \overline{K_A} = \bigoplus_{e<\omega} \overline{\Gamma_e(A)}$.

It is straightforward to check that $A \leq_e B$ implies $\overline{K_A} \leq_1 \overline{K_B}$, and so the skip operator on sets induces an operator on degrees: $\deg_e(A)^\diamond = \deg_e(A^\diamond)$. There are several ways in which it can be argued that the skip operator is the more natural analog of the Turing jump operator for the structure $\mathcal{D}_e$. Andrews, et al. [1] prove that:

1. $A \leq_e B$ if and only if $A^\diamond \leq_1 B^\diamond$, but there are sets A and B such that $A' \leq_1 B'$ but $A \not\leq_e B$.

2. If $\mathbf{b} \geq_e \mathbf{0}_e'$ then there is a degree $\mathbf{a}$ such that $\mathbf{a}^\diamond = \mathbf{b}$. In fact, $\mathbf{a}$ can be chosen to be quasiminimal.

3. On total enumeration degrees the skip and the jump coincide.

On the other hand, there are ways in which the skip behaves differently: even though the skip of a degree is never below that degree it can be, and most often is, to the side of it, so $\mathbf{a}$ and $\mathbf{a}^\diamond$ are usually incomparable degrees. Andrews et al. [1] push this to the extreme with the following theorem:

Theorem 3.22 (Andrews, et al.). *There are degrees* $\mathbf{a}$ *and* $\mathbf{b}$ *that form a* skip 2-cycle, *i.e.* $\mathbf{a}^\diamond = \mathbf{b}$ *and* $\mathbf{b}^\diamond = \mathbf{a}$. *Such degrees* $\mathbf{a}$ *and* $\mathbf{b}$ *must be above every hyperarithmetic degree.*

There is much more to investigate about the skip operator and its structural behavior, in particular, the authors leave open:

Question 3.23. Is the skip operator first order definable in $\mathcal{D}_e$?

The class of degrees on which the skip behaves just like the jump operator is the class of cototal enumeration degrees:

Definition 3.24. A set A is *cototal* if $A \leq_e \overline{A}$. An enumeration degree is *cototal* if and only if it contains a cototal set.

Clearly, every total degree is cototal: $A \oplus \overline{A} \equiv_1 \overline{A \oplus \overline{A}} = \overline{A} \oplus A$. Andrews et al. [1], also show that every Σ^0_2 enumeration degree and every continuous degree is cototal. On the other hand, sufficiently generic degrees are not cototal. Thus we have a proper superclass of the continuous degrees.

Motivation for the study of the cototal enumeration degrees came from symbolic dynamics. Jeandel and his group were studying the spectrum of a *minimal subshift.*

Definition 3.25. A set $S \subseteq 2^\omega$ is called a *subshift* if S is topologically closed and closed under the shift operator that maps $\alpha(0)\alpha(1)\alpha(2)\dots$ to $\alpha(1)\alpha(2)\dots$.

The subshift is *minimal* if it has no proper nonempty subset that is also a subshift.

The spectrum of a subshift S is the set of Turing degrees that compute a member of S.

Jeandel [17] had noticed that a Turing degree computes a member of a given minimal subshift S if and only if it can enumerate the set L_S, the *language of* S, consisting of all finite subwords of elements of S. Thus, the spectrum of a minimal subshift S is exactly characterized by the enumeration degree of L_S. He also noticed that the set L_S can be uniformly enumerated given any enumeration of its complement (the set of forbidden words in S), i.e. L_S is cototal. McCarthy [34] proved that every cototal degree contains the set L_S for some minimal subshift S and so we get a characterization of the cototal degrees.

Theorem 3.26 (Jeandel, McCarthy). *An enumeration degree is cototal if and only if it contains the language of a minimal subshift.*

It turns out that cototal degrees have numerous characterization arising in all kinds of areas of effective mathematics. The cototal degrees are:

1. The degrees on which the skip and the jump operator coincide.

2. The degrees of complements of maximal independent sets in computable graphs on ω.

3. The degrees of complements of maximal antichains in $\omega^{<\omega}$.

4. The enumeration degrees such that the set of Turing degrees above them is the spectrum of a structure and the the upward closure of an F_σ subset of ω^ω.

5. The degrees of sets with good approximations.

6. The degrees of points in computable G_δ topological spaces.

(1) and (2) are proved by Andrews et al. [1]. (3) is proved by McCarthy [34]. (4) was proved by Montalbán [38] and McCarthy [34]. (5) was proved by Miller and Soskova [37]. They used this characterization to prove that the cototal enumeration degrees are dense. (6) is proved by Kihara, Ng, and Pauly [22].

And so we come to a second open question related to definability:

Question 3.27. Are the cototal enumeration degrees first order definable in $\mathcal{D}_e$?

The last characterization of the cototal degrees is part of a more general program, initiated by Kihara and Pauly [23] and extended in Kihara, Ng, and Pauly [22], to transfer topological spaces and topological properties to the enumeration degrees.

Definition 3.28. A *represented space* is a pair of a second countable topological space X and listing of an open base $B^X = \{B_i\}_{i<\omega}$.
A *name* for a point $x \in X$ is an enumeration of the set $N_x = \{i \mid x \in B_i\}$.
For $x \in X$ and $y \in Y$, where X and Y are (possibly different) represented spaces, we say that $x \leq y$ if and only if every name for y uniformly computes a name for x.

Thus a represented space X gives rise to a class of enumeration degrees $\mathcal{D}_X \subseteq \mathcal{D}_e$. For example:

1. $\mathcal{D}_{S^\infty} = \mathcal{D}_e$, where S is the Sierpinski topology $\{\emptyset, \{1\}, \{0,1\}\}$.

2. $\mathcal{D}_{2^\omega} = \mathcal{D}_\mathbb{R}$ is the class of all total enumeration degrees.

3. $\mathcal{D}_{[0,1]^\omega}$ is the class of the continuous degrees.

4. $\mathcal{D}_{\mathbb{R}^<}$, where $\mathbb{R}^<$ denotes the reals equipped with the lower topology which is generated by $\{(q,\infty)\}_{q\in\mathbb{Q}}$, is the class of all semi-computable degrees.

Kihara, Ng, and Pauly [22] further investigate $\mathcal{D}_X$, where X is the ω-power of: the cofinite topology on ω, the telophase space, the double origin space, the quasi-Polish Roy space, the irregular lattice space. Thus we have many more classes whose first order definability in $\mathcal{D}_e$ can be pursued.

4 Automorphisms and automorphism bases

4.1 Global automorphisms

In this section we discuss the implications of definability for the automorphism group of the enumeration degrees.

Definition 4.1. Let $\mathfrak{A}$ be a structure with domain A. A set $B \subseteq A$ is an *automorphism base* for $\mathfrak{A}$ if any two automorphisms that agree on B coincide.

Equivalently, B is a base if the only automorphism that fixes all members of B is the identity.

Let us take a look at some highlights in Slaman and Woodin's [53] automorphism analysis:

Theorem 4.2 (Slaman, Woodin). *The Turing degrees have at most countably many automorphisms.*

There is a single degree $\mathbf{g} \leq \mathbf{0}_T^{(5)}$ *that is an automorphism base for* $\mathcal{D}_T$.

Relations on $\mathcal{D}_T$ *induced by definable relations in* Z_2 *are first order definable in* $\mathcal{D}_T$ *with such a parameter* $\mathbf{g}$.

Relations on $\mathcal{D}_T$ *induced by definable relations in* Z_2 *that are furthermore invariant under automorphisms are first order definable in* $\mathcal{D}_T$ *(without parameters).*

We will extract from this theorem a lot of information about the automorphisms of the enumeration degrees using the definability of the total degrees and the following old result of Selman [45].

Theorem 4.3 (Selman). $\mathbf{a} \leq \mathbf{b}$ *if and only if every total degree above* $\mathbf{b}$ *is also above* $\mathbf{a}$.

Thus, the total enumeration degrees are a *definable automorphism base* for $\mathcal{D}_e$. Definability implies that every automorphism of the enumeration degrees π induces an automorphism of the Turing degrees $\pi^*(\mathbf{a}) = \iota^{-1}(\pi(\iota(\mathbf{a})))$. The fact that the total degrees form an automorphism base tells us that this mapping is injective, and, in particular, a nontrivial automorphism of $\mathcal{D}_e$ gives rise to a nontrivial automorphism of $\mathcal{D}_T$. As promised we get the following:

Corollary 4.4. $\mathcal{D}_e$ *has at most countably many automorphisms. Furthermore, a single total degree below* $\mathbf{0}_e^{(5)}$ *is an automorphism base for* $\mathcal{D}_e$.

The most pressing open question here is therefore, whether the reverse relationship holds.

Question 4.5. Does every automorphism of $\mathcal{D}_T$ extend to an automorphism of $\mathcal{D}_e$?

A positive answer to the question above would give us that the two automorphism groups are isomorphic. It would also imply that automorphisms of the Turing degrees preserve the relations c.e. in and PA above, as they both have definable images in $\mathcal{D}_e$. By Theorem 4.2 this yields their first order definability in $\mathcal{D}_T$. If on the other hand we can rule out the existence of nontrivial automorphisms of $\mathcal{D}_T$ that preserve these relations then we would get that $\mathcal{D}_e$ is rigid. Our hope is that by proceeding in this fashion and uncovering more definable classes of total enumeration degrees we will put more and more restrictions on the possible extendable nontrivial automorphisms of $\mathcal{D}_T$ to eventually get rigidity.

4.2 Local and global structural interactions

The local structure of the enumeration degrees $\mathcal{D}_e(\leq \mathbf{0}'_e)$ consists of the initial interval bounded by $\mathbf{0}'_e$. Every degree in that interval consists entirely of Σ^0_2 sets. The local structure has been studied extensively as well. Cooper [7] proved that it is a dense structure and, as we mentioned earlier, Bianchini [4] extended this result to prove that every countable partial order can be embedded densely in $\mathcal{D}_e(\leq \mathbf{0}'_e)$. This gives the decidability of the the existential theory of $\mathcal{D}_e(\leq \mathbf{0}'_e)$. The two quantifier theory is much more difficult to analyze. Density prevents us from using the initial segment embedding method that Lerman and Shore [32] use for $\mathcal{D}_T(\leq \mathbf{0}'_T)$. Nevertheless, there are partial results: Lempp and Sorbi [29] show that every finite lattice can be embedded in $\mathcal{D}_e(\leq \mathbf{0}'_e)$ preserving least and greatest element and Lempp, Slaman, and Sorbi [27] prove that the extension of embeddings problem is decidable. Kent [20] showed that the three quantifier theory of $\mathcal{D}_e(\leq \mathbf{0}'_e)$ is not decidable. So the first open problem we have for $\mathcal{D}_e(\leq \mathbf{0}'_e)$ matches the one we have for $\mathcal{D}_e$:

Question 4.6. Is the two quantifier theory of $\mathcal{D}_e(\leq \mathbf{0}'_e)$ decidable?

The full theory was shown to be computably isomorphic to first order arithmetic by Ganchev and Soskova [13]. Their proof relies on a local version of the Coding Theorem that was already established by Slaman and Woodin [52] and the local definability of $\mathcal{K}$-pairs [12], which was not previously known. The local definability of $\mathcal{K}$-pairs unlocked a series of other first order definability results, proved in a series of papers by Ganchev and Soskova [12, 14, 15]:

Theorem 4.7 (Ganchev, Soskova). *The following classes have first order definitions in $\mathcal{D}_e$:*

1. *The downwards properly Σ^0_2 degrees, degrees that bound no nonzero Δ^0_2 degree.*

2. *The upwards properly Σ_2^0 degrees, degrees that are not bounded by any incomplete Δ_2^0 degree.*

3. *The Δ_2^0 total enumeration degrees.*

4. *The low enumeration degrees, degrees with $\mathbf{a}' = \mathbf{0}_e'$.*

5. *All members of the jump hierarchy: the low$_n$ and the high$_n$ degrees for $n \geq 1$.*

The local structure $\mathcal{D}_e(\leq \mathbf{0}_e')$ relates to first order arithmetic in a similar way as the global structure $\mathcal{D}_e$ relates to second order arithmetic. We can formulate a biinterpretability conjecture for the local structure as well. The Σ_2^0 sets form a countable class that can be naturally indexed. For example we can set $U_e = \Gamma_e(\emptyset')$, where $\{\Gamma_e\}_{e<\omega}$ list all enumeration operators.

Question 4.8. The biinterpretability conjecture for $\mathcal{D}_e(\leq \mathbf{0}_e')$ is that there is a definable coded model of first order arithmetic $\mathcal{M} = (\mathbb{N}^{\mathcal{M}}, +^{\mathcal{M}}, *^{\mathcal{M}})$ and a definable function $\varphi : \mathbb{N}^{\mathcal{M}} \to \mathcal{D}_e(\leq \mathbf{0}_e')$ such that $\varphi(e^{\mathcal{M}}) = \deg_e(U_e)$. Is it true?

The function φ above is called *an indexing* of the degrees in $\mathcal{D}_e(\leq \mathbf{0}_e')$. Slaman and Soskova [50] prove that a similar biinterpretability conjecture for $\mathcal{D}_T(\leq \mathbf{0}_T')$ is true modulo the use of finitely many parameters. Their approach transferred to the enumeration degrees does not lead to a similar result, so even biinterpretability for $\mathcal{D}_e(\leq \mathbf{0}_e')$ with parameters remains open. It does, however, allow them to uncover an important relationship between local and global structure. The starting point in both approaches is the following theorem of Slaman and Woodin [51], a consequence of the local coding theorem:

Theorem 4.9 (Slaman, Woodin). *There is an indexing of the c.e. Turing degrees that is definable from Δ_2^0 parameters in the local structure $\mathcal{D}_T(\leq \mathbf{0}_T')$.*

Slaman and Soskova [49] start with the result above transferred to $\mathcal{D}_e(\leq \mathbf{0}_e')$ via the standard embedding ι. They use the local definability of the total degrees, the low enumeration degrees, as well as several technical priority constructions, to prove that if a set of finitely many parameters defines an indexing of the image of the c.e. Turing degrees then the same set of parameters defines an indexing of the image of the Δ_2^0 Turing degrees. Next, using the definability of the jump operator and the image of the relation c.e. in, they show that every set of parameters that defines an indexing of the image of the Δ_2^0 Turing degrees also defines an indexing of the image of the degrees that are c.e. in and above some Δ_2^0 Turing degree. Next, using properties of sufficiently generic sets, they show that every set of parameters that defines an indexing as above, also defines an indexing of the image of all Turing degrees bounded by $\mathbf{0}''_T$. The last two steps can now be iterated any finite number of times to show that:

Theorem 4.10 (Slaman, Soskova). *Any set of parameters that defines an indexing of the image of the c.e. Turing degrees also defines an indexing of the image of the Turing degrees bounded by* $\mathbf{0}_T^{(n)}$ *for every natural number* n.

This theorem combines well with what we know about the automorphism group of $\mathcal{D}_e$, in particular, the fact that there is a single total degree below $\mathbf{0}_e^{(5)}$ that forms an automorphism base for $\mathcal{D}_e$. If a set of parameters defines an indexing of the image of the Turing degrees below $\mathbf{0}_T^{(5)}$ and an automorphism fixes these parameters then it is not difficult to see that the automorphism must fix all elements in the range of the definable indexing. As this includes the degree that by itself is an automorphism base for all of $\mathcal{D}_e$, the automorphism must be the identity.

Theorem 4.11 (Slaman, Soskova). *There is a finite set of* Δ^0_2 *total degrees that forms an automorphism base for the global structure* $\mathcal{D}_e$.

Similar arguments lead Slaman and Soskova [49] to the following consequence of their theorem:

Corollary 4.12 (Slaman, Soskova). *If* $\mathcal{D}_e$ *has a nontrivial automorphism then so does:*

1. *The local structure* $\mathcal{D}_e(\leq \mathbf{0}'_e)$.
2. *The structure of the* Δ^0_2 *Turing degrees* $\mathcal{D}_T(\leq \mathbf{0}'_T)$.
3. *The structure of the c.e. Turing degrees.*

Naturally, we wonder:

Question 4.13. Do the automorphisms of any of these structures extend to automorphisms of $\mathcal{D}_e$?

References

[1] Uri Andrews, Hristo A. Ganchev, Rutger Kuyper, Steffen Lempp, Joseph S. Miller, Alexandra A. Soskova, and Mariya I. Soskova *On cototality and the skip operator in the enumeration degrees*, Transactions of the American Mathematical Society 372 (3), 2019, pp. 1631–1670.

[2] Uri Andrews, Gregory Igusa, Joseph S. Miller, and Mariya I. Soskova *Characterizing the continuous degrees*, Israel Journal of Mathematics 234, 2019, pp. 743–767.

[3] Marat M. Arslanov, S. Barry Cooper, and Iskander Sh. Kalimullin *Splitting properties of total enumeration degrees*, Algebra and Logic 42 (1), 2003), pp. 1–13.

[4] Caterina Bianchini Bounding Enumeration Degrees, PhD Dissertation, University of Siena, 2000.

[5] Mingzhong Cai, Hristo A. Ganchev, Steffen Lempp, Joseph S. Miller, and Mariya I. Soskova *Defining totality in the enumeration degrees*, Journal of the American Mathematical Society 29 (4), 2016, pp. 1051–1067.

[6] William C. Calhoun, and Theodore A. Slaman *The Π^0_2 enumeration degrees are not dense*, The Journal of Symbolic Logic 61 (4) 1996, pp. 1364–1379.

[7] S. Barry Cooper *Partial degrees and the density problem II, The enumeration degrees of the Σ_2 sets are dense*, The Journal of Symbolic Logic 49 (2), 1984, pp. 503–513.

[8] — *Enumeration reducibility, nondeterministic computations and relative computability of partial functions*, in: K. Ambos-Spies, G. H. Müller, and G. E. Sacks (eds.) Recursion Theory Week (Oberwolfach, 1989), Springer, Berlin & Heidelberg 1990 [*Lecture Notes in Mathematics 1432*], pp. 57–110.

[9] Adam R. Day, and Joseph S. Miller *Randomness for non-computable measures*, Transactions of the American Mathematical Society 365 (7), 2013, pp. 3575–3591.

[10] Richard M. Friedberg and Hartley Rogers, Jr. *Reducibility and completeness for sets of integers*, Zeitschrift für mathematische Logik und Grundlagen der Mathematik 5, 1959, pp. 117–125.

[11] Hristo A. Ganchev, Iskander Sh. Kalimullin, Joseph S. Miller, and Mariya I. Soskova *A structural dichotomy in the enumeration degrees*, The Journal of Symbolic Logic, 2020 [to appear].

[12] Hristo A. Ganchev, and Mariya I. Soskova *Cupping and definability in the local structure of the enumeration degrees*, The Journal of Symbolic Logic 77 (1), 2012, pp. 133–158.

[13] — *Interpreting true arithmetic in the local structure of the enumeration degrees*, The Journal of Symbolic Logic 77 (4), 2012, pp. 1184–1194.

[14] — *Definability via Kalimullin pairs in the structure of the enumeration degrees*, Transactions of the American Mathematical Society 367 (7), 2015, pp. 4873–4893.

[15] — *The jump hierarchy in the enumeration degrees*, Computability 7 (2–3), 2018, pp. 179–188.

[16] Lance Gutteridge Some Results on Enumeration Reducibility, PhD Dissertation, Simon Fraser University, 1971.

[17] Emmanuel Jeandel *Enumeration in closure spaces with applications to algebra*, CoRR abs/1505.07578 2015.

[18] Carl G. Jockusch Jr. *Semirecursive sets and positive reducibility*, Transactions of the American Mathematical Society 131, 1968, pp. 420–436.

[19] Iskander Sh. Kalimullin *Definability of the jump operator in the enumeration degrees*, Journal of Mathematical Logic 3 (2), 2003, pp. 257–267.

[20] Thomas F. Kent *The Π_3-theory of the Σ^0_2-enumeration degrees is undecidable*, The

Journal of Symbolic Logic 71 (4), 2006, pp. 1284–1302.

[21] Thomas F. Kent, Andrew E. M. Lewis-Pye, and Andrea Sorbi *Empty intervals in the enumeration degrees*, Annals of Pure and Applied Logic 163 (5), 2012, pp. 567–574.

[22] Takayuki Kihara, Keng Meng Ng, and Arno Pauly *Enumeration degrees and non-metrizable topology*, Preprint, 2017.

[23] Takayuki Kihara, and Arno Pauly *Point degree spectra of represented spaces*, Submitted, 2015.

[24] Alistair H. Lachlan, and Richard A. Shore *The n-rea enumeration degrees are dense*, Archive for Mathematical Logic 31 (4), 1992, pp. 277–285.

[25] Daniel Lacombe *Quelques procédés de définition en topologie recursive*, in: Arend Heyting (ed.) Constructivity in Mathematics (Proceedings of the colloquium held at Amsterdam, 1957), North-Holland Publishing Company, Amsterdam 1959 [*Studies in Logic and the Foundations of Mathematics*], pp. 129–158.

[26] Jay Lagemann Embedding Theorems in the Reducibility Ordering of the Partial Degrees, PhD Dissertation, MIT, 1972.

[27] Steffen Lempp, Theodore A. Slaman, and Andrea Sorbi *On extensions of embeddings into the enumeration degrees of the Σ^0_2-sets*, Journal of Mathematical Logic 5 (2), 2005, pp. 247–298.

[28] Steffen Lempp, Theodore A. Slaman, and Mariya I. Soskova *Fragments of the theory of the enumeration degrees*, in preparation.

[29] Steffen Lempp, and Andrea Sorbi *Embedding finite lattices into the Σ^0_2 enumeration degrees*, The Journal of Symbolic Logic 67 (1), 2002, pp. 69–90.

[30] Manuel Lerman *Initial segments of the degrees of unsolvability*, Annals of Mathematics 93 (2), 1971, pp. 365–389.

[31] — Degrees of Unsolvability, Springer-Verlag, Berlin 1983 [*Perspectives in Mathematical Logic*].

[32] Manuel Lerman, and Richard A. Shore *Decidability and invariant classes for degree structures*, Transactions of the American Mathematical Society 310 (2), 1988, pp. 669–692.

[33] Leonid A. Levin *Uniform tests for randomness*, Doklady Akademii Nauk SSSR 227 (1), 1976, pp. 33–35.

[34] Ethan McCarthy *Cototal enumeration degrees and their applications to effective mathematics*, Proceedings of the American Mathematical Society 146 (8), 2018, pp. 3541–3552.

[35] Yurii T. Medvedev *Degrees of difficulty of the mass problem*, Doklady Akademii Nauk SSSR 104 1955, pp. 501–504.

[36] Joseph S. Miller *Degrees of unsolvability of continuous functions*, The Journal of Symbolic Logic 69 (2), 2004, pp. 555–584.

[37] Joseph S. Miller, and Mariya I. Soskova *Density of the cototal enumeration degrees*, Annals of Pure and Applied Logic 69 (5), 2018, pp. 450–462.

[38] Antonio Montalban *Computable structure theory: Within the arithmetic*, Preprint,

2018.

[39] John Myhill *Note on degrees of partial functions*, Proceedings of the American Mathematical Society 12, 1961, pp. 519–521.

[40] Andre Nies *Undecidable fragments of elementary theories*, Algebra Universalis 35 (1), 1996, pp. 8–33.

[41] David B. Posner, and Robert W. Robinson *Degrees joining to* $\mathbf{0}'$, The Journal of Symbolic Logic 46 (4), 1981, pp. 714–722.

[42] Hartley Rogers, Jr. *Some problems of definability in recursive function theory*, in: John N. Crossley (ed.) Sets, Models and Recursion Theory (Proceedings of the Summer School in Mathematical Logic and Tenth Logic Colloquium, Leicester, August-September 1965), North-Holland Publishing House, Amsterdam 1967 [*Studies in Logic and the Foundations Mathematics 46*], pp. 183–201.

[43] Gerald E. Sacks *On the degrees less than* $0'$, Annals of Mathematics (2) 77 (2), 1963, pp. 211–231.

[44] Mathias Schröder *Effective metrization of regular spaces*, in: Ker-l Ko, Anil Nerode, Marian Boykan Pour-El, Klaus Weihrauch, andJirí Wiedermann (eds.), Computability and Complexity in Analysis, [*Informatik Berichte 235*] (FernUniversität Hagen, Hagen, August 1998; Proceedings of the CCA Workshop, Brno, Czech Republic), pp. 63–80.

[45] Alan L. Selman *Arithmetical reducibilities* I, Zeitschrift für mathematische Logik und Grundlagen der Mathematik 17, 1971, pp. 335–350.

[46] Richard A. Shore *On the* $\forall\exists$*-sentences of* α*-recursion theory*, in: Jens Erik Fenstad (ed.) Generalized Recursion Theory, II (Proceedingds of the Second Symposium, University of Oslo, 1977), North-Holland Publishing Company, Amsterdam & New York 1978 [*Studies in Logic and the Foundations Mathematics 94*], pp. 331–353.

[47] Richard A. Shore, and Theodore A. Slaman *Defining the Turing jump*, Mathematical Research Letters 6 (5–6), 1999, pp. 711–722.

[48] Theodore A. Slaman, and Andrea Sorbi *A note on initial segments of the enumeration degrees*, The Journal of Symbolic Logic 79 (2), 2014, pp. 633–643.

[49] Theodore A. Slaman, and Mariya I. Soskova *The enumeration degrees: local and global structural interactions*, in: Foundations of mathematics, American Mathematical Society, Providence RI 2017 [*Contemporary Mathematics 690*], pp. 31–67.

[50] Theodore A. Slaman, and Mariya I. Soskova *The* Δ^0_2 *Turing degrees: automorphisms and definability*, Transactions of the American Mathematical Society 370 (2), 2018, pp. 1351–1375.

[51] Theodore A. Slaman, and W. Hugh Woodin *Definability in the Turing degrees*, Illinois Journal of Mathematics 30 (2), 1986, pp. 320–334.

[52] — *Definability in the enumeration degrees*, Archive for Mathematical Logic 36 (4-5), 1997, pp. 255–267.

[53] — *Definability in degree structures*, Preprint, 2005.

[54] Mariya I. Soskova *The automorphism group of the enumeration degrees*, Annals of Pure and Applied Logic 167 (10), 2016, pp. 982–999.

[55] Mariya I. Soskova, and Ivan N. Soskov *Embedding countable partial orderings in the enumeration degrees and the ω-enumeration degrees*, The Journal of Logic and Computation 22 (4), 2012, pp. 927–952.

www.ingramcontent.com/pod-product-compliance
Lightning Source LLC
Chambersburg PA
CBHW051122230426
43670CB00007B/644
* 9 7 8 1 8 4 8 9 0 3 4 0 1 *